Nanoelectronic
DEVICE
Applications Handbook

Devices, Circuits, and Systems

Series Editor

Krzysztof Iniewski

CMOS Emerging Technologies Inc., Vancouver, British Columbia, Canada

FORTHCOMING TITLES:

3 D Circuit and System Design: Multicore Architecture, Thermal Management, and Reliability
Rohit Sharma and Krzysztof Iniewski

Building Sensor Networks: From Design to Applications
Ioanis Nikolaidis and Krzysztof Iniewski

CMOS: Front-End Electronics for Radiation Sensors
Angelo Rivetti

Embedded and Networking Systems: Design, Software, and Implementation
Gul N. Khan and Krzysztof Iniewski

Energy Harvesting with Functional Materials and Microsystems
Madhu Bhaskaran, Sharath Sriram, and Krzysztof Iniewski

High Frequency Communication and Sensing: Traveling-Wave Techniques
Ahmet Tekin and Ahmed Emira

High-Speed Devices and Circuits with THz Applications
Jung Han Choi and Krzysztof Iniewski

Integrated Power Devices and TCAD Simulation
Yue Fu, Zhanming Li, Wai Tung Ng, and Johnny K.O. Sin

Labs-on-Chip: Physics, Design and Technology
Eugenio Iannone

Medical Imaging: Technology and Applications
Troy Farncombe and Krzysztof Iniewski

Metallic Spintronic Devices
Xiaobin Wang

Microfluidics and Nanotechnology: Biosensing to the Single Molecule Limit
Eric Lagally and Krzysztof Iniewski

MIMO Power Line Communications: Narrow and Broadband Standards, EMC, and Advanced Processing
Lars Torsten Berger, Andreas Schwager, Pascal Pagani, and Daniel Schneider

Mobile Point-of-Care Monitors and Diagnostic Device Design
Walter Karlen and Krzysztof Iniewski

Nanoelectronics: Devices, Circuits, and Systems
Nikos Konofaos

Nanomaterials: A Guide to Fabrication and Applications
Gordon Harling and Krzysztof Iniewski

Nanopatterning and Nanoscale Devices for Biological Applications
Krzysztof Iniewski and Seila Selimovic

Nanoplasmonics: Advanced Device Applications
James W. M. Chon and Krzysztof Iniewski

Nanoscale Semiconductor Memories: Technology and Applications
Santosh K. Kurinec and Krzysztof Iniewski

FORTHCOMING TITLES:

Nanoelectronic
D E V I C E
Applications Handbook

Edited by
James E. Morris
Krzysztof Iniewski

CRC Press
Taylor & Francis Group
Boca Raton   London   New York

CRC Press is an imprint of the
Taylor & Francis Group, an **informa** business

CRC Press
Taylor & Francis Group
6000 Broken Sound Parkway NW, Suite 300
Boca Raton, FL 33487-2742

First issued in paperback 2017

© 2013 by Taylor & Francis Group, LLC
CRC Press is an imprint of Taylor & Francis Group, an Informa business

No claim to original U.S. Government works

Version Date: 20130422

ISBN 13: 978-1-138-07259-6 (pbk)
ISBN 13: 978-1-4665-6523-4 (hbk)

Library of Congress Cataloging-in-Publication Data

IEEE International Conference on Nanotechnology (11th : 2011 : Portland, Ore.)
    Nanoelectronic device applications handbook / editors, James E. Morris, Krzysztof Iniewski.
    pages cm. -- (Devices, circuits, and systems)
    Selected papers from the 11th IEEE International Conference on Nanotechnology held in Portland, Oregon, 15-19th August, 2011.
    Summary: "The book has two purposes. First, it assembles the latest research in the field of nanoelectronics device technology in one place. Second, it exposes the reader to myriad applications that nanoelectronics devices technology has enabled. The book is meant for advanced graduate research work or for academicians and researchers. Contributed by world-renowned experts from academia and industry from around the globe, nanoelectronics devices will find widespread application range"-- Provided by publisher.
    Includes bibliographical references and index.
    ISBN 978-1-4665-6523-4 (hardback)
    1. Nanoelectronics--Congresses. 2. Nanostructured materials--Congresses. I. Morris, James E. II. Title.

TK7874.84.I3545 2013
620.1'15--dc23

2013013444

Visit the Taylor & Francis Web site at
http://www.taylorandfrancis.com

and the CRC Press Web site at
http://www.crcpress.com

Contents

SECTION I   Nano-CMOS Modeling

SECTION II   Nano-CMOS Technology

SECTION III   Nano Capacitors

SECTION IV   Terahertz Systems and Devices

SECTION V   Single-Electron Transistors and Electron Tunneling Devices

SECTION VI Quantum Cellular Automata

SECTION VII Memristors, Resistive Switches, and Memory

SECTION VIII   Graphene Preparation and Properties

SECTION IX   Graphene Devices

SECTION X   Carbon Nanotube Applications

SECTION XI Carbon Nanotube Transistor Modeling

SECTION XII Carbon Nanotube Transistor Fabrication

SECTION XIII Random CNT Network Transistors

SECTION XIV   Nano-Redundant Systems

SECTION XV   Nanowire Fabrication

SECTION XVI   Nanowire Applications

SECTION XVII Nanowire Transistors

SECTION XVIII Nanomagnetic Logic

SECTION XIX   Spintronics

SECTION XX   Nanodevice Modeling

Foreword

Nanotechnology, by its name, refers to technology at small scales, literally at nanometer scale dimensions. The fact that nanoscale features occur in one form or another across many different fields has led to "nanotechnology" being regarded as a somewhat broad umbrella encompassing a host of scientific and engineering disciplines. One of the most important areas within nanotechnology is that of *nanoelectronics*, which is concerned with nanometer scale devices, circuits, and architectures impacting the continued scaling of information processing systems, including communication and sensor systems, as well as providing an interface between the electronic and biological worlds. The present attention on nanotechnology and nanoelectronics has been driven from the top down by the continued scaling of semiconductor device dimensions into the nanometer scale regime, as well as the bottom-up growth of self-assembled materials based on semiconductor and carbon nanostructures. As nanotechnology has evolved from being defined by a set of fundamental discoveries and molecular scale capabilities, to becoming increasingly pervasive in all aspects of technology, it has become a ubiquitous presence in almost every conceivable technology of current importance to society.

The IEEE Nanotechnology Council has a mission to promote and support the theory, design, and development of nanotechnology and its scientific, engineering, and industrial applications. The flagship conference of the IEEE Nanotechnology Council is the IEEE International Conference on Nanotechnology (IEEE NANO), and we are especially pleased to support the publication of the current *Nanoelectronic Device Applications Handbook,* whose inspiration arose out of the results and ideas presented at the IEEE Nano 2011 conference held in Portland, Oregon. The particular focus of this volume on nanoelectronic devices is especially timely and relevant, given the evolution of the semiconductor industry into what is conceivably the world's largest nanotechnology industry. This volume is a unique assembly of topics ranging from present-day nanodevices used commercially today, new nanomaterials such as semiconductor nanowires to carbon-based electronics, and to potential future technologies based on entirely new concepts of information processing such as single electron devices, phase change devices, nanomagnetics, spintronics, and quantum computing. The reader should find a wealth of current topics in nanoelectronics within this handbook, which will serve as an invaluable reference on this ubiquitous technology.

Stephen M. Goodnick
IEEE Nanotechnology Council President

Preface

This book has its genesis in the 11th International Conference on Nanotechnology, the IEEE Nanotechnology Council's flagship conference, held in Portland, Oregon, 15–19 August 2011. This is first and foremost an engineering conference, with the research papers presented having more of an applied bias than is typical at most contemporary nanoscience and nanotechnology conferences which generally still focus on the science of this nascent field. The initial proposition was to bring the broadest possible range of this material to the attention of the wider nanotechnology community, but with nearly 400 presentations that was clearly impractical in one book. Nanomaterials was the largest single track, but an analysis and characterization of all the papers presented revealed two other threads running through most of them: either devices or applications or both. So the decision was made to limit the content to nanodevice applications, albeit liberally interpreted in some cases, eliminating all of the nanometrology, nanopackaging, nanophotonics, nanorobotics, quantum computing, nanotechnology education and commercialization, EHS issues, and most of the nanomaterials, nanotechnology in energy, and nano-biomedicine from consideration for inclusion, but providing a more coherent focus on the broad range of nanoelectronics. There is a wealth of information contained in these other areas, and interested readers are referred to the proceedings archive in IEEE Xplore at: http://ieeexplore.ieee.org/xpl/mostRecentIssue.jsp?reload=true&punum ber=6125891.

The response to invitations to authors to contribute was overwhelming, and what was initially envisaged as a standard 20-chapter monograph rapidly escalated into the much more comprehensive 68-chapter work. The book has been divided into sections for ease of navigation, and the Contents reveals both the breadth of topic coverage of the broad nanoelectronics field and depth in the inclusion of a number of chapters in each section, which reveal the diversity of research in each topic area. It must be emphasized that, notwithstanding the history above, the chapters are NOT the conference papers. Most have been expanded by at least 50% by the addition of extra material, typically additional new data or expanded applications context, but some report totally new work in the same area. So if a chapter herein whets the reader's appetite for more information, it would be worth checking out the conference paper in IEEE Xplore before looking further afield.

The first two sections are appropriately enough devoted to nanoscale advances in current MOSFET/CMOS (metal–oxide–semiconductor field-effect transistor/complementary metal–oxide–semiconductor) technology, the first focused on modeling and simulation. Chapter 1 (Rodriguez and Huq) describes an application of the online nanodevice simulation tool at nano-HUB.org, while Chapters 2 (Ashraf et al.) and 3 (Li) adopt Monte Carlo approaches to simulate the effects of channel interface defect trapping on MOSFET threshold voltages. In the second section, Chapters 4 (Khaderbad and Rao) and 5 (Srivastava and Malhotra) consider self-assembled monolayer techniques and the deposition of La_2O_3/HfO_2 gates, respectively, for nano-MOSFET performance enhancements, and Chapter 6 (Beg et al.) shows (by simulation) how varying device sizes in full adders can improve reliability (as determined by the static noise margin).

The next section contains four chapters on nano capacitors for four different applications. Chapter 7 (Sharma et al.) compares three technologies for electronics packaging. Chapter 8 (Arepalli) examines the effects of radiation on CNT composites and graphene for supercapacitors and field emission displays. Chapter 9 (Sayyad et al.) examines nanoparticle dispersions of an organic semiconductor for capacitive-type humidity sensors, and Chapter 10 (Li et al.) compares charge accumulation in SiO_2 and Si_3N_4/SiO_2 dielectric stacks for the control of stiction in capacitive MEMS/NEMS membrane devices.

Chapters 11 and 12 describe the fabrication of THz-capable devices. The nano-antenna arrays and MOM diodes for energy conversion described in Chapter 11 (Bareiβ et al.) can be fabricated at the requisite nanometer scale by nano-imprinting, and an InGaAs/InAlAs heterostructure is used for the novel Ballistic Deflection Transistor logic gates described in Chapter 12 (Wolpert and Ampadu).

Quantum–mechanical electron tunneling is the quintessential nanoscale phenomenon and drives much of the nanoelectronics field, whether intended or unintended. The first two chapters here, Chapters 13 (Ito et al.) and 14 (Karre et al.), describe different approaches to controlling the tunneling gap in single-electron transistor (SET) fabrication. Chapter 15 (Beiu et al.) uses SETs in its analysis of low-power axon-like communication, and Chapter 16 (D'Aloia et al.) describes the development of a sensitive mechanical stress sensor based on internal tunneling within a graphite nano-platelet composite.

The quantum cellular automata (QCA) chapters are natural successors to the SET section since the device state is also set by nanodot charging. However, Chapter 17 (Ottavi et al.) leads off by describing a magnetic variant of the QCA concept and demonstrates an HDL model for it. Chapter 18 (Kim and Swartzlander) presents a restoring divider circuit using more conventional QCAs, and Chapter 19 (Hook and Lee) describes a lattice implementation of QCA circuits at the molecular level for reliable room temperature operation. Chapter 20 (Wang et al.) returns to the problem of automated QCA circuit synthesis and the minimization of 4-variable logic functions.

Memristors offer considerable potential as both resistive switches and memory elements, and the next section is focused on memristors but also includes other examples of both. Chapter 21 (Delgado) presents a generalized treatment of the dynamic behavior of nonlinear nanodevices, but focuses primarily on memristors. Chapters 22 (Chen) and 23 (Linn et al.) are both concerned with the operation of crossbar memories, with Chapter 22 focused on the use of nonlinear resistive switches for sneak path reduction and Chapter 23 proposing and modeling antiserially connected memristors for the same purpose. Chapter 24 (Junsangsri and Lombardi) proposes and analyzes a memristor-based memory cell with ambipolar transistor controls. Chapter 25 (Cantley et al.) uses a memristor as the synapse in a neural learning circuit with nano-crystalline and nanoparticle thin-film transistors (TFTs). Finally, Chapter 26 (Maghsoudi and Martin) introduces a new memory concept, based on thermally actuated buckling of a nanoscale beam.

Graphene is the "material of the moment," with enormous potential for nanoelectronics applications, and not only for devices. Consequently, the next section contains chapters devoted to the fabrication and properties of graphene as necessary precursors to device development. A stress-free process for the transfer of graphene from its growth substrate to its intended application is reported in Chapter 27 (Chen et al.) along with sheet resistance control by plasma treatments. Chapter 28 (Li et al.) describes an alternative dielectrophoretic transfer technique for precise positioning on a substrate and a pH sensor application. Two more transfer techniques are described in Chapter 29 (Rahman and Norton).

Chapter 30 (Nayfeh et al.) reports on the fabrication and performance of a GHz RF graphene MOSFET, while Chapter 31 (Yu et al.) demonstrates that breakdown currents in such devices can be significantly increased by replacing the usual SiO_2 substrate layer with a more thermally conductive synthetic diamond. For active nanoelectronic devices, graphene must be converted from a semi-metal to a semiconductor, and Chapter 32 (Plachinda et al.) reports that band gaps of up to 0.81 eV are achievable with metal-bis-arene chemical treatments.

Carbon nanotubes (CNTs) are definitely moving beyond research into development. The first CNT section is devoted to general applications, before moving on to CNT transistors. There is a great deal of interest in vertical CNT device interconnect for 3D system integration and Chapter 33 (Vollebregt et al.) examines top-down and bottom-up CNT fabrication at relatively low temperatures. Next, Chapters 34 (Chen et al.) and 35 (Aria et al.) present an output interface design for CNT infrared sensors and the exploitation of the CNT surface-to-volume ratio in electrolytic capacitor electrodes. Finally, Chapter 36 (Abdellah et al.) describes a spray deposition technique for the fabrication of CNT gas (NH_3) sensors, and Chapter 37 (Aasmundtveit et al.) describes control of the synthesis temperature by Joule heating and of the growth direction by electric fields for CMOS/MEMS applications.

There are three sections devoted to CNT transistors, with the first two covering modeling and fabrication, in that order. Chapter 38 (Torres and Huq) extends a theoretical treatment of the bandgap variation with CNT diameter and temperature to its effects on transistor characteristics, and Chapter 39 (Chen et al.) uses time-dependent quantum mechanics to analyze CNT response to THz signals. Chapters 40 (Kato et al.) and 41 (Numata et al.) both address CNT TFT fabrication issues, with Chapter 40 reporting the stabilization of n-type CNT TFTs by Cs plasma irradiation and encapsulation, and Chapter 41 demonstrating printed CNT TFTs on flexible substrates.

The next two sections deal with the problems presented by a high rate of defective devices, first in CNT transistor devices (e.g., due to the mix of metallic and semiconducting CNTs) and then generalized to device-independent architectures. Chapter 42 (Gong et al.) first presents the characteristics of a CNT network TFT, and then goes on to percolation modeling of random effects. Chapter 43 (Ashraf et al.) similarly considers the effects of CNT diameter and misalignment and unwanted metallic CNTs on device yield, and Chapter 44 (Zhang and Delgado-Frias) analyzes the performance of an 8-transistor CNT SRAM array with faulty cell tolerance, for example, also with metallic CNTs. Chapters 45 (Zawodniok and Kundaikar) and 46 (Aymerich et al.) describe, respectively, a novel built-in self-test for nanofabric architectures and an adaptive fault-tolerant redundant architecture to accommodate defective devices by weighted voting.

Nanowires are also covered in three sections: fabrication, applications, and transistors. Epitaxial GaAs nanowire growth on silicon substrates, as demonstrated in Chapter 47 (Kang et al.), permits the integration of III–V optoelectronics with Si-microelectronics. Fabrication of both p- and n-type single-crystal tin oxide nanowires is described in Chapter 48 (Tran and Rananavare) for a ppb chlorine sensor and other applications, and similarly Chapter 49 (Ng et al.) describes the fabrication of single-crystal copper silicide nanowires for Li-ion battery anode terminals. Finally, Graf et al. (Chapter 50) demonstrate the effects of pulsed electrodeposition on the fabrication of metallic nanowires for interconnect.

In the applications area, Chapter 51 (Gupta et al.) characterizes ZnO nanowires for biosensing, Park et al. (Chapter 52) have fabricated PN heterojunctions from both p- and n-type ZnO nanorods on porous silicon, and Dar et al. (Chapter 53) demonstrate the use of GaN nanowires coated with silver nanoparticles as surface enhancement Raman spectroscopy biosensors.

Next, Bayraktaroglu and Leedy (Chapter 54) examine the properties of transparent columnar ZnO TFTs, and Kyogoku et al. (Chapter 55) examine the effects of Si-nanowire shapes and roughness on the energy band gap, which will impact transistor applications. Chapters 56 (Hossain et al.) and 57 (Martinez et al.) continue with simulation studies of the impacts of a channel trap and a channel donor, respectively, on the ON currents in nanowire transistors, while Chapters 58 (Huang et al.) and 59 (Jeong et al.) both present the characteristics of "gate-all-around" devices, the former addressing minimization of performance variations and the latter, the extraction of parasitic parameters.

Nanomagnetic logic (NML) is receiving a great deal of attention for its potential in nonvolatile memories. Chapter 60 (Das et al.) proposes an integration of NML with magnetoresistive RAM to create a new logic-in-memory architecture, Chapter 61 (Varga et al.) implements a full-adder circuit in NML, Chapter 62 (Breitkreutz et al.) models a simple NML 1-bit adder with irradiation reduced switching levels, and in Chapter 63, Pulecio et al. present an alternative concept of magnetic field-based computing with an image processing example.

In the last section, the remaining chapters (except for one) deal with spintronics. Chapter 64 (Rakheja and Naeemi) confronts the interconnection challenges facing all-spin logic and considers the possibilities for graphene interconnect. In Chapter 65, Wan et al. consider the impact of scattering centers on conductance anomalies in spin contacts, and Wijesundara and Stinaff look for electron-spin control mechanisms in quantum-dot molecules, which are candidates for quantum computing systems, in Chapter 66. In Chapter 67, Munira et al. close out the spintronics section with an examination of memory materials.

In closing with Chapter 68, Blom and Stokbro call for a new generation of simulation tools to handle nanoscale mechanisms in realistic nanodevice geometries.

We hope you enjoy the book, and that it serves as a continuing reference and source of ideas for years to come.

James E. Morris
Portland, Oregon

Krzysztof Iniewski
Vancouver, British Columbia, Canada

Editors

James (Jim) E. Morris is an ECE professor at Portland State University, Oregon, and professor emeritus at SUNY-Binghamton, having served as department chair at both. He has BSc and MSc degrees in physics from the University of Auckland, NZ, a PhD in electrical engineering from the University of Saskatchewan, Canada, and is an IEEE Fellow. Jim has served the IEEE Components Packaging and Manufacturing Technology (CPMT) Society as a treasurer (1991–1997), BoG member (1996–1998, 2011–2013), VP for conferences (1998–2003), distinguished lecturer (2000–), CPMT-Transactions associate editor (1998–), IEEE Nanotechnology Council representative (2007–) and co-chair of the CPMT Technical Committee (TC) on nanotechnology, and was recognized with the 2005 CPMT David Feldman Outstanding Contribution Award. On the Nanotechnology Council (NTC), he established the Nanopackaging TC, serves as the NTC Awards chair (2010–), and VP for conferences (2013–2014), and contributes to the *IEEE Nanotechnology Magazine*. He has been General Conference chair of four IEEE conferences, including IEEE NANO 2011, and program chair or treasurer for others. His current research spans electronics packaging, nanotechnology, and nanopackaging. He is actively involved in nanotechnology education and has chaired both the Education Society and CPMT Chapters at the local IEEE Oregon level. He has edited four books on electronics packaging, including one on nanopackaging, coauthored another, and is writing a sixth. This is his second in the nanotechnology field, both coedited with Kris Iniewski.

Krzysztof (Kris) Iniewski is managing R&D at Redlen Technologies Inc, a start-up company in Vancouver, Canada. Redlen's revolutionary production process for advanced semiconductor materials enables a new generation of more accurate, all-digital, radiation-based imaging solutions. Kris is also a president of CMOS Emerging Technologies (www.cmoset.com), an organization of high-tech events covering communications, microsystems, optoelectronics, and sensors.

In his career, Dr. Iniewski has held numerous faculty and management positions at the University of Toronto, University of Alberta, Simon Fraser University, and PMC-Sierra Inc. He has published over 100 research papers in international journals and conferences. He holds 18 international patents granted in the United States, Canada, France, Germany, and Japan. He is a frequently invited speaker and has consulted for multiple organizations internationally. He has written and edited several books for IEEE Press, Wiley, CRC Press, McGraw-Hill, Artech House, and Springer. His personal goal is to contribute to healthy living and sustainability through innovative engineering solutions. In his leisure time, Kris can be found hiking, sailing, skiing, or biking in beautiful British Columbia. He can be reached at kris.iniewski@gmail.com.

Contributors

Knut E. Aasmundtveit
Department of Micro and Nano Systems
    Technology
HiVe—Vestfold University College
Tønsberg, Norway

Alaa Abdellah
Institute for Nanoelectronics
Technische Universität München
Munich, Germany

Manoranjan Acharya
Intel Corp.
Hillsboro, Oregon

Zubair Ahmad
Low Dimensional Materials Research
    Center (LDMRC)
University of Malaya
Kuala Lumpur, Malaysia

Shunsuke Akimoto
Department of Electrical and Electronic
    Engineering
Tokyo University of Agriculture and Technology
Tokyo, Japan

Syed M. Alam
Everspin Technology
Chandler, Arizona

Edgar Albert
Institute for Nanoelectronics
Technische Universität München
Munich, Germany

Manuel Aldegunde
Electronics System Design Centre
Swansea University
Wales, United Kingdom

Paul Ampadu
Department of Electrical and Computer
    Engineering
University of Rochester
Rochester, New York

Sivaram Arepalli
Department of Energy Science
Sungkyunkwan University
Suwon, Korea

Adrianus I. Aria
Charyk Laboratory of Bioinspired Design
Graduate Aeronautical Laboratories
California Institute of Technology
Pasadena, California

Nabil Ashraf
School of Electrical, Computer and Energy
    Engineering
Arizona State University
Tempe, Arizona

Rehman Ashraf
Department of Electrical and Computer
    Engineering
Portland State University
Portland, Oregon

Abdullah Mohamed Asiri
Chemistry Department and the Center
    of Excellence for Advanced Materials
    Research
King Abdulaziz University
Jeddah, Saudi Arabia

Nivard Aymerich
Department of Electronic Engineering
Universitat Politècnica de Catalunya
Catalonia, Spain

Rock-Hyun Baek
POSTECH-Pohang
University of Science and Technology
Pohang, Republic of Korea

Alexander A. Balandin
Department of Electrical Engineering and
    Materials Science and Engineering
    Program
University of California
Riverside, California

Mario Bareiß
Institute for Nanoelectronics
Technische Universität München
Munich, Germany

Burhan Bayraktaroglu
Air Force Research Laboratory
Wright Patterson AFB, Ohio

Markus Becherer
Lehrstuhl für Technische Elektronik
Technische Universität München
Munich, Germany

C. I. M. Beenakker
Delft Institute of Microsystems and
    Nanoelectronics
Delft University of Technology
CT Delft, the Netherlands

Azam Beg
College of Information Technology
United Arab Emirates University
Al Ain, United Arab Emirates

Valeriu Beiu
College of Information Technology
United Arab Emirates University
Al Ain, United Arab Emirates

Paul L. Bergstrom
Department of Electrical and Computer
    Engineering
Michigan Technological University
Houghton, Michigan

G. H. Bernstein
Department of Electrical Engineering
    and Center for NanoScience and
    Technology
University of Notre Dame
Notre Dame, Indiana

N. Bhandari
School of Electronics and Computing
    Systems
University of Cincinnati
Cincinnati, Ohio

Sanjukta Bhanja
Department of Electrical Engineering
University of South Florida
Tampa, Florida

Anders Blom
QuantumWise A/S
Copenhagen, Denmark

U. Böttger
Peter Grünberg Institut
and
JARA Jülich-Aachen Research Alliance–
    Fundamentals for Future Information
    Technology
Jülich, Germany

Christina Brantley
U.S. Army
Redstone Arsenal, Alabama

Stephan Breitkreutz
Lehrstuhl für Technische Elektronik
Technische Universität München
Munich, Germany

R. Bruchhaus
Peter Grünberg Institut
and
JARA Jülich-Aachen Research Alliance–
    Fundamentals for Future Information
    Technology
Jülich, Germany

M. Cahay
School of Electronics and Computing
    Systems
University of Cincinnati
Cincinnati, Ohio

Kurtis D. Cantley
Department of Materials Science and
Engineering
University of Texas at Dallas
Richardson, Texas

Parthasarathi Chakraborti
Packaging Research Center
Georgia Institute of Technology
Atlanta, Georgia

Mary B. Chan-Park
School of Chemical and Biomedical
Engineering
Nanyang Technological University
Singapore

J. Charles
School of Electronics and Computing Systems
University of Cincinnati
Cincinnati, Ohio

Jamil Anwar Chaudry
Institute of Chemistry
University of the Punjab
Lahore, Pakistan

Daw Don Cheam
Institute of Microelectronics
Singapore

An Chen
Exploratory Research
Sunnyvale, California

C. C. Chen
Department of Materials Science and
Engineering
National Chiao Tung University
Hsinchu, Taiwan

Hongzhi Chen
Department of Electrical and Computer
Engineering
Michigan State University
East Lansing, Michigan

In-Gann Chen
Center for Micro/Nano Science and
Technology
National Cheng Kung University
Tainan, Taiwan

Liangliang Chen
Department of Electrical and Computer
Engineering
Michigan State University
East Lansing, Michigan

Lu-An Chen
Department of Materials Science and
Engineering
National Chiao Tung University
Hsinchu, Taiwan

Waileong Chen
Institute of Microelectronics
National Cheng Kung University
Tainan, Taiwan

Xuyuan Chen
Department of Micro and Nano Systems
Technology
HiVe—Vestfold University College
Tønsberg, Norway

Zuojing Chen
Department of Electrical and Computer
Engineering
University of Massachusetts
Amherst, Massachusetts

Malgorzata Chrzanowska-Jeske
Department of Electrical and Computer
Engineering
Portland State University
Portland, Oregon

Sorin Cotofana
Computer Engineering Laboratory
Delft University of Technology
CD Delft, the Netherlands

György Csaba
Department of Electrical Engineering
University of Notre Dame
Notre Dame, Indiana

Alessandro Giuseppe D'Aloia
Department of Astronautics, Electrical and
Energetical Engineering (DIAEE)
and
Research Center for Nanotechnology Applied
to Engineering (CNIS)
Sapienza University of Rome
Rome, Italy

Nitzan Dar
Department of Material Science and
    Engineering
National Cheng Kung University
Tainan, Taiwan

Jayita Das
Department of Electrical Engineering
University of South Florida
Tampa, Florida

P. P. Das
School of Electronics and Computing
    Systems
University of Cincinnati
Cincinnati, Ohio

Giovanni De Bellis
Department of Astronautics, Electrical
    and Energetical Engineering (DIAEE)
and
Research Center for Nanotechnology Applied
    to Engineering (CNIS)
Sapienza University of Rome
Rome, Italy

Alberto Delgado
Department of Electronic and Electrical
    Engineering
National University of Colombia at
    Bogota
Bogota, Colombia

José G. Delgado-Frias
School of Electrical Engineering and
    Computer Science
Washington State University
Pullman, Washington

Jaber Derakhshandeh
Delft Institute of Microsystems and
    Nanoelectronics
Delft University of Technology
CT Delft, the Netherlands

Eugene Edwards
U.S. Army
Redstone Arsenal, Alabama

Irina Eichwald
Lehrstuhl für Technische Elektronik
Technische Universität München
Munich, Germany

David R. Evans
Department of Physics
Portland State University
Portland, Oregon

Alexander Eychmüller
Technische Universität Dresden
Physical Chemistry and Electrochemistry
Dresden, Germany

Bernhard Fabel
Institute for Nanoelectronics
Technische Universität München
Munich, Germany

Melodie A. Fickenscher
Department of Physics
University of Cincinnati
Cincinnati, Ohio

Qiang Gao
Department of Electronic Materials Engineering
The Australian National University
Canberra, Australia

Morteza Gharib
California Institute of Technology
Pasadena, California

Avik W. Ghosh
Charles L. Brown Department of Electrical
    and Computer Engineering
University of Virginia
Charlottesville, Virginia

Qingqing Gong
Institute for Nanoelectronics
Technische Universität München
Munich, Germany

Stephen M. Goodnick
School of Electrical, Computer and Energy
    Engineering
Arizona State University
Tempe, Arizona

Matthias Graf
Department of Electronics Packaging
    Laboratory
Technische Universität Dresden
Dresden, Germany

Mélanie Guittet
Graduate Aeronautical Laboratories
California Institute of Technology
Pasadena, California

Yanan Guo
Materials Engineering and Centre for
    Microscopy and Microanalysis
University of Queensland
Queensland, Australia

Anurag Gupta
Department of Electrical and Computer
    Engineering
University of Alabama
Tuscaloosa, Alabama

Einar Halvorsen
Department of Micro and Nano Systems
    Technology
HiVe—Vestfold University College
Tønsberg, Norway

Ulrik Hanke
Department of Micro and Nano Systems
    Technology
HiVe—Vestfold University College
Tønsberg, Norway

Rikizo Hatakeyama
Department of Electronic Engineering
Tohoku University
Sendai, Japan

Andreas Hochmeister
Institute for Nanoelectronics
Technische Universität München
Munich, Germany

Nils Hoivik
Department of Micro and Nano Systems
    Technology
HiVe—Vestfold University College
Tønsberg, Norway

Loyd R. Hook IV
Department of Electrical and Computer
    Engineering
University of Oklahoma
Norman, Oklahoma

Arif Hossain
School of Electrical, Computer and Energy
    Engineering
Arizona State University
Tempe, Arizona

Po-Chun Huang
Department of Materials Science and
    Engineering
National Chiao Tung University
Hsinchu, Taiwan

Hasina F. Huq
Department of Electrical Engineering
The University of Texas-Pan American
Edinburg, Texas

Walid Ibrahim
College of Information Technology
United Arab Emirates University
Al Ain, United Arab Emirates

Kazuki Ihara
Technology Research Association for Single
    Wall Carbon Nanotubes (TASC)
and
NEC Corporation
Tsukuba, Japan

Ryoichi Ishihara
Delft Institute of Microsystems and
    Nanoelectronics
Delft University of Technology
Delft, the Netherlands

Mitsuki Ito
Department of Electrical and Electronic
    Engineering
Tokyo University of Agriculture and
    Technology
Tokyo, Japan

Jun-Ichi Iwata
Department of Applied Physics
The University of Tokyo
Tokyo, Japan

Howard E. Jackson
Department of Physics
University of Cincinnati
Cincinnati, Ohio

Chennupati Jagadish
Department of Electronic Materials
    Engineering
The Australian National University
Canberra, Australia

Gunther Jegert
Institute for Nanoelectronics
Technische Universität München
Munich, Germany

Yoon-Ha Jeong
The Department of Electrical Engineering,
    POSTECH
University of Science and Technology
Pohang, Republic of Korea

Hannah J. Joyce
Department of Physics
University of Oxford
Oxford, United Kingdom

Xueming Ju
Lehrstuhl für Nanoelektronik
Technische Universität München
Munich, Germany

Pilin Junsangsri
Department of Electrical and Computer
    Engineering
Northeastern University
Boston, Massachusetts

Karol Kalna
College of Engineering
Swansea University
Wales, United Kingdom

Jung-Hyun Kang
Department of Electronic Materials
    Engineering
The Australian National University
Canberra, Australia

P. Santosh Kumar Karre
Intel Corporation
Hillsboro, Oregon

Toshiaki Kato
Department of Electronic Engineering
Tohoku University
Sendai, Japan

Mrunal A. Khaderbad
Department of Electrical Engineering
IIT Bombay
Mumbai, India

Josef Kiermaier
Lehrstuhl für Technische Elektronik
Technische Universität München
Munich, Germany

Bruce C. Kim
Department of Electrical and Computer
    Engineering
University of Alabama-Tuscaloosa
Tuscaloosa, Alabama

Dae Mann Kim
National Center for Nanomaterials Technology
Pohang, Republic of Korea

Dong-Won Kim
Samsung Electronics
Republic of Korea

Ki Kang Kim
Department of Electrical Engineering and
    Computer Science
Massachusetts Institute of Technology
Cambridge, Massachusetts

Seong-Wan Kim
Quantum Intellectual Property Services
Austin, Texas

Ye-Ram Kim
The Department of Electrical Engineering,
    POSTECH
University of Science and Technology
Pohang, Republic of Korea

Yong Kim
Department of Physics
Dong-A University
Busan, Korea

Hagen Klauk
Max Planck Institute for Solid State Research
Stuttgart, Germany

Gregor Koblmüller
Walter Schottky Institute
Technische Universität München
Garching, Germany

Jing Kong
Department of Electrical Engineering and
    Computer Science
Massachusetts Institute of Technology
Cambridge, Massachusetts

Sambhav Kundaikar
Department of Electrical and Computer
    Engineering
Missouri University of Science and Technology
Rolla, Missouri

Shinya Kyogoku
Department of Applied Physics
The University of Tokyo
Tokyo, Japan

King Wai Chiu Lai
Department of Mechanical and Biomedical
    Engineering
City University of Hong Kong
Kowloon, Hong Kong

Donghwan Lee
Department of Chemistry and
    Research Institute for Natural
    Science
Hanyang University
Seoul, Korea

Jeong-Soo Lee
The Department of Electrical Engineering,
    POSTECH
University of Science and Technology
Pohang, Republic of Korea

Jongtaek Lee
Department of Chemistry and Research
    Institute for Natural Science
Hanyang University
Seoul, Korea

Kuo-Hao Lee
Department of Material Science and
    Engineering
National Cheng Kung University
Tainan, Taiwan

Jungwoo Lee
LG Chem
Daejeon, Korea

Samuel C. Lee
Department of Electrical and Computer
    Engineering
University of Oklahoma
Norman, Oklahoma

Sang-Hyun Lee
The Department of Electrical Engineering,
    POSTECH
University of Science and Technology
Pohang, Republic of Korea

Kevin Leedy
Air Force Research Laboratory
Wright Patterson AFB, Ohio

Nan Lei
School of Engineering and Computer
    Science
Washington State University
Vancouver, Washington

Dawen Li
Department of Electrical and Computer
    Engineering
University of Alabama-Tuscaloosa
Tuscaloosa, Alabama

Gang Li
Department of Micro and Nano Systems
    Technology
HiVe—Vestfold University College
Tønsberg, Norway

Pengfei Li
School of Engineering and Computer
    Science
Washington State University
Vancouver, Washington

Yiming Li
Department of Electrical and Computer
    Engineering
National Chiao Tung University
Hsinchu, Taiwan

Keng-Chih Liang
Department of Electro-Optical Engineering
National Cheng Kung University
Tainan, Taiwan

Carmen Maria Lilley
Department of Mechanical and Industrial
    Engineering
University of Illinois at Chicago
Chicago, Illinois

E. Linn
Institute of Materials in Electrical
    Engineering and Information
    Technology II (IWE II)
RWTH Aachen University
and
Jülich-Aachen Research Alliance (JARA)
    – Fundamentals for Future Information
    Technology
Aachen, Germany

Chih-Yi Liu
Department of Electro-Optical
    Engineering
National Cheng Kung University
Tainan, Taiwan

Chuan-Pu Liu
Department of Materials Science and
    Engineering
National Cheng Kung University
Tainan, Taiwan

Guanxiong Liu
Department of Electrical Engineering and
    Materials Science and Engineering Program
University of California
Riverside, California

Fabrizio Lombardi
Department of Electrical and Computer
    Engineering
Northeastern University
Boston, Massachusetts

Paolo Lugli
Institute for Nanoelectronics
Technische Universität München
Munich, Germany

Elham Maghsoudi
Department of Mechanical Engineering
Louisiana State University
Baton Rouge, Louisiana

Y. Malhotra
Department of Physics and Astrophysics
University of Delhi
Delhi, India

Michael James Martin
Department of Mechanical Engineering
Louisiana State University
Baton Rouge, Louisiana

Antonio Martinez
College of Engineering
Swansea University
Wales, United Kingdom

S. Menzel
Institute of Materials in Electrical Engineering
    and Information Technology II (IWE II)
RWTH Aachen University
and
JARA Jülich-Aachen Research Alliance–
    Fundamentals for Future Information
    Technology
Aachen, Germany

Munawar Ali Munawar
Institute of Chemistry
University of the Punjab
Lahore, Pakistan

Kamaram Munira
Charles L. Brown Department of Electrical
    and Computer Engineering
University of Virginia
Charlottesville, Virginia

Azad Naeemi
School of Electrical and Computer
    Engineering
Georgia Institute of Technology
Atlanta, Georgia

Siva G. Narendra
Tyfone, Inc.
Portland, Oregon

Osama M. Nayfeh
U.S. Army Research Laboratory
and
University of Illinois
Urbana-Champaign, Illinois

R. S. Newrock
Department of Physics
University of Cincinnati
Cincinnati, Ohio

Poh Keong Ng
Department of Electrical and Computer
  Engineering
University of Illinois at Chicago
Chicago, Illinois

Mohammed Niamat
Department of Electrical Engineering and
  Computer Science
University of Toledo
Toledo, Ohio

Yung-Tang Nien
Center for Micro/Nano Science and
  Technology
National Cheng Kung University
Tainan, Taiwan

Fumiyuki Nihey
Technology Research Association for
  Single Wall Carbon Nanotubes
  (TASC)
and
NEC Corporation
Tsukuba, Japan

Michael L. Norton
Department of Chemistry
Marshall University
Huntington, West Virginia

Hideaki Numata
Technology Research Association for Single
  Wall Carbon Nanotubes (TASC)
and
NEC Corporation
Tsukuba, Japan

Yosuke Osanai
Department of Electronic Engineering
Tohoku University
Sendai, Japan

Atsushi Oshiyama
Department of Applied Physics
The University of Tokyo
Tokyo, Japan

Marco Ottavi
Department of Electronic Engineering
University of Rome "Tor Vergata"
Rome, Italy

Eunkyung Park
LG Hausys
Gyeonggi-do, Korea

Taehee Park
Department of Chemistry and Research
  Institute for Natural Science
Hanyang University
Seoul, Korea

Paul Plachinda
Department of Physics
Portland State University
Portland, Oregon

Eric Polizzi
Department of Electrical and Computer
  Engineering
University of Massachusetts
Amherst, Massachusetts

Salvatore Pontarelli
Department of Electronic Engineering
University of Rome "Tor Vergata"
Rome, Italy

Wolfgang Porod
Department of Electrical Engineering and
  Center for NanoScience and Technology
University of Notre Dame
Notre Dame, Indiana

Javier Pulecio
Brookhaven National Laboratory
Upton, New York

Masudur Rahman
Department of Chemistry
Marshall University
Huntington, West Virginia

P. Markondeya Raj
Packaging Research Center
Georgia Institute of Technology
Atlanta, Georgia

Shaloo Rakheja
School of Electrical and Computer Engineering
Georgia Institute of Technology
Atlanta, Georgia

Katerina Raleva
University "Ss Cyril and Methodius" Skopje
Skopje, Macedonia

Shankar B. Rananavare
Department of Chemistry
Portland State University
Portland, Oregon

V. Ramgopal Rao
Department of Electrical Engineering
IIT Bombay
Mumbai, India

Alejandro Rodriguez
The University of Texas-Pan American
Edinburg, Texas

R. Rosezin
Peter Grünberg Institut
ForschungszentrumJülich GmbH
and
JARA Jülich-Aachen Research Alliance–
Fundamentals for Future Information
Technology
Jülich, Germany

Antonio Rubio
Department of Electronic Engineering
UniversitatPolitècnica de Catalunya
Catalonia, Spain

Paul Ruffin
U.S. Army
RDECOM/AMRDEC
Redstone Arsenal, Alabama

Takeshi Saito
National Institute of Advanced Industrial
Science and Technology (AIST)
and
Technology Research Association for Single
Wall Carbon Nanotubes (TASC)
Tsukuba, Japan

Adelio Salsano
Department of Electronic Engineering
University of Rome "Tor Vergata"
Rome, Italy

Fabrizio Sarasini
Department of Chemical Engineering
Materials Environment (DICMA)
Sapienza University of Rome
Rome, Italy

Sudeep Sarkar
Department of Computer Science and
Engineering
University of South Florida
Tampa, Florida

Maria Sabrina Sarto
Department of Astronautics, Electrical and
Energetical Engineering (DIAEE)
and
Research Center for Nanotechnology Applied
to Engineering (CNIS)
Sapienza University of Rome
Rome, Italy

Muhammad Hassan Sayyad
GIK Institute of Engineering Sciences and
Technology
Topi, Pakistan

Giuseppe Scarpa
Institute for Nanoelectronics
Technische Universität München
Munich, Germany

Hugo Schellevis
Delft Institute of Microsystems and
Nanoelectronics
Delft University of Technology,
Delft, the Netherlands

Doris Schmitt-Landsiedel
Lehrstuhl für Technische Elektronik
Technische Universität München
Munich, Germany

Reza Shahbazian-Yassar
Department of Mechanical Engineering
Michigan Technological University
Houghton, Michigan

Muhammad Shahid
Institute of Chemistry
University of the Punjab
Lahore, Pakistan

Himani Sharma
Packaging Research Center
Georgia Institute of Technology
Atlanta, Georgia

Jeng-Tzong Sheu
Institute of Nanotechnology/
    Department of Materials Science
    and Engineering
National Chiao Tung University
Hsinchu, Taiwan

Jun-Ichi Shirakashi
Department of Electrical and Electronic
    Engineering
Tokyo University of Agriculture and
    Technology
Tokyo, Japan

Leigh M. Smith
Department of Physics
University of Cincinnati
Cincinnati, Ohio

William A. Soffa
Department of Materials Science and
    Engineering
University of Virginia
Charlottesville, Virginia

Raj Solanki
Department of Physics
Portland State University
Portland, Oregon

Purushothaman Srinivasan
Texas Instruments
Dallas, Texas

A. Srivastava
Electronics and Communication Engineering
Indian Institute for Information Technology-
    Design and Manufacturing
Jabalpur, India

Eric A. Stinaff
Department of Physics and Astronomy
Ohio University
Athens, Ohio

Kurt Stokbro
QuantumWise A/S
Copenhagen, Denmark

Anand Subramaniam
Department of Electrical Engineering
University of Texas at Dallas
Richardson, Texas

Ryutaro Suda
Department of Electrical and Electronic
    Engineering
Tokyo University of Agriculture and
    Technology
Tokyo, Japan

Khaulah Sulaiman
Low Dimensional Materials Research
    Center (LDMRC)
University of Malaya
Kuala Lumpur, Malaysia

Mawahib Hussein Sulieman
United Arab Emirates University
Abu Dhabi, United Arab Emirates

Anirudha V. Sumant
Center for Nanoscale Materials
Argonne National Laboratory
DuPage County, Illinois

Earl E. Swartzlander Jr.
Department of Electrical and Computer
    Engineering
University of Texas at Austin
Austin, Texas

Bao Quoc Ta
Department of Micro and Nano Systems
    Technology
HiVe—Vestfold University College
Tønsberg, Norway

Mihai Tache
United Arab Emirates University
Abu Dhabi, United Arab Emirates

Alessio Tamburrano
Department of Astronautics, Electrical and
    Energetical Engineering (DIAEE)
and
Research Center for Nanotechnology Applied
    to Engineering (CNIS)
Sapienza University of Rome
Rome, Italy

Hark Hoe Tan
Department of Electronic Materials
    Engineering
The Australian National University
Canberra, Australia

Jesus Torres
The University of Texas—Pan American
Edinburg, Texas

Hoang A. Tran
Department of Chemistry
Portland State University
Portland, Oregon

Jacopo Tirillò
Department of Chemical Engineering
    Materials Environment (DICMA)
Sapienza University of Rome
Rome, Italy

Chia-Hao Tu
Department of Materials Science and
    Engineering
National Cheng Kung University
Tainan, Taiwan

Rao Tummala
Packaging Research Center
Georgia Institute of Technology
Atlanta, Georgia

Yonhua Tzeng
Institute of Microelectronics and Advanced
    Optoelectronics Technology Center
National Cheng Kung University
Tainan, Taiwan

Johan van der Cingel
Delft Institute of Microsystems and
    Nanoelectronics
Delft University of Technology
Delft, the Netherlands

Edit Varga
Department of Electrical Engineering and
    Center for Nano Science and Technology
University of Notre Dame
Notre Dame, Indiana

Dragica Vasileska
School of Electrical, Computer and Energy
    Engineering
Arizona State University
Tempe, Arizona

Srinivasa Vemuru
Department of Electrical and Computer
    Engineering and Computer Science
Ohio Northern University
Ada, Ohio

Eric M. Vogel
Department of Electrical Engineering
University of Texas at Dallas
Richardson, Texas

Sten Vollebregt
Delft Institute of Microsystems and
    Nanoelectronics
Delft University of Technology
Delft, the Netherlands

Fazal Wahab
GIK Institute of Engineering Sciences and
    Technology
Topi, Pakistan

J. Wan
School of Electronics and Computing Systems
University of Cincinnati
Cincinnati, Ohio

Peng Wang
Department of Electrical Engineering and
Computer Science
University of Toledo
Toledo, Ohio

Wen-Jing Wang
Department of Material Science and Engineering
National Cheng Kung University
Tainan, Taiwan

Yushu Wang
Packaging Research Center
Georgia Institute of Technology
Atlanta, Georgia

R. Waser
Institute of Materials in Electrical Engineering
and Information Technology II (IWE II)
RWTH Aachen University
Aachen, Germany
and
Peter Grünberg Institut
Forschungszentrum Jülich GmbH,
and
The JARA Jülich-Aachen Research
Alliance—Fundamentals for Future
Information Technology
Jülich, Germany

Kushal C. Wijesundara
Department of Physics and Astronomy
Ohio University
Athens, Ohio

Gilson Wirth
Department of Electrical Engineering
Universidade Federal do Rio Grande do Sul
(UFRGS)
Porto Alegre, Brazil

David Wolpert
IBM
Poughkeepsie, New York

Klaus-Jürgen Wolter
Electronics Packaging Laboratory
Technische Universität Dresden
Dresden, Germany

Ning Xi
Department of Electrical and Computer
Engineering
Michigan State University
East Lansing, Michigan

Hongyi Xu
Department of Materials Engineering
University of Queensland
Queensland, Australia

Jie Xu
Department of Mechanical Engineering
Washington State University
Vancouver, Washington

Wei Xue
Department of Mechanical Engineering
Washington State University
Vancouver, Washington

Jan M. Yarrison-Rice
Department of Physics
Miami University
Oxford, Ohio

Whikun Yi
Department of Chemistry and Research
Institute for Natural Science
Hanyang University
Seoul, Korea

Sigfrid Yngvesson
Department of Electrical and Computer
Engineering
University of Massachusetts—Amherst
Amherst, Massachusetts

Jie Yu
Department of Electrical Engineering
and Materials Science and
Engineering Program
University of California
Riverside, California

Maciej Zawodniok
Department of Electrical and Computer
Engineering
Missouri University of Science and
Technology
Rolla, Missouri

Liren Zhang
College of Information Technology
United Arab Emirates University
Al Ain, United Arab Emirates

Zhe Zhang
School of Electrical Engineering and
    Computer Science
Washington State University
Pullman, Washington

Jin Zou
Department of Materials Engineering
University of Queensland
Queensland, Australia

Ute Zschieschang
Max Planck Institute for Solid State
    Research
Stuttgart, Germany

Section I

Nano-CMOS Modeling

1 Validation of Nano-CMOS Predictive Technology Model Tool on NanoHUB.org

Alejandro Rodriguez and Hasina F. Huq

CONTENTS

1.1 INTRODUCTION

Next-generation electronic systems will need to adopt novel nano-electronic solutions to keep pace with Moore's law. Nano-CMOS (complementary metal–oxide–semiconductor) technology is among the most promising of nanotechnologies. Although most new nano-electronic technologies are still in their infancy, however, they present the potential for unprecedented levels of integration, low-power computing, signal processing, information storage, and possibly higher operating frequency with cost-effective system solution [1,2]. Advanced MOSFET (metal–oxide–semiconductor field-effect transistor) structures show a promise for scaling CMOS technology to gate lengths below 10 nm to enable continued improvements in integrated circuit (IC) cost and performance (e.g., higher operating speed) for at least 10 more years. Over the past three decades, by reducing the transistor gate lengths with each new generation of manufacturing technology, steady improvements in circuit performance (speed) and cost per function have been reported [2]. However, continued transistor scaling will not be as straightforward in the future as it has been in the past because the fundamental materials and process limits are rapidly being approached.

The challenge of extending Moore's law beyond the physical and economic barriers of the present semiconductor technologies calls for novel nano-electronic solutions. Silicon-based CMOS technology can be scaled down into the nanometer regime. High-performance, planar, ultrathin-body devices fabricated on silicon-on-insulator substrates have been demonstrated down to 15-nm gate lengths [1,2]. The effects of crystal orientation and roughness on carrier mobility, gate work function engineering, circuit performance, and sensitivity to process-induced variations are the critical factors to design nanoscale devices [3,4]. In this research, nano-CMOS models are investigated to analyze the CMOS behavior at the nanoscale. The importance of physical correlations among the parameters and the impact of process variations have been evaluated. An HSPICE (Simulation Program with Integrated Circuit Emphasis) model file is converted into a PSPICE format. Different models and parameters are investigated using BSIM 4 and BSIM 3 models. Various CMOS gates in a nanoscale are simulated and the results are analyzed. To predict the effect of temperature, the team simulated 32 and 65 nm CMOS gates. The temperature varied from 0°C to 1000°C in 10° increments. Predictive technology model (PTM) files estimate the behavior of submicron transistors. The files are generated by scaling down from larger known technology nodes. To facilitate the extraction of PTMs, online tools such as the nano-CMOS tool in NanoHUB.org are available for free to the public. This tool allows a user to automatically generate a model card by adjusting 10 parameters. Modifying the temperature parameter gives a new look into how submicron transistors will behave at high temperatures. Extensive research has been carried out into the design of PTM files for transistors in the submicron range [5]. Since PTM files can be used in conjunction with CAD (computer-aided design) programs, circuit designers are able to characterize nano-CMOS circuits even before the technology is available for manufacturing. Therefore, it is very important that the model files are as accurate as possible. An important aspect of circuit design is the knowledge on how a circuit will perform in high temperatures; yet, the extraction of PTM files at different temperatures other than room temperature is an area that has not been fully explored.

As mentioned above, one of the tools available to the nano-community is the nano-CMOS tool available in NanoHUB.org [6]. This tool allows a user to automatically generate a model card by adjusting 10 parameters. These parameters are physical gate length (L_{EFF}), L_{EFF} variation percentage, threshold voltage (V_{TH}), V_{TH} variation, power supply voltage (V_{DD}), equivalent electrical oxide thickness (T_{OX}), total source and drain series resistance (R_{dsw}), number of bias points in V_{gs} and V_{ds}, and temperature (T_{NOM}).

In this research, the main parameter of interest is T_{NOM}. The ability to modify T_{NOM} is a new feature added to the nano-CMOS tool on NanoHUB.org. This feature is not available in the original nano-CMOS tool [7]. T_{NOM} corresponds to the temperature at which the model card parameters are extracted. To test the effects of modifying the T_{NOM} parameter, various CMOS gates such as nine-stage inverter, NAND, NOR, XOR are designed in SPICE to observe the effects of temperature when: (1) only the T_{NOM} parameter is varied, (2) only the circuit simulation temperature ($T_{CIRCUIT}$) is varied, and (3) both T_{NOM} and $T_{CIRCUIT}$ are set to match accordingly. Measurements are taken for all three trials at 27°C and 125°C.

The chapter is organized as follows: Section 1.2 gives an overview of the tool used including its features and limitations. Section 1.3 explains about the model cards generated along with the I–V curves extracted using the tool at both 27°C and 125°C for various transistor sizes used in the simulations. Section 1.4 summarizes the methodology used for research and Section 1.5 draws some conclusions and future research.

1.2 NANO-CMOS MODEL

Nano-CMOS is an online tool developed by the Nanoscale Integration and Modeling (NIMO) Group at Arizona State University. It is designed to provide the users an easy method to customize the model file to whatever technology specifications they desire. The user is limited to the following constraints when using the tool:

- Designed only for bulk-CMOS.
- Able to simulate for physical gate length down to 10 nm and up to 145 nm.
- Temperature ranges input can be from 70 to 450 K (−203–177°C).
- Total source and drain series resistance input can be between 90 and 300 Ω-µm.
- Equivalent electrical oxide thickness can be from 0.8 to 1.8 nm.
- Power supply voltage can be between 0.7 and 1.8 V.
- Threshold voltage can be between −0.4 and 0.4 V.
- Allowable variation of threshold voltage can be between −100 and 100 mV.
- Model files are designed to work with SPICE programs compatible with BSIM4.

TABLE 1.1

Model Parameters for 32 and 65 nm Technology Nodes

Parameters	32 nm		65 nm	
	NMOS	PMOS	NMOS	PMOS
L_{eff} (nm)	12.6	12.6	24.5	24.5
L_{eff} Variation (%)	10	10	10	10
V_{th} (V)	0.16	−0.16	0.18	−0.18
V_{th} Variation (mV)	30	30	30	30
V_{dd} (V)	0.9	0.9	1.1	1.1
T_{ox} (nm)	1	1	1.2	1.2
R_{dsw} ($\Omega\,\mu$m)	150	150	165	165
Temperature (°C)	27/125	27/125	27/125	27/125
N_V_{gs}	10	10	10	10
N_V_{os}	20	20	20	20

Source: W. Zhao, and Y. Cao, New generation of predictive technology model for sub-45 nm design exploration, *Proceedings of the IEEE International Symposium on Quality Electronic Design, San Jose, CA,* pp. 585—590 © (March 2006) IEEE. With permission; W. Zhao, and Y. Cao, *ACM Journal on Emerging Technologies in Computing Systems (JETC),* 3(1): 1–17, April 2007. With permission.

The preconfigured bulk-CMOS model files include the nominal, slow–slow corner, and fast–fast corner model files for 32 nm PMOS (p-channel metal-oxide-semiconductor) and NMOS (n-channel metal-oxide-semiconductor), 45 nm PMOS and NMOS, 65 nm PMOS and NMOS, 90 nm PMOS and NMOS, and 130 nm PMOS and NMOS. Table 1.1 shows the parameters used to generate the model files.

A basic CMOS inverter, NAND, NOR, and XOR gates are designed in conjunction with the model files. All the CMOS gates are simulated using HSPICE. Figure 1.1 shows a nine-stage inverter circuit. The input voltage starts at 0 V and then increases to 1.8 V.

It is expected that the slow–slow corner is run at a slower speed, but has the best performance; the fast–fast corner is run at a higher speed, but shows the worst performance, and the nominal shows balanced performance between the two in terms of speed and performance. It is also observed that the larger technology nodes (90 and 130 nm) showed better performance in terms of rise time and fall time when compared to the small technology nodes (32 and 45 nm). The above-mentioned model parameters are included in PSPICE simulation and the temperature-dependent critical parameters are identified for simulation. Once the user adjusts the parameters to the desired technology node and simulates, the tool can output the *IV* curves of the nominal transistor in the linear and saturation region, the *IV* curves in the subthreshold region, and the HSPICE code for the nominal transistor model, the slow–slow corner transistor model, and the fast–fast corner transistor model. The generated code can be modified for any SPICE software that supports BSIM4.

1.3   NANO-CMOS CHARACTERISTICS

The model files used in the simulation are generated by using the parameters listed in Table 1.1. Those values are determined through a review of the literature [5]. They are also the default values set on the nano-CMOS tool for their respective technology nodes. For each of the four transistors (32 nm NMOS and PMOS, 65 nm NMOS and PMOS), the *IV* curves are generated to get a look at the current versus the voltage at 27°C and 125°C. The *IV* curves show the MOSFET operations in the triode and saturation regions. In addition, the *IV* curves in the logarithmic scales are shown for each MOSFET. The *IV* curves are shown in Figures 1.2 through 1.9. It is observed from the graphs that

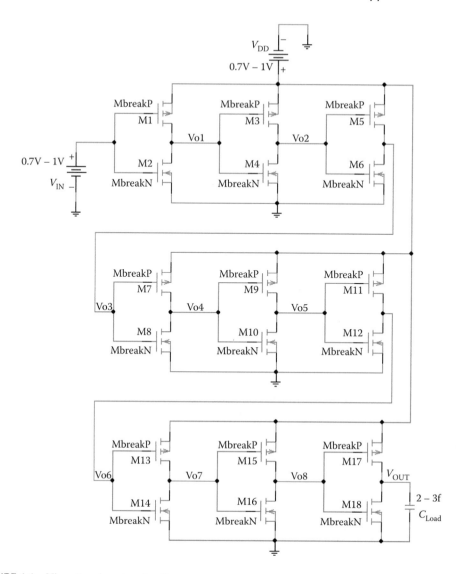

FIGURE 1.1   Nine-stage inverter circuit.

the NMOS transistor's current is lower in the triode region due to an increase of R_{ds} at a higher temperature. Eventually, the current at higher temperatures increases even at higher voltage (V_{ds}) levels.

Alternately, the *IV* curves for the PMOS transistors show that the current increases as the voltage increases, even at higher temperatures. For all instances of the *IV* curves in the subthreshold region, we observe that the currents at the higher and lower temperatures are approximately the same.

1.4 METHODOLOGY

Model cards are extracted for different temperatures. This is easily accomplished using the nano-CMOS tool. Then the four logic gates are designed in a SPICE program to measure the propagation delay of the output when the temperature is set at 27°C and 125°C. Three different configurations are used to add the effects of temperature in a circuit.

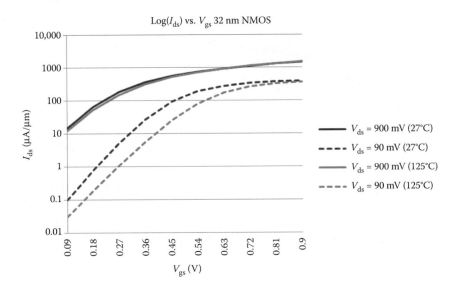

FIGURE 1.2 I_D–V_G curves for 32 nm NMOS. (A. Rodriguez and H. F. Huq, Validation of Nano-CMOS Predictive Technology Model Tool on NanoHUB.org., *IEEE International Conference on Nanotechnology*, Portland, OR, pp. 469–472, August 15–18, © (2011) IEEE. With permission.)

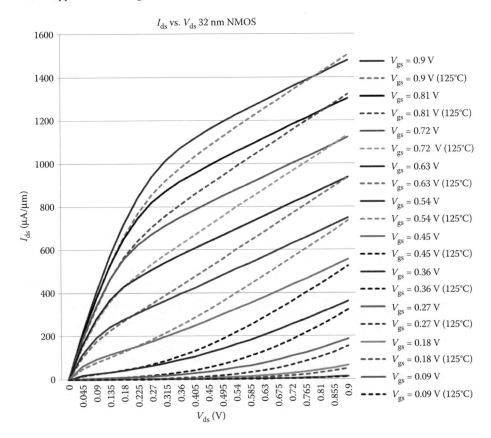

FIGURE 1.3 I_D–V_D curves for 32 nm NMOS. (A. Rodriguez and H. F. Huq, Validation of Nano-CMOS Predictive Technology Model Tool on NanoHUB.org., *IEEE International Conference on Nanotechnology*, Portland, OR, pp. 469–472, August 15–18, © (2011) IEEE. With permission.)

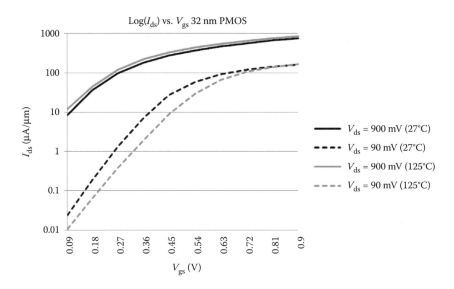

FIGURE 1.4 I_D–V_G curves for 32 nm PMOS. (A. Rodriguez and H. F. Huq, Validation of Nano-CMOS Predictive Technology Model Tool on NanoHUB.org., *IEEE International Conference on Nanotechnology*, Portland, OR, pp. 469–472, August 15–18, © (2011) IEEE. With permission.)

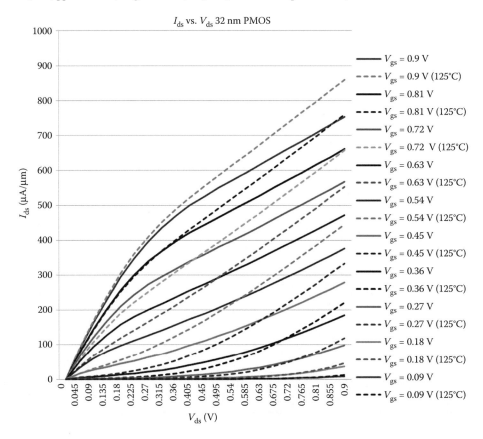

FIGURE 1.5 I_D–V_D curves for 32 nm PMOS. (A. Rodriguez and H. F. Huq, Validation of Nano-CMOS Predictive Technology Model Tool on NanoHUB.org., *IEEE International Conference on Nanotechnology*, Portland, OR, pp. 469–472, August 15–18, © (2011) IEEE. With permission.)

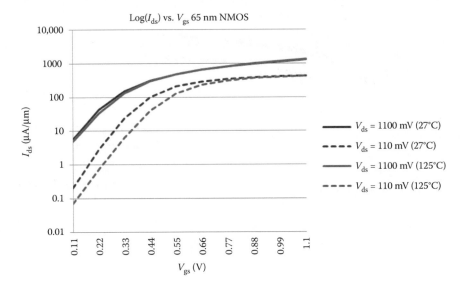

FIGURE 1.6　Log *IV* curves for 65 nm NMOS. (A. Rodriguez and H. F. Huq, Validation of Nano-CMOS Predictive Technology Model Tool on NanoHUB.org., *IEEE International Conference on Nanotechnology*, Portland, OR, pp. 469–472, August 15–18, © (2011) IEEE. With permission.)

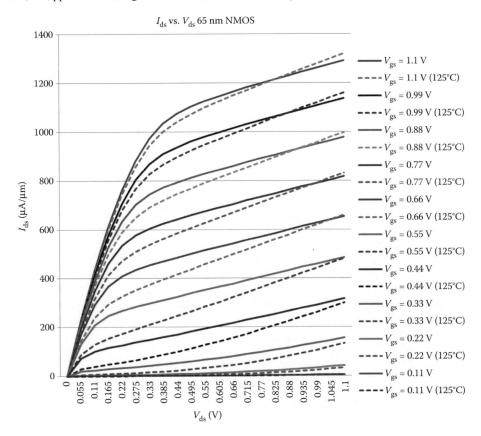

FIGURE 1.7　*IV* curves for 65 nm NMOS. (A. Rodriguez and H. F. Huq, Validation of Nano-CMOS Predictive Technology Model Tool on NanoHUB.org., *IEEE International Conference on Nanotechnology*, Portland, OR, pp. 469–472, August 15–18, © (2011) IEEE. With permission.)

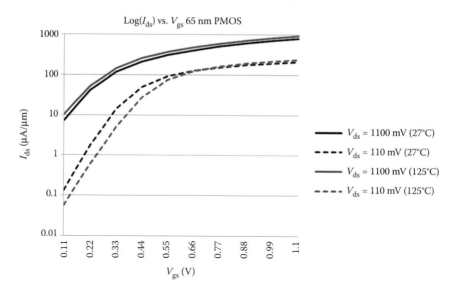

FIGURE 1.8 Log *IV* curves for 65 nm PMOS. (A. Rodriguez and H. F. Huq, Validation of Nano-CMOS Predictive Technology Model Tool on NanoHUB.org., *IEEE International Conference on Nanotechnology*, Portland, OR, pp. 469–472, August 15–18, © (2011) IEEE. With permission.)

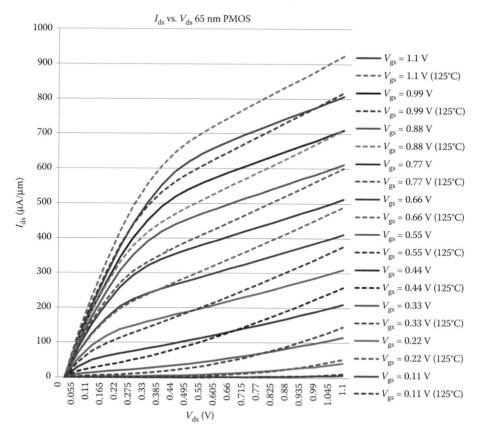

FIGURE 1.9 *IV* curves for 65 nm PMOS. (A. Rodriguez and H. F. Huq, Validation of Nano-CMOS Predictive Technology Model Tool on NanoHUB.org., *IEEE International Conference on Nanotechnology*, Portland, OR, pp. 469–472, August 15–18, © (2011) IEEE. With permission.)

- In the first configuration, the T_{NOM} parameter from nano-CMOS is varied to observe the effects of the performance of the circuits. When running the simulation, $T_{CIRCUIT}$ is left at room temperature (27°C).
- The second configuration used the model files extracted at room temperature and varied the $T_{CIRCUIT}$ parameter of the SPICE program.
- The third setup is a combination of the previous scenarios where both T_{NOM} and $T_{CIRCUIT}$ are set to 27°C, then again at 125°C.

It is expected that a transistor should perform slower at a higher temperature [8]. The characteristics of the *IV* curves show that once the voltage (V_{ds}) gets close to its maximum value, the performance of the transistors at different temperatures is only slightly different. Therefore, it is not expected for the performance degradation at the higher temperature to be of a great amount. We hypothesize that the third configuration will generate the most accurate trials in these simulations. The results from the simulations are shown in Tables 1.2 through 1.5. After analyzing the results, we observe a consistent

TABLE 1.2

Propagation Delays for CMOS NOT Simulation Results

		Nine-Stage Inverter		
Temperature Change	PTM (nm)	Average Propagation Delay(s)—27°C	Average Propagation Delay(s)—125°C	Variation %
T_{nom} (Nano-CMOS)	32	1.326E−10	1.226E−10	7.57918552
	65	1.658E−10	1.359E−10	18.03377563
$T_{circuit}$ (SPICE)	32	1.335E−10	1.870E−10	−40.1273885
	65	1.646E−10	2.420E−10	−47.0677606
T_{nom} and $T_{circuit}$	32	1.335E−10	1.552E−10	−16.2607718
	65	1.646E−10	1.625E−10	1.245821939

Source: A. Rodriguez and H. F. Huq, Validation of Nano-CMOS Predictive Technology Model Tool on NanoHUB.org., *IEEE International Conference on Nanotechnology*, Portland, OR, pp. 469–472, August 15–18, © (2011) IEEE. With permission.

TABLE 1.3

Propagation Delays for CMOS NAND Simulation Results

		NAND		
Temperature Change	PTM (nm)	Average Propagation Delay(s)—27°C	Average Propagation Delay(s)—125°C	Variation %
T_{nom} (Nano-CMOS)	32	1.397E−10	1.315E−10	5.871822413
	65	1.478E−10	1.257E−10	14.92385787
$T_{circuit}$ (SPICE)	32	1.404E−10	1.919E−10	−36.6809117
	65	1.485E−10	2.042E−10	−37.5084175
T_{nom} and $T_{circuit}$	32	1.404E−10	1.639E−10	−16.7378917
	65	1.485E−10	1.604E−10	−7.97979798

Source: A. Rodriguez and H. F. Huq, Validation of Nano-CMOS Predictive Technology Model Tool on NanoHUB.org., *IEEE International Conference on Nanotechnology*, Portland, OR, pp. 469–472, August 15–18, © (2011) IEEE. With permission.

TABLE 1.4

Propagation Delays for CMOS NOR

		NOR		
Temperature Change	PTM (nm)	Average Propagation Delay(s)—27°C	Average Propagation Delay(s)—125°C	Variation %
T_{nom} (Nano-CMOS)	32	1.519E−10	1.351E−10	11.05990783
	65	1.499E−10	1.217E−10	18.81881882
$T_{circuit}$ (SPICE)	32	1.535E−10	2.333E−10	−52.036494
	65	1.510E−10	2.276E−10	−50.6953642
T_{nom} and $T_{circuit}$	32	1.535E−10	1.784E−10	−16.2593679
	65	1.510E−10	1.527E−10	−1.09271523

Source: A. Rodriguez and H. F. Huq, Validation of Nano-CMOS Predictive Technology Model Tool on NanoHUB.org., *IEEE International Conference on Nanotechnology*, Portland, OR, pp. 469–472, August 15–18, © (2011) IEEE. With permission.

TABLE 1.5

Propagation Delays for CMOS XOR

		XOR		
Temperature Change	PTM (nm)	Average Propagation Delay(s)—27°C	Average Propagation Delay(s)—125°C	Variation %
T_{nom} (Nano-CMOS)	32	1.811E−10	1.647E−10	9.083379348
	65	1.790E−10	1.423E−10	20.50852193
$T_{circuit}$ (SPICE)	32	1.822E−10	2.781E−10	−52.6344676
	65	1.802E−10	2.708E−10	−50.2774695
T_{nom} and $T_{circuit}$	32	1.822E−10	2.210E−10	−21.2952799
	65	1.807E−10	1.833E−10	−1.41117875

Source: A. Rodriguez and H. F. Huq Validation of Nano-CMOS Predictive Technology Model Tool on NanoHUB.org., *IEEE International Conference on Nanotechnology*, Portland, OR, pp. 469–472, August 15–18, © (2011) IEEE. With permission.)

trend among all the simulation results. When only T_{NOM} is varied, the results show a decrease in propagation delay as the temperature increases. This goes against the standard behavior of propagation delay with temperature, but this type of outcome is suggested as a possibility in Ref. [9].

When only the temperature is modified in the SPICE file, we see that there is a drastic reduction in the performance at a higher temperature concurrent with the behavior observed in Ref. [10]. The last variation has not been tested previously. Results show a moderate and slight degradation in the performance at a higher temperature for the 32 and 65 nm nodes, respectively.

1.5   CONCLUSION

As stated in Refs. [10,11], temperature-dependent parameters within the model card are the temperature coefficient for threshold voltage (KT1), channel length dependence of the temperature coefficient for threshold voltage (KT1 L), body-bias coefficient of threshold voltage temperature (KT2), temperature coefficient for UA (UA1), temperature coefficient for UB

(UB1), temperature coefficient for UC (UC1), temperature coefficient for saturation velocity (AT), temperature coefficient for R_{dsw} (PRT), emission coefficients of junction for source and drain junctions, respectively (NJS, NJD), junction current temperature exponents for source and drain junctions, respectively (XTIS, XTID), and temperature at which parameters are extracted (TNOM). We have noticed that when modifying T_{NOM}, four additional parameters are affected. These are long-channel threshold voltage at $V_{bs} = 0$ (VTH0), first-order body bias coefficient (K1), low-field mobility (U0), and doping concentration in the channel (NDEP). Further research needs to be performed to determine if the addition of the temperature modification field on the nano-CMOS tool leads to more accurate results at higher temperatures based on the data obtained from the T_{NOM} and $T_{CIRCUIT}$ trials. In addition, the significance of the additional four parameters that change with T_{NOM} needs to be studied.

ACKNOWLEDGMENT

The authors wish to thank the NIMO group at Arizona State University for the valuable insight on the nano-CMOS tool.

REFERENCES

1. L. Chang, Y.-K. Choi, D. Ha, P. Ranade, S. Xiong, J. Boker, C. Hu, and T.-J. King, Extremely scaled silicon nano-CMOS devices, invited paper, *Proceedings of the IEEE*, 91(11), Nov. 2003.
2. M. M. Ziegler, and M. R. Stan, CMOS/nano co-design for crossbar-based molecular electronic systems, *IEEE Transactions on Nanotechnology*, 2(4), 217–230, Dec. 2003.
3. X. Liu, C. Lee, and C. Zhou, Carbon nanotube field-effect inverters, *Applied Physics Letters*, 79(20), 3329–3331, Nov. 2001.
4. Y. Cui, Z. Zhong, D. Wang, W. U. Wang, and C. M. Lieber, High performance silicon nanowire field effect transistors, *Nano Letters*, 3(2), 149–152, 2002.
5. W. Zhao, and Y. Cao, New generation of predictive technology model for sub-45 nm design exploration, *Proceedings of the IEEE International Symposium on Quality Electronic Design,* San Jose, CA, pp. 585–590, March 2006.
6. W. Zhao, and Y. Cao, Predictive technology model for nano-CMOS design exploration, *ACM Journal on Emerging Technologies in Computing Systems (JETC)*, 3(1), 1–17, April 2007.
7. Predictive Technology Model (PTM), http://www.eas.asu.edu/~ptm/
8. S. Sedra, and K.C. Smith, *Microelectronic Circuits* (5th edition). New York: Oxford, 2004.
9. R. Kumar, and V. Kursun, Reversed temperature-dependent propagation delay characteristics in nanometer CMOS circuits. *IEEE Transactions on Circuits and Systems*, 53(10):1078–1082, Oct. 2006.
10. R. Kumar, and V. Kursun, Modeling of temperature effects on nano-CMOS devices with the predictive technologies. *50th Midwest Symposium on Circuits and Systems, MWSCAS Conference*, Montreal, Canada, August 2007.
11. X. Xi et al., *BSIM4.3.0 MOSFET Model-User Manual*. Berkeley, CA: Dept. Elect. Comput. Eng., Univ. California, 2003.
12. A. Rodriguez and H. F. Huq, Validation of Nano-CMOS predictive technology model tool on NanoHUB. org., *IEEE International Conference on Nanotechnology*, August 15–18, 2011, Portland, OR, pp. 469–472.

2 Comparative Analysis of Mobility and Dopant Number Fluctuation Models for the Threshold Voltage Fluctuation Estimation in 45 nm Channel Length MOSFET Device

Nabil Ashraf, Dragica Vasileska, Gilson Wirth, and Purushothaman Srinivasan

CONTENTS

2.1 INTRODUCTION

In very small electronic devices, the alternate capture and emission of carriers at an individual defect site located at the interface of $Si:SiO_2$ of a MOSFET (metal–oxide–semiconductor field-effect transistor) generates discrete switching in the device conductance referred to as a random telegraph signal (RTS) or random telegraph noise (RTN). Accurate and physical models for random telegraph noise fluctuations (RTF) [1] are essential to predict and optimize circuit performance during the design stage [2]. Currently, such models are not available for circuit simulation. The compound between RTF and other sources of variation, such as random dopant fluctuations (RDF), further complicates the situation especially in extremely scaled CMOS (complementary metal–oxide–semiconductor) design. In the vicinity of a trap site, the electrostatic short-range Coulomb forces that exist between trap-inversion electrons–depletion ions modify the electrostatic surface potential in the channel from source to drain in a spatially random and discrete manner. Accurate replication of these multiple peaks and valleys of the surface potential is critical to be accounted for by the analytical models for inversion conditions and when spatial inhomogeneity exists due to the interface trap, inversion carriers, and depletion region dopant ions [3]. This aspect is not presently accounted for by most analytical device models, including the dopant number fluctuation [4] and the percolation theory model [5]. It will be shown that these two analytical models fail to account for the large threshold voltage fluctuations that are observed for source-side trap positions in the channel by 3D EMC device simulation [6] of a 45 nm MOSFET. Therefore, a new model is proposed. This new model, for the first time, highlights the carrier mobility fluctuations resulting from source side trap positions with the spatially variant short-range interaction force causing potential inhomogeneities and random spikes in the surface potential barrier near the source.

This chapter is organized as follows. In Section 2.2, a brief description of the theoretical model is given. A description of the analytical models is presented in Section 2.3. The results and summary of this work are discussed in Section 2.4.

2.2 THEORETICAL MODEL

In particle-based device simulation schemes, one couples the Monte Carlo transport kernel with a Poisson equation solver as shown diagrammatically in Figure 2.1 [7]. Briefly, after the free-flight scatter sequence, particle–mesh (PM) coupling takes place, which is followed by a Poisson equation solution for the electrostatic potential and the electric field needed in the subsequent free-flight scatter sequence.

The Poisson equation is solved on a mesh that is determined by the Debye criterion. Specifically, in critical device regions, the mesh has to be smaller than the extrinsic Debye length [8]. If the mesh is infinitely small, then the Coulomb potential is completely resolved. However, that would typically require a large number of node points. As in silicon devices, to get accurate results one has to solve the 2D/3D Poisson equation every 0.1 fs and the total simulation time is of the order of 5–10 ps. This means that the Poisson equation solution, which is a bottleneck for 3D simulations, has to be solved many times, which in turn requires the use of very efficient Poisson equation solvers. The time to solve the Poisson equation limits the number of node points that has to be used in the Poisson mesh, as the mesh has to be coarser, which in turn reduces the amount of the short-range Coulomb interaction that is accounted for via the solution of the 3D Poisson equation.

The short-range portion of the Coulomb interaction is typically accounted for by considering Coulomb scattering as an additional scattering mechanism in the k-space portion of the Monte Carlo transport kernel [9]. The proper calculation of electron–electron scattering and electron–ion scattering requires a proper screening model. Screening requires the evaluation of a distribution function, which is typically noisy and a time-consuming task [10]. The extent to which short-range and long-range coulomb interactions are taken into account is not really known and some overestimation or underestimation of the interaction usually occurs. Also, multiple scattering processes and dynamical screening are typically almost impossible to be accounted for.

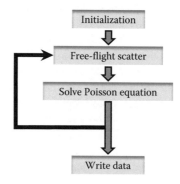

FIGURE 2.1   Typical flowchart of a particle-based device simulator.

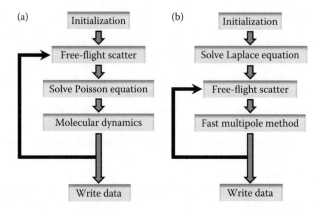

FIGURE 2.2 Philosophy behind the (a) corrected Coulomb approach, where correction force is used in the molecular dynamics routine, and (b) the fast multipole method, where the full Coulomb interaction is being considered to get the force on the electrons in the free-flight portion of the Monte Carlo transport kernel.

To avoid the problem with the k-space treatment of the Coulomb interaction, a real-space approach has been proposed by Lugli and Ferry [11] in which the electron–electron and the electron–ion interactions are accounted for via real-space molecular dynamics routine. It is important to note that direct application of the real-space molecular dynamics can be used for bulk systems only where it is not required to solve the Poisson equation. Hence, an approach is needed that correctly accounts for the full Coulomb interaction in particle-based device simulators. The group from Arizona State University (ASU) has, in that sense, been a pioneer in this field and in our simulation modules, we have currently implemented three approaches:

1. The corrected Coulomb approach—an approach that we have introduced [12].
2. The particle–particle–particle–mesh coupling method due to Hockney and Eastwood [13].
3. Fast multipole method [14].

The corrected Coulomb approach and the particle–particle–particle–mesh coupling method are similar in philosophy. Specifically, a correction force is calculated, given the mesh, and it is that correction force that is used in the molecular dynamics routine. The fast multipole method is completely different in philosophy because the Laplace equation is solved to account for the charges at the ohmic contacts, and later, only the fast multipole method is used to account for the full Coulomb interactions between the electrons and electrons and ions. The difference between these two ideologies is graphically shown in Figure 2.2.

2.3 DESCRIPTION OF THE ANALYTICAL MODELS

2.3.1 Number Fluctuation-Based Model

To properly account for the random number and position of the dopant ions in the depletion region of the channel, the model presented in ref. [4] accomplishes this in the following manner. The simulation domain is divided into small boxes by discretizing the channel length and width into small square cells of dimension l with a volume assisted by the depth of X that extends from the channel to ideally the maximum depletion width (depth). The random dopant ions are positioned in each of these cell volumes in random number and assortment that is based on the uniform nominal doping density. Thus, the number of dopant ions that can reside in a volume cell is dependent on the dimensions of the cell (square dimension l and depth dimension X). If this number is too small and close to unity, the calculated V_T will be too small and almost invariant from cell to cell. Therefore, the volume of the cell of dimension l and X for this research is chosen so that a dopant number variation up to a maximum of 4 can be expected to reside in the cell for different random dopant distribution configurations. The required device parameters for the 45 nm physical gate length MOSFET are L (gate length) = 45 nm, W (gate width) = 50 nm, t_{ox} (oxide thickness) = 0.9 nm, and N_A (substrate doping) = 8.9×10^{24} m^{-3}. We have shown earlier that V_T variation for a typical channel random dopant configuration correlates to a few nanometers in depth from the interface [15]. With the above information, the designed cell dimensions are computed to be $l = 10$ nm and $X = 4.5$ nm retrenched from the maximum possible depletion depth of $W_{max} = 12.34$ nm. After calculating the local cell threshold voltage from its dopant number value, all the threshold voltage values from all the cells in a 2D array have been averaged to extract the final value of the threshold voltage for a particular random dopant configuration.

Once the reference V_T values for different random dopant types and distributions are extracted using the procedure described in ref. [4], we have used the results from ref. [16] to calculate the threshold voltage V_T fluctuation percentage for an interface trap positioned along the channel from source to drain of an effective 32 nm channel length nMOSFET. Since a single interface trap is taken to be at locations that are 2 nm apart, the length l of the cell is now reduced to 2 nm, whereas the width l is kept at 10 nm. This places a maximum of 2 atoms per cell for a particular trap to interact with. Since the random interface trap is located at the middle of the gate width, the cell left edge in the width direction is at 20 nm and the right edge is at 30 nm (where an interface trap can be found) to accommodate the required 25 nm gate width point with a margin of ±5 nm.

2.3.2 Percolation-Based Analytical Model

Percolation theory-based conduction through the channel of a MOSFET, as proposed by Keyes [5], takes into account the random position and distribution of channel dopants along with the interface trap underneath the active gate region, as well as the rapid surface potential fluctuations out of their local inversion conditions. In a cellular arrangement along the width, length, and depth distribution of the active channel volume, this leads to certain cells being nonconducting and certain cells being conducting. Percolation paths through conducting cells that are adjacent to one another determine the current conduction probability from source to drain along the length direction at threshold condition. The relevant set of equations as detailed in Ref. [5] is used to determine the conduction probability Q for a set of 20 random dopant-type configurations. The key term (dQ/dV) is evaluated for a particular random dopant configuration for each trap position along the source and drain (length direction) and at the middle of the gate width (width direction) of the MOSFET and at each trap position. An average is made over all 20 random dopants to arrive at the threshold voltage value for the percolation model. For the reference threshold voltage needed in the evaluation of the dV term, analytical number-based threshold voltage expressions are utilized. The percolation model threshold voltage values in the presence of random dopants and random interface traps do not deviate more than 3% from number fluctuation-based threshold voltage values. This is due to the fact that there are long arrays of cells from source to drain of a MOSFET. Thus, Q is very small (~10^{-4} – 10^{-5} range) and does not deviate appreciably as a result of the trap's introduction and interaction with spatially inhomogeneous channel dopants and inversion carriers.

2.3.3 Newly Proposed Mobility Fluctuation-Based Analytical Model with Inherent Number Fluctuation from Dopants

The models discussed in Sections 2.3.1 and 2.3.2 are inadequate in a proper replication of barrier-limited short-range interactions between a source-side trap and neighboring channel dopants and inversion carrier electrons that have been simulated by the ensemble Monte Carlo (EMC) device simulation method. From a knowledge of carrier mobility fluctuations resulting from surface potential barrier spikes at source-side trap interactions with carriers and depletion ions, the authors propose a new analytical model that incorporates low-field (threshold condition) channel mobility fluctuations in the presence of random dopants and random interface traps. In the new model, the governing equations are taken from ref. [17] with the key equation that extracts the mobility values being referenced below:

$$g_{ds} = \mu_{eff} C_{ox} \frac{W}{L} \left(V_{gs} - V_T - \frac{A_b V_{ds}}{2} \right) \tag{2.1}$$

From Equation 2.1, the threshold voltage V_T can be determined from the known values of g_{ds} and μ_{eff}. For the calculation of g_{ds}, drain voltage bias is swept in 0.02 V increments from 0.25 to 0.45 V at a high enough gate bias of 0.8 V to cause sufficient inversion and maintain a linear region low-field approximation. The g_{ds} data are extracted from the EMC simulation for the above bias condition and the measurement of drain current. Also extracted from the EMC simulation is the surface-inverted charge Q_n for the above gate voltage. The channel mobility is then calculated using ref. [18]:

$$\mu_{eff} = \frac{g_{ds} L}{W Q_n} \tag{2.2}$$

The value of μ_{eff} as calculated from Equation 2.2 is fed back into Equation 2.1 to calculate V_T. The V_{ds} term in Equation 2.1 is fixed at 0.4 V to reduce the DIBL (drain-induced barrier lowering) effect and excessive drain-induced charge sharing to roll off V_T further.

2.4 SIMULATION RESULTS

In this section, we first present the extraction of V_T for different random dopant configurations and its fluctuations induced by a single interface trap. Figure 2.3 depicts the extracted threshold voltage distribution for different random dopant configurations taking account of the two existing analytical model-based computations and EMC device simulation. Figure 2.4 shows the fluctuations in threshold voltage for different interface trap positions from source to drain of an effective 32 nm

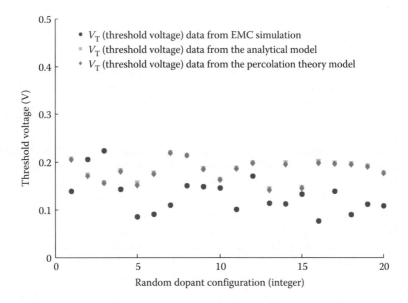

FIGURE 2.3 Threshold voltage as a function of different discrete random dopant configuration in the channel region when no interface random trap is present.

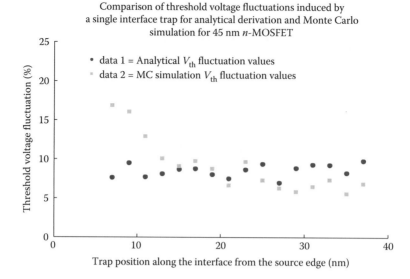

FIGURE 2.4   Threshold voltage fluctuations computed by the additional analytical model adjusted for the random interface trap's interactions with channel electrons and the EMC simulation method.

gate length MOSFET. The fluctuation trend of threshold voltage shown in this figure, for analytical model-based derivations, is compared with 3-D EMC device simulation. The results are presented in Figure 2.3. For the case of the EMC simulation method, a constant drain bias of 0.5 V is used for all threshold voltage extractions in the presence of random interface trap. Figure 2.4 illustrates the threshold voltage fluctuation extracted from an analytical model [4,6], adjusted for the trap's interactions with channel electrons inverted at threshold conditions, and also for EMC device simulation where usual short-range interactions between the trap to an electron–electron and a trap–electron–ion are accounted for. Deviations in V_T fluctuation values are noticed for the EMC simulation model in comparison with the analytical model due to the requirements of proper treatment of surface potential, mobility and inversion electron, and dopant number fluctuations through 3D short-range Coulomb force corrections. The analytical models thus exhibit inconsistencies in accurately replicating transport mechanisms existing in the vicinity of a nearby trap in the presence of random dopant ions and inversion electrons. Traps near the source end of the channel have the largest influence since they are major obstacles to the electrons because of the large input barrier experienced there. As the traps are positioned near the drain, due to the larger drift velocity and carrier excitation energy, the trap's interaction with channel carriers is minimal and the fluctuation deviation trend is more or less within the tolerance limit.

Figure 2.5 shows the expected deviations from mean values (error-bar) for threshold voltage fluctuation percentage observed in the presence of a single interface trap positioned from source to drain for the cases of (i) an EMC-based device simulation model, (ii) an analytical number fluctuation-based model, and (iii) an EMC-based simulation method with no short-range electron–electron and an electron–ion–trap interaction force. The results presented in Figure 2.5 show the importance of consideration of a short-range electron–electron and an electron–ion–trap interaction force for a proper estimation of large fluctuation values of threshold voltage in the presence of source-side trap positions. These results highlight the importance of a short-range electron–electron and an electron–ion–trap Coulomb interaction correction to the conventional particle–mesh coupling (PM) long-range Coulomb interactions. In addition, it is evident that the traps near the source end of the channel can cause significant mobility fluctuations apart from surface potential fluctuations impeding the electron flow and enhancing the local threshold voltage variations. So,

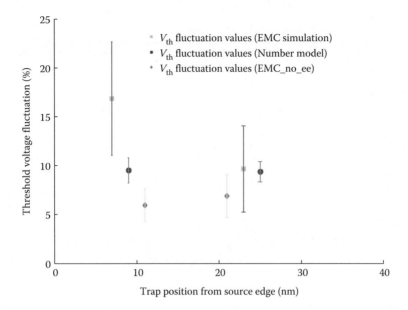

FIGURE 2.5 The error bar plot of threshold voltage fluctuation percentage for interface trap positions near the source and away from the source along the channel for the cases of (i) EMC simulation method, (ii) analytical model 1, and (iii) EMC simulation method with no short-range electron–electron and electron–ion–trap force consideration.

a significant mobility fluctuation needs to be added to the carrier number fluctuation for carrier electrons trapped near the source side by a random interface trap. The figure also demonstrates that as the traps are moved away from the source toward the drain end of the channel, the fluctuation pattern is within a few percentage tolerances between the analytical model and the EMC device simulation model.

The newly proposed mobility fluctuation-based analytical model with its inherent incorporation of dopant number fluctuation in the channel underneath the depletion region addresses the deficiency of the number and percolation theory models, as shown in Figure 2.6. Figure 2.7 shows the statistical set of reference effective channel mobility values for the different designated types and distributions of random dopants in the channel and bulk region. Figure 2.8 shows the percentage average mobility fluctuations over a set of 20 random dopant types for the case of a single interface trap when the trap is moved from the source junction edge to the drain junction edge of the MOSFET. The variation and scatter in mobility values are due to spatially inhomogeneous channel thickness, dopant number variations (position and number in close proximity to a carrier electron), and short-range spatial electron–electron and electron–trap–ion interactions. Figure 2.9 shows the threshold voltage values for the random dopant types considered for all three analytical models and the EMC device simulation environment. Figure 2.10 shows the percentage threshold voltage fluctuations for analytical model 1 based on dopant number fluctuations in the channel [4,16], EMC device simulation method [6], and new analytical model 3 (newly proposed), respectively. The new analytical model [17,18] correctly computes channel mobility fluctuations added to dopant number fluctuations. The figure clearly reveals that model 3 (newly proposed) is more compliant to the EMC device simulation results of threshold voltage fluctuations in the vicinity of the source of an effective 32 nm channel length MOSFET. The figure also demonstrates one important observation that the dopant number fluctuations and mobility fluctuation effects cannot be considered as an additive to give rise to the actual V_T fluctuation values predicted by a new analytical model. Figure 2.11 shows the error bar

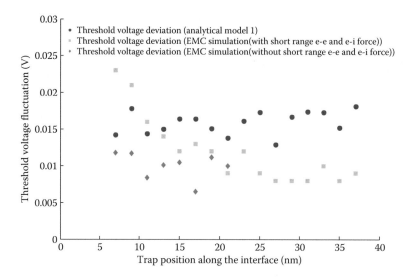

FIGURE 2.6   Threshold voltage fluctuation as a function of trap position at the interface in the channel region of the MOSFET for the first of the two analytical models and EMC simulation models with and without short-range Coulomb force corrections.

plot of extracted threshold voltage values for different random dopant configuration types for the cases of (i) an EMC simulation method, (ii) an analytical model 1, and (iii) a new mobility fluctuation-based model 3. Figure 2.12 further depicts the error bar plot for threshold voltage fluctuation percentage values in the presence of interface traps (source side, middle, and near drain) for all the above cases.

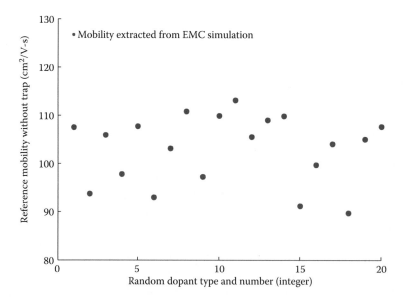

FIGURE 2.7   Effective channel mobility values for a different statistical set of random dopants. (N. Ashraf et al., Comparative analysis of mobility and dopant number fluctuations based models for the threshold voltage fluctuations estimation in 45 nm channel length MOSFET devices in the presence of random traps and random dopants, *IEEE International Conference on Nanotechnology*, Portland, OR, pp. 492–495, August 15–18, © (2011) IEEE. With permission.)

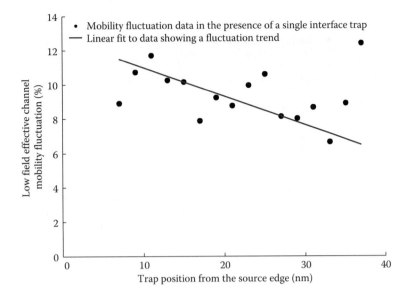

FIGURE 2.8 Effective channel mobility fluctuation as a function of trap position at the interface between source and drain junctions in the channel of the MOSFET. (N. Ashraf et al., Comparative analysis of mobility and dopant number fluctuations based models for the threshold voltage fluctuations estimation in 45 nm channel length MOSFET devices in the presence of random traps and random dopants, *IEEE International Conference on Nanotechnology*, Portland, OR, pp. 492–495, August 15–18, © (2011) IEEE. With permission.)

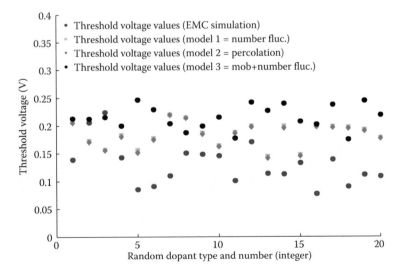

FIGURE 2.9 Threshold voltage values extracted for the two existing models in the literature along with the new analytical model and EMC device simulation for a statistical set of random dopant types designated as integer numbers. (N. Ashraf et al., Comparative analysis of mobility and dopant number fluctuations based models for the threshold voltage fluctuations estimation in 45 nm channel length MOSFET devices in the presence of random traps and random dopants, *IEEE International Conference on Nanotechnology*, Portland, OR, pp. 492–495, August 15–18, © (2011) IEEE. With permission.)

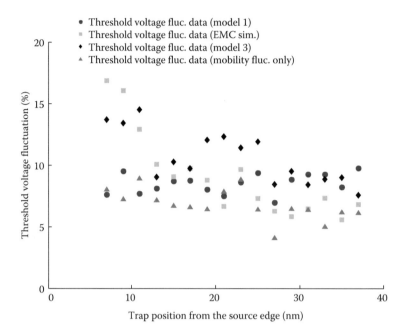

FIGURE 2.10 Percentage of averaged threshold voltage fluctuation values extracted for the two existing models in the literature along with the new analytical model and the EMC device simulation method for a single random interface trap positioned in the channel from source to drain of the MOSFET. (N. Ashraf et al., Comparative analysis of mobility and dopant number fluctuations based models for the threshold voltage fluctuations estimation in 45 nm channel length MOSFET devices in the presence of random traps and random dopants, *IEEE International Conference on Nanotechnology*, Portland, OR, pp. 492–495, August 15–18, © (2011) IEEE. With permission.)

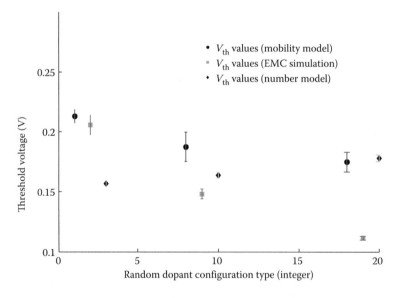

FIGURE 2.11 Threshold voltage distribution with expected deviations for different random dopant configuration types when the trap's effect is considered.

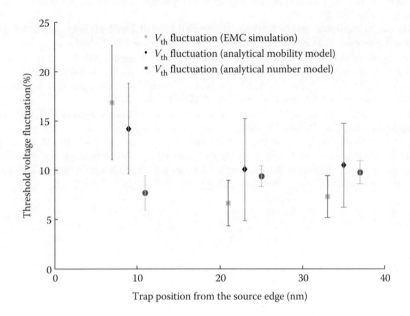

FIGURE 2.12 Threshold voltage fluctuation error bar plot considering different trap positions for the (i) EMC simulation method, (ii) analytical mobility fluctuation-based model, and (iii) analytical number fluctuation-based model.

REFERENCES

1. K. A. Ralls, W. J. Skocpol, L. D. Jackel, R. E. Howard, L. A. Fetter, R. W. Epworth, and D. M. Tennant, Discrete resistance switching in submicron silicon inversion layers: Individual interface traps and low frequency (1/f) noise, *Physical Review Letters*, 52, 228–231, 1984.
2. L. Brusamarello, G. I. Wirth, and R. da Silva, Statistical RTS model for digital circuits, *Microelectronics Reliability*, 49(9–11), 1064–1069, 2009.
3. D. Vasileska, H. R. Khan, S. S. Ahmed, C. Ringhoffer, and C. Heitzinger, Quantum and Coulomb effects in nanodevices, *International Journal of Nanoscience*, 4(3), 305–361, 2005.
4. X. Tang, V. K. De, and J. D. Meindl, Intrinsic MOSFET parameter fluctuations due to random dopant placement, *IEEE Transactions on Very Large Scale Integration (VLSI) Systems*, 5(4), 369–376, 1997.
5. R. W. Keyes, The effect of randomness in the distribution of impurity atoms on FET thresholds, *Appl. Phys.* 8, 251–259, 1975.
6. W. J. Gross, D. Vasileska, and D. K. Ferry, A novel approach for introducing the electron–electron and electron–impurity interactions in particle-based simulations, *IEEE Electron Device Letters* 20(9), 463–465, 1999.
7. W. J. Gross, D. Vasileska, and D. K. Ferry, 3D simulations of ultra-small MOSFETs with real-space treatment of the electron–electron and electron–ion interactions, *VLSI Design*, 10, 437–452, 2000.
8. D. Vasileska, S. M. Goodnick, and G. Klimeck, *Computational Electronics: Semiclassical and Quantum Transport Modeling*, Boca Raton, FL: Taylor & Francis, June 2010.
9. M. V. Fischetti and S. E. Laux, Long-range Coulomb interactions in small Si devices, Part I: Performance and reliability, *Journal of Applied Physics*, 89, 1205–1231, 2001.
10. D. K. Ferry, *Semiconductors*, Macmillan, USA September 1, 1991.
11. P. Lugli and D. K. Ferry, Degeneracy in the ensemble Monte Carlo method for high-field transport in semiconductors, *IEEE Transactions on Electron Devices*, 32(11), 2431–2437, 1985.
12. W. J. Gross, D. Vasileska, and D. K. Ferry, Ultrasmall MOSFETs: The importance of the full Coulomb interaction on device characteristics, *IEEE Transactions on Electron Devices*, 47(10), 1831–1837, 2000.
13. R. W. Hockney and J. W. Eastwood, *Computer Simulation Using Particles*, New York, NY: McGraw-Hill, 1981.

14. L. Greengard and V. Rokhlin, A fast algorithm for particle simulations, *Journal of Computational Physics*, 135(2), 280–292, 1997.
15. W. J. Gross, D. Vasileska, and D. K. Ferry, Ultra-small MOSFETs: The importance of the full Coulomb interaction on device characteristics, *7th International Workshop IWCE*, Glasgow, pp. 32–33, 2000.
16. K. Sonoda, K. Ishikawa, T. Eimori, and O. Tsuchiya, Discrete dopant effects on statistical variation of random telegraph signal magnitude, *IEEE Trans. Electron Dev.*, 54(8), 1918–1925, 2007.
17. X. Zhou and K. Y. Lim, Unified MOSFET compact *I–V* model formulation through physics-based effective transformation, *IEEE Trans. Electron Dev.*, 48(5), 887–896, 2001.
18. J. He, X. Zhang, Y. Wang, and R. Huang, New method for extraction of MOSFET parameters, *IEEE Electron Dev. Lett.*, 22(12), 597–59 2001
19. N. Ashraf, D. Vasileska, G. Wirth, and S. Purushothaman, Comparative analysis of mobility and dopant number fluctuations based models for the threshold voltage fluctuations estimation in 45 nm channel length MOSFET devices in the presence of random traps and random dopants, *IEEE International Conference on Nanotechnology*, Portland, OR, pp. 492–495, August 15–18, 2011.

3 Impact of Random Interface Traps on Asymmetric Characteristic Fluctuation of 16-nm-Gate MOSFET Devices

Yiming Li

CONTENTS

3.1 INTRODUCTION

Silicon device technology scaling and performance improvement require [1–3] not only overcoming a variety of fabrication challenges but also suppressing systematic variation and random effects [4–7]. Except process variation effect (PVE), random dopant fluctuation (RDF), as one of the known major intrinsic parameter fluctuations, complicates device manufacturing and degrades device characteristics in the nanometer scale complementary metal–oxide semiconductor (nano-CMOS) device era [8–30]. High-κ/metal-gate (HKMG) technology has been a key way to suppress RDF-induced variability and reduce leakage current [31–38]; however, HKMG may introduce random interface traps (ITs) at a high-κ/silicon interface and such IT fluctuation (ITF) degrades device characteristics considerably [39–47]. Various simulations of the device's variability induced by ITF were reported by using a one-dimensional (1D) IT's model for sub-65-nm CMOS devices [46], a 2D IT's model for 16-nm-gate HKMG devices [39], local interaction of the combined RDs and ITs [40,41], and full fluctuation among all random sources [4]. Recently, the asymmetric RDF on device characteristics was studied for 16-nm-gate HKMG metal–oxide–semiconductor field effect transistor (MOSFET) devices [48,49]. However, induced by random ITs, asymmetric physical and electrical characteristic fluctuations of 16-nm-gate HKMG MOSFETs have not been discussed yet.

In this chapter, the asymmetric characteristic fluctuation of the random ITs in 16-nm-gate HKMG MOSFETs is presented. Random ITs at the 2D interface of the hafnium oxide (HfO$_2$)/silicon of the 16-nm-gate HKMG N-MOSFETs are incorporated into an experimentally validated 3D device simulation [11] to quantify the random ITs-fluctuated characteristics. Large-scale statistical 3D device simulation is performed by solving a set of calibrated 3D electron–hole density-gradient equations coupling with the Poisson equation as well as electron–hole current continuity equations [50,51]. On the basis of the large-scale statistical 3D device simulation of random ITs-fluctuated samples, the random ITs-induced threshold voltage fluctuation ($\sigma_{V_{th}}$) and on-/off-state current fluctuation are estimated and compared with other significant fluctuation sources. In particular, an asymmetric drain-induced, barrier-lowering fluctuation σ_{DIBL} and a subthreshold swing fluctuation σ_{SS} [1] are studied by classifying the random ITs into the regions near the source side, around the middle of the channel, and near the drain side of the silicon channel, respectively. Our quantum mechanically corrected device simulation was compared with experimental data aiming at the greatest accuracy [11], which further enables us to explore both the individual and combined effects of randomly existing ITs and RDs on device characteristics in a unified way. The random 2D ITs are also solved with 3D RDs inside the silicon channel at the same time to assess the interaction effect of the combined ITs and RDs. For the studied 16-nm-gate HKMG N-MOSFETs with a device width of 16 nm, the high density of ITs (D_{it}) in the range of 3×10^{11}–3.3×10^{12} eV^{-1}cm^{-2} results in 26.3 mV fluctuation of threshold voltage ($\sigma_{V_{th,ITS}}$) and $\sigma_{V_{th,ITS}} = 10.2$ mV for the low D_{it} varying from 2.5×10^{10} to 3.8×10^{11} eV^{-1}cm^{-2}. Both the high and low D_{it}-induced $\sigma_{V_{th,ITs}}$ are smaller than that of $\sigma_{V_{th,RDs}} = 43$ mV. The largest asymmetric values of $\sigma_{V_{th}}$ and σ_{SS} are observed when random ITs with high D_{it} are near the source side, compared with the random ITs located at the middle of the channel and near the drain side. The engineering findings of this study indicate that both the ITs near the source side of the silicon channel and the RDs near the channel surface possess the largest characteristic fluctuations.

This chapter is organized as follows. In Section 3.2, we describe the simulation settings for random ITs-, RDs-, and the combined ITs and RDs (denoted as "ITs + RDs")-induced device and circuit's characteristic fluctuations. In Section 3.3, we discuss the findings of this study for the random ITs-fluctuated 16-nm-gate devices as well as the static random-access memory (SRAM). Finally, we draw conclusions and suggest future work.

3.2 THE STATISTICAL 3D DEVICE SIMULATION

The devices we examined are the 16-nm-gate titanium-nitride (TiN) gate MOSFETs with an amorphous-based TiN/HfO$_2$ gate stack which has an effective oxide thickness (EOT) of 0.8 nm, as shown in Figure 3.1a. The EOT varying from 0.4 to 1.2 nm will also be simulated for examining the impact of different high-κ gate stacks on fluctuation suppression. Note that the device's width is equal to the gate length of 16 nm, which is designed for the most critical assessment. The validated nominal device characteristics of the studied 16-nm-gate HKMG MOSFETs are calibrated, where the magnitude of the threshold voltages of the 16-nm-gate N- and P-MOSFETs is 250 mV. To examine the asymmetric characteristic fluctuation induced by random ITs, as shown in Figure 3.1a, we have partitioned the channel into three different regions: random ITs near the source side, around the middle of the channel, and near the drain side.

To carry out the statistical 3D device simulation with random ITs at HfO$_2$/Si interface, we first randomly generate 753 acceptor-like ITs (gray dots), as shown in Figure 3.1b, in a large 2D plane of $(224$ nm$)^2$ where the random ITs' concentration is around 1.5×10^{12} cm^{-2} in the large plane. Notably, this value is mainly for statistically generating the number of random ITs, which is not equal to the effective entire density of ITs (D_{it}). The total number of generated random ITs follows the Poisson distribution [1,42–47]. The large plane is then partitioned into many subplanes, where the size of each subplane is 16 nm^2. The number of random ITs in each subplane varies from 1 to 8 and the average number of ITs is 4, as shown in the histogram bar chart of Figure 3.1b. Each IT's energy on a subplane is randomly assigned according to the relation of D_{it} versus the energy [1,42–47], as shown in Figure 3.1b; consequently, each IT's density can be estimated according to its randomly assigned energy and the D_{it} may vary randomly from 1×10^{10} to 1×10^{12} eV^{-1} cm^{-2}, which quantitatively coincides with our experimental characterization for sub-20-nm HKMG CMOS devices. We repeat this process until all subregions are assigned. Therefore, 196 randomly generated 3D device samples with 2D random ITs at the HfO$_2$/Si interface are simulated to assess the influence of ITF.

For the RDF simulation, we mainly follow the simulation procedure reported in our recent work [4,8–13]. As shown in Figure 3.1c, the RDs in the 3D device channel region are statistically

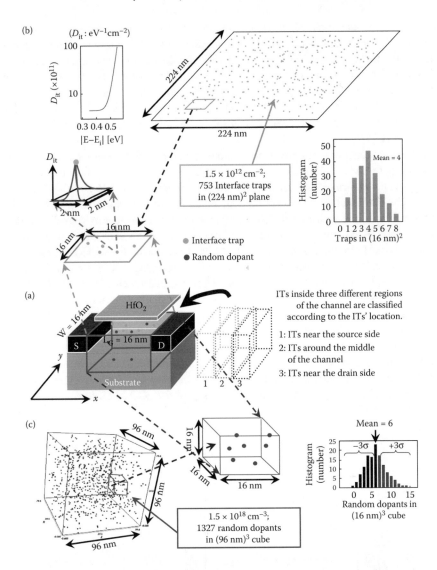

FIGURE 3.1 (a) Illustration of the studied device together with two sources of randomness: interface traps (gray dots) located at the HfO$_2$/Si interface and random dopants (dark dots) located at the silicon channel. Interface traps (ITs) inside three different regions of the channel are classified according to the ITs' location. They are, ITs near the source side, ITs around the middle of the channel and ITs near the drain side. Statistical 3D device simulation settings for the characteristic fluctuation resulting from random ITs and random dopants (RDs) are shown in (b) and (c). (b) We first generate 753 acceptor-like traps in a large plane for the 16-nm-gate N-MOSFET devices, where the IT's concentration at the large plane of $(224 \text{ nm})^2$ is around $1.5 \times 10^{12} \text{ cm}^{-2}$ and the total number of generated ITs follows the Poisson distribution. Each IT's energy on the plane is independently assigned according to the distribution of its density. Then, the entire plane is partitioned into many subplanes of $(16 \text{ nm})^2$ corresponding to the device's size, where the number of ITs in each subplane may vary from 1 to 8 and the average number is 4. Thus, the effective density of interface traps (D_{it}) varies from 3×10^{11} to 3.3×10^{12} eV^{-1}cm^{-2}. We call the ITs with high D_{it} for device D_{it} ranging from 3×10^{11} to 3.3×10^{12} eV^{-1} cm^{-2} and the ITs with low D_{it} for device D_{it} ranging from 2.5×10^{10} to 3.8×10^{11} eV^{-1}cm^{-2}. (c) For the setting of discrete dopants, impurities are randomly generated and distributed in a $(96 \text{ nm})^3$ cube with an average concentration of $1.5 \times 10^{18} \text{ cm}^{-3}$. There will be 1327 discrete dopants within the cube, and the number of discrete dopants varies from 0 to 14 (the average number is six) for all 216 subcubes of $(16 \text{ nm})^3$, which corresponds to the volume of the studied devices. Consequently, the total subcubes and subplanes can be mapped into the device's 3D channel and 2D surface for the characteristic fluctuation induced by the ITs, RDs, and combined ITs and RDs, respectively.

incorporated into the statistical 3D device simulation running on our parallel computing system [52]. Note that, for the best accuracy of our computational model, the implemented statistical 3D device simulation technique for estimating characteristic fluctuation was experimentally validated with silicon data for sub-20-nm devices in our earlier work [11], where the RDs-fluctuated mobility was validated with experimentally measured current–voltage (*I–V*) data. To compare with other significant fluctuation sources, such as PVE, oxide thickness fluctuation (OTF), and work function fluctuation (WKF) with different sized TiN grain, we follow the simulation procedures reported in our recent work [4].

The statistically generated ITs are further implemented for 16-nm-gate P-MOSFET devices so that the 16-nm-gate CMOS SRAM circuit can be simulated using coupled device–circuit solution methodology [13] to estimate the transfer characteristic fluctuation at the circuit level, as shown in Figure 3.2a. The computational flow of the coupled device–circuit simulation is shown in Figure 3.2b. Owing to the lack of well-established compact models for the 16-nm-gate CMOS devices, by using the coupled device–circuit simulation technique, the circuit-level fluctuations are estimated for the CMOS SRAM circuit, as shown in Figure 3.2a. To estimate the SRAM's static noise margin fluctuation σ_{SNM}, electrical characteristics of each randomly generated device in the tested circuit are first calculated by the 3D device simulation. The obtained result is then used as the devices' terminal characteristics in the coupled device–circuit steady-state simulation. The nodal equations of the tested SRAM circuit are formulated and then directly coupled to the device transport equations (in the form of a large matrix that contains both circuit and device equations), which are solved simultaneously to obtain the circuit transfer characteristics. The device characteristics obtained by device simulation, such as the distributions of potential and current density, are input in the SRAM circuit simulation through the device's contact terminals. Notably, to explore the influence of the intrinsic parameter fluctuation of σ_{SNM} of the SRAM circuit, the random samples are generated and performed, respectively.

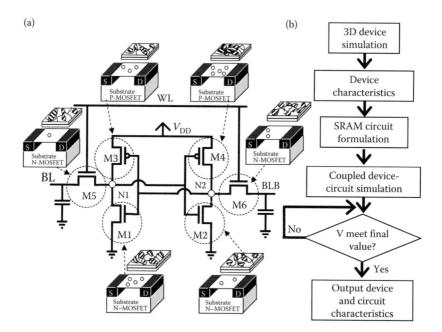

FIGURE 3.2 (a) To calculate the σ_{SNM}, the totally random generated N- and P-MOSFET devices with different random sources are assigned into the 6T SRAM circuit for the coupled device-circuit simulation. (b) The proposed flowchart for the coupled device-circuit simulation. Notably, the voltage of node $N_2(V)$ in the simulated 6T SRAM circuit is applied to 0.8 V. To assess the worst influences of ITs, RDs, and combined ITs and RDs on the 16-nm-gate SRAM's transfer characteristics, the cell ratio is set to 1 for all simulations.

3.3 RESULTS AND DISCUSSION

First, we explore the random ITs-fluctuated terminal current; Figure 3.3a shows the totally random ITs-induced fluctuations of drain current–gate voltage (I_D–V_G) curves of the 16-nm-gate N-MOSFETs with high D_{it}, where the line with circles indicates the nominal case and all dashed lines are the random ITs-fluctuated cases. The nominal case is the fresh device without random ITs. The random ITs located at the HfO$_2$/silicon interface may destroy the screening effect and the threshold voltage is simply raised. As shown in Figure 3.3b, the value of threshold voltage is determined from a constant current criterion when the drain current is greater than $10^{-7} \times (W/L)$ A, where L and W are the gate length and device width, respectively. The threshold voltage increases, and then the on-/off-state current (I_{on}/I_{off}) decreases accordingly, as the number of random ITs increases. The simulated $\sigma_{V_{th,ITS}}$ is 26.3 mV, which is smaller than the result of RDF ($\sigma_{V_{th,RDs}}$ = 43 mV). The random ITs-position induced different fluctuations of characteristics in spite of having the same number of ITs, as marked by an open bar in the inset of Figure 3.3b, where the magnitude of the spread characteristics of the threshold voltage increases as the number of random ITs increases. Figures 3.3b and c show the extracted I_{on} and I_{off} as functions of the number of random ITs, where each symbol shows each random IT-fluctuated result. The magnitude of each random ITs-fluctuated current decreases as the number of random ITs increases. Similarly, the random ITs-position induced rather different current fluctuations in spite of having the same number of random ITs, as marked by an open bar in insets of Figures 3.3c and d. To explore the impact of the random position of ITs on the device's physical characteristics, as shown in Figure 3.4, the random ITs-fluctuated surface potential and the electron

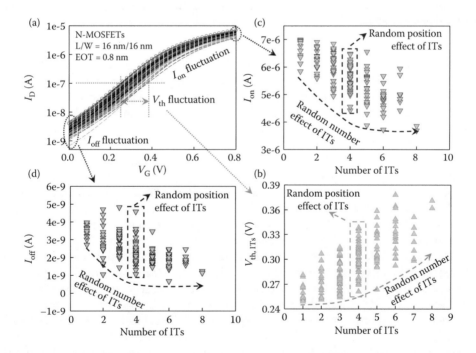

FIGURE 3.3 (a) The plot of ITs-fluctuated I_D–V_G curves shows the random ITs-induced DC characteristic fluctuations for the studied devices. The line with circles is the nominal case calibrated to V_{th} = 250 mV and others are the random ITs-fluctuated cases. From the I_D–V_G curve, we can extract the threshold voltage fluctuation, the off-state current fluctuation, and the on-state current fluctuation with respect to the number of ITs, as shown in (b), (c), and (d), respectively. We observe that the threshold voltage is increased, and thus the off-state current and the on-state current are decreased as the number of ITs is increased due to the relatively higher barriers generated. Except for the random number effect of ITs, the random position effect of ITs results in different threshold voltage for the same number of ITs.

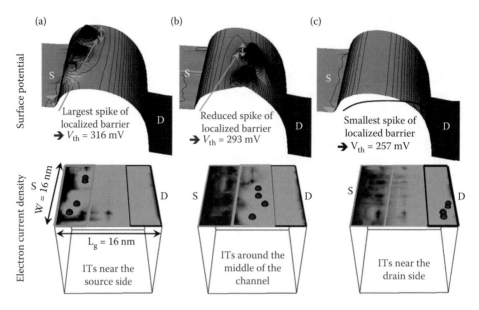

FIGURE 3.4 Plots of the ITs-fluctuated surface potential and the electron current density from the source side to the drain side for the device at the on-state condition. (a) Plots for the case of ITs near the source side. (b) Plots for the case of ITs around the middle of the channel. (c) Plots for the case of ITs near the drain side. The asymmetric impact of ITs on the surface potential results in different threshold voltage as well as a variance in the threshold voltage, as listed in Table 3.1.

current density from the source side to the drain side for the device at the on-state condition are further discussed. Three cases are selected among 196 simulations to demonstrate the random position associated local repulsive Coulomb field and disturbed electron current conducting path at the HfO$_2$/Si interface. Figure 3.4a shows the plots for the case of the random 4 ITs near the source side. Compared with the random ITs appearing in the other two regions, as shown in Figures 3.4b and c, the distribution of electron current density is seriously destroyed because of the random 4 ITs near the source side locally, which results in a relatively higher local spike of potential barriers and thus raises the threshold voltage ($V_{th} = 316$ mV). Figures 3.4b and c show plots for the case of the random 4 ITs around the middle of the channel and for the case of the random 4 ITs near the drain side. These random 4 ITs positioning away from the source side have relatively weakened local spikes of potential barriers which still effectively impede electron current conduction; however, they are weaker than that of the random 4 ITs appearing in the source side, as shown in Figure 3.4a. Thus, the device with random ITs of Figure 3.4c has minimal threshold voltage ($V_{th} = 257$ mV) and the largest area of electron current conduction, compared with the case of Figure 3.4a. The asymmetric impact of random ITs on the surface potential results in quite different threshold voltage as well as the variance of the threshold voltage, as listed in Table 3.1. The statistically simulated $I_D - V_G$ curves enable us to extract the $\sigma_{V_{th}}$, induced by different sources of fluctuations among different regions. As listed in Table 3.1, the ITF induced-$\sigma_{V_{th}}$ and σ_{SS} for the device with random ITs near the source side, around the middle of the channel, and near the drain side, respectively, are further calculated and classified. The studied 16-nm-gate N-MOSFETs with both the high and low D_{it} of ITs are simulated; significant values of $\sigma_{V_{th}}$ and σ_{SS} induced by random ITs with high D_{it} near the source side are observed, compared with the random ITs located at the middle of the channel and near the drain side. However, the simulation indicates that there is no differences in $\sigma_{V_{th}}$ and σ_{SS} for the cases of random ITs located at the middle of the channel and near the drain side. In particular, for a device with a low D_{it}, near the source side, the $\sigma_{V_{th}}$ is reduced from 28.7 to 10.2 mV; similarly, the σ_{SS} is reduced from 21.2 to 5.6 mV/Dec. The impact of RDs on the threshold voltage, SS, and DIBL, among the aforementioned three regions is

TABLE 3.1

The ITF Induced $\sigma_{V_{th}}$ and σ_{SS} for Device with ITs Near the Source Side, around the Middle of the Channel, and Near the Drain Side, Respectively

	$\sigma_{V_{th}}$ (mV)			σ_{SS} (mV/Dec)		
	Near Source Side	Around the Middle of the Channel	Near Drain Side	Near Source Side	Around the Middle of the Channel	Near Drain Side
High D_{it}	**28.7**	25.0	23.1	**21.2**	4.6	5.1
Low D_{it}	10.2	7.7	7.5	5.6	4.0	4.6

Note: The studied 16-nm-gate N-MOSFETs are with high and low D_{it} of ITs. Significant values of $\sigma_{V_{th}}$ and σ_{SS} induced by random ITs near the source side are observed, compared with the ITs located at the middle of the channel and near the drain side.

insignificant, but the impact of RDs near the channel surface is different from RDs located away from the depletion region of the channel [8–13,48,49].

The random ITs-fluctuated DIBL effect is pronounced for the 16-nm-gate N-MOSFETs, as shown in Figure 3.5a. For the device at the weak inversion region, there is a potential barrier at the channel region owing to a balance between the drift current and diffusion current. The barrier height decreases as the drain voltage increases; it thus results in an increased drain current, which is controlled not only by the gate voltage but also by the drain voltage. The DIBL effect could be observed through the $I_D - V_G$ curves of a device under the linear ($V_D = 0.05$ V) and saturated ($V_D = 0.8$ V) regions. It can be calculated by the lateral shift of threshold voltage divided by the difference of the drain voltage and is given in units (mV/V). Figure 3.5a shows the random ITs-fluctuated DIBL characteristic of the 16-nm-gate N-MOSFETs; the magnitude of DIBL decreases as the number of random ITs increases because the random ITs reduce the probability of electric-field lines penetrating from the drain side to the source side. The tendency of increasing fluctuation of DIBL follows the threshold voltage as the number of random ITs increases, as shown in Figure 3.3b. Figure 3.5b shows that the device with a high D_{it} of random ITs near the source side has a large DIBL at the same number of random ITs; Figure 3.5c shows the device with a high D_{it} of random ITs around the middle of the channel which has a small DIBL at the same number of random ITs. The random position effect of ITs on the V_{th}, DIBL, SS strongly depends on the random ITs near the source side. Such asymmetric fluctuation was also observed for devices with RDF [48,49].

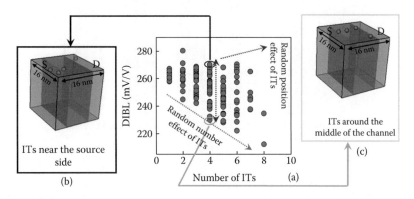

FIGURE 3.5 (a) The drain-induced barrier lowering (DIBL) versus the number of ITs. (b) Device with a high D_{it} of ITs near the source side has high DIBL at the same number of ITs. (c) Device with a high D_{it} of ITs around the middle of the channel has low DIBL at the same number of ITs.

We now turn to compare the ITs, RDs, and "ITs + RDs"-induced $\sigma_{V_{th}}$ of the studied 16-nm-gate N-MOSFETs. Figure 3.6 is the $\sigma_{V_{th}}$ as a function of EOT with respect to different fluctuation sources: ITs (with high D_{it}), RDs, and "ITs + RDs." The device with EOT = 0.8 nm exhibits $\sigma_{V_{th,RDs}}$ = 43 mV, $\sigma_{V_{th,ITs}}$ = 26.3 mV, and $\sigma_{V_{th,ITs+RDs}}$ = 45.4 mV. We note that $\sigma_{V_{th,ITs+RDs}}$ = 45.4 mV is smaller than the result calculated by $\sqrt{\sigma^2_{V_{th,ITs}} + \sigma^2_{V_{th,RDs}}}$ = 50.4 mV in which the random variables following statistically independent identical distribution (*iid*) is assumed. However, the *iid* assumption on the random variables of $V_{th,ITs}$ and $V_{th,RDs}$ is not always true owing to the local interaction of surface potentials at different extents between ITs and RDs concurrently existing in the surface/channel region of the N-MOSFETs. The relative error between $\sigma_{V_{th,ITs+RDs}}$ and $\sqrt{\sigma^2_{V_{th,ITs}} + \sigma^2_{V_{th,RDs}}}$ is about 11% overestimation, compared with the $\sigma_{V_{th,ITs+RDs}}$ of the N-MOSFETs; similarly, not shown here, the studied 16-nm-gate P-MOSFETs has $\sigma_{V_{th,ITs+RDs}}$ = 45.1 mV, which is smaller than the statistical sum $\sqrt{\sigma^2_{V_{th,ITs}} + \sigma^2_{V_{th,RDs}}}$ = 49.1 mV. Owing to the sizable threshold voltage fluctuation with respect to EOT nonlinearly, the statistical sums of the variances of two random variables induced by ITs and RDs disclose significant errors, compared with the statistical 3D device simulation together with the combined ITs and RDs simultaneously. This investigation shows that the local interaction of surface potentials owing to "ITs + RDs" could not be calculated independently by using the results of ITs and RDs. It explains why the *iid* assumption overestimates the threshold voltage fluctuations induced by the combined ITs and RDs. As shown in Figure 3.7, the threshold voltage fluctuation induced by various random sources for the 16-nm-gate N-MOSFETs, respectively, are compared. Among the five fluctuation sources, the RDF still dominates the $\sigma_{V_{th}}$ and the PVE plays a sizable fluctuation source; both should be minimized by process innovation continuously. The WKF-induced $\sigma_{V_{th}}$ is reduced from 36.7 to 11.1 mV when the minimal grain size of the TiN gate is reduced from (4 nm)^2 to (2 nm)^2. The ITF-induced $\sigma_{V_{th}}$ is reduced from 26.3 to 10.2 mV when the D_{it} is reduced from an order of 10^{12} to 10^{11} eV^{-1}cm^{-2}. We estimate the σ_{SNM} of the 6T SRAM circuit [53], as shown in Figure 3.2a. Butterfly curves fluctuation induced by RDF, PVE, WKF, and ITF of the tested 6T SRAM in read operation is first simulated, where the cell ratio CR = ((W/L)$_{\text{driver transistors:M1 and M2}}/(W/L)_{\text{access transistors:M5 and M6}}$) of the SRAM cell in this study is set as unitary, and the nominal value of SNM is 86 mV. They are then used to calculate σ_{SNM} with respect to different

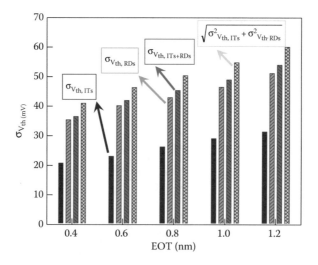

FIGURE 3.6 $\sigma_{V_{th}}$ as a function of EOT with respect to different fluctuation sources: ITs (with high D_{it}), RDs, and combined ITs and RDs (denoted as "ITs + RDs"). On the basis of the assumption of statistically independent identical distribution to the random variables of ITs and RDs, the $\sqrt{\sigma^2_{V_{th,ITs}} + \sigma^2_{V_{th,RDsp}}}$ is also calculated. Owing to the gate capacitance variation and surface potential interaction at different extents, resulting from combined ITs and RDs, the full 3D simulated $\sigma_{V_{th,ITs+RDs}}$ is smaller than the result of $\sqrt{\sigma^2_{V_{th,ITs}} + \sigma^2_{V_{th,RDsp}}}$.

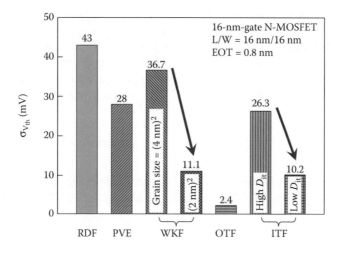

FIGURE 3.7 The threshold voltage fluctuation induced by various random sources for the 16-nm-gate N-MOSFETs. The WKF-induced $\sigma_{V_{th}}$ is reduced from 36.7 mV to 11.1 mV when the grain size of the metal gate is reduced from (4 nm)^2 to (2 nm)^2. The ITF-induced $\sigma_{V_{th}}$ is reduced from 26.3 mV to 10.2 mV when the effective density of the interface traps is reduced from high D_{it} to low D_{it}. Among fluctuations, the RDF dominates the $\sigma_{V_{th}}$ continuously and the PVE should be minimized.

TABLE 3.2

Summarized Static Noise Margin Fluctuation (σ_{SNM}) of 16-nm-gate SRAM Induced by RDF, PVE, WKF, and ITF, Respectively

Fluctuation Sources	RDF	PVE	WKF (Grain Size: 4 nm^2)	ITF (with High D_{it})
σ_{SNM} (mV)	27.4	20.8	25.3	19.1

Note: σ_{SNM} induced by WKF is with 4-nm metal grains and the σ_{SNM} induced by ITF is with high D_{it}.

fluctuation sources, as listed in Table 3.2. The relation between the device transconductance and SNM of SRAM can be viewed as $SNM \propto \sqrt{1 - (I_{nx}/g_{m,pmos})} - (I_{ax}/g_{m,nmos})$, where I_{nx} is the saturation drain current of the driver transistor of SRAM and I_{ax} is the saturation drain current of the access transistor. The σ_{SNM} is directly proportional to the fluctuation of transconductance and the fluctuation of transconductance is influenced by the $\sigma_{V_{th}}$. Thus, σ_{SNM} induced by random ITs with high D_{it} has a minimal value of 19.1 mV because the $\sigma_{V_{th,ITs}}$ is the lowest one among RDF, PVE, and WKF. However, the magnitude of σ_{SNM} for all RDF, PVE, WKF, and ITF is up to 20% of the nominal SNM, which may degrade the operation of 6T SRAM and should be improved by using multiple gate FETs or 8T SRAM cells [53].

3.4 CONCLUSIONS

In this chapter, the asymmetric characteristic fluctuation of the random ITs in 16-nm-gate HKMG MOSFETs has been presented. The random ITs-induced threshold voltage fluctuation and on-/off-state current fluctuation, drain-induced barrier-lowering fluctuation, and subthreshold swing fluctuation have been studied by classifying the random ITs into the regions near the source side, around the middle of the channel, and near the drain side of the silicon channel, respectively. For the studied 16-nm-gate HKMG N-MOSFETs with the device width of 16 nm, both high and low

D_{it}-induced $\sigma_{V_{th,ITs}}$ are smaller than that of $\sigma_{V_{th,RDs}}$ = 43 mV. The largest asymmetric values of $\sigma_{V_{th}}$ and σ_{SS} have been observed when random ITs with high D_{it} are near the source side, compared with the random ITs located at the middle of the channel and near the drain side. The engineering findings of this study indicate that both the ITs near the source side of the silicon channel and RDs near the channel surface possess the largest characteristic fluctuations. The random ITs have also exhibited sizable σ_{SNM} of the 16-nm-gate 6T SRAM circuit, compared with the σ_{SNM} induced by RDF, PVE, and WKF.

ACKNOWLEDGMENT

This work was supported in part by the National Science Council (NSC), Taiwan, under contract no. NSC 101-2221-E-009-092 and by TSMC, Hsinchu, Taiwan, under a 2011–2013 grant.

REFERENCES

1. S. M. Sze, *Physics of Semiconductor Devices*, 2nd Ed., New York, NY: John Wiley and Sons, 1981.
2. G. Moore, Progress in digital integrated electronics, in *IEDM Tech. Dig.*, 11–13, 1975.
3. International Technology Roadmap for Semiconductors [Online]. Available: http://www.itrs.net/
4. Y. Li, C.-H. Hwang, T.-Y. Li, and M.-H. Han, Process-variation effect, metal-gate work-function fluctuation, and random-dopant fluctuation in emerging CMOS technologies, *IEEE Trans. Electron Dev.*, 57(2), 437–447, 2010.
5. Y. Li, H.-W. Cheng, Y.-Y. Chiu, C.-Y. Yiu, and H.-W. Su, A unified 3D device simulation of random dopant, interface trap and work function fluctuations on high-κ/metal gate device, in *IEDM Tech. Dig.*, p. 5-5, 2011.
6. H.-W. Cheng, F.-H. Li, M.-H. Han, C.-Y. Yiu, C.-H. Yu, K.-F. Lee, and Y. Li, 3D device simulation of work-function and interface trap fluctuations on high-κ/metal gate devices, in *IEDM Tech. Dig.*, pp. 379–382, 2010.
7. K. J. Kuhn, M. D. Giles, D. Becher, P. Kolar, A. Kornfeld, R. Kotlyar, S. T. Ma, A. Maheshwari, and S. Mudanai, Process technology variation, *IEEE Trans. Electron Dev.*, 58(8), 2197–2208, 2011.
8. Y. Li, C.-H. Hwang, and T.-Y. Li, Random-dopant-induced device variability in nano-CMOS and digital circuits, *IEEE Trans. Electron Dev.*, 56(8), 1588–1597, 2009.
9. Y. Li, C.-H. Hwang, and T.-Y. Li, Discrete-dopant-induced timing fluctuation and suppression in nanoscale CMOS circuit, *IEEE Trans. Circuits and Systems Part II: Express Briefs*, 56(5), 379–383, 2009.
10. Y. Li, C.-H. Hwang, and T.-Y. Li, Random-dopant-induced variability in nano-CMOS devices and digital circuits, *IEEE Trans. Electron Dev.*, 56(8), 1588–1597, 2009.
11. Y. Li, S.-M. Yu, J.-R. Hwang, and F.-L. Yang, Discrete dopant fluctuated 20 nm/15 nm-gate planar CMOS, *IEEE Trans. Electron Dev.*, 55(6), 1449–1455, 2008.
12. Y. Li, C.-H. Hwang, and H.-M. Huang, Large-scale atomistic approach to discrete-dopant-induced characteristic fluctuations in silicon nanowire transistors, *Phys. Stat. Sol. (a)*, 205(6), 1505–1510, 2008.
13. Y. Li and C.-H Hwang, High-frequency characteristic fluctuations of nano-MOSFET circuit induced by random dopants, *IEEE Trans. Microwave Theory Tech.*, 56(12), 2726–2733, 2008.
14. J. Jaffari and M. Anis, Variability-aware bulk-MOS device design, *IEEE Trans. CAD Integrated Circuits Systems*, 27(2), 205–216, 2008.
15. Y. Li and C.-H Hwang, Discrete-dopant-induced characteristic fluctuations in 16 nm multiple-gate silicon-on-insulator devices, *J. Appl. Phys.*, 102(8), 084509, 2007.
16. Y. Li and S.-M. Yu, A coupled-simulation-and-optimization approach to nanodevice fabrication with minimization of electrical characteristics fluctuation, *IEEE Trans. Semi. Manuf.*, 20(4), 432–438, 2007.
17. N. Sano and M. Tomizawa, Random dopant model for three-dimensional drift-diffusion simulations in metal-oxide-semiconductor field-effect-transistors, *Appl. Phys. Lett.*, 79, 2267, 2007.
18. T. Ohtou, N. Sugii, and T. Hiramoto, Impact of parameter variations and random dopant fluctuations on short-channel fully depleted SOI MOSFETs with extremely thin BOX, *IEEE Electron Dev. Lett.*, 28(8), 740–742, 2007.
19. Y. Li and S.-M. Yu, Comparison of random-dopant-induced threshold voltage fluctuation in nanoscale single-, double-, and surrounding-gate field-effect transistors, *Jpn. J. Appl. Phys.*, 45(9A), 6860–6865, 2006.

20. C.L. Alexander, G. Roy, and A. Asenov, Random impurity scattering induced variability in conventional nano-scaled MOSFETs: Ab initio impurity scattering Monte Carlo simulation study, in *IEDM Tech. Dig.*, pp. 949–952, 2006.

21. H. Mahmoodi, S. Mukhopadhyay, and K. Roy, Estimation of delay variations due to random-dopant fluctuations in nanoscale CMOS circuits, *IEEE J. Solid-State Circuits*, 40(9), 1787–1796, 2005.

22. P. Dollfus, A. Bournel, S. Galdin, S. Barraud, and P. Hesto, Effect of discrete impurities on electron transport in ultrashort MOSFET using 3D MC simulation, *IEEE Trans. Electron Dev.*, 51(5), 749–756, May 2004.

23. A. Balasubramanian, P.R. Fleming, B.L. Bhuva, A.L. Sternberg, and L.W. Massengill, Implications of dopant-fluctuation-induced V_t variations on the radiation hardness of deep submicrometer CMOS SRAMs, *IEEE Trans. Dev. Mater. Reliab.*, 8(1), 135–144, 2003.

24. N. Sano, K. Matsuzawa, M. Mukai, and N. Nakayama, On discrete random dopant modeling in drift-diffusion simulations: Physical meaning of "atomistic" dopants, *Microelectron. Reliab.*, 42(2), 189–199, 2002.

25. N. Sano, K. Matsuzawa, M. Mukai, and N. Nakayama, Role of long-range and short-range Coulomb potentials in threshold characteristics under discrete dopants in sub-0.1 µm Si-MOSFETs, in *IEDM Tech. Dig.*, pp. 275–278, Dec. 2000.

26. H.-S. Wong, Y. Taur, and D. J. Frank, Discrete random dopant distribution effects in nanometer-scale MOSFETs, *Microelectron. Reliab.*, 38(9), 1447–1456, 1999.

27. P.A. Stolk, F.P. Widdershoven, and D.B.M. Klaassen, Modeling statistical dopant fluctuations in MOS transistors, *IEEE Trans. Electron Dev.*, 45(9), 1960–1971, 1998.

28. X.-H. Tang, V.K. De, and J.D. Meindl, Intrinsic MOSFET parameter fluctuations due to random dopant placement, *IEEE Trans. VLSI Systems*, 5(4), 369–376, 1997.

29. J.-R. Zhou and D.K. Ferry, 3D simulation of deep-submicron devices. How impurity atoms affect conductance, *IEEE Comput. Sci. Eng.*, 2(2), 30–37, 1995.

30. R. W. Keyes, Effect of randomness in distribution of impurity atoms on FET thresholds, *Appl. Phys.*, 8, 251–259, 1975.

31. Y. Li and H.-W. Cheng, Random work function induced threshold voltage fluctuation in metal-gate MOS devices by Monte Carlo simulation, *IEEE Trans. Semicond. Manuf.*, 25(2), 266–271, 2012.

32. Y. Li and H.-W. Cheng, Nanosized-metal-grain-induced characteristic fluctuation in 16-nm-gate complementary metal-oxide-semiconductor devices and digital circuits, *Jpn. J. Appl. Phys.*, 50(4), 04DC22, 2011.

33. K. Ohmori, T. Matsuki, D. Ishikawa, T. Morooka, T. Aminaka, Y. Sugita, T. Chikyow, K. Shiraishi, Y. Nara, and K. Yamada, Impact of additional factors in threshold voltage variability of metal/high-κ gate stacks and its reduction by controlling crystalline structure and grain size in the metal gates, in *IEDM Tech. Dig.*, pp. 1–4, Dec. 2008.

34. H. Dadgour, De Vivek, and K. Banerjee, Statistical modeling of metal-gate work-function variability in emerging device technologies and implications for circuit design, in *Proc. ICCAD*, pp. 270–277, 2008.

35. H. Daewon, H. Takeuchi, Y.-K. Choi, and T.-J. King, Molybdenum gate technology for ultrathin-body MOSFETs and FinFETs, *IEEE Trans. Electron Devices*, 51, 1989–1996, 2004.

36. A. Yagishita, T. Saito, K. Nakajima, S. Inumiya, K. Matsuo, T. Shibata, Y. Tsunashima, K. Suguro, and T. Arikado, Improvement of threshold voltage deviation in damascene metal-gate transistors, *IEEE Trans. Electron Dev.*, 48, 1604–1611, 2001.

37. J. L. He, Y. Setsuhara, I. Shimizu, and S. Miyake, Structure refinement and hardness enhancement of titanium nitride films by addition of copper, *Surf. Coat. Technol.*, 137, 38–42, 2001.

38. S. Berge, P. O. Gartland, and B. J. Slagsvold, Photoelectric work-function of a molybdenum single crystal for the (100), (110), (111), (112), (114), and (332) faces, *Surf. Sci.*, 43, 275–292, 1974.

39. Y. Li and H.-W. Cheng, Random interface-traps-induced electrical characteristic fluctuation in 16-nm-gate high-κ/metal gate complementary metal-oxide-semiconductor device and inverter circuit, *Jpn. J. Appl. Phys.*, 51(4), 04DC08, 2012.

40. Y. Li, H.-W. Cheng, and Y.-Y. Chiu, Interface traps and random dopants induced characteristic fluctuations in emerging MOSFETs, *Microelectron. Eng.*, 88(7), 1269–1271, 2011.

41. N. Ashraf, D. Vasileska, G. Wirth, and P. Srinivasan, Accurate model for the threshold voltage fluctuation estimation in 45-nm channel length MOSFET devices in the presence of random traps and random dopants, *IEEE Electron Dev. Lett.*, 32(8), 1044–1046, 2011.

42. O. Engström, Electron states in MOS systems, *ECS Trans.*, 35(4), 19–38, 2011.

43. M. Cassé1, K. Tachi1, S. Thiele1, and T. Ernst, Spectroscopic charge pumping in Si nanowire transistors with a high-κ/metal gate, *Appl. Phys. Lett.*, 96, 123506, 2010.

44. A. Appaswamy, P. Chakraborty, and J. Cressler, Influence of interface traps on the temperature sensitivity of MOSFET drain-current variations, *IEEE Electron Dev. Lett.*, 31(5), 387–389, 2010.
45. Md. Mahbub Satter and A. Haque, Modeling effects of interface trap states on the gate C-V characteristics of MOS devices on alternative high-mobility substrates, *Solid-State Electron.*, 54(6), 621–627, 2010.
46. P. Andricciola, H.P. Tuinhout, B. De Vries, N.A.H. Wils, A.J. Scholten, and D.B.M. Klaassen, Impact of interface states on MOS transistor mismatch, in *IEDM Tech. Dig.*, pp. 711–714, 2009.
47. P. K. Hurley, K. Cherkaoui, S. McDonnell, G. Hughes, and A.W. Groenland, Characterisation and passivation of interface defects in (100)-Si/SiO$_2$/HfO$_2$/TiN gate stacks, *Microelectron. Reliab.*, 47(8), 1195–1201, 2007.
48. Y. Li and K.-F. Lee, C.-Y. Yiu, Y.-Y. Chiu, and R.-W. Chang, Dual material gate approach to suppression of random-dopant-induced characteristic fluctuation in 16 nm MOSFE devices, *Jpn. J. Appl. Phys.*, 50(4), 04DC07, 2011.
49. K.-F. Lee, Y. Li, and C.-H. Hwang, Asymmetric gate capacitance and dynamic characteristic fluctuations in 16 nm bulk MOSFETs due to random distribution of discrete dopants, *Semicond. Sci. Technol.*, 25(4), 045006, 2010.
50. T.-W. Tang, X. Wang, and Y. Li, Discretization scheme for the density-gradient equations and effect of boundary conditions, *J. Comput. Electron.*, 1(3), 2002, 389–393.
51. S. Odanaka, Multidimensional discretization of the stationary quantum drift-diffusion model for ultrasmall MOSFET structures, *IEEE Trans. CAD Integr. Circuit Sys.*, 23(6), 837–842, 2004.
52. Y. Li, S. M. Sze, and T.-S. Chao, A practical implementation of parallel dynamic load balancing for adaptive computing in VLSI device simulation, *Eng. Comp.*, 18(2), 124–137, 2002.
53. Y. Li, H.-W. Cheng, and M.-H. Han, Statistical simulation of static noise margin variability in static random access memory, *IEEE Trans. Semicond. Manuf.*, 23(4), 509–516, 2010.

Section II

Nano-CMOS Technology

4 Bottom-Up Approaches for CMOS Scaling in the Nanoscale Era

Mrunal A. Khaderbad and V. Ramgopal Rao

CONTENTS

To date, feature sizes in complementary metal–oxide–semiconductor (CMOS) technologies have been scaled from 3 µm to sub-nanometers using the "top-down" scaling techniques [1]. In the sub-nanometers regime, miniaturization with the top-down techniques is not only becoming complex and expensive but also it is limited by the spatial resolution of lithography, variability, and longer fabrication turnaround times [2–4]. To overcome these issues, "bottom-up" nanotechnologies or the combination of bottom-up and top-down fabrication methodologies can be used to fabricate devices [5–8]. This approach provides a way to fabricate devices with feature sizes smaller than 10 nm, three-dimensional complex nanostructures and to make devices with novel functionalities (Figure 4.1).

In the bottom-up self-assembly, self-assembled monolayers (SAMs) of molecules are used to build well-ordered structures with novel electronic, optical, and magnetic properties [9,10]. SAMs show the feasibility to change surface chemical and physical properties at the molecular level. Thus, they have been widely used for modifying the surface wetting/adhesion properties, sensor applications, corrosion resistance, and molecular electronics [11,12]. SAMs with atomic thickness and spacing are used as ultra-thin resists and passivating layers [13]. For example, SAMs

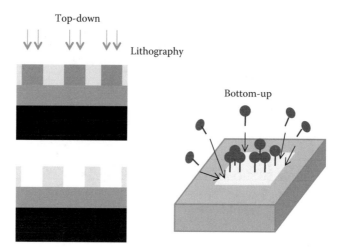

FIGURE 4.1 Top-down fabrication involving patterning, lithography, and etching versus bottom-up self-assembly process.

of 1H,1H,2H,2H-perfluorodecyltrichlorosilane [$CF_3(CF_2)_7(CH_2)_2SiCl_3$, FDTS] have been used as release and antistiction coatings in MEMS (microelectromechanical systems) technology [14].

This chapter describes the potential applications of metalloporphyrin SAMs in nano-CMOS scaling and in modifying electronic properties of graphene field effect transistors (FETs). First, an introduction to the SAMs and the formation of SAMs are discussed followed by porphyrins and their applications in nanoelectronics. Later, a few issues related to CMOS scaling are discussed. In this chapter, the applications of porphyrin SAMs as copper diffusion barriers, for gate electrode work function tuning, and in modifying the electronic properties of graphene FETs have been discussed.

4.1 SELF-ASSEMBLED MONOLAYERS

Self-assembly is the spontaneous organization of molecules into higher-ordered patterns or structures [15]. The molecules that form SAM are called *surfactants,* which comprising of a *head-group* which adsorbs on the substrate, an *end-group* that constitutes the outer surface of the film, and a *backbone* that connects the head- and the end-group. The idea of SAMs was first explained by Sagiv et al. [16] who show that the adsorption from organic solutions on polar solid surfaces can be used to produce homogeneous, compact monolayers. It was observed that the adsorption is always limited to the completion of a single monolayer, which is adsorbed directly on the surface of the substrate. Monolayer systems with many technological applications are the structures of silanes on hydroxylated surfaces (*siloxane chemistry*). SAM preparation involves the use of alkylchlorosilanes, alkylalkoxysilanes, and alkylaminosilanes to form monolayers [17,18]. Besides silanes (*R-SiX3, where X = Cl, OMe, OEt*), organometallics (*R-Li or R-MgX*) and alcohols (*R-OH*) are also used for the formation of SAMs (*where –R is the surface group*). Self-assembly of silanes takes place through surface silanol groups (–SiOH) via Si–O–Si bonds. In this process, silanol groups on hydroxylated surfaces act as a support for monolayer formation or convert to highly reactive sites such as Si–Cl or Si–Br for SAM development. In case of surfaces activated with Si–Cl bonds, monolayer formation happens via Si–O or Si–N linkages when exposed to alcohol or amine molecules [19]. Figure 4.2 depicts the monolayer formation on oxide surfaces using chlorine chemistry. Instead of the two-step chlorination/attachment process, direct R-OH condensation reaction with the –OH-terminated SiO_2 surface results in SAM deposition. Gao et al. [20] prepared highly ordered monolayers by the adsorption of octadecylphosphonic acid (ODPA) onto metal oxides such as zirconium oxide (ZrO_2), aluminum oxide (Al_2O_3), titanium dioxide (TiO_2), and

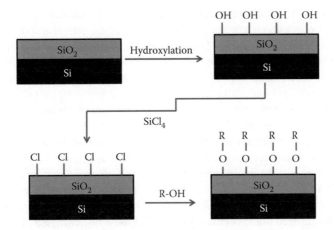

FIGURE 4.2 Formation of SAM on SiO_2 involving the initial chlorination of the surface followed by reaction with R-OH.

zirconated silica powders. Here, the covalent attachment occurs via the formation of P–O–Si ester linkage [21]. Using molecular self-assembly, Angst and Simmons [22], deposited monolayers of octadecyltrichlorosilane (OTS) and dimethyl octadecylchlorosilane (DMODCS) onto the thermal oxide of silicon. They found that close packed, high-quality monolayers were formed in the case of OTS on the hydrated oxide surface.

4.2 PORPHYRINS

Porphyrins are derived from the tetrapyrrole porphin molecule. They are aromatic conjugated molecules and can bind to all metals of the periodic table to form metalloporphyrins. In the real world, chlorophyll contains porphyrin with a magnesium central metal ion, and the Fe(II)porphyrin complex is a part of the hemoglobins and myoglobins, which transport oxygen in living beings [23]. Porphyin derivatives have numerous applications in pressure-sensitive paints, biosensors, gas sensors, and thin-film transistors [24–27]. Moreover, combining porphyrins with other organic semiconductors or devices interestingly led to novel concepts for sensing applications. For example, poly(3-hexylthiophene) (P3HT) and Cu(II)tetraphenyl porphyrin composite-based OFETs were demonstrated as sensors for the nitro-based explosive compounds [28]. Iron(III)porphyrin coated on an SU-8 micro cantilever probe with an integrated piezoresistive readout was shown as a carbon monoxide (CO) sensor [29]. Imahori and Fukuzumi [30] developed molecular photovoltaic devices by self-assembly of porphyrins and fullerenes on Au and indium tin oxide (ITO) electrodes. In these systems, porphyrins act as electron donors and fullerenes (C_{60}) as electron acceptors. In their work, thiol chemistry was used for porphyrin monolayer formation on gold and porphyrin-C_{60} dyads were combined with SAMs on ITO.

4.3 ISSUES RELATED TO CMOS SCALING

Miniaturization of CMOS devices has major advantages like increased processing power, higher transistor density, and reduced cost per transistor. But, the scaling also leads to short-channel effects (SCEs) like increased leakage currents, hot-carrier injections, and increased source–drain resistance [31,32]. Besides SCEs, in the nano-CMOS regime, there are challenges in the back-end-of-line (BEOL) processes such as increased line resistance. Though low-k dielectrics enable interconnect scaling, they pose leakage and reliability problems [33,34]. Coming to front-end-of-line (FEOL), for the 45 nm node and beyond, high-k/metal-gate technologies have been explored to reduce leakage currents and reliable high-speed operation [35]. For this, high-k materials, including hafnium oxide (HfO_2) and Al_2O_3, are investigated to replace the silicon dioxide gate dielectric. Concurrently, mid-gap metals and dual work function metals are suggested to eliminate effects such as poly-silicon depletion and poly-silicon dopant penetration. To achieve acceptable threshold voltage (V_t), gate electrodes must have appropriate work functions. In this scenario, metal gate work function tuning is an important process step to optimize device performance in nanoscale technologies.

New channel materials with improved transport properties (*heterogeneous integration*), such as SiGe, Ge, III–V semiconductors, and contemporary device structures, are being investigated to endorse CMOS scaling [36]. Graphene draws immense interest because of its electronic properties such as ballistic carrier transport and quantum Hall effect, thus making it a promising material and building block of future electronic devices and as a possible channel material for CMOS scaling. In spite of graphene's amazing properties, there are some hurdles to surmount before it could be considered as a viable candidate to replace silicon. The most important challenges with graphene are the absence of band gap and ambipolar conduction that must be addressed to facilitate graphene's use in logic devices and digital electronics [37].

In the following sections, we discuss the applications of metalloporphyrin SAMs as copper diffusion barriers, for work function tuning, and for tuning carrier injection in graphene FETs by forming zinc–porphyrin SAMs at the source–drain interface.

4.4 PORPHYRIN SAMs AS COPPER DIFFUSION BARRIERS

This section presents the application of zinc–porphyrin SAMs as Cu diffusion barriers for advanced back-end CMOS technologies. The SAM layers are integrated with various inter-layer dielectrics (ILDs) such as SiO_2, hydrogen silsesquioxane (HSQ), and black diamond (BD).

4.4.1 COPPER/LOW-*k* TECHNOLOGIES IN ULSI METALLIZATION

Resistor–capacitor (RC) parasitics play a prominent role in overall chip performance in ultra-large scale integration (ULSI) technologies. That is why, in current technologies, copper metallization is used due to its lower resistivity (1.8 $\mu\Omega$ cm) as compared with traditional Al metallization (3.3 $\mu\Omega$ cm) [38,39]. In addition, the low-*k* dielectric between interconnects results in a significant reduction in RC delay, extending the performance enhancement curve for at least one technology generation [40]. Among various low-*k* materials, silsesquioxane (*elementary unit* $(R–SiO_{3/2})_n$)-based dielectrics, with a *k* value less than that of SiO_2's relative permittivity, such as HSQ and methyl-silsesquioxane (CH_3–$SiO_{3/2}$ (*MSQ*)), have been widely explored [41,42]. Amorphous carbon and polymers have also been explored for BEOL purposes [43,44]. Lowering the *k* values in silica-based materials can be achieved by doping with fluorine or carbon or by introducing CH_3 groups [45]. Also, BD is a low-*k* film from *Applied Materials*, consisting of PECVD organosilicate material with a changeable organic phase [46]. Though Cu/low-*k* technologies provide solutions to CMOS scaling, there are reliability issues such as copper diffusion through ILD, copper drift, poor adhesion, and thermal stability [47,48].

Thus, Cu needs an appropriate diffusion barrier which is scalable along with other device dimensions [49,50].

4.4.2 Cu DIFFUSION BARRIERS IN ULSI METALLIZATION

Present CMOS technology node is 22 nm and following the 32 nm node. Future CMOS technology nodes are 15 nm, 14 nm etc., need ultrathin Cu diffusion barriers [51]. Sputtered TaN or ternary nitride alloys, such as W–Ge–N, Ta–Si–N, and Ta–W–N, deposited using reactive sputtering, can act as diffusion barriers [52–55], but we realize that thinner barriers are problematic with this kind of deposition. Barrier layers deposited by atomic layer deposition (ALD) are thinner as compared to sputtered films, but they tend to have high-defect densities and grain boundaries [56]. On the other hand, self-forming diffusion barriers reduce the overall Cu conductivity [57–59]. Alternately, SAMs with step coverage can be used as diffusion barriers. They are formed either by vapor-phase self-assembly or by wet chemical methods. Organosilane monolayers can inhibit Cu diffusion into SiO_2 and it was observed that the size and configuration of the terminal functional group and molecular chain length play an important role in the barrier properties. SAMs with long-chain lengths screen Cu atoms from the influence of the substrate and the aromatic rings sterically hinder Cu diffusion between the molecules through the SAM layer [60–63].

The following subsection presents the application of 5-(4-hydroxyphenyl)-10, 15, 20-tri (*p*-tolyl) zinc(II) porphyrin (Zn(II)TTPOH) SAMs as Cu diffusion barriers for advanced back-end CMOS technologies.

4.4.3 PORPHYRIN SAMS AS Cu DIFFUSION BARRIERS IN ULSI METALLIZATION

4.4.3.1 Bias–Temperature–Stress Analysis

Cu diffuses through ILDs under high bias–temperature conditions [64]. The diffused charge leads to a lateral shift of voltage in the capacitance–voltage (*CV*) curves. Bias-temperature stress (BTS) studies were carried out on Cu/SiO_2/*p*-Si and Cu/SAM/SiO_2/*p*-Si MOSCAP (metal–oxide–silicon capacitors) test structures to characterize the Cu diffusion (*device fabrication procedure, SAM formation can be found in ref. 65*). Figure 4.3 shows the schematic representation of MOSCAP used in this study and the Zn(II)TTPOH molecule used for SAM preparation.

Figure 4.4a describes the pre-stress and post-stress *C–V* characteristics for the MOS capacitors (EOT = 40 nm) with and without the porphyrin SAM, obtained at 50 kHz frequency, using the Agilent 4284-A precision LCR meter. In this study, a Cu/SiO_2/*p*-Si MOS capacitor was subjected to 2.5 MV/cm electric field stress at 100°C for 30 min, whereas the Cu/SAM/SiO_2/*p*-Si MOS capacitor was subjected to 4.5 MV/cm electric field stress at 100°C for 30 min. Comparing the *C–V* plots, it is evident that the *C–V* curve shift is less in the case of the Cu/SAM/SiO_2/Si MOS structure compared to that of the MOS structure without SAM [65]. This shows the effectiveness of Zn porphyrin SAM as a Cu diffusion barrier. Figure 4.4b shows the HFCV (hydrogen fuel-cell vehicle) characteristics of the Cu/HSQ/*p*-Si and Cu/SAM/HSQ/*p*-Si MIS test structures [66]. Comparing the CV plots in Figure 4.4b, it is clear that the *C–V* curve shift is less in the case of the Cu/SAM/HSQ/Si MIS structure compared to that of the MIS structure without SAM.

BTS studies (*to evaluate barrier performance*) were also performed on the MIS capacitor structures with BD as the dielectric. The devices were subjected to a stress of 1.25 MV/cm at 100°C. The CV curves obtained for devices without SAM and with SAM are shown in Figure 4.5. After stress, devices without SAM degraded to a higher extent as compared to devices with SAM [67].

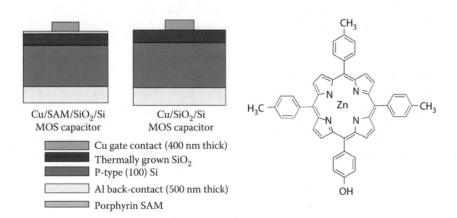

FIGURE 4.3 Device schematic and 5-(4-hydroxyphenyl)-10, 15, 20-tri (*p*-tolyl) zinc(II) porphyrin molecule.

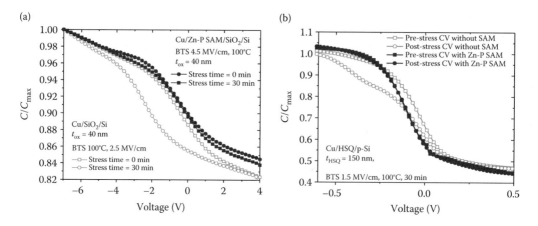

FIGURE 4.4 (a) Pre-stress and post-stress $C–V$ characteristics for the Cu MOS capacitors with and without the porphyrin SAM. (M. A. Khaderbad et al., Metallated porphyrin self assembled monolayers as Cu diffusion barriers for the nano-scale CMOS technologies, *8th IEEE Conference on Nanotechnology, 2008. NANO '08.* Arlington, Texas, pp. 167–170. © (2008) IEEE. With permission.) (b) Pre-stress and post-stress $C–V$ characteristics for the MIS capacitors with and without the porphyrin SAM. (U. Roy et al., Hydroxy-phenyl Zn(II) porphyrin self-assembled monolayer as a diffusion barrier for copperlow k interconnect technology, *Electron Devices and Semiconductor Technology, 2009. IEDST '09. 2nd International Workshop on,* pp. 1–5. © (2009) IEEE. With permission.)

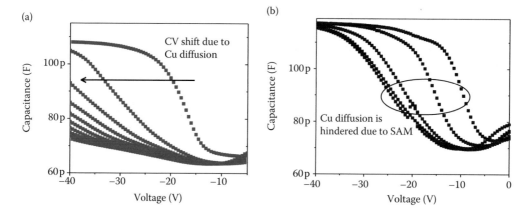

FIGURE 4.5 BTS $C–V$ characteristics (a) for the Cu/BD/Si MIS capacitors. (b) BTS CV with porphyrin SAM. (M. A. Khaderbad et al., Porphyrin Self-assembled monolayer as a copper diffusion barrier for advanced CMOS technologies, *Electron. Dev., IEEE Trans.,* 99, 1–7. © (2012) IEEE. With permission.)

4.4.3.2 Secondary Ion Mass Spectrometry

To observe the diffusion of Cu in ILD, the time-of-flight secondary ion mass spectrometry (ToFSIMS) was acquired using a TRIFT V nano-TOF instrument manufactured by Physical Electronics, MN, USA. Figure 4.6a shows the SIMS depth profile of Cu in the Cu/SiO$_2$/Si and Cu/SAM/SiO$_2$/Si samples, annealed at 400°C in N$_2$ atmosphere for 1 h. The ILD thickness is around 150 nm. It can be clearly observed that the Cu has penetrated more into the sample in which there is no SAM barrier [67]. This shows the effectiveness of the Zn(II)TTPOH monolayer as a good barrier layer to copper diffusion into SiO$_2$. Similar results are observed in samples with BD ILD (Figure 4.6b) samples, annealed at 300°C, for 30 min in N$_2$ atmosphere.

The observed BTS and SIMS analysis trends indicate that zinc porphyrin monolayers act as efficient copper diffusion barriers for porous low-k ILDs. Besides, no change in bulk permittivity or

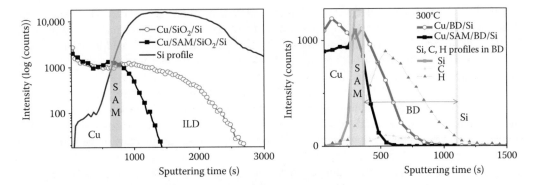

FIGURE 4.6 SIMS depth profile of copper is obtained with the Cu/SiO$_2$/Si and Cu/SAM/SiO$_2$/Si structures after annealing for 1 h in N$_2$ atmosphere. Si profile (Cu-ILD interface) can also be seen in the figure SIMS depth profile of copper obtained with the Cu/BD/Si and Cu/SAM/BD/Si structures after annealing for 1 h in N$_2$ atmosphere. (M. A. Khaderbad et al., Porphyrin Self-assembled monolayer as a copper diffusion barrier for advanced CMOS technologies, *Electron. Dev., IEEE Trans.*, 99, 1–7. © (2012) IEEE. With permission.)

the bulk mechanical properties of the ILD were observed after SAM integration. This work clearly shows that the bottom-up grown monolayers on ILDs improve the device reliability by hindering Cu diffusion without affecting the ILD film bulk properties [65–67].

The following section presents a technique for the tuning of the gate metal work function (Φ_{metal}) using SAMs of metalloporphyrins.

4.5 PORPHYRIN SAMs FOR METAL GATE WORK FUNCTION TUNING

Among various components, the MOSFET gate is an important entity in the overall transistor design and scaling. New technical challenges and SCEs arise with the scaling of feature sizes in the sub-nanometer technological regime [68,69]. The demand for gate engineering has been driven by various technical concerns [70]. For example, p-type metal-oxide-semiconductor (PMOS) and n-type metal-oxide-semiconductor (NMOS) need low and symmetrical threshold voltages (V_t) for proper switching, high-performance, and for low-power applications. In the case of polysilicon gates, this was achieved by different threshold-adjustment implants for n-type and p-type devices [71]. But, boron penetration, which is a side effect of gate doping, causes charge trapping at the gate–oxide interface, impurity diffusion into gate oxide, and uncontrollable transistor threshold voltages. Moreover, high doping levels in the polysilicon gate result in polydepletion effects for ultra-thin oxide MOSFETs due to the serial connection of the smaller polydepletion capacitance with oxide capacitance. This leads to the degradation of inversion gate capacitance and trans-conductance, and drive currents [72,73]. One alternate method to solve the above problems is to use metal gates instead of polysilicon [74,75]. Also, the thickness of the silicon dioxide (SiO$_2$) dielectric or its Si–O–N analogue is becoming sufficiently thin that the direct gate tunnelling currents through the dielectrics are posing reliability problems, increased power dissipation, and eventually deteriorate the device performance and circuit stability for VLSI circuits [76,77]. A solution to the problem is the replacement of SiO$_2$ by an alternative insulator with a higher dielectric constant, so that the physical thickness of the dielectric could be increased [78–80].

4.5.1 POTENTIAL GATE ELECTRODES FOR WORK FUNCTION ENGINEERING

The V_t of a MOSFET mainly depends on the gate work function. For bulk devices, the required metal work functions for replacing the conventional n- and p-polysilicon gates are about 4 and 5 eV, respectively, to control V_t swings. In other words, for NMOS and PMOS devices to have low and symmetric threshold voltages, two gate electrodes with different work functions are needed. This calls for the integration of multiple metals on a single substrate. In such processes, transition metals such as W, Ti, Ta, Mo, and Ru, and their metallic derivatives, WN, TiN, TaN, MoN, TaSiN, and MoSiN are used as metal gates. These transition metals are known to possess some desirable properties such as mid-gap work function, high melting point, low resistivity, and high thermal stability [81–83].

However, the use of two different metals for PMOS and NMOS devices may require a complicated, selective deposition and etching process for integration onto the same silicon substrate. However, in another approach, a single metal with a tunable work function can be used. Using a single metal is simpler from a process integration perspective. This tuning has been successfully implemented using various techniques, including alloying, metal inter-diffusion, dopant implantation, silicidation, and nitridation [84–86]. Also, SAMs of dipolar organic molecules at the metal electrode and gate dielectric interface can be used to selectively tune the work function. Molecules in SAM create an effective dipole at the interface, which enables the change in metal work function [87,88]. By means of first principles study, Heimel et al. [89] had shown that the local ionization potential and electron affinity, together with the interface dipole, determine the work function modification.

4.5.2 Porphyrin SAMs for Work Function Engineering

Here, TTPOH molecules with different central metal ions have been used to form SAMs on SiO_2, HfO_2, and Al_2O_3 gate dielectrics. Moreover, dipolar properties of porphyrin macrocycles can be tuned by incorporating various metal species in them or by using various subgroups. This allows work-function tuning for different technological applications. The Kelvin probe force microscopy (KPFM) characterizations of SAMs show local changes in the surface potential induced by the alignment of molecular dipole moments. Figure 4.7a depicts KPFM imaging of patterned Zn(II) TTPOH SAM on Si, showing higher potential with respect to Si substrate. Figure 4.7b shows the surface potential of Mg, Fe, Co, and Zn TTPOH SAMs on Si. Figure 4.7c shows the estimated dipole moments of metallo-TTPOH molecules using DFT (density functional theory) calculations (performed using the DMol3 module in the Accelrys Materials Studio [91]). It can be clearly seen that TTPOH with Co has less dipole moment as compared to that with Zn central metal.

4.5.2.1 HFCV Analysis

The chemical structure of TTPOH with different central metal ions for SAM preparation is shown in Figure 4.8a.

For work-function tuning experiments, $Al/SiO_2/Si$ and $Al/SAM/SiO_2/Si$ MOSCAP test structures were fabricated (*fabrication procedure can be found in ref. [92]*). HFCV technique was used to measure the flat-band voltage (V_{fb}), the work function difference (Φ_{ms}), and other important

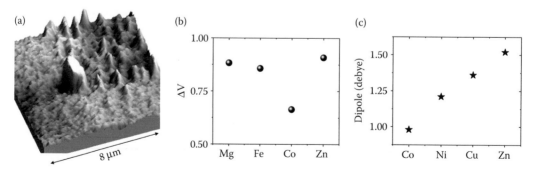

FIGURE 4.7 (a) KPFM image showing 30 mV potential corresponding to the SAM dipole [90]. (b) Surface potential measured for TTPOH SAMs with different transition metal atoms. (c) Dipole moment estimated from DFT calculations.

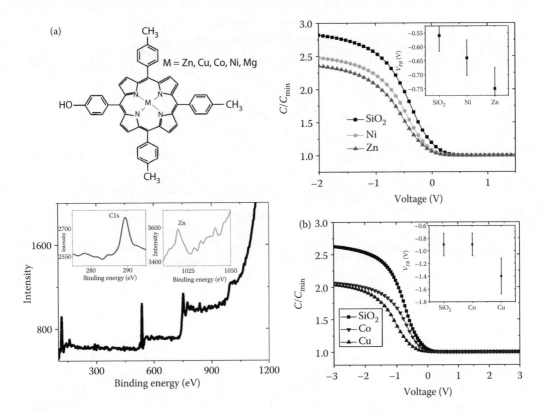

FIGURE 4.8 (a) 5-(4-Hydroxyphenyl)-10,15,20-tri(*p*-tolyl) porphyrin with different central metal ions and an XPS spectrum of Zn(II)TTPOH SAM on SiO$_2$. (b) HFCV plots Al/SiO$_2$/*p*-Si, Al/(Zn/Ni) porphyrin SAM/SiO$_2$/*p*-Si MOS CAPs (t_{ox} = 3 nm), and Al/SiO$_2$/*p*-Si, Al/(Cu, Co) porphyrin SAM/SiO$_2$/p-Si MOS CAPs (t_{ox} = 3.5 nm). (M. A. Khaderbad et al., Variable interface dipoles of metallated porphyrin self-assembled monolayers for metal-gate work function tuning in advanced CMOS technologies, *IEEE Trans. Nanotechnol.*, 9(3), 335–337. © (2010) IEEE. With permission.)

electrical parameters. The *C–V* curves for bare SiO$_2$ (3–4 nm thick) and the SiO$_2$ surface covered with TTPOH SAM with various metal derivatives are presented in Figure 4.8. The dipolar SAM layer at the metal–oxide interface modifies the potential, thereby changing the effective metal work function. This can be given as below

$$V_{fb} = \phi_{ms} - \frac{Q}{C_{ox}} \ \& \ \phi_{ms,mod} = \phi_{ms,withSAM} - \phi_{ms,withoutSAM} \qquad (4.1)$$

where $\Phi_{ms,mod}$ is the modified metal work function due to SAM.

For work-function tuning experiments on high-*k* dielectrics, HfO$_2$ was deposited on a p-type (100) Si wafer using an AMAT gate stack cluster tool with an integrated CVD chamber for MOCVD of high-*k* materials. Tetrakis(diethylamino)hafnium (TDEAH) was used as a precursor in thin-film deposition with oxygen as the oxidant gas [79,93].

Figure 4.9a illustrates the ground state UV–Vis reflection spectra of the Zn(II)TTPOH SAM on HfO$_2$, clearly showing the presence of Zn(II)TTPOH. Figure 4.9b shows the *C–V* curves of bare HfO$_2$ (EOT~5.46 nm) and the HfO$_2$ surface covered with Zn-TTPOH SAM (EOT~5.59 nm) [94]. Figure 4.9b also shows the hysteresis analysis on the above MOSCAPs.

Further, similar experiments were performed on devices with high-*k* Al$_2$O$_3$ [94]. Al$_2$O$_3$ deposition was done by reactive sputtering using a pulsed DC power supply [95,96]. Figure 4.10a shows

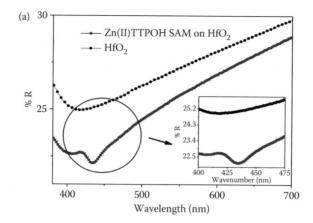

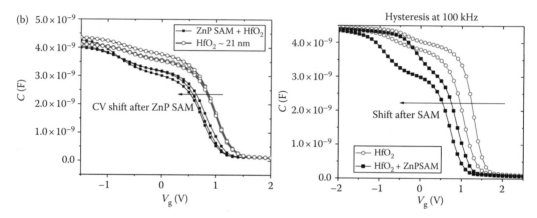

FIGURE 4.9 (a) UV reflection spectra SAM on HfO_2. (b) HFCV plots (t_{HfO2}~21 nm) showing a shift due to ZnP SAM formation on HfO_2 and hysteresis analysis on MOSCAPs (t_{HfO2}~21 nm) showing a shift due to ZnP SAM formation on HfO_2. (M. A. Khaderbad et al., Bottom-up method for work function tuning in high-k/metal gate stacks in advanced CMOS technologies, *11th IEEE Conference on, Nanotechnology, 2011. NANO '11.* Portland, Oregon, pp. 269–273. © (2011) IEEE. With permission.)

the XPS spectrum of Zn(II)TTPOH SAM on Al_2O_3 and in the UV–Vis spectra of SAM on Al_2O_3, and the peak around 429 nm clearly shows the presence of Zn(II)TTPOH porphyrin (Figure 4.10b).

Figure 4.11 shows the HFCV curves of bare Al_2O_3 (~15 nm thick and k~7) and Al_2O_3 with Cu and Ni TTPOH SAMs. As seen in the figure, the metal-gate work function was modified successfully by the adsorbed molecules and Cu(II)TTPOH SAM shows a higher shift as compared to Ni(II)TTPOH SAM. Figure 4.11b shows HFCV curves for the 5.22 nm Al_2O_3 MOSCAP with and without Cu(II) TTPOH SAM, with EOTs of 3.32 and 2.91 nm, respectively (shift~−0.79 V).

Figure 4.12a presents the flat-band voltage values for Al_2O_3 thicknesses. A constant shift in the presence of Cu(II)TTPOH SAM shows that the change in the magnitude of work function ($|\Delta\Phi_{ms}|$ ~ 0.8 V) arises from the dipole moment associated with the organic monolayer. To see the effect of temperature, the devices were exposed to annealing conditions of 400°C for an hour in ambient N_2. CV characteristics were obtained for these devices before and after annealing and are shown in Figure 4.12b. HFCV curves show that the devices are intact after exposure to these processing conditions of temperature. This shows that the above technique can be very effective in advanced CMOS technologies involving gate-last CMOS processes.

Thus, in porphyrins, central metal ion, substituents, their position, monolayer assembly, and intramolecular interactions have an effect on the value of dipole and subsequent electronic properties.

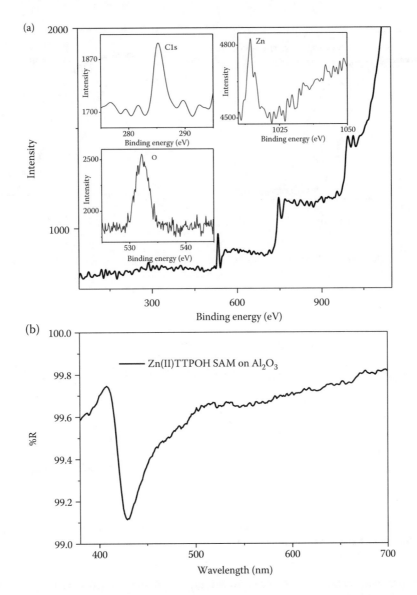

FIGURE 4.10 (a) XPS spectrum of the UV reflection spectra SAM on HfO$_2$. (b) UV reflection spectra of SAM on Al$_2$O$_3$. (M. A. Khaderbad et al., Bottom-up method for work function tuning in high-k/metal gate stacks in advanced CMOS technologies, *11th IEEE Conference on, Nanotechnology, 2011. NANO '11.* Portland, Oregon, pp. 269–273. © (2011) IEEE. With permission.)

4.6 UNIPOLAR GRAPHENE FIELD EFFECT TRANSISTORS BY MODIFYING SOURCE AND DRAIN ELECTRODE INTERFACES WITH Zn(II)TTPOH

This section refers to the unipolar operation of reduced graphene oxide (RGO) FETs by modification of source–drain (S–D) electrode interfaces with SAMs of Zn(II)TTPOH molecules.

4.6.1 RGO FETs: OVERVIEW

Graphene, a two-dimensional network of carbon atoms, has sparked interest in the research community because of its unique electrical and mechanical properties. It has a large specific surface

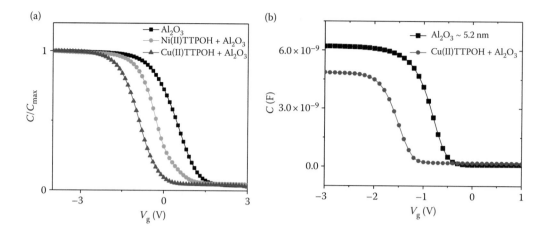

FIGURE 4.11   HFCV curves for Al_2O_3 as a gate dielectric with (a) thickness = 15 nm and (b) thickness = 5.22 nm. (M. A. Khaderbad et al., Bottom-up method for work function tuning in high-k/metal gate stacks in advanced CMOS technologies, *11th IEEE Conference on Nanotechnology, 2011. NANO '11.* Portland, Oregon, pp. 269–273. © (2011) IEEE. With permission.)

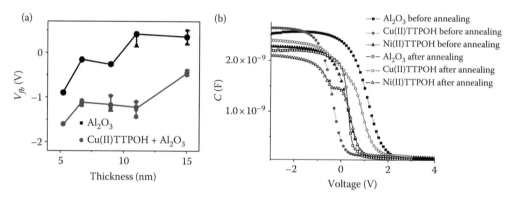

FIGURE 4.12   (a) Plot of flatband voltage versus the thickness with the corresponding error bars and (b) CV curves obtained before and after annealing at 400°C.

area, high intrinsic mobility, high Young's modulus, and thermal conductivity [95]. *Reduced graphene oxide (RGO)*, which is a solution processed form of graphene, is being considered in electrical, energy, and sensor applications [96]. The lower carrier mobilities in RGO FETs (in comparison to pristine graphene) [97] are due to the presence of defects and a disconnected network of π-delocalized regions in the carbon atom arrangement. The ambipolar conductance in RGO makes it unsuitable for fabricating logic gates or circuits. This is because the power consumption is higher in such circuits compared to unipolar logic. Previous attempts to achieve unipolar transport in graphene include nitrogen doping or the utilization of cobalt electrodes in graphene FETs, resulting in asymmetric electron hole currents in the devices [98,99].

4.6.2   Electrical Transport Modification in RGO FETs Using Porphyrin SAMs

Interface or surface modification through molecular self-assembly provides a way to modify the electronic properties of devices at lower costs [100]. In general, the hole and electron injection barrier heights are determined by the differences between the metal electrode work function (φ) and the highest occupied molecular orbital (HOMO) or the lowest unoccupied molecular orbital (LUMO) of the semiconductor. However, through the integration of dipoles at metal–semiconductor interfaces, the barrier heights can be modulated, significantly affecting the charge injection [101]. Chen et al. have used self-assembled functionalized aromatic thiols to tune the hole injection barrier (φ_h) of copper(II) phthalocyanine on Au(111) [102].

Here, controlling carrier injection in graphene FETs with the integration of Zn(II)TTPOH SAM at the electrode interfaces of RGO transistors has been reported. SAM and device fabrication have been reported elsewhere [103]. Figure 4.13 shows the SEM images of the fabricated devices.

Figure 4.14a shows the interface energy level diagram of RGO with Pt electrodes before and after SAM modification. As shown in this figure, barrier heights required for charge injection from a metal to the semiconductor with Zn(II)TTPOH interfacial modification can be expressed as [104]

$$\varphi_e = \varphi_M + \Delta\varphi_{Dipole} - \chi \tag{4.2}$$

$$\varphi_h = E_g - (\varphi_M + \Delta\varphi_{Dipole} - \chi) \tag{4.3}$$

where φ_M is the work function of the electrode, χ is the electron affinity (LUMO), and E_g is the band gap of the Zn-porphyrin; $\Delta\varphi_{dipole}$ is the barrier change due to the interface dipole.

FIGURE 4.13 SEM images of the fabricated devices. (Reprinted with permission from M. A. Khaderbad et al., Fabrication of unipolar graphene field effect transistors by modifying source and drain electrode interfaces with Zn-porphyrin, *ACS Appl. Mater. Interfaces*, 4(3), 1434–1439. Copyright 2012, American Chemical Society.)

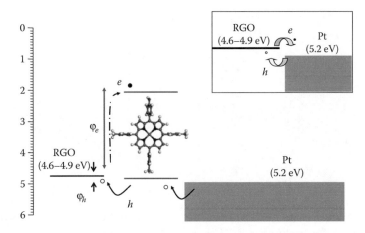

FIGURE 4.14 Schematic representation of the SAM-modified source (drain) interface (inset shows the RGO-Pt interface). (Reprinted with permission from M. A. Khaderbad et al., Fabrication of unipolar graphene field effect transistors by modifying source and drain electrode interfaces with Zn-porphyrin, *ACS Appl. Mater. Interfaces*, 4(3), 1434–1439. Copyright 2012, American Chemical Society.)

In this work, energy levels of Zn(II)TTPOH were measured using cyclic voltammetry. In a cyclic voltammogram, the onset of oxidation is related to the HOMO energy level and the LUMO energy level can be estimated from the reduction potential. For Zn(II)TTPOH, first oxidation occurs at 0.68 V and first reduction occurs at −1.41 V. These values correspond to the HOMO level of 5.1 eV and LUMO level of 2.97 eV, respectively. Zn(II)TTPOH dipole ($\Delta\varphi_{dipole}$) was estimated from the surface potential measurements of Zn(II)TTPOH SAM on RGO using KPFM. From the KPFM values and the HOMO–LUMO levels of Zn(II)TTPOH calculated from CV, φ_e and φ_h can be calculated (from Equations 4.2 and 4.3) as 2.2 and 0.11 eV, respectively, showing a higher barrier for electrons as compared to holes. At the metal–organic interfaces, the injection current I is described as follows [105]:

$$I = 4A\Psi^2 N_0 e_0\ \mu E\ \exp\left(-\frac{\varphi_B}{k_B T}\right)\exp(f^{1/2})\tag{4.4}$$

where A is the injecting area, Ψ is a slowly varying function of the electric field, N_0 is the density of the unoccupied sites in the semiconductor, μ is the bulk mobility of the injected charge carriers in the organic semiconductor, E is the electric field, φ_B is the height of the injection barrier, k_B is the Boltzmann's constant, and T is the temperature.

As injection current $I\ \alpha\ \exp(-\varphi_B/k_B T)$, a higher injection barrier for electrons as compared to that of holes is the source of unipolar behavior in Zn porphyrin-modified graphene FETs. This is

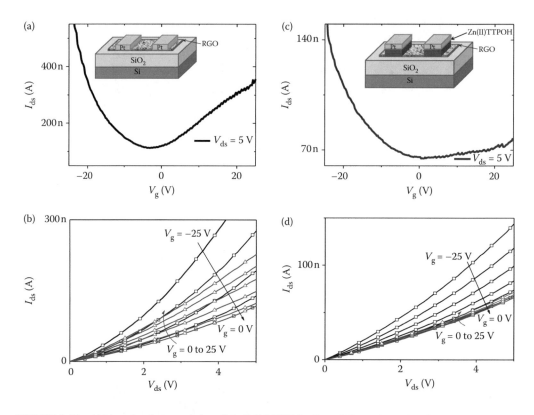

FIGURE 4.15 (a) Transfer characteristics of the RGO FET for $V_{ds} = 5$ V (a schematic of the device is shown in the inset). (b) Output characteristics of RGO FETs. (c) Transfer characteristics of the RGO-SAM FET for $V_{ds} = 5$ V. (d) Output characteristics of RGO-SAM FETs for different gate voltages. (Reprinted with permission from M. A. Khaderbad et al., Fabrication of unipolar graphene field effect transistors by modifying source and drain electrode interfaces with Zn-porphyrin, *ACS Applied Materials & Interfaces*, 4(3), 1434–1439. Copyright 2012, American Chemical Society.)

clearly observed in the *IV* characteristics of Zn porphyrin-modified graphene FETs (Figure 4.15). The fabrication of these devices is explained elsewhere [103].

Figure 4.15b shows the output characteristics of the graphene device showing the electron modulation and the hole transport by the applied gate voltage.

4.7 SUMMARY

This chapter explains the bottom-up approaches for CMOS scaling in the nanoscale era. In the BEOL processes, it has been shown that porphyrin SAMs can be used as effective copper diffusion barriers, thus improving the device performance and lifetime. BTS and SIMS studies confirm the effectiveness of zinc-porphyrin SAM as a good copper diffusion barrier. No change in permittivity of the ILD was observed with SAM integration. Nanoindentation measurements on BD films show that there is no effect on the bulk mechanical properties of BD because of SAM integration.

In the front-end-of-line (FEOL), a technique to modify the magnitude of metal work function is demonstrated by changing the central metal ion (Zn, Co, Cu, Ni) in the porphyrin SAM on SiO_2, HfO_2, and Al_2O_3 gate dielectrics. KPFM and DFT simulations of porphyrin SAMs and porphyrins show that the variation in dipole moment in porphyrins can be achieved by changing the central metal ion. HFCV analysis on MOSCAPs with different EOTs shows that the shift in flat-band voltage is due to the dipole moment associated with central metal ion in the porphyrin ring. Hence, this technique has a potential for applications in CMOS technologies involving the gate last processes.

Also, bottom-up fabrication methods can be used for interface engineering of nanoelectronic devices. Zn(II)TTPOH treatment of RGO at the source–drain interface inhibits electron injection by increasing the barrier height for electrons at the RGO–Pt interface. This work clearly demonstrates that relevant interface modifications using SAM provide a powerful approach to improve the performance of RGO FETs and are critical for applications such as logic gates and integrated circuitry.

REFERENCES

1. E. J. Nowak, Maintaining the benefits of CMOS scaling when scaling bogs down, *IBM J. Res. Dev.*, 46(2–3), 169–180, 2002.
2. Y. Ando, K. Miyake, A. Mizuno, A. Korenaga, M. Nakano, and H. Mano, Fabrication of nanostripe surface structure by multilayer film deposition combined with micropatterning, *Nanotechnology*, 21, 095304, 2010.
3. S. Samukawa, Ultimate top-down etching processes for future nanoscale devices: Advanced neutral-beam etching, *Jpn. J. Appl. Phys.*, 45(4A), 2395–2407, 2006.
4. J. Lian, L. Wang, X. Sun, Q. Yu, and R. C. Ewing, Patterning metallic nanostructures by ion-beam-induced dewetting and Rayleigh instability, *Nano Letters*, 6(5), 1047–1052, 2006.
5. W. Lu and C. M. Lieber, Nanoelectronics from the bottom-up, *Nat. Mater.*, 6, 841–850, 2007.
6. N. Nuraje, S. Mohammed, L. Yang, and H. Matsui, Biomineralization nanolithography: Combination of bottom-up and top-down fabrication to grow arrays of monodisperse gold nanoparticles along peptide lines, *Angew. Chem. Int. Ed.*, 48, 2546–2548, 2009.
7. T. Kamins, Beyond CMOS electronics: Self-assembled nanostructures, *Springer Ser. Mater. Sci.*, 106, Part III, 227–256, 2009.
8. B. Yu and M. Meyyappan, Nanotechnology: Role in emerging nanoelectronics, *Solid-State Electron.*, 50, 536–544, 2006.
9. Y. Huang, X. F. Duan, Y. Cui, L. J. Lauhon, K.-Ha Kim, and C. M. Lieber, Logic gates and computation from assembled nanowire building blocks, *Science*, 294(5545), 1313–1317, 2001.
10. M. A. Khaderbad, A. Kushagra, M. Ravikanth, and V. Ramgopal Rao, Bottom-up approaches for nanoelectronics, *Cutting Edge Nanotechnology*, D. Vasileska (Ed.), 2010, ISBN: 978-953-7619-93-0, INTECH, Croatia.
11. K. Nayak, P. Kulkarni, A. Deepu, V. Sitaraman, A. Punidha, A. Saha, M. Ravikanth, S. Mitra, S. Mukherji, and V. R. Rao, Patterned microfluidic channels using self-assembled hydroxy-phenyl porphyrin monolayer, *Proceedings of the 7th IEEE International Conference on Nanotechnology*, August 2–5, Hong Kong, 2007.
12. G. M. Whitesides and B. Grzybowski, Self-assembly at all scales, *Science,* 295, 5564, 2418, 2002.

13. Y. Xia, X. Zhao, and G. M. Whitesides, Pattern transfer: Self-assembled monolayers as ultrathin resists, *Microelectron. Eng.*, 32(1–4), *Nanotechnology*, 255–268, 1996.

14. U. Srinivasan, M. R. Houston, R. T. Howe, and R. Maboudian, Alkyltrichlorosilane based self-assembled monolayer films for stiction reduction in silicon micromachines, *J. Microelectromech. Syst*, 7, 252–60, 1998.

15. F. Schreiber, Structure and growth of self-assembling monolayers, *Progr. Surf. Sci.*, 65, 151–256, 2000.

16. J. Sagiv, Organized monolayers by adsorption. 1. Formation and structure of oleophobic mixed monolayers on solid surfaces, *J. Am. Chem. Soc.*, 102(1), 92–98, 1980.

17. A. Ulman, Formation and structure of self-assembled monolayers, *Chem. Rev.* 96, 1533–1554, 1996.

18. D. K. Aswal, S. Lenfant, D. Guerin, J. V. Yakhmi, and D. Vuillaume, Self assembled monolayers on silicon for molecular electronics, *Anal. Chim. Acta*, 568, 84–108, 2006.

19. R. C. Major and X.-Y. Zhu, Two-step approach to the formation of organic monolayers on the silicon oxide surface, *Langmuir*, 17(18), 5576–5580, 2001.

20. W. Gao, L. Dickinson, C. Grozinger, F. G. Morin, and L. Reven, Self-assembled monolayers of alkylphosphonic acids on metal oxides, *Langmuir*, 12(26), 6429–6435, 1996.

21. A. Raman, M. Dubey, I. Gouzman, and E. S. Gawalt, Formation of self-assembled monolayers of alkylphosphonic acid on the native oxide surface of SS316 L, *Langmuir*, 22(15), 6469–6472, 2006.

22. D. L. Angst and G. W. Simmons, Moisture absorption characteristics of organosiloxane self-assembled monolayers, *Langmuir*, 7(10), 2236–2242, 1991.

23. D. Dolphin, The porphyrins, *Physical Chemistry,* Part C, Academic Press, New York, 1978.

24. S. Grenoble, M. Gouterman, G. Khalil, J. Callis, and L. Dalton, Pressure-sensitive paint (PSP): Concentration quenching of platinum and magnesium porphyrin dyes in polymeric films, *Elsevier J. Lumin.* 113, 33–44, 2005.

25. D. B. Papkovsky, T. O'Riordan, and A. Soini, Phosphorescent porphyrin probes in biosensors and sensitive bioassays, *Biochem. Soc. Trans.*, 28, 74–77, 2000.

26. S. Tao and G. Li, Porphyrin-doped mesoporous silica films for rapid TNT detection, *Colloid Polym Sci.*, 285, 721–728, 2007.

27. P. Checcoli, G. Conte, S. Salvatori, R. Paolesse, A. Bolognesi, M. Berliocchi, F. Brunetti, A. D'Amico, A. Di Carlo, and P. Lugli, Tetra-phenyl porphyrin based thin film transistors, *Synth. Metals,* 138(1–2), 261–266, 2003.

28. R. S. Dudhe, S. P. Tiwari, H. N. Raval, M. A. Khaderbad, J. Sinha, M. Yedukondalu, M. Ravikanth, A. Kumar, and V. R. Rao, Explosive vapour sensor using poly (3-hexylthiophene) and Cu-tetraphenylporphyrin composite based organic field effect transistors, *Appl. Phys. Lett.*, 93, 263306, 2008.

29. C. Vijaya Bhaskar Reddy, M. A. Khaderbad, S. Gandhi, M. Kandpal, S. Patil, K. Narasaiah Chetty, K. Govinda Rajulu, P. C. K. Chary, M. Ravikanth and V. Ramgopal Rao, Piezoresistive SU-8 cantilever with Fe(III)porphyrin coating for CO sensing, *Nanotechnology, IEEE Transactions on*, PP(99), 1, 2012.

30. H. Imahori and S. Fukuzumi, Porphyrin- and fullerene-based molecular photovoltaic devices, *Adv. Funct. Mater.*, 14(6), 525–536, 2004.

31. D. J. Frank, R. H. Dennard, E. K. Nowak, P. M. Solomon, Y. Taur, and H-S. P. Philip Wong, Device scaling limits of Si MOSFETs and their application dependencies, *Proceedings of the IEEE*, 89(3), 259–287, 2001.

32. B. P. Wong, A. Mittal, Y. Cao, and G. Starr, *Nano-CMOS Circuit and Physical Design*, Chapter 1, Wiley, New Jersey, 2004, ISBN: 978-0-471-46610-9.

33. K. Maex, M. R. Baklanov, D. Shamiryan, F. Iacopi, S. H. Brongersma, and Z. S. Yanovitskaya, Low dielectric constant materials for microelectronics, *J. Appl. Phys.*, 93(11), 8793–8841, 2003.

34. C. D. Hartfield, E. T. Ogawa, Y-J. Park, T-C. Chiu, and H. Guo, Interface reliability assessments for copper/low-k products, *Device Mater. Reliab., IEEE Trans.*, 4(2), 129–141, 2004.

35. R. Chau, S. Datta, M. Doczy, B. Doyle, J. Kavalieros, and M. Metz, High-K/metal-gate stack and its MOSFET characteristics, *IEEE Electron Dev. Lett.*, 25(6), 408–410, June 2004.

36. T. Skotnicki, J. A. Hutchby, T.-J. King, H.-S. P. Wong, and F. Boeuf, Toward the introduction of new materials and structural changes to improve MOSFET performance, 2005, *Circuits Dev. Mag., IEEE,* 21(1), 16–26.

37. S. K. Banerjee, L. F. Register, E. Tutuc, D. Basu, S. Kim, D. Reddy, and A. H. MacDonald, Graphene for CMOS and beyond CMOS applications, *Proc. IEEE*, 98(12), 2032–2046, 2010.

38. H. Havemann and J. A. Hutchby, High-performance interconnects: An integration overview, *Proc. IEEE*, 89(5), 586–601, 2001.

39. S. P. Murarka, Multilevel interconnections for ULSI and GSI era, *Mater. Sci. Eng. R: Reports*, 19(3–4), 87–151, 1997.

40. B. Li, T. D. Sullivan, T. C. Lee, and D. Badami, Reliability challenges for copper interconnects, 2004, *Microelectron. Reliab.*, 44(3), 365–380.

41. E. S. Moyer, K. Chung, M. Spaulding, T. Deis, R. Boisvert, C. Saha, and J. Bremmer, Ultra low dielectric constant silsesquioxane based resin [ILDs], *Interconnect Technology, IEEE International Conference*, San Francisco, CA, pp. 196–197, 1999.

42. H-J. Lee, E. K. Lin, H. Wang, W. Wu, W. Chen, and E. S. Moyer, Structural comparison of hydrogen silsesquioxane based porous low-*k* thin films prepared with varying process conditions, *Chem. Mater.*, 14(4), 1845–1852, 2004.

43. D. T. Price, R. J. Gutmann, and S. P. Murarka, Damascene copper interconnects with polymer ILDs, *Thin Solid Films,* 308–309, 31, 523–528, 1997.

44. Jeremy A. Theil, Fluorinated amorphous carbon films for low permittivity interlevel dielectrics, *J. Vac. Sci. Technol. B,* 17, 2397, 1999.

45. K. J. Chao, P. H. Liu, A. T. Cho, K. Y. Huang, Y. R. Lee, and S. L. Chang, Preparation and characterization of low-k mesoporous silica films, *Stud. Surf. Sci. Catal., Elsevier,* 154, Part 1, 94–101, 2004.

46. W. H. Teh, C. F. Tsang, A. Trigg, K. W. Teoh, R. Kumar, N. Balasubramanian, D. L. Kwong, S. E. Ong, Farah Malik, and C. L. Gan, Adhesion studies of Ta/Low-k (black diamond) interface using thermocompressive wafer bonding and four-point bend, *J. Electrochem. Soc.,* 153(9), G795–G798, 2006.

47. Y. Shacham-Diamond, D. A. Hoffstetter, and W. G. Oldham, Reliability of copper metallization on silicon-dioxide, *VLSI Multilevel Interconnection Conference, IEEE*, Santa Clara, CA, pp. 109–115, 1991.

48. G. B. Alers, K. Jow, R. Shaviv, G. Kooi, and G. W. Ray, Interlevel dielectric failures in copper/low-k structures, *Dev. Mater. Reliab., IEEE Trans.,* 4(2), 148–152, June 2004.

49. Zs. Tokei, Y-L. Li, and G. P. Beyer, Reliability challenges for copper low-k dielectrics and copper diffusion barriers, *Microelectron. Reliab.*, 45(9–11), 1436–1442, September 2005.

50. A. E. Kaloyeros and E. Eisenbraun, Ultrathin diffusion barriers/liners for gigascale copper metallization, *Ann. Rev. Mater. Sci.*, 30, 363–385, 2000.

51. A. E. Kaloyerosa, E. T. Eisenbrauna, K. Dunna, and O. van der Straten, Zero thickness diffusion barriers and metallization liners for nanoscale device applications, *Chem. Eng. Commun.*, 198(11), 1453–1481, 2011.

52. S. R. Burgess, H. Donohue, K. Buchanan, N. Rimmer, and P. Rich, Evaluation of Ta and TaN-based Cu diffusion barriers deposited by Advanced Hi-Fill (AHF) sputtering onto blanket wafers and high aspect ratio structures, *Microelectron. Eng.,* 64(1–4), 307–313, October 2002.

53. J. Li, H-S. Lu, Y-W. Wang, X-P. Qu, Sputtered Ru-Ti, Ru-N, and Ru-Ti-N films as Cu diffusion barrier, *Microelectron. Eng.,* 88(5), 635–640, May 2011.

54. S. Rawal, D. P. Norton, T. J. Anderson, and L. McElwee-White, Properties of W–Ge–N as a diffusion barrier material for Cu, *Appl. Phys. Lett.,* 87, 111902, 2005.

55. M-A. Nicolet, Ternary amorphous metallic thin films as diffusion barriers for Cu metallization, *Appl. Surf. Sci.,* 91(1–4), 269–276, October 1995.

56. C. H. Peng, C. H. Hsieh, C. L. Huang, J. C. Lin, M. H. Tsai, M. W. Lin, C. L. Chang, W. S. Shue, and M. S. Liang, A 90 nm generation copper dual damascene technology with ALD TaN barrier, *IEDM*, 603–606, 2002.

57. J. Koike and M. Wada, Self-forming diffusion barrier layer in Cu-Mn alloy metallization, *Appl. Phys. Lett.*, 87, 041911, 2005.

58. D-C. Perng, K-C. Hsu, and J-B. Yeh, A 3 nm self-forming InO$_x$ diffusion barrier for advanced Cu/porous low-k interconnects, *Jpn. J. Appl. Phys.*, 49, 05FA04, 2010.

59. J. Koike, M. Haneda, J. Iijima, Y. Otsuka, H. Sako, and K. Neishi, Growth kinetics and thermal stability of a self-formed barrier layer at Cu-Mn/SiO2 interface, *J. Appl. Phys.*, 102, 043527, 2007.

60. A. Krishnamoorthy, K. Chanda, S. P. Murarka, G. Ramanath, and J. G. Ryan, Self-assembled near-zero-thickness molecular layers as diffusion barriers for Cu metallization,, *Appl. Phys. Lett.,* 78, 2467–2469, 2001.

61. A. M. Caro, G. Maes, G. Borghs, and C. M. Whelan, Screening self-assembled monolayers as Cu diffusion barriers, 2008, *Microelectron. Eng.*, 85(10), 239–42.

62. N. Mikami, N. Hata, T. Kikkawa, and H. Machida, Robust self-assembled monolayer as diffusion barrier for copper metallization, *Appl. Phys. Lett.* 83(25), 5181–5183, 2003.

63. P. G. Ganesan, A. P. Singh, and G. Ramanath, Diffusion barrier properties of carboxyl- and amine-terminated molecular nanolayers, *Appl. Phys. Lett.,* 85, 579–581, 2004.

64. J. Cluzel, F. Mondon, D. Blachier, Y. Morand, L. Martel, and G. Reimbold, Electrical characterization of copper penetration effects in silicon dioxide, *Annual International Reliability Physics Symposium, IEEE*, Dallas, Texas, pp. 431–432, 2002.

65. M. A. Khaderbad, K. Nayak, M. Yedukondalu, M. Ravikanth, S. Mukherji, and V. R. Rao, Metallated porphyrin self assembled monolayers as Cu diffusion barriers for the nano-scale CMOS technologies, *8th IEEE Conference on Nanotechnology, 2008. NANO '08.* Arlington, Texas, pp. 167–170.

66. U. Roy, M. A. Khaderbad, M. Yedukondalu, M. G. Walawalkar, M. Ravikanth, S. Mukherji, and V. R. Rao, Hydroxy-phenyl Zn(II) porphyrin self-assembled monolayer as a diffusion barrier for copper-low k interconnect technology, *Electron Devices and Semiconductor Technology, 2009. IEDST '09. 2nd International Workshop on*, pp. 1–5.

67. M. A. Khaderbad, R. Pandharipande, S. Madhu, M. Ravikanth, and V. R. Rao, Porphyrin Self-assembled monolayer as a copper diffusion barrier for advanced CMOS technologies, *Electron. Dev., IEEE Trans.*, 99, 1–7, 2012.

68. Y. Taur, D. A. Buchanan, W. Chen, D. J. Frank, K. E. Ismail, S-H. Lo, G. A. Halasz, R. G. Viswanathan, H. -J. C. Wann, S. J. Wind, and H-S. Wong, CMOS scaling into the nanometer regime, *Proc. IEEE*, 85(4), 486–504, 1997.

69. C. Hu, Future CMOS scaling and reliability , *Proc. IEEE*, 81(5), 682–689, 1993.

70. B. Yu, D-H. Ju, W-C. Lee, N. Kepler, T-J. King, and C. Hu, Gate engineering for deep-submicron CMOS transistors, *Electron. Dev., IEEE Trans.*, 45(6), 1253–1262, 1998.

71. G. J. Hu and R. H. Bruce, Design tradeoffs between surface and buried channel FET's, *IEEE Trans. Electron. Dev.*, 32(3), 584–588, 1985.

72. H. J. Oguey and B. Gerber, MOS voltage reference based on polysilicon gate work function difference, *Solid-State Circuits, IEEE J.*, 15(3), 264–269, 1980.

73. C-H. Choi, P. R. Chidambaram, R. Khamankar, C. F. Machala, Z. Yu, and R. W. Dutton, Dopant profile and gate geometric effects on polysilicon gate depletion in scaled MOS, *Electron. Dev., IEEE Trans.*, 49(7), 1227–1231, 2002.

74. Y-C. Yeo, Q. Lu, P. Ranade, H. Takeuchi, K. J. Yang, I. Polishchuk, T-J. King, C. Hu, S. C. Song, H. F. Luan, and D-L. Kwong, Dual-metal gate CMOS technology with ultrathin silicon nitride gate dielectric, *Electron. Dev. Lett., IEEE*, 22(5), 227–229.

75. I. Polishchuk, P. Ranade, T-J. King, and C. Hu, Dual work function metal gate CMOS technology using metal interdiffusion, *Electron. Dev. Lett., IEEE*, 22(9), 444–446.

76. D. J. DiMaria and E. Cartier, Mechanism for stress-induced leakage currents in thin silicon dioxide films, *J. Appl. Phys.*, 78(6), 3883–3894, 1995.

77. R. Moazzami and C. Hu, Stress-induced current in thin silicon dioxide films, 1992, *Electron Devices Meeting, IEDM*, 139–142.

78. N. R. Mohapatra, M. P. Desai, S. G. Narendra, and V. R. Rao, The effect of high-k gate dielectrics on deep submicrometer CMOS device and circuit performance, *Electron. Dev., IEEE Trans.*, 49(5), 826–831, 2002.

79. G. D. Wilk, R. M. Wallace, and J. M. Anthony, High-k gate dielectrics: Current status and materials properties considerations, *J. Appl. Phys.*, 89(10), 5243–5275, 2001.

80. B. Cheng, M. Cao, V. Ramgopal Rao, A. Inani, P. V. Voorde, W. Greene, Z. Yu, H. Stork, and J. C. S. Woo, The impact of high-k gate dielectrics and metal gate electrode on sub 100 nm MOSFETs, *Electron. Dev., IEEE Trans.*, 46, 1537–1544.

81. H. D. B. Gottlob, T. Echtermeyer, M. Schmidt, T. Mollenhauer, J. K. Efavi, T. Wahlbrink, M.C. Lemme et al., 0.86-nm CET gate stacks with epitaxial Gd_2O_3 High-*k* dielectrics and FUSI NiSi metal electrodes, *IEEE Electron. Dev. Lett.*, 27(10), 814–816, 2006.

82. T. Li, C. Hu, W. Ho, H. Wang, and C. Chang, Continuous and precise work function adjustment for integratable dual metal gate CMOS technology using Hf–Mo binary alloys, *IEEE Trans. Electron. Dev.*, 52, 1172–1179, 2005.

83. K. Shiraishi, K. Yamada, K. Torii, Y. Akasaka, K. Nakajima, M. Kohno, T. Chikyo, H. Kitajima, and T. Arikado, Physics in Fermi level pinning at the PolySi/Hf-based high-k oxide interface, *Tech. Dig.- Symp. VLSI Technol.*, 108–109.

84. X. P. Wang, A. Lim, M.-F. Li, C. Ren, W. Y. Loh, C. X. Zhu, A. Chin et al., Work function tunability of refractory metal nitrides by lanthanum or aluminum doping for advanced CMOS devices, *IEEE Trans. Electron Dev.*, 54(11), 2871–2877, 2007.

85. J. H. Sim, H. C. Wen, J. P. Lu, and D. L. Kwong, Work function tuning of fully silicided NiSi metal gates using a TiN capping layer, *Electron. Dev. Lett., IEEE*, 25(9), 610–612.

86. D. S. Yu, A. Chin, C. H. Wu, M. –F. Li, C. Zhu, S. J. Wang, W. J. Yoo, B. F. Hung, and S. P. McAlister, Lanthanide and IR-based dual metal-gate/HfAlON CMOS with large work-function difference, *Electron Devices Meeting, 2005. IEDM Technical Digest. IEEE International*, pp. 634–637, 2005.

87. B. de Boer, A. Hadipour, M. M. Mandoc, T. van Woudenbergh, and P. W. M. Blom, Tuning the metal work function with self-assembled monolayers, *Adv. Mater.,* 17(5), 621–625, 2005.

88. K. Asadi, F. Gholamrezaie, E. C. P. Smits, P. W. M. Blom, and B. Boer, Manipulation of charge carrier injection into organic field-effect transistors by self-assembled monolayers of alkanethiols, *J. Mater. Chem.*, 17, 1947–1953, 2007.

89. G. Heimel, L. Romaner, E. Zojer, and J.-L. Brédas, Toward control of the metal–organic interfacial electronic structure in molecular electronics: A first-principles study on self-assembled monolayers of π-conjugated molecules on noble metals, *Nano Letters,* 7(4), 932–940, 2007.

90. M. A. Khaderbad, M. Rao, K. B. Jinesh, R. Pandharipande, S. Madhu, M. Ravikanth, and V. R. Rao, Effect of central metal ion on molecular dipole in porphyrin self-assembled monolayers, *Nanosci. Nanotechnol. Lett.,* 4(7), 729–732(4).

91. B. Delley, From molecules to solids with the DMol3 approach, *J. Chem. Phys.,* 113, 7756–7764, 2000.

92. M. A. Khaderbad, U. Roy, M. Yedukondalu, M. Rajesh, M. Ravikanth, and V. R. Rao, Variable interface dipoles of metallated porphyrin self-assembled monolayers for metal-gate work function tuning in advanced CMOS technologies, *IEEE Trans. Nanotechnol.,* 9(3), 335–337, 2010.

93. S. V. Elshocht, R. Carter and M. Caymax et. al, Scalability of MOCVD-deposited hafnium oxide, *Mat. Res. Soc. Symp. Proc.,* 765, D2.7.1–D2.7.6, 2003.

94. M. A. Khaderbad, R. Pandharipande, A. Gautam, A. Mishra, M. Bhaisare, A. Kottantharayil, Y. Meesala, M. Ravikanth, and V. R. Rao, Bottom-up method for work function tuning in high-k/metal gate stacks in advanced CMOS technologies, *11th IEEE Conference on Nanotechnology, 2011. NANO '11.* Portland, Oregon, pp. 269–273.

95. J. C. Meyer, A. K. Geim, M. I. Katsnelson, K. S. Novoselov, T. J. Booth, and S. Roth, The structure of suspended graphene sheets, *Nature,* 446(7131), 60–63, 2007.

96. Y. Si and E. T. Samulski, Synthesis of water soluble graphene, *Nano Letters,* 8(6), 1679–1682, 2008.

97. K. A. Mkhoyan, A. W. Contryman, J. Silcox, D. A. Stewart, G. Eda, C. Mattevi et al., Atomic and electronic structure of graphene-oxide, *Nano Letters,* 9(3), 1058–1063, 2009.

98. X. Wang, X. Li, L. Zhang, Y. Yoon, P. K. Weber, H. Wang et al., N-doping of graphene through electrothermal reactions with ammonia, *Science,* 324(5928), 768–771, 2009.

99. R. Nouchi, M. Shiraishi, and Y. Suzuki, Transfer characteristics in graphene field-effect transistors with Co contacts, *Appl. Phys. Lett.,* 93(15), 152104–3, 2008.

100. Y. Selzer and D. Cahen, Fine tuning of Au/SiO$_2$/Si diodes by varying interfacial dipoles using molecular monolayers, *Adv. Mater.,* 13(7), 508–11, 2001.

101. X. Cheng, Y. Y. Noh, J. Wang, M. Tello, J. Frisch, and R. P. Blum et al., Controlling electron and hole charge injection in ambipolar organic field-effect transistors by self-assembled monolayers, *Adv. Funct. Mater.,* 19(15), 2407–15, 2009.

102. W. Chen, C.n Huang, X. Y. Gao, L. Wang, C. G. Zhen, D. Qi, S. Chen, H. L. Zhang, K. P. Loh, Z. K. Chen, et al., Tuning the hole injection barrier at the organic/metal interface with self-assembled functionalized aromatic thiols, *J. Phys. Chem. B,* 110(51), 26075–26080, 2006.

103. M. A. Khaderbad, V. Tjoa, M. Rao, R. Phandripande, S. Madhu, J. Wei, M. Ravikanth, N. Mathews, S. G. Mhaisalkar, and V. R. Rao, Fabrication of unipolar graphene field effect transistors by modifying source and drain electrode interfaces with Zn-porphyrin, *ACS Appl. Mater. Interfaces,* 4(3), 1434–1439, 2012.

104. H. Ishii, K. Sugiyama, E. Ito, and K. Seki, Energy level alignment and interfacial electronic structures at organic/metal and organic/organic interfaces, *Adv. Mater.,* 11(8), 605–25, 1999.

105. J. C. Scott and G. G. Malliaras, Charge injection and recombination at the metal–organic interface, *Chem. Phys. Lett.,* 299(2), 115–9, 1999.

5 Study of Lanthanum Incorporated HfO$_2$ Nanoscale Film Deposited as an MOS Device Structure Using a Dense Plasma Focus Device

A. Srivastava and Y. Malhotra

CONTENTS

5.1 INTRODUCTION

Scaling of a field effect transistor (FET) has been the major driving force behind the continuous improvements of the silicon-based semiconductor chips for achieving higher packaging density, higher speed, and lower power consumption/dissipation. With the mass scale production technology node reaching 45 nm and below, there was an urgent requirement of oxide thickness of less than 1.2 nm. Such an ultra-thin gate oxide led to direct tunneling resulting in an exponential increase of gate leakage current which was unacceptable due to the huge power consumption [1,2]. So, a hafnium-based high-κ dielectric constant was adopted, which replaced the traditional SiO$_2$ due to its relatively high dielectric constant and wideband gap. But hafnium-based gate dielectric materials posed new problems such as an increase in scattering for carriers, a lower effective mobility, and an increase in the interface-state density [4]. The continuous and aggressive scaling of the transistor further leads to a demand for new and better high-κ gate dielectric stacks to overcome the problem [1,2]. Lathanum incorporated in hafnium seems to be one of the promising new approaches to overcome the issues related to the hafnium-based high-κ dielectric stacks. There are many reports of attempts to incorporate La$_2$O$_3$ into HfO$_2$ film, and improvements in oxide-charge density (Q_{ox}) as well as D_{it} have been reported [4,5].

Another important problem with high-κ gate dielectric materials/stacks is the choice of technique used to deposit the dielectric film/stacks. This is very critical in determining the deposited film properties and the interface properties of gate dielectric with silicon substrate [1,3]. This chapter investigates the microstructure and electrical properties of HfO$_2$ and La$_2$O$_3$/HfO$_2$ gate stacks deposited using a novel deposition technique, namely the dense plasma focus (DPF) under optimized conditions. It further investigates other important parameters through electrical characterization like flat-band voltage (V_{fb}) and oxide-charge density (Q_{ox}) extracted from a high-frequency (1 MHz) $C–V$ curve.

5.2 EXPERIMENTAL DETAILS

HfO_2 and La_2O_3 targets having high purity (99.9% purity) and supplied by M/s Semiconductor Technology in the form of a disk are inserted at the top of the anode to deposit thin films using the DPF machine subsystem (schematic-[5]). P-type silicon (100) having resistivity around 10 Ω-cm was used as a substrate to deposit La_2O_3/HfO_2 gate stacks. The wafers were cleaned using the standard cleaning procedure to remove the organic and inorganic contaminations in which the wafers were first etched in dilute HF (1:20), then rinsed in deionized water, and finally dried in dry N_2 atmosphere immediately before loading in the vacuum chamber. The silicon substrate was then inserted from the top of the plasma chamber with the help of a brass rod and kept at an optimized distance from the top of the anode. The shutter is placed in between the anode top and the substrate to exclude all those ions hitting the substrates that are produced without good focusing. The background pressure of the vacuum chamber was kept at 10^{-3} mm Hg. The deposition is carried out in high-purity ambient gas, while the important DPF parameters like capacitor voltage as well as gas pressure in the plasma chamber are optimized for best focusing so that it can ionize the disk of material to be deposited. Throughout the deposition process of the gate dielectric material, the gas pressure is maintained at 80 Pa. The capacitor is charged to about 15 kV by a high-voltage charger as discharge takes place through the electrode assembly with the help of a spark gap arrangement. The gas breakdown starts between the anode and the cathode near the insulating sleeve. In the DPF device, plasma generation consists of three phases: (a) inverse pinch phase, (b) axial phase, and finally (c) radial collapse or the focus phase. In the focused phase, the plasma has a density of ~10^{26} m^{-3} and a temperature of ~1–2 keV. At this phase, approximately 10^{17} ions per burst are produced. Once the strong evidence of formation of focused plasma is achieved after a couple of shots, the shutter is opened. The high-temperature, high-density argon-focused plasma created at the top of the anode ionizes the hafnium oxide target to produce fully ionized ions which move upward in a fountain-like structure along with the argon ions in the postfocus phase and are deposited on the substrate by a single focus shot. Once the ions from the hafnium oxide target are deposited and the substrates cool down, the modified anode on which the hafnium oxide target was fixed is removed through a glass window and the modified anode fitted with lanthanum oxide target is grooved into the central anode. Subsequently, ions from the lanthanum oxide target are deposited using argon plasma produced in the DPF device using one focused shot on the top of the hafnium oxide deposited film which is above the p-silicon substrate. The film thickness is around 30 nm, which is measured using an ellipsometer (SENTECH SE850). The top electrodes in the form of aluminum (Al) dots on the as-deposited gate dielectric stacks were fabricated by a thermal evaporator using the shadow mask technique having a diameter of around 200 μm, and the back contact was fabricated again by the thermal evaporator to get a planar Al thin film having a thickness of about 200 nm. One of the major advantages of using DPF is that it can be used to deposit not only gate stacks using only two shots, but also for phase change of materials [7–10], which would lead to better electrical properties due to the lesser defects/trap both within the film and at the interface when films are treated under inert gas ambient inside the DPF chamber. Moreover, it may not require post-annealing processing steps on deposited gate stacks, which is a very essential step in the traditional/normal deposition techniques to reduce oxide-charge density, interface traps, and so on.

In this chapter, we have not used post-deposition annealing techniques to reduce oxide-charge density, interface traps, and others, which are detrimental the reliability and performance of the device. Instead, we have treated the film in inert gas such as argon or other gases like nitrogen and so on in order to reduce the oxide-charge density, interface traps, and so forth, thereby reducing the expensive fabrication steps and cost.

5.3 RESULTS

The surface morphology is studied using CP-II atomic force microscopy (AFM) in noncontact mode. Figure 5.1 shows the AFM image with a scan area of 1×1 μm^2 for the deposited La$_2$O$_3$/HfO$_2$ gate stacks on silicon substrates using two DPF shots. The AFM image shows the formation of nanostructures. The size of the nanostructures is found to lie around 10–20 nm. The maximum height of the nanostructures is found to be above 20 nm. The average height, rms roughness, and roughness average of nanostructures are found to be around 60, 33, and 26 nm, respectively.

The x-ray diffraction (XRD) studies that are used to examine the structural properties of the deposited gate stack are carried out using the Rigaku, Rotaflex rotating anode XRD instrument. Figure 5.2 shows the XRD spectra of the La$_2$O$_3$/HfO$_2$ gate stacks deposited by DPF. The peaks marked with #, *, and □ correspond to HfO$_2$, La$_2$HfO$_7$, and La$_2$O$_3$. The presence of La$_2$HfO$_7$ peaks confirms the incorporation of lanthanum oxide in the hafnium oxide-based gate dielectric fabricated over the silicon wafer.

The preliminary results of capacitance versus voltage and current versus voltage characteristics are measured using an Agilent (HP) mode 4284A LCR meter and a semiconductor characterization system KEITHLEY 4200-SCS, respectively. The variation of gate current with gate

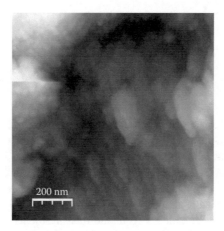

FIGURE 5.1   AFM characteristics of La$_2$O$_3$/HfO$_2$ thin film MOS capacitors.

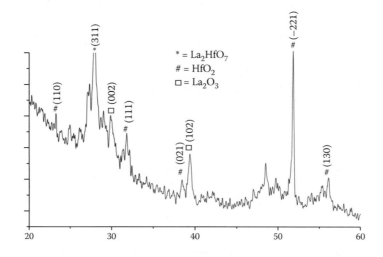

FIGURE 5.2   XRD spectra of La$_2$O$_3$/HfO$_2$ gate stacks deposited by DPF.

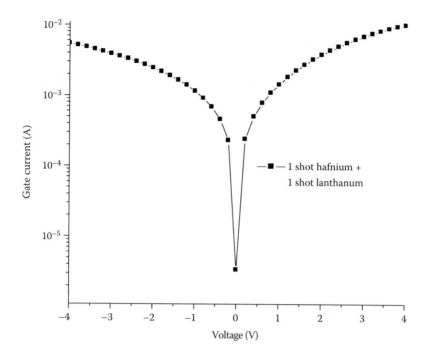

FIGURE 5.3   Current–voltage characteristics of pure La_2O_3/HfO_2 gate dielectric stacks deposited using the DPF system under optimized conditions.

voltage is plotted in Figure 5.3. The accumulation, depletion, and inversion can be clearly seen and the leakage current can be determined from the gate current versus the gate voltage graph. The variation of capacitance with gate voltage is plotted in Figure 5.4. The capacitance is measured for two DPF shots when one shot is of hafnium and the other is of lanthanum giving rise to the lanthanum-incorporated hafnium-based gate dielectric. The other important parameters such as the dielectric constant, electrical thickness, oxide capacitance, flat-band voltage, and the threshold voltage are determined and tabulated from the measurements performed at 1 MHz for the lanthanum-incorporated hafnium-based gate dielectric stacks MOS (metal–oxide–semiconductor) device deposited using two DPF shots.

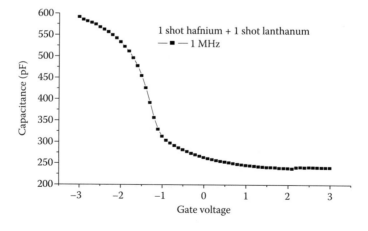

FIGURE 5.4   Capacitance–voltage characteristics of pure La_2O_3/HfO_2 gate dielectric stacks deposited using the DPF system under optimized conditions.

TABLE 5.1

Electrical Parameters of MOS Devices at 1 MHz

Electrical Parameters

Dielectric constant (ε)	26
Oxide capacitance (C_{ox} (pF))	590
Oxide thickness (T_{ox} (nm))	32
Flat-band capacitance (C_{FB} (pF))	385
Flat-band voltage (V_{FB} (V))	−1.28
Threshold voltage (V_{TH} (V))	−0.46

The dielectric constant value was determined to be 26, the oxide thickness value around 32 nm, the oxide capacitance around 590 pF, and the flat-band capacitance around 385 pF determined by measurements from a 1 MHz capacitive voltage curve. The flat-band voltage was determined to be −1.28 V, while the threshold voltage was determined to be −0.46 V for 1 MHz with a diameter of a dot around 200 μm as tabulated in Table 5.1.

5.4 CONCLUSION

Preliminary results show a successful deposition of gate dielectric stacks of La$_2$O$_3$/HfO$_2$ using only two DPF shots. The AFM image shows the formation of nanostructures of La$_2$O$_3$/HfO$_2$ gate stacks. The RMS roughness and average roughness of the dielectric film prepared was of the order of a few nanometers. The electrical results of the La$_2$O$_3$/HfO$_2$ gate stacks deposited using two DPF shots showed improved values of capacitance with lanthanum incorporation. The other important electrical parameters such as leakage current, oxide charges/capacitance, threshold voltage, and flat-band voltage with the incorporation of lanthanum in hafnium were successfully determined. We extend our study to find the relationship between the electrical and morphological properties of La$_2$O$_3$/HfO$_2$ gate dielectric stacks as an MOS structure using the DPF deposition technique. The use of a DPF machine with less processing steps for the deposition of gate dielectric stacks and avoiding the post deposition annealing steps has been well established.

ACKNOWLEDGMENTS

The authors would like to thank Professor M. P. Srivastava, Physics Department, Delhi University, for providing useful experimental support to deposit La$_2$O$_3$/HfO$_2$ gate dielectric stacks using two shots of the DPF device on silicon substrate. The authors also wish to thank Professor V. Ramgopal Rao, Electrical Engineering Department, Indian Institute of Technology, Mumbai, India, for his useful comments, suggestions, and experimental support to carry out electrical characterization under the INUP project titled "NANO SCALE MOSFETs—Scalability Issues and Possible Solutions" at IIT, Mumbai.

REFERENCES

1. H. Wong and H. Iwai, On the scaling issues and high-κ replacement of ultrathin gate dielectrics for nanoscale MOS transistors, *Microelectron. Eng.* 83, 1867, 2006.
2. H. Wong and H. Iwai, The road to miniaturization, *Phys. World*, 18, 40, 2005.
3. A.Srivastava, R. K. Nahar, and C. K. Sarkar, Study of the effect of thermal annealing on high k hafnium oxide thin film based MOS and MIM capacitor, *Journal of Materials Science: Materials in Electronics* (Springer), 22, 882, 2010.
4. H. Fujisawa, A. Srivastava, K. Kakushima, P. Ahmed, K. Tsutsui, N. Sugii, T.Hattori, C. K. Sarkar, and H. Iwai, Electrical characterization of W/HfO$_2$ MOSFETs with La$_2$O$_3$ incorporation, *ECS Trans.*, 18(1), 39, 2009.

5. A. Srivastava, R. K. Nahar, C. K. Sarkar, and Y. Malhotra, Study of hafnium oxide thin film deposited using dense plasma focus machine as a gate dielectric for a MOS device, *Microelectron. Reliab.*, 51(4), 751, 2011.

6. D. Zade, S. Sato, K. Kakushima, A. Srivastava,, P. Ahmet, K. Tsutsui, A. Nishiyama et al., Effects of La$_2$O$_3$ incorporation in HfO2 gated nMOSFETs on low-frequency noise, *Microelectron. Reliab.*, 51(4), 746. 2011.

7. R. S. Rawat, M. P. Srivastava, S. Tandon, and A. Mansingh, Crystallization of an amorphous lead zirconate titanate thin film with a dense-plasma-focus device, *Phys. Rev. B*, 47, 4858, 1993.

8. R. Sagar and M. P. Srivastava, Amorphization of thin flim of Cds due to ion irradiation by dense plasma focus, *Phys. Lett. A*, 183, 209, 1996.

9. P. Agarwala, S. Annapoorni, M. P. Srivastava, R. S. Rawat, and P. Chauhan, Magnetite phase due to energetic argon ion irradiation from dense plasma focus on hematite thin film, *Phys. Lett. A*, 231, 434, 1997.

10. P. Agarwal, M. P. Srivastava, P. N. Dheer, V. P. N. Padmabhan, and A. K. Gupta, Enhancement in Tc of superconducting BPSCCO thick films due to irradiation of energetic argon ions of dense plasma focus, *Physica C*, 313, 87, 1999.

6 Low-Power Reliable Nano Adders

Azam Beg, Mawahib Hussein Sulieman, Valeriu Beiu, and Walid Ibrahim

CONTENTS

6.1 INTRODUCTION

Addition is a common arithmetic operation in a wide variety of digital applications. Full adder (FA) cells are used in many arithmetic operations and are crucial in both central and floating-point units. They are also used extensively for cache as well as for memory address calculations. The soaring demand for mobile electronic devices such as portable computers, smart phones, and tablets necessitates power-efficient very-large-scale integration (VLSI) circuits. As FA cells are on the critical path, they determine the system's overall performance. That is why designing faster and low-power FAs was the main driving force behind many of the reported results [1–4].

For several decades, increasing the performance by reducing the propagation delay was the key design objective of the VLSI community. During that time, the FA's performance and power consumption have also improved significantly due to the relentless scaling of CMOS (complementary metal–oxide–semiconductor) devices. CMOS scaling was always used to implement faster, smaller, and cheaper integrated circuits, which were optimized for minimizing delay. However, with the massive scaling of the CMOS devices, both leakage and dynamic currents have increased exponentially. Simultaneously, with the constant increase in demand for battery-operated mobile devices (especially during the last decade), energy efficiency has become the most stringent design objective, replacing propagation delay as the main design objective.

Unfortunately, the continuous scaling of CMOS technology has also brought several new challenges. With the transistor channel length closing in on 10 nm, the manufacturers run into a number of fundamental limitations. This massive scaling of the CMOS transistors is introducing large parameter fluctuations, including the threshold voltage (V_{TH}) variations [5–7]. This has led to a constant reduction in the available reliability margins. However, most of the FA designs which have been proposed and analyzed assumed that the basic fabric of gates/transistors is reliable enough. That is why reliability was not considered as an optimization criterion. It is only very recently that the reliability of FAs has started to be thoroughly investigated [8–12]. Such papers have tried to evaluate and compare the reliability of the existing FAs, but have neglected power and delay. In Refs. [13,14], we have used a reverse sizing scheme (proposed in Ref. [15]) to simultaneously improve on both the reliability and the power efficiency of two FAs. The simulation results have shown that using the reverse sizing scheme allows for significantly improving the reliability and power efficiency at the expense of a slight increase in propagation delay.

This chapter presents yet another technique for sizing the transistors for increasing the reliability (measured here in terms of static noise margin, SNM) of FAs of different styles of design, specifically targeting advanced CMOS processes, that is, below 32 nm. The proposed sizing method takes advantage of the short channel effects (which become significant below 32 nm) by using them favorably to simultaneously increase reliability and reduce power. This chapter is organized as follows. Section 6.2 provides a brief review of low-power FAs. The effect of the transistor sizing on V_{TH} variations is presented in Section 6.3. The classical, reverse, and optimal sizing methods are detailed in Section 6.4, followed by simulation results and analyses in Section 6.5, and the concluding remarks in Section 6.6.

6.2 LOW-POWER FULL ADDERS

Many low-power FA designs based on pass-transistor logic have been proposed [16–19]. In general, the pass-transistor FAs have fewer gates/transistors and achieve lower power consumption [20]. However, they do suffer from degraded signal levels due to V_{TH} loss and charge sharing. One of the early low-power FA designs is the SERF (Static Energy Recovery Full adder) [16], which has two XNOR gates and one MUX (multiplexer). These three gates are implemented using 10 pass transistors.

A systematic approach for constructing 10-transistors FAs was proposed in Ref. [17]. This paper has detailed 41 different FAs based on XOR–XNOR gates. An in depth analysis has shown that three of the 41 proposed FAs consume less power than the SERF. Another FA design consists of six MUXs, each implemented by two pass transistors, for a total of 12 transistors [18]. An alternate FA design is the Complementary and Level Restoring Carry Logic (CLRCL) [19]. This FA consists of one XOR and one MUX. Both functions were implemented using pass-transistor logic. Although the CLRCL seems similar to the FAs described previously, it has the clear advantage of having only one V_{TH} loss when compared to the other FAs which exhibit double V_{TH} losses.

The CLRCL avoids multiple V_{TH} losses by restoring the degraded output at each stage using CMOS inverters. The single V_{TH} loss of the CLRCL FA design should enable it to function at lower supply voltages. Simulation results have shown that the CLRCL FA has the lowest energy consumption and lowest delay compared to the other adders in 35 nm technology [19]. Considering these advantages, we have decided to select this FA design and enhance it by replacing the pass transistors with transmission gates (TGs). These minimize signal degradation by limiting it only to charge sharing. The schematic representation of the CLRCL FA with TGs (TG FA) is shown in Figure 6.1. It should be mentioned that two inverters had to be added to the circuit presented in Ref. [19] for driving the TGs. One inverter generates the complementary input carry (C'_{in}), while the second one restores the voltage level of the *Sum* output.

6.3 EFFECTS OF THRESHOLD VOLTAGE VARIATIONS

One of the fundamental limitations of the bulk MOSFETs is their accuracy in reproducing V_{TH} over the transistors inside a chip (intradie variations). This problem arises from the random fluctuations of both *the number of dopants* and of *their physical locations*. Previous simulations [21] have suggested that V_{TH} variations could be approximated by a Gaussian distribution with standard deviation:

$$\sigma V_{TH\_RDF} = 3.19 \times 10^{-8} T_{ox} N_{ch}^{0.4} \sqrt{W_{eff} L_{eff}} \tag{6.1}$$

where t_{ox} is the oxide thickness, N_A is the channel doping, while L_{eff} and W_{eff} are the channel effective length and width, respectively.

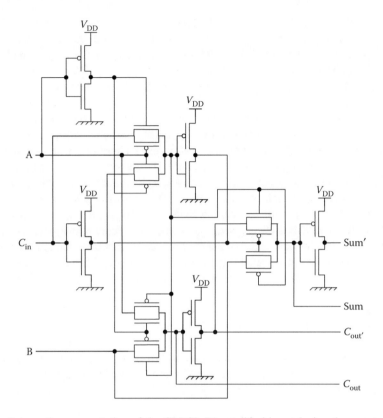

FIGURE 6.1 Schematic representation of the CLRCL FA modified by replacing the pass transistors with transmission gates (TG FA).

From Equation 6.1, it is clear that σ is inversely proportional to W_{eff} and L_{eff}. It follows that increasing size ($W_{\text{eff}} \times L_{\text{eff}}$) will reduce σ, and should improve the transistor's reliability. In Ref. [22], we have used σ to estimate the switching probability of failure of both nMOS and pMOS transistors. The simulation results have clearly shown that the probability of failure of the CMOS transistors depends on not only the transistor type (i.e., nMOS or pMOS) but also on the logic applied at the gate terminal (i.e., HI or LO), as well as on the variations of these voltages (which can be related to the static and dynamic noise margins of the gates). It was also shown that the switching probability of failure of a pMOS transistor is better than that of an equally sized nMOS one. The main conclusion which has emerged was that, in order to improve the reliability of CMOS gates, the reliability of the nMOS transistors has to also be improved, ideally matching the reliability of the pMOS transistors.

6.4 CLASSICAL, REVERSE, AND OPTIMAL TRANSISTOR SIZING

For classical CMOS transistors, the drive current when a transistor is "ON" is

$$I_{\text{ON}} \propto \frac{W_{\text{eff}}}{L_{\text{eff}}} \times \mu^* \times C_{\text{ox}} \times (V_{\text{DD}} - V_{\text{TH}})^2, \tag{6.2}$$

where μ^* is the carrier effective mobility. As mentioned before, maximizing the performance has for many years been the fundamental objective of VLSI designers. Therefore, they routinely adjusted the sizing of the nMOS and the pMOS transistors in order to maintain the same I_{ON} current when either the pull-up (I_{up}) or the pull-down (I_{down}) stacks are switched "ON." The aim of transistor sizing is to balance the rise and fall times. To maximize the performance while balancing the I_{up} and

I_{down} currents, VLSI designers were setting the length of the nMOS and pMOS transistors to the technological minimum ($L_{nMOS} = L_{pMOS} = min$). This minimizes both the resistances and the gate's capacitances. Afterward, the width of the nMOS was set to $W_{nMOS} = 2 \times L_{nMOS}$. Finally, W_{pMOS} was adjusted (increased) based on the gate topology such that $I_{up} = I_{down}$. This was required due to the difference in mobility between the holes (in case of pMOS) and the electrons (in case of nMOS).

Increasing W_{pMOS} to balance the rise and fall times does improve the reliability of the pMOS transistors (as increasing W_{pMOS} increases the area $W_{pMOS} \times L_{pMOS}$, reducing σ_{pMOS}). However, as mentioned in Section 6.3, this does not affect the reliability of the CMOS gate because the pMOS transistors are already more reliable than the nMOS ones. Therefore, when relying on the classical sizing method, improving the nMOS's reliability by increasing W_{nMOS} can be done only by making W_{pMOS} even larger. To overcome this constraint of the classical sizing, in Ref. [15], we have proposed a reverse sizing scheme. The aim of the reverse sizing scheme was to improve the reliability of the nMOS transistors by making their areas ($W_{nMOS} \times L_{nMOS}$) relatively larger than the area of the pMOS transistors ($W_{pMOS} \times L_{pMOS}$), while also reducing the dynamic power dissipation (as limiting I_{ON}). The reverse sizing scheme relies on the fact that changing the transistor's W and L dimensions affects both its reliability (see Equation 6.1) and the I_{ON} current (see Equation 6.2). The reverse sizing method starts by reducing W_{nMOS} and W_{pMOS} to the minimum technological limit ($W_{nMOS} = W_{pMOS} = min$), which implicitly limits I_{ON}, and setting $L_{pMOS} = 2 \times W_{pMOS}$. Afterward, L_{nMOS} is increased to compensate for the difference in mobility (between electrons and holes). This is exactly the opposite of classical sizing, which keeps L_{nMOS} and L_{pMOS} to the technological minimum and increases W_{pMOS} to compensate for the slower mobility of the holes.

It is worth mentioning that balancing the rise and fall times by selecting the proper transistor sizing depends on the transistor type, and on how the transistors are connected together. Transistors connected in series should be sized differently than the transistors connected in parallel or individual transistors.

The dynamic power dissipation is calculated as

$$P = a/2 \ C f V^2, \tag{6.3}$$

where a is the fraction of clock cycles in which the output transitions, C is the load capacitance, f is the switching frequency, and V_{DD} is the supply voltage. Obviously, the simplest and most effective way to reduce power dissipation is to reduce V_{DD}. Ultra low-power FA designs operating in sub-V_{TH} have been proposed [2,4]. However, operating in sub-V_{TH} raises two major concerns. The first one is that reducing V_{DD} reduces I_{ON}, which in turn degrades the performance. The second one is that reducing V_{DD} negatively affects the reliability [23–25] as the effects of variations (not only V_{TH} variations) become more pronounced.

To overcome the above constraint, an optimal sizing method is proposed here. Its aim is not only to balance the rise and fall times but also to maximize the allowed static noise margins (SNMs). Maximizing the allowed SNMs (as is commonly done for RAM cells) will push the limits where the circuit will still operate correctly, even if V_{DD} is reduced and the variations are high. This should allow reducing V_{DD} and the associated dynamic power dissipation. Instead of keeping one of the transistor dimensions to the technological minimum and changing the other dimension to balance the rise and fall time, the optimal sizing method works by allowing all the widths and lengths of the transistor's channels to be adjusted simultaneously. For maximizing the SNMs, the optimal sizing method starts by taking advantage of the short channel effects and adjusts the lengths of the CMOS transistors such as $V_{TH\_nMOS} = V_{TH\_nMOS} = V_{DD}/2$. Only afterward, the balancing of the rise and fall times is done by adjusting the widths of the transistors (for classical CMOS sizing).

For a specific channel length, V_{TH} can be calculated as

$$V_{TH} = VTH0 - \frac{0.5 \times (ETA0 + ETAB \times V_{bseff})}{\cosh(DSUB \times L_{eff}/l_t) - 1} \times V_{ds}, \tag{6.4}$$

where *VTH0* is the threshold voltage of a long channel device (10 μm) at $V_{bs} = 0$, *ETA0* and *ETAB* are, respectively, the drain-induced barrier lowering (DIBL) coefficient in the sub-threshold region and the body-bias coefficient for the sub-threshold DIBL effect, *DSUB* is the coefficient exponent in the sub-threshold region, L_{eff} is the effective channel length, V_{bs_eff} is the effective substrate bias voltage, V_{ds} is the drain-to-source voltage, and l_t is the characteristic length which depends on the depletion width X_{dep}, the oxide thickness t_{ox}, and the gate dielectric constant (see Ref. [26]).

It is clear from Equation 6.4 that V_{TH} can easily be adjusted by slightly modifying the channel length. We have used the BSIM4v4.7 level 54 [26] for finding the *optimum transistor length* (L_{opt}) for both the nMOS and the pMOS transistors such that $V_{TH\_nMOS} = V_{TH\_pMOS} = V_{DD}/2$. Afterward, we have used the classical sizing approach, that is, we set the width of the nMOS transistor to $W_{nMOS} = 2 \times L$ min (the minimum technological size), and the width of the pMOS channel is adjusted such that the *voltage transfer curve* (VTC) is symmetric (a sizing which also balances the rise and fall times).

6.5   RESULTS AND ANALYSES

Simulations were performed to study the effect of using the optimal sizing method on the performance, the power consumption, and the reliability of two FA cells: the standard mirrored 28T FA (Figure 6.2) and the TG FA (Figure 6.1). All the simulations were done using NGSpice (ver. 24) with 22 nm Predictive Technology Model v2.1 (metal gate, high-*k*, and strained-Si) [27,28]. Power and delay for the FAs have been measured using the test setup circuit shown in Figure 6.3.

To calculate the SNMs, the results from NGSpice were inputted into a MATLAB script which plots the voltage transfer characteristics (VTCs) and calculates V_{IL}, V_{IH}, V_{OL}, V_{OH}, and SNM. The MATLAB script is used to automatically locate the inflection points (slope = −1) of the VTCs. The script calculates $d(V_{out})/d(V_{in})$ and identifies the minimum and maximum points, which are the two points that correspond to the (V_{IL}–V_{IH}) and (V_{OL}–V_{OH}) of the VTC. The SNM is then easily calculated as $min((V_{IL}-V_{OL}), (V_{OH}-V_{IH}))$.

It is important to mention that, in case of multi-input–multi-output circuits such as the FA, the shapes of the VTCs (and their associated SNMs) depend upon the applied input vectors. The VTCs for the TG FA (because its SNMs are better than 28T FA) and the worst-case SNMs are shown in Figure 6.4.

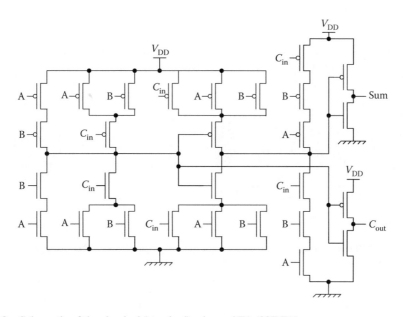

FIGURE 6.2   Schematic of the classical (standard) mirrored FA (28T FA).

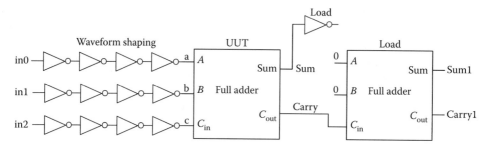

FIGURE 6.3 The test circuit used for measuring the power consumption and the delay of an FA.

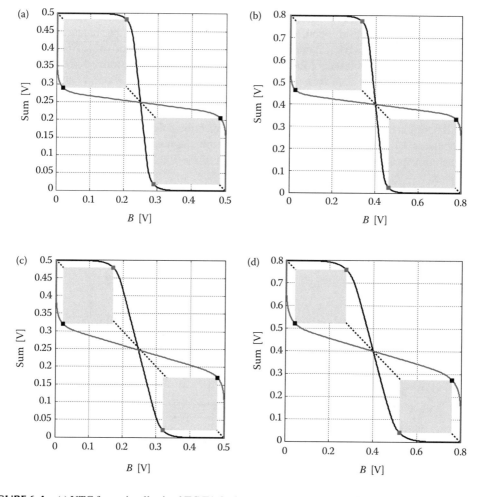

FIGURE 6.4 (a) VTC for optimally sized TG FA for input B and output Sum (when input $A = 0$; input $C_{in} = 1$; and $V_{DD} = 0.5$ V). (b) VTC for optimally sized TG FA for input B and output Sum (when input $A = 0$; input $C_{in} = 1$; and $V_{DD} = 0.8$ V). (c) VTC for classically sized TG FA for input B and output Sum (when input $A = 1$; input $C_{in} = 0$; and $V_{DD} = 0.5$ V). (d) VTC for classically sized TG FA for input B and output Sum (when input $A = 1$; input $C_{in} = 0$; and $V_{DD} = 0.8$ V).

The simulation results show that, for the 22 nm PTM HP technology, the L_{opt} for the pMOS nodes (L_{opt_pMOS}) is 29.4 nm, and for the nMOS nodes (L_{opt_nMOS}) it is 24.9 nm. This means that increasing the length of the nMOS channel by only 2.9 nm (13.2%) and the pMOS channel by 7.4 nm (35.9%) will improve the noise margin by setting V_{TH_nMOS} and V_{TH_pMOS} to $V_{DD}/2$. After resolving the optimal channel length for nMOS and pMOS nodes, W_{nMOS} is set to 44 nm and W_{pMOS} is adjusted such that the VTCs are as balanced as possible (equivalently, the rise and fall times are balanced). Table 6.1 shows the classical and the optimal sizing for the 28T and TG-FAs.

Table 6.2 shows the SNM, power, delay, and power-delay product (PDP) for the 28T FA and TG FA in case of a nominal ($V_{DD} = 0.8$ V) as well as a lower supply voltage ($V_{DD} = 0.5$ V). The power and delays have been measured with inputs of 50% duty cycles at a frequency of 100 MHz.

Table 6.2 shows that, in the case of nominal V_{DD} (0.8V), using the optimal sizing method improves the SNMs of the 28T FA and the TG FA by 35% and 33%, respectively, compared to the classical sizing method. The optimal sizing method maintains its efficiency even at lower voltages. Using the optimal sizing method at $V_{DD} = 0.5$V improves the SNMs of the 28T FA by 28% and of the TG FA by 27%.

A comparison between the classical and optimal-sized FAs at the same supply voltage shows that the optimally sized ones consume more power than the classical counterparts. However, such a comparison is misleading as, due to the significant improvement in SNMs, the supply voltage in case of the optimal FAs can be reduced while still achieving the same reliability as the classical designs at nominal V_{DD}. The optimal sized FAs consume less than the classical sized FAs at the same SNM (reliability). Table 6.2 shows that using the optimal sizing method at $V_{DD} = 0.5$ V will reduce the power consumption of the 28T FA by 79%, while increasing the delay by about 8× and

TABLE 6.1

Classical and Optimal Transistor Dimensions (nm) for 28T and TG FAs

		pMOS		nMOS	
Circuit	Sizing Method	W	L	W	L
28T FA	Optimal	205	29.4	44	24.9
	Classical	65	22	44	22
TG FA	Optimal	205	29.4	44	24.9
	Classical	65	22	44	22

TABLE 6.2

Comparison of 28T and TG FAs under Different Conditions

Type	Sizing Method	V_{DD} [V]	SNM for Sum [V]	Power [W]	Delay [s]	PDP [J]
28T FA	Optimal	0.5	0.189	2.83E-08	6.08E-10	1.72E-17
		0.8	0.307	7.83E-08	7.38E-11	5.78E-18
	Classical	0.5	0.148	1.29E-08	9.67E-11	1.25E-18
		0.8	0.227	5.06E-08	3.02E-11	1.53E-18
TG FA	Optimal	0.5	0.189	2.59E-08	8.97E-10	2.33E-17
		0.8	0.311	8.67E-08	9.45E-11	8.19E-18
	Classical	0.5	0.149	1.35E-08	2.89E-10	3.91E-18
		0.8	0.233	9.31E-08	4.03E-11	3.75E-18

the PDP by 3×. In the case of TG FA, using the optimal sizing method at $V_{DD} = 0.5$ V reduces the power consumption by 3.34× , while increasing both the delay by 9.5× and PDP by 2.8×.

6.6 CONCLUSIONS

A method for increasing the SNM (which is linked to reliability) of digital circuits has been introduced in this chapter. The method has been evaluated for two different FAs, the classical CMOS mirrored FA (28T FA) and the one using TGs (TG FA) when implemented in the 22 nm PTM HP. The method significantly enhances the SNMs at both the nominal as well as lower supply voltages. Additionally, the power is also reduced while the delay is increased when moving to larger channel lengths.

REFERENCES

1. M. Alioto and G. Palumbo, High-speed/low-power mixed full adder chains: Analysis and comparison versus technology, *Proc. ISCAS*, New Orleans, LA, USA, May 2007, pp. 2998–3001.
2. S. Aunet, B. Oelmann, T. S. Lande, and Y. Berg, Multifunction subthreshold gate used for a low power full adder, *Proc. NorChip*, Oslo, Norway, Nov. 2004, pp. 44–47.
3. S. Goel, A. Kumar, and M. A. Bayoumi, Design of robust, energy-efficient full adders for deep-submicrometer design using hybrid-CMOS logic style, *IEEE Trans. VLSI Syst.*, 14, Dec. 2006, 1309–1321.
4. K. Granhaug and S. Aunet, Six subthreshold full adder cells characterized in 90 nm CMOS technology, *Proc. DDECS*, Prague, Czech Republic, Apr. 2006, pp. 25–30.
5. SIA, *International Technology Roadmap for Semiconductors* (ITRS), 2011 [Online]. Available at: http://public.itrs.net
6. H. Iwai, Roadmap for 22 nm and beyond (invited), *Microelectr. Eng.*, 86, Jul.-Sep. 2009, 1520–1528.
7. C. Millar, D. Reid, G. Roy, S. Roy, and A. Asenov, Accurate statistical description of random dopant-induced threshold voltage variability, *IEEE Electr. Dev. Lett.*, 29, Aug. 2008, 946–948.
8. S. Purohit, M. Margala, M. Lanuzza, and P. Corsonello, New performance/power/area efficient, reliable full adder design, *Proc. GLSVLSI*, Boston, MA, USA, May 2009, pp. 493–498.
9. T. J. Dysart and P. M. Kogge, Analyzing the inherent reliability of moderately sized magnetic and electrostatic QCA circuits via probabilistic transfer matrices, *IEEE Trans. VLSI Syst.*, 17, Apr. 2009, 507–516.
10. H. B. Marr, J. George, D. V. Anderson, and P. Hasler, Increased energy efficiency and reliability of ultra-low power arithmetic, *Proc. MWSCAS*, Knoxville, TN, USA, Aug. 2008, pp. 366–369.
11. W. Ibrahim, V. Beiu, and M. H. Sulieman, On the reliability of majority gates full adders, *IEEE Trans. Nanotech.*, 7, Jan. 2008, 56–67.
12. W. Ibrahim and V. Beiu, Threshold voltage variations make full adders reliabilities similar, *IEEE Trans. Nanotech.*, 9, Nov. 2010, 664–667.
13. M. H. Sulieman and W. Ibrahim, Design of low-power and reliable nano adders, *Proc. IEEE-NANO*, Portland, USA, Aug. 2011, pp. 441–444.
14. W. Ibrahim, A. Beg and V. Beiu, Highly reliable and low-power full adder cell, *Proc. IEEE-NANO*, Portland, USA, Aug. 2011, pp. 500–503.
15. M. H. Sulieman, V. Beiu, and W. Ibrahim, Low-power and highly reliable logic gates: Transistor-level optimizations, *Proc. IEEE-NANO*, Seoul, Korea, Aug. 2010, pp. 254–257.
16. R. Shalem, E. John, and L. K. John, A novel low-power energy recovery full adder cell, *Proc. GLSVLSI*, Ann Arbor, USA, March 1999, pp. 380–383.
17. H. Bui, Y. Wang, and Y. Jiang, Design and analysis of low-power 10-transistor full adders using novel XOR-XNOR gates, *IEEE Trans. Circuits Systems—II*, 49, Jan. 2002, 25–30.
18. Y. Jiang, A. Alsheridah, Y. Wang, E. Shah, and J. Chung, A novel multiplexer-based low power full adder, *IEEE Trans. Circuits Systems—II*, 51, Jul. 2004, 45–48.
19. J. F. Lin, M. H. Sheu, and C. C. Ho, A novel high-speed and energy efficient 10-transistor full adder design, *IEEE Trans. Circuits Systems—II*, 54, May 2007, 1050–1059.
20. R. Zimmermann and W. Fichtner, Low-power logic styles: CMOS versus pass-transistor logic, *IEEE J. Solid-State Circ.*, 32, Jul. 1997, 1079–1090.
21. A. Asenov, A. R. Brown, J. H. Davies, S. Kaya, and G. Slavcheva, Simulation of intrinsic parameter fluctuations in decananometer and nanometer-scale MOSFETs, *IEEE Trans. Electr. Dev.*, 50, Sep. 2003, 1837–1852.

22. W. Ibrahim and V. Beiu, Using Bayesian networks to accurately calculate the reliability of complementary metal oxide semiconductor gates, *IEEE Trans. Reliab.*, 40, Jul. 2011, 538–549.

23. D. Bol, Robust and energy-efficient ultra-low-voltage circuit design under timing constraints in 65/45 nm CMOS, *J. Low Power Electr. Appl.*, 1, Jan. 2011, 1–19.

24. D. Bol, R. Ambroise, D. Flandre, and J.-D. Legat, Interests and limitations of technology scaling for subthreshold logic, *IEEE Trans. VLSI Syst.*, 17, Oct. 2009, 1508–1519.

25. M. Alioto, Understanding DC behavior of subthreshold CMOS logic through closed-form analysis, *IEEE Trans. Circ. & Syst. 1*, 57, Jul. 2010, 1597–1607.

26. *BSIM4v4.7 MOSFET Model, User's Manual*, Apr. 2011, Available at: http://www-device.eecs.berkeley.edu/bsim/Files/BSIM4/BSIM470/BSIM470_Manual.pdf

27. W. Zhao and Y. Cao, New generation of predictive technology model for sub-45 nm early design exploration, *IEEE Trans. Electr. Dev.*, 53, Nov. 2006, 2816–2823.

28. *Predictive Technology Model* [Online]. Available at: http://ptm.asu.edu

Section III

Nano Capacitors

7 Package-Compatible High-Density Nano-Scale Capacitors with Conformal Nano-Dielectrics

Himani Sharma, P. Markondeya Raj,
Parthasarathi Chakraborti, Yushu Wang, and Rao Tummala

CONTENTS

7.1 INTRODUCTION

Passive components support key electrical functions in any electronic system. They constitute ~80% of all components [1] and take up to ~50% of the printed wiring board area [2]. They, therefore, substantially influence the size, cost, and performance of the electronic system as a whole. Capacitors (C) are the most challenging among all the passives (R, L, and C). They perform a multitude of important functions such as decoupling, noise suppression, power storage conditioning and modulation, sensing, and signal processing. The never-ending demand for the high-performance, portable electronics has propelled the need for miniaturization of electronic components, including the capacitors. This demand for miniaturization entails the fabrication of thin power modules that are less than 200 µm in thickness to be embedded into 3D silicon and organic packages [3]. Additionally, these packages need to be reliably and safely operated over a broad temperature range [4]. So far, these passive components, especially the capacitors, have been major impediments on the road to system miniaturization and thickness reduction. Although the demand for high-density capacitors in integrated thin power modules has been increasing, the volumetric density of the available discrete capacitors has only gone through incremental changes over the past few decades. This is because of several fundamental limitations with existing capacitor technologies. Today's high-density capacitors face the imminent demand of ultrahigh volumetric densities, ultralow leakage currents, high-frequency operation, and faster charge–discharge speeds. The current capacitor technologies are stretching their limits to explore novel materials, processes, and integration methods to meet the demands, but suffer from several fundamental limitations. A new class of nanocapacitor technologies is needed to address these limitations, which is the main focus of this chapter.

Starting from the origins of the capacitor that can be traced back to the invention of the Leyden jar, a ceramic capacitor [5], the capacitor technology has taken a leap through its 250 years of technical evolution. The capacitor development process has been evolutionary rather than revolutionary. Early development was primarily driven by the need for banks of low-cost energy storage capacitors used in large-pulse power systems. The production of aluminum electrolytic capacitors in 1936 was a stepping stone toward the production of commercial capacitors and was widely used during the Second World War [6]. The discovery of barium titanate (BT) (with a relative permittivity of 1000) in 1941 [7] and its family of materials, thereafter, led to its use as a passive electronic component in a host of defense equipment. The use of fabrication technologies such as tape casting and the ceramic electrode cofiring process during the 1970s and 1980s led to the development of the multilayer ceramic capacitor (MLCC). The MLCC-based capacitors have demonstrated the potential to achieve capacitance densities in the range of 100 $\mu F/cm^2$. Dielectrics such as BT used in MLCCs have high dielectric constants. As of 2010, the global annual BT-based MLCC stands at 1 trillion units. However, they are incompatible with silicon because of their high-temperature processing requirements. Therefore, they cannot be integrated into silicon packages as thin films. The state-of-the-art tantalum-based capacitor technology with high volumetric densities serves as an alternative route to the above approach. There are also several issues, though, with the integration of existing tantalum-based capacitors into the silicon package. Tantalum processing with existing microparticles requires high-temperature processing in the range of more than 1500°C [8]. Moreover, the volumetric density of the system is further reduced in the final system when the packaging volume is taken into account. Trench capacitors address the limitations of achieving high capacitance density using thin films on the walls of the trench in silicon. However, the silicon trench capacitors have not exceeded capacitance densities of more than 40–50 $\mu F/cm^2$ and yet suffer from the need for high-cost tools [9]. In view of the above limitations, researchers around the globe are working on novel capacitor technologies to pursue high capacitance densities (>100 $\mu F/cm^2$) in thin-film high-density capacitors. A simple 2D graph in Figure 7.1 shows the effective surface area enhancement and the corresponding capacitance densities of the current and emerging technologies.

The basic capacitor technology meets its target capacitance by enhancing one or more of the three fundamental parameters—electrode surface area, dielectric permittivity, and thinner dielectric. Combining the three parameters in the best way to achieve the highest capacitance density, lowest leakage currents, and highest reliability remains the ultimate goal for most of the capacitor applications. While dielectric permittivity is a material-dependent parameter, the electrode surface area and the thickness of the deposited dielectric are both process-driven. Based on these critical, process-driven parameters, the state-of-the-art capacitor technologies are broadly categorized as planar and 3D high-density capacitors (shown in Figure 7.2).

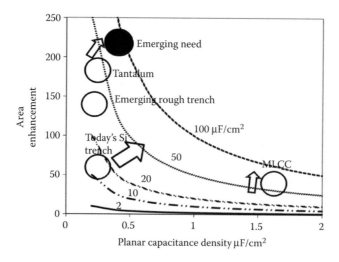

FIGURE 7.1 Volumetric densities of various state-of-the-art and emerging capacitor technologies.

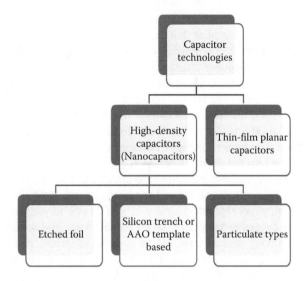

FIGURE 7.2 Classification scheme for various capacitor technologies.

The "planar" capacitors require planar (2D) electrode surfaces, but utilize ultrathin, high-permittivity dielectrics such as barium–strontium titanates (BST), strontium titanates (ST), or BT. On the other hand, the current "three-dimensional" or high-density capacitors utilize extremely high electrode surface areas with low-to-moderate permittivity dielectrics. These include a silicon trench [10], MLCCs [11], and particulate capacitors (tantalum based) [12]. More recently, there is an increasing focus on benefiting from all the three key capacitance parameters, viz., ultrahigh surface electrode area, thin dielectrics with high permittivity in the same component. This new class of capacitors is termed "nanocapacitors" and will be the focus for this chapter. For an easier and more comprehensive understanding, the chapter will be divided into etched foil nanocapacitors, silicon trench nanocapacitors, and nanoparticulate-based capacitors.

7.2 ETCHED-FOIL NANOCAPACITORS

Adding a new dimension to the planar electrodes is the key to increase the surface area and thereby the volumetric density of the capacitors. The most commonly used method to do so is by etching deep and porous channels inside the electrode. Silicon, aluminum, and titanium are some of the electrode systems that are commonly etched using electrochemical (wet-etching) or reactive plasma (dry-etching) processes [13–17] to yield high-aspect-ratio deep channels or trenches. Depositing conformal high-permittivity dielectrics over larger areas is critical to form the capacitor and usually requires vapor deposition techniques such as chemical vapor deposition (CVD) and atomic layer deposition (ALD) [18]. These deposition techniques have their set of drawbacks, which includes high cost, high deposition time, and low throughput. On the other hand, methods such as thermal oxidation or nitridation limit the dielectrics to lower permittivity values. To circumvent these issues, anodizable metals, also known as valve metals, such as Al, Ta, Ti, and Nb [19], have been widely used as nanoelectrode templates for the fabrication of nanocapacitors. Here, the dielectric formation is accomplished by oxidizing the electrode surface by immersing the etched metal into an electrolyte solution and applying a voltage bias to the electrode. This method, known as anodization, is widely used in the electrolytic capacitor and tantalum industry. The consistency, conformality, and thickness uniformity of the dielectric are monitored by the time, current, voltage, and the concentration of the electrolyte. The liquid electrolyte is capable of reaching the narrowest channel in the metal, which, on contact with the metal surface, oxidizes to form a dielectric oxide on the anode. The dielectric thickness increases with the anodizing voltage at a rate of ~1–2 nm/V [20]. It is known that out of all the available dielectric materials, oxides, nitrides, and oxynitrides have the highest breakdown voltages. This implies that these dielectric films can be easily thinned down to a few 10 s of nanometers without compromising on their reliability. The oxide layer formed during anodization shows a voltage-dependent resistance that causes the current to increase more steeply as the voltage increases during anodization. The rated anodizing, and the operating voltages are carefully designed to produce safe and reliable capacitors.

The counter electrode used for these capacitors is usually a liquid electrolyte. Aluminum electrolytic capacitors are also designated as "wet" or "nonsolid" capacitors due to the use of liquid electrolyte as the cathode. The liquid has the advantage of filling the deep-etch structures or channels, thus optimally fitting into the anode structure and accessing the high-surface-area electrode. A conventional aluminum electrolytic capacitor thus consists of a wound capacitor element that is impregnated with a liquid electrolyte, connected to terminals, and sealed in a can. The element is composed of an anode foil, paper separators saturated with an electrolyte, and a cathode foil. These capacitors routinely offer capacitance values from 0.1 μF to 3 F and voltage ratings from 5 to 700 V. They are polar devices, having distinct positive and negative terminals, and are offered in several design configurations. Though these conventional liquid aluminum electrolytic capacitors provide high capacitance densities required for various applications such as decoupling, power, noise suppression, and so on, there are several limitations associated with these electrolytic capacitors. Some of the major disadvantages are the high impedance and thermal instability, and liquid electrolyte leakage, owing to the use of a low-conductive (10^{-2}–10^{-3} S/cm) and thermally unstable ionic liquid electrolyte.

To cope with these limitations, solid aluminum capacitors using solid counter electrodes have been developed. The most prominent solid counter electrodes are broadly classified as organic and inorganic counter electrodes based on their chemical properties. Conducting polymers such as poly-ethylenedioxythiophene (PEDT), poly-3-hexylthiophene (P3HT), polyethylene oxide (PEO), polypyrrol, and complexes such as tetracyanoquinodimethane (TCNQ) are organic counter electrodes, while pyrolytic manganese dioxide and ruthenium oxide are categorized as inorganic counter electrodes. These solid counter electrode based capacitors have several advantages over the traditional liquid electrolytic capacitors. For example, PEDT polymer-based capacitors have lower equivalent series resistance (ESR) than the liquid electrolytic capacitors because of the higher conductivity of PEDT [21]. This permits a single polymer-type capacitor to replace several liquid aluminum electrolytic-type capacitors, resulting in a reduced total number of components on the board and more available real estate. Moreover, since the polymer is a solid, it also gives longer life, not following the classic Arrhenius formula. Instead of doubling life with every 10°C drop in temperature, the lifetime will be 10 times longer for every 20°C decrease in temperature. Another key advantage of the solid electrolyte over the liquid ones is their capability to heal the defects or pinholes in the dielectrics. Manganese dioxide (MnO_2) and PEDT are widely used in tantalum particulate capacitors as self-healable counter electrodes. The self-healable mechanism in Al-electrolytic capacitors with MnO_2 as the counter electrode is shown in Figure 7.3.

This self-healing process is an important factor in the steady-state reliability characteristics of tantalum capacitors, which are referenced as having "no wear-out mechanism." In MnO_2-based self-healable reactions, the conductive MnO_2 electrode is locally converted to highly resistive Mn_2O_3, as shown in Reaction 7.1:

$$2MnO_2 \rightarrow Mn_2O_3 + O^* \tag{7.1}$$

Aluminum oxide (Al_2O_3) has pinholes or defects or regions thinner than the surrounding dielectric. A large proportion of the capacitor current (charging, leakage) flows through the leaky site causing excessive localized heating. As the temperature at the defect site rises, Reaction 7.1 takes place, converting the conductive manganese dioxide (MnO_2), which has a resistivity of between 1 and 10 ohm/cm^2,

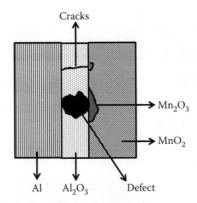

FIGURE 7.3 Self-healing mechanism of MnO_2 in Al/Al_2O_3 system.

to a less conductive form (Mn_2O_3) having a resistivity between 10^6 and 10^7 ohm/cm^2. Thus, the conduction site is effectively "plugged" or "capped," thereby decreasing the leakage current.

Though MnO_2 has shown several benefits as a counter electrode in Al- and Ta-based capacitors, it is unsafe during high-voltage operation since it forms an ignitable redox couple with metal at higher voltages [22]. Other limitations such as high ESR due to higher resistivity have led to the use of alternative cathode materials such as conducting polymers. Among the intrinsically conducting polymers (ICPs), thiophenes are most widely used in the capacitor industry due to numerous advantages such as higher conductivity, improved safety, and self-healing. PEDT is one classic ICP with much higher conductivity than MnO_2, thereby lowering the characteristic ESR in the capacitor [23]. Furthermore, it has improved safety aspects during the high-voltage operation. In addition, PEDT possesses self-healing ability, high-temperature stability, and outstanding electronic properties in the doped as well as undoped states. Even with those attractive merits, PEDT-based capacitors exhibit poorer electrical characteristics such as high leakage currents as the operating voltage of the Ta and Al electrolytic capacitors is increased [24]. The conducting polymer deposition method plays a critical role in determining the leakage properties of the fabricated capacitor. *In situ* deposition of conducting polymers such as PEDT has generally shown poor electrical characteristics due to the chemical interactions between the oxide dielectric and the residual oxidizers entrapped in the deep Al/Si trenches. Tantalum capacitors that use PEDT as the cathode electrode eliminate the residual oxidizers by carrying out several cycles of washing and reformation (reanodization to form the dielectric). More recently, prepolymerized PEDT polymer dispersions have also been introduced to tantalum technology, greatly improving the capacitor voltage ratings of 25 V and reliability [25].

Titanium nitride (TiN) deposited through ALD has recently gained momentum due to its high conductivity and conformality across deep trenches [26]. ALD, being self-limiting (deposition of one atomic or molecular layer in each reaction cycle) in behavior, is capable of sequentially depositing conformal thin films of materials (metals and dielectrics alike) onto substrates with complex, contoured surfaces. It has become the leading process to achieve such coatings, yielding a high degree of thickness control and conformality in the most demanding nanostructures. Banerjee et al. [27] have demonstrated metal–insulator–metal (MIM) nanocapacitor arrays with ALD inside AAO (anodized aluminum oxide) nanopores, making use of self-assembly and self-alignment. ALD titanium nitride formed the bottom as well as the top electrode, whereas ALD Al_2O_3 was sandwiched as the dielectric. The highly regular arrays in AAO have reported a capacitance density of 100 μF/cm^2 for 10-μm-thick anodic aluminum oxide.

The Packaging Research Center at the Georgia Institute of Technology has coupled the low-cost dielectric deposition technique with top electrode ALD to show a cost-effective approach to high-density nanocapacitors that has the potential to be integrated directly into silicon or glass

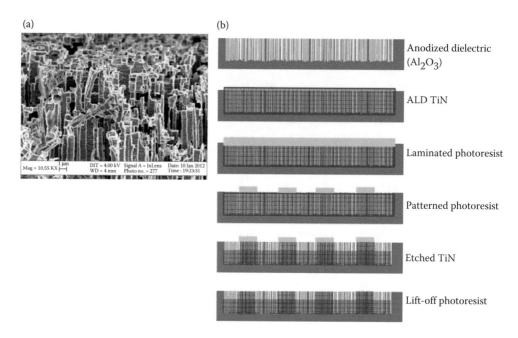

FIGURE 7.4 (a) SEM micrograph of the 75-μm-thick etched Al foil; (b) Step-by-step fabrication of the etched Al capacitor with ALD TiN as a top electrode.

substrates. Commercially available etched aluminum foils, capable of introducing 50–100 times more the surface area depending upon the foil thickness and the extent of etch, were utilized as the bottom electrode. These metal foils were anodized with respective chemistries to yield highly conformal, uniformly thick dielectrics. The anodized metal foils can be subsequently integrated into organic packages by traditional processes such as lamination and etching. Etched foils with 75–105-μm thickness, having channels with an aspect ratio of above 50, were used in this study. The SEM image of the 75-μm-thick etched Al foil is shown in Figure 7.4a. ALD TiN was used as the counter electrode of the capacitor because of its ability to conformally coat the higher surface area of the anodized etched foil. An approximately 20-nm-thick blanket TiN film deposited by ALD was patterned using dry photoresist as an etch mask to isolate smaller devices. Reactive ion etcher (RIE) was used for the nitride etching process (with the following parameters: SF_6—45 sccm, O_2—5 sccm, RF power: 350 W). A step-by-step illustration of the procedure for the fabrication of etched-metal nanocapacitors is shown in Figure 7.4b. Such low-temperature, low-cost, high-density nanocapacitors showed capacitance densities of above 6–7 $\mu F/cm^2$ at kHz frequencies, which are significantly higher than what are feasible with high-permittivity thin-film capacitors so far, demonstrated with high-temperature processes.

7.3 SILICON TRENCH CAPACITORS

Silicon trench capacitors were introduced to miniaturize the decoupling capacitors in high-performance processors. The idea of using high-value metal oxide semiconductor (MOS) capacitors made in silicon with enlarged surface using etched ridges in Si was first proposed by Rosenfeld [28] and then by Lehmann et al. who used arrays of deep trenches [29,30]. As the surface opening of the storage capacitor is scaled down, adequate capacitance is maintained by etching deeper trenches. Submicrometer opening trenches with aspect ratios of above 40 were reported by IBM in the late 1980s [31]. Equivalent capacitance densities of 10–20 $\mu F/cm^2$ were reported in 1991, but with micrometer-scale dimensions leading to fF capacitors. This technology is now implemented over larger areas to yield μF capacitors with millimeter or centimeter dimensions using deep silicon trenches over a large area with high K

dielectrics. The silicon trench formation processes can be classified into wet-etching and dry-etching processes, where dry etching uses inductive coupled plasma with fluorine chemistry, which then diffuses to the low-pressure etching chamber where silicon is processed. Wet etching, on the other hand, is a low-cost, fast-etching technique that is accomplished with photo electrochemical method.

The deposition of dielectrics in silicon trenches is usually achieved with reasonable conformality using CVD techniques [32]. Liquid-phase CVD and thermal oxynitridation are known to be conformal and compatible with active circuitry [33]. Therefore, these techniques are widely pursued as the oxynitride deposition route. Other alternate routes include thermal oxidation and nitridation. However, ALD [34] is usually the preferred route for uniform step coverage unlike CVD. Silicon oxide is the traditional dielectric choice because of its high breakdown voltage and compatibility with silicon. A further increase of capacitance was obtained by introducing a pure nitride dielectric, without oxide, where leakage is reduced by operation at a lower voltage (Vdd from 1.8 to 1.2 V) for dynamic random-access memory (DRAM) applications [35]. ALD is pursued as the most versatile technique to deposit high K materials with thickness uniformity, step coverage, and relatively low temperatures. Several higher-permittivity dielectrics such as Al_2O_3, HfO_2, TiO_2, and Ta_2O_5 have been deposited by ALD in deep trenches to enable high capacitance densities. State-of-the-art industrial trench capacitors in silicon reach capacitance densities up to 200 nF/mm^2 at 11 V breakdown voltage [36]. More recently, results on multilayer MIM capacitors with ALD Al_2O_3 dielectric layers [37] have shown a different way to increase the capacitance density as compared to earlier works on single-layer MIM capacitors [38,39]. Kammerer et al. reported the fabrication and characterization of multiple MIM capacitors by depositing $TiN/Al_2O_3/TiN/Al_2O_3/TiN$ stacks by ALD in deep trenches with an aspect ratio of 20, yielding a very high capacitance density of 440 nF/mm^2 at a breakdown voltage (BDV) >6 V [37,40].

Even with the advances and improvements in the ALD technique, the process of metal and dielectric deposition by ALD remains challenging due to the low-throughput vacuum tools, costly precursors, packaging-incompatible infrastructure, and scaling costs. Chemical solution deposition, including sol–gel and other solution-deposition techniques, offers a low-cost route to derive inorganic thin films with much higher K than the ALD oxides. Integration of thin-film sol–gel-derived ceramic capacitors has been demonstrated in both organic packages and silicon [41,42]. GT-PRC has shown an innovative and low-cost solution-coating technology to form conformal electrodes and dielectrics in silicon trench structures to form trench capacitors. An all-solution-derived capacitor with a PZT/LNO/SiO_2/Si structure was demonstrated [42]. Vacuum-infiltration was carried out to conformally coat the sol–gel precursors on the deep trenches by applying a controlled pressure gradient over the wafer that drove the solution down the trench surfaces [42]. A schematic 3D structure thus formed by the sol–gel vacuum infiltration technique is shown in Figure 7.5a. A SEM cross-section of the conformal sol–gel coating along the via wall is also shown in Figure 7.5b. Multiple coatings were applied to increase the coating thickness before thermally annealing the films in air at high temperatures (>700°C). A representative SEM image of the sol–gel-deposited (PZT/LNO/SiO_2/Si) is shown in Figures 7.6a and b [42]. The thin films showed a capacitance density of 2–3 µF/cm^2. In conjunction with the planar capacitors demonstrated, the feasibility of conformal solution coatings in trenches with an aspect ratio of 5–10 can potentially lead to a capacitance density of 10–30 µF/cm^2, which can meet the emerging needs for thin power supply capacitors in a silicon interposer. These thin capacitor components can then be embedded in the package to enable true system integration and miniaturization.

7.4 PARTICULATE CAPACITORS

The approach to a ultrahigh-surface-area bottom electrode is achieved by sintering valve metal particles on a metal wire. This ultrahigh surface area is capable of providing high volumetric density in capacitors. Tantalum capacitors are the most widely and commercially manufactured discrete capacitors. These are fabricated by sintering micrometer-sized tantalum powder into a pellet. The pellets are sintered with controlled open-pore volume. The particles create point contacts with each

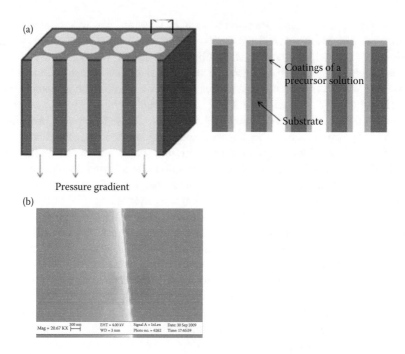

FIGURE 7.5 Schematic representation of a three-dimensional trench structure after sol–gel vacuum infiltration. Cross-section of a conformal ceramic coating along the via wall.

other in random fashion, leaving large gaps between most of the particles. Thus, the pressed tantalum pellet is a highly porous metal with channels or tunnels that are interconnected throughout the pellet and extends to the outside surface. Figure 7.7 shows the illustration of a tantalum capacitor. The volume enhancement for particulate electrodes is a function of the particle's packing fraction and the particle size. Tantalum requires very high temperatures for sintering (>1400°C), thus making it incompatible for silicon or organic packages.

Tantalum capacitors employ the anodization technique for dielectric deposition. Being a valve metal, tantalum forms an insulating oxide when a potential is applied in an electrolytic bath. For the capacitor application, pressed and sintered tantalum pellets are immersed into an electrolytic solution as an anode. The liquid electrolytes seep inside the narrow tunnels of the pellet, converting the metal into the oxide, Ta_2O_5. The depth and consistency of the Ta_2O_5 layer are usually determined by the time, current, and the applied voltage during the anodization process. The thickness of the dielectric is in the range of 18 Å/V [43]. The capacitors are polar. Therefore, the dielectric is more stable when the tantalum is used as the anode. Any reverse bias can destabilize the dielectric by depleting the oxygen in the dielectric. Thus, the Ta_2O_5 layer is generally reformed (posthealed) before the cathode (top electrode) is deposited. This is accomplished to prevent high initial leakage that will otherwise deteriorate the dielectric until it is completely damaged in that area and lead to shorting of the capacitor as a whole. Therefore, tantalum capacitors employ self-healable materials such as manganese dioxide (MnO_2) which can heal dielectric defects that can otherwise cause reliability issues. MnO_2 electrodes are formed by dipping the tantalum pellet in $Mn(NO_3)_2$, followed by pyrolyzing the salt to convert into the oxide. This process is repeated to fill the tantalum pellet. Typically, this conversion process takes place at a temperature range of 200–280°C. The MnO_2 now extends into the pellet and covers the outside surface of the pellet. The pellet is then dipped in a carbon solution, and subsequently in silver paints and cured. The silver paint surface is connected to a lead-frame element with a conductive epoxy and the tantalum riser wire is welded to another lead-frame element. The structure is then packaged with the lead frame extending out through the plastic. When normalized to 100-μm thickness, commercially available tantalum capacitors have shown a capacitance density of 140 μF/cm² with a voltage rating of 6 V. For a 10 V rating, 100 μF/cm²

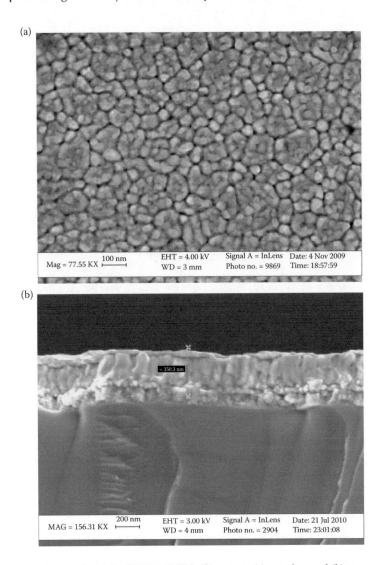

FIGURE 7.6 SEM micrographs of the PZT/LNO/SiO$_2$/Si system: (a) top view and (b) cross-sectional view.

is available, whereas only 15 µF/cm^2 is available for a voltage rating of above 20 V. Furthermore, the leakage currents are generally reported to be close to 0.1 µA/µF.

Besides their ultrahigh capacitance densities, tantalum capacitors have traditionally had two weaknesses—a susceptibility to ignition when they fail and higher ESR than that from other capacitor technologies. Moreover, the tantalum capacitors, owing to their high processing temperatures, remain silicon- and package-incompatible and are limited by the moderate K dielectric, Ta$_2$O$_5$. Though the safety and the ESR concerns are addressed by introducing conductive polymer (PEDT) as a replacement for the conventional manganese dioxide, tantalum processing temperatures and the restriction to use high K dielectrics remain a major bottleneck for tantalum capacitors.

To address the challenges posed by valve metals and to advance further in capacitance densities with integrable approaches, low-temperature sinterable metals such as copper are pursued. This also opens an avenue to deposit high K ferroelectrics as dielectrics, which is not feasible with tantalum particles. GT-PRC has shown the feasibility of high-density nanocapacitors, utilizing copper particulate electrodes, ALD alumina as a dielectric, and conducting polymer as the counter electrode [44]. These capacitors showed a capacitance density of 60 µF/cm^2 using low-temperature,

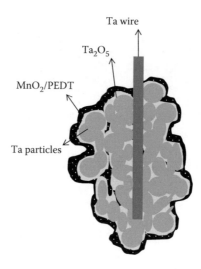

FIGURE 7.7   Cross-section of a discrete tantalum capacitor with Ta wire for interconnection.

silicon- or package-compatible processes. The cross-section of the capacitor is illustrated in Figure 7.8. The thicker electrode capacitors showed as high as 400–500 times surface area enhancement, corresponding to a much higher capacitance density. The capacitance versus voltage plots for copper nanoparticulate capacitors based on the thickness of the electrode are shown in Figure 7.9 [44]. The capacitance density corresponds to 60 $\mu F/cm^2$.

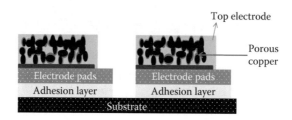

FIGURE 7.8   Integrated high-density nanocapacitor with a copper nano-electrode.

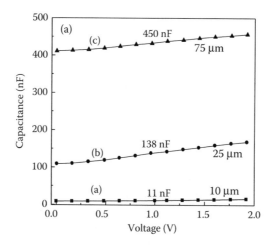

FIGURE 7.9   Capacitance vs. applied voltage curves of the copper nano-electrode capacitors. The electrode thickness is shown for each curve.

This novel particulate capacitor technology is unique compared to current technologies, such as MLCC, trench, and tantalum, in many ways: (1) There is direct deposition of particulate electrodes onto silicon as thin integrated passive device (IPD) or interposer architectures, unlike the bulky surface mount discretes (SMD). (2) It is a combination of high K films and high-surface-area electrode technologies. The highest capacitance density can thus be, obtained with leading edge particulate electrode technologies. (3) Capacitor arrays can be formed in a planar format with multiple independent terminals that are compatible with a thin silicon IC stack on the top. This unique nanoelectrode capacitor technology thus goes beyond the current technologies and has the potential of achieving higher volumetric capacitance density with superior leakage current, breakdown voltage, and ESR properties.

7.5 CONCLUSIONS

The need for miniaturized passive components arises from the increasing demand for lighter, higher-performance, and highly integrated ultrathin packages at reduced cost. This chapter discusses one aspect of these passive components, that is, the state-of-the-art high-density capacitor technologies, benchmarking them against their capacitance densities, silicon and package compatibility, low-cost manufacturable processes, and electrical properties such as leakage current and breakdown voltages. The demand for high capacitance densities in microscaled device dimensions has provided the motivation to develop nanocapacitors with unique electrode structures and designs, as well as superior material properties and processes. Three classes of nanocapacitors, viz., etched foils, silicon trench, and particulate types, were discussed in this chapter. A consolidated understanding of the role of high-surface-area electrodes, along with high-permittivity dielectrics in fabricating nanocapacitors with ultrahigh capacitance densities, has been provided. Novel high-surface-area nanoporous electrode technologies with a conformal dielectric for integration of high-density capacitors as thin films on silicon and glass with densities of 100 μF/cm^2 using 100-μm-thick films are discussed.

REFERENCES

1. Raj PM, Sharma H, Sethi K, Wang Y, Sundaram V, Tummala R. Thin film power components with nanoscale electrodes and conformal dielectrics. In: *IEEE International Conference on Nanotechnology* 2011: 106–110.
2. Tummala RR. *Fundamentals of Microsystems Packaging*. McGraw-Hill, New York 2001, **7**:2.
3. Raj PM, Sharma H, Mishra D, Chakraborti P, Lei S, Yang I, Tummala R. Silicon- and glass-integrated thin film power components. In: *CARTS International*. Las Vegas 2012.
4. Brennecka GL, Ihlefeld JF, Maria JP, Tuttle BA, Clem PG. Processing technologies for high-permittivity thin films in capacitor applications. *Journal of the American Ceramic Society* 2010, **93**(12):3935–3954.
5. Ho J, Jow TR, Boggs S. Historical introduction to capacitor technology. In *IEEE: Electrical Insulation Magazine* 2010, **26**(1):20–25.
6. Niwa S, Taketani Y. Development of new series of aluminium solid capacitors with organic semiconductive electrolyte (OS-CON). *Journal of Power Sources* 1996, **60**(2):165–171.
7. Mansbridge GF. Patent 19,451. UK; 1900.
8. Balaji T, Govindaiah R, Sharma M, Purushotham Y, Kumar A, Prakash T. Sintering and electrical properties of tantalum anodes for capacitor applications. *Materials Letters* 2002, **56**(4):560–563.
9. Hoogeland DJ, Jinesh K, Roozeboom F, Besling W, Lamy Y, Van de Sanden M, Kessels W. Plasma-assisted ALD of TiN/AlO stacks for MIMIM trench capacitor applications. *ECS Transactions* 2009, **25**(4):389–397.
10. Katsumata R, Tsuda N, Idebuchi J, Kondo M, Aoki N, Ito S, Yahashi K, Satonaka T, Morikado M, Kito M. Fin-array-FET on bulk silicon for sub-100 nm trench capacitor DRAM. In: *IEEE Symposium on VLSI Technology* 2003: 61–62.
11. Pithan C, Hennings D, Waser R. Progress in the synthesis of nanocrystalline BaTiO3 powders for MLCC. *International Journal of Applied Ceramic Technology* 2005, **2**(1):1–14.
12. Larmat F, Reynolds JR, Qiu YJ. Polypyrrole as a solid electrolyte for tantalum capacitors. *Synthetic Metals* 1996, **79**(3):229–233.

13. Blauw MA, van Lankvelt P, Roozeboom F, van de Sanden M, Kessels W. High-rate anisotropic silicon etching with the expanding thermal plasma technique. *Electrochemical and Solid-State Letters* 2007, **10**(10):H309–H312.

14. Diesinger H, Bsiesy A, Herino R. Nano-structuring of silicon and porous silicon by photo-etching using near field optics. *Physica Status Solidi (a)* 2003, **197**(2):561–565.

15. Kemell M, Ritala M, Leskelä M, Ossei-Wusu E, Carstensen J, Föll H. Si/Al2O3/ZnO: Al capacitor arrays formed in electrochemically etched porous Si by atomic layer deposition. *Microelectronic Engineering* 2007, **84**(2):313–318.

16. Smela E, Kallenbach M, Holdenried J. Electrochemically driven polypyrrole bilayers for moving and positioning bulk micromachined silicon plates. *Journal of Microelectromechanical Systems* 1999, **8**(4):373–383.

17. Van Den Meerakker J, Elfrink R, Weeda W, Roozeboom F. Anodic silicon etching; the formation of uniform arrays of macropores or nanowires. *Physica Status Solidi (a)* 2003, **197**(1):57–60.

18. Elam J, Routkevitch D, Mardilovich P, George S. Conformal coating on ultrahigh-aspect-ratio nanopores of anodic alumina by atomic layer deposition. *Chemistry of Materials* 2003, **15**(18):3507–3517.

19. Kukli K, Ritala M, Leskelä M. Development of dielectric properties of niobium oxide, tantalum oxide, and aluminum oxide based nanolayered materials. *Journal of the Electrochemical Society* 2001, **148**:F35.

20. Velten D, Biehl V, Aubertin F, Valeske B, Possart W, Breme J. Preparation of TiO_2 layers on cp-Ti and Ti6Al4V by thermal and anodic oxidation and by sol-gel coating techniques and their characterization. *Journal of Biomedical Materials Research* 2002, **59**(1):18–28.

21. Kirchmeyer S, Reuter K. Scientific importance, properties and growing applications of poly (3,4-ethylenedioxythiophene). *Journal of Materials Chemistry* 2005, **15**(21):2077–2088.

22. Freeman Y, Harrell W, Luzinov I, Holman B, Lessner P. Electrical characterization of tantalum capacitors with poly(3,4-ethylenedioxythiophene) counter electrodes. *Journal of the Electrochemical Society* 2009, **156**:G65–G70.

23. Prymak JD. Replacing MnO2 with conductive polymers in Tantalum capacitors. In: *CARTS. Europe* 2009.

24. Alapatt GF, Harrell WR, Freeman Y, Lessner P. Observation of the Poole-Frenkel effect in tantalum polymer capacitors. In: *IEEE SoutheastCon* 2010: 498–501.

25. Reed E, Haddog G. Reliability of high-voltage tantalum polymer capacitors, *CARTS 2011 Proceedings*, March 28–31, Jacksonville, FL, pp. 1–13.

26. Blomberg T, Wenger C, Kaynak CB, Ruhl G, Baumann P. ALD grown NbTaOx based MIM capacitors. *Microelectronic Engineering* 2011, **88**(8):2447–2451.

27. Banerjee P, Perez I, Henn-Lecordier L, Lee SB, Rubloff GW. Nanotubular metal–insulator–metal capacitor arrays for energy storage. *Nature Nanotechnology* 2009, **4**(5):292–296.

28. Rosenfeld R, Bean K. Fabrication of topographical ridge guides on silicon for VHF operation. In: *IEEE Ultrasonics Symposium* 1972: 186–189.

29. Foell H, Lehmann V. Etching method for generating apertured openings or trenches in layers or substrates composed of n-doped silicon. Patent 4874484. US; 1989.

30. Lehmann V, Föll H. Formation mechanism and properties of electrochemically etched trenches in n-type silicon. *Journal of the Electrochemical Society* 1990, **137**:653–659.

31. Rajeevakumar TV, Lii T, Weinberg ZA, Bronner GB, McFarland P, Coane P, Kwietniak K, Megdanis A, Stein KJ, Cohen S. Trench storage capacitors for high density DRAMs. *Electron Devices Meeting, Technical Digest, International* 1991: 835–838.

32. IslamRaja MM, Cappelli M, McVittie J, Saraswat K. A 3-dimensional model for low-pressure chemical-vapor-deposition step coverage in trenches and circular vias. *Journal of Applied Physics* 1991, **70**(11):7137–7140.

33. Lin CC, Wei RC. Nitrogen-plasma treatment of carbon nanotubes and chemical liquid phase deposition of alumina for electrodes of aluminum electrolytic capacitors. *Journal of the Electrochemical Society* 2012, **159**(5):A664–A668.

34. Lee B, Park SY, Kim HC, Cho KJ, Vogel EM, Kim MJ, Wallace RM, Kim J. Conformal Al_2O_3 dielectric layer deposited by atomic layer deposition for graphene-based nanoelectronics. *Applied Physics Letters* 2008, **92**(20):203102–203103.

35. Liu CH, Wu JM. Improvement of electrical properties and reliability of (BaSr) TiO thin films on Si substrates by NO prenitridation. *Electrochemical and Solid-State Letters* 2006, **9**(2):G23–G26.

36. Roozeboom F, Klootwijk J, Verhoeven J, Heuvel FC, Dekkers W, Heil S, Hemmen JL, Sanden MCM, Kessels W, LeCornec F. ALD options for Si-integrated ultrahigh-density decoupling capacitors in pore and trench designs. *ECS Transactions* 2007, **3**(15):173.

37. Klootwijk J, Jinesh K, Dekkers W, Verhoeven J, Van Den Heuvel F, Kim HD, Blin D, Verheijen M, Weemaes R, Kaiser M. Ultrahigh capacitance density for multiple ALD-grown MIM capacitor stacks in 3-D silicon. *IEEE Electron Device Letters* 2008, **29**(7):740–742.
38. Bajolet A, Giraudin J, Rossato C, Pinzelli L, Bruyère S, Crémer S, Jagueneau T, Delpech P, Montès L, Ghibaudo G. Three-dimensional 35 nF/mm2 MIM capacitors integrated in BiCMOS technology. In: *IEEE Solid-State Device Research Conference* 2005: 121–124.
39. Gutsche M, Seidl H, Luetzen J, Birner A, Hecht T, Jakschik S, Kerber M, Leonhardt M, Moll P, Pompl T. Capacitance enhancement techniques for sub-100 nm trench DRAMs. In: *IEEE Electron Devices Meeting* 2001: 18.16.11–18.16.14.
40. Matters-Kammerer MK, Jinesh KB, Rijks TGSM, Roozeboom F, Klootwijk JH. Characterization and modeling of atomic layer deposited high-density trench capacitors in silicon. *IEEE Transactions On Semiconductor Manufacturing* 2012, **25**(2): 247–254.
41. Abothu IR, Raj PM, Hwang JH, Kumar M, Iyer M, Yamamoto H, Tummala R. Processing, properties and electrical reliability of embedded ultra-thin film ceramic capacitors in organic packages. In: *IEEE Electronic Components Technology Conference* 2007: 1014–1018.
42. Wang Y, Xiang S, Raj P, Sharma H, Williams B, Tummala R. Solution-derived electrodes and dielectrics for low-cost and high-capacitance trench and Through-Silicon-Via (TSV) capacitors. In: *IEEE Electronics component Technology Conference* 2011: 1987–1991.
43. Klerer J. Determination of the density and dielectric constant of thin TaO films. *Journal of the Electrochemical Society* 1965, **112**:896.
44. Sharma H, Sethi K, Raj PM, Tummala R. Fabrication and characterization of novel silicon-compatible high-density capacitors. *Journal of Materials Science: Materials in Electronics* 2012, **23**(2):528–535.

8 Modified Carbon Nanostructures for Display and Energy Storage

Sivaram Arepalli

CONTENTS

8.1 INTRODUCTION

Carbon nanostructures have been gaining considerable interest over the last three decades after the discovery of fullerenes in 1985 showed the possibility of stable curvatures of carbon forms [1]. Elongated forms of fullerenes in the shape of carbon nanotubes (CNTs) and nanocones have become the focus of intense research over the last two decades [2]. The last few years have seen a resurgence of interest in the flat carbon nanostructure called graphene [3]. All the above-mentioned carbon nanostructures exhibit very unique and superior mechanical, electrical, thermal, optical, and chemical properties, resulting in diverse applications ranging from nanoelectronics and nanocomposites, to nanomedicine. The large surface areas and porosities of these carbon nanostructures also helped their applications in energy production, storage, and transmission [4–7]. In recent years, one-dimensional (1D) nanostructure field emitters (e.g., nanowires) have attracted great interest due to their potential application as field emission (FE) flat-panel displays [8,9]. The incorporation of defects in CNTs has been shown to enhance the surface areas of the CNTs, and subsequently increase the electric double-layer capacitance (EDLC). The ion irradiation of CNTs generates fundamental as well as technological interest since it has the potential to introduce a wide range of defects in a controlled manner for tailoring the material properties [10]. We have investigated the influence of Ga^+ ion irradiation on the EDLC of multiwalled carbon nanotubes (MWCNTs). Cyclic voltammetry (CV) was employed to analyze the capacitive performance of MWCNTs before and after irradiation at varying cumulative ion doses. Also, CNTs in layered transition metal oxides are very attractive materials for high-energy-density supercapacitor applications. These materials overcome the drawbacks of conventional supercapacitors, but their commercial realization has been hindered by the high cost and poor electrical conductivity, which results in low power density, a narrow operation voltage window, and sluggish faradaic redox kinetics [11–15]. The randomly entangled mesoporous network of CNTs provides the electrical conducting pathways that allow easy diffusion of ions to the active surface area of nanosized metal oxides, which facilitates the faradaic processes across the interface. We found that our MoO_3–MWCNT nanocomposites-based electrochemical supercapacitor in terms of energy storage mechanism is operated in dual modes: (i) pseudocapacitance mode due to MoO_3 and (ii) EDLC mode due to MWCNTs. The results showed that the presence of MWCNTs in the nanocomposites improved the electronic conductivity, homogeneous electrochemical accessibility, and high ionic conductivity by avoiding an agglomerative binder, which makes them a promising material for the fabrication of electrochemical energy storage devices Recently, a few researchers have reported the electron field emission property of graphene film and few-layer graphene prepared by different techniques [16–20]. Chemically exfoliated single-layer graphene by an electrophoretic deposition technique, screen-printed graphene, and vertically oriented graphene grown by plasma-enhanced chemical vapor deposition (PECVD) techniques have been used to investigate its field emission properties. Palnitkar et al. used boron and nitrogen doping of graphene (produced by an arc discharge method) to tailor the turn-on field [20]. We explored the effect of morphological disorder on the field emission property of graphene synthesized by a thermal chemical vapor deposition technique (TCVD).

8.2 EXPERIMENTAL METHODS

Conventional CVD synthesis of MWCNT samples was carried out at 750°C using iron on silicon substrates for the first set of experiments. The precursor gas mixture composed of C_2H_4, Ar, and H_2 was used to produce vertically aligned MWCNT arrays. The MWCNT carpets were irradiated uniformly with 30 keV Ga^+ ions in a focused ion beam instrument (FIB, SMI3050TB). The structural changes in MWCNTs were investigated by HRTEM (FETEM, JEM2100F) and by Raman spectroscopy (Renishaw, RM1000). The electrode performance was characterized by CV using a multichannel potentiostat and a galvanostat (Bio-Logic Science Instruments, VMP3) with MWCNTs as a working electrode, Pt as a counter electrode, and Ag/AgCl as a reference electrode. The charge–discharge behavior was probed in a 0.1–0.9 V voltage range using an H_2SO_4 (2 M) electrolyte.

The synthesis of the intertwined MoO_3–MWCNT composites was carried out using a hydrothermal method by mixing the appropriate amount of modified MWCNTs and MoO_2 powder in H_2O_2 under ultrasonication for 20 min [21,22]. The pH of the resulting solution was maintained at 2 by adding dilute HCl. After stirring for 2 h, the mixture was transferred to a 20-mL Teflon autoclave and heated to 180°C for 6 h. The resulting black and white precipitates of the composite and MoO_3 nanowires were filtered and washed several times with deionized water and ethanol. They were then dried at 120°C for 12 h on a hot plate. The crystal structures of the MoO_3 nanowires and intertwined MoO_3–MWCNT nanocomposites were examined by x-ray diffraction (XRD; Rigaku Rotaflex D/Max) with Cu Kα radiation ($\lambda = 1.5418$ Å). The morphology was observed by field emission scanning electron microscopy (FESEM, JEOL JSM-7401F). High resolution transmission electron microscopy (HRTEM, JEM 2100F) and selected area electron diffraction (SAED) were carried out at 200 kV. The surface electronic states were examined by x-ray photoelectron spectroscopy (XPS, Perkin-Elmer PHI 660) and the binding energies were calibrated using C1s 284.6 eV as a reference.

Large-area monolayer graphene was synthesized on Cu foil by the CVD method reported earlier [23]. Copper foil of dimensions 70 cm × 35 cm × 70 μm was rolled into a cylindrical CVD quartz chamber and annealed in H_2 atmosphere for 1 h at 950°C. Graphene was synthesized at 950°C under gas flow of H_2 and CH_4 (80 sccm and 250 sccm). The Cu foil was cut in equal small pieces (1 cm × 1 cm), coated with the polymer poly(methyl methacrylate) (PMMA), and immersed in a copper etchant ($FeCl_3$) solution. Mechanical stressing of the copper foil resulted in nonplanar graphene sheets. The PMMA-coated graphene sheets were transferred onto a p-type Si(100). PMMA was then removed by acetone and the samples were annealed at 300°C for 1 h to improve the contact between the Si and graphene. The morphological investigation of the samples was done by FESEM and structural studies were carried out by Raman spectroscopy (Renishaw, RM-1000 Invia). The field emission characteristics of the planar and agglomerated single-layer graphene samples were measured in a high vacuum chamber with a parallel diode-type configuration at a base pressure of the order of 10^{-7} Torr. The field emission current was measured at different voltages using an automatically controlled Keithley 2001 electrometer and power supply (Fug Power HCN 700-3500).

8.3 RESULTS AND DISCUSSIONS

8.3.1 IRRADIATED MWCNTs

The unirradiated and irradiated MWCNT samples were examined using SEM, TEM, and Raman spectroscopy. Figure 8.1 shows micrographs of pristine and irradiated MWCNTs. The damage to the graphitic wall structure is evident at two higher ion doses, as it becomes broken and curly in Figure 8.1c and d.

The Raman spectra of pristine and irradiated MWCNTs recorded in the 1000–3200 cm^{-1} range are shown in Figure 8.2. The D-band is the disorder-induced band in the graphite lattice, the G-band is due to the tangential vibration of carbon atoms, the D'-band at ~1610 cm^{-1} is associated with the maximum in the graphene two-dimensional (2D) phonon density of states, and the second harmonic of the D-band in the 2600–2800 cm^{-1} range is denoted by G'. The second-order G'-band is an intrinsic property of a 2D graphene lattice. The analysis of the Raman data indicates several interesting features. First, there is a significant upshift in the position of all the Raman bands with increased ion dose as noted in Table 8.1. Second, a significant increase in the D- and G-band intensity ratio, that is, I_D/I_G, as well as the FWHM of the corresponding band after Ga^+ ion irradiation indicates an increase in the density of structural defects. Moreover, the ratio is related to the size of the disordered graphite clusters during the amorphization process of CNTs induced by ion bombardment during irradiation [24]. A further increase of disorder can destroy the aromatic ring structure and the I_D/I_G ratio is found to decrease at higher doses, which could be the transformation of graphite to tetrahedral amorphous carbon.

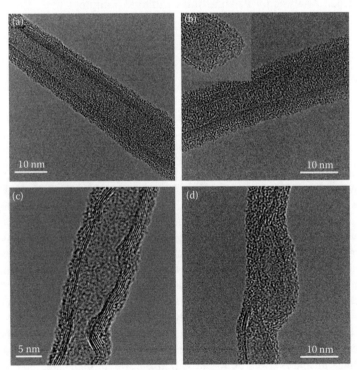

FIGURE 8.1 TEM micrographs of (a) pristine MWCNT sample and samples irradiated with 30 keV Ga ions at doses of (b) 2×10^{13}, (c) 2×10^{14}, and (d) 4×10^{14} ions/cm². The inset of (b) shows the opened tip of MWCNT. (Reprinted with permission from A. C. Ferrari and J. Robertson, *Phys. Rev. B* **61**, 14095, 2000, Copyright 2011, American Institute of Physics.)

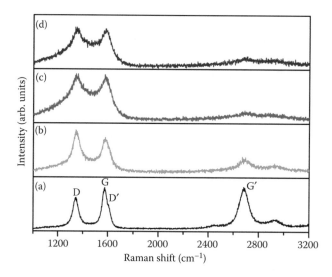

FIGURE 8.2 Raman spectra of (a) pristine MWCNT sample and irradiated samples at doses of (b) 2×10^{13}, (c) 2×10^{14}, and (d) 4×10^{14} ions/cm². The designations are given in the text. (Reprinted with permission from A. C. Ferrari and J. Robertson, *Phys. Rev. B***61**, 14095, 2000, Copyright 2011, American Institute of Physics.)

TABLE 8.1
Summary of Raman Analysis for Pristine and Irradiated Samples

Dose (ions/cm²)	D-Band (cm⁻¹)	G-Band (cm⁻¹)	D′-Band (cm⁻¹)	G′-Band (cm⁻¹)	I_D/I_G	$I_{G'}/I_G$
0	1344 (51)	1575 (45)	1610 (26)	2684 (91)	0.75	1.03
2×10^{13}	1349 (91)	1579 (80)	—	2690 (97)	1.17	0.30
2×10^{14}	1352 (260)	1581 (98)	—	2700 (141)	1.11	0.10
4×10^{14}	1355 (255)	1581 (101)	—	2700 (138)	1.09	0.11

Note: The values shown in parentheses are the full width at half maximum (FWHM) of the corresponding peak.

The upshift in the G-band is caused by stress generated in MWCNTs after Ga⁺ ion irradiation [25]. The D-band position also shows an upshift, implying that the defects created due to Ga⁺ ion irradiation have a different energy compared to the as-grown MWCNTs. Third, the presence of a 2D graphene structure in the sample can be monitored by the nature of the G′-band in the Raman spectra [26]. A well-defined G′-band is observable in the case of a pristine sample. As the irradiation dose increases, the intensity of the G′-band decreases and disappears at a dose of 2×10^{14} ions/cm². The absence of a D′-band and the decreased intensity of the G′-band indicate the amorphous nature of the samples irradiated at a dose of 2×10^{14} ions/cm² and higher.

The electrochemical performance of unirradiated and irradiated samples is shown in Figure 8.3. The double-layer capacitive behavior is clearly observed with near rectangular-shaped voltam-mograms for all samples. The peak at 0.55 V in the pristine sample can be attributed to oxygen containing functional groups contributing as pseudocapacitance to the overall capacitance. The specific capacitance increases (from 50 to 115 F/g) at 2×10^{13} ions/cm² and then starts decreasing exponentially with increasing irradiation dose and saturates (at 75 F/g) after 2×10^{14} ions/cm² as shown in Figure 8.4. The enhanced specific capacitance can be attributed to the defects creation and tip opening of MWCNTs by ion irradiation. The inner cavities of opened-tip nanotubes provide an additional surface to form the EDLC [27]. The HRTEM image shown in the inset of Figure 8.1b

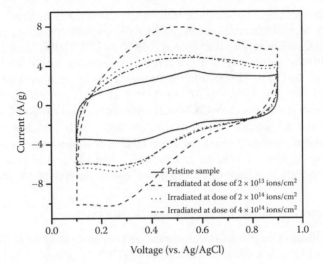

FIGURE 8.3 Cyclic voltammograms recorded at a 50 mV/s for the pristine and Ga ions irradiated MWCNT samples. (Reprinted with permission from A. C. Ferrari and J. Robertson, *Phys. Rev. B***61**, 14095, 2000, Copyright 2011, American Institute of Physics.)

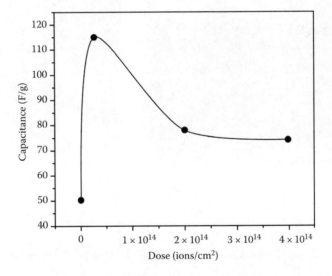

FIGURE 8.4 Variation of specific capacitance of MWCNT samples with ions dose. The specific capacitance decreases with ion dose and shows saturation behavior at higher doses. (Reprinted with permission from A. C. Ferrari and J. Robertson, *Phys. Rev. B***61**, 14095, 2000, Copyright 2011, American Institute of Physics.)

confirmed that the inner cavity of the opened-tip MWCNTs can be easily filled with electrolyte. Therefore, the opened-tip nanotubes play a positive role for charging the double layer. HRTEM and Raman results show that after Ga$^+$ ion irradiation, there is an increase of defects in MWCNTs.

The formation of the defects on the walls of MWCNTs by ion irradiation is very favorable for charging the double-layer capacitance of CNTs, which further improves the capacitive behavior of nanotube electrodes. The defects creation mechanism in MWCNTs can be explained on the basis of electronic and nuclear energy loss in a graphite target by Ga$^+$ ion irradiation.

The electronic and nuclear energy losses of 30 keV Ga$^+$ ion in a carbon target with the density of graphite are estimated to be 0.22 and 1.42 keV/nm, respectively. The dominant mechanism for defect creation on CNTs by irradiation of charged particles is the knock-on atomic displacement due to kinetic energy transfer [28]. The structural transformations in CNTs by irradiation are due to the defects mainly in the form of vacancies and interstitials. For graphitic structures, the threshold energy (T_d) required to produce a Frenkel pair is estimated to be 20 eV [28]. Those recoiled atoms with energy slightly above T_d can travel more distance in CNTs than in other solids due to the open structure inside the graphitic walls of the CNTs. Controlled defect creation and tip opening of MWCNTs is responsible for peak capacitance obtained at 2×10^{13} ions/cm^2. The decrease of capacitance at higher doses is attributed to the amorphization of MWCNTs, which lowers the conductivity of nanotubes [29]. The simulation showed that the damage created in the nanotube quickly increases with the irradiation dose. However, at higher irradiation doses, the defect number saturates toward a constant value. This is because at a certain irradiation dose, the system has reached a large degree of disorder so that an incoming energetic ion will not change the internal total coordination numbers substantially.

8.3.2 INTERCALATION OF MoO$_3$–MWCNT NANOCOMPOSITES

Figure 8.5a shows FESEM images of the as-synthesized intertwined MoO$_3$–MWCNT composite. The MWCNTs and MoO$_3$ nanowires in the MoO$_3$–MWCNT composites had mean diameters of

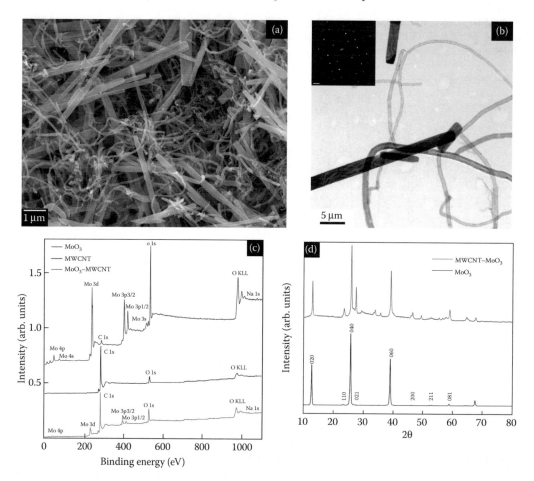

FIGURE 8.5   FESEM image of the MoO$_3$–MWCNT composite synthesized at 180°C for 6 h, (b) HRTEM images MoO$_3$–MWCNT nanocomposites (inset SAED), (c) XPS survey spectrum of MoO$_3$ (top), MWCNT (middle) and MoO$_3$–MWCNT nanocomposite (bottom), and (d) XRD patterns of the α-MoO$_3$ nanowires (bottom) and MoO$_3$–MWCNT composite (top) synthesized at 180°C for 6 h; the MoO$_3$ nanowires show an orthorhombic crystal structure.

~40 nm and ~50 nm, respectively, with a length in the range of tens of micrometers. The FESEM images showed that the MoO_3 nanowires were intertwined in a randomly entangled porous network of MWCNTs. The crystal structure of MoO_3 in the composites was examined by TEM. Figure 8.5b presents the TEM images with SAED patterns showing clear diffraction spots in the insets. This pattern is typical for a single crystalline system and the lattice spacing (2.4 Å) is in good agreement with the orthorhombic MoO_3. Figure 8.5c shows the XRD patterns of the MoO_3 nanowires and MoO_3–MWCNT composites. All XRD peaks can be indexed to the orthorhombic structure of the MoO_3 phase (JCPDS card no. 05-0508) and crystalline MWCNT phase. As also indicated by electron diffraction, the XRD pattern also showed sharp and well-defined peaks for the MoO_3 nanowires, indicating good crystallinity with an orthorhombic phase in the space group Pnma (with lattice constants $a = 13.85$ Å, $b = 3.696$ Å, and $c = 3.966$ Å). The slight decrease (0.08 nm) in the d value for the MoO_3–MWCNT composites suggests a decrease in the lattice volume that forms an interconnected network.

The detail of the chemical composition of the composite was measured by XPS. Figure 8.5d shows the XP spectra of molybdenum, oxygen, and carbon of MoO_3, MWCNTs, and MoO_3–MWCNT nanocomposite. The main peak at 284.6 eV in the MWCNT spectrum was attributed to sp^2-hybridized carbon in the graphitic layers of the MWCNTs [30–33] and the peak at approximately 285.7 eV was attributed to surface hydroxyl groups [34–37]. On the other hand, this C 1s region in the spectrum of the MoO_3–MWCNT composite could be deconvoluted into three peaks. The graphite-like carbon peak was found shifted to a higher binding energy due to the formation of Mo–C bonds [38] and the peaks at approximately 286.8 and 288.7 eV were assigned to the C–N and C = O groups, respectively. The Mo 3d5/2 peak positions in the composite are shifted toward a higher binding energy compared to that in MoO_3, suggesting the formation of some Mo–C bonds [35–38]. Overall, XPS data suggest that MWCNTs and MoO_3 in the composite interact via surface oxygen functionality and via some Mo–C bonding. The estimated Mo/carbon ratio was 1:2, which is in agreement with the starting stoichiometry.

The charge storage properties of the MoO_3 and MoO_3–MWCNT composites were measured by CV in 1 M NaOH and 1 M H_2SO_4 as shown in Figure 8.6. The charge storage in the layered structure of MoO_3 nanowires occurred through the following steps: reaction of cations with an electroactive material followed by a redox reaction, electrochemical adsorption of cations on the electroactive material through a charge transfer process, and the intercalation of ions into the van der Waals gaps of the layered structure of MoO_3 [11,39]. In addition to these mechanisms, the presence of the MWCNT matrix serves as an electrical conducting pathway leading to the active sites of MoO_3, which contributes to the total capacitance through an electrostatic charge. The enhancement in the capacitance of MoO_3 when intertwined with MWCNTs is clearly evident by the increase in current. The inset in Figure 8.6a shows the typical response of the MoO_3 nanowires where the electrolyte cation cannot fully utilize the active surface area because of its poor conductivity.

The cathodic half-cycle constitutes the adsorption of the hydrogen ion, while proton desorption occurs in the reverse cycle. The hydrogen ion forms a double layer along the interface and also diffuses into the crystal lattice of MoO_3, causing intercalation. As can be seen from Figure 8.6b and d, the cyclic voltammograms show sharp peaks for the first few initial cycles exhibiting the redox behavior of the system. After continuous cycling, there is a smooth electron transition, which can be seen from the diminishing peaks of the voltammograms. Hydrogen evolution is accompanied by the decrease in the capacitance values. The changes in the acid concentration not only enhance the accumulation of charges but also introduce hydrogen evolution at less negative potentials.

The cycle-dependent electrochemical responses of the nanocomposite were examined to better understand the contribution from double-layer and pseudocapacitance. Figure 8.6b and c shows the cyclic voltammograms of an MoO_3–MWCNT composite recorded after 5, 10, 50, and 100 cycles at 20 mV/s scan rates. The charge stored in the composite electrode was calculated by integrating the voltammetry current according to the following equation:

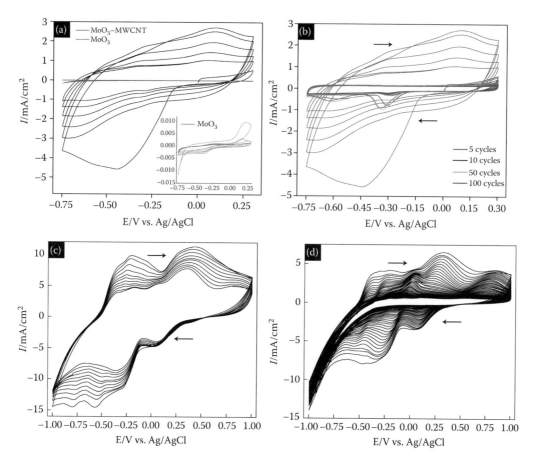

FIGURE 8.6   (a) Cyclic voltammograms of the MoO$_3$–MWCNT nanocomposite and MoO$_3$ nanowires (inset) in 1 M NaOH at 20 mV/s vs. Ag/AgCl as a reference electrode, first five cycles, (b) cyclic behavior of the MoO$_3$–MWCNT nanocomposite in 1 M NaOH at 20 mV/s vs. Ag/AgCl as a reference electrode, (c) cyclic voltammograms of the MoO$_3$–MWCNT nanocomposites in 1 M H$_2$SO$_4$ at 20 mV/s vs. Ag/AgCl as a reference electrode, first 5 cycles, and (d) cyclic behavior of the MoO$_3$–MWCNT nanocomposites in 1 M H$_2$SO$_4$ at 20 mV/s vs. Ag/AgCl as a reference electrode.

$$q = \frac{1}{v} \int_{E_c}^{E_a} |i| \, dE \tag{8.1}$$

where E is the electrode potential, i is the measured current density, v is the scan rate, and E_c and E_a are the cathodic and anodic potential limits, respectively. Figure 8.7a and b shows the specific capacitance with respect to the cycle number. The initial capacitance was 210 F/g, 178 F/g, which decreased rapidly, leveling off at 110 F/g, 98 F/g after approximately 100 cycles for 1 M H$_2$SO$_4$ and NaOH, respectively. The magnitude of the stored charge for the initial cycles is quite large as the cations can access the active sites of MoO$_3$ through the conducting MWCNT matrix to form an adsorption layer through a charge-transfer process, resulting in redox pseudocapacitance. The MoO$_3$ nanowires have a highly crystalline layered structure and the MWCNTs provide facile conducting pathways for the cations to enter into the van der Waals gaps (as mentioned earlier), giving rise to an intercalation pseudocapacitance mechanism. The contribution from double-layer capacitance was minor as indicated by the charge-storage magnitude that was strongly dependent on the scan rate.

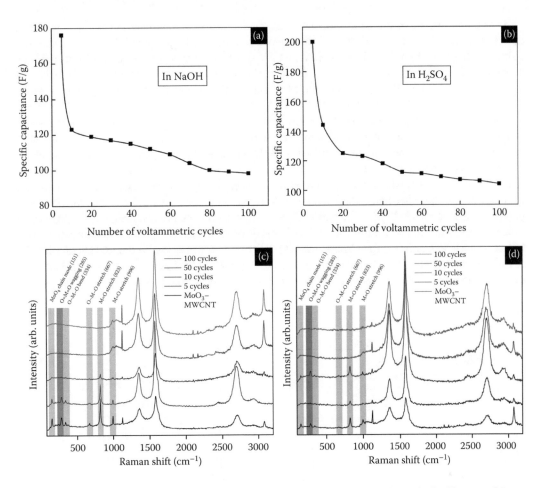

FIGURE 8.7 (a) Specific capacitance decreases upon cycling in 1 M NaOH vs. Ag/AgCl as a reference electrode, (b) specific capacitance decreases upon cycling in 1 M H_2SO_4 vs. Ag/AgCl as a reference electrode, (c) Raman spectra of MoO_3–MWCNTs recorded at 514-nm excitation after 5, 10, 50, and 100 (top) voltammetric cycles in 1 M NaOH vs. Ag/AgCl as a reference electrode, and (d) Raman spectra of MoO_3–MWCNTs recorded at 514-nm excitation after 5, 10, 50, and 100 (top) voltammetric cycles in 1 M H_2SO_4 vs. Ag/AgCl as a reference electrode.

The cycling-induced structural and morphological changes were measured by Raman spectroscopy. Figure 8.7c and d shows the Raman spectra of the MoO_3–MWCNT composites before and after cycling (which appear as superpositions of the MWCNT and MoO_3 peaks). The Raman peaks observed for MoO_3 in the pristine MoO_3–MWCNT composite are characteristic of a single-crystal α-MoO_3 system, which is a thermodynamically stable phase with a layered structure. Raman bands at 996 (A_g, υ_s M = O stretch), 823 (A_g, υ_s M = O stretch), 667 (B_{2g}, B_{3g} υ_s O–M–O stretch), 473 (A_g, υ_s O–M–O stretch and bend), 380 (B_{1g}, δ O–M–O scissor), 376 (B_{1g}), 366 (A_{1g}, δ O–M–O scissor), 334 (A_g, B_{1g}, δ O–M–O bend), 293 (B_{3g}, δ O = M = O wagging), 285 (B_{2g}, δ O = M = O wagging), 247 (B_{3g}, τ O = Mo = O twist), 216 (A_g, rotational rigid MoO_4 chain mode, R_c), 197 (B_{2g}, τ O = Mo = O twist), 159 (A_g/B_{1g}, translational rigid MoO_4 chain mode, T_b), 129 (B_{3g}, translational rigid MoO_4 chain mode, T_c), 116 (B_{2g}, translational rigid MoO_4 chain mode, T_c), 100 (B_{2g}, translational rigid MoO_4 chain mode, T_a), and 89 cm^{-1} (A_g, translational rigid MoO_4 chain mode, T_a) closely resemble the single-crystal bands for α-MoO_3, a thermodynamically stable phase with layered structure. There was no β-MoO_3 phase present in the sample as the peak at 902 cm^{-1} was absent.

The Raman spectra of the nanocomposites recorded after five voltammetric cycles in 1 M NaOH and H_2SO_4 show the structural distortions introduced by the intercalation of Na and H ions decreases, indicating the loss of structural disorder. When Na and H and electrons are inserted into α-MoO_3 nanowires, the electrons reduce Mo^{6+} ions to Mo^{5+}, thus creating $Mo^{5+}=O$ and $Mo^{5+}-O$ bonds. The intercalation of Na and H ions induces the Mo^{5+} states by reducing the Mo^{6+} states. The intensity of the Raman bands of MoO_3 in the MoO_3–MWCNT composite decreased after five voltammetric cycles, and almost disappeared after 50 cycles, indicating a gradual loss of crystallinity of the MoO_3 nanowires, which is most likely due to some degree of irreversible intercalation, as shown in Figure 8.8. This loss of the Raman signature of the α-MoO_3 phase was accompanied by a decrease in the charge storage capacity measured by CV. The Raman signature of α-MoO_3 disappeared completely after 50 and 100 cycles, which is the region where the specific capacitance becomes constant (Figure 8.8a and b). On the other hand, the Raman spectra also show the structural changes in MWCNTs upon cycling as shown in Figure 8.7c and d. Three distinct peaks can be seen from the spectra: (1) the D-band in the range of 1300–1400 cm^{-1}, which is the disorder-induced band in the graphite lattice that reflects defects in CNTs; (2) the G-band close to 1600 cm^{-1}, which is due to the tangential vibration of carbon atoms; and (3) the D*-band in the range of 2600–2800 cm^{-1}, which is the second harmonic of the D-band. The structural integrity of the nanotubes can be estimated qualitatively by calculating

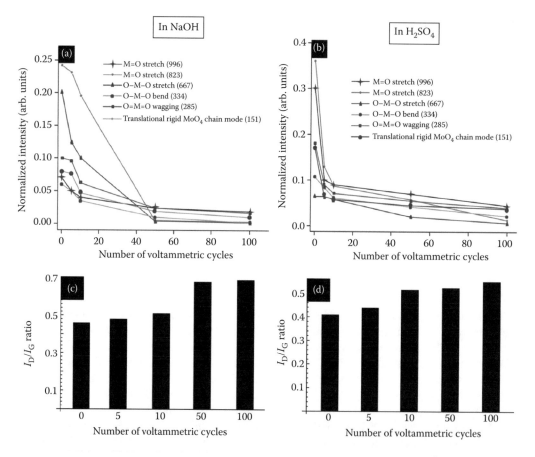

FIGURE 8.8 Decrease of the MoO_3 Raman bands in the MoO_3–MWCNT nanocomposite upon cycling, indicating structural changes. The peak positions were calculated after fitting with the Lorentzian function and intensities were normalized with respect to the MWCNT G-band (a) 1 M NaOH, (b) 1 M H_2SO_4. Increase of the I_D/I_G ratio for MWCNTs with the number of cycles, indicating the structural changes (c) 1 M NaOH and (d) 1 M H_2SO_4.

the peak intensity ratio of the D- and G-bands (I_D/I_G). Figure 8.8c and d shows the I_D/I_G ratios of the MWCNTs in the composite, which were calculated after 5, 10, 50, and 100 voltammetric cycles.

The observed gradual increase in the I_D/I_G ratio is consistent with the increase in defect density in the MWCNT walls, which may result from charge centers appearing on the MWCNTs as a result of faradaic reactions and/or intercalation of the electrolyte into the MWCNT structure. Thus, the MWCNTs act not only as a support matrix that provides electrical conducting pathways to the active sites of the metal oxide but also they contribute to the overall capacitance of the system through faradaic and/or intercalation mechanisms (as well as double-layer formation). The nanocomposite was examined by HRTEM to further visualize the structural and morphological changes. Figure 8.5 shows the HRTEM images and electron diffraction pattern of the MoO_3–MWCNT composite after 10 and 100 cycles in 1 M H_2SO_4 and NaOH. The crystallinity of the MoO_3 nanowires changed with increasing cycles, and the diffraction pattern exhibited amorphous rings rather than sharp diffraction spots (Figure 8.5b). This supports the idea that the MoO_3 crystal structure is modified irreversibly upon cycling, which is most likely through irreversible cation intercalation.

Electrochemical impedance analysis was performed to gain an understanding of the double-layer and pseudocapacitance of the composite system. Figure 8.6 shows the impedance spectra of the MoO_3–MWCNT nanocomposites recorded after 10, 50, and 100 voltammetry cycles in 1 M NaOH and 1 M H_2SO_4. The impedance plot revealed R_s values (solution resistance or initial resistance) of 0.028 Ω (after 10 cycles), 0.022 Ω (after 50 cycles), and 0.025 Ω (after 100 cycles) in 1 M NaOH and 0.046 Ω (after 10 cycles), 0.048 Ω (after 50 cycles), and 0.039 Ω (after 100 cycles) in 1 M H_2SO_4.

This shows that the low equivalent series resistance is relatively independent of cycling due to the good electrical conductivity of the MWCNT support network. Initially, redox processes were dominant, resulting in the larger semicircle at higher frequencies. At lower frequencies, diffusion behavior was dominant. The insertion of hydrogen and sodium ions into the interiors of composite materials results in the loss of crystallinity of oxide material and introduces defects on the sidewalls of CNTs. The plots can be further divided into two regions for further analysis: a higher-frequency region that shows the total resistance of the system, and a lower-frequency region dominated by the diffusion behavior. The impedance plots after 10 and 50 cycles show the lower-frequency regions dominated by diffusion phenomena. On the other hand, the impedance plot after 100 cycles shows linearity, which is almost parallel to the imaginary axis, indicating steady capacitive behavior with little diffusion limitation. The decrease in the diffusion with increased cycle number suggested a loss of a slow, diffusion-controlled charge storage process. This can be interpreted as a decrease in the storage mechanism related to intercalation into MoO_3. In an acidic medium, the impedance spectra are dominated by a noncapacitive faradaic reaction, probably hydrogen evolution. With a number of cycles, the material exhibits amorphous behavior and the capacitance contribution is mainly due to the formation of a double layer.

The redox kinetics and cycling behavior of the nanoarchitectures of MoO_3–MWCNTs can be modified to produce high capacitance values when supported on a conducting support matrix of MWCNT. The electrochemical charge storage in the MoO_3–MWCNT composite involves the contribution from double-layer, faradaic, and intercalation mechanisms. CNTs not only offer mechanical integrity to MoO_3 but also provide electrical conducting pathways to its active sites. The initial storage capacity was 178 F/g in an basic medium and 190 F/g in an acidic medium, and decreased rapidly with repeated cycling, reaching a steady value of 98 F/g (basic) and 102 F/g (acidic) after approximately 80 cycles. The loss of the MoO_3 structure observed by Raman spectroscopy and TEM, as well as impedance spectroscopy, argues toward gradual loss of the intercalation charge storage mechanism upon repeated cycling due to irreversible intercalation.

8.3.3 Disordered Graphene

The effects of morphological disorder of graphene on electron field emission properties are reported below. Figure 8.9a and b shows the morphological distribution of the planar single-layer graphene

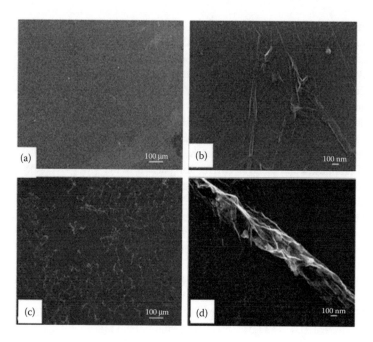

FIGURE 8.9   SEM images of planar (a and b) and aggregated (c and d) graphene at different magnifications. (P. Rai et al., Modified carbon nanostructures for energy and display applications, *Proc. 11th IEEE International Conference on Nanotechnology*, August 15–18, Portland, Oregon, pp. 76–79, © (2011) IEEE. With permission.)

(PSLG) that seems to be a continuous layer of graphene. Figure 8.9c and d shows the SEM images at different magnifications of agglomerated single-layer graphene (ASLG).

The structural properties and qualities of these two types of samples can be deduced by Raman spectroscopy. Raman spectra of the planar and agglomerated single-layer graphene are shown in Figure 8.10. Both the Raman spectra show the basic feature of a graphene, that is, a G-band that originates from doubly degenerate (iTO and LO) phonon mode (E_{2g} symmetry) at the Brillouin zone center and a 2D band that originates from a second-order process, involving two iTO phonons

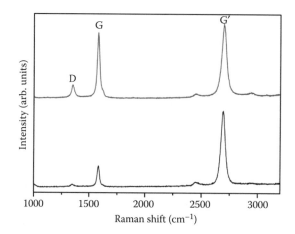

FIGURE 8.10   Raman spectra of planar (lower curve) and agglomerated graphene (upper curve). (P. Rai et al., Modified carbon nanostructures for energy and display applications, *Proc. 11th IEEE International Conference on Nanotechnology*, August 15–18, Portland, Oregon, pp. 76–79, © (2011) IEEE. With permission.)

near the K point [40,41]. The calculated values of the G- to 2D band (I_G/I_{2D}) intensity ratio and the D- to G-band (I_D/I_G) intensity ratio were found to be 0.28 and 0.10, respectively. On the basis of the above data obtained by the Raman spectrometer, we can conclude that our planar graphene shows an excellent agreement with the features of continuous monolayer graphene.

The upper plot of Figure 8.10 shows the Raman spectrum of the ASLG. The increased value of the I_D/I_G ratio implies the highly disordered state and increased number of the defects/edges in the agglomerated graphene, and consequently this sample should have more number of sites for field emission. Moreover, after agglomeration of the single-layer graphene, there is a possibility of the formation of the double, triple, or multilayer graphene structure, which this will increase the I_G/I_{2D} ratio. Thus, the comparative analysis of the PSLG and ASLG shows that after the agglomeration of monolayer graphene, the emission sites may increase significantly.

Figure 8.11 shows the plot of current density against the applied electric field for PSLG and ASLG, obtained from the field emission characterization technique as described in the experimental section. The turn-on field E_{to} corresponding to 10 μA/cm^2 for the PSLG was found to be 4.31 V/μm, whereas turn-on value E_{to} corresponding to ASLG was found to be 2.26 V/μm. This enhancement in the field emission properties in ASLG is mainly due to the random shape of the graphene.

If we compare this work with the other graphene-based field emission, we find that pristine ASLG gives similar or better performance without any treatment. For example, Qi et al. [42] reported that the plasma etching treatment for 3 min on the few layers of graphene sheets decreases the turn-on electric field from 3.91 to 2.23 V/μm, and increases the maximum emission current density, drawn at a field of 4.4 V/μm, from 0.033 to 1.330 mA/cm^2. But by using our simple method, the turn-on values were found to decrease from 4.31 to 2.26 V/μm and increase in maximum current density from 0.007 to 3.525 mA/cm^2 at 4.26 V/μm (Table 8.2 and Figure 8.11). Three very interesting observations can be made from the results obtained. (1) There was considerable reduction in the "turn-on" field for morphologically disordered graphene. (2) β_L did not show much difference (%) while the value of β_H changed significantly (%). (3) The surface of the planar monolayer graphene has a considerable field enhancement factor.

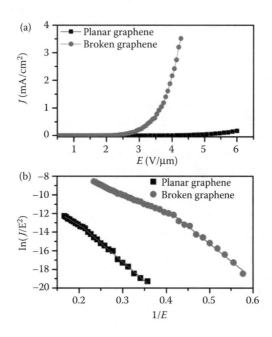

FIGURE 8.11 (a) Plots of the electron-emission current density (J) as a function of applied electric field (E) for the CVD grown planar and broken graphene. (b) Corresponding F–N plots. (Nanotechnology (IEEE-NANO), 2011 11th IEEE Conference, pp. 76–79, © (2011) IEEE.)

TABLE 8.2
Different Parameters Obtained from Field Emission Measurements

Sample	E_{turnon} (V/μm)	$β_L$	$β_H$	J_m (mA/cm²)
Planar	4.31	1870	2690	0.007
Agglomerated/broken	2.26	2060	3870	3.525

Source: Nanotechnology (IEEE-NANO), 2011 11th IEEE Conference, pp. 76–79, © (2011) IEEE.

We believe that the reduction in the turn-on field is due to a change in morphology, which gives rise to additional emission sites in the form of defects and irregular nanostructure shapes. Only a slight change in the lower slope of the F–N plot implies the verification of a double barrier model of field emission. Moreover, planar monolayer graphene showed a significant field emission property. We suggest that field emission from planar monolayer graphene come mainly from its grain boundaries. Further investigations are needed to have a better control on the stresses used on the copper foil and then to correlate the extent of morphological disorder to the field emission properties.

8.4 CONCLUSIONS

In conclusion, we observed that a moderate level of ion irradiation can produce considerable changes in the nanostructures such as MWCNT. Higher doses will result in amorphization, reducing the usefulness of the defect creation. The irradiation levels used in this study are routinely used while carrying out ion etching, and therefore one has to be cognizant of these limitations as well. The electrochemical charge storage in the MoO_3–MWCNT composite involves the contribution from double-layer, faradaic, and intercalation mechanisms. The loss of charge storage capacity upon repeated cycling can be explained as due to irreversible intercalation of the nanocomposite. Minute mechanical stresses present while transferring single-layer graphene resulted in improved field emission characteristics. Controlled modifications of carbon nanostructures such as MWCNT and graphene can be used in applications such as supercapacitors and field emission displays.

ACKNOWLEDGMENTS

The support from the Korean Ministry of Education, Science and Technology under the grant R31-2008-10029 (World Class University Program) is greatly appreciated. Technical assistance from the present and past members of the NMESG (NanoMaterials for Energy Science Group) of Sungkyunkwan University is acknowledged. Collaborative work from group members of "Nanoscale Devices and Materials Physics Laboratory" helped to carry out the metal oxide nanocomposite work.

REFERENCES

1. K. M. Kadish and R. S. Ruoff (Eds), *Fullerenes: Chemistry, Physics and Technology*, John Wiley & Sons Inc., New York, 2000.
2. M. Meyyappan (Ed), *Carbon Nanotubes: Science and Applications*, CRC Press LLC, Boca Raton, 2005.
3. S. Pati, T. Enoki, and C.N.R. Rao (Eds), *Graphene and Its Fascinating Attributes*, World Scientific Publishing Co Pvt. Ltd., Singapore, 2011.
4. J. R. Miller and P. Simon, Electrochemical capacitors for energy management, *Science*, **321**, 651, 2008.
5. T. Brezesinski, J. Wang, S. H. Tolbert, and B. Dunn, Ordered mesoporous α-MoO_3 with iso-oriented nanocrystalline walls for thin-film pseudocapacitors, *Nat. Materials*, **9**, 146, 2010.
6. V. L. Pushparaj, M. M. Shaijumon, A. Kumar, S. Murugesan, L. Ci, Robert Vajtai, R. J. Linhardt, O. Nalamasu, and P. M. Ajayan, Flexible energy storage devices based on nanocomposite paper, *Proc.Nat.l Acad.Sci.*, **104**, 13574, 2007.

7. P. Simon and Y. Gogotsi, Materials for electrochemical capacitors, *Nature Materials*, **7**, 845, 2008.
8. Z. W. Pan, Z. R. Dai, and Z. L. Wang, Nanobelts of semiconducting oxides, *Science*, **291**, 1947, 2001.
9. Y. Zhu, Y. Bando, L.Yin, and D. Golberg, Field nanoemitters: Ultrathin BN nanosheets protruding from Si_3N_4 nanowires, *Nano Lett.* **6**, 2982, 2006.
10. M. Hoefer and P. R. Bandaru, Determination and enhancement of the capacitance contributions in carbon nanotube based electrode systems, *Appl. Phys. Lett.* **95**, 183108, 2009.
11. T. Brezesinski, J. Wang, S.H. Tolbert, and B. Dunn, Ordered mesoporous α-MoO_3 with iso-oriented nanocrystalline walls for thin-film pseudocapacitors, *Nat. Mater.* **9**, 146, 2010.
12. H.Y. Lee and J.B. Goodenough, Ideal supercapacitor behavior of amorphous $V_2O_5.nH_2O$ in potassium chloride (KCl) aqueous solution, *J. Solid State Chem.* **148**, 81–84, 1999.
13. A.S. Aricò, P. Bruce, B. Scrosati, J.M. Tarascon, and W. Van Schalkwijk, Nanostructured materials for advanced energy conversion and storage devices, *Nat. Mater.* **4** 366–377, 2005.
14. P. Simon and Y. Gogotsi, Materials for electrochemical capacitors, *Nat. Mater.* **7**, 845, 2008.
15. J. Wang, J. Polleux, J. Lim, and B. Dunn, Pseudocapacitive Contributions to Electrochemical Energy Storage in TiO2 (Anatase) Nanoparticles, *J. Phys. Chem. C.* **111**, 14925, 2007.
16. Z.-S. Wu, S. Pei, W. Ren, D. Tang, L. Gao, B. Liu, F. Li, C. Liu, and H.-M. Cheng, Field emission of single-layer graphene films prepared by electrophoretic deposition, *Adv. Mater.* **21**, 1756, 2009.
17. M. Qian, T. Feng, H. Ding, L. Lin, H. Li, Y. Chen, and Z. Sun, Electron field emission from screen-printed graphene films, *Nanotechnology* **20**, 425702, 2009.
18. A. Malesevic, R. Kemps, A. Vanhulsel, M. P. Chowdhury, A. Volodin, and C. V. Haesendonck, Field emission from vertically aligned few-layer graphene, *J. Appl. Phys.* **104**, 084301, 2008.
19. J. L. Qi, X. Wang, W. T. Zheng, H. W. Tian, C. Q. Hu, and Y. S. Peng, Ar plasma treatment on few layer graphene sheets for enhancing their field emission properties, *J. Phys. D: Appl. Phys.* **43**, 055302, 2010.
20. U. A. Palnitkar, Ranjit V. Kashid, Mahendra A. More, Dilip S. Joag, L. S. Panchakarla, and C. N. R. Rao, Remarkably low turn-on field emission in undoped, nitrogen-doped, and boron-doped grapheme, *Appl. Phys. Lett.* **97**, 063102, 2010.
21. S.R. Sivakkumar, J.M. Ko, D.Y. Kim, B.C. Kim, and G.G. Wallace, Performance evaluation of CNT/polypyrrole/MnO_2 composite electrodes for electrochemical capacitors, *Electrochimica Acta* **52**, 7377, 2007.
22. M. Jayalakshmi, M.M. Rao, N. Venugopal, and K.B. Kim, Hydrothermal synthesis of SnO_2–V_2O_5 mixed oxide and electrochemical screening of carbon nano-tubes (CNT), V_2O_5, V_2O_5–CNT, and SnO_2–V_2O_5–CNT electrodes for supercapacitor applications, *J. Power Sources* **166**, 578, 2007.
23. F. Gunes, H.-J. Shin, C. Biswas, G. H. Han, E. S. Kim, S. J. Chae, J.-Y. Choi, and Y. H. Lee, Layer-by-layer doping of few-layer graphene film, *ACS Nano* **4**, 4595, 2010.
24. A. C. Ferrari and J. Robertson, Interpretation of Raman spectra of disordered and amorphous carbon, *Phys. Rev. B* **61**, 14095, 2000.
25. A. Misra, P. K. Tyagi, P. Rai, D. R. Mahopatra, J. Ghatak, P. V. Satyam, D. K. Avasthi, and D. S. Misra, Axial buckling and compressive behavior of nickel-encapsulated multiwalled carbon nanotubes, *Phys Rev. B* **76**, 014108, 2007.
26. A. R. Adhikari, M. Huang, H. Bakhru, R. Vajtai, C. Y. Ryu, and P. M. Ajayan, Stability of ion implanted single-walled carbon nanotubes: Thermogravimetric and Raman analysis, *J. Appl. Phys.* **100**, 64315, 2006.
27. Y. Yamada, O. Kimizuka, K. Machida, S. Suematsu, K. Tamamitsu, S. Saeki, Y. Yamada et al., Hole Opening of Carbon Nanotubes and Their Capacitor Performance, *Energy Fuels* **24**, 3373, 2010.
28. A. V. Krasheninnikov and K. Nordlund, Irradiation effects in carbon nanotubes, *Nucl. Instr. and Meth. B* **216**, 355, 2004.
29. Y. Yamada, O. Kimizuka, O. Tanaike, K. Machida, S. Suematsu, K. Tamamitsu, S. Saeki, Y. Yamada, and H. Hatori, Capacitor properties and pore structure of single- and double-walled carbon nanotubes, *Electrochemical and Solid-State Letters* **12**, K14, 2009.
30. P. Li, X. Lim, Y. Zhu, T. Yu, C.K. Ong, Z. Shen, A.T.S. Wee, and C.H. Sow, Tailoring wettability change on aligned and patterned carbon nanotube films for selective assembly, *J. Phys. Chem. B* **111**, 1672, 2007.
31. N.I. Kovtyukhova, T.E. Mallouk, L. Pan, and E.C. Dickey, Individual single walled nanotubes and hydro-gels made by oxidative exfoliation of carbon nanotube ropes, *J. Am. Chem. Soc.* **125**, 9761, 2003.
32. D. Yu and L. Dai, Self-Assembled graphene/carbon nanotube hybrid films for supercapacitor, *J. Phys. Chem. Lett.* **1**, 467, 2010.
33. C. H. Chen, H. C. Su, S. C. Chuang, S. J. Yen, Y. C. Chen, Y. T. Lee, H. Chen et al., Hydrophilic modi-fication of neural microelectrode arrays based on multi-walled carbon nanotubes, *Nanotechnology* **21**, 485501, 2010.

34. C.C. Lin, B.T.T. Chu, G. Tobias, S. Sahakalkan, S. Roth, M.L.H. Green, and S.Y. Chen, Electron transport behavior of individual zinc oxide coated single-walled carbon nanotubes, *Nanotechnology* **20**, 105703, 2009.
35. T. Wei, G. Luo, Z. Fan, C. Zheng, J. Yan, C. Yao, W. Li, and C. Zhang, Preparation of graphene nanosheet/polymer composites using in situ reduction–extractive dispersion, *Carbon* **47**, 2296, 2009.
36. S. Stankovich, D.A. Dikin, R.D. Piner, K.A. Kohlhaas, A. Kleinhammes, Y. Jia, Y. Wu, S.T. Nguyen, and R.S. Ruoff, Synthesis of graphene-based nanosheets via chemical reduction of exfoliated graphite oxide, *Carbon* **45**, 1558, 2007.
37. S. Park, J. An, R.D. Piner, I. Jung, D. Yang, A. Velamakanni, S.T. Nguyen, and R.S. Ruoff, Aqueous suspension and characterization of chemically modified graphene sheets, *Chem. Mater.* **20**, 6592, 2008.
38. L. Fu, Z. Liu, Y. Liu, B. Han, P. Hu, L. Cao, and D. Zhu, Beaded cobalt oxide nanoparticles along carbon nanotubes: Towards more highly integrated electronic devices, *Adv. Mater.* **17**, 217, 2005.
39. I. Shakir, M. Shahid, H.W. Yang, and D.J. Kang, Structural and electrochemical characterization of α-MoO$_3$ nanorod-based electrochemical energy storage devices, *Electrochim Acta* **56**, 376, 2010.
40. L.M. Malard, M.A. Pimenta, G. Dresselhaus, and M.S. Dresselhaus, Raman spectroscopy in graphene, *Physics Reports*, **473**, 51, 2009.
41. A. Reina, X. Jia, J. Ho, D. Nezich, H. Son, V. Bulovic, M. S. Dresselhaus, and J. Kong, Large area, few-layer graphene films on arbitrary substrates by chemical vapor deposition, *Nano Lett.* **9**, 30, 2009.
42. J. L. Qi, X. Wang, W. T. Zheng, H. W. Tian, C. Q. Hu, and Y. S. Peng, Ar plasma treatment on few layer graphene sheets for enhancing their field emission properties, *J. Phys. D: Appl. Phys.* **43**, 055302–08, 2010.
43. P. Rai, S. Pandey, G. Arabale, P. Nikolaev, and S. Arepalli, Modified carbon nano structures for energy and display applications, *Proc. 11th IEEE International Conference on Nanotechnology*, August 15–18, Portland, Oregon, pp. 76–79, 2011.

9 Production and Characterization of Nanoparticle Dispersions of Organic Semiconductors for Potential Applications in Organic Electronics

Muhammad Hassan Sayyad, Fazal Wahab,
Munawar Ali Munawar, Muhammad Shahid,
Jamil Anwar Chaudry, Khaulah Sulaiman, Zubair Ahmad,
and Abdullah Mohamed Asiri

CONTENTS

9.1 INTRODUCTION

The synthesis and characterization of organic semiconductors are receiving great attention because of their potential use in a wide range of electronic and photonic devices, such as junction diodes, organic field effect transistors, memories, solar cells, organic light-emitting diodes, optical power-limiting, radio frequency identification tags (RFIDs), sensors, and so on [1–7]. Most of the devices are based on the thin films of organic semiconductors, and expensive equipment is required for depositing their thin films because of the poor solubility of many important families of organic semiconductors in common organic solvents. Solution-processable organic semiconductors are potential candidates for the low-cost, large-area, and large-scale production of many electronic and photonic devices. Using pulsed laser ablation of organic semiconductors in polar solvents, dispersions of the nanoparticles of these materials can be produced [8] and their thin films can be deposited employing the low-cost deposition techniques such as drop casting, dipping, spin casting, inkjet printing, and so on.

The synthesis and characterization of nanoscaled materials have received much more attention because of their unique properties and applications. Most of these studies have been devoted to the nanoparticles of metals and inorganic semiconductors. A large number of organic semiconductors and their derivatives have been synthesized and many are being synthesized every year but little is known about the organic semiconducting nanomaterials and their potential applications. Devices fabricated using nanoparticles have exhibited better performance. Using chemically synthesized nanoparticles of cobalt phthalocyanine, a highly responsive glucose biosensor has been reported [9]. It is expected that the other devices fabricated using nanoparticles of organic semiconductors will exhibit better performance.

A wide variety of materials are used as sensors for capturing physical, chemical, or biological stimuli and converting them to measurable output signals. Nanomaterials bring many advantages to sensor technology and can be used to overcome many limitations of bulk material-based sensors. They are attractive because of their small characteristic size and correspondingly large surface-to-volume ratio. Nanomaterials increase a sensor's active surface area and generate novel interfaces, thus improving the sensitivity, response, and recovery times. In recent years, there has been great progress in the application of nanomaterials in sensor technologies. In these applications, nanomaterials such as metal nanoparticles, nanostructured metal oxides, and carbon nanotubes have been been actively investigated.

Humidity measurement and control are of great importance for many industrial, medical, agricultural, and domestic applications [10 and references therein]. Therefore, many relative humidity (RH) sensors using various technologies and active materials have been fabricated and investigated. Recently, we have investigated the sensing potential of bulk organic semiconducting thin films [11–13]. Organic semiconductors have been observed to show a large change in their electrical properties with variation in RH. In this chapter, a brief introduction of the organic semiconductors is given. Methods of generation of the nanoparticles of organic semiconductors are mentioned and the PLA technique for the generation of the nanoparticle dispersion of organic semiconducting materials is briefly described. Bulk and nanoscaled organic semiconductor-based sensors are discussed. The potential of the nanoparticle dispersion of organic semiconductors for humidity sensor applications is demonstrated. The fabrication, study, and results of a capacitive-type humidity sensor fabricated using nanoparticles of organic semiconductors are presented.

9.2 ORGANIC SEMICONDUCTORS

Organic semiconductors are multifunctional materials because of their capability of exhibiting a variety of properties, such as charge transport, photoconductivity, electroluminescence, and so on. The electrical and optical properties of organic semiconductors are highly dependent on ambient conditions, which makes them very promising for the development of various types of sensors, such as temperature, light, humidity, and so on. Phthalocyanines belong to a class of organic semiconductors and have been extensively employed in the fabrication of organic and organic–inorganic devices [14–16]. They exhibit good thermal and chemical stability [17,18]. Phthalocyanines can be easily deposited as thin films by thermal evaporation without dissociation. Therefore, many studies have been carried out on phthalocyanine-based thin-film devices [19–21].

Organic semiconductors are π-conjugated systems and can be classified into two broad categories of low-molecular-weight compounds and long-chain polymers. Low-molecular-weight compounds comprise the families of azo dyes, perylenes, phthalocyanines, porphyrins, and so on. Numerous organic devices have been fabricated and studied using these materials. Recently, the devices fabricated using nanomaterials of low-molecular-weight organic semiconductors have shown better performance; therefore, it is expected that, in the future, there will be more activities in the synthesis and characterization of low-molecular-weight organic semiconducting nanomaterials.

9.3   GENERATION OF NANOPARTICLES OF ORGANIC SEMICONDUCTORS

The formation of nanoscaled organic semiconductors has recently attracted a great deal of interest due to the possibility of tailoring the specific material function depending on the particle size [8], their role in the modification of optical and electrooptical properties [22], their potential impact in several attractive economic fields [23], and so on. Various methods such as reprecipitation, miniemulsion, direct condensation of organic vapor, precipitation by solvent displacement with or without subsequent microwave irradiation, supercritical fluids, vapor deposition, laser ablation, template-based approaches, and so on have been developed to prepare the nanoparticles of organic semiconductors [22–25].

Laser ablation in liquids has received much attention as a very simple nanoparticle production technique with some remarkable features, such as the synthesis of nanoparticle dispersions in different liquids, stability of the dispersions, and control of the shape and size. Using laser ablation in a liquid, nanoparticles of a wide range of materials, such as metals, metal oxides, organic/inorganic semiconductors, and so on, are obtainable in different liquids. Nanoparticle synthesis using this technique involves a large number of parameters, such as laser wavelength, laser energy, pulse width, repetition rate, liquid type, and liquid composition [26]. By varying these parameters, the size, shape, and morphology of the nanostructured materials can be controlled. In this method, a pulsed laser beam with enough energy falls onto a solid target in a transparent liquid, and a high-temperature (about 6000 K), high-pressure (about 1 GPa), and short-lived plasma plume comprising nanoparticles of the target material is produced on the solid–liquid interface [27]. Many studies regarding the generation and evolution of the laser plasma plume have been carried out. This plasma plume quickly cools and a dispersion of the nanoparticles of the target material is formed.

9.4   ORGANIC SEMICONDUCTOR-BASED SENSORS

Organic sensors offer great potential for the applications in security, industry manufacturing, environmental monitoring, and medical science. Besides the advantages of easy processing, low cost, and compatibility with plastic substrates, another important advantage of organic-based sensors is that the organic compounds can also be chemically functionalized. This property makes them more favorable, thereby improving the selectivity of the sensor. Compared to the traditional detection methods, organic sensors have higher selectivity, sensitivity, and lower cost [28,29]. Pure polymer, polymer blends, and polymer–inorganic composites have also been studied for the purposes, resulting in different degrees of advancements in this area [30–33]. In organic sensors, analyte molecules diffuse into the organic film and interact with the organic material. The analyte molecules can have various effects on organic film, such as doping of carrier density, dipole-induced trapping, and retardation of charges [34]. Those interactions alter the conductivity, threshold voltages, and mobility of devices and hence change the current flow through the thin films between electrodes. Several mechanisms of sensing such as capacitance [35–37], electrical resistance [29], and surface acoustic wave (SAW) [38,39] have been proposed.

Capacitive sensors work at either change in dielectric permittivity or induced polarization in the organic film. For instance, water molecules are bound at suitable sites in organic materials during the adsorption and desorption processes in capacitive humidity sensors. Since there exists a big difference between the dielectric constant of water (~80) and organic materials (~5), the water molecules adsorbed in the film influence the dielectric constant greatly. If the electrodes are similar to the parallel plates, the principle of operation of the capacitive-type humidity sensor is based on a familiar expression defined as $C = (\varepsilon A)/d$, where C is the capacitance, A is the electrode area, d is the distance between the two electrodes, and ε is the permittivity of the sensing material, which is a function of RH. With the adsorption of humidity into the film or the desorption of moisture from the film, the dielectric constant of the sensing organic semiconductor changes; therefore, a variation of capacitance as a result of a change in the dielectric constant of the sensing film is measured.

Phthalocyanines represent a large family of organic semiconducting materials with high chemical and thermal stability. Using these compounds, numerous organic electronic devices have been fabricated.

The interaction of various analytes with the surfaces of the thin films of these compounds produces reversible changes in their physical properties. By measuring these changes using different technologies, such as electrical, optical, and so on, their potential for numerous sensors has been studied.

9.5 HUMIDITY SENSING APPLICATIONS OF NANOPARTICLES OF ORGANIC SEMICONDUCTORS

To investigate the sensing potential of the nanoparticles of organic semiconductors, the nanoparticle dispersion of organic semiconductor CoPc was generated using PLA. Using this dispersion, a surface-type humidity sensor was fabricated and its humidity-dependent capacitance was measured.

9.5.1 GENERATION OF NANOPARTICLES OF CoPc

The molecular structure of the organic semiconductor cobalt phthalocyanine (CoPc) is shown in Figure 9.1. It was prepared in accordance with the method of Sakamoto and Ohno, with minor modifications [40].

The synthesized CoPc was dried for 24 h under a 200 W tungsten filament lamp and a pellet was formed using a press. A nanoparticle dispersion of CoPc was produced using the PLA technique. A schematic representation of the experimental setup used is shown in Figure 9.2. In this method,

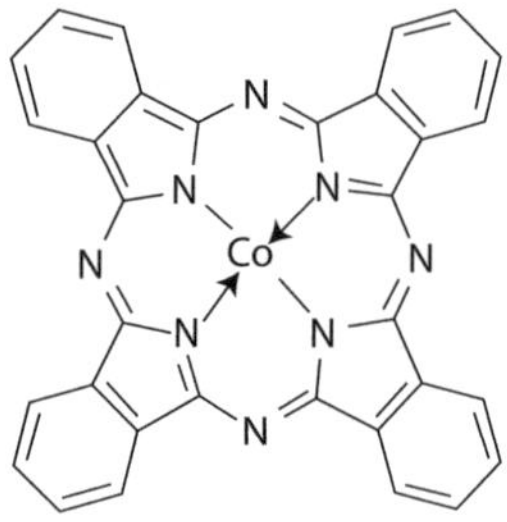

FIGURE 9.1 Molecular structure of CoPc. (M.H. Sayyad et al., Production and characterization of nanoparticle dispersions of organic semiconductors for potential applications in organic electronics, *Proceedings of the 11th IEEE International Conference on Nanotechnology*, August 15–18, Portland, Oregon, pp. 193–196, © (2011) IEEE. With permission.)

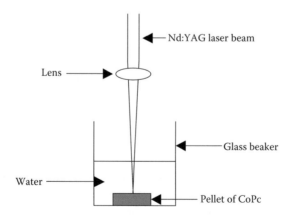

FIGURE 9.2 Setup for the synthesis of nanoparticles of solid targets using pulsed laser ablation in liquid medium. (M.H. Sayyad et al., Production and characterization of nanoparticle dispersions of organic semiconductors for potential applications in organic electronics, *Proceedings of the 11th IEEE International Conference on Nanotechnology*, August 15–18, Portland, Oregon, pp. 193–196, © (2011) IEEE. With permission.)

a pellet of CoPc was placed under a thin layer of deionized water in a beaker and exposed to a nanosecond Nd:YAG laser radiation working at 532 nm and 10 Hz. Operating the laser for 30 min, a colloidal solution of the nanoparticles of CoPc was obtained.

9.5.2 DEVICE FABRICATION AND CHARACTERIZATION

The Ag/NanoCoPc/Ag-type capacitive humidity sensor was fabricated by drop casting the nanoparticle dispersion of CoPc. The 100-nm-thick silver electrodes were deposited on a glass substrate with a gap of 50 μm by the thermal vacuum evaporation technique using a mask. The electrodes were deposited at a rate of 0.2 nm/s and a vacuum of 10^{-5} mbar. The structure of the fabricated device is shown in Figure 9.3.

The humidity-dependent capacitance was investigated using a homemade experimental setup shown in Figure 9.4. It consists of a humidifier and a measurement chamber. The RH in the chamber was obtained by bubbling dry nitrogen in the humidifier containing distilled water. The percentage of RH inside the measurement chamber was controlled by two valves. The RH and capacitance were measured using commercial digital meters. During capacitive measurements, the frequency was kept at 1 kHz.

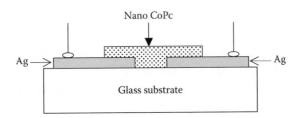

FIGURE 9.3 Cross-sectional view of the Ag/NanoCoPc/Ag surface-type sensor. (M.H. Sayyad et al., Production and characterization of nanoparticle dispersions of organic semiconductors for potential applications in organic electronics, *Proceedings of the 11th IEEE International Conference on Nanotechnology*, August 15–18, Portland, Oregon, pp. 193–196, © (2011) IEEE. With permission.)

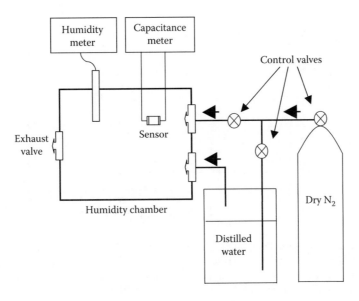

FIGURE 9.4 Experimental setup for humidity-dependent capacitance measurements.

9.5.3 Humidity-Dependent Capacitance

Humidity can affect human comfort, numerous material properties, and industrial processes. It is used to represent the amount of water vapor in the air and is defined in terms of RH, specific humidity, and absolute humidity.

Absolute humidity measures the water vapor content in absolute terms, that is, how much water vapor is present, without any concern about the maximum amount the air could carry. The RH is the type mostly used in conventional humidity meters. The RH is the ratio of partial water vapor pressure (p_w) in an air–water mixture at any temperature to the saturated vapor pressure (p_s) of water at the same temperature. RH as a percentage is defined as [41]

$$RH = 100 \times \frac{p_w}{p_s} \qquad (9.1)$$

Figure 9.5 shows the variation in the capacitance of the Ag/NanoCoPc/Ag sensor with an RH at room temperature and under dark conditions. It is observed that in the range 0–55% RH, the capacitance of the sensor does not change but above 55% RH, it increases significantly with a corresponding increase in humidity. Similar behavior has been reported in the case of a polypyrrole-based sensor [42] and our earlier studies of porphyrin-based sensors [29,43].

This type of capacitance variation with the increase in humidity can be explained on the basis of water adsorption and dielectric physics theory [42]. According to the adsorption theory, the adsorption of water molecules on the sensing material takes place during the two stages of chemisorption and physisorption. When the sensor is exposed to humidity, the formation of a chemisorbed layer starts. The capacitance of the sensor does not change until the complete chemisorbed layer is formed. The threshold RH at which the chemisorbed layer is completed has been observed to be dependent upon the material type, film surface morphology, and film thickness. When the chemisorbed layer is completed with an increase in further humidity, many more physisorbed layers are

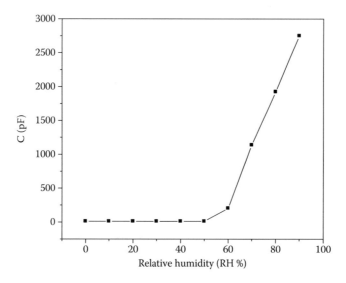

FIGURE 9.5 Capacitance–humidity relationship for the Ag/NanoCoPc/Ag sensor. (M.H. Sayyad et al., Production and characterization of nanoparticle dispersions of organic semiconductors for potential applications in organic electronics, *Proceedings of the 11th IEEE International Conference on Nanotechnology*, August 15–18, Portland, Oregon, pp. 193–196, © (2011) IEEE. With permission.)

formed; owing to the formation of these additional layers on the surface of a sensitive material, the capacitance of the sensor increases.

According to the dielectric physics theory, the capacitance of an ideal capacitor is independent on the frequency [42]. At low RH, the capacitor exhibits ideal behavior. This shows that the amount of water vapor adsorbed by the sensing material is not large enough. At higher RH, as more water molecules are adsorbed, the leak conduction appears in the sensing material; therefore, the expression of the capacitance of the sensor could be written as [42]

$$C = \varepsilon^* C_0 = \left(\varepsilon_r - i \frac{\gamma}{\omega \varepsilon_0} \right) C_0. \tag{9.2}$$

where C_0 is the capacitance of the ideal capacitor, ε^* is the complex dielectric constant, ε_r is the relative dielectric constant of the ideal capacitor, ε_0 is the vacuum capacitance constant, ω is the frequency, and γ is the leak conductance. Using this equation, the increase in capacitance of the sensor with the increase in RH can be explained. The leak conductance γ is directly proportional to the RH, so the capacitance of the sensor is directly proportional to the RH. Thus, the capacitance of the sensor increases with an increase in humidity. We also see from the expression that the capacitance is inversely proportional to ω. Therefore, the sensor is expected to show better performance at low frequencies. At low frequencies, the carriers have enough time to respond and the capacitance is greatly affected. At higher frequencies, the change of the electric field becomes too fast for the carriers to follow and the capacitance is not affected significantly by RH.

Response and recovery times are the most important parameters for all humidity sensors. Response/recovery time is defined as the time taken by a capacitive sensor to attain 90% of the total capacitance change. The response and recovery times were measured using the method described in Ref. [43]. The response time of the sensor was measured from 30% to 90% RH at room temperature. In this method, the response time is measured by quickly moving the sensor from an RH level of 30% to 90% and the recovery time is measured by moving very quickly from a humidity level of 90% RH, maintained in a closed chamber, to an open atmosphere at a humidity level of 30% RH. Both the response and recovery times for the Ag/NanoCoPc/Ag sensor were observed for approximately less than 5s, which is much less than the times reported for the capacitive humidity sensors based on other nanostructured materials [44,45] and many commercially available humidity meters. Therefore, this sensor can be used for the fabrication of a fast hygrometer for the measurement of high RH where the commercially available humidity sensors are either too slow or incapable of working [44].

9.6   SUMMARY

Using laser ablation, nanoparticle dispersions of almost all the organic semiconductors and conjugated polymers can be synthesized. Their morphological, spectroscopic, and electrical characterization can help in the identification of their potential applications in numerous areas. Using the technique of PLA, the nanoparticle dispersion of an organic semiconductor CoPc is produced in deionized water and the potential of these nanoparticle dispersions for the fabrication of an organic nanoparticle-based sensor is demonstrated. The sensor has been observed to be very sensitive and quicker in response than the commercially available low-cost sensors. This study demonstrates the great potential of the nanodispersion of CoPc in water, prepared using laser ablation, for the fabrication of a low-cost, very sensitive, and fast humidity sensor for the measurement of high RH. Owing to the high surface-to-volume ratio, the nanostructured materials are potential candidates for gas sensors. Therefore, it is proposed that the nanoparticle dispersion of CoPc and other active organic semiconductors may also be used as active materials in gas sensors.

REFERENCES

1. Z. Ahmad and M. H. Sayyad, Extraction of electronic parameters of Schottky diode based on an organic semiconductor methyl-red, *Physica E,* 41, 631–634, 2009.
2. S. M. Khan, M. H. Sayyad, and K. S. Karimov, Investigation of temperature-dependent electrical properties of p-VOPc/n-si heterojunction under dark conditions, *Ionics,* 17, 307–313, 2011.
3. M. H. Sayyad, Z. Ahmad, Kh. S. Karimov, M. Yaseen, and M. Ali, Photo organic field effect transistor based on a metallo-porphyrin, *Journal of Physics D (Applied Physics),* 42, 105112, 2009.
4. Z. Ahmad, P. C. Ooi, K. C. Aw, and M. H. Sayyad, Electrical characteristics of poly(methylsilsesquioxane) thin films for non-volatile memory, *Solid State Communications,* 151, 297–300, 2011.
5. S. M. Khan, M. Kaur, J. R. Heflin, and M. H. Sayyad, Fabrication and characterization of ZnTPP:PCBM bulk heterojunction (BHJ) solar cells, *Journal of Physics and Chemistry of Solids,* 72, 1430–1435, 2011.
6. Z. Ahmad, M. H. Sayyad, M. Yaseen, K. C. Aw, M. M. Tahir, and M. Ali, Potential of 5,10,15,20-Tetrakis(3/,5/-di-tertbutylphenyl)porphyrinatocopper(II) for a multi-functional sensor, *Sensors and Actuators B: Chemical,* 155, 81–85, 2011.
7. F. Aziz, M. H. Sayyad, K. Sulaiman, B. H. Majlis, Khassan S. Karimov, Z. Ahmad, and G. Sugandi, Influence of humidity conditions on the capacitive and resistive response of an Al/VOPc/Pt co-planar humidity sensor, *Measurement Science and Technology,* 23, 014001, 2012.
8. J. Hobley, T. Nakamori, S. Kajimoto, M. Kasuya, K. Hatanaka, H. Fukumura, and S. Nishio, Formation of 3,4,9,10-perylenetetracarboxylicdianhydride nanoparticles with perylene and polyyne byproducts by 355 nm nanosecond pulsed laser ablation of microcrystal suspensions, *Journal of Photochemistry and Photobiology A: Chemistry,* 189, 105–113, 2007.
9. K. Wang, J.-J. Xu, and H.-Y. Chen, A novel glucose biosensor based on the nanoscaled cobalt phthalocyanine–glucose oxidase biocomposite, *Biosensors and Bioelectronics,* 20, 1388–1396, 2005.
10. Z. Chen and C. Lu, Humidity sensors: A review of materials and mechanisms, *Sensor Letters,* 3, 274–295, 2005.
11. M. Saleem, M. H. Sayyad, K. S. Karimov, M. Yaseen, and M. Ali, Synthesis and application of Ni(II) 5,10,15,20-tetrakis(4′-isopropylphenyl) porphyrin in a surface-type multifunctional sensor, *Physica Scripta,* 82, 015703–015708, 2010.
12. M. H. Sayyad, F. Aziz, K. S. Karimov, M. Saleem, Z. Ahmad, and S. M. Khan, Characterization of vanadylphthalocyanine based surface-type capacitive humidity sensors, *Journal of Semiconductors,* 31, 114002, 2010.
13. M. H. Sayyad, Z. Ahmad, M. Yaseen, K. C. Aw, M. M. Tahir, and M. Ali, Potential of 5,10,15,20-Tetrakis(3/,5/-di-tertbutylphenyl)porphyrinatocopper(II) for a multifunctional sensor, *Sensors and Actuators B: Chemical,* 155, 81–85, 2011.
14. M. Biancardo and F. C. Krebs, Battery effects in organic photovoltaics based on polybithiophene, *Solar Energy Materials and Solar Cells,* 92, 353–356, 2008.
15. G. Liang, T. Cui, and K. Varahramyan, Electrical characteristics of diodes fabricated with organic semiconductors, *Microelectronic Engineering,* 65, 279–284, 2003.
16. J. Portier, J. H. Choy, and M. A. Subramanian, Inorganic–organic-hybrids as precursors to functional materials, *International Journal of Inorganic Materials,* 3, 581–592, 2001.
17. A. S. Riad, S. M. Khalil, and S. Darwish, Effect of temperature on photoconduction and low frequency capacitance measurements on β-CuPc photovoltaic cells, *Thin Solid Films,* 249, 219–223, 1994.
18. T. G. Abdel-Malik and G. A. Cox Charge transport in nickel phthalocyanine crystals. I. Ohmic and space-charge-limited currents in vacuum ambient *Journal of Physics C: Solid State Physics,* 10, 63–74, 1997.
19. M. M. El-Nahass, K. F. Abd-El-Rahman, and A. A. A. Darwish, Fabrication and electrical characterization of p-NiPc/n-Si heterojunction, *Microelectronics Journal,* 38, 91–95, 2007.
20. A. C. Varghese and C. S. Menon, Electrical properties of nickel phthalocyanine thin films using gold and lead electrodes, *Journal of Materials Science: Materials in Electronics,* 17, 149–153, 2006.
21. O. V. Molodtsova, T. Schwieger, and M. Knupfer, Electronic properties of the organic semiconductor hetero-interface CuPc/C_{60}, *Applied Surface Science,* 252, 143–147, 2005.
22. Z. Jia, D. Xiao, W. Yang, Y. Ma, J. Yao, and Z. Liu, Preparation of perylene nanoparticles with a membrane mixer, *Journal of Membrane Science,* 241, 387–392, 2004.
23. D. Oliveira, K. Baba, J. Mori, Y. Miyashita, H. Kasai, H. Oikawa, and H. Nakanishi, Using an organic additive to manipulate sizes of perylene nanoparticles, *Journal of Crystal Growth,* 312, 431–436, 2010.
24. X. Xu and L. Li, Organic semiconductor nanoparticle film: Preparation and application smart nanoparticles technology, A. A. Hashim, Ed., In Tech *Janeza Trdine* 9, 51000 Rijeka, Croatia, 2012.

25. Y. S. Zhao, H. Fu, A. Peng, Y. Ma, D. Xiao, and J. Yao, Low-dimensional nanomaterials based on small organic molecules: Preparation and optoelectronic properties, *Advanced Materials,* 20, 2859–2876, 2008.

26. F. Barreca, N. Acacia, E. Barletta, D. Spadaro, G. Currò, and F. Neri, Small size TiO_2 nanoparticles prepared by laser ablation in water, *Applied Surface Science*, 256(21), 6408–6412, 2010.

27. P. Liu, W. Cai, L. Wan, M. Shi, X. Luo, and W. Jing, Fabrication and characteristics of rutile TiO_2 nanoparticles induced by laser ablation, *Transactions of Nonferrous Metals Society of China*, 19, Supplement 3(0), s743–s747, 2009.

28. M. H. Sayyad, M. Shah, K. S. Karimov, Z. Ahmad, M. Saleem, and M. M. Tahir, Fabrication and study of NiPc thin film based surface type photocapacitors, *Journal of Optoelectronics and Advanced Materials*, 10, 2805–2810, 2008.

29. M. Saleem, M. H. Sayyad, Kh. S. Karimov, M. Yaseen, and M. Ali, Cu(II) 5,10,15,20-tetrakis(4-isopropylphenyl) porphyrin based surface-type resistive–capacitive multifunctional sensor, *Sensors and Actuators B*, 137(2), 442–446, 2009.

30. M. H. Sayyad, K. ul Hasan, M. Saleem, Kh. S. Karimov, F. A. Khalid, M. Karieva, Kh. Zakaullah, and Z. Ahmad, Electrical properties of poly-N-epoxypropylcarbazole/vanadium pentoxide composite, *Eurasian Chemico-Technological Journal*, 9(1), 57–62, 2007.

31. K. S. Karimov, K. Akhmedov, I. Qazi, and T. A. Khan, Poly-N-epoxypropylcarbazole complexes photocapacitive detectors, *Journal of Optoelectronics and Advanced Materials*, 9, 2867–2872, 2007.

32. N. Parvatikar, S. Jain, C. M. Kanamadi, B. K. Chougule, S. V. Bhoraskar, and M. V. N. A. Prasad, Humidity sensing and electrical properties of polyaniline/cobalt oxide composites, *Journal of Applied Polymer Science*, 103(2), 653–658, 2006.

33. S. Jain, S. Chakane, A. B. Samui, V. N. Krishnamurthy, and S. V. Bhoraskar, Humidity sensing with weak acid-doped polyaniline and its composites, *Sensors and Actuators B: Chemical*, 96(12), 124–129, 2003.

34. H. E. Katz and J. Huang, Thin-film organic electronic devices, *Annual Review of Materials Research*, 39(1), 71–92, 2009.

35. C. Roman, O. Bodea, N. Prodan, A. Levi, E. Cordos, and I. Manoviciu, A capacitive-type humidity sensor using crosslinked poly(methyl methacrylate-co-(2 hydroxypropyl)-methacrylate), *Sensors and Actuators B: Chemical*, 25(13), 710–713, 1995.

36. M. Matsuguchi, S. Umeda, Y. Sadaoka, and Y. Sakai, Characterization of polymers for a capacitive-type humidity sensor based on water sorption behavior, *Sensors and Actuators B: Chemical*, 49(3), 179–185, 1998.

37. Z. Ahmad, M. H. Sayyad, M. Yaseen, K. C. Aw, M. M. Tahir, and M. Ali, *Sensors and Actuators B: Chemical,* 155(2), 81–85, 2011.

38. A. E. Hoyt, A. J. Ricco, J. W. Bartholomew, and G. C. Osbourn, SAW sensors for the room-temperature measurement of CO_2 and relative humidity, *Analytical Chemistry*, 70(10), 2137–2145, 1998.

39. C. Caliendo, I. Fratoddi, M. V. Russo, and C. Lo Sterzo, Response of a Pt-polyyne membrane in surface acoustic wave sensors: Experimental and theoretical approach, *Journal of Applied Physics*, 93(12), 10071–10077, 2003.

40. K. Sakamoto and E. Ohno. Synthesis of cobalt phthalocyanine derivatives and their cyclic voltammograms, *Dyes and Pigments*, 35(4), 375–386, 1997.

41. J. Fraden, *Handbook of Modern Sensors: Physics, Designs, and Applications*, 4th ed., Springer, New York, 2010, pp. 445–459.

42. T. Zhang et al., Analysis of dc and ac properties of humidity sensor based on polypyrrole materials, *Sensors and Actuators B*, 131(2), 687–691, 2008.

43. M. H. Sayyad et al., Synthesis of Zn(II) 5,10,15,20-tetrakis(4-isopropylphenyl) porphyrin and its use as a thin film sensor, *Applied Physics A*, 98(1), 103–109, 2010.

44. W.-P. Chen, Z.-G. Zhao, X.-W. Liu, Z.-X. Zhang, and C.-G. Suo, A capacitive humidity sensor based on multi-wall carbo nanotubes (MWCNTs), *Sensors and Actuators B*, 149(1), 136–142, 2010.

45. L. Xiaowei, Z. Zhengang, L. Tuo, and W. Xin, Novel capacitance-type humidity sensor based on multiwall carbon nanotube/SiO_2 composite films, *Journal of Semiconductors*, 32(3), 034006–034011, 2011.

46. M. H. Sayyad, F. Wahab, Z. Ahmad, M. Shahid, J. A. Chaudry, and M. A. Munawar, Production and characterization of nanoparticle dispersions of organic semiconductors for potential applications in organic electronics, *Proceedings of the 11th IEEE International Conference on Nanotechnology*, August 15–18, Portland, Oregon, pp. 193–196, 2011.

10 Investigation of Charge Accumulation in Si_3N_4/SiO_2 Dielectric Stacks for Electrostatically Actuated NEMS/MEMS Reliability

Gang Li, Ulrik Hanke, and Xuyuan Chen

CONTENTS

10.1 INTRODUCTION

Dielectric charging, understood to mean the accumulation of electric charge in the dielectric, constitutes a major failure mechanism that inhibits the commercialization of various electrostatically actuated micro- and nanoelectromechanical systems (MEMS and NEMS) [1]. In such NEMS/MEMS devices containing dielectric materials, dielectric charging always occurs when there is a strong electric field. Stiction induced by dielectric charging is considered the main failure mechanism in electrostatic NEMS/MEMS devices especially for electrostatic capacitive MEMS switches. The accumulated charge in the dielectric of a capacitive MEMS switch will cause the metal bridge to be partially or fully pulled down, degrading the on–off ratio of the switch [2–5]. Charge accumulation in the dielectric must be avoided for high reliability in such devices. Though the topic has been intensively investigated, little information is available in the literature providing a fundamental solution to this problem; even the charging processes are still not thoroughly understood. Therefore, thoroughly understanding the charge accumulation mechanisms to develop a strategy to control dielectric charging is still crucial for minimizing charge accumulation in the dielectric.

It is well known that the deposited dielectric films used in NEMS/MEMS devices, typically SiO_2 and Si_3N_4, contain a large density of traps associated with dangling bonds. However, SiO_2 and Si_3N_4 are the two most common dielectric materials used in radio frequency (RF) MEMS capacitive switches [6,7]. Silicon dioxide has a lower trap density than silicon nitride, which implies that devices made with silicon dioxide dielectric layers should be less prone to charge trapping, that is, having a longer lifetime. Plasma-enhanced chemical vapor deposition (PECVD) silicon dioxide, however, has a lower dielectric constant, 4.1–4.2, when compared to PECVD silicon nitride, 6–9, which leads to a decrease in the down-state capacitance [7–9].

This work investigates how charge accumulates in a double-layer Si_3N_4/SiO_2 dielectric in comparison with that in a single-layer SiO_2 dielectric with the aim of controlling charge accumulation in the dielectric. The double-layer dielectric structure makes it possible to modify the charge accumulation compared with the single-layer dielectric structure because several key parameters of charge injection will be changed for the Si_3N_4/SiO_2 stacked dielectric due to the heterojunction formation, for example, the effective tunneling masses, injection barriers, and the electric field across the dielectric. Furthermore, one benefit from the Si_3N_4/SiO_2 stacked dielectric is that the effective dielectric constant will increase relative to a single SiO_2 dielectric, potentially improving the isolation performance of a capacitive MEMS switch. The dielectric charging behavior can be characterized with a metal–insulator–semiconductor (MIS) structure. By investigating the $C–V$ curves of the MIS structures, the kinetics of charge injection in the dielectrics can be analyzed qualitatively and quantitatively.

10.2 EXPERIMENTS

10.2.1 SAMPLE PREPARATION

In our experiment, the charge accumulation behavior in different dielectrics was investigated by using the MIS capacitor structure instead of an actual NEMS/MEMS membrane device. Three MIS capacitor devices have been designed as shown in Figure 10.1. The dimension of the test MIS structure is designed according to the measurement range of the measurement setup. For each device, both the silicon dioxide and silicon nitride films were fabricated by using the PECVD technique. The total dielectric thickness is about 300 nm in all cases. Then the Al metal electrodes the with a circular area of 2.25×10^{-4} cm^2 were sputtered and patterned by means of photolithography.

10.2.2 C–V MEASUREMENTS

It is well known that charge accumulation in dielectric of an MIS structure can be evaluated by $C–V$ measurement. The flat band voltage (V_{FB}) of the $C–V$ curve, which is a function of the total amount of the space charge in the dielectric, can provide information on how many charges are present in the dielectric of the MIS structure. The shift of the $C–V$ curve after applying the DC bias toward the left or right indicates that the net positive or negative charge is injected into the dielectric.

In the experiments, the silicon substrate is placed on the probe station chuck and all the samples are measured in ambient air at room temperature. First, an initial $C–V$ measurement before applying

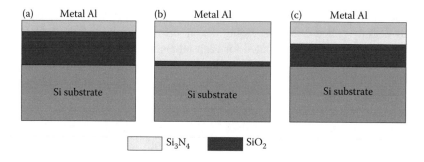

FIGURE 10.1 Schematic cross-sectional views of the three kinds of MIS structures: (a) MOS device: Al-SiO$_2$(300 nm)-n-Si, (b) thick-oxide metal-nitride-oxide-semiconductor (MNOS) device: Al-SiO$_2$ (200 nm)-Si$_3$N$_4$ (100 nm)-n-Si, and (c) thin-oxide MNOS device: Al-SiO$_2$ (50 nm)-Si$_3$N$_4$ (250 nm)-n-Si.

a DC bias stress was performed. Then all devices were electrically stressed with high DC biases, so that charges can be injected into the dielectric films. By comparing the measured $C–V$ curves before and after charge injection, the behavior of charge accumulation can be obtained.

The amount of trapped charge calculated from the change in the flat band voltage is just an approximation of the charge located in the bulk dielectric because the position of the trapped charges in the dielectric will affect the result. The charge near the Si/dielectric interface will contribute a larger $C–V$ shift than the charges near the metal electrode due to there being more mirror charges in the semiconductor [10]. Even then, this method is still a good way to evaluate how charge accumulates in different dielectrics.

10.3 RESULTS AND DISCUSSION

10.3.1 Charge Injection Results

The $C–V$ curves were measured for all the samples by performing a voltage sweep from −20 to 20 V and applying a small-signal AC voltage with a frequency of 100 kHz. Before applying a DC bias to the MIS device, $C–V$ measurement on a virgin MIS device was carried out twice. This is to make sure that the sweep voltage does not influence the charge accumulation in the dielectric.

In all our experiments, to inject space charges into the dielectrics, the samples are electrically stressed by applying a DC bias for 1 min to the metal electrode of the MIS device. Then, to determine the charge accumulation before relaxation, $C–V$ measurement is immediately performed after the DC bias is removed. In Figure 10.2a and c, the measured $C–V$ curves for the three fabricated MIS devices before and after DC stress are shown.

For all the three samples shown in Figure 10.2, when the metal electrode is biased with low positive voltages (say 28 and 30 V), there was no obvious shift between the two $C–V$ curves measured before and after DC stress. This indicates that the net injected charge in the dielectric of the sample device is negligible. In contrast, when the applied voltage increases to 32 V or more, the obtained $C–V$ curve shifts indicate that the number of charges injected into the dielectric of the three devices is considerable and cannot be neglected. The results can be explained as follows.

For a thick dielectric layer, there exists a critical electric field that can determine whether significant charge tunneling will take place or not. If the applied electric field is lower than the critical value, the dominant conduction mechanism is ascribed to temperature-dependent Schottky emission, which takes place significantly only at high temperature [11] and is negligible at room temperature [12]. Therefore, there was no significant $C–V$ curve shift for the three devices at low bias voltage such as 28 and 30 V because the measurement is performed at room temperature. Once the electric field is over the critical value, Fowler–Nordheim (FN) tunneling through a triangular barrier begins to dominate the charge injection [13]. The tunneling probability depends exponentially on the tunneling barrier height, the effective mass of tunneling carriers, and the electric field across the dielectric. This conduction mechanism dominates over Schottky emission at room temperature under a high electric field. Therefore, $C–V$ curve shifts were observed when the applied voltages increased to 32 V.

Moreover, the measured $C–V$ curves for the three samples shift in different directions. For the MOS device shown in Figure 10.2a, the shift is to the right side, indicating that net negative charges were injected into the silicon oxide film. A similar result is also found for the thin oxide MNOS device. In this case, the amount of net charges and their polarity is mainly attributed to the injection of charges with the same polarity as the potential of the Si substrate.

However, for the thick oxide MNOS device, as shown in Figure 10.2b, the $C–V$ curve shifted to the left side when the metal electrode is biased with high positive bias voltage. This result indicates that net positive charges were accumulated in the dielectric of the thick oxide MNOS device. In this case, the amount of net charges and their polarity is mainly attributed to the injection of charges with the same polarity as the potential of the Al electrode.

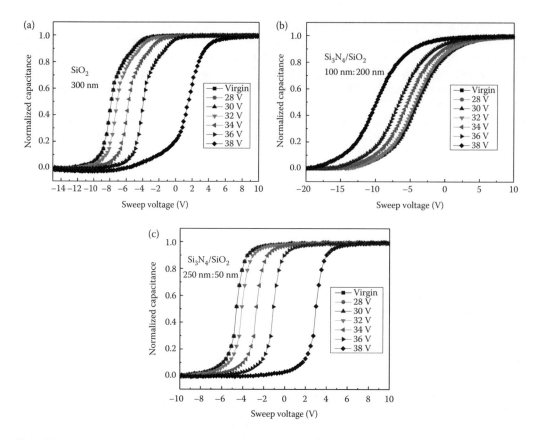

FIGURE 10.2 Measured *C–V* curves for the three MIS devices before and after the charge injection: (a) MOS device, $SiO_2 = 300$ nm, (b) thick oxide MNOS device, $Si_3N_4/SiO_2 = 100$ nm:200 nm, and (c) thin oxide MNOS device, $Si_3N_4/SiO_2 = 250$ nm:50 nm. Positive DC bias voltages of 28, 30, 32, 34, 36, and 38 V at the metal electrode of MIS devices are applied for 1 min, respectively.

The experimental results in the MOS and the thin oxide MNOS devices show that an electron injection from the silicon substrate dominates the charge injection at high positive bias voltage. When a large positive bias is applied on the metal electrode of the MIS structure, the *C–V* curve shift benefits from both the electron injection from the semiconductor and a hole injection from the metal electrode into the dielectric, but the electron injection is the dominant injection mechanism that leads to negative charge accumulation in the dielectric [13,14]. However, this explanation cannot apply to the result obtained for the thick oxide MNOS device, in which the net positive charge accumulation was observed after the positive bias stress.

10.3.2 CHARGE INJECTION MODEL

The experimental results suggest that the effects of both hole injection and electron injection should be taken into account when we analyze charge accumulation. To better understand the above-mentioned charge injection behavior, we have to investigate the charge transport mechanism in a general MIS capacitor structure under high bias voltage. In fact, charge transport in the insulator of an MIS capacitor structure under an electric field depends not only on the insulator itself but also on the interaction of the insulator with its contact electrodes [15,16]. Consequently, carrier conduction processes can be classified into two groups: (i) bulk-limited (Poole–Frenkel, ohmic, and space-charge-limited) conduction and (ii) electrode-limited (Schottky and Fowler–Nordheim) conduction [17].

The electrode-limited conduction corresponds to the charge injection into the dielectric from the contact interfaces. Schottky emission, as shown in Figure 10.3, is the temperature-induced flow of charge carriers that passes over a potential-energy barrier into the conduction band. This takes place significantly if the temperature is high enough and the electric field is lower than about 10^8 Vm^{-1}. FN tunneling is known as a temperature-independent charge transfer mechanism, which is applied to a triangular barrier and takes place when the barrier thickness is relatively large [18]. In our case, the FN tunneling mechanism is important as an electrode-related process because the dielectric layer is relatively thick (about 300 nm) and the electric field in the dielectric is relatively high.

The bulk-limited conduction is related to the charge-trapping process in the dielectric after charge injection. Poole–Frenkel conduction was observed to be the dominant conduction mechanism [19]. The electrode-limited conduction happens before the subsequent bulk-limited conduction, and charge conduction in the bulk dielectric is very difficult to control, which involves the modification of the electrical properties of the dielectric materials, while the electrode-limited conduction can be easily modified, such as by changing the work function of the electrode. We focus on the electrode-limited conduction for the purpose of modifying the charge accumulation in the dielectric in this work.

In Figure 10.3, and later figures, the tunneling barriers are shown as exactly triangular. In reality, the barriers are rounded and reduced in strength by image-force effects [26]. In the simplified mathematical analysis that follows, we treat the barriers as exactly triangular. This does not affect the qualitative validity of our results. More generally, we simplify the influence of bulk-limited conduction on charge-trapping process, and assume the bulk of dielectric as a black box where charge trapping happens during the process of the charge carriers tunneling through the insulator. Specifically, it is assumed that any differences in the rates of electrons and holes supply to the interface may be neglected, and that the charge trapped in the dielectric as a result of a particular injection process is proportional to the typical tunneling probability for that injection process.

According to the above arguments, the processes of charge injection and their trapping in the SiN film of MIS structure under high stress voltage can be described by a modified FN tunneling charge injection model (see Figure 10.4). In the model, we assume that both the electron injection from the grounded electrode and the hole injection from the positively biased electrode contribute to the charge injection. Thus, in this model, the amount of trapped charge relates to the FN tunneling probabilities. The FN tunneling probability for a triangular barrier is given by the well-known Wentzel–Kramer–Brillouin (WBK) approximation [20,21].

$$P = \exp\left(-\frac{4}{3} \frac{\sqrt{2qm^*}}{\hbar} \frac{\phi^{3/2}}{E} \right) \tag{10.1}$$

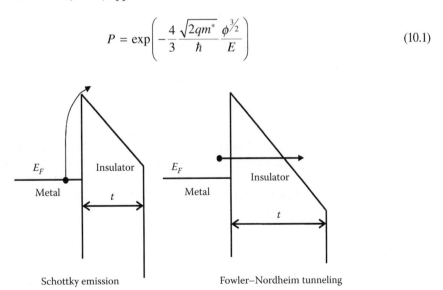

FIGURE 10.3 Energy diagram of Schottky emission and Fowler–Nordheim tunneling.

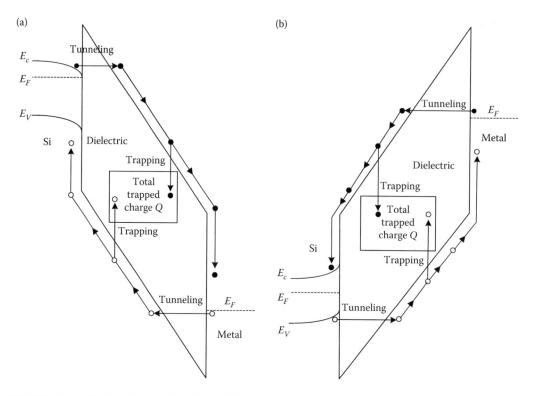

FIGURE 10.4 Modified Fowler–Nordheim (FN) tunneling energy diagram of a general Si/dielectric/metal structure: (a) with a positive stress voltage at the metal electrode, and (b) with a negative stress voltage at the metal electrode.

where $\hbar$ is the Planck's constant, m^* is the effective mass of the tunneling carrier, q is the charge of the electron, ϕ is the barrier for the tunneling carrier, and E is the effective electric field inside the dielectric. From this equation, the effective masses of the tunneling carriers, injection barrier, and the electric field are important parameters that will influence charge injection in the dielectric under high applied voltage.

Based on Equation 10.1, for positive stress voltage (voltage on Al is positive relative to Si), the amount of trapping holes is proportional to the tunneling probability of holes from the Fermi level of metal to the dielectric and can be expressed as

$$N_e \propto P_e = e^{-4\sqrt{2m_e^*\phi_e^3}/3q\hbar E_e} \tag{10.2}$$

where N_e is the number of trapped electrons in the dielectric, P_e is the tunneling probability of electrons, m_e^* is the effective mass of the tunneling electron, E_e is the electric field strength in the dielectric in which electrons tunnel into, and ϕ_e is the barrier height for electron tunneling. Similarly, the tunneling probability of electrons from the valence band of the silicon semiconductor to the dielectric can be expressed as

$$N_h \propto P_h = e^{-4\sqrt{2m_h^*\phi_h^3}/3q\hbar E_h} \tag{10.3}$$

where N_h is the number of trapped holes in the dielectric, P_h is the tunneling probability of electrons, m_h^* is the effective mass of tunneling holes, E_h is the electric field strength in the dielectric in which holes tunnel into, and ϕ_h is the barrier height for hole tunneling.

10.3.3 ANALYSIS OF RESULTS AND DISCUSSION

Based on Equation 10.1, the charge accumulation behavior at high bias voltage will now be analyzed. The key parameters are the effective masses of the tunneling carriers m^*, the injection barrier ϕ, and the electric field strength E in the dielectric. Using Equations 10.2 and 10.3, the ratio of tunneling probabilities may be formed.

$$\frac{P_e}{P_h} = e^{-A(r-1)}, \quad A = \frac{4\sqrt{2m_h^*\phi_h^3}}{3q\hbar E_h}, \quad r = \frac{E_h}{E_e}\sqrt{\frac{m_e^*\phi_e^3}{m_h^*\phi_h^3}} \tag{10.4}$$

From Equation 10.4, we see that $P_e < P_h$ if $r > 1$ and $P_e > P_h$ when $r < 1$. Thus we can use the parameter r to predict the polarity of the accumulated charge. For the MOS device, the electric field strength is the same for charge injection of both electrons and holes. For the MNOS device, the electric field strengths are different for electron and hole charge injections due to the different dielectric constants of SiO₂ and Si₃N₄. To calculate the field strengths E_O in SiO₂ and E_N in Si₃N₄, we model the double-layer dielectric as two parallel capacitors in series. Charge conservation then gives

$$C_{ox} V_{ox} = C_N V_N \tag{10.5}$$

where C_{ox} and C_N are the capacitances formed by SiO₂ and Si₃N₄ films in the MNOS structure, and V_{ox} and V_N are the voltage drops across the SiO₂ and Si₃N₄ films, respectively. Then Equation 10.5 gives

$$\frac{V_{ox}}{V_N} = \frac{C_N}{C_{ox}} = \frac{\varepsilon_N/d_N}{\varepsilon_{ox}/d_{ox}} = \frac{\varepsilon_N d_{ox}}{\varepsilon_{ox} d_N} \tag{10.6}$$

where ε_{ox} and ε_N are the effective dielectric constants of SiO₂ and Si₃N₄ films, and d_{ox} and d_N are the thickness of SiO₂ and Si₃N₄ dielectric films, respectively. The ratio of the electrical field strengths for hole and electron tunneling is then

$$\frac{E_h}{E_e} = \frac{E_N}{E_O} = \frac{V_N d_O}{V_O d_N} = \frac{\varepsilon_{ox}}{\varepsilon_N} \tag{10.7}$$

The parameter r is then conveniently given by known material properties as

$$r = \frac{\varepsilon_{ox}}{\varepsilon_N}\sqrt{\frac{m_e^*\phi_e^3}{m_h^*\phi_h^3}} \tag{10.8}$$

We use the typical dielectric constant of PECVD SiO₂ $\varepsilon_{ox} = 4.2\varepsilon_0$ and Si₃N₄ $\varepsilon_N = 7.0\varepsilon_0$. Then, based on Equation 10.8, we can make Table 10.1.

In the case of the MOS device, the band gap E_g of silicon dioxide is 8.1 eV as shown in Figure 10.4. When the metal electrode is positively biased, the potential barrier of silicon dioxide for the hole injection from the Al electrode is higher than the barrier for the electron injection from the silicon substrate.

Thus, from the estimation given in Table 10.1, we get $P_e > P_h$, which concludes that the electron injection from the silicon substrate will dominate the charge injection. This result complies with the experimental result in Figure 10.2a, that is, electrons accumulate in the SiO₂ film.

The tunneling barrier heights in the MNOS device are shown in Figure 10.5 when the metal electrode is positively biased.

TABLE 10.1

Estimated Tunneling Parameters

	ϕ_e (eV)	ϕ_h (eV)	m_e^*/m_0	m_h^*/m_0	r	
SiO$_2$ 300 nm	3.15 [22]	4.9 [22]	0.42 [23]	0.33 [23]	0.35	$P_e > P_h$
Si$_3$N$_4$/SiO$_2$ 250 nm/50 nm	3.15 [22]	2.4 [22]	0.42 [24]	0.3 [25–27]	1.07	$P_e < P_h$
Si$_3$N$_4$/SiO$_2$ 100 nm/200 nm	3.15	2.4	0.42	0.3	1.07	$P_e < P_h$

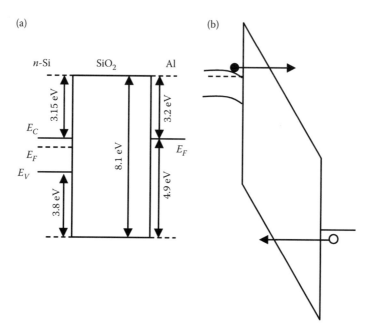

FIGURE 10.5 (a) Energy diagram of metal-SiO$_2$ (200 nm)-Si$_3$N$_4$ (100 nm)-n-Si (thick oxide MNOS) device without bias voltage at the metal electrode. (b) Modified Fowler–Nordheim tunneling when the metal electrode is positively biased.

For the thick oxide MNOS device shown in Figure 10.6, the estimates in Table 10.1 give $P_h > P_e$, which concludes that the hole injection from the Al electrode will dominate the charge injection. This result also complies with the experimental result in Figure 10.2a.

Similarly, for the thin oxide MNOS device shown in Figure 10.7, we still get $P_h > P_e$, which also indicates that the hole injection from the Al electrode will dominate the charge injection in the thin oxide MNOS device. However, the experimental result shows that the electron injection from the silicon substrate dominates the charge injection. The predicted result based on the simple model estimate is in conflict with the experimental result in Figure 10.2c. This suggests that the charge injection mechanism in the double-layer dielectric should be more complicated.

A possible explanation for this discrepancy can be as follows: It is well known that the SiO$_2$/Si$_3$N$_4$ interface has a much higher trap density than the bulk of SiO$_2$ or Si$_3$N$_4$. These interface traps are more easily charged or discharged if the distance from this interface to the substrate is small [27]. Therefore, the thickness of SiO$_2$ between the silicon nitride and the silicon surface will strongly affect the charge accumulation behavior in the MNOS structure. If the thickness of the SiO$_2$ is thin, the electron injection will be enhanced due to the small distance from the SiO$_2$/Si$_3$N$_4$ interface to

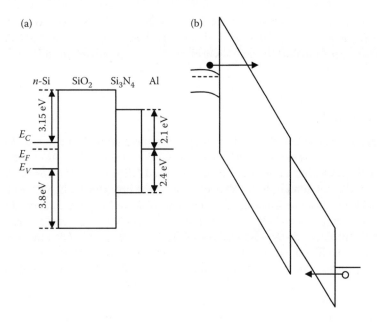

FIGURE 10.6 (a) Energy diagram of metal-SiO$_2$ (200 nm)-Si$_3$N$_4$ (100 nm)-n-Si (thin oxide MNOS) device without bias voltage at the metal electrode. (b) Modified Fowler–Nordheim tunneling when the metal electrode is positively biased.

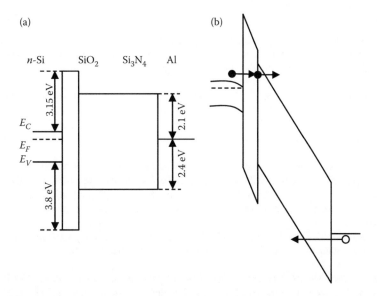

FIGURE 10.7 (a) Energy diagram of metal-SiO$_2$ (50 nm)-Si$_3$N$_4$ (250 nm)-n-Si (thin oxide MNOS) device without bias voltage at the metal electrode. (b) Modified Fowler–Nordheim tunneling when the metal electrode is positively biased.

the electron injection interface, as shown in Figure 10.7. This can be understood as a kind of trap-assisted tunneling [28]. In such a case, the SiO$_2$/Si$_3$N$_4$ interface traps have a high possibility to capture the injected electrons instead of holes because this interface is too far to be reached for the hole injection from the Al electrode. As a result, the electron injection dominates the charge injection.

From the above experimental and modeling results, it follows that the tunneling barrier heights for electrons and holes in a given MIS structure are very important parameters to evaluate charge injection behavior, which will change as the dielectric changes. The effects of SiO_2 thickness on charge accumulation in the Si_3N_4/SiO_2 stacked dielectric may make it possible to balance the number of injected charges from the top and bottom electrodes by optimizing the thickness ratio of Si_3N_4 to SiO_2.

10.4   CONCLUSIONS

This work demonstrates an investigation of charge accumulation in Si_3N_4/SiO_2 dielectric stacks in comparison with that in a single-layer SiO_2 dielectric for the purpose of controlling dielectric charging to improve the reliability of electrostatically actuated NEMS/MEMS membrane devices. The behavior of charge accumulation in the dielectric has been investigated by performing $C–V$ measurements on an MIS capacitor structure. To explain the observed experimental results, we basically proposed a balance of holes and electrons from a modified FN limited charge injection model in combination with the interface trap-assisted tunneling. It is demonstrated that the sign of the charge accumulated in the MNOS devices can be controlled by the SiO_2 thickness between the Si_3N_4 and the Si substrates. For the thin oxide MNOS structure, we attributed the negative charge accumulation to the enhancement of the electron injection from the silicon substrate, which is caused by the Si_3N_4/SiO_2 interface trap-assisted tunneling due to lower distance from this interface to the electron injection interface. Conversely, the positive charge accumulation in the thick oxide MNOS structure was assumed to come from the suppression of the electron injection because the oxide is too thick to allow the interface trap-assisted tunneling.

REFERENCES

1. U. Zaghloul, B. Bhushan, G. Papaioannou, F. Coccetti, P. Pons, and R. Plana, Nanotribology-based novel characterization techniques for the dielectric charging failure mechanism in electrostatically actuated NEMS/MEMS devices using force-distance curve measurements, *J. Colloid Interface Sci.*, 365, 236, 2012.
2. H. San, Z. Deng, Y. Yu, G. Li, and X. Chen, Study on dielectric charging in low-stress silicon nitride with the MIS structure for reliable MEMS applications, *J. Micromech. Microeng.*, 21, 125019, 2011.
3. C. L. Goldsmith, J. Ehmke, A. Malczewski, B. Pillans, S. Eshelman, Z. Yao, J. Brank, and M. Eberly, Lifetime characterization of capacitive RF MEMS switches, IEEEMTT-S Int. Microwave Symp. Digest, p. 227, 2001.
4. W. M. van Spengen, R. Puers, R. Mertens, and I. DeWolf, A comprehensive model to predict the charging and reliability of capacitive RF MEMS switches, *J. Micromech. Microeng.*, 14, 514, 2004.
5. M. Exarchos, V. Theonas, P. Pons, G. J. Papaioannou, S. Melle, D. Dubuc, F. Cocetti, and R. Plana, Investigation of charging mechanisms in metal-insulator-metal structures, *Microelectron. Reliab.*, 45, 1782, 2005.
6. G. J. Papaioannou, M. Exarchos, and V. Theona, Effect of space charge polarization in radio frequency microelectromechanical system capacitive switch dielectric charging, *Appl. Phys. Lett.*, 89, 103512, 2006.
7. K. Seshan, *Handbook of Thin-Film Deposition Processes and Techniques: Principles, Methods, Equipment and Applications* (2nd ed), Noyes Publications/William Andrew Publishing, New York, 2002.
8. M. J. Madu, *Fundamentals of Microfabrication: The Science of Miniaturization* (2nd ed), CRC Press, Boca Raton, FL, 2002.
9. R. W. Herfst, P. G. Steeneken, and J. Schmitz, Time and voltage dependence of dielectric charging in RF MEMS capacitive switches, 45th Annual International Reliability, p. 417, 2007.
10. H. Krause, Electron injection and trap filling in insulating layers, *Phys. Status Solidi A*, 36, 705, 1976.
11. K. H. Ahn and S. Baik, Significant suppression of leakage current in $(Ba,Sr)TiO_3$ thin films by Ni or Mn doping, *J. Appl. Phys.*, 92, 5, 2002.
12. K. L. Ngai and Y. Hsia, Empirical study of the metal-nitride-oxide-semiconductor device characteristics deduced from a microscopic model of memory traps, *Appl. Phys. Lett.*, 4, 159, 1982.
13. K Lehovec, C. H. Chen, and A. Fedotowsky, MNOS charge versus centroid determination by staircase charging, *IEEE Trans. Electron Devices*, 25, 1030, 1978.

14. A. K. Agarwal and M. H. White, New results on electron injection, hole injection, and trapping in MONOS nonvolatile memory devices, *IEEE Trans. Electron Devices*, 32, 941, 1985.
15. T. C. McGill, S. Kurtin, L. Fishbone, and C. A. Mead, Contact-limited currents in metal-insulator-metal structures, *J. Appl. Phys.* 41, 3831, 1970.
16. J. G. Simmons, Conduction in thin dielectric films, *J. Phys. D Appl. Phys.*, 4, 613, 1971.
17. A. I. K. Choudhury, M. R. R. Mazumder, K. Z. Ahmed, and Q. D. M. Khosru, Explanation for reduced Fowler–Nordheim tunneling current in ultra-thin silicon nitride gate dielectric, 3rd ICECE, Dhaka, Bangladesh, p. 2714, 2004.
18. J. G. Simmons, Poole–Frenkel effect and Schottky effect in metal-insulator-metal systems, *Phys. Rev.*, 155, 657, 1967.
19. E. H. Rhoderick, *Metal-Semiconductor Contacts*, Clarendon, Oxford, 1978.
20. M. H. White and C. C. Chao, Statistics of deep-level amphoteric traps in insulators and at interfaces, *J. Appl. Phys.*, 57, 2318, 1985.
21. D. Bohm, *Quantum Theory*, Prentice-Hall, Inc., Englewood Cliffs, 1951.
22. A. V. Vishnyakov, Y. N. Novikov, V. A. Gritsenko, and K. A. Nasyrov, The charge transport mechanism in silicon nitride: Multi-phonon trap ionization, *Solid-State Electron.*, 53, 251, 2009.
23. M. I. Vexler, S. E. Tyaginov, and A. F. Shulekin, Determination of the hole effective mass in thin silicon dioxide film by means of an analysis of characteristics of a MOS tunnel emitter transistor, *J. Phys. Condens. Matter*, 17, 8057, 2005.
24. B. Brar, G. D. Wilk, and A. C. Seabaugh, Direct extraction of the electron tunneling effective mass in ultrathin SiO₂, *Appl. Phys. Lett.*, 69, 2728, 1996.
25. T. Maruyama and R. Shirota, The low electric field conduction mechanism of silicon oxide-silicon nitride-silicon oxide interpoly-Si dielectrics, *J. Appl. Phys.*, 78, 3912, 1995.
26. S. Makram-Ebeid and M. Lannoo, Quantum model for phonon-assisted tunnel ionization of deep levels in a semiconductor, *Phys. Rev. B*, 25, 6406, 1982.
27. H. Castán, S. Dueñas, H. García, A. Gómez, L. Bailón, M. Toledano-Luque, A. del Prado, I. Mártil, and G. González-Díaz, Effect of interlayer trapping and detrapping on the determination of interface state densities on high-k dielectric stacks, *J. Appl. Phys.*, 107, 114104, 2010.
28. K. A. Nasyrov, S. S. Shaimeev, and V. A. Gritsenko, Trap-assisted tunneling hole injection in SiO₂: Experiment and theory, *J. Exp. Theor. Phys.*, 109, 786, 2009.

Section IV

Terahertz Systems and Devices

11 Nano Antennas for Energy Conversion

*Mario Bareiß, Andreas Hochmeister, Gunther Jegert,
Gregor Koblmüller, Ute Zschieschang, Hagen Klauk,
Bernhard Fabel, Giuseppe Scarpa, Wolfgang Porod,
and Paolo Lugli*

CONTENTS

11.1 INTRODUCTION

Renewable energy sources, like hydroelectric power, wind power, tidal power, and photovoltaics, have great importance for satisfying the global demand for energy in the future, since the resources of fossil fuels are limited and nuclear power plants can be hazardous radiation sources for thousands of years if the outer shell is damaged. Photovoltaics is one of the most promising energy sources because the solar power density that reaches the earth can be as large as 136 mW/cm^2 [1] and will continue to be available for millions of years. The state-of-the-art commercial Si solar cells are autarkic power plants, but they have the drawbacks of a relatively small conversion efficiency, the need for a high-temperature manufacturing process, and the fact that the use of strong chemicals during the fabrication process cannot be avoided. Further, the absorption range of Si solar cells is limited by the large bandgap of Si, so that Si solar cells cannot convert the infrared radiation produced by the sun [3]. These drawbacks can be overcome by utilizing the arrays of nano antennas, either as individual devices for energy harvesting or in combination with organic or inorganic solar cells as hybrid photovoltaic devices to extend the absorption spectrum and thus increase the total efficiency [4–6]. The antenna length is usually set to several micrometers corresponding to the infrared regime, so that the antennas can be easily fabricated by optical lithography. The main challenge in the fabrication process of nano antenna arrays is therefore the implementation of a large number of terahertz rectifying devices, which are required to convert the terahertz alternating current induced in the antennas by the infrared radiation into a direct current. Efficient rectifiers, such as pn-junctions or Schottky diodes, cannot be used in this application, since the cutoff frequency of these devices is usually in the megahertz or at the best in the gigahertz range [7]. Recently, Schottky diodes with a cutoff frequency of a few terahertz have been reported [8], but only in the form of individual diodes that cannot be scaled to arrays consisting of millions of such devices. More promising are the metal–oxide–metal (MOM) junctions featuring a thin oxide layer with a thickness in the range of a few nanometers and with metals of different work functions. Such MOM junctions work as tunneling diodes, since the junctions exhibit asymmetric I–V characteristics with respect to the polarities of terahertz (THZ) alternating currents, owing to the fact that electrons tunnel through the thin barrier within femtoseconds [9,10]. However, this is true only if the area of the MOM junction is in the nanometer range, and therefore, simple conventional optical lithography is not applicable for the fabrication of nano antenna arrays, including MOM tunneling diodes. So far, single antenna-coupled MOM diodes (ACMOMDs) fabricated by electron-beam lithography and liftoff techniques have been reported and have shown a promising performances [11,12]. However, these devices have two main drawbacks. The first drawback is that the tunneling dielectric is obtained by the natural oxidation of an aluminum layer, resulting in an insulator with a thickness of 2–3 nm. The advantage of this fabrication method is its simplicity, but this oxide has poor reproducibility and poor electrical stability, resulting in a large fraction of shorted diodes. As an alternative to natural oxidation, a plasma-induced oxide growth has the advantage that it is far more reproducible, that the thickness of the oxide is limited to 3.6 nm, and that the physical properties of this oxide, such as its compactness, result in a dielectric that is electrically stable at high electrical fields. The second drawback of fabricating MOM diodes by electron-beam lithography is that this process is very time-consuming, which is problematic when fabricating arrays of millions of nano antennas. A more promising fabrication technique is the direct transfer-printing of complete arrays of nano antennas and MOM-rectifying diodes. Stacks of several metals are deposited onto a pretreated stamp having nanometer-size structures and are then transfer-printed onto a target substrate, which may consist of Si, SiO$_2$, glass, or other materials [2]. Ultrathin dielectrics with a thickness of a few nanometers can also be fabricated and transferred in this way. The stamp can be either a soft stamp (e.g., polydimethylsiloxane, PDMS) or a hard stamp (e.g., silicon). Hard stamps have the advantages of providing higher resolution of the printed structures and the possibility of reusing the stamps several tens of times, provided the stamps are properly cleaned after each transfer step.

11.2 NANOTRANSFER PRINTING

The basic process steps of the nanotransfer-printing process are given here. The main challenge that has to be overcome is the adhesion force. Since the materials that are to be transferred onto the target substrate have to be easily delaminated from the stamp, the surface energy of the stamp has to be reduced as much as possible. In contrast, the adhesion between the surface of the materials to be transferred and the target substrate has to be as strong as possible to guarantee a successful transfer. The surface energy of the stamp can be reduced by applying a hydrophobic molecular self-assembled monolayer (SAM). A material with a poor adhesion, ideally a noble metal, such as gold or platinum, providing very weak adhesion due to the small reactivity of noble metals, is deposited onto the hydrophobic stamp surface so that the adhesion between the stamp and the material stack is further decreased. After that, the MOM stack of the rectifying tunneling diodes is deposited, including the thin insulator. The last layer deposited onto the stamp should be an adhesion promoter to achieve a large adhesion force between the material stack to be transferred and the surface of the target substrate. Here, chromium or titanium is the material of choice. They can be further activated by a plasma treatment, which results in the formation of hydrophilic hydroxyl groups on the metal surface. Prior to the transfer-printing process, the target substrate is activated depending on the material it is made of. For example, the density of hydroxyl groups on the surface of a silicon substrate can be greatly increased by an RCA 2 (at 75°C) or a piranha clean. A plasma treatment results in a hydrophilic surface as well [13]. The density of hydroxyl groups on the activated silicon surface increases up to 5 OH/nm^2 [14]. A silicon substrate covered with a thermally grown silicon dioxide layer needs a 1 min RCA 1 clean

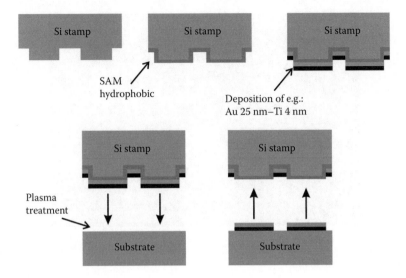

FIGURE 11.1   Basic steps of a nanotransfer-printing process: A Si stamp is covered by a hydrophobic self-assembled monolayer (SAM). After that, a noble metal (here Au) is deposited as the initial metal layer. To provide a good adhesion, the last metal layer consists of an adhesion promoter, which is Ti here. After a plasma treatment, the stamp and the substrate are brought into contact and the metal stacks on the elevated structures on the stamp are transferred onto the target substrate.

prior to a plasma treatment to achieve a hydrophilic surface [15]. Glass can be activated by a plasma treatment. After activating the target substrate, the stamp is placed on top of the target substrate and placed into an imprinting machine. Typical pressures that are applied are in the range of several tens of bars up to 1 kbar, depending on the printing machine. After demolding, the material stacks located on the elevated regions of the stamp are transfer-printed onto the target substrate and the stamp can be "recycled" by removing the metals remaining on the stamp by wet etching. In this way, hard stamps can be used several tens of times. An overview of the printing process is shown in Figure 11.1.

11.3   FABRICATION

11.3.1   NANO ANTENNA

Two main fabrication methods exist for producing hard stamps, namely, molecular beam epitaxy (MBE) and e-beam lithography, both of which are described here.

Under economic aspects, the fabrication of MBE stamps lacks in cost-effectiveness, even if the stamps are "recycled." However, stamp feature sizes down to a few nanometers can be achieved by MBE, since the epitaxially grown layer thicknesses can be controlled on the scale of a single atomic layer. A superlattice of AlGaAs and GaAs is grown lattice-matched on a 2 in GaAs wafer, which has a thickness of 300 μm. The Al content in the superlattice must be around 80% with respect to the other group III element to achieve a sufficient etching selectivity between GaAs and AlGaAs. A 5 mm × 8 mm piece is cut out of the wafer, and then the vertical side edges of the wafer piece are selectively etched using citric acid or HF. Citric acid selectively removes GaAs with respect to AlGaAs, thus producing a side edge with three-dimensional features corresponding to the periodicity of the AlGaAs/GaAs superlattice, as shown in Figure 11.2. When etching the MBE stamp with HF (5%), AlGaAs is selectively etched, resulting in small gaps on the stamp. The etched MBE stamps are then either left in clean-room conditions for 1 day to allow the surface to oxidize or covered with a hydrophobic SAM featuring a thiol group, such as n-octanethiol. When depositing metals onto an MBE stamp, this hydrophobic layer serves as a delaminating layer, so that the adhesion between the deposited metals and the stamp is weak; thus, the metals can be removed easily from the stamp. A gold layer with a thickness

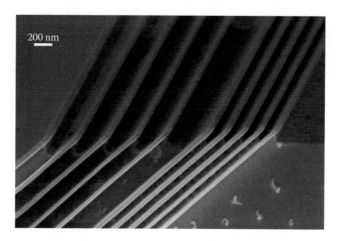

FIGURE 11.2 Scanning electron microscopy (SEM) image of a stamp produced by molecular beam epitaxy with feature sizes as small as 30 nm.

of 15–45 nm is followed by a Ti layer with a thickness of 4 nm, which serves as an adhesion promoter, then deposited by thermal evaporation onto the patterned stamp. A layer of TiO_2 is formed by allowing the surface of the Ti layer to oxidize. In general, TiO_2 does not feature a highly hydrophilic surface; therefore, an oxygen-plasma treatment for several minutes destroys the titanium oxide bonds, and by exposing the stamp to air, hydrophilic hydroxyl groups are formed on the Ti/TiO_2 surface. Silanols are formed on the Si target substrate (either with or without a layer of SiO_2). The hydrophilic character of a bare Si surface is stable for several hours; however, in the case of an SiO_2 surface, a reactivation of the surface is necessary after approximately half an hour. After activating the Ti on the stamp and the SiO_2 on the substrate, both the stamp and the substrate were brought into contact and the 200-nm-thick Au/Ti lines were transferred onto the Si substrate. The custom-made MBE printing machine, as well as the nanotransfer process, is described elsewhere [16]. MBE stamps with structures down to 30 nm have been fabricated, leading to transferred Au lines with widths down to 43 nm. Other metals like aluminum have also been transferred. In a subsequent process step, the printed lines can be adjusted to the antenna length of several micrometers via a focused ion beam (FIB).

So far, we have presented an unconventional method for the fabrication of hard stamps with small feature sizes and MBE resolution. However, to cover large areas with structures in a single printing step, planar Si stamps must be considered more useful than stamps that consist only of the 300 µm wide edge of a wafer piece. Using electron-beam lithography followed by dry etching, the surface of an Si stamp can be patterned into any desired features with a resolution of approximately 45 nm over areas up to several square centimeters. To render the surface of the Si stamp hydrophobic and obtain low adhesion between the stamp surface and the deposited metals, the stamp is covered with an organic silane-based SAM, for example, 1*H*,1*H*,2*H*,2*H*-perfluorooctyltrichlorosilane [17]. The subsequent steps of covering the stamp with materials to be transferred to the target substrate are analogous to the MBE stamp process.

11.3.2 MOM Tunneling Diodes

As discussed above, the microscale and nanoscale MOM tunneling diodes have been fabricated in a nanotransfer-printing process. For the microscale diodes [18], a planar Si wafer comprising a silane-based hydrophobic SAM served as the stamp. A stack of 10 nm thick Au, acting as a delamination layer between the diodes and the stamp, followed by a 20 nm thick Al layer, was deposited by thermal evaporation onto the stamp through a shadow mask, resulting in parallel metal lines on the substrate. The width of the lines ranged from 10 to 200 µm. After the metal deposition, a natural oxygen layer formed spontaneously on the Al surface upon exposure to air. As discussed above, the spontaneously grown aluminum oxide layer does not provide a compact and pure insulator. Therefore, an oxygen-plasma treatment was performed to increase the thickness of the aluminum oxide from ~1.6 nm to ~3.6 nm, resulting in a pure and compact AlO_x layer [19,20]. After that, a 30 nm thick layer of Au, followed by 4 nm thick Ti acting as an adhesion layer between the MOM diodes and the target substrate, was evaporated through a second shadow mask, resulting in metal lines arranged perpendicular to the first set of metal lines. The active area of the microscale MOM diodes defined at the intersections of the two metal stacks ranges from 10^{-6} cm^2 to 10^{-4} cm^2.

Nanoscale MOM tunneling diodes were also fabricated using a flat Si stamp with pillars having a diameter of 50 nm [2]. The first layer that was deposited onto the stamp was 15 nm of AuPd (delamination layer). AuPd was chosen since it provides a very small surface roughness because of its small grain size. This is important in the nanometer regime because materials with rough surfaces are more difficult to transfer, since the contact area between the material to be transferred and the target substrate is very small. Then a 25 nm thick layer of Al (first electrode) was deposited and its surface oxidized by an oxygen-plasma treatment to obtain the tunneling dielectric, followed by the deposition of a 20 nm thick layer of AuPd as the second electrode. 4 nm of Ti serves again as an adhesion promoter. After activating the stamp and the target substrate to create hydrophilic surfaces, the completed microscale diodes are then transfer-printed onto an Si wafer covered with a 200 nm thick SiO_2 layer, whereas the nanoscale diodes were transfer-printed onto an Si wafer covered with a conductive layer of 25 nm of AuPd, which has a layer of 4 nm of Ti on top to provide good adhesion to the materials to be transferred. The printing process has been carried out on an NIL-2.5 Nanoimprinter from Obducat. The pressure during transfer was 50 bar and it was held for 5 min. The temperature was set to 200°C, which is necessary to remove the water formed during the transfer process. The titanol surface groups from the stamp and the silanol surface groups from the target substrate react to a titansiloxane group and water is released. As this reaction is reversible, the removal of water is important. This is one reason why hard stamps, such as GaAs or Si, are more suitable than soft PDMS stamps [21–23], which cannot withstand high temperatures, although it has also been shown in the literature that temperatures around 50°C improve the transfer process when using soft stamps [24].

11.4 RESULTS

11.4.1 Nano Antennas

A scanning electron microscope (SEM) and a probe station were used to characterize transfer-printed antenna structures fabricated with an MBE stamp. We have been able to show that the antenna structures are continuous over a length of almost 2 mm. By optical lithography and a liftoff process, the resistivity of the ensemble of transferred lines consisting of aluminum was determined to have a value of 2.3×10^{-8} Ω m, which is similar to the bulk resistivity of aluminum [25].

Antenna structures fabricated in a transfer-printing process using a flat Si-stamp were also characterized by SEM. The Au lines are continuous over an area of 0.6 mm × 0.6 mm, but larger areas can also be covered. The transfer yield is almost 100%, which is defined by the percentage of successfully transferred structures with respect to the structures on the stamp on which the materials were initially deposited. The morphology of the lines is excellent and even more complex structures like grid-like structures can be transferred when tailoring the printing parameters. Further processes on the transferred antenna structures can include optical lithography and lift-off, for example, to define contact pads, or also more complex processes like FIB. With respect to liftoff processes, we want to point out that the adhesion of the Au lines to the substrate is high due to the Ti layer.

11.4.2 MOM Tunneling Diodes

For the microscale transfer-printed MOM tunneling diodes, the transfer yield is also almost 100% [26]. Atomic force microscope (AFM) measurements of the MOM diodes on the stamp before printing and on the substrate after printing show that the heights are similar, which indicates that the morphology of the metal layers was not affected by the transfer process. To show that the ultrathin AlO_x dielectric still works as an insulating layer after transfer, electrical measurements were performed using a probe station. Since the MOM diode represents the overlap of an $Au/Al/AlO_x$ line and a Au/ Ti line, the electrical characterization could easily be done by contacting the two electrodes without touching or affecting the active area and especially the ultrathin dielectric. The *I–V* characteristics of the printed MOM tunneling diodes are presented in Figure 11.3a (symbols). The applied bias was varied on the Au electrode. Asymmetric *I–V* characteristics with respect to the polarities, as determined in the measurements, are essential; otherwise, the MOM junction would not represent an unbiased rectifying device. To show that direct (and Fowler–Nordheim) tunneling instead of, for example, trap-assisted tunneling, is the main current mechanism occurring in the MOM diodes, simulations were carried out using a kinetic Monte Carlo model [27,28] and numerical simulations [29]. Using these models, we have been able to not only validate this hypothesis but also show that transport in the diodes is dominated by electron tunneling. The current density can be described by the Tsu-Esaki formula [30]. For low voltages below 3.3 and 4.3 V for negative and positive biases, respectively, direct tunneling was observed and for higher voltages, Fowler–Nordheim tunneling was observed. However, in the initial Monte Carlo model, planar metal layers without any surface roughness were assumed. For currents below 5×10^{-4} A/cm^2, the simulation fits the experimental data very well. However, for larger currents, the simulation does not match the experimental results. Here, the effects of the surface roughness of the insulator/electrode interface are important, especially for the electron injection into

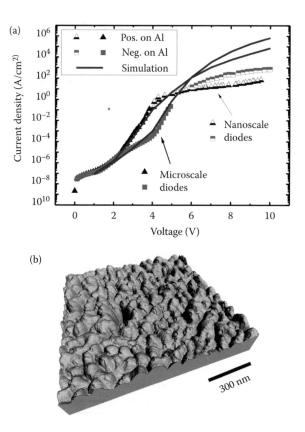

FIGURE 11.3 (a) Electrical characterization and simulations of a printed MOM tunneling diode. (Reprinted with permission from M. Barciß et al., High-yield transfer printing of metal–insulator–metal nanodiodes, *ACS Nano*, 6, 2853–2859, 2012/03/27. Copyright 2012, American Chemical Society.) (b) Roughness of the Al/AlO_x layer determined by AFM.

the aluminum electrode. The nonideal inhomogeneous distribution of the applied electric field was found to cause this derivation. By measuring the surface roughness of the Au/Al/AlO$_x$ layer with an AFM after transfer-printing (see Figure 11.3b), an root mean square (RMS) value of 0.5 nm over an interval of 20 nm was determined. The electric field at local height peaks on the surface is larger than on a planar metal layer. By including these effects in the Monte Carlo simulation, a better effect, especially for higher voltages, was achieved. A further parameter besides the tunneling barrier height of 3.3 eV for the Al/AlO$_x$ interface and 4.3 eV of the Au/AlO$_x$ interface that was assumed when taking the surface roughness of the interface into account is an effective electron tunneling mass of 0.35 m$_0$, which is reasonable [31]. Since aluminum oxide is known to have a work function of around 1 eV, the measured values of the barrier heights are in good agreement with the theoretical values [18,26].

The morphological characterization of the MOM tunneling nanodiodes was achieved by SEM. Here, the transfer yield was also found to be excellent at 98%. The diameter of the transferred MOM pillars was found to increase from 50 to 94 nm, which results from the fact that during evaporation, the material was accumulated on the sidewalls of the previously deposited materials on the stamp. However, the electrical characterization of the nanodiodes is more challenging than that the microdiodes because, unlike the microdiodes, the nanodiodes cannot be contacted outside the active area. Therefore, for the characterization of the nanodiodes, a conductive atomic force microscope (C-AFM) was employed to make soft electrical contact to the top electrode, while the bottom electrode was connected via a clamp that was attached to the conductive substrate (see Figure 11.4).

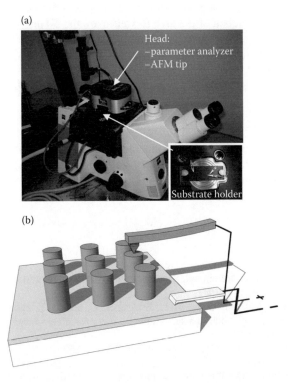

FIGURE 11.4 (a) A photograph of the conductive AFM setup: The parameter analyzer and the AFM tip are implemented in the head, and the substrate holder comprises a ceramic plate and gold clamps for connecting the sample substrate. (b) A schematic view of the AFM setup: The MIM pillars (vertical pillars) that are transfer-printed onto the Si target substrate coated with a conductive AuPd layer are contacted from the top via a conductive AFM tip (here positive potential). The AuPd layer on the substrate is connected via a gold clamp (here negative potential). (Reprinted with permission from M. Bareiß et al., High-yield transfer printing of metal–insulator–metal nanodiodes, *ACS Nano*, 6, 2853–2859, 2012/03/27. Copyright 2012, American Chemical Society.)

The *I-V* characteristics were determined only for voltages above 5 V since the noise introduced by the C-AFM setup did not permit the measurement of smaller voltages. The bias-dependent current densities measured in the nanodiodes are perfectly consistent with those measured previously in the microdiodes, which confirms the reliability of the materials and the fact that downscaling does not change the diode's characteristics (Figure 11.3a). Note that very large electric fields can be applied to the nanodiodes without causing damage, which was not possible with the microdiodes. We believe that the reason is the fact that the number of potentially catastrophic defects scales with the diode area. Since all fabrication was carried out under clean-room conditions, the absolute defect number may even be practically zero over sufficiently small areas [2].

11.5 CONCLUSION

In summary, we demonstrated the transfer-printing of arrays of nano antenna structures using either an MBE stamp or a patterned Si stamp featuring an excellent morphology and reliable electrical resistivity. Microscale and nanoscale MOM tunneling diodes with an ultrathin dielectric of 3.6 nm of AlO_x fabricated in an oxygen-plasma process were transfer-printed in a single printing step. The transfer yield is almost 100% and the electrical characteristics are in line with the Monte Carlo and numerical simulations.

The main challenge that we want to overcome in our future work is the implementation of a MOM nanodiode in an array of transfer-printed antennas on even larger areas to be suitable for energy-harvesting applications. Since the cutoff frequency of a MOM diode is dependent on its area, a further goal is the reduction of the area to values around 50 nm. Here, MBE stamps might be particularly well suited, since we already showed a successful transfer of Au lines having a width of 43 nm. A second, perpendicular transfer comprising Al and a thin layer of AlO_x would create a nanoscale diode with a smaller area. This would expand the possible absorption spectrum into the high THz regime and could lead to higher energy absorption from sunlight.

ACKNOWLEDGMENTS

The authors thank Rosi Heilmann for her invaluable help. The authors acknowledge the financial support from the German Research Funding (DFG) through the TUM International Graduate School of Science and Engineering (IGSSE) and the Institute for Advanced Studies (IAS), the focus group "Nanoimprint and Nanotransfer," and the DFG Excellence Cluster "Nanosystem Initiative Munich."

REFERENCES

1. M. P. Thekaekara and A. J. Drummond, Standard values for the solar constant and its spectral components, *Nat. Phys. Sci.*, 229, 6–9, 1971.
2. M. Bareiß, F. Ante, D. Kälblein, G. Jegert, C. Jirauschek, G. Scarpa, B. Fabel, E. M. Nelson, G. Timp, U. Zschieschang, et al., High-yield transfer printing of metal–insulator–metal nanodiodes, *ACS Nano*, 6, 2853–2859, 2012/03/27 2012.
3. S. S. Iyer and Y. H. Xie, Light emission from silicon, *Science*, 260, 40–46, 1993.
4. C. Fumeaux, W. Herrmann, F. K. Kneubühl, and H. Rothuizen, Nanometer thin-film Ni-NiO-Ni diodes for detection and mixing of 30 THz radiation, *Infrared Phys. Technol.*, 39, 123–183, 1998.
5. C. Fumeaux, W. Herrmann, H. Rothuizen, P. De Natale, and F. K. Kneubühl, Mixing of 30 THz laser radiation with nanometer thin-film Ni-NiO-Ni diodes and integrated bow-tie antennas, *Appl. Phys. B Lasers Opt.*, 63, 135–140, 1996.
6. I. Wilke, Y. Opplinger, W. Herrmann, and F. K. Kneubuhl, Nanometer thin-film Ni-NiO-Ni diodes for 30 THz radiation, *Appl. Phys. A*, 58, 329–341, 1994.
7. M. Bareiß, A. Hochmeister, G. Jegert, U. Zschieschang, H. Klauk, B. Fabel, G. Scarpa, W. Porod, and P. Lugli, Quantum carrier dynamics in ultra-thin MIM tunneling diodes, *17th International Conference on Electron Dynamics in Semiconductors, Optoelectronics and Nanostructures, EDISON 17*, 2011.

8. S. Barbieri, J. Alton, C. Baker, T. Lo, H. Beere, and D. Ritchie, Imaging with THz quantum cascade lasers using a Schottky diode mixer, *Opt. Express,* 13, 6497–6503, 2005.

9. J. A. Bean, B. Tiwari, G. H. Bernstein, P. Fay, and W. Porod, Thermal infrared detection using dipole antenna-coupled metal-oxide-metal diodes, *J. Vac. Sci. Technol. B,* 27, 11–14, 2009.

10. J. A. Bean, A. Weeks, and G. D. Boreman, Performance optimization of antenna-coupled Al/AlO$_x$/Pt tunnel diode infrared detectors, *IEEE J. Quantum Electron.,* 47, 126–135, 2011.

11. B. A. Slovick, J. A. Bean, P. M. Krenz, and G. D. Boreman, Directional control of infrared antenna-coupled tunnel diodes, *Opt. Express,* 18, 20960–20967, 2010.

12. B. Tiwari, J. A. Bean, G. Szakmany, G. H. Bernstein, P. Fay, and W. Porod, Controlled etching and regrowth of tunnel oxide for antenna-coupled metal-oxide-metal diodes, *J. Vac. Sci. Technol. B,* 27, 2153–2160, 2009.

13. M. Bareiß, M. A. Imtaar, B. Fabel, G. Scarpa, and P. Lugli, Temperature enhanced large area nano transfer printing on Si/SiO$_2$ substrates using Si wafer stamps, *J. Adhes.,* 87, 893–901, 2011.

14. R. K. Iler, *The Chemistry of Silica,* Wiley, New York, 1979.

15. T. Suni, K. Henttinen, I. Suni, and J. Mäkinen, Effects of plasma activation on hydrophilic bonding of Si and SiO$_2$, *J. Electrochem. Soc.,* 149, G348–G351, 2002.

16. S. Harrer, S. Strobel, G. Scarpa, G. Abstreiter, M. Tornow, and P. Lugli, Room temperature nanoimprint lithography using molds fabricated by molecular beam epitaxy, *IEEE Trans. Nanotechnol.,* 7, 363–370, 2008.

17. S. A. Kulinich and M. Farzaneh, Hydrophobic properties of surfaces coated with fluoroalkylsiloxane and alkylsiloxane monolayers, *Surf. Sci.,* 573, 379–390, 2004.

18. M. Bareiß, A. Hochmeister, G. Jegert, U. Zschieschang, H. Klauk, R. Huber, D. Grundler et al., Printed array of thin-dielectric metal-oxide-metal (MOM) tunneling diodes, *J. Appl. Phys.,* 110, 044316, 2011.

19. H. Ryu et al., Logic circuits based on individual semiconducting and metallic carbon-nanotube devices, *Nanotechnology,* 21, 475207, 2010.

20. U. Zschieschang, F. Ante, M. Schlörholz, M. Schmidt, K. Kern, and H. Klauk, Mixed self-assembled monolayer gate dielectrics for continuous threshold voltage control in organic transistors and circuits, *Adv. Mat.,* 22, 4489–4493, 2010.

21. Y.-L. Loo, D. V. Lang, J. A. Rogers, and J. W. P. Hsu, Electrical contacts to molecular layers by nano-transfer printing, *Nano Lett.,* 3, 913–917, 2003.

22. Y.-L. Loo, R. L. Willett, K. W. Baldwin, and J. A. Rogers, Interfacial chemistries for nanoscale transfer printing, *J. Am. Chem. Soc.,* 124, 7654–7655, 2002.

23. J. Zaumseil, M. A. Meitl, J. W. P. Hsu, B. R. Acharya, K. W. Baldwin, Y.-L. Loo, and J. A. Rogers, Three-dimensional and multilayer nanostructures formed by nanotransfer printing, *Nano Lett.,* 3, 1223–1227, 2003.

24. J.-h. Choi, K.-H. Kim, S.-J. Choi, and H. H. Lee, Whole device printing for full colour displays with organic light emitting diodes, *Nanotechnology,* 17, 2246–2249, 2006.

25. M. Bareiß, A. Hochmeister, G. Jegert, G. Koblmüller, U. Zschieschang, H. Klauk, B. Fabel, G. Scarpa, W. Porod, and P. Lugli, Energy Harvesting using Nano Antenna Array, *Nanotechnology (IEEE-NANO), 2011 11th IEEE Conference on,* pp. 218–221, 2011.

26. M. Bareiß, B. N. Tiwari, A. Hochmeister, G. Jegert, U. Zschieschang, H. Klauk, B. Fabel et al., Nano antenna array for terahertz detection, *IEEE Trans. Microwave Theory Tech.,* 59, 2751–2757, 2011.

27. G. Jegert, A. Kersch, W. Weinreich, and P. Lugli, Monte Carlo simulation of leakage currents in TiN/ZrO$_2$/TiN capacitors, *IEEE Trans. Electron Devices,* 58, 327–334, 2011.

28. G. Jegert, A. Kersch, W. Weinreich, U. Schroder, and P. Lugli, Modeling of leakage currents in high-kappa dielectrics: Three-dimensional approach via kinetic Monte Carlo, *Appl. Phys. Lett.,* 96, 062113, 2010.

29. C. Jirauschek, Accuracy of transfer matrix approaches for solving the effective mass Schrödinger equation, *IEEE J. Quantum Electron.,* 45, 1059–1067, 2009.

30. R. Tsu and L. Esaki, Tunneling in a finite superlattice, *Appl. Phys. Lett.,* 22, 562–564, 1973.

31. W. Y. Ching and Y.-N. Xu, First-principles calculation of electronic, optical and structural properties of alpha-Al$_2$O$_3$, *J. Am. Ceram. Soc.,* 77, 404–411, 1994.

12 Ballistic Transistor Logic for Circuit Applications

David Wolpert and Paul Ampadu

CONTENTS

12.1 INTRODUCTION

As complementary metal-oxide-semiconductor (CMOS) transistors near atomic-scale limitations, researchers are searching for novel devices to complement or supplant conventional silicon electronics. Several alternative devices have been proposed as potential successors to conventional Si CMOS. Unlike the transition from bipolar devices to CMOS, however, a clear successor to the metal-oxide-semiconductor field-effect transistor (MOSFET) has not yet emerged. Some devices offer improved speed, some promise unmatched power efficiency, and still others offer versatility and robustness through runtime configurability. Because no device is superior in *every* metric, future computing systems may offer heterogeneous integration of CMOS and other technologies, providing a judicious balance among the devices optimized for the particular performance metric desired.

This chapter describes a recent entrant into the emerging devices landscape, the ballistic deflection transistor (BDT) [1], which promises terahertz speeds. In Section 12.2, we provide a brief overview of nanoscale devices and their circuit applications. Section 12.3 describes the operation of the BDT, including ballistic transport in two-dimensional electron gas (2DEG) heterostructures. In Section 12.4, we present examples of BDT logic gates—an inverter, a two-input NAND, and a two-input general-purpose gate (GPG) used to construct random logic. BDT circuit design methodology, a full adder implementation case study, and a discussion of applications suited for BDTs are presented in Section 12.5. Future work on voltage-mode BDT logic and potential limitations are discussed in Section 12.6, and conclusions are presented in Section 12.7.

12.2 NANOSCALE DEVICES AND CIRCUIT APPLICATIONS

12.2.1 CONVENTIONAL DEVICES AND CIRCUITS

Conventional device improvements include new material structures, fabrication processes, or integration methods that have properties similar to standard MOSFETs. Surveys of nanoscale devices emphasizing these traditional improvements are available, for example, Refs. [2,3]. For the continued scaling of conventional MOSFETs beyond the 14 nm node, new lithography techniques such as extreme ultraviolet lithography, double patterning, and immersion lithography are being explored [4]. III–V compound semiconductors, such as GaAs, have been used in electronic and optical applications, resulting in operating frequencies of over 250 GHz in a GaAs heterojunction bipolar transistor (HBT) [5]. Channel properties can be adjusted through the use of Ge, SiGe, or strained Si approaches [6]. An excellent discussion of these techniques is available in Ref. [7], with mobility improvements of up to 10× reported. Metal gates with high-κ dielectrics, multigate FETs [8], and finFETs [9] have been employed to reduce leakage current and improve I_{on}/I_{off} ratios. Silicon-on-insulator (SoI) technology enables large performance improvements and a tighter control of noise, coupling, and substrate leakage [2]. Ultrathin SoI substrates have facilitated the creation of 3D integrated circuits [10], which reduce the wire length, improve the latency and system performance, and enable the integration of different substrates [11]. As device dimensions reach the electron mean free path at room temperature in some materials, ballistic transport can be harnessed, promising terahertz-range frequencies [12–14] (ballistic devices will be discussed in more detail in Section 12.3).

12.2.2 NONCONVENTIONAL DEVICES AND CIRCUITS

Nonconventional devices exhibit properties fundamentally different from conventional CMOS. Molecular structures have been used to create both passive and active devices [15,16]. For example, nanowires can be used with active molecular switches to create logic crossbar structures, such as memristor crossbar latches [17]; nanowires are also finding use as gate channels [18]. Graphene sheets, nanoribbons, and carbon nanotubes can exhibit either metallic or semiconducting properties depending on their chirality; these structures have some of the highest carrier mobilities of any known device, and should enable operating frequencies well into the terahertz range [19,20]. In addition, nanotube devices can utilize band-to-band tunneling to achieve very low power consumption [21,22]. These features have encouraged a great deal of research on nanotube circuit modeling [23] and implementations such as frequency doublers and mixers [20]. Single-electron transistors (SETs) [24,25] are deep nanoscale devices that make use of the Coulomb barrier tunneling to provide logic. This "one electron at a time" transport can be regarded as the ultimate limit in low-power nanoelectronics. SET logic has been explored using both a traditional voltage-mode operation and a novel charge state logic where the state is determined by whether or not an electron is present in a node [26]. Spintronics [27,28] uses the spin of an electron or a group of electrons to encode logic information, and can be reconfigured at runtime by adjusting the device magnetization. Ferromagnetic semiconductors are increasingly implemented as spin-valve devices [27] and offer promise for large, nonvolatile random-access memories. An excellent survey of nontraditional devices can be found in Ref. [29].

Research is also ongoing in the areas of DNA computing [30,31], peptide computing [32], and, most notably, quantum computing [33]. These technologies promise immense computational power for parallel computing problems such as database searches and large number factorization. Finally, nonlinear optical devices such as semiconductor all-optical switches and amplifiers have recently been used to create Boolean logic functions [34]. These devices will be immensely useful in fiber optic systems, circumventing the bandwidth limitations of electro-opto conversions.

12.3 BALLISTIC DEFLECTION TRANSISTORS

One promising group of devices utilizes ballistic transport, with near-terahertz results already reported [12,13]. Various III–V compound heterostructures, such as AlGaAs/InGaAs or InGaAs/InAlAs systems, can be used to create a 2DEG layer with very high electron mobilities [35]. In 2DEG structures with dimensions smaller than the electron mean free path l_e, electrons can travel without any scattering events. This is referred to as ballistic transport and can be observed even at room temperature, for example, Refs. [36,37]. One-dimensional electron gas transistor channels have been used to create high-electron-mobility transistors (HEMTs) [38], which can exhibit up to terahertz performance.

Ballistic transport can be explained using Figure 12.1, which is derived from an experiment by Hirayama and Tarucha [39]. The black lines are boundaries, electrons are ejected from terminal 1, and terminal 2 is positively charged. In the ordinary drift-diffusive transport regime, scattering eliminates the initial electron momentum before electrons enter the central region. This loss of momentum is defined by the momentum relaxation time

$$\tau_m = m^* \mu / e \tag{12.1}$$

where m^* is the electron effective mass, μ is the electron mobility, and e is an elementary charge. Scattered electrons drift toward the positively charged terminal 2, resulting in the electron flow shown in Figure 12.1a. As transport becomes more ballistic, τ_m increases and electrons maintain their momentum for a longer period, resulting in a larger percentage of carriers reaching terminal 3 instead of terminal 2. In the ballistic limit (where τ_m is much larger than the travel time between terminals 1 and 3), electron paths follow a billiard model [40], and most electrons travel from terminal 1 to terminal 3, shown in Figure 12.1b. In the region between ballistic and drift-diffusive transport, electrons behave quasi-ballistically and can be guided by applied potentials and geometrical boundaries. In semiconducting 2DEG systems, τ_m and l_e strongly depend on temperature, so devices that are ballistic at liquid nitrogen or liquid helium temperatures can become drift-diffusive at room temperature. A number of novel devices have been proposed using 2DEG ballistic transport [36,41–45]. In this chapter, we focus on the BDT.

The BDT [14] is a six-terminal coplanar structure etched into a 2DEG. The device consists of a grounded electron source, left and right gates, and three biased drains. The center drain is a constant pull-up potential used to accelerate the electrons from the source toward the central deflector; the left and right drains are outputs. The dimensions and materials allow the electrons to travel in a quasi-ballistic manner at room temperature, guided by the deflector and lateral gate potentials as shown in Figure 12.2. A SEM image of a BDT is shown in Figure 12.3. The steering voltage is much smaller than that required for gate pinch-off (e.g., in the results shown, differential input voltages of ±150 mV are used). The low capacitance of the 2DEG features (~0.2 fF) [36] the results in an estimated f_T in the terahertz range. Mateos et al. [12] have shown a 1 THz response in a similar structure, the ballistic rectifier.

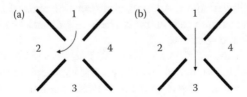

FIGURE 12.1   (a) Drift-diffusive transport versus (b) ballistic transport.

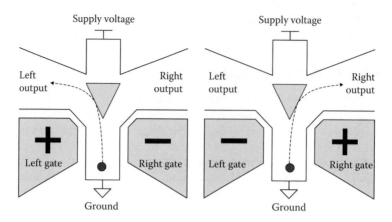

FIGURE 12.2 Schematic of BDT shown under two gate bias conditions.

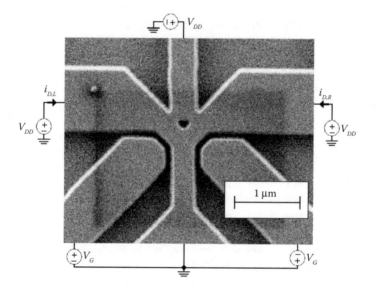

FIGURE 12.3 SEM image of a fabricated BDT.

The BDT has been fabricated in an $In_{0.53}Ga_{0.47}As–In_{0.52}Al_{0.48}As$ heterostructure with an InP sub-strate, using mask layers defined by electron beam lithography, with reactive ion etching used to create the raised mesas shown in the SEM image of Figure 12.3. The 9 mm thick 2DEG occurs about 60 nm beneath the surface. The minimum etch width is 70 nm (the gate–channel spacing) and the etch depth is approximately 130 nm. The electron beam lithography process is less susceptible to orientation changes, diffraction effects, or other photolithography issues; however, in large designs, beam drift can still result in variations. Additional information about the process used is available in Ref. [46]. The electron mean free path in the 2DEG is calculated to be 120 nm [46]; still ballistic effects are observed at room temperature.

The room temperature measured response of the BDT across a range of differential gate voltages is shown in Figure 12.4, with a grounded source and 1 V bias applied to each of the three drains. The x-axis is the left gate voltage. The asymmetry between the left and right outputs is caused by process variation and a slight offset in contact placement. A positive left gate and negative right gate voltage results in current gain through the left output branch; current gain is seen through the right

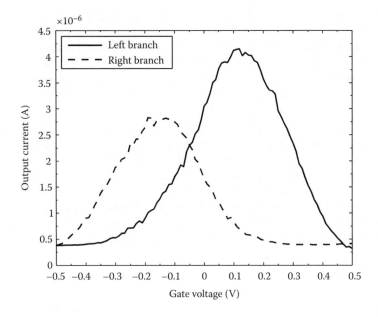

FIGURE 12.4 Measured output response of the BDT.

output branch when the voltages are reversed. The overlapping response in Figure 12.4 is a result of electrons scattering away from the intended output.

A major challenge of cascading logic with BDTs is converting the output *current* of one device into the *input* voltage of the next device. In the BDT, an accumulated gate charge is not dissipated by switching the driving gate input; instead, the lack of driving current creates a high resistance path to V_{DD} and ground through the driving device. One solution is to use a string of resistors between V_{DD} (e.g., 1 V) and V_{SS} (e.g., −1 V) as a current-to-voltage converter, shown outside the dashed box in Figure 12.5. The output current from the BDT affects the voltage division between V_{DD} and V_{SS}, and the resistor values can be calculated such that the output nodes V_{out} match the range of input gate voltages (i.e., if $V_{g,left}$ and $V_{g,right}$ equal −0.15 and 0.15 V, respectively, $V_{out,left}$ and $V_{out,right}$ will equal 0.15 and −0.15 V, respectively). More electrons on a node result in a lower voltage; thus, a flow of electrons represents logic "0."

The empirical model in Figure 12.5 uses two voltage-controlled current sources (VCCS) to recreate the output behavior from Figure 12.4. The model in Figure 12.5 assumes that the two gate voltages are differential; for nondifferential gate voltages, each VCCS instance must be replaced by a two-input polynomial VCCS (PVCCS) instance, with both gate voltages connected to each PVCCS.

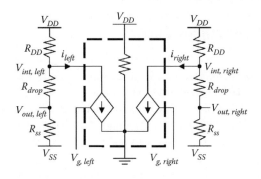

FIGURE 12.5 Empirical model of the BDT.

The six-terminal behavioral model is completed with a leakage path between ground and the supply voltage, represented by a resistor. To simplify the model, output symmetry is forced by inverting the measured right channel response of about 0 V to create the left channel response.

The BDT has been successfully tested at 20 GHz (the maximum operating frequency of the available equipment); however, the low capacitance of the 2DEG gate structures should yield operating frequencies in the terahertz range. Future experiments include designing a terahertz-capable test environment to determine the maximum operating frequency.

There are four sets of current paths in the BDT that consume power—(1) a path between the top drain and the source (~0.8 μA over a 1 V drop), (2) a path between V_{DD} and V_{SS} in the current-to-voltage converter (two ~2 μA paths over a 2 V drop), (3) a path between V_{DD} in the current-to-voltage converter and the BDT source (the total current through both output drains is 4.6 μA over a 1 V drop), and (4) the average gate leakage current (two gates with leakage 0.4 μA over an average 1 V drop). This results in a net power consumption of 14.2 μW and does not include the switching of the ~2 fF gate capacitance, which is likely to be very small compared to the static currents. In contrast, a 0.5 μm NMOSFET from a commercial 65 nm technology consumes an average of 174 μW. In addition, a single BDT provides the same functionality as a dual-rail CMOS inverter, providing more functionality for lower equivalent power consumption.

12.4 LOGIC DESIGN WITH BALLISTIC DEFLECTION TRANSISTORS

In this section, we explore logic gate design with ballistic deflection transistors while providing three different types of results. The first proof of concepts is evaluated using a Monte Carlo particle simulator based on classical mechanics [47], building upon the billiard model proposed by Beenakker and van Houten [40]. Screenshots from this simulator provide an intuitive picture of predicted electron flows through the device. In addition to the Monte Carlo simulator, we have created an empirical model of the BDT based on measurements of a single device. This empirical model is extrapolated to provide a gate-level functional analysis of the designs that includes the various nonidealities of the fabricated devices. Finally, the measurement results of each logic gate are provided to prove the functionality of each presented logic gate.

12.4.1 INVERTING BDT

The BDT provides complementary outputs; thus, a single BDT can act as both an inverter and a buffer. For example, if we examine Figure 12.4, an input of 0.15 V (logic "1") produces a large current in the left output branch and a small current in the right output branch. A large output current reduces the output voltage of that branch (the resulting output voltage is a function of the output current and the current-to-voltage converter on that output), creating a logic "0"—the inverted input. On the opposite output terminal, the small output current has little effect on the output voltage, maintaining a logic "1"—the same value as the input.

12.4.2 TWO-INPUT NAND GATE WITH BDTS

The two-input NAND gate can be used to construct any arbitrarily complex logic function; thus, achieving this function in novel devices is an important milestone if they are to be used for general-purpose computation. The NAND gate design [48] shown in Figure 12.6 (with the SEM image in Figure 12.7) functions as follows: the source is shown as the arrow entering the bottom channel of the BDT labeled "1." The differential gate inputs A and $\overline{A}$ guide electrons into the channel labeled $A = 1$ when gate A is high and gate $\overline{A}$ is low, and into channel $A = 0$ in the opposite case. The differential gate inputs B and $\overline{B}$ guide electrons in the central region to the channels labeled $B = 0$ or $B = 1$ similar to gate A. This results in a flow of electrons (representing a "0") at the output F only when both A and B are high. For each other input combination, electrons are diverted to either the

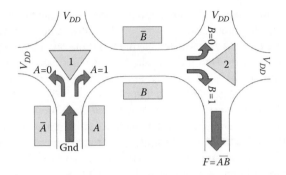

FIGURE 12.6   BDT two-input NAND gate schematic.

FIGURE 12.7   SEM image of the BDT NAND gate.

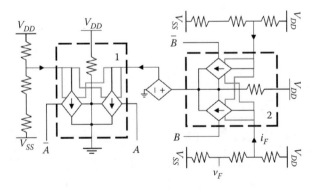

FIGURE 12.8   Circuit model of the BDT NAND gate.

left output channel or the top right output channel. This behavior results in the logic function *A* NAND *B*.

The empirical model in Figure 12.8 is similar to the schematic in Figure 12.6, with a current-controlled voltage source (CCVS) allowing the VCCS instances in BDT 2 to react to the current exiting BDT 1. The CCVS has no physical equivalent in the gate, but is needed in the circuit model to convert the output signal from BDT 1 into a voltage to adjust the VCCS in BDT 2.

Here, we describe an important improvement to the model presented in Ref. [48]. In the previous model, it was assumed that the drain potentials in both BDTs were properly balanced, which would result in the $A = 0/B = 1$ and $A = 1/B = 0$ states having the same output voltage. Unfortunately, that is not the case, as shown by the measurement results from the fabricated BDT NAND gate in Figure 12.9b. The uneven drain potential increases the amount of incorrectly steered current in BDT 1, such that even when $A = 0$, a large current flows into BDT 2; when $B = 1$, that undesired current flow reduces the output voltage (see $A = 0/B = 1$ in Figure 12.9b), limiting the gate's I_{on}/I_{off} ratio.

To consider this effect in our empirical model, we have added a new parameter that allows us to tune the switching behavior of BDT 1 by adjusting the current of the $A = 0$ case (V_{Low}). The adjustment is made by connecting the drains of each device to that device's VCCS instances, which take the difference between the drain potentials and scale the results according to the empirical data. As shown in Figure 12.9a, adjusting V_{Low} allows us to closely match the reduced $A = 0/B = 1$ state from the measured results. To tune the model, we performed an additional experiment on a single BDT in which we varied the left output drain bias while keeping the right output drain bias fixed. Based on the empirical data, we found that when the left drain bias is just 33% of the right drain bias (a loose approximation of three biased drains to the right of BDT1 and one biased drain to the left of BDT1), the switching ratio is reduced to just 54% of the ratio when the output drains are equal. $V_{Low} = 54\%$ correlates very closely with our model in Figure 12.9a, which shows that the $A = 0/B = 1$ state in the measured results falls within the range of $V_{Low} = 40\%$ and $V_{Low} = 60\%$. All transient waveforms in this chapter were taken at low frequency, using an arbitrary clock cycle of 1 ns for simulations and representing DC measurement results in transient format for presentation purposes.

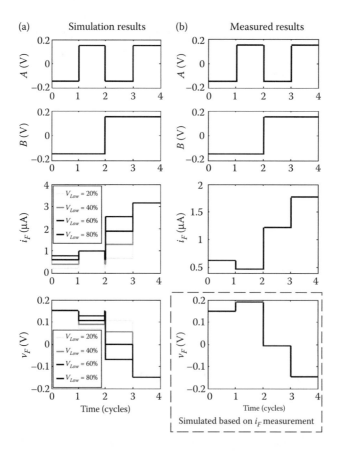

FIGURE 12.9 BDT NAND gate waveforms. (a) Simulated. (b) Measured.

Note that the fabricated gate did not include the current-to-voltage converters; the measured output currents i_F in Figure 12.9b are passed through a simulated current-to-voltage converter to provide the voltage response v_F in the dashed box. The empirical model is generated from a different chip than the fabricated NAND gate; thus, although the relative behaviors of the model and measurements are very similar, the magnitudes of the output currents are unequal. This is corrected in the voltage outputs by adjusting the current-to-voltage converters.

12.4.3 General-Purpose Gate with BDTs

To improve the utility of BDTs in digital logic applications, we have designed a new type of gate structure that is symmetric, reducing the problem of uneven drain potentials that limited the NAND gate response. The design consists of three BDTs [49] with the sources of two rotated BDTs connected to the left and right drains of a center BDT, shown in Figure 12.10 (with the SEM image in Figure 12.11). Electrons are injected through the source, S, and steered to the left or right BDT by gate inputs A and $\bar{A}$. As the electrons are drawn to the pull-up potentials on either side, gate inputs B and $\bar{B}$ further steer the electrons to one of four output drains.

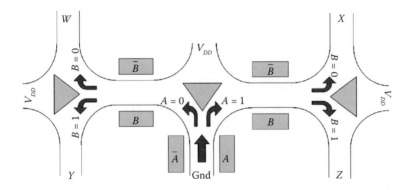

FIGURE 12.10   BDT general-purpose gate (GPG) schematic. (D. Wolpert et al., General purpose logic gate using ballistic nanotransistors, *Proc. 11th IEEE Conf. Nanotechnol.*, Portland, OR, p. 1, © (August 2011) IEEE. With permission.)

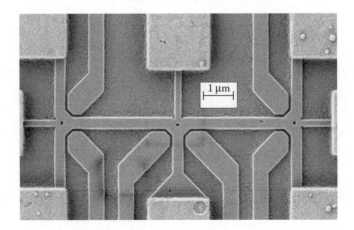

FIGURE 12.11   SEM image of the BDT general-purpose gate (GPG). (D. Wolpert et al., General purpose logic gate using ballistic nanotransistors, *Proc. 11th IEEE Conf. Nanotechnol.*, Portland, OR, p. 1, © (August 2011) IEEE. With permission.)

The novelty of this gate structure is that there are four gate outputs, each corresponding to a specific combination of inputs A and B. We label these outputs W, X, Y, and Z, and they correspond to the input values $A = B = 0$, $A = 1$ & $B = 0$, $A = 0$ & $B = 1$, and $A = B = 1$. The logic function of the GPG is defined by how these four drains are connected to output terminals F and $\bar{F}$, as shown in Table 12.1. This structure is capable of producing any two-input function and its complement; thus, we name the structure as a two-input GPG.

If a drain is connected to terminal F, only one of the four input vectors (e.g., $A = B = 0$ for drain W) will result in a flow of electrons through that drain, reducing the voltage at F; thus, connecting a drain to output terminal F sets that table entry to logic low, while connecting a drain to output terminal $\bar{F}$ sets that table entry to logic high. For example, the BDT NAND gate shown in Figure 12.8 is very similar to the combination of the center and right BDTs in Figure 12.12. In the NAND gate, the function A NAND B is achieved by connecting the lower right drain to an output F. In the GPG, the function A NAND B is also achieved by connecting the lower right drain (Z) to F. Indeed, the circuit model for the GPG, shown in Figure 12.12, is the same for the center and right-side BDTs as the circuit model for the NAND gate from Figure 12.8; however, the simulation results shown for the GPG in the NAND configuration, discussed momentarily, do not have the same asymmetry issues as the two-BDT NAND gate, resulting in a much larger I_{on}/I_{off} ratio. The improved asymmetric circuit model presented in the previous subsection is not needed for the GPG (aside from process variations) because of the improved device symmetry. A list of output terminal connections for other common logic functions is provided in Table 12.2.

TABLE 12.1
General-Purpose Gate Function Definition

A	B	F
0	0	W
0	1	Y
1	0	X
1	1	Z

If $W|X|Y|Z$ connected to $\bar{F}$, truth table entry is 1
If $W|X|Y|Z$ connected to F, truth table entry is 0

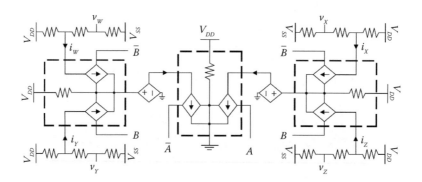

FIGURE 12.12 Circuit model of the BDT general-purpose gate (GPG). (D. Wolpert et al., General purpose logic gate using ballistic nanotransistors, Proc. 11th IEEE Conf. Nanotechnol., Portland, OR, p. 2, © (August 2011) IEEE. With permission.)

TABLE 12.2
GPG Connections for Common Logic Functions

Function	Terminal Connections			
	W	**X**	**Y**	**Z**
NAND	$\overline{F}$	$\overline{F}$	$\overline{F}$	F
AND	F	F	F	$\overline{F}$
NOR	$\overline{F}$	F	F	F
OR	F	$\overline{F}$	$\overline{F}$	$\overline{F}$
XNOR	$\overline{F}$	F	F	$\overline{F}$
XOR	F	$\overline{F}$	$\overline{F}$	F

In Figure 12.13, we present the measurement results of a fabricated GPG connected in XOR and XNOR configurations. In Figure 12.13b and c, the measured current response i_F represents the current flowing out through the terminals connected to F. The measured output currents were passed through simulated current-to-voltage converters to generate the output voltages v_F, which could then be applied to successive BDT logic stages. In this chip, input voltages of only ±100 mV are sufficient to switch between logic states, resulting in a noise margin of approximately 7 μA in the XNOR configuration and 5 μA in the XOR configuration. The peak current in the XNOR configuration is 24 μA, resulting in an I_{on}/I_{off} ratio of 1.6. The I_{on}/I_{off} ratio is expected to improve dramatically as the geometries and process steps are further refined. Gate leakage was particularly problematic in the fabricated test chip—the large potential difference between the gate and the 1V drain biases causes a gate leakage of ~20 μA in the fabricated GPG.

In Figure 12.14, we present the XOR and XNOR simulation results corresponding to the measurements in Figure 12.13. The waveforms match very closely with the exception of small asymmetries in the $A = 0/B = 1$ and $A = 1/B = 0$ cases resulting from process variations. In Figure 12.15, we present simulation results from the GPG configurations of other common logic functions, demonstrating the versatility of the GPG for circuit design. In each configuration, the current-to-voltage converters need to be modified depending on how many BDT drains are connected to the internal voltage node (V_{INT} from Figure 12.5) of terminals F and $\overline{F}$. As shown, each output response is capable of closely reproducing the input voltage range, resulting in a straightforward connection to the gate input of the next stage of the GPG or other devices. An additional advantage of the GPG is its potential for use as a device fabric [50]; the slower e-beam processing could be used to print dies full of unconnected GPGs, and a conventional metallization process could be used to add the resistors and connect the gates into the desired functionality. While this could also be done with the NAND gate, the GPG is capable of creating any two-input logic function using only three BDTs, while multiple NAND gates are required to create most functions. For example, to create an XNOR function, five NAND gates are required (corresponding to 10 BDTs), while the GPG can create that functionality by using only three BDTs. Work is currently underway to improve the speed of e-beam processing the using arrays of beams working simultaneously; this method has the added benefit of eliminating the costly process of fabricating nanoscale masks [51].

12.5 CIRCUIT DESIGN WITH BALLISTIC DEFLECTION TRANSISTORS

In this section, we present a circuit design methodology for integrating discrete BDT logic gates into a larger system. First, we examine a logic cell, and then we use this cell in a case study—a full adder design consisting of five GPCs. We also discuss a number of design considerations for BDT circuits using our empirical circuit model, including necessary modifications to the concept of noise

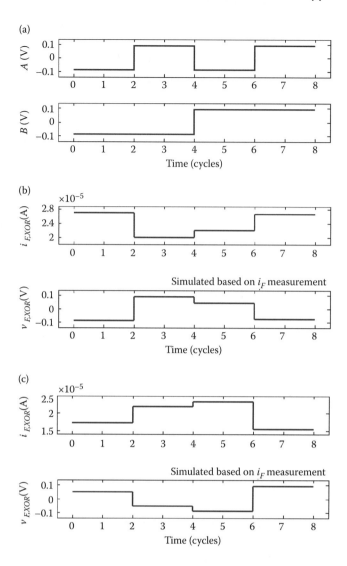

FIGURE 12.13 Measured current response of fabricated GPG. (a) Input voltages, (b) XOR configuration, and (c) XNOR configuration, with simulated current-to-voltage conversion shown for each case. (D. Wolpert et al., General purpose logic gate using ballistic nanotransistors, *Proc. 11th IEEE Conf. Nanotechnol.*, Portland, OR, p. 2, © (August 2011) IEEE. With permission.)

margins and constraints when cascading devices. Finally, we describe potential system applications for BDT logic.

12.5.1 BDT Circuit Design Methodology

Circuit design with BDTs is somewhat similar to the traditional dual-rail logic design; each logic gate requires complementary gate inputs. Unlike the traditional logic design, the outputs F and $\overline{F}$ are generated by connecting the BDT drains to current-to-voltage converters, which then produce the desired output voltage. These additional internal connections complicate circuit design, so we begin with a simple case of one logic cell; in the next subsection, we will show how to combine these logic cells to create more complex functionality.

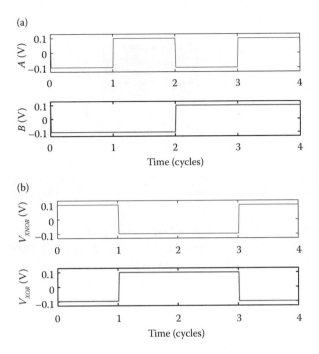

FIGURE 12.14 Simulation results from the BDT general-purpose gate model. (a) Input vectors. (b) GPG XNOR/XOR response. (D. Wolpert et al., General purpose logic gate using ballistic nanotransistors, *Proc. 11th IEEE Conf. Nanotechnol.*, Portland, OR, p. 3, © (August 2011) IEEE. With permission.)

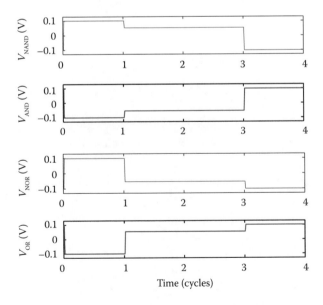

FIGURE 12.15 Simulation results from the other BDT GPG configurations using the inputs in Figure 12.16a. (D. Wolpert et al., General purpose logic gate using ballistic nanotransistors, *Proc. 11th IEEE Conf. Nanotechnol.*, Portland, OR, p. 3, © (August 2011) IEEE. With permission.)

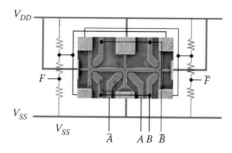

FIGURE 12.16 A BDT XOR/XNOR logic cell. (D. Wolpert et al., General purpose logic gate using ballistic nanotransistors, *Proc. 11th IEEE Conf. Nanotechnol.*, Portland, OR, p. 3, © (August 2011). With permission.)

The logic cell consists of a GPG, two current-to-voltage converters, and power and ground rails. The current-to-voltage converters provide the outputs F and $\bar{F}$, defined by the GPG output drain connections as shown in Table 12.2. An example of an XOR/XNOR logic cell is shown in Figure 12.16. The voltage rail connections and current-to-voltage converters are indicated by the lighter-colored lines. Inputs A and $\bar{A}$ are connected to the central BDT, while inputs B and $\bar{B}$ are connected to both the left and right BDTs. The GPG outputs are connected such that the left current-to-voltage converter provides $F = A$ XOR B, while the right current-to-voltage converter provides $\bar{F} = A$ XNOR B.

Defining multicell circuit functionality (an example of a multicell circuit is provided in Figure 12.17a) is a three-step process. First, we create a grid of "unconfigured" logic cells (meaning

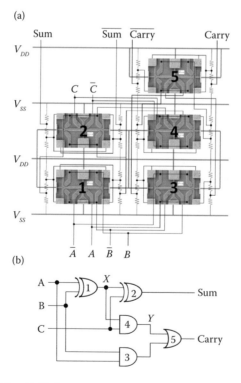

FIGURE 12.17 BDT full adder. (a) GPG-based design and (b) corresponding gate schematic. (D. Wolpert et al., General purpose logic gate using ballistic nanotransistors *Proc. 11th IEEE Conf. Nanotechnol.*, Portland, OR, p. 4, © (August 2011). With permission.)

the GPG inputs and outputs of each cell are left floating), one cell per logic gate in the circuit. Then we analyze the circuit and group any interconnected gates together to attempt to minimize inter-connect distances. Once we have found an appropriate logic gate mapping, we define each GPG by connecting the appropriate output drains to the current-to-voltage converters in that cell with metal wires. Finally, we connect the logic cells by connecting each output F and $\overline{F}$ to the appropriate inputs in the following cell using metal wires.

During this connection process, we can also pay attention to the output load of each cell. As shown in Figure 12.16, input A is only connected to a single gate, while input B is connected to two gates. To balance the load when connecting up multiple logic cells, cells that drive multiple other cells should be routed first so that they can be connected to the single input (A) whenever possible, rather than the double input load (B).

12.5.2   CASE STUDY: BDT FULL ADDER

The use of the GPG logic cells makes a circuit layout very regular (aside from signal routing), as shown in Figure 12.17a for a 5-GPG full adder. In Figure 12.17a, the voltage rail connections and current-to-voltage converters are indicated by the lighter-colored lines, which form a simple grid pattern. The gate schematic in Figure 12.17b is provided to make the logic connections in Figure 12.17a easier to follow. Note that GPG 2 and GPG 4 are inverted to simplify connection with the power and ground rails.

The output waveforms in Figure 12.18 indicate that the circuit functions correctly, although the voltage difference between logic high and logic low in the carry input is reduced. To explore why the sum behavior is in full swing and the carry behavior is not, we have included waveforms for the nodes labeled X and Y in Figure 12.17b. Node X corresponds to the output A XOR B. Following the probability analysis presented in Ref. [48] and the simulation results in Figure 12.14, we see that the $A = 1/B = 0$ and $A = 0/B = 1$ input states in the XOR gate outputs are in full swing; thus, the output remains a full swing after passing through the XOR-configured GPG. The full-swing output of GPG 1 also enables a full-swing output in GPG 2, resulting in the desirable *Sum* wave-form shown in Figure 12.18. Unfortunately, the carry generation circuit requires AND and OR GPG configurations, which do have reduced-swing states for the $A = 1/B = 0$ and $A = 0/B = 1$ input cases. These reduced-swing states results are clearly seen for certain input combinations in the Y waveform in Figure 12.18 (we will discuss the solid and dashed lines momentarily). The nonideal output from GPG 4 (node Y) exacerbates the swing reduction in GPG 5, although there is still a clear distinction between logic low (<0 V) and logic high (>0 V) in the *Carry* waveform.

One way to reduce the impact of the intermediate states in AND and OR configurations is to adjust the current-to-voltage converters. The current-to-voltage converters are designed to provide a full-swing output (either +100 mV or −100 mV) for the input conditions where $A = 0/B = 0$ and $A = 1/B = 1$. These current-to-voltage converters result in the solid lines in Figure 12.18. Instead of designing the current-to-voltage converters to swing between ±100 mV, we can design them to improve the output swing of cascaded gates by setting the output voltage to the average of the reduced-swing state and the nearest (in terms of voltage) full-swing state. For example, let us con-sider the NAND gate waveform in Figure 12.15. Instead of setting the $A = 0/B = 0$ case to 100 mV, we can average the $A = 0/B = 0$ case with the $A = 1/B = 0$ case. With this approach, the maximum distance of any output voltage from the ±100 mV peak voltages will only be 25 mV, whereas the reduced-swing states shown in Figure 12.15 have a 50 mV offset. As shown in the *Carry* waveform in Figure 12.18, this increases the difference between the output and the switching threshold from −30.7 mV to −47.1 mV, an improvement of 53%. The impact of the reduced-swing state will be fur-ther reduced as the device I_{on}/I_{off} ratio increases.

When describing the difference between the on and off states in the BDT, the term "noise mar-gin" is insufficient—the output current in the BDT degrades both when the input voltage drops below a certain threshold *and* when the input voltage *increases* beyond a certain threshold. Thus,

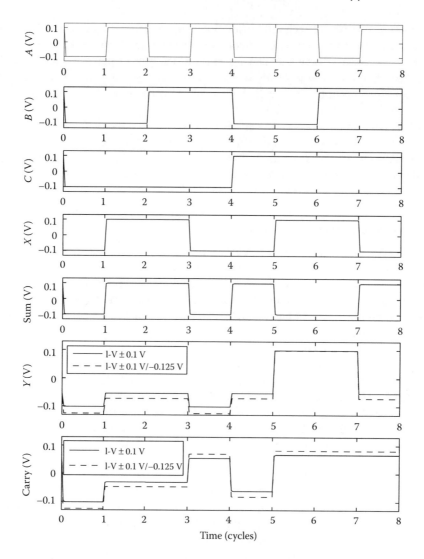

FIGURE 12.18 Simulated output response of the GPG-based full adder. (D. Wolpert et al., General purpose logic gate using ballistic nanotransistors, *Proc. 11th IEEE Conf. Nanotechnol.*, Portland, OR, p. 4, © (August 2011) IEEE. With permission.)

in a circuit design with BDT, we do not refer to the difference between logic high and logic low as a noise margin, but as a noise window. We define the noise window as the input voltage range over which the output voltage exceeds the switching voltage (or is below the switching voltage for logic low); the switching voltage is the average of the output states; the BDT operates at ±0.1 V, and thus the switching voltage is 0 V. In a single BDT, the input voltage range exceeding the switching voltage is between −20 and 318 mV, indicating that the *Carry* output waveform in Figure 12.18 is still well within the noise window ($V_{out,low} = -47.1$ mV and $V_{out,high} = 69.7$ mV).

12.5.3 BDT APPLICATIONS

The BDT is proposed as a high-speed supplement to conventional Si MOSFET systems. As the BDT process matures and new logic structures and circuits are created, the remaining challenges are those of integration. One possible way of combining the BDT with silicon CMOS would be through

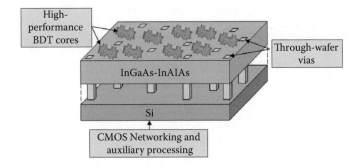

FIGURE 12.19 Heterogeneous integration of BDT cores and conventional Si CMOS.

three-dimensional integration of heterogeneous wafers, which have been previously proposed to link optical I/O and CMOS logic [11]. One example of this heterogeneous integration is shown in Figure 12.19, with high-performance BDT cores interfacing with a conventional Si CMOS chip using through-wafer vias. To integrate BDTs and MOSFETs, the voltage ranges must be shifted from ±0.15 V to a full-swing CMOS range, which will require a more complex circuit than the conventional level shifters commonly used in CMOS technologies.

Using this type of configuration, BDTs could provide terahertz-frequency accelerator cores for existing chip-multiprocessors. Further into the future, the BDT may take on the role of general-purpose computation, using the higher-density conventional CMOS devices purely for networking (and even this may eventually be replaced by optical I/O) and other tasks with lower-frequency requirements.

12.6 LIMITATIONS AND POTENTIAL SOLUTIONS

The BDT, like all nanoscale devices, is susceptible to a variety of variations in fabrication. In addition, the mode of operation presented in this chapter requires pull-up potentials to ensure current flow. These issues, and potential solutions to these issues, are discussed in this section.

12.6.1 VARIABILITY

Device variations have large impacts on leakage current and switching behavior. This chapter reports the results from three chips—a BDT device chip (Figure 12.4), a chip for the two-BDT NAND gates (Figure 12.9), and a chip for the GPGs (Figure 12.13).

Among the three chips, leakage currents vary between 0.4 and 20 μA with only up to 10% of this leakage current between the source and the top drain. Gate leakage in these devices is related to the etch depth of the trench between the gate and the channel; if the etch depth is too shallow, high-energy electrons can escape the 2DEG layer and tunnel underneath the trench, resulting in gate leakage. Unfortunately, the steepness of the trench walls is limited by the fabrication process; thus, if the trenches are made too deep, it will increase the gate–channel separation, reducing the steering capability of the device.

The I_{on}/I_{off} ratios in the three chips vary between 1.4 and 10. The switching capability of the BDT may be improved by adjusting the geometry parameters such as the triangle location and the corner radius in the central area of the BDT [47].

BDT circuits are also susceptible to variations in resistance values in the current-to-voltage converters. Figure 12.20 shows the impact of systematic resistance variations of up to ±20% on the output voltages for logic high ($V_{out,high}$) and logic low ($V_{out,low}$). The horizontal dashed lines indicate the ideal values of the output voltages, showing that resistance variations can have a significant impact on the BDT noise window. The impact of a 20% variation in resistance can impact the output voltage by as much as 60%.

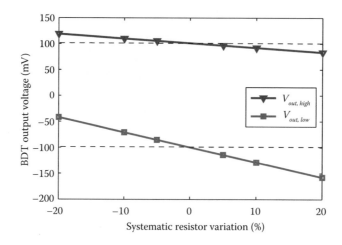

FIGURE 12.20   Impact of systematic resistor variations on the BDT output voltage.

12.6.2   ALTERNATIVE BDT CONNECTION METHODS

One potential solution to the resistor variation issue is to use an alternative method of connecting the gates together, for example, operating the BDT in a voltage mode where the outputs are directly connected to the next logic stage rather than by using current-to-voltage converters. This would eliminate the need for all the resistors in Figure 12.17a, greatly reducing the design complexity and area. An example of a BDT operating in the voltage mode is shown in Figure 12.21. Thus far, the differential output voltage generated is only 15% of the differential input voltage (over the range of input voltages between ±50 and ±300 mV); however, we expect the output response to improve significantly as device designs are optimized for this mode. An additional alternative of connecting the BDTs is to use complementary structures such as those presented in Ref. [52].

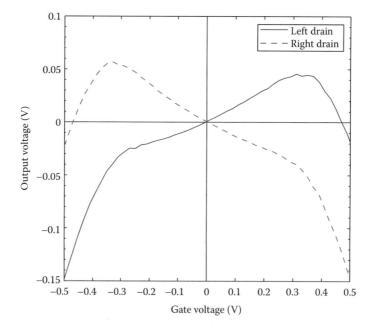

FIGURE 12.21   Voltage-mode operation of a BDT showing output voltages generated by a ±0.5 V range of differential input voltages.

12.6.3 FUTURE CIRCUIT IMPROVEMENTS

At the circuit level, the two main areas of improvement we are currently researching are the new gate structures and the new methods of connecting BDTs. The GPG concept could be extended to a 7-BDT three-gate input GPG (with the additional BDTs connected to the drains of the outer two BDTs in Figure 12.10); however, the number of devices that may be cascaded with a single 2DEG channel is limited by the mean free path, as described in Section 12.3. This 7-BDT structure would be able to perform the half-adder operation using two gates instead of five (although the BDT count would only be reduced by one); more importantly, these larger structures would dramatically reduce the wiring complexity of BDT circuits.

12.7 CONCLUSIONS

This chapter has described the circuit applications of a newly developed nanodevice, termed the ballistic deflection transistor (BDT), which promises terahertz performance at moderate voltages. With the successful fabrication of a set of BDT logic gates and a preliminary circuit design methodology, the BDT has the potential for consideration in the next generation of nanoscale devices. In particular, the GPG may provide a new integration opportunity for mass fabrication. A full adder has been presented as a case study on using multiple GPGs to develop circuits based on BDTs.

The presented overview of nanoscale devices reveals a number of exciting options with the potential of replacing conventional CMOS when further scaling becomes too costly. This wide variety of options may in fact introduce unexpected system complexities, as there may not be a single "predominant" device in which the resources are pooled, as was the case when CMOS emerged as the electronics standard of sorts. A number of conventional improvements may be combined, such as strained-Si channels with finFET gates, to extend the reach of conventional devices; however, these improvements do not offer the same potential performance enhancement as nonconventional ideas such as molecular or optical computing. The ballistic deflection transistor provides an interesting middle ground, pairing a mature semiconductor technology with nonconventional device and circuit design. The presented circuit design methodology points out a few complexities that must be addressed prior to adoption (such as logic depth and signal degradation), but overall the potential performance benefits of the BDT make it a strong candidate for use in the next era of computing—heterogeneous integration of nanoelectronics and CMOS. Continued work on improving the device, along with the BDT logic and circuit design, should help to further differentiate it from other alternatives.

ACKNOWLEDGMENT

The authors would like to thank other members of the BDT group at the University of Rochester and the University of Massachusetts at Lowell. BDT fabrication was performed at the Cornell Nanofabrication Facility (CNF), a member of the NSF-sponsored NNI network.

REFERENCES

1. D. Wolpert, H. Irie, R. Sobolewski, P. Ampadu, Q. Diduck, and M. Margala, Ballistic deflection transistors and the emerging nanoscale era, *Proc. IEEE Int. Symp. Circuits Syst.*, Taipei, Taiwan, pp. 61–64, May 2009.
2. W. Haensch et al., Silicon CMOS devices beyond scaling, *IBM J. Res. Dev.*, 50(4/5), 339–361, Jul./Sept. 2006.
3. T. Skotnicki, J. Hutchby, T.-J. King, H.-S. Wong, and F. Boeuf, The end of CMOS scaling: Toward the introduction of new materials and structural changes to improve MOSFET performance, *IEEE Circuits Devices Mag.*, 21(1), 16–26, Jan./Feb. 2005.
4. K. Ronse et al., Lithography options for the 32 nm half pitch node and beyond, *IEEE Trans. Circuits Syst.-I: Regular Papers*, 56(8), 1883–1890, Aug. 2009.

5. T. Oka et al., Advanced performance of small-scaled InGaP/GaAs HBT's with f_T over 150 GHz and f_{max} over 250 GHz, *Int. Electron Devices Mtg. (IEDM'98)*, San Francisco, CA, pp. 653–656, Dec. 1998.

6. M. Bohr, The evolution of scaling from the homogeneous era to the heterogeneous era, *IEEE Int. Elec. Dev. Mtg. (IEDM'11)*, Washington, DC, 1.1.1–1.1.6, Dec. 2011.

7. M. Lee, E. Fitzgerald, M. Bulsara, M. Currie, and A. Lochtefeld, Strained Si, SiGe, and Ge channels for high-mobility metal-oxide-semiconductor field-effect transistors, *J. Appl. Phys.*, 97 011101-1–011101-28, 2005.

8. I. Ferain, C. Colinge, and J.P. Colinge, Multigate transistors as the future of classical metal-oxide-semiconductor field-effect transistors, *Nature* 479, 310–316, Nov. 2011.

9. S. Mujumdar, Layout-dependent strain optimization for p-channel trigate transistors, *IEEE Trans. Elec. Dev.*, 59(1), 72–78, Jan. 2012.

10. V. F. Pavlidis and E. G. Friedman, *Three-Dimensional Integrated Circuit Design*, Burlington, MA: Morgan Kaufmann, 2008.

11. K. Banerjee, S. J. Souri, P. Kapur, and K. C. Saraswat, 3-D ICs: A novel chip design for improving deep-submicrometer interconnect performance and systems-on-chip integration, *Proc. IEEE*, 89(5), 602–633, May 2001.

12. J. Mateos et al., Microscopic modeling of nonlinear transport in ballistic nanodevices, *IEEE Trans. Elec. Dev.*, 50(9), 1897–1905, Sept. 2003.

13. H. Irie and R. Sobolewski, Picosecond electric pulse excitation of three-branch ballistic nanodevices, *J. Phys.: Conf. Series*, 193(012097), 1–4, 2009.

14. Q. Diduck, M. Margala, and M. J. Feldman, Terahertz transistor based on geometrical deflection of ballistic current, *Proc. IEEE MTT-S Int. Microwave Symp.*, San Francisco, CA, pp. 345–347, Jun. 2006.

15. M. Stan, P. Franzon, S. Goldstein, J. Lach, and M. Ziegler, Molecular electronics: From devices and interconnect to circuits and architecture, *Proc. IEEE*, 91(11), 1940–1957, Nov. 2003.

16. A. E. Kaloyeros et al., Conformational molecular switches for post-CMOS nanoelectronics, *IEEE Trans. Circuits Syst.-I: Regular Papers*, 54(11), 2345–2352, Nov. 2007.

17. P. J. Kuekes, Molecular crossbar latch, U.S. Patent No. 6,586,965, Jul. 2003.

18. T. Palacios, Applied physics: Nanowire electronics comes of age, *Nature* 481, 152–153, Jan. 2012.

19. A. K. Geim and K. S. and Novoselov, The rise of graphene, *Nat. Mater.*, 6 183–191, 2007.

20. D. Akinwande, Y. Nishi, and H.-S. P. Wong, Carbon nanotube quantum capacitance for nonlinear terahertz circuits, *IEEE Trans. Nanotechnol.*, 8(1), 31–36, Jan. 2009.

21. S. O. Koswatta, D. E. Nikonov, and M. S. Lundstrom, Computational study of carbon nanotube p-i-n tunnel FETs, *Int. Elec. Dev. Mtg. (IEDM'05)*, Washington, DC, pp. 525–528, Dec. 2005.

22. A. Raychowdhury and K. Roy, Carbon nanotube electronics: Design of high-performance and low-power digital circuits, *IEEE Trans. Circuits Syst.-I: Regular Papers*, 54(11), 2391–2401, Nov. 2007.

23. J. Zhang, N. P. Patil, and S. Mitra, Design guidelines for metallic-carbon-nanotube-tolerant digital logic circuits, *Proc. Design Automation Test Europe (DATE'08)*, Munich, Germany, pp. 1009–1014, Mar. 2008.

24. M. Kastner, The single electron transistor and artificial atoms, *Ann. Phys.*, 9(11–12), 885–894, Nov. 2000.

25. G. Zardalidis and I. G. Karafyllidis, SECS: A new single-electronic-circuit simulator, *IEEE Trans. Circuits Syst.-I: Regular Papers*, 55(9), 2774–2784, Oct. 2008.

26. K. K. Likharev, Single-electron devices and their applications, *Proc. IEEE*, 87(4), 606–632, Apr. 1999.

27. S. Sarma, Ferromagnetic semiconductors: A giant appears in spintronics, *Nat. Mater.*, 2, 292–294, May 2003.

28. S. A. Wolf, A. Y. Chtchelkanova, and D. M. Treger, Spintronics–a retrospective and perspective, *IBM J. Res. Dev.*, 50(1), 101–110, Jan. 2006.

29. G. Bourianoff, J. E. Brewer, R. Cavin, J. A. Hutchby, and V. Zhirnov, Boolean logic and alternative information-processing devices, *Computer*, 41(5), 38–46, May 2008.

30. L. M. Adleman, Molecular computation of solutions to combinatorial problems, *Science*, 266(11), 1021–1024, Nov. 1994.

31. S. M. R. Hasan, A novel mixed-signal integrated circuit model for DNA-protein regulatory genetic circuits and genetic state machines, *IEEE Trans. Circuits Syst.-I: Regular Papers*, 55(5), 1185–1196, Jun. 2008.

32. H. Hug and R. Schuler, Strategies for the development of a peptide computer, *Bioinformatics*, 17(4), 364–368, 2001.

33. D. P. DiVincenzo, Quantum computation, *Science*, 270(5234), 255–261, Oct. 1995.

34. J. H. Kim et al., All-optical logic gates using semiconductor optical-amplifier-based devices and their applications, *J. Korean Phys. Soc.*, 45(5), 1158–1161, Nov. 2004.

35. S. Datta, *Electronic Transport in Mesoscopic Systems*, Cambridge, UK: Cambridge University Press, 2005.

36. A. Song, Room temperature ballistic nanodevices, *Encycl. Nanosci. Nanotechnol.*, 9, 371–389, 2004.

37. C. W. J. Beenakker and H. Van Houten, Quantum transport in semiconductor nanostructures, *Solid State Phys.*, 44, 1–111, 1991.

38. T. Mimura, S. Hiyamizu, T. Fujii, and K. Nanbu, A new field-effect transistor with selectively doped GaA/n-Al$_x$Ga$_{1-x}$As heterojunctions, *Jpn. J. Appl. Phys.*, 19(L225-7), 1980.

39. Y. Hirayama and S. Tarucha, High temperature ballistic transport observed in AlGaAs/InGaAs/GaAs small four-terminal structures, *Appl. Phys. Lett.*, 63(17), 2013–2017, 1993.

40. C. W. J. Beenakker and H. Van Houten, Billiard model of a ballistic multiprobe conductor, *Phys. Rev. Lett.*, 63(17), 1857–1860, 1989.

41. J. S. Galloo et al., Ballistic GaInAs/AlInAs devices technology and characterization at room temperature, *Proc. 4th IEEE Conf. Nanotechnol. (NANO'04)*, Munich, Germany, 98–100, Aug. 2004.

42. T. Palm and L. Thylen, Analysis of an electron-wave Y-branch switch, *Appl. Phys. Lett.*, 60(2), 237–239, Jan. 1992.

43. S. Reitzenstein, L. Worschech, and A. Forchel, A novel half-adder circuit based on nonometric ballistic Y-branched junctions, *IEEE Elec. Dev. Lett.*, 24(10), 625–627, 2003.

44. H. Q. Xu, Electrical properties of three-terminal ballistic junctions, *Appl. Phys. Lett.*, 78, 2064–2066, 2001.

45. R. Fleischmann and T. Geisel, Mesoscopic rectifiers based on ballistic transport, *Phys. Rev. Lett.*, 89, 016804-1–016804-4, Jun. 2002.

46. H. Irie, Q. Diduck, M. Margala, R. Sobolewski, and M. J. Feldman, Nonlinear characteristics of T-branch junctions: Transition from ballistic to diffusive regime, *Appl. Phys. Lett.*, 93, 053502, 2008.

47. D. Huo, Q. Yu, D. Wolpert, and P. Ampadu, A simulator for ballistic nanostructures in a 2-D electron gas, *ACM J. Emerging Technol. Computing Syst. (JETC)*, 5(1), 5.1–5.21, Jan. 2009.

48. D. Wolpert, Q. Diduck, and P. Ampadu, NAND gate design for ballistic deflection transistors, *IEEE Trans. Nanotechnol.*, 10(1), 150–154, Jan. 2011.

49. D. Wolpert, I. Iniguez-de-la-Torre, V. Kaushakl, M. Margala, and P. Ampadu, General purpose logic gate using ballistic nanotransistors, *Proc. 11th IEEE Conf. Nanotechnol.*, Portland, OR, pp. 1171–1176, Aug. 2011.

50. T. Jhaveri et al., Maximization of layout printability/manufacturability by extreme layout regularity, *J. Micro/Nanolith. MEMS MOEMS*, 6(031011), Sept. 2007.

51. B. J. Lin, Future of multiple e-beam direct-write systems, *Proc. 4th SPIE Conf. Alternative Lithographic Technologies*, San Jose, CA, 832302, Feb. 2012.

52. T. Palm and L. Thylen, Designing logic functions using an electron waveguide Y-branch switch, *J. Appl. Phys.*, 79(10), 8076–8081, May 1996.

Section V

Single-Electron Transistors and Electron Tunneling Devices

13 Simultaneously Controlled Tuning of Tunneling Properties of Integrated Nanogaps Using Field-Emission-Induced Electromigration

Mitsuki Ito, Shunsuke Akimoto, Ryutaro Suda, and Jun-Ichi Shirakashi

CONTENTS

13.1 INTRODUCTION

The fabrication of large-scale, reproducible nanogap electrodes is a key issue for the integration of nanoscale devices such as single-electron transistors (SETs). SETs have been studied extensively because of their potential advantages in high integration density and low power consumption. There are many experimental reports on the fabrication techniques of metallic SETs, based on nanogaps with randomly placed nanoparticles [1–5]. However, it is almost impossible to control the position of islands of SETs using these fabrication techniques. On the other hand, highly elaborate shadow evaporation [6,7] and scanning probe microscope (SPM)-based oxidation [8,9] have also been reported, and the fabrication methods have achieved a high-temperature operation of SETs. However, it is known that these approaches are complicated and require special techniques and systems.

Recently, the electromigration method, based on the electrical current-induced diffusion of metal atoms in a thin metal film, has been extensively investigated as a fabrication method of nanogaps (i.e., nanoscale electrodes with nanometer separation) [10–14]. From the point of view of the integration of nanogaps, parallel fabrication of multiple nanogaps based on the electromigration method has already been reported by other researchers [15,16]. However, the electrical properties of individual nanogaps are not in situ measured in the method. In addition, the nanogap electrodes fabricated by the electromigration method generally have a tendency to exhibit high tunnel resistance.

Hence, we consider that it is difficult to utilize this conventional electromigration procedure for the fabrication/integration of nanogaps with well-controlled tunnel resistance.

Over the past few decades, field emission phenomena have been widely utilized in vacuum microelectronics and extensively studied theoretically [17]. The current flowing through nanogaps is dominated by an electron-tunneling process leading to a high electric field. Such a field emission current also has a tendency to reconstruct the atomic geometry at the tip of the nanogap electrodes or even destroy the tip surface [18]. By utilizing such phenomena, we have reported a wide range of control over tunnel resistance of nanogaps by field-emission-induced electromigration, a technique called "activation" [19,20]. "Activation" is based on moving atoms induced by the Fowler–Nordheim (F–N) field emission current passing through the nanogaps. Using the activation procedure, we can control the electrical properties of the nanogaps by adjusting the magnitude of the preset current of the activation and easily obtain nanogap-based SETs [21]. Furthermore, we have also demonstrated the integration method of two SETs with similar electrical properties by using the activation procedure, and the charging energy of the two SETs can be simultaneously controlled by the preset current in the activation [22,23]. Consequently, it is considered that the electrical properties of tunneling devices based on nanogaps can be tuned by adjusting the magnitude of the preset current during the activation procedure. In this chapter, we present a remarkably simple and easy technique for the simultaneous tuning of tunnel resistance of three nanogaps connected in series using the activation procedure. Furthermore, a reproducible method based on the activation scheme is also reported in detail for the fabrication/integration of planar-type nanogap-based SETs.

13.2 EXPERIMENTAL DETAILS

13.2.1 Activation Procedure for Series-Connected Nanogaps

The fabrication and activation procedures of the samples are composed of two main steps. First, we fabricated the initial Ni nanogaps by electron-beam (EB) lithography, EB deposition, and lift-off process on thermally oxidized silicon substrates. In this study, we use multiple initial Ni nanogaps connected in series. The tip shape of the nanogaps is sharpened in order to produce a field emission current caused by electric field concentration at the tip of the nanogap electrode.

In the next step, the activation experiments were carried out in a vacuum chamber using a semi-conductor parameter analyzer. During the activation, the field emission current I was simultaneously applied to the initial nanogaps connected in series in current source mode. The activation process is performed as follows (Figures 13.1a, 13.1b, and 13.1c). First, we set the preset current I_S. Then F–N field emission current is induced in the series-connected Ni nanogaps. Since the three nanogaps are connected in series (Figure 13.1a), an identical field emission current passes through each nanogap device during the activation procedure. It should be noted that the activated atoms induced by the F–N field emission current, which is completely controlled by the current source, are just moved from the source to the drain electrode when the voltage drop across the nanogap is suddenly decreased. Thus, the sample with wider initial gap separation is preferentially and immediately narrowed by the field emission current with a higher voltage drop (Figure 13.1b), and the applied current I is slowly ramped up until it reaches the preset current I_S. As a result, it is considered that the electrical properties of the activated nanogaps become similar to each other, with the progress of the activation (Figure 13.1c). In other words, the gaps of the series combination of the three Ni nanogap electrodes can self-regulate with final spacing set by the applied current. After performing the activation, I_D–V_D properties of each nanogap were measured individually. This procedure was continuously repeated while increasing the preset current I_S from 1 nA to 30 μA. The advantage of this fabrication method is that nanogaps having similar electrical properties can be simultaneously formed because the magnitude of the field emission current becomes the same in each nanogap. Hence, it is strongly suggested that the simultaneous tuning of nanogaps can be achieved simply and easily by just passing a field emission current through a large number of nanogaps connected in series.

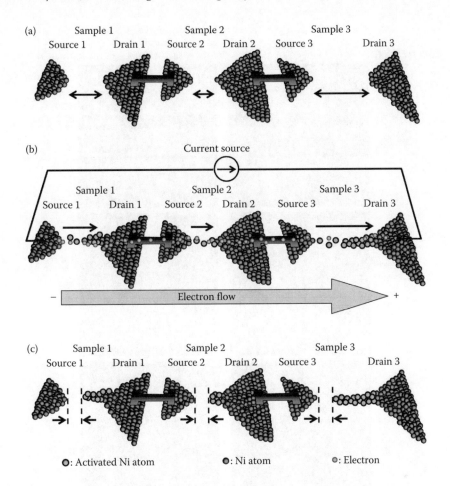

FIGURE 13.1 Schematic representation of the integration process during activation. (a) Three initial nanogaps with a separation of a few tens of nanometers are defined by conventional electron-beam lithography and lift-off process. (b) An identical field emission current passes through each device during the activation procedure. Then the sample with the wider initial gap separation is preferentially and immediately activated by the field emission current with a higher voltage drop. (c) Nanogaps having similar electrical properties can be simultaneously formed, because the gaps self-regulate with final spacing set by the applied current.

13.3 RESULTS AND DISCUSSION

13.3.1 Tuning of Structural and Electrical Properties of Integrated Nanogaps by Activation

The activation with a current source was performed to simultaneously control the multiple planar-type series-connected Ni nanogaps at room temperature. Scanning electron microscopy (SEM) images of samples A-1, A-2, and A-3 before performing the activation are shown in Figures 13.2a–c, respectively. In the SEM images, initial gap separation of the samples A-1, A-2, and A-3 is estimated to be approximately 114, 80, and 108 nm, respectively. The initial nanogaps showed high tunnel resistance above 100 TΩ, which clearly reflects the initial gap separation of the samples. On the other hand, after performing the activation with the preset current I_S of 30 μA, the separation of the gaps was reduced to less than 10 nm, which is due to the accumulation of Ni atoms at the tip of each drain electrode, as shown in Figures 13.2d–f.

Figure 13.3a shows the I_D–V_D properties of series-connected nanogaps (A-1, A-2, and A-3) during the total activation steps with the preset current I_S ranging from 1 nA (1st activation step) to

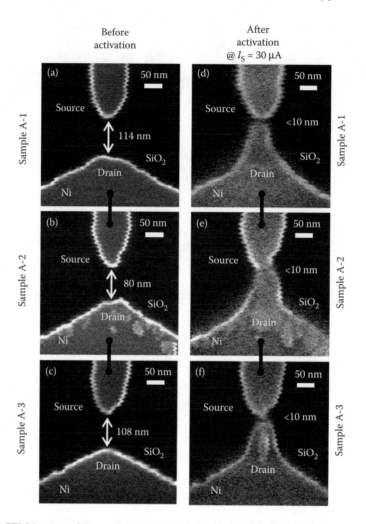

FIGURE 13.2 SEM images of three series-connected Ni nanogaps before (samples (a) A-1, (b) A-2, and (c) A-3) and after (samples (d) A-1, (e) A-2, and (f) A-3) performing the activation with the final preset current $I_S = 30\ \mu A$.

30 μA (8th activation step). From the figure, the remarkable decrease of the drain voltage V_D was clearly observed at the final stage of each activation step. This result indicates that at the final stage of activation, the activated atoms simultaneously moved from the source to the drain electrode in each sample along the electron flow and then caused the decrease of the separation of the gaps. The voltage V_S is defined as the voltage V_D at which the current I_D is the same as the preset current I_S, as shown in Figure 13.3b, representing the I_D–V_D properties of the 6th activation step with the preset current I_S of 3 μA.

Furthermore, Figure 13.4a shows the drain current I_D as a function of the duration of activation T_P in the total activation steps with the preset current I_S ranging from 1 nA (1st) to 30 μA (8th). The drain current I_D is seen to increase exponentially, while increasing the duration of each activation step, as shown in Figure 13.4a. Figure 13.4b also exhibits the relation between the resistance R_D, given by V_D/I_D, and the duration of activation T_P with the preset current I_S ranging from 1 nA (1st) to 30 μA (8th). It is observed that R_D slightly increases at the initial stage of each activation step as the duration of activation T_P increases. On the other hand, the value of the maximum resistance R_D in each activation step decreases with the progress of activation. This marked decrease in the

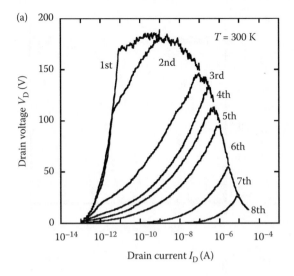

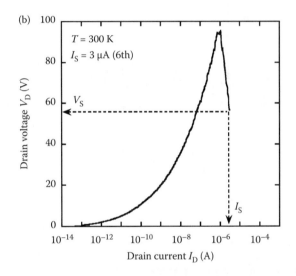

FIGURE 13.3 (a) I_D–V_D curves of series-connected three samples A-1, A-2, and A-3 during the total activation steps. (b) I_D–V_D properties of samples A-1, A-2, and A-3 during the 6th activation with the preset current $I_S = 3\ \mu A$. Voltage V_S is defined as drain voltage V_D at which drain current I_D is the same as the preset current I_S.

resistance R_D suggests that the formation of a hillock by the accumulation of Ni atoms is caused at the tip of the drain electrode, and then the gap separation is further reduced during the activation with each preset current.

The I_D–V_D characteristics of each sample after performing the activation with different preset currents $I_S = 1$ nA (1st), 500 nA (4th), and 30 μA (8th) at room temperature are displayed in Figure 13.5. When the preset current I_S was set to 1 nA, the I_D–V_D characteristics showed insulating-like I_D–V_D properties. On the other hand, the I_D–V_D curves of the simultaneously activated samples with the preset current $I_S = 500$ nA displayed nonlinear properties. From the figure, although the nanogaps before the activation (or even in the activation with $I_S = 1$ nA) showed insulating properties due to the wide separation of the initial gaps, the nonlinear I_D–V_D characteristics were obtained after the activation with the preset current $I_S = 500$ nA, which is strongly suggestive of the tunneling

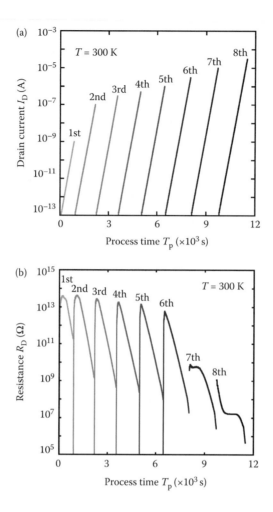

FIGURE 13.4 (a) Drain current I_D, passing through samples A-1, A-2, and A-3, as a function of the duration of activation T_P in the total activation steps with the preset current I_S ranging from 1 nA (1st) to 30 μA (8th). (b) Relation between the resistance R_D (= V_D/I_D) and the duration of activation T_P with the preset current ranging from 1 nA (1st) to 30 μA (8th) at room temperature.

properties of the simultaneously activated nanogaps. Additionally, the linear I_D–V_D properties were also observed when the preset current I_S was set to 30 μA. Consequently, the I_D–V_D properties of the three samples were simultaneously varied from "insulating" to "metallic" through "tunneling" properties by increasing the preset current. These results suggest that the electrical properties of the activated three nanogaps become similar to each other with the progress of the activation, and the accumulation of atoms within the gaps is significantly increased due to the higher preset currents.

Several kinds of nanogap resistances of the samples as a function of the preset current I_S during the activation are displayed in Figure 13.6. The tunnel resistances in the low-voltage regime of the three simultaneously activated nanogaps were defined as R_{A-1}, R_{A-2}, and R_{A-3}. The tunnel resistance of each nanogap was individually measured after the activation at room temperature, and decreases simultaneously ranging from 100 TΩ to 100 kΩ while increasing the preset current I_S from 1 nA (1st) to 30 μA (8th). This result indicates that the preset current I_S is a dominant parameter for simultaneous control of the tunnel resistance of the nanogaps. Moreover, a marked decrease in the tunnel resistance is clearly observed in the range of 100 nA < I_S < 1 μA. It is interpreted as a transition of the electrical properties of the nanogaps from the insulating regime to the tunneling regime.

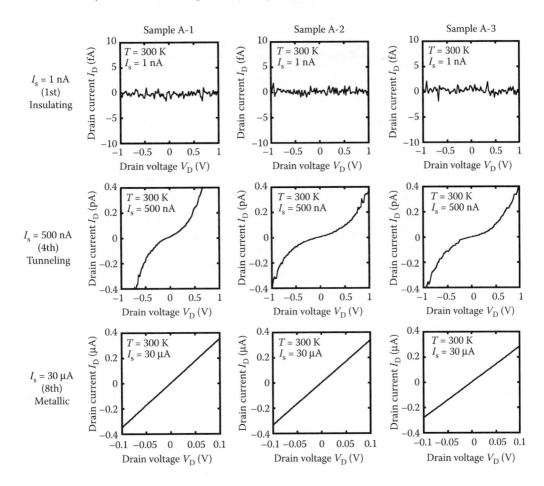

FIGURE 13.5 $I_D–V_D$ characteristics of samples A-1, A-2, and A-3 after performing the activation with different preset currents $I_S = 1$ nA (1st), 500 nA (4th), and 30 µA (8th) at room temperature.

These results suggest that the wide range of control of the electrical properties of the nanogaps connected in series is achieved by the activation procedure. Additionally, a resistance R_s given by V_S/I_S and a differential resistance obtained by dV_S/dI_S were determined during the activation at room temperature and clearly decreased by increasing the preset current I_S from 1 nA (1st) to 30 µA (8th), showing a linear relation with a slope of approximately −1. In this figure, the tunnel resistance of the nanogaps after the activation can be successfully predicted by a relation between "R_{A-1}, R_{A-2}, and R_{A-3} and R_s" or "R_{A-1}, R_{A-2}, and R_{A-3} and dV_S/dI_S." In other words, if we know the R_S or the dV_S/dI_S during the activation procedure, one can easily estimate the tunnel resistance of the nanogaps. This tendency is quite similar to that of individually activated nanogaps [19,20]. Furthermore, it should be noted that the tunnel resistance of simultaneously activated nanogaps was almost the same at each preset current. These results indicate that the simultaneous control of the electrical properties of multiple nanogaps connected in series is successfully achieved by the activation procedure, despite the differences in the samples having different initial nanogap separation distances.

13.3.2 Integration of SETs Using Activation

The integration method of two SETs based on nanogaps X–1 and X–2 using the activation process is illustrated in Figures 13.7a–e. In the first step, two initial Ni nanogaps with a separation of a few tens of nanometers are fabricated by using conventional EB lithography and lift-off techniques on

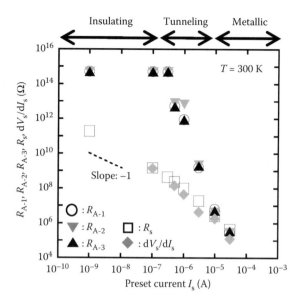

FIGURE 13.6 Dependence of resistances of the nanogaps on preset current I_S. The low-bias voltage tunnel resistances R_{A-1}, R_{A-2}, and R_{A-3}, the resistance $R_S(= V_S/I_S)$, and the differential resistance dV_S/dI_S were estimated from the activation measurements.

an SiO_2/Si substrate (Figure 13.7a). Next, the F–N field emission current is applied to the initial nanogaps connected in series, as shown in Figure 13.7b. At each nanogap, the migration of Ni atoms from the source (cathode) to the drain (anode) electrode is caused along the direction of the electron flow (Figure 13.7c). The accumulated Ni atoms within the gaps play a dual role of both the reduction of gap width and the formation of islands of SETs. Finally, two SETs are simultaneously integrated using the activation procedure (Figure 13.7d). The advantage of this fabrication method is that the SETs with similar electrical properties could be simultaneously formed because the magnitude of the field emission current becomes the same in each nanogap. Hence, it is strongly suggested that the integration of SETs could be easily achieved by passing a field emission current through many nanogaps connected in series, as shown in Figure 13.7e.

Figures 13.8a and b show the SEM images of samples B-1 and B-2, connected in series, before performing the activation, respectively. In these SEM images, initial gap separation of the samples B-1 and B-2 is estimated to be approximately 66 and 50 nm, respectively. Both of the initial nanogaps exhibited high tunnel resistance above 100 TΩ, reflecting the initial gap separation of the nanogap electrodes. On the other hand, the SEM images of both the simultaneously activated samples B-1 and B-2 with the final preset current $I_S = 10$ μA are shown in Figures 13.8c and d, respectively. From the images, the Ni atoms are accumulated in the gap between the source and the drain electrodes at each device, suggesting the formation of the Ni dot/cluster structures (islands) within the gap.

Figures 13.9a and b represent the I_D–V_D characteristics of the samples B-1 and B-2, respectively, which are simultaneously activated with the preset current $I_S = 1.5$ μA. For clarity, each I_D–V_D curve is shifted by 50 fA. In both devices, the gate voltage V_G was swept from −14 to 14 V by a 0.5 V step at 16 K. A Coulomb blockade voltage of each device was obviously modulated by the gate voltage V_G quasi-periodically. These characteristics suggest that the SETs consist of multiple islands. Modulation properties of the Coulomb blockade voltage (stability diagram) can be more clearly seen in contour plots of the drain current I_D of the devices as a function of the drain voltage V_D and the gate voltage V_G, as shown in Figures 13.9c and d. The charging energy and electrical current levels of both SETs were almost the same. Hence, the two SETs with similar electrical properties were successfully integrated by the activation, despite the difference of initial gap separation between the

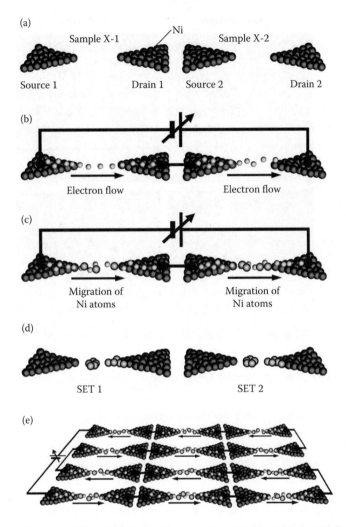

FIGURE 13.7   Sketched description of the integration process using the activation. (a) Two initial nanogaps with separation of a few tens of nanometers are defined by conventional electron beam lithography and lift-off process. (b) Field emission current passes through the two nanogaps connected in series. (c) Ni atoms at the tip of source electrodes, which are activated by the field emission current, are simultaneously migrated from the source (cathode) to the drain (anode). (d) Two SETs are simultaneously formed and integrated by the accumulation of the Ni atoms within the gap. (e) The concept of the large-scale integration of planar-type nanogap-based SETs using the activation procedure.

samples (66 nm in B-1 and 50 nm in B-2). As described above, the preset current during the activation procedure can tune the electrical properties of the devices. In particular, the tunnel resistance of the nanogaps and the charging energy of the SETs show the same tendency on the preset current of the activation in spite of the variation of the initial gap separation of individually activated nanogaps [19–21]. In other words, the tunnel resistance and the charging energy of the devices strongly depend on the preset current of the activation but hardly relate to the initial gap separation of the nanogaps [20,21]. Since the two nanogaps are connected in series, the identical field emission current passes through both devices during the activation procedure. Thus, the sample with wider initial gap separation is preferentially and immediately activated by the field emission current with a higher voltage drop. As a result, it is considered that the electrical properties of both SETs become similar to each other with the progress of the activation.

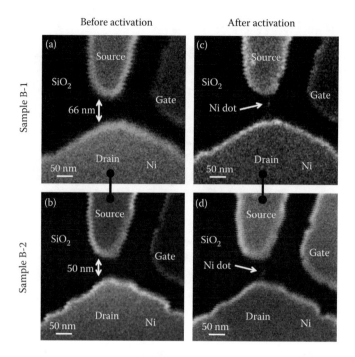

FIGURE 13.8 SEM images of two nanogaps connected in series before (samples (a) B-1 and (b) B-2) and after (samples (c) B-1 and (d) B-2) performing the activation with the final preset current $I_S = 10\ \mu A$.

Figure 13.10 represents the relation between the charging energy E_C of the two simultaneously activated SETs and the preset current I_S during the activation procedure. Opened circles and triangles indicate the samples B-1 and B-2, respectively. The charging effects of each device were estimated from the Coulomb blockade voltages of the I_D–V_D curves. As seen in Figure 13.10, the charging energy E_C of both simultaneously activated SETs was almost the same at each preset current I_S, ranging from 500 nA to 5 μA, where the samples show SET properties. As discussed above, since the preset current I_S in the activation procedure is a dominant parameter for controlling the electrical properties of the devices, the integration of the two SETs with similar electrical properties is achieved by passing an identical field emission current through the serially connected nanogaps. Furthermore, as the preset current I_S increases from 500 nA to 5 μA, the charging energy E_C of both SETs uniformly decreases from approximately 1000 to 150 meV. This tendency is quite similar to that of the individually activated SETs [21]. Therefore, if the activation procedure is applied to the series-connected nanogaps with an appropriate preset current, one can easily obtain the integrated SETs having similar electrical properties. Since the charging energy of the integrated SETs strongly relates to the preset current of the activation, the islands and tunnel barriers placed within the gap between the source and the drain electrodes are simultaneously tuned by the magnitude of the preset current. Hence, it is suggested that the simultaneous control of the charging energy of the integrated SETs is possible with the preset current of the activation. The results imply that the activation method is useful for the integration of nanogap-based SETs with similar electrical properties.

13.4 CONCLUSION

We performed simultaneous tuning of tunnel resistance of multiple nanogaps by a simple and easy method. This method is based on field-emission-induced electromigration and is called "activation." The control of the tunnel resistance of three nanogaps was simultaneously achieved by performing

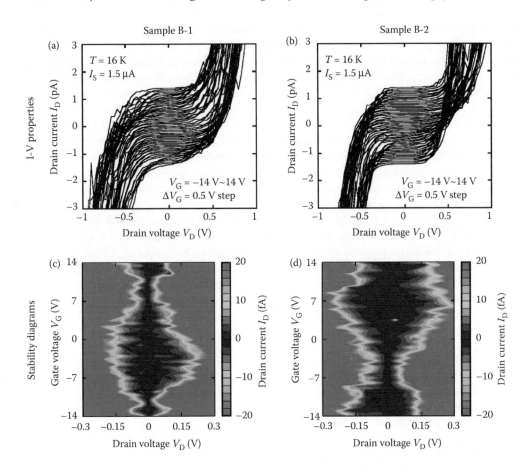

FIGURE 13.9 Current–voltage (I_D–V_D) characteristics (samples (a) B-1 and (b) B-2) and stability diagrams (samples (c) B-1 and (d) B-2) of two simultaneously activated devices (samples B-1 and B-2) with the preset current $I_S = 1.5\ \mu A$. In I_D–V_D properties, each curve is shifted by 50 fA for clarity, and gate voltage V_G is varied from −14 V (bottom) up to 14 V (top) with a step of 0.5 V.

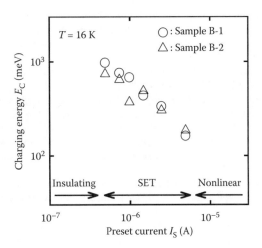

FIGURE 13.10 Charging energy E_C of two simultaneously activated SETs (samples B-1 and B-2) as a function of preset current I_S at 16 K.

the activation process for the series-connected nanogaps. The I_D–V_D properties of the nanogaps were simultaneously varied while increasing the preset current of the activation, indicating that the three nanogaps having similar electrical properties were successfully integrated by the activation process. Furthermore, the tunnel resistance of the simultaneously activated nanogaps decreased, ranging in the order of 100 TΩ to 100 kΩ, while increasing the preset current I_S from 1 nA to 30 μA. The gaps of the series combination of the three Ni nanogap electrodes can self-regulate with the final spacing set by the applied current. Consequently, the preset current in the activation procedure can simultaneously tune the tunnel resistance of the integrated nanogaps, in spite of the variation of the initial gap separation of the nanogap electrodes connected in series.

The integration of the SETs was achieved by simultaneously performing the activation procedure for the series-connected nanogaps. The I_D–V_D curves of the two simultaneously activated devices displayed Coulomb blockade properties. Additionally, the Coulomb blockade voltage of each device was clearly modulated by gate voltage V_G. The charging energy and electrical current levels of both SETs were almost the same. As a result, the two SETs with similar electrical properties were successfully integrated by the activation procedure. Moreover, the charging energy of both SETs decreased uniformly with an increase in the preset current. Hence, it is suggested that the simultaneous control of the charging energy of the integrated SETs is possible with the preset current of the activation. These results indicate that the activation procedure is suitable for the simultaneous control of the electrical properties of multiple nanogaps and allows us to integrate planar-type nanogap-based SETs simply and easily.

REFERENCES

1. Y. Yasutake, K. Kono, M. Kanehara, T. Teranishi, M. R. Buitelaar, C. G. Smith, and Y. Majima, Simultaneous fabrication of nanogap gold electrodes by electroless gold plating using a common medical liquid, *Appl. Phys. Lett.*, 91, 203107-1-3, 2007.
2. Y. Azuma, Y. Yasutake, K. Kono, M. Kanehara, T. Teranishi, and Y. Majima, Single-electron transistor fabricated by two bottom-up processes of electroless Au plating and chemisorption of Au nanoparticle, *Jpn. J. Appl. Phys.*, 49, 090206-1-3, 2010.
3. D. L. Klein, P. L. McEuen, J. E. B. Katari, R. Roth, and A. P. Alivisatos, An approach to electrical studies of single nanocrystals, *Appl. Phys. Lett.*, 68, 2574–2576, 1996.
4. W. Chen, H. Ahmed, and K. Nakazato, Coulomb blockade at 77 K in nanoscale metallic islands in a lateral nanostructure, *Appl. Phys. Lett.*, 66, 3383–3384, 1995.
5. E. M. Ford and H. Ahmed, Control of Coulomb blockade characteristics with dot size and density in planar metallic multiple tunnel junctions, *Appl. Phys. Lett.*, 75, 421–423, 1999.
6. Y. Nakamura, D. L. Klein, and J. S. Tsai, $Al/Al_2O_3/Al$ single electron transistors operable up to 30 K utilizing anodization controlled miniaturization enhancement, *Appl. Phys. Lett.*, 68, 275–277, 1996.
7. Y. Nakamura, C. D. Chen, and J. S. Tsai, 100-K operation of Al-based single-electron transistors, *Jpn. J. Appl. Phys.*, 35, L1465–L1467, 1996.
8. J. Shirakashi, K. Matsumoto, N. Miura, and M. Konagai, Room temperature Nb-based single-electron transistors, *Jpn. J. Appl. Phys.*, 37, 1594–1598, 1998.
9. J. Shirakashi, K. Matsumoto, N. Miura, and M. Konagai, Single-electron charging effects in Nb/Nb oxide-based single-electron transistors at room temperature, *Appl. Phys. Lett.*, 72, 1893–1895, 1998.
10. H. Park, A. K. L. Lim, A. P. Alivisatos, J. Park, and P. L. McEuen, Fabrication of metallic electrodes with nanometer separation by electromigration, *Appl. Phys. Lett.*, 75, 301–303, 1999.
11. S. I. Khondaker and Z. Yao, Fabrication of nanometer-spaced electrodes using gold nanoparticles, *Appl. Phys. Lett.*, 81, 4613–4615, 2002.
12. K. Tsukagoshi, E. Watanabe, I. Yagi, and Y. Aoyagi, The formation of nanometer-scale gaps by electrical degradation and their application to C_{60} transport measurements, *Microelectron. Eng.*, 73, 686–688, 2004.
13. K. I. Bolotin, F. Kuemmeth, A. N. Pasupathy, and D. C. Ralph, Metal-nanoparticle single-electron transistors fabricated using electromigration, *Appl. Phys. Lett.*, 84, 3154–3156, 2004.
14. K. Luo, D.-H. Chae, and Z. Yao, Room-temperature single-electron transistors using alkanedithiols, *Nanotechnology*, 18, 465203, 2007.

15. D. E. Johnston, D. R. Strachan, and A. T. Charlie Johnson, Parallel fabrication of nanogap electrodes, *Nano Letters* 7, 2774–2777, 2007.
16. S. L. Johnson, D. P. Hunley, A. Sundararajan, A. T. Charlie Johnson, and D. R. Strachan, High-throughput nanogap formation using single ramp feedback control, *IEEE Trans. Nanotechnol.*, 10, 806–809, 2011.
17. M. Araidai and K. Watanabe, Ab initio calculation of surface atom evaporation in electron field emission, *e-J. Surf. Sci. Nanotechnol.*, 5, 106–109, 2007.
18. P. Kumar and K. Sangeeth, Electric breakdown between nanogap separated platinum electrodes, *Nanosci. Nanotechnol. Lett.*, 1, 194–198, 2009.
19. S. Kayashima, K. Takahashi, M. Motoyama, and J. Shirakashi, Control of tunnel resistance of nanogaps by field-emission-induced electromigration, *Jpn. J. Appl. Phys.*, 46, L907–L909, 2007.
20. Y. Tomoda, K. Takahashi, M. Hanada, W. Kume, and J. Shirakashi, Fabrication of nanogap electrodes by field-emission-induced electromigration, *J. Vac. Sci. Technol. B*, 27, 813–816, 2009.
21. W. Kume, Y. Tomoda, M. Hanada, and J. Shirakashi, Fabrication of single-electron transistors using field-emission-induced electromigration, *J. Nanosci. Nanotechnol.*, 10, 7239–7243, 2010.
22. S. Ueno, Y. Tomoda, W. Kume, M. Hanada, K. Takiya, and J. Shirakashi, Integration of single-electron transistors using field-emission-induced electromigration, *J. Nanosci. Nanotechnol.*, 11, 6258–6261, 2011.
23. S. Ueno, Y. Tomoda, W. Kume, M. Hanada, K. Takiya, and J. Shirakashi, Field-emission-induced electromigration method for the integration of single-electron transistors, *Appl. Surf. Sci.*, 258, 2153–2156, 2012.

14 High-Resistive Tunnel Junctions for Room-Temperature-Operating Single-Electron Transistors Fabricated Using Chemical Oxidation of Tungsten Nanoparticles

P. Santosh Kumar Karre, Daw Don Cheam,
Manoranjan Acharya, and Paul L. Bergstrom

CONTENTS

14.1 INTRODUCTION

Single-electron transistor (SET) devices are of immense interest because of their potential for low-power operation and high integration density. The basic building blocks of a SET device are tunnel junctions with very small capacitance that are connected in series with a central conducting island. Charge conduction occurs through two tunnel junctions that are connected to the source and drain terminals of the device. As the dimensions of the conducting island reduce to a few nanometers, the electronic wave functions in the conducting island become discrete. The discrete charge on the conducting island is accessed by tunneling through source and drain terminals. The central conducting island is capacitively coupled to a gate terminal, which is used to control the charge on the island. SET devices operate on the principle of a Coulomb blockade, which is the suppression of electron tunneling below a threshold voltage; the capacitance of the tunnel junction system dictates the threshold voltage, whereas a Coulomb blockade is observed when the device dimensions are reduced to a few nanometers [1]. SET devices can be realized with semiconducting or metallic systems. This chapter looks at metallic systems where tungsten and tungsten oxide are used. A pictorial view of a double-junction SET device is shown in Figure 14.1, where the conducting island is

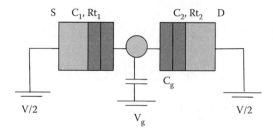

FIGURE 14.1   Pictorial view of the double-junction SET device.

sandwiched between the two tunnel junctions that are connected to the source and drain terminals. One tunnel junction is characterized by the tunnel resistance Rt_1 and capacitance C_1, whereas the other tunnel junction is characterized by the resistance Rt_2 and capacitance C_2, and the capacitance of the gate coupled to the central conducting island is C_g. The energy required to add an electron to the conducting island is the charging energy, and the external applied bias provides energy for tunneling. If the applied bias is greater than the charging energy, it results in electron tunneling.

The ambient thermal temperature also provides energy for tunneling; if the charging energy is less than the thermal noise, the Coulomb blockade events are dissipated. For room-temperature operations of SET devices, the charging energy should be greater than the thermal noise. The Coulomb blockade events are observable at room temperature if discrete energy levels can be maintained in the central conducting island; the high tunnel resistance of the tunnel junctions helps to confine the electronic wave function and results in the room-temperature operation of these devices. There are two important constraints on SET device operations: the capacitance associated with the central island should be very small and the tunnel junction resistance should be very large [2]. For the room-temperature operation of SET devices, the confinement of electrons to the central conducting island is a necessary requirement. For the room-temperature operation of SET devices, the conducting island dimensions should be below 10 nm for the capacitance to be favorable. Technologically, the fabrication of sub-10 nm features is a challenge. Many approaches have been used in the past for the realization of SET devices; we present the fabrication of SET devices based on focused ion beam (FIB) techniques. This work looks at a multijunction SET system with tungsten as a conducting island and tungsten oxide as a tunnel junction. An FIB system is used for the fabrication of a SET device, and both FIB etching and deposition have been investigated. The fabricated multijunction SET devices show a room-temperature Coulomb blockade. The key to the realization of room-temperature-operating SET devices is the fabrication of sub-10 nm nanoparticles in tungsten using FIB deposition and the chemical oxidation of tungsten in peracetic acid to realize high-resistive tunnel junctions. Based on the oxide thickness of the tungsten oxide, the barrier height of the fabricated tunnel junctions were in the range of ~1.5–2.5 eV. The device characteristics for the multijunction SET system are presented showing a clear Coulomb blockade with the charging energies close to 160.0 meV. The SET device is a very sensitive electrometer; the charge on the gate electrode of the SET device can be used to modulate the device characteristics; thus, the SET device can be used as a sensor. The charge on the gate electrode can be modulated using a chemical, optical, electrical, or any other transduction mechanism. The current work looks at a SET device as a gas sensor.

14.2   FOCUSED ION BEAM-BASED DEVICE FABRICATION

FIB technologies have been very useful in the realization of nanostructures for nanoelectron devices [3,4]. An FIB system is very similar to a scanning electron microscope (SEM) in operation but uses a focused beam of Ga^+ ions. It was used in this work for the selective modification of the sample surface. FIB systems can be used for etching and deposition; both the techniques have been investigated for SET device fabrication. FIB etching resulted in a minimum island size of 40 nm. Conducting

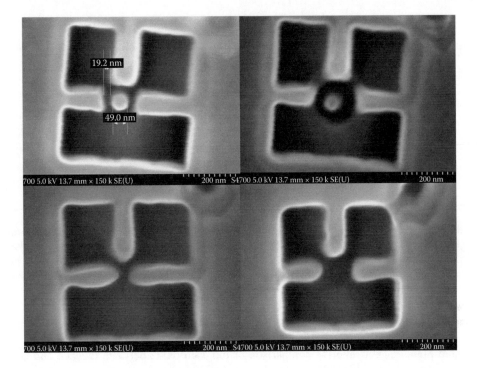

FIGURE 14.2 Minimum island size with FIB etching.

islands below 40 nm were not feasible to be fabricated using FIB etching, and the interaction of beam energies during FIB etching for islands below 40 nm resulted in the evaporation of the island's material being etched as shown in Figure 14.2 [5]. FIB deposition uses a precursor gas that is chemically adsorbed on the sample surface, and the energy for the chemical reaction is provided by the FIB beam energy, resulting in the desired material being deposited on the sample surface and the other gases, the by-products, being removed by the FIB vacuum system. FIB deposition was used for the successful deposition of nanoislands of sub-10 nm dimensions for the realization of the SET devices.

A silicon substrate having a 300 nm-thick Al_2O_3 passivating film deposited using a Perkin-Elmer 2400-8J parallel plate radio frequency (rf) sputtering system was used. Nanoislands were randomly deposited on the sample surface. The tungsten nanoislands were fabricated using FIB deposition in a Hitachi FB-2000A system [6]. The nanoislands were deposited randomly in an area of 16 μm × 16 μm on an Al_2O_3 thin film. The average diameter of the tungsten nanoislands was 8 nm with an average height of 3.3 nm. To reduce the size of the islands further, the sample was subjected to a brief chemical etch in a 1:1 sulfuric peroxide solution (piranha etch). The samples were rinsed in deionized water and then introduced into a solution of peracetic acid, which is a mixture of glacial acetic acid and hydrogen peroxide in a 1:1 volume ratio. Peracetic acid, being a strong oxidizing agent, was considered for the oxidation of tungsten. A pictorial view of the deposited nanoislands and the chemically oxidized tungsten islands is shown in Figure 14.3.

Thermal oxidation of tungsten nanoislands was considered but the higher temperatures involved in the thermal oxidation of the tungsten nanoislands were found to increase the nanoisland size, thus degrading the room-temperature performance of the SET device. The initial oxidation of tungsten nanoislands was performed by thermal oxidation; the high temperature of ~900°C in the process led to an increase in the island size. An increase in the island size is not desirable for room-temperature operation of a SET device; the capacitance of the SET device increases with an increase in the island size. The oxidation of tungsten nanoislands needs to be performed at lower temperatures, which help retain a lower island size. The chemical oxidation

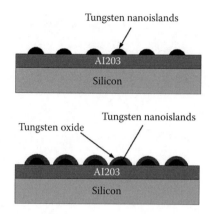

FIGURE 14.3 Pictorial view of the fabricated tungsten nanoislands and the chemically oxidized tungsten nanoislands forming tungsten oxide tunnel junctions. (P. S. K. Karre et al., High resistive tunnel junctions for the room temperature operating single-electron transistor fabricated using chemical oxidation of tungsten nano particles, *Digest IEEE Int'l Conference on Nanotechnology (Nano 2011)*, August 13–18, Portland, OR, pp. 1456–1459, © (2011) IEEE. With permission.)

of tungsten using peracetic acid was the best alternative in maintaining the critical dimensions of the tungsten nanoislands to realize the room-temperature operation of the SET devices. The fabricated tunnel islands and the comparison of tungsten oxide between thermal and chemical oxidation can be seen in Figure 14.4.

Tungsten oxide acts as tunnel junctions for the device. The connecting nanowires from the source, drain, and gate terminals were also fabricated using FIB deposition, and after the formation of tunnel junctions, the nanowires were deposited in tungsten connecting the active area of the device to the source, drain, and gate terminals. To estimate the thickness of tungsten oxide formed around the tungsten nanoislands during the chemical oxidation process, a cross section of the sample was polished using electromechanical polishing down to 1 μm smoothness. The nanoislands were deposited close to the edge of the polished wafer surface. The oxidation time of the as-deposited tungsten nanoislands was varied from 1 to 30 min. The samples of tungsten nanoislands deposited using FIB deposition were introduced into peracetic acid after a quick dip in piranha solution and then oxidized in peracetic acid for different time intervals from 1 to 30 min. The samples were rinsed in deionized water and dried in a nitrogen ambient. The cross section of the sample was milled using FIB etching and observed using a field emission scanning electron microscope (FE-SEM) to estimate the thickness of the tungsten oxide. A cross section of one of the samples is shown in the SEM micrograph in Figure 14.5, showing the FIB milled cross section of the sample with tungsten nanoislands and the deposited tungsten oxide in the series of SEM micrographs.

The SEM micrograph in Figure 14.6 shows the source, drain, and gate terminals of the fabricated multijunction SET device. The dimensions of the source drain and gate terminals are $100 \times 100 \ \mu m^2$ and the connecting leads are 50 μm in length and 250 nm in width. All the connecting nanostructures for the device are fabricated using FIB deposition of tungsten. The deposition parameters were different for the connecting nanoleads, nanoislands and the source, drain, and gate terminals with different beam currents, and beam apertures to realize the differences in the nanostructures in the as-deposited films. The conducting nanoislands of tungsten were deposited in an area of $16 \ \mu m \times 16 \ \mu m$ on the Al_2O_3 thin film at the center of the device with a beam having a beam current of 2.0–3.5 nA as shown in the exploded view of the SEM micrograph. The nanoislands are randomly deposited between the source and drain terminals. The geometry of the device allows 10 nanoislands between the source and drain terminals along the source and drain terminals, thus making the device a multijunction SET system. The average height of the deposited nanoislands is 3.3 nm as shown in the atomic force microscopy (AFM) micrograph of the as-deposited nanoislands

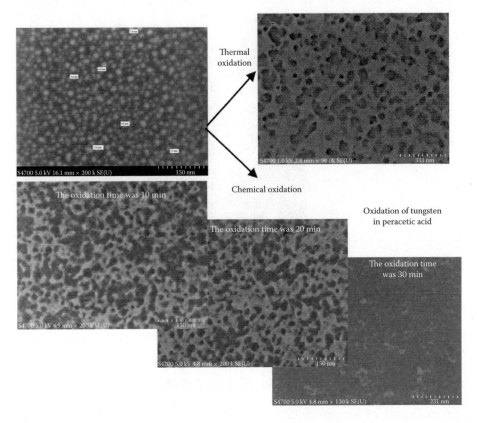

FIGURE 14.4 Oxidation of tungsten nanoislands using chemical oxidation in peracetic acid. (P. S. K. Karre et al., High resistive tunnel junctions for theroom temperature operating single-electron transistor fabricated using chemical oxidation of tungsten nano particles, *Digest IEEE Int'l Conference on Nanotechnology (Nano 2011)*, August 13–18, Portland, OR, pp. 1456–1459, © (2011) IEEE. With permission.)

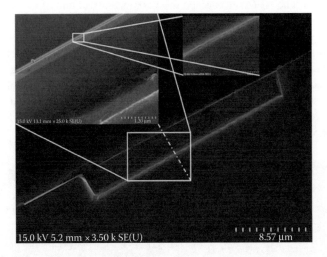

FIGURE 14.5 Cross section of one of the samples showing the tungsten nanoislands and the tungsten oxide formed after the chemical oxidation in peracetic acid. (P. S. K. Karre et al., High resistive tunnel junctions for theroom temperature operating single-electron transistor fabricated using chemical oxidation of tungsten nano particles, *Digest IEEE Int'l Conference on Nanotechnology (Nano 2011)*, August 13–18, Portland, OR, pp. 1456–1459, © (2011) IEEE. With permission.)

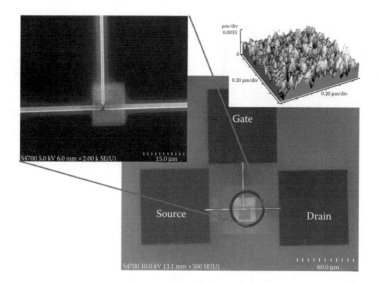

FIGURE 14.6 SEM micrograph of the fabricated STE device, active area, and the as-deposited tungsten nanoislands. (P. S. K. Karre et al., High resistive tunnel junctions for theroom temperature operating single-electron transistor fabricated using chemical oxidation of tungsten nano particles, *Digest IEEE Int'l Conference on Nanotechnology (Nano 2011)*, August 13–18, Portland, OR, pp. 1456–1459, © (2011) IEEE. With permission.)

of tungsten using FIB deposition. The nanoislands were deposited last after the deposition of probing pads and the connecting nanoleads of the device. The fabricated SET structure was oxidized in peracetic acid to form the tungsten oxide; the tungsten oxide was formed on top of the nanoislands, reducing the height and size of the as-deposited tungsten nanoislands as well as the entire device. A blanket thin film of 30 nm Al_2O_3 was deposited as a passivating layer on top of the fabricated SET device; FIB etching was used to open up the underlying source, drain, and gate terminals by timed FIB etching of the passivating Al_2O_3 layer using a beam having a beam current of 4.0–8.0 nA for an etch time of 10 min. The AFM micrograph shows the nanostructure of the as-deposited tungsten nanoislands using FIB deposition in Figure 14.6. The SET device fabrication with the FIB deposition process is a slow process, and to demonstrate a progression toward large-scale production of these devices, step and flash imprint lithography (SFIL) was successfully utilized for the fabrication of the connecting nanowires from the source, drain, and gate terminals coming into the active area of the device. The active area of the device is fabricated using FIB deposition of the nanoislands in tungsten. The fabricated SET devices using SFIL and FIB deposition show clear Coulomb blockade properties at room temperature. The micrograph in Figure 14.7 clearly shows the SFIL fabricated structures with the SET devices [8].

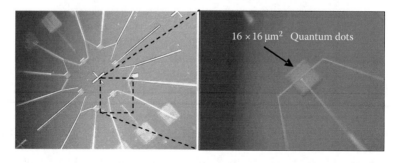

FIGURE 14.7 Micrograph showing the SFIL and FIB deposited SET devices.

14.3 RESULTS AND DISCUSSION

The tungsten oxide thickness for different oxidation times was estimated using the FE-SEM measurements from a cross section of the sample. The EDS and AFM methods were also used to characterize the tungsten oxide thin films. The thickness of the tungsten oxide increased from 2.3 nm for a 60 s oxidation to 9.6 nm for an oxidation time of 360 s and decreased thereafter. The thickness of the tungsten oxide increased with the oxidation time initially; as the tungsten became more hydrated and soluble, it was found to dissolve in the hydrogen peroxide present in the peracetic acid [9,10], thus reducing the thickness of the tungsten oxide and the tungsten after 360 s. The tungsten nanoislands were completely oxidized and dissolved for an oxidation time of 20 min and more. All the samples that were oxidized had a starting average island size of ~8 nm.

The tungsten oxide tunnel junctions were characterized by fabricating the SET devices and extracting the junction parameters from the device characteristics. The device characteristics of the room-temperature-operating SET device were obtained using the Keithley 4200-SCS semiconductor characterization system coupled with a Micromanipulator 8000 series probe station. The device characteristics showed a clear Coulomb blockade and Coulomb oscillations in the differential conductance

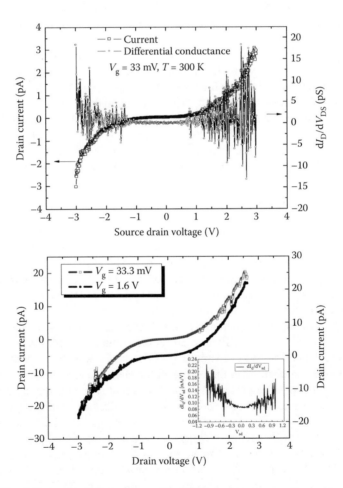

FIGURE 14.8 The *I–V* characteristics of a working SET device at room temperature showing a clear Coulomb blockade for different gate voltages. (P. S. K. Karre et al., High resistive tunnel junctions for the-room temperature operating single-electron transistor fabricated using chemical oxidation of tungsten nano particles, *Digest IEEE Int'l Conference on Nanotechnology (Nano 2011)*, August 13–18, Portland, OR, pp. 1456–1459, © (2011) IEEE. With permission.)

of the device as shown in Figure 14.8. The characteristics for the two different devices are shown where the source and drain characteristics for different gate voltages are investigated. The measurements were performed at a room temperature of 23°C. The source drain voltage of the device was varied from –3.0 to 3.0 V for a fixed gate voltage of 33.0 mV and 1.6 V. The drain current of the SET device was found to be of the order of a few pA, showing a clear Coulomb blockade and Coulomb oscillations in the differential conductance of the device; the single-electron effects were observed in multiple devices showing that the SET device characteristics were reproducible across multiple devices. The drain current shows a variation in the blockade voltage based on an applied gate bias. A fine structure in the nonlinear device characteristics was seen at higher source and drain voltages due to the differences in the actual island size of the individual conducting islands and the oxide thickness resulting from these variations, thus making the tunnel junctions asymmetrical in nature.

It is estimated from the geometry of the device that there are 10 nanoislands participating in the conduction process, thus making the device a 20-junction SET system. Assuming that all the tunnel junctions are similar and have the same capacitance and resistance, the charging energy of the device can be calculated using the Orthodox theory [11,12] that is applied to a multijunction SET system. The device parameters were extracted from the (I–V) characteristics of the device, and the resistance of the tunnel junctions was found to be ~29 GΩ, which is much larger than the quantum of resistance (~26 KΩ). The effective capacitance of the device was found to be 0.499 aF. The capacitance of the individual junction was estimated to be 0.947 aF, and the charging energy of the device was found to be 160.0 meV. The threshold voltage of the device was estimated on the basis of the effective capacitance of the device ~0.4 V, but the actual threshold voltage of the device can be seen to be ~0.7–1.5 V. This difference in the threshold voltage is due to the fact that all the conducting islands are nonuniform and contribute to the differences in the actual threshold voltage and the estimated threshold voltage, where all the junctions were considered to be identical for the device parameter extraction. The barrier height of the tunnel junctions was extracted from the conductance of the device by using the Brinkmann fit method [13]. The barrier height of the tunnel junctions was found to vary from ~2.3 to ~1.3 eV for thicknesses of ~2 to ~9 nm, respectively.

14.4 SET AS GAS SENSOR

The SET device has been demonstrated to be an N_2O gas sensor, where the sensing mechanism is based on the charge modulation on the gate electrode due to the interaction of the incoming gas species with the surface complexes on the gate electrode. The gas sensing is expected to be mediated through ionosorption [14] of the gas molecules on the surface of the gate terminal. Figure 14.9 shows the proposed sensing mechanism of the SET gas sensor, where the incoming gas molecules are adsorbed on the gate surface, facilitating charge transfer [15].

The gate terminal is made of tungsten using the FIB deposition process. The nonlinear Coulomb blockade along with the Coulomb oscillations are shown in Figure 14.10; the blockade length (the voltage range where the current is not increasing) of the device was ~6.0 V without the introduction of N_2 gas into the chamber. With the N_2 gas in the chamber, the blockade length reduced to ~2.0 V. The reduction in the blockade length of the device is attributed to the increase in the overall capacitance of the device due to the adsorption of the gas molecules on the gate surface. The drain current of the device was monitored as a steady stream of N_2 gas was injected into the flow cell; after the injection of the steady stream of N_2 gas, a constant flow of N_2O gas was injected, which reduced the drain current of the device. The presence of the incoming gas molecules can be sensed by the reduction of the drain current of the device and also by the reduction of the Coulomb blockade length of the device. The sensing mechanism was through ionosorption; the native amorphous tungsten oxide surface of the gate pad has an interatomic distance of 1.79 Å between tungsten and oxygen [16]. The number of available tungsten oxide sites for the ionosorption of the incoming gas on a 100 μm × 100 μm pad was calculated to be approximately 7.8×10^{10}. The stable isotopic configuration expected for N_2O is N–N–O [17], and the number of electrons transferred between N_2O

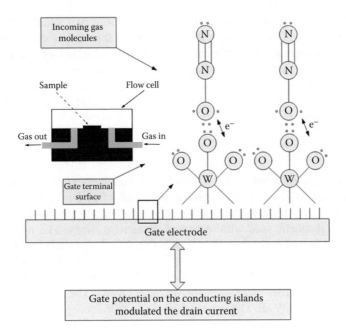

FIGURE 14.9 Proposed ionosorption mechanisms resulting in charge transfer in the SET device.

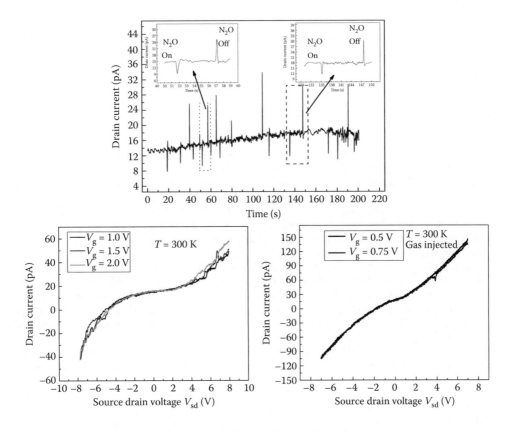

FIGURE 14.10 Coulomb blockade changes with and without gas injection.

and the WO_3 surface was calculated to be approximately 1.56×10^{11}, which is in close agreement with the measured drop in the drain current. Depending on the surface state of the gate terminal and the incoming gas molecule, the electrons can be transferred to or from the gate terminal. The removal of electrons from the gate terminal induces a positive bias on the gate, resulting in a downward spike in the drain current, and vice versa. The SET device characteristics are modulated on the basis of the charge transfer on the gate terminal. The SET device has also been demonstrated to be a photondetector using bacteriorhodopsin [18].

14.5 CONCLUSIONS

A novel approach was developed in the fabrication of high-resistive tunnel junctions for room-temperature-operating SETs using the chemical oxidation of tungsten nanoislands in peracetic acid. The tunnel junction thickness was found to increase and then decrease due to the dissolution of tungsten in peracetic acid with increased dissolution times. The tunnel junctions were estimated to have a tunnel resistance of ~29 GΩ; the SET devices fabricated with these large tunnel-resistant tunnel junctions showed a clear Coulomb blockade and Coulomb oscillations at room temperature. The fabricated tunnel junctions had an estimated barrier height in the range of ~2.3 to ~1.3 eV for thicknesses of ~2 to ~9 nm, respectively. The effective capacitance of the device was found to be 0.499 aF. The capacitance of the individual junction was estimated to be 0.947 aF, and the charging energy of the device was found to be 160.0 meV. The SET device was used as a gas sensor demonstrating the sensing of N_2O gas with the SET device and an optical sensor using bacteriorhodopsin.

ACKNOWLEDGMENT

The research reported in this document was performed in connection with contract DAAD17-03-C-0115 with the U.S. Army Research Laboratory.

REFERENCES

1. H. Grabert and M. H. Devaret, *Single Charge Tunneling, Coulomb Blockade Phenomena in Nanostructures*, Plenum, New York, 1992.
2. K. K. Likharev, Single-electron devices and their applications, *Proc. IEEE 87*, 606, 1999.
3. D. Petit aid, D. Wood, R. P. Cowburn, C. C. Faulkner, and S. Johnstone, Nanometer scale patterning using focused ion beam milling, *Rev. Sci. Instr.*, 76, 026105, 2005.
4. M. Nakayama, F. Wakaya, J. Yanagisawa, and K. Gamo, Focused ion beam etching of resist/Ni multilayer films and applications to metal island structure formation, *Journal of Vacuum Science & Technology B: Microelectronics and Nanometer Structures*, 16(4), 2511, 2514, Jul 1998. doi: 10.1116/1.590200.
5. P. S. K. Karre, P. L. Bergstrom, G. Mallick, and S. P. Karna, Fabrication of quantum islands for single electron transistors using focused ion beam technology, *Proc. IWPSD'05: Thirteenth Int. Workshop on the Physics of Semiconductor Devices*, 2005.
6. P. S. K. Karre, P. L. Bergstrom, G. Mallick, and S. P. Karna, Room temperature operational single electron transistor fabricated by focused ion beam deposition, *J. Appl. Phys.*, 102, 024316, 2007.
7. P. S. K. Karre, D. D. Cheam, M. Acharya, and P. L. Bergstrom. High resistive tunnel junctions for the room temperature operating single electron transistor fabricated using chemical oxidation oftungsten nano particles, *Digest IEEE Int'l Conference on Nanotechnology (Nano 2011)*, 13–18 August 2011, Portland, OR, pp. 1456–1459.
8. D. D. Cheam, P. S. K. Karre, M. Palard, and P. L. Bergstrom, Step and flash imprint lithography for quantum dots based room temperature single electron transistor fabrication, *Microelectron. Eng.*, 86(4–6), 646–649, 2009.
9. G. S. Kanner and D. P. Butt, Raman and electrochemical probes of the dissolution kinetics of tungsten in hydrogen peroxide, *J. Phys. Chem. B*, 102, 9501, 1998.
10. B. Reichman and A. J. Bard, The electrochromic process at WO_3 electrodes prepared by vacuum evaporation and anodic oxidation of W, *J. Electrochem. Soc.*, 126, 583, 1979.

11. J. P. Pekola, K. P. Hirvi, J. P. Kauppinen, and M. A. Paalanen, Thermometry by arrays of tunnel junctions, *Phys. Rev. Lett.*, 73, 2903, 1994.
12. K. P. Hirvi, J. P. Kauppinen, A. N. Korotkov, M. A. Paalanen, and J. P. Pekola, Arrays of normal metal tunnel junctions in weak Coulomb blockade regime, *Appl. Phys. Lett.*, 67, 2096, 1995.
13. W. F. Brinkman, R. C. Dynes, and J. M. Rowell, Tunneling conductance of asymmetrical barriers, *J. Appl. Phys.*, 41(5), 1970.
14. M. J. Madou and S. R. Morrison, *Chemical Sensing with Solid State Devices*, Academic Press Inc, Boston, MA, 1989.
15. P. S. K. Karre, M. Acharya, W. R. Knudsen, and P. L. Bergstrom, Single electron transistor based gas sensing with tungsten nano particles at room temperature, *IEEE Sensors J.*, 8(6), 797–802, 2008.
16. F. Wang and R. D Harcourt, Electronic structure study of the N_2O isomers using post-Hartree-Fock and density functional theory calculations, *J. Phys. Chem. A.* (Article) 104(6), 1304–1310, 2000.
17. T. Pauporte, Y. Soldo-Olivier, and R. Faure, XAS study of amorphous WO_3 formation from a peroxo-tungstate solution, *J. Phys. Chem. B.* (Article) 107(34), 8861–8867, 2003.
18. K. A. Walczak, M. Acharya. D. Lueking, P. L. Bergstrom, and C. Friedrich, Integration of the bion-anomaterial bacteriorhodopsin and single electron transistors, *Proceedings of the 26th Army Science Conference*, Paper #MP-06, December 2008, Orlando, FL, USA. 4 p.

15 Axon-Inspired Communication Systems

Valeriu Beiu, Liren Zhang, Azam Beg, Walid Ibrahim, and Mihai Tache

CONTENTS

15.1 INTRODUCTION

The latest ITRS (International Technology Roadmap for Semiconductors) [1] has identified power consumption as a grand challenge for nanoelectronics. A related daunting task is represented by the passive wires (interconnects), see [2–8]. They are the ones driving the power/energy consumption, as their number increases exponentially (each device has to be connected by several wires to other devices, and the number of devices is following the exponential Moore's law), while their scaling is limited (as their parasitic capacitances and RC-delays are not scaling in sync with the devices). In fact, the huge processing potential offered by parallel architectures will not be realized satisfactorily without addressing the interconnection challenge which is intimately entangled with power and energy consumption. During the 2011 International Solid-State Circuits Conference, power consumption has been a leitmotif [9]. Dr. Oh-Hyun Kwon, the president of Samsung Electronics, has stated the need *"to reduce power consumption by a compound annual rate of 20% over the next decade,"* while Professor Asad Abidi (University of California, Los Angeles) has mentioned that the power efficiency challenge *"requires radical changes in the way we do things."* One of the radical approaches—advocated by Dr. Jack Sun, Vice President for Research and Development and Chief Technology Officer of TSMC, to solve the challenges associated with power constraints—the chip industry could envisage is to *"get some inspiration from the amazing human brain."* On the one hand, structures in the brain are characterized by massive interconnections, but, contrary to common thought, most of these are local with an amazingly sparse global interconnect network. On the other hand, cortical neurons possess an average of 8000 (up to a maximum of 100,000) synaptic connections, which differentiate them significantly from present-day CMOS (complementary metal–oxide–semiconductor) gates having an average of four inputs only. With on the order of 10^{10} neurons in the human cortex, and about 10^{14} synaptic connections, the brain is inspirational for the future of interconnects (supporting massively parallel multi-/many-cores) due to its ultra-low-power consumption (about 20 W) and extreme reliability. This is because the neurons themselves are able to communicate (dendritic and axonal communications) at quite large distances on a very limited power budget.

If we consider an older 1.0 μm technology (using an Al and SiO$_2$ dielectric), the transistor delay was 20 ps, and the RC delay of a 1 mm line was 1 ps. In a 32 nm technology (using Cu and low-k dielectric), the transistor delay is about 1 ps, and the RC delay of a 1 mm line is around 250 ps. In addition, up to 80% of microprocessor power is being consumed (wasted) by interconnects [10]. This communication challenge [11] is only now starting to be fully appreciated as it is bridging the power and the reliability challenges, as on-chip communications are getting more and more power hungry and less and less reliable (i.e., more and more sensitive to noise and variations). Just like the wires connecting the devices inside an integrated chip, the connections between neurons occupy a large fraction of the total volume, and the "wires" (dendrites and axons) consume energy during signaling. In fact, although the human brain represents only 2% of the total body weight, it consumes 20% (going up to 40% for kids) of its resting energy. Impressive advances in brain scanning have evolved from the early positron emission tomography—PET (developed about 20 years ago), to magnetic resonance imaging—MRI (developed about 10 years ago). Variations include functional MRI (fMRI), blood oxygen level-dependent (BOLD) MRI, and diffusion MRI among others. These have led to a very precise mapping of the brain's functional areas and of their connectivities (see Figure 15.1). The rapidly developing diffusion tensor MRI (DT-MRI) offers the means of probing the structural arrangement of tissue at the cellular level, and has already been used to estimate the brain connectivity. Such techniques can be used to map the path of (white matter) tracts in the brain *in vivo* noninvasively, using the so-called tractography methods.

All of these results are supporting the scarcity of the long interconnects in the brain. The detailed map of the full set of neurons and synapses within the nervous system of an organism is known as a connectome (http://en.wikipedia.org/wiki/Connectome), and the National Institutes of Health is supporting the Human Connectome Project (http://www.humanconnectomeproject.org/), which started in 2011. In fact, the first comprehensive attempt to reverse-engineer the mammalian brain was started in 2005 as the EPFL/Blue Brain Project (http://bluebrain.epfl.ch/), and impressive simulation results of neocortical columns (about 10,000 biologically accurate individual neurons) have already been obtained (see Figure 15.2). The expectation is that the Blue Brain Project will be expanded and continued under the EU's FET Flagship program as the Human Brain Project (http://www.humanbrainproject.eu/), a fact which has just been announced in January 2013 (http://www.nature.com/news/brain-simulation-and-graphene-projects-win-billion-euro-competition-1.12291).

Transferring information inside the brain is in fact achieved by both electrical and chemical processes. Inside one neuron (see Figure 15.3a), information is transmitted along the dendrites and axons by electrical pulses supported by the active membrane—and in particular by the voltage-gated ion channels. The gaps between neurons are known as synaptic clefts, and represent obstacles for these pulses. Chemicals called neurotransmitters are used to convey information over these gaps.

Neurons themselves are isolated by semipermeable membranes which act as barriers to particular types of ions. The movement of ions through a membrane is controlled by many different ion

(a) (b)

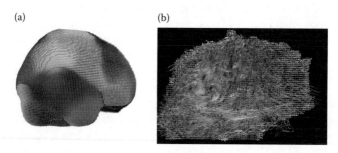

FIGURE 15.1 The brain: (a) Tensor visualization with MedINRIA (from http://www-sop.inria.fr/asclepios/ software/MedINRIA/screenshoots/); (b) HARDI (high angular resolution diffusion imaging) tracks showing information transfer between the two hemispheres (from http://www.humanconnectomeproject.org/wp-content/uploads/2011/06/cing7-e1307569111895-720x476.jpg).

(a)   (b)

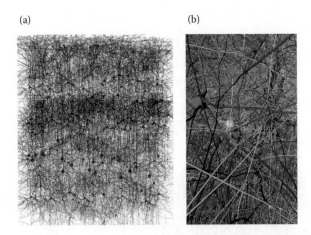

FIGURE 15.2   A cortical column: (a) Layers and their connectivity (©BBP/EPFL, from http://bluebrain.epfl.
ch/); (b) Simulated view inside a cortical column (©BBP/EPFL, from http://bluebrain.epfl.ch/).

channels. It is this movement that leads to different ion concentrations inside and outside a neuron,
which in turn creates a difference of potential called membrane potential. Normally, the concentration of Na^+ is higher outside a neuron, while the concentration of K^+ is higher inside the neuron.
Besides ion channels, which are normally closed most of the time and open only very briefly, the
Na^+/K^+ pumps are the ones responsible for maintaining these ion concentrations.

Trying to find inspiration from the neurons in general, and from the way they communicate in
particular, this chapter will focus on the axon-inspired communication (i.e., at long distances and on
a limited power budget), analyzing the dense and locally connected arrays of voltage-gated (nonlinear) ion channels [12–15]. The resulting arrays of the voltage-gated ion channels will be simulated
at the logical level using single-electron technology/transistors (SETs) [16]. Such an approach should
lead to practical power/energy lower bounds for nanoelectronics.

15.2   AXON COMMUNICATION

In the early 1950s, scientists were starting to understand the propagation of the action potential as
the basis for the propagation of nerve impulses [12,13]. That theory became a cornerstone of cellular
biophysics, and set an entire field of scientists to work on excitable membranes. Inside one neuron
(Figure 15.3a), information is transmitted by electrical pulses [17] along the dendrites and axons.
The pulses are shaped and propagated by the active membrane, which contains many voltage-gated
ion channels. Ion channels are pore-forming proteins (or an assembly of several proteins) [18] which
are present in the membranes (lipid bilayers) of all biological cells [19,20]. The typical ion channel
is about one or two atoms wide and it is both very selective (ion specificity being about 1/10,000) as
well as fast (allowing the passage of about 10^{7-8} ions/s). Ions like, for example, Na^+ and K^+ can move
across the cell membrane based on their gradients. The ions move through an ion channel in a row
[21,22] conceptually similar to, but physically very different from, the tunneling events of single-
electron transistors [23]. Some ion channels are controlled by a "gate" which may open or close
selectively, performing exactly like a switch. In particular, voltage-gated ion channels open or close
depending on the voltage gradient across the membrane [24,25], and behave very similarly to the
well-known transistors. In a recursive manner, these voltage-gated ion channels allow the flow of
certain (selected) ions down their electrochemical gradient, and hence they are themselves creating,
maintaining, and propagating the voltage gradients across the membrane.

When a neuron is stimulated, the voltage-gated Na^+ channels open and the Na^+ ions start flowing into the cell. This causes a local increase of the membrane potential, which in turn activates
neighboring voltage-gated Na^+ channels, which also open. The resulting local depolarization also

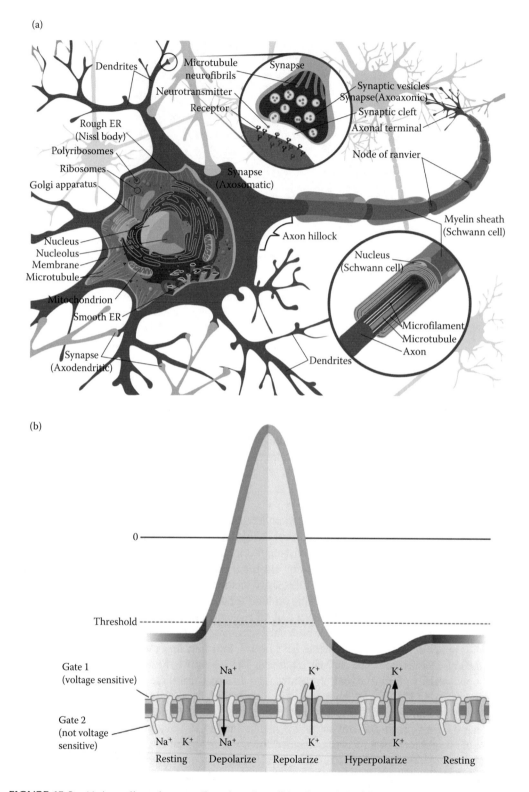

FIGURE 15.3   (a) A myelinated neuron (from http://en.wikipedia.org/wiki/File:Complete_neuron_cell_diagram_en.svg); (b) Action potential (theoretical approximation of the membrane potential). (From http://www.mhhe.com/biosci/esp/2001_gbio/folder_structure/an/m2/s4/assets/images/anm2s4_1.jpg.)

stimulates nearby voltage-gated K$^+$ channels to also open. Therefore, K$^+$ ions start flowing out of the cell in a process called repolarization. Still, even before the Na$^+$ and K$^+$ ions across the membrane equilibrate, both the Na$^+$ and K$^+$ channels close automatically after a very brief period of time (resting). The main result of these timely orchestrated actions of large numbers of voltage-gated ion channels is a local and moving reversal of the membrane potential, which is known as an action potential (Figure 15.3b). In the end, K$^+$ ions move outside and restore the resting potential. The resulting "spikes" are the ones representing and propagating information along an axon. After an action potential has taken place, there is a period of time (known as the refractory period) during which the membrane cannot be stimulated again.

A detailed understanding of the energy consumption associated with an action potential is still in its infancy, but it is a hot topic very actively pursued. For details, the interested reader should consult Refs. [27–43]. In this chapter, we are not going to analyze such aspects; our aim is to show how borrowing some inspiration from the way axons communicate could be translated to another implementation medium—nanoelectronics.

15.3 OPTIMAL ARRAYS FOR MINIMIZING POWER

Obviously, as our major aim is to reduce power, it seems only normal that we would want to minimize the number of ion channels. It follows that the neighboring ion channels (i.e., the ion channels around a given ion channel) should be positioned in some optimal way. This can be better appreciated by looking at Figure 15.4a, which—although not an array of ion channels—endorses nature's appetite for arrays at the nanoscale. Additionally, one would also want to reduce the time the ion channels are open while also maximizing the distances they can communicate (also known as broadcast coverage area), but unfortunately these two aims are conflicting.

Here, we present an upper bound on the broadcast coverage area with the condition that each voltage-gated ion channel (which is in fact both "a sensor" and "a broadcasting node") is assumed to be surrounded by the same number of neighboring voltage-gated ion channels (which means that here we consider only regular arrays). All the voltage-gated ion channels are used just for rebroadcasting the information, that is, for propagating the action potential further. This interpretation of the membrane of an axon is equivalent to a sensor network used for broadcasting a message. That is why it can be modeled using the unit disk graph [44]. Each and every voltage-gated ion channels will be considered to communicate/broadcast over a disk of radius r (which is related to the time the voltage-gated ion channel is open). As shown in Figure 15.4b, X is defined as the originating node (there is normally more than one originating node). X broadcasts information to its neighboring ion channels Y_i, located inside a circle as $|XY_i| \leq r$, for further rebroadcasting. Let $TA_{k,n}$ be the total broadcast coverage area of $m(k,n)$ ion channels as in a k-hop broadcasting scheme, and $AS_{k,n}$ be the broadcast coverage area of each ion channel involved in this broadcasting process. Here, k represents the number of broadcasting hops from ion channel X, while n is the number of neighboring ion channels.

We consider simplistically that information (i.e., the flow of ions) from X travels in a "single hop" to the neighboring ion channels Y_i ($i = 1, 2, \ldots, n$). The total "coverage area" will be determined by the distance $|XY_i|$ and the forwarding angles $\alpha_i = \angle Y_i X Y_j$, where $j = (i + 1) \bmod n$.

Theorem: If an ion channel X has n ($n \geq 3$) neighboring ion channels Y_i ($i = 1, 2, \ldots, n$) for broadcasting relay, the total broadcasting coverage area is optimized when all forwarding neighboring ion channels Y_i are symmetrically located on the border of X's transmission range.

Proof: The total broadcasting coverage area $TA_{1,1}(x)$ is given by

$$TA_{1,1}(x) = \pi + 2\arcsin(x/2) + r\sqrt{1 - (x/2)^2}, \quad 0 \leq x \leq r, \tag{15.1}$$

(a)

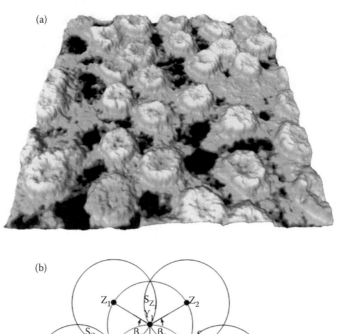

(b)

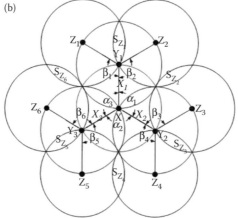

FIGURE 15.4 (a) Nuclear envelope of *Xenopus laevis* oocyte, a "biological nano-array," imaged with atomic force microscopy (area is about 600 nm × 600 nm; from http://celldynamics.uni-muenster.de/Projects/ProA6txt.html); (b) multi-hop broadcast with three forwarding neighboring nodes (i.e., three voltage-gated ion channels surrounding each voltage-gated ion channel).

where x is the distance between X and Y_i. Differentiating, we obtain:

$$dTA_{1,1}(x)/dx = 2\sqrt{1 - (x/2)^2} > 0, \quad 0 \leq x \leq r. \tag{15.2}$$

It follows that $TA_{1,1}(x)$ is maximized when $x = r$:

$$Max(TA_{1,1}) = 4\pi/3 + \sqrt{3}/2 \approx 1.609\pi, \quad \text{where } r = 1. \tag{15.3}$$

Hence, when $x_i = r = 1$, the total broadcast coverage area is:

$$TA_{1,n} = \pi + n \times (\pi/3 + \sqrt{3}/2) - \sum_{i=1}^{n} SV_i(\alpha_i), \tag{15.4}$$

with $\Sigma_{i=1}^{n}\alpha_i = 2\pi$ and α_i the angle between X and Y_i (as defined above), while SV_i is the overlapping area between two adjacent ion channels. When $\alpha_i \geq 2\pi/3$, $SV_i(\alpha_i) = 0$. From Equation 15.4, we can infer that $TA_{1,n}$ is maximized when $\Sigma_{i=1}^{n}SV_i(\alpha_i)$ is minimized. Using Lagrange relaxation, let

$$g(\alpha_1,\ldots,\alpha_n) = \sum_{i=1}^{n}\alpha_i - 2\pi. \tag{15.5}$$

Define:

$$F = \sum_{i=1}^{n} SV_i(\alpha_i) - \beta\, g(\alpha_1,\ldots,\alpha_n),$$
$$= \sum_{i=1}^{n} SV_i(\alpha_i) - \beta\sum_{i=1}^{n}\alpha_i - 2\pi \tag{15.6}$$

with β the Lagrange multiplier. F is minimized if:

$$\begin{cases} \partial F/\partial\alpha_1 = \partial SV_1(\alpha_1)/\partial\alpha_1 + \beta = 0 \\ \qquad\cdots \\ \partial F/\partial\alpha_n = \partial SV_n(\alpha_n)/\partial\alpha_n + \beta = 0 \\ \partial F/\partial\beta = \sum_{i=1}^{n}\alpha_i - 2\pi = 0 \end{cases} \tag{15.7}$$

Hence:

$$\partial SV_1(\alpha_1)/\partial\alpha_1 = \partial SV_2(\alpha_2)/\partial\alpha_2 = \cdots = \partial SV_n(\alpha_n)/\partial\alpha_n.$$

The solution for Equation 15.7 is $\alpha_i = 2\pi/n$ ($i = 1,2,\ldots n$); therefore, $Max(TA_{1,n})$ can be obtained if and only if:

$$\begin{cases} \alpha_i = 2\pi/n \\ x_i = 1 \end{cases}, \quad i = 1,2,\ldots,n.$$

This implies that $TA_{1,n}$ is maximized when Y_i are symmetrically located on the border of X's transmission range.

As an example, let us consider a single-hop broadcast relay scheme from X via three neighboring ion channels Y_i ($i = 1, 2, 3$) and assume that $x = r = 1$. $TA_{1,3}$ is maximized when Y_i ($i = 1, 2, 3$) are located on the border of X's transmission range with forwarding angles $\alpha_i = 2\pi/3$:

$$Max(TA_{1,3}) = \pi + 3 \times \left(\frac{\pi}{3} + \frac{\sqrt{3}}{2}\right) \approx 2.827\pi. \tag{15.8}$$

The optimum coverage area follows as:

$$Max(AS_{1,3}) = Max(TA_{1,3})/4 = 0.706\pi. \tag{15.9}$$

If the number of neighboring ion channels is increased to $n = 4$, the neighboring ion channels Y_i ($i = 1, 2, 3, 4$) should be located on the border of X's transmission range at forwarding angles $\alpha_i = \pi/2$, which leads to:

$$Min\left[\sum_{i=1}^{4} SV_i(\pi_i)\right] = 4 \times SV(\pi/2) = 4 \times \left(\frac{\pi}{12} + \frac{\sqrt{3}}{2} - 1\right), \tag{15.10}$$

and the maximum $TA_{1,4}$ is:

$$Max(TA_{1,4}) = \pi + 4 \times \left(\frac{\pi}{3} + \frac{\sqrt{3}}{2}\right) - Min\left[\sum_{i=1}^{4} SV_i(\alpha_i)\right] = 2\pi + 4 \approx 3.273\pi, \tag{15.11}$$

while the corresponding value for the coverage area becomes:

$$Max(AS_{1,4}) = Max(TA_{1,4})/5 \approx 0.655\pi. \tag{15.12}$$

From Equations 15.6, 15.9, and 15.12, it is clear that the maximum of $AS_{1,n}$ decreases when n increases, which is due to the fact that the overlapping area starts increasing for $n > 3$. Still, for $n = 2$, broadcasting generates interstices which do degrade the broadcasting capabilities. It follows that the broadcasting coverage area reaches an upper bound when $n = 3$ (i.e., $\alpha_i = 2\pi/3$).

Figure 15.4b shows a 2-hop broadcast relay scheme, where there are $n = 3$ neighboring ion channels around each forwarding voltage-gated ion channel. All voltage-gated ion channels are assumed to have the same coverage range of radius $r = 1$ [44]. Ion channel X broadcasts the information to its neighboring ion channels Y_i ($i = 1, 2, 3$) in the first hop. Upon receiving the information from Y_i ($i = 1, 2, 3$), voltage-gated ion channels Z_j ($j = 1, 2, \ldots, 6$) rebroadcast the information to their neighbors in the second hop. A loop of broadcasting occurs if rebroadcasting back to X, which in practice is supporting the spiking process. Let S_{Zm} ($m = 1, 2, \ldots, 6$) denote the area overlapped by two neighboring voltage-gated ion channels Z_j ($j = 1, 2, \ldots, 6$), and β_j denote the forward angles between voltage-gated ion channels Y_i ($i = 1, 2, 3$) and Z_j. From the previous theorem, the optimum broadcasting coverage area is obtained when all the neighboring voltage-gated ion channels are uniformly located on the border of the transmission range of the broadcasting voltage-gated ion channels:

$$Max(TA_{2,3}) = Max(TA_{1,3}) + 6 \times \left(\frac{\pi}{3} + \frac{\sqrt{3}}{2}\right) - Min\left[\sum_{m=1}^{6} S_{Zm}\right]. \tag{15.13}$$

By using the Lagrange relaxation technique again, the minimum of $\sum_{m=1}^{6} S_{Zm}$ is obtained for forwarding angles $\beta_j = 2\pi/3$ ($j = 1, 2, \ldots, 6$):

$$Min\left[\sum_{m=1}^{6} S_{Zm}\right] = 3 \times 2\int_0^1\left[\sqrt{1 - (y - 1/2)^2} - \sqrt{3}/2\right]dy = \pi - 3\sqrt{3}/2 \tag{15.14}$$

and

$$Max(TA_{2,3}) = \left[\pi + 3\left(\frac{\pi}{3} + \frac{\sqrt{3}}{2}\right)\right] + \left[6\left(\frac{\pi}{3} + \frac{\sqrt{3}}{2}\right)\right] - \left[\pi - 3\sqrt{3}/2\right] \approx 6.308\pi. \tag{15.15}$$

If $k \geq 2$, $n = 3$, and $r = 1$, the maximum of $TA_{k,3}$ is reached when the forwarding neighboring voltage-gated ion channels are symmetrically located on the border of the broadcasting voltage-gated ion channels' transmission range:

$$Max(TA_{k,3}) = Max(TA_{k-1,3}) + 3k\sqrt{3}. \tag{15.16}$$

Using Equation 15.15, this becomes:

$$Max(TA_{k,3}) = Max(TA_{1,3}) + \sum_{w=2}^{k} 3\sqrt{3} \cdot w = (2\pi + 3\sqrt{3}/2) + \frac{3\sqrt{3}}{2}(k^2 + k - 1), \quad k > 1. \tag{15.17}$$

The total number of forwarding voltage-gated ion channels can now be easily calculated as:

$$m_{k,3} = 1 + 3 + 3 \times 2 + \cdots + 3k = (3k^2 + 3k + 2)/2 \tag{15.18}$$

and the optimum coverage area follows:

$$Max(AS_{k,3}) = \frac{Max(TA_{k,3})}{m_{k,3}} = \frac{4\pi + 3\sqrt{3}(k^2 + k - 1)}{3k^2 + 3k - 2}. \tag{15.19}$$

For $k \geq 2$ and $n = 4$, the maximum value of the total broadcasting coverage area is

$$Max(TA_{k,4}) = Max(TA_{1,4}) + \sum_{w=2}^{k} (\pi + 4w) = 2k^2 + (\pi + 2)k + \pi, \tag{15.20}$$

and in this case the total number of forwarding nodes is:

$$m_{k,4} = 1 + 4 + 4 \times 2 + \cdots + 4k = 2k^2 + 2k + 1,$$

hence the maximum $AS_{k,4}$ becomes

$$Max(AS_{k,4}) = \frac{Max(TA_{k,4})}{m_{k,4}} = \frac{2k^2 + (\pi + 2)k + \pi}{2k^2 + 2k + 1}. \tag{15.21}$$

The main conclusion is that *the number of ion channels is minimized when the forwarding nodes are symmetrically located for n = 3* (as the average area reaches its upper bound). Voltage-gated ion channels cannot be distributed regularly, and certainly can never be located on such ideal positions. Still, the broadcast coverage area upper bound should lead to *a lower bound on the power required for communicating as minimizing the number of ion channels* [26–28]. Such an array (see Figure 15.5a) is easy to understand, while the mapping should in fact be on a cylinder in 3D as the simplest approximation of an axon (see Figure 15.5b).

15.4 AXON-INSPIRED COMMUNICATION IN SINGLE ELECTRON TECHNOLOGY

Single-electron technology (SET) is considered to be the ultimate low-power nanoelectronic technology [23]. This is why we have decided to use SETs for implementing the elementary logical blocks which could emulate the functioning of voltage-gated ion channels in nanoelectronics. Once

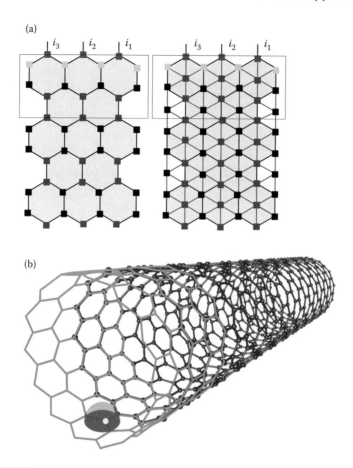

FIGURE 15.5   (a) 2D hexagonal and triangular arrays with three inputs. (A. Beg and V. Beiu, Ultra low power/energy SET-based Axon-inspired communication, *11th International Conference on Nanotechnology (IEEE Nano 2011)*, Portland, OR, pp. 1183–1186. © (2011) IEEE. With permission.); (b) 3D hexagonal communication array showing one voltage-gated ion channel (light dot) broadcasting (surrounding hemisphere) to the nearby voltage-gated ion channels. (V. Beiu et al., Minimizing communication power using near-neighbor axon-inspired lattices, *Proc. IEEE-NANO*, Portland, OR, 426–430. © (2011) IEEE. With permission.)

these are done, we shall estimate the resulting power and energy consumptions of the resulting axon-inspired transmission schemes.

SETs are quantum-based devices relying on a Coulomb blockade [23]. Over the last 20 years, researchers have been looking into various logic gates built with SETs, while there has been very limited work on investigating the power and energy characteristics of such gates [45–48], and even less when going to the circuit and system levels [49]. Sulieman and Beiu [48] have successfully compared different SET-based full adders, including not only the delay and design complexity but also the sensitivity to variations using SIMON and MATLAB. SIMON is the most well-known simulator for SET devices and circuits [50]. It simulates co-tunneling using Monte Carlo and master equations, and has also been used for the simulations presented in this chapter.

The hexagonal and triangular arrays are presented in Figure 15.5a. Such arrays were also suggested in Refs. [14, 51], and preliminary results were reported in Refs. [15,16]. Based on the theorem presented in Section 15.3, the hexagonal array is expected to be more power- and energy-efficient than the triangular one. For the hexagonal array, the basic building block equivalent to the functioning of a voltage-gated ion channel is a two-input OR (OR-2) gate. Obviously, such an implementation will not emulate the self-timed shut-down mechanism of voltage-gated ion channels, and could end up wasting more energy as electrons are still going to be thrown away, while ions are recirculated. Similarly, for the triangular array, we will use a three-input OR gate (OR-3). We have designed and simulated both these gates using SIMON [50]. Both gates were built using an NOR gate followed by an INV, as can be seen in Figure 15.6. The OR-2 and OR-3 gates map directly onto the nodes (voltage-gated ion channels) of the arrays presented in Figure 15.5a.

For mimicking an action potential propagating on an axon, we have decided to use pulses (spikes) that are 1 μs wide, spaced at 10 ms intervals. The sequence used for the input vectors for the OR-2 was as follows: 00 at 0 ms, 11 at 10 ms, 00 at 20 ms, 10 at 30 ms, and 00 at 40 ms. Similarly, the

(a)

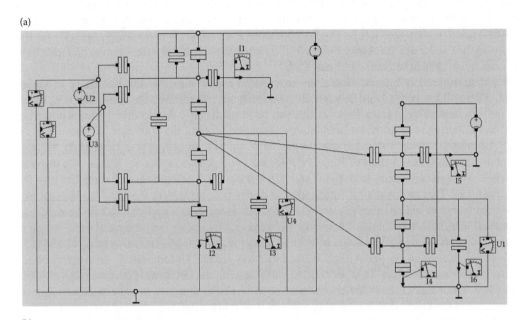

(b)

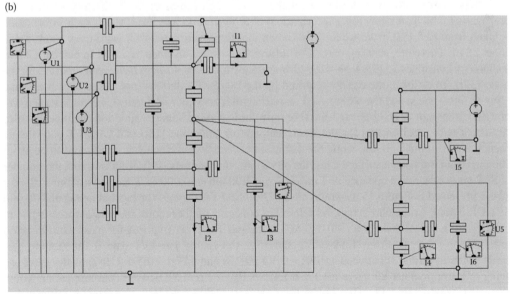

FIGURE 15.6 SET-based OR gates (using SIMON): (a) OR-2 as a NOR-2 followed by an INV; (b) OR-3 as a NOR-3 followed by an INV. (A. Beg and V. Beiu, Ultra low power/energy SET-based Axon-inspired communication, *11th International Conference on Nanotechnology (IEEE Nano 2011)*, Portland, OR, USA, pp. 1183–1186. © (2011) IEEE. With permission.)

sequence for OR-3 was: 000 at 0 ms, 111 at 10 ms, 000 at 20 ms, 110 at 30 ms, 000 at 40 ms, 100 at 50 ms, and 000 at 60 ms.

After performing the simulations, we have estimated the power and energy as

$$P_{avg} = \frac{V_{DD}}{T} \int_{t=0}^{T} i(t)\,dt \quad \text{and} \quad E = V_{DD} \int_{t=0}^{T} i(t)\,dt, \tag{15.22}$$

where the supply voltage is $V_{DD} = 6.5$ mV, $i(t)$ is the instantaneous current (obtained from simulations), P_{avg} is the average power, and E is the energy. The total time for simulations was $T = 40$ ms for OR-2 and $T = 60$ ms for OR-3. The instantaneous P and E (as "power-delay-product") for the OR-2 and the OR-3 gates are shown in Figure 15.7, while the numerical values for the different time intervals are presented in Tables 15.1 and 15.2.

An axon-inspired communication is an array-based communication link like those in Figure 15.5a, where all the nodes (voltage-gated ion channels) are transmitting the same data (i.e., either "0" or "1"). Interestingly, from Table 15.1, it can be immediately inferred that the transition from "00" to "11" exhibits lower power and energy consumption than the "00" to "10." Similarly, for OR-3, we found the lowest P and E was for the "000" to "111" transition (see Table 15.2). We also found that the maximum values of P and PDP for the OR-3 gate are about $6 \times$ higher than the ones for OR-2, while the minimum values of P and PDP for the OR-3 gate are about $5 \times$ smaller than the ones for OR-2! This means that, if functioning perfectly (i.e., no errors), the OR-3 (to be used in a triangular array) would be more power and energy efficiency than a perfect OR-2 (to be used in a hexagonal array). This is counterintuitive as we were expecting that the hexagonal array which is using OR-2 gates would achieve lower power and energy than the triangular array using OR-3 gates, while our simulations show that it might be the other way around. Unfortunately, when errors occur, that is, transitions other than "000" to "111" to "000" appear, the OR-2 could become more efficient than OR-3. These results show that errors could be much dearer than one would expect, and suggest that a detailed combined power-reliability analysis is absolutely needed.

Finally, we are now in a position to make some rough estimates on the energy required for propagating data (communicating information) at a given distance. For CMOS, such information is taken from Ref. [52], from where it is known that transmitting a 64-bit word over 1 mm needs about 25 pJ. Therefore, sending one bit of information over a distance of 1 mm in current CMOS technology consumes 25 pJ/64-bit = 0.4 pJ/bit/mm (estimates of 4 nJ/cm have been reported for the brain [37]). To estimate the energy required by the two SET solutions (one based on a hexagonal array of OR-2 gates, and the other based on a triangular array of OR-3 gates), we consider the size of a SET transistor to be of about 1 nm (not only similar to the voltage-gated ion channels but also consistent with requirements for the room temperature operation [23]). An OR SET gate should occupy about 10 nm × 10 nm, while the distance between two SET gates is considered to be about 100 nm (from the known surface densities of various ion channels [19]). It follows that the number of SET gates covering a distance of 1 mm is 1 mm/100 nm = 10,000 SET gates. Structures like the arrays presented in Figure 15.5a would have 47,000 SET OR-2 gates (hexagonal array) and 70,000 SET OR-3 gates (triangular array). As reliability concerns might require more than three inputs in parallel, we estimate that about 100,000 SET gates per 1 mm are required for these axon-inspired transmission schemes. From Tables 15.1 and 15.2, the energy per SET gate if functioning correctly (without errors) is between $1.013E - 019$ J (OR-3) and $4.77E - 019$ J (OR-2) as the gates are going to transmit either all 0's or all 1's. It follows that a 1 mm SET-based communication would need $100,000 \times 4.77E - 19 = 0.047$ pJ/bit/mm (in fact, $47,000 \times 4.77E - 19 = 0.022$ pJ/bit/mm and $70,000 \times 1.01E - 19 = 0.007$ pJ/bit/mm, respectively). The main conclusion is that an axon-inspired communication like those presented in this chapter could be about $10 \times$ (0.047 pJ/bit/mm vs 0.4 pJ/bit/mm) to $50 \times$ (0.007 pJ/bit/mm vs 0.4 pJ/bit/mm) more energy-efficient than the current CMOS solutions, but depending strongly on reliability.

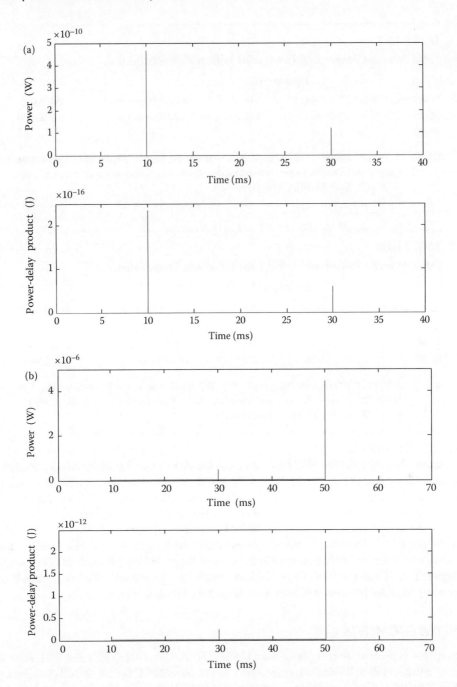

FIGURE 15.7 Power and energy (power-delay-product, or PDP) for the test sequence: (a) OR-2; (b) OR-3. (A. Beg and V. Beiu, Ultra low power/energy SET-based Axon-inspired communication, *11th International Conference on Nanotechnology (IEEE Nano 2011)*, Portland, OR, USA, pp. 1183–1186. © (2011) IEEE. With permission.)

15.5 CONCLUSIONS

This chapter aimed to identify the practical limits for ultra-low power/energy communications when using electronic principles. To this end, it has started by presenting a theoretical analysis supporting the fact that hexagonal arrays would be optimal to minimize communication power/energy consumption, because they are minimizing the number of nodes while maximizing the coverage

TABLE 15.1

OR-2 Average Power and Energy for Different Transitions

Time Interval (ms)	Input Vector		Average Power (W)	Energy (J)
	From	To		
10–20	00	11	4.777E–017	4.777E–019
30–40	00	10	5.937E–016	5.937E–018

Source: A. Beg and V. Beiu, Ultra low power/energy SET-based Axon-inspired communication, *11th International Conference on Nanotechnology (IEEE Nano 2011)*, Portland, OR, USA, pp. 1183–1186. © (2011) IEEE. With permission.

TABLE 15.2

OR-3 Average Power and Energy for Different Transitions

Time Interval (ms)	Input Vector		Average Power (W)	Energy a (J)
	From	To		
10–20	000	111	1.013E–017	1.013E–019
30–40	000	110	2.245E–015	2.245E–017
50–60	000	100	3.882E–015	3.882E–017

Source: A. Beg and V. Beiu, Ultra low power/energy SET-based Axon-inspired communication, *11th International Conference on Nanotechnology (IEEE Nano 2011)*, Portland, OR, USA, pp. 1183–1186. © (2011) IEEE. With permission.

area. Later, we have considered SET-based gates as the elementary building blocks for emulating axon-inspired locally connected arrays of voltage-gated ion channels. On the basis of simulations, we have found that axon-inspired SET-based solutions could offer energy savings in the range of 10 × to 50 × (over current CMOS solutions), while the particular logic gate used seems to play a significant role as the arrays themselves, with reliability being an open topic for further research. Finally, neither voltage-gated ion channels nor electronic nanodevices can be positioned very precisely; hence, they can never be located on the ideal positions suggested by the ideal arrays presented in this chapter. That is why the results reported here should be seen more like "a lower bound" on the power/energy required for nanoelectronic axon-inspired communications.

ACKNOWLEDGMENTS

This work was supported by two grants from UAEU (NRF RSA 1108-00329 and NRF RSA 1108-00451), by a British Council Prime Minister Initiative to Connect PMI2 grant (RC GS 271), by a grant from SRC (2011-RJ-2150G), and by a grant from Intel (2011-05-24G). This document is an output from the PMI2 Project funded by the UK Department for Innovations, Universities and Skills (DIUS) for the benefit of the United Arab Emirates Higher Education Sector and the UK Higher Education Sector. The views expressed are not necessarily those of the DIUS, or of the British Council.

REFERENCES

1. *International Technology Roadmap for Semiconductors (ITRS)*, SEMATECH, New York, NY, USA, 2011. Available at: public.itrs.net

2. J. D. Meindl, Q. Chen, and J. A. Davis, Limits on silicon nanoelectronics for Terascale integration, *Science*, 293, Sept. 14, 2001, 2044–2049.

3. V. Beiu, U. Rückert, S. Roy, and J. Nyathi, On nanoelectronic architectural challenges and solutions, *Proc. IEEE-NANO*, Munich, Germany, Aug. 2004, 628–631.

4. S. Rakheja and A. Naeemi, Interconnects for novel state variables, *IEEE Trans. Electr. Dev.*, 57, Oct. 2010, 2711–2718.

5. R. W. Keyes, The wire-limited logic chip, *IEEE J. Solid-State Circ.*, 17, 1982, 1232–1233.

6. R. W. Keyes, The power of connections, *IEEE Circ. Dev. Mag.*, 7, 1991, 32–35.

7. J. A. Davis, R. Venkatesan, A. Kaloyeros, M. Beylansky, S. J. Souri, K. Banerjee, K. C. Saraswat, A. Rahman, R. Reif, and J. D. Meindl, Interconnect limits on Gigascale integration in the 21st century, *Proc. IEEE*, 89, 2001, 305–324.

8. V. Beiu, W. Ibrahim, and R. Z. Makki, On wires at low electron densities, *Proc. IEEE-NANO*, Genoa, Italy, Jul. 2009, pp. 703–706.

9. J. M. Rabaey, Beyond the horizon: The next 10× reduction in power—Challenges and solutions *Proc. ISSCC'11*, San Francisco, CA, USA, Feb. 2011. Plenary panel available at: http://isscc.org/doc/2011/ISSCC2011_PlenarySession4_slides.pdf and also http://isscc.org/videos/2011_plenary.html#session4

10. N. Magen, A. Kolodny, U. Weiser, and N. Shamir, Interconnect-power dissipation in a microprocessor, *Proc. SLIP'04*, Paris, France, Feb. 2004, pp. 7–13.

11. J. D. Meindl, Beyond Moore's law: The interconnect era, *Comp. Sci. Eng.*, 5, Jan. 2003, 20–24.

12. A. L. Hodgkin, and A. F. Huxley, A quantitative description of membrane current and its application to conduction and excitation in nerve, *J. Physiol.*, 117, Aug. 1952, 500–544.

13. A. L. Hodgkin and R. D. Keynes, The potassium permeability of a giant nerve fibre, *J. Physiol.*, 128, Apr. 1955, 61–88.

14. V. Beiu, Brain-inspired information processors, *Intl. Forum on Minimum Energy Electr. Syst. (MEES2020)*, Abu Dhabi, UAE, May 2010. Available at http://grc.src.org/member/event/e003960/E003960_SessionV_Beiu.pdf

15. V. Beiu, L. Zhang, W. Ibrahim, and M. Tache, Minimizing communication power using near-neighbor axon-inspired lattices, *Proc. IEEE-NANO*, Portland, OR, USA, Aug. 2011, 426–430.

16. A. Beg and V. Beiu, Ultra low power/energy SET-based axon-inspired communication, *Proc. IEEE-NANO*, Portland, OR, USA, Aug. 2011, 1183–1186.

17. S. B. Laughlin and T. J. Sejnowski, Communication in neural networks, *Science*, 301, Sept. 2003, 1870–1874.

18. T. Albrecht, J. B. Edel, and M. Winterhalter, Special issue on new developments in nanopore research, *J. Phys. Cond. Mat.*, 22, Nov. 17, 2010.

19. M. Hilt and W. Zimmermann, Hexagonal, square, and stripe patterns of the ion channel density in bio-membranes, *Phys. Rev. E*, 75, 2007, art. 016202.

20. R. Phillips, T. Ursell, P. Wiggins, and P. Sens, Emerging roles for lipids in shaping membrane-protein function, *Nature*, 459, May 21, 2009, 379–385.

21. P. H. Nelson, A permeation theory for single-file ion channels: One- and two-step models, *J. Chem. Phys.*, 134(16), Apr. 25, 2011, art. 165102.

22. M. Ø. Jensen, D. W. Borhani, K. Lindorff-Larsen, P. Maragakis, V. Jogini, M. P. Eastwood, R. O. Dror, and D. E. Shaw, Principles of conduction and hydrophobic gating in K^+ channels, *Proc. Natl. Acad. Sci.*, 107, Mar. 30, 2010, 5833–5838.

23. K. K. Likharev, Single-electron devices and their applications, *Proc. IEEE*, 87, Apr. 1999, 606–632.

24. W. A. Catterall, Ion channel voltage sensors: Structure, function, and pathophysiology, *Neuron*, 67, Sept. 2010, 915–928.

25. R. E. Dempski, K. Hartung, T. Friedrich, and E. Bamberg, Fluorometric measurements of intermolecular distances between the α- and β-subunits of the Na^+/K^+-ATPase, *J. Biol. Chem.*, 281(47), Nov. 24, 2006, 36338–36346.

26. A. A. Faisal, and J. E. Niven, A simple method to simultaneously track the numbers of expressed channel proteins in a neuron, *Proc. CompLife*, Springer, LNBI 4216, 2006, pp. 257–267.

27. D. Attwell, and S. B. Laughlin, An energy budget for signaling in the grey matter of the brain, *J. Cereb. Blood Flow Metab.*, 21, Oct. 2001, 1133–1145.

28. P. Lennie, The cost of cortical computation, *Curr. Biol.*, 13, Mar. 18, 2003, 493–497.

29. K. Koch, J. McLean, R. Segev, M. A. Freed, M. J. Berry II, V. Balasubramanian, and P. Sterling, How much the eye tells the brain, *Curr. Biol.*, 16(14), Jul. 25, 2006, 1428–1434.

30. H. Alle, A. Roth, and J. R. P. Geiger, Energy-efficient action potentials in hippocampal mossy fibers, *Science*, 325, Sept. 11, 2009, 1404–1408.

31. S. Bernèche, and B. Roux, Energetics of ion conduction through the K^+ channel, *Nature*, 414, Nov. 1, 2001, 73–77.

32. E. Schneidman, I. Segev, and N. Tishby, Information capacity and robustness of stochastic neuron models, in S. A. Solla, T. K. Leen and K.-R. Müller (Eds.): *Advances in Neural Information Processing Systems (NIPS'99)*, MIT Press, Cambridge, MA, 2000, pp. 178–184.

33. J. A. White, J. T. Rubinstein, and A. R. Kay, Channel noise in neurons, *Trends Neurosci.*, 23(3), Mar. 2000, 131–137.

34. M. E. Raichle and D. A. Gusnard, Appraising the brain's energy budget, *Proc. Natl. Acad. Sci.*, 99(16), Aug. 6, 2002, 10237–10239.

35. S. Schreiber, C. K. Machens, A. V. M. Herz, and S. B. Laughlin, Energy-efficient coding with discrete stochastic events, *Neural Comp.*, 14(6), Jun. 2002, 1323–1346.

36. N. Cohen, From ionics to energetics in the nervous system, *Solid State Ionics*, 176(19–22), Jun. 2005, 1661–1666.

37. P. Crotty, T. Sangrey, and W. B. Levy, Metabolic energy cost of action potential velocity, *J. Neurophysiol.*, 96(3), Sept. 2006, 1237–1246.

38. B. C. Carter and B. P. Bean, Sodium entry during action potentials of mammalian neurons: Incomplete inactivation and reduced metabolic efficiency in fast-spiking neurons, *Neuron*, 64, Dec. 24, 2009, 898–909.

39. A. Hasenstaub, S. Otte, E. Callaway, and T. J. Sejnowski, Metabolic cost as a unifying principle governing neuronal biophysics, *Proc. Natl. Acad. Sci.*, 107(27), Jul. 6, 2010, 12329–12334.

40. C. Howarth, C. M. Peppiatt-Wildman, and D. Attwell, The energy use associated with neural computation in the cerebellum, *J. Cereb. Blood Flow Metab.*, 30(2), Feb. 2010, 403–414.

41. B. Sengupta, M. Stemmler, S. B. Laughlin, and J. E. Niven, Action potential energy efficiency varies among neuron types in vertebrates and invertebrates, *PLoS Comput. Biol.*, 6(7), Jul. 2010, art. e1000840.

42. A. Singh, R. Jolivet, P. J. Magistretti, and B. Weber, Sodium entry efficiency during action potentials: A novel single-parameter family of Hodgkin–Huxley models, in J. Lafferty, C. K. I. Williams, J. Shawe-Taylor, R. S. Zemel, and A. Culotta (Eds.): *Advances in Neural Information Processing Systems (Proc. NIPS'10)*, MIT Press, Cambridge, MA, 2011, pp. 2173–2180.

43. A. Moujahid, A. d'Anjou, F. J. Torrealdea, and F. Torrealdea, Energy and information in Hodgkin–Huxley neurons, *Phys. Rev. E*, 83, 2011, art. 031912.

44. B. N. Clark, C. J. Colbourn, and D. S. Johnson, Unit disk graphs, *Discrete Math.*, 86, Dec. 1990, 165–177.

45. C. Choi, J. Lee, S. Park, I.-Y. Chung, C.-J. Kim, B.-G. Park, D. M. Kim, and D. H. Kim, Comparative study on the energy efficiency of logic gates based on single-electron transistor technology, *Semicond. Sci. Technol.*, 24, May 2009, art. 065007.

46. V. Saripalli, V. Narayanan, and S. Datta, Analyzing energy-delay behavior in room temperature single electron transistors, *Proc. VLSID*, Bangalore, India, Jan. 2010, 399–404.

47. J. Lee, J. H. Lee, I. -Y. Chung, C. -J. Kim, B. -G. Park, D. M. Kim, and D, H. Kim, Comparative study on energy-efficiencies of single-electron transistor-based binary full adders including nonideal effects, *IEEE Trans. Nanotech.*, 10, Sept. 2011, 1180–1190.

48. M. H. Sulieman, and V. Beiu, On single-electron technology full adders, *IEEE Trans. Nanotech.*, 4, Nov. 2005, 669–681.

49. M. H. Sulieman, and V. Beiu, Characterization of a 16-bit threshold logic single electron technology adder, *Proc. ISCAS*, Vancouver, Canada, May 2004, pp. 681–684.

50. C. Wasshuber, H. Kosina, and S. Selberherr, SIMON—A simulator for single-electron tunnel devices and circuits, *IEEE Trans. CAD IC Syst.*, 16, Sept. 1997, 937–942.

51. V. Beiu, The trustworthy wings of the mysterious butterfly, *Intl. Nanotech. Conf. Comm. & Coop. INC6*, Grenoble, France, May 2010 [Online]. Available at http://www.inc6.eu/presentations/We1-1_Beiu.pdf

52. W. J. Dally, The end of denial architecture and the rise of throughput computing (keynote), *Proc. ASYNC*, Chapel Hill, USA, May 2009 [Online]. Available: http://asyncsymposium.org/async2009/slides/dally-async2009.pdf (also presented at *HiPC'09* and *DAC'09*; see http://videos.dac.com/46th/wedkey/dally.html).

16 Electromechanical Modeling of GNP Nanocomposites for Integrated Stress Monitoring of Electronic Devices

Alessandro Giuseppe D'Aloia, Alessio Tamburrano, Giovanni De Bellis, Jacopo Tirillò, Fabrizio Sarasini, and Maria Sabrina Sarto

CONTENTS

16.1 INTRODUCTION

Since it was first obtained as a single sheet in 2004 by Geim et al. [1], graphene has attracted tremendous interest in the scientific community thanks to its outstanding mechanical [2–4], thermal [5], and especially electrical and transport properties [6]. The interest for graphene has been particularly pushed by the difficulties in exploiting and controlling the properties of carbon nanotubes (CNTs) throughout the last decade, as well as the high cost of the latter structures. Along with the interest in carbon-based nanostructures, composites filled with CNTs [7–10], graphite nanoplatelets (GNPs) [11–17], or combined systems [18–21] have also recently drawn much attention from the scientific community, thanks to the very low loadings required to reach considerable improvement in functional properties, traditionally only achievable with high concentrations of conventional fillers. These new composites are continuously finding applications in various fields such as electromagnetic shielding, aeronautics, photovoltaics, actuators, and sensors. For instance, several studies have focused on the piezoresistive behavior of polymers reinforced with single- or multiwalled CNTs [22,23], expanded graphite, and GNPs [24–26].

An important upcoming application of CNT-based composites is their use in high-density electronics to replace common commercial grease in thermal interface materials (TIMs) [27–29]. TIMs, when applied between heat sources and heat sinks, are essential ingredients in thermal management, especially in the future generations of integrated circuits or ultrafast high-power density communication systems. Nevertheless, the outcomes of experiments involving CNT-based TIMs were controversial, thus providing strong motivation for further research on alternative fillers, such as GNPs [30–33].

Hung et al. [30] studied heat transport in polymer nanocomposites reinforced with GNPs using high-precision thermal conductivity measurements. They found that the resistance to heat conduction across interfaces between GNPs and the polymeric matrix has a strong effect on heat transport in nanocomposites.

Yu et al. [31] reported an efficient process for converting natural graphite into few-layer GNPs. When embedded in a polymeric matrix, GNPs cause a remarkable enhancement of the thermal conductivity of the nanocomposite at low-volume loadings and significantly outperform the CNT-based fillers.

In a more recent publication, the same group [32] showed that by combining single-walled CNTs and GNPs, it is possible to achieve a synergistic effect in the thermal conductivity enhancement of epoxy-based composites.

Finally, in a very recent publication [33], Shahil et al. developed a synthesis process of TIMs made of a graphene–multilayer graphene nanocomposite polymer and demonstrated extremely high thermal coefficient enhancement at low filler loadings.

Owing to their multifunctional properties, GNP-based composites could replace common TIMs in electronic devices, combining thermal management utilities with integrated stress–strain monitoring capabilities of the packaging. In other words, it could be possible to realize smart TIMs able to detect local stress–strain states due to the delamination phenomena, for example. For this purpose, the availability of simulation models capable of predicting the effective electrical conductivity of the composite, as a function of the type, volume fraction, and mechanical stress state, is of paramount importance for design purposes.

In this chapter, we present a full model for the evaluation of the electromechanical response of a GNP-based nanocomposite for possible application as a smart TIM in electronic/nanoelectronics. Section 16.2 describes the synthesis and fabrication processes of GNPs and GNP-based nanocomposites developed in the CNIS Lab of Sapienza University. A complete experimental characterization of the dc conducting properties of both GNPs and nanocomposites with increasing filler concentration is also performed. Next, the electromechanical model of the nanocomposite material is presented in Section 16.3. This model is based on the assumption that GNPs are not deformable particles randomly distributed inside the polymeric matrix. The tunneling-based conducting model proposed in Refs. [34–37] is applied to define the relation between the dc effective electrical conductivity of the nanocomposite and the average probabilistic distance among the nanoparticles in the matrix. The model is calibrated using the electrical conductivity values of the filler and of the composites measured in Section 16.2. Finally, in Section 16.4, the developed model is applied to predict the response of a GNP-based TIM. The results of the numerical simulations demonstrate the validity of the proposed approach and can provide an indications on the optimum design of GNP-filled nanocomposites with enhanced electromechanical response.

16.2 EXPERIMENTAL

16.2.1 GNP Synthesis and Characterization

GNPs were synthesized through thermal expansion in the CNIS labs starting from a commercially available graphite intercalation compound (GIC), namely Grafguard 160-50N (intercalated expandable graphite flakes) as described in Refs. [17,21]. After thermal treatment at 1150°C, leading to thermally expanded graphite oxide (TEGO, with the typical worm-like structure of Figure 16.1a), the expanded graphite was tip sonicated in acetone, thus obtaining GNPs having thicknesses in the range 3–15 nm and lateral dimensions around 1–10 μm, as shown in Figure 16.1b.

A critical issue concerns the characterization of the GNP electrical conductivity, which strongly depends on the set of parameters used during the thermal expansion of GICs. Thick films made of

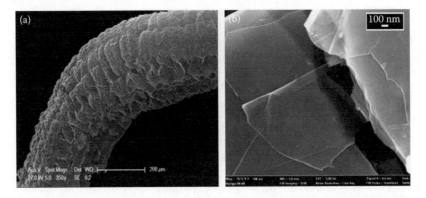

FIGURE 16.1   Micrographs of TEGO (a) and of ultrasonication (b). SEM analyses were performed at DICMA (a) and at SNN-Lab by Zeiss Auriga (b).

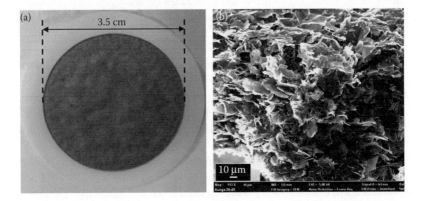

FIGURE 16.2   GNP film fabricated for dc electrical conductivity measurement (a) and a corresponding SEM image of a fracture section (b). (A.G. D'Aloia et al. Electromechanical modeling of GNP nanocomposites for stress sensor applications, *IEEE NANO 2011*, Portland, Oregon, USA, pp. 1648–1651, © (2011) IEEE. With permission.)

GNPs were then fabricated by vacuum filtration (using nanopore-sized AAO membranes as filters) to estimate the effective dc conductivity of the GNPs as described in Refs. [16,38].

The typical route used to produce the thick film is as follows: 40 mg TEGO was dispersed in 125 mL acetone through an ultrasonic probe operating in pulse mode for 40 min. The resulting GNP suspension was then subjected to vacuum filtration to obtain a compact film with thickness around 200 µm. Figure 16.2a and b shows, respectively, a photograph of the film obtained after vacuum filtration and a SEM micrograph highlighting the layered structure.

The sheet electrical resistance R_s of the GNP film was obtained through the four-point probe method [39] at room temperature, using a Signatone S301 stand, collinear tungsten carbide tips, and a Keithley 6221 dc/ac current source connected to a Keithley 2182A nano-voltmeter.

Then the dc electrical conductivity value of 10.5 kS/m was calculated from the measured R_s considering the average thickness of the GNP film.

16.2.2   Production and Characterization of Nanocomposites

Nanocomposites were prepared using a commercially available epoxy-based vinyl ester matrix, kindly provided by Reichhold (DION 9102). According to the preparation method described in Ref. [17], the filler was dispersed in acetone. The so-obtained suspension was mixed with the polymeric matrix and then poured into rectangular-shaped molds. Finally, the composite was cured in air for

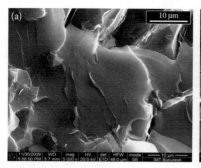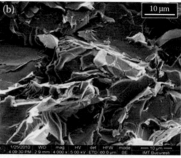

FIGURE 16.3 SEM micrographs showing the fracture surface of 0.5% wt. GNP (a) and 2% wt. GNP (b) nanocomposites. SEM analyses were performed at IMT Bucharest.

24 h and postcured for a further 24 h at 70°C. Nanocomposites loaded with GNPs at four different weight fractions (0.25%, 0.5%, 1%, and 2%) were prepared accordingly.

Extensive SEM analyses were performed to assess the uniformity of filler dispersion. As an example, Figure 16.3a and b shows the HR-SEM images of the fracture surfaces of 0.5% wt. and 2% wt. GNP composites [17].

The dc electrical conductivity of nanocomposites with different concentrations of GNPs was measured applying the four-wire volt-amperometric method. The test samples have been prepared as follows. At first, the two opposite faces of each parallelepiped-shaped specimen have been coated with silver paint and dried at 60°C for 10 min. Afterward, tin-coated copper wires have been bonded to the aforementioned faces through a conductive silver-loaded epoxy adhesive. This step was followed by a curing phase at 120°C for 10 min.

The electrical measurements were performed in delta mode using a Keithley 6221 dc/ac current source connected to a Keithley 2182A nano-voltmeter, controlled by a PC for data acquisition and analysis. The dc electrical conductivity γ was evaluated from the measured resistance values. The results obtained for the different GNP-based nanocomposites are shown in Figure 16.4.

Finally, the fabricated GNP nanocomposites were subjected to tensile testing according to Ref. [40]. Three dog-bone-shaped flat composite specimens were tested for each weight percentage.

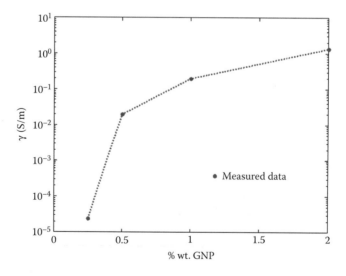

FIGURE 16.4 Measured dc effective conductivity of nanocomposites filled with different weight fractions of GNPs. (A.G. D'Aloia et al. Electromechanical modeling of GNP nanocomposites for stress sensor applications, *IEEE NANO 2011*, Portland, Oregon, USA, pp. 1648–1651, © (2011) IEEE. With permission.)

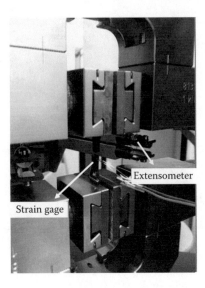

FIGURE 16.5   Experimental setup for tensile tests of nanocomposites.

A Zwick/Roell Z010 universal testing machine equipped with a 10 kN load cell was used to perform the mechanical tests. The tensile tests were performed at a cross-head speed of 0.5 mm/min. The specimens were gripped between flat mechanical wedge grips. The longitudinal strain was measured using an extensometer with a 10 mm gage length. Transverse strain was measured by a strain gage, as shown in Figure 16.5. This allowed us to measure the elastic constants of the composites, namely, Young's modulus E and Poisson's ratio υ, together with maximum strength and elongation at failure. Before mechanical property measurements, the specimen surfaces were mechanically polished to minimize the influence of the surface flaws. Figure 16.6 shows a specimen before and after the test.

The tensile properties of nanocomposites filled at 1% wt. and 2% wt. are summarized in Table 16.1, while Figure 16.7 shows the measured tensile stress σ versus strain ε.

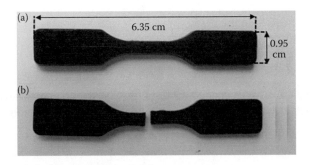

FIGURE 16.6   Dog-bone specimens before (a) and after (b) the tensile test.

TABLE 16.1

Nanocomposite Tensile Properties Obtained from Experimental Tests

GNP (% wt.)	Young's Modulus (GPa)	Maximum Strength (MPa)	Strain at Failure (%)	Poisson's Ratio
1	3.19 ± 0.01	63.79 ± 2.50	2.62 ± 0.08	0.34 ± 0.01
2	3.16 ± 0.03	40.88 ± 2.81	1.76 ± 0.16	0.33 ± 0.02

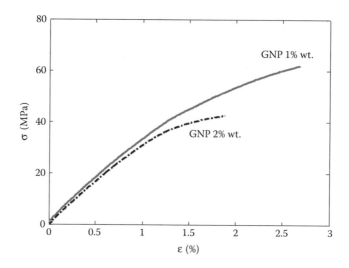

FIGURE 16.7 Measured tensile stress–strain characteristic of GNP-based nanocomposites at 1% wt. and 2% wt.

16.3 STRESS SENSOR MODELING

The stress sensor sketched in Figure 16.8a consists of a brick-shaped specimen of a GNP-based nanocomposite, having thickness t, and lateral dimensions L_x and L_y, with $L_x \geq L_y \gg t$, and it is supplied by an ideal current source I_s, which is injected into the L_y-long side-face of the specimen uniformly, and extracted from the corresponding opposite side.

The current I_s is assumed to flow in the (x, y)-plane of the sensor and to be uniformly distributed across the specimen thickness. Therefore, the distributions of the electric potential $V(x, y)$ and of the current density components $J_x(x, y)$ and $J_y(x, y)$ over the sensor surface are obtained from the solution of the following 2D differential equations:

$$-\frac{\partial V}{\partial x} = \frac{1}{\gamma_x} J_x \tag{16.1}$$

$$-\frac{\partial V}{\partial y} = \frac{1}{\gamma_y} J_y \tag{16.2}$$

where γ_x and γ_y are the effective dc conductivities of the GNP-based composite realizing the sensor along the x- and y-axes, respectively. Without mechanical stress the application results in $\gamma_x = \gamma_y = \gamma$

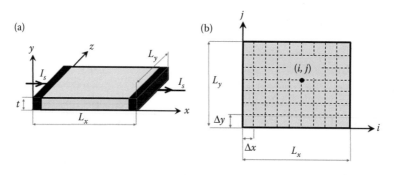

FIGURE 16.8 Stress sensor supplied by current I_s (a) and a 2D-equivalent discrete model (b).

because GNPs are assumed to be distributed uniformly and randomly into the matrix. Otherwise, if there is an applied stress producing the deformations ε_x and ε_y along the x- and y-axes, it results in $\gamma_x \neq \gamma_y$. The variations of γ_x and γ_y due to the deformations of the sensor are predicted applying the model reported in the next section. The deformations ε_x and ε_y corresponding to the applied stress are computed in Section 16.4 as well. Equations 16.1 and 16.2 are solved numerically by applying a leap-frog finite different scheme. For this purpose, it is set at $x = i\Delta x$ and $y = j\Delta y$, with $i \in [0,n]$, $j \in [0,m]$, and $\Delta x \ll L_x$, $\Delta y \ll L_y$. Therefore, the brick-shaped specimen is discretized in $(n \times m)$ cells, each having a thickness t and lateral dimensions Δx and Δy, as shown in Figure 16.8b.

Equations 16.1 and 16.2 are expressed in the following discrete form:

$$V_{(i,j)} - V_{(i+1,j)} = \frac{\Delta x}{\gamma_x} J_{x(i+1/2,j)} \tag{16.3}$$

$$V_{(i,j)} - V_{(i,j+1)} = \frac{\Delta y}{\gamma_y} J_{y(i,j+1/2)} \tag{16.4}$$

where $V_{(i,j)}, V_{(i+1,j)}, V_{(i,j+1)}, J_{x\,(i+1/2,j)}$, and $J_{y(i,j+1/2)}$ are the discretized electric potential and the current density components in the (i,j), $(i+1,j)$, $(i,j+1)$, $(i+1/2,j)$, and $(i,j+1/2)$ nodes, respectively, as sketched in Figure 16.9a.

Referring to the sample cell of Figure 16.9b, the currents flowing in the x- and y-directions, I_x and I_y, are defined by the following zero-order approximated expression:

$$I_{x\,(i+1/2,j)} = J_{x\,(i+1/2,j)}\,\Delta yt \tag{16.5}$$

$$I_{y\,(i,j+1/2)} = J_{y\,(i,j+1/2)}\,\Delta xt \tag{16.6}$$

which are obtained in the hypothesis that the density current components assume the constant value $J_{x(i+1/2,j)}$ and $J_{y(i,j+1/2)}$ over the cell lateral surface $\Sigma_x = \Delta xt$ and $\Sigma_y = \Delta yt$, respectively.

The voltage drop between the (i, j) and $(i + 1, j)$ nodes can be expressed as a function of the current $I_{x(i+1/2,j)}$ flowing through the effective conductance G_x of the cell along the x-direction:

$$V_{(i,j)} - V_{(i+1,j)} = I_{x(i+1/2,j)}/G_x \tag{16.7}$$

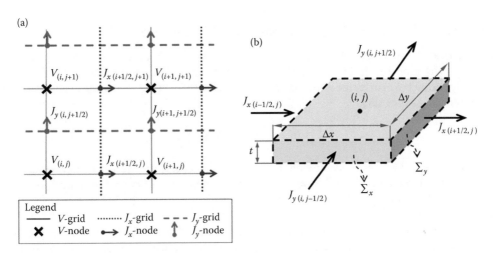

FIGURE 16.9 Discretized electric potential and current density components appearing in Equations 16.3 and 16.4 (a) and elementary cell centered in the (i,j)-node (b).

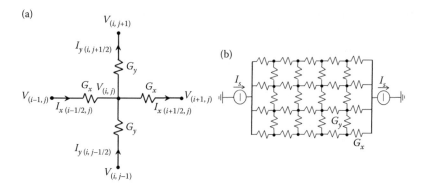

FIGURE 16.10 Circuit model of a single cell (a) and an equivalent network model of the stress sensor with current injection and extraction (b).

A similar expression defines the voltage drop between the (i,j) and $(i,j + 1)$ nodes:

$$V_{(i,j)} - V_{(i,j+1)} = I_{y(i,j+1/2)}/G_y \tag{16.8}$$

with G_y being the effective conductance of the cell along the y-direction.

The expressions of G_x and G_y are obtained comparing Equations 16.5, 16.7 with Equation 16.3, and Equations 16.6, 16.8 with Equation 16.4. It results in

$$G_x = \gamma_x \frac{\Delta y t}{\Delta x} \tag{16.9}$$

$$G_y = \gamma_y \frac{\Delta x t}{\Delta y} \tag{16.10}$$

The equivalent circuit representation of Equations 16.7 and 16.8 at the (i,j) node is shown in Figure 16.10a and the equivalent network model of the stress sensor is reported in Figure 16.10b, where the current injection and extraction are represented by the ideal generator I_s.

16.4 NANOCOMPOSITE ELECTROMECHANICAL MODELING

The electrical transport in GNP-based nanocomposites is described applying the tunneling percolation model proposed in [34–37]. In the hypothesis that GNPs are distributed randomly and uniformly within the matrix, the effective electrical conductivity of the composite can be described by the following expression:

$$\gamma(\theta) = \gamma_0 \exp \left[-2\,\delta(\theta)/\xi\right] \tag{16.11}$$

where θ is the GNP weight concentration in the composite, $\delta(\theta)$ is the average distance among the GNPs dispersed in the matrix at the weight concentration θ, $\xi = 9.22$ nm is the characteristic tunneling length [35], and γ_0 is the limit value of the electrical conductivity in the hypothesis that δ approaches to zero.

It is assumed that γ_0 is equal to the measured conductivity of the GNP film shown in Figure 16.4. The average distance $\delta(\theta)$ among the GNPs in a nanocomposite with filler concentration θ is then evaluated solving Equation 16.11 for $\delta(\theta)$, considering the corresponding measured value of the dc electrical conductivity $\gamma(\theta)$ reported in Figure 16.4. The results are shown in Figure 16.11.

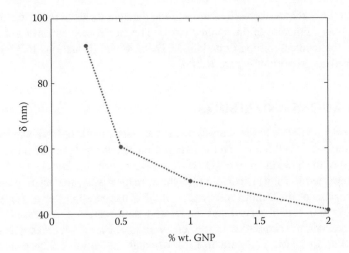

FIGURE 16.11 Average distance among the GNPs in the matrix as a function of the weight ratio. (A.G. D'Aloia et al. Electromechanical modeling of GNP nanocomposites for stress sensor applications, *IEEE NANO 2011*, Portland, Oregon, USA, pp. 1648–1651, © (2011) IEEE. With permissi.)

The deformations of the sensor along the *x*- and *y*-axes, ε_x and ε_y, are computed in the hypothesis that the stress is applied in the (x, y)-plane with components σ_x, σ_y, and that shear stress is not present. In the elastic range, it results in

$$\begin{bmatrix} \varepsilon_x \\ \varepsilon_y \end{bmatrix} = \frac{1}{E} \begin{bmatrix} 1 & -v \\ -v & 1 \end{bmatrix} \begin{bmatrix} \sigma_x \\ \sigma_y \end{bmatrix} \tag{16.12}$$

where E is the measured Young's modulus of the nanocomposite and v is the corresponding Poisson coefficient.

As a consequence, owing to the applied stress condition, the dimensions L_x and L_y of the strain sensor become L_x' and L_y':

$$L_x' = L_x (1 + \varepsilon_x) \tag{16.13}$$

$$L_y' = L_y (1 + \varepsilon_y) \tag{16.14}$$

This implies that, in the hypothesis that the GNPs are nondeformable particles, the average distances among the geometrical centers of the GNPs dispersed in the matrix along the *x*- and *y*-axes, that is, δ_x and δ_y, respectively, become δ_x' and δ_y':

$$\delta_x' = \delta_x (1 + \varepsilon_x) \tag{16.15}$$

$$\delta_y' = \delta_y (1 + \varepsilon_y) \tag{16.16}$$

Therefore, the effective electrical conductivity of the nanocomposite assumes different values γ_x and γ_y along the *x*- and *y*-axes, respectively, as it appears in Equations 16.9 and 16.10. These conductivities are obtained for a given GNP concentration θ replacing in Equation 16.11 γ with γ_x or γ_y, and δ with δ_x' or δ_y'.

Moreover, combining the resulting expressions with Equations 16.15, 16.16, and 16.12, the result is that γ_x, γ_y can be formulated explicitly as functions of the applied stress components σ_x and σ_y:

$$\gamma_x(\theta) = \gamma_0 \exp \{-2\delta_x (\theta) [1 + E^{-1} (\sigma_x - v\sigma_y)]/\xi\} \tag{16.17}$$

$$\gamma_y(\theta) = \gamma_0 \exp \{-2\delta_y (\theta)[1 + E^{-1} (\sigma_y - v\sigma_x)]/\xi\} \tag{16.18}$$

It is noted that the previous expressions are valid in the hypothesis of linearity, that is, that E, and ν are not varying as a function of the applied stress. The electromechanical response of the strain sensor is finally obtained combining Equations 16.17, 16.18 with Equations 16.7 through 16.10, and solving the equivalent network of Figure 16.10b.

16.5 APPLICATIONS AND RESULTS

The proposed model is applied to the analysis of a stress sensor made of GNP nanocomposites at 1% or 2% wt. The thickness t is 0.5 mm and the lateral dimensions are $L_x = 14.7$ mm and $L_y = 8.2$ mm, which are typical of the TIM in an Intel® Dual Core/GT2 Ivy Bridge Die.

The curves reported in Figure 16.12a and b show, respectively, the relative deformations and the lateral dimensions of the strain sensor made of GNP nanocomposites at 1% wt. as a function of σ_x, in case of $\sigma_y = 0$. It is observed that the relative deformation ε_x increases with σ_x, whereas ε_y decreases. Moreover, the dimension of the sensor along the x-axis increases nearly 3 mm when σ_x varies from 0 up to 30 MPa (Figure 16.12b), whereas the length of the y-side decreases less than 2 mm.

The average distances among the particles is then estimated using Equations 16.15 and 16.16 for the two concentrations of 1% wt. and 2% wt. It is observed that as a stress σ_x is applied, δ_x' increases exponentially, while δ_y' decreases, as shown in Figure 16.13a and c. On the other hand, the electrical conductivity along the x-axis γ_x decreases, whereas γ_y increases, as shown in Figure 16.13b and d.

Figure 16.14a shows the voltage across the sensor supplied by the current $I_s = 1$ μA. The corresponding gage factor, shown in Figure 16.14b, is given by

$$G = \frac{(\Delta R)/R_0}{(\Delta L)/L_0} \tag{16.19}$$

where ΔR and ΔL are the changes in electrical resistance and in the length, respectively, due to the applied stress; R_0 and L_0 are the initial electrical resistance and length of the stress sensor.

It is noted that the voltage is strongly dependent on the applied stress for both GNP concentrations. This means that the GNP-based nanocomposites developed in this study can be very sensitive even to small variations of the applied stress, and are also suitable to be used as a vibration sensor.

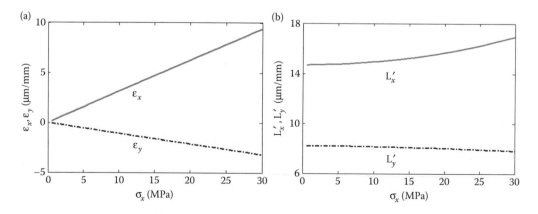

FIGURE 16.12 Relative deformations ε_x, ε_y, (a), and dimensions L_x' and L_y' (b) as a function of the stress σ_x, for a stress sensor made of GNP-nanocomposites at 1% wt.

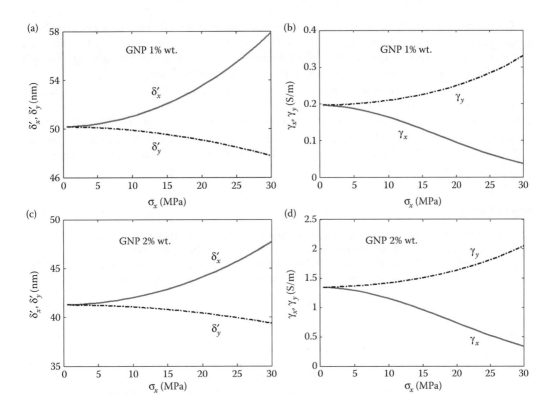

FIGURE 16.13 Average distance δ'_x or δ'_y among the GNPs in nanocomposites with different concentrations (a), (c), and corresponding electrical conductivities γ_x and γ_y (b), (d) as functions of the stress σ_x.

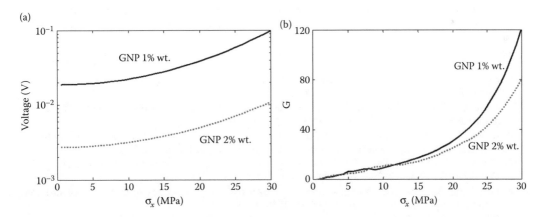

FIGURE 16.14 Voltage across a sensor made of GNP nanocomposites at 1% wt. or 2% wt. (a) of GNPs, and corresponding gage factor G (b) as a function of the applied stress σ_x.

16.6 CONCLUSIONS

In this study, we propose a full electromechanical model of a TIM made of a GNP nanocomposite for integrated stress monitoring of electronic devices. The proposed simulation approach is based on the tunneling percolation model of the dc electrical conductivity of the nanocomposite. Moreover, it is assumed that GNPs are not deformable particles, so that the average distance among them in the composites can be directly correlated to the applied stress condition.

To calibrate the developed model for quantitative analysis, thick GNP films have been produced to measure the dc effective conductivity of the GNPs. Moreover, nanocomposite samples filled with different concentrations of GNPs have been fabricated to estimate the dc conductivity, Young's module, and Poisson coefficient of the composite as functions of the filler concentration.

The obtained results demonstrate that strain sensors containing GNP at 1% wt. or 2% wt. are characterized by an electromechanical characteristic that is very sensitive to small variations of the applied stress. Quantitative experimental validations of the complete electromechanical model of the sensor are the objectives of a future study. Moreover, the combined optimization of the thermal and electromechanical performances of a smart TIM made of a GNP composite in a real device configuration will be carried out.

REFERENCES

1. K.S. Novoselov, A.K. Geim, S.V. Morozov, D. Jiang, Y. Zhang, S.V. Dubonos, I.V. Grigorieva, A.A. Firsov, Electric field effect in atomically thin carbon films, *Science*, 306(5696), 666–669, 2004.
2. H.C. Schniepp, K.N. Kudin, J.L. Li, R.K. Prud'homme, R. Car, D.A. Saville, I.A. Aksay, Bending properties of single functionalized graphene sheets probed by atomic force microscopy, *ACS Nano*, 2(12), 2577–2584, 2008.
3. C. Lee, X. Wei, J.W. Kysar, J. Hone, Measurement of the elastic properties and intrinsic strength of monolayer graphene, *Science*, 321(5887), 385–388, 2008.
4. I.W. Frank, D.M. Tanenbaum, A.M. van der Zande, P.L. McEuen, Mechanical properties of suspended graphene sheets, *J. Vac. Sci. Technol. B*, 25(6), 2558–2561, 2007.
5. A.A. Balandin, S. Ghosh, W. Bao, I. Calizo, D. Teweldebrhan, F. Miao, Superior thermal conductivity of single-layer graphene, *Nano Lett.*, 8(3), 902–907, 2008.
6. X. Du, I. Skachko, A. Barker, E.Y. Andrei, Approaching ballistic transport in suspended graphene, *Nat. Nanotechnol.*, 3(491), 1–5, 2008.
7. Z. Liu, G. Bai, Y. Huang, Y. Ma, F. Du, F. Li, Reflection and absorption contributions to the electromagnetic interference shielding of single-walled carbon nanotube/polyurethane composites, *Carbon*, 45, 821–827, 2007.
8. S.H. Park, P. Thielemann, P. Asbeck, P.R. Bandaru, Enhanced dielectric constants and shielding effectiveness of, uniformly dispersed, functionalized carbon nanotube composites, *Appl. Phys. Lett.*, 94(24), 243111-1-3 2009.
9. L.L. Wang, B.K. Tay, K.Y. See, Z. Sun, L.K. Tan, D. Lua, Electromagnetic interference shielding effectiveness of carbon-based materials prepared by screen printing, *Carbon*, 7(8), 1905–1910, 2009.
10. M.H. Al-Saleh, U. Sundararaj, Electromagnetic interference shielding mechanisms of CNT/polymer composites, *Carbon*, 47(7), 1738–1746, 2009.
11. J. Liang, Y. Wang, Y. Huang, M. Yanfeng, Z. Liu, J. Cai et al. Electromagnetic interference shielding of graphene/epoxy composites, *Carbon*, 47(3), 922–925, 2009.
12. Z. Mo, Y. Sun, H. Chen, P. Zhang, D. Zuo, Y. Liu et al. Preparation and characterization of a PMMA/Ce(OH)3, Pr2O3/graphite nanosheet composite, *Polymer*, 46(26), 12670–12676, 2005.
13. G. Chen, W. Wenig, D. Wu, C. Wu, PMMA/graphite nanosheets composite and its conducting properties *Eur. Polym. J.*, 39(12), 2329–2335, 2003.
14. L.N. Song, M. Xiao, Y.Z. Meng, Electrically conductive nanocomposites of aromatic polydisulfide/expanded graphite, *Comp. Sci. Tech.*, 66(13), 2156–2162, 2006.
15. B.F. He, S. Lau, H.L. Chan, J. Fan, High dielectric permittivity and low percolation threshold in nanocomposites based on poly(vinylidene fluoride) and exfoliated graphite nanoplates, *Adv. Mater.*, 21(6), 710–715, 2009.
16. M.S. Sarto, A.G. D'Aloia, A. Tamburrano, G. De Bellis, Synthesis, modeling, and experimental characterization of graphite nanoplatelet-based composites for EMC applications, *IEEE Trans. EMC*, 54(1), 17–27, 2012.
17. G. De Bellis, A. Tamburrano, A. Dinescu, M.L. Santarelli, M.S. Sarto, Electromagnetic properties of composites containing graphite nanoplatelets at radio frequency, *Carbon*, 49(13), 4291–4300, 2011.
18. G. De Bellis, I.M. De Rosa, A. Dinescu, M.S. Sarto, A. Tamburrano, Electromagnetic properties of carbon-based nanocomposites: The effect of filler and resin characteristics, Int. Symp. IEEE NANO 2010, Seoul, Korea, 486–489, 2010.

19. I.M. De Rosa, F. Sarasini, M.S. Sarto, A. Tamburrano, EMC impact of advanced carbon fiber/carbon nanotube reinforced composites for next-generation aerospace applications, *IEEE Trans. EMC*, 50(3), 556–563, 2008.
20. I.M. De Rosa, A. Dinescu, F. Sarasini, M.S. Sarto, A. Tamburrano, Effect of short carbon fibers and CNTs on microwave absorbing properties of polyester composites containing nickel-coated carbon fibers, *Composites Sci. Technol.*, 70(1), 102–109, 2010.
21. G. De Bellis, I.M. De Rosa, A. Dinescu, M.S. Sarto, A. Tamburrano, Electromagnetic absorbing nanocomposites including carbon fibers, nanotubes and graphene nanoplatelets, IEEE Int. Symp. EMC, Fort Lauderdale, USA, 202–207, 2010.
22. I. Kang, M.J. Schulz, J.H. Kim, V. Shanov, D. Shi. A carbon nanotube strain sensor for structural health monitoring, *Composites Sci. Technol.*, 15(6), 737–-748, 2006.
23. Z.M. Dang, M.J. Jiang, D. Xie, S.H. Yao, L.Q. Zhang, Supersensitive linear piezoresistive property in carbon nanotubes/silicon rubber nanocomposites, *J. Appl. Phys.*, 104(2), 024114-1-6, 2008.
24. Y.-J. Kim, J.Y. Cha, H. Ham, H. Huh, D.-S. So, I. Kang, Preparation of piezoresistive nano smart hybrid material based on graphene, *Curr. Appl. Phys.*, 11(1), S350–S352, 2011.
25. F.R. Al-solamy, A.A. Al-Ghamdi, W.E. Mahmoud, Piezoresistive behavior of graphite nanoplatelets based rubber nanocomposites, *Polym. Adv. Technol.*, 23(3), 478–482, 2011.
26. A.G. D'Aloia, A. Tamburrano, G. De Bellis, M.S. Sarto, Electromechanical modeling of GNP nanocomposites for stress sensor applications, IEEE NANO 2011, Portland, Oregon, USA, pp. 1648–1651, 2011.
27. H. Huang, C. Liu, Y. Wu, Aligned carbon nanotube composite films for thermal management, *Adv. Mater.*, 17(13), 1652–1656, 2002.
28. A.M. Marconnet, N. Yamamoto, M. Panzer, B. Wardle, K.E. Goodson, Thermal conduction in aligned carbon nanotube-polymer nanocomposites with high packing density, *ACSNANO*, 5(6), 4818–4825, 2011.
29. W. Lin, K.S. Moon, C.P. Wong, A combined process of in situ functionalization and microwave treatment to achieve ultrasmall thermal expansion of aligned carbon nanotube-polymer nanocomposites: Toward applications as thermal interface materials, *Adv. Mater.*, 21(3), 2421–2424, 2009.
30. M.T. Hung, O. Choi, Y.S. Ju, T. Hahn, Heat conduction in graphite-nanoplatelet-reinforced polymer nanocomposites, *Appl. Phys. Lett.*, 89(2), 023117-1-6, 2006.
31. A. Yu, P. Ramesh, X. Su, E. Bekyarova, M.E. Itkis, R.C. Haddon, Graphite nanoplatelet-epoxy composites thermal interface materials, *J. Phys. Chem. C*, 111(21), 7565–7569, 2007.
32. A. Yu, P. Ramesh, X. Su, E. Bekyarova, M.E. Itkis, R.C. Haddon, Enhanced thermal conductivity in hybrid graphite nanoplatelet-carbon nanotube filler for epoxy composites, *Adv. Mater.*, 20(24), 4740–4744, 2008.
33. K.M.F. Shahil, A.A. Baladin, Graphene-multilayer graphene nanocomposites as highly efficient thermal interface materials, *Nano Lett.*, 12(2), 861–867, 2012.
34. J. Hicks, A. Behnam, A. Ural, A computational study of tunneling-percolation electrical transport in graphene-based nanocomposites, *Appl. Phys. Lett.*, 95(21), 213103-1-3, 2009.
35. G. Ambrosetti, C. Grimaldi, I. Balberg, T. Maeder, A. Danani, P. Ryse, Solution of the tunneling-percolation problem in the nanocomposite regime, *Phys. Rev. B*, 81(15), 155434-1-12, 2010.
36. J.R. Macdonald, On the mean separation of particles of finite size in one to three dimensions, *Mol. Phys.*, 44(5), 1043–1049, 1981.
37. G. Ambrosetti, N. Johner, C. Grimaldi, A. Danani, P. Ryse, Percolative properties of hard oblate ellipsoids of revolution with a soft shell, *Phys. Rev. E*, 78(6), 061126-1-11, 2008.
38. G. De Bellis, F. Ruggeri, A. Broggi, A. Tamburrano, M.L. Santarelli, M.S. Sarto, Effect of the synthesis parameters on the dc resistance of graphite nanoplatelets thick films, GraphITA, L'Aquila, Italy, 2011.
39. ASTM F390-98, Standard test method for sheet resistance of thin metallic films with collinear four-probe array.
40. ASTM D638-10, Standard test method for tensile properties of plastics.

Section VI

Quantum Cellular Automata

17 An HDL Model of Magnetic Quantum-Dot Cellular Automata Devices and Circuits

Marco Ottavi, Salvatore Pontarelli, Adelio Salsano, and Fabrizio Lombardi

CONTENTS

17.1 INTRODUCTION

Among the disparate emerging technologies that have been proposed to overcome the limitations of "end-of-the-roadmap" CMOS (complementary metal–oxide–semiconductor), quantum-dot cellular automata (QCA) shows promising features to achieve both high computational throughput and low-power dissipation. The QCA computational paradigm [1,2] introduces highly pipelined architectures with extremely high speed (of the order of *THz*), while radically departing from the switch-based operation of CMOS. QCA manufacturability has been demonstrated both for metal-dot QCA [3] and molecular scale allowing room temperature operation. Recently, magnetic QCA (MQCA) based on Co nanomagnets has been analyzed [4–8]. The use of nanomagnets is very attractive because MQCA can operate at room temperature, and has been shown to be easier than the molecular implementation of an electrostatic QCA. Moreover, MQCA can also be integrated with other emerging technologies such as magnetic RAM for memory design. The clocking mechanism of MQCA is similar to electrostatic QCA; the use of abrupt switching in electrostatic QCA is unreliable [2] due to the possible generation of metastable states, so a quasi-adiabatic clocking scheme has been proposed to overcome the kink probability in QCA circuits [2]. For MQCA, a three-phase snake clock has also been proposed [9]. Finally, a technology-based solution has been proposed in Ref. [7] to stabilize the magnetization state of nanomagnets by adding biaxial anisotropy. This arrangement modifies the framework in which MQCA circuits can be designed, thus requiring further investigation into mechanisms (also at circuit level) to leverage the newly introduced functionalities.

In addition to the advances in cell manufacturing and fabrication, research at the higher circuit and system levels has been pursued for QCA. Various QCA architectural solutions have been proposed, such as memories [10,11] and microprocessors [12]. As for tools, QCADesigner [13] has been widely utilized by manually placing the electrostatic QCA cells on a two-dimensional layout and simulating their behavior. This is accomplished by solving the quantum equations describing the QCA circuit. However, owing to its low-level model, QCADesigner incurs high computational penalties and is not suitable to design or simulate logic circuits of even medium complexity; therefore, new environments suitable for CAD implementation must be devised for circuit-level QCA design. Hence, both a SPICE level model [14] and a simple VHDL level model [15] have been proposed for nanomagnet-based devices.

Ottavi et al. [16] have introduced an HDL-based design tool (HDLQ) that overcomes the limitations of simulators like QCADesigner with respect to a circuit-level evaluation for electrostatic QCA. In this chapter, an HDL framework (and associated tool) based on Ref. [7] is proposed as a viable support for the design of systems based on the MQCA nanodevices. The models of the proposed framework are compatible with the HDLQ framework and also provide a support to simulate the behavior of MQCA. The modeling proposed in this work is based on Ref. [7] and is valid until the underlying physical assumptions will be proven viable for the actual manufacturing of these devices. This framework utilizes different and novel models by which MQCA cells can be simulated and a circuit-level assessment can be pursued at reduced computational complexity compared with other (physically based) simulators, such as OOMMF. HDL models for a MQCA cell as well as building blocks are proposed to ensure magnetization, clocking, and signal propagation; functions and test benches are also presented for the proposed CAD tool (denoted by HDLM). Moreover, the lazy AND and the dictator gates are modeled to ensure correct MQCA operation. The effectiveness of the proposed tool is further evidenced by its application to the novel design presented in this chapter for an n-input AND gate.

This chapter is organized as follows: Section 17.2 provides an overview of MQCA; in Section 17.3, the basic principles of the proposed framework and tool (denoted as HDLM) are described. In Section 17.4, the HDL model of a MQCA cell is introduced in detail. Section 17.5 presents the functional model used in the HDL simulation for the MQCA building blocks. Some of these blocks (wire, majority voter) are similar to the electrostatic implementation, while others (the lazy AND and the dictator gate) are specific to an MQCA implementation. Section 17.6 introduces a novel n-input AND gate that exploits the characteristics and functionality of MQCA; this also shows the effectiveness of the proposed tool to investigate new gates while establishing at a functional level its operational features. The conclusion is provided in the last section.

17.2   REVIEW

QCA operates on a computational paradigm based on the interactions of a set of bistable cells. The two stable states of a cell lead to a straightforward correspondence with the logic (boolean) values of zero and one.

The initial proposal of electrical quantum-dot cellular automata (EQCA or just QCA) was based on the Coulombic interaction of arrays of cells or QCA cells. Each cell contains four quantum dots which trap two extra electrons. Coulombic interaction within the QCA cells is such that only two stable states are possible, that is, those in which the extra electrons occupy diagonally positioned quantum dots (Figure 17.1a). Through Coulombic interactions and an externally imposed clocking E-field, adjoining QCA cells are then able to transfer and process boolean operations or transfer information in Figure 17.1b in which the EQCA binary wire (i.e., the combination of cells that permits data transfer), inverter (i.e., the combination of cells that permits the inversion of the boolean value), and the majority voter (i.e., the combination of cells that permits one to perform the majority

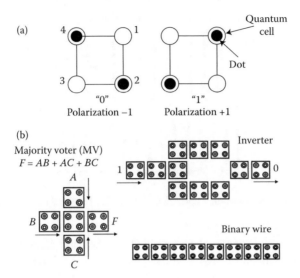

FIGURE 17.1   (a) Bistable feature of an EQCA cell (b) EQCA gates.

function $F = AB + AC + BC$) are depicted. EQCA cells, moreover, can be put in a metastable state through the E-field clock to process consecutive operations on the same cells; the possibility of having an inversion of the value stored on the cells (the so-called kink) is a function of the dimension of the clocking zones [17].

Recently, the QCA functional paradigm has raised attention when associated to magnetic devices in which the information is encoded in the magnetization state of the ferromagnetic devices. These devices (normally used for data storage) have shown to be particularly well suited also for logic functionality when included in architectures comparable to those used in QCA.

Cowburn and Welland [4] have realized MQCA operation by manufacturing chains of 110 nm diameter disk-shaped magnetic particles that manifest collective behavior. The magnetic anisotropy of the chains determines the preferred magnetization direction of disks and consequently the processing of binary information. Imre et al. [18] introduce a further shape-induced anisotropy with the creation of narrow nanomagnets on the scale of 10–100 nm. They can be assumed as single-domain magnets while still above the superparamagnetic limit. These elongated micromagnets are extremely stable given their shape-induced anisotropy. The magnetization points always to their long axis with two possible verses when no external magnetic field is induced. MQCA devices based on these magnets are expected to operate at room temperature and yield great promise for circuit design.

Similar to the EQCA functional paradigm, Figure 17.2a shows the two stable states of an MQCA cell (also referred to as a nanomagnet). An MQCA cell can also assume a metastable null state: while in the logic (zero and one) states, the magnetization is aligned to the vertical axis (which has a stable energy level), the magnetic field in the metastable state is aligned horizontally and does not interact with the neighboring cells (therefore, corresponding to a functionally null state).

Logic operation and signal propagation are performed in two steps in MQCA [7]. In the first step, all nanomagnets are aligned along their magnetically hard (lithographically short) axes by applying a global external magnetic field. In the second step, the external field is removed. If an input is imposed, then the dipole field alignment between the neighbors pushes them out of their metastable state and induces an antiparallel magnetization state. Therefore, the behavior of a horizontal line of MQCA cells can be viewed as a chain of inverters propagating the signal by successive operations (inversion) of the input value. Figure 17.2b shows the state of a chain of nanomagnets (binary wire) in its stable state.

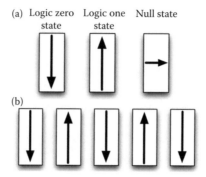

FIGURE 17.2 (a) Bistable feature of an MQCA cell (b) MQCA binary wire. (Copyright 2011 IEEE, with permission.)

Similar to the kink occurrence in electrostatic QCA, the cascade propagation in a horizontal wire may fail when the number of MQCA cells is increased [5]. To overcome this limitation, one of the most commonly used solutions requires the partition of an MQCA circuit into small zones driven by a suitable clock circuitry [8]. Carlton et al. [7] have proposed a solution to this problem; it introduces a hard axis stability by adding a biaxial anisotropy term to the net magnetization energy of each nanomagnet. Simulation performed using the OOMMF simulator [19] has shown that this technique allows a signal to correctly propagate up to 30 nanomagnets. When simulated, the input is transferred through 30 nanomagnets in 3 ns, corresponding to a propagation time through a single MQCA cell of approximately 100 ps. While a horizontal wire of nanomagnets tends to align in an antiparallel configuration, the vertical wire tends to align in parallel. Therefore, the horizontal wires invert the signal, while the vertical wires perform no inversion.

The basic logic gate for MQCA is still the majority voter and works in a similar fashion as for electrostatic QCA, that is, the output cell assumes the configuration of the majority of the inputs (Figure 17.3a). Together with the inversion provided by the antiparallel magnetization, this forms a functionally complete gate set.

Carlton et al. [7] have addressed the behavior of majority gates with legs of unequal length. The input that arrives earlier at the majority gate of Figure 17.3b imposes the magnetization on the center cell (crossing) of a majority gate. This may generate the wrong output and propagate an erroneous result also toward the other legs of the crossing. Figure 17.3b shows the propagation of information from the input nearest to the crossing toward the other inputs and to the output. A metastable (null)

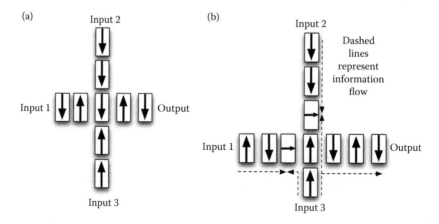

FIGURE 17.3 (a) MQCA majority gate (b) MQCA majority gate with uneven legs. (Copyright 2011 IEEE, with permission.)

A	B	output
0	0	0
0	1	0
1	0	0
1	1	z

FIGURE 17.4 Truth table of lazy AND gate. (Copyright 2011 IEEE, with permission.)

state is still present in those nanomagnets that have an equal distance from the inputs. To address the issues related to this race condition in the signals of the majority gate, Carlton et al. [7] have proposed two functional blocks:

1. *Lazy AND:* this gate acts as an AND gate when the output is supposed to be a logic zero. When the gate should output a logic 1, it generates no output, that is, the value of its output cell is in the metastable (null) state.
2. *Dictator (majority) gate:* this is a modified majority gate, in which the two vertical inputs (i.e., labeled 1 and 3) have a weaker coupling to the center nanomagnet (the nanomagnets are separated by a longer distance). It can only change the output provided all inputs agree; otherwise, the output toggles only once the value coming from the input labeled 2 reaches the center nanomagnet.

The lazy AND gate is physically realized by adding some extra nanomagnets (orthogonally placed with respect to the original direction) to block the magnetization corresponding to the logic one value. The truth table of the lazy AND is given in Figure 17.4. If one of the inputs is zero, then the output is zero; otherwise, the output remains in the metastable state (denoted as z, in analogy with the high impedance state of CMOS technology). By the truth table, the arrival order of the inputs is irrelevant. For example, assume that the A input signal arrives first: if A is zero, then the output is zero regardless of the value of B. However, if A is one, the gate will wait for the arrival of the input signal on B to decide whether the output must toggle to zero or remain in the metastable state.

For the dictator majority gate, the order in which the inputs arrive is important. Assume that the two vertical inputs and the single horizontal input are available; let the horizontal input be defined as *dominant*, that is, the distance of the horizontal nanomagnet from the center nano-magnet is smaller than the distance for the vertical nanomagnets. However, the magnetization of only one vertical input is not sufficient to toggle the majority gate. The vertical inputs can impose the value on the majority gate only if they arrive first and are identical. If only one vertical input reaches the center nanomagnet, or the vertical inputs are in disagreement (different values), then the gate remains in the z state. The horizontal input is capable of imposing the magnetization on the gate if the gate is still in the metastable state when the horizontal signal arrives. Using the lazy and the dictator gates, new and different logic gates can be designed; these gates are insensitive to the propagation delay of the signals. Figure 17.5 shows an AND gate that is insensitive to the order of arrival at its inputs.

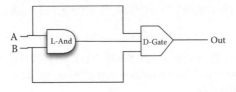

FIGURE 17.5   AND gate realized with dictator and lazy gates. (Copyright 2011 IEEE, with permission.)

17.3   HDL FRAMEWORK

The complexity of low-level physical simulations introduces more usable simulation models such as already proposed at SPICE-like level in Ref. [14] or at VHDL level in Ref. [15] in which a VHDL model for domain-wall-based devices has been proposed). This chapter proposes a novel HDL-based framework (with associated tool) by which MQCA can be assessed by simulation. The tool of the proposed HDL framework is referred to as HDLM due to its compatibility with a previous tool proposed by the same authors and applicable to electrostatic QCA design [16]. Modeling in HDLM relies on two components.

- A model for the MQCA cell as related to the unique features of this technology (such as magnetization) and its interaction with the immediate neighboring cells.
- The models for some basic building blocks for designing MQCA circuits. These building blocks include the majority voter as well as a few gates that are specifically used to alleviate some of the problems incurred in signal propagation by MQCA.

HDLM is compatible with HDLQ [16] because it uses similar principles and data structures; this permits primitives to be utilized as models for characterizing the cell and circuits, while changing the mode of operation from Coulombic interactions (for electrostatic QCA) to magnetization (for MQCA).

Similar to Ref. [16], the model is realized by using a Verilog HDL description of MQCA. The comparative advantages of such an approach are as follows:

1. OOMMF is not suitable to simulate circuits (even of modest size) due to the complexity of the equations involved in the MQCA cells.
2. The overall design process is highly simplified when an HDL description is used. Moreover, as HDL is widely used in the digital design community, many tools are available and compatible with this language and description.

Therefore, a Verilog description can be used to model the basic MQCA cell, and the structures described in the previous section, that is, the lazy AND and the dictator gates. HDL modeling allows the presence of an event-driven simulation to be leveraged, so, for example, the occurrence of transitions on a neighboring cell induces an event, and the evaluation of the next state in the cells. Consequently, by introducing a delay in updating the output of a cell, a cascade of switching events, as expected in a nanomagnet array, can be generated to model MQCA.

17.4   HDL MODEL OF AN MQCA CELL

This section describes the model for the behavior of an MQCA cell as introduced in the previous section. In HDLM, an MQCA cell requires a different characterization from the electrostatic cell of HDLQ. In particular, different functions must be utilized in the model to capture the magnetic properties of MQCA.

17.4.1   I/O INTERFACE

For an MQCA cell, we must define an I/O interface, that is, a model by which cell interactions occur among neighboring cells in the layout. In HDLM, this interface is characterized by the following features.

- Four inputs corresponding to the North, South, East, and West (N,S,E,W) directions.
- The input (called status) corresponding to the application of the external magnetic field that provides the metastable state.
- An output corresponding to the value assumed by the cell itself.

```
module MQCAcell (
                // directions
                N,
                E,
                S,
                W,
                //output
                value,
                //clock status
                //status
);
        //status 00 = relax, 01 = switch, 10 = hold, 11 = release
        parameter reset = 2'b00;
        parameter switch = 2'b01;
        parameter hold = 2'b10;
        input N, E, S, W;
        input [1:0] status;
        output value;
        reg value;
```

FIGURE 17.6   Interface of the Verilog code for an MQCA device.

While the N and S directions contribute to cell magnetization in a parallel manner, the E and W directions contribute in an antiparallel manner. Directional inputs and the output can assume three values, corresponding to the logic (zero and one) and to the metastable states (i.e., the z value).

Finally, the external magnetization field is considered to act as a clock signal and can assume the reset, switch, or hold values. If the clock signal is reset, the output of the cell is z, regardless of the value of the other inputs; otherwise, the Verilog model evaluates the magnetization of the cell by using a magnetization function based on the values seen at the directional inputs N, S, E, W.

The I/O interface of the Verilog module is presented in Figure 17.6. The module declares the above-mentioned inputs and outputs, and the parameters correspond to the possible states (*reset*, *switch*, *hold*) of the clock signal.

17.4.2 MAGNETIZATION

To evaluate the magnetization of the cell, a function converting the binary value to a magnetization value is used. The function *bin2mag* is given by

$$bin2mag = \begin{cases} +1 & \text{if } x = 1 \\ 0 & \text{if } x = z \\ -1 & \text{if } x = 0 \end{cases} \qquad (17.1)$$

The function *bin2mag* permits the magnetization of a cell to be evaluated by adding the magnetization values of the vertical inputs and subtracting the magnetization values of the horizontal inputs using

$$output(N,S,E,W) = bin2mag(N) + bin2mag(S) \\ - (bin2mag(E) + bin2mag(W)) \qquad (17.2)$$

The last step of this computation is the reverse conversion from the magnetization to the binary representation that follows from Equation 17.2. In Figure 17.7, the Verilog functions implementing the above two equations have been reported.

```
function integer bintomag; // conversion function from binary to
polarization
input a; // input argument port
        case (a)
        1'b0: bintomag=-1;
        1'b1: bintomag=10;
        1'bz: bintomag=0;
        1'bx: bintomag=0;
endcase
endfunction
function compute; // function definition
input N, E, S, W;
integer polarization;
        begin
        polarization=bintomag(N)+bintomag(S)-1*(bintomag(E)+bintomag(W));
        // reverse conversion
        if (polarization>0)
           compute=1'b1;
        else if (polarization<0)
           compute=1'b0;
        else
           compute=1'bz;
        end
  endfunction
```

FIGURE 17.7   Verilog code for the magnetization functions.

17.4.3 PROPAGATION AND CELL PLACEMENT

To emulate the propagation delay through the nanomagnets, the output receives the computed value within a specified delay (set to a default value of 100 ps). The use of the magnetization function closely resembles the physics behavior. It has a high level of flexibility because it can be used to provide weights to the inputs (as discussed in a later section when the model of the dictator gate will be described). Figure 17.8 shows the Verilog code describing the propagation delay in the MQCA device.

Placement and connection between cells are performed as follows: the nanomagnets are placed on a grid layout, such that each magnet can have at most four neighbors, one for each direction. The directional inputs (N,S,E,W) are connected to the output of the corresponding neighbor, if present. If no cell is present in that direction, then the corresponding input is connected to a fixed z value. Finally, all cells are connected to a specific clock signal. The use of a clock signal is utilized to define the so-called clocking zones, similar to Ref. [9].

17.5 HDL MODELS OF MQCA BUILDING BLOCKS

HDL simulation models are presented in this section; MQCA building blocks, such as a wire, the majority gate, and the specific structures that have been proposed for MQCA (lazy AND and dominant majority gate), are assessed and evaluated.

```
always @(status,N,E,S,W)
  begin
  if (status == reset) //reset
     value <=1'bz;
  else if (value ==1'bz) //compute
     value <= #1 compute(N,E,S,W);
  end
```

FIGURE 17.8   Verilog code for evaluation of the magnetization of an MQCA device.

17.5.1 MQCA Binary Wire

The binary wire is a well-known block of the QCA functional paradigm. It is composed of a series of adjacent MQCA cells and allows the propagation of the information through the nanomagnets. Figure 17.9 shows the Verilog code of a horizontal MQCA wire. The MQCA cells making up the wire have been indexed with x and y coordinates in the two-dimensional grid layout. The signal connecting the output of a cell to the input of its neighbor is defined as a bidimensional wire. The output of the cell (with coordinates (x,y)) is connected to the signal $v[x][y]$, while the west inputs are connected to $v[x-1][y]$ and the east input is connected to the $v[x+1][y]$ signal. The simulation starts by initially imposing an external magnetic field, corresponding to clock = "1," thereby, forcing the z value to all cells. After the clock is lowered, the signal starts propagating in the west–east direction by the concatenation of events as triggered by the change in state of the neighboring cell located in the west (W) direction. Figure 17.10 shows its waveforms. Each output change occurs with the allowed switching time of a nanomagnet; currently, this is set to the default value of 100 ps delay, as reported in Ref. [7]. The wire behaves like the cascading effect of a domino chain; after all the cells are magnetized, the wire remains in a steady state until a new clock rising event erases the attained state.

17.5.2 MQCA Majority Gate

A further building block that is modeled in HDLM is the majority gate. Figure 17.11 shows its MQCA layout; its inputs are positioned in the North, South, and West directions, while the output is in the East direction. The vertical inputs are noninverting, that is, only the horizontal input is inverting. In Figure 17.11, all inputs have the same length, and therefore they have the same delay. The waveform in Figure 17.11 shows the inputs, the values of the MQCA cells (for the input values to the majority gate), and the MQCA cells making up the wire that moves the computed value to the output. The signals arrive at the cell labeled Q22 at the same time, as expected for a correct functionality of the majority gate. Moreover, similar to the binary wire case, the waveforms show that for the majority voter also, a change in the inputs will not affect the computed value after the cells attain their final value.

```
MQCAcell Q02(nocell, v[1][2], nocell, in,      v[0][2], clock); //input 1
MQCAcell Q12(nocell, v[2][2], nocell, v[0][2], v[1][2], clock); ///wire
MQCAcell Q22(nocell, v[3][2], nocell, v[1][2], v[2][2], clock); ///wire
MQCAcell Q32(nocell, v[4][2], nocell, v[2][2], v[3][2], clock); ///wire
MQCAcell Q42(nocell, v[5][2], nocell, v[3][2], v[4][2], clock); ///wire
MQCAcell Q52(nocell, nocell,  nocell, v[4][2], out,     clock); //output
```

FIGURE 17.9 Verilog code of an MQCA wire. (Copyright 2011 IEEE, with permission.)

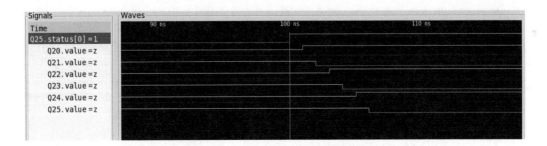

FIGURE 17.10 Waveform of an MQCA wire. (Copyright 2011 IEEE, with permission.)

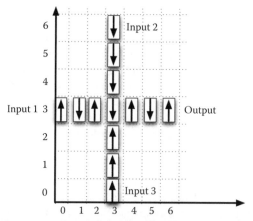

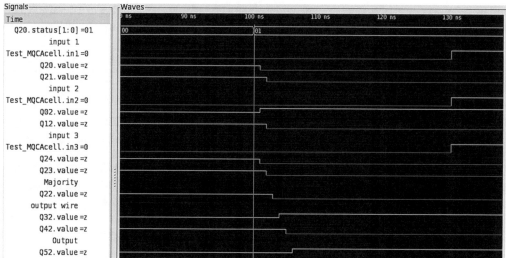

FIGURE 17.11 Layout and waveform of an MQCA majority gate. (Copyright 2011 IEEE, with permission.)

17.5.3 MQCA-SPECIFIC BLOCKS

The functionality of the proposed simulation model in HDLM is extended to the MQCA-specific blocks (the lazy AND and the dictator gates).

1. *Lazy AND gate:* The lazy AND gate can be constructed from the majority gate by employing two modifications. The first modification consists of setting one of the inputs to the logic zero value to achieve an AND function with the remaining two inputs. The other modification is related to the computation of the so-called *mag2bin* function. As per the implementation of the lazy gate, the nanomagnet that computes the AND function can assume only a zero or z value, so the *mag2bin* function must be modified as per the following equation:

$$mag2bin = \begin{cases} 0 & \text{if } M < 0 \\ z & \text{if } M \geq 0 \end{cases} \qquad (17.3)$$

The lazy AND gate can therefore be constructed as an MQCA cell with an input whose value is fixed at 0 and the *mag2bin* function is modified as in the above equation.

2. *Dictator gate:* The dictator gate can be considered as a modification of the majority gate. This gate is a majority gate that has vertical inputs at a distance further away than expected. Therefore, the interaction between the vertical inputs and the center of the majority gate is weaker. This behavior can be simulated by adding a weight in the magnetization function as given previously in Equation 17.2. As in accordance with Ref. [7], only when the vertical inputs have the same logic value, it is possible to impose a value at the output of the gate. When only one of the inputs is defined, or the two inputs have different values, then the output of the gate is in the z state. The modified magnetization function is given as follows:

$$output(N, S, E, W) = 0.5 \cdot bin2mag(N) + 0.5 \cdot bin2mag(S)$$
$$- (bin2mag(E) + bin2mag(W)) \tag{17.4}$$

This function produces a value with magnitude greater than 0.5 when either the horizontal input is set, or the two vertical inputs have the same value. The corresponding *mag2bin* function can be expressed as follows:

$$mag2bin = \begin{cases} 0 & \text{if } M < -0.5 \\ z & \text{if } -0.5 \le M \le 0.5 \\ 1 & \text{if } M > 0.5 \end{cases} \tag{17.5}$$

The above-described models have been implemented in Verilog and therefore can be used to simulate circuits as well as different gates as presented in the next section.

17.6   A NOVEL N-INPUT AND GATE IN MQCA

In this section, it is shown that the MQCA-specific building blocks introduced in the previous section together with HDLM can be used to design a novel MQCA gate, namely, an n-input AND gate. By leveraging the functional characteristics of MQCA with the lazy AND gate, a compact implementation of an n-input AND gate can be designed. This gate consists of two blocks.

- A block made of multiple lazy AND gates
- A block that resolves the output magnetization when all inputs are 1

The first block works similar to a wired AND gate, in which multiple lazy AND outputs are connected onto a single output wire (Figure 17.12 for $n = 7$). This makes the layout similar to a wired AND; when at least one input is 0, then the correspondent lazy AND output will dominate the magnetization on the output wire (because the other outputs will be in the so-called z state). When all inputs are equal to 1, the wired output remains in the z state; this condition is resolved by the second block referred to as the resolution block. The resolution block has the multiple lazy AND gate output and an additional input as inputs (denoted as "the longest input wire" in Figure 17.12). It generates a 1 as output if the additional input is 1 and the other input is z, 0 otherwise. This block is therefore made of the longest input wire and the output wire. According to the order of arrival of the inputs, the n-input lazy AND gate has no constraint (being composed of 2-input lazy AND gates), whereas the second block has to be designed such that the signal propagating on the longest input wire must always arrive after the result of the lazy AND wired function. This constraint can be accomplished by using a snake-shaped wire such as the one depicted in Figure 17.12. All inputs are placed in the

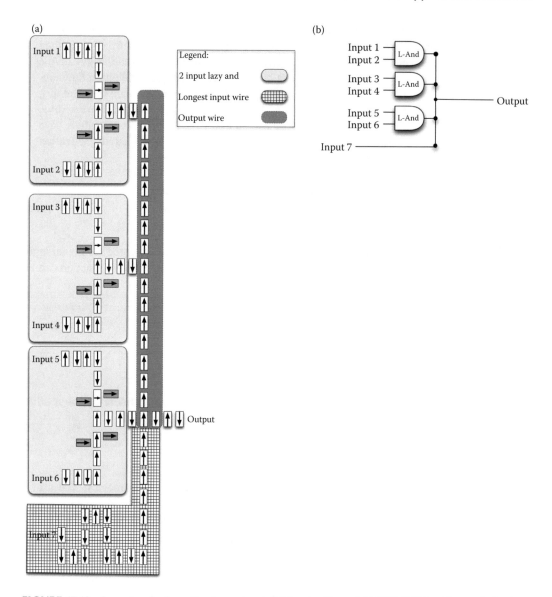

FIGURE 17.12 Layout and schematic of an *n*-input AND gate. (Copyright 2011 IEEE, with permission.)

W direction, while the output is in the E direction. The proposed AND gate has been described using the previously presented Verilog models and simulated to verify its correctness. The simulation results prove the functionalities of the proposed *n*-AND.

17.7 CONCLUSION

This chapter has presented a new framework for analyzing MQCA devices and circuits; this framework has been implemented in HDL and a tool (HDLM) has been designed. In HDLM, the Verilog description has been implemented at the nanomagnet level to leverage the event-driven simulation engine ability of modeling a cascading propagation effect. Different weights are possible in HDLM to model the interaction between the neighboring magnets, thus allowing the operation of MQCA-specific functionalities (such as the lazy AND and the dictator gates) to be evaluated. With the proposed tool, it is possible to model and simulate not only the typical QCA paradigm building blocks but also the specific gates that have been introduced for MQCA. Finally, a novel MQCA gate (i.e., an n-input AND) has been introduced in this chapter; the operation of this gate has been simulated by utilizing HDLM. The proposed n-input AND gate exploits the novel characteristics of the MQCA functional paradigm and has been assessed using the proposed tool.

ACKNOWLEDGMENT

This research was partially funded by the Italian Ministry for University and Research; Program "Incentivazione alla mobilità di studiosi stranieri e italiani residenti all'estero," D.M. n.96, 23.04.2001.

REFERENCES

1. C.S. Lent, P.D. Tougaw, W. Porod, and G.H. Bernstein, Quantum cellular automata., *Nanotechnology*, 4(1), 49–57, 1993.
2. C.S. Lent and P.D. Tougaw, A device architecture for computing with quantum dots, in *Proc. of the IEEE*, 85, 541–557, 1997.
3. A.O. Orlov, I. Amlani, G.H. Bernstein, C.S. Lent, and G.L. Snider, Realization of a functional cell for quantum-dot cellular automata, *Science*, 277, 928–931, 1997.
4. R.P. Cowburn and M.E. Welland, Room temperature magnetic quantum cellular automata, *Science*, 287(5457), 1466–1468, 2000.
5. A. Imre, G. Csaba, G.H. Bernstein, W. Porod, and V. Metlushko, Investigation of shape-dependent switching of coupled nanomagnets, *Superlattices and Microstructures*, 34(3–6), 513–518, 2003, *Proc. of the 6th Int. Conf. on New Phenomena in Mesoscopic Structures*, Hawaii.
6. M.T. Alam, J. DeAngelis, M. Putney, X.S. Hu, W. Porod, M. Niemier, and G.H. Bernstein, Clocking scheme for nanomagnet QCA, in *IEEE-NANO 2007*, Hong Kong, Aug. 2007, pp. 403–408.
7. D. Carlton, N. Emley, E. Tuchfeld, and J. Bokor, Simulation studies of nanomagnet-based architecture, *Nano Letters*, 8(12), 4173–4178, 2008.
8. A. Kumari and S. Bhanja, Landauer clocking for magnetic cellular automata (mca) arrays, *Very Large Scale Integration (VLSI) Systems, IEEE Transactions on*, 99, 1–4, 2010.
9. M. Graziano, A. Chiolerio, and M. Zamboni, A technology aware magnetic QCA NCL-HDL architecture, in *9th IEEE Conference on Nanotechnology, 2009. IEEE-NANO 2009*, Genoa, Italy, July 2009, pp. 763–766.
10. M. Ottavi, V. Vankamamidi, F. Lombardi, S. Pontarelli, and A. Salsano, Design of a QCA memory with parallel read/serial write, in *VLSI, 2005. Proc. IEEE Comp. Soc. Annual Symp. on*, May 2005, pp. 292–294.
11. M. Ottavi, V. Vankamamidi, F. Lombardi, and S. Pontarelli, Novel memory designs for QCA implementation, in *5th IEEE Conference on Nanotechnology, 2005*, July 2005, pp. 545–548 vol. 2.
12. M.T. Niemier and P.M. Kogge, Logic in wire: Using quantum dots to implement a microprocessor, in *The 6th IEEE International Conference on Electronics, Circuits and Systems, 1999. Proceedings of ICECS '99*, Pafos, Cyprus, 1999, Vol. 3, pp. 1211–1215.
13. K. Walus, T.J. Dysart, G.A. Jullien, and R.A. Budiman, QCAdesigner: A rapid design and simulation tool for quantum-dot cellular automata, *IEEE Trans. Nanotechnology*, 3(1), 26–31, 2004.
14. G. Csaba, A. Imre, G.H. Bernstein, W. Porod, and V. Metlushko, Nanocomputing by field-coupled nanomagnets, *Nanotechnology, IEEE Transactions on*, 1(4), 209–213, 2002.
15. J.-O. Klein, E. Belhaire, C. Chappert, R.P. Cowburn, D. Petit, and D. Read, VHDL simulation of magnetic domain wall logic, *Magnetics, IEEE Transactions on*, 42(10), 2754–2756, 2006.
16. M. Ottavi, L. Schiano, F. Lombardi, and D. Tougaw, HDLQ: A HDL environment for QCA design, *J. Emerg. Technol. Comput. Syst.*, 2(4), 243–261, 2006.
17. V. Vankamamidi, M. Ottavi, and F. Lombardi, Two-dimensional schemes for clocking/timing of QCA circuits, *Computer-Aided Design of Integrated Circuits and Systems, IEEE Transactions on*, 27(1), 34–44, 2008.
18. A. Imre, G. Csaba, L. Ji, A. Orlov, GH Bernstein, and W. Porod, Majority logic gate for magnetic quantum-dot cellular automata, *Science*, 311(5758), 205–208, 2006.
19. M. J. Donahue and D. Porter, OOMMF User's guide, *Interagency Report NISTIR,*, no. 6376, 1999, National Institute of Standards and Technology, Gaithersburg.

18 Restoring Divider Design for Quantum-Dot Cellular Automata

Seong-Wan Kim and Earl E. Swartzlander, Jr.

CONTENTS

18.1 INTRODUCTION

Quantum-dot cellular automata (QCA) [1,2] is a promising emerging nanotechnology that may mitigate the problems that are anticipated for CMOS (complementary metal–oxide–semiconductor) due to the continued reduction of feature sizes. Since QCA operate according to different principles from CMOS technology, they require different design methods. Numerous nanoelectronic devices are being investigated and many experimental devices have been developed. DNA origami may serve as the interface between QCA circuits and CMOS systems on silicon substrates [3]. An architecture for data input into a molecular QCA circuit from an external CMOS circuit has been proposed [4]. The controlled formation and occupation of a new form of quantum-dot assemblies at room temperature have been demonstrated [5]. Furthermore, there is an attempt to integrate CMOS, single-electron transistors, and QCA [6].

High-level circuit design is needed to keep pace with the changing physical characteristics. Arithmetic units, especially adders, multipliers, and dividers, play an important role in the design of digital processors and application-specific systems.

Iterative computational circuit designs for QCA are difficult to build with conventional sequential circuit design methods that are based on state machines. State machines for QCA have problems due to long delays between the state machine and the units to be controlled. Even a simple 4-bit microprocessor that has been implemented with QCA [7] was made without using a state machine. Owing to the difficulty of designing sequential circuits, there has been little research into using QCA to realize iterative computational units, such as dividers. Most previous research has focused on simpler arithmetic unit designs, such as adders and multipliers.

This chapter presents a digit recurrent restoring binary divider. To use pipelining, an array structure is implemented in QCA. In Section 18.2, QCA cells and their operation as logic components are explained. In Section 18.3, the restoring binary divider design is presented. In Section 18.4, an implementation of the array restoring dividers using the proposed method is reviewed in detail. Finally, the simulation results and analysis of the designs and the summary are presented in Sections 18.5 and 18.6.

18.2 BACKGROUND

18.2.1 QCA CELL

A semiconductor QCA cell has four quantum dots and two electrons that are trapped inside the cell. The electrons reside in the dots as shown in Figure 18.1. Binary information is encoded by the positions of the electrons, and a QCA cell allows two available polarizations, $P = \pm 1$. Since the quantum dots are coupled by tunnel barriers, the two electrons can change their positions freely by controlling the potential barriers using a clocking mechanism. The computation is performed by interactions based on Coulombic forces between neighboring QCA cells. Since the basic principle of operation is very different from CMOS, QCA "circuits" have many unique characteristics [8–10].

18.2.2 LOGIC GATES IN QCA

The basic circuit elements in QCA are inverters and 3-input majority gates. All the other logic gates, such as AND gates and OR gates, can be realized using these basic elements. A conventional QCA inverter and its symbol are shown in Figure 18.2. Since the conventional inverter is large, several variations of the inverter have been developed as shown in Figure 18.3. The various inverters are used for the implementations in this chapter since their different relative positions of the input cells and the output cells are useful for circuit optimization.

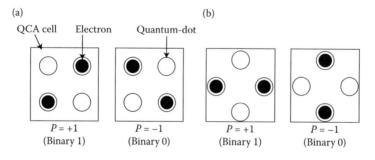

FIGURE 18.1 Basic QCA cells with two possible polarizations. (a) Regular cells, (b) 45° rotated cells. (S.-W. Kim and E. E. Swartzlander, Jr., Restoring divider design for quantum-dot cellular automata, *IEEE 11th Conference on Nanotechnology (NANO-2011)*, Portland, OR, August 15–19, pp. 1295–1300. © (2011) IEEE. With permission.)

The 3-input majority gate in QCA is configured as shown in Figure 18.4. Its function is expressed using the following logic equation: $M(A,B,C) = AB + BC + CA$.

2-Input AND gates and OR gates are implemented by fixing one input of the majority gate as follows:

$$A \cdot B = M(A,B,0)$$

$$A + B = M(A,B,1)$$

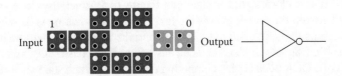

FIGURE 18.2 Layout and schematic symbol of a conventional inverter in QCA. (S.-W. Kim and E. E. Swartzlander, Jr., Restoring divider design for quantum-dot cellular automata, *IEEE 11th Conference on Nanotechnology (NANO-2011)*, Portland, OR, August 15–19, pp. 1295–1300. © (2011) IEEE. With permission.)

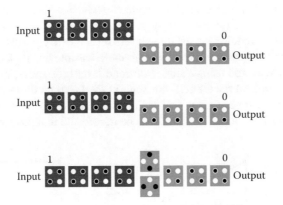

FIGURE 18.3 Layout of various inverters in QCA. (S.-W. Kim and E. E. Swartzlander, Jr., Restoring divider design for quantum-dot cellular automata, *IEEE 11th Conference on Nanotechnology (NANO-2011)*, Portland, OR, August 15–19, pp. 1295–1300. © (2011) IEEE. With permission.)

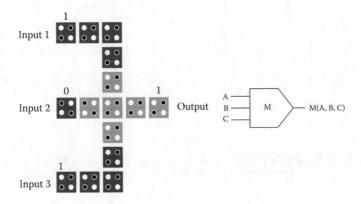

FIGURE 18.4 Layout and schematic symbol of a 3-input majority gate in QCA. (S.-W. Kim and E. E. Swartzlander, Jr., Restoring divider design for quantum-dot cellular automata, *IEEE 11th Conference on Nanotechnology (NANO-2011)*, Portland, OR, August 15–19, pp. 1295–1300. © (2011) IEEE. With permission.)

While the logic optimization methods for CMOS can be used for QCA circuits, majority logic reduction methods specialized for QCA [11,12] are especially useful for circuits using XOR gates. For example, a full adder circuit can be implemented using only three majority gates and a few inverters.

18.2.3 Clock Zones

All QCA cells are organized into clock zones. The designs in this chapter use four clock zones, and the computations are performed sequentially in the same order as that of clock zones. Each clock zone has a different clock signal as shown in Figure 18.5. When the clock signal is high, the potential barriers between the quantum dots are low and the polarization is 0. When the clock signal is low, the electrons in a QCA cell are localized and the polarization will be held as ±1. Using these 90° phase-shifted signals, each clock zone has one of four phase states among Switch, Hold, Release, and Relax. A QCA cells begins computing during the Switch state and holds the polarization during the Hold state. The QCA cell prepares for the next computing during the Release and Relax states. A QCA wire transfers information using clock zones as shown in Figure 18.6.

18.2.4 Coplanar Wire Crossings

There are two kinds of wire crossings in QCA: coplanar wire "crossings" and multilayer crossovers. Coplanar wire crossings are implemented using only one QCA layer, which is an advantage. In contrast, multilayer crossovers require at least three layers for the wire crossovers and via interconnections. Since coplanar wire crossings currently seem more feasible for implementation, the dividers in this chapter are implemented using them. With multilayer crossovers, the full subtractor, multiplexer, and controlled full subtractor cells are slightly smaller than those designed with coplanar wire crossings. Coplanar wire crossings [13] are implemented using regular cells and 45° rotated cells as shown in Figure 18.7. Signals A and B can be transferred independently.

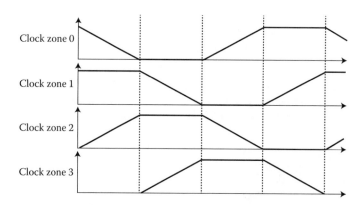

FIGURE 18.5 QCA clock signals for four clock zones. (S.-W. Kim and E. E. Swartzlander, Jr., Restoring divider design for quantum-dot cellular automata, *IEEE 11th Conference on Nanotechnology (NANO-2011)*, Portland, OR, August 15–19, pp. 1295–1300. © (2011) IEEE. With permission.)

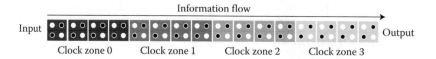

FIGURE 18.6 A QCA wire with clock zones. (S.-W. Kim and E. E. Swartzlander, Jr., Restoring divider design for quantum-dot cellular automata, *IEEE 11th Conference on Nanotechnology (NANO-2011)*, Portland, OR, August 15–19, pp. 1295–1300. © (2011) IEEE. With permission.)

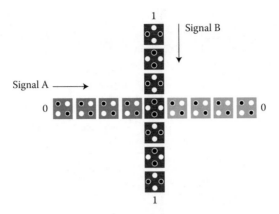

FIGURE 18.7 Example of a coplanar wire crossing. (S.-W. Kim and E. E. Swartzlander, Jr., Restoring divider design for quantum-dot cellular automata, *IEEE 11th Conference on Nanotechnology (NANO-2011)*, Portland, OR, August 15–19, pp. 1295–1300. © (2011) IEEE. With permission.)

For signal B in Figure 18.7, the explanation is as follows. Empty dots (shown in white) have a charge of +½. Populated dots (in black) have a charge of +½ for the dot added to −1 for the electron, giving a total charge of −½. The input for the first cell in the path is 1. The lowermost quantum dot of the first QCA cell is empty, so it has a +½ charge; thus, an electron of the second QCA cell is attracted to the uppermost quantum dot of the second QCA cell. If the input is changed from 1 to 0, the lowermost quantum dot of the first QCA cell has −½ charge, so an electron at the uppermost quantum dot of the second QCA cell is repulsed. As a result, the state of the second QCA cell will be changed from 0 to 1, and so on.

Coplanar wire crossings are susceptible to sneak noise from neighbor cells and are very sensitive to geometric misalignment. To avoid such problems, careful implementation based on design guidelines for robust operation is required.

18.3 RESTORING BINARY DIVIDER

18.3.1 CONVENTIONAL RESTORING BINARY DIVIDER ARCHITECTURE

The division operations [14–16] are performed in a bit sequential manner by the following formula:

$$P^{(i+1)} = r \times P^{(i)} - q_{i+1} \times D$$

where, $i = 0, 1, 2, \ldots, i - 1$ is the recursion index, $P^{(i)}$ is the partial remainder in the ith iteration. D is the divisor, and r is the radix. In this chapter, $r = 2$. The initial partial remainder $P^{(0)}$ equals the dividend, and $P^{(n)}$ is the final remainder. Restoring division operates on fixed point fractional numbers and is based on the following assumptions:

$0 \leq P^{(i+1)} < D, D < 1$, the allowable quotient digits q_{i+1} are from the digit set $\{0,1\}$.

The quotient digit is selected by performing a sequence of subtractions and shifts. Each time, D is subtracted from the partial remainder $r \times P^{(i)}$ until the difference becomes negative. Then, D is added back to the negative difference, which is called restoring. Since $r = 2$, the quotient digit is determined as follows:

$$q_{i+1} = \begin{cases} 0 & \text{if } 2P^{(i)} < D \\ 1 & \text{if } 2P^{(i)} \geq D \end{cases}$$

The partial remainder is obtained by one left shift of $P^{(i)}$ and by one subtraction such that $P^{(i+1)} = 2P^{(i)} - D$. If it is positive, then $q_{i+1} = 1$, else $q_{i+1} = 0$ and one restoring addition is to be

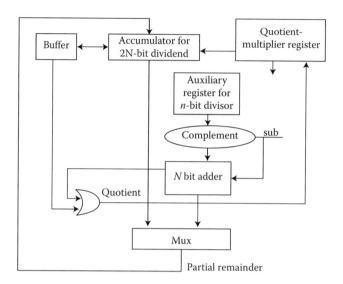

FIGURE 18.8 Restoring divider block diagram. (S.-W. Kim and E. E. Swartzlander, Jr., Restoring divider design for quantum-dot cellular automata, *IEEE 11th Conference on Nanotechnology (NANO-2011)*, Portland, OR, August 15–19, pp. 1295–1300. © (2011) IEEE. With permission.)

performed. The addition is needed to restore the correct partial remainder, $P^{(i+1)} = P^{(i+1)} + D = 2P^{(i)}$. Thus, performing binary restoring division requires at least one subtraction and may require one addition to determine each quotient bit. However, nonperforming restoring binary division needs only one subtraction per quotient bit because it uses a register to restore the partial remainder. Figure 18.8 shows a block diagram of a conventional restoring divider. This restoring divider starts with the dividend in the partial remainder register. If the correct quotient digit is a 0, the remainder is corrected by performing a compensating addition step. This scheme is not well suited to QCA implementation due to the delays of QCA wires.

18.3.2 RESTORING ARRAY DIVIDER FOR QCA

The design of high-speed iterative QCA arrays for parallel division uses a large amount of replicated units (for the comparison of the partial remainder and divisor) to reduce the number and size of QCA wires. The shift is realized by pipelined QCA wires. To improve the throughput of the array divider, pipelining can be used by inserting latches on the output of each row of cells. In general, a $k \times k$ restoring array divider receives a $2k$-bit dividend and k-bit divisor, and produces a k-bit quotient and $2k$-bit remainder including k leading 0s. Consider the division of a $2k \times k$ restoring array divider with $k = 3$:

$$\text{Dividend } z = .z_1z_2z_3z_4z_5z_6, \quad \text{Divisor } d = .d_1d_2d_3,$$

$$\text{Quotient } q = .q_1q_2q_3, \quad \text{and} \quad \text{Remainder } s = .000s_4s_5s_6.$$

Figure 18.9 shows an example of a binary restoring divider. A block diagram of a 3 bit × 3 bit restoring array divider with controlled subtractor cells is shown in Figure 18.10. Each cell consists of a full subtractor and a two-input multiplexer that is the basic element, which is replicated n^2 times to construct an $n \times n$ restoring array divider. When the control input $P = 1$, divider input d is subtracted from the partial remainder and the difference cell$_{\text{out}}$ is passed down to find the difference between the previous partial remainder and the divisor. Otherwise, when the control input $P = 0$, which triggers the multiplexer in a row cell, the partial remainder input bits are passed down unchanged.

```
                    1   1   0  ←——— Quotient
      .1  1  0 | .1  0   0   1   1   1        Sub/restoring
                  .1  1   0            ←——— (mux)

                    .1  1   1                 Sub/restoring
                    .1  1   0         ←——— (mux)

                      .0  1   1               Sub/restoring
                      .0  0   0       ←——— (mux)

                        0   1   1     ←——— Final remainder
```

FIGURE 18.9 Example of a 3 bit × 3 bit binary restoring divider. (S.-W. Kim and E. E. Swartzlander, Jr., Restoring divider design for quantum-dot cellular automata, *IEEE 11th Conference on Nanotechnology (NANO-2011)*, Portland, OR, August 15–19, pp. 1295–1300. © (2011) IEEE. With permission.)

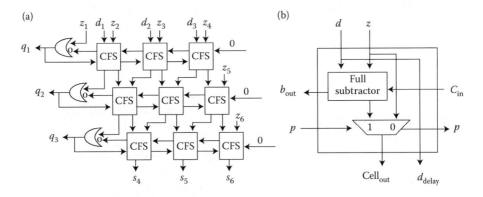

FIGURE 18.10 A 3 bit × 3 bit binary restoring divider. (a) Block diagram, (b) controlled full subtractor (CFS) cell. (Modified from S.-W. Kim and E. E. Swartzlander, Jr., Restoring divider design for quantum-dot cellular automata, *IEEE 11th Conference on Nanotechnology (NANO-2011)*, Portland, OR, August 15–19, pp. 1295–1300 © (2011) IEEE.)

The difference $\text{cell}_{out} = z\bar{P} + zdc_{in} + \bar{z}\bar{d}\,c_{in}P + z\bar{d}\,\bar{c}_{in} + \bar{z}d\,\bar{c}_{in}P$, that is

$$\text{cell}_{out} = \begin{cases} z \oplus d \oplus c_{in} & \text{if } P = 1 \text{ (Subtraction)} \\ z & \text{if } P = 0 \text{ No Operation} \end{cases}$$

Quotients q_i are obtained from the left of each row. $q_i = \bar{b}_{out} + z_i = P$, where z_i is the bit-shifted output in the $(i-1)$th iteration. If $z_i = 1$ or $b_{out} = 0$, then the subtraction is performed.

18.4 IMPLEMENTATION DETAILS

18.4.1 DESIGN GUIDELINES FOR ROBUST QCA CIRCUITS

To design a robust circuit, the divider has been designed using coplanar wire crossings with the design guidelines suggested in refs. [17,18]. Coplanar wire crossings are used for this research since a physical implementation of multilayer crossovers has not been demonstrated yet. If multilayer crossovers are available for a design, the design can be implemented more efficiently. Robust operation of a majority of gates is attained by limiting the minimum and maximum number of cells that are driven by the output, which is verified using the coherence vector method. The maximum cell number for each circuit component in a clock zone is determined by simulations with sneak noise sources. For example, the maximum cell number for a simple wire is 14 and the minimum is 2.

18.4.2 Implementation of the Restoring Divider

18.4.2.1 Basic Elements for Restoring Array Divider

The restoring array divider is composed of controlled full subtractor cells that have one full subtractor and one two-input multiplexer. The full subtractor takes three inputs (A_i, B_i, and C_i) and produces two outputs (D_i and C_{i+1}), where $D_i = A_i - B_i - C_i$ is the difference and C_{i+1} is the borrow signal that carries from one cell to the next. A simplified majority expression for a subtractor design in QCA is

$$D_i = A_i B_i C_i + \bar{A}_i \bar{B}_i C_i + A_i \bar{B}_i \bar{C}_i + \bar{A}_i B_i \bar{C}_i$$
$$= C_i(\bar{A}_i \bar{B}_i + A_i B_i) + \bar{C}_i(\bar{A}_i B_i + A_i \bar{B}_i)$$
$$= (A_i C_i + \bar{B}_i C_i)(\bar{A}_i C_i + B_i C_i) + \bar{C}_i(\bar{A}_i B_i + A_i \bar{B}_i + A_i C_i + \bar{A}_i C_i + B_i C_i + \bar{B}_i C_i)$$

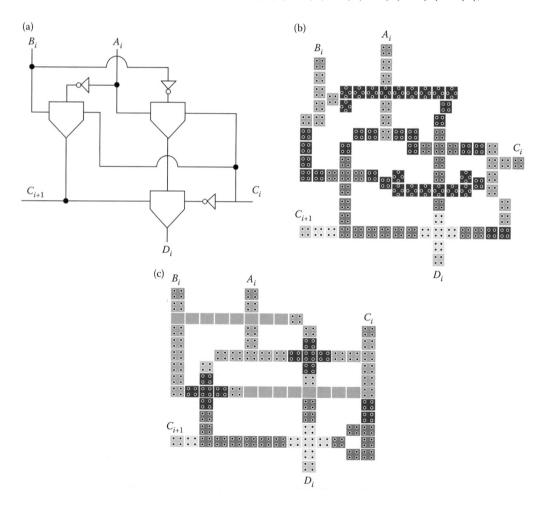

FIGURE 18.11 A full subtractor. (a) Schematic. (b) Coplanar crossing layout. ((a and b) Modified from S.-W. Kim and E. E. Swartzlander, Jr., Restoring divider design for quantum-dot cellular automata, *IEEE* 11th *Conference on Nanotechnology (NANO-2011)*, Portland, OR, August 15–19, pp. 1295–1300 © (2011) IEEE.), (c) Multilayer crossover layout.

$$= (A_iC_i + \bar{B}_iC_i + A_i\bar{B}_i)(\bar{A}_iC_i + B_iC_i + \bar{A}_iB_i) + \bar{C}_i(\bar{A}_iB_i + B_iC_i + \bar{A}_iC_i)$$
$$+ \bar{C}_i(A_i\bar{B}_i + \bar{B}_iC_i + A_iC_i)$$
$$= M(A_i,\bar{B}_i,C_i)M(\bar{A}_i,B_i,C_i) + \bar{C}_iM(\bar{A}_i,B_i,C_i) + \bar{C}_iM(A_i,\bar{B}_i,C_i)$$
$$= M(M(\bar{A}_i,B_i,C_i),M(A_i,\bar{B}_i,C_i),\bar{C}_i)$$

$$C_{i+1} = \bar{A}_iB_i + \bar{A}_iC_i + B_iC_i$$
$$= M(\bar{A}_i,B_i,C_i)$$

A full subtractor can be designed using only three majority gates and three inverters. Figure 18.11 shows the full subtractor schematic and layout. Both outputs of the full subtractor (i.e., the difference and borrow signals) have one clock latency with single and multilayer wire crossings. Simulation results are shown in Figure 18.12.

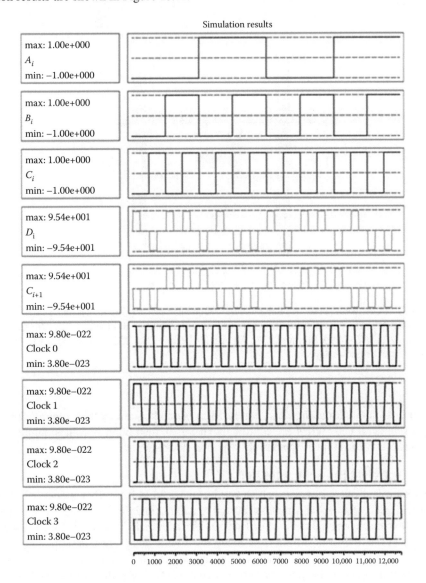

FIGURE 18.12 Simulation results of the full subtractor.

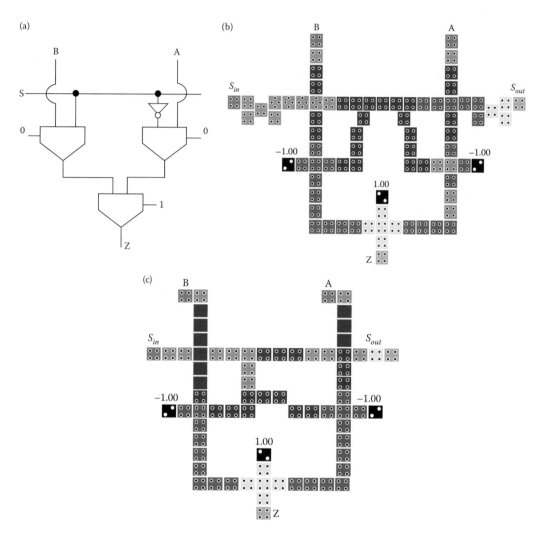

FIGURE 18.13 A 2:1 Multiplexer. (a) Schematic. (b) Coplanar crossing layout. ((a and b) S.-W. Kim and E. E. Swartzlander, Jr., Restoring divider design for quantum-dot cellular automata, *IEEE 11th Conference on Nanotechnology (NANO-2011)*, Portland, OR, August 15–19, pp. 1295–1300. © (2011) IEEE. With permission.) (c) Multilayer crossover layout.

The two input multiplexer shown in Figure 18.13 is another cell element required for the controlled full subtractor cell. When the control signal S is 0, then input A comes out and when the control signal S is 1, then input B comes out with one clock latency.

18.4.2.2 Controlled Full Subtractor Cell

The CFS (controlled full subtractor) cell shown in Figure 18.14 basically conducts $z - d$ if $P_{in} = 1$ to find the difference between the previous partial remainder and divisor. The borrow signal comes out after a two clock latency. Instead of shifting the partial remainder left to form $2P^{(i)}$, the remainder is fixed and the divisor is shifted right along the divisor delay lines, and the quotient bits are obtained from the left of each row. This cell has the difference D_i and borrow out C_{i+1} with three clock latency and two clock latency, respectively. To make the array divider, this cell unit is modified for synchronization.

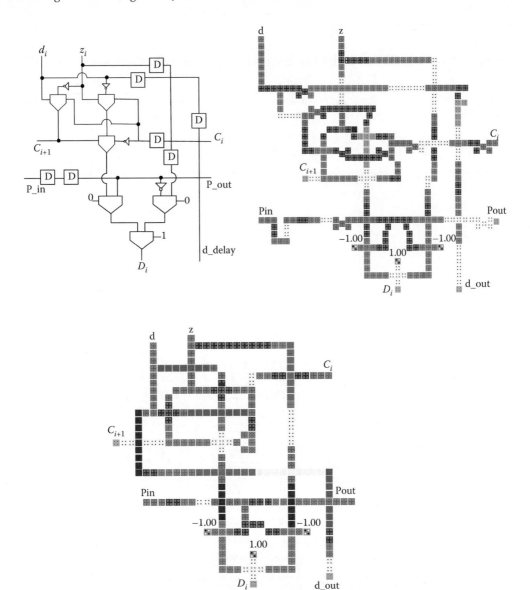

FIGURE 18.14 Controlled full subtractor cell. (a) Schematic. (b) Coplanar crossing layout. ((a and b) Modified from S.-W. Kim and E. E. Swartzlander, Jr., Restoring divider design for quantum-dot cellular automata, *IEEE* 11th *Conference on Nanotechnology (NANO-2011)*, Portland, OR, August 15–19, pp. 1295–1300, © (2011) IEEE.) (c) Multilayer crossover layout.

18.4.2.3 Comparison of Coplanar and Multilayer Wire Crossings

Coplanar wire crossings are unique to QCA technology. With two cell types (regular and rotated), coplanar crossings can be realized. Many clock zones between the regular cells across the rotated cells are needed so the area will be large. Coplanar crossings are the current wiring method. Multilayer crossovers have some potential advantages such as latency and area minimization, but physical feasibility, assigning clock phases, and placement and routing are currently uncertain. Table 18.1 shows a comparison of the basic elements for a restoring divider with coplanar and multilayer wire crossings. With multilayer crossovers, each component has fewer cells and less area. Small elements such as the full subtractor and multiplexer have the same latency with both types

TABLE 18.1

Comparison of Basic Elements with Coplanar and Multilayer Wire Crossings

Elements	Coplanar			Multilayer		
	Area (μm^2)	Latency	# of Cells	Area (μm^2)	Latency	# of Cells
Full subtractor	0.11	1	90	0.08	1	82
Multiplexer	0.13	1	79	0.11	1	71
Controlled full subtractor cell	0.58	D_i C_{i+1}	324	0.45	D_i C_{i+1}	255
		3 2			2 2	

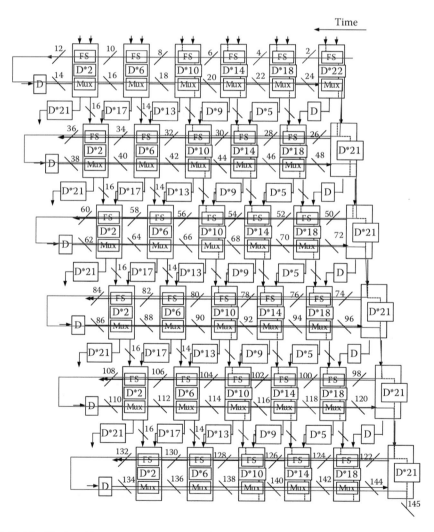

FIGURE 18.15 Timing block diagram for a 6 bit × 6 bit restoring array divider. (S.-W. Kim and E. E. Swartzlander, Jr., Restoring divider design for quantum-dot cellular automata, *IEEE 11th Conference on Nanotechnology (NANO-2011)*, Portland, OR, August 15–19, pp. 1295–1300. © (2011) IEEE. With permission.)

of crossings, but the controlled full subtractor cell unit has slightly less latency when realized with multilayer wire crossings.

18.4.2.4 Timing Analysis

The restoring array divider resembles an array multiplier, but it has different delay characteristics due to borrow signal propagation along the row and quotient signal feedback to each cell. From the rightmost column, CFS cells have 22, 18, 14, 10, 6, and 2 latches between the full subtractor and the multiplexer to implement the pipelining. Each cell has 21 delay latches for the divisor to pass down to the next row. A timing block diagram for a 6 bit × 6 bit restoring array divider is shown in Figure 18.15.

18.5 SIMULATION RESULTS

18.5.1 RESTORING DIVIDER RESULTS

Simulations were done with QCADesignerv2.0.3 [19] assuming coplanar wire crossings. The size of the basic quantum cell was set at 18 nm × 18 nm with 5 nm diameter quantum dots. The center-to-center distance is set at 20 nm for adjacent cells. The following parameters are used for a bistable approximation: 25,600 samples for the 3 bit × 3 bit restoring array divider, 156,000 samples for the 6 bit × 6 bit restoring array divider, 0.001 convergence tolerance, 65 nm radius effect, 12.9 relative permittivity, 9.8e–22 clock high, 3.8e–23 clock low, 1 clock amplitude factor, 11.5 layer separation, 100 maximum iterations per sample. The layout of the 3 bit × 3 bit restoring array divider is shown in Figure 18.16. Each cell with different delays is tested exhaustively, and the full integration is verified with input vectors: Dividend {39, 48, 42, 33, 36}, Divisor {6, 7, 6, 5, 5}. Correct output data come out from 37 clock latency as shown in Figure 18.17 such that Quotient {6, 6, 7, 6, 7} and Remainder {3, 6, 0, 3, 1}. The results are shown as integers to make it easier to understand even though they are fractions. The first consecutive division inputs are {100111, 110}, which are {39, 6} in decimal. .10011/.110 is expressed as the integer 0.609375/0.75. 0.609375 equals 0.75 times 0.75 (Quotient) plus 0.046875 (Remainder) and it is .110 and .000011 as the quotient and remainder fixed fractional numbers, respectively.

As the input word size increases, the number of RAS cells grows along the column lines in proportion.

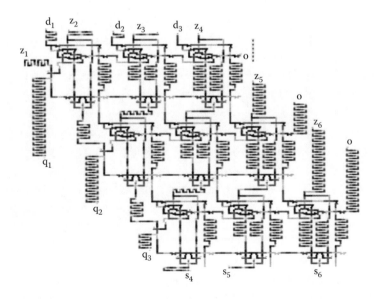

FIGURE 18.16 Layout of the 3 bit × 3 bit restoring array divider.

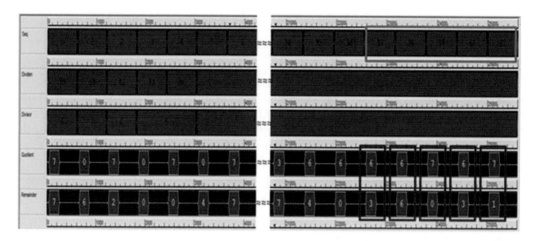

FIGURE 18.17 Simulation results of the 3 bit × 3 bit restoring array divider.

Figure 18.18 shows the layout of the 6 bit × 6 bit restoring array divider. The output waveform with 145 clock latency is shown in Figure 18.19. Six consecutive input vectors and outputs are:

Dividend {2672, 3335, 2228, 2346, 3670, 2328}, Divisor {51, 56, 52, 50, 62, 44},

Quotient {52, 59, 42, 46, 59, 52}, and Remainder {20, 31, 44, 46, 12, 40}.

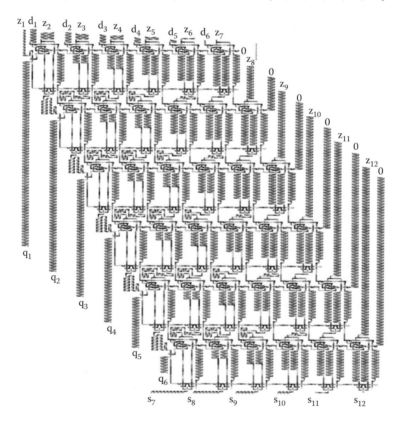

FIGURE 18.18 Layout of the 6 bit × 6 bit restoring array divider. (S.-W. Kim and E. E. Swartzlander, Jr., Restoring divider design for quantum-dot cellular automata, *IEEE 11th Conference on Nanotechnology (NANO-2011)*, Portland, OR, August 15–19, pp. 1295–1300. © (2011) IEEE. With permission.)

FIGURE 18.19 Simulation results of the 6 bit × 6 bit restoring array divider. (S.-W. Kim and E. E. Swartzlander, Jr., Restoring divider design for quantum-dot cellular automata, *IEEE 11th Conference on Nanotechnology (NANO-2011)*, Portland, OR, August 15–19, pp. 1295–1300. © (2011) IEEE. With permission.)

TABLE 18.2
Comparison of QCA Dividers

Dividers	Latency	Area (μm²)	# of Cells	Throughput
3 bit × 3 bit RAD	37	15.05	6451	1
6 bit × 6 bit RAD	145	86.22	42,236	1
12 bit × 12 bit RAD	577	740.44	301,395	1

Source: S.-W. Kim and E. E. Swartzlander, Jr., Restoring divider design for quantum-dot cellular automata, *IEEE 11th Conference on Nanotechnology (NANO-2011)*, Portland, OR, August 15–19, pp. 1295–1300. © (2011) IEEE. With permission.

18.5.2 COMPARISON ANALYSIS OF DIVIDERS

Table 18.2 shows a comparison of the 3 bit × 3 bit, 6 bit × 6 bit, and 12 bit × 12 bit QCA restoring array dividers. The dividers get much larger as the operand size increases. Also, the latency increases by almost a factor of 4 as the word size is doubled. Full simulation time took 42 h for the 12-bit restoring array divider. The restoring array divider has good throughput since an output is produced on every clock cycle after the 37th clock in the case of the 3 bit × 3 bit restoring array divider.

18.6 CONCLUSION

A digit recurrence restoring the binary divider design for QCA is presented in detail. It is a conventional design that uses controlled full subtractor cells to produce a relatively simple and efficient implementation. The design has been optimized for realization with QCA technology. The restoring array divider is implemented with CFS cell blocks. It can be easily enlarged by irregular block cells without long data connections, but it is large and has a long latency due to the restoring algorithm (recursive division). However, by using pipelining, it has a good throughput.

ACKNOWLEDGMENT

This chapter is an expanded version of Ref. [20]. Earl Swartzlander is supported in part by a grant from AMD, Inc.

REFERENCES

1. C. S. Lent, P. D. Tougaw, W. Porod, and G. H. Bernstein, Quantum cellular automata, *Nanotechnology*, 4, 49–57, 1993.
2. C. S. Lent, P. D. Tougaw, and W. Porod, Quantum cellular automata: The physics of computing with arrays of quantum dot molecules, *Proceedings of Workshop on Physics and Computation*, Dallas, Texas, pp. 5–13, 1994.
3. M. Lieberman, DNA origami as circuitboards for QCA, *1st International Workshop on Quantum-Dot Cellular Automata (1WQCA)*, Vancouver, p. 8, Aug. 2009.
4. K. Walus, F. Karim, and A. Ivano, Architecture for an external input into a molecular QCA circuit, *Journal of Computational Electronics*, 8, 35–42, 2009.
5. M. B. Haider, J. L. Pitters, G. A. DiLabio, L. Livadaru, J. Y. Mutus, and R. A. Wolkow, Controlled coupling and occupation of silicon atomic quantum dots at room temperature, *Physical Review Letters*, 102, 046805, 2009.
6. A. A. Prager, A. O. Orlov, and G. L. Snider, Integration of CMOS, single electron transistors, and quantum-dot cellular automata, *IEEE Nanotechnology Materials and Devices Conference*, pp. 54–58, June 2009.
7. K. Walus, M. Mazur, G. Schulhof, and G. A. Jullien, Simple 4-bit processor based on quantum-dot cellular automata (QCA), *16th International Conference on Application-Specific Systems, Architecture and Processors*, Samos, Greece, pp. 288–293, 2005.
8. C. S. Lent and P. D. Tougaw, A device architecture for computing with quantum dots, *Proceedings of the IEEE*, 85, 541–557, 1997.
9. M. T. Niemier and P. M. Kogge, Logic in wire: Using quantum dots to implement a microprocessor, *The 6th IEEE International Conference on Electronics, Circuits and Systems*, pp. 1211–1215, 1999.
10. M. T. Niemier and P. M. Kogge, Problems in designing with QCAs: Layout = timing, *International Journal of Circuit Theory and Applications*, 29, 49–62, 2001.
11. R. Zhang, K. Walus, W. Wang, and G. A. Jullien, A method of majority logic reduction for quantum cellular automata, *IEEE Transactions on Nanotechnology*, 3, 443–450, 2004.
12. W. Wang, K. Walus, and G. A. Jullien, Quantum-dot cellular automata adders, *Third IEEE Conference on Nanotechnology*, San Francisco, California, vol. 1, pp. 461–464, 2003.
13. D. Tougaw and C. S. Lent, Logical devices implemented using quantum cellular automata, *Journal of Applied Physics*, 75, 1818–1825, 1994.
14. A. B. Gardiner and J. Hont, Comparison of restoring and nonrestoring cellular-array dividers, *Electronics Letters*, 7, 172–173, 1971.
15. B. Parhami, *Computer Arithmetic: Algorithms and Hardware Designs*, 2nd Edition, New York, NY: Oxford University Press, Inc., 2010.
16. M. D. Ercegovac and T. Lang, *Division and Square Root: Digit-Recurrence Algorithms and Implementations*, Boston: Kluwer Academic Publishers, 1994.
17. K. Kim, K. Wu, and R. Karri, Towards designing robust QCA architectures in the presence of sneak noise paths, *Proceedings of the Conference on Design, Automation and Test in Europe*, Munich, Germany, pp. 1214–1219, 2005.
18. K. Kim, K. Wu, and R. Karri, The robust QCA adder designs using composable QCA building blocks, *IEEE Transactions on Computer-aided Design of Integrated Circuits and Systems*, 26, 176–183, 2007.
19. K. Walus, T. J. Dysart, G. A. Jullien, and R. A. Budiman, QCADesigner: A rapid design and simulation tool for quantum-dot cellular automata, *IEEE Transactions on Nanotechnology*, 3, 26–31, 2004.
20. S.-W. Kim and E. E. Swartzlander, Jr., Restoring divider design for quantum-dot cellular automata, *IEEE 11th Conference on Nanotechnology (NANO-2011)*, Portland, OR, August 15–19, pp. 1295–1300, 2011.

19 LINA-QCA
Theory, Design, and Viable Implementation Strategies

Loyd R. Hook IV and Samuel C. Lee

CONTENTS

19.1 INTRODUCTION

Researchers have been predicting limitations to the size scaling for CMOS (complementary metal–oxide–semiconductor)-integrated circuitry for many years. Often, these limits have been overcome and extended to smaller and smaller sizes [1]. However, as the CMOS minimum feature size has continued to shrink over the past decade, other concerns which are less fundamental and more practical have arisen. A particularly well-known example of this fact can be seen with limitations on the CPU clock speed due to power dissipation challenges. In response to this effect, CPU designers have resorted to techniques such as integrating multiple core CPUs in attempting to maintain the exponential computing performance growth which has driven the multibillion dollar industry for several decades. However, multi-core performance advantages also have limitations stemming from the ability to parallelize applications which will limit their effectiveness to increase practical performance [2]. In response to issues such as these, academia and industry have been researching alternative technologies to extend and replace CMOS designs in the coming years. However, these technologies have suffered from their own theoretical and practical challenges and thus a clear alternative for CMOS circuitry has not yet emerged.

One of the leading contenders to eventually replace CMOS is the quantum-dot cellular automata (QCA) nanoelectronics paradigm [3]. QCA have a relatively well-developed theoretical background in which THz switching of single nanometer molecular devices in a highly pipelined circuit architecture has been shown to be possible [4]. Additionally, alternate computing structures such as reversible logic [5] or systolic architectures [6,7] have been shown to be particularly well suited to be developed in the QCA framework. Advantages such as these provide the basis for continuing research into QCA as a potential replacement for CMOS designs; however, other pressing concerns must be addressed before large-scale integration and development can proceed.

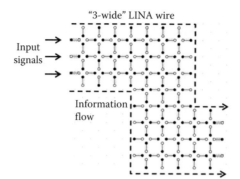

FIGURE 19.1   LINA wire (here shown with two 90° turns) with three "integrated signals."

In particular, as with many other nanoelectronic proposals, reliability issues in the presence of room temperature thermal noise are a major concern due to the relatively small energies utilized to store and communicate data. Design alternatives which focus on increasing the reliability have tended to create additional problems for implementation. For instance, molecular QCA cells with sizes and cell spacing of ~1 nm have been proposed to raise state separation energies above the room temperature thermal noise and thus create reliable computing constructs [8]. However, at this ultra-small-size scale, practical difficulties arise, such as the layout of clocking wires necessary for the correct operation and functional pipelining of VLSI level QCA circuitry or power dissipation of circuitry with speeds, which create a reasonable alternative to current CMOS that will hamper near-term implementation prospects. Additionally, at this size, scale deposition and patterning of QCA logic structures will be very difficult and will require a significant advancement of current technologies before large-scale circuits could be generated [9]. Until issues such as these can be resolved, experimentation and demonstration of room temperature large-scale QCA circuitry may not be possible.

These issues have prompted research into other methods of increasing circuit reliabilities that are amenable to larger cells and structures required of currently available or near-term materials and circuit layout technologies. The lattice-based integrated-signal nanocellular automata (LINA)-type QCA [10] (shown in Figure 19.1), which offer potential solutions to many of these challenges are a result of this research. LINA is a lattice-based design that enables a mixture of self-assembly and patterning techniques to overcome deposition and/or positioning challenges and is flexible enough to incorporate error-prone fabrication and various clocking wire minimum pattern sizes. LINA designs were developed to enable utilization of self-assembled materials with relatively large inter-/intracellular spacings into reliable designs, thereby opening up a large area of currently implementable materials for circuit fabrication and room temperature operation. This chapter presents the theory and design of LINA circuitry along with simulation results to verify its design. Section 19.2 describes the underlying theory and advantages of LINA, Section 19.3 introduces logical structures and circuits, Section 19.4 presents design principles and simulation results, Section 19.5 offers a design example in which LINA is used, and the chapter concludes with Section 19.6.

19.2   THEORY AND ADVANTAGES OF LINA

One of the most pressing challenges facing the implementation of room temperature electric QCA in large reliable circuits is the precise deposition and patterning of single nanometer-sized molecules into structures which perform useful communication and logic. Single nanometer-sized molecules have been considered to be required to increase energy separation between ground and excited cell states above error-producing thermal noise. Proposals involving a mixture of self-assembly and patterning have offered hope to solve this challenge, but may have issues at the required size scales and integration levels [9]. In general, techniques to deposit and pattern molecules on the required size scale and with the precise placement of traditional QCA designs remain very difficult in the near term.

One reason for this difficulty lies in the traditional QCA cellular designs, and the resultant circuit layouts, which utilize some circuit design methods developed for other technologies. Lattice-based designs have recently been proposed [11], which attempt to solve the deposition issue by laying out cells based on common self-assembly formations, such as those shown in the nanoparticles of Ref. [12], thereby producing a method in which the relative cellular layout can be achieved. However, the ability to pattern large-scale complex circuit layouts on this small cellular size scale is not available for most large-scale lithography techniques. Additionally, the previous lattice-based designs do not account for common fabrication defects such as cell deletion, misalignment, or other fault mechanisms, commonly found with any self-assembly process. LINA attempts to advance the transition to lattice-based designs by incorporating additional "integrated signals" in communication and logic structures. This has the effect of increasing reliability by taking advantage of the inherent majority property of QCA to self-correct logical errors due to the previously mentioned error modes. The increased reliability also allows for larger cell sizes and cell spacings (possibly greater than 20 nm), which are able to be utilized for reliable room-temperature operation. This may allow for other larger nanoparticles, which can be more easily controlled, to be utilized in the role of a LINA QCA cell. Another added benefit of the LINA designs is the flexibility to adjust structure widths based on large-scale patterning technologies, thereby potentially allowing currently available patterning techniques to be used in LINA QCA circuits.

The integrated signals of the LINA designs carry common information much like the redundant designs would [13], but advance the concept by restoring and repairing signals continually and allowing for multiple paths for each individual signal. Additionally, owing to the unique geometries of these integrated signal designs, they do not require the condensation of multiple signals into a single wire or cell for input to logic. This is possible because LINA gates are able to accept each of the fully integrated-signal wires as separate inputs. This allows for robust communication and logic structures and removes weak points which are susceptible to logical faults.

The structure of the integrated signal wires can be seen in Figure 19.2, where three different possible wire widths are shown. The term "wire width" or "n-wide" is in reference to the number of integrated signals which are contained in the wire and not to its geometrical width. For example, the wire of Figure 19.2b has a width of 3 (making it a "3-wide" wire), even though the geometrical width is 7 times the lattice spacing. In general, LINA wires can be made into any odd integer number width providing design flexibility for increased reliability or to fit with clocking or patterning technologies.

One major difference between LINA and traditional QCA is that the information carried by LINA wires must be understood based not only on the cell state but also on the geometric position. Another added complexity arises from the fact that LINA utilizes two cellular orientations in the lattice arrangement. Therefore, in order to be able to quantify and study the operation of LINA wires and gates, a convention is developed based on the geometrical layout shown in Figure 19.3. In this convention, the binary state of the cell is determined by the position of the free electron (shown in the figure as the filled-in black dot location). When the electron is in a positive location, according to the axis parallel to the cell's orientation, the cell is given a binary value of 1 and a polarization of +1. When the electron is in a negative direction, the cell is given a binary value of 0 and a polarization of −1.

The convention also dictates that the origin of the coordinate plane (0,0) is taken as the center cell of the input plane of a wire, regardless of the width of the wire. From there, the axes must be positioned along the directions of the two cellular orientations. This convention allows for the wire shown in the figure to be given the binary value of 1, due to the value of the cell at the coordinate origin. The scale of the axes is based on a common cellular lattice spacing (L), which directly corresponds to the eventual intercellular spacing, and along with the intracellular-dot spacing (D), provides the position of the cells' component parts.

With this convention in hand, the expected binary value of the remaining cells of the wire can be obtained using Equation 19.1, where $V_{x,y}$ is the binary value of the cell at position (x,y). For

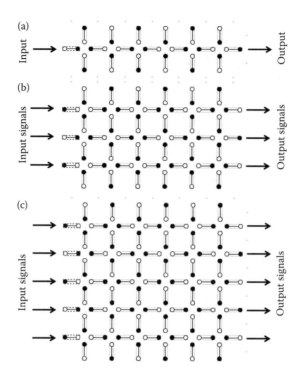

FIGURE 19.2 (a) 1-wide LINA wire, (b) 3-wide LINA wire, and (c) 5-wide LINA wire.

example, in Figure 19.3, the cell at position (0,0) has a binary value of 1 ($V_{0,0} = 1$). Therefore, using Equation 19.1, we can surmise the expected binary value of the cell at (2, −2) to be ($V_{2,-2} = V_{0,0} = 1$) because $(-2 - 2)_{\mathrm{mod4}}$ is equal to 0. Likewise, for the cell at (3,5), the expected binary value ($V_{3,1} = \mathrm{not}(V_{0,0}) = 0$) because $(1 - 3)_{\mathrm{mod4}}$ is equal to 2.

$$V_{x,y} = \begin{cases} V_{0,0}, & (y - x)\,\mathrm{mod}\,4 == 0 \\ \overline{V_{0,0}}, & (y - x)\,\mathrm{mod}\,4 == 2 \end{cases} \tag{19.1}$$

In addition, the correct binary value of any true output plane of a LINA circuit will be the expected value of the center cell on the output plane. Furthermore, the total probability of correct logical

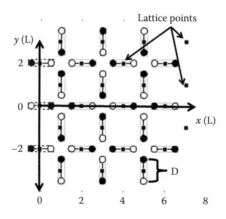

FIGURE 19.3 Geometrical layout convention used for LINA designs. The numbers of the scale are associated with integer multiples of the lattice spacing (L).

output (PCLO) of this output can be obtained with knowledge of the individual PCLO of the cells on the output plane. The individual PCLOs are combined through the use of the probabilistic ensemble majority voting methods, such as shown the for a 3-wide wire in Equation 19.2 where "P_i" is the PCLO for Cell$_i$ (here Cell$_1$, Cell$_2$, and Cell$_3$ are the output cells) and P_{Total} is the total output PCLO. For example, if the PCLOs of the three output signals in the wire in Figure 19.2b are 0.95, 0.97, and 0.95, then P_{Total} would be equal to 0.99465 according to

$$P_{Total} = \sum_{i=1,(j \neq k \neq i)}^{3} ((1 - P_i) P_j P_k) + P_1 P_2 P_3 \qquad (19.2)$$

Though there is an increase in the LINA design and/or layout complexity, as opposed to the traditional QCA, this is offset by the significant increase in reliabilities of LINA. An example of this reliability increase can be seen in Figure 19.4, which shows reliability numbers for two wires obtained by utilizing coherence vector simulation methods. As can be seen from the figure, a 300 nm wire created from cells with an inter-dot distance (D) equal to 10 nm and an intercellular lattice space

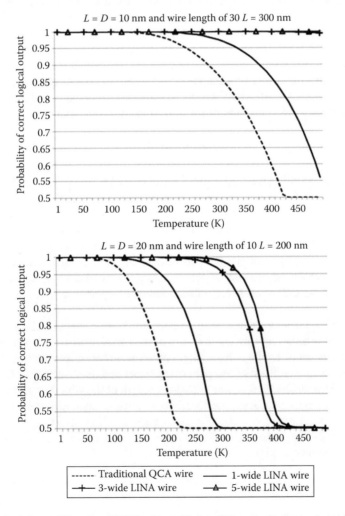

FIGURE 19.4 PCLO for traditional and LINA wires utilizing different cell sizes and spacings. (The simulations were performed with a relative permittivity of 1, a high clock level of 9.8e–20 J, and a low clock level of 3.8e–23 J.)

(*L*) distance also equal to 10 nm was simulated. This wire shows a PCLO of 0.86 for the traditional QCA designs at a room temperature of 300 K. PCLO for the LINA designs are much improved at this temperature with the 1-wide LINA PCLO of 0.97, 3-wide PCLO of four 9's (0.9999), and 5-wide PCLO of six 9's.

The predicted increase in room-temperature reliability is consistent for different *L* and *D* values (as well as for other simulation methods). PCLO for the room temperature operation of wires with *L* = *D* = 20 nm and overall length of 200 nm demonstrate this fact. Traditional QCA designs are nearly completely random (PCLO ~0.5) at this temperature. The 1-wide LINA PCLO is also nearly random here; however, the 3-wide LINA PCLO is 0.96 and the 5-wide PCLO is 0.98. The increase in reliability, especially for the larger spacings, allows for a greater number of potential implementation technologies compared to traditional QCA.

19.3 LINA LOGIC STRUCTURES AND CIRCUITRY

19.3.1 LINA Logic Structures

Traditional QCA logic design is centered on the three input majority voter gate, which is typically made up of a single QCA cell. This gate, along with the ability to invert a signal, gives the traditional QCA design logical completeness. The LINA designs are also based on the three input majority gate; however, the implementation of this logic gate requires a fully populated $n \times n$ area of the lattice map to perform computations (where the input and output wires are n-wide). The LINA majority gate performs the function in Equation 19.3 where the right input is being inverted in the majority gate of Figure 19.5. All LINA widths provide this same majority function as long as the center cell in the input wires is taken to be the logical value of the input.

$$Output = Maj\,(A, B, \bar{C}) \tag{19.3}$$

The same cellular structure is also used to provide a planer wire crossing for the LINA architecture. Also shown in Figure 19.5, the crossing structure requires a change in the clocking scheme for correct operation. This is accomplished with the application of an additional nontraditional clock signal to the cells in the crossing structure. The new clock only allows for signal pass-through and does not lend itself to a hold phase in which other cells are driven or a null phase in which other cells are unaffected. (The clock phases can be seen in the full adder design example of Section 19.5.) Additionally, for a correct functioning of the crossing, the two incoming wires must be 2 clock phases out of sync with each other. More information on this planer wire crossing can be found in Ref. [10].

Also required for logic completeness is an inverter structure—which in LINA can come in many forms. Owing to the inherent inverter chain nature of the LINA design, the simplest inverter can be achieved by carefully choosing the correct length of wire. However, this method of inverting a signal will be intolerant to even the slightest patterning errors and thus may not be the optimal choice for this function. Also, the method of controlling the length of the wire does not scale well to large circuitry where wires must meet at precise locations to perform logic with some of the inputs being inverted and some not. Instead, a dedicated LINA logic block to invert signals is preferred. The inverter block is created by removing a column in an otherwise normal LINA wire. An example of this technique can be seen in Figure 19.5.

19.3.2 Layout of LINA Blocks

Owing to the regular structure of the LINA lattice map and other clocking considerations, it is convenient (and in many ways necessary) to group LINA cells together when laying out circuitry. The majority gate and planer wire crossing of the previous section are examples of this grouping;

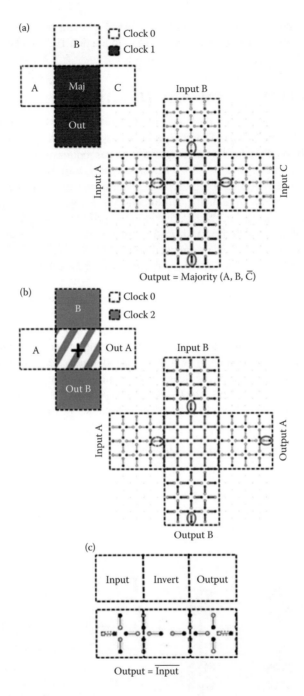

FIGURE 19.5 (a) 3-wide LINA majority gate and associated block diagram. (b) 3-Wide planer wire crossing and associated block diagram. (Clocking scheme for planer wire x-ing can be found in Ref. [10].) (c) 1-wide LINA inverter and associated block diagram. *Note*: Logical values of the input and outputs are taken at the cells circled.

however, wires should also be grouped in this way. This provides the advantage of allowing a simplified layout and design process as groups, or "blocks" of cells are used. Blocks are defined as squares of cells that have a length equal to the LINA design width. It should be pointed out that the width of the blocks and therefore the LINA design should be directly related to the layout technologies for the clocking structures as well as the cells themselves; therefore, intra-block cells should utilize the same clock phase. The first example of using blocks can be seen in Figure 19.5, and an example of using blocks to design a circuit will be seen in Section 19.5 in which a full adder design is produced.

19.4 LINA DESIGN SPACE: POWER, SPEED, RELIABILITY

It has already been shown how LINA designs can improve reliability by adding width to gates and wires. However, it is common to nanoelectronics proposals and QCA specifically that an increase in reliability is usually accompanied by an increase in power requirements. Certainly, adding additional cells to circuitry, in the form of LINA wires and gates, will create additional power which must be dissipated to the environment. However, LINA designs actually generate less heat for fixed desired reliabilities as opposed to reducing cell dimensions and spacing for the most relevant conditions. To prove this statement, formulas derived in Ref. [14] (and utilized in Ref. [15]) for power dissipation are utilized along with the LINA design tools. Concurrent evaluation of power, speed, and reliability is complicated by their interrelated nature and the fact that QCA cells both process and communicate information. This provides a rather difficult task in developing suitable metrics for quantifying the results of this evaluation.

Beginning by assuming a reasonable upper limit of power dissipation of 100 W/cm², and that the average per cell power dissipation is evenly distributed throughout the cell, we are able to formulate a reasonable metric defined as the "effective QCA half-pitch." This value is based on the distance between the basic QCA majority and fan-out gates (which utilize the same cellular structures and are elemental QCA logic components) as shown in Figure 19.7a. This quantity is not directly related to the CMOS half-pitch (which is based on the transistor as the elemental structure), but attempts to mimic the simplicity in understanding the CMOS metric coveys. The QCA effective pitch (F_p) is calculated in part by equating the power density of a single array block to our predefined limit of 100 W/cm² as in

$$N_c * P_c / F_p^2 = \frac{100\ W}{cm^2} \tag{19.4}$$

with the length from the elemental logic device to the next along the side of a square being F_p (the "full" pitch), N_c being the number of QCA cells in the square of the full pitch, and P_c being the power per cell dissipated at a given clock frequency.

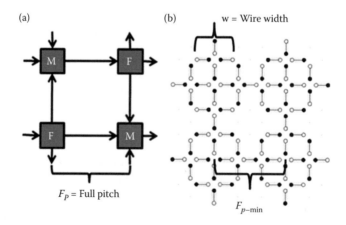

FIGURE 19.6 (a) Section of array of the majority (M) and fan-out (F) gates and associated pitch (F_p) to provide a metric for density, power, and speed considerations. (b) Example of the smallest 1-wide LINA array of elements. (L. R. Hook IV and S. C. Lee. Lattice-based integrated-signal nanocellular automata (LINA) for the future of QCA-based nanoelectronics, *2011 IEEE Nanotechnol. Conf. (IEEE-NANO)* © (2011) IEEE. With permission.)

Lower limits to the pitch are calculated based on the smallest possible array block layout for any given cell spacing. Figure 19.6 shows an example of the smallest functional layout for a 1-wide LINA design. These two values give insight into the LINA design by revealing two areas of design space for a given cell and dot spacings and desired clock frequency in which the circuit density is either limited by power dissipation or by cell geometries. This space is shown in Figure 19.7b as the two segments of the line connecting the half-pitch values in all of the wire design lines.

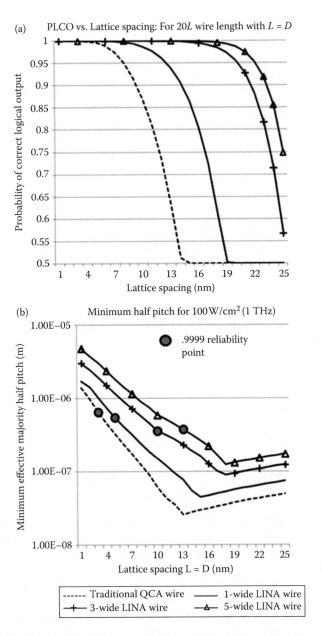

FIGURE 19.7 (a) PCLO for room temperature operation with increasing lattice spacing from 1 to 25 nm for traditional and 1,3,5-wide LINA wire designs. Wire length is 20L. (b) Minimum "effective majority" half-pitch for four traditional and LINA wire circuit design along with the 4 9's reliability point at 1 THz. Note that the minimum half-pitch is seen with the 3-wide LINA design at this reliability. The simulations feeding these data were run with a relative permittivity of 1, a high clock level of 9.8e–20 J, and a low clock level of 3.8e–23 J.

TABLE 19.1

Minimum Half-Pitch and Wire Design for Various Clock Speeds and Reliabilities

Clock Speed	Reliability >6 9's	Reliability >5 9's	Reliability >4 9's
1 THz	590 nm	517 nm	360 nm
	$L = 12$ nm	$L = 13$ nm	$L = 10$ nm
	5-wide	5-wide	3-wide
	LINA	LINA	LINA
500 GHz	180 nm	154 nm	91 nm
	$L = 10$ nm	$L = 11$ nm	$L = 13$ nm
	5-wide	5-wide	5-wide
	LINA	LINA	LINA
100 GHz	15 nm	12 nm	9 nm
	$L = 3$ nm	$L = 4$ nm	$L = 3$ nm
	1-wide	1-wide	Traditional
	LINA	LINA	QCA

Figure 19.7, as a whole, depicts data which are used in the process for determining the minimum half-pitch values based on reliability, power, speed, lattice spacing, and wire design. Figure 19.7a shows the reliabilities for various lattice spacings for four different wire designs (traditional, 1-wide LINA, 3-wide LINA, and 5-wide LINA) for a wire with length equal to 20 lattice spacings. This reliability is used to delineate various points along the half-pitch lines of Figure 19.7b to give minimum half-pitch for any desired reliability. Table 19.1 shows the calculated values for the half-pitch for various reliabilities based on this process. The table shows clear advantages for the LINA designs at high frequencies and with high reliability requirements. The data also show that the traditional QCA designs are advantageous in circuits where the desired frequency or reliabilities are low. (Further data reiterating these claims can be found in Ref. [10].)

It is interesting to see in Table 19.1 that the 1 nm cell spacing does not provide a minimum half-pitch for any of the entries. This is due in part to the number of cells and thus switching events that must be accounted for by using such relatively small cells and the large power per cell values which scale inversely with cell spacing due to higher kink energies. It is also interesting to note that simulations involving QCA designs do not yield a particular small circuit area if 1 THz switching with high reliabilities is desired; instead, a trade-off between circuit area, clock frequency, and reliability will have to be made. These types of trades are common for IC designers even today, and as technologies improve, these trades are sure to grow more favorable.

19.5 DESIGN EXAMPLE: DESIGN AND SIMULATION OF A ROOM TEMPERATURE, RELIABILITY >0.99999 FULL ADDER

As an example of LINA designs and how the results found in the previous sections may be used, consider the design of a full adder with greater than 5 9's reliability at 100 GHz clock speed. Table 19.1 shows that for this speed and reliability, 1-wide LINA provides the greatest space/power efficiency using a 4 nm spaced LINA circuit. Many circuits will be restricted by the material implementation and thus will not be able to precisely choose lattice spacing. Additionally, owing to the low reliabilities for the smaller width designs with spacings larger than a few nanometers, such as traditional QCA or 1-wide LINA, larger width designs may be necessary. However, for this example, the ability to precisely define spacings will be assumed to demonstrate the use of Table 19.1.

The full adder design chosen is based on the simplified full adder sum and carry outputs from Ref. [16]. The logical schematic of the full adder design which was chosen can be seen in Figure 19.8a. Figure 19.8b shows the LINA blocks design which can be used to implement the full adder with any width of the LINA circuit as has been discussed. The blocks are labeled with either the input

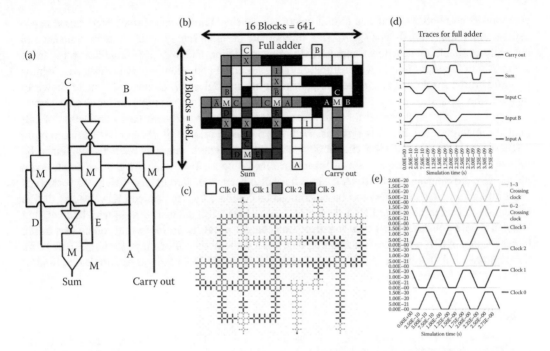

FIGURE 19.8   (a) Logical schematic of the full adder cell with majority logic reduction. (b) LINA blocks the layout of a full adder. (c) 1-wide LINA layout of a full adder circuit. (d) Traces for the inputs and outputs of a full adder circuit. (e) Clock traces used for the full adder.

(A, B, or C) or the block function (M for majority, X for crossing, or I for inversion). Additionally, the circuit uses a simple design rule which dictates a single block spacing between adjacent wires, although it may need to be changed based on the specifics of the implementation technologies. Figure 19.8c shows the 1-wide 4 nm LINA layout for the circuitry corresponding to the blocks design. The sum and carry-out outputs are taken from the bottom of the circuit and are valid for one complete clock cycle after the corresponding inputs are latched in, as can be seen by the simulation results of Figure 19.8d. Another important feature of this particular full adder design is the use of five LINA coplanar wire crossing blocks and both LINA wire crossing clock patterns. As can be seen in the diagrams, the coplanar crossings are able to pass signals which are half a clock out of phase with each other (clocks 0 and 2 or clocks 1 and 3). The clock signals which allow this are shown along with the traces for all of the other clock signals in Figure 19.8e. Overall, the proposed full adder circuit takes up an area equal to 16×14 blocks. The actual area per block is dependent on the lattice spacing used as well as the width of the LINA wire. Overall, the 4 nm 1-wide full adder uses 91 blocks and 553 LINA cells and encompasses an area of 256 nm $\times$ 192 nm = 49152 nm^2 and one full clock cycle for complete operation.

As has been briefly mentioned, the blocks design for this full adder can be implemented with any width of LINA wire. To accomplish this, each block must encompass a completely populated $n \times n$ square area of the LINA lattice map with the length of the block equal to the width of the LINA wire. Therefore, LINA designs are able to be directly scalable as implementation technologies improve.

19.6   CONCLUSIONS

If the QCA paradigm is to compete as a suitable replacement for CMOS-integrated circuitry in the coming years, implementation strategies for large-scale room-temperature operation must be developed. Traditional QCA designs make this difficult due to size restrictions for highly reliable components and both cellular and circuit layout geometries. Lattice-based integrated-signal nano-cellular automata (LINA)-type QCA offer an alternate design strategy which is more amenable to currently available or near-term nanoparticle implementation technologies by adhering to lattice maps (which fit more directly with self-assembled materials) and allowing for flexible wire widths to adjust to the resolution of large-scale patterning technologies. LINA have also been shown to increase the reliability in the presence of thermal excitations and assembly or patterning errors, thereby increasing the potential for room-temperature operation and near term fabrication of cellular automata-based computing systems. With these facts in mind, this chapter has begun to build a foundation upon which LINA logical devices can be designed. This has been accomplished by developing a logically complete set of primitives, including a LINA majority gate, inverter, and a planer wire crossing structure. LINA designs have also been evaluated in terms of the power/speed/reliability/area trades and shown to be advantageous compared to traditional QCA designs in this respect. As an example of the LINA design process, a LINA full adder was designed, laid out, and simulated with good results. Taken together, this chapter provides an important foundation to the logical design of LINA devices, thus adding depth to a strategy that offers hope for large-scale room temperature implementation and continued viability of the QCA paradigm to supplement and/or replace CMOS in the near future.

REFERENCES

1. International Technology Roadmap for Semiconductors (ITRS) 2010 Update, http://www.itrs.net/Links/2010ITRS/Home2010.htm, 2010.
2. H. Sutter, The free lunch is over: A fundamental turn toward concurrency in software, *Dr. Dobb's Journal*, 30(3), 202–210, 2005.
3. P. D. Tougaw and C. S. Lent, Logical devices implemented using quantum cellular automata, *J. Appl. Phys.*, 75, 1818–1825, 1994.
4. J. Timler and C. S. Lent, Power gain and dissipation in quantum-dot cellular automata, *J. Appl. Phys.*, 91(2), 823–831, 2002.
5. H. Thapliyal and N. Ranganathan, Reversible logic-based concurrently testable latches for molecular QCA, *IEEE Trans. Nanotechnol.*, 962–69, 2010.
6. A. Fijany, B. Toomarian, and M. Spotnitz, Novel highly parallel and systolic architectures using quantum dot-based hardware, *Parallel Comput.*, 484–492, 1999.
7. L. Lu, W. Liu, M. O'Neill, and E. E. Swartzlander Jr., QCA systolic matrix multiplier, *2010 IEEE Annual Symposium on VLSI*, Belfast, UK, pp.149–154, 2010.
8. Y. Wang and M. Lieberman, Thermodynamic behavior of molecular-scale quantum-dot cellular automata (QCA) wires and logic devices, *IEEE Trans. Nanotechnol.*, 3, 368–376, 2004.
9. W. Hu, K. Sarveswaran, M. Lieberman, and G. H. Bernstein, High-resolution electron beam lithography and DNA nano-patterning for molecular QCA, *IEEE Trans. Nanotechnol.*, 4, 312–316, 2005.
10. L. R. Hook IV and S. C. Lee, Lattice-based integrated-signal nanocellular automata (LINA) for the future of QCA-based nanoelectronics, *2011 IEEE Nanotechnol. Conf. (IEEE-NANO)*, 2011.
11. L. R. Hook IV and S. C. Lee, Design and simulation of 2-D 2-dot quantum-dot cellular automata logic, *IEEE Trans. Nanotechnol.*, 10, 996–1003, 2011.
12. E. V. Shevchenko, D. V. Talapin, N. A. Kotov, S. O'Brien, and C. B. Murray, Structural diversity in binary nanoparticle superlattices, *Nature*, 439, 55–59, 2006.
13. T. Dysart and P. Kogge, Reliability impact of N-modular redundancy in QCA, *IEEE Trans. Nanotechnol.*, 10, 1015–1022, 2011.
14. J. Timler and C. S. Lent, Power gain and dissipation in quantum-dot cellular automata, *J. Appl. Phys.*, 91(2), 823–831, 2002.
15. S. Srivastava, S. Sarkar, and S. Bhanja, Estimation of upper bound of power dissipation in QCA circuits, *IEEE Trans. Nanotechnol.*, 8, 116–127, 2009.
16. R. Zhang, K. Walus, W. Wang, and G. A. Jullien, A method of majority logic reduction for quantum cellular automata, *IEEE Trans. Nanotechnol.*, 3, 443–450, 2004.

20 Minimal Majority Gate Mapping of Four-Variable Functions for Quantum-Dot Cellular Automata

Peng Wang, Mohammed Niamat, and Srinivasa Vemuru

CONTENTS

20.1 INTRODUCTION

Complementary metal–oxide–semiconductor (CMOS) scaling faces many serious difficulties due to the approaching fundamental device physics limits. The quantum effects will dominate the device performance as the dimension approaches the sub-10 nm range due to an increase in the gate leakage current, as well as in capacitive coupling, doping, and lithography fluctuations [1]. Many technologies are proposed for the replacement of CMOS technology, such as quantum-dot cellular automata (QCA) [2–4], single electron tunneling (SET) [5], and tunneling phase logic (TPL) [6]. QCA, one of the viable technologies for the implementation of future digital systems, will be the focus of this chapter. QCA technology has excellent features for nanoelectronic integrated circuit implementations, such as extremely high packing densities (10^{12} devices/cm^2), simple interconnections, small signal delays, and low-power consumption. The basic circuit unit for QCA circuits is a three-input majority gate. In addition to developing logic circuits, QCA are also used to create interconnects. Thus, large QCA digital circuit architectures can be built using simple structures such as wires, majority gates, and inverters. The focus of this chapter is to present logic synthesis approaches for QCA applications based on majority gates.

Research in majority logic synthesis could be traced back to the 1960s. Reduced-unitized-table [7], Karnaugh-map (K-map) [8], and Shannon's decomposition principles [9] were employed for majority logic synthesis. However, these methods were only applicable to small networks since they were solving the problem manually. Recent efforts to synthesize the majority logic for QCA circuits involved developing algebraic expressions based on a geometric interpretation of Boolean functions. This resulted in 13 standard functions with corresponding majority gate implementations [10]. However, these methods were limited to the three-variable Boolean logic functions. The approaches that are used to handle logic circuits with more than three variables are proposed in Refs. [11–15]. In these methods, logic synthesis tools such as SIS [16] are initially used to decompose the problem into three-feasible nodes. The three-feasible nodes are then synthesized to obtain majority expressions. In this chapter, we use four-feasible nodes as the starting point. With this approach, 143 different functions are identified for possible implementation using majority gates. The architecture implementations using this approach result in fewer majority gates and logic levels than prior methods.

20.2 BACKGROUND MATERIAL

Lent et al. proposed a QCA model that uses the electron position in a cell to determine binary values [17,18]. The propagation of binary information in a QCA circuit is based on the Coulomb repulsion between electrons rather than on the actual flow of electrical current.

20.2.1 QCA Cells

A standard QCA cell consists of four quantum dots which are confined by the cell boundary. Two electrons placed within this cell can move between the quantum dots through electron tunneling. The tunneling is controlled by potential barriers that are raised or lowered and by the interaction with neighboring cells. The two free electrons in a cell will occupy diagonal sites because of the Coulomb repulsion that pushes them away from each other. Thus, there are two energetically stable states of the two electrons in the QCA cell representing Logic "0" and Logic "1" as shown in Figures 20.1a and 20.1b, respectively.

20.2.2 QCA Devices

A majority of the gates and inverters are the basic logic gates in QCA technology. A QCA majority gate can perform a three-input logic function shown in Equation 20.1. A simple layout of a QCA majority gate is shown in Figure 20.2a and that of a QCA inverter is shown in Figure 20.2b.

$$M(A, B, C) = AB + AC + BC \tag{20.1}$$

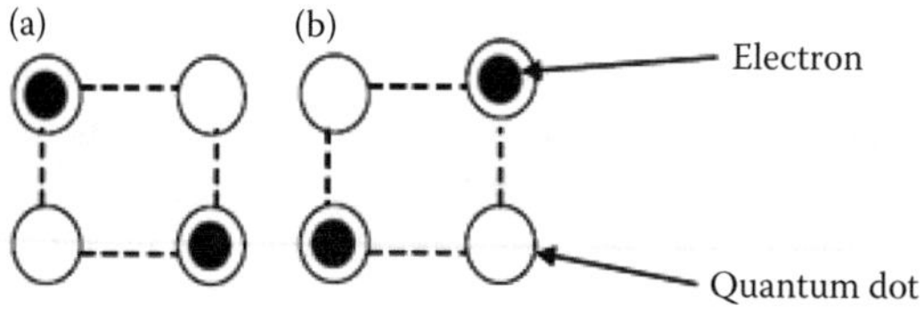

FIGURE 20.1 (a) QCA Cell Logic '0'; (b) QCA Cell Logic '1'. (P. Wang et al., Minimal majority gate mapping of 4-variable functions for quantum cellular automata, *Proceedings of the IEEE NANO*, Portland, OR, pp. 1307–1312. © (August 2011) IEEE. With permission.)

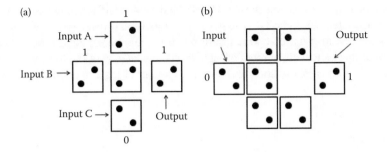

FIGURE 20.2 (a) QCA majority gate (b) QCA inverter. (P. Wang et al., Minimal majority gate mapping of 4-variable functions for quantum cellular automata, *Proceedings of the IEEE NANO*, Portland, OR, pp. 1307–1312. © (August 2011) IEEE. With permission.)

By forcing one of the three inputs of the majority gate to a constant logic "0" or a "1," a two-input AND or a two-input OR gate can be realized as shown below.

$$M(A, B, 0) = AB \tag{20.2}$$

$$M(A, B, 1) = A + B \tag{20.3}$$

20.2.3 QCA CLOCK

A QCA clock, comprising four distinct periodic phases [19], is needed for both combinational and sequential circuits. The QCA clock performs three main functions. First, the application of clock phases decides the direction of the information flow in the circuit. Second, it provides a mechanism to simultaneously propagate the information through the circuit. Third, it provides the power needed to activate the circuit. The clock controls the tunnel barrier for electrons. If the barrier is low, then the electrons are free to travel through the tunnel; thereby, the QCA cell can be easily polarized by its neighbor cells. If the barrier is high, the tunneling of electrons is not feasible and the QCA cell will hold its polarization state.

The QCA clock works in a pipeline fashion in four clock phases, Switch, Hold, Release, and Relax, as shown in Figure 20.3 [20]. In the Switch phase, the barrier begins to rise slowly and the polarization of the cell is determined by its neighbor cells (which are in the Hold phase). When the cell goes into the Hold phase, the polarization is formed and the barrier reaches its peak. The cell will affect its neighbors to pass on the information. Then, in the Release phase, the barrier begins to lower and the cell may lose its polarization since the binary information has already been passed. Finally, in the Relax phase, the tunneling barrier is in the lowest state, and as a result the QCA cell will remain unpolarized. All the four clock phases constitute one clock cycle, in which one bit information is passed. As the clock continues, the information is propagated through the QCA circuit.

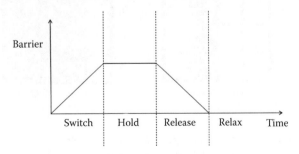

FIGURE 20.3 Four phases of the clock for the QCA circuit.

Using universal logic based on the AND, OR, and NOT functions, any Boolean logic function can be constructed using the QCA majority logic gates and inverters. However, these implementations in general may not be optimal in terms of the number of logic gates or the number of logic levels. With the increasing number of logic levels, the latency of the logic structure increases. Therefore, there is a strong motivation to reduce the number of logic gates and logic levels.

20.3 TERMINOLOGY

20.3.1 PRIMITIVES

The Boolean logic functions that can be represented by a single majority gate are called primitives. The primitives are the basic units that are used to form all other majority expressions. There are five types of primitives based on the number of input variables to a majority gate. The first type is the trivial case where there are no inputs, that is, a logic "1" or "0," referred to as a constant-type primitive ("C" type). The second type has a single input variable such as a or c' and is called a single-type primitive ("S" type). The third and the fourth types of primitives both have two input variables, which are the "AND" type and the "OR" type, respectively. Note that the input variable can appear in the AND and OR expressions as itself or in its complement form. The last type of primitives has all the three variable inputs and is called the "T" type primitive. All these primitives can be realized with one majority gate and their K-map implementations are shown in Figure 20.4.

Since any logic function can be constructed by using primitives, it is important to identify the possible primitives. Let us consider the three-variable Boolean functions. For the first two types, there are eight primitives, logic "1," logic "0," single inputs a, b, and c, and their complementary

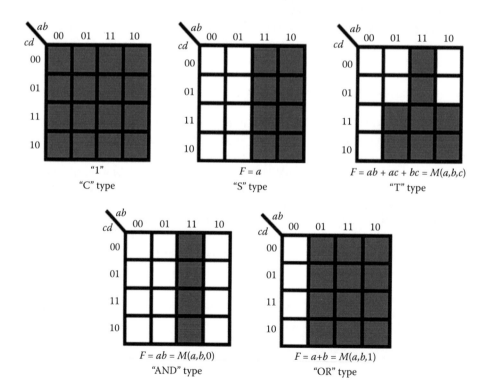

FIGURE 20.4 K-map representation of the five primitive types. (P. Wang et al., Minimal majority gate mapping of 4-variable functions for quantum cellular automata, *Proceedings of the IEEE NANO*, Portland, OR, pp. 1307–1312. © (August 2011) IEEE. With permission.)

literals a', b', and c'. The "AND/OR" type primitives are all based on the two-input majority gates. Therefore, any combination of two inputs from "a," "b," "c," "(a')," "(b')," "(c')" can be used. The combinations that are not allowed are the three pairs, "$(a\ a')$," "$(b\ b')$," and "$(c\ c')$." Each of these combinations appears once for the AND primitive and once for the OR primitive. So, the total number of primitives is given as

$$(C_2^6 - 3)*2 = 24 \tag{20.4}$$

The "T"-type primitives are based on the three-input majority gates and use all three variables "a," "b," and "c." Each input to the majority gate can appear as itself or its complement, thus giving a total of 2^3 "T" type primitives. Therefore, the total number of primitives for the three-variable problems is shown as

$$8(C_2^6 - 3)*2 + 2^3 = 40 \tag{20.5}$$

Similarly, 90 primitives are possible for the four-variable problem and are shown as

$$10 + (C_2^8 - 4)*2\,C_3^4* = 90 \tag{20.6}$$

A partial list of the primitives of the four-variable functions are shown in Table 20.1.

TABLE 20.1
A Partial List of 90 Possible Four-Variable Primitives

No.	Regular Logic	Majority Logic	Type	No.	Regular Logic	Majority Logic	Type
1	0	0	C	...	...	...	
2	1	1	C	74	$a'b' + a'd + b'd$	$M(a',b',d)$	T
3	a	a	S	75	$ac + cd + ad$	$M(a,c,d)$	T
4	a'	a'	S	76	$a'c' + c'd' + a'd'$	$M(a,c,d)'$	T
5	b	b	S	77	$a'c + cd + a'd$	$M(a',c,d)$	T
6	b'	b'	S	78	$ac' + c'd' + ad'$	$M(a,c',d')$	T
7	c	c	S	79	$ac' + c'd + ad$	$M(a,c',d)$	T
8	c'	c'	S	80	$a'c + cd' + a'd'$	$M(a',c,d')$	T
9	d	d	S	81	$ac + cd' + ad'$	$M(a,c,d')$	T
10	d'	d'	S	82	$a'c' + c'd + a'd$	$M(a',c',d)$	T
11	ab	$M(a,0,b)$	And	83	$bc + cd + bd$	$M(b,c,d)$	T
12	ac	$M(a,0,c)$	And	84	$b'c' + c'd' + b'd'$	$M(b,c,d)'$	T
13	ad	$M(a,0,d)$	And	85	$b'c + cd + b'd$	$M(b',c,d)$	T
14	bc	$M(b,0,c)$	And	86	$bc' + c'd' + bd'$	$M(b,c',d')$	T
15	bd	$M(b,0,d)$	And	87	$bc' + c'd + bd$	$M(b,c',d)$	T
16	cd	$M(c,0,d)$	And	88	$b'c + cd' + b'd'$	$M(b',c,d')$	T
17	$a' + b'$	$M(a,0,b)'$	Or	89	$bc + cd' + bd'$	$M(b,c,d')$	T
...	...	...	...	90	$b'c' + c'd + b'd$	$M(b',c',d)$	T

Source: P. Wang et al., Minimal majority gate mapping of 4-variable functions for quantum cellular automata, *Proceedings of the IEEE NANO,* Portland, OR, pp. 1307–1312. © (August 2011) IEEE. With permission.

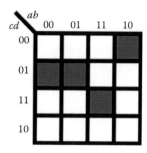

FIGURE 20.5 K-map implementation.

20.3.2 MAJORITY EXPRESSION

The K-map-based geometric interpretation of the majority gate synthesis method using primitives will be illustrated in this section. Any four-variable Boolean logic function can be mapped to a four-variable K-map. The function, $f = ab'c'd' + abcd + a'c'd$, and its K-map shown in Figure 20.5 will be used as an example to illustrate the synthesis process.

The synthesis solution targeting majority gates will be a tree structure with each of its nodes containing at most three branches representing the three inputs to the majority function. By selecting proper primitives, the top node of the tree gives the synthesis result of the function. Heuristic rules are used in the selection of primitives. The objective of the solution is to cover all the on-set cells using as few primitives as possible. The first primitive should cover as many on-set squares and as few off-set squares as possible. If several primitives can satisfy this rule, one of them is used as the starting solution. The second primitive is selected such that it covers the on-set squares that are not part of the first primitive and does not cover any off-set cells covered by the first primitive. It is also preferred that the second primitive covers as many on-set squares as possible. Once the first two primitives are selected, there are few choices for the third one. It must cover the on-set squares that are covered only once by the first two primitives and it should not cover the off-set squares that are covered once by the first two primitives. In summary, all the on-set cells are covered by at least two of the three primitives and all the off-set cells are covered at most by one primitive. This process yields a solution for the function involving at most four majority gates.

However, not all the four-variable functions could be represented by only three primitives. After the first two primitives are selected, it may not be possible to find a primitive that "will cover the on-set squares that are covered only once by the first two primitives and do not cover the off-set squares that are covered once by the first two squares." In this case, the required cell coverage should be expanded to another level of majority functions that could be realized with three primitives. As this process is continued, it will lead to a tree-like structure solution for the four-variable logic functions.

Applying this principle to $f = ab'c'd' + abcd + a'c'd$, primitive 79 from Table 20.1 ($M(a, c', d)$) is selected first as it covers all the on-set squares and four off-set squares. Then, primitive 4 of Table 20.1, a' is selected as the second primitive. There is no single primitive that provides the needed pattern cover, shown in the middle K-map of Figure 20.6. Therefore, another level with three primitives is needed to obtain the necessary coverage. A possible solution to obtain the cover for the needed primitive is found as shown in the rightmost three K-maps in Figure 20.6. The solution for the function as shown in Figure 20.6 is $f = M(M(a, c', d), a', M(M(a, b', c), M(c',d',0), M(b, d, 0)))$. There may be more than one solution for a given four-variable function. Most of the four-variable solutions can be obtained in three levels of majority gates and all of them can be obtained in four levels.

20.3.3 FOUR-DIMENSIONAL CUBE AND FOUR-VARIABLE K-MAP

A four-variable Boolean function can be represented by a four-dimensional (4-D) cube as shown in Figure 20.7 [21]. Each variable of the function, a, b, c, and d, represents a dimension of a 4-D

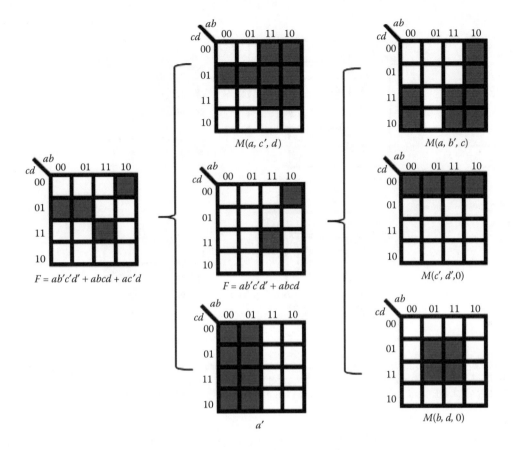

FIGURE 20.6 Solution based on the K-map. (P. Wang et al., Minimal majority gate mapping of 4-variable functions for quantum cellular automata, *Proceedings of the IEEE NANO*, Portland, OR, pp. 1307–1312. © (August 2011) IEEE. With permission.)

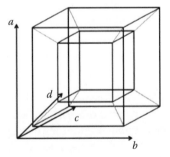

FIGURE 20.7 A 4-D cube. (P. Wang et al., Minimal majority gate mapping of 4-variable functions for quantum cellular automata, *Proceedings of the IEEE NANO*, Portland, OR, pp. 1307–1312. © (August 2011) IEEE. With permission.)

cube. The 16 different minterms are represented by the 16 vertices in a 4-D cube or by the 16 cells in a four-variable K-map. For example, the minterm $abc'd$ can be represented by a vertex $(1, 1, 0, 1)$ and also as an on-set square in the K-map. For a four-variable function, all the on-set vertices represented by minterms in a 4-D cube can be considered as a certain spatial structure which could be rotated, flipped, or mirrored to represent several other four-variable functions.

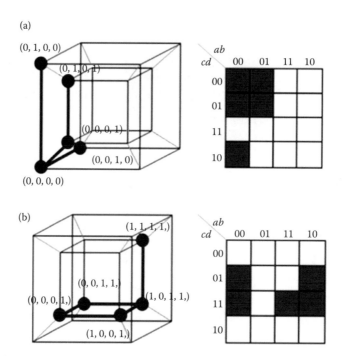

FIGURE 20.8 (a) Interpretation of $f = b'd + acd$. (b) Interpretation of $f = a'c' + a'b'd'$. (P. Wang et al., Minimal majority gate mapping of 4-variable functions for quantum cellular automata, *Proceedings of the IEEE NANO*, Portland, OR, pp. 1307–1312. © (August 2011) IEEE. With permission.)

As an example, consider two logic functions $f = a'c' + a'b'd'$ and $f = b'd + acd$. After mapping these two logic functions to a 4-D cube and the corresponding K-map, as shown in Figures 20.8a and b, it can be clearly seen that the spatial structures formed by their on-set vertices are exactly the same (one face and another vertex above it), and the only difference is that they are in different locations. If one can find all possible spatial on-set structures for the four-variable functions and determine the simplified majority expressions for each structure, any four-variable problem could be targeted to these solutions to obtain the minimal majority expressions. These structures are referred to as standard functions.

20.3.4 HAMMING DISTANCE

Hamming distance between two minterms is the number of different literals in the two minterms. It is easy to visualize on the K-map as the number of horizontal and vertical steps needed to move from cells representing the minterms. The maximum Hamming distance between two minterms in a four-variable K-map is four. For example, the distance between $a'bc'd$ and $ab'c'd$ is two since the literals involving a and b are different in the two terms. Since the K-map is foldable in both the horizontal and vertical directions, one must be careful in computing the Hamming distance from the K-map.

20.3.5 A METHOD TO FIND STANDARD FUNCTIONS

Hamming distance is utilized to identify all possible standard functions of the four-variable problems. If the Hamming distances between all pairs of vertices in a Boolean logic function are exactly the same as the Hamming distances between all pairs of vertices in another Boolean logic function, then these two functions will have the same structure and can be represented by the same standard function.

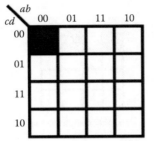

$$F = a'b'c'd' = M'\,(M(a\ b\ 1),\, M(c\ d\ 1),\, 1)$$

FIGURE 20.9 The standard function with one minterm.

The standard functions are searched based on the number of minterms and the search is limited to at most eight-minterm problems, because Boolean functions with 9–15 minterms can be represented by the complementary function comprising seven to one minterms. The search begins with functions with only one minterm. There are 16 positions for this single vertex in a 4-D cube. No matter which minterm is selected, its spatial structure remains the same on a single vertex. Since the spatial structures are all the same, any one of them could be a standard function. For example, the solution for $a'b'c'd'$ is shown in Figure 20.9 as a standard function. Other standard functions with one minterm will have similar solutions except that the inputs to the majority gates will be different.

For Boolean functions with two minterms, there are 15 possible locations where the second minterm can be placed in the K-map as shown in Figure 20.10. Deletion of the structures with similar Hamming distance yields four unique combinations with Hamming distances of one to four as shown in Figure 20.11.

The concept of a distance vector is introduced before continuing to the cases involving three or more minterms. For the four-variable case, the distance vector is a 4-tuple. The first element number represents the number of minterm pairs whose hamming distance is one, the second element represents the number of pairs whose distance is two, the third number represents the number of pairs whose distance is three, and the fourth number represents the number pairs whose distance is four. The distance vectors for the four two-minterm standard function are shown in Figure 20.11. In the figure, the left column denotes the Hamming distances going from one to four, and the right column denotes the distance vector. Since the functions have one pair of minterms, only one of 4-tuples have a value of 1 with the others being 0.

If two functions have the same distance vectors, they are represented by the same standard function. One hundred and forty-three distinct structures are obtained for the 4-variable functions resulting in 143 standard functions. For brevity, a partial list of standard functions is given in Table 20.2. The second column gives the Boolean expression of the function. The third column gives an optimal majority gate implementation. The last column specifies the distance vector for the corresponding standard functions.

20.4 LOGIC SYNTHESIS METHODOLOGY AND IMPLEMENTATION

The flowchart for the proposed majority logic synthesis method is shown in Figure 20.12. The input is an arbitrary Boolean logic function F and the output is an optimal synthesized majority network based on the four-feasible implementations, G_m. The first step of the method is simplifying and decomposing the Boolean function F. A minimized 4-feasible logic network G in which each node contains four or less inputs will be generated after this step. Each node in G is then converted to the corresponding optimal majority expression using the four-variable standard functions and primitives. A preliminary solution is obtained when all the nodes in the 4-feasible network have been converted to majority expression. The solution is then searched to eliminate repeated terms and doubled inversions resulting in a final solution of the majority gate representation, G_m, of the function F.

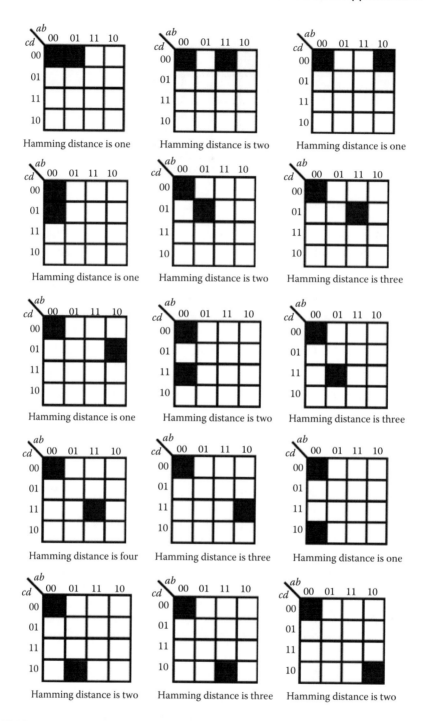

FIGURE 20.10 Process for searching standard functions for two minterm functions.

20.4.1 SIMPLIFYING AND DECOMPOSING

The effectiveness of the proposed synthesis solution depends on the pre-processing of the function F prior to decomposition into the 4-feasible nodes. Pre-processing involves Boolean simplification and algebraic factorization and is performed using the sis tool. The sis pre-processing commands used are given in Figure 20.13. After pre-processing, the network needs to be decomposed to a

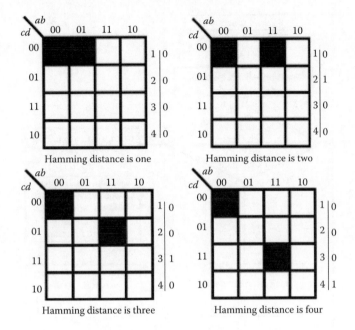

FIGURE 20.11 Four standard functions for two-minterm functions with distance vectors.

TABLE 20.2
Partial List of Standard Functions of Four-Variable Functions

No.	Standard Functions	Majority Expressions	Distance Vector
1	$F = a'b'c'd'$	$F = M(M(a'\,b'\,0), M(c'\,d'\,0), 0)$	None
2	$F = a''c'd'$	$F = M(a', M(c'\,d'\,0), 0)$	1,0,0,0
3	$F = a'b'c'd' + abc'd'$	$F = M(M(a'\,b'\,0), M(c'\,d'\,0), M(a\,b\,0))$	0,1,0,0
4	$F = a'b'c'd' + abc'd$	$F = M(M(a'\,b\,c'), M(b'\,d'\,0), M(a\,d\,0))$	0,0,1,0
5	$F = a'b'c'd' + abcd$	$F = M\{M[M(a'\,b'\,0), d', 0], M[M(a\,b\,0), c, 0], M'\,(c, d'\,0)\}$	0,0,0,1
...	...	...	...
78	$F = a'b'c'd' + a'bc'd' + abc'd' + ab'c'd'$ $+ a'b'c'd + a'bc'd + abc'd$	$F = M[M(a'\,b\,c'), M(c'\,d'\,0), c']$	9,9,3,0
79	$F = a'b'c'd' + a'bc'd' + abc'd' + ab'c'd'$ $+ a'b'c'd + a'bc'd + a'b'cd$	$F = M\{M(a'\,b'\,0), M[M(b'\,c'\,d), a', 0]\,1\}$	8,8,4,1
80	$F = a'b'c'd' + a'bc'd' + abc'd' + ab'c'd'$ $+ a'b'c'd + a'bc'd + abcd$	$F = M\{M(a'\,b'\,0), M[M(a'\,c\,d), M(a'\,c'\,0), M(a\,b\,0)], 1\}$	7,8,5,1
81	$F = a'b'c'd' + a'bc'd' + abc'd' + ab'c'd'$ $+ a'b'c'd + a'bc'd + a'b'cd'$	$F = M[M(a'\,c'\,d'), M(b'\,c\,0), c']$	8,9,4,0
...	...	...	...
141	$F = a'b'c'd' + a'bc'd' + abc'd' + a'bc'd +$ $a'b'cd + abcd + ab'cd + ab'cd'$	$F = M\{M[M(a'\,b\,d'), c', 0], M(a\,b'\,d), c, 0], 1\}$	6,12,6,4
142	$F = a'b'c'd' + a'bc'd' + abc'd' + abc'd +$ $ab'c'd + a'b'cd + a'bcd + abcd' + ab'cd'$	$F = M[M(a\,c'\,d'), M(a'\,c'\,d), M(a\,c\,d)]$	4,12,12,0
143	$F = a'b'c'd' + abc'd' + a'bc'd + ab'c'd +$ $a'b'cd + abcd + a'bcd' + ab'cd'$	$F = M(M\{M[M(a\,b\,d), M(a'\,b'\,d), M(a'\,b\,d')], c', 0\}, M\{M[M(a\,b\,d), M(a'\,b\,d'), M(a\,b'\,d')], c, 0\}, 1)$	0,24,0,4

Source: P. Wang et al., Minimal majority gate mapping of 4-variable functions for quantum cellular automata, *Proceedings of the IEEE NANO*, Portland, OR, pp. 1307–1312. © (August 2011) IEEE. With permission.

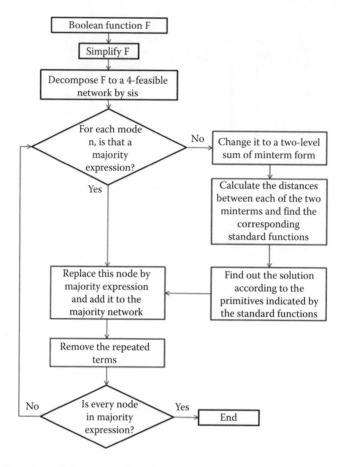

FIGURE 20.12 Flowchart of the proposed method. (P. Wang et al., Minimal majority gate mapping of 4-variable functions for quantum cellular automata, *Proceedings of the IEEE NANO*, Portland, OR, pp. 1307–1312. © (August 2011) IEEE. With permission.)

1. collapse	13. resub -a -d
2. sweep	14. sweep
3. eliminate 5	15. gcx -bt 10
4. simplify -m nocomp -d	16. resub -a -d
5. resub -a -d	17. sweep
6. gkx -abt 30	18. gkx -ab
7. resub -a -d	19. resub -a -d
8. sweep	20. sweep
9. gcx -bt 30	21. gcx -b
10. resub -a -d	22. resub -a -d
11. sweep	23. sweep
12. gkx -abt 10	24. eliminate

FIGURE 20.13 Commands for Pre-processing. (P. Wang et al., Minimal majority gate mapping of 4-variable functions for quantum cellular automata, *Proceedings of the IEEE NANO*, Portland, OR, pp. 1307–1312. © (August 2011) IEEE. With permission.)

1. xl_part_coll -n4 -m -g2	5. xl_imp -n4
2. xl_coll_ck -n4	6. xl_partition -n4 -t
3. xl_partition -n4 -m	7. xi_cover -n4 -e30 -u200
4. full_simplify	8.xl_coll_ck -n4 -k

FIGURE 20.14 Commands for the first approach.

1. xl_part_coll -n4 -m -g2	10. fx
2. xl_coll_ck -n4	11. resub -a; sweep
3. xl_partition -n4 -m	12. eliminate -1; sweep
4. sweep; eliminate -1	13. full_simplify -m nocon
5. simplify -m nocomp	14. xl_imp -n4
6. eliminate -1	15. xl_partition -n4 -t
7. sweep; eliminate 5	16. xl_cover -n4 -e30 -u20
8. simplify -m nocomp	17. xl_cool_ck -n4 -k
9. resub -a	

FIGURE 20.15 Commands for the second approach.

4-feasible network. Two approaches are used with a different sequence of sis commands as shown in Figures 20.14 and 20.15. Depending on the function F, one of these two methods results in a 4-feasible function that is better suited to the majority gate synthesis. Therefore, both methods are tried out and given as input to the majority gate converter.

20.4.2 Conversion to Majority Gates

After a 4-feasible network is generated, each node in this network is converted to majority expression. A detailed explanation of the conversion is demonstrated using the implementation of the function $f = b(a'c'd + acd')$.

Step 1: Simplify the function to the minterm form

This step is illustrated using the K-map and the minterms are mapped to two cells using the following equation:

$$f = b(a'c'd + acd') = a'bc'd + abcd' \tag{20.7}$$

Step 2: Calculate the distance of the resulting minterms

The Hamming distance between the two minterms is identified as 3 and the resulting distance vector is (0, 0, 1, 0).

Step 3: Find the corresponding standard function

The standard function 4 in Table 20.1 is found to have the same distance vector as this function, which means they have the same spatial structure and the same type of solution. Therefore, the solution for this problem consists of one "T" type primitive and two "AND" type primitives.

Step 4: Solution targeting the majority gates

As can be seen from Figure 20.16, the "T" type primitive covers both minterms while the two "AND" type primitives cover only one minterm each. The resulting implementation is given as $f = M(M(b\ c\ d), M(a'\ c'\ 0), M(a\ d'\ 0))$ and its logic diagram is shown in Figure 20.17a. The logic implementation of the same function using MALS described in Ref. [13] is shown in Figure 20.17b.

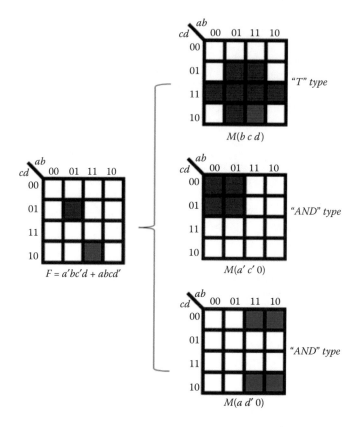

FIGURE 20.16 Solution for $f = b(a'c'd + acd')$. (P. Wang et al., Minimal majority gate mapping of 4-variable functions for quantum cellular automata, *Proceedings of the IEEE NANO*, Portland, OR, pp. 1307–1312. © (August 2011) IEEE. With permission.)

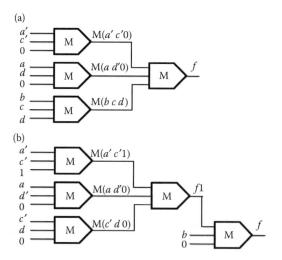

FIGURE 20.17 Majority gate solution using (a) proposed method, and (b) MALS [13]. (P. Wang et al., Minimal majority gate mapping of 4-variable functions for quantum cellular automata, *Proceedings of the IEEE NANO*, Portland, OR, pp. 1307–1312. © (August 2011) IEEE. With permission.)

20.4.3 REMOVE REPEATED TERMS

The Boolean logic function has already been converted to its majority gate expression. The result can be further simplified by removing redundancies which include repeated terms and doubled inverters. The process of removing is illustrated by an example given below.

$$\left.\begin{array}{l} f = abcd \\ g = abc' + abd' \end{array}\right\} \tag{20.8}$$

This multi-output Boolean logic network is already a 4-feasible network. The logic functions can be directly converted into majority expression yielding the solution shown below.

$$\left.\begin{array}{l} f = abcd = M(M(a\,b\,0), M(c\,d\,0), 0) \\ g = abc' + abd' = M(M(a\,b\,0), M'(c\,d\,0), 0) \end{array}\right\} \tag{20.9}$$

This solution contains two levels of logic and six majority gates, but it can be seen that the terms $M(a\,b\,0)$ and $M(c\,d\,0)$ appear in both functions "f" and "g." Removing the redundant terms, a simpler solution that uses the four gates is shown in

$$\left.\begin{array}{l} f = abcd = M((1),(2),0) \\ g = abc' + abd' = M((1),(2)',0) \\ (1) = M(a\,b\,0) \\ (2) = M(a\,b\,0) \end{array}\right\} \tag{20.10}$$

20.5 RESULTS AND COMPARISON

The proposed method and MALS are applied to an MCNC/SIGDA benchmark circuit "majority." The generated results from the two methods are compared, including logic equations, circuit diagrams, and QCA layouts. The logic function of the benchmark circuit "majority" is given below.

$$\left.\begin{array}{l} f = (h)' \\ (h) = a'b'd' + a'c'd' + a'd'e' + b'c'd' + b'd'e' + c'd'e' \end{array}\right\} \tag{20.11}$$

The SIS tool is used to simplify and decompose the circuit to generate a 4-feasible network shown below.

$$\left.\begin{array}{l} f = (x) + d \\ (x) = abc + abe + ace + bce \end{array}\right\} \tag{20.12}$$

The proposed method is used to convert each node in the function to its corresponding majority expression. Node f contains two variables, so one-to-one mapping is performed on it. Node (x) contains four variables, so the four-variable standard functions are used to convert this node. Its distance vector is given as $(4, 6, 0, 0)$ and standard function No. 35 is selected. Based on the majority expression for the standard function, the solution for this node contains three levels and five majority

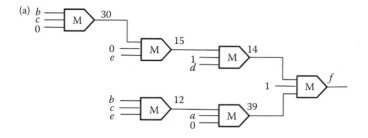

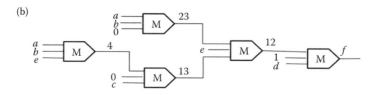

FIGURE 20.18   (a) MALS implementation and (b) implementation of the proposed method.

gates. The third level is constructed by one "T" type primitive; the second level is constructed by three "AND" type primitives; the first level is also constructed by one "T" type primitive. After all the redundant terms are removed, the final result is shown in Equation 20.13.

$$
\left.
\begin{aligned}
f &= (12) + d = M((12)\,d\,0) \\
(12) &= (23)(13) + (23)e + (13)e = M((23)\,(13)\,e) \\
(23) &= ab = M(a\,b\,0) \\
(13) &= (4)c = M((4)\,c\,0) \\
(4) &= ab + ae + be = M(a\,b\,e)
\end{aligned}
\right\}
\tag{20.13}
$$

The result of the same benchmark circuit benchmark generated by MALS is shown below.

$$f = (14) + (39) = M((14)\,(39)\,1)$$

$$(12) = bc + be + ce = M(b\,c\,e)$$

$$(14) = (15) + d = M((15)\,d\,1)$$

$$(15) = (30)e = M((30)\,e\,0)$$

$$(30) = bc = M(b\,c\,0)$$

$$(39) = (12)a = M((12)\,a\,0)$$

The circuit diagram for the proposed method and MALS are also shown in Figures 20.18a and b, respectively. The proposed method takes one less gate than the existing method, which is a 16.7% reduction.

TABLE 20.3
MCNC/SIGDA Benchmark Circuit Synthesis Comparisons

	MALS		Our Method		Reduction %	
	Level	Gate	Level	Gate	Level	Gate
b1	3	9	2	8	33.3%	11.1%
cm42a	2	21	2	17	0	19.0%
decod	3	28	3	26	0	7.1%
cm82a	4	16	3	10	25%	37.5%
majority	4	6	4	5	0	16.7%
z4 ml	4	27	4	15	0	44.44%
cm152a	6	21	6	21	0	0
tcon	2	24	2	24	0	0
cmb	4	44	4	27	0	38.6%
cc	5	44	5	41	0	6.8%
cht	4	120	4	119	0	0.83%
unreg	5	84	5	84	0	0
cm85a	7	34	6	22	14.3%	35.3%
cm151a	7	42	7	21	0	50%
cm162a	7	46	7	40	0	13.0%
cu	7	46	7	39	0	15.2%
cm163a	7	42	7	33	0	21.4%
pm1	6	45	6	34	0	24.4%
mux	9	46	7	38	22.2%	17.4%
total	96	745	91	624	5.21%	16.2%

Although the majority gate implementations of all 143 functions are developed, they are not guaranteed to be in the simplest form as the current implementations are based on heuristics developed in the implementations of the standard functions. Still, the proposed method gives better results than the current approaches when applied to several MCNC benchmarks. This synthesis method is currently being integrated with the sis logic synthesis tool. Recently, Kong et al. [15] improved on the 3-feasible function approach of Ref. [13] by performing better decomposition in sis. These approaches will be extended to the 4-feasible functions when the proposed method is integrated into sis.

A comparison of the gate counts and the number of levels for various MCNC benchmarks using the proposed method and the method described in Ref. [13] is shown in Table 20.3. From the table, it can be seen that on average, there is a total reduction of 16.2% in the number of gates and 5.21% in the number of levels using the proposed method as compared to MALS.

20.6 CONCLUSIONS

As progress is made in nanoelectronic devices, corresponding developments must be made to use these devices in systems applications. Automated synthesis is a key step in this process. This chapter addresses the majority gate logic synthesis targeting QCA applications. A method is proposed for the first time that can handle 4-feasible Boolean networks. A total of 143 four-variable standard functions and their majority gate implementations are presented. This majority gate-based synthesis method is currently being integrated with the sis logic synthesis tool. The primitives are separated into different catalogs, thereby providing a heuristic in searching them rather than using an exhaustive search.

ACKNOWLEDGMENT

This work was supported in part by the National Science Foundation under awards 0958298 and 0958355.

REFERENCES

1. International technology roadmap for semiconductors (ITRS). [Online]. Available at: http://www.itrs.net
2. G. L. Snider, A. O. Orlov, I. Amlani, X. Zuo, G. Bernstein, C. Lent, J. Merz, and W. Porod, Quantum-dot cellular automata: Review and recent experiments, *Journal of Applied Physics*, 85, 8, 1999.
3. C. S. Lent and P. D. Tougaw, A device architecture for computing with quantum dots, *Proceedings of the IEEE*, 85, 541, 1997.
4. K. G. Walus, A. Jullien, and V. S. Dimitrov, Computer arithmetic structures for quantum cellular automata, *Conference Record of the 37th Asilomar Conference on Signals, Systems and Computers*, Monterey, CA, pp. 1435–1439, 2004.
5. T. Oya, T. Asai, T. Fukui, and Y. Amemiya. A majority-logic nanodevice using an irreversible single-electron boxes, *IEEE Transactions on Nanotechnology*, 2, 15–22, 2003.
6. H. A. Fahmy and R. A. Kiehl, Complete logic family using tunneling-phase-logic devices, *Proceedings of the International Conference on Microelectronics*, State of Kuwait, pp. 22–24, Nov. 1999.
7. S. B. Akers, Synthesis of combinational logic using three-input majority gates, *Proceedings of the 3rd Annual Symposium on Switching Circuit Theory and Logic*, Chicago, IL, pp. 149–157, 1962.
8. H. S. Miller and R. O. Winder, Majority logic synthesis by geometric methods, *IRE Transactions on Electronic Computers*, EC-11, 89–90, 1962.
9. S. Muroga, *Threshold Logic and Its Applications*. New York, NY: Wiley, 1971.
10. R. Zhang, K. Walus, W. Wang, and G. A. Jullien, A method of majority logic reduction for quantum cellular automata, *IEEE Transactions on Nanotechnology*, 3, 443–450, 2004.
11. M. Bonyadi, S. Azghadi, N. Rad, K. Navi, and E. Afjei, Logic optimization for majority gate-based nanoelectronic circuits based on genetic algorithm, *International Conference on Electrical Engineering*, San Francisco, USA, pp. 1–5, 2007.
12. S. Rai, Majority gate based design for combinational quantum cellular automata (QCA) circuits, in *Proceedings of the 40th Southeastern Symposium on System Theory*, New Orleans, LA, pp. 222–224, 2008.
13. R. Zhang, P. Gupta, and N. K. Jha, Majority and minority network synthesis with application to QCA-, SET-, and TPL-based nanotechnologies, *IEEE Transactions on Computer-Aided Design*, 26, 1233–1245, 2007.
14. Z. Huo, Q. Zhang, S. Haruehanroengra, and W. Wang, Logic optimization for majority gate- based nanoelectronic circuits, *International Symposium on Circuits and Systems*, Island of Kos, Greece, pp.1307–1310, 2006.
15. K. Kong, Y. Shang, and R. Lu, An optimized majority logic synthesis methodology for quantum-dot cellular automata, *IEEE Transactions on Nanotechnology*, 9, 170–183, 2010.
16. E. M. Sentovich, K. J. Singh, L. Lavagno, C. Moon, R. Murgai, A. Saldanha, H. Savoj, P. R. Stephan, R. K. Brayton, and A. Sangiovanni-Vincentelli, SIS: A system for sequential circuit synthesis, Technical Report, University of California, Berkeley, 1992.
17. C. S. Lent, P. D. Tougaw, W. Porod, and G. H. Bernstein Quantum cellular automata, *Nanotechnology*, 4, 49–57, 1993.
18. C. S. Lent and B. Isaksen, Clocked molecular quantum-dot cellular automata, *IEEE Transactions on Electron Devices*, 50, 1890–1896, 2003.
19. K. Hennessy and C. S. Lent, Clocking of molecular quantum-dot cellular automata, *Journal of Vacuum Science and Technology*, 19(5), 1752–1755, 2001.
20. E. P. Blair and C. S. Lent, Quantum-dot cellular automata: An architecture for molecular computing, *International Conference on Simulation of Semiconductor and Devices*, Cambridge, MA, pp. 14–18, 2003.
21. R. K. Brayton, C. McMullen, G. D. Hachtel, and A. Sangiovanni-Vincentelli, *Logic Minimization Algorithms for VLSI Synthesis*. Norwell, MA: Kluwer, 1984.
22. P. Wang, M. Niamat, and S. Vemuru, Minimal majority gate mapping of 4-variable functions for quantum cellular automata, *Proceedings of the IEEE NANO*, Portland, OR, pp. 1307–1312, Aug. 2011.

Section VII

Memristors, Resistive Switches, and Memory

21 Nanodevices
Describing Function and Liénard Equation

Alberto Delgado

CONTENTS

21.1 INTRODUCTION

Nanotechnology is a promising research area with applications in different disciplines. In electrical and electronic engineering, nanodevices are new building blocks for circuits that increase the designer possibilities beyond traditional elements. In this chapter, two methods to study the behavior of nonlinear circuits and systems are illustrated for memristors, and nanodevices based on deoxyribonucleic acid (DNA) or carbon nanotubes (CNTs).

The two points of view are: (i) frequency analysis with the describing function (DF) [1,2], and (ii) time analysis with the Liénard equation [3]. The purpose of these tools is to predict the behavior of circuits and systems with nonlinear nanodevices or even to guide the design of nanodevices to synthesize any desired dynamics. Also, the tunable nanodevices could increase the complexity and applications of new circuits.

In the literature, it is common to find experimental $v-i$ characteristics for nanodevices; these plots show nonlinear behavior that can be described by piecewise linear equations. Some typical nanodevices are based on oxide films [4–6], DNA [7], or CNTs [8]. In other nanodevices, the nonlinear $v-i$ relationship can be expressed as formulas, $v = \varphi(i)$, $\phi = f(q)$; the current is related to the charge, $i = dq/dt$, and the voltage is related to the flux, $v = d\phi/dt$.

This chapter is divided as follows. Section 21.2 presents the definition of DF and its calculation, step by step, for a general nonlinearity. Then, the general formula is applied to memristors, and DNA or CNTs-based nanodevices. Section 21.3 introduces the Liénard equation for series RLC circuits with nanodevices; two types of nonlinearities are considered $v = \varphi(i)$ and $\phi = f(q)$; also, five conditions for the existence of limit cycles are formulated. Section 21.4 presents the simulations to support the theoretical predictions for self-sustained oscillations. Finally, Section 21.5 includes some conclusions and future work.

21.2 DESCRIBING FUNCTION

In nonlinear control theory, the DF is used to predict the behavior of a nonlinear device in closed loop with a linear system, see Figure 21.1. The characteristic polynomial (denominator) of the closed-loop transfer function produces two equations; the solution yields the frequency and amplitude (ω, a) of the limit cycle or self-sustained oscillation [1]. This approach is called *analysis in the frequency domain*; the input voltage is periodic with known amplitude.

The closed-loop transfer function obeys the expression,

$$\frac{Y(j\omega)}{R(j\omega)} = \frac{\Psi(a)G(j\omega)}{1 + \Psi(a)G(j\omega)} \tag{21.1}$$

where $\Psi(a)$ is the DF of the nonlinear nanodevice and $G(j\omega)$ is the linear system transfer function in the frequency domain. The closed loop has a limit cycle when,

$$1 + \Psi(a)G(j\omega) = 0 \tag{21.2}$$

The function $G(j\omega)$ is a complex quantity with real and imaginary parts; if the DF is real, there are two equations,

$$\Psi(a) . \operatorname{Re}[G(j\omega)] = -1$$
$$\operatorname{Re}[G(j\omega)] = \frac{-1}{\Psi(a)} \tag{21.3}$$

$$\Psi(a) . \operatorname{Im}[G(j\omega)] = 0$$
$$\operatorname{Im}[G(j\omega)] = 0 \tag{21.4}$$

Solving Equation 21.4 produces the limit cycle frequency and Equation 21.3 yields the oscillation amplitude [2]. When a closed-loop system like Figure 21.1 is given, in a practical situation, the first step is to find the DF $\Psi(a)$ for the nonlinearity.

The definition of a real DF $\Psi(a)$ for odd, time invariant, memoryless nonlinearities is the following [1]:

$$\Psi(a) = \frac{2}{\pi a} \int_0^\pi i(\theta) \sin\theta \, d\theta \tag{21.5}$$

This analysis is valid for the input voltage,

$$v(\theta) = a \sin\theta \tag{21.6}$$

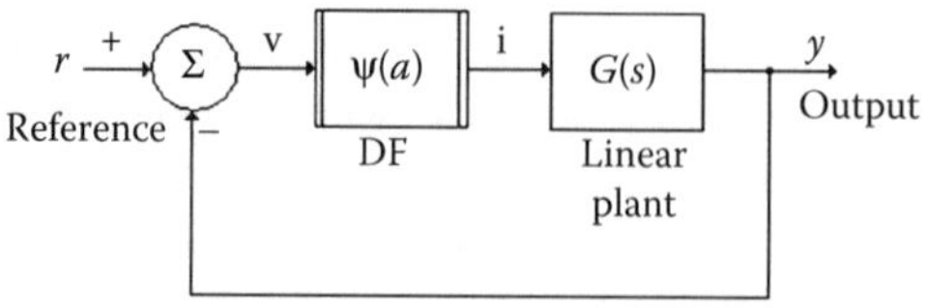

FIGURE 21.1 Linear plant in closed loop with a nanodevice, s is the Laplace operator.

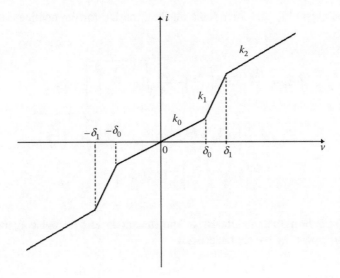

FIGURE 21.2   Nonlinear static v–i characteristic for a general nanodevice.

Figure 21.2 shows a voltage–current characteristic for a general nanodevice where (k_0, k_1, k_2) are slopes; this v–i function can be studied using the DF as follows.

To evaluate integral (21.5), it is necessary to find the current $i(\theta)$ for different intervals following the input voltage (21.6),

$$
\begin{aligned}
0 \le \theta < \beta_0; \quad & i = k_0.v \\
\beta_0 \le \theta < \beta_1; \quad & i = \delta_0(k_0 - k_1) + k_1.v \\
\beta_1 \le \theta < \pi - \beta_1; \quad & i = \delta_0 k_0 + (\delta_1 - \delta_0)k_1 + k_2.(v - \delta_1) \\
\pi - \beta_1 \le \theta < \pi - \beta_0; \quad & i = \delta_0(k_0 - k_1) + k_1.v \\
\pi - \beta_0 \le \theta \le \pi; \quad & i = k_0.v
\end{aligned}
\tag{21.7a}
$$

where

$$
\beta_0 = \sin^{-1}\left(\frac{\delta_0}{a}\right); \quad \beta_1 = \sin^{-1}\left(\frac{\delta_1}{a}\right);
\tag{21.7b}
$$

Writing again Equation 21.5 and using Equation 21.7 yields,

$$
\begin{aligned}
\Psi(a) = \ & \frac{2}{\pi a} \int_0^{\beta_0} k_0 a \sin^2 \theta \, d\theta \\
& + \frac{2}{\pi a} \int_{\pi - \beta_0}^{\pi} k_0 a \sin^2 \theta \, d\theta \\
& + \frac{2}{\pi a} \int_{\beta_0}^{\beta_1} \left[\delta_0(k_0 - k_1) + k_1 a \sin \theta\right] \sin \theta \, d\theta \\
& + \frac{2}{\pi a} \int_{\pi - \beta_0}^{\pi - \beta_1} \left[\delta_0(k_0 - k_1) + k_1 a \sin \theta\right] \sin \theta \, d\theta \\
& + \frac{2}{\pi a} \int_{\beta_1}^{\pi - \beta_1} \left[\delta_0 k_0 + (\delta_1 - \delta_0)k_1 + k_2 a \sin \theta - \delta_1 k_2\right] \sin \theta \, d\theta
\end{aligned}
\tag{21.8}
$$

Integrating each term [9], and after some algebra, the DF for the nonlinearity of Figure 21.2 obeys,

$$\Psi(a) = \frac{2k_0}{\pi}\left[\sin^{-1}\left(\frac{\delta_0}{a}\right) - \frac{\delta_0}{a}\sqrt{1 - \left(\frac{\delta_0}{a}\right)^2}\right]$$

$$+ \frac{4}{\pi a}\left[\delta_0 k_0 + (\delta_1 - \delta_0)k_1 - \frac{\delta_1 k_2}{2}\right]\sqrt{1 - \left(\frac{\delta_1}{a}\right)^2}$$

$$- \frac{2k_2}{\pi}\sin^{-1}\left(\frac{\delta_1}{a}\right) + k_2; |a| > \delta_1 \tag{21.9}$$

Formula (21.9) can be used to find the DF of known nanodevices by replacing the value δ_0, δ_1, k_0, k_1, k_2 from the corresponding v–i characteristics.

21.2.1 MEMRISTOR

Figure 21.3 shows the v–i response for a memristor [4–6].
To reduce Figure 21.2 to Figure 21.3,

$$k_1 = k_2, \quad \delta_0 = \delta_1$$

Replacing these relationships in Equation 21.9,

$$\Psi(a) = \frac{2(k_0 - k_1)}{\pi}\left[\sin^{-1}\left(\frac{\delta_0}{a}\right) + \frac{\delta_0}{a}\sqrt{1 - \left(\frac{\delta_0}{a}\right)^2}\right] + k_1; \quad |a| > \delta_0 \tag{21.10}$$

Equation 21.10 is the memristor DF when the frequency of the input voltage is constant.

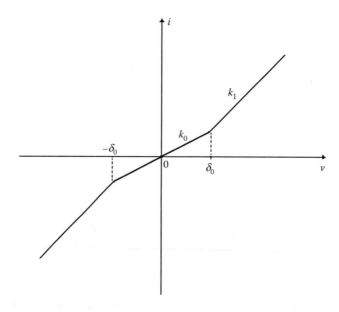

FIGURE 21.3 Nonlinear v–i characteristic for a memristor with fixed frequency.

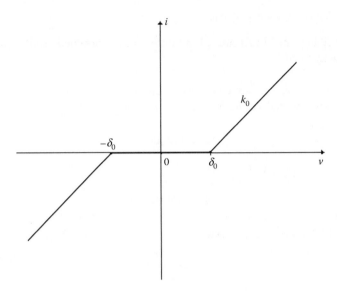

FIGURE 21.4 Nonlinear v–i characteristic of DNA or CNT-based nanodevice.

21.2.2 NANODEVICES BASED ON DNA OR CNTS

Other types of nanodevices use the molecule of life (DNA) or CNTs between a pair of electrodes [7,8]. For these electronic devices, the general DF (21.9) can be applied to the v–i characteristic, see Figure 21.4.

To reduce Figure 21.2 to Figure 21.4,

$$k_0 = 0; \quad k_1 = k_2 = k_0; \quad \delta_0 = \delta_1$$

Replacing these parameters in Equation 21.9,

$$\Psi(a) = k_0 - \frac{2k_0}{\pi}\left[\sin^{-1}\left(\frac{\delta_0}{a}\right) - \frac{\delta_0}{a}\sqrt{1 - \left(\frac{\delta_0}{a}\right)^2}\right]; \quad |a| > \delta_0 \qquad (21.11)$$

Equation 21.11 is the DF for the typical nonlinearity of DNA or CNT-based nanodevices.

21.3 THE LIÉNARD EQUATION

The second approach to study circuits and systems with nonlinear nanodevices is known as *time domain analysis* and involves nonlinear differential equations; here, numerical solutions are required when the differential equation is not standard.

The forced Liénard equation [3] has the structure,

$$\frac{d^2 x}{dt^2} + a_1(x)\frac{dx}{dt} + a_0(x) = u \qquad (21.12)$$

where $a_1(x)$; $a_0(x)$ are nonlinear functions; u is the forcing input.

21.3.1 CONDITIONS FOR A LIMIT CYCLE

The solution of Equation 21.12 presents a limit cycle or self-sustained oscillation if the following five conditions are satisfied,

1. The coefficients $a_1(x)$; $a_0(x)$ are continuously differentiable for all x.
2. Coefficient $a_0(x)$ has odd symmetry,

$$a_0(x) = -a_0(-x) \ \forall x \tag{21.13}$$

3. Coefficient $a_0(x)$ is positive for x positive,

$$a_0(x) > 0 \ \forall x > 0 \tag{21.14}$$

4. Coefficient $a_1(x)$ has even symmetry,

$$a_1(x) = a_1(-x) \ \forall x \tag{21.15}$$

5. The function $F(x)$ has one positive zero at $x = a$,

$$F(x) = \int_0^x a_1(\lambda) \, d\lambda \tag{21.16}$$

$F(x) < 0$, for $0 < x < a$;
$F(x) > 0$ and nondecreasing for $x > a$;
$F(x) \to \infty$ as $x \to \infty$;

The Liénard equation 21.12 appears naturally in series RLC circuits with memristors or nanodevices with nonlinear $v–i$ characteristics.

21.3.2 QUADRATIC MEMRISTOR

Consider the circuit shown in Figure 21.5 with a memristor $\phi = f(q)$.
    The corresponding differential equation,

$$R.i + L\frac{di}{dt} + \frac{1}{C}\int_0^t i(\tau) \, d\tau + \frac{df(q)}{dq}i = u \tag{21.17}$$

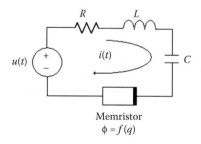

FIGURE 21.5   RLC circuit with a memristor $\phi = f(q)$.

Here, the memristance $M = df(q)/dq$ is quadratic in relation with the charge,

$$M(q) = \frac{\partial f(q)}{\partial q} = k(q^2 - \alpha) \tag{21.18}$$

where (k, α) are constants; replacing Equation 21.18 by Equation 21.17 with $i = dq/dt$ yields,

$$\frac{d^2q}{dt^2} + \left(\frac{R + k(q^2 - \alpha)}{L}\right)\frac{dq}{dt} + \frac{1}{LC}q = u \tag{21.19}$$

This Liénard equation, for the memristive circuit shown in Figure 21.5, has coefficients $a_0(x)$ and $a_1(x)$ that satisfy the limit cycle conditions 1–4. Equation 21.19 can be written as Equation 21.12 with coefficients (21.20).

$$a_0(x) = \frac{1}{LC}x$$
$$a_1(x) = \left(\frac{R + k(x^2 - \alpha)}{L}\right) \tag{21.20}$$

Condition five for a limit cycle (21.16) requires the integration of $a_1(x)$,

$$F(x) = \int_0^x \frac{1}{L}\left[R + k(\lambda^2 - \alpha)\right]d\lambda = \frac{1}{L}\left[(R - k\alpha)x + \frac{k}{3}x^3\right] \tag{21.21}$$

Finding the positive zero,

$$\frac{1}{L}\left[(R - k\alpha)x + \frac{k}{3}x^3\right] = 0$$
$$x = \sqrt{\frac{3(k\alpha - R)}{k}} \tag{21.22}$$

Formula (21.22) is dependent on the memristor parameters (k, α) and the resistive part R of the circuit, that is, the elements that damp oscillations or dissipative elements.

21.3.3 Nonlinear Nanodevice

Figure 21.6 is a series RLC circuit with a general nonlinear nanodevice $v = \varphi(i)$.

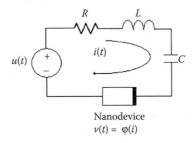

FIGURE 21.6   RLC circuit with a nonlinear nanodevice $v = \varphi(i)$.

The differential equation in this case includes the function $\varphi(i)$,

$$R.i + L\frac{di}{dt} + \frac{1}{C}\int_0^t i(\tau)\,d\tau + \varphi(i) = u \tag{21.23}$$

If the input u is constant, then the derivative with respect to time of Equation 21.23 follows the expression,

$$\frac{d^2i}{dt^2} + \frac{1}{L}\left[R + \Gamma(i)\right]\frac{di}{dt} + \frac{1}{LC}i = 0$$

$$\Gamma(i) = \frac{\partial\varphi(i)}{\partial i} \tag{21.24}$$

In some cases, from the nanodevice experimental v–i characteristic, it could be possible to interpolate a continuous function $v = \varphi(i)$.

21.4 SIMULATIONS

In this section, two series RLC circuits with quadratic memristors are simulated [10] to verify the presence of limit cycles if the five conditions of the Liénard equation are satisfied [3]. The parameters for the simulation are: $R = 4\ \Omega$, $L = 2$ H, and $C = 0.5$ F.

21.4.1 Limit Cycle

In the first case, the constants for $M(q)$ are $(k, \alpha) = (10, 0.5)$; the positive zero is located at $x = 0.55$. Integral (21.16) is the cubic equation (21.25) shown in Figure 21.7; notice that $F(x)$ satisfies condition five for a limit cycle.

$$F(x) = -0.5x + \frac{5}{3}x^3 \tag{21.25}$$

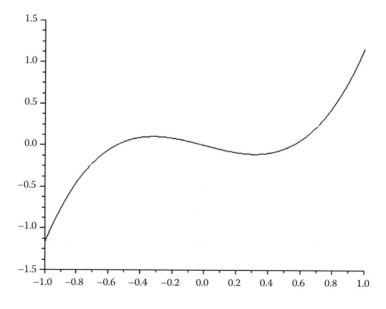

FIGURE 21.7 $F(x)$ for Liénard equation with a limit cycle.

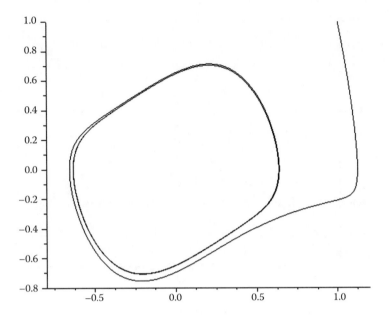

FIGURE 21.8 Limit cycle from an initial condition (1,1).

Writing the unforced Liénard equation for this memristive circuit $(k,\alpha) = (10, 0.5)$ with initial conditions (21.26),

$$\frac{d^2x}{dt^2} + (-0.5 + 5x^2)\frac{dx}{dt} + x = 0$$

$$\frac{dx(0)}{dt} = 1; \quad x(0) = 1$$

(21.26)

Figure 21.8 is the solution of Equation 21.26; there is a limit cycle because $a_1(x)$; $a_0(x)$ satisfy all the required conditions.

21.4.2 STABLE FIXED POINT

In the second case, the constants for $M(q)$ are $(k, \alpha) = (2, 0.5)$; there is no positive zero; instead, $x = \pm$ j 2.12. Integral (21.16) is the cubic equation (21.27) shown in Figure 21.9; notice that $F(x)$ does not satisfy condition five for a limit cycle.

$$F(x) = 1.5x + \frac{1}{3}x^3$$

(21.27)

Writing the unforced Liénard equation for this memristive circuit $(k,\alpha) = (2, 0.5)$ with initial conditions (21.28),

$$\frac{d^2x}{dt^2} + (1.5 + x^2)\frac{dx}{dt} + x = 0$$

$$\frac{dx(0)}{dt} = 1; \quad x(0) = 1$$

(21.28)

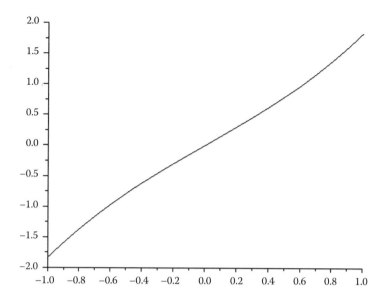

FIGURE 21.9   *F(x)* does not satisfy the conditions for a limit cycle.

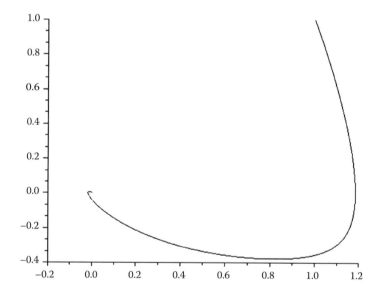

FIGURE 21.10   Stable fixed point for an initial condition (1,1).

Figure 21.10 is the solution of Equation 21.28; there is no limit cycle because $a_1(x)$; $a_0(x)$ do not satisfy the required conditions.

21.5   CONCLUSIONS

This chapter has presented two methods, *frequency-domain analysis and time-domain analysis*, to study circuits and systems with nonlinear nanodevices; the corresponding mathematical concepts are the *DF* and the *Liénard equation*, respectively.

The DF for a general *v–i* characteristic was calculated and the final formula, with particular parameters, was used to derive the DFs for a memristor and for DNA and CNTs-based nanodevices.

In the time domain, the Liénard equation was formulated for RLC series circuits with nanodevices; two types of nonlinearities were considered, $v = \varphi(i)$ and $\phi = f(q)$. Conditions for the existence of a limit cycle were formulated and the theory was illustrated with simulations.

Future work could consider a time-varying DF to improve the modeling of frequency-dependent nanodevices such as the memristor [5]; also, the design of nanodevices to synthesize particular $v-i$ characteristics is an open problem.

REFERENCES

1. Khalil, H.K. *Nonlinear Systems*. Macmillan Publishing, New York, NY, 1992.
2. Delgado, A. The memristor as controller. *IEEE Nanotechnology Materials and Devices Conference*, Monterey, California, Oct. 12–15, 2010.
3. Strogatz, S.H. *Nonlinear Dynamics and Chaos*. Perseus Books Publishing, Cambridge, 1994.
4. Chua, L.O. Memristor—The missing circuit element. *IEEE Transactions on Circuit Theory*, 18(5), 507–519, 1971.
5. Strukov, D.B., Snider, G.S., Stewart, D.R., and Williams, R.S. The missing memristor found. *Nature*, 453, 80–83, 2008.
6. Jeong, H. J., Lee, J. Y., Ryu, M. K., and Choi, S. Y. Bipolar resistive switching in amorphous titanium oxide thin film. *Physica Status Solidi RRL* 4(1–2), 28–30, 2010.
7. Cohen, H., Nogues, C., Naaman, R., and Porath, D. Direct measurement of electrical transport through single DNA molecules of complex sequence. *Proceedings of the National Academy of S ciences, USA*, 102(33), 11589–11593, 2005.
8. Kaun, C.-C., Larade, B., Mehrez, H., Taylor, J., and Guo, H. Current—voltage characteristics of carbon nanotubes with substitutional nitrogen. *Physical Review B*, 65, 205416 (1–5), 2002.
9. Wolfram Alpha LLC. 2012. Wolfram|Alpha. http://www.wolframalpha.com/input/?i = integrate (access May 22, 2012). The use of Wolfram|Alpha does not imply in any way that Wolfram endorses the particular substance or general quality of this work.
10. Consortium Scilab–Digiteo 2011. Scilab: Free and Open Source software for numerical computation (OS, Version 5.XX) [Software]. Available from: http://www.scilab.org.

22 Sensing and Writing Operations of Nano-Crossbar Memory Arrays

An Chen

CONTENTS

22.1 INTRODUCTION

Many nanoarchitectures are based on the crossbar arrays where active devices are built at the junctions between access lines laid out in orthogonal arrangement [1]. NanoFabrics [2], nanoscale programmable logic array (NanoPLA) [3], and nanoscale application-specific integrated circuit (NASIC) [4] are some examples of the two-dimensional (2D) crossbar-based nanoarchitectures. Crossbar arrays can also be built on top of CMOS (complementary metal–oxide–semiconductor) to implement three-dimensional (3D) architectures such as CMOS/nanowire/molecular (CMOL) [5] or field-programmable nanowire interconnect (FPNI) [6]. These hybrid nanoarchitectures can utilize both the computation power of CMOS and the high-density data storage and signal-routing capabilities of crossbar arrays. Crossbar array architectures attract great attention in the development of molecular electronics [7–10]. Some molecule-based devices have different resistance states and can be electrically switching between them. The resistive network of these devices can be built as crossbar arrays to implement logic and memory functions. The simple structure of crossbar arrays may enable the bottom-up self-assembly process to achieve low fabrication cost [11].

With the CMOS-based flash memories quickly approaching the scaling limit, several novel memory concepts have emerged as potential candidates for the next-generation nonvolatile memories [12]. Most of these emerging memories are simple two-terminal devices, which are also compatible with crossbar array architectures. By building memory devices at the crossing points of horizontal and vertical access lines, a footprint as small as $4F^2$ (F: feature size of a technology node) can be achieved. In recent years, there has been a strong demand for high-density and low-cost solid-state memories driven by data-intensive consumer applications. Crossbar memory arrays and their 3D stacks have been considered as promising solutions in this direction. An eight-layer one-time-programmable memory was demonstrated using the crossbar arrays of antifuse/diode memory devices [13]. More recently, multilayer-programmable memory was also demonstrated on oxide-based nonlinear resistive switching memories [14]. Phase-change memory has also been integrated in crossbar arrays with two-terminal select devices [15–17].

A well-known challenge with crossbar arrays is the large amount of parasitic leakage paths (i.e., sneak paths). When accessing a selected device in the array, the sneak paths formed by unselected devices degrade the sensing margin and the effective voltage/current delivered to the selected device. At the same time, the unselected devices are also exposed to the operation voltages which may cause an unexpected disturbance. Select devices with rectifying behaviors can help to reduce parasitic leakage and disturbance. However, two-terminal select devices (e.g., diodes) compatible with both crossbar architectures and the switching requirements are still under research. Low processing temperature is also required on these devices for 3D memory architectures, which constrains the available select device options. Passive crossbar arrays without select devices provide a baseline for analysis and benchmark [18–20]. Therefore, this chapter discusses the reading and writing accessibility of passive crossbar arrays, using both an analytical approach and a statistical simulation method, to discuss the effects of key device and array parameters on crossbar array operations.

The reading operation of a crossbar array needs to maximize the sensing margin defined as the output signal difference between different states of a selected device. The writing operation requires a sufficiently high voltage and current to be delivered to a selected device. At the same time, both operations need to minimize the disturbance on unselected devices, as well as to maximize the speed and power efficiency. The accessibility of a crossbar array is determined by the feasibility and efficiency of the reading and writing operations.

This chapter is organized into the following sections. Section 22.2 provides a detailed analysis on the reading operation of crossbar arrays based on circuit simplification and analytical solutions. It will show that the voltage configurations and sensing method determine the accessibility of crossbar arrays. Section 22.3 introduces a matrix-based solution applicable to general resistance distributions in crossbar arrays. Statistical analysis can be developed based on the large number of random resistance patterns. Section 22.4 discusses the feasibility of improving the accessibility of crossbar arrays using nonlinear memory device characteristics. The writing operation of crossbar arrays will be analyzed in Section 22.5. The last Section 22.6 will summarize this chapter.

22.2   READING OPERATION OF CROSSBAR ARRAYS

Many emerging memory devices exhibit resistive switching characteristics, that is, their resistance can be electrically switched between a high-resistance state (HRS, or off-state) and a low-resistance state (LRS, or on-state). The switching from HRS to LRS is usually called the "set" and that from LRS to HRS the "reset." Logic information is encoded in these different resistance states (e.g., HRS for logic "0" and LRS for logic "1"). The analysis in this chapter starts with simple linear devices with constant HRS and LRS resistances, that is, R_{off} and R_{on}, respectively. The on/off ratio (R_{off}/R_{on}) is an important parameter for an array sensing margin. It will be shown later that when appropriate nonlinearity exists in the current–voltage ($I–V$) characteristics in HRS and LRS, the sensing margin may be improved.

Figure 22.1 illustrates a crossbar array with m wordlines (WLs) and n bitlines (BLs). For simplicity, the WL and BL resistance is ignored in this analysis. Take the Cu interconnect as an example. It is estimated that for the feature size down to 22 nm, the line resistance between neighboring devices is ~0.2 Ω [21]. For memory devices with LRS resistance above 20 kΩ and WL/BL lines with less than 1000 devices, the total line resistance is below 1% of memory device resistance and may be ignored in the first-order analysis. However, for memory devices with lower operating resistance, interconnect lines with higher resistivity, and larger arrays, line resistance may not be negligible in accurate analysis. Quantitative solutions including line resistance are more complex.

Under the assumption of negligible line resistance, the array symmetry allows the selected device to be placed anywhere in the array without affecting the analysis. In Figure 22.1, the selected device (R_j) is assumed to be between WL1 and BL1. Notice that the unselected devices

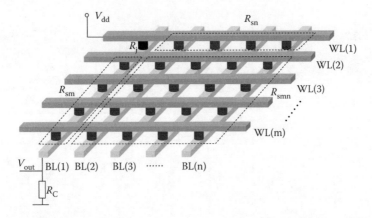

FIGURE 22.1 An illustration of an $m \times n$ crossbar array with a voltage-divider sensing scheme. The selected device is assumed to be at the upper left corner (R_j).

can be divided into three groups: (1) devices sharing WL with R_j (R_{sn}); (2) devices sharing BL with R_j (R_{sm}); (3) devices sharing no lines with R_j (R_{smn}). The unselected devices in group R_{sn} and R_{sm} are also called "half-selected" devices because they are partially exposed to the voltage bias applied on the selected device.

A voltage-divider sensing scheme is used here to read R_j. The selected WL (i.e., WL1) is biased to V_{dd} and the selected BL (i.e., BL1) is grounded through a reference resistor (R_C). The state of R_j (i.e., LRS or HRS) is determined by reading the current flowing from V_{dd} (WL1) to ground (BL1) through R_j. This current is measured from the readout voltage (V_{out}) through R_C. R_j in LRS (HRS) leads to high (low) current and high (low) V_{out}. If there were no sneak paths caused by unselected lines and devices, R_j and R_C form a voltage divider and V_{out} is given by $V_{dd} \cdot R_C/(R_j + R_C)$. Therefore, R_C can be optimized to enlarge the sensing margin ΔV_{out} (i.e., $V_{out}|_{Rj \text{ in LRS}} - V_{out}|_{Rj \text{ in HRS}}$). It is straightforward to show that ΔV_{out} maximizes at $R_C = \sqrt{R_{on} \cdot R_{off}}$.

22.2.1 Voltage Configurations for Reading Operation

In reality, a large number of sneak paths exist owing to the unselected lines and devices. The bias schemes of the unselected WLs/BLs are important for the sensing margin during the reading operation. There are three bias options for the unselected WLs/BLs (V_{dd}, ground, or floating), which give nine possible voltage configurations for reading (Table 22.1). Notice that it is also possible to bias unselected WLs/BLs to a fraction of V_{dd} (partial bias), similar to the partial bias schemes used for the writing of crossbar arrays. However, as will be shown later, partial voltage schemes may not

---

TABLE 22.1

Summary of Possible Voltage Configurations for the Reading Operation and the Crossbar Array Accessibility for Each Configuration

Configuration	1	2	3	4	5	6	7	8	9
Unsel. WLs	F	F	F	V_{dd}	V_{dd}	V_{dd}	0	0	0
Unsel. BLs	F	V_{dd}	0	F	V_{dd}	0	F	V_{dd}	0
Accessibility	No	No	Yes	No	No	No	Yes	Yes	Yes

Source: A. Chen, The accessibility of nano-crossbar arrays of resistive switching devices, *11th IEEE Conference on Nanotechnology (IEEE-NANO)*, Portland, OR, pp. 1767–1771, © (2011) IEEE. With permission.

Note: Labels: "0", grounded; "F", floating.

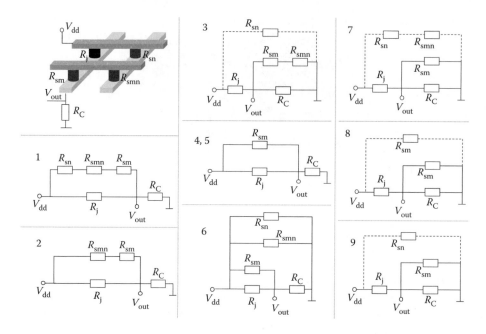

FIGURE 22.2 Equivalent circuits for the different reading voltage configurations.

effectively improve the sensing margin, because they do not eliminate or reduce the parasitic leakage current through these sneak paths.

For simplicity, it is also assumed that there is no parasitic access resistance from the voltage sources to the WLs and BLs. Therefore, these lines can be accurately biased to V_{dd} or ground. With the line resistance being ignored, the same voltage is passed along each WL and BL. Circuit symmetry allows unselected lines biased to the same voltage to be lumped together. As a result, large crossbar arrays can be simplified to simple 2×2 arrays with unselected devices lumped into equivalent resistances R_{sn}, R_{sm}, and R_{smn}. Crossbar arrays with different voltage configurations can be converted to equivalent circuits shown in Figure 22.2 and solved analytically. These equivalent circuits also help to identify whether crossbar arrays can be accessed in each voltage configuration. Essentially, if leakage paths formed in parallel with R_j between V_{dd} and V_{out} (configurations 1, 2, 4–6), the state of R_j cannot be accurately identified because V_{out} is largely determined by the random resistance patterns of these parallel paths instead of R_j. In the other four voltage configurations (3 and 7–9), some unselected devices form parallel paths with the reference resistor R_C between V_{out} and ground, which only affects the actual reference resistance. V_{out} is still mainly determined by R_j, that is, V_{out}'s for $R_j = R_{on}$ and $R_j = R_{off}$ can be distinguished, although it is no longer effective to optimize the sensing margin by choosing R_C. Parallel leakage paths also form between V_{dd} and ground (dashed lines), which does not affect the sensing signals but reduces the power efficiency of the reading operation.

The last row of Table 22.1 summarizes the accessibility of each voltage configuration in reading operations.

22.2.2 Sensing Margin

In accessible voltage configurations (3 and 7–9), V_{dd} is divided between R_j and an effective reference resistor formed by R_C and parallel resistors. The range of V_{out} and the sensing margin (ΔV_{out}) can be derived analytically. Since V_{out} is higher when R_j is in LRS than when it is in HRS, V_{out} is equal to $V_{out}(R_j = R_{on}) - V_{out}(R_j = R_{off})$. For both $R_j = R_{on}$ and $R_j = R_{off}$, V_{out} reaches the maximum (minimum) when unselected devices in parallel with R_C are all in HRS (LRS). Let $V_{on}^{min}(V_{on}^{max})$ represent the minimum (maximum) value of V_{out} for $R_j = R_{on}$, and V_{off}^{min} (V_{off}^{max}) represent the minimum (maximum)

TABLE 22.2

Minimum and Maximum Values of V_{out} for $R_j = R_{on}$ and $R_j = R_{off}$

	V_{on}^{min}/V_{dd}	V_{on}^{max}/V_{dd}	V_{off}^{min}/V_{dd}	V_{off}^{max}/V_{dd}
Configuration 3	$\dfrac{n}{\dfrac{n}{r_C}+(m-1)(n-1)+n}$	$\dfrac{nk}{\dfrac{nk}{r_C}+(m-1)(n-1)+nk}$	$\dfrac{n}{\dfrac{nk}{r_C}+k(m-1)(n-1)+n}$	$\dfrac{n}{\dfrac{nk}{r_C}+(m-1)(n-1)+n}$
Configurations 7–9	$\dfrac{1}{\dfrac{1}{r_C}+m}$	$\dfrac{k}{\dfrac{k}{r_C}+(m-1)+k}$	$\dfrac{1}{\dfrac{k}{r_C}+k(m-1)+1}$	$\dfrac{1}{\dfrac{k}{r_C}+m}$

Source: A. Chen, The accessibility of nano-crossbar arrays of resistive switching devices, *11th IEEE Conference on Nanotechnology (IEEE-NANO)*, Portland, OR, pp. 1767–1771, © (2011) IEEE. With permission.

Note: $r_C = Rd/R_{on}$; $k = R_{off}/R_{on}$; m and n are the WL and BL numbers.

value of V_{out} for $R_j = R_{off}$. These parameters are calculated as shown in Table 22.2. All these boundary values increase with rising R_C and decrease with a larger array size. For large arrays ($n, m \rightarrow \infty$), the different expressions in Table 22.2 for configuration 3 and configurations 7–9 converge to the same results. The *worst-scenario sensing margin* is given by the difference between V_{on}^{min} and V_{off}^{max}, defined as a normalized value of $(V_{on}^{min} - V_{off}^{max})/V_{dd}$. Since V_{on}^{min} is independent of the on/off ratio k and V_{off}^{max} is lower for larger k, a higher on/off ratio helps to increase the sensing margin.

Figure 22.3a plots the worst-scenario sensing margin as a function of R_C (normalized to R_{on}) and array size for a square crossbar array ($m = n$) with the on/off ratio $k = 10$. This level of the on/off ratio is typical for resistive switching memory devices. The sensing margin decreases monotonically with the increase of array size from slightly above 10% (a 16-bit array of 4×4) to around 0.1% (a 64-kbit array of 256×256). This result shows that the sensing margin of crossbar arrays with linear resistive switching devices is very low even for kilobit arrays, which limits the feasible crossbar array size. To illustrate the R_C effects, the dependence of sensing margin on R_C is also plotted for 8×8 and 256×256 arrays in Figure 22.3b and c. There exists an optimal R_C value for the maximum sensing margin. Both the optimal R_C and the maximum sensing margin decrease with increasing array size.

22.2.3 Disturbance and Power Efficiency

In addition to the sensing margin, *reading disturbance* on unselected devices is another key performance measure. It can be seen from Figure 22.2 that the unselected devices along parallel leakage paths (dashed lines) are biased to V_{dd} during reading. Although V_{dd} for reading can be made low enough to not switch these devices, the voltage stress on the unselected devices may still degrade their reading endurance. Take configuration 3 as an example; the reading of R_j exposes all the half-selected devices connected in parallel in group R_{sn} to the reading voltage. If R_j is continuously read to the end of its reading endurance, all these devices in R_{sn} also experience similar reading stress. In configuration 7, V_{dd} is divided between R_{sn} and R_{smn} along the parallel paths. However, R_{smn} is much smaller than R_{sn} in large arrays, so devices in the R_{sn} group still experience the voltage stress close to V_{dd}. This constant exposure of unselected devices to voltage stress during reading operations may significantly accelerate the reading endurance failure of crossbar arrays.

The parallel leakage paths (dashed lines) also waste power, resulting in low *power efficiency* in the reading operation. Take configuration 9 as an example. In the worst case, $R_j = R_{off}$ and unselected devices in group R_{sn} are all R_{on}. It can be shown that the power efficiency of the reading operation is lower than $1/[(n-1)k]$, which is ~0.1% for a 100×100 array with an on/off ratio of 10. The best scenario for power efficiency arises when $R_j = R_{on}$ and unselected devices in R_{sn} are all R_{off}. Calculation

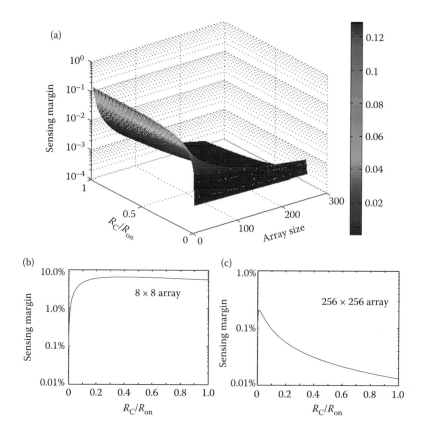

FIGURE 22.3   (a) The worst-scenario sensing margin of a square crossbar array as a function of R_C and array size (number of WLs/BLs) for voltage configurations 7–9; (b) and (c) show the dependence of the sensing margin on R_C for 8×8 and 256×256 arrays. The on/off ratio k is assumed to be 10.

shows that the best power efficiency approaches $k/(n-1)$. With $k = 10$, the efficiency is still below 10% for arrays larger than 100×100. Therefore, most power is wasted during the reading of a crossbar array. The analysis above also shows that the on/off ratio may or may not help power efficiency, depending on the state of R_j and resistance patterns of the crossbar array. Power efficiency always decays with increasing array size because of more leakage paths.

Leaving unselected WLs and BLs floating (configuration 1) may cause less disturbance and waste less power. This is because unselected devices are serially connected in each sneak path and therefore each of them is exposed to lower voltage in average. However, in the worst scenario, the state of R_j cannot be identified because of the sneak paths in parallel with it. Notice that the worst scenario for a sensing margin when unselected lines are floating is different from when they are grounded. Simply applying the same worst scenarios on different voltage configurations will lead to erroneous conclusions.

22.3   A CROSSBAR ARRAY SOLUTION BASED ON MATRIX ALGEBRA

The analysis above is based on simplified crossbar array equivalent circuits and analytical solutions. A general solution for crossbar arrays can be developed based on matrix algebra. The general solution provides detailed voltage/and current distribution in crossbar arrays and reading output voltage for any resistance patterns.

With the line resistance being ignored, the same voltage exists along each WL and BL. The WL voltages ($V_{WL1}, \ldots, V_{WLm}$) and BL voltages ($V_{BL1}, \ldots, V_{BLn}$) define the voltage distribution in crossbar

TABLE 22.3

Variables Used in the Matrix-Based Crossbar Array Solution

Variables	Definitions
$V_{\mathrm{WL}i}$ $(1 \le i \le m)$, $V_{\mathrm{BL}j}$ $(1 \le j \le n)$	WL and BL voltages
R_{ij} $(1 \le i \le m, 1 \le j \le n)$	Resistance at junction (i, j), that is, R_{on} or R_{off}
R_R^S, V_R^S	Access resistance and applied voltage on the selected WL (row)
R_R^U, V_R^U	Access resistance and applied voltage on the unselected WL (row)
R_C^S, V_C^S	Access resistance and applied voltage on the selected BL (column)
R_C^U, V_C^U	Access resistance and applied voltage on the unselected BL (column)

arrays. Based on these $(m + n)$ variables, all junction voltages, current distributions, and output voltage can be easily derived. For an array with a given resistance pattern $\{R_{ij}\}$ $(1 \le i \le m, 1 \le j \le n)$, these unknown variables $\{V_{\mathrm{WL}i}, V_{\mathrm{BL}j}\}$ can be solved based on matrix algebra. Table 22.3 shows the definition of the variables involved in the matrix-based solution. Access resistance is defined as the resistance between voltage sources and WLs/BLs. Notice that the analysis in the previous section ignored access resistance. An extremely high access resistance simulates floating lines. Kirchhoff's law defines $(m + n)$ equations for these unknown voltages and can be written in a matrix format as shown in Figure 22.4. All WL/BL voltages can be solved together from this matrix equation.

The general solution can be applied on any resistance patterns. A large number of random resistance patterns can be generated to calculate the distribution of crossbar array parameters. Figure 22.5 shows an example of the calculated distributions of the readout voltage (V_{out}) for 1000 randomly generated resistance patterns for a 64×64 array. The gap between the V_{out} distributions for R_j in LRS and HRS represents the sensing margin. As shown in this figure, the sensing margin is only ~1%. It should be pointed out that for a $(n \times m)$ array, there are $2^{n \times m}$ possible resistance patterns because each junction has two possible resistance values (R_{on} or R_{off}). So even a small 8×8 array has $>10^{19}$ (2^{64}) possible resistance patterns and it is impossible to simulate even a small portion of these possibilities. It has been shown that many of these resistance patterns are equivalent. Independent patterns of an $(n \times n)$ array are between n^n and n^{2n} [23]. This is significantly less than $2^{n \times n}$ possible patterns, but still a very large number. Therefore, for larger arrays, the simulation based on a small number of samples becomes less representative.

$$
\begin{pmatrix}
1+\sum_{j=1}^{n}\frac{R_R^S}{R(1,j)} & 0 & \cdots & 0 & -\frac{R_R^S}{R(1,1)} & -\frac{R_R^S}{R(1,2)} & \cdots & -\frac{R_R^S}{R(1,n)} \\
0 & 1+\sum_{j=1}^{n}\frac{R_R^U}{R(2,j)} & \cdots & 0 & -\frac{R_R^U}{R(2,1)} & \frac{R_R^U}{R(2,2)} & \cdots & \frac{R_R^U}{R(2,n)} \\
\vdots & \vdots & \ddots & \vdots & \vdots & \vdots & \ddots & \vdots \\
0 & 0 & \cdots & 1+\sum_{j=1}^{n}\frac{R_R^U}{R(m,j)} & -\frac{R_R^U}{R(m,1)} & -\frac{R_R^U}{R(m,2)} & \cdots & \frac{R_R^U}{R(m,n)} \\
\frac{R_C^S}{R(1,1)} & \frac{R_C^S}{R(2,1)} & \cdots & \frac{R_C^S}{R(m,1)} & -1-\sum_{i=1}^{m}\frac{R_C^S}{R(i,1)} & 0 & \cdots & 0 \\
\frac{R_C^U}{R(1,2)} & \frac{R_C^U}{R(2,2)} & \cdots & \frac{R_C^U}{R(m,2)} & 0 & -1-\sum_{i=1}^{m}\frac{R_C^U}{R(i,3)} & \cdots & 0 \\
\vdots & \vdots & \ddots & \vdots & \vdots & \vdots & \ddots & \vdots \\
\frac{R_C^U}{R(1,n)} & \frac{R_C^U}{R(2,n)} & \cdots & \frac{R_C^U}{R(m,n)} & 0 & 0 & \cdots & -1-\sum_{i=1}^{m}\frac{R_C^U}{R(i,n)}
\end{pmatrix}
\begin{pmatrix}
V_{\mathrm{WL}1} \\ V_{\mathrm{WL}2} \\ \vdots \\ V_{\mathrm{WL}m} \\ V_{\mathrm{BL}1} \\ V_{\mathrm{BL}2} \\ \vdots \\ V_{\mathrm{BL}n}
\end{pmatrix}
=
\begin{pmatrix}
V_R^S \\ V_R^U \\ \vdots \\ V_R^U \\ -V_C^S \\ -V_C^U \\ \vdots \\ -V_C^U
\end{pmatrix}
$$

FIGURE 22.4 The matrix-based crossbar array equation with WL/BL voltages as unknown variables.

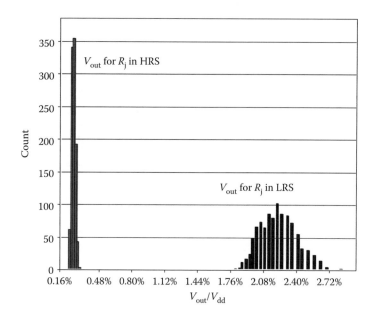

FIGURE 22.5 Distribution of readout voltage (V_{out}) for R_j in HRS and LRS based on 1000 random resistance patterns of a 64 × 64 crossbar array. The gap between these two distributions represents the sensing margin. It is assumed that the on/off ratio is 10, R_C/R_{on} is 0.1, and access resistance is negligible ($10^{-5} \cdot R_{on}$).

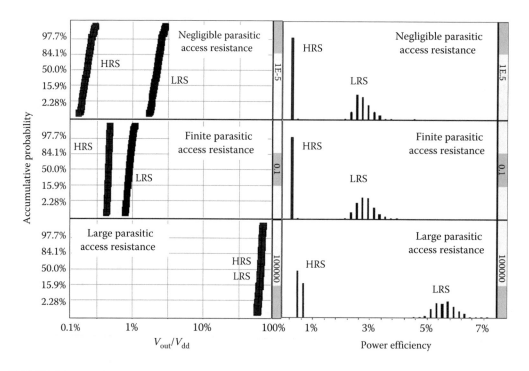

FIGURE 22.6 Left panels: accumulative probability of V_{out} for R_j in LRS and HRS; right panels: histogram of power efficiency of the reading operation. Three access resistance values are simulated: negligible ($10^{-5} \cdot R_{on}$, top panels), finite ($0.1 \cdot R_{on}$, middle panels), and large ($10^5 \cdot R_{on}$, bottom panels).

Figure 22.6 shows the distributions of V_{out} normalized to V_{dd} (left panels) and power efficiency (right panels) calculated from this matrix-based statistical analysis. The top, middle, and bottom rows assume negligible access resistance (= $10^{-5} \cdot R_{on}$), finite access resistance (= $0.1 \cdot R_{on}$), and very high-access resistance ($10^5 \cdot R_{on}$), respectively. The top row is similar to the simplified analysis in Section 22.2 and the bottom row simulates floating unselected WLs and BLs. The results in Figure 22.6 indicate that the sensing margin (the gap between HRS and LRS V_{out}'s) decreases with increasing access resistance. Notice that floating unselected WLs/BLs helps to improve the power efficiency (especially for R_j in LRS); however, the sensing margin diminishes to 0. The power efficiency of reading an LRS device is higher than that of reading an HRS device, because a relatively higher percentage of current passes through LRS devices than HRS devices for the same unselected device resistance pattern. However, overall power efficiency is still very low because of the power wasted in the sneak paths.

22.4 EFFECT OF NONLINEAR MEMORY CHARACTERISTICS

Many RRAM devices demonstrate nonlinear I–V characteristics, which may improve the sensing margin. There are many forms of nonlinear I–V characteristics. Figure 22.7 shows a simple parabolic shape nonlinearity in I–V curves: both LRS and HRS can be described as $I = a \cdot V + b \cdot V^2$, where a and b are constants with different values for LRS and HRS. A space–charge-limited conduction (SCLC) transport mechanism may give this type of I–V characteristics. Tunneling-based transport mechanisms (e.g., trap-assisted tunneling) may provide exponential I–V dependence and stronger nonlinearity. The nonlinearity ratio (η_{LRS} and η_{HRS}) here is defined as the ratio of the resistance at 0 V over that at V_{dd}. This ratio is defined for both LRS and HRS. Notice that different definitions of the nonlinearity ratio may be used in different papers, leading to different conclusions. By definition, the higher the nonlinearity ratio, the larger the changes of resistance over the operation voltage range. Nonlinearity in Figure 22.7 essentially describes the voltage-dependent resistance, that is, higher resistance at lower voltage. For nonlinear devices, the on/off ratio is also voltage dependent and should be defined at a certain voltage. If not specified, the on/off ratio of nonlinear devices discussed in this chapter is defined at V_{dd}.

To explain the nonlinearity effect, Figure 22.8 reconstructs the crossbar array in Figure 22.1 to highlight the composition of sneak paths. $R(i,j)$ at junction (i,j) is the selected device. WL$_i$ is biased to V_{dd} and BL$_j$ is biased to ground through R_C. Unselected WLs/BLs are assumed to be floating. For the current to flow from WLi to BLj, it can go directly through $R(i,j)$ (the selected path) or through parallel leakage paths formed by three unselected devices (sneak paths). Current direction along any sneak path is always a combination of WL-to-BL, BL-to-WL, and WL-to-BL; therefore, one of the three unselected devices is always reverse biased along any sneak path. This explains why rectifying devices (e.g., diode) can improve the accessibility of crossbar arrays because their high reverse resistance reduces the leakage through sneak paths. Notice that the voltage drop between WLi and BLj is divided by the three unselected devices along any sneak path, that is, voltage of these unselected devices is relatively smaller than the voltage on the selected device. Therefore, the

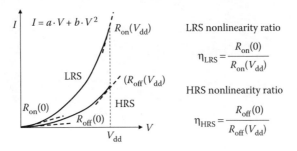

FIGURE 22.7 An illustration of nonlinear I–V characteristics and the definition of nonlinearity ratio.

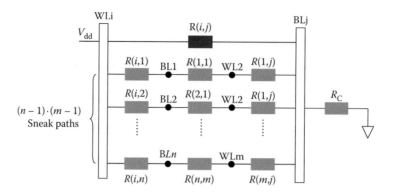

FIGURE 22.8 Reconstruct a crossbar array to highlight the composition of sneak paths.

nonlinearity-induced resistance increase is more prominent on unselected devices than on the selected device. This also helps to reduce the leakage through sneak paths and improve the sensing margin.

To quantify the nonlinearity effect on the sensing margin, let us use configuration 9 in Figure 22.2 (unselected WLs and BLs are all grounded) as an example. The worst-scenario sensing margin $(V_{on}^{min} - V_{off}^{max})/V_{dd}$. The minimum V_{out} for $R_j = R_{on}$ (i.e., V_{on}^{min}) arises when unselected devices in the R_{sm} group are all in LRS, that is, $R_{sm} = R_{on}/(m-1)$. Because the voltage on R_j and R_{sm} is below V_{dd}, both R_j and R_{sm} increase from their values at V_{dd} when $V_{out} < V_{dd}/2$, $V(R_j)$ is larger than $V(R_{sm})$ and therefore R_j increases less than R_{sm}, which helps to increase V_{on}^{min} and the sensing margin. Similarly, nonlinearity in HRS also increases V_{off}^{max}, which, however, reduces the sensing margin. Therefore, the change in the sensing margin depends on the nonlinearity ratios of LRS and HRS.

Figure 22.9 shows the calculated sensing margin for a 64-kbit array (256 × 256) as a function of the nonlinearity ratios of LRS and HRS. As expected, the sensing margin increases with the LRS nonlinearity, but decreases with the HRS nonlinearity. With appropriate nonlinearity ratios in LRS and HRS, sensing margin close to 5% can be achieved in a 256 × 256 array, significantly better than the sensing margin of <0.2% for linear devices.

Figure 22.10 shows the calculated worst-scenario sensing margin as a function of R_C and array size for nonlinear devices with $\eta_{LRS} = 100$ and $\eta_{HRS} = 1$. The overall shape is very similar to

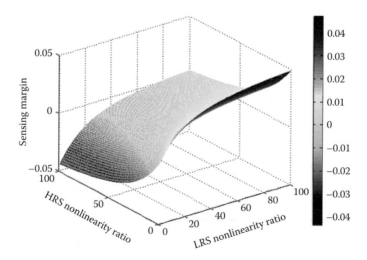

FIGURE 22.9 Worst-scenario sensing margin versus LRS and HRS nonlinearity ratios for a 256 × 256 array with an on/off ratio of 10 defined at V_{dd}.

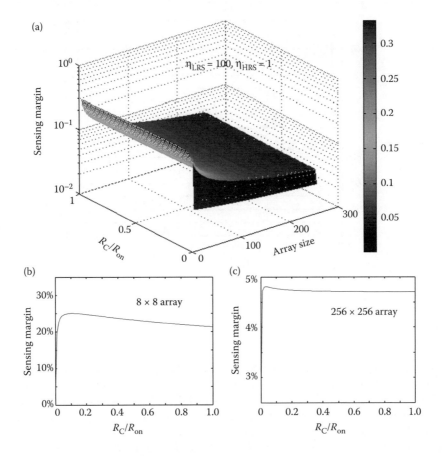

FIGURE 22.10   (a) The worst-scenario sensing margin of a square crossbar array with nonlinear RRAM devices as a function of R_C and array size; (b) and (c) show the dependence of the sensing margin on R_C for 8×8 and 256×256 arrays.

Figure 22.3 for linear devices; however, the sensing margin is significantly higher. Figure 22.10b and c plots the sensing margin dependence on R_C for two example array sizes. At optimal R_C, the maximum sensing margin is ~25% for a small 8×8 array and is slightly below 5% for a 256×256 array. It is expected that stronger nonlinearity (e.g., exponential I–V characteristics) may improve the sensing margin even more and enable larger crossbar array sizes.

The matrix-based method in Section 22.3 can also be used to simulate the statistical distribution of crossbar arrays of a nonlinear device. Figure 22.11 compares the calculated V_{out} distributions for linear and nonlinear devices in a 64×64 array. The sensing margin increases significantly owing to the nonlinearity in the memory device characteristics.

22.5 WRITING OPERATION OF CROSSBAR ARRAYS

The writing operation of crossbar arrays has different requirements from the reading operation. Sufficiently high voltage and current need to be delivered to the selected device and, at the same time, disturbance on unselected devices needs to be minimized. The parasitic leakage paths affect the switching operation by degrading the voltage delivery to the selected device and reducing power efficiency. The WL and BL resistance is important for the writing operation because line resistance degrades the voltage delivery along the lines. Since many leakage paths exist along a line, to drive a certain level of current to the end of a line, significantly higher current may be required at the beginning of a line. This high current requirement may cause reliability concerns.

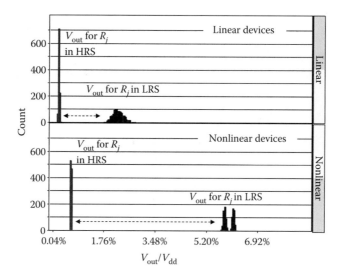

FIGURE 22.11 Compare the readout voltage distribution of 64×64 crossbar arrays with linear and nonlinear devices, showing the improvement of the sensing margin owing to nonlinearity. The nonlinearity ratios are assumed to be $\eta_{LRS} = 100$ and $\eta_{HRS} = 1$.

Resistive switching devices can be unipolar (set and reset happen in the same voltage polarity direction) or bipolar (set and reset in the opposite directions). The switching operation of a crossbar array of bipolar switching devices is more complicated than that of unipolar devices, because the reversal of the switching voltage may have different effects on the selected device depending on its location inside of the array. Select devices can help to improve the writing operation, but it is challenging to develop functional select devices for bipolar switching devices. Two terminal devices that provide both two-direction conduction and rectifying characteristics are difficult to find.

During the writing operation, the selected WL is biased to V_{dd} and the selected BL is grounded. Two partial bias schemes are commonly used for the unselected WLs/BLs: (1) the "1/2 bias scheme" and (2) the "1/3 bias scheme." In the "1/2 bias scheme," all unselected WLs and BLs are biased to $V_{dd}/2$. If all line resistance and access resistance are ignored, unselected devices in the R_{smn} group are zero biased and unselected devices in the R_{sm} and R_{sn} groups are all biased to $V_{dd}/2$. In the "1/3 bias scheme," the unselected WLs and BLs are biased to $V_{dd}/3$ and $2V_{dd}/3$, respectively. As a result, all the unselected devices are biased to either $+V_{dd}/3$ or $-V_{dd}/3$. Figure 22.12 illustrates these two bias schemes and the junction voltages. Therefore, the "1/2 bias scheme" minimizes the number of biased unselected devices because most unselected devices (group R_{smn}) are zero biased. However,

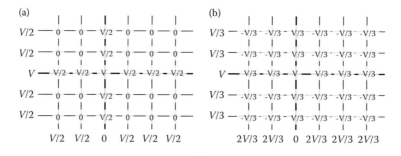

FIGURE 22.12 Partial bias schemes for the writing of crossbar arrays: (a) 1/2 bias scheme; (b) 1/3 bias scheme. The values at the junction points are the voltage drop at the junctions.

the half-selected devices may experience up to $V_{dd}/2$ voltage disturbance. The "1/3 bias scheme" minimizes the maximum voltage disturbance because the maximum voltage on unselected devices is $V_{dd}/3$. However, all the unselected devices are exposed to a certain voltage bias. So the "1/2 bias scheme" is preferred for leakage reduction and the "1/3 bias scheme" is favored for disturbance minimization.

The voltage distribution during the writing operation can also be solved using the same matrix algebra in Section 22.3. Figure 22.13 shows the distribution of the selected device voltage and the maximum voltage on unselected devices, based on the randomly generated resistance patterns in crossbar arrays. The "1/2 bias scheme" is used in this calculation. Two levels of access resistance ($10^{-5} \cdot R_{on}$ and $10^{-2} \cdot R_{on}$) are simulated and two array sizes are considered, 256×256 (on the left) and 64×64 (on the right). With negligible access resistance, V_{dd} is accurately delivered to the selected device and maximum disturbance voltage on unselected devices is $V_{dd}/2$. Both distributions focused narrowly on these values. These results do not change with the increase of array size. With finite access resistance, the voltage delivered to the selected device degrades significantly from the V_{dd} by more than 50% in large arrays. To compensate for this degradation, higher external voltage needs to be applied to ensure the switching of a selected device; however, this also increases the change of disturbance on unselected devices. Although the maximum disturbance voltage is also decreasing, the gap between the selected device voltage and the maximum disturbance voltage on unselected devices decreases in large arrays. Both distributions also show finite slope, indicating increasing dependence of voltage distributions on the resistance patterns of the crossbar array.

The simulation in Figure 22.13 only considered the effects of access resistance and still neglected the line resistance; therefore, the same voltage is passed along all the lines no matter how large the array is. Line resistance has an important impact on the writing operation because the resistance-induced voltage degradation may cause large variation on device voltages at the beginning and the end of the lines. This issue is more prominent for crossbar arrays of resistive switching devices owing to the high writing voltage and leakage in the arrays. In comparison, flash memory arrays have a much lower writing current (owing to the tunneling-based writing mechanism) and lower leakage (owing to the transistors as select devices). Figure 22.14 calculated the voltage decay along an access line up to 512 devices along the line for different line resistance R_L (defined as the resistance

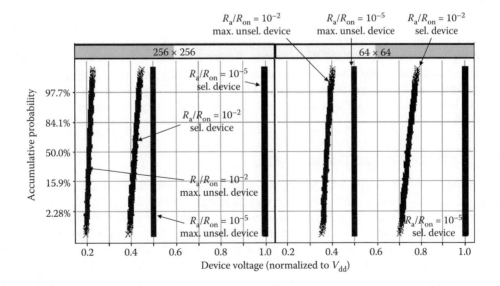

FIGURE 22.13 Calculated distribution of the selected device voltage and the maximum voltage of unselected devices for the "1/2 bias" scheme. Two different access resistance R_a ($= 10^{-5}R_{on}$ or $10^{-2}R_{on}$) and two different array sizes (256×256 and 64×64) are simulated.

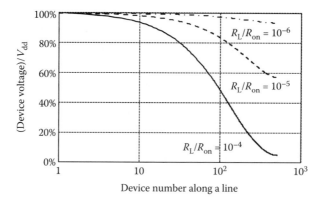

FIGURE 22.14 Calculated voltage decay along a line for different line resistance (R_L) of 10^{-6}, 10^{-5}, and 10^{-4} (normalized to R_{on}).

between two neighboring devices). Assuming a Cu-interconnect and resistivity of 3.5 μΩ cm at the 22 nm technology node [21], an interconnect line aspect ratio of 2 will give R_L~1–2 Ω. For resistive switching memories with R_{on} of 10–100 kΩ, this corresponds to R_L/R_{on} between 10^{-4} and 10^{-5} in the figure. The voltage passing along the line may decay by 20%–50% just after 100 devices. This voltage degradation may constrain realistic crossbar arrays to a relative small size. Select devices can help to reduce array leakage and voltage degradation, which may enable a larger array size. Nonlinear device characteristics have similar effects. However, it is still a great challenge and may be impossible to build crossbar array at the size similar to flash memories, because of the intrinsically higher leakage and writing current. This issue becomes more critical for highly scaled devices, because both array leakage current and line resistance increases when device size scales down.

Another important issue rarely discussed for the writing operation of crossbar arrays is the current delivery. Resistive switching memories usually require 10–100 μA or even a higher switching current [24]. Calculation shows that even for a small 1 k (32 × 32) array with R_L/R_{on} of 10^{-4}, the current delivered to the end of a line may be as small as 2% of the initial value from the power sources. This is due to the large amount of sneak paths that diverge the current while it travels along a line. This inefficient current delivery in crossbar arrays requires much higher current from voltage sources, presenting both design and reliability challenges. Resistive switching devices with low-switching current (e.g., sub-1 μA operation) have been demonstrated; however, the reliable operation of these devices with array statistics is still lacking [25]. Similar to the switching voltage, the current degradation also limits a realistic crossbar array size and imposes stringent requirements on memory device characteristics.

22.6 DISCUSSIONS AND SUMMARY

Crossbar memory arrays have a broad range of applications in nanoelectronics. Baseline analysis in this chapter shows that sneak paths and sensing margin degradation limit practical crossbar memory array size. With the linear resistive switching memory devices, it is expected that the crossbar array size is limited to below several kilobits. Nonlinearity in memory device characteristics can help to reduce leakage through sneak paths and may enable crossbar array size up to several hundred kilobits. However, stable and repeatable resistive switching memory devices with sufficiently strong nonlinearity are still under research. Voltage and current delivery requirements during writing operations impose more constraints on crossbar array size. Line resistance needs to be considered in realistic analysis of crossbar array switching operations because of the resistance-induced voltage and current degradation. Owing to these constraints, it is impractical for crossbar arrays to compete with flash-based memory arrays in terms of array size. However, small crossbar arrays can be combined with transistor access devices where each small crossbar array is addressed by one transistor. The array size needs to be chosen appropriately to optimize both operation performance and array efficiency.

Emerging oxide-based resistive switching devices are still facing severe challenges, among which variability is considered an intrinsic limitation owing to the stochastic switching mechanisms. In a transistor-based memory array, the isolation provided by transistors can help to minimize the effect of each device on the overall array operation. However, in crossbar arrays without such isolation components, the variability of each device is well reflected in the overall array performance. The sensing margin may shrink owing to the spreading of the readout voltage distribution. The writing operation may be affected by the unpredicted voltage/current variation owing to the device instability. Although nonlinear memory devices are considered a promising solution for self-selecting crossbar arrays without select devices, large variability may also exist in nonlinear resistive switching devices and limit crossbar array designs.

Two-terminal select devices are ultimately required to build functional crossbar arrays. Select devices need to not only isolate devices but also allow sufficient voltage and current to be passed to devices. Diodes with rectifying characteristics are considered a promising solution; however, diodes with a high rectifying ratio and drive current are difficult to find. Contact resistance is an important practical limitation on the diode drive current, which becomes an even bigger challenge when devices scale down. In addition, diodes usually work only for unipolar switching devices. Some volatile switching devices have also been considered for select devices in crossbar arrays [26]; however, their feasibility has not been fully proven either.

In summary, crossbar memory arrays provide promising opportunities for nanoelectronics. Their high density and low cost may enable unprecedented applications. To realize these potential opportunities, many challenges in the sensing and writing of crossbar arrays still need to be addressed by both device innovation and creative circuit designs.

REFERENCES

1. N.Z. Haron and S. Hamdioui, Emerging crossbar-based hybrid nanoarchitectures for future computing systems, *International Conference on Signals, Circuits and Systems*, Nabeul, Tunisia, 2008.
2. S.C. Goldstein et al., Nanofabrics: Spatial computing using molecular electronics, *Proceedings of the 28th Annual International Symposium on Computer Architecture (ISCA'01)*, Gothenburg, Sweden, 2001, pp. 178–191.
3. A. DeHon, Array-based architecture for FET-based nanoscale electronics, *IEEE Transactions on Nanotechnology*, **2**, 2003, 23–32.
4. T. Wang et al., Opportunities and challenges in application-tuned circuits and architecture based on nanodevices, *First ACM Conference on Computer Frontier*, Ischia, Italy, 2004, pp. 503–511.
5. K.K. Likharev, et al., Afterlife for silicon: CMOL circuit architectures, *Proceedings of 2005 5th IEEE Conference on Nanotechnology*, 2005, 175–178.
6. G.S. Sinder and R.S. Williams, Nano/CMOS architecture using a field-programmable nanowire interconnect, *Nanotechnology* **18**, 2007, 1–11.
7. Y. Chen, et al, Nanoscale molecular-switch crossbar circuits, *Nanotechnology* **14**, 2003, 462–468.
8. M.M. Ziegler and M.R. Stan, Design and analysis of crossbar circuits for molecular nanoelectronics, *Proceedings of the 2002 2nd IEEE Conference on Nanotechnology*, Arlington, VA, 2002, pp. 323–327.
9. G.S. Rose et al., Design approaches for hybrid CMOS/molecular memory based on experimental device data, *Proceedings of the 16th ACM Great Lakes Symposium on VLSI*, Philadelphia, PA, 2006, pp. 2–7.
10. M.R. Stan, et al, Molecular electronics: From devices and interconnect to circuits and architecture, *Proc. IEEE* **91**, 2003, 1940–1957.
11. W. Liu and C.M. Lieber, Nanoelectronics from the bottom up, *Nature Mater.* **6**, 2007, 841–850.
12. Emerging Research Devices, International Technology Roadmap of Semiconductors (ITRS) report, 2011.
13. M. Crowley, et al, 512Mb PROM with 8 layers of antifuse/diode cells, *IEEE International Solid-State Circuits Conference*, 2003, 16.4.

14. C.J. Chevallier, et al, A 0.13 μm 64 Mb multi-layered conductive metal-oxide memory, *IEEE International Solid-State Circuits Conference*, 2010, 14.3.
15. Y.C. Chen, et al, An access-transistor-free (0T/1R) non-volatile resistance random access memory (RRAM) using a novel threshold switching, self-rectifying, *IEDM Tech. Dig.*, 2003, pp. 905–908.
16. D. Kau, et al, A stackable cross point phase change memory, *IEDM Tech. Dig.*, 2009, pp. 617–620.
17. Y. Sasago, et al, Cross-point phase change memory with $4F^2$ cell size driven by low-contact-resistivity poly-Si diode, *Symposium on VLSI Technology*, 2009, pp. 24–25.
18. A. Flocke and T. G. Noll, Fundamental analysis of resistive nano-crossbars for the use in hybrid nano/CMOS-memory, *33rd European Solid-State Circuits Conference*, Munich, 2007, pp. 328–331.
19. A. Flocke et al., A fundamental analysis of nano-crossbars with non-linear switching materials and its impact on TiO_2 as a resistive layer, *8th IEEE Conference on Nanotechnology*, Arlington, TX, 2008, pp. 319–322.
20. G. Csaba and P. Luigi, Read-out design rules for molecular crossbar architectures, *IEEE Trans. Nanotech.* **9**, 2009, 369–374.
21. E. Linn, R. Rosezin, C. Kügeler, and R. Waser, Complementary resistive switches for passive nanocrossbar memories, *Nat. Mater.* **9**, 2010, 403–406.
22. A. Chen, The accessibility of nano-crossbar arrays of resistive switching devices, *11th IEEE Conference on Nanotechnology (IEEE-NANO)*, Portland, OR, 2011, pp. 1767–1771.
23. P.P. Sotiriadis, Information storage capacity of crossbar switching networks, *Proceedings of the 13th ACM Great Lakes Symposium on VLSI*, Washington, DC, 2003, pp. 45–49.
24. A. Chen, et al, Non-volatile resistive switching for advanced memory applications, *International Electron Device Meeting (IEDM) Technical Digest*, 2005, pp. 765–768.
25. C.H. Cheng, A. Chin, and F. S. Yeh, Ultralow switching energy $Ni/GeO_x/HfON/TaN$ RRAM, *IEEE Electron. Lett.* **32**, 2011, 366–368.
26. M.J. Lee, et al, Two series oxide resistors applicable to high speed and high density nonvolatile memory, *Adv. Mater.* **19**, 2007, 3919–3923.

23 Modeling of Complementary Resistive Switches

E. Linn, S. Menzel, R. Rosezin, U. Böttger, R. Bruchhaus, and R. Waser

CONTENTS

23.1 INTRODUCTION

Resistive switches [1] are two-terminal devices that offer a nonvolatile switching behavior when applying voltage pulses. Although individual devices have a low-power requirement [2], passive crossbar arrays show significant power losses due to parasitic current sneak paths [3]. This so-called "sneak-path problem" does not only lead to increased power consumption, but it also complicates or even prevents proper array read operations. In Figure 23.1, a section of a passive crossbar array is depicted, illustrating the read operation of the lower left element. In case 1, the element is considered to be in the high-resistive state (HRS), while in case 2 the element is considered to be in the

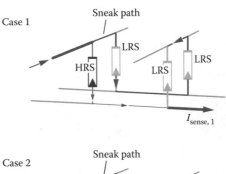

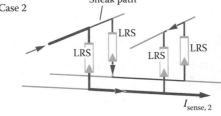

FIGURE 23.1 Sneak-path problem in passive crossbar arrays.

low-resistive state (LRS). Since neighboring cells are low resistive in both cases, the read currents are large in both cases, making a proper reading difficult. To solve this problem, the implementation of a rectifying diode-like selection device was suggested so far [4]. However, this approach is difficult to realize for bipolar resistive switches, since high current densities are required. In an alternative approach, antiserially connected bipolar resistive switches are applied to form a complementary resistive switch (CRS) [5,6]. Complementary resistive switches alleviate size limitations for passive crossbar array memory devices by the elimination of sneak paths because of high-resistive storing states. In Figure 23.2, the basic behavior of a CRS cell is illustrated. If, for example, element B is in the LRS and element A is in the HRS, almost all voltage drops at element A until $V_{th,1}$ is reached. At this point (A), element A switches to the LRS and element B remains in the LRS, because the potential drop at A is far below $V_{th,RESET}$. The CRS state is defined as "ON" with both elements how being low resistive and having an equal voltage drop. If the voltage reaches $V_{th,2}$(B), element B becomes high resistive, because this is equivalent to a voltage drop of $V_{th,RESET}$ at element B. This state is defined as "0." For all applied voltages larger than $V_{th,3}$, element B stays high resistive and element A low resistive. If a potential V comes into the range $V_{th,4} < V < V_{th,3}$(C), the high-resistive element B switches to the LRS and both elements in the CRS are in the LRS (state "ON"). If the negative potential exceeds $V_{th,4}$, element A switches back to HRS (D) and the resulting state is "1." In Table 23.1, the CRS states are depicted. The "OFF" state is only found in uninitialized cells and is not considered further in the following. The states "0" and "1" are the logical storage states and the state "ON" occurs only on a reading of the memory state. The internal memory states "0" and "1" of a CRS cell are indistinguishable at low voltages because state "0" as well as state "1" show a high resistance. Therefore, no parasitic current paths, which can only be induced by low-resistive cells in the crossbar, can arise. To read the stored information ("0" or "1") of a single CRS cell, a read voltage must be applied to the CRS cell. If the CRS cell is in state "1," the cell switches to the "ON," state and if the cell is in state "0," the cell remains in state "0." By this selective switching

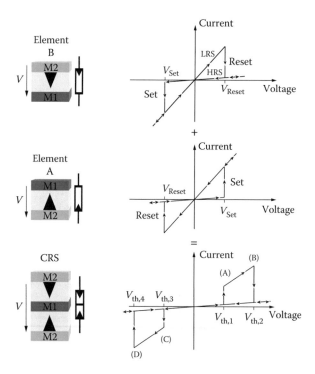

FIGURE 23.2 A complementary resistive switch results from the antiserial connection of two elements A and B. For example, M1 is Cu and M2 is Pt.

TABLE 23.1

CRS States

CRS	Element A	Element B	Overall Resistance
0	HRS	LRS	≈ HRS
1	LRS	HRS	≈ HRS
ON	LRS	LRS	LRS + LRS
OFF	HRS	HRS	>> HRS; only initial state

to the "ON," state, the stored information is destroyed (destructive read). An alternative destructive read approach uses a voltage $V > V_{th,2}$ resulting in a current spike when HRS/LRS applies while no current spike occurs for LRS/HRS. In ref. [7], this mode is used for logic applications. Note that for a destructive readout, it is necessary to write back the previous state of the cell after the reading. This requirement is not present when using a nondestructive readout scheme as proposed in ref. [8]. In general, the writing of state "1" requires a negative voltage ($V < V_{th,4}$), and for writing a "0" a positive voltage $V > V_{th,2}$ is required.

Since the CRS cells consist of two antiserially connected bipolar resistive elements, for example, electrochemical metallization (ECM) elements [1], it is straightforward to use their corresponding compact simulation models for circuit simulation. First, a basic model (using a fixed threshold voltage) is applied. In the next step, dynamical models (the so-called memristive models [9,10]) are applied.

23.2 BASIC SIMULATION MODELS

In Figure 23.3a, a simulation of a bipolar resistive switch is depicted when applying a triangular voltage sweep. Fixed set and reset voltages are assumed. The corresponding simulation of two antiserially connected elements is depicted in Figure 23.3b. From this simulation result, the basic functionality of CRS can be understood in detail. Initially, element A is high resistive (HRS) and element B is low resistive (LRS), which corresponds to the HRS/LRS state. At $V_{th,1}$(A), element A switches, resulting in LRS/LRS. If the voltage exceeds $V_{th,2}$, the cell switches to LRS/HRS (B). For voltages smaller than $V_{th,3}$ the element B switches to LRS, and hence the cell is again in the LRS/LRS state (C). When the voltage exceeds $V_{th,4}$, the cell switches back to HRS/LRS (D). To read the device, a voltage in the range $V_{th,1} < V < V_{th,2}$ is applied, which switches the cell to LRS/LRS if the previous state was HRS/LRS (read a "1"). If the previous state was LRS/HRS, the CRS cell does not switch and stays in the HRS (read a "0"). After a readout, a write-back step is performed.

23.3 MEMRISTIVE MODELS

To simulate a large memory or logic structures along with the CMOS periphery, more realistic dynamical compact circuit models are needed. While these models should be as simple as possible, they have to provide the correct input–output behavior. Since bipolar resistive switches show a pinched hysteresis loop, they can be considered as memristive systems or memristors [9–11]. A memristive system is a special case of a dynamical system and is defined by a state equation (23.1) and a readout equation (23.2):

$$\dot{\mathbf{x}} = f(\mathbf{x}, u), \tag{23.1}$$

$$y = h(\mathbf{x}, u) \cdot u. \tag{23.2}$$

u is the current I and y the voltage V, or vice versa, while $\mathbf{x}$ is the inner state variable which is multidimensional in general. For $u = I$, h is the resistance R, whereas h is the conductance G for $u = V$. $h(\mathbf{x},0) \neq \infty$ must hold to result in a pinched hysteresis loop [10]. Additionally, for a nonvolatile

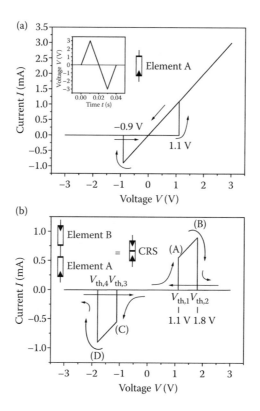

FIGURE 23.3 Spice simulation of a resistive switch (a) and CRS cell (b) with fixed threshold voltages. (From E. Linn et al., Modeling complementary resistive switches by nonlinear memristive system, *Proceedings of the 11th IEEE Conference on Nanotechnology*, Portland, OR, pp. 1474–1478, © (2011) IEEE. With permission.)

memory device, $f(\mathbf{x},u) = 0$ for $u = 0$ must hold, since no change of state should occur without external excitation. For modeling resistive switches as memristive systems, it is crucial to identify the inner state variables. At least one state variable describing a structural change is needed, for example, the length of a filament in electrochemical metallization cells (ECM) [13].

23.3.1 Linear Memristive Model

Several models in the literature are based on the basic memristive approach by the HP group [11]. Here, an adapted model implementation from ref. [14] for a memristive element A (Figure 23.4a) is applied. In this model, $\sigma(\bullet)$ is the step function, w_A is the state variable which corresponds to the filament length in an ECM device, d is the thickness of the active layer, C_1 is a fitting constant, R_{LRS} is the low resistance value, and R_{HRS} is the high resistance value:

$$\dot{w}_A = f(w_A, I_A) = C_1 \cdot I_A \cdot \left(\sigma((d - w_A) \cdot I_A) + \sigma(-w_A \cdot I_A) \right). \tag{23.3}$$

$$V_A = R(w_A) \cdot I_A, \tag{23.4}$$

$$\text{with } R_A(w_A) = (R_{LRS} - R_{HRS}) \cdot w_A/d + R_{HRS}). \tag{23.5}$$

In Equation 23.3, there is a range limitation, allowing only filament lengths in the range of $0 < w_A < d$. After reaching the range limit, the state change is set to zero until the current polarity changes. Within the range $0 < w_A < d$, Equation 23.3 reduces to

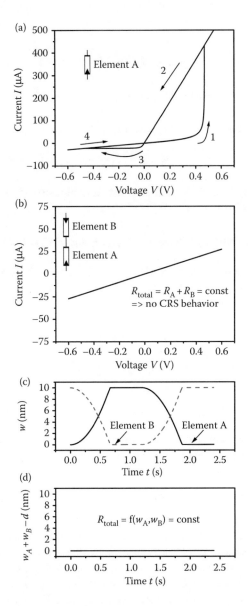

FIGURE 23.4 Spice simulation for linear model. (From E. Linn et al., Modeling complementary resistive switches by nonlinear memristive system, *Proceedings of the 11th IEEE Conference on Nanotechnology,* Portland, OR, pp. 1474–1478, © (2011) IEEE. With permission.)

$$\dot{w}_A = C_1 \cdot I_A. \tag{23.6}$$

For an element B, index A is replaced by B in Equations 23.3 through 23.6. If we consider two memristive elements A and B connected antiserially (which means $I_B = -I_A$) and with initial values $w_A(t = 0) = 0$ and $w_B(t = 0) = d$, which corresponds to HRS/LRS, we obtain an $I-V$ characteristic not showing any CRS behavior (cf. Figures 23.4b and 23.3b). This is a direct consequence of the linear dependency of R on the state variable w and the fact that $\dot{w}$ is an odd function of the input current I. This can be seen when calculating the total resistance of the CRS cell

$$R_{total} = R(w_A) + R(w_B) \tag{23.7}$$

for the given system

$$\dot{w} = C_1 \cdot I. \tag{23.8}$$

$$R(w) = ((R_{LRS} - R_{HRS}) \cdot w/d + R_{HRS}). \tag{23.9}$$

By integration of Equation 23.8 and the use of initial conditions, we can calculate

$$w_A = \int I_A \, dt, \; w_B = d - \int I_A \, dt \; \text{and} \; w_A + w_B = d. \tag{23.10}$$

Since w_A and w_B even up (Figure 23.4c,d), a constant total resistance R_{total} (Equation 23.7) results from Equations 23.9 and 23.10:

$$R_{total} = (R_{LRS} - R_{HRS}) + 2 \cdot R_{HRS} = R_{HRS} + R_{LRS} = \text{const.} \tag{23.11}$$

Since the total resistance is constant, no CRS behavior can result from such a linear memristive model. The equations modeling a single element (Equations 23.8 and 23.9) can also be transformed to a second-order linear system:

$$\dot{w}_1 = C_1 \cdot I, \tag{23.12}$$

$$\dot{w}_2 = C_1 \cdot I, \tag{23.13}$$

$$R(w_1, w_2) = R_{LRS} \cdot w_1/d + R_{HRS} \cdot w_2/d \; \text{with} \; w_1 = d - w_2 = w. \tag{23.14}$$

Thus, such a linear memristive model does not provide correct simulation results for CRS cells. Subsequently, this model is too simple to render correct bipolar resistive switch behavior.

23.3.2 NON-LINEAR MEMRISTIVE MODEL

To overcome the limitation of the linear model, a nonlinear model must be implemented. Here, an electron-transfer mechanism at the interfaces is assumed which is described by the Butler–Volmer equation. This was simulated in ref. [15]. Figure 23.5a shows the applied equivalent circuit. The main circuit consists of a voltage source, the memristive element A, and a series resistor. An auxiliary circuit is used for a state variable calculation with $I_w = \dot{w}_A$ and $V_w = w_A$ implementing the same range limitation $0 < w_A < d$ as for the linear model (Figure 23.5a, right). The initial memristive model (Figure 23.5b) includes two paths, one for the ionic I_{ion} and the other for the electronic current I_{el}. In this model implementation, $I_{el} \gg I_{ion}$ holds for the whole voltage range considered. Since we assume two identical interfaces, both interfaces can be described by one current source (I_{ion} in Figure 23.5b) using the Butler–Volmer equation [15,16]. This results in

$$I_{ion} = C_2 \cdot \sinh\left(\frac{V_1}{2 \cdot V_T}\right). \tag{23.15}$$

In the gap between a filament and an electrode, the ionic resistance and the electronic resistance are given as

$$R_{ion,gap} = \left(1 - \frac{w_A}{d}\right) \cdot R_{ion,0} \tag{23.16}$$

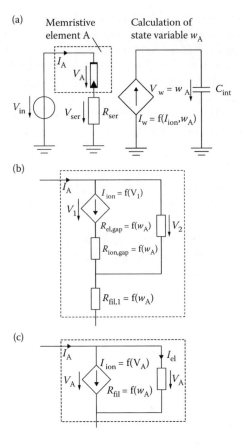

FIGURE 23.5 Memristive model of an ECM element based on an electron-transfer mechanism. (From E. Linn et al., Modeling complementary resistive switches by nonlinear memristive system, *Proceedings of the 11th IEEE Conference on Nanotechnology*, Portland, OR, pp. 1474–1478, © (2011) IEEE. With permission.)

and

$$R_{el,gap} = \left(1 - \frac{w_A}{d}\right) \cdot R_{fil,max},\tag{23.17}$$

respectively.

Both resistances are dependent on the state variable w_A, as is the resistance of the filament

$$R_{fil,1} = \frac{w_A}{d} \cdot R_{fil,0}.\tag{23.18}$$

To satisfy Equation 23.2, the current I_A must be an explicit function of V_A and w_A. This is not the case for the initial model in Figure 23.5b, since there are implicit dependencies:

$$I_A = I_{ion} + I_{el} = C_2 \cdot \sinh\left(\frac{V_1}{2 \cdot V_T}\right) + \frac{V_2}{R_{el,gap}}\tag{23.19}$$

with $V_1 = V_A - I_{ion} \cdot R_{ion,gap} - I_A \cdot R_{fil,1}$ and $V_2 = V_A - I_A \cdot R_{fil,1}$.

By making two simplifications, an explicit formulation as a memristive system is possible. The first one concerns the ionic current I_{ion}, which can be defined only by the Butler–Volmer equation, neglecting the small voltage drop at the ionic resistance of the gap $R_{ion,gap}$. Note that for larger voltages V_{in}, this resistance must be considered. The second simplification is also related to the ionic current I_{ion}, which is small compared to the electronic current I_{el}, and thus the voltage drop of the ionic current at the filament ($R_{fil,1}$) is neglected as well. This simplification is valid for the small-to-moderate voltages which are considered in this work. The corresponding circuit model is depicted in Figure 23.5c. The resistance R_{fil} is the sum of $R_{fil,1}$ and $R_{el,gap}$, which is equivalent to Equation 23.9:

$$R_{fil} = R_{fil,1} + R_{el,gap} = \frac{w_A}{d} \cdot R_{fil,0} + \left(1 - \frac{w_A}{d}\right) \cdot R_{fil,max}. \tag{23.20}$$

$R_{fil,0}$ (=1 kΩ) is the minimum filament resistance value and $R_{fil,max}$ (=1 MΩ) is the maximum filament resistance value. Owing to the applied simplifications, $V_1 = V_A$ and $V_2 = V_A$ result, leading to the final memristive system (Figure 23.5c)

$$\dot{w}_A = f(w_A, V_A) = C_1 \cdot I_{ion} \cdot \left(\sigma((d - w_A) \cdot I_{ion}) + \sigma(-w_A \cdot I_{ion})\right),$$

$$\text{with } I_{ion} = C_2 \cdot \sinh\left(\frac{V_A}{2 \cdot V_T}\right). \tag{23.21}$$

$$I_A = G(w_A, V_A) \cdot V_A = C_2 \cdot \sinh\left(\frac{V_A}{2 \cdot V_T}\right) + \frac{V_A}{R_{fil}}$$

$$= \left(\frac{C_2}{V_A} \cdot \sinh\left(\frac{V_A}{2 \cdot V_T}\right) + \frac{1}{\frac{w_A}{d} \cdot R_{fil,0} + \left(1 - \frac{w_A}{d}\right) \cdot R_{fil,max}}\right) \cdot V_A. \tag{23.22}$$

This system is in accordance with Equations 23.1 and 23.2 and satisfies $f(w_A, V_A) = 0$ for $V_A = 0$, since $\sinh(0) = 0$. Because $\sinh(0)/0 = 1$ also, $G(w_A, 0) \neq \infty$ holds. Note that the change of the state variable (Equation 23.21) is controlled by the ionic current and the readout current (Equation 23.22) is dominated by the electronic current. Figure 23.6 shows the simulation results. In Figure 23.6a, a typical I–V characteristic of element A and a series resistor R_{ser} is shown. If we now consider two memristive elements A and B connected antiserially, we observe a CRS behavior (Figure 23.6b). This I–V characteristic is similar to the characteristic shown in Figure 23.3b. The characteristics of the inner state variables w_A and w_B, which correspond to the filament length in each element, are shown in Figure 23.6c. In contrast to the linear model, the state variables of the elements A and B do not even up for the nonlinear model (Figure 23.6d).

The different dynamic can be understood by having a closer look at the voltage divider which applies for the CRS. First, almost all voltage drops at element A, which is in HRS. Since a state change is a nonlinear function of V_A, the filament growth velocity increases with applied voltage V_{in}. When the filament reaches the counter electrode ($w_A = d$), element A switches to LRS. Now, both elements A and B are in the LRS, taking an equal voltage drop which may additionally be reduced by a series resistor. With a further increase of the applied voltage V_{in}, the voltage V_B becomes large enough to induce a fast decrease of the filament, thus switching element B to HRS. The influence of a series resistor on the I–V characteristics of single elements as well as CRS cells is shown in Figure 23.7. For small R_{ser}, the very asymmetric I–V characteristic of an ECM

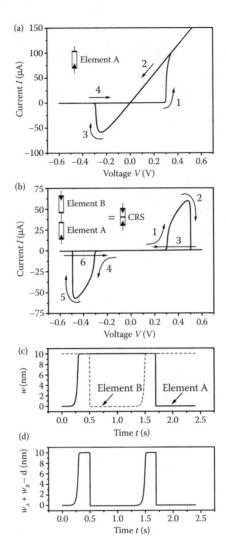

FIGURE 23.6 Spice simulation for nonlinear model. (From E. Linn et al., Modeling complementary resistive switches by nonlinear memristive system, *Proceedings of the 11th IEEE Conference on Nanotechnology*, Portland, OR, pp. 1474–1478, © (2011) IEEE. With permission.)

element (Figure 23.7a) only results in a current spike in the CRS configuration (Figure 23.7b). For larger R_{ser}, a stable LRS/LRS state is obtained for the CRS cells built of asymmetric ECM cells, as stated in ref. [5].

23.4 CONCLUSION

For realistic simulations, physics-based nonlinear memristive models must be applied. By applying a physics-based nonlinear model for ECM elements, it is possible to simulate the correct CRS behavior for antiserially combined elements. However, the results show that it is not feasible to model a CRS cell by two antiserially connected linear current-controlled memristive elements. Therefore, simple linear memristive models, which are often used in the literature, are inapplicable for simulation. Since CRS behavior should result for all bipolar resistive switches connected antiserially, the occurrence of CRS behavior is a good indicator for model consistency.

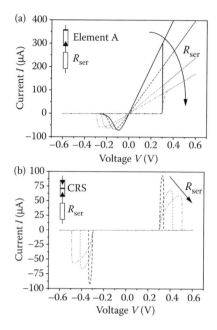

FIGURE 23.7 Variation of series resistance for a resistive switch (a) and CRS (b), simulated with the physical-based model from Figure 23.5. (From E. Linn et al., Modeling complementary resistive switches by nonlinear memristive system, *Proceedings of the 11th IEEE Conference on Nanotechnology*, Portland, OR, pp. 1474–1478, © (2011) IEEE. With permission.)

REFERENCES

1. R. Waser, S. Menzel, and R. Bruchhaus, *Nanoelectronics and Information Technology* (3rd edition), Wiley-VCH, Weinheim, Germany, 2012.
2. V. V. Zhirnov, R. Meade, R. K. Cavin, and G. Sandhu, Scaling limits of resistive memories, *Nanotechnology*, 22, 254027/1–21, 2011.
3. A. Flocke and T. G. Noll, Fundamental analysis of resistive nano-crossbars for the use in hybrid Nano/CMOS-memory, *Proceedings of the 33rd European Solid-State Circuits Conference*, Munich, Germany, 2007, pp. 328–331.
4. J.-J. Huang, Y.-M. Tseng, C.-W. Hsu, and T.-H. Hou, Bipolar nonlinear Ni/TiO$_2$/Ni selector for 1S1R crossbar array applications, *IEEE Electron Device Lett.*, 32, 1427–1429, 2011.
5. E. Linn, R. Rosezin, C. Kügeler, and R. Waser, Complementary resistive switches for passive nanocrossbar memories, *Nat. Mater.*, 9, 403–406, 2010.
6. R. Rosezin, E. Linn, L. Nielen, C. Kügeler, R. Bruchhaus, and R. Waser, Integrated complementary resistive switches for passive high-density nanocrossbar arrays, *IEEE Electron Device Lett.*, 32, 191–193, 2011.
7. R. Rosezin, E. Linn, C. Kügeler, R. Bruchhaus, and R. Waser, Crossbar logic using bipolar and complementary resistive switches, *IEEE Electron Device Lett.*, 32, 710–712, 2011.
8. S. Tappertzhofen, E. Linn, L. Nielen, R. Rosezin, F. Lentz, R. Bruchhaus, I. Valov, U. Böttger, and R. Waser, Capacity based nondestructive readout for complementary resistive switches, *Nanotechnology*, 22, 395203/1–7, 2011.
9. L. O. Chua and S. M. Kang, Memristive devices and systems, *Proc. IEEE*, 64, 209–223, 1976.
10. L. O. Chua, Resistance switching memories are memristors, *Appl. Phys. A-Mater. Sci. Process.*, 102, 765–783, 2011.
11. D. B. Strukov, G. S. Snider, D. R. Stewart, and R. S. Williams, The missing memristor found, *Nature*, 453, 80–83, 2008.
12. E. Linn, S. Menzel, R. Rosezin, U. Böttger, R. Bruchhaus, and R. Waser, Modeling complementary resistive switches by nonlinear memristive system, *Proceedings of the 11th IEEE Conference on Nanotechnology*, Portland, OR, 2011, pp. 1474–1478.

13. I. Valov, R. Waser, J. R. Jameson, and M. N. Kozicki, Electrochemical metallization memories—Fundamentals, applications, prospects, *Nanotechnology*, 22, 254003/1–22, 2011.
14. S. Shin, K. Kim, and S. M. Kang, Compact models for memristors based on charge-flux constitutive relationships, *IEEE Trans. Comput-Aided Des. Integr. Circuits Sys.*, 29, 590–598, 2010.
15. S. Menzel, B. Klopstra, C. Kügeler, U. Böttger, G. Staikov, and R. Waser, A simulation model of resistive switching in electrochemical metallization memory cells, *Mater. Res. Soc. Symp. Proc.*, 1160, 101–106, 2009.
16. S. Menzel, U. Böttger, and R. Waser, Simulation of multilevel switching in electrochemical metallization memory cells, *J. Appl. Phys.*, 111, 014501, 2012.

24 Hybrid Design of a Memory Cell Using a Memristor and Ambipolar Transistors

Pilin Junsangsri and Fabrizio Lombardi

CONTENTS

24.1 INTRODUCTION

With the scaling of complementary metal oxide semiconductor (CMOS) in the nano ranges, the technology roadmap predicted by Moore's law is becoming difficult to meet. The so-called emerging technologies have been widely reported to supersede or complement CMOS. Integration of significantly different technologies such as spintronics [1], carbon nanotube field effect transistor [2], metananomaterial-based optical circuits [3], and, more recently, the memristor [4] have gained attention, thus creating new possibilities for designing innovative circuits and systems. This type of design style is commonly referred to as "hybrid" because it exploits different characteristics of emerging technologies (provided they show compatible features, inclusive of manufacturing and fabrication). A hybrid approach relies on partially utilizing CMOS, while introducing emerging technologies as needed for performance improvement. This is very attractive for memories in which the modular cell-based organization of these systems is well suited to new technologies and innovative paradigms for design.

In this chapter, the memristor is utilized as a storage element; the memristor shows many advantageous features for memory design, such as nonvolatility, linearity, lower power dissipation, and good scalability. The hybrid cell proposed in this chapter also utilizes ambipolar transistors; the ambipolar (p and n) characteristics of this type of transistor allow for the fast and accurate control of the memristance in the proposed memory cell. The hybrid memory cell is analyzed with respect to the two memory operations (READ and WRITE) and the characteristics of the memristor range for its on/off states. It is shown that as the voltage across the memristor is low, a refresh operation could be required for multiple consecutive READ operations. This operational feature is also related to the substantial difference in the READ and WRITE times (nearly two orders of magnitude) and the memristance range. Extensive simulation results using HSPICE are provided to substantiate the performance of the proposed memory cell; metrics (READ time, WRITE time, power dissipation) are assessed under different operating conditions (such as by varying the feature size and supply voltage).

This chapter is organized as follows. Section 24.2 provides a brief review of the memristor while Section 24.3 discusses the operational features of an ambipolar transistor and introduces the macroscopic model used in this manuscript for simulation. Section 24.4 presents the operational features of the proposed memory cell inclusive of the WRITE and READ operations. In Section 24.5, extensive simulation results are presented to assess the performance of the proposed memory cell. In Section 24.6, a comparative discussion is pursued with respect to other memristor-based memory cells. Section 24.7 concludes this chapter.

24.2 MEMRISTOR

In circuit theory, the memristor (or memory resistor) is the fourth fundamental element that utilizes for its operation the relationship between flux and electric charge. This element was postulated by Chua in 1971 [5] based on the concept of symmetry with other circuit elements, such as the resistor, inductor, and capacitor (Figure 24.1). However, it remained of theoretical interest for more than 30 years till HP Labs provided a physical implementation [4] based on a nano-scale thin film of titanium dioxide for fabrication. The relationship between the flux and the electric charge of a memristor is given by [6]

$$d\phi = M*dq \tag{24.1}$$

where M is the memristance or memristor value (in Ω), ϕ is the flux through the magnetic field, and q is the electric charge, that is, the electric charge moving through the memristor is proportional to the flux of the magnetic field that flows through the material. Therefore, the magnetic flux between the terminals is a function of the amount of charge (i.e., q) that flows through the device. Equation 24.1 is equivalent to V = MI, where V and I are the voltage and current across the memristor, respectively [6].

A memristor operates as a *variable resistor* whose value depends on the direction of the current or the voltage across it, that is, if there is a positive voltage across the memristor, its memristance

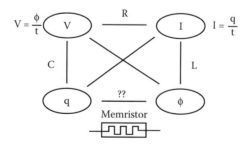

FIGURE 24.1 Relationship between fundamental circuit elements. (P. Junsangsri, F. Lombardi, Design of a hybrid memory cell using memristance and ambipolarity, *IEEE Transactions on Nanotechnology*, 12(1), 71–80. © (2013) IEEE. With permission.)

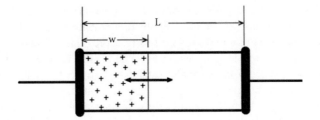

FIGURE 24.2 TiO_2 film sandwiched between two Pt electrodes. (P. Junsangsri, F. Lombardi, Design of a hybrid memory cell using memristance and ambipolarity, *IEEE Transactions on Nanotechnology*, 12(1), 71–80. © (2013) IEEE. With permission.)

reduces to a small value (given by R_{ON}); if there is a negative voltage across the memristor, its memristance increases to a high value (given by R_{OFF}). Hereafter, the memristor is considered as a switching resistance device; as applicable to the HP Labs implementation [4], the rate of change for the memristance is usually linear provided its value is not close to the extreme values. If the memristance value is close to the extreme values, nonlinearity is likely to occur for its rate of change. As physical implementation of a memristor, HP Labs has fabricated a device based on a titanium dioxide film sandwiched between two platinum electrodes (Figure 24.2) [8]. As shown in Figure 24.2, the memristor consists of two parts (or *regions*), the doped region and the undoped region. The widths of the doped region (w) and the undoped region (L−w) change depending on the direction of the current or voltage across it. Let R_{ON} be the resistance for a completely doped memristor and R_{OFF} be the resistance for a completely undoped memristor; then the current–voltage relationship of a memristor is given as follows:

$$v(t) = \left\{ R_{ON} \frac{w(t)}{L} + R_{OFF} \left(1 - \frac{w(t)}{L}\right) \right\} i(t) \tag{24.2}$$

where w(t) is the width of the doped region and L is the TiO_2 thickness [4]. As a function of time, the width of the doped region is given by

$$w(t) = \mu_v \frac{R_{ON}}{L} q(t) \tag{24.3}$$

where μ_v represents the average dopant mobility ($\sim 10^{-10}$ cm²/s/V). By differentiating w(t) in (24.3) with respect to time, the rate of change for the width of the doped region is given by

$$\frac{dw(t)}{dt} = \mu_v \frac{R_{ON}}{L} i(t) \tag{24.4}$$

To model the characteristics of a memristor, different HSPICE models have been proposed in the technical literature [9–11]. These HSPICE models are based on various window functions to simulate behaviorally and macroscopically a memristor according to its physical model. In this chapter, the memristor model of Ref. [9] is used because it has been shown to closely resemble the HP Labs memristor parameters and operation [5].

The memristor has received considerable attention over the last few years. For circuit design, the memristor has been widely advocated as a memory element, mostly for multilevel storage operation [12]. This is accomplished by using a reference (resistive) array, whose resistance values are predetermined and fixed. In this design, a comparator is used to compare the resistance of the memristor with a resistor in the array. As a memristor can attain a very high resistance, the required resistive array may cause a large power dissipation; moreover, the provision of a comparator in the cell may result in a considerable area overhead. In Ref. [13], the fundamental electrical properties of memristors are encapsulated into a set of compact closed-form expressions for characterizing nonvolatile memory operation. Moreover, the design, basic (READ and WRITE) operations, data integrity, and noise tolerance of these memory circuits are also established. In the design of Ref. [13], a memristor is read or written by directly forcing the input voltage source into the memristor itself and comparing it with the reference; so this memory requires three voltage levels (0, V_{DD}, and $V_{DD}/2$), and therefore an additional voltage line is needed. Area and power dissipation are considerably increased. In Ref. [6], a memristor-based content addressable memory (MCAM) is proposed; also, in this case, two levels of supply voltage (0, V_{DD} and $V_{DD}/2$) are required, thus incurring the same disadvantages in terms of area and power dissipation as in Ref. [13].

24.3 AMBIPOLAR TRANSISTOR

As opposed to a traditional (unipolar silicon CMOS) device whose behavior (either p-type or n-type) is determined at fabrication, ambipolar devices can be operated in a switched mode (from p-type ton-type, or vice versa) by changing the gate bias [14,15]. Ambipolar conduction is characterized by the superposition of electron and hole currents; this behavior has been experimentally reported in different emerging technologies, such as carbon nanotubes [16], graphene [17], silicon nanowires [14,18], organic single crystals [19], and organic semiconductor heterostructures [20]. An ambipolar transistor can be used to control the direction of the current based on the voltage at the so-called polarity gate. In this chapter, a four-terminal ambipolar transistor (double gate MOSFET or DG-FET) is utilized. The second gate (referred to as the polarity gate, PG) controls its polarity, that is, when PG is "0," the ambipolar transistor behaves like an N-channel metal oxide semiconductor (NMOS); when PG is "1," it behaves like a P-channel metal oxide semiconductor (PMOS) [21]. The symbol and the modes of operation of the ambipolar transistor used in this chapter are shown in Figure 24.3.

In the technical literature and to the best knowledge of the authors, there is no HSPICE-compatible model to simulate the behavior of an ambipolar transistor; therefore, in this chapter, the model of Figure 24.4 is utilized at a macroscopic level for simulating the characteristics of an

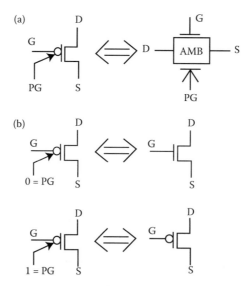

FIGURE 24.3 Ambipolar transistor. (a) Symbol. (b) Characteristic. (P. Junsangsri, F. Lombardi, Design of a hybrid memory cell using memristance and ambipolarity, *IEEE Transactions on Nanotechnology*, 12(1), 71–80. © (2013) IEEE. With permission.)

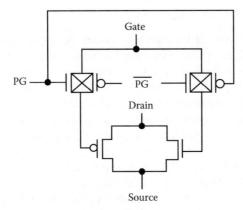

FIGURE 24.4 Model of an ambipolar transistor. (P. Junsangsri, F. Lombardi, Design of a hybrid memory cell using memristance and ambipolarity, *IEEE Transactions on Nanotechnology*, 12(1), 71–80. © (2013) IEEE. With permission.)

ambipolar transistor by using the transmission gates and MOSFETs. From this model, the characteristics of an ambipolar transistor that behaves like an NMOS or PMOS (based on the voltage at PG) are simulated. Figure 24.5 shows the relationship between the ambipolar current (I_D) and the voltage across the drain and source of the ambipolar transistor (V_{DS}) of Figure 24.4 as found by simulation at 32 nm feature size.

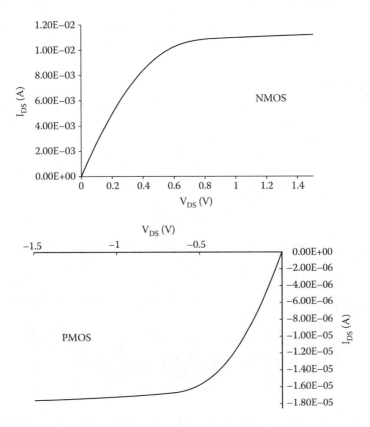

FIGURE 24.5 I_D and V_{DS} of an ambipolar transistor using the model of Figure 24.4 when biased as PMOS and NMOS.

24.4 PROPOSED MEMORY CELL

By exploiting the characteristics of a memristor, a novel design of a memory cell is proposed in this section; as shown in Figure 24.6, a memristor is used as a storage element, while the ambipolar transistors are used as control elements. The new memory cell is obtained by connecting the memristor and two ambipolar transistors in series (Figure 24.6). In the proposed memory cell, data are sent through the bit line (BL) and the inverse bit line (BL′), while the word line (WL) is used for line selection. When the memory cell is selected, the voltage at WL is set to V_{DD} and data are sent through BL and BL′. BL′ and L2 are connected by the transistor NT2; L2 is also connected to the PG of the ambipolar transistors. Therefore, in the proposed design, BL′ is used for polarity selection during the READ/WRITE operation. When BL′ is "0," the ambipolar transistors behave as NMOS, and the current flows from node A to B (Figure 24.6), or from the drain side to the source side of the ambipolar transistors (node A is the source of AMB1 and node B is the drain of AMB2). If BL′ is "1," the ambipolar transistors behave as PMOS. The current flows from node B to node A, or from the source to the drain of the ambipolar transistors. Hence, the memristor is written along both directions, that is, the WRITE operation is bidirectional. To understand the characteristics of the circuit in Figure 24.6, assume that WL is high (logic "1") when selecting the memory cell. The memristance is equal to R_{ON} when the boundary of the TiO_2 film in the memristor moves to the right side; this is accomplished by forward biasing the voltage across the memristor. For R_{OFF} memristance, it is accomplished by reverse biasing the voltage across the memristor to move the boundary of the TiO_2 film in the memristor to the left side. The READ and WRITE operations of the proposed memory cell are as follows.

24.4.1 WRITE OPERATION

In the proposed cell, the memristor is written with the data on BL and BL′. Owing to the symmetric conductance of the n and p types of the ambipolar transistors, the currents that flow in and out of the memristor are equal.

24.4.1.1 WRITE a "1"

To WRITE a "1," the memristance must be biased to R_{ON}, that is, BL and BL′ are set to V_{DD} and ground (GND), respectively. The NMOS transistors (NT1 and NT2) can pass the GND value with no significant voltage drop across them, so the GND voltage from BL′ is passed to line L2. L2 is connected to the polarity gate of the ambipolar transistors; so when the voltage at L2 is GND, both ambipolar transistors behave as NMOS. For BL, owing to the voltage drop across the transistor NT1, the voltage at line L1 (V_{L1}) is given by

$$V_{L1} = V_{DD} - V_{NT1} \tag{24.5}$$

where V_{DD} is the supply voltage and V_{NT1} is the threshold voltage of the transistor NT1.

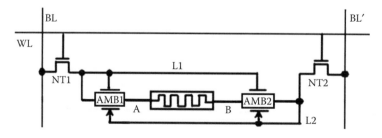

FIGURE 24.6 Proposed memory cell design. (P. Junsangsri, F. Lombardi, Design of a hybrid memory cell using memristance and ambipolarity, *IEEE Transactions on Nanotechnology*, 12(1), 71–80. © (2013) IEEE. With permission.)

Consider the ambipolar transistor AMB1; in Figure 24.6, the gate and drain of AMB1 are connected together by L1, and V_{GS} and V_{DS} of AMB1 are equal; therefore, AMB1 operates in the saturation region ($V_{DS} \geq V_{GS} - V_T$ and $V_{GS} > V_T$). The voltage at node A of the memory cell in Figure 24.6 (V_A) is

$$V_A = V_{L1} - V_{DS1} \tag{24.6}$$

where V_{DS1} is the voltage between the drain and the source of AMB1.

Next, consider the ambipolar transistor AMB2; assume the memristor holds as data a "0" (R_{OFF}) prior to the WRITE "1" operation. The memristance R_{OFF} is relatively high, so the voltage drop across the memristor (V_{mem}) is also high, thus resulting in a high voltage difference between nodes A and B, where V_A is given in Equation 24.6 and V_B is nearly GND. For AMB2, V_{GS2} is also relatively high (nearly V_{L1}), while V_{DS2} is relatively low (nearly zero); so AMB2 operates in the linear region. Owing to the voltage drop across the memristor, the boundary of the memristor changes and its memristance is reduced during the WRITE "1" operation. When the memristance is reduced, V_{mem} is also reduced. The voltage at node B increases when reducing the voltage difference across the memristor. The voltage at node B increases until AMB2 operates in the saturation region ($V_{DS2} \geq V_{GS2} - V_{T2}$), where V_{DS2} is the voltage difference between node B and L2, V_{GS2} is the voltage difference between L1 and L2, and V_{T2} is the threshold voltage of AMB2. So the total current that passes through AMB1, AMB2, and the memristor suddenly increases; then V_B (V_A) increases (decreases) at a higher rate as shown in Figure 24.7. Figure 24.7 shows the plot of the voltage at nodes A and B, and the memristance of the memory cell versus time (ns). As explained previously, the voltage at node A (B) slightly decreases (increases) first; when the memristance reduces till $V_{DS2} \geq V_{GS2} - V_{T2}$, AMB2 operates in the saturation region, and the voltage at node A (B) suddenly decreases (increases).

24.4.1.2 WRITE a "0"

The process for writing a "0" is similar to the one for writing a "1," but with the inverted logic value. For writing a "0," the memristor must be in the R_{OFF} state. BL is at GND (i.e., 0), while BL′ is at V_{DD} (i.e., 1). As BL′ is at V_{DD}, the ambipolar transistors behave as PMOS and are ON (BL is "0"). Then the memristor is in the R_{OFF} state because the voltage at node B is higher than at A (or the current flows from L2 to the drain of AMB2).

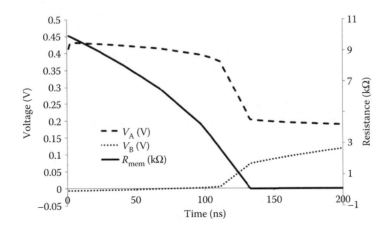

FIGURE 24.7   Plot of voltage, resistance (y-axis), and time (x-axis) for the proposed memory cell.

24.4.2 READ OPERATION

Consider next the READ operation. Recall that the memristor will change its memristance value if there is a current or voltage across it. To prevent this from happening, the READ operation must be fast; in particular, it should be faster than the time required for reaching the threshold time of the memristor (given by 12 ns [6]). In the proposed cell, the READ operation occurs by precharging the bit lines (BL and BL′) to V_{DD} and GND, respectively, and WL is then set. Similar to the WRITE "1" operation, both AMB1 and AMB2 are ON during the first part of the READ operation. The voltages of the bit lines are precharged to their appropriate value; when connected through the ambipolar transistors and memristor, the voltages of the bit lines tend to balance their values, that is, the voltage of BL is transferred to BL′. When the voltage of BL′ is increased up to a value higher than the threshold voltage of the ambipolar transistors, the ambipolar transistors behave as PMOS, that is, L1 and L2 are disconnected. So the voltage difference between BL and BL′ is dependent on the data in the memory cell, that is, the memristance value. When the memristance of the memory cell is R_{ON}, the voltage across the memristor is very low. The voltage difference between BL and BL′ is lower than when the memristance is R_{OFF} because the voltage from BL is better transferred to BL′ and the voltage difference between the bit lines for R_{ON} is less than for R_{OFF}. Consider the voltage difference across the memristor during a READ operation; the ambipolar transistors are OFF, so the voltage difference across the memristor drops to zero. As both ambipolar transistors are OFF, the voltage difference between the bit lines does not affect the memristance of the memory cell.

24.5 SIMULATION RESULTS

HSPICE [22] has been used to simulate the characteristics of the proposed memory cell; the memristor model from Ref. [8] (with a memristance range of 100–19 kΩ) is employed. The macroscopic model of Figure 24.4 is utilized for an ambipolar transistor; the transistor sizes are then adjusted to generate the symmetric conduction between the PMOS and NMOS behaviors. In this chapter, the NMOS and PMOS transistors of the macroscopic model of Figure 24.4 have a feature size of 32 nm [23]. The circuit is then designed by setting $L_{eff} = 12.6$ nm, $V_{th} = 0.16$ V (NMOS) or -0.16V (PMOS), $V_{DD} = 0.9$ V, and $T_{ox} = 1$ nm. The WRITE driver of Figure 24.8 [24] is utilized for the two memory operations (WRITE and READ).

24.5.1 WRITE OPERATION

In this section, the WRITE operation is simulated. Recall that the memristor retains its value when the timing of the voltage difference across the memristor is less than 12 ns [6]. To simulate the

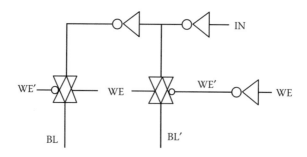

FIGURE 24.8 Driver circuit for WRITE operation. (P. Junsangsri, F. Lombardi, Design of a hybrid memory cell using memristance and ambipolarity, *IEEE Transactions on Nanotechnology*, 12(1), 71–80. © (2013) IEEE. With permission.)

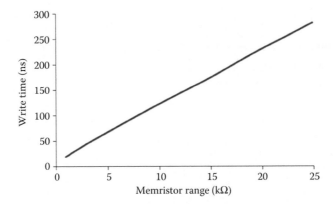

FIGURE 24.9 Plot of WRITE time (ns) versus memristance range (kΩ).

WRITE operation, the WRITE time of a memristor can be found by considering the memristance of the memory cell. The WRITE time is the time to fully bias the memristor to its desired state, so this is a function of the value of the memristance. Figure 24.9 shows the plot of the memristance range (*x*-axis) versus the WRITE time (*y*-axis) of the proposed memory cell when using the model of Ref. [9]. Figure 24.9 shows that by increasing the memristance range, the WRITE time is increased too; the WRITE time is 219 ns for the considered memristor (with a memristance range of 100–19 kΩ).

24.5.2 READ Operation

Using the driver circuit in Figure 24.8, when WE is ON, BL and BL' are precharged to V_{DD} and GND, respectively, prior to the READ operation. So the voltage at node IN in Figure 24.8 is V_{DD}. Next, WE is OFF to isolate the input voltage from node IN and the bit lines (BL and BL'). So WL is ON to start the READ process. Simulation of the READ operation shows that the voltage differences between BL and BL' for the "1" (R_{ON}) and "0" (R_{OFF}) states are not the same; for a 1 ns READ operation, the voltage difference between BL and BL' in the "1" state (R_{ON}) is about 0.1205 V, while it is 0.5817 V for the "0" state (R_{OFF}). Moreover, the proposed memory cell is similar in behavior to an SRAM because the voltage is kept within L1 and L2 when there is no READ/WRITE operation. By opening only WL, the voltages from L1 and L2 are sent to BL1 and BL2, respectively; then the memristor is read without disturbing its current state.

Even though the memristance of the memory cell does not cause a state change, in some cases (such as if selecting the wrong threshold voltage for the ambipolar transistor when it is operating as a PMOS), the memristance may slightly change its value during a read "0" (R_{OFF}) operation. Following the precharging of BL and BL', every time the memory cell is read, it is biased to the R_{ON} state, that is, if it is already in the R_{ON} state, its value must be unchanged. The READ time is established by considering the rate of change in memristance from R_{OFF} to R_{ON}. Hence, multiple WRITE operations with alternating values are simulated. As shown in Figure 24.10, two WRITE operations are used, that is, the memristor changes from R_{OFF} to R_{ON} (write "1"), and from R_{ON} to R_{OFF} (write "0").

As shown in Figure 24.11, the shaded region corresponds to the WRITE time (T_W) when the memristor is changed from R_{OFF} to R_{ON}. By considering the slight memristance rate of change in value every time a READ operation occurs, a state change (i.e., from R_{OFF} to R_{ON}) can occur after multiple and consecutive READ operations. Therefore, a refresh operation is required. Let the memristance value (given by $R_{OFF}/2$) be the threshold level for the READ operation; when this value

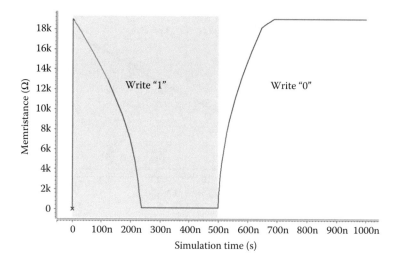

FIGURE 24.10 Memristance of the proposed memory cell (*y*-axis) versus time (*x*-axis) for consecutive WRITE "1" and "0" operations.

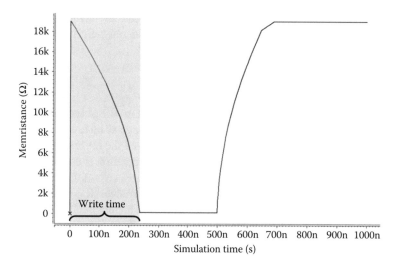

FIGURE 24.11 Memristance of the proposed memory cell (*y*-axis) versus time (*x*-axis) for consecutive WRITE "1" and "0" operations.

is reached, a refresh operation must take place to prevent the memristor from changing its state. So the READ time is given by

$$T_R = \frac{T_W}{2N} \tag{24.7}$$

where T_R is the READ time, T_W is the WRITE time, and N is the number of consecutive READ operations. By considering $R_{OFF}/2$ as the threshold level of the memristance, half of the WRITE time ($T_W/2$) corresponds to the time to reach the threshold level, that is, when the memristance of the proposed memory cell reaches this threshold level, a refresh operation is required.

24.5.3 Refresh

As mentioned previously in the R_{OFF} state, the READ operation slightly changes the memristance value up to a possible change of state, that is, from R_{OFF} to R_{ON}. To avoid this state change from occurring, a refresh operation is required. For detecting the voltage difference between BL and BL′ (V_{DIFF}), a comparison circuit is needed; if the memristor is in the R_{OFF} state and V_{DIFF} is lower than the voltage difference between the bit lines (when the memristance is equal to half of R_{OFF}), the refresh operation must start by rewriting "0" to the memory cell. In the proposed memory cell, the READ operation is affected only for "0" (i.e., the R_{OFF} state), because the voltage difference between the bit lines biases the memristor to R_{ON}.

24.5.4 Transistor Sizing

Consider next the size of the transistors; there are some trade-offs between the transistor size (NT1, NT2, and the ambipolar transistors in Figure 24.6), the WRITE time, and the number of READ operations for assessing the impact of the change in memristance. For a large transistor size, the voltage from BL or BL′ will be transferred faster to the memristor, thus resulting in a low WRITE time. However, the number of READ operations is also small, because there is a high voltage across the memristor, that is, a high rate of change is applicable to the memristance.

Figure 24.12 shows the plots of the WRITE and READ times (ns) versus the transistor size (nm). This graph shows that if the transistor size increases, the WRITE time decreases because the NMOS transistors allow a higher bias from the bit lines (BL and BL′) to the memristor. As for the READ time, it is more than 200 times less than the WRITE time, but it shows the same dependency versus the transistor size. Moreover, for the READ time, by increasing the transistor size, the voltage across the memristor will increase, and the rate of change for the memristance will also be high. Therefore, by considering that the memory cell must be read nearly 100 times consecutively before changing its state, the utilization of small-sized transistors will result in a slower WRITE time. By having a fast READ time, the data in the memory cell can be read more times prior to the point at which the memristor changes its state.

24.5.5 Comparison

In this section, the proposed memory cell is analyzed with respect to its CMOS feature size; designs using 32, 45, and 65 nm [23] have been simulated. Two values for the supply voltage (i.e., 0.9 and

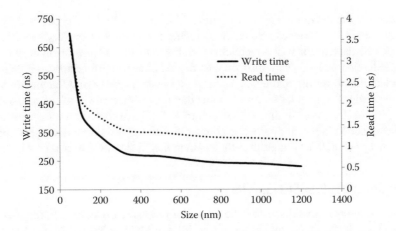

FIGURE 24.12 Plot of WRITE and READ times versus transistor size (NMOS).

TABLE 24.1
WRITE/READ Times at Different Supply Voltages and Feature Sizes

	32 nm		45 nm		65 nm	
V_{DD} (V)	0.9 V	1 V	0.9 V	1 V	0.9 V	1 V
WRITE time (ns)	219	210	260	245	290	275
READ time (ns)	1.095	1.05	1.30	1.225	1.45	1.375

1.0 V) have been used. The results are given in Table 24.1. Table 24.1 shows that when the feature size is reduced, the WRITE and READ times decrease significantly too; moreover, the supply voltage significantly affects the WRITE and READ times. In Table 24.1, the READ time corresponds to the maximum value found over 100 consecutive READ operations (so without incurring in a change of state). By increasing the supply voltage, the WRITE time decreases; at a lower supply voltage, the voltage across the memristor is also lower and the change of memristance state occurs at a slower rate. So the WRITE time is also slower. The READ time reported in Table 24.1 corresponds to the maximum value for a READ operation as performed at least 100 times prior to reaching the threshold and changing the state of the memristance; hence, if the READ time is faster than the value given in Table 24.1, then the number of consecutive READ operations for state change (i.e., from R_{OFF} to R_{ON}) is also larger.

24.5.6 POWER

Next, consider the power dissipation of the proposed memory cell circuit; the memory cell consists of a small number of transistors (i.e., two NMOS transistors and two ambipolar transistors, so 4T). Based on the number of transistors, the capacitance of the proposed memory cell is less than a traditional RAM cell (requiring 6T). Also, the proposed circuit does not require standby power to retain the memristance value, and there is no direct path connecting V_{DD} to GND. However, leakage power due to switching is still present, thus causing dynamic dissipation. This type of power dissipation is dependent on the clock frequency and is given by the well-known equation

$$P_{Dyn} = \frac{1}{2} CV^2 f_{clk} \tag{24.8}$$

where P_{Dyn} denotes the dynamic power dissipation, V is the supply voltage (denoted by V_{DD}), C is the capacitance, and f_{clk} is the clock frequency. To reduce power dissipation, V_{DD} as well as the clock frequency must be decreased. While technology scaling (such as 32 nm) operates at a low supply voltage, the reduction in frequency also implies degradation in performance. HSPICE has been used to simulate the power dissipation of the proposed memory cell using the macroscopic model presented previously. The simulation results show that the average power of the proposed memory cell is only 5.9% of the power in the MCAM cell of Ref. [6] also based on a memristor; the main reason for the reduction in power dissipation compared with Ref. [6] is that the supply voltage of the proposed memory cell (0.9 V) is significantly less than that of the MCAM cell (3 V).

24.6 COMPARATIVE DISCUSSION

In this section, a comparative discussion between the proposed memory cell and the MCAM of Ref. [6] is pursued. It should be noted that in Ref. [6], the MCAM has a different functionality as the search data operation is available to compare the stored data during the READ operation, that is, if the stored data is matched with the search data, the match line will be discharged, or else the original state will be preserved. As the proposed memory cell is binary, then for a READ operation, the stored value shows as a voltage difference between the two bit lines.

Ref. [6] requires two voltage sources, V_{DD} and $V_{DD}/2$ (and a V_{DD} of 3 V). This is significantly higher than the supply voltage of the proposed cell (based on CMOS scaling). Simulation also shows that at the same supply voltage (i.e., 0.9 V), the proposed memory cell requires only 33.037% of the power of Ref. [6]. This is a direct result of the nonstatic power dissipation source in the proposed memory cell and the lower number of components in the circuit, that is, four transistors and one memristor, while Ref. [6] uses seven transistors and two memristors for a 7T NOR type, and five transistors and two memristors for the 5T NOR type.

As for the memory operations, simulation results show that the WRITE time of the proposed memory cell is slower than that of Ref. [6] because the voltage across the memristor of Ref. [6] (1.5 V) is higher than that of the proposed memory cell (0.9 V), that is, the rate of change of the memristor in the MCAM is faster than in the proposed memory cell. However, at the same supply voltage (0.9 V), the simulation results show that for WRITE, the proposed memory cell is slightly slower than the MCAM [6]. This occurs because in the proposed memory cell, the voltage drop across the transistors during a WRITE operation is high; so the voltage across the memristor in the proposed memory cell is less than half of the supply voltage. In the MCAM, the supply voltage ($V_{DD}/2$) is directly forced to the memristor for a WRITE operation; so the WRITE time of the MCAM [6] is not affected by the voltage drop across the transistor. By reducing the voltage drop across the transistors, the WRITE time of the proposed memory cell can be improved with respect to the MCAM of Ref. [6]. As for the READ time, this operation is significantly faster in the proposed cell (1 ns) than in Ref. [6] (12 ns) because in Ref. [6] the search time is dependent on the discharging process of the match line.

24.7 CONCLUSION

This chapter presented a novel memory cell whose circuit consists of a memristor, two NMOS, and two ambipolar transistors, and hence its hybrid nature. In this cell, the memristor is utilized as a storage element due to the excellent features such as nonvolatility, linearity, low power consumption, and good scalability. The proposed hybrid cell also utilizes ambipolar transistors for the control of the memristance in the operation of the memory cell. The hybrid memory cell has been analyzed with respect to the two memory operations (READ and WRITE) and the characteristics of the memristor range for its on/off states.

Macroscopic models are utilized to characterize the nonvolatile feature of the memory cell. In the proposed memory cell, during the READ operation, the voltage across the memristor is low; so a refresh operation may be required when multiple consecutive READ operations occur. This operational feature is also related to the substantial difference in the READ and WRITE times (nearly two orders of magnitude) and the memristance range. Extensive simulation results using HSPICE have been provided to substantiate the performance of the proposed memory cell; different metrics (READ time, WRITE time, power dissipation) have also been assessed under different operating conditions (such as by varying the feature size and supply voltage).

REFERENCES

1. G.I. Bourainoff, P.A. Gargini, D.E. Nikonov, Research directions beyond CMOS computing, *Solid-State Electronics*, 51(11–12), 1426–1431, 2007.
2. D. Akinwande, S. Yasuda, B. Paul, S. Fujita, G. Close, H.S.P. Wong, Monolithic integration of CMOS VLSI and CNT for hybrid nanotechnology applications, *Proceedings of the 38th European Solid-State Device Research Conference, ESSDERC'08*, pp. 91–94, 2008.
3. N. Engheta, Circuit with light at nanoscales: Optical nanocircuits inspired by metamaterials, *Science*, 317(5845), 1698–1702, 2007.

4. D.B. Strukov, G.S. Snider, D.R. Stewart, R.S. Williams, The missing memristor found, *Nature*, 453, 80–83, 2008.

5. L.O. Chua, Memristor—The missing circuit element, *IEEE Transactions on Circuit Theory*, CT-18(5), 507–519, 1971.

6. K. Eshraghian, K.R. Cho, O. Kavehei, S.K. Kang, D. Abbott, S.M. Steve Kang, Memristor MOS content addressable memory (MCAM): Hybrid architecture for future high performance search engines, *IEEE Transactions on VLSI Systems*, 19(8), 1407–1417, 2011.

7. P. Junsangsri, F. Lombardi, Design of a hybrid memory cell using memristance and ambipolarity, *IEEE Transactions on Nanotechnology*, 12(1), 71–80, 2013.

8. S. Williams, How we found the missing memristor, *IEEE Spectrum*, 45(12), 28–35, 2008.

9. D. Batas, H. Fiedler, A memristor SPICE implementation and a new approach for magnetic flux controlled memristor modeling, *IEEE Transactions on Nanotechnology*, 99, 1–1, 2009.

10. Z. Biolek, D. Biolek, V. Biolova, SPICE model of memristor with nonlinear dopant drift, *Radioengineering*, 18(2), pt. 2, 210–214, 2009.

11. A. Rak, G. Cserey, Macromodeling of the memristor in sPICE, *IEEE Transactions on Computer-Aided Design of Integrated Circuits and Systems*, 29(4), 632–636, 2010.

12. H. Kim, M. Pd. Sah, C. Yang, L.O. Chua, Memristor-based multilevel memory, *12th International Workshop on Cellular Nanoscale Networks and Their Applications (CNNA)*, 2010.

13. Y. Ho, G.M. Huang, P. Li, Nonvolatile memristor memory: Device characteristics and design implications, *ICCAD'09*, November 2–5, 2009, San Jose, California, USA.

14. S. Koo et al., Enhanced channel modulation in dual-gated silicon nanowire transistors, *Nano Letters*, 5(12), 2519–2523, 2005.

15. Y.-M. Lin et al., High-performance carbon nanotube field-effect transistor with tunable polarities, *IEEE Transactions on Nanotechnology*, 4, 481–489, 2005.

16. S. Heinze et al., Unexpected scaling of the performance of carbon nanotube Schottky-barrier transistors, *Physical Review B*, 68, 235418, 2003.

17. K.S. Novoselov et al., Electric field effect in atomically thin carbon films, *Science*, 306(5696), 666–669, 2004.

18. A. Colli et al., Top-gated silicon nanowire transistors in a single fabrication step, *ACS Nano*, 3(6), 1587–1593, 2009.

19. A. Dodabalapur et al., Organic heterostructure field-effect transistors, *Science*, 269(5230), 1560–1562, 1995.

20. J.H. Schön et al., Ambipolar pentacene field-effect transistors and inverters, *Science*, 287(5455), 1022–1023, 2000.

21. M.H. Ben Jamaa et al., Novel library of logic gates with ambipolar CNTFETs: Opportunities for multilevel logic synthesis, *DATE2009*, 2009, pp. 622–627.

22. Star-Hspice User Guide, Avant! Corporation, Release 2002. 2 June 2002.

23. http://ptm.asu.edu/.

24. CMOSSRAM Circuit Design and Parametric Test in Nano-Scaled Technologies Frontiers in Electronic Testing, 2008, Volume 40, 13–38.

25 Spike Timing-Dependent Plasticity Using Memristors and Nano-Crystalline Silicon TFT Memories

Kurtis D. Cantley, Anand Subramaniam, and Eric M. Vogel

CONTENTS

25.1 INTRODUCTION

Interest in the possibility of using memristive devices as synapses in artificial neural circuits was sparked by the demonstration of TiO_2 resistive switches by HP Labs in 2008 [1]. A great deal of the resulting research has been centered on implementing spike timing-dependent plasticity (STDP), which is a synaptic learning mechanism based on timing differences between action potentials [2,3]. However, time scales of biological inter-spike intervals (ISIs) are on the order of tens of milliseconds, much longer than the typical electronic phenomena. This makes STDP a difficult scheme to implement efficiently using electronics. Proposed solutions have involved pulse width or height modulation [4–6] or pulse shaping [7] and would require somewhat extensive circuitry for each neuron. Additionally, the reports do not explain the learning characteristics beyond pair-based trials. Experiments on biological synapses indicate a much more complex reality, in that the exact mechanisms of synaptic learning cannot be explained by pair-based STDP alone [8,9]. Specifically, asymmetric temporal integration of the synaptic weight changes has been demonstrated in spike triplet and quadruplet, as well as frequency-dependent experiments [10–12]. Progress continues on developing models that explain the observed effects more thoroughly [13,14].

Although the exact mechanisms of biological synaptic modification are not known, the consequences of STDP in large networks lead to many potential applications. For example, it is well known that STDP learning rules assist in the recognition of spatio-temporal spike patterns within a

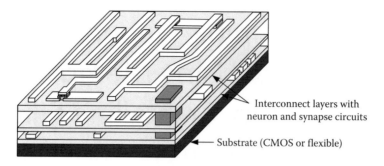

Interconnect layers with
neuron and synapse circuits

Substrate (CMOS or flexible)

FIGURE 25.1 An illustration of arbitrarily connected neural circuits. The low-temperature processing would enable three-dimensional fabrication on top of a CMOS core, or over a large area on a flexible substrate.

large population [15–17]. Such behavior could be useful for recognizing recurring patterns in real-time data from large sensor streams. In addition, STDP has the ability to detect phase differences in spike patterns, which is useful for sound source localization. Finally, STDP can be used for complex visual pattern extraction. Section 25.5 discusses these examples in detail.

Other sections of this chapter describe circuits which realize a compact physical implementation of the STDP learning mechanism with biologically realistic action potentials. Simulation results of spike triplet and frequency dependence are also shown and compared to measured biological data. Complex synaptic behavior is a natural outcome of using ambipolar nano-crystalline silicon (nc-Si) thin-film transistor (TFT) memory devices in conjunction with memristive devices. As described in Section 25.2, the former can be fabricated using gold nanoparticles embedded in the gate dielectric, and the latter could be based on any number of material systems. Metal oxides such as TiO_2 or HfO_2 that can be deposited by atomic layer deposition (ALD) or sputtering would be preferable, however. The reason is that together with the low-temperature, large-area nc-Si deposition, these processes could enable the fabrication of a three-dimensional system with physical structure similar to the human neocortex, as illustrated in Figure 25.1. Distribution directly on top of a silicon CMOS core could be one possibility, or the rigid crystalline silicon could be completely eliminated in favor of flexible substrates.

25.2 DEVICE MODELS

Several reports of micron-scale ambipolar nc-Si TFTs exist [18,19], but the simulations in this chapter make use of the SPICE model provided in Ref. [20]. This is different from previous works [21,22] in that the model is more thoroughly vetted for submicron channel geometries. The spiking neuron circuit schematic used for the simulations is shown in Figure 25.2a, and is a heavily modified version of that originally proposed by Mead [23]. Additions of a transistor at the input and different charge leakage paths during resting or spiking (controlled by V_{leak} and V_{reset}, respectively) help to stabilize the action potential width. A more detailed description of neuron circuit operation is contained in Ref. [24]. In that work, each synapse consisted of one TFT and one memristive device, whereas here the synapses are comprised of only a single ideal memristor. The state variable w/D of the memristive device (definition provided in Ref. [1]) represents the synaptic weight. A path for potentiation current through the memristor is provided by a nanoparticle memory TFT (NP-TFT). One of these devices is placed at the output of each neuron circuit such that it would drive a large number of synapses.

Memory characteristics of the NP-TFT are modeled by adjusting the actual gate voltage of the intrinsic ambipolar nc-Si TFT using a dependent voltage source ($V_{G,adj}$) to account for charge trapped in the nanoparticles. The total amount of trapped charge Q_{np} is calculated by integrating the oxide current I_{ox} across a large capacitor. The magnitude of the oxide current is calculated based on the Poole–Frenkel equation [25] and has been closely matched to experimental data. Charge leaks

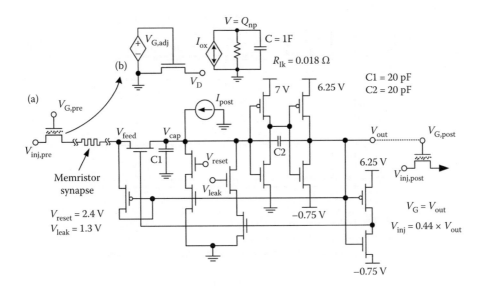

FIGURE 25.2 (a) Schematic diagram of the circuit used to investigate STDP learning. The NP-TFT for the postsynaptic neuron is actually left unconnected for these simulations. (b) The SPICE model subcircuit for the NP-TFT consists of two parts: the intrinsic transistor with adjusted gate voltage $V_{G,adj}$ to account for the nanoparticle charge Q_{np}, and an auxiliary circuit used for the trapped charge calculation.

off the capacitor through a small resistor, the value of which determines the time constant of retention. A schematic of the subcircuit is shown in Figure 25.2b. Testing the NP-TFT is performed via simulation as follows. The gate voltage is first quickly ramped up to the programming voltage of 4.5 V and back down, and the drain current is measured during this sweep (square symbol in Figure 25.3). Five milliseconds later, a 1 ms programming pulse (4.5 V) is applied (see Figure 25.3c), causing electrons to flow through the oxide and be trapped in the nanoparticles near the semiconductor surface. Another probing ramp 5 ms after that measures the total threshold voltage shift (circle in Figure 25.3), followed by a final probing ramp 40 ms after the programming pulse. Most of the charge has leaked off the nanoparticles by this time (triangle in Figure 25.3).

Transfer characteristics of the TFT device during each of the three ramps are shown in Figure 25.4, with symbols that correspond to those in Figure 25.3. The maximum threshold voltage shift for this device is slightly above 1 V, and charge constantly leaks off (via R_{lk} in Figure 25.2b) such that the retention time is in the 50–100 ms range (time constant is approximately 10 ms). The programming pulse used is similar in height and duration to the action potential output of the neuron circuit in Figure 25.2a. The SPICE model used here is based on the characteristics of recently fabricated NP-TFT devices. Depending on the gate dielectric thickness, the chosen material (SiO$_2$ or HfO$_2$), and the location of the nanoparticles within the dielectric, as well as the programming voltage and retention time, can be finely tuned. The intrinsic TFT used for the fabrication is the same as those of Ref. [20], though the model has a slight threshold voltage shift to generate more symmetric learning as described in the following sections. Other similar devices have also been demonstrated using organic materials [26] and amorphous silicon [27].

25.3 SYNAPTIC LEARNING MECHANISM

25.3.1 ACTION POTENTIAL PAIRS

Changes in synaptic weight as a function of spike timing differences were measured using 100 ms transient simulations of the circuit shown in Figure 25.2a. The results of two individual simulations

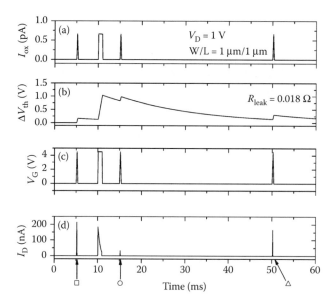

FIGURE 25.3 Traces of (a) current through the gate dielectric I_{ox}, (b) the threshold voltage shift as a function of time, (c) applied gate voltage V_G, and (d) drain current I_D through the NP-TFT. A fast gate voltage sweep is used to probe I_D–V_G immediately before and after the programming pulse (which occurs at $t = 10$ ms) then again at 40 ms later after the charge has leaked from the nanoparticles. The three labeled probing regions correspond to the symbols in Figure 25.4.

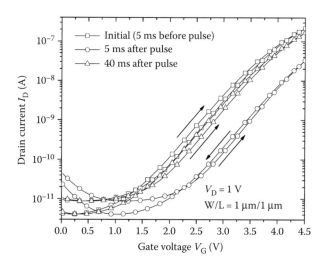

FIGURE 25.4 Transfer characteristics of the NP-TFT immediately prior to, 5 ms after, and 40 ms after the programming pulse (presynaptic action potential V_G). These curves correspond to the drain current measured during the three gate voltage probing sweeps shown in Figure 25.3.

for pre–post and post–pre pairs are shown in Figure 25.5. The presynaptic action potential ($V_{gate,pre}$) was always applied at $t = 50$ ms, while the timing of the postsynaptic action potential was varied to obtain different $\Delta t = t_{post} - t_{pre}$. A dependent voltage source provides the injection voltage V_{inj}, which is defined to be the output voltage times a multiplication factor of 0.44. Thus, V_{inj} will normally be near ground, but spike to ~2 V during an action potential. Timing of the postsynaptic action potential at V_{out} was varied using a very short 100 nA current pulse (I_{post}) that was injected directly onto

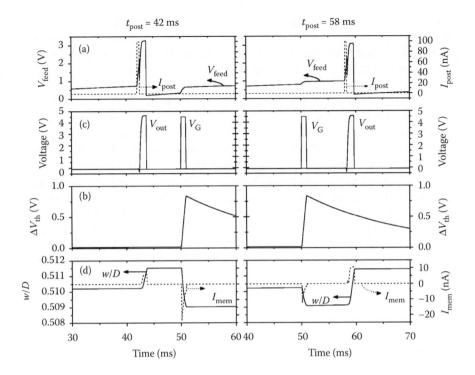

FIGURE 25.5 Pair-based STDP testing procedure. The left column is a post–pre pair resulting in overall synaptic depression, whereas the right-hand column is a potentiating pre–post pair. (a) The voltage at the input node of the postsynaptic neuron V_{feed}, along with the injection current I_{post} required to generate an action potential. (b) The output voltage V_{out} of the postsynaptic neuron (see Figure 25.2a along with the gate voltage applied to the NP-TFT. This voltage V_G induces trapped charge in the nanoparticles, resulting in the threshold voltage shifts shown in (c). (d) The current through the memristive device I_{mem}, the magnitude of which relates to the change in the state variable w/D, and thus the synaptic weight.

the node V_{cap} (see Figure 25.5a). Owing to the large magnitude of the current, the neuron reaches the threshold of 1.9 V almost immediately. However, it must be turned off at the instant V_{feed} reaches 1.9 V to avoid inducing errors in the synaptic weight measurement, and the time at which it turns off is recorded as t_{post}. Also, the current source must be placed at V_{cap} and injected directly onto the "membrane" capacitor C1. If it is placed at the V_{feed} node, too much reverse current will flow when the feed transistor starts to turn off, which will artificially increase potentiation, resulting in an erroneous STDP curve. Careful observation of the circuit operation is therefore necessary when evaluating the performance of the system and subsequent synaptic weight changes.

In addition to the current pulse injected into the postsynaptic neuron to initiate spikes at precise times, a constant DC current $I_{post,DC}$ was added to imitate the excitation of the neuron by many other afferents. This current causes the voltage V_{feed} to build up and be greater than zero, but is set less than the leakage current so it does not cause the neuron to fire. Under realistic operating conditions, the summation of many different excitatory postsynaptic potentials (EPSPs) will generate approximately the same effect [24]. During a presynaptic action potential, the memory TFT is on and forward current flows through the memristor to the postsynaptic neuron, decreasing the synaptic weight. The magnitude of the decrease depends not only on the magnitude of the injection voltage ($V_{inj} \approx 2$ V) but also on the instantaneous value of V_{feed}. If the postsynaptic neuron is nearly at threshold (~1.9 V), there is almost no voltage drop across the memristor and thus very little depression. Dependence of the weight change on the postsynaptic voltage is also a characteristic of biological synapses [13,28].

Upon termination of the presynaptic action potential, the memory transistor channel remains on for an extended period while the trapped charge leaks away. The length of time between the initial presynaptic pulse and the postsynaptic neuron action potential controls the reverse current through the memristor, and thus the potentiation (Figure 25.5d). In other words, the magnitude of the weight increase varies with the instantaneous conductance of the TFT channel at the time of the postsynaptic spike. Although somewhat difficult to see in Figure 25.3b, a continuous application of presynaptic pulses will shift the threshold voltage beyond 1 V. Therefore, the potentiation is not necessarily dependent only on the nearest-neighbor spike interaction, but will in fact be stronger if several presynaptic action potentials occur immediately prior to a post spike [29].

The results of performing many trials of post–pre (left column of Figure 25.5) and pre–post (right column of Figure 25.5) are shown in Figure 25.6. Different DC injection currents of 0.5, 1, and 1.5 nA are shown in Figure 25.6a, b, and c, respectively, each with different values of the initial synaptic weight (w/D = 0.2, 0.5, and 0.8). The similarity to biological measurements such as those reported in Ref. [2] is striking, especially for the case that $I_{post,DC}$ = 1 nA. Dependence of the weight change on the initial weight is also observed in biology, but a more detailed analysis is required to determine if the trends are the same in these circuits [30]. The value of V_{feed} in these simulations varies depending on how long the postsynaptic action potential occurred before that of the presynaptic neuron because V_{feed} is reset after the spike. It then slowly builds up to different levels at t_{pre} = 50 ms.

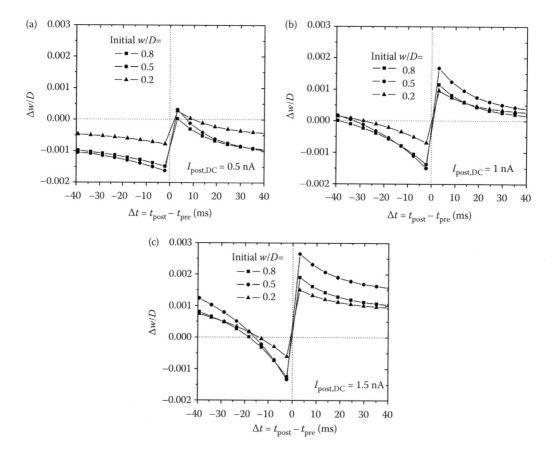

FIGURE 25.6 Pair-based STDP curves for various initial synaptic weight values (w/D) and DC postsynaptic injection currents $I_{post,DC}$. The different values of (a) 0.5 nA, (b) 1 nA, and (c) 1.5 nA vary the voltage V_{feed} at the time of the first spike. Each data point is obtained from individual simulations like those shown in Figure 25.5.

In simulations of the pre–post pair, V_{feed} is always the same value at the t_{post} because it is not reset until that point (see the right-hand column of Figure 25.5a).

25.3.2 Spike Triplets

In a manner similar to the previous section with action potential pairs, the synaptic weight changes resulting from spike triplets were obtained. Temporal plots of the main circuit variables analogous to Figure 25.5 are shown in Figure 25.7. In all of the cases, a 1 nA DC injection current was used. Data for a full set of triplets are shown in Figure 25.8, with a direct comparison to biological measurements [11] and the model of Clopath and Gerstner [13]. Although there is some difference in the actual levels of potentiation, the overall trend for potentiation and depression is very well-matched.

25.3.3 Frequency Dependence

A final experiment involves testing the changes in synaptic weight as a function of the frequency of applied action potentials. With Δt set to specific positive values, the pairs are repeated 10 times at different frequencies. The periods are kept larger than Δt so that the spike train is always pre–post alternating. Depending on the frequency, the postsynaptic spike may occur at varying times during the duty cycle of t_{pre}, as shown in the insets of Figure 25.9. Specifically, in the upper inset, the majority of the pairs look like post–pre sequences, whereas at a lower frequency for the same Δt, they are clearly pre–post (lower inset). The weight change is positive in the former case because the NP-TFT channel essentially stays on for the duration of the stimulus, resulting in synaptic potentiation. In the latter case, depression dominates because the DC injection current is set to 0 nA in this simulation. If it is larger, potentiation occurs for all frequencies when the pairs are pre–post alternating.

Simulations using trains of post–pre pairs were also performed (not shown). The curves generated were similar in shape to those of Figure 25.9, but with more dominant depression at low frequencies. At high frequencies, however, there is still strong potentiation. From a Hebbian perspective, the relationship makes sense because concurrent high-frequency activity should strengthen the neuronal

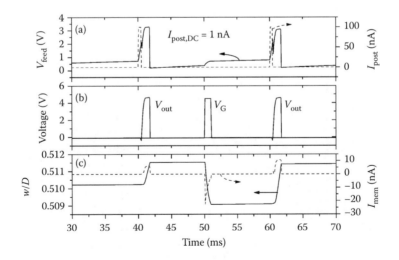

FIGURE 25.7 Plots of different circuit variables as a function of time during a (10,–10) post–pre–post triplet test. (a) The injection current pulses I_{post} quickly increase V_{feed} to the threshold, causing the postsynaptic neuron to fire at $t = 40$ and 60 ms. These action potentials are shown together with the presynaptic spike V_G in (b). (c) Current through the memristive device I_{mem} alters the synaptic weight w/D as a function of time by an amount proportional to the polarity and the magnitude. This triplet results in overall synaptic potentiation.

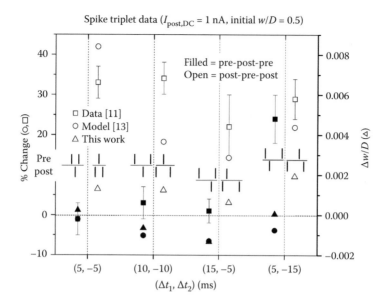

FIGURE 25.8 A direct comparison of synaptic weight changes in spike triplet experiments between Refs. [11] and [13] and this work. The biological data taken in the hippocampus cells indicate dominance of synaptic potentiation.

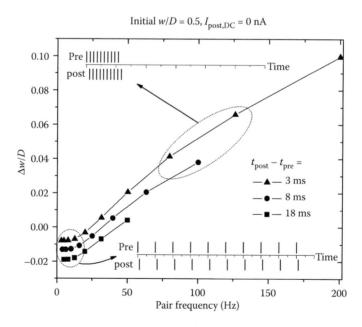

FIGURE 25.9 Frequency dependence of the synaptic weight change for different Δt values. Ten action potential pairs were applied at different frequencies with the given pre–post offset, and the total weight change was measured. The insets show approximately what the pre- and postsynaptic firing looks like for different regions of the graph.

connection regardless of the exact timing. Very similar behavior and curve shapes have been widely reported in biological measurements [10,31,32].

25.4 DISCUSSION

Further analysis is warranted regarding some issues related to the devices and circuits described in this work. The first calculates the approximate power dissipation of the individual components and the system, and compares that to the human brain. It can be seen in Figure 25.7c that the current flow through the memristor during any action potential (pre- or postsynaptic) is on the order of 10 nA given the voltages used. Also, the memristor state variable in this case was $w/D = 0.5$, which corresponds to $R_{mem} = 25$ MΩ. At this value, the average energy per synapse per spike (with 1 ms duration) is ~2.5 pJ. There is no power dissipation of the synapse when neither the pre- nor postsynaptic neurons are in the nonfiring state. Measurements made on a fabricated neuron circuit indicate that the total power consumption, including in standby mode (nonfiring), is ~12 nW at 100 Hz. Now imagine a system with the same scale as that of the human neocortex (10^{10} neurons [33] and 10^{14} synapses [34]). Assuming that all the synapses are at the same weight of $w/D = 0.5$ and an overall average firing rate of 100 Hz, the total power consumption approaches 25 kW, most of which is in the synapses. This value is clearly far too large for any realistic electronic system, but could easily be reduced by a factor of 100 or more with some simple design alterations. This would put the total power requirement at approximately the same magnitude as the human brain [35].

Reduction of the injection voltage V_{inj} would be the main factor that significantly diminishes the amount of current flowing through the memristive devices. It is currently set to approximately 2 V ($0.44 \times V_{out}$) because the value must be greater than the threshold of the input inverter (~1.9 V) in order for the postsynaptic neuron to fire. In other words, the maximum value that V_{cap} could ever obtain is the same as V_{inj} (see Figure 25.2a). Reducing the value of V_{inj} to 0.2 V or less would be ideal from the power reduction perspective as well as a fan-in and density viewpoint. With reduced V_{inj}, a much smaller input capacitor C1 could be used to obtain the same voltage at V_{cap}. One possible method to obtain such low-voltage operation could be the addition of a front-end amplifier to translate ~100 mV at V_{cap} into the ~1.9 V necessary to trip the inverter. Along the same lines, reducing the device threshold voltage [36] and optimizing the inverter transfer curve would allow much lower power supply voltages and switching currents. Another point to be made is that even after significant improvements, the required power may still be too much for a small chip, but could easily be dissipated over larger areas.

Bias degradation of the nc-Si silicon TFT devices is an additional matter of concern [37]. There is some research that addresses the circuit-level impact of TFT degradation [38], but none specifically targeting neuromorphic circuits and systems. However, increased deposition temperatures will produce more stable nc-Si films, leading to reduced threshold voltage shift over time [39]. In turn, the neuron circuit operation should remain relatively stable, and an advantage of this architecture would be the ability to adapt to changes in the firing rate, pulse width, and so on. Future measurements should confirm the expectation that device degradation will not significantly impact on the overall operation of these systems.

25.5 APPLICATIONS

Although pair-based STDP fails to explain all the effects observed in biology, it remains a fundamental first-order approximation of true learning rules. Potential applications for a hardware implementation of the STDP learning rule are outlined in this section. Fundamental system-level consequences of STDP are explained first, followed by the applications. In the first case, a methodology for performing spatio-temporal pattern recognition is shown, having potential use for real-time processing of large amounts of input data. Described next is sound source localization via detection of interaural time differences, which is a capability of animals that can be attributed to

STDP learning rules. Finally, the extraction of features such as edges from visual scenes is discussed. Other applications that have been investigated previously but are not explicitly included here are pattern recall [40] and dataset classification [41].

25.5.1 SPATIO-TEMPORAL PATTERN RECOGNITION

It has been recognized for some time that when implemented as single layer perceptron networks, STDP learning rules enable the identification of recurring spike patterns [15–17]. Because the patterns are time-resolved and distributed within a subset of the afferent neurons located in space, this is generally referred to as spatio-temporal pattern recognition. Theoretical studies consist of randomly generated afferent spike trains with Poisson-distributed ISI. At specific times, a fraction of the population of input neurons simultaneously produces unique patterns, and the output neuron learns over time to respond only to these presentations. The recurring patterns are the same each time they occur for a particular neuron, but different between neurons. They are also generated such that the firing rate remains approximately constant so as not to be distinguished in this way. At first, the output neuron responds randomly to the input spike trains and is not selective to the recurring patterns. After some time (or a certain number of pattern presentations), the synaptic weights are altered such that the output responds quickly to the start of the pattern. The STDP learning rule controls the weight change and there is no supervision or training of the network. Successful identification of the simultaneous patterns increases with the frequency of their presentation, the proportion of afferents involved, and the high precision firing times within the pattern (low jitter) [17].

In a single layer, the spatio-temporal pattern recognition is interesting but not particularly useful. Connected with multiple layers in a hierarchical structure, far more complex patterns, or even the absence of patterns, could be detected. These systems would be useful for any application involving the real-time processing of large amounts of data that contain recurring temporal patterns. Examples might include large sensor networks connected to electronic systems or networks to detect errors or intrusions, power grids to proactively prevent faults, or to traffic monitoring systems to efficiently route cars or emergency vehicles.

25.5.2 SOUND LOCALIZATION

The ability to identify the spatial position of a sound source is an important aspect of animal perception. At frequencies in which the wavelength is shorter than the spacing of the ears, interaural intensity difference is typically used to locate sounds. When the wavelength is longer, the interaural time difference (ITD) is detected to precisely locate the source on the azimuthal plane. Coincidence detection properties of neurons and the STDP learning rule are both utilized in the auditory system to perform sound localization [42–44]. The neuron circuits in this work function as integrators or coincidence detectors depending on the action potential frequency [24,45].

In owls, the ability to detect ITDs of down to a few microseconds corresponds to angular precision of a few degrees, helping these nocturnal birds locate prey by sound alone. Other animals such as bats and dolphins use echolocation for navigation and to detect and avoid nearby objects. Essentially the same mechanisms can be used in engineered systems such as robots and autonomous vehicles to assist in navigation. A distinct advantage to using neural networks to perform this task is not only the efficiency, but also the possibility of integrating these data with that from other sensor systems (for instance, an artificial visual cortex as described in Section 25.5.3) to more fully comprehend the contents of the surrounding environment. In addition, one could imagine modules for various electronic systems that incorporate this localization technology, helping microphones to pick out the location of a voice when there is a noisy background. Again, the potential to efficiently integrate this system directly with another capability such as speech recognition is a distinct advantage of these networks.

25.5.3 VISUAL DATA PROCESSING

The visual cortex is the most densely connected area of the human cortex [46]. It performs many complex processing operations. One known function is orientation sensitivity, and the ability to detect edges of objects within the visual field. The emergence of this capability is a consequence of STDP [47–49], and several studies have been devoted to emulating it using electronic devices [50–52]. Toward the overall goal of building an artificial visual cortex that can process complete visual scenes and return information in real time, interim applications could include the recognition and classification of specific images. Recognizing handwriting [52], faces [51], and the locations of nearby cars traveling down a highway [50] are examples of tasks that digital systems have difficulty in performing successfully. Artificial neural networks promise to provide the ability to efficiently differentiate between similar letters or distinguish faces with subtle variations in the shapes of noses or eyes. They will also be able to classify cars, trucks, and motorcycles into distinct categories. And perhaps the most important feature inherent in these systems will be no requirement for programming or explicitly training the network to perform such a wide variety of functions.

25.6 CONCLUSIONS

Neuron circuits comprising of ambipolar nc-TFTs and NP-TFTs are simulated using SPICE. Single memristive devices are used for the synapses, and are driven by the NP-TFTs connected to the presynaptic neuron outputs. Using this compact configuration, the circuits exhibit timing-dependent synaptic learning with biologically realistic action potentials (no changes in the height or width, and no pulse shaping is required). The striking similarity to biological synapses is demonstrated through pair- and triplet-based experiments, as well as in the frequency dependence of the weight changes. The significance of these results is highlighted by the use of materials and devices which could conceivably be used to build dense, large-scale systems with 3-D physical structure similar to the neocortex.

Beyond the investigation of the learning mechanisms, considerable discussion was devoted to other important issues. These include improving the circuit performance for use in large networks, reducing power consumption, and mitigating concerns about device and circuit reliability. Three potential applications of these circuits were also presented in detail. One consequence of STDP, the ability of the network to respond to recurring spatio-temporal patterns, may be very useful for systems that must process large amounts of input data streaming in real time. Sound localization can also be accomplished by taking advantage of the timing-dependence of the learning to detect phase differences in aural signals. An artificial visual cortex capable of extracting and processing visual information such as edges in multiple receptive fields is the last of the applications discussed. With such a rich set of capabilities and applications, it should be clear that artificial neural networks employing STDP learning have great potential for future electronic systems that could help provide humans with more detailed and useful information about the complex environment around them.

REFERENCES

1. D. B. Strukov, G. S. Snider, D. R. Stewart, and R. S. Williams, The missing memristor found, *Nature*, 453(7191), 80–83, 2008.
2. G.-Q. Bi and M.-M. Poo, Synaptic modification by correlated activity: Hebb's postulate revisited, *Annual Review of Neuroscience*, 24, 139–166, 2001.
3. S. Song, K. D. Miller, and L. F. Abbott, Competitive Hebbian learning through spike-timing-dependent synaptic plasticity, *Nature Neuroscience*, 3(9), 919–926, 2000.
4. G. S. Snider, Spike-timing-dependent learning in memristive nanodevices, in *IEEE International Symposium on Nanoscale Architectures (NANOARCH)* Anaheim, CA, 2008, pp. 85–92.
5. S. H. Jo, T. Chang, I. Ebong, B. B. Bhadviya, P. Mazumder, and W. Lu, Nanoscale memristor device as synapse in neuromorphic systems, *Nano Letters*, 10(4), 1297–1301, 2010.

6. D. Kuzum, R. G. D. Jeyasingh, B. Lee, and H. S. P. Wong, Nanoelectronic programmable synapses based on phase change materials for brain-inspired computing, *Nano Letters,* 12(5), 2179–2186, 2011.
7. B. Linares-Barranco and T. Serrano-Gotarredona, Exploiting memristance in adaptive asynchronous spiking neuromorphic nanotechnology systems, in *9th IEEE Conference on Nanotechnology (IEEE NANO),* Genoa, Italy, 2009, pp. 601–604.
8. J. Lisman and N. Spruston, Questions about STDP as a general model of synaptic plasticity, *Frontiers in Synaptic Neuroscience,* 2(140), 2010. DOI: 10.3389/fnsyn.2010.00140.
9. H. Z. Shouval, S. S. H. Wang, and G. M. Wittenberg, Spike timing dependent plasticity: A consequence of more fundamental learning rules, *Frontiers in Computational Neuroscience,* 4(19), 2010. DOI: 10.3389/fncom.2010.00019.
10. H. Markram, J. Lubke, M. Frotscher, and B. Sakmann, Regulation of synaptic efficacy by coincidence of postsynaptic APs and EPSPs, *Science,* 275(5297), 213–215, 1997.
11. H.-X. Wang, R. C. Gerkin, D. W. Nauen, and G.-Q. Bi, Coactivation and timing-dependent integration of synaptic potentiation and depression, *Nature Neuroscience,* 8, 187–193, 2005.
12. R. C. Froemke and Y. Dan, Spike-timing-dependent synaptic modification induced by natural spike trains, *Nature,* 416, 433–438, 2002.
13. C. Clopath and W. Gerstner, Voltage and spike timing interact in STDP—A unified model, *Frontiers in Synaptic Neuroscience,* 3(25), 2010. DOI: 10.3389/fnsyn.2010.00025.
14. C. G. Mayr and J. Partzsch, Rate and pulse based plasticity governed by local synaptic state variables, *Frontiers in Synaptic Neuroscience,* 2(33), 2010. DOI: 10.3389/fnsyn.2010.00033.
15. W. Gerstner, R. Ritz, and J. van Hemmen, Why spikes? Hebbian learning and retrieval of time-resolved excitation patterns, *Biological Cybernetics,* 69(5), 503–515, 1993.
16. R. Guyonneau, R. VanRullen, and S. J. Thorpe, Neurons tune to the earliest spikes through STDP, *Neural Computation,* 17(4), 859–879, 2005.
17. T. Masquelier, R. Guyonneau, and S. J. Thorpe, Spike timing dependent plasticity finds the start of repeating patterns in continuous spike trains, *PLoS ONE,* 3(1), e1377, 2008. DOI:10.1371/journal.pone.0001377.
18. C.-H. Lee, A. Sazonov, M. R. E. Rad, G. R. Chaji, and A. Nathan, Ambiploar thin-film transistors fabricated by PECVD nanocrystalline silicon, in *Materials Research Society Spring Meeting Proceedings,* San Francisco, CA, 2006.
19. K.-Y. Chan, J. Kirchhoff, A. Gordijn, D. Knipp, and H. Stiebig, Ambipolar microcrystalline silicon thin-film transistors, *Thin Solid Films,* 517(23), 6383–6385, 2009.
20. A. Subramaniam, K. D. Cantley, H. J. Stiegler, R. A. Chapman, and E. M. Vogel, Submicron ambipolar nanocrystalline silicon thin-film transistors and inverters, *IEEE Transactions on Electron Devices,* 59(2), 359–366, 2011.
21. K. D. Cantley, A. Subramaniam, H. J. Stiegler, R. A. Chapman, and E. M. Vogel, SPICE simulation of nanoscale non-crystalline silicon TFTs in spiking neuron circuits, in *53rd IEEE International Midwest Symposium on Circuits and Systems (MWSCAS),* Seattle, WA, 2010, pp. 1202–1205.
22. K. D. Cantley, A. Subramaniam, H. J. Stiegler, R. A. Chapman, and E. M. Vogel, Hebbian learning in spiking neural networks with nano-crystalline silicon TFTs and memristive synapses, *IEEE Transactions on Nanotechnology,* 10(5), 1066–1073, 2011.
23. C. Mead, *Analog VLSI and Neural Systems.* New York: Addison-Wesley, 1989.
24. K. D. Cantley, A. Subramaniam, H. J. Stiegler, R. A. Chapman, and E. M. Vogel, Neural learning circuits utilizing nano-crystalline silicon transistors and memristors, *IEEE Transactions on Neural Networks and Learning Systems,* 23(4), 565–573, 2012.
25. S. M. Sze, *Physics of Semiconductor Devices,* 3rd ed. Hoboken, NJ: John Wiley and Sons, 2007.
26. F. Alibart, S. Pleutin, D. Guérin, C. Novembre, S. Lenfant, K. Lmimouni, C. Gamrat, and D. Vuillaume, An organic nanoparticle transistor behaving as a biological spiking synapse, *Advanced Functional Materials,* 20(2), 330–337, 2010.
27. Y. Kuo and H. Nominanda, Nonvolatile hydrogenated-amorphous-silicon thin-film-transistor memory devices, *Applied Physics Letters,* 89(173503), 2006. DOI: 10.1063/1.2356313.
28. J. M. Brader, W. Senn, and S. Fusi, Learning real-world stimuli in a neural network with spike-driven synaptic dynamics, *Neural Computation,* 19(11), 2881–2912, 2007.
29. A. Morrison, M. Diesmann, and W. Gerstner, Phenomenological models of synaptic plasticity based on spike timing, *Biological Cybernetics,* 98(6), 459–478, 2008.
30. M. C. W. van Rossum, G. Q. Bi, and G. G. Turrigiano, Stable Hebbian learning from spike timing-dependent plasticity, *The Journal of Neuroscience,* 20(23), 8812–8821, 2000.

31. J-P. Pfister and W. Gerstner, Triplets of spikes in a model of spike timing-dependent plasticity, *The Journal of Neuroscience,* 26(38), 9673–9682, 2006.
32. P. J. Sjostrom, G. G. Turrigiano, and S. B. Nelson, Rate, timing, and cooperativity jointly determine cortical synaptic plasticity, *Neuron,* 32(6), 1149–1164, 2001.
33. B. Pakkenberg and H. J. G. Gundersen, Neocortical neuron number in humans: Effect of sex and age, *The Journal of Comparative Neurology,* 384(2), 312–320, 1997.
34. Y. Tang, J. R. Nyengaard, D. M. G. DeGroot, and H. J. G. Gundersen, Total regional and global number of synapses in the human brain neocortex, *Synapse,* 41(3), 258–273, 2001.
35. D. Attwell and S. B. Laughlin, An energy budget for signaling in the grey matter of the brain, *Journal of Cerebral Blood Flow and Metabolism,* 21(10), 1133–1145, 2001.
36. A. Subramaniam, K. D. Cantley, R. A. Chapman, H. J. Stiegler, and E. M. Vogel, Submicron ambipolar nanocrystalline-silicon TFTs with high-κ gate dielectrics, in *International Semiconductor Device Research Symposium (ISDRS),* College Park, MD, 2011.
37. M. J. Powell, S. C. Deane, and W. I. Milne, Bias-stress-induced creation and removal of dangling-bond states in amorphous silicon thin-film transistors, *Applied Physics Letters,* 60(2), 207–209, 1992.
38. D. R. Allee, L. T. Clark, B. D. Vogt, R. Shringarpure, S. M. Venugopal, S. G. Uppili, K. Kaftanoglu, H. Shivalingaiah, Z. P. Li, J. J. Ravindra Fernando, E. J. Bawolek, and S. M. O'Rourke, Circuit-level impact of a-Si:H thin-film-transistor degradation effects, *IEEE Transactions on Electron Devices,* 56(6), 1166–1176, 2009.
39. A. Subramaniam, K. D. Cantley, R. A. Chapman, B. Chakrabarti, and E. M. Vogel, Ambipolar nanocrystalline-silicon TFTs with submicron dimensions and reduced threshold voltage shift, in *69th Annual Device Research Conference (DRC) Digest,* Santa Barbara, CA, 2011.
40. J. Arthur and K. Boahen, Learning in silicon: Timing is everything, *Advances in Neural Information Processing Systems,* 18, 75–82, 2006.
41. L. Bako, Real-time classification of datasets with hardware embedded neuromorphic neural networks, *Briefings in Bioinformatics,* 11(3), 348–363, 2010.
42. C. E. Carr and M. Konishi, A circuit for detection of interaural time differences in the brain stem of the barn owl, *The Journal of Neuroscience,* 10(10), 3227–3246, 1990.
43. B. Glackin, J. A. Wall, T. M. McGinnity, L. P. Maguire, and L. J. McDaid, A spiking neural network model of the medial superior olive using spike timing dependent plasticity for sound localization, *Frontiers in Computational Neuroscience,* 4(18), 2010. DOI: 10.3389/fncom.2010.00018.
44. W. Maass and C. M. Bishop, *Pulsed Neural Networks.* Cambridge, MA: The MIT Press, 1999.
45. P. König, A. K. Engel, and W. Singer, Integrator or coincidence detector? The role of the cortical neuron revisited, *Trends in Neurosciences,* 19(4), 130–137, 1996.
46. E. R. Kandel, J. H. Schwartz, and T. M. Jessell, *Principles of Neural Science,* Fourth ed. New York: McGraw-Hill, 2000.
47. A. Delorme, L. Perrinet, and S. J. Thorpe, Networks of integrate-and-fire neurons using Rank Order Coding B: Spike timing dependent plasticity and emergence of orientation selectivity, *Neurocomputing,* 38(40), 539–545, 2001.
48. T. Masquelier and S. J. Thorpe, Unsupervised learning of visual features through spike timing dependent plasticity, *PLoS Computational Biology,* 3(2), e31, 2007.
49. S. Thorpe, A. Delorme, and R. Van Rullen, Spike-based strategies for rapid processing, *Neural Networks,* 14(6–7), 715–725, 2001.
50. M. Suri, O. Bichler, D. Querlioz, O. Cueto, L. Perniola, V. Sousa, D. Vuillaume, C. Gamrat, and B. DeSalvo, Phase change memory as synapse for ultra-dense neuromorphic systems: Application to complex visual pattern extraction, in *International Electron Devices Meeting (IEDM),* Washington, DC, 2011, pp. 11–79–11–82.
51. B. Linares-Barranco, T. Serrano-Gotarredona, L. A. Camuas-Mesa, J. A. Perez-Carrasco, C. Zamarreo-Ramos, and T. Masquelier, On spike-timing-dependent-plasticity, memristive devices, and building a self-learning visual cortex, *Frontiers in Neuroscience,* 5(26), 2011. DOI: 10.3389/fnins.2011.00026.
52. G. S. Snider, Self-organized computation with unreliable, memristive nanodevices, *Nanotechnology,* 18(365202), 2007.

26 Thermally Actuated Nanoelectromechanical Memory

A New Memory Concept for Spacecraft Application

Elham Maghsoudi and Michael James Martin

CONTENTS

26.1 INTRODUCTION

Probes used in planetary exploration place greater demands on electronics than terrestrial, or even orbital, systems. The most challenging missions are those that go beyond the orbit of Mars. Because missions to Jupiter's moons, as well as Saturn and its moons, have been identified as high-priority missions by the National Research Council's Decadal Survey [1], the development of electronics for these applications is part of NASA's technology development roadmap [2].

There are three technical challenges in engineering electronics for these applications: low available power, high required reliability, and the need for the configuration to be radiation-resistant. Because missions to the outer planets operate where the solar energy available is much lower than that available on Earth, spacecraft generally rely on radioisotope thermoelectric generators (RTGs). A typical system provides around 500 W, which must satisfy all of the spacecraft's power needs, including avionics, thermal management, propulsion, and communications [3].

These missions are also of long duration. The journey from Earth to the outer planets requires several years, during which time electronic components may be used in communications with Earth, monitoring spacecraft health, and acquiring data from secondary scientific objectives. Other components may not be used until the spacecraft arrives at its final destination. In both cases, the components must meet higher standards of reliability than terrestrial systems to avoid endangering the success of the mission [4].

The most formidable challenge for spacecraft electronics is the high-radiation environment of space. Deep-space missions remain exposed to high levels of radiation for years. When the spacecraft enters the Jovian environment, the radiation environment becomes even more severe [5], limiting the operational life of a spacecraft in these conditions [6]. There are three major failure modes associated with radiation: total dose damage, latch-up, and single-event upsets. Total dose damage is the deterioration due to the cumulative radiation damage to the material of the device. This can be mitigated by shielding and by materials selection, but still limits the overall life of the system.

Latch-up is a catastrophic failure mode caused by a single high-energy ion causing a thermal runaway in the chip. This can be mitigated by measures such as current sensing and alternate circuit configurations. Because this failure mode is only an issue in powered circuits, it is not the primary concern in nonvolatile memory systems.

The single-event upsets pose the greatest challenge to reliable data storage. A high-energy particle can "flip" a bit that is stored by semiconductor memory, changing a 1 to a 0, or vice versa. This will affect both volatile and nonvolatile memories. Because the spacecraft depends on volatile memories for short-term operations and nonvolatile memories for storing both scientific data and the operating commands of the software, compromised data can affect the ability of the spacecraft to return data and the survival of the spacecraft itself.

In volatile memories, this is usually dealt with by storing data in triplicate and designing the circuit to "vote" on the correct choice. In nonvolatile memories, this is resolved through the frequent use of error codes and rewriting of data. This complication affects the complexity, energy costs, and reliability of the system. These problems can potentially be avoided by using nanomechanical memory as a substitute for conventional systems.

26.2   NANOMECHANICAL MEMORY

Previous researchers have designed nonvolatile electromechanical memories and volatile mechanical memories operating on the basis of the displacement of bridge structures. The concept of mechanical bistability of a doubly clamped bridge was used to implement these devices [7,8]. A potentially simpler nonvolatile memory device is the buckled-beam nanomechanical memory [9–11]. These devices have been electrostatically actuated and successfully demonstrated in laboratory experiments [10], but have not been the subject of extensive performance or reliability analysis.

The optimal design, power requirements, and writing time of buckling-beam-based memories are still not well characterized. These are scientific goals that are necessary for both space-based and terrestrial applications.

This chapter presents a preliminary analysis of a thermally actuated nanoelectromechanical memory mechanism. The mechanism combines thermal actuation with the buckling-beam concept. The simulation of a unit bridge of an array of buckling-beam memory using a finite-difference method gives an overview of how geometry variation affects the power requirements for thermal actuation and writing time of the device.

26.3   BUCKLING MECHANISM OF THERMAL MEMORY

Figure 26.1 illustrates the buckling-beam memory that uses an array of nanofabricated beams and thermal buckling of these devices. Each beam represents a single bit of memory. The data storage density is inversely proportional to the area occupied by each beam, so the ideal array would use the smallest beams possible. The data are then "written" by buckling the beams. A buckled down beam is a "1" and a buckled up beam is a "0."

Beam (a) is in the initial, unbuckled state. It is being heated, through either a thin-film resistor or a heated tip designed as a writing device, at a constant rate.

After a time t_b, the thermal stresses in the axial direction will cause the force in the beam to exceed the force required for buckling from the column stability theory, causing the beam to buckle down as shown in (b). If the heat is added to the bottom of the beam, it will buckle up as (c), which is corresponding to a "0."

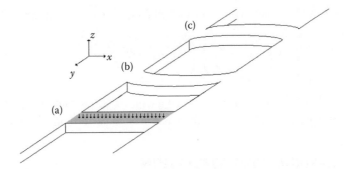

FIGURE 26.1 Nanomechanical memory actuation in a memory array.

The critical buckling load is defined as follows [12]:

$$F_b = EI \cdot \left(\frac{\pi}{\kappa l} \right)^2,$$ (26.1)

where E is the modulus of elasticity, I is the moment of inertia, l is the length of the bridge, and κ is the column effective length factor that is equal to 0.5 for both end-fixed bridges.

The thermal force due to thermal stress is as follows [13]:

$$F_b = \iint\limits_{\text{Area}} \alpha E \cdot (T - T_w) \cdot dy \cdot dz,$$ (26.2)

where α is the thermal expansion coefficient, T is the temperature distribution, and T_w is the wall temperature (Figure 26.2).

In an ideal case, where there are no losses to the surrounding material or the gas, and the temperature inside the bridge is uniform, Equation 26.2 is simplified as

$$F_b = A_c \alpha \cdot E \cdot (T_b - T_w),$$ (26.3)

where T_b is the total required buckling temperature, and A_c is the cross section of the beam. T_b can be found by combining the formulation for thermal stress inside the beam (Equation 26.3) with the critical buckling load (Equation 26.1):

$$T_b = \left(\frac{\pi d}{l} \right)^2 \cdot \left(\frac{1}{3\alpha} \right) + T_w .$$ (26.4)

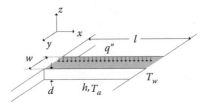

FIGURE 26.2 Simulation geometry.

Using Equation 26.4, the required buckling energy for the ideal case with no losses is obtained as

$$Q = \left(\frac{\pi^2}{3}\right) \cdot \left(\frac{\rho \cdot c_p}{\alpha}\right) \cdot \left(\frac{w \cdot d^3}{l}\right), \tag{26.5}$$

where ρ is the density, c_p is the specific heat, w is the width, d is the thickness, and l is the length of the bridge. To investigate the nonideal case, there is a need to perform the simulations.

26.4 THERMAL–STRUCTURAL SIMULATION

A simplified model of a beam undergoing thermal buckling is shown in Figure 26.2. The beam is of length l, width w, and depth d, and is at an initial temperature T_w. Depending on how the device is packaged, it may be surrounded by gas molecules at an ambient temperature T_a, and have a convective heat transfer coefficient h. For thermal actuation, the heat load is assumed to be applied on the top of the beam, at a rate of q'' (in W/m^2).

As the temperature of the beam increases, the thermal stresses in the beam will increase, and the beam will buckle. To capture this process, the transient heat conduction equation inside the beam must be solved:

$$\rho \cdot c_p \cdot \frac{\partial T}{\partial t} = k \cdot \left(\frac{\partial^2 T}{\partial x^2} + \frac{\partial^2 T}{\partial y^2} + \frac{\partial^2 T}{\partial z^2}\right), \tag{26.6}$$

where k is the thermal conductivity of the material [14].

The boundary conditions are based on the thermal conditions at the surface and ends of the beam. There are three different types of boundary condition in this model (Figure 26.2). The Dirichlet boundary condition at both ends is expressed as a constant wall temperature T_w. There are two different types of Nuemann boundary conditions that are applied to heated and unheated surfaces. The boundary condition for the heated top surface includes constant heat flux as well as convective heat transfer while the boundary condition for unheated surfaces only includes the convective heat transfer.

The heat transfer coefficient, h, is calculated based on the free molecular theory for an ideal gas [15]:

$$h = \sigma_T n_i \cdot \left(\frac{\gamma + 1}{\gamma - 1}\right) \cdot \sqrt{\frac{k_b^3 T_a}{8\pi m}}, \tag{26.7}$$

where k_b is the Boltzmann's constant, σ_T is the thermal accommodation coefficient, γ is the specific heat ratio of the gas, m is the molecular weight of the gas, and n_i is the number density of the gas. For an ideal gas, n_i can be calculated using

$$n_i = \frac{P_{gas}}{mRT_a}, \tag{26.8}$$

where P_{gas} is the pressure of the gas and R is the ideal gas constant.

The time-dependent temperature field will be used to find the thermal stresses in the beam. When the axial load on the beam is equal to the critical buckling load, the simulated beam is assumed to buckle [16]. The axial load is calculated using Equation 26.2, where T is the nodal temperature distribution.

A finite-difference solver that allows the temperature, materials, ambient conditions, heat load, and geometry to be changed is used to solve these equations. Using these results, a buckling time t_b and a total energy for buckling Q for any given geometry and ambient conditions are computed:

$$Q = t_b \cdot q^{\bullet}, \tag{26.9}$$

where $q^{\bullet}$ is the total heating rate added to the top:

$$q^{\bullet} = wlq''. \tag{26.10}$$

26.5 GEOMETRY EFFECTS ON WRITING TIME AND POWER CONSUMPTION

The need to balance the mechanical and thermal properties of both the material and the geometry suggests that this will be an optimization problem. The ideal array would use the smallest beams possible to maximize data storage density. Figure 26.3 shows the buckling time variations versus the dimension variations for silicon bridges. The total heating rate added to the top surface is constant, equal to 99.2 mW. Buckling time variations by the bridge length show that longer structures buckle faster. As the thickness of the bridge increases, the energy consumption increases due to an increase in the moment of inertia (Equation 26.1), leading to an increase in the buckling time. Similarly, an increase in the width leads to an increase in the moment of inertia. As a result, the same behavior as thickness variation is expected to lead to the buckling time increase by the width increase. However, Figure 26.3b shows that the width variations have a less significant effect on buckling time in comparison with the length and thickness variations. Decreasing the thickness and width reduces both

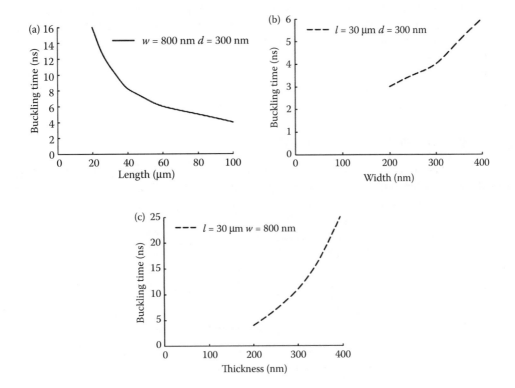

FIGURE 26.3 Buckling time variations by (a) length, (b) width, and (c) thickness variations.

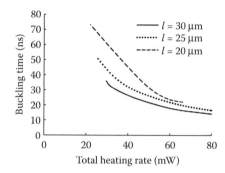

FIGURE 26.4 Buckling time variations by the total heat added to the top surface.

energy consumption and buckling time, improving the system's performance. However, decreasing the length has the opposite effect: increasing energy consumption and buckling time.

Figure 26.4 shows the buckling time variations versus the total heating rate at the top of the bridge. The thickness d and the width w are constant, equal to 300 and 800 nm, respectively. The buckling time decreases as the total heat added to the top increases. Data storage density is a trade-off. To balance these constraints, a length of 20 μm and the smallest possible thickness and width to fabricate are suggested. The buckling time changes relatively with the length and inversely with the total heating rate.

Figure 26.5 shows the required energy variation with the total heating rate at the top. Longer structures will buckle at lower temperatures, and will require less energy to actuate. They will also have higher convective losses, but lower conduction losses.

Calculations show that for a bridge with a length of 20 μm, a width of 800 nm, and a thickness of 300 nm, the least total energy of 1.58 nJ is required to buckle the bridge. This value is smaller than the estimated one calculated using Equation 26.5 (3.16 nJ). If a ratio of estimated energy divided by calculated energy is defined, it remains constant around 2 for varying heating rates as long as the thickness and the width are constant. Although the required energy to buckle for a beam with constant dimensions is expected to be constant, it decreases slightly as the total heating rate at the top increases. The results show that it is more efficient to increase the heating rate and make the buckling faster. The faster buckling does not increase the losses.

26.6 HIGH-ENERGY PARTICLE COLLISION

These structures will also be subjected to unintentional heat addition. This represents the mechanical equivalent of the single-event upset for conventional semiconductor devices. In the highly

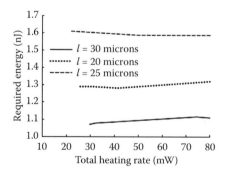

FIGURE 26.5 Required energy versus total heating rate.

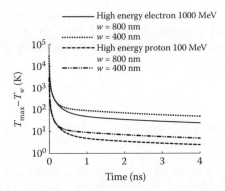

FIGURE 26.6 Heat dissipation after high-energy electron and proton collision for a silicon bridge with a length of 20 μm and a thickness of 300 nm.

radioactive environment of space, high-energy protons and electrons may strike the structure. These particles can be a result of solar events or the planetary environment [3]. To avoid accidental resetting of the spacecraft memory, the beam must be sized so that the energy of the particles does not cause buckling.

The trapped electron and proton energy spectra were previously obtained using the radiation belt models for the Europa orbit of the Jupiter. It shows that the highest energy and the most probable trapped electron and trapped protons carry 1000 and 100 MeV, respectively [5]. If these high-energy particles hit the device and cause undesired buckling, the data will be scrambled. The bridge must be sized to avoid undesired buckling after high-energy particle collision. The worst scenario is that only one particle hits the center of the top surface. The particle is small enough to add instantaneous heat to the central node at the top surface. The amount of this heat depends on the energy the particle carries. The high temperature is calculated assuming that the total energy is transferred to a node.

$$T_{\text{hep}} = \frac{E_{\text{hep}}}{\rho \cdot c_p \cdot \delta x \cdot \delta y \cdot \delta z} + T_w \tag{26.11}$$

where E_{hep} is the energy of the corresponding high-energy particle, and δx, δy, and δz are the space discretizations in x, y, and z, respectively. The same finite-difference solver is used to investigate the probability of the buckling at the time of collision. The rate of heat dissipation is also calculated.

Figure 26.6 shows the heat dissipation of a bit with a length of 20 μm, widths of 400 and 800 nm, and a thickness of 300 nm, which is hit by the highest-energy electron and proton at the top surface. If buckling occurs, it must be at the very beginning of the collision. The simulation results show that buckling does not occur. In addition, the heat due to the collision dissipates in less than 6 ns, which is much less than the quickest buckling time for the same dimensions, 11 ns (Figure 26.4). Although the rate of heat dissipation decreases by decreasing the width of the bridge, high-energy particle collision will not lead to an undesired buckling in the silicon memory bit.

26.7 CONCLUSIONS

Thermally actuated nanoelectromechanical memory offers low consumption and rapid write times with a data storage density comparable to existing technologies. The data storage density is a trade-off. Decreasing the thickness and the width is desired for both energy consumption and buckling time. However, decreasing the length has inverse effects. To balance these constraints, a length of 20 μm and the smallest possible thickness and width to fabricate are suggested. The buckling time decreases with increasing the length and the total heating rate.

High-energy particle collision will not lead to undesired buckling in the memory bit. The heat due to the collision dissipates in less than 10 ns, which shows that this memory is highly protective in extreme radiation environment. The resistance to accidental buckling from radiation makes this technology appropriate for spacecraft applications.

REFERENCES

1. Committee on the Planetary Science Decadal Survey, Vision and voyages for planetary science in the decade 2013–2022, National Research Council 2011.
2. M. A. Meador, B. Files, J. Li, H. Manohara, D. Powell, and E. J. Siochi, Draft nanotechnology roadmap-technology area 10, National Aeronautics and Space Administration, November 2010.
3. P. Fortescue, J. Stark, and G. Swinerd, *Spacecraft Systems Engineering* (Wiley, England, 2003).
4. M. D. Griffin and J. R. French, *Space Vehicle Design* (AIAA, NJ, 2004).
5. I. Jun and H. B. Garrett, Comparison of high-energy particle environments at the Earth and Jupiter, *Radiation Protection Dosimetry*, 116(1–4): 50–54, 2005.
6. D. Brown, D. Agle, M. Martinez, and G. Napier, Juno Launch, National Aeronautics and Space Administration, Press Kit, August 2011.
7. T. Nagami, Y. Tsuchiya, K. Uchida, H. Mizuta, and S. Oda, Scaling analysis of nanoelectromechanical memory devices, *Japanese Journal of Applied Physics*, 49, 044304-1-5, 2010.
8. R. L. Badzey, G. Zolfagharkhani, A. Gaidarzhy, and P. Mohanty, A controllable nanomechanical memory element, *Applied Physics Letters*, 85, 3587, 2004.
9. B. Hälg, On a micro-electro-mechanical nonvolatile memory cell, *IEEE Transactions on Electron Devices*, 37, 2230–2236, 1990.
10. D. Roodenburg, J. W. Spronk, H. S. J. van der Zant, and W. J. Venstra, Buckling beam micromechanical memory with on-chip readout, *Applied Physics Letters*, 94, 183501-1-3, 2009.
11. B. Charlot, W. Sun, K. Yamashita, H. Fujita, and H. Toshiyoshi, Bistable nanowire for micromechanical memory, *Journal of Micromechanics and Microengineering*, 18, 045005, 7, 2008.
12. S. P. Timoshenko and J. N. Goodier, *Theory of Elasticity* (Wiley, NY, 1951).
13. B. A. Boley and J. H. Weiner, *Theory of Thermal Stresses* (Wiley, NY, 1960).
14. F. P. Incropera, D. P. DeWitt, T. L Bergman, and A. S. Lavine, *Introduction to Heat Transfer* (Wiley, NY, 2007).
15. M. J. Martin and B. H. Houston, Frequency-dependent free-molecular heat transfer of vibrating cantilever and bridges, *Physics of Fluids*, 21, 017101, 2009.
16. M. Chiao and L. Lin, Self-buckling of micromachined beams under resistive heating, *Journal of Microelectromechanical Systems*, 9, 146–151, 2000.

Section VIII

Graphene Preparation and Properties

27 Low-Stress Transfer of Graphene and Its Tunable Resistance by Remote Plasma Treatments in Hydrogen

Waileong Chen, Chia-Hao Tu, Keng-Chih Liang, Chih-Yi Liu, Chuan-Pu Liu, and Yonhua Tzeng

CONTENTS

27.1 INTRODUCTION

In this chapter, a novel, low-stress process for transferring thermal chemical vapor deposition (CVD) single-layer graphene from copper foils to destination substrates is demonstrated. An efficient transfer of graphene by directly fishing up the graphene with the target substrate rather than by using the "poly(methyl methacrylate) (PMMA)-based transfer" method is reported. Electrical and optical characteristics of the as-transferred graphene and the hydrogen remote-plasma-modified graphene are presented. Although graphene is mechanically very strong considering its atomically thin structure, a large-area, single-layer graphene is practically very fragile, especially during handling and transfer from one substrate to another. Handling of a large-area freestanding graphene is even more challenging. The combination of effective transfer and surface treatment of graphene by hydrogenation allows the fine-tuning of its electrical resistivity for practical applications.

Since the isolation and confirmation of a stable, atomically thin graphene of a 2D graphite-like carbon network, many means of synthesizing and a wide variety of applications of graphene and modified graphene have been proposed or demonstrated. Graphene film has been attracting much attention because of its exceptional properties, such as extremely high electron mobility [1], low resistivity [2], high thermal conductivity [3,4], and high mechanical strength [5]. Graphene is a promising material for advanced device applications in the future.

Graphene film can be synthesized on catalyst layers such as copper [6,7] and nickel foil [8] by using CVD. For future applications of graphene, the large-scale growth and transfer of graphene from the catalyst are required, and the graphene film must be exfoliated from the catalyst to be integrated to the appropriate surface for further applications. Graphene film might need to be placed on a substrate that is not compatible with the synthesis conditions for graphene. Therefore, an effective transfer process is needed to remove graphene from where it is grown and transfer it to where it will be used. The extremely thin, single-layer graphene of one atom thickness makes a damage-free transfer process even more challenging. Besides, the removal of substance between the transferred graphene and the destination substrate must start from the center toward the edges of the graphene. Otherwise, the trapping of undesired substance between the transferred graphene and the destination substrate will be an issue. For some applications, it is also desirable for more than one graphene film to be transferred to the same destination substrate to form a stacked structure of graphene films of the same or different characteristics.

A few transfer processes have been reported. Generally, coatings on graphene, such as PMMA, have been used as a mechanical support to transfer graphene after the underlying catalyst is etched away with acid solution (e.g., HCl, NHO_3, and $FeCl_3$) [9–11] for several hours. PMMA is then etched away after the graphene is transferred and placed on top of the destination substrate. Stress is induced in the graphene during the transfer and curing process of PMMA. A second coating of PMMA was found to dissolve the first coating and help relax the stress on the graphene. The suggested multiple coatings of PMMA and the need to etch it for exposing the graphene add three more process steps and consume additional chemicals. For applications, that require the transfer of multiple graphene films of the same or different properties, the process is even more complicated and tedious because PMMA needs to be coated and etched repetitively. Furthermore, the graphene film showed strong n-type doping after using PMMA to support and transfer to substrate. Perhaps the residual of PMMA remaining on the graphene, which induced the charge transfer, evidently occurs at the interface between graphene and PMMA, the residual still left on the graphene surface even when the PMMA has been removed by acid solution [12]. However, a "direct transfer method" has been reported by the Ruoff group [7,13], and the graphene process method not only shows more easier transfers and short-time consumption but also it reduces the chemical damage of graphene from the solvent.

As reported in previous research works [14–16], when graphene is fully terminated by hydrogen, the electrically conductive graphene is converted into electrically insulating graphane, which has direct band gaps with Eg = 3.5 eV for the chair form and with Eg = 3.7 eV for the boat form. The calculated graphane C–C bond length of 1.52 Å is similar to the sp^3 bond length of 1.53 Å in diamond and is much greater than the 1.42 Å characteristics of sp^2 carbon in graphene [17]. So the atomic structure is also different from that of graphene. Graphene has a 2D flat plane structure, whereas graphane has an extended 2D covalently bonded structure [17]. Moreover, what if only one side of the graphene carbon atoms is terminated with hydrogen atoms? It is called graphone, which has been calculated to have an indirect band gap of 0.46 eV. And it is ferromagnetic while graphene and graphane are nonmagnetic [18]. Therefore, it is possible to tune the electronic and magnetic properties of graphene by hydrogenation. The possibility of band gap engineering with hybrid graphane–graphene nanostructures is desirable and promising [19]. To achieve this, a novel handling and transfer process for synthesized graphene is necessary.

Here, we report a transfer process that is easy and of low stress to graphene; the number of layers and quality of the as-grown graphene and hydrogen-terminated graphene were evaluated by Raman spectroscopy. We also show that the resistance of graphene can be tuned by exposure to remote hydrogen plasma for different periods of time. The Raman spectra of as-transferred and modified graphene are compared to show the structural modification of graphene by the hydrogen plasma.

27.2 EXPERIMENTAL

Graphene was grown by thermal CVD on copper foils of 50 μm thickness in a gas mixture of methane and hydrogen at a temperature about 1000°C at 1 Torr gas pressure or lower. A transfer process with minimized stress to graphene without requiring the coating of PMMA on graphene has been developed. Because the graphene film is one carbon atom thick, the thickness of graphene is only ~0.34 nm; to protect and locate the graphene, we stuck a Teflon tape to the edge of the graphene before we carried out the chemical etching process. The schematic and flowchart for the transfer process are shown in Figure 27.1a and b. Instead of using PMMA coatings, the graphene on copper remains floating on the surface of the aqueous chemicals and deionized (DI) water until the copper is completely etched away

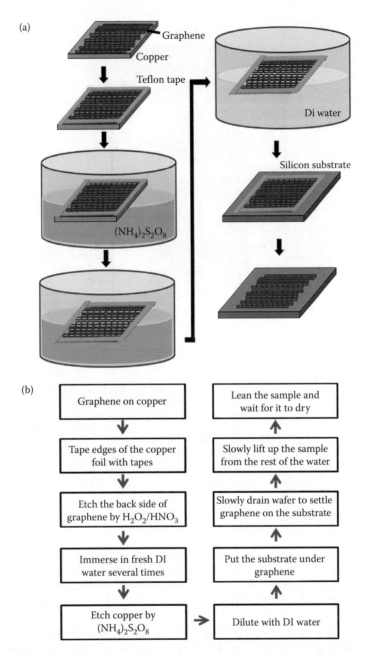

FIGURE 27.1 (a) Schematic diagrams, and (b) a flow chart representing a low-stress graphene transfer process.

and the graphene is placed on top of a destination substrate. During the entire process, graphene is not picked up by any stressful process, but instead remains floating on the surface of the liquids. The destination substrate is placed in the liquid to accept the graphene when the liquid is drained and the graphene is lowered toward the substrates. A variety of modified transferring processes allows the graphene to be transferred in different kinds of supporting substrates, including various kinds of grids. Graphene with both sides exposed can be modified before being placed on another substrate.

Transferred graphene was subjected to remote microwave plasma generated in hydrogen at 1 Torr gas pressure as shown in Figure 27.2a. The distance between graphene and the plasma ball

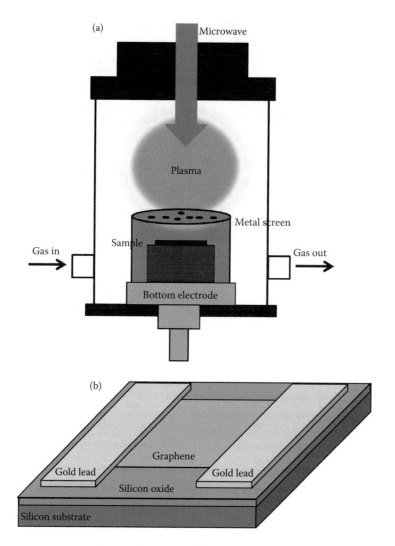

FIGURE 27.2 (a) Schematic diagram showing the remote hydrogenation plasma; and (b) Schematic diagram showing the geometry of graphene and electrodes for resistance measurements. (W. Chen et al., Low-stress transfer of graphene and its tuneable resistance by remote plasma treatments in hydrogen, *11th IEEE International Conference on Nanotechnology*, Portland, Oregon, August 15–18. © (2011) IEEE. With permission.)

was about 10 cm with a metal screen between the plasma and the graphene. After exposing the remote plasma species for 10 min, the sample was taken out to measure its resistance. The electrical resistance of graphene was measured by depositing gold contacts on two sides of a nearly square-shaped graphene sheet. The geometry is shown in Figure 27.2. An ohmmeter was used to measure the resistance between these two gold contacts as shown in Figure 27.2b at a different time of the hydrogenation process. All experiments were executed at room temperature. Graphene and hydrogenated graphene were characterized by resistance measurements, optical microscopy, scanning electron microscopy (SEM), and Raman spectroscopy.

27.3 RESULTS AND DISCUSSION

Figure 27.3a shows an optical photo of a single-layer graphene floating on DI water. There is a circle on the graphene. This graphene is successfully transferred from a copper foil to a silicon

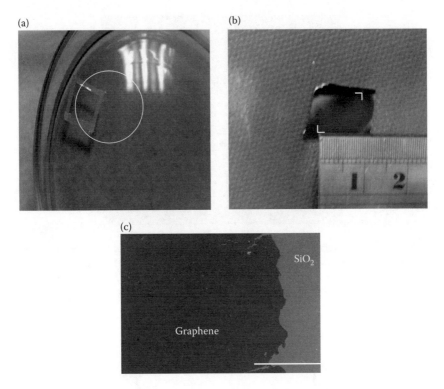

FIGURE 27.3 (a) Optical image showing graphene marked by a circle floating on DI water, (b) Optical photo of a transferred graphene on a destination substrate. (c) SEM image of a transferred graphene on an SiO$_2$/Si substrate. The scale bar is 300 μm.

oxide/silicon substrate shown in Figure 27.3b. The graphene floating on liquids was in a relaxed state with minimum applied stress during the whole transferring process. Neither the protective coatings nor the mechanical handling tools were in touch with the graphene. This novel process makes it very useful for the graphene to be safely transferred from one substrate to another. It also allows further modification of the graphene in the middle of the transferring process. Therefore, stacking of multiple layers of graphene and modified graphene can be implemented effectively for a variety of innovative applications. Figure 27.3c is an SEM image showing the graphene after transfer.

Figure 27.4 shows one example of the hydrogen-plasma-treated single-layer graphene. The optical micrograph shows the boundary between graphene and hydrogen-plasma-modified graphene. The different colors of graphene and hydrogen-plasma-modified graphene are due to the interference of these films of different atomic structure and thickness (graphene is about 0.34 nm thick, and when fully hydrogenated, graphane is about 2.62 nm thick). Part of the graphene was covered by an aluminum foil when the sample was treated by remote hydrogen plasma.

The resistance shown in Figure 27.5a was measured immediately after every hydrogen plasma treatment was completed for a fixed time period. At first, the resistance increased slowly from 0.90 kΩ. It increased to 200 kΩ abruptly at the last treatment. It can be speculated that when one hydrogen atom attacks one carbon atom to break the π bond, one unpaired electron will be released simultaneously. Owing to the π-bonding network breaking and the p-electron associated with the unhydrogenated carbon atoms being localized and unpaired, the energy of hydrogen bonding to the rest of the carbon atoms is lowered. As the amount of the broken π-bonding network increased, the trend is accelerated, and the resistance increases abruptly until carbon sites available for bonding to hydrogen reach saturation.

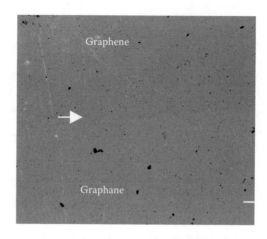

FIGURE 27.4 Optical micrograph showing the boundary between as-grown graphene and hydrogenated graphene as pointed by an arrow. The scale bar is 100 μm. (W. Chen et al., Low-stress transfer of graphene and its tuneable resistance by remote plasma treatments in hydrogen, *11th IEEE International Conference on Nanotechnology*, Portland, Oregon, August 15–18. © (2011) IEEE. With permission.)

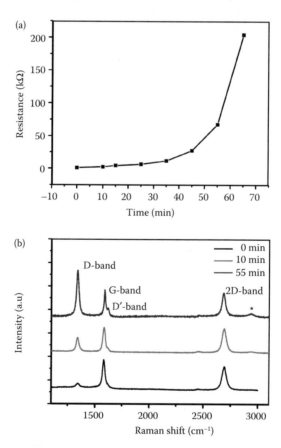

FIGURE 27.5 (a) Electrical resistance of a single-layer graphene as a function of exposure time to a remote hydrogen plasma. (b) The evolution of Raman spectra (633 nm laser excitation) with increasing exposure time to a remote microwave plasma in hydrogen. (0 min bottom; 55 min top.) (W. Chen et al., Low-stress transfer of graphene and its tuneable resistance by remote plasma treatments in hydrogen, *11th IEEE International Conference on Nanotechnology*, Portland, Oregon, August 15–18. © (2011) IEEE. With permission.)

To investigate the transformation of graphene to graphane, the evolution of Raman spectra at different stages of hydrogen plasma treatment was monitored. These spectra are characterized by the presence of three typical bands in graphene and graphane. They are the disorder-induced D band at around 1344 cm^{-1}, the radial breathing mode G band at around 1580 cm^{-1}, and the second order of zone boundary phonons 2D band at around 2680 cm^{-1}. As the plasma treatment time increased, the intensities of the two bands rose. A shoulder peak of the G band, called the D' band, originating from an intervalley double resonance Raman process, where the associated phonons have small wave vectors in the vicinity of Γ point, can be observed. Another featured band assigned to the combination of the D band and the G band can be found at around 2924 cm^{-1}, as pointed out by a star mark in Figure 27.5b. The appearance of the D' band and the D + G peak indicates that graphene is turning into graphane by hydrogen termination. An increase in the I_D/I_G ratio was also examined. These are all related to an increase in the amount of sp^3 bonding. Thus, from the above data, we can know that the increase in resistance is a result of the hydrogenation of graphene.

27.4 CONCLUSION

A novel process for transferring thermal CVD-grown graphene on copper foils to destination substrates for practical applications is reported. The transferred graphene is subjected to reactions with remote microwave plasma in hydrogen. The electrical resistance of modified graphene as a function of plasma treatment parameters is reported.

ACKNOWLEDGMENTS

We gratefully acknowledge the financial support by the Ministry of Education and the National Science Council in Taiwan via grants 99-2120-M-006-004 and 99-2911-I-006-504.

REFERENCES

1. X. Du, I. Skachko, A. Barker, and E. Y. Andrei, Approaching ballistic transport in suspended graphene, *Nature Nanotechnology* 3, 491–495, 2008.
2. J. H. Chen, C. Jang, S. Xiao, M. Ishigami, and M. S. Fuhrer, Intrinsic and extrinsic performance limits of graphene devices on SiO$_2$, *Nature Nanotechnology* 3, 206–209, 2008.
3. W. W. Cai, A. L. Moore, Y. Zhu, X. Li, S. Chen, L. Shi, and R. S. Ruoff, Thermal transport in suspended and supported monolayer graphene grown by chemical vapor deposition, *Nano Letters* 10, 1645–1651, 2010.
4. S. Chen, L. Brown, M. Levendorf, W. Cai, S.-Y. Ju, J. Edgeworth, X. Li et al., Oxidation resistance of graphene-coated Cu and Cu/Ni alloy, *ACS Nano* 5, 1321–1327, 2011.
5. F. Schedin, A. K. Geim, S. V. Morozov, E. W. Hill, P. Blake, M. I. Katsnelson, and K. S. Novoselov, Detection of individual gas molecules adsorbed on graphene, *Nature Materials* 6, 652–655, 2007.
6. X. Li, W. Cai, I. H. Jung, J. H. An, D. Yang, A. Velamakanni, R. Piner, L. Colombo, and R. S. Ruoff, Synthesis, characterization and properties of large-area graphene films, *ECS Transactions* 19, 41–52, 2009.
7. X. Li, W. Cai, L. Colombo, and R. S. Ruoff, Evolution of graphene growth on Ni and Cu by carbon isotope labelling, *Nano Letters* 9, 4268, 2009.
8. K. S. Kim, Y. Zhao, H. Jang, S. Y. Lee, J. M. Kim, K. S. Kim, J.-H. Ahn, P. Kim, J.-Y. Choi, and B. H. Hong, Large-scale pattern growth of graphene films for stretchable transparent electrodes, *Nature Nanotechnology* 457, 706–710, 2009.
9. X. Li, Y. Zhu, W. Cai, M. Borysiak, B. Han, D. Chen, R. D. Piner, L. Colombo, and R. S. Ruoff, Transfer of large-area graphene films for high-performance transparent conductive electrodes, *Nano Letters* 9, 4359–4363, 2009.
10. W. Regan, N. Alem, B. Aleman, B. Geng, C. Girit, L. Maserati, F. Wang, M. Crommie, and A. Zettl, A direct transfer of layer-area graphene, *Applied Physics Letters* 96, 113102, 2010.
11. X. Li, Y. Zhu, W. Cai, M. Borysiak, B. Han, D. Chen, R. D. Piner, L. Colombo, and R. S. Ruoff, Transfer of large-area graphene films for high-performance transparent conductive electrodes, *Nano Letters* 9(12), 4359–4363, 2009.

12. Geringer, D. Subramaniam, A. K. Michel, B. Szafranek, D. Schall, A. Georgi, T. Mashoff, D. Neumaier, M. Liebmann, and M. Morgenstern, Electrical transport and low-temperature scanning tunneling microscopy of microsoldered graphene, *Applied Physics Letters* 96, 082114, 2010.
13. X. Li, W. Cai, J. An, S. Kim, J. Nah, D. Yang, R. Piner et al., Large-area synthesis of high-quality and uniform graphene films on copper foils, *Science* 324, 1312, 2009.
14. D. C. Elias, R. R. Nair, T. M. G. Mohiuddin, S. V. Morozov, P. Blake, M. P. Halsall, A. C. Ferrari et al., Control of graphene's properties by reversible hydrogenation: Evidence for graphane, *Science* 323, 610, 2008.
15. A. Gupta, G. Chen, P. Joshi, S. Tadigadapa, and P. C. Eklund, Raman scattering from high-frequency phonons in supported n-graphene layer films, *Nano Letters* 6(12), 2667–2673, 2006.
16. M. E. Kompan and D. S. Krylov, Detecting graphene–graphane reconstruction in hydrogenated nanoporous carbon by Raman spectroscopy, *Technical Physics Letters* 36, 1140–1142, 2010.
17. J. O. Sofo, A. S. Chaudhari, and G. D. Barber, Graphane: A two-dimensional hydrocarbon, *Physical Review B* 75, 153401, 2007.
18. J. Zhou, Q. Wang, Q. Sun, X. S. Chen, Y. Kawazoe, and P. Jena, Ferromagnetism in semihydrogenated graphene sheet, *Nano Letters* 9, 3867–3870, 2009.
19. Y. H. Lu and Y. P. Feng, Band-gap engineering with hybrid graphane–graphene nanoribbons, *The Journal of Physical Chemistry* C113, 20841–20844, 2009.
20. W. Chen, C.-H. Tu, K.-C. Liang, C.-Y. Liu, C.-P. Liu, and Y. Tzeng, Low-stress transfer of graphene and its tuneable resistance by remote plasma treatments in hydrogen, *11th IEEE International Conference on Nanotechnology*, Portland, Oregon, August 15–18, 2011.

28 High-Yield Dielectrophoretic Deposition and Ion Sensitivity of Graphene

Pengfei Li, Nan Lei, Jie Xu, and Wei Xue

CONTENTS

28.1 INTRODUCTION

Graphene research has attracted enormous attention since the discovery of the material in 2004 [1]. Graphene can be viewed as a two-dimensional (2D) honeycomb lattice that is constructed with a single layer of carbon atoms. It not only provides a unique structure for fundamental research but is also an ideal material for future nanoelectronics and sensor applications [2,3]. Graphene has demonstrated fascinating chemical, mechanical, and electrical properties. For example, when exposed to different gases, graphene can be chemically doped to either p-type or n-type [4]. A nanoindentation measurement of a freestanding monolayer graphene membrane using atomic force microscopy (AFM) shows that the material has an ultrahigh Young's modulus of 1.0 TPa, which makes graphene one of the strongest materials ever measured [5]. Depending on the number of layers in the sheet, the electrical properties of graphene alter accordingly, changing it from a semiconductor to a metal, or vice versa [6].

However, owing to the single-atom thickness, the handling of graphene has become a major bottleneck in this field, especially in the development of practical devices. The most common method to generate graphene is mechanical exfoliation of bulk graphite [1,7]. Single- or double-atomic-layer graphene can be obtained after repeatedly peeling layers off the highly oriented pyrolytic graphite using scotch tapes. However, the process is time-consuming, unscalable, and has low yield and low repeatability. Consequently, this technique has been used primarily to create individual devices for the study of novel physics of graphene; it is not suitable for large-scale device fabrication. Alternatively, graphene can be directly grown on substrates with chemical vapor deposition (CVD) and epitaxy methods [8,9]. However, these methods have a limited choice of substrates, which are not entirely compatible with traditional semiconductors and insulators in electronic device fabrication. Recently, chemical methods have been developed to produce stable, homogeneous graphene solutions in large quantities [10]. Depositing graphene from the solutions provides an effective approach to integrate graphene into device fabrication.

Owing to the small dimensions of graphene, the fabrication of graphene-based devices usually relies on expensive and time-consuming electron-beam lithography systems. The series-style fabrication and time-consuming processes used in the electron-beam lithography are not suitable for mass production of such devices. The CVD- and epitaxy-grown graphene can be processed using the traditional lithography-based techniques. However, further modifications are needed to enhance the compatibility of the related materials to the standard substrates used in the semiconductor industry. By contrast, dielectrophoresis, a simple electrical approach using alternating current (AC) signals can not only deposit nanoscale materials inexpensively, but also it can achieve large-scale assembly of materials over large areas [11]. This method has been successfully used in the controlled assembly of nanowires, carbon nanotubes, and graphene [11–13]. Furthermore, it is a solution-based method that can fully utilize the latest advances in the production of graphene solutions.

Here, we demonstrate the controlled assembly of graphene sheets on a substrate with high precision and high yield. Prefabricated electrodes are used to generate strong electric fields in the solution. The dispersed graphene sheets are attracted by the fields and land on the substrate to bridge the electrodes. A thermal annealing step is carried out to enhance the adhesion between the graphene and the substrate. Electrical characterization of the deposited graphene is conducted by using it as a semiconducting material in a liquid-gated field-effect transistor (FET). The assembled graphene sheets demonstrate high sensitivity toward pH values in an aqueous environment with excellent reversibility and repeatability. The ion sensing mechanism of the graphene-based device is described and discussed in the chapter. Our study demonstrates that the dielectrophoresis process is an effective approach to fabricate graphene-based electronics and sensors for practical applications.

28.2 DEVICE DESIGN AND FABRICATION

The electrodes used for graphene assembly are fabricated on a 4-inch silicon wafer with a 200-nm-thick surface thermal oxide layer. They are fabricated with traditional optical lithography and metal wet etching to maintain low costs and scalability. The detailed fabrication process can be found in our previous reports [14–17]. Figure 28.1a illustrates an electrode design that contains arrays of tips on both sides. Each electrode contains 21 comb fingers on the edge. The designed width and length of each finger are $w = 3$ μm and $l = 10$ μm, respectively. The gap between the two opposite fingers is $g = 3$ μm. The comb-shaped electrodes enable strong and directional electric fields between the tips. Figure 28.1b shows a scanning electron microscopy (SEM) image of a pair of fabricated electrodes. The scale bar shown in the image is 100 μm.

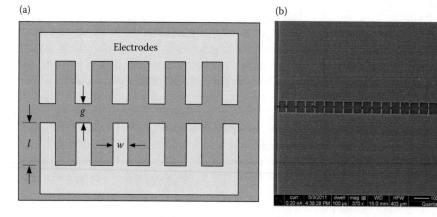

FIGURE 28.1 (a) Schematic illustration of an electrode design. (b) SEM image of a pair of fabricated electrodes.

The graphene solution is obtained from NanoIntegris (Skokie, IL, USA) with the single- to triple-atomic-layer graphene making up approximately 56% of the content. The sample has a concentration of 0.05 mg/mL and it is utilized as is without further chemical or physical modifications. During the dielectrophoresis process, a small drop of the graphene solution is placed on the substrate to cover the electrodes. The process takes 30–60 s with an instant voltage drop of 1–2 V observed from the oscilloscope during the graphene-capturing event. The dielectrophoretic assembly of graphene is based on the spatial redistribution of charges inside the material. The polarized graphene is subject to a net force and can be moved to follow the electric field direction. After removing the dispersion with a pipette, the deposited graphene is bonded to the substrate based on van der Waals forces. The use of electrodes in the graphene deposition process produces ready-to-use devices, and no further processing steps for electrical contacts are needed.

28.3 DIELECTROPHORESIS DEPOSITION OF GRAPHENE

The schematic diagram of the dielectrophoresis process is shown in Figure 28.2. A function generator is used as the AC source and an oscilloscope is used to monitor the voltage change during the graphene deposition process. The frequency and the peak-to-peak voltage of the AC signal are set as 5 MHz and 10 V, respectively. These parameters are selected based on our previous experimental studies on carbon nanotubes and analytical results from other groups [14–16]. Because the capturing event of graphene on the device can cause a slight voltage drop across the electrodes, the voltage change displayed on the oscilloscope indicates the landing moment of the material.

The dielectrophoretically assembled graphene sheets are inspected with SEM. Figure 28.3 shows the assembled graphene sheets on the electrodes. Multiple sheets are deposited between the electrodes to form conductive channels. Such graphene sheets can only be observed in between the electrode tips where the electric field has the highest magnitude during the dielectrophoresis process. This proves that dielectrophoresis is an "active" approach to assemble graphene and is suitable for device integration. On this particular device, we find that 13 out of 21 pairs of fingers are connected by graphene (marked as ellipses) with no graphene observed on the remaining 8 pairs (marked as X). This is because the graphene deposited in the early stage decreases the potential across the electrodes, changing the electric field distribution in the solution. Consequently, the dielectrophoretic forces on the dispersed graphene are weakened to a level where they are not able to drag the graphene sheets to the electrodes for further deposition. Nevertheless, the number of deposited

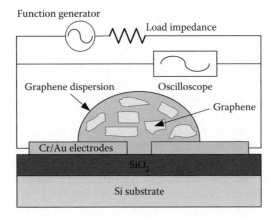

FIGURE 28.2   The experimental system for the dielectrophoretic deposition of graphene. (P. Li et al., High-yield dielectrophoretic deposition and ion sensitivity of graphene, in *2011 11th IEEE International Conference on Nanotechnology*, Portland, OR, pp. 1327–1330. © (2011) IEEE. With permission.)

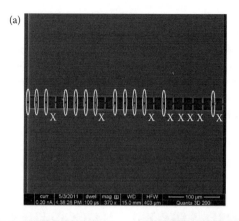

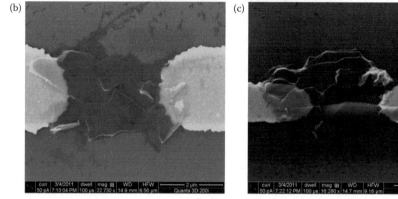

FIGURE 28.3 SEM images of the dielectrophoretically deposited graphene sheets. (a) Top view of a device with 21 electrode fingers. (b) Top view and (c) tilted view of the deposited graphene. (P. Li et al., High-yield dielectrophoretic deposition and ion sensitivity of graphene, in *2011 11th IEEE International Conference on Nanotechnology*, Portland, OR, pp. 1327–1330. © (2011) IEEE. With permission.)

graphene sheets is adequate for the graphene-based device to be functional. We believe that the yield of the graphene deposition on the electrode fingers can be improved by further optimization of the electrode design.

The deposited graphene sheets are then treated with a thermal annealing step at 200°C for 10 min on a hot plate. This step serves two purposes. First, it can reduce the contact resistance between the graphene and the electrode, resulting in an improved contact [11]. The electrical characteristics of the graphene sheets before and after the annealing step are illustrated in Figure 28.4. The resistance of this particular device decreases by 41%, from 345.13 to 202.17 Ω. Although not all the devices demonstrate the same trend of decreased resistance, previous investigations by other groups have proven that the effective electrical contact can be improved [18,19]. Second, we discover that the thermal annealing step can dramatically increase the adhesion between the graphene and the substrate. In our pH sensing experiments that involve intensive solution placing and removing steps, the thermally annealed devices demonstrate much higher repeatability than the nonannealed counterparts. Based on the fact that the nonannealed devices often show sudden, irreversible increase in resistance, we believe that the loosely bonded graphene sheets can be washed away in the solution removing steps. Although similar phenomena have occasionally been observed in thermally annealed devices, they are rare and the corresponding resistance decrease shows much smaller values. Therefore, we believe that the discrepancy between the two groups of devices originates from the bonding strength of the graphene and the substrate.

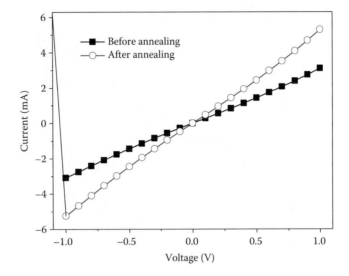

FIGURE 28.4   Current–voltage (*I–V*) characteristics of deposited graphene before and after thermal annealing at 200°C for 10 min. (P. Li et al., High-yield dielectrophoretic deposition and ion sensitivity of graphene, in *2011 11th IEEE International Conference on Nanotechnology*, Portland, OR, pp. 1327–1330. © (2011) IEEE. With permission.)

28.4   ELECTRICAL CHARACTERISTICS OF GRAPHENE

The deposited graphene is electrically characterized using a liquid-gated FET configuration. The measurement system setup for the graphene-based transistor is shown in Figure 28.5. The two electrodes, previously used for the nanomaterial deposition, now serve as source and drain terminals. A droplet of distilled water is place on the substrate to cover the graphene. It is used as the liquid gating medium, as suggested by previous research reports [20,21]. A standard Ag/AgCl reference probe, serving as the gate electrode, is immersed in the water droplet. The suspended end is approximately 2 mm away from the graphene. A semiconductor device analyzer is used to apply the drain (V_d) and gate (V_g) voltage, and record the induced drain current (I_d) at the same time.

The output characteristics (I_d–V_d) of the graphene-based device are shown in Figure 28.6a. The drain voltage V_d is swept from −1 to 1 V with a 0.01 V step and the gate voltage is swept from −0.4 to 0.4 V with a 0.2 V step. The relationship between the drain current and drain voltage is highly linear in the measured range of −1 V to 1 V. This range is selected to protect the device from

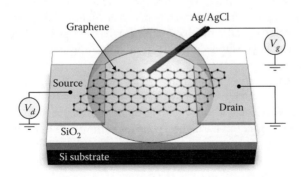

FIGURE 28.5   Schematic of a graphene-based field-effect transistor with an Ag/AgCl reference probe as the gate electrode.

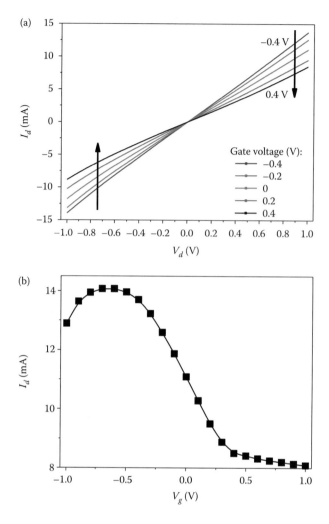

FIGURE 28.6 (a) Output characteristics of a graphene device with the reference gate voltage ranging from −0.4 to 0.4 V in 0.2 V steps. (b) Transfer characteristics of the same device with the gate voltage ranging from −1 to 1 V in 0.1 V steps. The drain current is fixed at −1 V.

burning out by high current. Although drain current saturation is not observed, the field effect is apparent from the figure. Furthermore, we notice that when the gate voltage is shifted from a positive region to a negative region, the drain current becomes higher. This phenomenon indicates that graphene is a p-type material and a majority of its carriers are positively charged holes. During operation, the negative voltage can accumulate more holes into the channel area, leading to the higher drain current.

The transfer characteristics (I_d–V_g) of the same device are shown in Figure 28.6b. The drain voltage is fixed at −1 V and the gate voltage is swept from −1 to 1 V with a 0.1 V step. The smallest current obtained in this curve is 8.08 mA at $V_g = 1$ V and the highest current of 14.07 mA is observed at a gate voltage of −0.7 V. On the right side of the highest point (or V_g from −0.7 to 1.0 V), the current decreases as the gate voltage increases. This means that fewer holes are involved in the current conductance at a higher voltage, proving the p-type characteristics of the deposited graphene. On the other side of this point, as V_g decreases from −0.7 to −1 V, the current is decreased. This can be explained by the depletion of the minority carriers, electrons, from the graphene channel by the decreased negative gate voltage. In addition, when V_g is in the high, positive range of 0.5–1 V, the current plot is relatively flat with a saturation current of 8.1–8.3 mA. This indicates that graphene is a highly conductive material, which cannot be completely switched off using the current device configuration.

28.5 ION SENSITIVITY OF GRAPHENE

To investigate its potential in sensing applications, the deposited graphene is characterized for its response toward pH values in solutions. The electrodes initially used for dielectrophoresis are now used as electrical contacts. Figure 28.7 shows the system layout for the ion sensitivity measurement. The electrodes initially used for dielectrophoresis are now used as electrical contacts. A bias voltage V_0 is applied across the electrode pair and the current I_0 is monitored with a data acquisition system. Solutions with various pH values, ranging from 5 to 9 (standard buffer solutions from Fisher Scientific, Pittsburgh, PA, USA), are sequentially dropped on the device to cover the graphene. The assembled graphene is used as a two-terminal pH-sensitive resistor and its real-time resistance R in different solutions can be measured.

The graphene device shows a stable response toward pH values in liquid. Figure 28.8a shows the recorded data in a resistance–time (R–t) plot from the real-time data acquisition. The resistance of graphene is highly sensitive to pH values with the highest resistance (~475 Ω) at pH 5 and the lowest resistance (~405 Ω) at pH 9. A near-equal amount of decrement in the resistance of graphene can be observed when the pH value is changed from 5 to 9, and a similar amount of increment can be observed with increasing pH values. In addition, the graphene-based pH sensors show high repeatability among devices. The measurement of three different devices demonstrates that all sensors show similar electrical and ion-sensitive characteristics, as illustrated in Figure 28.8b. The normalized resistance, defined as

$$\Delta R/R_r = (R - R_{min})/(R_{max} - R_{min}), \tag{28.1}$$

is used to evaluate the sensor performance, where ΔR, R_r, R_{max}, and R_{min} are the real-time sensor resistances relative to its lowest value, the range of resistance in the entire measurement, the highest resistance, and the lowest resistance, respectively. Our measurement shows that the normalized resistance and the pH value have a relatively linear relationship, which can be expressed as

$$\Delta R/R_r = -0.24 \times pH + 2.2. \tag{28.2}$$

As a result, the assembled graphene demonstrates reversible and repeatable pH sensitivity, enabling its potential in chemical- and bio-sensing applications.

The pH sensing mechanism of graphene can be explained by the electrolyte-induced capacitive gating principles that involve the adsorption of ions and the creation of an electrical double layer at the liquid–solid interface [22,23]. The hydroxonium ions (H_3O^+) and hydroxyl ions (OH^-) from the

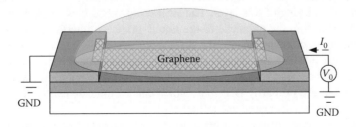

FIGURE 28.7 Testing scheme of deposited graphene in an aqueous environment. (P. Li et al., High-yield dielectrophoretic deposition and ion sensitivity of graphene, in *2011 11th IEEE International Conference on Nanotechnology*, Portland, OR, pp. 1327–1330. © (2011) IEEE. With permission.)

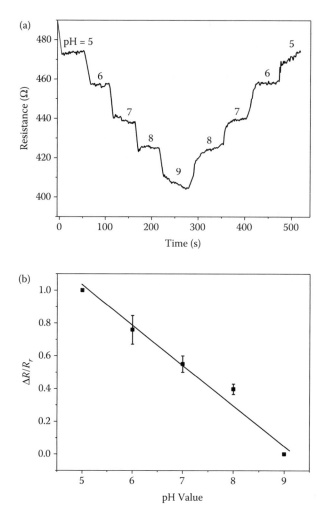

FIGURE 28.8 (a) Recorded real-time pH response of the assembled graphene. (b) Relationship between the normalized resistance and pH values. (P. Li et al., High-yield dielectrophoretic deposition and ion sensitivity of graphene, in *2011 11th IEEE International Conference on Nanotechnology*, Portland, OR, pp. 1327–1330. © (2011) IEEE. With permission.)

pH buffer solution can be adsorbed on the graphene surface and create an electric double layer at the liquid–graphene interface. It is a nonfaradaic (or capacitive) process, which means the ions cannot transmit through the electric double layer directly. Instead, the energy transfer is achieved through the charging and discharging of this layer. The surface-adsorbed ions can attract mobile carriers with opposite charges from the graphene to the liquid–graphene interface, changing the conductance of the material. In our experiments, the surface-adsorbed ions in acidic solutions are primarily H_3O^+ ions, which can partially deplete the holes—the majority carriers—from the graphene. As a result, the resistance has a relatively high value in the pH 5 solution. In alkaline solutions, the surface-adsorbed ions are primarily OH^- ions, which can accumulate extra holes in the graphene. Therefore, the resistance is decreased accordingly.

Our research investigates the inherent pH sensitivity of graphene because the tested devices are configured as two-terminal resistors. The devices can also be configured as FET sensors with an additional terminal added to the device as a gate electrode. With the ability to control the amount of charge carriers in graphene through accumulation or depletion, the sensors may show enhanced performance. Two recent reports investigated the pH sensitivity of graphene using a liquid-gated transistor configuration [24,25]. Although their research was based on CVD- and epitaxy-grown graphene samples, the measurement technology can be easily extended to dielectrophoretically assembled graphene. Their results demonstrated enhanced pH sensitivity with nonzero gate voltages. One interesting effect presented in both reports is that the resistance–pH (R–pH) relationship switches its trend at a gate voltage of approximately −0.1 V. We expect that a similar phenomenon can be observed in the devices using dielectrophoretically assembled graphene. Additional experiments will be conducted to investigate the performance of the graphene transistor-based pH sensors.

28.6 CONCLUSIONS

In conclusion, we report a high-yield method to actively deposit graphene on prefabricated electrodes using dielectrophoresis. The electrode design enables highly concentrated electric fields to apply strong dielectrophoretic forces on graphene sheets in dispersion. The graphene sheets were precisely deposited on a substrate at the desired locations. Thermal annealing was used to improve the contact between the graphene and the electrodes as well as to enhance the adhesion between the graphene and the substrate. The use of electrodes in the deposition process enables the direct development of ready-to-use graphene devices, and no further processing steps for electrical contacts are needed. The deposited graphene demonstrated p-type characteristics when measured in a liquid-gated FET. The deposited graphene devices were used as sensors and they demonstrated satisfactory performance toward pH sensing. The electrical double layer at the liquid–graphene interface created by the capacitive gating effect was considered as the primary sensing mechanism. Our investigation shows that dielectrophoresis is a low-cost, simple, and fast method to deposit graphene and has great potential for device integration. We believe that the graphene-based sensors fabricated with this approach can be developed and applied to a wide range of areas in the near future.

REFERENCES

1. K. S. Novoselov, A. K. Geim, S. V. Morozov, D. Jiang, Y. Zhang, S. V. Dubonos, G. IV, and A. A. Firsov, Electric field effect in atomically thin carbon films, *Science,* 306, 666–669, 2004.
2. P. Avouris, Graphene: Electronic and photonic properties and devices, *Nano Lett.,* 10, 4285–4294, 2010.
3. A. K. Geim and K. S. Novoselov, The rise of graphene, *Nat. Mater.,* 6, 183–191, 2007.
4. X. Wang, X. Li, L. Zhang, Y. Yoon, P. K. Weber, H. Wang, J. Guo, and H. Dai, N-doping of graphene through electrothermal reactions with ammonia, *Science,* 324, 768–771, 2009.
5. C. Lee, X. Wei, J. W. Kysar, and J. Hone, Measurement of the elastic properties and intrinsic strength of monolayer graphene, *Science,* 321, 385–388, 2008.
6. A. H. Castro Neto, F. Guinea, N. M. R. Peres, K. S. Novoselov, and A. K. Geim, The electronic properties of graphene, *Rev. Mod. Phys.,* 81, 109–162, 2009.
7. N. Lei, P. Li, W. Xue, and J. Xu, Simple graphene chemiresistors as pH sensors: Fabrication and characterization, *Meas. Sci. Technol.,* 22, 107002, 2011.
8. X. Li, W. Cai, J. An, S. Kim, J. Nah, D. Yang, R. Piner et al., Large-area synthesis of high-quality and uniform graphene films on copper foils, *Science,* 324, 1312–1314, 2009.
9. P. W. Sutter, J.-I. Flege, and E. A. Sutter, Epitaxial graphene on ruthenium, *Nat. Mater.,* 7, 406–411, 2008.
10. S. Park and R. S. Ruoff, Chemical methods for the production of graphenes, *Nat. Nanotechnol.,* 4, 217–224, 2009.
11. A. Vijayaraghavan, S. Blatt, D. Weissenberger, M. Oron-Carl, F. Hennrich, D. Gerthsen, H. Hahn, and R. Krupke, Ultra-large-scale directed assembly of single-walled carbon nanotube devices, *Nano Lett.,* 7, 1556–1560, 2007.
12. A. Vijayaraghavan, C. Sciascia, S. Dehm, A. Lombardo, A. Bonetti, A. C. Ferrari, and R. Krupke, Dielectrophoretic assembly of high-density arrays of individual graphene devices for rapid screening, *ACS Nano,* 3, 1729–1734, 2009.
13. B. R. Burg, J. Schneider, S. Maurer, N. C. Schirmer, and D. Poulikakos, Dielectrophoretic integration of single- and few-layer graphenes, *J. Appl. Phys.,* 107, 034302, 2010.

14. P. Li and W. Xue, Selective deposition and alignment of single-walled carbon nanotubes assisted by dielectrophoresis: From thin films to individual nanotubes, *Nanoscale Res. Lett.,* 5, 1072–1078, 2010.

15. P. Li, C. M. Martin, K. K. Yeung, and W. Xue, Dielectrophoresis aligned single-walled carbon nanotubes as pH sensors, *Biosensors,* 1, 23–35, 2011.

16. P. Li, N. Lei, J. Xu, and W. Xue, High-yield fabrication of graphene chemiresistors with dielectrophoresis, *IEEE Trans. Nanotechnol.,* 11, 751–759, 2012.

17. P. Li, N. Lei, J. Xu, and W. Xue, High-yield dielectrophoretic deposition and ion sensitivity of graphene, in *2011 11th IEEE International Conference on Nanotechnology,* Portland, OR, 2011, pp. 1327–1330.

18. D. Joung, A. Chunder, L. Zhai, and S. I. Khondaker, High yield fabrication of chemically reduced graphene oxide field effect transistors by dielectrophoresis, *Nanotechnology,* 21, 165202, 2010.

19. B. R. Burg, F. Lutolf, J. Schneider, N. C. Schirmer, T. Schwamb, and D. Poulikakos, High-yield dielectrophoretic assembly of two-dimensional graphene nanostructures, *Appl. Phys. Lett.,* 94, 053110, 2009.

20. S. Rosenblatt, Y. Yaish, J. Park, J. Gore, V. Sazonova, and P. L. McEuen, High performance electrolyte gated carbon nanotube transistors, *Nano Lett.,* 2, 869–872, 2002.

21. D. W. Kim, G. S. Choe, S. M. Seo, J. H. Cheon, H. Kim, J. W. Ko, I. Y. Chung, and Y. J. Park, Self-gating effects in carbon nanotube network based liquid gate field effect transistors, *Appl. Phys. Lett.,* 93, 243115, 2008.

22. A. B. Artyukhin, M. Stadermann, R. W. Friddle, P. Stroeve, O. Bakajin, and A. Noy, Controlled electrostatic gating of carbon nanotube FET devices, *Nano Lett.,* 6, 2080–2085, 2006.

23. J. H. Back and M. Shim, pH-dependent electron-transport properties of carbon nanotubes, *J. Chem. Phys. B,* 110, 23736–23741, 2006.

24. P. K. Ang, W. Chen, A. T. S. Wee, and K. P. Loh, Solution-gated epitaxial graphene as pH sensor, *J. Am. Chem. Soc.,* 130, 14392–14393, 2008.

25. I. Heller, S. Chatoor, J. Männik, M. A. G. Zevenbergen, C. Dekker, and S. G. Lemay, Influence of electrolyte composition on liquid-gated carbon nanotube and graphene transistors, *J. Am. Chem. Soc.,* 132, 17149–17156, 2010.

29 Multilayer Graphene Grid and Nanowire Fabrication and Printing

Masudur Rahman and Michael L. Norton

CONTENTS

29.1 INTRODUCTION

Molecular sensor development requires the fabrication of structures with nanometer precision. Recently, graphene or few-layer graphene (FLG) has been proposed as a material for making advanced electronic devices. The transport properties of graphene, a single-atom-thick layer of graphite (~0.35 nm) [1], are influenced by atomic-scale defects, and, more importantly from a sensor perspective, by adsorbates [2,3] and the local electronic environment. Highly oriented pyrolytic graphite (HOPG) is one of the best precursors for generating high-quality, crystalline graphene [1], which fortuitously is also an ideal substrate material for high-resolution atomic force microscope (AFM) studies [4]. Several approaches have been used to produce graphene for large-area electronics, including epitaxial growth [5–8], transfer-printing [9,10], electrostatic deposition [1,11], and solution-based deposition [12,13]. At the same time, efforts have been made to tailor graphene sheets into nanoscale features [9,14–17]. Keun et al. developed chemical vapor deposition (CVD) to grow a graphene layer on nickel films and transferred them to polydimethylsiloxane (PDMS) using an etching method [18,19]. However, the incompatibility between PDMS and the mechanical properties of graphene usually causes breaks during the fabrication process, especially during the etching step [20]. For large-scale production, CVD is a promising technique; however, the quality of the graphene layers (roughness) requires improvement. Novoselov et al. described a very simple sticky tape method [21]. First, press the adhesive tape onto a sample of graphite and pull. Then, repeatedly stick the carbon-covered tape against itself and peel away. Thereby, the first carbon flake breaks up further into thin, hundred-micron-wide fragments. Then, press this carbon-coated tape onto an Si surface and carefully remove the tape. Some single- or few-layer graphene will adhere to the Si substrate. However, one cannot mass-produce graphene with the sticky tape method. Liu et al. used a chemically modified silicon wafer to covalently attach graphene [22]. For graphene-based electronics, fabrication on chemically modified silicon will be difficult to control. To prepare a few hundred microns long graphene wires and to avoid the wet etching or chemical step, in this chapter, the authors demonstrate a different process for patterning HOPG, producing ordered features at the micron and ~100 nanometer scales. Two different methods are used to transfer the patterned graphene onto substrates for characterization, a thermal tape (T-tape) method to transfer onto glass and direct transfer onto a PDMS substrate, both performed without applying any electrostatic force [11].

29.2   PATTERN GENERATION ON HOPG

29.2.1   PREPARATION OF A MICRON-SCALE PATTERN

Ni transmission electron microscope (TEM) grids (SPI Inc., Mesh 2000 lines/inch, pitch 12.5 μm, bar width 5 μm, hole width 7.5 μm) were used as physical masks for the oxygen plasma lithography [21]. The TEM grids were placed on the HOPG (SPI Inc., ZYA grade) surface. Next, an O_2 plasma was used to etch the graphite through the apertures in the TEM grid. A modified Harrick plasma etcher was used to produce the O_2-based plasma employed for the 40 min periods with a pressure of 100 mTorr O_2.

The scanning electron microscope (SEM) (JEOL 5310LV) imaging in the inset of Figure 29.1a (right upper corner) shows that the oxygen plasma lithography successfully replicates the TEM grid in the HOPG. The higher-resolution SEM image (Figure 29.1b) indicates that the grid pattern was uniformly etched. To determine the etching rate, an AFM topography analysis was performed for a 52×52 μm area (Figure 29.1c). The line profile analysis confirmed an average etch rate of 1.3 nm/min (Figure 29.1d). A Nano-R (Pacific Nanotechnology) AFM microscope was used to collect images of samples in noncontact mode using TM300-A (SensaProbes) AFM probes.

29.2.2   PREPARATION OF A NANOSCALE PATTERN

To produce patterns with finer features, we used phase-shift lithography [23], reducing the critical dimension in the product from 5 micron bars to ~100 nm wide (full width half maximum (FWHM)) lines. Freshly cleaved HOPG substrates were coated with 30% S1813 (MicroChem Corp.) photoresist (diluted with Thinner P from MicroChem Corp.), spun at 800 rpm for 10 s and 3000 rpm for 30 s, followed by a soft bake of 1 min at 116°C, exposed to UV for 30 s, and developed with 351 developer (Rohm and Haas) for 40 s. A PDMS mask was used for the phase mask [23]. After development, the features were etched into the resist patterned HOPG substrate using an O_2-based plasma for 10 min at an etching rate of 1.3 nm/min (see above). Finally, the photoresist was removed by soaking/rinsing the HOPG substrate in acetone for 10 min.

Figure 29.2a presents the multilayer graphene (MLG) line pattern generated on the HOPG substrate after 10 min oxygen plasma etching. The topography line scan indicates $Z = 10.2$ nm, which is equivalent to ~29 graphene layers and ~100 nm width. The etching rate is consistent with the oxygen plasma etch rate indicated in Figure 29.1d. The 3D image of the high-resolution line pattern on HOPG is shown in Figure 29.2b. It indicates that the lines were continuous and were not interrupted by the graphene layers.

29.3   PRINTING OF THE PATTERNED GRAPHENE

29.3.1   THERMAL TAPE METHOD TO PRINT ONTO GLASS

T-tape from Nitto Denko Inc. was used to transfer the multilayer grid patterned graphene onto glass coverslips. First, T-tape was used to peel off thick MLG flakes, which contained the TEM grid pattern. Figure 29.3 shows a schematic illustration of this printing process. T-tape removes flakes that are thicker than the etched, patterned depth. Therefore, extra graphene layers were

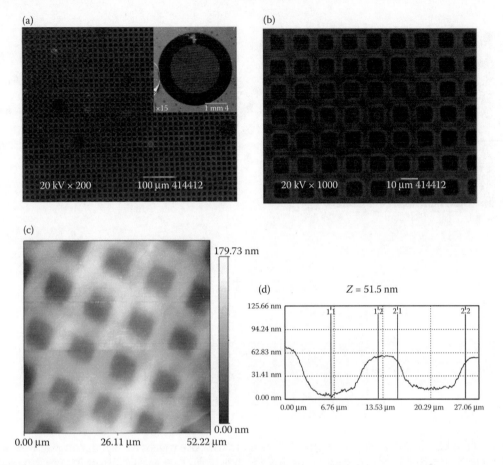

FIGURE 29.1 (a) SEM image of TEM grid pattern on HOPG produced using oxygen plasma lithography; inset (right upper corner) shows the replica of TEM grid on the HOPG substrate. (b) Higher-resolution SEM image of TEM grid pattern. ((a,b) Adapted from M. Rahman and M. Norton. Widefield optical and AFM analysis of few layer graphene nanowires functionalized with DNA, Extended abstract of a paper presented at Microscopy and Microanalysis, Nashville, Tennessee, August 7–11, 2011.) (c) AFM image of features produced. (d) Line profile analysis indicating $Z = 51.5$ nm/40 min etching rate. ((c,d) M. Rahman and M.L. Norton, *11th IEEE Conference on Nanotechnology (IEEE-NANO)*, Portland, OR, August 15–18, pp. 592–595. © (2011) IEEE. With permission.)

peeled off using scotch tape until the grid pattern was observable using optical microscopy. The T-tape with the MLG was placed side down on a plasma cleaned glass coverslip and thermal release of the MLG was accomplished by removing the tape after incubation at 120°C for 5 min.

Figure 29.3 shows the schematic illustration of the process for transferring the patterned graphene onto glass slides using T-tape. During release, the T-tape also leaves some residue on the top of the patterned MLG. To remove this residue, we washed the transferred MLG with acetone for 5 min with sonication and finally rinsed with isopropanol and used N_2 to blow the sample dry. Figure 29.4a shows a light microscopy (LM) image of the printed, patterned MLG on a glass coverslip. Darker regions mean that extra graphene layers, still remain. Although if we peel off more graphene layers using scotch tape, we could remove these extra graphene layers, this will also increase the loss of patterned graphene. AFM imaging was performed at the same printed region (Figure 29.4b). The arrow indicates a defect located in both the LM and AFM images. Line profile analysis (Figure 29.4c) confirmed that ~140 layers of graphene were printed onto the glass. The AFM result shows that some roughness remains and we believe that the automated process

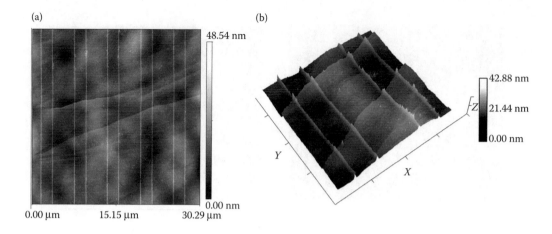

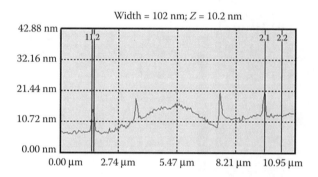

FIGURE 29.2 (a) AFM image of graphene ~100 nm FWHM line pattern on HOPG. (Adapted from M. Rahman and M. Norton. Widefield optical and AFM analysis of few layer graphene nanowires functionalized with DNA, Extended abstract of a paper presented at Microscopy and Microanalysis, Nashville, Tennessee, August 7–11, 2011.) (b) 3D image of high-resolution line pattern on HOPG. (M. Rahman and M.L. Norton, *11th IEEE Conference on Nanotechnology (IEEE-NANO)*, Portland, OR, August 15–18, pp. 592–595. © (2011) IEEE. With permission.)

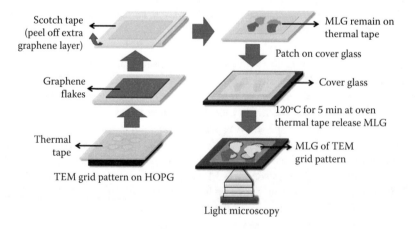

FIGURE 29.3 Schematic illustration of the process for printing the patterned graphene onto glass slides using thermal tape. (M. Rahman and M.L. Norton, *11th IEEE Conference on Nanotechnology (IEEE-NANO)*, Portland, OR, August 15–18, pp. 592–595. © (2011) IEEE. With permission.)

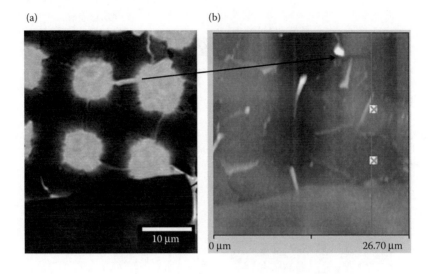

(a) (b)

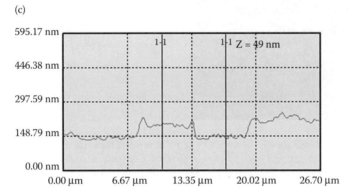

(c)

FIGURE 29.4 (a) Transmitted light microcopy image of MLG pattern on glass slide. (b) AFM image of the same region of the printed patterned MLG graphene on the glass slide. (c) Line profile indicating $Z = 49$ nm. (M. Rahman and M.L. Norton, *11th IEEE Conference on Nanotechnology (IEEE-NANO)*, Portland, OR, August 15–18, pp. 592–595. © (2011) IEEE. With permission.)

could not only decrease the roughness but also would provide more controllability of the number of graphene layers printed.

29.3.2 DIRECT PRINTING OF GRAPHENE NANOWIRES ONTO PDMS

Freshly prepared flat PDMS stamps (1 cm × 1 cm) were washed with toluene, then with ethanol, then oven dried for 2 h at 60°C, and finally plasma cleaned for 2 min before printing. The plasma cleaned PDMS stamps were then placed on the patterned HOPG surface. PDMS placement was started at one edge of the graphite sample and proceeded to the other edge with the application of gentle pressure using curved-style forceps. At the end of this process, the PDMS was securely attached to the HOPG. We then gently removed the printed FLG adhering to the PDMS from the HOPG surface, using an inverse peeling process. For optical microscopy analysis/imaging, the PDMS/FLG assembly was placed face down on a plasma cleaned glass coverslip.

The ~100 nm wide (FWHM) graphite nanowire was printed onto PDMS using soft lithography as illustrated in Figure 29.5. Figure 29.6a presents the LM image of the fourth print of FLG lines (or nanowires) onto PDMS. The graphene wire profile analysis result (Figure 29.6b) indicates that ~10

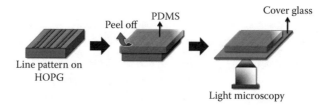

FIGURE 29.5 Schematic illustration of the method for printing graphene nanowires onto PDMS. (M. Rahman and M.L. Norton, *11th IEEE Conference on Nanotechnology (IEEE-NANO)*, Portland, OR, August 15–18, pp. 592–595. © (2011) IEEE. With permission.)

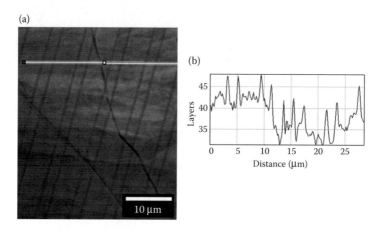

FIGURE 29.6 (a) Light microcopy image of FLG pattern on PDMS. (b) Plot profile of the printed graphene nanowires. (M. Rahman and M.L. Norton, *11th IEEE Conference on Nanotechnology (IEEE-NANO)*, Portland, OR, August 15–18, pp. 592–595. © (2011) IEEE. With permission.)

graphene layers were printed onto the PDMS. The profile was analyzed using ImageJ 1.43u NIH software and calibrated using the absorbance value provided by Nair et al. [26]. This result also indicates the transfer of patterned graphene nanowires along with additional graphene layers onto the PDMS substrate.

29.4 CONCLUSION

We have developed a method for printing graphene nanowires onto PDMS substrates without using any wet chemical processing. The potential for simplicity, speed, and reproducibility anticipated for lithographic patterning via contact printing on a transparent, biocompatible, and flexible surface makes this combination of methods and materials a strong candidate for nanofabrication of platforms supporting sensing nanoarchitectures. The process described in this report increases the potential of graphene as an improved substrate requiring simple chemical modification for the preparation of active sites for the potential localization of biomolecules such as DNA and for controlling the electrical conductivity of graphene wires.

ACKNOWLEDGMENTS

The authors would like to express their sincere thanks to David Neff, Molecular and Biological Imaging Center, Marshall University, for the helpful discussion of ImageJ profile analysis. This work was supported financially by grants from NSF EPSCoR 0554328 and from the ARO G: W911NF-08-1-0109 and W911NF-11-1-0024.

REFERENCES

1. A. N. Sidorov, M. M. Yazdanpanah, R. Jalilian, P. J. Ouseph, R. W. Cohn, and G. U. Sumanasekera, Electrostatic deposition of graphene, *Nanotechnology,* 18, 135301–135304, 2007.
2. F. Schedin, A. K. Geim, S. V. Morozov, E. W. Hill, P. Blake, M. I. Katsnelson, and K. S. Novoselov, Detection of individual gas molecules adsorbed on graphene, *Nat Mater,* 6, 652–655, 2007.
3. C. Berger, Z. Song, T. Li, X. Li, A. Y. Ogbazghi, R. Feng, Z. Dai et al., Ultrathin epitaxial graphite: 2D electron gas properties and a route toward graphene-based nanoelectronics, *J Phys Chem B,* 108, 9912–19916, 2004.
4. M. Rahman and M. L. Norton, Two-dimensional materials as substrates for the development of origami-based bionanosensors, *IEEE Trans Nanotechnol,* 9, 539–542, 2010.
5. C. Berger, Z. Song, X. Li, X. Wu, N. Brown, C. Naud, D. Mayou et al., Electronic confinement and coherence in patterned epitaxial graphene, *Science,* 312, 1191–1196, 26 2006.
6. S. Y. Zhou, G. H. Gweon, A. V. Fedorov, P. N. First, W. A. de Heer, D. H. Lee, F. Guinea, A. H. Castro Neto, and A. Lanzara, Substrate-induced bandgap opening in epitaxial graphene, *Nat Mater,* 6, 770–775, 2007.
7. J. Coraux, A. T. N'Diaye, C. Busse, and T. Michely, Structural coherency of graphene on Ir(111), *Nano Lett,* 8, 565–570, 2008.
8. L. Song, L. Ci, W. Gao, and P. M. Ajayan, Transfer printing of graphene using gold film, *ACS Nano,* 3, 1353–1356, 23 2009.
9. X. Liang, Z. Fu, and S. Y. Chou, Graphene transistors fabricated via transfer-printing in device active-areas on large wafer, *Nano Lett,* 7, 3840–3844, 2007.
10. J. H. Chen, M. Ishigami, C. Jang, D. R. Hines, M. S. Fuhrer, and E. D. Williams, printed graphene circuits, *Adv Mater,* 19, 3623–3627, 2007.
11. X. Liang, A. S. Chang, Y. Zhang, B. D. Harteneck, H. Choo, D. L. Olynick, and S. Cabrini, Electrostatic force assisted exfoliation of prepatterned few-layer graphenes into device sites, *Nano Lett,* 9, 467–472, 2009.
12. G. Eda, G. Fanchini, and M. Chhowalla, Large-area ultrathin films of reduced graphene oxide as a transparent and flexible electronic material, *Nat Nanotechnol,* 3, 270–274, 2008.
13. S. Stankovicha, D. A. Dikina, R. D. Pinera, K. A. Kohlhaasa, A. Kleinhammesc, Y. Jiac, Y. Wuc, S. T. Nguyenb, and R. S. Ruoff, Synthesis of graphene-based nanosheets via chemical reduction of exfoliated oxide, *Carbon,* 45, 1558–1565, 2007.
14. X. Wang, Y. Ouyang, X. Li, H. Wang, J. Guo, and H. Dai, Room-temperature all-semiconducting sub-10-nm graphene nanoribbon field-effect transistors, *Phys Rev Lett,* 100, 206803, 2008.
15. Y. Ouyang, Y. Yoon, J. K. Fodor, and J. Guo, Comparison of performance limits for carbon nanoribbon and carbon nanotube transistors, *Appl Phys Lett,* 89, 203107, 2006.
16. Y. Ouyang, Y. Yoon, and J. Guo, Scaling behaviors of graphene nanoribbon FETs: A three-dimensional quantum simulation study, *IEEE Trans Electron Devices,* 54, 2223–2231, 2007.
17. L. Tapaszto, G. Dobrik, P. Lambin, and L. P. Biro, Tailoring the atomic structure of graphene nanoribbons by scanning tunnelling microscope lithography, *Nat Nanotechnol,* 3, 397–401, 2008.
18. K. S. Kim, Y. Zhao, H. Jang, S. Y. Lee, J. M. Kim, J. H. Ahn, P. Kim, J. Y. Choi, and B. H. Hong, Large-scale pattern growth of graphene films for stretchable transparent electrodes, *Nature,* 457, 706–710, 5 2009.
19. S. Bae, H. Kim, Y. Lee, X. Xu, J. S. Park, Y. Zheng, J. Balakrishnan et al., Roll-to-roll production of 30-inch graphene films for transparent electrodes, *Nat Nanotechnol,* 5,574–578, 2010.
20. M. Rahman, H. A. Howells, and L. M. Norton, Patterning and transfer of graphene onto substrates for addressing biomolecules, in *ACS 239th National Meeting*, San Francisco, CA, 2010, p. COLL 20452.
21. K. S. Novoselov, A. K. Geim, S. V. Morozov, D. Jiang, Y. Zhang, S. V. Dubonos, I. V. Grigorieva, and A. A. Firsov, Electric field effect in atomically thin carbon films, *Science,* 306, 666–669, 2004.
22. L. H. Liu and M. Yan, Simple method for the covalent immobilization of graphene, *Nano Lett,* 9, 3375–3378, 2009.
23. M. Rahman and L. M. Norton, Hierarchical lithography for generating molecular testbeds, *IEEE Sens J,* 10, 498–502, 2010.
24. M. Rahman and M.L. Norton, *11th IEEE Conference on Nanotechnology (IEEE-NANO)*, 15–18 August 2011, pp. 592–595.
25. M. Rahman and M. Norton, Widefield optical and AFM analysis of few layer graphene nanowires functionalized with DNA, Extended abstract of a paper presented at Microscopy and Microanalysis, Nashville, Tennessee, August 7–11, 2011
26. R. R. Nair, P. Blake, A. N. Grigorenko, K. S. Novoselov, T. J. Booth, T. Stauber, N. M. R. Peres, and A. K. Geim, Fine structure constant defines visual transparency of graphene, *Science,* 320, 1308–1308, 2008.

Section IX

Graphene Devices

30 Nanotransistors Using Graphene Interfaced with Advanced Dielectrics for High-Speed Communication

Osama M. Nayfeh, Ki Kang Kim, and Jing Kong

CONTENTS

30.1 INTRODUCTION

Graphene is emerging as a wonder material for future radiofrequency electronics and is mechanically compatible with the ubiquitous integration on arbitrary substrates—rigid [1,2], flexible [3], stretchable [4], and transparent [5]. Several studies have been performed demonstrating GHz unity-gain frequency (f_t) in graphene transistors [2]; however, there are minimal reports that also achieve GHz maximum oscillation frequency (f_{max}) [6]. Achieving power gain requires high-performance device characteristics in addition to well-designed device layouts in low-loss configurations. Achieving both high-frequency f_t and f_{max} is critical for the realization of graphene-based advanced active radio-frequency (RF) circuits such as amplifiers and oscillators. In this chapter, transistors are constructed and examined based on chemical vapor deposited (CVD) graphene. Transistors incorporate scaled (~10 nm) plasma-assisted atomic-layer-deposited (ALD) gate dielectrics and use an RF layout designed for targeting improved RF performance. An extrinsic f_t and f_{max} in the GHz regime are both achieved under suitable bias conditions that favor large transconductance and low output resistance due to the onset of an observed saturation-like behavior. The dependence of the gain on the applied gate and drain voltages is measured and examined. Increasing f_t and f_{max} are consistent with bias conditions that correspond to increased g_m and I_d. Simple, small-signal RF models are used to extract the small-signal capacitances and to determine the performance-limiting factors.

30.2 GRAPHENE PREPARATION AND DEVICE FABRICATION

Top-gated, gate-last RF transistor structures in the G–S–G layout configuration are constructed using the process flows reported in Refs. [2,7]. Starting substrates are Cu-catalyzed CVD graphene transferred to 300 nm SiO_2/n+ Si [1,2,5]. The starting materials for graphene growth are Alfa Aesar Cu foils. The foils are annealed in hydrogen at 975°C/325 mtorr for 35 min. Next, the foils are exposed to 18 sccm of methane at a total pressure of 1.5 torr. The wafers are cooled for 10–15 min

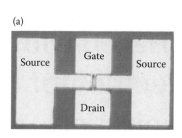

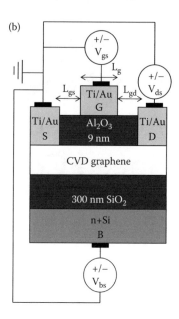

FIGURE 30.1 (a) RF transistor device micrograph in the G–S–G configuration. (b) Cross-sectional schematic and biasing conditions. $L_g = L_{sd} = 1.5$ mm and W = 75 mm.

at 325 mtorr with 12 sccm of hydrogen. The graphene/Cu foil is coated with polymethyl methacrylate (PMMA) (8%). The back-side graphene on Cu is etched with a direct O_2 plasma (20-sccm O_2/70 W). The Cu foil is etched using a Transene Inc. ferric-chloride-based Cu etchant. An HCl etch is used to remove the remnant iron particles. The PMMA/graphene film is placed on the SiO_2/n+ Si substrates; the PMMA is dissolved in acetone, and a 475°C Ar/H_2 anneal is used to remove the residual PMMA. SiO_2 is produced by dry/wet/dry oxidation, and the breakdown voltage is ~150 V. The Raman signatures of the graphene are characteristic of largely mono/bilayer graphene.

Graphene on SiO_2 substrates are cleaned with acetone/methanol/isopropanol, followed by a deionized water rinse and N_2 dry. Samples are baked for 5 min at 125°C to accelerate the removal of solvent. AZ5214 photoresist is coated, and source/drain contact regions are patterned by image reversal. Then, the 5-nm Ti/100-nm Au is evaporated, and the substrates are soaked for 24 h in acetone, followed by liftoff using rinses in acetone/methanol/IPA/H_2O. To minimize damage, liftoff is performed with gentle pipette rinsing and no ultrasonication. Following source/drain formation, graphene channels are patterned with AZ5214 and defined using a direct plasma etch (20-sccm O_2/70 W). Substrates are solvent-cleaned, and an ~9 nm gate dielectric is deposited by plasma-assisted atomic layer deposition as reported in Ref. [7] directly into the graphene. The film thicknesses are determined from the spectroscopic ellipsometry and capacitance–voltage measurements of the identical films deposited on Si. Top gates are formed identical to the process for a source/drain. Figure 30.1a shows a top-down optical image of a constructed two gate-finger field-effect-transistor (FET) and Figure 30.1b shows a cross-sectional schematic of the device.

30.3   DEVICE ELECTRICAL MEASUREMENTS

The DC characteristics of constructed devices with a gate length $L_g = 1.5$ μm, gate-source/drain length $L_{sd} = 1.5$ μm, and width W = 75 μm are measured using a Keithley 4200 semiconductor parametric analyzer. Figure 30.2a shows representative transfer (I_d vs. V_{gs}) characteristics, and Figure 30.2b shows output characteristics (I_d vs. V_{ds}). The characteristics are typical of devices produced using this CVD graphene and that incorporate the plasma-assisted Al_2O_3 gate dielectric, producing with a positive dirac point voltage V_{dp} that corresponds to an effectively heavily p-type

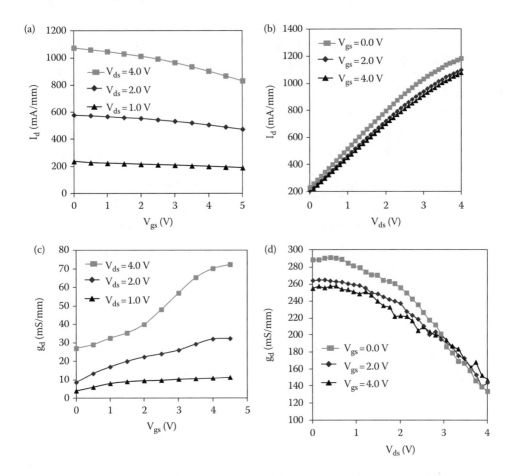

FIGURE 30.2 Measured (a) transfer I_d versus V_{gs} characteristics representative of heavily p-type doped graphene due to the PA-ALD gate-dielectric process and (b) output (I_d vs. V_{ds}) characteristics showing signs of saturation with $V_{ds} > 3$ V. (c) Extracted transconductance g_m versus V_{gs} with weak modulation for $0 < V_{gs} < 2$ V. (d) Conductance g_d that increases with V_{ds} due to onset of saturation-like effects.

"oxygen doped" film [7]. The transconductance $g_m = dI_d/dV_{gs}$ and is shown in Figure 30.2c along with the total conductance $g_d = dI_d/dV_{ds}$ in Figure 30.2d. The FETs operate normally on "depletion mode," and with increasing V_{gs}, there is gate-controlled modulation of the Fermi level that reduces the effective channel charge (hole) density and the overall I_d. Owing to a density-dependent carrier mobility [2,3] observed for graphene synthesized by this method and using this dielectric [7], g_m also increases with increasing V_{gs} and is 28 mS/mm with $V_{gs} = 0$ V/$V_{ds} = 4$ V to 72 mS/mm with $V_{gs} = 6$ V/$V_{ds} = 4$ V. With low V_{ds}, the output characteristics are near ohmic due to good quality contacts. With increasing V_{ds}, g_{ds} decreases from 260 to 160 mS/mm. The reduction in g_{ds} is due to the onset of saturation-like behavior. The origin of this behavior is still under examination and could be due to the slight band-gap opening with this dielectric. Strong saturation is required to significantly decrease g_{ds} and obtain significant intrinsic voltage gain for the devices [6].

Two-port RF measurements were taken in the ground–signal–ground (G-S-G) configuration and collected using an Agilent PNA sampling in the 10 MHz–50 GHz range. The current amplification h_{21}, unilateral power gain (U), and Mason's power gain (MSG) versus frequency are extracted from the collected S parameters for stable conditions and are shown in Figure 30.3a. The h_{21} and U characteristics follow a near-ideal −20 dB/dec behavior expected for good-quality ideal RF transistors. f_t/f_{max} are ~3.2 GHz/1.8 GHz for $V_{gs} = 5$ V/$V_{ds} = 5$ V, demonstrating both GHz current gain

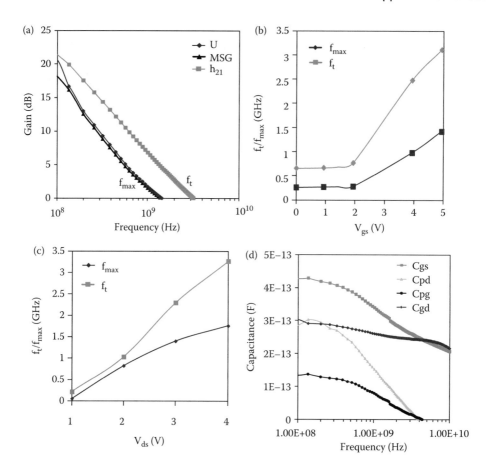

FIGURE 30.3 Measured (a) Mason's unilateral gain (U), transducer gain (MAG), and current gain (h_{21}) characteristics versus frequency. (b) The extracted f_t and f_{max} as a function of V_{gs} in agreement with the g_m characteristics and nearly constant over a 2 V range. (c) f_t and f_{max} as a function of V_{ds}. (d) Relevant extracted small-signal capacitances.

and maximum oscillation frequency. The increased frequency performance with increased V_{gs} is consistent with the increased g_m observed in the DC measurements.

Figure 30.3b shows the dependence of f_t and f_{max} as a function of V_{gs}. The behavior is consistent with the g_m behavior. While the f_t and f_{max} are not completely linear across the full 6 V window, there is not an apparent g_m collapse that severely diminishes the frequency response due to graphene being able to sustain a significant charge density >10^{13} cm²/Vs. In fact, we find near-linear characteristics over the 2–3 V range, which could be favorable for linear RF circuits. Also, with reduced V_{ds}, there is a more linear g_m, which corresponds to less modulation of f_t/f_{max} with V_{gs}.

30.4 SMALL-SIGNAL ANALYSIS AND DISCUSSION

The f_t and f_{max} can depend significantly on the capacitances of both the intrinsic and extrinsic (including parasitic) parts of the transistor. To obtain further insight into performance-limiting factors and how to achieve an increased frequency performance device, conventional transistor small-signal models are used to extract the small-signal parameters of the intrinsic components C_{gs}, C_{gd}, as well as the extrinsic (parasitic) C_{pg}, C_{pd}. The small-signal g_m, g_d, and R_{ds} are also extracted for comparison with the DC data. The capacitances are presented as a function of the frequency and are shown in Figure 30.3d. The parasitic elements are extracted using the admittance (Y) parameters obtained by converting from the scattering (S) parameters. Using the Y parameters, the following expressions are used that are derived from the equivalent conventional FET small-signal circuit [8]. As can be seen, the dominant capacitance is the intrinsic C_{gs}; however, the parasitic gate/drain pad capacitances are still large enough to limit the overall frequency/power performance.

30.5   CONCLUSION

Radiofrequency transistors based on CVD graphene and incorporating plasma-assisted atomic layer-deposited gate dielectric are constructed and examined. The DC characteristics for this process are representative of a heavily p-type doped graphene. The RF extrinsic f_t (unity-gain frequency) and f_{max} (maximum oscillation frequency) are both in the GHz regime. The dependency of f_t/f_{max} as a function of V_{gs} and V_{ds} are consistent with the g_m and I_d dependencies. f_t/f_{max} are near linear over a 2 V V_{gs} range, consistent with a near-constant g_m for this operating region. The small-signal capacitances are extracted from conventional FET circuit models to determine the performance-limiting factors.

REFERENCES

1. R. Piner, I. Velamakanni, E. Jung, E. Tutuc, S.K. Banerjee, L. Colombo, and R.S. Ruoff, Large-area synthesis of high-quality and uniform graphene films on copper foils, *Science Express Reports*, 324(5932), 1312–1314, 2009.
2. O.M. Nayfeh, Radio-frequency transistors using chemical-vapor-deposited monolayer graphene: Performance, doping, and transport effects, *IEEE Transactions on Electron Devices*, 58(9), 2847–2853, 2011.
3. O.M. Nayfeh, Graphene transistors on mechanically flexible polyimide incorporating atomic-layer-deposited gate dielectric, *IEEE Electron Device Letters*, 32(10), 1349–1351, 2011.
4. S-Ki Lee, B.J. Kim, H. Jang, S.C. Yoon, C. Lee, B.H. Hong, J.A. Rogers, J.Ho Cho, and J-H. Ahn, Stretchable graphene transistors with printed dielectrics and gate electrodes, *Nano Letters*, 11(11), 4642–4646, 2011.
5. S. Bae, H.K. Kim, Y.B. Lee, X.F. Xu, J.S. Park, Y. Zheng, J. Balakrishnan et al., Roll-to-roll production of 30-inch graphene films for transparent electrodes, *Nature Nanotechnology*, 5, 574–578, 2010.
6. K. Kim, J-Y. Choi, T. Kim, S-H. Cho, and H-J. Chung, A role for graphene in silicon-based semiconductor devices, *Nature*, 479, 338–344, 2011.
7. O.M. Nayfeh, T. Marr, and M. Dubey, Impact of plasma-assisted atomic-layer-deposited gate dielectric on graphene transistors, *IEEE Electron Device Letters*, 32(4), 473–475, 2011.
8. J.G. Dambrine, A. Cappy, D. Heliodore, and E. Playez, A new method for determining the FET small-signal equivalent circuit, *IEEE Transactions on Microwave Theory*, 36(7), 1151–1159, 1988.

31 Graphene-on-Diamond Devices and Interconnects
Carbon sp²-on-sp³ Technology

Jie Yu, Guanxiong Liu, Alexander A. Balandin,
and Anirudha V. Sumant

CONTENTS

31.1 INTRODUCTION

Graphene is a promising material for future electronics owing to its extremely high carrier mobility [1–3], thermal conductivity [4,5], saturation velocity [2,3], and ability to integrate with almost any substrate [6]. The most feasible are applications that do not require a bandgap but can capitalize on graphene's superior current-carrying capacity. Graphene field-effect transistors (FETs) and interconnects built on SiO_2/Si substrates reveal the breakdown current density, J_{BR}, of ~1 $\mu A/nm^2$ [7–9], which is ~100 × larger than the fundamental electromigration limit for the metals [10]. However, the current-carrying capacity of graphene-on-SiO_2/Si devices is still smaller than the maximum achieved in carbon nanotubes (CNTs) [11–13]. In this chapter, we outline the graphene-on-diamond technology, which enables the fabrication of graphene-on-diamond devices and interconnects with a substantially enhanced breakdown current density. The discussion in this chapter follows our report of a systematic study of the current-induced breakdown in graphene-on-diamond devices [14]. The study demonstrated that by replacing SiO_2 with a synthetic diamond, one can solve the early-graphene-device-failure problem and increase J_{BR} of graphene by an order of magnitude to above ~10 $\mu A/nm^2$ [14]. We used recent advances in the chemical vapor deposition (CVD) and processing of diamond for fabricating >40 graphene devices on ultra-nanocrystalline diamond (UNCD) and single-crystal diamond (SCD) substrates with a surface roughness below $\delta H \approx 1$ nm. It was found that not only SCD but also UNCD with a grain size $D \sim 5$–10 nm can improve J_{BR}, owing to the increased thermal conductivity of UNCD at higher temperatures. The obtained results are important for graphene applications in interconnects [7,15] and radio-frequency transistors [16], and can lead to the new planar sp²-on-sp³ carbon-on-carbon technology.

31.2 SELECTION OF SUBSTRATE MATERIALS

Graphene devices are commonly fabricated on Si/SiO_2 substrates with an SiO_2 thickness of $H \approx 300$ nm [1–3]. Owing to optical interference, graphene becomes visible on Si/SiO_2 (300-nm) substrates, which facilitates its identification. It was discovered that graphene has excellent heat conduction properties with the intrinsic thermal conductivity, K, exceeding 2000 W/mK at room temperature (RT) [4,5]. However, in typical device structures, for example, FETs or interconnects, most of the heat propagates directly below the graphene channel in the direction of the heat sink, that is, the bottom of the Si wafer [17,18]. For this reason, the highly thermally resistive SiO_2 layers act as the thermal bottleneck, not allowing one to capitalize on graphene's excellent intrinsic thermal properties. Theoretical considerations suggest that the breakdown mechanism in sp^2-bonded graphene should be similar to that in sp^2-bonded CNTs. Unlike in metals, the breakdown in CNTs was attributed to the resistive heating or local oxidation, assisted by defects [11–14]. Thermal conductivity of SiO_2, $K = 0.5$–1.4 W/mK at RT [19] is more than 1000-times smaller than that of Si, $K = 145$ W/mK, which indicates that the use of materials with higher K, directly below graphene, can improve graphene's J_{BR} and reach the maximum values observed for CNTs.

Synthetic diamond is a natural candidate for use as a bottom dielectric in graphene devices, which can perform the function of a heat spreader. Recently, there was a major progress in CVD diamond growth performed at low temperature, T, compatible with Si complementary metal–oxide–semiconductor (CMOS) technology [20–22]. There are other potential benefits of utilizing diamond thin films as substrates for graphene devices instead of SiO_2. The energy of the optical phonons in diamond, $E_p = 165$ meV, is much larger than that in SiO_2, $E_p = 59$ meV. The latter can improve the saturation velocity in graphene when it is limited by the surface electron—phonon scattering [23]. The lower trap density achievable in diamond, compared to SiO_2, indicates a possibility of reduction of the $1/f$ noise in graphene-on-diamond devices [24], which is essential for applications in r.f. transistors and interconnects. It was also demonstrated that replacing SiO_2 with diamond-like carbon (DLC) helps to substantially improve the radio frequency characteristics of the graphene transistors [16]. However, DLC is an amorphous material with $K = 0.2$–3.5 W/mK at RT [25], which is a low value even when compared to SiO_2. Depending on the H content, as-deposited DLC films have high internal stress, which needs to be released by annealing at higher $T \sim 600°$C [26]. These facts provide strong motivations for the search of other carbon materials, which can be used as substrates for graphene devices.

31.3 SYNTHETIC DIAMOND GROWTH

Synthetic diamond can be grown in a variety of forms from UNCD films with the small grain size, D, and, correspondingly low K, to SCD, with the highest K among all bulk solids. Microcrystalline diamond (MCD) has larger D than UNCD but suffers from unacceptable surface roughness, δH, and high thermal boundary resistance, R_B [5]. Up to date despite attempts in many groups to fabricate graphene devices on diamond with acceptable characteristics, no breakthrough was reported. The major stumbling blocks for the development of viable graphene-on-diamond sp^2-on-sp^3 technology are the high δH of synthetic diamond, difficulty of visualization of graphene on diamond, and problems with the top-gate fabrication—no bottom gates are possible on SCD substrates. The study reported in Ref. [14] used the most recent advances in CVD diamond growth and polishing as well as our experience of graphene device fabrication to prepare a large number of test structures, and study the current-carrying and thermal characteristics of graphene-on-diamond devices in the practically relevant ambient conditions. In this chapter, we discuss two main forms of diamond—UNCD and SCD—which represent two extreme cases in terms of D and K.

The examined UNCD films were grown on Si substrates in the microwave plasma chemical vapor deposition (MPCVD) system at the Argonne National Laboratory [14]. Figure 31.1a,b shows the MPCVD system used for the growth inside a clean room and schematic of the process, respectively. The growth conditions were altered to obtain a larger D, in the range 5–10 nm, instead of the

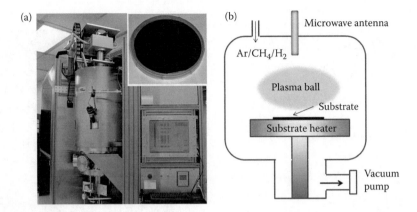

FIGURE 31.1 Synthetic diamond growth and characterization. (a) Large-area MPCVD system used for the synthetic diamond growth. The inset shows a 100-mm Si/UNCD wafer. (b) Schematics describing the UNCD growth in the MPCVD system.

typical grain sizes $D \approx 2–5$ nm in UNCDs. This was done to increase K of UNCD without strongly increasing the surface roughness. In our investigation, we intentionally did not increase D beyond 10 nm or used MCD in order to keep δH in the range suitable for polishing. The inset shows a 100-mm UNCD/Si wafer.

The surface roughness of the synthetic diamond substrate plays an important role in reducing electron scattering at the graphene—diamond interface and increasing the electron mobility, μ. We performed the chemical mechanical polishing (CMP) to reduce the as-grown surface roughness from $\delta H \approx 4–7$ nm to below $\delta H \approx 1$ nm, which resulted in a corresponding reduction of the thickness, H, from the as-grown $H \approx 1$ μm to ~700 nm. The H value was selected keeping in mind conditions for graphene visualization on UNCD, together with the thermal management requirements. The SCD substrates were type IIb (100) grown epitaxially on a seed diamond crystal and then laser cut from the seed. For graphene devices fabrication, the SCD substrates were acid washed, solvent cleaned, and put through the hydrogen termination process [27]. The near-edge x-ray absorption fine-structure spectroscopy (NEXAFS) of the grown UNCD film confirms its high sp^3 content and quality (Figure 31.2a). The strong reduction of δH is evident from the atomic force microscopy (AFM) images of the as-grown UNCD and UNCD after CMP presented in Figure 31.2b and c, respectively.

31.4 GRAPHENE PREPARATION AND CHARACTERIZATION

Graphene and few-layer graphene (FLG) were prepared by exfoliation from the bulk highly oriented pyrolytic graphite (HOPG) to ensure quality and uniformity in thickness. We selected flakes of the rectangular-ribbon shape with the width $W \geq 1$ μm, which is larger than the phonon mean free path (MFP) Λ in graphene [5]. The condition $W > \Lambda$ ensured that K does not undergo additional degradation due to the phonon-edge scattering. The length, L, of graphene ribbons was in the range 10–60 μm. We selected ribbons with the small aspect ratio $\gamma = W/L \sim 0.03–0.1$ to imitate interconnects. Raman spectroscopy was used for determining the number of atomic planes, n, in FLG, although the presence of sp^2 carbon at the grain boundaries in UNCD made the spectrum analysis more difficult. Figure 31.3a shows spectra of the graphene-on-UNCD/Si and UNCD/Si substrate. One can see the 1332 cm^{-1} peak, which corresponds to the optical vibrations in the diamond crystal structure. The peak is broadened due to the small D in UNCD. The bands at ~1170, 1500, and 1460 cm^{-1} are associated with the presence of *trans*-poly-acetylene and sp^2 phase at grain boundaries [28,29]. The graphene G peak at 1582 cm^{-1} and $2D$ band at ~2700 cm^{-1} are clearly recognizable. Figure 31.3b presents spectra of the graphene-on-SCD and SCD substrate and the difference between the two. The intensity and width of the 1332 cm^{-1} peak confirms that we have a single-crystal diamond.

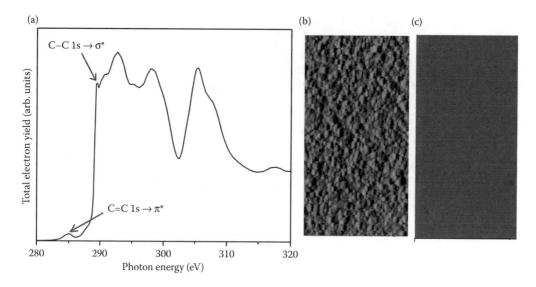

FIGURE 31.2 Material characterization of a synthetic diamond. (a) NEXAFS data for a deposited UNCD thin film revealing its high sp^3 content and quality. The exciton peak at ~289.3 eV corresponds to 1s → σ* resonance from sp^3 carbon. The peak at ~285 eV corresponds to 1s → π* resonance from sp^2 carbon at grain boundaries. The revealed sp^2 fraction is 2%, which is lower than the typical 5% sp^2 content, owing to larger D in our UNCD. (b) and (c) AFM images of the as-grown and chemical–mechanical polished UNCD, respectively.

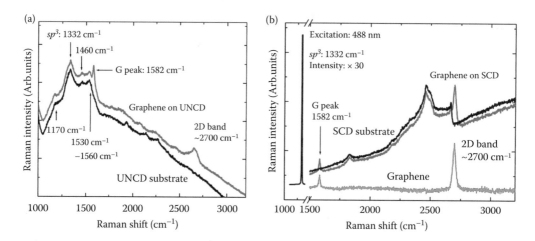

FIGURE 31.3 Micro-Raman spectroscopic analysis of graphene-on-diamond samples. (a) Raman spectra of graphene-on-UNCD (upper curve) and UNCD substrate (lower curve). (b) Raman spectra of graphene-on-SCD and SCD substrate. The difference in spectra (lower curve) was used to determine the number of atomic planes, n. The specific example shows a single-layer graphene. The data indicate that micro-Raman spectroscopy can be used for identification of graphene on synthetic diamond samples, which contain sp^2 phase on grain boundaries.

31.5 GRAPHENE DEVICE FABRICATION

We focused our discussion on devices made of FLG with $n \leq 5$. FLG supported on substrates or embedded between dielectrics preserves its transport properties better than single-layer graphene. Two-terminal (i.e., interconnects) and three-terminal (i.e., FETs) devices were fabricated on both UNCD/Si and SCD substrates. The electron-beam lithography (EBL) was used to define the source, drain

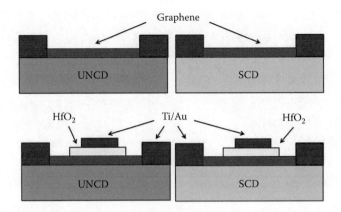

FIGURE 31.4 Schematic of the graphene-on-diamond devices showing the two-terminal and three-terminal devices fabricated for testing on UNCD/Si and SCD substrates.

contacts, and gate electrodes. The contacts consisted of a thin Ti film covered by a thicker Au film. The top-gate HfO$_2$ dielectric was grown by the atomic layer deposition (ALD). The novelty in our design, as compared to the graphene-on-SiO$_2$/Si devices, was the fact that the gate electrode and pad were completely separated by the HfO$_2$ layer to avoid oxide lift-off sharp edges, which can affect the connection of the gate electrode.

Figure 31.4 shows schematics of the fabricated devices. For testing the breakdown current density in FLG, we used two-terminal devices in order to minimize extrinsic effects on the current and heat conduction. Three-terminal devices were utilized for μ measurements. We also fabricated conventional graphene-on-SiO$_2$/Si devices as references. Figure 31.5 is an optical microscopy image of two-terminal graphene-on-SCD devices. Figures 31.6a and b show the scanning electron microscopy (SEM) images of the two-terminal and three-terminal graphene-on-UNCD devices, respectively. We electrically characterized >40 graphene-on-diamond devices and >10 graphene-on-SiO$_2$/Si reference devices [14]. To understand the origin of the breakdown, we correlated J_{BR} values with the thermal resistances of the substrates. We measured the effective K of the substrates and determined their thermal resistance as $R_T = H_S/K$, where H_S is the substrate thickness.

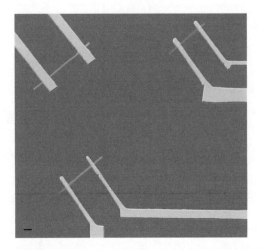

FIGURE 31.5 Optical microscopy image of the two-terminal graphene devices—prototype interconnects—on a single-crystal synthetic diamond.

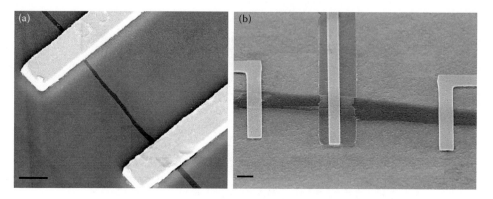

FIGURE 31.6 SEM images of the two-terminal (i.e., interconnects) and three-terminal (i.e., transistors) graphene-on-UNCD/Si devices. (a) The two-terminal devices were used for the breakdown current density testing. (b) The three-terminal devices were utilized to measure the mobility. The scale bar is 2 μm.

31.6 CHARACTERISTICS OF GRAPHENE-ON-DIAMOND DEVICES

Figure 31.7a shows R_T for the UNCD/Si and Si/SiO$_2$ (300 nm) substrates as a function of T. Note that R_T for Si increases approximately linearly with T, which is expected because the intrinsic thermal conductivity of crystalline materials decreases as $K \sim 1/T$ for T above RT. The T dependence of R_T for UNCD/Si is completely different, which results from the interplay of heat conduction in UNCD and Si. In UNCD, K grows with T owing to the increasing inter-grain transparency for the acoustic phonons that carry heat [5]. UNCD/Si substrates, despite being more thermally resistive than Si wafers at RT, can become less thermally resistive at high T. The R_T value for SCD substrate is ~0.25 × 10^{-6} m^{2}K/W, which is more than an order of magnitude smaller than that of Si at RT. The thermal interface resistance, R_B, between FLG and the substrates is $R_B \approx 10^{-8}$ m^{2}K/W, and it does not depend strongly on either n or the substrate material [5]. For this reason, R_B does not affect the R_T trends.

Figure 31.7b shows the current–voltage (I–V) characteristics of graphene-on-SCD FET at low source–drain voltages for different top-gate, V_{TG}, bias. The inset demonstrates a high quality of the HfO$_2$ dielectric and metal gate deposited on top of the graphene channel. The linearity of I–Vs

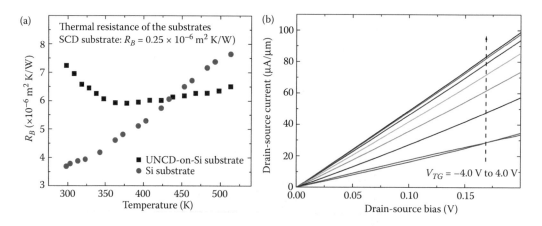

FIGURE 31.7 Electrical and thermal characteristics of graphene-on-diamond. (a) Thermal resistance of a UNCD/Si substrate and a reference Si wafer. Note that the thermal resistance of the composite UNCD/Si substrate decreases at the high temperature. (b) Low-field I–V characteristics of top-gate graphene-on-SCD devices. (The figure is based on the experimental data reported in Yu J. et al., *Nano Lett.*, 12, 1603, 2012.)

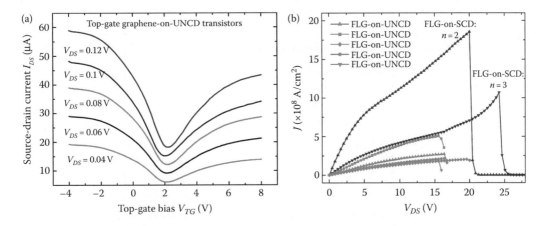

FIGURE 31.8 Breakdown current testing of graphene-on-diamond devices. (a) The source–drain current in the three-terminal graphene-on-UNCD devices as a function of the top-gate bias. (b) The breakdown current density in the two-terminal graphene-on-UNCD and graphene-on-SCD devices. Note an order of magnitude improvement in the current-carrying ability of graphene devices and interconnects fabricated on the single-crystal synthetic diamond. (The figure is based on the experimental data reported in Yu J. et al., *Nano Lett.*, 12, 1603, 2012.)

confirms that the contacts are Ohmic. Figure 31.8a presents the source–drain, I_{SD}, current as a function of V_{TG} for graphene-on-UNCD FET. In the good top-gate graphene-on-diamond devices, the extracted μ was ~1520 cm²/V/s for the electrons and ~2590 cm²/V/s for the holes. These mobility values are acceptable for applications in downscaled electronics. In Figure 31.8b, we show results of the breakdown testing. For graphene-on-UNCD, we obtained $J_{BR} \approx 5 \times 10^8$ A/cm² as the highest value, while the majority of devices broke at $J_{BR} \approx 2 \times 10^8$ A/cm². The reference graphene-on-SiO₂/Si had $J_{BR} \approx 10^8$ A/cm², which is consistent with the literature [7–9]. The maximum achieved for graphene-on-SCD was as high as $J_{BR} \approx 1.8 \times 10^9$ A/cm². This is an important result, which shows that via improved heat removal from graphene channel one can reach, and even exceed, the maximum current-carrying capacity of ~10 μA/nm² (= 1 × 10⁹ A/cm²) reported for CNTs [11–14]. The surprising improvement in J_{BR} for graphene-on-UNCD is explained by the reduced R_T at high T where the failure occurs. At this temperature, R_T of UNCD/Si can be lower than that of Si/SiO₂.

31.7   THERMAL BREAKDOWN IN GRAPHENE DEVICES

The location of the current-induced failure spot and J_{BR} dependence on electrical resistivity, ρ, and length, L, can shed light on the physical mechanism of the breakdown. The failures in the middle of CNTs and $J_{BR} \sim 1/\rho$ were interpreted as signatures of the electron diffusive transport, which resulted in the highest Joule heating in the middle [11–13]. The failures at the CNT-metal contact were attributed to the electron ballistic transport through CNT and energy release at the contact. There is a difference in contacting CNT with the diameter $d \sim 1$ nm and graphene ribbons with $W \geq 1$ μm. It is easier to break CNT-metal than the graphene-metal contact thermally. In our study, we observed the failures both in the middle and near the contact regions (see Figure 31.9). The difference between these two types was less pronounced than that in CNTs. The failures occurred not exactly at the graphene–metal interface but at some distance, which varied from sample to sample. We attributed it to the width variations in graphene ribbons leading to breakdowns in the narrowest regions, or in the regions with defects, which are distributed randomly. We did not observe scaling of J_{BR} with ρ as we did in the case of CNTs.

The breakdown current density, J_{BR}, for graphene scaled well with ρL. Figure 31.9 shows data for graphene-on-UNCD with a similar aspect ratio. From the fit to the experimental data, we obtained

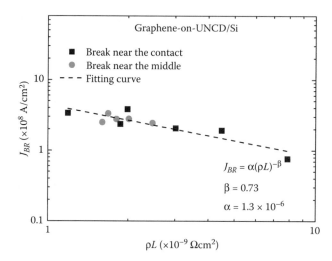

FIGURE 31.9 Scaling of the breakdown current density. J_{BR} as a function of the electrical resistance and length of graphene interconnects. The device failures close to the middle of the graphene channel and to the graphene—metal contact are indicated with the circles and rectangles, respectively. (The figure is based on the experimental data reported in Yu J. et al., *Nano Lett.*, 12, 1603, 2012.)

$J_{BR} = \alpha(\rho L)^{-\beta}$, where $\alpha = 1.3 \times 10^{-6}$ and $\beta = 0.73$. For graphene-on-SCD, the slope is $\beta = 0.51$. Previously, the scaling with $(\rho L)^{-\beta}$ (where $\beta = 0.6$–0.7) was observed in carbon nanofibers (CNF) [30], which had a similar aspect ratio. Such $J_{BR}(\rho L)$ dependence was explained from the solution of the heat-diffusion equation, which included thermal coupling to the substrate. However, the thermally induced J_{BR} for CNF was ~10^6 A/cm^2—much smaller than the record $J_{BR} \approx 1.8 \times 10^9$ A/cm^2 we obtained for graphene-on-SCD.

31.8 CONCLUSIONS

We described a possibility of a substantial increase in the current-carrying capacity of graphene devices and interconnects via their fabrication on synthetic diamond substrates. The obtained results are important for the proposed graphene applications in interconnects and radiofrequency transistors, and can eventually lead to the new planar sp^2-on-sp^3 carbon-on-carbon technology.

ACKNOWLEDGMENTS

The work at the University of California at Riverside was supported by the Office of Naval Research (ONR) through award N00014-10-1-0224, Semiconductor Research Corporation (SRC) and Defense Advanced Research Project Agency (DARPA) through the FCRP Center on Functional Engineered Nano Architectonics (FENA), and DARPA Defense Microelectronics Activity (DMEA) under agreement number H94003-10-2-1003. The work at the Argonne National Laboratory was supported by the U.S. Department of Energy (DOE), Office of Science and Office of Basic Energy Sciences under Contract DE-AC02-06CH11357. NEXAFS studies were performed at the University of Wisconsin Synchrotron Radiation Center.

REFERENCES

1. Novoselov, K. S. et al. Electric field effect in atomically thin carbon films. *Science* 306, 666–669, 2004.
2. Novoselov, K. S. et al. Two-dimensional gas of massless Dirac fermions in graphene. *Nature* 438, 197–200, 2005.

3. Zhang, Y. B., Tan, Y. W., Stormer, H. L., and Kim, P. Experimental observation of the quantum Hall effect and Berry's phase in graphene. *Nature* 438, 201–204, 2005.

4. Ghosh, S. et al. Dimensional crossover of thermal transport in few-layer graphene. *Nat. Mat.* 9, 555, 2010.

5. Balandin, A. A. Thermal properties of graphene and nanostructured carbon materials. *Nat. Mat.* 10, 569–581, 2011.

6. Palacios, T. Graphene electronics: Thinking outside the silicon box. *Nat. Nano.* 6, 464–465, 2011.

7. Murali, R. et al. Breakdown current density of graphene nanoribbons. *Appl. Phys. Lett.* 94, 243114, 2009.

8. Yu, T. et al. Bilayer graphene system: Current-induced reliability limit. *IEEE Electron Device Lett.* 31, 1155–1157, 2010.

9. Lee, K. J., Chandrakasan, A. P., and Kong, J. Breakdown current density of CVD-grown multilayer graphene interconnects. *IEEE Electron Device Lett.* 32, 557–559, 2011.

10. Christou, A. *Electromigration and Electronic Device Degradation* (Wiley-Interscience, New York, NY, 1994).

11. Collins, P. G., Hersam, M., Arnold, M., Martel, R., and Avouris, P. Current saturation and electrical breakdown in multiwalled carbon nanotubes. *Phys. Rev. Lett.* 86, 3128, 2001.

12. Tsutsui, M. et al. Electrical breakdown of short multiwalled carbon nanotubes. *J. Appl. Phys.* 100, 094302, 2006.

13. Huang, J. H. et al. Atomic scale imaging of wall-by-wall breakdown and concurrent transport measurements in multiwall carbon nanotubes. *Phys. Rev. Lett.* 94, 236802, 2005.

14. Yu, J., Liu, G., Sumant, A. V., Goyal, V., and Balandin, A. A. Graphene-on-diamond devices with increased current-carrying capacity: Carbon sp^2-on-sp^3 technology. *Nano Lett.*, 12, 1603, 2012.

15. Shao, Q., Liu, G., Teweldebrhan, D., and Balandin, A. A. High-temperature quenching of electrical resistance in graphene interconnects. *Appl. Phys. Lett.* 92, 202108, 2008.

16. Wu, Y. et al. High-frequency scaled graphene transistors on diamond-like carbon. *Nature.* 472, 74–78, 2011.

17. Freitag, M. et al. Energy dissipation in graphene field-effect transistors. *Nano Lett.* 9, 1883–1888, 2009.

18. Subrina, S., Kotchetkov, D., and Balandin, A. A. Heat removal in silicon-on-insulator integrated circuits with graphene lateral heat spreaders. *IEEE Electron Device Lett.* 30, 1281, 2009.

19. Yamane, T. et al. Measurement of thermal conductivity of silicon dioxide thin films using a 3ω method. *J. Appl. Phys.* 91, 9772, 2002.

20. Sumant, A. V. et al. Ultrananocrystalline and nanocrystalline diamond thin films for MEMS/NEMS applications. *MRS Bull.* 35, 281, 2010.

21. Sumant, A. V. et al. Large area low temperature ultrananocrystalline diamond films and integration with CMOS devices for monolithically integrated MEMS/NEMS-CMOS systems. *Proc. SPIE* 7318, 17, 2009.

22. Goldsmith, C. et al. Charging characteristics of ultrananocrystalline diamond in RF-MEMS capacitive switches. *IEEE Intl. Microwave Symp. Dig.* 1246–1249, 2010. DOI: 10.1109/MWSYM.2010.5517781.

23. Meric, I. et al. Current saturation in zero-bandgap, top-gated graphene field-effect transistors. *Nat. Nanotechnol.* 3, 654–659, 2008.

24. Liu, G. et al. Low-frequency electronic noise in the double-gate single-layer graphene transistors. *Appl. Phys. Lett.* 95, 033103, 2009.

25. Shamsa, M. et al. Thermal conduction in diamond-like carbon thin films. *Appl. Phys. Lett.* 89, 161921, 2006.

26. Friedmann et al. Thick stress-free amorphous tetrahedral carbon films with hardness near that of diamond. *Appl. Phys. Lett.* 71, 3820, 1997.

27. Sumant, A. V. et al. Correlation between surface chemistry and nanotribology for ultrananocrystalline diamond, and application to micro- and nanomechanical systems. *Phys. Rev. B* 76, 235429, 2007.

28. Ferrari, A. C. and Robertson, J. Origin of the 1150-cm⁻¹ Raman mode in nanocrystalline diamond. *Phys. Rev. B* 63, 121405, 2001.

29. Shamsa, M. et al. Thermal conductivity of nitrogenated ultrananocrystalline diamond films on silicon. *J. Appl. Phys.* 103, 083538, 2008.

30. Suzuki, M. et al. Current-induced breakdown of carbon nanofibers. *J. Appl. Phys.* 101, 114307, 2007.

32 Graphene Band Gap Modification via Functionalization with Metal-Bis-Arene Molecules

Paul Plachinda, David R. Evans, and Raj Solanki

CONTENTS

32.1 INTRODUCTION

Graphene continues to draw immense interest because of its unusual electronic and spin properties resulting from a simple structure composed of a single layer of carbon atoms arranged in a two-dimensional honeycomb pattern [1,2]. These properties, including the ballistic carrier transport and quantum Hall effect, make it a promising candidate as a building block of future nanoelectronic devices and as a possible replacement for silicon [3,4]. In spite of graphene's amazing properties, there are some obstacles that need to be overcome before it can be considered as a viable candidate to replace silicon. The main barrier is the absence of a band gap. Therefore, producing a band gap is probably one of the most important challenges that must be addressed before graphene can ultimately enable practical applications ranging from digital electronics to infrared nanophotonics.

A number of possible solutions have been proposed and demonstrated for producing band gaps in single- and double-layered graphene. One of the more straightforward methods involves the growth of epitaxial graphene on a lattice-matched (SiC) substrate to induce a stress and, as a result, open up a band gap of about 0.26 eV [5]. Alternatively, a somewhat more successful method utilizes quantum confinement to open a band gap in graphene by the fabrication of nanoscale structures, that is, quantum dots and nanoribbons, where the band gap varies inversely with the nanoscale structure dimension [6–8]. However, the nonuniform edges of these structures play a major role in degrading their electrical properties [9]. For example, with graphene nanoribbon of widths less than 10 nm, a band gap of about 0.4 eV has been reported; however, associated electron motilities are between 100 and 200 $cm^2/V.s$. Mobility is believed to be degraded by edge scattering [10].

A graphene bilayer shares many of the interesting properties of a single-layered film and provides a richer band structure, albeit without a band gap. Theoretical studies have predicted that a

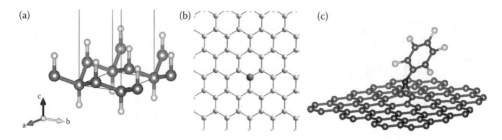

FIGURE 32.1 View of the molecular structures of functionalized graphene. (a) Hydrogenated graphene. (Adapted from D. K. Samarakoon and X.-Q. Wang, *ACS Nano,* 4, 4126–30, 2010.) (b) With metal adatom (Fe). (Adapted from A. V. Krasheninnikov et al., *Physical Review Letters,* 102, 126807, 2009.) (c) Perfluorophenylazide (PFPA)-functionalized graphene. (Adapted from K. Suggs, D. Reuven and X.-Q. Wang, *Journal of Physical Chemistry C,* 115, 3313–3317, 2011.)

significant band gap could be induced by breaking the inversion of the two layers through the application of a perpendicular electric field [11–13]. Using a single or double (top and bottom) gates to apply a strong electric field, a tunable band gap of up to about 250 meV has been created in graphene field effect transistors [14,15].

Chemical modification of graphene by covalently functionalizing its surface potentially allows a wider flexibility in engineering electronic structure, in particular the local density of states of the carbon atoms bound to the modifier that can result in the opening of the band gap. Such binding can involve covalent hydrogenation of graphene to modify hybridization of carbon atoms from sp^2 to sp^3 geometry as shown in Figure 32.1a [16–18]. Methods have also been developed to functionalize graphene covalently with molecular species and adatoms (Figure 32.1b) [19–24]. Among these, perfluorophenylazide (PFPA) functionalization of graphene is well developed using a nitrene intermediate (Figure 32.1c). We have utilized films of this molecule to act as adhesion layers to produce long ribbons of exfoliated graphene [21–24].

To minimize the effects of edge scattering of carriers, wider strips of graphene would be required for the fabrication of devices; however, these wider strips will be semimetallic. Therefore, our objective is to identify molecules that, when covalently bonded to graphene, can break its conical band structure and open up an energy gap. To achieve this goal, we have examined the electronic structure of metal-arene (MA)-functionalized graphene and report below our results based on the first-principles density-functional calculations of the band gap-functionalized graphene (MA-Gr). It is shown that the MA covalently binds at the π-conjugation of graphene and changes the electronic properties from metallic to semiconducting. We also show that the energy gap can be tuned by adjusting the number of bound MA adducts.

During our previous investigation of utilizing PFPA as an adhesion layer, it was determined that PFPA was covalently bonded to graphene [21]. Hence, we first simulated the covalent binding energy and the resulting induced band gap in graphene of PFPA and its derivatives. (For the calculation methods, see below.) Derivatives that were considered were obtained by substituting the nitrogen in the nitrene radical by the element listed in Table 32.1. It can be seen that the widest gap

TABLE 32.1

Band Gaps of PFPA-Functionalized Graphene

Element	B	C	N	O	P	S	Cl	As
Gap (eV)	~0[a]	0.11	0.24	0	0.22	~0[a]	0	~0[a]

[a] Nonzero values below 0.001 Ha (~0.027 eV) cannot be calculated exactly and are therefore set to approximate zero.

of 0.24 eV induced in graphene is when it is functionalized with a single PFPA molecule per 6×6 graphene supercell. Functionalization with two molecules slightly increases the gap to 0.28 eV. This gap is not wide enough for the fabrication of practical devices. Therefore, we next focused on MA-functionalized graphene.

32.2 METHODS

The calculations were conducted within the framework of the density functional theory (DFT) as implemented in the $DMol_3$ package [27]. The generalized gradient approximation (GGA) in BLYP [28,29] exchange-correlation parameterization was used for both final geometry optimization and band structure calculation. Initial geometry optimization was performed using the local density approximation (LDA) with the Vosko–Wilk–Nusair (VWN) [30] correlation function. A 6×6 graphene supercell with a vacuum space of 11.5 Å normal to the graphene plane was used. The geometry optimization convergence criterion was satisfied when the total energy change was less than 3×10^{-5} Ha. Only one k-point (gamma) was used throughout the calculations since the distance between neighboring k-points was only 0.077 1/Å due to a large supercell choice. For the band structure computation, the k-path selected was Γ–M–K–Γ, with 24, 20, and 40 k-points on each segment correspondingly. Although the GGA approach systematically underestimates the band gaps, we are primarily interested in the mechanism of a gap opening. For that purpose, the GGA approach is expected to provide qualitatively correct information. A more precise GW approach is very costly with this system, consisting of a total of 94 atoms. The $DMol_3$ package utilizes a numerical orbital basis set for the radial part of the wave function centered on the atoms. This allows one to include a thick vacuum layer into the model without increase of the computation time.

To pursue the effect of adduct concentration on the electronic structures, we have considered two configurations by adding one or four MA molecules onto a 6×6 rhombus cell, respectively. The cell constitutes 72 carbon atoms of graphene and 1 metal, 6 carbon, and 6 hydrogen atoms of each MA molecule.

32.3 RESULTS AND DISCUSSION

Various studies conducted on adatoms of transition metals on graphene and CNTs have demonstrated a broad potential for modifying the electronic structure of graphene [25,26,31,32]. Partially filled d-shells can play the role of electron donors [33], disturbing the π-conjugated system and thus possibly opening a gap. However, it was found that a single transition element adatom produces an insignificant charge transfer of ~0.01 electrons per carbon atom from the transition metal to the graphene sheet, and thus hardly changes the band structure [34]. Even so, "sandwich compounds" such as metallocenes and bis-arenes have long been known and are of great interest in inorganic chemistry. Moreover, it is well known that ligand aromaticity is preserved in these compounds; therefore, replacing one of the aromatic ligands with a graphene sheet would seem to be an attractive method for functionalization since

1. Metallocenes and metal-bis-arene compounds are known to be good electronic donors [35] and graphene shows the strongest interaction with electron donor and acceptor molecules via molecular charge transfer [36].
2. The geometrical structure of η^6 compounds is similar to the honeycomb structure of graphene.

Based on these observations, we undertook a study of MA-functionalized graphene for the following 3d metals: Ti, V, Cr, Mn, and Fe (Zn and Cu have a closed 3d shell and are therefore ignored) in two different configurations—one and four MA molecules per 6×6 graphene supercell (Figure 32.2).

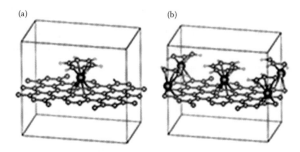

FIGURE 32.2 Ball-and-stick presentation of optimized structures of MA-functionalized graphene (top view) with one (a) and four (b) MA adducts per 6×6 graphene supercell.

32.3.1 Binding Energy

MA produces very strong bonds with the graphene sheet. The binding energy of Cr-MA was found by extrapolating to 0 K of the binding energies obtained by Hess's law for different values of thermal occupancy smearing. A quite high value of this energy of −4.72 eV (455.41 kJ/mol), which is about 25 kJ/mol per electron, indicates strong bonding and presents a solid evidence for the real existence of such compounds.

32.3.2 Possible Synthesis Strategy

MA complexes are known to exist for all transition metals [37], and their structure and chemistry resemble that of metallocenes. One of the many possible ways of synthesis is condensation of metal vapors on graphene, followed by aromatic molecules such as benzene at low temperature and pressure. Elschebroich and Kündig [38,39] have successfully applied this strategy to polyaromatics, which leads us to believe that this conventional synthesis approach may allow us to produce MA-functionalized graphene sheets. Varying the concentration of metal vapor and aromatic molecules could provide a means to achieve a different degree of functionalization.

32.3.3 Geometrical Properties

We summarize in Table 32.2 the optimized configuration of MA-functionalized graphene.

TABLE 32.2
Geometrical Parameters of Free MA Molecules and MA Molecules Bound to the Graphene Sheet

Metal Atom	Ti	V	Cr	Mn	Fe
$\angle$(C–M–C) (°) (MA)	103.477	101.397	99.76	101.285	104.353
M–C(a) (A)	2.312	2.255	2.211	2.244	2.314
$\angle$(C–M–C) (°) (1 mol)	105.411	101.433	101.183	99.748	97.475
M–C(a)/M–C(g) (A)	2.361/2.368	2.295/2.310	2.235/2.266	2.171/2.266	2.186/2.358
$\angle$(C–M–C) (°) (4 mol)	105.346	102.886	100.991	100.198	102.835
M–C(a)/M–C(g) (A)	2.335/2.391	2.26/2.334	2.216/2.274	2.161/2.298	2.198/2.394

Source: (P. Plachinda, D. Evans, and R. Solanki, Modification of graphene band structure by haptic functionalization, *2011 11th IEEE Conference on Nanotechnology (IEEE-NANO)*, Aug. 15–18, pp.1187–1192. © (2011) IEEE. With permission.)

TABLE 32.3

Electronic Configuration of the Metal Atoms in the MA, and the Corresponding Energy Gap Opening in the MA-Gr as a Result of Functionalization

Metal Atom	Ti	V	Cr	Mn	Fe
Number of valence electrons	16	17	18	19	20
Electronic configuration:					
e^*_{1g} (yz, xz)				↑	↑↑
a'_{1g} (z^2)		↑	↓↑	↓↑	↓↑
e_{2g} (x^2–y^2, xy)	↓↑ ↓↑	↓↑ ↓↑	↓↑ ↓↑	↓↑ ↓↑	↓↑ ↓↑
Number of unpaired electrons	0	1	0	1	2
Eg (1 molecule) (eV)	0.40815	0.10884	0.38094	0.29931	0
Eg (4 molecule) (eV)	0.32652	0.78909	0.8163	0	0.48978

Source: (P. Plachinda, D. Evans, and R. Solanki, Modification of graphene band structure by haptic functionalization, *2011 11th IEEE Conference on Nanotechnology (IEEE-NANO)*, Aug. 15–18, pp.1187–1192. © (2011) IEEE. With permission.)

As can be seen from Table 32.2, the bonding to an "infinite" graphene sheet changes the structural features with respect to the free $M[\eta^6\text{-(arene)}_2]$ molecule. General trends in the bonding lengths demonstrate the following features: the M–C (graphene) bond lengths remain about 3% longer than the one in the free molecule; the M–C (arene) bonds, however, remain almost unchanged with respect to the free molecules for Ti–Cr metals and become about 3% shorter for Mn and Fe. The reason for the extreme behavior of the bond lengths can be explained by considering a molecular orbital representation of the MAs. The usage of molecular orbitals (instead of Wannier functions as for a periodic system) is justified because we assume little interaction between the molecules from the neighboring cells. In the $Cr[\eta^6\text{-(arene)}_2]$, electrons fully occupy the a_{1g} binding orbital, whereas adding additional electrons, as happens in Mn and Fe, leads to partial population of the antibonding, twice degenerate e_{1g}^* orbital, which is composed of the 4p and 4s atomic orbitals of the metal and antibonding π^*-orbitals of graphene and arene (see Table 32.3).

The conjugated π-system of the graphene sheet can effectively redistribute additional electron density donated by the metal atom, and thus decrease the number of electrons on the M–C (graphene) bond, thus weakening it. Deviation from the single-molecule behavior for the M–C (arene) bond (Figure 32.3a) is hardly observed for Ti, V, and Cr. Mn and Fe compounds, as mentioned above, demonstrate about 3% shortening of the bond. We relate this distortion to the Jahn–Teller effect: unpaired electrons in the Mn and Fe compounds occupy a doubly degenerate e_{1g}^* level, and therefore the Mn and Fe compounds undergo geometrical distortion that removes degeneracy. This asymmetry in bond length also leads to the difference in the C (arene)–M–C (graphene) angles: for Ti–Cr, they exceed those for the free molecule, but for Mn–Fe, they are less. This phenomenon results in less mixing of the localized atomic d-orbitals and leads to the creation of narrow bands in the band structure of MA-Gr, decreasing the band gap compared to the corresponding Ti–Cr compounds.

32.3.4 Electronic Properties

The graphene–metal interaction in haptic functionalization has direct consequences on the electronic properties of graphene. As previously reported, the functionalization of graphene with radical (primarily hydrogen, epoxide, and nitrene) groups locally disrupt the planarity of the graphene sheet, changing the local hybridization from sp^2 to sp^3 geometry [16–18], which induces an sp^3-type defect-like state near the Fermi level. (Hereinafter, under the Fermi level in insulators, we understand the top of the valence band, where it is conventionally put by the majority of the DFT programs.) In

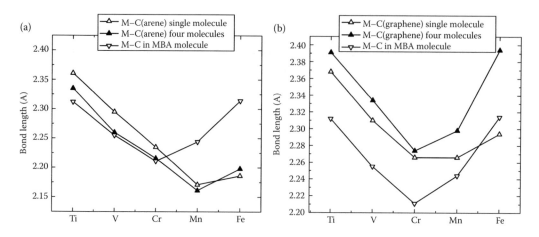

FIGURE 32.3 Geometrical parameters of the MA-Gr. (a) M–C (arene) bond length. (b) M–C (graphene) bond length. (P. Plachinda, D. Evans, and R. Solanki, Modification of graphene band structure by haptic functionalization, *2011 11th IEEE Conference on Nanotechnology (IEEE-NANO)*, Aug. 15–18, pp. 1187–1192. © (2011) IEEE. With permission.)

our case, however, the graphene sheet is not distorted in the z-direction and thus rehybridization of carbon atoms does not occur. The local bonding configuration is, however, significantly affected by the electronic structure of the functionalizing atom, and especially its d-electrons that were found to lie close to the Fermi level. This is similar to the situation with sp^3-type "impurity" states for radical functionalization. Partially occupied, highly localized d-orbitals near the Fermi level cause repulsion of the π-bands, causing the energy band of pristine graphene to be shifted away from the Fermi level due to the π–d interaction.

The calculated band structures for CrBA-functionalized graphene are compared with pristine graphene in Figure 32.4. It is readily observable that after haptic functionalization, the linear dispersion law of pristine graphene at the Dirac point is entirely broken. Since the calculations were conducted using a 6×6 supercell, the K point maps to the Γ point, due to the folding of the reciprocal space. The π- and π*-bands in the Γ direction in pristine graphene have a separation of about

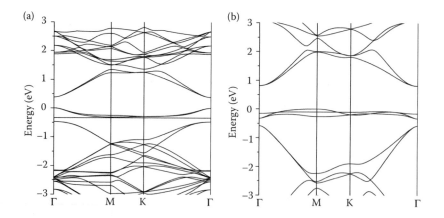

FIGURE 32.4 Calculated band structures for one single-molecule CrBA-functionalized graphene (a) and four-molecule CrBA-functionalized graphene (b). (P. Plachinda, D. Evans, and R. Solanki, Modification of graphene band structure by haptic functionalization, *2011 11th IEEE Conference on Nanotechnology (IEEE-NANO)*, Aug. 15–18, pp. 1187–1192. © (2011) IEEE. With permission.)

11.26 eV. The bands that were previously intersecting at the Dirac point are now shifted together at the new Γ point of the supercell. The π- and π^*-bands preserve their arrangement in the functionalized graphene; however, the distance between them grows from about 1 eV in 1-MA-Gr to 1.25 eV in 4-MA-Gr. (This growth is attributed to the shift of the π^*-band by +0.25 eV relative to the 1-MA-Gr.) As the analysis of the density of states calculated for different atoms and projected on different angular momenta demonstrates, a system of pure (with no π-admixture) localized d-bands of the metal is now located between the π- and π^*-bands of graphene, preventing them from crossing. These d-bands cause a strong repulsion and are responsible for the opening of the gap now between the bands produced by the metal. These electronic properties of MA-functionalized products are in contrast to the sp^3 rehybridization and loss of π-electrons found upon the addition of acceptor chemical groups or metals in other functionalization schemes [15,25,40]. The fact that the band gap strongly depends on the nature of the functionalizing metal atom confirms our idea about the importance of the number of d-electrons or the modification of the band structure. Both occupied and empty d-levels of the metal form flat bands close to the Fermi level. These "impurity" states are probably responsible for bringing the strongest contribution to the band repulsion, more than the d-admixture of the former pure π- and π^*-bands. Additional flat d-bands produced by the localized electrons of the metal atoms in the MA-Gr can be successfully utilized to mimic dopant levels of conventional semiconductors.

An important property of the MA-induced perturbation of the band structure is that alteration in the electronic structure of graphene increases with the increasing MA functionalization concentration. We have investigated the functionalization of graphene at a higher adduct concentration by including three more MA functional groups in the unit cell (see Figure 32.2). Owing to the limitations imposed by the nature of the DFT calculations, we cannot study the variation of the electronic properties imposed by continuously changing the concentration of functionalizing molecules. This would require dealing with huge supercells. Two aforementioned geometries correspond to one functionalizing molecule per 6 × 6 and 3 × 3 graphene supercells. An important idea is to demonstrate the ability to tweak the band gap by varying the concentration, possibly even in the broader limits than discussed here. This corresponds to one MA molecule per 3 × 3 graphene supercell. The exact positions of functionalizing molecules inside the unit cell are not important since there are many ways to redefine the lattice. This geometry is further labeled as 4-MA-Gr unlike 1-MA-Gr with only one MA per 6 × 6 graphene supercell. As the concentration of functionalizing molecules increases (i.e., by transition from 1-MA-Gr to 4-MA-Gr), repulsion between the π-bands (graphene) increases as well, leading to a wider band opening. The extracted energy gap is 0.44 and 0.98 eV for one and four CrBA adducts (i.e., CrBA-Gr and 4-GrBA-Gr), respectively, on a graphene unit cell consisting of 72 graphene-carbon atoms. A higher number of d-bands complicates the picture. The distance between the π-bands in 1-VBA-Gr at the Γ-point is about 0.8 eV but "impurity" levels decrease it 10 times to 0.08 eV. The distance between the d-levels in 1-MnBA-Gr is about 2.7 eV, which is much more than the distance between the π-bands; thus, the highest occupied molecular orbital (HOMO) and the lowest unoccupied molecular level (LUMO) of 1-MnBA-Gr line up with the π- and π^*-bands of graphene. As the number of electrons in the system increases, the Fermi level drifts up, causing a transition from a semiconducting to a metallic state. The band alignments of 1-MnBA-Gr and 1-FeBA-Gr are almost the same; however, owing to the extra electron of Fe, the Fermi level becomes coincident with the former π^*-band, making the iron compound semimetallic. The same trend is observed in transition from 4-MnBA-Gr to 4-FeBA-Gr. The latter becomes metallic for the very same reason: since the Fermi level is located higher in the iron compound, it becomes metallic despite its very close similarity to the band alignment between 4-MnBA-Gr and 4-FeBA-Gr. The band diagrams of all substances under consideration are presented in Figure 32.5.

A closer analysis of band alignment demonstrates that a gap opening can be primarily attributed to the interaction of the d-electrons with the π-conjugated system. Although the carbon atoms on graphene connecting to MA essentially retain a flat band configuration corresponding to sp^2 hybridization, additional π–d interaction is nevertheless present. Local modification of the original

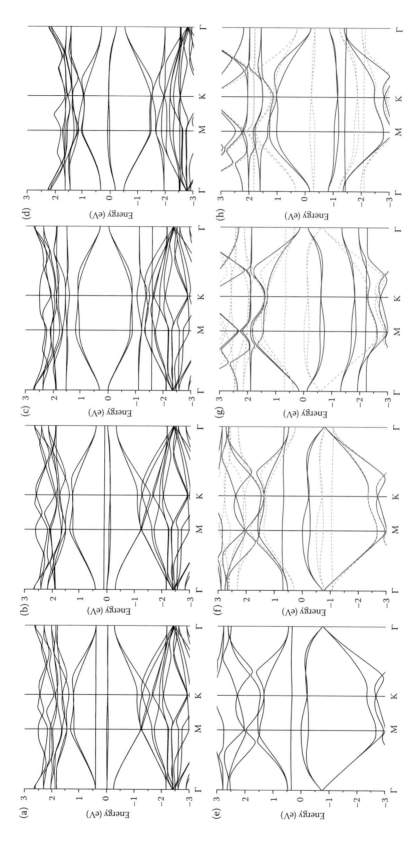

FIGURE 32.5 Band structures of 1-MA-Gr (a–d) and 4-MA-Gr (e–h), where M = Ti (a,e), V (b,f), Mn (c,g), and Fe (d,h). Solid and dashed lines correspond to the spin-up and spin-down bands. Energy reference level coincides with the position of the Fermi level. (P. Plachinda, D. Evans, and R. Solanki, Modification of graphene band structure by haptic functionalization, *2011 11th IEEE Conference on Nanotechnology (IEEE-NANO)*, Aug. 15–18, pp. 1187–1192. © (2011) IEEE. With permission.)

π-conjugation in the vicinity of the metal atom is manifested by rehybridization, that is, the HOMO and the LUMO of MA-Gr now are formed by the π-backbonding [41] mechanism. This rehybridization, however, is in contrast to conventional $sp^2 \rightarrow sp^3$ rehybridization because it occurs without a major geometrical distortion of the underlying graphene sheet. Accordingly, carrier scattering can be substantially regarded as being due to electrostatic interaction similar to that observed for ionized dopant impurities in conventional semiconductors rather than due to localized defect states. As a consequence, it is to be expected that mobility degradation will be much less in MA-Gr than in covalently functionalized graphene for which significant nonplanarity of the graphene sheet is unavoidable. Indeed, this is a crucial difference and was the original motivation for considering this type of functionalization since it seems rather obvious that the preservation of aromaticity, viz., sp^2 hybridization, should result in less degradation of carrier transport properties. Of course, this must be confirmed experimentally and work in that direction is in progress.

The charge densities of the corresponding HOMO/LUMO at the band center (the Γ point) are shown in Figure 32.5. Different atoms demonstrate different mixing of atomic orbitals that take part in the formation of HOMO and LUMO. The HOMO of 1-CrBA-Gr and 1-MnBA-Gr (not shown) are constructed by the σ-type donation mechanism (empty d-orbital is interacting with the filled π-orbital). The LUMO of these compounds and 1-VBA-Gr, in turn, demonstrate π-type back-donation behavior (filled d-orbital interacts with an empty π*-orbital). The HOMO of 1-VBA-Gr is entirely represented by the d_z^2 orbital of the metal atom. The MO picture of the 1-iron is somewhat different from that of the other atoms: π-type back-donation for HOMO and the σ-type donation for LUMO. This is to be contrasted with 4-MA functionalization. As can be seen in Figure 32.6, an increase of adduct concentration impedes the donation mechanism causing the HOMOs of 4-Cr to consist of the unhybridized d_z^2 orbitals of the metal atom and the LUMO of the π-type back-donation MOs. Molecular orbitals of other MA-Gr structures are not shown here due to space limitations. However,

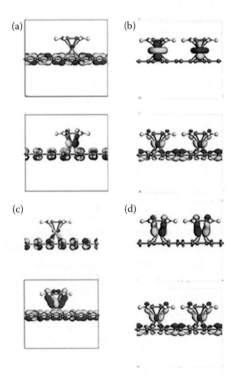

FIGURE 32.6 Molecular orbitals (HOMO—top row, LUMO—bottom row) for 1-CrBA-Gr (a), 4-CrBA-Gr (b), 1-FeBA-Gr (c), and 4-FeBA-Gr (d).

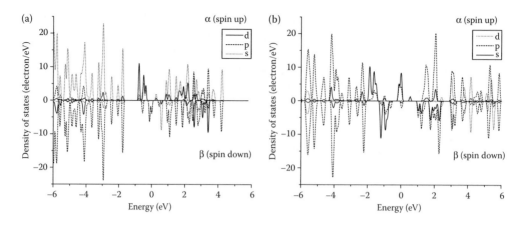

FIGURE 32.7 (a) Spin-resolved density of states in 4-VMB-Gr. (b) Spin-resolved density of states in 4-FeMB-Gr.

the donation mechanism changes depending on the number and energy of the filled d-orbitals of the central atom. A proportional increase of functionalizing molecules causes a change in the amount of π-conjugated bonds broken due to the backbonding mechanism. This correlates with the associated increase of the band gap and thus provides support for the suggested scenario of the d-level-induced band gap opening.

As was previously demonstrated, graphene functionalized with transition metal atoms can demonstrate profound magnetic properties [25]. Band structures of 4-VBA-Gr and 4-FeBA-Gr (Figures 32.7a and 32.7b) demonstrate strong spin polarization. Spin-unrestricted calculations reveal significant differences between the spin-up (α) and spin-down (β) densities of states (DoS) pointing to ferromagnetic behavior of 4-VMB-Gr and 4-FeMB-Gr.

Iron demonstrates an even stronger difference in the DoS for the spin-up and spin-down electrons. Band structures for compounds with different central metal atoms having paired electrons obviously do not demonstrate any ferromagnetic behavior. However, owing to the presence of unpaired electrons, the manganese compound can exist in both ferromagnetic and antiferromagnetic states. A closer analysis of the magnetic properties of the MA-Gr will be published separately. The presence of a nontrivial magnetic structure suggests wide possible applications of the MA-functionalized graphene in spintronics.

32.4 CONCLUSIONS

We have studied the electronic characteristics of MA-functionalized graphene and have shown that the MA adducts mainly preserve the sp^2 hybridization network of the carbons on graphene away from the functionalizing groups. However, the π-conjugation of graphene near the Fermi level is greatly modified by the π-backbonding process, caused by d-orbitals of the metal atom in the functionalizing molecule, which leads to the opening of a substantial band gap dependent upon the adduct concentration and number of the occupied d-orbitals. Band gaps vary from 0.81 eV for 4-CrBA-Gr to 0.11 eV for 1-VBA-Gr, and to zero for 1-FeBA-Gr and 4-MnBA-Gr. Moreover, the electronic structure of the functionalizing metal in the MA molecules allows the possibility of controlled modification of both the band gap itself and the position of the Fermi level with respect to the "d-impurity" levels and native graphene bands. Some functionalizing molecules result in a ferromagnetic behavior along with the opening of the band gap. Such dependence of the electronic properties on the type of functionalizing metal suggests a novel tunable approach for the "band engineering" of graphene. Our findings on the nature of an MA-functionalization-induced band gap provide useful guidelines for enabling the flexibility and optimization of graphene-based nanodevices.

REFERENCES

1. A. K. Geim and K. S. Novoselov, The rise of graphene, *Nature Materials,* 6, 183–91, 2007.
2. M. J. Allen, V. C. Tung, and R. B. Kaner, Honeycomb carbon: A review of graphene, *Chemical Reviews,* 110, 132–145, 2010.
3. C. Rao, A. Sood, K. Subrahmanyam, and A. Govindaraj, Graphene: The new two-dimensional nanomaterial, *Angewandte Chemie International Edition,* 48, 7752–7777, 2009.
4. A. H. C. Neto, F. Guinea, N. M. R. Peres, K. S. Novoselov, and A. K. Geim, The electronic properties of graphene, *Review of Modern Physics,* 81, 109–162, 2009.
5. V. M. Pereira, A. H. Castro Neto, and N. M. R. Peres, Tight-binding approach to uniaxial strain in graphene, *Physical Review B,* 80, 045401, 2009.
6. B. Trauzettel, D. V. Bulaev, D. Loss, and G. Burkard, Spin qubits in graphene quantum dots, *Nature Physics,* 3, 192–196, 2007.
7. K. Nakada, M. Fujita, G. Dresselhaus, and M. S. Dresselhaus, Edge state in graphene ribbons: Nanometer size effect and edge shape dependence, *Physical Review B,* 54, 17954, 1996.
8. L. Brey and H. A. Fertig, Electronic states of graphene nanoribbons studied with the Dirac equation, *Physical Review B,* 73, 235411, 2006.
9. K. A. Ritter and J. W. Lyding, The influence of edge structure on the electronic properties of graphene quantum dots and nanoribbons, *Nature Materials,* 8, 235–242, 2009.
10. X. Li, X. Wang, L. Zhang, S. Lee, and H. Dai, Chemically derived, ultrasmooth graphene nanoribbon semiconductors, *Science,* 319, 1229–1232, 2008.
11. E. V. Castro, K. S. Novoselov, S. V. Morozov, N. M. R. Peres, J. M. B. L. dosSantos, J. Nilsson, F. Guinea, A. K. Geim, and A. H. CastroNeto, Biased bilayer graphene: Semiconductor with a gap tunable by the electric field effect, *Physical Review Letters,* 99, 216802, 2007.
12. E. McCann, Asymmetry gap in the electronic band structure of bilayer graphene, *Physical Review B,* 74, 161403, 2006.
13. Y. Zhang, T.-T. Tang, C. Girit, Z. Hao, M. C. Martin, A. Zettl, M. F. Crommie, Y. R. Shen, and F. Wang, Direct observation of a widely tunable bandgap in bilayer graphene, *Nature,* 459, 820–823, 2009.
14. K. F. Mak, C. H. Lui, J. Shan, and T. F. Heinz, Observation of an electric-field-induced band gap in bilayer graphene by infrared spectroscopy, *Physical Review Letters,* 102, 256405, 2009.
15. D. K. Samarakoon and X.-Q. Wang, Tunable band gap in hydrogenated bilayer graphene, *ACS Nano,* 4, 4126–30, 2010.
16. D. C. Elias, R. R. Nair, T. M. G. Mohiuddin, S. V. Morozov, P. Blake, M. P. Halsall, A. C. Ferrari et al., Control of graphene's properties by reversible hydrogenation: Evidence for graphane, *Science,* 323, 610–613, 2009.
17. O. Leenaerts, B. Partoens, and F. M. Peeters, Hydrogenation of bilayer graphene and the formation of bilayer graphane from first principles, *Physical Review B,* 80, 245422, 2009.
18. M. Z. S. Flores et al., Graphene to graphane: A theoretical study, *Nanotechnology,* 20, 465704, 2009.
19. J. Choi, K.-j. Kim, B. Kim, H. Lee, and S. Kim, Covalent functionalization of epitaxial graphene by azidotrimethylsilane, *The Journal of Physical Chemistry C,* 113, 9433–9435, 2009.
20. M. Quintana, K. Spyrou, M. Grzelczak, W. R. Browne, P. Rudolf, and M. Prato, Functionalization of graphene via 1,3-dipolar cycloaddition, *ACS Nano,* 4, 3527–3533.
21. L.-H. Liu and M. Yan, Simple method for the covalent immobilization of graphene, *Nano Letters,* 9, 3375–3378, 2009.
22. L.-H. Liu, G. Zorn, D. G. Castner, R. Solanki, M. M. Lerner, and M. Yan, A simple and scalable route to wafer-size patterned graphene, *Journal of Materials Chemistry,* 20, 5041–5046, 2010.
23. L.-H. Liu, M. M. Lerner, and M. Yan, Derivitization of pristine graphene with well-defined chemical functionalities, *Nano Letters,* 10, 3754–3756.
24. L.-H. Liu, G. Nandamuri, R. Solanki, and M. Yan, Electrical properties of covalently immobilized single-layer graphene devices, *Journal of Nanoscience and Nanotechnology,* 11, 1288–1292, 2011.
25. A. V. Krasheninnikov, P. O. Lehtinen, A. S. Foster, P. Pyykko, and R. M. Nieminen, Embedding transition-metal atoms in graphene: Structure, bonding, and magnetism, *Physical Review Letters,* 102, 126807, 2009.
26. K. Suggs, D. Reuven, and X.-Q. Wang, Electronic properties of cycloaddition-functionalized graphene, *Journal of Physical Chemistry C,* 115, 3313–3317, 2011.
27. DMol3. San Diego, CA: Accelrys Software Inc., 2010.
28. A. D. Becke, Density-functional exchange-energy approximation with correct asymptotic behavior, *Physical Review A,* 38, 3098, 1988.

29. C. Lee, W. Yang, and R. G. Parr, Development of the Colle-Salvetti correlation-energy formula into a functional of the electron density, *Physical Review B,* 37, 785, 1988.

30. S. H. Vosko, L. Wilk, and M. Nusair, Accurate spin-dependent electron liquid correlation energies for local spin density calculations: A critical analysis *Canadian Journal of Physics,* 58, 1200–1211, 1980.

31. Ishii, M. Yamamoto, H. Asano, and K. Fujiwara, DFT calculation for adatom adsorption on graphene sheet as a prototype of carbon nanotube functionalization, *Journal of Physics: Conference Series,* 100, 052087, 2008.

32. K. Suggs, D. Reuven, and X.-Q. Wang, Electronic properties of cycloaddition functionalized graphene, *The Journal of Physical Chemistry C,* 115(8), 3313–3317, 2011.

33. V. Zólyomi, Á. Rusznyák, J. Koltai, J. Kürti, and C. J. Lambert, Functionalization of graphene with transition metals, *Physica Status Solidi (B),* 247(11–12), 2920–2923, 2010.

34. O. Leenaerts, B. Partoens, and F. M. Peeters, Paramagnetic adsorbates on graphene: A charge transfer analysis, *Applied Physics Letters,* 92, 243125, 2008.

35. R. L. Brandon, J. H. Osiecki, and A. Ottenberg, The reactions of metallocenes with electron acceptors 1a, *The Journal of Organic Chemistry,* 31, 1214–1217, 1966.

36. C. N. R. Rao, K. S. Subrahmanyam, H. S. S. Ramakrishna Matte, B. Abdulhakeem, A. Govindaraj, B. Das, P. Kumar, A. Ghosh, and D. J. Late, A study of the synthetic methods and properties of graphenes, *Science and Technology of Advanced Materials,* 11, 054502, 2010.

37. D. Astruc, Chapter 11: Metallocenes and sandwich complexes, in *Organometallic Chemistry and Catalysis* Berlin: Springer, 2007, pp. 251–288.

38. C. Elschenbroich and R. Möckel, Bis(η6-naphthalin)chrom(0), *Angewandte Chemie,* 89, 908–909, 1977.

39. E. P. Kundig and P. L. Timms, Metal atom preparation and ligand displacement reactions of bisnaphthalenechromium and related compounds, *Journal of the Chemical Society, Chemical Communications,* pp. 912–913, 1977.

40. S. M.-M. Dubois, Z. Zanolli, X. Declerck, and J.-C. Charlier, Electronic properties and quantum transport in Graphene-based nanostructures, *The European Physical Journal B,* 72, 1–24, 2009.

41. A. D. McNaught and A. Wilkinson, *IUPAC. Compendium of Chemical Terminology, 2nd ed. (The Gold Book).* Oxford: Blackwell Scientific Publications, 2006.

42. P. Plachinda, D. Evans, and R. Solanki, Modification of graphene band structure by haptic functionalization, *2011 11th IEEE Conference on Nanotechnology (IEEE-NANO),* Aug. 15–18, pp. 1187–1192.

Section X

Carbon Nanotube Applications

33 Integrating Low-Temperature Carbon Nanotubes as Vertical Interconnects in Si Technology

Sten Vollebregt, Ryoichi Ishihara, Jaber Derakhshandeh,
Johan van der Cingel, Hugo Schellevis, and C. I. M. Beenakker

CONTENTS

33.1 INTRODUCTION

Since their discovery, many applications have been proposed for vertically aligned carbon nanotubes (CNT). Some potential areas for the use of CNT are supercapacitors [1], microelectromechanical systems (MEMS) [2], and displays [3]. For these applications, the unique high-aspect-ratio (HAR) features of CNT and their large surface area compared to their volume can be exploited. One promising application within microelectronics that has received a lot of attention is the use of aligned CNT as vertical interconnects [4].

CNT are attractive for interconnects as the currently used interconnect materials like Cu are approaching their physical limitations due to the continued downscaling in the semiconductor manufacturing process. In the roadmap for the semiconductor industry made by the International Technology Roadmap for Semiconductors (ITRS), it can be found that around 2015, for the highest-performance integrated circuits, the electrical current density in the interconnects surpasses the maximum current density of Cu, with no manufacturable known solutions available, as shown in Figure 33.1 [5]. CNT, on the other hand, have been demonstrated to be able to carry current densities up to 10^9 A/cm^2 [6]. Besides that, CNT have been demonstrated to be able to transport heat very efficiently, with thermal conductivities up to 3500 W/mK being demonstrated at room temperature (in comparison, the thermal conductivity of Cu is about 343 W/mK) [7]. This could aid in decreasing the temperature of the interconnect stack, again aiding in improving reliability [8].

Besides the clear reliability advantage, several other issues exist within the interconnect technology. For local interconnects, which form the connections between logic blocks like adders, the interconnect sizes are being scaled with the same trend as the transistor size. Owing to this, the cross section of the interconnect has been pushed well into the nanometer regime. Due to grain boundary and surface scattering, the resistivity of Cu increases by 2–5 times the bulk value [9].

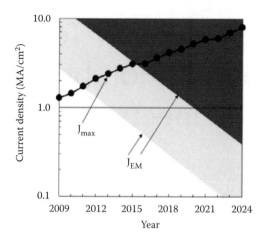

FIGURE 33.1 Progression of J_{max} over the years according to the 2010 update of the ITRS roadmap. (From the International Technology Roadmap for Semiconductors, 2010, http://public.itrs.net/.)

This effect becomes more severe when the interconnect size is further reduced. Owing to their one-dimensional (1D) nature, CNT do not suffer from these scaling effects. However, this 1D nature does introduce a quantum resistance of 6.45 kΩ per single-walled nanotube (SWCNT), while the situation is slightly more complex for multiwalled nanotubes (MWCNT) [10]. This basically means that many parallel conducting tubes are required to achieve an electrical conductivity close to copper. Simulations have shown that dense, vertically aligned SWCNT bundles can indeed compete with Cu vias, while dense MWCNT bundles display a higher resistance [11,12]. Still, as via resistance is only a small part of the total interconnect resistance, a small increase can be acceptable if other properties like electrical reliability are superior to that of Cu.

For vertical interconnects (vias), a key requirement is thus selective growth of vertically aligned CNT directly on top of electrically conductive layers with high tube density. Without the direct growth on a conductive layer, difficult transfer techniques would be necessary to allow electrical contact to the CNT. For the best electrical performance, the CNT density should be high, as bundle resistance decreases if the number of tubes is increased as was mentioned before. Low-temperature deposition (preferably below 500°C, or 400°C for modern low-k dielectrics) is important to allow integration with already-fabricated devices (e.g., CMOS transistors). If growth temperature can be brought back down to 350°C, it will even allow growth on certain flexible substrates.

Two approaches can be defined to integrate interconnects: the traditional top-down approach and the bottom-up approach as presented by Li et al. [13]. Both methods are shown in Figure 33.2. In the top-down approach, a contact opening is etched using a plasma through the dielectric between the metal layers, followed by CNT growth inside the opening and subsequent metallization. In case CNT density is low and/or the CNT height is longer than the oxide thickness, planarization might be required. In case of the bottom-up approach, CNT bundles are first grown at the desired location, covered by a dielectric, planarized, and finally covered by the next metallization. The distinct advantage of the bottom-up method is that it allows the creation of HAR vias, without the need for the etching of, and metal deposition in, HAR openings. The bottom-up approach is also attractive for HAR MEMS [14].

Few publications exist that use mainstream silicon process technology to integrate low-temperature CNT growth, and combine this with electrical characterization of the as-grown CNT bundles. In this chapter, we demonstrate vertically aligned high-density growth of CNT on sputtered TiN layers at temperatures as low as 500°C. We found that specific processing steps used in silicon technology on an exposed TiN diffusion barrier can have a large impact on the CNT growth at low temperature. We will discuss the processing steps we found to be harmful to CNT growth, and a

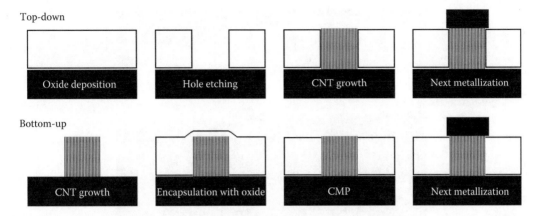

FIGURE 33.2   Graphical representation of top-down and bottom-up integration process.

method to prevent damage. Using this, we demonstrate integrated top-down CNT via structures to measure the bundle resistance of the vertically aligned CNT grown at 500°C. Finally, we demonstrate that it is possible to use plasma-enhanced chemical vapor deposited (PECVD) silicon oxide and nitride to cover CNT for bottom-up integration.

33.2   EXPERIMENTAL

In this section, we describe the process steps employed to create structures to electrically characterize the vertically grown CNT. Previously, we demonstrated that resist residues remaining after standard photolithography and wet etching of the catalyst during cleaning can negatively impact growth [15]. To prevent damage to the catalyst layer, we changed to a lift-off process using pure negative resist to pattern the Fe catalyst layer.

In case of a lift-off, a thin layer is deposited on top of a photoresist layer, and subsequently removed by dissolving the photoresist and "lifting off" the layer on top. This in contrast to general lithography, where a layer is removed by selective etching through a (photo-resist) mask. The minimum feature sizes obtained recently with this method are 0.8 μm holes and 0.5 μm lines, which is limited by our ASML PAS 5500/80 waferstepper. Another advantage of lift-off is that it allows the patterning of materials that are difficult to etch, like Pd, which we recently demonstrated to allow low-temperature, high-density vertically aligned growth [16].

CNTs are grown using a commercially available AIXTRON BlackMagic Pro 4″ chemical vapor deposition (CVD) reactor. We used 100 mm p-type Si wafers as substrates. For the measurement samples, we sputtered 500 nm of Ti and 50 nm of TiN on the Si wafers using an SPTS Sigma sputter coater. Then, 1.5 μm of tetraethyl orthosilicate (TEOS) is deposited using a Novellus Concept One PECVD reactor. After this, the wafer is coated with 1.5 μm AZ nLOF2020 negative resist, exposed, and developed. This is followed by dry etching of the oxide using an LAM Drytek 384T etcher with fluorine chemistry. The resist is not stripped but used to perform a lift-off, thus enabling automatic alignment. Before this, 5 nm of Fe is evaporated using a Solution CHA e-beam evaporator. Lift-off is performed with n-methyl-2-pyrrolidone (NMP) at 70°C. Normally, no ultrasonic treatment is required to completely remove the resist, making the method attractive for fabrication.

The catalyst layer is activated by annealing the sample inside the CNT reactor at 500°C with 700 sccm H_2 flowing. During this step, the Fe layer is calcined and break up into a cluster of nanoparticles. From these nanoparticles, CNT can be grown using CVD. After 3 min, 50 sccm of C_2H_2 is added and CNT are grown for 5 min to reach the required height of 1.5 μm. During both steps, the pressure is regulated to 80 mbar. The CNT are covered by a second metallization of 100 nm of Ti and 3 μm of Al using sputtering. Finally, the Al and Ti layers are patterned using lithography and

FIGURE 33.3 SEM cross section of CNT via fabrication using a described process with materials and thicknesses as indicated. Note that for this wafer 2 μm of TEOS was used.

wet etching. In Figure 33.3, a scanning electron microscope (SEM) cross section of a CNT via can be found fabricated using this process with the different layer thicknesses indicated.

To analyze the samples, we used an FEI/Philips XL50 SEM, Renishaw inVia Raman spectroscope with 633 nm laser, NT-MDT nTegra atomic force microscope (AFM), and an Agilent 4156C parameter analyzer in combination with a semiautomatic probe station. In case of Raman spectroscopy, three measurements are performed for each sample, which are averaged to increase accuracy.

33.3 RESULTS AND DISCUSSION

In this section, we discuss the effect of different surface treatments on the TiN support layer and the resulting low- and high-temperature CNT growth on that layer. Afterward, using a sacrificial layer to protect the TiN, we employed the top-down approach to fabricate and subsequently electrically characterize the as-grown CNT bundles. Finally, we investigate the change in crystallinity of CNT bundles covered with PECVD oxide or nitride, which is required for bottom-up integration.

33.3.1 SURFACE TREATMENT OF TiN

We use TiN as a diffusion barrier for our electrical measurement structures, since it is not as sensitive to oxidation in air as Ti (and thus allows transfer of wafers between machines) and it is known to form a good electrical contact to CNT [17]. Besides that, TiN is a well-known diffusion barrier in semiconductor technology. It was found that certain common process steps used in semiconductor technology can induce microscopic changes in the TiN layer, preventing the low-temperature, self-aligned growth of CNT. We investigated the impact of chemical solutions and plasma treatment on the TiN and the subsequent low- and high-temperature CNT growth.

First, we investigate which treatments prevent low-temperature growth. For this, we created five different samples with the same TiN layer and exposed this layer to different conditions (see also Table 33.1): A: 10 min. 99.9% HNO_3 (our standard metal clean); B: 1 min. 0.55% HF (used to etch SiO_2 and Ti); C: 1 kW oxygen plasma (used for resist stripping); D: 100 W fluorine chemistry soft landing step used for oxide etching; E: pristine TiN. After this, we evaporated Fe and grew CNT at 500°C (low temperature) and 650°C (high temperature).

In Figure 33.4, the SEM images taken from the resulting CNT growth on the five different samples at low and high temperatures can be found. As can be directly observed, both plasma-treated samples display no self-aligned vertical growth at low temperature. The other three samples grow

TABLE 33.1

Overview of Surface Treatments Used on Different Samples

Sample	Treatment
A	10 min HNO_3 (99.9%)
B	1 min HF (0.55%)
C	5 min 1 kW oxygen plasma
D	1 min 100 W fluorine plasma
E	No treatment

Source: S. Vollebregt et al., Integrating low temperature aligned carbon nanotubes as vertical interconnects in Si technology, *11th IEEE Conference on Nanotechnology*, Portland, Oregon, August 15–19, pp. 985–990. © (2011) IEEE. With permission.

self-aligned CNT with a height of several microns. Interestingly, sample B displays the highest CNT (3.7 μm), followed by sample A (3.3 μm) and E (2.5 μm). We are unsure if this is caused by slight differences in the treated surface (HF might passivate the TiN surface) or due to small temperature differences in the reactor (as growth rate is highly temperature dependent). We approximate the tube density to be in the order of 10^{11} tubes/cm^2 for all aligned samples, and using transmission electron microscopy, we determined the tube diameter to be in the range of 10–15 nm (not shown here).

The high-temperature samples again display a difference between the samples A, B, E and C, D. Although self-aligned growth is now possible in all samples, the CNT height of sample C and D is approximately half of that of the other samples. Also, density and alignment suffer from plasma treatment. At high temperature, the length of samples A and B is found to be the same (31 μm), while E is slightly longer (35 μm), which is in contrast to the situation at low temperature.

Using Raman spectroscopy, we investigated if the treatment of the support layer influenced the crystal quality of the CNT growth. Raman spectroscopy is a nondestructive and fast method to investigate the crystallinity of CNT as several bands in the spectrum have been shown to be sensitive to defects [18]. Here, we investigated the D, D′, and G′ bands, and determined their intensity to the graphite reference G band. As is known from the literature, the $I_{D/G}$ and $I_{D'/G}$ ratios normally show a decline for increasing crystallinity, while $I_{G'/G}$ increases [18,19]. A decreasing band full width at half maximum (FWHM) is also associated with increasing crystallinity [18,20]. In Table 33.2, the intensity ratios and FWHM of the different Raman bands are listed.

Table 33.2 clearly shows that CNT grown at higher temperatures have higher crystallinity. The band ratios and widths of samples A, B, and E are close to each other, indicating that the chemical treatment of the TiN surface with either HNO_3 or HF has no profound impact on CNT crystallinity. On the other hand, samples C and D display a different behavior. The $I_{D/G}$ ratio is significantly higher for the low-temperature C sample, and lower for the high-temperature C and D samples. This could indicate lower and higher crystallinity for the low- and high-temperature samples, respectively. On the other hand, the FWHM of the low-temperature growth on sample C suggests a higher crystallinity compared to samples A, B, D, and E. The changes in $I_{D'/G}$ and $I_{G'/G}$ are less profound. We can thus conclude that the plasma treatment appears to improve CNT crystallinity for the high-temperature samples, while for low temperature, the results are inconclusive.

To investigate the potential causes for the change in low- and high-temperature CNT growth on plasma-treated surfaces, we measured the sample surface after the treatment with AFM, which are shown in Figure 33.5. As can be seen, no significant difference exists between samples A and E (and B, which is not displayed here). Sample C, on the other hand, appears to have a more smooth surface (i.e., less sharp edges between the different TiN grains). The change in surface roughness is only minor and unlikely to be the cause for the change in growth. Sample D displays the largest change.

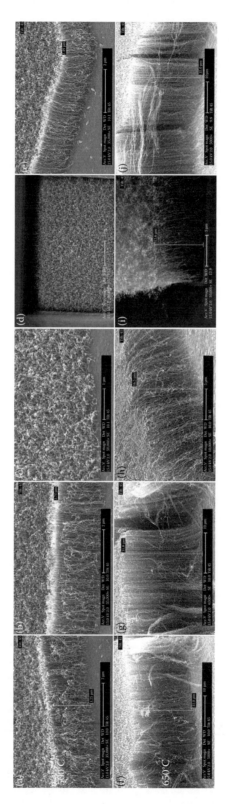

FIGURE 33.4 SEM images of CNT grown on differently treated TiN substrates at 500 and 650°C: (a)–(e) Samples A through E at 500°C; (f)–(j) Samples A through E at 650°C. See also Table 33.1 for an overview of treatment conditions for samples A through E. (S. Vollebregt et al., Integrating low temperature aligned carbon nanotubes as vertical interconnects in Si technology, *11th IEEE Conference on Nanotechnology*, Portland, Oregon, August 15–19, pp. 985–990. © (2011) IEEE. With permission.)

TABLE 33.2

Raman Data Obtained from CNT Grown on Different Samples

Sample	$I_{D/G}$	$I_{D'/G}$	$I_{G'/G}$	FWHM (cm^{-1}) D	G	D'	G'
A: 500°C	2.84	1.24	0.30	81	63	44	148
650°C	2.22	0.81	0.88	54	47	39	92
B: 500°C	2.80	1.18	0.27	81	62	43	151
650°C	2.28	0.75	0.63	52	47	37	90
C: 500°C	3.13	1.30	0.44	69	59	44	116
650°C	1.56	0.57	0.78	50	42	33	75
D: 500°C	2.66	1.29	0.37	85	63	45	136
650°C	1.63	0.73	0.79	57	45	38	81
E: 500°C	2.74	1.14	0.30	80	62	43	148
650°C	2.13	0.68	0.60	53	46	36	93

Source: S. Vollebregt et al., Integrating low temperature aligned carbon nanotubes as vertical interconnects in Si technology, *11th IEEE Conference on Nanotechnology*, Portland, Oregon, August 15–19, pp. 985–990. © (2011) IEEE. With permission.

It appears that the TiN layer was partly sputtered by the fluorine plasma, forming clusters of small particles. The surface morphology of sample D is changed extensively, which could account for the observed change in CNT growth on these samples.

To further examine the influence of the treatment on the TiN layer and subsequent CNT growth, we also performed AFM on samples on which the Fe catalyst was deposited and activated for CNT growth. The samples were placed in the CNT reactor after Fe evaporation, followed by 3 min of annealing at 500°C, while H_2 was flowing. This is the default activation step for both low- and high-temperature CNT growth. After this, the samples were cooled down in a N_2 environment. Figure 33.6 displays the AFM results. As can be seen, both samples A and E display small nanoparticles of similar size (the bigger bright spots on sample A are most likely particles deposited on the wafer during HNO_3 treatment). The same holds for sample B (not shown here). On samples C and D, however, besides the small nanoparticles, a significant amount of larger nanoparticles can be found. This will result in CNT growth with large diameter distribution and, most likely, lower density. This can explain the absence of self-alignment at low growth temperatures for samples C and D. It can also explain the higher crystal quality observed by Raman spectroscopy on those samples. As found by Antunes et al. [20], CNT grown from larger nanoparticles at the same temperature show a lower $I_{D/G}$, $I_{D'/G}$, and FWHM, and a higher $I_{G'/G}$. This, however, does not account for the large $I_{D/G}$ ratio observed for the low-temperature C sample. Sample D has low-temperature Raman data that match closer with that obtained from sample E. Indeed, the nanoparticle distribution of sample D is, with respect to sample C, more similar to that of sample E.

Increased surface roughness of sample D is most likely the cause of the observed broader particle deposition. For sample C, we believe that the oxidation of the TiN surface due to the oxygen plasma (in case of our Ti/TiN stack, sheet resistance increases from 1.158 Ω/sq to 1.165 Ω/sq) alters the surface properties in such a way that activation of the catalyst layer becomes more difficult.

33.3.2 *I–V* Characterization of Top-Down Integrated CNT

We created four-point probe vertical interconnect measurement structures using a top-down approach as described in the experimental section and shown in the inset of Figure 33.8. To protect

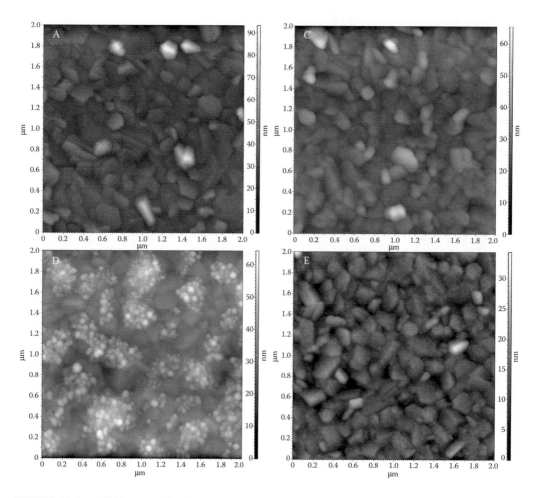

FIGURE 33.5 AFM images taken from samples A, C, D, and E displaying the effect of the surface treatments on the TiN surface. (S. Vollebregt et al., Integrating low temperature aligned carbon nanotubes as vertical interconnects in Si technology, *11th IEEE Conference on Nanotechnology*, Portland, Oregon, August 15–19, pp. 985–990. © (2011) IEEE. With permission.)

the TiN layer from damage during plasma etching, we sputtered an additional 100 nm of Ti on top of this layer before TEOS deposition. During the contact opening etch, we stop on this layer without completely removing it (the etch rate of Ti is low in our fluorine etcher). Before Fe evaporation, we remove the sacrificial Ti layer by a short 60 s etch in 0.55% HF. Using this method, we are able to grow perfectly aligned high-density bundles of CNT within the contact openings at 500°C; see also Figure 33.7a. As the CNT height was controlled to be the same as the depth of the opening, no planarization is required for the second metallization. As can be seen in Figure 33.7b, good contact is achieved, although some irregularities exist between the bulk metal and the metal on the CNT due to the small spacing between the oxide opening and the CNT bundle.

We measured the resistivity of the square and round CNT bundles with different diameters. The resulting *I–V* characteristics of the square structures can be found in Figure 33.8. As can be seen, all structures display good linearity, indicating good metal–CNT contact without a significant Schottky barrier.

The resistivities of the structures can be found in Figure 33.9. For comparison, we also added values of several publications found in the literature [21–24]. As can be seen, smaller structures

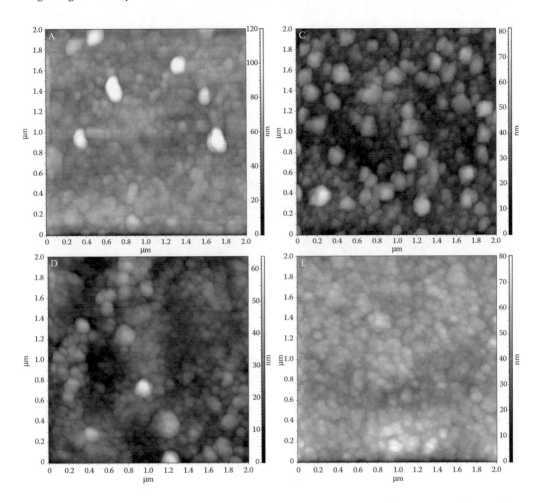

FIGURE 33.6   AFM images taken from samples A, C, D, and E displaying the catalyst nanoparticles after activation. (S. Vollebregt et al., Integrating low temperature aligned carbon nanotubes as vertical interconnects in Si technology, *11th IEEE Conference on Nanotechnology*, Portland, Oregon, August 15–19, pp. 985–990. © (2011) IEEE. With permission.)

FIGURE 33.7   SEM images taken from CNT growth in contact openings (a) before and (b) after metallization. (S. Vollebregt et al., Integrating low temperature aligned carbon nanotubes as vertical interconnects in Si technology, *11th IEEE Conference on Nanotechnology*, Portland, Oregon, August 15–19, pp. 985–990. © (2011) IEEE. With permission.)

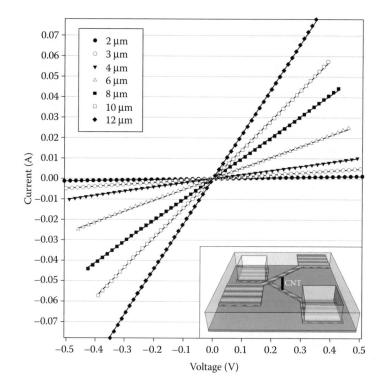

FIGURE 33.8 *I–V* measurements of structures with different bundle sizes. Inset: Four-point probe measurement structure.

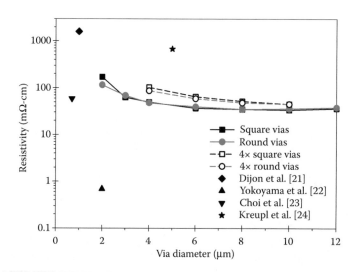

FIGURE 33.9 Extracted resistivity of structures with different diameter and morphology compared to other values found in the literature. (S. Vollebregt et al., Integrating low temperature aligned carbon nanotubes as vertical interconnects in Si technology, *11th IEEE Conference on Nanotechnology*, Portland, Oregon, August 15–19, pp. 985–990. © (2011) IEEE. With permission.)

suffer from an increase in resistivity, probably caused by a combination of a not-yet optimized lithographic process that slightly undersizes the smaller structures. A good match can be found between the square and round structures, indicating that for this size, there is no preferential morphology for creating CNT vertical interconnect bundles, as expected. We also measured the resistivity of structures consisting of four smaller bundles spaced by 3 μm with a total area of a larger bundle (e.g., 4 × 4 μm to get the same area as a single 8 μm bundle). As can be seen from Figure 33.9, the resistivity of a single large bundle is always lower than that of several smaller bundles.

Compared to the other values found in the literature, our average resistivity of 35 mΩ-cm is between the lowest and highest found. It must be noted that the value from Yokoyama et al. [22] is achieved with structures fabricated with very specific equipment after a long optimization process, while our reported values are not yet optimized. Our growth temperature (500°C) is relatively low compared to the temperatures used by Dijon et al. (590°C, [21]), Choi et al. (600°C, [23]), and Kreupl et al. (700°C, [24]). We believe our resistivity can still be reduced by at least one order of magnitude by further optimizing our process.

As the CNT all have diameters between 10 and 15 nm, the grown CNT are multiwalled, which implies that they are all (semi)metallic [25]. This is confirmed by the Raman spectra (not shown here), which do not show any radial breathing modes, the fingerprints of SWCNT [26]. To accurately determine the average tube resistivity, we require the contact resistance, which is nonneglectable for CNT. In a recent publication, we investigated the CNT resistance versus length to determine the contact resistance, and found that this resistance is low compared to the CNT bundle resistance [27]. We attribute this to embedding of the CNT tips by the sputtered Ti layer used for the second metallization.

33.3.3 COVERING CNT FOR BOTTOM-UP

To allow low-temperature bottom-up integration, CNT have to be covered by a low-temperature dielectric, in contrast to the high-temperature LPCVD TEOS used by Li et al. [13]. For low-temperature dielectric deposition, PECVD is the preferred method in the semiconductor industry. However, it is likely that the plasma might damage the grown CNT by ion bombardment or oxidation. To investigate this, we grew freestanding bundles of CNT with a height of approximately 5 μm on TiN substrates (Figure 33.10a) and covered them with PECVD TEOS, silicon oxide, and nitride at temperatures of 400, and 350°C for the TEOS deposition.

Figure 33.10b shows an SEM image taken of an array of 5 × 5 2-μm-wide CNT bundles covered by 1 μm TEOS (deposited using TEOS and O_2). As can be seen, good step coverage is achieved, although deposition on the sidewalls of the CNT is lower compared to the total thickness of the TEOS layer (approximately 600 nm). In Figure 33.10c, the same array is completely covered by 5 μm of PECVD oxide (deposited from SiH_4 and N_2O). Planarization will be necessary to remove the excess oxide. Finally, Figure 33.10d shows the array covered by 200 nm silicon nitride (from SiH_4 and NH_3). As can be seen, for unknown reasons, the deposition is significantly less smooth compared to the oxide depositions.

Using Raman spectroscopy, we investigated the CNT crystallinity before and after PECVD dielectric deposition of 1 μm of TEOS or oxide and 200 nm nitride. As can be seen in Figure 33.11, only minor changes to the crystallinity can be observed. The TEOS deposition appears to induce the least amount of damage, with only the width of the D band (around 1330 cm^{-1}) increasing slightly. Oxide deposition from silane and nitrous oxide, on the other hand, display an increase of the D band, indicating less crystallinity. The width of the D band is similar to that of the D band after TEOS deposition. Finally, nitride deposition has a D band intensity in between that of CNT covered by TEOS and oxide. Again, the width of the D band is similar to that of the TEOS-covered CNT. In the second-order band region, changes are even smaller (not shown here). We can thus conclude that PECVD deposition does not induce a significant amount of defects in the CNT bundles, opening

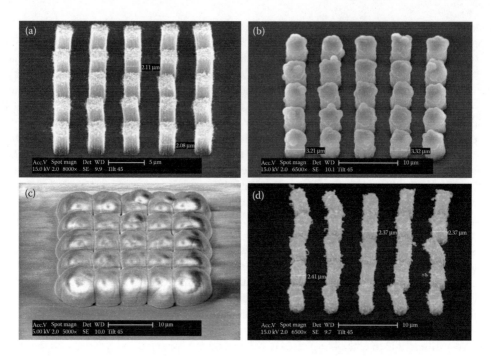

FIGURE 33.10 SEM images of (a) freestanding bundles grown on TiN, (b) bundles covered by 1 μm TEOS, (c) bundles covered by 5 μm SiO_2, and (d) bundles covered by 200 nm Si_3N_4. (S. Vollebregt et al., Integrating low temperature aligned carbon nanotubes as vertical interconnects in Si technology, *11th IEEE Conference on Nanotechnology*, Portland, Oregon, August 15–19, pp. 985–990. © (2011) IEEE. With permission.)

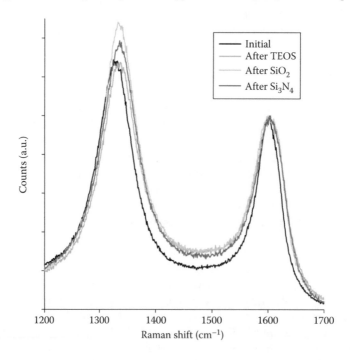

FIGURE 33.11 Raman spectra of CNT before and after covering with PECVD dielectrics. The intensity is normalized to the G-band at 1580 cm^{-1}. (S. Vollebregt et al., Integrating low temperature aligned carbon nanotubes as vertical interconnects in Si technology, *11th IEEE Conference on Nanotechnology*, Portland, Oregon, August 15–19, pp. 985–990. © (2011) IEEE. With permission.)

up the possibility to use this method for low-temperature, bottom-up integration. TEOS deposition appears to be an especially attractive candidate. In a recent publication, we measured CNT structures fabricated using this bottom-up approach and demonstrated that the resistivity is in the same order of magnitude as the results displayed here [28].

33.4   CONCLUSIONS

We demonstrated the high-density, low-temperature, vertically aligned growth of CNT on conductive TiN layers deposited on Si substrates. It was found that plasma bombardment can damage the TiN diffusion barrier, while wet chemical cleaning has no apparent impact. From the AFM measurements, it was observed that the plasma-treated samples display a broader distribution of catalyst nanoparticles after activation. This can account for the lack of self-aligned vertical growth at low temperature. With a sacrificial Ti layer, damage can be prevented.

Using this approach, we successfully created CNT via measurement structures, displaying a relative low resistivity of 35 mΩ-cm and linear I–V characteristics, indicating good ohmic contact between the metal contacts and CNT. Finally, we demonstrate that it is possible to use PECVD dielectrics to cover CNT without inducing significant damage to the CNT, which can be used for bottom-up processes. TEOS appears to be an especially attractive candidate, introducing hardly any change in CNT crystallinity.

Both methods are simple and effective to fully integrate CNT as vertical interconnects in silicon process technology. Although the resistivity is still high compared to that of Cu, further process and CNT growth optimization should be able to decrease its value.

ACKNOWLEDGMENTS

The authors would like to thank M. R. Tajari Mofrad for assistance during the electrical measurements and the Dimes Technology Centre staff for all fabrication support. A part of the work has been performed in the project JEMSiP_3D, which is funded by the Public Authorities in France, Germany, Hungary, the Netherlands, Norway, and Sweden, as well as by the ENIAC Joint Undertaking.

REFERENCES

1. D. N. Futaba, K. Hata, T. Yamada, T. Hiraoka, Y. Hayamizu, Y. Kakudate, O. Tanaike, H. Hatori, M. Yumura, and S. Iijima, Shape-engineerable and highly densely packed single-walled carbon nanotubes and their application as super-capacitor electrodes, *Nature Materials*, 5, 987–994, 2006.
2. Y. Hayamizu, T. Yamada, K. Mizuno, R. C. Davis, D. N. Futaba, M. Yumura, and K. Hata, Integrated three-dimensional microelectromechanical devices from processable carbon nanotube wafers, *Nature Nanotechnology*, 3, 289–294, 2008.
3. S. Fan, M. G. Chapline, N. R. Franklin, T. W. Tombler, A. M. Cassell, and H. Dai, Self-oriented regular arrays of carbon nanotubes and their field emission properties, *Science*, 283, 512–514, 1999.
4. J. Robertson, Growth of nanotubes for electronics, *Materials Today*, 10(1–2), 36–43, 2007.
5. International Technology Roadmap for Semiconductors, 2010, http://public.itrs.net/.
6. B. Q. Wei, R. Vajtai, and P. M. Ajayan, Reliability and current carrying capacity of carbon nanotubes, *Applied Physics Letters*, 79(8), 1172–1174, 2001.
7. E. Pop, D. Mann, Q. Wang, K. Goodson, and H. Dai, Thermal conductance of an individual single-wall carbon nanotube above room temperature, *Nano Letters*, 6(1), 96–100, 2006.
8. N. Srivastava, H. Li, F. Kreupl, and K. Banerjee, On the applicability of single-walled carbon nanotubes as VLSI interconnects, *IEEE Transactions on Nanotechnology*, 8(4), 542–559, 2009.
9. S. M. Rossnagel, R. Wisnieff, D. Edelstein, and T. S. Kuan, Interconnect issues post 45 nm, in *International Electron Devices Meeting (IEDM)*, 2005, pp. 89–91.
10. A. Naeemi and J. D. Meindl, *Carbon Nanotube Electronics*, ser. Series on Integrated Circuits and Systems. Springer US, 2009, ch. Performance Modeling for Carbon Nanotube Interconnects, pp. 163–190.

11. H. Li, N. Srivastava, J.-F. Mao, W.-Y. Yin, and K. Banerjee, Carbon nanotube vias: A reality check, in *International Electron Devices Meeting (IEDM)*, 2007, pp. 207–210.

12. H. Li, N. Srivastava, J.-F. Mao, W.-Y. Yin, and K. Banerjee, Carbon nanotube vias: Does ballistic electron-phonon transport imply improved performance and reliability, *IEEE Transactions on Nanotechnology*, 58(8), 2689–2701, 2011.

13. J. Li, Q. Ye, A. Cassell, H. T. Ng, R. Stevens, J. Han, and M. Meyyappan, Bottom-up approach for carbon nanotube interconnects, *Applied Physics Letters*, 82(15), 2491–2493, 2003.

14. D. N. Hutchison, N. B. Morrill, Q. Aten, B. W. Turner, B. D. Jensen, L. L. Howell, R. R. Vanfleet, and R. C. Davis, Carbon nanotubes as a framework for high-aspect-ratio MEMS fabrication, *Journal Of Microelectromechanical Systems*, 19(1), 75–82, 2010.

15. S. Vollebregt, R. Ishihara, J. Derakhshandeh, W. H. A. Wien, J. van der Cingel, and C. I. M. Beenakker, Patterned growth of carbon nanotubes for vertical interconnect in 3D integrated circuits, in *Proceedings of The Annual Workshop on Semiconductor Advances for Future Electronics and Sensors*, 2010, pp. 184–187.

16. S. Vollebregt, J. Derakhshandeh, R. Ishihara, M. Y. Wu, and C. I. M. Beenakker, Growth of high-density self-aligned carbon nanotubes and nanofibers using palladium catalyst, *Journal of Electronic Materials*, 39(4), 371–375, 2010.

17. S. Sato, M. Nihei, A. Mimura, A. Kawabata, D. Kondo, H. Shioya, T. Iwai, M. Mishima, M. Ohfuti, and Y. Awano, Novel approach to fabricating carbon nanotube via interconnects using size-controlled catalyst nanoparticles, in *IEEE International Interconnect Technology Conference*, 2006, 230–232.

18. S. Vollebregt, R. Ishihara, F. D. Tichelaar, Y. Hou, and C. I. M. Beenakker, Influence of the growth temperature on the first and second-order raman band ratios and widths of carbon nanotubes and fibers, *Carbon*, 50, 3542–3554, 2012.

19. Y.-J. Lee, The second order raman spectroscopy in carbon crystallinity, *Journal of Nuclear Materials*, 325, 174–179, 2004.

20. E. F. Antunes, A. O. Lobo, E. J. Corat, and V. J. Trava-Airoldi, Influence of diameter in the Raman spectra of aligned multi-walled carbon nanotubes, *Carbon*, 45, 913–921, 2007.

21. J. Dijon, H. Okuno, M. Fayolle, T. Vo, J. Pontcharra, D. Acquaviva, D. Bouvet et al., Ultra-high density carbon nanotubes on Al-Cu for advanced vias, in *International Electron Devices Meeting (IEDM)*, 2010, p. 33.4.1.

22. D. Yokoyama, T. Iwasaki, K. Ishimaru, S. Sato, T. Hyakushima, M. Nihei, Y. Awano, and H. Kawarada, Electrical properties of carbon nanotubes grown at a low temperature for use as interconnects, *Japanese Journal of Applied Physics*, 47(4), 1985–1990, 2008.

23. Y.-M. Choi, S. Lee, H. S. Yoon, M.-S. Lee, H. Kim, I. Han, Y. Son, I.-S. Yeo, U.-I. Chung, and J.-T. Moon, Integration and electrical properties of carbon nanotube array for interconnect applications, in *Nanotechnology, 2006. IEEE-NANO 2006. Sixth IEEE Conference on*, 2006, 262–265.

24. F. Kreupl, A. P. Graham, G. S. Duesberg, W. Steinhögl, M. Liebau, E. Unger, and W. Hönlein, Carbon nanotubes in interconnect applications, *Microelectronic Engineering*, 64, 399–408, 2002.

25. A. Naeemi and J. D. Meindl, Compact physical models for multiwall carbon-nanotube interconnects, *IEEE Electron Device Letters*, 27(5), 338–340, 2006.

26. M. S. Dresselhaus, G. Dresselhaus, R. Saito, and A. Jorio, Raman spectroscopy of carbon nanotubes, *Physics Reports*, 409, 47–99, 2005.

27. S. Vollebregt, R. Ishihara, F. D. Tichelaar, J. van der Cingel, and K. Beenakker, Electrical characterization of carbon nanotube vertical interconnects with different lengths and widths, in *IEEE International Interconnect Technology Conference*, 2012.

28. S. Vollebregt, R. Ishihara, J. van der Cingel, and K. Beenakker, Low-temperature bottom-up integration of carbon nanotubes for vertical interconnects in monolithic 3D integrated circuits, in *Proceedings of the 3rd IEEE International 3D System Integration Conference*, 2012, 1–4.

29. S. Vollebregt, R. Ishihara, J. Derakhshandeh, J. van der Cingel, H. Schellevis, and C. I. M. Beenakker, Integrating low temperature aligned carbon nanotubes as vertical interconnects in Si technology, *11th IEEE Conference on Nanotechnology*, Portland, Oregon, August 15–19, 2011, pp. 985–990.

34 Readout Circuit Design for MWCNT Infrared Sensors

*Liangliang Chen, Ning Xi, Hongzhi Chen,
and King Wai Chiu Lai*

CONTENTS

34.1 INTRODUCTION

An optoelectronic device includes all studies and applications on electronic devices that source, detect, and control light. Through the detectors, light signal will be converted to an electrical signal and then be processed by a voltage monitoring or current monitoring circuit. In wide bandwidth light, infrared (IR) light was extensively used from military to nonmilitary purposes. There are thousands of commercialized applications in industrial, scientific, and medical areas, including hyperspectral imaging in biological and mineralogical measurements; target acquisition and tracking, night vision in military applications [1]; IR data communications by standards published by IrDA; IR telescope in astronomy; environment monitoring in meteorology, and so on. It is also closely related to the human body, because humans at normal temperature can radiate around 12 µm wavelength IR light based on Wien's displacement law.

In all IR imaging systems, it was mainly composed of optical components, including lens, mirrors, and IR sensors. The typical IR sensors, including the FPA, QD, and QW structures [2,3], are temperature dependent and work under very low temperatures in order to maintain high efficiency. Experimentally, it is shown that carbon nanotube (CNT)-based IR sensors perform well at room temperature [4]. Our group studied the noncryogenic cooled CNT-based IR imaging system in detail.

However, the photocurrent is from a pico-ampere to a nano-ampere in a CNT IR sensor [5,6] and it is bias dependent. To read photocurrent in the CNT IR sensor and to integrate the sensor into the imaging system, a low-noise, high-gain readout circuit is needed. As shown in Figure 34.1, a fully

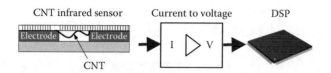

FIGURE 34.1 Current readout system for the CNT IR sensor.

current readout system for the CNT-based IR sensor includes a sensor, a current-to-voltage module, and a digital signal processing module (DSP).

To distinguish the voltage difference on the photodetector, it must generate at least a millivolt-scale voltage when light irradiates on the detector. While the photovoltaic effect is not as great as a silicon photodiode in a CNT-based IR detector [6-8], it can generate a nanoampere-scale photocurrent without voltage changes. However, in current monitoring circuits, for the purpose of reducing the bias voltage influence on photodetectors, it requires that readout circuits implement zero input impedance. In a in CNT-based IR detector especially, the photocurrent characteristic greatly depends on the applied bias voltage [7,8]. There are two requirements of readout circuits of a CNT-based IR detector: one relates to the high resolutions to the picoamperescale, the other is no bias application on the detector. To get a good resolution, a high-performance current-to-voltage amplifier (also named the trans-impedance amplifier) becomes the most important module in the readout system.

34.2 CURRENT-TO-VOLTAGE CONVERSION METHOD

There are two types of CNT detector signal monitoring, including voltage monitoring and current monitoring. Voltage monitoring requires a great voltage difference (at least mV scale) on the detector, when light irradiates on the detector. In CNT detector experiments [9], the CNT photovoltaic effect is not like a silicon photodiode; it generates a nanoampere scale photocurrent while the voltage does not change much. However, current monitoring requires that the readout circuits present zero input impedance to the detector, and the readout circuit absorbs the detector's current without producing a voltage across the detector, especially for a CNT-based detector, in which the photocurrent characteristic of a CNT detector depends on the applied bias voltage. It is the best way to use zero bias for current monitoring on the CNT detector.

34.2.1 RESISTOR-BASED CURRENT READOUT METHOD

To convert pA/nA photocurrent to voltage, a current-to-voltage amplifier (also named a trans-impedance amplifier) becomes the most important module in a readout system; then a microprocessor can be used to process the photocurrent. The basic principle of a current-to-voltage converter (IV converter) is to use photocurrent (I_p) multiplied by a resistor. These circuits, shown in Figure 34.2, need

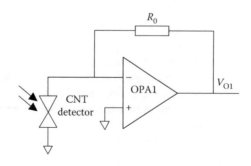

FIGURE 34.2 Schematic of the R type IV converter.

very large (approximately GΩ) and precise resistors to detect pA currents [10]. The thermal noise of a large resistor will also limit the resolution of the circuit. The deep negative feedback of the amplifier (OPA1) makes the negative node (−) and positive node (+) work in the same potential, so the CNT detector is at zero bias in this condition.

34.2.2 Capacitor-Based Current Readout Method

A capacitive trans-impedance amplifier (CTIA) works as a current-to-voltage converter (*IV* converter) and it also can be designed to have high charge-to-voltage conversion ratios and low read noise [10,11]. Figure 34.3 is the basic circuit of a CTIA using a switch capacitor as the adjust resistor. There are two working phases in Figure 34.3. In phase 1, switch S is high, a OPA2 works like a voltage follower. The CNT detector forms a circular loop with current source in OPA2 and the photocurrent is a partial current of the current source. In phase 2, switch S is low in T_{off} seconds, the photocurrent (I_p) will charge a negative node (−) of OPA2, and the charge will redistribute in it. $Q = I_p*T_{off}$. $V_{O2} = Q/C_0$ (T_{off} represents switch S off time). This circuit can detect as low current as possible based on the noise process in the input and was used in our readout system. It also achieves a large power supply rejection ratio (PSRR), a high open-loop gain, and a large dynamic range. The CTIA gain is given by $G = Q/C_0$, QIN being the input charge integrated on the CTIA's feedback capacitor.

34.3 DESIGN OF SENSOR READOUT SYSTEM

34.3.1 Zero Bias Current-to-Voltage Conversion

There are two requirements of readout circuits of a CNT-based IR detector: one is the high resolution to a picoampere scale; the other is no bias or bias modulation on the detector. To reduce noise and read such low current of an MWCNT detector, a capacitive transimpedance amplifier, shown in Figure 34.3, was designed in this IR readout system, because CTIA have high charge-to-voltage conversion ratios and low read noise [11].

The amplifier works in deep negative feedback control, which will make bias on a CNT detector modulated by the positive input of OPA2 as shown in Figure 34.3. The circuit can detect low currents because of the input noise reduction.

34.3.2 High-Resolution ADC

In high-resolution ADC, Delta-sigma (ΔΣ; or sigma-delta, ΣΔ) modulation is a method for encoding high-resolution signals into lower-resolution signals using pulse-density modulation. A high

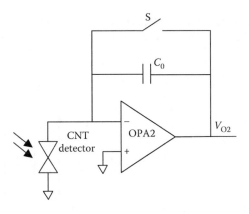

FIGURE 34.3 Schematic of the C type *IV* converter. (L. Chen et al., Readout system design for MWCNT infrared sensors, *IEEE Nano Conference*, Portland, Oregon, pp. 1635–1638. © (2011) IEEE. With permission.)

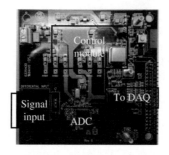

FIGURE 34.4 ADC PCB board in a readout system. (L. Chen et al., Readout system design for MWCNT infrared sensors, *IEEE Nano Conference*, Portland, Oregon, pp. 1635–1638. © (2011) IEEE. With permission.)

performance, 24-bit Σ-Δ analog-to-digital converter (ADC) was used in our experiments. It combines wide input bandwidth and high speed with the benefits of Σ-Δ conversion with a performance of 106 dB SNR at 625 kSPS, making it ideal for high-speed data acquisition. A wide dynamic range, combined with significantly reduced anti-aliasing requirements, simplifies the design process. In addition, the device offers programmable decimation rates, and the digital FIR filter can be adjusted if the default characteristics are not appropriate to the application. It is ideal for applications demanding high SNR without a complex front-end signal processing design. Figure 34.4 shows the ADC part in a readout system, including the ADC control part, ADC chip, input, and out interface [9].

34.4 READOUT SYSTEM TESTING

34.4.1 HARDWARE SETUP OF TESTING SYSTEM

In this section, the hardware setup and some experiment results of ROIC are presented. For the sake of reducing the noise after CTIA, a low-pass filter (LPF) was used to limit the bandwidth and optimize the thermal noise contribution on CTIA. Meanwhile, the LPF can weaken all high-frequency noise, including the power frequency noise. The readout system also includes ADC circuits, which were used to convert analog current to digital values. Figure 34.5 is the experimental PCB of a readout system.

34.4.2 TEST RESULTS

On the basis of this readout system, a group of experiments were conducted to verify the performance of stability. In Figure 34.6, the x-axis corresponds to laser intensity. The digital output

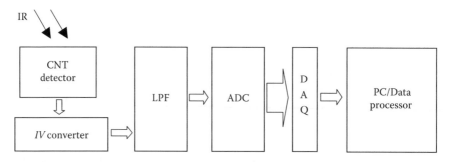

FIGURE 34.5 Diagram of a readout system.

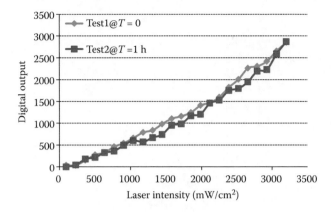

FIGURE 34.6   Readout test on CNT-based IR detector.

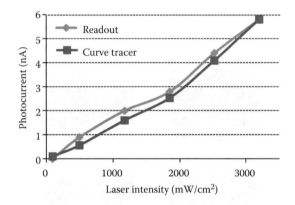

FIGURE 34.7   Compare readout system with Curve Tracer. (L. Chen et al., Readout system design for MWCNT infrared sensors, *IEEE Nano Conference*, Portland, Oregon, pp. 1635–1638. © (2011) IEEE. With permission)

at $T = 1$ h maintains almost the same value at $T = 0$ with a different laser input. Meanwhile, the readout was also compared with a Curve Tracer (Agilent 4155C Semiconductor Analyzer) measurement in Figure 34.7. The results show that the readout system can reach pA resolution and has good stability performance [9].

34.5   APPLICATION OF READOUT SYSTEM

34.5.1   Readout in Single Pixel Camera

At last the readout system was also integrated in an MWCNT-based single pixel imaging system [12]. The hardware setup was shown in Figure 34.8, IR irradiation was controlled by DMD, and a lens was used to focus all IR light on the detector [9].

Based on the hardware, three groups of experiments were conducted. A rectangle bar was moved from the top to the bottom, where the recovery image can follow an objective image. The result shows that the lower left corner is destructive. From these results, we find that the readout system can work in the MWCNT IR single imaging system (Figure 34.9).

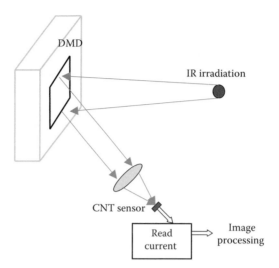

FIGURE 34.8   CNT-based single pixel IR imaging system.

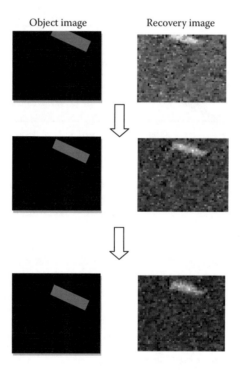

FIGURE 34.9   Recovery image based on a single MWCNT detector.

34.6 CONCLUSION

The design and experimental results of an MWCNT-based IR detector ROIC system are presented in this study. By this system, we can use it to test the low current to a sub-nanoampere in the CNT-based IR detector; it also works in the CNT-based single pixel IR imaging system. Although in our readout system it can work at a hundred Hertz, it is not enough for high-resolution image recovery. Meanwhile, an ultra-fast readout method is also a challenge project in nanoelectronics.

ACKNOWLEDGMENT

This research study is partially supported under NSF Grants IIS-0713346 and ONR Grants N00014-04-1-0799 and N00014-07-1-0935.

REFERENCES

1. F. Szmulowicz, G. J. Brown, H. C. Liu et al., GaAs/AlGaAs p-type multiple quantum wells for infrared detection at normal incidence: Model and experiment, *Opto Electronics Review,* 9, 164–173, 2001.
2. M. Kolahdouz, A. A. Farniya, L. D. Benedetto, and H. H. Radamson. Improvement of infrared detection using Ge quantum dots multilayer structure, *Applied Physics Letters*, 96, 2135161–2135163, 2010.
3. A. Dehe, H. L. Hartnagel, D. Pavlidis, K. Hong, and E. Kuphal. Properties of InGaAs/InP thermoelectric and surface bulk micro machined infrared sensors, *Applied Physics Letters,* 69, 3039–3041, 1996.
4. I. A. Levitsky and W. B. Euler, Photoconductivity of single-wall carbon nanotubes under continuous-wave near infrared illumination, *Applied Physics Letters*, 83, 1857–1859, 2003.
5. J., N. Xi, H. Chen, K. W. C. Lai, G. Li, and U. C. Wejinya, Design, manufacturing, and testing of single carbon nanotube based infrared sensors, *IEEE Transactions on Nanotechnology*, 8(2), 245–251, 2009.
6. J. Zhang, N. Xi, K. W. C. Lai, H. Chen, and Y. Luo, Single carbon nanotube based photodiodes for infrared detection, *Proceedings of the 7th IEEE International Conference on Nanotechnology*, 8(2), pp. 1156–1160, 2009.
7. J. Zhang, N. Xi, H. Chen, K. W. C. Lai, G. Li, and U. Wejinya, Photovoltaic effect in single carbon nanotube-based Schottky diodes, *International Journal of Nanoparticles*, 1(2), 108–118, 2008.
8. H. Chen, N. Xi, K. W. C. Lai, C. K. M. Fung, and R. Yang, Development of infrared detectors using single carbon nanotubes based field effect transistors, *IEEE Transaction on Nanotechnology*, 5(9), 582–589, 2010.
9. L. Chen, N. Xi, H. Chen, and K. W. C. Lai, Readout system design for MWCNT infrared sensors, *IEEE Nano Conference*, Portland, Oregon, pp. 1635–1638, 2011.
10. J. G. Graeme, *Photodiode Amplifiers, OP Amp Solutions*, McGraw-Hill Press, New York, pp. 23–30, 1995.
11. B. Fowler, J., D. How, and M. Godfrey. Low FPN high gain capacitive transimpedance amplifier for low noise CMOS image sensors, *Proceedings of the SPIE*, Vol. 4306, San Jose, CA.
12. M. F. Duarte, M. A. Davenport, D. Takhar et al., Single pixel imaging via compressive sampling, *IEEE Signal Processing Magazine,* pp. 83–92, 2008.

35 Use of Vertically Aligned Carbon Nanotubes for Electrochemical Double-Layer Capacitors

Adrianus I. Aria, Mélanie Guittet, and Morteza Gharib

CONTENTS

35.1 INTRODUCTION

The demand for, and consumption of, global energy is continually increasing across a world whose population will grow to nearly 9 billion people in the next couple of years. It is poised to continue its steady increase in the next several decades. The climate change and the decreasing availability of fossil fuels have hastened the development of satisfying sustainable and renewable energy technologies. Since electrical energy storage systems such as batteries and capacitors are considered as the most critical link between the ever-increasing renewable energy supply and demand, a lot of work has been done to improve their efficiency and performance. Further, the advancement of micro/nano-electro-mechanical devices for telecommunication and biomedical applications initiates the demand for reliable lightweight electrical energy storage systems with small form factors. To fulfill all these requirements, the development of new materials with an exceptional specific surface area and excellent electrochemical properties has become the focus of much research effort.

Electrical energy can be stored either chemically, for example, batteries, or physically, for example, capacitors. Chemically, the electrical energy is converted into chemical energy via Faradaic reduction and oxidation (redox) reactions. Although the energy conversion from electrical to chemical and vice versa is thermodynamically reversible in principle, in practice, such conversion often involves some degree of irreversibility due to the electrode–electrolyte interphase changes during the charging and discharging processes [1]. Therefore, the charge/discharge rates of batteries are very low and their lifecycle is limited to only several thousand cycles. However, depending on the types of batteries, they can have a very high gravimetric energy density of up to 200 Wh/kg [2].

In contrast to batteries, capacitors store electrical energy physically via non-Faradaic electrostatic process. Owing to the absence of phase changes due to redox reactions, the charging and discharging processes are highly reversible. Therefore, capacitors in general have an exceptional cyclability, which leads to a very stable behavior over a very long lifecycle and very high charge/

discharge rates. However, the gravimetric energy density of capacitors is typically very low, because it is limited by the accessible surface area of the electrodes.

Great efforts have been made in the last few decades to combine the high energy-storage capability of conventional batteries and the high-power-delivery capability of conventional solid-state capacitors, resulting in the invention of electrochemical double-layer capacitors (EDLCs). EDLCs are electrical energy storage devices that utilize the highly reversible electrostatic accumulation of ions of the electrolytes on the surface of the active electrode materials. When an EDLC is charged, cations accumulate on the surface of the negatively polarized electrode, creating a capacitor-like electrical double-layer separation. Similarly, an electric double layer is also formed on the surface of the positively polarized electrode due to the accumulation of anions. When an EDLC is discharged, cations and anions of the electrolytes can be transported back from the surface of the electrodes in a very fast response time (Figure 35.1). Therefore, EDLCs are capable of providing short high-energy bursts that neither batteries nor solid-state capacitors can provide efficiently, with a remarkably long lifecycle [1,3]. Such capability has been used to provide high-power pulses for a few seconds, for a wide range of applications such as electric transportation technology, emergency backup power, consumer electronics, such as laptops or cell phones, medical electronics, and military devices, including communication devices, spacecraft probes, and missile systems [4,5].

The gravimetric-specific double-layer capacitance (C_G) of an EDLC is dictated by the dielectric constant of the electrolyte (ε_e) and the accessible surface area of the electrodes (A) according to the following relation:

$$C_G = \frac{\varepsilon_e \varepsilon_0 A}{d_{dl} m}$$

(35.1)

where ε_0 is the vacuum dielectric constant, d_{dl} is the electrical double-layer separation thickness, and m is the mass of active electrode materials. Obviously, active electrode materials with a high-specific surface area are the key component to achieve a high-specific capacitance. To date, activated carbons (ACs) have been the most widely used active electrode materials in commercially available EDLCs (Figure 35.2a and b). ACs are known to have a very high-specific surface area, up to 2000 m²/g [3,6], due to their tremendous network of micropores (pore size: <2 nm). However, the electrolyte accessibility of the micropores dominated network is considerably poor such that an electric double layer cannot be formed on most of the micropores. Thus, the ACs-based EDLCs usually exhibit unsatisfactory capacitance, which is about 10–20% of the theoretical values [2].

Compared to ACs, carbon nanotubes (CNTs) have a relatively lower specific surface area since their porous network is dominated by mesopores (pore size: 2–50 nm) and macropores (pore size: >50 nm). However, these mesopores and macropores are large enough to be accessed freely by ions

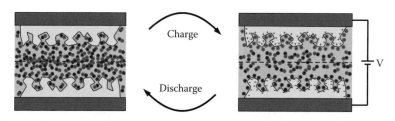

FIGURE 35.1 Schematic of an EDLC in charged and discharged states. In an EDLC, charge is stored electrostatically in the form of reversible ion adsorption on the surface of the active materials. When the EDLC is charged, cations accumulate on the surface of the negatively polarized electrode creating a capacitor-like electrical double-layer separation. Similarly, an electric double layer is also formed on the surface of the positively polarized electrode due to the accumulation of anions.

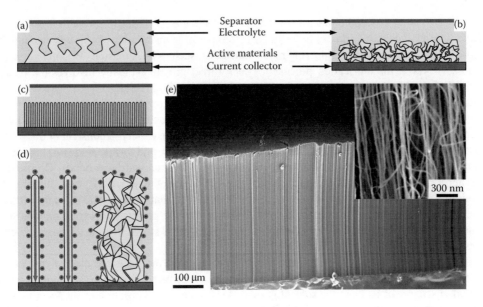

FIGURE 35.2 Schematic construction of an EC, which typically comprises of active electrode materials attached to current collectors, electrolytes, and a separator. Compared to bulk graphitic carbons (a), randomly oriented microporous ACs (b) and VACNTs (c) are more preferably used as active electrode materials because of their high specific surface area and optimal pore size distribution. Between these two high specific surface area materials, CNTs have a higher electrical conductivity which may lead to a smaller IR drop (d). Low-magnification (e) and high-magnification (inset) electron microscopy images of a vertically aligned carbon nanotube array on a current collector.

of the electrolyte such that their intrinsic capacitance is higher than that of ACs (Figure 35.2c). CNTs are basically 1D graphitic structure made of rolled-up graphene sheets. They possess an exceptionally high surface-to-volume ratio, that is, an extremely high surface area available for chemisorptions or physisorptions in a small form factor, with an aspect ratio of up to 320 million. Owing to their sp^2 carbon hybridization, the electronic transport along their main axis is almost perfectly ballistic, which means that the electrons flow directly without being scattered (Figure 35.2d). Therefore, the electrical conductivity of CNTs is truly exceptional and much superior to the currently available electronic materials [7]. CNTs are also known to have an excellent electrochemical stability, as well as exceptional thermal and mechanical properties.

Among diverse types of CNTs, vertically aligned CNTs (VACNTs) oriented perpendicularly to the current collectors are considered to be well suited to be used as active electrode materials in EDLCs (Figure 35.2e). Since the physical properties of VACNTs, such as the packing density, number of walls, and length, can be varied by changing the growth conditions, the porosity of VACNTs can be tuned such that their porous network is dominated by mesopores while maximizing their specific surface area. In addition, a low contact resistance between the VACNTs and the current collector can be maintained such that the overall IR drop can be lowered to achieve an even higher power density than that of the ACs-based EDLCs.

VACNTs are normally grown using thermal chemical vapor deposition (CVD) on silicon substrates with carbon-containing gas and hydrogen as the precursor gases at an elevated temperature. Typically, the substrates are precoated with a thin-layer transition metal catalyst of iron (Fe), nickel (Ni), or cobalt (Co), along with a thin supporting layer of aluminum oxide (Al_2O_3) serving as a buffer layer between the catalyst and the substrate. Methane (CH_4), acetylene (C_2H_2), or ethylene (C_2H_4) is usually used as the carbon containing a precursor gas. Alternatively, a floating catalyst method may also be used for growing VACNTs. Here, the substrate is exposed to a mixture of catalyst vapor and carbon-containing precursor gas, which is typically a mixture of ferrocene ($Fe(C_5H_5)_2$) and xylene (C_8H_{10}) at an elevated temperature. These VACNTs are then transferred onto metal current collectors using conductive epoxies or low melting temperature metal alloys [8,9]. It is also possible to grow VACNTs directly on conducting metal substrates, such as Ni, tungsten (W), and aluminum (Al) foil. However, the fabrication process for such direct growth typically involves the plasma-enhanced CVD (PECVD) or low-pressure CVD (LPCVD) method [10,11].

According to Equation 35.1, the dielectric constant of the electrolyte is another key component to achieve the high-specific capacitance of EDLCs. Since it is assumed that all ions are in the solvated state, the dielectric constant of the electrolyte is dominated by the dielectric constant of the solvent. Typically, protic solvents such as water, hydrogen cyanide (HCN), and formic acid (FA) have a very high dielectric constant due to the existence of strongly structured hydrogen bonds. On the other hand, aprotic solvents such as propylene carbonate (PC), acetonitrile (ACN), and dimethyl formamide (DMF) typically have a lower dielectric constant than that of the protic solvents, although it is still higher than that of the unstructured solvents due to the presence of strong dipole–dipole interactions. Water, for instance, has a dielectric constant of 80°C at 25°C, while PC has a dielectric constant of 64.

From a capacitance point of view, the use of aqueous alkaline or acidic electrolytes usually leads to a higher specific capacitance than that of organic electrolytes. However, at a high operating potential, discharge of H_2 from the solvent is very likely to happen. Thus, to prevent decomposition due to the electrolysis of water ($E° = 1.23$ V), the operating potential for aqueous electrolytes is limited to 1 V [2,3,12]. From an energy density and power density perspective, the use of nonaqueous solvents is more favored due to the lack of electrochemically active H atoms such that a higher operating potential of 2.5–3.5 V is achievable. The gravimetric energy density (E_G) stored by EDLCs is given by the following relation:

$$E_G = \frac{1}{2} C_G \left(\frac{V}{2} \right)^2 \tag{35.2}$$

where V is the operating potential of EDLCs. The gravimetric power density, P_G, of EDLCs is given by the following relation:

$$P_G = \frac{1}{mR} \left(\frac{V}{2} \right)^2 \tag{35.3}$$

where R is the equivalent series resistance (ESR) measured from the IR drop of EDLCs. Therefore, a threefold increase in operating potential achieved using nonaqueous electrolytes will result in an order of magnitude increase in stored energy for the same capacitance value. However, it is important to note that the resistivity of most nonaqueous electrolytes is substantially larger than that of aqueous electrolytes, which results in a higher ESR.

Among a few available nonaqueous electrolytes that can be reliably used to obtain a high value of specific capacitance as well as energy density and power density, a mixture of tetraalkylammonium-tetrafluoroborate (Et_4NBF_4) in PC is the most common choice. Although Et_4NBF_4/PC electrolytes are currently more expensive than their aqueous counterparts, they offer a better stability at a higher operating voltage and at a larger temperature range. Et_4NBF_4/PC electrolytes are also considered to be relatively safe due to their high boiling point and low toxicity.

35.2 CAPACITIVE BEHAVIOR OF VACNTs

Cyclic voltammetry is a very common and widely used electrochemical characterization technique, because of its ability and effectiveness to quickly observe an electrochemical behavior over a wide potential range. It is basically a cyclic measurement of potential-dependent current when the potential is varied linearly across the potential window at a constant scan rate. Since, in principle, capacitors are free from Faradaic redox reactions, the plot of current versus potential (called voltammogram) of ideal capacitors is perfectly rectangular. The gravimetric specific capacitance (C_G) of an ideal capacitor is given by the following relation:

$$C_G = \frac{I}{(\mathrm{d}V/\mathrm{d}t)m} \tag{35.4}$$

where I is the recorded response current and $\mathrm{d}V/\mathrm{d}t$ is the scan rate. Since, in reality, Faradaic processes are always involved and the ESR is always nonzero, the voltammogram shape of an EDLC is trapezoidal with nonconstant current during the linear potential sweep. Hence, the gravimetric specific capacitance of an EDLC is given by the following relation:

$$C_G = \frac{\displaystyle\int_{V_1}^{V_2}\{I_C(V) - I_D(V)\}\mathrm{d}V}{2(V_2 - V_1)(\mathrm{d}V/\mathrm{d}t)m} \tag{35.5}$$

where I_C and I_D are the recorded response current during the charge and discharge state, respectively, and V_1 and V_2 are the lower and upper limits of the potential window, respectively.

The cyclic voltammograms of VACNTs in 1 M Et$_4$NBF$_4$/PC in a two electrodes configuration at various scan rates show a smooth and symmetrical shape over a potential range of 0–2.5 V (Figure 35.3a). Such featureless voltammograms indicate the absence of Faradaic reaction during the charge and discharge cycles. A linear increase of response current during the linear potential sweep implies a nonzero finite ESR of the VACNT-based EDLCs. A relatively sharp transient response when the potential sweep changes sign indicates a rapid charge storage and delivery kinetic of the VACNTs. As per the calculations in Equation 35.5, the average gravimetric specific capacitance of VACNTs at a scan rate of 100 mV/s is about 31 F/g.

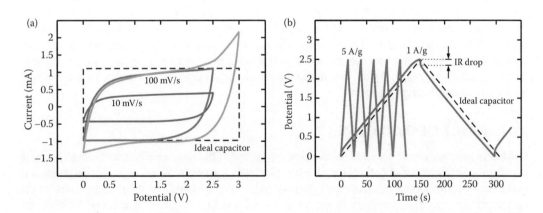

FIGURE 35.3 Cyclic voltammograms of VACNTs in 1 M Et$_4$NBF$_4$/PC electrolyte at scan rates of 10 and 100 mV/s in two electrodes configuration (a). The smooth, symmetrical shape of the voltammograms indicates the absence of a Faradaic behavior during the charge and discharge cycles. Ideally, the Et4NBF4/PC electrolyte has a voltage window of 3 V. However, a significant increase of current at a potential higher than 2.5 V indicates undesirable Faradaic losses. Galvanostatic charge/discharge curves of a vertically aligned carbon nanotubes array in a 1 M Et$_4$NBF$_4$/PC electrolyte at a gravimetric current density of 1 and 5 A/g in two electrodes configuration (b). Although the shape of the charge/discharge curves is almost perfectly triangular, the presence of IR drop caused by undesirable Faradaic losses is observable.

At a larger potential range of 0–3 V, the Faradaic behavior of VACNTs becomes more pronounced. At such a high potential, the response current increases exponentially, suggesting the occurrence of potential-dependent redox reactions where the relative quantity of the active redox species is dependent on the potential. Since, in principle, the Et_4NBF_4/PC electrolyte is stable at a potential of up to 3 V, the observed potential-dependent redox reactions may be caused by the discharge of electroactive species from the VACNTs.

The cyclic voltammograms of VACNTs also show a dependency of the response current on the scan rate. An increase of response current with an increasing scan rate implies that the electrochemical characteristic of VACNTs is dominated by their capacitive behavior [13,14]. Note that the relation between the specific capacitance of VACNTs and the scan rate follows a power law, where C_G scales as $dV/dt^{-1/2}$. This suggests a large variation in the diffusion path length of the ions in the electrolyte, which may be caused by the presence of a complex porous network with various pore sizes in the VACNTs.

The capacitive behavior of active electrode materials can also be probed using galvanostatic charge/discharge analysis at a constant current density, where the response potential is recorded and plotted against the time. For an ideal capacitor, the shape of the potential versus time curve is perfectly triangular with constant slopes showing a perfect capacitive behavior. Since, in reality, the ESR is always nonzero, the charge/discharge curves of an EDLC are typically curved. An IR drop can be obviously seen from the charge/discharge curves when the ESR is significantly high.

It is possible for a VACNT-based EDLC to have a high ESR. Factors that influence the ESR of an EDLC include the internal resistance of active electrode materials, contact resistance between active electrode materials and current collectors, wettability of active electrode materials by the electrolytes, and conductivity of the electrolyte. Since the internal resistance of VACNTs is typically very low, such a high ESR may be attributed to a poor wettability by the electrolyte and a poor contact to the current collector. The slightly curved and nontriangular shape of galvanostatic charge/discharge curves of VANCTs shows a nonzero ESR (Figure 35.3b). The ESR of VACNTs is measured to be ~10 Ω, which is considered reasonable, although it is relatively higher than that reported by previous studies [8,15,16].

Gravimetric specific capacitance can be calculated directly from the charge/discharge curves using Equation 35.4, where the scan rate, dV/dt, is measured from the slope of the discharge curves. At a current density of 5 A/g, the gravimetric specific capacitance of VACNTs is about 25 F/g. A much higher gravimetric specific capacitance of about 61 F/g can be reached at a lower current density of 1 A/g. Clearly, the gravimetric specific capacitance of VACNTs increases as the applied current density decreases, and the relation between these two parameters follows an inverse law, where C_G scales as $(I/m)^{-1}$. Again, this finding suggests the presence of a complex porous network with a large number of micropores.

35.3 EFFECT OF OXIDATION

Oxidation processes have been known for decades as one of the most credible methods to increase the specific surface area of carbon-based active electrode materials. When an oxidation process is performed on carbon-based active electrode materials, oxygen atoms attack the defect sites on the surface of the materials, resulting in the removal of impurities and modification of their porous structures. Oxygen is readily physisorbed by carbons as molecular O_2 whenever they are exposed to air or any oxygen containing gas. In addition, chemisorptions of oxygen during the oxidation process result in the presence of oxygenated groups on the surface of the materials. ACs are typically made by selective oxidation of carbon-rich organic precursors [3], resulting in a substantial increase of the surface area and a better pore size distribution.

Carbon-based active electrode materials, including VACNTs, can be oxidized using several known methods such as the liquid-phase or gas-phase oxidation and plasma treatments. Liquid-phase oxidations are carried out by exposing these materials to strong acids and oxidants, including potassium chlorate ($KClO_3$), potassium permanganate ($KMnO_4$), sodium nitrate ($NaNO_3$), phosphoric acid (H_3PO_4), nitric acid (HNO_3), and sulfuric acid (H_2SO_4). It is important to note that in addition to the highly corrosive nature of these chemicals, the oxidation processes usually involve the generation of toxic and explosive gases [17]. Liquid-phase oxidations are also considered impractical for oxidizing VACNTs, because the capillary forces induced by these liquid chemicals are very likely to destroy the vertical alignment of the VACNTs. Hence, the gas-phase oxidations and plasma treatments are more favorably used for oxidizing VACNTs.

Gas-phase oxidations using oxygen (O_2), ozone (O_3) or carbon dioxide (CO_2) gas, as well as water vapor (H_2O) at an elevated temperature are considered a safer, more convenient and more practical way compared to their liquid-phase counterparts, especially in large industrial-scale production.

Compared to both the aforementioned methods, oxygen plasma treatments are able to oxidize carbon-based active electrode materials in a much faster way without sacrificing safety, convenience, and practicality. However, oxygen plasma treatments are more commonly used in small-scale production.

The presence of oxygenated groups on the surface of ACs or oxidized VACNTs should influence their electrical properties and electrochemical characteristics. Since pristine VACNTs are technically nonpolar, the presence of polar oxygenated groups on their surfaces should improve their wettability in highly polar electrolytes. On the other hand, the presence of such groups may also negatively affect the ESR, the self-discharge and cyclability characteristics as well as the ion adsorption behavior of the VACNTs.

Previously reported studies suggest that the specific capacitance of the VACNTs can be increased dramatically by the presence of oxygenated groups [18,19]. The presence of such oxygenated groups improves VACNTs' wettability in highly polar electrolytes, especially in aqueous electrolytes, by increasing the number of contact sites with ions of the electrolytes [20]. This, in turn, improves the pore access and increases the wetted surface area of the VACNT–electrolyte interface, which ultimately amplifies the total area of the electric double layer. Therefore, the average gravimetric specific capacitance of VACNT-based EDLCs caused by the presence of the electric double layer increases as the average oxygen/carbon atomic ratio (O/C ratio) of the VACNTs increases. Note that the O/C ratio represents the surface concentration of oxygenated groups on the VACNTs and is measured by elemental analysis using energy-dispersive x-ray spectroscopy (EDX), x-ray photoelectron spectroscopy (XPS), or electron energy loss spectroscopy (EELS).

There is no doubt that the presence of oxygenated groups improves VACNTs' wettability, which in turn contributes to the increase of their capacitance. Nevertheless, such a significant increase in capacitance may also be attributed to the occurrence of fast Faradaic redox reactions. Basically, oxygenated groups, whether acidic, basic, neutral, or amphoteric oxides, may act as pseudocapacitive materials attached to the surface of VACNTs. Like in batteries, Faradaic redox reactions allow a much higher capacitance due to interphase changes during the charging and discharging processes. Oxygenated groups, such as hydroxyl (C—OH) and carbonyl (C=O), are known to be electroactive and may involve in redox reactions of oxygen as follows [18,21–23]:

$$C{-}OH \Leftrightarrow C{=}O + H^+ + e^- \tag{35.6}$$

$$C{=}O + e^- \Leftrightarrow C{-}O^- \tag{35.7}$$

For pristine VACNTs, their capacitive behavior is dominated by the non-Faradaic electric double-layer behavior. However, a nonzero O/C ratio of the pristine VACNTs also means that Faradaic redox reactions are also involved during the charging and discharging cycles, although their contribution is limited. As the O/C ratio of the VACNTs increases due to a prolonged oxidation process, the contribution of Faradaic redox reactions relative to the non-Faradaic electric double layer to the overall capacitance is also expected to increase. Indeed, the specific capacitance of VACNTs increases almost linearly from about 29 F/g to about 83 F/g as the average

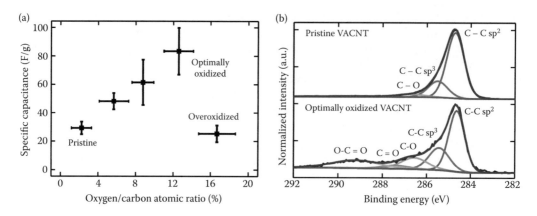

FIGURE 35.4 Specific capacitance of VACNT-based supercapacitors in 1 M Et₄NBF₄/PC as a function of their oxygen/carbon atomic ratio. Specific capacitance is obtained from galvanostatic charge/discharge analysis at a current density of 5 A/g. X-ray photoelectron spectroscopy spectra of the pristine and optimally oxidized VACNTs showing a significant increase in surface concentration of C–O, C=O, and O–C=O bonds (b).

oxygen/carbon atomic ratio (O/C ratio) increases from about 2% for the pristine VACNTs to about 13% for the well oxidized ones (Figure 35.4a). Note that these capacitances are measured at a current density of 5 A/g.

XPS analysis before and after the oxidation process shows the evolution of oxygenated groups attached to the VACNTs (Figure 35.4b). Deconvolution of the C 1s XPS spectra of both pristine and oxidized VACNTs shows five distinct peaks associated with C–C sp² (284.5 ± 0.1 eV, FWHM 0.9 eV), C–C sp³ (285.5 ± 0.1 eV, FWHM 1.3 eV), C–O (286.5 ± 0.1 eV, FWHM 1.4 eV), C=O (287.4 ± 0.2 eV, FWHM 1.4 eV), and O–C=O (289.1 ± 0.4 eV, FWHM 1.7 eV) [20,24,25]. For pristine VACNTs, the presence of a strong C–C sp² peak and a mild C–C sp³ peak indicates a relatively high-quality graphitic structure with limited defects and disorders. In addition, a very weak C–O peak indicates the presence of hydroxyl or epoxide groups at the basal planes of CNTs. Note that a pristine VACNT is not free from the hydroxyl or epoxide groups, although their concentration is extremely low.

As expected, the surface concentration of the hydroxyl or epoxide groups increases as VACNTs undergo a prolonged oxidation process. An increase in the intensity of the C=O peak can also be observed, suggesting an increase in the surface concentration of carbonyl groups at the edge boundaries and defect sites of VACNTs. Similarly, an increase in the intensity of the O–C=O peak indicates an increase in the surface concentration of the ester or carboxyl groups at the edge boundaries and defect sites of VACNTs. Further, an increase in the intensity of the C–C sp³ peak indicates a considerable increase of defect density induced by oxygen uptake, producing a strong disruption to the π-bond network of the VACNTs.

VACNTs with an excessive surface concentration of the oxygenated groups due to over-oxidation are most likely to exhibit high rates of self-discharge [26]. During the charging process, dissociation of the –OH groups may occur, leading to the formation of –H or –O free radicals. These free radicals may generate molecular H₂, O₂, or H₂O₂, which are typically involved in self-discharge processes. Moreover, these oxygenated groups are also known to be active catalysts for the electrochemical decomposition of electrolytes [21,26], resulting in a high leakage current, a poor cyclability, and a very short lifecycle. Furthermore, over-oxidized VACNTs are expected to have a lower electrical conductivity than the pristine ones due to the disappearance of their π-bond network. All the above-mentioned factors contribute to a significant deterioration in capacitance for over-oxidized VACNTs, which is proven by a decrease of specific capacitance of VACNT-based EDLCs from more than 80 F/g to about 25 F/g as the O/C ratio by increases about 17% (Figure 35.4).

35.4 LIFETIME AND PERFORMANCE

As mentioned earlier, capacitors, including EDLCs, in principle, have an excellent stability to be used for millions of cycles at a very high charge/discharge rate. Such capability arises from the absence of Faradaic redox reactions that eliminates the electrode–electrolyte interphase changes. Although, in reality, limited Faradaic redox reactions are involved, the pristine VACNT-based EDLCs are able to withstand more than 120,000 charge/discharge cycles at a high current density of 10 A/g without any significant degradation in capacitance. Note that there exists a transient condition such that they need to be cycled for several charge/discharge cycles before reaching their maximum capacitance.

The optimally oxidized VACNTs are able to withstand about the same number of charge/discharge cycles at a moderate current density of 10 A/g while only losing 10% of their original capacitance. On the other hand, the over-oxidized VACNTs are only capable of withstanding less than 25,000 charge/discharge cycles before losing more than 15% of their original capacitance. Such fast degradation in capacitance is actually expected from VACNTs that have been exposed to a prolonged oxidation process. These VACNTs have an excessive surface concentration of oxygenated groups involved in Faradaic redox reactions such that they exhibit high rates of self-discharge, high ESR, and poor cyclability. Nonetheless, their lifecycle is still much longer than that of typical batteries.

As mentioned earlier, the performance of VACNT-based EDLCs can be improved by an effective utilization of the oxidation process. Basically, when VACNTs have the right surface concentration of oxygenated groups, they have a relatively high gravimetric specific capacitance in nonaqueous electrolytes. Using Equations 35.2 and 35.3, such capacitance can be translated into gravimetric energy density and power density. Optimally oxidized VACNT-based EDLCs typically exhibit a gravimetric energy density of 16 Wh/kg at a gravimetric power density of 6 kW/kg, or a gravimetric energy density of 7 Wh/kg at a gravimetric power density of 12 kW/kg (Figure 35.5). These findings exceed the gravimetric energy density of currently available AC-based EDLCs [27,28], or even other CNT-based EDLCs [10,15,26,29,30].

The performance of VACNT-based EDLCs can be further improved by incorporating pseudocapacitive materials onto the CNTs. Pseudocapacitive materials, including metal oxides and conducting polymers, have been extensively studied in the past to increase the gravimetric specific

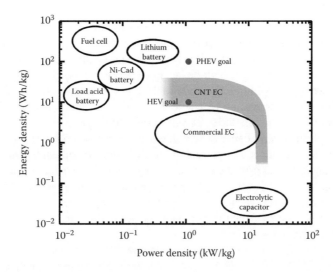

FIGURE 35.5 Ragone plot for CNT-based ECs compared to currently available energy storage systems and performance targets of hybrid electric vehicles (HEV) and plug-in hybrid electric vehicles [1]. The energy density and power density of both CNT-based ECs are calculated based on the mass of active electrode materials only.

capacitance of ACs. Ruthenium oxides (RuO_2), cobalt oxides (Co_3O_4), manganese oxides (MnO_2), and titanium dioxide (TiO_2) are the most common metal oxides to be used along with VACNTs due to their unique oxidation states that are accessible at relatively low potential and simplicity in the fabrication process [3,22]. Conducting polymers such as polyaniline (PANI), polypyrrole (PPy), and polythiophene (PT) have been used in many electrochemical applications because of their compatibility with CNTs and inherent high specific capacitance [3,22]. However, owing to the presence of Faradaic redox reactions, both metal oxides and conductive polymers cannot be cycled fast enough and often suffer from lack of stability. Gravimetric energy densities of about 32 and 22 Wh/kg have been achieved in the past using the $Ni(OH)_2$/CNTs composite and PANI/CNTs composite, respectively [31,32].

Improvement to the performance of VACNT-based EDLCs can also be achieved by incorporating other microporous carbons onto the CNTs and using ionic liquid electrolytes. A gravimetric energy density of about 50 Wh/kg at a power density of about 22 kW/kg has been achieved using AC-CNT composites in the ionic liquid electrolyte [6]. Ionic liquid electrolytes, such as the 1-ethyl-3-methylimidazoliumbis(trifluoromethylsulfonyl)imide (EMIM-Tf_2N), are nonvolatile room-temperature solvent-free electrolytes with a large electrochemical window. A further improvement to the performance of VACNT-based EDLCs can be achieved by optimizing their physical properties, especially their packing density. A gravimetric energy density of about 70 Wh/kg at a power density of about 24 kW/kg has been achieved in the past using compacted VACNTs [16]. These compacted VACNTs have a packing density 10 times higher than that of the common as-grown VACNTs.

35.5   APPLICATIONS AND CONCLUDING REMARKS

Although the gravimetric energy density of VACNTs-based EDLCs is relatively higher than that of commercially available EDLCs, it is still much lower when compared with that of batteries, especially lithium-ion batteries [23,25]. Therefore, it is unlikely for EDLCs to substitute batteries as the main electrical energy storage devices. However, since EDLCs in general have a much better lifecycle and gravimetric power density, they can complement the operation of batteries in hybrid systems [4]. In such systems, an EDLC acts as a buffer to uptake and deliver high power on demand that certainly cannot be handled by batteries. Thus, the battery's energy can be delivered at the capacitor's rate [1].

These EDLC–battery hybrid systems are useful in many modern applications, particularly for small mobile or wireless devices such as mobile phones, wireless sensors, portable computers, and wearable drug delivery systems. These modern devices do not require a lot of energy to be delivered continuously but, instead, very short high-power low-energy bursts. Using batteries for repeated delivery of high-power bursts will quickly degrade the lifecycle of the batteries. Furthermore, most users do not want to wait for hours to fully charge these devices. Hence, hybridization with EDLCs will improve the charging rate of batteries as well as their lifecycle.

EDLC–battery hybrid systems can also be utilized in green transportation and renewable energy applications. Using these systems, as opposed to purely battery-based energy storage systems, the charging/discharging time of midsized electric cars may be cut down from several hours to less than an hour. In addition, faster acceleration, higher top speed, and better regenerative braking can all be accommodated by these hybrid systems. EDLC–battery hybrid systems will also allow an efficient production of renewable energy from the sun, wind, or ocean tides and currents, by stabilizing the generated power before being stored.

At the moment, the biggest limiting factor in producing VACNT-based EDLCs is their extremely high price. However, CNTs and VACNTs in particular are expected to be much cheaper in the near future due to a significant increase in supplies. In addition, since no corrosive liquids need to be used as electrolytes, cheap lightweight materials such as aluminum, polypropylene, and polyimide can be used for current collectors, separators, and cases, respectively. The small ecological footprint of VACNT-based EDLCs and their lack of toxic materials should also be highlighted. Finally,

VACNT-based EDLCs are relatively safe since they do not have reactivity issues associated with lithium, such as leakages or even explosions.

ACKNOWLEDGMENT

This work was supported by The Office of Naval Research under Grant number N00014-11-1-0031 and The Fletcher-Jones Foundation under Grant number 9900600. The authors gratefully acknowledge support and infrastructure provided for this work by the Charyk Laboratory for Bioinspired Design, the Kavli Nanoscience Institute (KNI), and the Molecular Materials Research Center (MMRC) at the California Institute of Technology.

REFERENCES

1. Conway, B.E. *Electrochemical Supercapacitors: Scientific Fundamentals and Technological Applications* (Kluwer Academics/Plenum Publisher, New York, 1999).
2. Liu, C., Yu, Z., Neff, D., Zhamu, A., and Jang, B.Z. Graphene-based supercapacitor with an ultrahigh energy density, *Nano Letters*, 2010, 10(12), 4863–4868.
3. Simon, P. and Gogotsi, Y. Materials for electrochemical capacitors, *Nature Materials*, 2008, 7(11), 845–854.
4. Miller, J.R. and Simon, P. Electrochemical capacitors for energy management, *Science*, 2008, 321(5889), 651–652.
5. Miller, J.R. and Burke, A.F. Electrochemical capacitors: Challenges and opportunities for real-world applications, *Electrochemical Society Interface*, 2008, 17, 53–57.
6. Lu, W., Hartman, R., Qu, L., and Dai, L. Nanocomposite electrodes for high-performance supercapacitors, *The Journal of Physical Chemistry Letters*, 2011, 2(6), 655–660.
7. Jorio, A., Dresselhaus, G., and Dresselhaus, M. Carbon nanotubes: Advanced topics in the synthesis, structure, properties and applications (Springer, Berlin, 2007).
8. Kumar, A. Contact transfer of aligned carbon nanotube arrays onto conducting substrates, *Applied Physics Letters*, 2006, 89(16), 163120.
9. Fu, Y., Qin, Y., Wang, T., Chen, S., and Liu, J. Ultrafast transfer of metal-enhanced carbon nanotubes at low temperature for large-scale electronics assembly, *Advanced Materials*, 2010, 22(44), 5039–5042.
10. Yoon, B-J., Jeong, S-H., Lee, K-H., Seok Kim, H., Gyung Park, C., and Hun Han, J. Electrical properties of electrical double layer capacitors with integrated carbon nanotube electrodes, *Chemical Physics Letters*, 2004, 388(1–3), 170–174.
11. Signorelli, R., Ku, D.C., Kassakian, J.G., and Schindall, J.E. Electrochemical double-layer capacitors using carbon nanotube electrode structures, *Proceedings of the IEEE*, 2009, 97(11), 1837–1847.
12. Stoller, M.D., Park, S., Zhu, Y., An, J., and Ruoff, R.S. Graphene-based ultracapacitors, *Nano Letters*, 2008, 8(10), 3498–3502.
13. Lufrano, F. and Staiti, P. Conductivity and capacitance properties of a supercapacitor based on Nafion electrolyte in a nonaqueous system, *Electrochemical and Solid-State Letters*, 2004, 7(11), A447–A450.
14. Chen, J.H., Li, W.Z., Wang, D.Z., Yang, S.X., Wen, J.G., and Ren, Z.F. Electrochemical characterization of carbon nanotubes as electrode in electrochemical double-layer capacitors, *Carbon*, 2002, 40(8), 1193–1197.
15. Niu, C. High power electrochemical capacitors based on carbon nanotube electrodes, *Applied Physics Letters*, 1997, 70(11), 1480.
16. Futaba, D., Hata, K., Yamada, T., Hiraoka, T., Hayamizu, Y., and Kakudate, Y. Shape-engineerable and highly densely packed single-walled carbon nanotubes and their application as super-capacitor electrodes, *Nat. Mater.*, 2006, 5(12), 987–994.
17. Marcano, D.C., Kosynkin, D.V., Berlin, J.M., Sinitskii, A., Sun, Z., Slesarev, A., Alemany, L.B., Lu, W., and Tour, J.M. Improved synthesis of graphene oxide, *ACS Nano*, 2010, 4(8), 4806–4814.
18. Lee, S.W., Yabuuchi, N., Gallant, B.M., Chen, S., Kim, B-S., Hammond, P.T., and Shao-Horn, Y. High-power lithium batteries from functionalized carbon-nanotube electrodes, *Nat. Nano*, 2010, 5(7), 531–537.
19. Hirsch, A. Functionalization of single-walled carbon nanotubes, *Angewandte Chem. Int. Ed.*, 2002, 41(11), 1853–1859.
20. Aria, A.I. and Gharib, M. Reversible tuning of the wettability of carbon nanotube arrays: The effect of ultraviolet/ozone and vacuum pyrolysis treatments, *Langmuir*, 2011, 27(14), 9005–9011.
21. Hsieh, C-T. and Teng, H. Influence of oxygen treatment on electric double-layer capacitance of activated carbon fabrics, *Carbon*, 2002, 40(5), 667–674.

22. Pan, H., Li, J.Y., and Feng, Y.P. Carbon nanotubes for supercapacitor, *Nanoscale Res. Lett.*, 2010, 5(3), 654–668.

23. Lee, S.W., Gallant, B.M., Lee, Y., Yoshida, N., Kim, D.Y., Yamada, Y., Noda, S., Yamada, A., and Shao-Horn, Y. Self-standing positive electrodes of oxidized few-walled carbon nanotubes for light-weight and high-power lithium batteries, *Energy Environ. Sci.*, 2012, 5(1), 5437–5444.

24. Yang, D., Velamakanni, A., Bozoklu, G., Park, S., and Stoller, M. Chemical analysis of graphene oxide films after heat and chemical treatments by X-ray photoelectron and Micro-Raman spectroscopy, *Carbon*, 2009, 47(1), 145–152.

25. Byon, H.R., Gallant, B.M., Lee, S.W., and Shao-Horn, Y. Role of oxygen functional groups in carbon nanotube/graphene freestanding electrodes for high performance lithium batteries, *Adv. Funct. Mater.*, 2012 (published online).

26. Pandolfo, A. and Hollenkamp, A. Carbon properties and their role in supercapacitors, *J. Power Sources*, 2006, 157(1), 11–27.

27. Fernandez, J., Arulepp, M., Leis, J., Stoeckli, F., and Centeno, T. EDLC performance of carbide-derived carbons in aprotic and acidic electrolytes, *Electrochim. Acta*, 2008, 53(24), 7111–7116.

28. Simon, P. and Burke, A. Nanostructured carbons: Double-layer capacitance and more, *Electrochem. Soc. Interface*, 2008, 17(1), 38–43.

29. Du, C., Yeh, J., and Pan, N. High power density supercapacitors using locally aligned carbon nanotube electrodes, *Nanotechnology*, 2005, 16(4), 350.

30. Shah, R., Zhang, X., and Talapatra, S. Electrochemical double layer capacitor electrodes using aligned carbon nanotubes grown directly on metals, *Nanotechnology*, 2009, 20(39), 395202.

31. Nam, K-W., Kim, K-H., Lee, E-S., Yoon, W-S., Yang, X-Q., and Kim, K-B. Pseudocapacitive properties of electrochemically prepared nickel oxides on 3-dimensional carbon nanotube film substrates, *J. Power Sources*, 2008, 182(2), 642–652.

32. Mi, H., Zhang, X., An, S., Ye, X., and Yang, S. Microwave-assisted synthesis and electrochemical capacitance of polyaniline/multi-wall carbon nanotubes composite, *Electrochem. Commun.*, 2007, 9(12), 2859–2862.

36 Spray Deposition of Carbon Nanotube Thin Films

Alaa Abdellah, Paolo Lugli, and Giuseppe Scarpa

CONTENTS

36.1 INTRODUCTION

Random carbon nanotube (CNT) networks have raised a continuously increasing interest among a broad and multidisciplinary community of researchers over the last decade. The remarkable and concurrently diverse properties of such networks have rendered them suitable for a wide range of applications in science and engineering. Among the most promising applications are transparent conductive electrodes [1–3], thin-film transistors and circuits [4,5], and mechanical and chemical sensors [6–8]. The desired functionality of the CNT films produced for different applications depends primarily on the choice of the appropriate raw material. Nevertheless, it is strongly influenced by major process-specific aspects. A random CNT network can either be directly grown by chemical vapor deposition (CVD) or processed from a solution of well-dispersed nanotubes. The CVD process involves catalyst nanoparticles acting as seeds for the growth of the CNTs. Although this method leads to films with individually separated tubes and better inter-tube junctions, it is a high vacuum/temperature process that is not suitable for the emerging field of low-cost flexible electronics. Solution-based methods, on the other hand, have several advantages. In general, relying on this kind of low-temperature processes overrides most constraints on the choice of the substrate material, enabling fabrication even on plastic substrates. The omission of a high-vacuum process further reduces costs significantly.

A significant step toward the commercialization of novel CNT electronics involves the development of large-area and high-throughput processes for the fabrication of high-quality CNT films. Among all the different fabrication techniques available, spray deposition is considered to be one of the most competitive in terms of the cost–performance ratio. It offers the means for depositing CNT films with high uniformity and low roughness over large areas in a high-throughput (inline) process.

The successful adoption of this technique to CNT film fabrication for various applications has been previously reported in the literature [3,9].

In this chapter, we demonstrate a reliable and reproducible spray deposition process for the fabrication of CNT films exhibiting state-of-the-art performance. The convenient control of major process parameters enables a fine and accurate tuning of film characteristics, hence rendering this process suitable for a wide range of device applications with different requirements. Films that are fabricated using this process are further applied as resistive networks in gas-sensing devices. We demonstrate that such devices, incorporating highly uniform pristine CNT films, exhibit a high sensitivity toward ammonia (NH_3) without the need for any further functionalization of the CNT film, thereby maintaining a simple fabrication process.

36.2 MATERIALS AND METHODS

Spray technology is used in a wide variety of applications such as surface coating, humidification, combustion, and others. Each application area requires different spray nozzles to fulfill the specified criteria and achieve the desired results. The layers deposited here were sprayed by an air atomizing nozzle. This type of nozzles provides, along with ultrasonic spray nozzles, the finest degree of atomization. It is worth mentioning that ultrasonic spray nozzles have also been successfully applied to the deposition of high-uniformity CNT films [3]. For our static test system, a commercially available automatic air atomizing spray gun (Krautzberger GmbH, Germany) was used. The spray gun contains an internal pneumatic control system that is activated by an electromechanical 3/2-way valve connected to a timer for precise spray time adjustment. Other major spray parameters to be adjusted for obtaining the desired spray characteristics are the material flow rate, atomizing gas (N_2) pressure, nozzle-to-sample distance, and substrate temperature. For air-assisted nozzles, the diameter of the orifice is one of the most significant dimensions for atomization. Here, a nozzle with a 0.5 mm orifice diameter was chosen. The atomizing gas (N_2) pressure was kept below 1 bar throughout all experiments in order to obtain the desired spray characteristics and a uniform spray pattern across the area of interest. We use an approach introduced previously for the spray deposition of active polymers in organic photodiodes [10,11], in which we operate within the wet spraying regime while heating up the substrate in order to speed up the drying of wet droplets arriving at the substrate. This approach allows the formation of dry layers with good thickness control over time, enhancing at the same time reproducibility. Figure 36.1a shows a schematic drawing of the test setup, indicating the vertical arrangement of a nozzle and a substrate.

36.2.1 SOLUTION PREPARATION

To disperse CNTs in an aqueous solution, a high molecular weight cellulose derivative, sodium carboxymethyl cellulose (CMC), is used. This kind of dispersant has been reported previously as an excellent agent for dispersing single-walled nanotubes (SWNTs) in water [12]. As a first step, an adequate quantity of CMC is added to Millipore DI water so as to obtain a 0.5 wt% aqueous solution of CMC. This solution is then stirred for ≥12 h at room temperature to uniformly dissolve the surfactant in water. The desired amount of SWNTs can then be added to the previously prepared CMC stock solution. Typically, a concentration between 0.03 and 0.05 wt% of SWNTs is used. Actual dispersion of the CNTs is achieved by sonication of the complete solution for 90 min in a bath sonicator. For dispersions prepared by probe sonication, the duration is reduced to 15 min. The solution is then centrifuged at 15,000 rpm for 2 h and the final solution is obtained by decanting the top 80% of the supernatant.

36.2.2 BASIC CNT FILM FABRICATION

For the fabrication of highly uniform and smooth CNT films, the final CNT solution is spray deposited onto plain glass substrates. Prior to film deposition, the substrates are thoroughly cleaned by

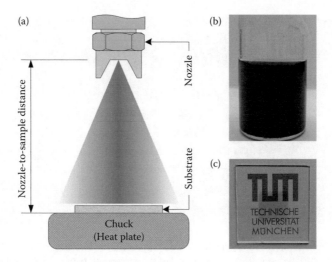

FIGURE 36.1 (a) Schematic drawing of the test setup, indicating the vertical arrangement of a nozzle and a substrate. (b) Vial containing a CNT/surfactant aqueous dispersion after sonication and centrifugation, ready for deposition. (c) An image of a CNT film on glass, uniformly deposited over an area of 1 cm². (A. Abdellah et al., IEEE-NANO 2011, *11th IEEE Conference on Nanotechnology 2011*, Portland, OR, pp.1118–1123. © (2011) IEEE. With permission.)

sonication in acetone and isopropanol. The substrates are then exposed to oxygen plasma for further cleaning and surface treatment. An increased wettability of the glass surface after plasma treatment is beneficial for achieving uniform surface coating during spray deposition of the water-based CNT solution. Chemical post-deposition treatment is necessary to remove the CMC-matrix embedding the CNTs, thus changing the film behavior from insulating to conductive. For this purpose, samples are immersed in dilute HNO_3 (13–23%) for $\geq$24 h at room temperature. Some samples are additionally immersed in concentrated HNO_3 (65%) for 1 h to further increase conductivity via stronger doping of CNTs [3,9]. Samples are finally rinsed in DI water and subsequently dried. Note that all steps involved in CNT film fabrication are performed entirely in ambient conditions.

36.2.3 BASIC CNT FILM CHARACTERIZATION

The surface characterization and thickness measurements of fabricated CNT films were performed with an atomic force microscope (AFM) operated in a tapping mode. Information on CNT length distribution was extracted from AFM images of low-density films using image processing software (ImageJ). Sheet resistances were measured using a four-point collinear probe method on a Keithley 4200 semiconductor parameter analyzer. Optical transmission was measured using a chopped 300 W xenon arc research source passing through an Oriel Cornerstone 260 1/4 m monochromator and a calibrated silicon photodiode with a preamplifier connected to an Oriel Merlin digital lock-in radiometry system.

36.3 RESULTS AND DISCUSSION

36.3.1 BASIC FILM CHARACTERISTICS

Prior to the adoption of CNT films in any specific application, it is necessary to optimize and evaluate the elementary film performance on the basis of some primary figures of merit and their dependence on different process parameters. Among the most important figures are sheet resistance or conductivity, optical transmittance or absorption spectra, mechanical stability, CNT distribution, and surface characteristics. For this reason, we first present an overview of basic film performance

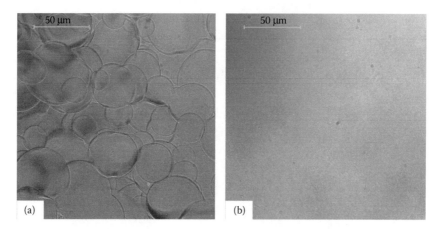

FIGURE 36.2 (a) Optical microscopy image of an as-prepared film, showing disk-like features originating from spray deposition. (b) The same film after removal of the surfactant matrix. The final film of directly interconnected CNTs is formed and exhibits a conductive behavior. (A. Abdellah et al., IEEE-NANO 2011, *11th IEEE Conference on Nanotechnology 2011*, Portland, OR, pp.1118–1123. © (2011) IEEE. With permission.)

and the means for controlling the same before discussing the device-specific aspects of gas sensors as one possible application of such CNT films.

As already mentioned in the previous section, we use CMC as a surfactant to aid dispersion of the nanotubes in water. This implies that the surfactant will still be present in all deposited films before performing further post-deposition treatments. As-prepared films are therefore composed of a CNT network embedded into a CMC-matrix. Single CNTs and CNT-bundles are suspended within the insulating matrix, and hence the film does not exhibit any conductive behavior. Figure 36.2a shows an optical microscopy image of an as-prepared film. Note the clearly distinguishable disk-like structures typical for spray-deposited layers. Droplets are deformed upon impact onto the substrate and dry rapidly creating the features observed. After removal of the CMC-matrix, the suspended CNT network embedded in the matrix collapses, thereby forming the final film of directly interconnected CNTs and exhibiting conductive behavior. This is achieved by chemical post-deposition treatment as described in the experimental section. An optical microscopy image of such a pristine CNT film is shown in Figure 36.2b. It is clear from the image that the features previously observed in as-prepared films completely vanish after CMC removal. Moreover, the film seems highly uniform with no agglomerates or other clear particles visible at an optical scale.

An accurate examination of the surface characteristics at the nano-scale is required for a detailed evaluation of the pristine CNT film quality. This involves the characterization of uniformity, surface roughness, and CNT length distribution. Surface roughness is of especially great significance as it can become a crucial factor in certain applications, for example, when considering these films as transparent electrodes for optoelectronic devices. The CNT length distribution, on the other hand, is of relevance for the evaluation of the dispersion quality and an understanding of the contribution of different process steps in altering the physical properties of the raw CNT material. Sonication during the solution preparation was shown to shorten the CNTs and has therefore to be carefully optimized in terms of power and duration [13]. Figure 36.3 shows two AFM images of the film introduced in Figure 36.1c, scanned over a large area of $10 \times 10 \ \mu m^2$ (left image) as well as a zoom-in of a smaller area of $2 \times 2 \ \mu m^2$ (right image). For the $10 \times 10 \ \mu m^2$ scan, the values measured for the average and RMS surface roughness are 5.79 and 7.41 nm, respectively. Very similar values (5.64 and 7.13 nm) are obtained from the $2 \times 2 \ \mu m^2$ scan, which demonstrates a high uniformity of the films independent of the total area under investigation.

Another major figure of merit necessary toward a full characterization of CNT films is the sheet resistance (R_{sh}) or DC conductivity (σ_{dc}) and its dependence on the film thickness (d) or density.

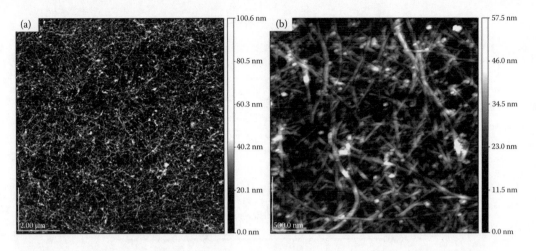

FIGURE 36.3 AFM images of a CNT film scanned over a large area of $10 \times 10 \ \mu m^2$ (a) as well as a zoom-in of a smaller area of $2 \times 2 \ \mu m^2$ (b). The values measured for the average and RMS surface roughness are 5.79 and 7.41 nm, respectively. (A. Abdellah et al., IEEE-NANO 2011, *11th IEEE Conference on Nanotechnology 2011*, Portland, OR, pp.1118–1123. © (2011) IEEE. With permission.)

Owing to the percolating nature of the CNT films, R_{sh} is expected to vary inversely with film density. Since CNT density is correlated to film thickness, R_{sh} is also expected to scale inversely with film thickness. Figure 36.4a plots the relation between R_{sh}, as determined by four-point probe measurements, and d. One can clearly observe the reduction in sheet resistance with increasing film thickness or density, reaching as low as 60 Ω/sq at approx. 42 nm. DC conductivity can be calculated in a straightforward manner from the sheet resistance and thickness through $\sigma_{dc} = 1/(R_{sh} \cdot d)$. Figure 36.4b shows the calculated values of σ_{dc} as a function of d. As opposed to a material-specific constant value for the conductivity, as expected from continuous metallic or semiconducting films, σ_{dc} for a CNT thin film shows a strong dependence on thickness. The conductivities calculated for different films vary by a factor of almost 4 in a range of thicknesses between 10 and 50 nm. As more CNTs are deposited, more conduction paths are created through the film and conductivity continues to increase until a thickness of approx. 50 nm is reached, when the film conductivity approaches saturation. This behavior is consistent with the results reported previously by other groups examining the same CNT material used here [14]. However, the saturation conductivity reached there at 50 nm was 400 S/cm, whereas for our films conductivity approaches 4000 S/cm at 45 nm, thereby being one order of magnitude higher. Besides the choice of a raw material with a given purity and ratio of metallic to semiconducting tubes, numerous other factors influence the conductivity of the films including the length and diameter distribution of tubes and bundles, presence of residual surfactants, and doping during post-deposition treatments.

For certain applications of CNT films, evaluation of the electrical properties solely is not sufficient. The optical characterization of fabricated films can give insight into the influence and degree of doping/dedoping through chemical and thermal treatments. This can be achieved by the examination of metallic and semiconducting transitions in the absorption spectrum of films processed differently, as discussed previously. Moreover, in applications requiring optically transparent CNT films, the transmittance in the visible spectral range is of special interest. For this purpose, the transmittance at a wavelength of 550 nm is typically used as a representative value for comparison. As the optical losses increase naturally with increasing film thickness while the electrical losses decrease, a trade-off must be found for applications requiring both transparency and conductivity. Hence, a common approach for the evaluation of the electro-optical performance of such films is to plot sheet resistance as a function of the transmittance at 550 nm. This relation is plotted in

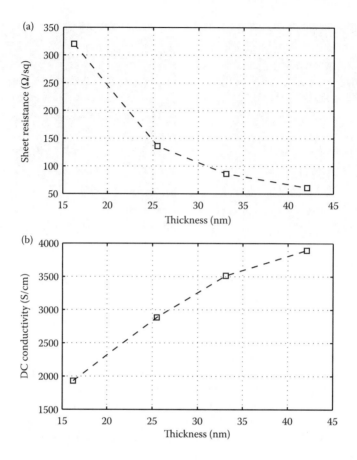

FIGURE 36.4 (a) Relation between sheet resistance (R_{sh}), as determined by four-point probe measurements, and film thickness. (b) The calculated values of DC conductivity (σ_{dc}) as a function of film thickness. (A. Abdellah et al., IEEE-NANO 2011, *11th IEEE Conference on Nanotechnology 2011*, Portland, OR, pp.1118–1123. © (2011) IEEE. With permission.)

Figure 36.5 for the fabricated CNT films. A good trade-off between sheet resistance and transmittance is achieved with films having 200 Ω/sq at 80% or 105 Ω/sq at 65% transmittance. These values closely match the state-of-the-art performance of films fabricated using more common and well-established deposition methods, such as spin coating or even vacuum filtration. Another method to quantify and compare transparent conductive thin films is by using the formula defined in Equation 36.1, where σ_{op} is the optical conductivity and $Z_0 = 377 \ \Omega$ is the impedance of free space [15].

$$T = \left(1 + \frac{Z_0}{2R_{sh}} \cdot \frac{\sigma_{op}}{\sigma_{dc}}\right)^{-2} \tag{36.1}$$

This formula is originally applied to characterize thin metal films, where the absorption of the material is significantly lower than the reflectance and the thickness of the film is much smaller than the wavelength of interest. Typically, the value for the optical conductivity of CNT films at 550 nm is chosen to be 200 S/cm, as determined by Ruzicka et al. [16]. Since optical conductivity is independent of the doping levels, this value is assumed to be constant even among differently processed films. By fitting the data plotted in Figure 36.5 to this formula, a value of 7.5 is obtained for the ratio $\gamma = \sigma_{dc}/\sigma_{op}$, which is often used as a figure for a quantitative comparison of CNT thin-film

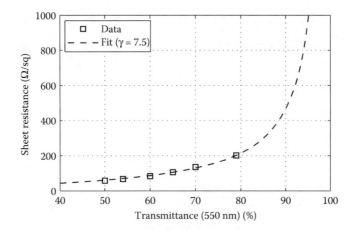

FIGURE 36.5 Plot of sheet resistance (R_{sh}) as a function of optical transmittance at 550 nm. A good trade-off between sheet resistance and transmittance is achieved with films having 200 Ω/sq at 80% or 105 Ω/sq at 65% transmittance.

performance. This figure should however be considered with caution and may not be suitable for straightforward comparison, since sheet resistance also depends on the type of SWNTs used and the level of doping. Tube diameter, average bundle length, and film roughness are detrimental factors for transmittance as well. The DC to optical conductivity ratio (γ), extracted from the relation between sheet resistance and transmittance, is therefore best suitable for the comparison of different films prepared using the same method and under similar conditions. Nevertheless, the values determined above are comparable to those reported for CNT films incorporating the same type of nanotubes (P3-SWNT, Carbon Solutions Inc), although they are fabricated using a different method [17].

All the figures discussed above demonstrate that the readily scalable air-atomizing spray deposition process presented here yields high-quality CNT thin films with a competitive electro-optical performance suitable for different device applications.

36.3.2 Effect of Process Parameter

The previous section was dedicated to a general description of the materials and methods involved in the CNT thin-film technology developed here. Note that a careful optimization of the different process parameters and processing conditions is crucial for achieving reliable and reproducible results using spray deposition. Several process-specific issues have to be addressed for the spray deposition and post-deposition treatment of CNT thin films based on different types of surfactants.

36.3.2.1 Sonication Time and CNT Distribution

The effect of sonication power and duration on the quality of the CNT dispersion used for nanotube deposition from a solution is of major significance for the performance of the resulting CNT thin film. It is necessary to differentiate between the different sources used for ultrasonic-assisted CNT dispersion into the solution. The two types typically used in such an application are bath and probe sonicators. While in a bath sonicator the ultrasonic power is generated in a water bath containing a flask with the CNT solution, a probe sonicator is directly immersed into the flask where the ultrasonic power is generated locally. This renders probe sonicators more powerful, hence requiring shorter sonication durations for solution preparation. The results discussed here to demonstrate the effect of sonication time on the characteristics of the final CNT film refer to the dispersions prepared by bath sonication. Figure 36.6 compares the relation between sheet resistance and transmittance for a variety of spray-deposited films from two CNT solutions prepared with different sonication durations. The data plotted show a clear shift toward higher values of sheet resistance at any given

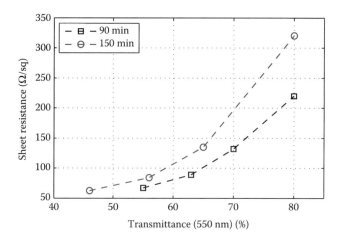

FIGURE 36.6 Variation of sheet resistance as a function of transmittance at 550 nm for SWNT films deposited from solutions with different sonication times. All films were chemically treated in concentrated HNO$_3$ (65%) for 3 h following CMC removal.

transmittance. Since excessive sonication introduces tube cutting and hence results in reduction of the average tube length [18], the electro-optical characteristics of the final CNT film suffer. This is a direct consequence of the percolation nature of such a CNT network, being composed of a random arrangement of particles with high-aspect ratios, where the percolation threshold scales inversely with the average tube length. A reduction of the average tube length therefore leads to an increase in the percolation threshold, with shorter sonication allowing for lower sheet resistance at a given optical transmittance or tube density [19].

To be able to evaluate the quality of a given CNT dispersion, an analysis of the CNT bundle distribution within the resulting film has to be performed. When considering films with very low CNT density or surface coverage, it is possible to extract information about the CNT length distribution by means of image processing. For this purpose, the CNT solution under investigation is highly diluted and used to spray a CNT network with ultra-low density onto a Si substrate. The sample is then processed further, according to the usual procedure, and its topography is characterized by AFM. Analysis of CNT bundles and their length distribution is finally performed using the image processing software ImageJ. Figure 36.7 shows the original AFM image obtained for such a sample along with the processed image used to generate the corresponding bundle length distribution. A histogram containing the bundle length distribution obtained for a sample sprayed with the CNT solution of 90 min sonication is plotted in Figure 36.8. A total of 79 CNTs are counted and considered in this distribution, hence allowing an acceptable statistical analysis. The histogram shows the relative frequency of occurrence for different tube lengths. It is then possible to fit the given tube length data with a log-normal distribution, as represented by the solid red line in the same figure. This is in good agreement with the CNT length distributions typically found in the literature [3]. For our film, the best fit is obtained using a log-normal distribution with a mean of 744 nm and a standard deviation of 95 nm. Longer SWNT bundles result in lower sheet resistances, due to less intertube junctions, which prevail in determining the overall film conductivity.

36.3.2.2 Chemical Treatment with Concentrated HNO$_3$

Chemical post-deposition treatment with dilute HNO$_3$ is needed for surfactant (CMC) removal as described above. At the same time, however, this treatment introduces an intentional p-type doping of the SWNT film. Treatment with concentrated HNO$_3$ is found to be effective when a strong doping of the film is desired. We analyzed the reduction in sheet resistance of SWNT films with different thicknesses due to concentrated HNO$_3$ (65%) treatment. The different samples were directly

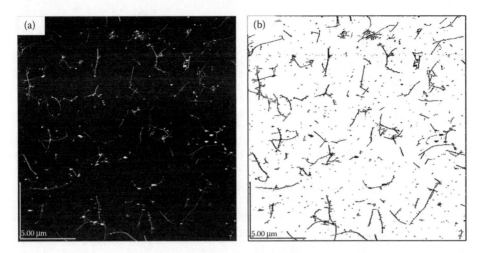

FIGURE 36.7 (a) AFM image of a CNT film with an extremely low tube density scanned over an area of $20 \times 20 \ \mu m^2$. (b) The processed image used for bundle length analysis.

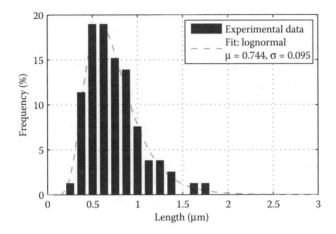

FIGURE 36.8 Histogram showing the relative frequency of occurrence for different tube lengths. It is possible to fit the given tube length data with a log-normal distribution with a mean of $\mu = 744$ nm and a standard deviation of $\sigma = 95$ nm. (A. Abdellah et al., IEEE-NANO 2011, *11th IEEE Conference on Nanotechnology 2011*, Portland, OR, pp.1118–1123. © (2011) IEEE. With permission.)

immersed into the HNO_3 and the sheet resistance was recorded for varying treatment intervals. A graph containing the collected data is shown in Figure 36.9a. It is found that treatment with concentrated HNO_3 for 1 h seems to be sufficient to approach the lowest value of sheet resistance for any given film, after which only slight reduction is observed. For all films within the thickness range examined here, a stabilization of sheet resistance occurs within 3 h of treatment. Moreover, the absolute change in sheet resistance for thinner films is larger than for thicker films. This is assumed to be due to the fact that thicker films, that is, ones with a higher tube density, contain a larger number of metallic tubes and hence more metallic paths, suppressing the effect of doping on the conductivity of semiconducting tubes. The relative change in sheet resistance is however similar for all samples, with a reduction by factors of 1.3–1.5 after treatment.

36.3.2.3 Optical Characteristics

In a discussion about the optical properties of CNT thin films, optical transmittance is of particular interest for organic optoelectronic device applications. As semitransparent electrodes in OLEDs

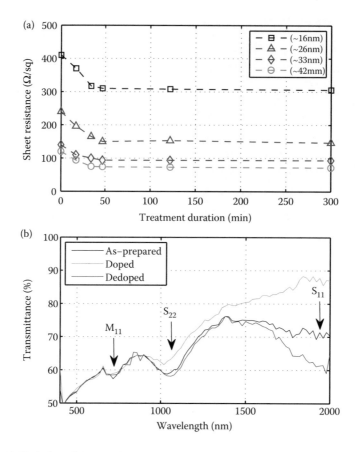

FIGURE 36.9 (a) Variation of sheet resistance as a function of concentrated HNO$_3$ (65%) treatment duration for four films of different thicknesses. All films were previously subject to identical treatment in dilute HNO$_3$ (13–23%) for CMC removal. (b) Optical transmittance spectra of an as-prepared, doped, and dedoped SWNT film.

or OPDs, an optimum of sheet resistance and transmittance within the visible range of the optical spectrum has to be carefully determined. Additionally, fundamental information on specific properties of the nanotubes incorporated in a film, such as electronic transitions and doping of nanotubes, can be obtained by the analysis of a broader spectral range extending beyond the visible wavelengths.

The measured optical transmittance decreases with increasing CNT film thickness and is wavelength dependent. It is therefore necessary to define a representative value for the transmittance of any given CNT thin film. While in some cases this measure is defined as the average transmittance over a certain spectral range, usually including the visible and infrared range (400–1800 nm) [3], a more common convention is to use the transmittance at a fixed wavelength of 550 nm as a representative value. The latter yields moderate values and results in more realistic figures within the context of optoelectronic applications operated in the visible spectral range. Note that this figure can be considered to represent a lower limit, since film transmittance increases considerably in the infrared range. Depending on the application under development, a trade-off between optical and electrical losses must be found so as to fulfill the specifications required.

The characteristic absorption peaks attributed to electronic transitions at van Hove singularities can be observed in the optical transmittance spectrum of SWNT thin films, thereby deducing information about the extent of doping/dedoping through chemical and thermal treatments. To demonstrate this, an SWNT film was fabricated by spray deposition, which was then characterized once as-prepared and again after performing different post-deposition treatments. Figure 36.9b shows the wavelength-dependent optical transmittance curve of the as-prepared, spray-deposited film, where the different transitions are evident as absorption peaks. Interband energy transitions in semiconducting SWNTs are designated by S_{11} and S_{22} while the transition in metallic SWNTs is designated by M_{11}, as typically found in the literature [20]. As photons are dependent on SWNTs, peak absorption occurs between energies with local maxima of electron density (van Hove singularities). Owing to p-type doping, electrons in the valence band are withdrawn and therefore no longer contribute to interband energy transitions. The withdrawal of electrons can be equivalently viewed as the injection of holes. Effects of p-type doping through chemical treatment of the same SWNT film by concentrated HNO_3 (65%) can be seen in Figure 36.9b (doped). A nearly complete bleaching of the S_{11} transition and significant reduction of the S_{22} transition can be observed for the doped film. The preceding observation clearly confirms the occurrence of electron withdrawal from the valance band of the semiconducting SWNTs upon doping with HNO_3. A similar effect is observed when films are chemically treated with $SOCl_2$. The p-type doping of SWNTs can be reversed by a thermal treatment. Here, the film is annealed on a hot plate at 200°C for 2 h under ambient conditions. A substantial increase of the absorption peak corresponding to the S_{11} energy transition is observed, as seen in the transmittance curve (dedoped) shown in Figure 36.9b. Owing to the partial removal of the previously adsorbed oxygen species after thermal annealing of the film, electrons are returned to their initial state in the first van Hove singularity. Consequently, photon energies equal to the S_{11} electronic interband transition are absorbed as indicated by a dip in the transmittance curve located at a wavelength of approx. 1900 nm. The return of electrons to the p-type nanotubes causes a reduction in the excess charge carriers, which in turn leads to a reduction of the overall film conductivity. Sheet resistance is hence decreased from 100 Ω/sq for the HNO_3 doped film to 420 Ω/sq by thermally induced dedoping. From the preceding example, we conclude that a qualitative evaluation of the extent of film doping is possible by a simple analysis of the optical absorption spectrum in the near-infrared range.

36.3.3 CNT NETWORKS FOR GAS SENSING

The inherently high surface-to-volume ratio of CNTs makes them an ideal candidate for environmental gas-sensing applications [6,7,21–24]. A common implementation for CNT gas sensors consists of an interdigitated electrode structure (IDES) forming the two contact terminals of the simple resistive architecture, above or beneath which a uniformly distributed resistive random network of CNTs is deposited. It is worth mentioning that the CNT film incorporated in our gas sensors has a significantly lower density/thickness than the films introduced above. The utilization of low-density films leads to better sensor response and higher sensitivities. This is considered to be a consequence of the more direct exposure of a greater portion of CNTs contributing to current conduction through the film, thereby enhancing the change of the overall film conductance when exposed to a given concentration of the test gas. A sensor readout is performed by applying a sensing current and measuring the voltage drop across the two terminals of the device, thus monitoring the relative change in resistance/conductance with respect to different controlled environmental conditions. Before sensor response toward the test gas is evaluated, the initial resistance of the network is monitored for a specified time in order to determine a baseline for the device under test. Baseline characterization was performed with a sensing current of 10 μA under a constant carrier (N_2) gas flux of 200 mL/min at room temperature. A continuous increase in the resistance of the network can be observed during the first hour before a stable state is approached. In our opinion, this effect is attributed to a continuation of the dedoping process due to the flow of a relatively high-current density through the network. After reaching a stable initial resistance, the sensor response is investigated by exposing the sensor module inside a gas chamber to different concentrations of ammonia. The main figure used for the evaluation of sensor performance is the sensitivity, defined in Equation 36.2 as the relative change in resistance, where R_i and R_f are the initial and final resistance values of an exposure cycle, respectively.

$$S = \frac{R_f - R_i}{R_i} \times 100 \tag{36.2}$$

Figure 36.10a depicts the measured sensor resistance over time. The graph is divided into four segments, based on four exposure/recovery cycles at different NH_3 concentrations of 10, 25, 50, and 100 ppm. Each cycle is composed of an exposure interval followed by a recovery interval. During exposure, NH_3 is allowed to enter the chamber along with the carrier gas at room temperature and a constant flux of 200 mL/min for a 100 s time interval. Recovery is then introduced by heating the sensor module to 80°C and increasing the N_2 flux to 1000 mL/min for 300 s, after which heating is stopped again. The high flux is maintained for another 300 s to accelerate cooling of the sensor and to purge any residual test gas molecules out of the chamber. Recovery is then completed by a final 300 s interval under sensing conditions to restore the initial resistance. The increase in thermal energy during the heating cycle is necessary to enhance desorption of gas molecules attached to the CNT film and to enable a complete recovery. One of the most prominent features observed in the response is the clear and immediate change in resistance at the start and end of NH_3 exposure.

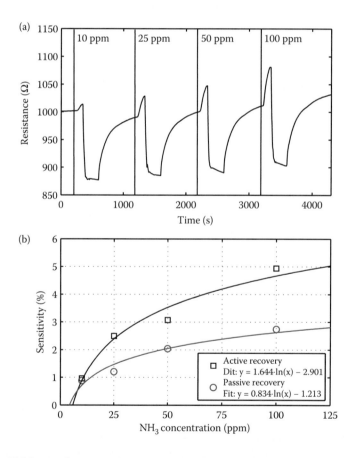

FIGURE 36.10 (a) Measured sensor resistance over time. The graph is divided into four segments, based on four exposure/recovery cycles at different NH_3 concentrations of 10, 25, 50, and 100 ppm. (b) Plot displaying the sensitivity as a function of NH_3 concentration. Sensitivities presented here correspond to the relative resistance changes in an exposure interval as short as 60 s with values reaching 5% at a concentration of 100 ppm. (A. Abdellah et al., IEEE-NANO 2011, *11th IEEE Conference on Nanotechnology 2011*, Portland, OR, pp. 1118–1123. © (2011) IEEE. With permission.)

Taking a look at the magnitude of change in resistance after exposure to different concentration reveals a clear increase in sensitivity with increasing concentration, at least up to a concentration of 100 ppm. This becomes more evident when displaying the sensitivity as a function of NH_3 concentration, as plotted in Figure 36.10b. To avoid inaccuracies in sensitivity calculations due to delays in the gas system, R_i is recorded at the instance of abrupt change in resistance and R_f is defined as the resistance of the network 60 s later. This means that the sensitivities presented here correspond to an exposure interval of 60 s, which is to our knowledge shorter than the usual exposure intervals reported in the literature. Sensitivity is found to have a logarithmic dependence on the concentration, with sensitivities of 1% for concentrations as low as 10 ppm and reaching 5% at a concentration of 100 ppm. We consider these values for sensitivity to be very competitive taking into account the short exposure intervals of 60 s [8]. The response of the same sensor with passive recovery, that is, no heating or increased flux is applied during the recovery interval, is also shown in Figure 36.10b. It can be seen that the sensitivity is significantly lower due to incomplete recovery of the CNT film. Active recovery can therefore enhance sensitivity by a factor of 2, using the same sensor architecture in addition to a simple heating element.

36.4 SUMMARY

In conclusion, we presented a reliable and reproducible spray deposition process for the fabrication of highly uniform CNT films exhibiting a state-of-the-art performance. The convenient control of major process parameters enables a fine and accurate tuning of film characteristics. Average surface roughness was shown to be 5.79 nm, measured over an area of $10 \times 10\ \mu m^2$. The calculated conductivities were shown to vary with film thickness due to the percolating nature of CNT films. Conductivity increases to 4000 S/cm at 45 nm and approaches saturation beyond this thickness. A good trade-off between sheet resistance and transmittance is achieved with the best films having 200 Ω/sq at 80%.

Further, we report the successful implementation of CNT-based gas sensors with exceptionally high sensitivities and a fast response to various test gases. Highly uniform CNT thin films, prepared using a reliable and reproducible, low-cost spray deposition process, are utilized as resistive networks for gas detection. The sensitivity toward NH_3 is examined along with passive and active recovery routines. Sensors show a clear and immediate change in resistance as a response to test gas exposure. Sensitivities of 1% for concentrations as low as 10 ppm and reaching 5% at a concentration of 100 ppm can be achieved in case of NH_3. We consider these values for NH_3 sensitivity to be competitive, taking into account the short exposure intervals of 60 s considered in our sensitivity calculations.

REFERENCES

1. D. Zhang et al., Transparent, conductive, and flexible carbon nanotube films and their application in organic light-emitting diodes, *Nano Letters*, 6(9), 1880–1886, 2006.
2. Y.-M. Chien, F. Lefevre, I. Shih, and R. Izquierdo, A solution processed top emission OLED with transparent carbon nanotube electrodes, *Nanotechnology*, 21(13), 134020, 2010.
3. R. C. Tenent et al., Ultrasmooth, large-area, high-uniformity, conductive transparent single-walledcarbon-nanotube films for photovoltaics produced by ultrasonic spraying, *Advanced Materials*, 21(31), 3210–3216, 2009.
4. M. Ha et al., Printed, sub-3V digital circuits on plastic from aqueous carbon nanotube inks, *ACS Nano*, 4(8), 4388–95, Aug. 2010.
5. C. Wang, J. Zhang, and C. Zhou, Macroelectronic integrated circuits using high-performance separated carbon Nanotube thin-film transistors, *ACS Nano*, 4(12), 7123–7132, Nov. 2010.
6. J. Li, Y. Lu, Q. Ye, M. Cinke, J. Han, and M. Meyyappan, Carbon nanotube sensors for gas and organic vapor detection, *Nano Letters*, 3(7), 929–933, Jul. 2003.
7. E. Bekyarova et al., Chemically functionalized single-walled carbon nanotubes as ammonia sensors, *The Journal of Physical Chemistry B*, 108(51), 19717–19720, Dec. 2004.

8. M. Penza, R. Rossi, M. Alvisi, G. Cassano, and E. Serra, Functional characterization of carbon nanotube networked films functionalized with tuned loading of Au nanoclusters for gas sensing applications, *Sensors and Actuators B: Chemical*, 140(1), 176–184, Jun. 2009.

9. S. Kim, J. Yim, X. Wang, D. D. C. Bradley, S. Lee, and J. C. DeMello, Spin- and spray-deposited single-walled carbon-nanotube electrodes for organic solar cells, *Advanced Functional Materials*, 20(14), 2310–2316, 2010.

10. A. Abdellah, D. Baierl, B. Fabel, P. Lugli, and G. Scarpa, Exploring spray technology for the fabrication of organic devices based on poly(3-hexylthiophene). *IEEE-NANO 2009. 9th IEEE Conference on Nanotechnology*, Genoa, pp. 831–934, 2009.

11. A. Abdellah, B. Fabel, P. Lugli, and G. Scarpa, Spray deposition of organic semiconducting thin-films: Towards the fabrication of arbitrary shaped organic electronic devices, *Organic Electronics*, 11(6), 1031–1038, Jun. 2010.

12. T. Takahashi, K. Tsunoda, H. Yajima, and T. Ishii, Dispersion and purification of single-wall carbon nanotubes using carboxymethylcellulose, *Japanese Journal of Applied Physics*, 43(6A), 3636–3639, Jun. 2004.

13. D. Hecht, L. Hu, and G. Grüner, Conductivity scaling with bundle length and diameter in single walled carbon nanotube networks, *Applied Physics Letters*, 89(13), 133112, Sep. 2006.

14. E. Bekyarova et al., Electronic properties of single-walled carbon nanotube networks, *Journal of the American Chemical Society*, 127(16), 5990–5995, Apr. 2005.

15. L. Hu, D. Hecht, and G. Grüner, Percolation in transparent and conducting carbon nanotube networks, *Nano Lett*, 4(12), 2513–2517, Dec. 2004.

16. B. Ruzicka and L. Degiorgi, Optical and dc conductivity study of potassium-doped single-walled carbon nanotube films, *Physical Review B*, 61(4), R2468–R2471, Jan. 2000.

17. Y. Zhou, L. Hu, and G. Grüner, A method of printing carbon nanotube thin films, *Applied Physics Letters*, 88(12), 123109, Mar. 2006.

18. M. F. Islam, E. Rojas, D. M. Bergey, A. T. Johnson, and A. G. Yodh, High weight fraction surfactant solubilization of single-wall carbon nanotubes in water, *Nano Letters*, 3(2), 269–273, Feb. 2003.

19. D. Simien, J. A. Fagan, W. Luo, J. F. Douglas, K. Migler, and J. Obrzut, Influence of nanotube length on the optical and conductivity properties of thin single-wall carbon nanotube networks, *ACS Nano*, 2(9), 1879–84, Sep. 2008.

20. S. Niyogi et al., Chemistry of single-walled carbon nanotubes, *Accounts of Chemical Research*, 35(12), 1105–1113, Dec. 2002.

21. J. Kong et al., Nanotube molecular wires as chemical sensors, *Science*, 287(5453), 622–625, Jan. 2000.

22. O. K. Varghese, P. D. Kichambre, D. Gong, K. G. Ong, E. C. Dickey, and C. A. Grimes, Gas sensing characteristics of multi-wall carbon nanotubes, *Sensors and Actuators B: Chemical*, 81(1), 32–41, Dec. 2001.

23. Y.-T. Jang, S.-I. Moon, J.-H. Ahn, Y.-H. Lee, and B.-K. Ju, A simple approach in fabricating chemical sensor using laterally grown multi-walled carbon nanotubes, *Sensors and Actuators B: Chemical*, 99(1), 118–122, Apr. 2004.

24. M. Arab, F. Berger, F. Picaud, C. Ramseyer, J. Glory, and M. Maynelhermite, Direct growth of the multi-walled carbon nanotubes as a tool to detect ammonia at room temperature, *Chemical Physics Letters*, 433(1–3), 175–181, Dec. 2006.

25. A. Abdellah et al., IEEE-NANO 2011, *11th IEEE Conference on Nanotechnology*, Portland, OR, pp. 1118–1123, 2011.

37 Electrical Control of Synthesis Conditions for Locally Grown CNTs on a Polysilicon Microstructure

Knut E. Aasmundtveit, Bao Quoc Ta, Nils Hoivik, and Einar Halvorsen

CONTENTS

37.1 INTRODUCTION

The integration of carbon nanotubes (CNTs) with micro-electromechanical systems (MEMS) or nano-electromechanical systems (NEMS) has attracted increasing interest over the last decade [1–7]. Recent semiconductor advancements further demonstrate the feasibility of using CNTs to construct nano-devices [3–7]. The possibility of combining CNTs with on-chip integrated circuits (ICs) and MEMS enables many applications of CNTs in devices, including physical, chemical, and biological sensors, as well as nanotube-based transistors. Combining the three technologies on a single platform will allow for true integration of micro–nano systems. CNTs with their huge surface-to-volume ratio give ultra-sensitive sensors, ICs/CMOS provide signal processing capability, and MEMS give actuation opportunities as well as further sensor options.

To combine CNTs with MEMS and IC technologies, a desirable approach is to construct the on-chip microelectronics first using readily available foundry services and synthesize or assemble CNTs selectively at desired locations afterward. CNTs are normally synthesized in bulk volumes by arc discharge, laser ablation, or chemical vapor deposition at high temperatures, above 700°C, which is far higher than the acceptable post-processing temperatures for circuits from standard foundry services. Processed CMOS/MEMS devices should normally not be exposed to temperatures above 300–400°C, depending on the duration of high-temperature exposure. *In situ* manufacturing of CNTs on an IC (CMOS) device is therefore not compatible with these high-temperature CNT manufacturing techniques. Alternatively, CNTs can be synthesized separately and deposited on the microsystem, but it is difficult to control the deposition locations of CNTs, and to obtain a reliable and low-cost assembly process. A number of ways to integrate CNTs in microsystems and ICs have been proposed and demonstrated:

- Manual transfer of CNTs by micro-manipulators, and welding the terminals of the CNTs to MEMS structures using electron or ion-beam-induced metal deposition in an SEM and/ or FIB [8]. Whereas this has indeed been successfully demonstrated on a research laboratory scale, the technique relies on a serial process occupying high-cost equipment, and is therefore not scalable to a low-cost industrial manufacturing process.
- Dielectrophoresis allows the alignment of CNTs between the given electrodes [9]. However, this process will also rely on a separate process for electrical connection of the CNTs to the microstructure, such as the above-mentioned electron-beam or ion-beam metal deposition. Similar to the above, this implies that the process is not compatible with manufacturing in an industrial setting.
- Processing from a solution of CNTs, low-temperature and low-cost deposition of CNTs is possible [10]. This technique will, however, result in random CNT networks not utilizing the properties of single CNTs as the CNT-to-CNT interconnects are likely to dominate the performance. Furthermore, additional patterning is needed to locate the CNT network in the desired area, which is normally performed through a photolithography process.
- CNTs grown on a separate wafer can be transferred to the active wafer [11]. The cited process requires a number of process steps, including Au evaporation on top of the CNTs, thermal release tape transfer of Au + CNT, and Au etching. Also, CNT patterning as well as metal connection deposition is needed in this case.

Heading for a low-cost CMOS-compatible process of CNT–microsystem integration, a local synthesis technique that uses local Joule heating has been studied and developed by Christensen et al. [1,2] on SOI structures, thus demonstrating the direct integration of CNTs on a Si microsystem at room temperature. Similar approaches have been addressed by other groups, using SOI [12], or metal [13,14] growth structures.

Selected properties of the methods described above are summarized in Table 37.1, focusing on the relevance for low-cost, industrially relevant processes for CNT–microsystem integration.

This study presents the local synthesis and direct integration of CNTs on polysilicon microstructures fabricated from a commercial process (PolyMUMPS), thus investigating the applicability for the CNT direct integration on a realistic device platform. Furthermore, we present a method to automatically control the entire growth process, in particular the important local synthesis temperature. This control method uses only the measurement of electrical resistance of the structure; therefore, it is more cost-effective and less time consuming than other methods that require additional optical equipment to determine optimum growth conditions. Importantly, it is scalable to wafer-level processing and well adapted to be implemented in an industrial setting.

37.2 METHODOLOGY

To demonstrate the concept, we use a microstructure, shown in Figure 37.1, consisting of two suspended polysilicon microbridges, 5 μm wide and 160 μm long. The two microbridges are separated by 15 μm, and they are suspended 3.5 μm above the substrate. The suspended bridge structure is obtained through the etching of a 3.5 μm sacrificial layer in the PolyMUMPS process. A dual layer of 3 nm Fe and 2 nm Ni is evaporated to serve as the catalyst. Such thin layers do not result in a short-circuit over the microstructure. The catalyst film is not patterned, since the areas for CNT growth will be selected by the localized resistive heating. The synthesis is accomplished using resistive heating on the growth structure while a local electric field is established between the two bridges to guide the growth direction of CNTs [1,2]. The electrical arrangement is also illustrated

TABLE 37.1

Comparison of Different Processes for CNT–Microsystem Integration

CNT–Microsystem Integration Technique	Low-Temperature Process/CMOS Compatible	Serial or Parallel Process	Advanced or Low-Cost Equipment	Additional Post-Patterning Required	Ability to Connect Individual Nanotubes	Comments
CVD in furnace	No	Parallel	Medium	Catalyst pre-patterning	Yes	Not compatible with CMOS
Manual transfer and metal deposition	Yes	Serial	Advanced	No	Yes	Highest control, but highest cost
Dielectrophoresis	Yes	Serial	Advanced	No	Yes	Advanced equipment and serial process for ion-beam induced metal deposition
Solution processing	Yes	Parallel	Low cost	Yes	Random network	
Wafer transfer	Yes	Parallel	Medium	Yes	Yes, but difficult to control # of CNTs	Several process steps required
Local synthesis/direct integration	Yes	Parallel	Low cost	No	Yes	Presented in more detail in this chapter

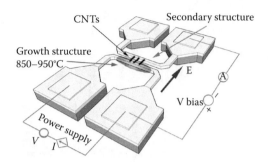

FIGURE 37.1 Polysilicon microstructures and electrical arrangement. The growth structure is heated by Joule heating. CNTs grow from the growth structure, and connect to the secondary structure under the assistance of an electric field.

in the figure. The synthesis is conducted in a room temperature chamber with a pressure of 0.4 bar, acetylene (C_2H_2) gas flow of 30 ccm (cubic centimeter per minute) after the growth structure is heated to the desired temperature (850–950°C). When CNTs grown connect the two microbridges, individual connections are detected electrically, with the current between the two microbridges increasing in steps every time a new CNT connection is made.

To get high-quality CNTs, it is very important to control the synthesis temperature [15], which in this case is the temperature of the growth structure. A technique that analyzes the color of the emitted light has been applied in previous studies [1,2]. Although that technique can potentially be adapted to an industrial process, it requires additional optical equipment, as well as complex signal processing to enable automation. Here, we demonstrate a simpler technique not requiring any special or additional equipment. Furthermore, this technique makes the synthesis process easily automated, as it only needs the measurement of the electrical resistance of the growth structure during the heating process, in which the supplied power increases gradually with a step of 1 mW/100 ms.

37.3 CHARACTERIZATION OF CNT GROWTH STRUCTURE

37.3.1 CALCULATION AND SIMULATION

The technique referred to is to measure the resistance of the polysilicon microbridge to determine the temperature. Therefore, the relationship between the measured resistance and the temperature profile of the polysilicon microbridge needs to be understood and characterized.

Electrical conduction in polysilicon can be described in terms of three mechanisms: conduction within the grain, carrier transport across the grain boundary, and conduction in the regions between grains. Hence, the resistivity of the grain (ρ_g), the resistivity of the barrier (ρ_B), and the region between grains (ρ_{gb}) all contribute to the resulting resistivity, as formulated below [16].

$$\rho = \left(1 - \frac{2W + \delta}{L}\right)\rho_g + \left(\frac{2W}{L}\right)\rho_B + \frac{\delta}{L}\rho_{gb} \tag{37.1}$$

where L is the grain length, W is the width of the depletion region in the grain, and δ is the width of the region between grains (the amorphous grain boundary). Whereas the resistivity of the grain (ρ_g) (being equivalent to the resistivity of single crystalline silicon) is well known, the resistivities ρ_B and ρ_{gb} are higher than ρ_g, but the numerical values are less certain. The quantitative analysis below will therefore focus on the term containing ρ_g. In the limiting case where $W \ll L$ and $\delta \ll L$, Equation 37.1 indeed simplifies to $\rho \approx \rho_g$.

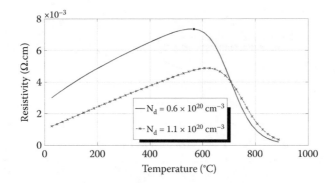

FIGURE 37.2 Calculated resistivity of silicon versus temperature, the two curves representing the upper and lower specification limits for the doping concentration of the PolyMUMPS process. The resistance of the polysilicon microbridge (Figure 37.1) would be proportional to these curves if the whole bridge could be at the same temperature. (Bao Quoc Ta et al., Electrical control of synthesis conditions for locally grown CNTs on polysilicon microstructure, *11th IEEE International Conference on Nanotechnology*, August 15–18, 2011, Portland, Oregon, pp. 374–377. © (2011) IEEE. With permission.)

The individual grain resistivity, equivalent to the resistivity of single crystalline silicon, depends strongly on the temperature and the doping concentration [17–19], as shown in Figure 37.2.

As the temperature increases, the resistivity increases due to the decrease in carrier mobility (thermal scattering), the density of charge carriers being fairly constant (equal to the doping density). However, when the temperature increases sufficiently, the number of charge carriers increases due to the thermal excitation of the electron–hole pairs. Above 550°C, the effect of increased conductivity due to these intrinsically excited charge carriers will dominate the effect of reduced mobility, causing a dramatic decrease in the resistivity as shown in Figure 37.2.

The temperature distribution of the microbridge when subject to resistive Joule heating, as obtained by a finite element method (FEM) simulation, is shown in Figure 37.3. The FEM model considers, as a simplification, only heat loss by conduction through the polysilicon microbridge to the substrate, neglecting other heat loss mechanisms (notably through convection). The doping level is assumed to be at the lower specified limit (cf. Figure 37.2). The parabolic-like distribution shows

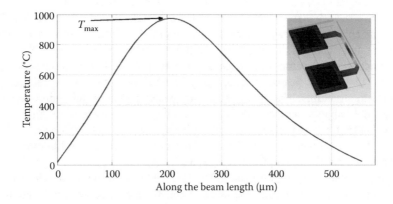

FIGURE 37.3 Simulation of temperature distribution of a polysilicon microbridge under Joule heating with Thomson effect. For a simplification, the simulation considers the heat loss only by conduction through the polysilicon microbridge to the anchors. (Bao Quoc Ta et al., Electrical control of synthesis conditions for locally grown CNTs on polysilicon microstructure, *11th IEEE International Conference on Nanotechnology*, August 15–18, 2011, Portland, Oregon, pp. 374–377. © (2011) IEEE. With permission.)

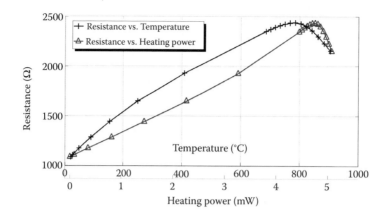

FIGURE 37.4 Total resistance of the bridge versus the highest temperature generated by Joule heating, and versus the heating power (COMSOL simulation). (Bao Quoc Ta et al., Electrical control of synthesis conditions for locally grown CNTs on polysilicon microstructure, *11th IEEE International Conference on Nanotechnology*, August 15–18, 2011, Portland, Oregon, pp. 374–377. © (2011) IEEE. With permission.)

a temperature maximum (T_{max}) near the center of the bridge, somewhat shifted away from the center due to the Thomson effect [20,21]. As CNTs are required to grow at the bridge center region, the highest temperature (T_{max}) needs to be controlled.

The resulting resistance of the microbridges under investigation was calculated by FEM (COMSOL). Figure 37.4 shows this resistance as a function of T_{max}, and as a function of the corresponding heating power. The simulation considers only the ρ_g part in Equation 37.1, and takes into account the temperature-dependent resistivity (Figure 37.2) and the simulated temperature distribution caused by Joule heating (Figure 37.3).

37.3.2 EXPERIMENTAL CHARACTERIZATION

Figure 37.5 shows the measured resistance versus the applied heating power. The shape of the curve is qualitatively in accordance with the calculated curve in Figure 37.4. The higher power supply needed in the experiment, compared with the simulation, can be attributed to convection playing a role in the heat transfer from the microbridge. The relationship between resistance and temperature of the polysilicon structure is repeatable for every heating cycle after the first one. The first heating cycle serves as an annealing process that causes an increased grain size, reduced grain boundaries, and thus reduces and stabilizes the resistivity [21–23]. Referring to Equation 37.1, this means that the system further approaches the limiting case where $W \ll L$ and $\delta \ll L$, where the simplification $\rho \approx \rho_g$ is valid. This implies that the released polysilicon bridges should be preheated to become predictable growth structures. The growth structure (center region of the microbridge) repeatedly reaches the desired temperature (850–950°C) when the overall resistance is about 1.5× the room temperature value (R_T), whereas the maximum resistance value (1.63 × R_T) occurs at a somewhat lower temperature. The temperature was determined experimentally, as a calibration, by using temperature-indicating paint (Tempilaq from Tempil, NJ, USA).

The desired synthesis temperature is at 900°C [1–3]. This is in the range where the resistance decreases with increasing temperature (Figures 37.4 and 37.5). This causes challenges if the heating process is to be controlled using a voltage source, with stepwise increase in a controlled constant voltage. An increase of supplied voltage (V) would cause an additional increase in the Joule heating V^2/R due to the simultaneous reduction of the resistance; hence, there is a positive feedback that eventually has been shown to destabilize the temperature.

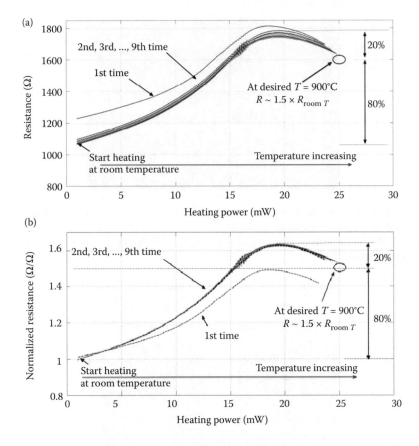

FIGURE 37.5 The total measured resistance (a) and normalized resistance (b) of the bridge versus the heating power. (Bao Quoc Ta et al., Electrical control of synthesis conditions for locally grown CNTs on polysilicon microstructure, *11th IEEE International Conference on Nanotechnology*, August 15–18, 2011, Portland, Oregon, pp. 374–377. © (2011) IEEE. With permission.)

In this study, we controlled the heating process by supplied power instead of by supplied voltage. During the process, the measured resistance (R) is used to control the electric current (I) to keep the power (I^2R), and thus the temperature at the desired value. This feedback control is done automatically using a LabView program. By using this method, we obtained stable temperature for the growth of CNTs. An artifact of this control method is shown in Figure 37.5: the resistance oscillation at the supplied heating power is ~16 mW. This is caused by a time delay in the heating power control, and the inherent strong increase in resistance at these temperatures. Since this artifact occurs at temperatures well below the desired growth temperature, it has no effect on the performance of our growth structure.

37.4 RESULTS AND DISCUSSION

Figure 37.6 shows a scanning electron microscopy (SEM) image of a locally synthesized CNT connecting the two microbridges. CNTs grown appear straight across a gap of 15 μm, although the electric field in this study is 0.2 V/μm—about 10 times smaller than that discussed in Refs. [1,2].

We note that CNTs also grow in the opposite direction from the left side of the growth structure, shown in Figure 37.7, possibly influenced by the fringing electric field, as illustrated in Figure 37.8.

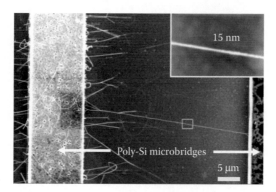

FIGURE 37.6 Locally synthesized CNT connecting two polysilicon microstructures. (Bao Quoc Ta et al., Electrical control of synthesis conditions for locally grown CNTs on polysilicon microstructure, *11th IEEE International Conference on Nanotechnology*, August 15–18, 2011, Portland, Oregon, pp. 374–377. © (2011) IEEE. With permission.)

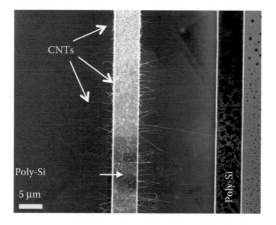

FIGURE 37.7 CNTs grown in the opposite direction, possibly due to the fringing electric field. (Bao Quoc Ta et al., Electrical control of synthesis conditions for locally grown CNTs on polysilicon microstructure, *11th IEEE International Conference on Nanotechnology*, August 15–18, 2011, Portland, Oregon, pp. 374–377. © (2011) IEEE. With permission.)

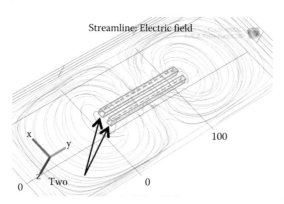

FIGURE 37.8 Simulation of the electric field between two microbridges (by COMSOL Multiphysics). (Bao Quoc Ta et al., Electrical control of synthesis conditions for locally grown CNTs on polysilicon microstructure, *11th IEEE International Conference on Nanotechnology*, August 15–18, 2011, Portland, Oregon, pp. 374–377. © (2011) IEEE. With permission.)

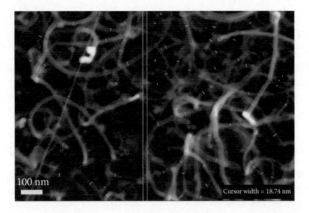

FIGURE 37.9 Measurement of the diameter of synthesized CNTs. (Bao Quoc Ta et al., Electrical control of synthesis conditions for locally grown CNTs on polysilicon microstructure, *11th IEEE International Conference on Nanotechnology*, August 15–18, 2011, Portland, Oregon, pp. 374–377. © (2011) IEEE. With permission.)

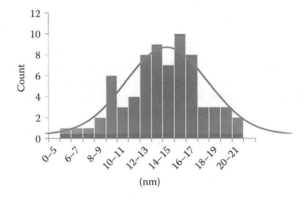

FIGURE 37.10 Distribution of the synthesized CNT diameters, for a random selection of CNTs. The solid line is a normal distribution of the same average and standard deviation. (Bao Quoc Ta et al., Electrical control of synthesis conditions for locally grown CNTs on polysilicon microstructure, *11th IEEE International Conference on Nanotechnology*, August 15–18, 2011, Portland, Oregon, pp. 374–377. © (2011) IEEE. With permission.)

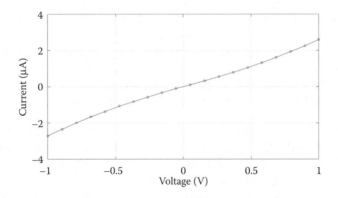

FIGURE 37.11 *I–V* curve of the bridging CNTs, shown in Figure 37.6, measured between two microbridges at room temperature, corresponding to an overall resistance of 400 kΩ. (Bao Quoc Ta et al., Electrical control of synthesis conditions for locally grown CNTs on polysilicon microstructure, *11th IEEE International Conference on Nanotechnology*, August 15–18, 2011, Portland, Oregon, pp. 374–377. © (2011) IEEE. With permission.)

On SEM inspection, Figures 37.9 and 37.10 show that the synthesized CNTs have a uniform and narrow distribution of CNT diameters (~80% in the range of 9–17 nm); and very high aspect ratios up to 1000:1 (length/diameter).

Current (I)–voltage (V) measurement is taken between the two microbridges. Figure 37.11 is the I–V curve of the sample in Figure 37.6 showing a fairly ohmic behavior, and an overall resistance of 400 kΩ. The present result is comparable with the previous results from refs. [1,2] (from 400 kΩ up to 6 MΩ).

37.5   OUTLOOK AND POTENTIAL INDUSTRIAL IMPLEMENTATION

There is presently a very large interest in the integration of nanomaterials, and CNTs, in particular, in microsystems and microelectronics. This arises, on the one hand, from the unique properties of these nanomaterials:

- The nanoscale dimensions open up for miniaturization beyond what traditional micro-engineering can obtain, and beyond what is commonly regarded as physical limits for Si miniaturization.
- The huge surface-to-volume ratio of nanomaterials enables unsurpassed sensitivity in gas/chemical/biochemical sensing devices.

The other important reason for the importance of nanomaterial integration is that any device needs to interface with the macroscopic world. Microsystems and microelectronics can provide exactly this bridge between the nanomaterial and the macroscopic world, giving signal processing functionality, microfluid flow handling, actuation, and additional sensing capabilities on established manufacturing platforms.

Different methods for CNT integration in Si microsystems have been proposed and demonstrated, as outlined in the Introduction. The main benefits of the technique of direct integration by localized resistive heating is

- Localized CNT positions, integrating the CNTs where they are needed in active circuitry
- Low process temperature, being CMOS compatible
- Low cost
- Batch fabrication possibility

The goal will be to develop the process, so that it is fully compatible with the established industrial microengineering processes. As the industry heads toward heterogeneous integration, substantial effort is put into wafer-level integration of different technologies, in particular the integration of MEMS and CMOS technologies. Our demonstrated process may then be the last process step to add nano-functionality to the combined MEMS + CMOS wafers, potentially even including other types of technologies such as GaAs, SiC, and so on. We therefore regard this process as an important step toward realizing complete, integrated micro and nanosystems.

The main contributions from the work presented in this chapter are

- The demonstration of the direct integration process on a commercially available polysilicon platform (PolyMUMPS): This demonstrates that the process is compatible with standardized Si manufacturing processes, and that it does not rely on using more specialized and expensive technology platforms such as SOI.
- The technique of monitoring the temperature of the CNT growth structure through resistance measurements is important for the implementation in an industrial process. Previous techniques have involved analysis of the color of emitted light, requiring transparent windows in reactors and optical readouts that are not readily compatible with

parallel batch manufacturing. The presently proposed technique using only electrical measurements is easily integrated in existing manufacturing platforms, allowing simultaneous monitoring of a large number of growth structures. It also opens up for easy automation of the process.

- The technique of controlling the temperature of the CNT growth structure through the temperature monitoring described above, and through the control of the supplied power, allows for a predictable and reliable control of the growth temperature, also in the temperature regions where the Si microbridge resistance decreases with the temperature. A reliable prediction and control of the synthesis temperature is essential for providing CNTs with the predictable properties and quality. Importantly, a temperature control based only on electrical measurements as outlined above is crucial for industrial implementation. Simultaneous control of the processing parameters for a large number of CNT growth structures can be realized without the need for expensive, dedicated equipment.

The work presented in this chapter therefore demonstrates the feasibility of the CNT direct integration process to be scaled up to an industrial compatible, wafer-level batch process at low cost, compatible with pre-processed wafers.

Our current work heads for an improved characterization and understanding of the growth process, where we aim at optimizing the process conditions for predictable and well-defined direct integration of CNTs in microsystems, as well as demonstration of sensor concepts.

37.6 CONCLUSION

We have demonstrated a process suitable for locally synthesizing and integrating CNTs on polysilicon structures at ambient temperature, thus showing compatibility with CMOS/MEMS technology. Our technique of monitoring the CNT synthesis temperature, which needs only the measurement of the growth structure electrical resistance, brings up a potential for an automated and mass production of CNT-integrated CMOS or MEMS devices. The method of controlling the synthesis temperature, by a feedback control to keep the supplied electrical power constant, tolerates the change in resistivity of polysilicon at elevated temperatures, thus keeping the synthesis process stable and well controlled. Our developments bring up the potential for the commercialization of synthesis and integration of CNTs into CMOS/MEMS devices.

ACKNOWLEDGMENTS

The authors are grateful to Egil Erichsen and Bodil Holst, University of Bergen, Norway, for SEM imaging of samples.

REFERENCES

1. D. Christensen, O. Englander, K. Jongbaeg, and L. Liwei, Room temperature local synthesis of carbon nanotubes, in *Nanotechnology, 2003. IEEE-NANO 2003. 2003 Third IEEE Conference on*, San Francisco, CA, August 12–14, 2003, vol. 2, pp. 581–584.
2. H. Chiamori, W. Xiaoming, G. Xishan, T. Bao Quoc, and L. Liwei, Annealing nano-to-micro contacts for improved contact resistance, in *Nano/Micro Engineered and Molecular Systems (NEMS), 2010 5th IEEE International Conference on*, Xiamen, China, January 20–23, 2010, pp. 666–670.
3. A. Jungen, C. Stampfer, J. Hoetzel, V. M. Bright, and C. Hierold, Process integration of carbon nanotubes into microelectromechanical systems, *Sensors and Actuators A: Physical*, 130–131, 588–594, 2006.
4. Z. Chen, J. Appenzeller, Y.-M. Lin, J. Sippel-Oakley, A. G. Rinzler, J. Tang, S. J. Wind, P. M. Solomon, and P. Avouris, An integrated logic circuit assembled on a single carbon nanotube, *Science*, 311, 1735, 2006.

5. R. F. Smith, T. Rueckes, S. Konsek, J. W. Ward, D. K. Brock, and B. M. Segal, Carbon nanotube based memory using CMOS production techniques, in *Compound Semiconductor Integrated Circuit Symposium, 2006. CSIC 2006. IEEE*, San Antonio, TX, November 12–15, 2006, pp. 47–50.

6. J. E. Jang, S. N. Cha, Y. Choi, G. A. J. Amaratunga, D. J. Kang, D. G. Hasko, J. E. Jung, and J. M. Kim, Nanoelectromechanical switches with vertically aligned carbon nanotubes, *Applied Physics Letters*, 87, 163114-3, 2005.

7. A. Star, K. Bradley, J.-C. P. Gabriel, and G. Gruner, Nano-electronic sensors: Chemical detection using carbon nanotubes, *Polymeric Materials: Science & Engineering*, 89, 204, 2003.

8. J. J. Brown, J. W. Suk, G. Singh, A. I. Baca, D. A. Dikin, R. S. Ruoff, and V. M. Bright, Microsystem for nanofiber electromechanical measurements, *Sensors and Actuators A: Physical*, 155, 1–7, 2009.

9. M. A. Cullinan and M. L. Culpepper, Design and fabrication of single chirality carbon nanotube-based sensors, in *Nanotechnology (IEEE-NANO), 2011 11th IEEE Conference on*, Portland, OR, August 15–19, 2011, pp. 26–29.

10. G. Qingqing, E. Albert, B. Fabel, A. Abdellah, P. Lugli, M. B. Chan-Park, and G. Scarpa, Solution-processable random carbon nanotube networks for thin-film transistors, in *Nanotechnology (IEEE-NANO), 2011 11th IEEE Conference on*, Portland, OR, August 15–19, 2011, pp. 378–381.

11. N. Patil, A. Lin, E. R. Myers, R. Koungmin, A. Badmaev, Z. Chongwu, H. S. P. Wong, and S. Mitra, Wafer-scale growth and transfer of aligned single-walled carbon nanotubes, *Nanotechnology, IEEE Transactions on*, 8, 498–504, 2009.

12. D. S. Engstrøm, N. L. Rupesinghe, K. B. K. Teo, W. I. Milne, and P. Bøgild, Vertically aligned CNT growth on a microfabricated silicon heater with integrated temperature control—Determination of the activation energy from a continuous thermal gradient, *Journal of Micromechanics and Microengineering*, 21, 015004, 2011.

13. S. Dittmer, S. Mudgal, O. A. Nerushev, and E. E. B. Campbell, Local heating method for growth of aligned carbon nanotubes at low ambient temperature, *Low Temperature Physics*, 34, 834–837, 2008.

14. S. Dittmer, O. A. Nerushev, and E. E. B. Campbell, Low ambient temperature CVD growth of carbon nanotubes, *Applied Physics A: Materials Science & Processing*, 84, 243–246, 2006.

15. F. Ding, K. Bolton, and A. Rosén, Nucleation and growth of single-walled carbon nanotubes: A molecular dynamics study, *The Journal of Physical Chemistry B*, 108, 17369–17377, 2004.

16. D. M. Kim, A. N. Khondker, S. S. Ahmed, and R. R. Shah, Theory of conduction in polysilicon: Drift-diffusion approach in crystalline-amorphous-crystalline semiconductor system. Part I: Small signal theory, *Electron Devices, IEEE Transactions on*, 31, 480–493, 1984.

17. G. W. Ludwig and R. L. Watters, Drift and conductivity mobility in silicon, *Physical Review*, 101, 1699–1701, 1956.

18. E. M. Conwell, Properties of silicon and germanium, *Proceedings of the IRE*, 40, 1327–1337, 1952.

19. J. E. Suarez, B. E. Johnson, and B. El-Kareh, Thermal stability of polysilicon resistors, *Components, Hybrids, and Manufacturing Technology, IEEE Transactions on*, 15, 386–392, 1992.

20. A. W. Van Herwaarden and P. M. Sarro, Thermal sensors based on the Seebeck effect, *Sensors and Actuators*, 10, 321–346, 1986.

21. G. R. Lahiji and K. D. Wise, A batch-fabricated silicon thermopile infrared detector, *Electron Devices, IEEE Transactions on*, 29, 14–22, 1982.

22. M. M. Mandurah, K. C. Saraswat, C. R. Helms, and T. I. Kamins, Dopant segregation in polycrystalline silicon, *Journal of Applied Physics*, 51, 5755–5763, 1980.

23. N. C. C. Lu, L. Gerzberg, and J. D. Meindl, A quantitative model of the effect of grain size on the resistivity of polycrystalline silicon resistors, *Electron Device Letters, IEEE*, 1, 38–41, 1980.

24. Bao Quoc Ta, N. Hoivik, E. Halvorsen, and K. E. Aasmundtveit, Electrical control of synthesis conditions for locally grown CNTs on polysilicon microstructure, *11th IEEE International Conference on Nanotechnology*, August 15–18, 2011, Portland, Oregon, pp. 374–377.

Section XI

Carbon Nanotube Transistor Modeling

38 A Qualitative Comparison of Energy Band Gap Equations with a Focus on Temperature and Its Effect on CNTFETs

Jesus Torres and Hasina F. Huq

CONTENTS

38.1 INTRODUCTION

Carbon nanotubes (CNTs) have attracted the most attention due to their extraordinary electrical, mechanical, and optical properties [1–3]. In addition to the efforts of developing new electronic devices, direct bandgap one-dimensional (1D) nanostructures are playing an important role because of the desire to integrate both electronic and optoelectronic technologies on the same material [2]. For example, CNT field-effect transistors (CNTFETs) have generated considerable attention over the past few years because of their semiconductive properties and have reached a high level of performance [4–7]. Meanwhile, several major technology-related questions still need to be addressed. The research on temperature-dependent CNTFET model is not adequate; which is one of the critical and dominant factors on electronics device performance.

The role of temperature has been established in semiconductors as one factor affecting the band gap. In CNTFETs, temperature has been neglected due to its small contribution of affecting the band gap, at room temperature, as opposed to how temperature affects silicon base semiconductors as in Ref. [8]. A qualitative comparison of various energy band gap equations is taken to view the effects of temperature outside room temperature.

Three band gap equations are used in this research; one of them is dependent mainly on the diameter, another on temperature and defect, and the last one has the use of temperature with parameters depending on chirality as in Refs. [8–10]. The three equations are derived from single-wall nanotubes (SWNTs) and applied to coaxial-type CNTFETs. However, the temperature in CNTFETs is said to be negligible and can be ignored as stated in Ref. [8]. Thus, an investigation on how the temperature would affect these devices is conducted by using different equations with a set of various parameters.

The implementation of the band gap equation is followed by using the I_D equation derived from ballistic nanotransistors as in Ref. [11], and the *I–V* curves can be simulated by using the MATLAB source codes found in the simulation tool named FETToy available at Nanohub.org. Temperature is the main parameter that is varied during simulation. The chirality vector is chosen for the small band gap of CNTs, such as $m - n = 3$ reported in Ref. [10]. Results have been obtained solely from simulations. For the next phase, other parameters would be included such as how temperature affects the Fermi levels at the channel, charge density, valley degeneracy, electron concentration, quantum capacitance, and mobile charge. The final step would be to compare the analytical model with the experimental results of a CNTFET.

38.2 METHODOLOGY

38.2.1 Band Gap with Diameter

The theory behind CNTs begins with an understanding of graphene. That is because SWNTs are considered to be rolled up graphene sheets. To solve the band structure of the graphene π orbitals, the steps are taken from [8] where the first step is to solve the Schrödinger equation

$$H\Psi = E\Psi, \tag{38.1}$$

where H is defined as the Hamiltonian, Ψ is the wave function, and E is the energy of the electrons in the π orbitals. The final solution to the energy is shown as

$$E = E_0 \mp \gamma_0 \left(1 + 4\cos\left(\frac{\sqrt{3}k_x a}{2}\right)\cos\left(\frac{k_y a}{2}\right) + 4\cos^2\left(\frac{k_y a}{2}\right) \right)^{1/2}. \tag{38.2}$$

Equation 38.2 demonstrates the valence bonds of graphene using the negative sign, and the conduction band vice versa where γ_0 is the tight-bonding integral. Using Equation 38.2 the density of states (DOS) can be found by differentiating it. For a detailed look into solving the Schrödinger equation, the reader is asked to refer to Ref. [12]. Finally, by using Equation 38.2, the band gap equation is shown as

$$E_g = 2 \times \left(\frac{\partial E}{\partial k}\right) \times \frac{2}{3d} \approx \frac{0.7 \text{ eV}}{d(\text{nm})}. \tag{38.3}$$

Equation 38.3 does not have temperature dependence as the next two band gap equations to be shown, but is good to analyze with others to show how much the energy band gap differs.

38.2.2 Band Gap with Defect and Temperature

As mentioned before, SWNTs are made up from rolled up sheets of graphene and different electrical properties can be achieved depending on their chiral vector described as

$$C = n_1 a_1 + m a_2, \tag{38.4}$$

where (m, n) are integers and (a_1, a_2). Thus, by using Equation 38.4 if $m = 0$, the SWNT is thought to be zigzag and if $n = m$, it is said to be armchair; otherwise it is simply called chiral.

The next band gap equation can be found in more detail in Ref. [9], but basically the band gap equation is found by using the ambipolar random telegraph signal. The SWNTs used in this reference have small band gaps, where (m, n) will be in accordance with $m - n = 3$. They define the ratio of defect emission and capture times as

$$\frac{\tau_c}{\tau_e} = g \exp\left(\frac{E_t - E_f}{k_B T}\right), \tag{38.5}$$

where g is the trap degeneracy, E_t is the defect energy, E_f is the Fermi energy, and $k_B T$ is the thermal energy. By using Equation 38.5, the band gap equation is found as

$$E_g = k_B T \left[\ln\left(\frac{\tau_e}{\tau_c}\right)_{ne} + \ln\left(\frac{\tau_e}{\tau_c}\right)_{po} \right]. \tag{38.6}$$

The temperature parameter in Equation 38.6 varied from 4.2 K up to 115 K from reviewing [9].

38.2.3 BAND GAP WITH TEMPERATURE

The last band gap equation comes from Ref. [10] and is a bit complicated compared to the other two. The main idea is that there are two parts of the band gap equation, which consists of harmonic and anharmonic parts. The harmonic portion comes from the ground-state geometry and the anharmonic part is from thermal expansion. Thus, the band gap equation can be shown as

$$\Delta E_g(T) = \frac{\alpha_1 \Theta_1}{e^{\Theta_1/T} - 1} + \frac{\alpha_2 \Theta_2}{e^{\Theta_2/T} - 1}. \tag{38.7}$$

The temperature in Equation 38.7 is varied such that $T < 400$ K and the chirality of the SWNTs used must follow the form of

$$U = (n - m) \bmod 3 = 2 \tag{38.8}$$

$$d = \sqrt{n^2 + m^2 + nm}, \tag{38.9}$$

where Equation 38.9 is the diameter of the SWNT and must remain dimensionless. Equation 38.7 also has some parameters as functions which are

$$\Theta_1 = \frac{A}{d^2} \tag{38.10}$$

$$\xi = (-1)^\upsilon \cos(3\theta) \tag{38.11}$$

$$f_\eta(\xi) = \gamma_1^\eta \xi + \gamma_2^\eta \xi^2 \tag{38.12}$$

$$\alpha_1 = \alpha_1^0 + f_{\alpha_1}(\xi)d \tag{38.13}$$

$$\Theta_2 = \Theta_2^\infty + \frac{f_{\Theta_2}(\xi)}{d} \tag{38.14}$$

$$\alpha_2 = \frac{1}{d}\left(B + \frac{f_{\alpha_2}(\xi)}{d}\right). \tag{38.15}$$

The remaining parameters for the previous set of equations are shown in Table 38.1.

TABLE 38.1
Parameters for Previous Equations

A	9.45×10^3 K
α_1^0	-1.70×10^{-5} eV/K
$\gamma_1^{\alpha_1}$	1.68×10^{-6} eV/K
$\gamma_2^{\alpha_1}$	6.47×10^{-7} eV/K
Θ_2^∞	470 K
$\gamma_1^{\Theta_2}$	1.06×10^3 K
$\gamma_2^{\Theta_2}$	-5.94×10^{-2} K
B	-4.54×10^{-4} eV/K
$\gamma_1^{\alpha_2}$	-2.68×10^{-3} eV/K
$\gamma_2^{\alpha_2}$	-2.23×10^{-5} eV/K

Source: From R. B. Capaz et al., *Physics Review Letters*, 94, 036801, 2005.
            With permission.

38.3 RESULTS

38.3.1 BAND GAP WITH TEMPERATURE

The band gap equations are varied in different ways since each has specific things to observe; but the main feature that is noticed in the research is the temperature. At first, the diameter in Equation 38.3 is varied from 0.8 to 3 nm. It is mentioned in Ref. [12] that the energy band gap from Equation 38.3 will vary from 0.2 to 0.9 eV. A MATLAB plot of Equation 38.3 is shown in Figure 38.1; note that this is the only equation without temperature and can be used to compare how much the others will vary.

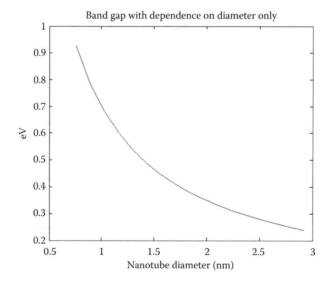

FIGURE 38.1   The diameter in Equation 38.3 is varied from 0.8 to 3 nm.

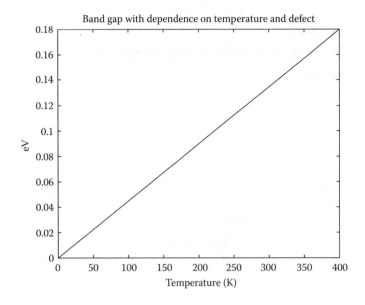

FIGURE 38.2   The band gap using Equation 38.6 behaves in a linear fashion with temperature.

Equation 38.6 shows a difference; there is no diameter parameter, but the measurements taken in Ref. [9] are done by having SWNTs in the range of 1–3 nm. The defect ratios are kept as $(\tau_e/\tau_c)_{ne} = 15$ and $(\tau_e/\tau_c)_{po} = 12$. The band gap using Equation 38.6 behaves in a linear fashion as shown in Figure 38.2.

Finally, the band gap equation used in Equation 38.7 is varied by temperature and is also shown with different diameter sizes of 1, 2.1, and 3 nm, as shown in Figure 38.3.

The band gap equations are simulated at temperatures of 50, 300, and 400 K with diameters of 1, 2.1, and 3 nm, respectively. The temperature-dependent energy band gap values are listed in Table 38.2. As it is shown, the energy band gaps vary tremendously depending on some critical parameters such as temperature, diameter, defect ratio, and so on.

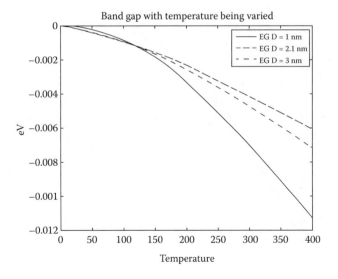

FIGURE 38.3   Temperature dependent band-gap behavior using the Equation 38.7

TABLE 38.2
Band Gaps at Different Diameters and Temperatures

Equation	50 K	300 K	400 K
	Diameter = 1 nm		
(38.3)	0.8317	0.8317	0.8317
(38.6)	0.02238	0.1343	0.1791
(38.7)	−0.00236	−0.007038	−0.01128
	Diameter = 2.1 nm		
(38.3)	0.4045	0.4045	0.4045
(38.6)	0.02238	0.1343	0.1791
(38.7)	−0.00039	−0.0004779	−0.00072
	Diameter = 3 nm		
(38.3)	0.2789	0.2789	0.2789
(38.6)	0.2238	0.1343	0.1791
(38.7)	−0.00043	−0.0004287	−0.0061

38.3.2 I–V Characteristics

The output I–V characteristics are also analyzed. The diameters used in the simulation are 1, 2.1, and 3 nm, respectively. The I–V characteristics are simulated at 50, 300, and 400 K. Figures 38.4 through 38.7 show the effects of temperature. The maximum current is affected as temperature increases, but the linear region seems not to be affected much by the temperature from Vds voltages of 0–0.2 V. It is also observed that the current of the CNTFET is a function of energy gap, diameter, defect of CNT, chiral vectors, chiral angel as well as temperature. It is observed from the figures that the drain current in the saturation region at higher temperatures is higher than at the lower temperature. However, for the lower drain–source voltages, the difference between the higher and lower temperature current regions remains almost constant.

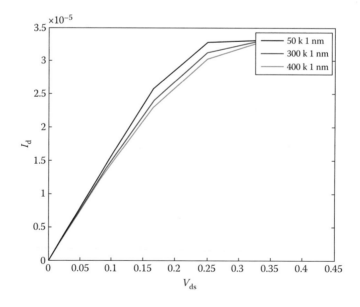

FIGURE 38.4 I–V curves at different temperatures.

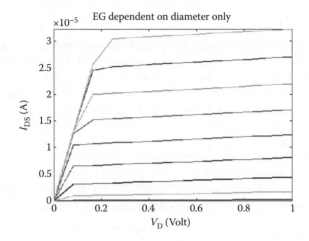

FIGURE 38.5 *I–V* using bandgap equation 38.3. Diameter 1 nm, temperature 50 K.

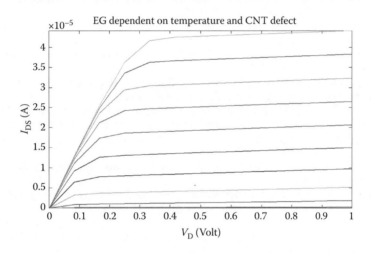

FIGURE 38.6 *I–V* using bandgap equation 38.6. Diameter 1 nm, temperature 300 K.

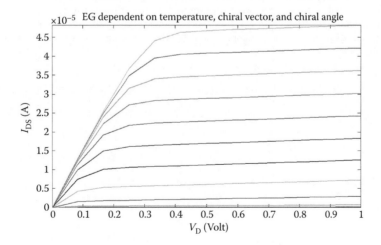

FIGURE 38.7 *I–V* using bandgap equation 38.7. Diameter 3 nm, temperature 400 K.

38.4 CONCLUSION

The research is specifically focused on the current–voltage characteristics of CNTFET considering the temperature variations. The results are in good accordance with their references. Temperature does show that it has some impact on the energy band gap and I–V characteristics. However, since each equation has specific variables, it is also important to note which parameters would be more important to go along with temperature. The effect of varying temperature on CNTFET also influences the on current, off current, on–off current ratio, transconductance, drain conductance characteristics, and the subthreshold swing.

Temperature definitely plays a significant role in the performance of all types of transistors. The need of CNTFET arises from the fact that most of the transistors design encounters a lot of problems during its operation in a very low-power supply [13,14]. Therefore, CNTFET can be a good solution, which provides more reliability even during aggressive scaling.

REFERENCES

1. J. Guo, S. Goasguen, M. Lundstrom, and S. Datta, Metal–insulator–semiconductor electrostatics of carbon nanotubes. *Applied Physics Letters*, 81(8), 2002.
2. P. Avouris and J. Chen, Nanotube electronics and optoelectronics, *Materials Today*, 9(10), 46–54, 2006.
3. S. Iijima, Helical microtubules of graphitic carbon, *Nature*, 354(6348), 56–58, 1991.
4. S. J. Tans, A. R. M. Verschueren, and C. Dekker, Room-temperature transistor based on a single carbon nanotube, *Nature*, 393, 49–52, 1998.
5. R. Martel, T. Schmidt, H. R. Shea, T. Hertel, and P. Avouris, Single- and multi-wall carbon nanotube field-effect transistors, *Appl. Phys. Lett.*, 73, 2447–2449, 1998.
6. J. J. Guo, D. B. Farmer, Q. Wang, E. Yenilmez, R. G. Gordon, M. Lundstrom, and H. J. Dai, Self-aligned ballistic molecular transistors and electrically parallel nanotube arrays, *Nano Lett.*, 4, 1319–1322, 2004.
7. P. L. McEuen, M. S. Fuhrer, and H. Park, Single-walled carbon nanotube electronics, *IEEE Transactions on Nanotechnology*, 1(1), 78–85, 2002.
8. F. Leonard, *The Physics of Carbon Nanotube Devices*, 13 Eaton Ave. Norwich, NY 13815: William Andrew Inc., 2009, p. 20.
9. F. Liu, M. Bao, and K. L. Wang, Determination of the small band gap of carbon nanotubes using the ambipolar random telegraph signal, *Nano Letters*, 5(7), 1333–1336, 2005.
10. R. B. Capaz, C. D. Spataru, P. Tangney, M. L. Cohen, and S. G. Louie, Temperature dependence of the band gap of semiconducting carbon nanotubes, *Physics Review Letters*, 94, 036801, 2005.
11. A. Rahman, J. Guo, and S. Datta, Theory of ballistic nanotransistors, *IEEE Transactions on Electron Devices*, 50(9), 1853–1864, 2003.
12. J. J. Kong, *Carbon Nanotube Electronics*, New York: Springer, 2009, p. 1–9
13. Q. Lu, D. Park, and A. Kalnitsky et al., Leakage current comparison between ultra-thin Ta_2O_5 films and conventional gate dielectrics, *IEEE Electron Device Letters*, 19(9), 341–342, 1998.
14. K. Natori, Ballistic metal-oxide-semiconductor field effect transistor, *Journal of Applied Physics*, 76(8), 4879–4890, 1994.

39 Real-Time Quantum Simulation of Terahertz Response in Single-Walled Carbon Nanotube

Zuojing Chen, Eric Polizzi, and Sigfrid Yngvesson

CONTENTS

39.1 INTRODUCTION

Recently, a terahertz response of single-walled carbon nanotubes has been successfully observed and measured [1,2]. Carbon nanotubes (CNTs) have emerged as one of the most active areas of nanoscale science and technology research [3,4]. They are truly nanoscale materials with typical diameters of 1–2 nm, and can be regarded as being close to ideal 1-D conductors with an unprecedented mean free path of about 1 μm at 300 K. Their cross-sectional dimensions are about an order of magnitude smaller than the limiting dimensions at which complementary metal oxide semiconductor (CMOS) technology is likely to encounter insuperable scaling limits in a few years. Not surprisingly, it has turned out to be quite difficult to realize actual working applications by using materials at such a radically different scale; this is still a largely unmet challenge! For example, CNT field-effect transistors (CNT-FETs) have been predicted to perform up to terahertz frequencies [5,6], but experiments show switching speeds [4] (digital circuits) or cutoff frequencies (analog circuits) that are at least a couple of orders of magnitude below that predicted intrinsic performance. Measurements on CNTs have been extended to about 60 GHz, and much higher frequencies (up to several THz) are required to verify the properties of CNTs as predicted by existing theories. These theories predict plasmon waves propagating along the tubes at slow speeds of roughly 0.01c (c is the speed of light), a unique feature of 1-D conductors [7]. The transmission line model for a metallic single-walled CNT (m-SWCNT) introduced in Ref. [7] includes a unique kinetic inductance element that is about 1000 times greater than the magnetic inductance usually considered for macroscopic conductors. It also incorporates a quantum capacitance. Based on this model, one predicts that plasmon resonances should occur at terahertz frequencies for CNTs as short as 1 μm. It is a current, so far unmet, challenge for experimentalists to measure and take advantage of these resonances. The unusual antenna properties of CNTs have been predicted in Ref. [8]. Surface wave types of fields near the tubes are predicted to be enhanced by hundreds of times [9], and this has also not yet been demonstrated experimentally. Such surface waves are promising for many applications, similar to those that are presently under intense development in the optical/NIR range, employing surface plasmon polaritons (SPPs) [10]. The utilization of the plasmon phenomena shows potential for shrinking circuit sizes to a small fraction of a wavelength, at THz as well as in the visible/NIR. Many unique applications of CNTs in the terahertz frequency range thus appear possible when these types of plasmon phenomena are well understood.

In this chapter, we present large-scale quantum atomistic time-domain simulations to gain an in-depth picture of electron transport phenomena at very high frequencies in CNTs. These numerical models are expected to provide fundamental insights for understanding plasmonic and other many-body excitations and then enhance the reliability in designing tunable carbon-based electronic devices. The simulations are performed with short time steps to be able to correctly represent phenomena up to over 100 THz. We especially emphasize an investigation of any resonances, as well as the kinetic inductance.

For large systems under time-dependent external perturbations such as electromagnetic (EM) fields, pulsed lasers, AC signals, particle scattering, and so on, a full quantum treatment of the problem is still considered to be very challenging. Reliable modeling approaches in the time domain are often limited in terms of trade-off between robustness and performances [11,12]. In our work, effective modeling and propagation schemes are carried out to simulate the CNT THz response. Our propagation schemes consist of performing a direct integration of the time-ordered evolution operator. The numerical treatment of time-ordered evolution operators often gives rise to the matrix exponential. The most obvious way to address this numerical problem would be to directly diagonalize the Hamiltonian while selecting the relevant number of modes needed to accurately expand the solutions. Direct diagonalization techniques, however, have been known to be very computationally demanding, especially for large systems. Consequently, the mainstream in time-dependent simulations uses traditional approximations such as split operator techniques or a perturbation theory. Here, we rather perform exact diagonalizations by taking advantage of our new linear scaling eigenvalue solver FEAST [13]. By using FEAST, the solution of the eigenvalue problem is reformulated into solving a set of well-defined, independent linear systems along a complex energy contour. Additionally, obtaining the spectral decomposition of the matrix exponential becomes a suitable alternative to PDE-based techniques, such as the Crank–Nicolson schemes [14], and can also potentially be performed using a time-domain parallelism.

39.2 MODEL AND METHODOLOGY

In time-dependent quantum systems, the electrons obey the time-dependent Schrödinger equation:

$$i\hbar \frac{\partial}{\partial t} \Psi(t) = \hat{H}\Psi(t). \tag{39.1}$$

Besides appropriate boundary conditions, the time-dependent Schrödinger equation requires an initial value condition $\Psi(t=0) = \Psi_0$ that completely determines the dynamics of the system.

Using a single electron picture, and in the time-dependent density functional theory (TDDFT) framework [15], the solutions of the stationary Kohn–Sham Schrödinger-type Equation 39.2 are taken as initial wave functions $\Psi_0 = \{\psi_1^{(0)}, \psi_2^{(0)}, \ldots, \psi_{N_e}^{(0)}\}$ and will be propagated over time.

$$\left[-\frac{\hbar^2}{2m} + v_{KS}[n](\mathbf{r}) \right] \psi_j^{(0)}(\mathbf{r}) = E_j^{(0)} \psi_j^{(0)}(\mathbf{r}). \tag{39.2}$$

For a system of interest that is composed of N_e electrons, the electron dynamics can be described by a set of one-body equations, the Kohn–Sham equations. It has the same form as the Schrödinger equation, where $\Psi = \{\psi_1, \psi_2, \dots, \psi_{N_e}\}$ and with ψ_j the solution of

$$i\hbar \frac{\partial}{\partial t} \psi_j(\mathbf{r}, t) = \left[-\frac{\hbar^2}{2m} \nabla^2 + v_{KS}[n](\mathbf{r}, t) \right] \psi_j(\mathbf{r}, t). \tag{39.3}$$

The Kohn–Sham potential v_{KS} is a functional of the time-dependent density and it is conventionally separated in the following way:

$$v_{KS}[n](\mathbf{r}, t) = v_{ext}(\mathbf{r}, t) + v_H[n](\mathbf{r}, t) + v_{xc}[n](\mathbf{r}, t), \tag{39.4}$$

where the first term represents the external potential, the second term is the Hartree potential that accounts for the electrostatic interaction between the electrons, and the last term is defined as the exchange-correlation potential that accounts for all the nontrivial many-body effects. The density of the interacting system can be obtained from the time-dependent Kohn–Sham wave functions

$$n(\mathbf{r}, t) = \sum_{j=1}^{N_e} |\psi_j(\mathbf{r}, t)|^2. \tag{39.5}$$

TDDFT can indeed be viewed as a reformulation of time-dependent quantum mechanics where the basic variable is no longer the many-body wave function, but the time-dependent electron density $n(\mathbf{r}, t)$. For any fixed, initial many-body state, the Runge–Gross theorem [15] shows that there is one-to-one correspondence between densities and the potential, which means that the external potential uniquely determines the density. The Kohn–Sham approach chooses a noninteracting system, which has a density that is equal to the interacting system.

Formally, the solution of Equation 39.1 can be written as

$$\Psi(t) = \hat{U}(t, 0)\Psi_0 = T \exp\left\{ -\frac{i}{\hbar} \int_0^t d\tau \hat{H}(\tau) \right\} \Psi_0, \tag{39.6}$$

where the evolution operator $\hat{U}$ is unitary and can be represented using a time-ordered exponential $T\exp$, which is a nontrivial mathematical object. In most cases, the problem is addressed using very small time steps and the time-independent Hamiltonian approximation within the intervals. Intermediate physical solutions are computed in addition to the final solution $\Psi(t)$ to describe the evolution of the system over $[0,t]$. This can be accomplished by dividing $[0,t]$ into smaller time intervals since using the intrinsic properties of the evolution operator, one can apply the following decomposition:

$$\hat{U}(t, 0) = \hat{U}(t_n, t_{n-1})\hat{U}(t_{n-1}, t_{n-2})\dots\hat{U}(t_2, t_1)\hat{U}(t_1, t_0). \tag{39.7}$$

If the Δ_t chosen is very small, it is reasonable to consider the $\hat{H}(\tau)$ constant within the time interval $[t, t + \Delta_t]$, leading to

$$\Psi(t + \Delta_t) = \hat{U}(t + \Delta_t, t)\Psi(t) = \exp\left\{ -\frac{i}{\hbar} \Delta_t \hat{H}(t) \right\} \Psi(t). \tag{39.8}$$

Equation 39.8 requires the solution of eigenvalue problems, while the exact Hamiltonian diago-nalization is often considered to be computationally challenging. For large systems, approximations such as the Crank–Nicolson or split operator are commonly made since the direct factorization of the evolution operator is not practical using conventional eigenvalue solvers. In our previous work [16], we have proposed new effective and direct numerical propagation schemes that go beyond the per-turbation theory and linear response. We also perform the exact diagonalizations of Hamiltonians by taking advantage of a new linear scaling eigenvalue solver, FEAST.

Denoting $\mathbf{H}$ the $N \times N$ Hamiltonian matrix obtained after the discretization of $\hat{H}$ at a given time t and where N could represent the number of basis functions (or number of nodes using real-space mesh techniques), $\mathbf{H}$ can then be diagonalized as follows:

$$\mathbf{D} = \mathbf{P}^T \mathbf{H} \mathbf{P}, \tag{39.9}$$

where the columns of the matrix $\mathbf{P} = \{\mathbf{p_1}, \mathbf{p_2}, ..., \mathbf{p_M}\}$ represent the eigenvectors of $\mathbf{H}$ associated with the M lowest eigenvalues regrouped within the diagonal matrix $\mathbf{D} = \{d_1, d_2, ..., d_M\}$. Now, we can get the resulting matrix form of the time propagation equation, which is given by

$$\Psi(t + \Delta_t) = \mathbf{P} \exp\left\{-\frac{i}{\hbar}\Delta_t \mathbf{D}\right\} \mathbf{P}^T \mathbf{S} \Psi(t). \tag{39.10}$$

Using the property (39.7), the solution $\Psi(t)$ can finally be obtained as a function of Ψ_0:

$$\Psi(t) = \mathcal{T}\left\{\prod_i \left[\mathbf{P}_i \exp\left(-\frac{i}{\hbar}\Delta_t \mathbf{D}_i\right)\mathbf{P}_i^T \mathbf{S}\right]\right\}\Psi_0, \tag{39.11}$$

where $\mathbf{D}_i = \mathbf{P}_i^T \mathbf{H}(t_i)\mathbf{P}_i$. $\mathbf{S}$ is a symmetric positive-definite matrix that satisfies $\mathbf{P}^T \mathbf{S} \mathbf{P} = \mathbf{I}$.

In addition to the (basic) direct propagation scheme above, two highly efficient propagation schemes using larger time intervals have also been proposed in Ref. [16].

39.3   SIMULATION RESULTS

Based on the modeling strategies presented here, we have recently developed a highly efficient numerical framework for performing first-principle TDDFT using all-electron calculations and 3-D finite element discretization. While this framework is already capable of reproducing optical absorp-tion spectra of molecules (from H2 to C60) [17], it is still currently being optimized to effectively address THz responses for much larger-scale systems running on high-end parallel architectures.

In this section, however, we present the preliminary results obtained using our 3-D time-depen-dent, numerical framework, making use of an atomistic, empirical pseudopotential, which has allowed use to capture the THz response of metallic CNT and accurately reproduce experimental results on the Fermi velocity and kinetic inductance. In our model, we use an empirical pseudopo-tential [18], and propose to compute the response of a time-dependent, external perturbation applied to the system (EM radiation). The evolution of all $t = 0$ wave functions (i.e., all electrons present in the system) are considered with the time-dependent Hamiltonian. Our model also uses real-space mesh techniques for discretization (finite element method), and a time-dependent version of the atomistic mode approach described in Ref. [19]. Moreover, if the empirical pseudopotential U_{eps} is supposed to be time independent, the total atomistic potential can then be decomposed as follows:

$$U(x, y, z, t) = U_{eps}(x, y, z) + U_{ext}(x)\sin(\omega t), \tag{39.12a}$$

$$U_{ext}(x) = \frac{2x - L}{L}U_0, \tag{39.12b}$$

where x is the longitudinal direction of the tube, L represents the distance between contacts ($x \in [0,L]$), and $\omega = 2\pi f$, with f being the corresponding frequency of the AC signal and U_0 its amplitude. The time-dependent external potential applied to the CNT then maintains zero in the middle of the tube but oscillates at both ends alternatively to the $\pm U_0$ values.

39.3.1 Electron Resonances

We apply the direct propagation scheme to an isolated (5,5) metallic carbon nanotube (see Figure 39.1). In the noninteracting Kohn–Sham system, the N ground-state Kohn–Sham orbitals are taken as the initial states and are propagated over time. A very short pulse (time domain) is injected into the device. We calculate the current density in the middle of the tube, and then Fourier transform the current density to get the responses of the CNT.

We studied CNTs with different lengths, which are described in Table 39.1 (the CNT length in the table represents the entire computational domain).

Since we have the wavefunction at any time, we could get density and current at any time t. The probability current density $\vec{j}$ of the wave function Ψ is defined as

$$\vec{j} = \frac{\hbar}{2mi}\left(\Psi^*\vec{\nabla}\Psi - \Psi\vec{\nabla}\Psi^*\right) = \frac{\hbar}{m}\mathrm{Im}(\Psi^*\vec{\nabla}\Psi). \tag{39.13}$$

After integrating over the cross section, we have the current in the middle of the tube versus time.

Figure 39.2 shows the current versus time at the middle of the tube for 12 unit cell and 24 unit cell SWCNTs, where the width of the pulse is 0.04 fs, and the total propagation time is 1.25×10^{-13} s.

Then, we Fourier transform the current:

$$X_k = \sum_{n=0}^{N-1} I_n e^{-\frac{2\pi i}{N}kn} \quad k = 0,\dots,N-1$$

with the results shown in Figure 39.3.

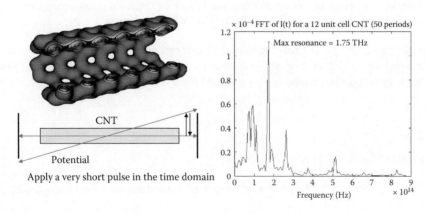

FIGURE 39.1 Schematic representation of the simulation setup. (Chen, Z. et al., Real-time quantum simulation of terahertz response in single wall carbon nanotube, *2011 11th IEEE Conference on Nanotechnology (IEEE-NANO)*, Portland, OR, pp. 1339–1342. © (2011) IEEE. With permission.)

TABLE 39.1
(5,5) CNTs of Different Lengths

Unit Cell	Number of Atoms	CNT Length (nm)	Size of System Matrix
6	120	1.62	19,050
12	240	3.25	37,050
24	480	6.5	73,050
48	960	13	145,050

Source: Chen, Z. et al., Real-time quantum simulation of terahertz response in single wall carbon nanotube, *2011 11th IEEE Conference on Nanotechnology (IEEE-NANO)*, Portland, OR, pp. 1339–1342. © (2011) IEEE. With permission.

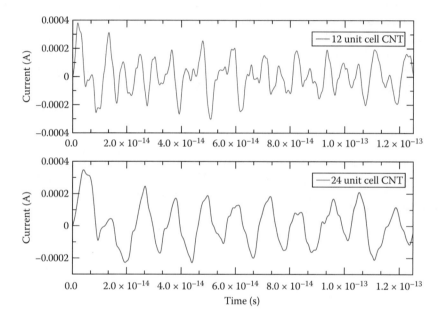

FIGURE 39.2 Current density in the middle of nanotube versus time. (Chen, Z. et al., Real-time quantum simulation of terahertz response in single wall carbon nanotube, *2011 11th IEEE Conference on Nanotechnology (IEEE-NANO)*, Portland, OR, pp. 1339–1342. © (2011) IEEE. With permission.)

The frequencies of the maximum response of 12 unit cell and 24 unit cell SWCNTs are 175 and 88 THz, respectively, which correspond to the resonance frequency. We know that the lowest frequency for a Fabry–Perot resonance can be obtained from

$$L = \frac{\lambda}{2}, \tag{39.14}$$

so the phase velocity for the 12 unit cell CNT is

$$v = f * \lambda = f * 2L = 1.13 \times 10^8 \text{ cm/s}.$$

And, similarly, the phase velocity for the 24 unit cell CNT is

$$v = 1.14 \times 10^8 \text{ cm/s}.$$

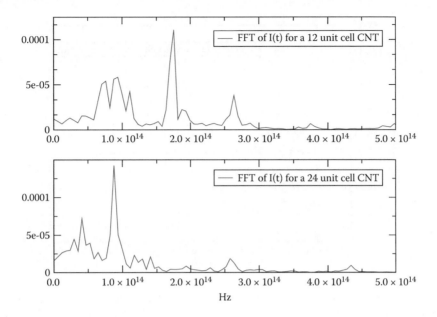

FIGURE 39.3 Fourier transform of the current density. (Chen, Z. et al., Real-time quantum simulation of terahertz response in single wall carbon nanotube, *2011 11th IEEE Conference on Nanotechnology (IEEE-NANO)*, Portland, OR, pp. 1339–1342. © (2011) IEEE. With permission.)

We can see that the phase velocity is constant, showing that the high-frequency electron response is dominated by single-particle excitations rather than collective plasmon modes. Burke [7] and Hanson et al. [8] have found the plasmon velocity of SWCNT to be approximately 3×10^8 and 6×10^8 cm/s. Our simulations show that we have obtained a phase velocity that is consistent with the Fermi velocity 8×10^7 cm/s, which was also measured in a recent experiment [2]. We have also performed simulations in which the applied potential was varied sinusoidally at the resonant frequencies given above. In this case, we find responses from our simulations such as the peaks shown in Figure 39.3 below the resonances. These responses are consistent with the electron excitations due to the HOMO-LUMO gap in the system. For the 12 unit cell case, if one assumes the lower response at 100 THz, the corresponding energy is 0.414 eV, and in our simulation, we find that the HOMO-LUMO gap for this system is 0.416 eV.

39.3.2 KINETIC INDUCTANCE

From the solutions for the wave functions, one can now investigate many properties of the CNT. In particular, within the real-space mesh framework, the electron density is given by

$$\mathbf{n}(t) = \sum_{j=1}^{N_e} |\Psi_j(t)|^2. \tag{39.15}$$

Kinetic inductance is an important property of the CNT. The total kinetic energy can be expressed in terms of the current I and equivalence to the inductance used:

$$E_k = \frac{mI^2}{2nA^2q^2} = \frac{1}{2}L_kI^2. \tag{39.16}$$

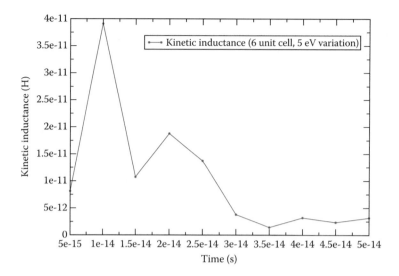

FIGURE 39.4 Simulation result of the kinetic inductance versus time. (Chen, Z. et al., Real-time quantum simulation of terahertz response in single wall carbon nanotube, *2011 11th IEEE Conference on Nanotechnology (IEEE-NANO)*, Portland, OR, pp. 1339–1342. © (2011) IEEE. With permission.)

The probability current $\vec{j}$ of the wave function Ψ is defined as

$$\vec{j} = \frac{\hbar}{2mi}\left(\Psi^*\vec{\nabla}\Psi - \Psi\vec{\nabla}\Psi^*\right) = \frac{\hbar}{m}\operatorname{Im}(\Psi^*\vec{\nabla}\Psi). \tag{39.17}$$

Integrating over the cross section ($I = \int_{y,z} \vec{j}\,dy\,dz$), we have a probability current in the middle of the CNT. Kinetic energy can also be calculated from wave function using the formula:

$$E_k = \frac{\hbar^2}{2m}\left|\nabla\Psi\right|^2. \tag{39.18}$$

Because $L_k = \frac{2E_k}{I^2}$ and the current I could be zero at some time, we then take an average of E_k and I over each period to calculate the average kinetic inductance.

After plotting the kinetic inductance (Figure 39.4) and performing cubic least square fitting, we obtain a kinetic inductance of 2.5 pH, as the actual length of CNT is around 1.4 nm, so we have a unit kinetic inductance of 3.57 pH/nm (considering spin). One theoretical estimate for a unit kinetic inductance of SWCNT is 6.7 pH/nm [20], which is consistent with Léonard's book [21]. Burke has a unit kinetic inductance of 4 pH/nm by using a nanotransmission line model [7]. A measured kinetic inductance result is 7.8 pH/nm (15 parallel tubes) [22]. Our result is then consistent with other theoretical estimates and measured ones.

In summary, our simulation results of electron resonances are congruent with the ballistic electron resonance model. The electron velocity is found to be constant over different CNTs and equal to the Fermi velocity, which means that the THz electron response is dominated by single-particle excitations rather than collective plasmon modes. In addition, our estimated kinetic inductance of SWCNT agrees with other theoretical estimates and experimental measures.

ACKNOWLEDGMENTS

This chapter is based upon work supported by the National Science Foundation: Grants No. ECCS 1028510 and ECCS 0846457.

REFERENCES

1. K. Fu, R. Zannoni, C. Chan, S. Adams, J. Nicholson, E. Polizzi, and K. Yngvesson, Terahertz detection in single wall carbon nanotubes, *Applied Physics Letters*, 92, 033105, 2008.
2. Z. Zhong, N. Gabor, J. Sharping, A. Gaeta, and P. McEuen, Terahertz time-domain measurement of ballistic electron resonance in a single-walled carbon nanotube, *Nature Nanotechnology*, 3(4), 201–205, 2008.
3. P. Avouris, Z. Chen, and V. Perebeinos, Carbon-based electronics, *Nature Nanotechnology*, 2, 605–615, 2007.
4. J. Appenzeller, Carbon nanotubes for high-performance electronics progress and prospect, *Proceedings of IEEE*, 96, 201–211, 2008.
5. K. Alam and R. Lake, Performance metrics of a 5 nm, planar, top gate, carbon nanotube on insulator (COI) transistor, *IEEE Transactions on Nanotechnology*, 6(2), 186–190, 2007.
6. D. Kienle, Terahertz response of carbon nanotube transistors, *Physical Review Letters*, 103, 026601, 2009.
7. P. Burke, Luttinger liquid theory as a model of the gigahertz electrical properties of carbon nanotubes, *IEEE Transactions on Nanotechnology*, 1(3), 129–144, 2002.
8. G. Hanson, Current on an infinitely-long carbon nanotube antenna excited by a gap generator, *IEEE Transactions on Antennas and Propagation*, 54(1), 76–81, 2006.
9. M. V. Shuba, S. A. Maksimenko, and G. Ya. Slepyan, Absorption cross-section and near-field enhancement in finite-length carbon nanotubes in the terahertz-to-optical range, *Journal of Computational Theoretical Nanoscience*, 6, 2016–2023, 2009.
10. Heber J., Surfing the wave, *Nature*, 461, 720, 2009.
11. T. Iitaka, Solving the time-dependent Schrödinger equation numerically, *Physical Review E*, 49, 4684–4690, 1994.
12. A. Castro, M. Marques, and A. Rubio, Propagators for the time-dependent Kohn-Sham equations, *Journal of Chemical Physics*, 121(8), 3425–3433, 2004.
13. E. Polizzi, Density-matrix-based algorithm for solving eigenvalue problems, *Physical Review B*, 79(11), 115112, 2009.
14. J. Crank and P. Nicolson, A practical method for numerical evaluation of solutions of partial differential equations of the heat-conduction type, *Advances in Computational Mathematics*, 6(1), 207–226, 1996.
15. E. Runge and E. Gross, Density-functional theory for time-dependent systems, *Physical Review Letters*, 52(12), 997–1000, 1984.
16. Z. Chen and E. Polizzi, Spectral-based propagation schemes for time-dependent quantum systems with application to carbon nanotubes, *Physical Review B*, 82, 205410, 2010.
17. Z. Chen and E. Polizzi unpublished.
18. A. Mayer, Band structure and transport properties of carbon nanotubes using a local pseudopotential and a transfer-matrix technique, *Carbon*, 42(10), 2057–2066, 2004.
19. D. Zhang and E. Polizzi, Efficient modeling techniques for atomistic-based electronic density calculations, *Journal of Computational Electronics*, 7(3), 427–431, 2008.
20. D. Kienle and F. M. C. Léonard, Terahertz response of carbon nanotube transistors, *Physical Review Letters*, 103, 026601, 2009.
21. F. Léonard, *The Physics of Carbon Nanotube Devices*. William Andrew: New York, 2008.
22. M. Zhang, X. Huo, P. Chan, Q. Liang, and Z. Tang, Radio-frequency transmission properties of carbon nanotubes in a field-effect transistor configuration, *IEEE Electron Device Letters*, 27(8), 668–670, 2006.
23. Chen, Z., Yngvesson, S., and Polizzi, E., Real-time quantum simulation of Terahertz response in single wall carbon nanotube, *2011 11th IEEE Conference on Nanotechnology (IEEE-NANO)*, Portland, OR, pp. 1339–1342, 2011.

Section XII

Carbon Nanotube Transistor Fabrication

40 Fabrication of Stable n-Type Thin-Film Transistor with Cs Encapsulated Single-Walled Carbon Nanotubes

Toshiaki Kato, Rikizo Hatakeyama, and Yosuke Osanai

CONTENTS

40.1 INTRODUCTION

Thin-film transistors (TFTs) are one of the most promising practical applications of single-walled carbon nanotubes (SWNTs) due to their flexible filament-like structure and high carrier mobility [1]. For the fabrication of industrial electrical devices, it is an inevitable issue to utilize both p- and n-type transistors as basic components of the electrical circuits. Since oxygen and water molecules adsorbing on the surface of SWNTs are known to play a role as an electron acceptor against SWNTs, SWNTs–TFTs have p-type semiconducting features. To date, there are several reports on the fabrication of n-type SWNTs–TFTs by functionalizing the outside surface of SWNTs [2]. However, the operation of n-type SWNTs–TFTs is limited only under the specific condition and the fabrication of stable n-type SWNTs–TFTs under the various environmental conditions has not been realized.

Here, we report on the successful fabrication of very stable n-type SWNTs–TFTs by encapsulating Cs atoms in SWNTs with a plasma ion irradiation method [3–11]. Since the graphitic network of carbon cells protects the inside Cs atoms from other reactive molecules that exist outside SWNTs, the n-type features are found to be very stable even after soaking for a long time in water.

40.2 EXPERIMENTAL

The Cs atom encapsulation is carried out by a plasma ion irradiation method. $Cs^{+}-e^{-}$ plasmas are generated by a thermal contact ionization method (Figure 40.1a) [3–9]. The typical plasma parameters are measured by a Langmuir probe and can be estimated from this I_p-V_p curve as follows: electron density $(n_e) \approx 10^9 \text{ cm}^{-3}$, electron temperature $(T_e) \approx 0.2 \text{ eV}$, and space potential $(\varphi_s) \approx -3 \text{ V}$. SWNTs are deposited on a SiO_2 (300 nm)/Si substrate. To promote the adsorption of SWNTs, the SiO_2 substrate surface is functionalized by 3-aminopropyltriethoxysilane (APTES) prior to the SWNTs deposition [12]. The pairs of Au electrodes are fabricated by a conventional photolithography

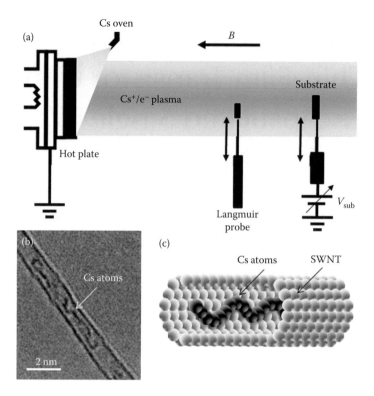

FIGURE 40.1 (a) Schematic illustration of Cs @ SWNTs formation process. (b, c) A typical TEM image and model of Cs @ SWNT.

technique. SWNTs–TFTs are put on a SUS plate, which is inserted into the plasma region. After the Cs irradiation, Cs-irradiated SWNTs–TFTs are carefully rinsed by purified water to remove the Cs atoms adsorbing on the outside of the SWNTs. The transport properties of SWNTs–TFTs are measured by a vacuum probe station and semiconductor parameter analyzer under a field-effect transistor (FET) configuration. The structure of Cs-irradiated SWNTs is also observed by transmission electron microscopy (TEM).

40.3 RESULTS AND DISCUSSION

Figures 40.1b and c show a typical TEM image and a model of Cs encapsulated (@) SWNTs. The one-dimensional chain-like structures of Cs atoms are often and clearly observed inside SWNTs by TEM observation. The typical source–drain current (I_{ds}) versus gate bias voltage (V_{gs}) curves of the same device after the Cs plasma irradiation are investigated as a function of Cs irradiation time (Figure 40.2a). The source–drain voltage (V_{ds}) is fixed at 1 V. The saturated source–drain current (I_{ds}) ratio of n- to p-channel [$(I_{onn}-I_{off})/(I_{onp}-I_{off})$] is utilized as a guidepost to estimate the conducting type of SWNTs–TFTs (Figure 40.2b), where I_{off} denotes the off current corresponding with the current from metallic SWNTs. The value of $(I_{onn}-I_{off})/(I_{onp}-I_{off})$ clearly increases with an increase in the Cs irradiation time. A similar tendency is also observed from the plot of the threshold gate bias voltage (V_{gth}). V_{gth} shifts to the negative V_{gs} direction with an increase in the Cs irradiation time. This indicates that the electrical transport property of SWNTs–TFTs can be precisely controlled by tuning the doping density of Cs inside SWNTs.

To confirm the stability of Cs @ SWNTs–TFTs, the transistors are soaked in purified water for a certain period. Figure 40.3 gives the soaking time dependence of I_{ds}–V_{gs} curves for Cs @ SWNTs–TFTs. Although the threshold of the gate bias voltage slightly shifts to the positive V_{gs} direction

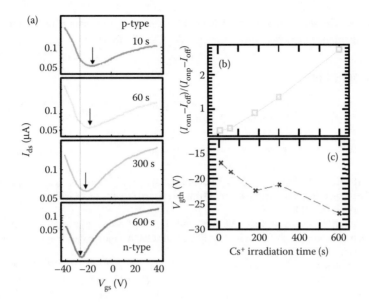

FIGURE 40.2 (a) I_{ds}–V_{gs} curves of Cs @ SWNTs-TFTs for different Cs plasma irradiation times. (b) (I_{onn}–I_{off})/(I_{onp}–I_{off}) and (c) V_{gth} plot as a function of Cs plasma irradiation time.

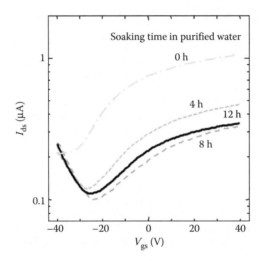

FIGURE 40.3 The soaking time dependence of Cs @ SWNTs-TFTs features in the purified water.

between 0 and 4 h soaking time, the change in the I_{ds}–V_{gs} feature is almost saturated by 8 h. This indicates that Cs atoms adsorbing on the outside of SWNTs should be washed out during the 0–4 h. Since the I_{ds}–V_{gs} does not show an obvious change after an 8 h soak and the n-type feature is maintained, the transition from a p- to an n-type property should originate from the encapsulated Cs atoms inside SWNTs, which remain very stable even under water.

40.4 CONCLUSION

We have succeeded in fabricating very stable n-type SWNTs–TFTs with Cs@SWNTs and found that the transport property of SWNT–TFTs clearly changes from a p- to an n-type after Cs plasma irradiation. On the basis of systematic investigation, the n-type property is very stable under the long soak in water. These stable n-type SWNTs–TFTs are very important for the industrial fabrication of high-performance electrical circuits with SWNTs–TFTs.

ACKNOWLEDGMENT

This work was supported by a grant-in-aid for scientific research from the Ministry of Education, Culture, Sports, Science and Technology, Japan.

REFERENCES

1. Q. Cao, H.-S. Kim, N. Pimparkar, J. P. Kulkarni, C. Wang, M. Shim, K. Roy, M. A. Alam, and J. A. Rogers, Medium-scale carbon nanotube thin-film integrated circuits on flexible plastic substrates, *Nature*, 454(7203), 495–500, 2008.
2. P. Qi, O. Vermesh, M. Grecu, A. Javey, Q. Wang, and H. Dai, Toward large arrays of multiplex functionalized carbon nanotube sensors for highly sensitive and selective molecular detection, *Nano Letters*, 3(3), 347–351, 2003.
3. T. Hirata, R. Hatakeyama, T. Mieno, and N. Sato, Production and control of K–C_{60} plasma for material processing, *Journal of Vacuum Science Technology A*, 14(2), 615–618, 1995.
4. G.-H. Jeong, T. Hirata, R. Hatakeyama, K. Tohji, and K. Motomiya, C_{60} encapsulation inside single-walled carbon nanotubes using alkali-fullerene plasma method, *Carbon*, 40(12), 2247–2253, 2002.
5. G.-H. Jeong, R. Hatakeyama, T. Hirata, K. Tohji, K. Motomiya, T. Yaguchi, and Y. Kawazoe, Formation and structural observation of cesium encapsulated single-walled carbon nanotubes, *Chemical Communications*, (1), 152–153, 2003.
6. G.-H. Jeong, A. A. Farajian, R. Hatakeyama, T. Hirata, T. Yaguchi, K. Tohji, H. Mizuseki, and Y. Kawazoe, Cesium encapsulation in single-walled carbon nanotubes via plasma ion irradiation: Application to junction formation and ab initio investigation, *Physical Review B*, 68(7), 075410-1-6, 2003.
7. T. Izumida, R. Hatakeyama, Y. Neo, H. Mimura, K. Omote, and Y. Kasama, Electronic transport properties of Cs-encapsulated single-walled carbon nanotubes created by plasma ion irradiation, *Applied Physics Letters*, 89(9), 093121-1-3, 2006.
8. Y. F. Li, R. Hatakeyama, T. Kaneko, T. Izumida, T. Okada, and T. Kato, Electronic transport properties of Cs-encapsulated double-walled carbon nanotubes, *Applied Physics Letters*, 89(9), 093110-1-3, 2006.
9. R. Hatakeyama and Y. F. Li, Synthesis and electronic-property control of Cs-encapsulated single- and double-walled carbon nanotubes by plasma ion irradiation, *Journal of Applied Physics*, 102(3), 034309-1-7, 2007.
10. T. Kato, R. Hatakeyama, J. Shishido, W. Oohara, and K. Tohji, p–n Junction with donor and acceptor encapsulated single-walled carbon nanotubes, *Applied Physics Letters*, 95(8), 083109-1-3, 2009.
11. Y. F. Li, R. Hatakeyama, W. Oohara, and T. Kaneko, Formation of p–n junction in double-walled carbon nanotubes based on heteromaterial encapsulation, *Applied Physics Express*, 2(9), 095005-1-3, 2009.
12. L. Zhang, S. Zaric, X. Tu, X. Wang, W. Zhao, and H. Dai, Assessment of chemically separated carbon nanotubes for nanoelectronics, *Journal of the American Chemical Society*, 130(8), 2686–2691, 2007.

41 Printing Technology and Advantage of Purified Semiconducting Carbon Nanotubes for Thin Film Transistors

Hideaki Numata, Kazuki Ihara, Takeshi Saito, and Fumiyuki Nihey

CONTENTS

41.1 INTRODUCTION

Single-wall carbon nanotubes (CNTs) [1] have a large intrinsic mobility [2] and are expected to be useful as a new electronic material. However, their electronic structures depend on their diameters and chirality, where one-third of the pristine CNTs are metallic (m-) and the other two-thirds are semiconducting (s-) [3]; thus, they intrinsically contain metallic contamination. Therefore, it is crucial to mitigate the effect of m-CNTs when making use of CNTs as semiconducting materials. There are two approaches, increasing s-CNT purity and decreasing the influence of m-CNT. Using a CNT random network as a thin film transistor (TFT) channel [4] is one of the best ideas to prevent device shorts caused by m-CNTs. According to the simulation work, a CNT random network is tolerant to m-CNT contamination [5]. Furthermore, CNT random networks are compatible with solution processes such as s-CNT purification [6] and printing fabrication. The device yield and performance of a CNT random network TFT can be increased by the use of purified s-CNTs.

On the other hand, printed electronics enable large-area and flexible electronics on plastic substrates [7]. The key feature of print fabrication is simultaneous material deposition, and pattern definition by fully additive processes. This feature reduces the number of process steps and the amount of waste material during fabrication, and leads to low-cost and eco-friendly fabrication from a minimum of materials. Furthermore, maskless printing enables on-demand production by decreasing the lead time with offering flexibility in the circuit design. Therefore, printed electronics are attractive even though the resolution of the printing methods is three orders of magnitude larger than that of photolithography. Actually, it is difficult to fabricate high-density and high-functioning circuits by printing. Therefore, printed electronics are rather suited to implementing devices such as large-area sensing sheets, lightweight human interfaces, RF devices with multiple antennas, and so on. It is expected that such de novo electronics fields will be open by printed electronics.

The key issue is successful fabrication of printed TFTs in order to realize printed electronics, and the performance of the TFTs strongly affects the application field. The performance of printed TFTs is expected to increase by adopting CNT channels with high carrier mobility. A high-mobility TFT with a solution-processed CNT network was fabricated on an Si wafer [8]. It showed the excellent potential of a CNT network channel, but it was solid and its gate electrode was not individually fashioned. A flexible and high-mobility TFT was also fabricated on a polymeric substrate [9]; however, it required an additional patterning process for device separation. Inkjet printing is a promising candidate for the direct formation of a CNT network, but because the performance of the current TFTs does not appear competitive [10,11], technological development for printed CNT–TFTs is desired.

We developed a fabrication technology for printed CNT–TFTs. All electrodes, insulators, and CNT channels were directly printed without additional patterning processes. In this fabrication, surface-free energy was controlled for precise pattern definition. In addition, a mechanism for adsorbing CNTs was applied to channel formation in order to achieve homogeneous dispersion of the CNT random networks. We also developed an s-CNT purification method and fabricated the CNT–TFTs on the plastic films. The TFT characteristics clearly demonstrated the advantages of the purified s-CNT ink. The on-current of the printed CNT–TFT was increased without deterioration in the on/off ratio. A large field-effect mobility of 3.6 cm^2/Vs was obtained for the printed CNT–TFTs with an on/off ratio of about 1000.

41.2 CNT INK PREPARATION

In this study, we used two types of CNT inks, CNT-1 and CNT-2. For CNT-1 preparation, commercially available CNTs (CoMoCAT SG65, Southwest Nano Technologies) were dispersed into a mixed organic solvent of xylene and dichloroethane by ultrasonication. CNT-1 has low surface tension and dries quickly on the substrates. It contains only volatile materials except for CNTs; however, it may contain about 33% m-CNTs. CNT-2 was an aqueous suspension and has higher surface tension than CNT-1. CNTs grown by chemical vapor deposition [12] were dispersed into heavy water (D$_2$O) with polyoxyethylene alkyl ether, a nonionic surfactant, and s-CNT was purified by the electric-field-induced layer formation (ELF) method [13]. The micelles containing s-CNT were extracted to the anode region by applying a direct current (DC) electric field for a few tens of hours (Figure 41.1a). This extraction was repeated twice, and we estimated from the absorption spectra (Figure 41.1b) that the purity of s-CNT had increased more than 95%. According to thermogravimetric analysis (TGA), most parts of the surfactants can be removed by heat treatment at 180°C in air (Figure 41.2). Both types of CNT inks can be handled in ambient conditions.

41.3 PRINTING FABRICATION TECHNOLOGY

41.3.1 Device Fabrication Flow

Device fabrication flow is shown in Figure 41.3. All device elements were patterned by maskless printing methods with a minimum of materials. First, gate electrodes were printed on a polyimide (PI) film by use of an inkjet printer with nano-Ag ink (NPS-J, Harima Chemicals) and sintered in air. Next, gate insulators were formed by an ink dispenser with PI ink (CT4112, Kyocera Chemical)

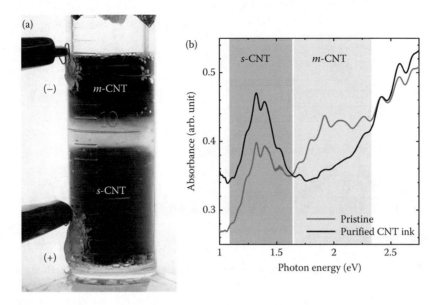

FIGURE 41.1 (a) *s*-CNT purification by the electric-field-induced layer formation (ELF) method [13]. (b) Absorption spectra of purified *s*-CNT ink and pristine CNT suspension. (H. Numata et al., Printing technology and advantage of purified semiconducting carbon nanotubes for thin film transistor fabrication on plastic films, *Proc. Int. Conf. lEEE Nano 2011*, Portland, OR, pp. 1000–1005. © (2011) IEEE. With permission.)

and cured in an N_2 oven. For some TFT samples, a 30-nm-thick SiO_2 film was sputter deposited. This film was not for insulation but acted as a seed layer for a 3-aminopropyltriethoxysilane (APTES) monolayer, which was formed just before CNT channel formation. Then the source and drain electrodes were printed with nano-Ag ink and sintered. Finally, CNT ink was cast on the channel regions by a dispenser. When CNT-2 ink was used for the fabrication, the surfactants were removed from the channel regions by posttreatment with a combination of heat and wet processes. The maximum temperature was 200°C during the fabrication. In this study, we used a sputtered SiO_2 layer as a seed for APTES modification. However, reliable insulation is not required for the layer, so it can be replaced by printable materials such as hydrogen silsesquioxane.

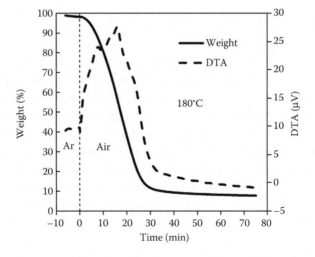

FIGURE 41.2 TGA chart of the surfactants used for CNT-2.

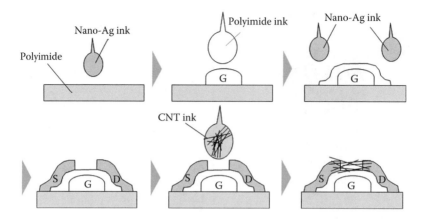

FIGURE 41.3 Print fabrication flow.

In this fabrication process, all device geometries were defined by printing methods without an additional patterning process. Therefore, it is important to control the wettability between the material inks and the underlying surfaces. In addition, there are two keys in the fabrication process. One is the reliability and uniformity of the printed gate insulators in order to successfully form such layered device structures. The other is the homogeneous dispersion of the CNT networks to improve device quality and performance. The details are described later.

41.3.2 WETTABILITY

Wetting tests were performed and surface free energy was quantitatively evaluated with Kaelble's model [14], which provides dispersive (γ_S^d) and polar (γ_S^p) components of the surface free energy separately. The contact angles (θ) were measured with three probe liquids, water, methylene iodide, and formamide, whose polar (γ_L^p) and dispersive (γ_L^d) surface tension components are known. According to Kaelble's model, the relation of the parameters is described as

$$\frac{(\gamma_L^p + \gamma_L^d)(1 + \cos\theta)}{2\sqrt{\gamma_L^d}} = \frac{\sqrt{\gamma_L^p}}{\sqrt{\gamma_L^d}}\sqrt{\gamma_S^p} + \sqrt{\gamma_S^d}. \tag{41.1}$$

All measured data were plotted, and the surface free energy components γ_S^d and γ_S^p were estimated from the squares of the intercept and slope of the linear regression line, respectively (Figure 41.4a).

We used a corona discharge scanner to change the surface free energy because it enables simple and rapid treatment in the atmosphere. Figure 41.4b shows the surface energy of the PI film treated by the corona discharge method. In Figure 41.4b, the horizontal axis indicates the treatment time for a unit film length (the reciprocal of the scanning speed). The film had large γ_S^d and small γ_S^p before treatment. After treatment, γ_S^d decreased from the initial value, and γ_S^p increased with the treatment time. In Figure 41.4b, the total free energy represents the summation of γ_S^d and γ_S^p.

Figures 41.5a–c show the Ag dot patterns printed on the PI surfaces with various surface free energy components. The printed Ag dots were enlarged when γ_S^d and γ_S^p were 37.0 and 1.9 mJ/m^2, respectively. This is a hydrophobic surface, and the dispersive component was too large to print fine nano-Ag patterns precisely (Figure 41.5a). The fine Ag dots were obtained when γ_S^d and γ_S^p were 24.3 and 18.8 mJ/m^2, respectively (Figure 41.5b). It is interesting that the total surface energy was larger than in the former case. In the case of a rather hydrophilic surface, γ_S^d and γ_S^p were 28.1 and 34.7 mJ/m^2, respectively; the printed Ag ink was spread on the surface; and the dots could not be separated (Figure 41.5c). Thus, it is very important to control the surface free energy for precise pattern definition in printing.

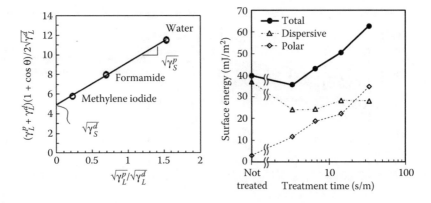

FIGURE 41.4 (a) Surface free energy estimation based on Kaelble's model [14], and (b) estimated surface energy for the PI surface treated by the corona discharge method. (H. Numata et al., Printing technology and advantage of purified semiconducting carbon nanotubes for thin film transistor fabrication on plastic films, *Proc. Int. Conf. IEEE Nano 2011*, Portland, OR, pp.1000–1005. © (2011) IEEE. With permission.)

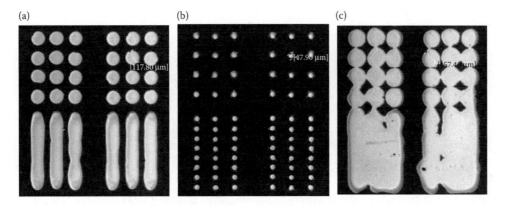

FIGURE 41.5 Ag dot patterns printed on the PI surfaces with various surface free energies. (a) $\gamma_S^d = 37.0$ mJ/m^2 and $\gamma_S^p = 1.9$ mJ/m^2, (b) $\gamma_S^d = 24.3$ mJ/m^2 and $\gamma_S^p = 18.8$ mJ/m^2, and (c) $\gamma_S^d = 28.1$ mJ/m^2 and $\gamma_S^p = 34.7$ mJ/m^2.

41.3.3 Gate Insulators

The electric properties of the insulators were evaluated to assess the printed Ag/PI/Ag stacked structures. The thickness of the printed films was controlled by surface free energy and printing conditions such as ink volume and solute concentration. Figures 41.6a and b show the current–voltage (*I–V*) and capacitance–voltage (*C–V*) characteristics for the printed PI layers whose thickness was about 600 nm. The printed Ag electrodes have much rougher surfaces because the grains grow up to about 100 nm during sintering. However, the printed PI layers have very small leakage currents for a 40 V bias (Figure 41.6a). The capacitances were independent of the bias, and the thickness of the printed PI was estimated from the capacitance value. A relative dielectric constant (k_{GI}) of 2.59 was estimated in advance from reference samples, which had a 500-nm-thick, spin-coated PI layer sandwiched between the upper and lower Au electrodes, which were deposited by E-gun evaporation. In Figure 41.7, the horizontal and vertical axes show the estimated thickness and the normalized conductance at 1 MHz. We evaluated 91 samples, and an average thickness (t_{GI}) of 630 nm with a standard deviation (σ) of 4.9% was obtained. Consequently, reliable and uniform printed insulators were obtained.

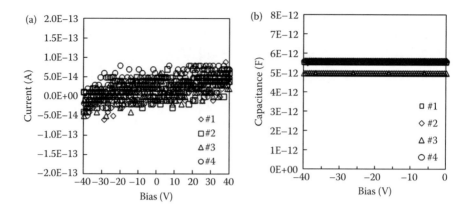

FIGURE 41.6   (a) *I–V* characteristics, and (b) *C–V* characteristics of the printed PI films.

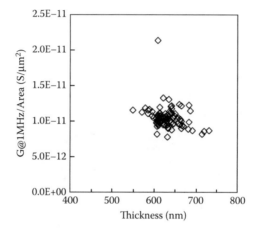

FIGURE 41.7   Estimated thickness and normalized conductance of the printed PI films at 1 MHz.

41.3.4 CNT RANDOM NETWORK CHANNEL

Homogeneous dispersion is required for a CNT network because aggregation and bundles emphasize the effect of *m*-CNTs on the device characteristics. CNT-1 ink has small surface tension and dries rapidly. Thus, it is expected that the droplet spreads out immediately and is parched before aggregation on the solid surface. However, it is difficult to precisely control the size of the channel regions because the droplet edge is dynamic. CNT-2 ink has a larger surface tension and its droplet is relatively stable. In this case, CNTs tend to be deposited at the edge of the droplet and form a ring-shaped condensed region. It is a well-known phenomenon, the so-called "coffee stain." In the drying droplet, the contact line is pinned, and the evaporating solvent from the edge is replenished from the interior; consequently, an outward capillary flow occurs and carries the solute to the edge [15]. In such a condensed region, the *m*-CNT density would exceed the percolation threshold [16] locally and result in leakage current. To prevent "coffee stain" formation at the edge, CNTs need to be actively deposited in the middle of the droplet against the outward flow. Thus, a mechanism to adsorb CNTs should be appended to the surface to be printed.

Many surface treatment methods were investigated for effective CNT adsorption. An APTES-modified SiO_2 surface provided an excellent result. An amino-silanized SiO_2 surface was used to adsorb CNTs surrounded by ionic surfactants of sodium dodecylsulfate (SDS) [17]. An APTES-modified surface was also used for the suspensions of CNTs in amide solvents [18]. We carried out an adsorption

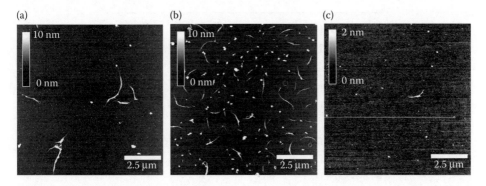

FIGURE 41.8 Results of the adsorption test for CNT ink. AFM height images of the CNTs on a (a) bare SiO_2 surface, (b) APTES-modified SiO_2 surface, and (c) HMDS-modified SiO_2 surface.

test with an organic CNT suspension that was similar to CNT-1 ink. Although no surfactant was used, a high boiling solvent mixture of dodecane and triglyme (triethylenglykol-dimethylether) was used. A puddle of the suspension was formed on the surface for 60 min and blown away by N_2 gas. CNTs adsorbed on the surface were observed with an atomic force microscope (AFM). Figures 41.8a–c indicate AFM images of the CNT adsorption test. A few CNTs were observed on the bare SiO_2 surface (Figure 41.8b). On the APTES-modified SiO_2 surface, many CNTs were adsorbed uniformly (Figure 41.8b). Conversely, they were rarely adsorbed on the hexamethyldisilazane (HMDS)-modified SiO_2 surface (Figure 41.8c). The APTES and HMDS functionalize the SiO_2 surface with the amino and methyl groups, respectively. It was confirmed that CNTs are effectively adsorbed by the amino groups. It was also confirmed that the APTES-modified SiO_2 surface actively adsorbed the micelles containing s-CNT in CNT-2 ink [19]. The APTES layer chemically couples with the SiO_2 surface, so a 30-nm-thick sputtered SiO_2 layer was used as a seed layer in the device fabrication. The layer can be replaced by printable materials such as hydrogen silsesquioxane because reliable insulation is not required.

41.4 PRINTED CNT THIN FILM TRANSISTORS

41.4.1 TFT Fabrication

A 10×10 TFT array was printed on a 48.7 mm $\times$ 42.3 mm PI film substrate (Figure 41.9a). The fabrication conditions were determined by a combination of two types of CNT inks and three types of surfaces on which the CNT channels were printed. The fabricated samples are listed in Table 41.1 with symbols used in the following figures. The open and gray-colored symbols correspond to the samples with

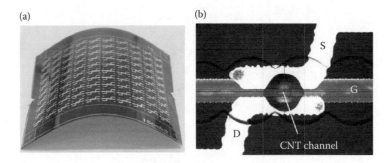

FIGURE 41.9 (a) Fabricated 10×10 array sample, and (b) micrograph of a printed CNT–TFT on a plastic film. (H. Numata et al., Printing technology and advantage of purified semiconducting carbon nanotubes for thin film transistor fabrication on plastic films, *Proc. Int. Conf. IEEE Nano 2011*, Portland, OR, pp. 1000–1005. © (2011) IEEE. With permission.)

TABLE 41.1
Sample List

No.	CNT link	Surface	Symbol	Remarks
1	CNT-1	CT4112	○	
2	CNT-1	SiO$_2$	◇	
3	CNT-1	APTES	△	
4	CNT-1	CT4112	□	Excessive shots
5	CNT-2	CT4112	◉	
6	CNT-2	SiO$_2$	◆	
7	CNT-2	APTES	▲	
8	CNT-2	APTES	▣	Post-treatment

CNT-1 and -2 inks, respectively. The symbol shapes of circles, diamonds, and triangles correspond to the surfaces of CT4112, bare SiO$_2$, and APTES-modified SiO$_2$, respectively. Each sample has 100 TFTs, and the amount of CNT ink was varied in the samples. A small droplet of 10–20 nL CNT ink was cast 1–16 times to form a TFT channel. In Table 41.1, the square symbols represent the additional samples. Sample 4 was fabricated under conditions similar to those of sample 1, but the "excessive shots" were further cast to the CNT channel. Sample 8 had conditions similar to those of sample 7, but "posttreatment" to remove surfactants from the CNT network was different. Figure 41.9b represents the micrograph of a printed CNT–TFT immediately after a droplet of CNT-2 ink was cast between the source and drain electrodes. The channel width (W) and length (L) were about 610 and 100 μm, respectively.

41.4.2 RESULTS AND DISCUSSION

Figure 41.10 shows typical transfer characteristics of CNT–TFTs with a gate voltage that varied from −20 to 20 V at a drain voltage (V_D) of −2 V. They indicate p-type characteristics and their on/off ratio was greater than 10,000. The printed CNT–TFTs had excellent quality. The relationship between on-current (I_{on}) and on/off ratio for all devices was studied (Figure 41.11). In Figure 41.11,

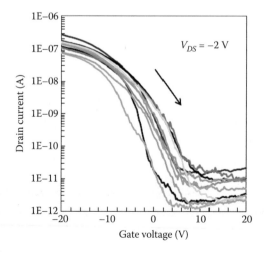

FIGURE 41.10 Transfer characteristics of printed CNT–TFTs. (H. Numata et al., Printing technology and advantage of purified semiconducting carbon nanotubes for thin film transistor fabrication on plastic films, *Proc. Int. Conf. IEEE Nano 2011*, Portland, OR, pp. 1000–1005. © (2011) IEEE. With permission.)

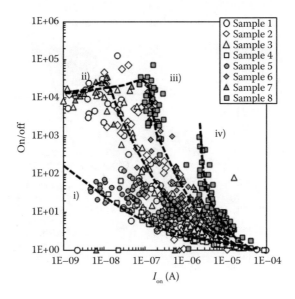

FIGURE 41.11 I_{on} dependence of on/off ratio for printed CNT–TFTs. (H. Numata et al., Printing technology and advantage of purified semiconducting carbon nanotubes for thin film transistor fabrication on plastic films, *Proc. Int. Conf. IEEE Nano 2011*, Portland, OR, pp. 1000–1005. © (2011) IEEE. With permission.)

there are four trend curves, (i)–(iv). Curve (i) corresponds to TFTs with low on/off ratios. For curves (ii) and (iii), the maximum on/off ratio exceeded 10,000. However, there was a trade-off between I_{on} and the on/off ratios observed in higher I_{on} regions. The critical values of I_{on} for which the on/off ratio decreased drastically were different for curves (ii) and (iii). Curve (iv) is a group of TFTs that has a large I_{on} with a rather high on/off ratio.

The I_{on} dependence of the off-current (I_{off}) is also plotted in Figure 41.12 in order to understand the deterioration in the on/off ratio in the high I_{on} region. For the CNT random network TFTs, I_{on} represents the total number of CNTs in the channel, and I_{off} is strongly affected by m-CNTs because all of the current paths consist only of m-CNTs.

Curve (i) mainly consists of TFTs of samples 4 and 5, and the TFTs have large values of I_{off} even in the small I_{on} region. When I_{on} increases, I_{off} also increases, and the on/off ratio approaches 1. In the preliminary experiments for sample 4, when excessive shots were cast into the CNT channel, the cast ink washed away the formerly deposited CNTs in the channel and a ripple-like aggregation was formed (Figure 41.13a). For sample 5, stable droplets of CNT ink were formed and the surface had no mechanism for the adsorption of CNTs. Therefore, CNTs tended to aggregate like "coffee stains" as shown in Figure 41.13b. For both samples, it is conceived that stripes of CNT aggregations formed across the channel area. In such aggregations, the density of m-CNT exceeds the percolation threshold locally and caused large values of I_{off}, even in the small I_{on} region.

Curves (ii) and (iii) consist of TFTs with CNT-1 and -2 inks, respectively. In the small I_{on} region, the values of I_{off} were very small and the on/off ratios were approximately 10,000. As I_{on} increased, I_{off} increased, and each curve has a critical value of I_{on} for which I_{off} increased drastically. It is thought that the m-CNT density exceeded the percolation threshold and the leakage current increased above the critical I_{on}. The critical I_{on} values are about 3×10^{-8} A and 2×10^{-7} A for curves (ii) and (iii), respectively. The difference between them clearly shows the advantage of employing s-CNT purification for CNT-2. CNT-2 ink contains an order of magnitude fewer m-CNTs than CNT-1 ink. Therefore, the critical I_{on} for the TFTs with CNT-2 ink increased than that for the TFTs with CNT-1 ink by up to an order of magnitude. The I_{on} values increased without deterioration in device quality for the printed CNT–TFT by adopting purified s-CNT ink.

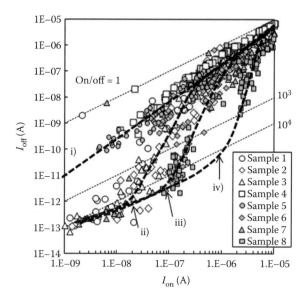

FIGURE 41.12 I_{on} dependence of I_{off} for printed CNT–TFTs. (H. Numata et al., Printing technology and advantage of purified semiconducting carbon nanotubes for thin film transistor fabrication on plastic films, *Proc. Int. Conf. IEEE Nano 2011*, Portland, OR, pp. 1000–1005. © (2011) IEEE. With permission.)

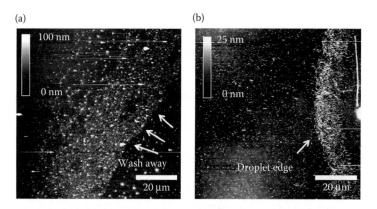

FIGURE 41.13 Results of the morphology test in the preliminary experiments. (a) When excessive shots were cast into the CNT channel, the cast ink washed away the formerly deposited CNTs. (b) "Coffee stains" appeared when a stable droplet was formed and the surface had no mechanism to adsorb CNTs.

In curve (iv), large I_{on} TFTs with a rather high on/off ratio were obtained. We used a simple parallel plate model for the gate capacitance and calculated field-effect mobility (μ) using the formula

$$\mu = \frac{L}{W} \frac{dI_D}{dV_G} \frac{t_{GI}}{k_{GI}\varepsilon_0} \frac{1}{V_D}, \tag{41.2}$$

where L is the channel length, W is the channel width, t_{GI} is the gate insulator thickness, k_{GI} is the relative dielectric constant of the gate insulator, V_D is the drain voltage, and dI_D/dV_G is the mutual conductance estimated from the transfer characteristics. The relationship between field-effect mobility and on/off ratio for printed CNT–TFTs was plotted in Figure 41.14. In curve

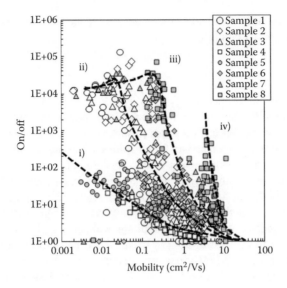

FIGURE 41.14 Relationship between field-effect mobility and on/off ratio for printed CNT–TFTs.

(iv), we obtained a CNT–TFT that has a high mobility of 3.6 cm²/Vs with an on/off ratio of about 1,000.

Curve (iv) consists of a part of the TFTs in sample 8. Sample 8 was fabricated with CNT-2 ink and APTES-modified SiO_2. Figure 41.15 indicates the AFM phase image of the CNT channel in sample 8. The CNTs are indicated as dark and narrow lines. A dense and uniform CNT random network was observed. Therefore, we can obtain a large I_{on} for TFTs with a rather high on/off ratio. The TFTs of sample 8 belong to curves (iii) and (iv). On the contrary, sample 7 has no TFTs belonging to curve (iv). Samples 7 and 8 were fabricated under the same conditions, except for the posttreatment to remove the surfactants from the channels. Sample 7 was baked first and then wet processed. Sample 8 was soaked in isopropyl alcohol and rinsed with deionized water before heat treatment. In the preliminary experiment, we found that the CNTs aggregated slightly when the device was soaked in a liquid before heat treatment. It is thought that the micelles migrated and reconstructed on the surface in the liquid environment. Consequently, the CNT morphology was modified by the posttreatment. It is expected that the slight aggregation decreased the CNT–CNT contact resistance and increased the mobility for curve (iv). It was also

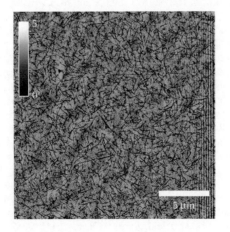

FIGURE 41.15 AFM phase image of printed CNT channel in sample 8.

reported that the Y-type junctions had low contact resistance between the CNTs, and it led to increased mobility of CNT–TFTs [20].

41.5 SUMMARY

A fabrication technology for printed CNT–TFTs was developed. All electrodes, insulators, and CNT channels were directly patterned without additional patterning processes such as photolithography and etching. In the fabrication process, surface free energy was controlled for precise pattern definition. In addition, a mechanism for adsorbing CNTs was applied to channel formation in order to achieve homogeneous dispersion of the CNT random networks. The CNT–TFTs were fabricated on the plastic films with a minimum of materials. A purified *s*-CNT ink was used, and its advantage was clearly demonstrated by the TFT characteristics. The on-current of the printed CNT–TFT increased without deterioration in the on/off ratio. A high mobility of 3.6 cm^2/Vs was obtained with an on/off ratio of about 1000. In addition, we successfully fabricated high-uniformity printed CNT–TFTs recently with a σ of 30% by use of this print fabrication technology. The drain current distributions for the on- and off-states were entirely separated, and this result indicates that the printed CNT–TFTs are capable of select switching in a two-dimensional matrix [19]. These results are definitely promising for the realization of printed electronics integrated with CNT–TFTs.

ACKNOWLEDGMENT

The authors would like to thank Shinichi Yorozu and Hiroyuki Endoh for their helpful discussions. A part of this work was supported by New Energy and Industrial Technology Development Organization (NEDO).

REFERENCES

1. S. Iijima and T. Ichihashi, Single-shell carbon nanotubes of 1-nm diameter, *Nature*, 363, 603–605, 1993.
2. T. Dürkop, S. A. Getty, E. Cobas, and M. S. Fuhrer, Extraordinary mobility in semiconducting carbon nanotubes, *Nano Lett.*, 4, 35–39, 2004.
3. R. Saito, M. Fujita, G. Dresselhaus, and M. S. Dresselhaus, Electronic structure of chiral graphene tubules, *Appl. Phys. Lett.* 60, 2204–2206, 1992.
4. E. S. Snow, J. P. Novak, P. M. Campbell, and D. Park, Random networks of carbon nanotubes as an electronic material, *Appl. Phys. Lett.*, 82, 2145–2147, 2003.
5. M. Ishida and F. Nihey, Estimating the yield and characteristics of random network carbon nanotube transistors, *Appl. Phys. Lett.*, 92, 163507, 2008.
6. M. S. Arnold, A. A. Green, J. F. Hulvat, S. I. Stupp, and M. C. Hersam, Sorting carbon nanotubes by electronic structure using density differentiation, *Nat. Nanotech.*, 1, 60–65, 2006.
7. T. Sekitani, M. Takamiya, Y. Noguchi, S. Nakano, Y. Kato, K. Hizu, H. Kawaguchi, T. Sakurai, and T. Someya, A large-area flexible wireless power transmission sheet using printed plastic MEMS switches and organic field-effect transistors, *IEDM Tech. Dig.* 2006, pp. 287–290, San Francisco.
8. M. C. LeMieux, M. Roberts, S. Barman, Y. W. Jin, J. M. Kim, and Z. Bao, Self-sorted, aligned nanotube networks for thin-film transistors, *Science*, 321, 101–104, 2008.
9. E. S. Snow, P. M. Campbell, M. G. Ancona, and J. P. Novak High-mobility carbon-nanotube thin-film transistors on a polymeric substrate, *Appl. Phys. Lett.*, 86, 033105, 2005.
10. P. Beecher, P. Servati, A. Rozhin, A. Colli, V. Scardaci, S. Pisana, T. Hasan et al., Ink-jet printing of carbon nanotube thin film transistors, *J. Appl. Phys.*, 102, 043710, 2007.
11. T. Takenobu, N. Miura, S. Y. Lu, H. Okimoto, T. Asano, M. Shiraishi, and Y. Iwasa, Ink-jet printing of carbon nanotube thin-film transistors on flexible plastic substrates, *Appl. Phys. Exp.*, 2, 025005, 2009.
12. T. Saito, S. Ohshima, T. Okazaki, S. Ohmori, M. Yumura, and S. Iijima, Selective diameter control of single-walled carbon nanotubes in the gas-phase synthesis, *J. Nanosci. Nanotech.*, 8, 6153–6157, 2008.
13. K. Ihara, H. Endoh, T. Saito, and F. Nihey, Separation of metallic and semiconducting single-wall carbon nanotube solution by vertical electric field, *J. Phys. Chem. C*, 115, 22827–22832, 2011.
14. D. H. Kaelble, *Physical Chemistry of Adhesion*, John Wiley & Sons, USA, 1971.

15. R. D. Deegan, O. Bakajin, T. F. Dupont, G. Huber, S. R. Nagel, and T. A. Witten, Capillary flow as the cause of ring stains from dried liquid drops, *Nature*, 389, 827–829, 1997.
16. C. H. Seager and G. E. Pike, Percolation and conductivity: A computer study. II, *Phys. Rev. B* 10, 1435–1446, 1974.
17. M. Burghard, G. S. Duesberg, G. Philipp, J. Muster, and S. Roth, Controlled adsorption of carbon nanotubes on chemically modified electrode arrays, *Adv. Mater.*, 10, 584–588, 1998.
18. J. Liu, M. J. Casavant, M. Cox, D. A. Walters, P. Boul, W. Lu, A. J. Rimberg, K. A. Smith, D. T. Colbert, and R. E. Smalley, Controlled deposition of individual single-walled carbon nanotubes on chemically functionalized templates, *Chem. Phys. Lett.*, 303, 125–129, 1999.
19. H. Numata, K. Ihara, T. Saito, H. Endoh, and F. Nihey, Highly uniform thin-film transistors printed on flexible plastic films with morphology-controlled carbon nanotube network channels, *Appl. Phys. Exp.*, 5, 055102, 2012.
20. D. M. Sun, M. Y. Timmermans, Y. Tian, A. G. Nasibulin, E. I. Kauppinen, S. Kishimoto, T. Mizutani, and Y. Ohno, Flexible high-performance carbon nanotube integrated circuits, *Nat. Nanotech.*, 6, 156–161, 2011.
21. H. Numata, K. Ihara, T. Saito, and F. Nihey, Printing technology and advantage of purified semiconducting carbon nanotubes for thin film transistor fabrication on plastic films, *Proc. Int. Conf. IEEE Nano 2011*, Portland, OR, pp. 1000–1005, 2011.

Section XIII

Random CNT Network Transistors

42 Solution-Processed Random Carbon Nanotube Networks Used in a Thin-Film Transistor

Qingqing Gong, Edgar Albert, Bernhard Fabel, Alaa Abdellah, Paolo Lugli, Giuseppe Scarpa, and Mary B. Chan-Park

CONTENTS

42.1 CARBON NANOTUBE NETWORKS AS ELECTRONIC MATERIAL

The single-walled carbon nanotube (CNT) was discovered in 1993 [1], with a rolled graphene character [2]. Individual CNTs have a low density, high stiffness, and high axial strength [3]. CNT thin films exhibit superior visible and infrared optical transmittance comparable with the commercial indium tin oxide layer [4]. Depending on the chirality, single-walled CNTs can be either metallic or semiconducting [5]. As-grown CNTs contain both semiconducting and metallic species [6,7]. The metallic content can be eliminated by electrical breakdown [8] or by using density gradient sorting [9]. Semiconducting CNTs used in field-effect transistors exhibit near ballistic transport and high mobility [10,11]. The superior mechanical, optical, and electronic properties of CNTs make them attractive as emerging research material. In recent years, random CNT networks have been applied in high-frequency technique [12–14], chemical and biosensing [15–17], and in flexible and stretchable logic circuits [18–22].

42.2 RANDOM CNT NETWORK TRANSISTORS

42.2.1 FABRICATION OF RANDOM NETWORK TRANSISTORS

Single-walled CNTs can be synthesized by arc-discharge, laser ablation, and chemical vapor deposition methods [23–25]. Commercial CNTs are available in solid or liquid form from companies such as NanoIntegris and SWeNT. NanoIntegris CNTs are produced via arc-discharge and then purified with the density-gradient ultracentrifugation (DGU) method [9]. Separated CNTs with a semiconducting purity of up to 99% are available. The average tube diameter of NanoIntegris CNTs is ca.1.4 nm. The SWeNT CNTs are synthesized via an optimized chemical vapor deposition process, the CoMoCAT process, which provides CNTs with narrow chirality distribution peaked at

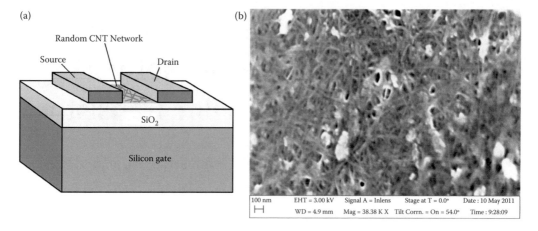

FIGURE 42.1 (a) SEM image of a CNT network deposited on a silicon substrate (SWeNT 90%-sc CNTs dissolved in NMP with a concentration of 10 μg/mL, scale bar: 100 nm). (b) Schematic diagram of a back-gated CNT network transistor. CNTs from the active channel between the source and drain contacts. The silicon substrate with an oxide layer serves as the back-gate. (From Q. Gong et al., *11th Proc. IEEE-Nano*, Portland, pp. 378–381. Copyright 2011 IEEE, with permission.)

(6,5) and (7,5) species [26]. As-grown nanotubes are purified by the DGU method with semiconducting content of around 90%. The average diameter of SWeNT CNTs is ca. 0.8 nm.

The solubility of single-walled CNTs in organic solvents is generally low. For instance, single-walled CNTs have room-temperature solubility of 10 μg/mL in *N*-methylpyrrolidinone (NMP) and 95 μg/mL in 1,2-dichlorobenzene (which is the highest solubility reported in Ref. [27]). Dispersion of single-walled CNTs in aqueous solution can be achieved with surfactants or biomolecules [28,29]. Surfactants can alter the electronic properties of CNTs, and therefore must be removed after deposition of nanotubes from solution to substrate. Commonly used surfactants include sodium dodecyl sulfate (SDS) and sodium cholate (SC) [9].

Random CNT networks can be deposited or transferred onto various substrates such as silicon wafer, quartz, glass, or flexible polymeric film. A monolayer such as 3-aminopropyltriethoxy-silane (APTES) can be applied to the substrate surface for improved adhesion and uniform distribution of the deposited nanotubes [30,31]. There are various coating techniques [32] for depositing CNTs from solution, such as dip coating [14], drop casting [33], spin coating [34], spray coating [35,36], inkjet printing [37], roll-to-roll printing [19], and transfer printing via polydimethysiloxane (PDMS) stamp or thermally activated adhesive tape [38,39]. Figure 42.1 shows an SEM image of a random CNT network drop-cast on an Si/SiO$_2$ substrate.

When used as a conducting channel in thin-film transistors, the CNT networks must be effectively connected to the source and drain electrodes. For instance, conductive materials with similar composition, such as graphene [39] or metallic CNT films [19], have been reported for contacting semiconductor-enriched nanotube networks. Among metal contacts, palladium (Pd) is considered to be suitable due to its good wetting interaction combined with a high work function $\Phi_{Pd} = 5.1$ eV [10,40]. A schematic diagram of a back-gated random network transistor is illustrated in Figure 42.1. CNTs are distributed between and beneath the source/drain contacts. The silicon substrate with its thermal oxide layer serves as a global back gate. To achieve efficient gate control, a separate top-gate needs to be built upon or under the network channel.

42.2.2 Characterization of Random Network Transistors

Figure 42.2 shows the *I–V* characteristics of a back-gated random network transistor. The output characteristics I_D–V_{DS} were measured at V_{GS} from −4 V to +4 V in 2 V steps. The saturation of the

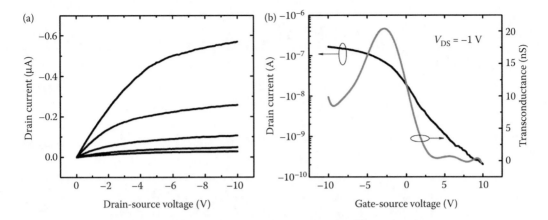

FIGURE 42.2 *I–V* characteristics of a back-gated random network transistor on a silicon wafer with 200 nm SiO$_2$, contacted with 10 nm Pd/30 nm Au electrodes. The channel geometry is $L = 100$ μm, $W = 1000$ μm. NanoIntegris 98%-sc CNTs was dissolved in NMP with a concentration of 2.5 μg/mL). (a) Output characteristics $I_D – V_{DS}$, measured at $V_{GS} = -4$ V to $+ 4$ V in 2 V steps (from top to bottom). (b) Transfer characteristic $I_D – V_{GS}$ and the transconductance g_m, measured at fixed drain-source voltage $V_{DS} = -1$ V.

drain currents at high drain voltage can be seen in the left subfigure. The drain conductance g_D can be calculated from the output curves using Equation 42.1. The effective mobility μ_{eff} is determined by g_D with Equation 42.2, where L is the channel length, W is the channel width, V_{th} is the threshold voltage, and C_{ox} is the oxide capacitance [41].

Figure 42.2 also shows the drain current I_D and transconductance g_m as a function of the gate-source voltage V_{GS}, measured at $V_{DS} = -1$ V. From the transfer curve, the on- and off-currents are defined as the drain current at the on- and off-states of the transistor, and the on/off ratio is the ratio of these. For the random CNT network transistors, the maximum on/off ratio has been reported to exceed 10^5 [22], which is smaller than the ratio of transistors with individual CNTs as an active channel. The ratio of random network transistors depends on the channel material and on the channel length, and is limited by the presence of metallic tubes in the network, which results in a high off-current.

The transconductance g_m is defined by Equation 42.3. Using Equation 42.4, the field-effect mobility μ_{FE} can be derived from g_m [41,42]. The carrier mobility determines the carrier velocity, and hence the switching speed [11]. Generally, the field-effect mobility μ_{FE} is smaller than the effective mobility μ_{eff} for the same device, although μ_{FE} is commonly used [41]. The gate capacitance per unit area C_{ox} can be derived from two models. The parallel plate model treats the active channel material as a uniform sheet and depends only on the dielectric material (oxide in this case) thickness t_{ox}, and the dielectric constant ε_{ox}, as given in Equation 42.5. For SiO$_2$, the oxide dielectric constant $\varepsilon_{ox} = 3.9$, so that the gate capacitance of the 200 nm SiO$_2$ layer is calculated to be $C_{ox} = 17.26$ nF/cm^2. Equation 42.6 is a modified array model, which includes the influence of the average spacing between nanotubes, defined as Λ_0 [43]. The quantum capacitance of CNT is $C_Q = 4.0 \times 10^{-10}$ F/m [21]. R is the mean radius of the CNTs. As the nanotube density increases, the modified array model approaches the parallel plate model. At sufficient high nanotube density, the difference between the two models is negligible.

$$g_d = \left. \frac{\partial I_D}{\partial V_{DS}} \right|_{V_{GS}=\text{const}} \tag{42.1}$$

$$\mu_{eff} = \frac{L \cdot g_d}{W \cdot C_{ox} \cdot (V_{GS} - V_{th})} \tag{42.2}$$

$$g_\mathrm{m} = \left.\frac{\partial I_\mathrm{D}}{\partial V_\mathrm{GS}}\right|_{V_\mathrm{DS}=\mathrm{const}} \tag{42.3}$$

$$\mu_\mathrm{FE} = \frac{L \cdot g_\mathrm{m}}{W \cdot V_\mathrm{DS} \cdot C_\mathrm{ox}} \tag{42.4}$$

$$C_\mathrm{ox} = \frac{\varepsilon_0 \cdot \varepsilon_\mathrm{ox}}{t_\mathrm{ox}} \tag{42.5}$$

$$C_\mathrm{ox} = \left\{ C_\mathrm{Q}^{-1} + \frac{1}{2\pi \cdot \varepsilon_0 \cdot \varepsilon_\mathrm{ox}} \cdot \ln\left[\frac{\Lambda_0}{R} \cdot \frac{\sinh\left(2\pi \cdot t_\mathrm{ox}/\Lambda_0\right)}{\pi}\right] \right\}^{-1} \cdot \Lambda_0^{-1} \tag{42.6}$$

42.2.3 COMPARING PERFORMANCE OF RANDOM NETWORK TRANSISTORS

In our previous study [33], we built various kinds of random CNT network transistors and studied the effect on performance metrics such as the on/off ratio and field-effect mobility of device parameters such as the network density, the band gap, and the degree of semiconductor enrichment. Single-walled CNTs were purchased from SWeNT (90%-sc, mean diameter 0.8 nm) and from NanoIntegris (98%-sc, mean diameter 1.4 nm). CNTs were dissolved in NMP without additional surfactants. Random nanotube networks were then constructed by drop-casting nanotube suspension on the Si/SiO$_2$ substrate. The network density was controlled by varying the suspension concentration from 1.25 to 10 μg/mL, as well as by multiple casting of the 10 μg/mL suspension to achieve high network density.

Figure 42.3 summarizes the results of the performance comparison. Generally, the on/off ratio decreases with increasing network density due to the increase of metallic content in the nanotube network, which leads to a more rapid rise of the off-current than the on-current. The field-effect mobility increases with increasing network density, although this trend saturates above a certain density threshold, as shown in Figure 42.3. This is presumably due to the preferential charge transport through the metallic nanotube network at high density, which dilutes the contribution of the

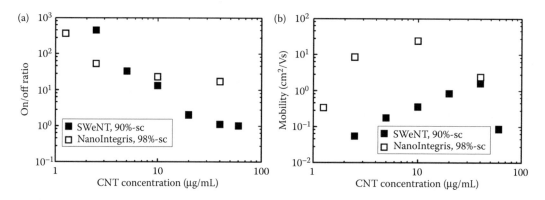

FIGURE 42.3 Performance (a) on/off ratio; (b) field-effect mobility (μ_FE) of the SWeNT (90%-sc) and NanoIntegris (98%-sc) CNTs-based field-effect transistors versus network density. The network density is parameterized by the concentration of the CNT/NMP suspension, from which the CNT networks were deposited. (From Q. Gong et al., *11th Proc. IEEE-Nano*, Portland, pp. 378–381. Copyright 2011 IEEE, with permission.)

semiconducting network to the channel conductance, with a corresponding reduction in the trans-conductance of the channel and in the field-effect mobility.

The band gap of semiconducting CNTs scales inversely with the tube diameter [44], so that the SWeNT CNTs have a larger band gap than the NanoIntegris nanotubes. The large band gap results in a lower conductivity of the semiconducting nanotubes and, consequently, a lower charge mobility, as shown in Figure 42.3. Owing to the much higher metallic content (10% metallic), the SWeNT nanotube-based network transistors have a substantially lower on/off ratio than the NanoIntegris (2% metallic) nanotube-based network transistors for the high network density devices. At a low network density, semiconducting nanotubes dominate the charge transport of both sets of devices. The higher on/off ratio of the low-density SWeNT devices than NanoIntegris devices is due to the larger band gap of the SWeNT nanotubes.

42.2.4 Simulating Nanotube Networks with a Percolation Model

To understand the correlation between the electrical characteristics and the geometrical properties of CNT networks, a Monte Carlo-based method can be applied to simulate the networks of randomly distributed single-walled CNT networks [45,46]. The generated thin films are confined to a 3D box of dimensions $L \times W \times H$, where $L, W,$ and H are the length, width, and height. The CNTs within this volume are modeled as cylinders with length L_{CNT} and diameter d_{CNT}. The electrodes are represented by metallic tubes positioned in the two opposite defined regions of the 3D box (Figure 42.4).

The orientation of CNTs inside the defined volume is determined by two angles $\alpha \in [0, \pi]$, the angle between the nanotube axis and the plane of the active channel, and $\beta \in [0, \pi]$, the angle between the nanotube axis and the direction of current flow. The length of each CNT as well as its diameter are sampled from a log-normal distribution [47–49]. The CNTs are generated and placed randomly inside the given volume so that at least one end of each nanotube lies inside the 3D box. CNTs are randomly generated until a given total tube number or a desired density is reached. For each network, a finite range of chiral indices (n,m) can be specified for the generated CNTs. CNTs are either metallic or semiconducting, depending on the chiral indices. After network generation, redundant nanotubes which do not contribute to a percolation path (a continuous chain of nanotubes) between the electrodes are removed. Next, the tube–tube junctions are calculated. The junction points between two CNTs can be metallic–metallic (MM), semiconducting–semiconducting (SS), or metallic–semiconducting (MS). The resistance of conducting CNTs is about 6 kΩ/μm of nanotube length. The resistance of metallic nanotubes can be calculated using Equation 42.7 with a mean tube length $L_{CNT} = 1$ μm [47].

$$R_{CNT} = \frac{h}{4e^2}(1 + L_{CNT}) \tag{42.7}$$

The resistances of intertube junctions are considered to follow a normal distribution. The homogeneous junctions (MM- or SS-type) have a resistance that ranges between a minimum value of 180 kΩ and a maximum value of 420 kΩ. The resistances of heterogeneous junctions (MS-type) are considered to be a hundred times larger than homogeneous junctions, between 18 and 42 MΩ, as reported in Ref. [48]. After generation of the nanotube network and determination of the junction resistances, the network is converted to a simulation program with integrated circuit emphasis (SPICE)-compatiable circuit file and the total current across the thin film is calculated with HSPICE.

Figure 42.4 shows the conductance of simulated nanotube networks as a function of network density. The channel length L was fixed and two channel widths, $W = L$ and $W = 2L$, were modeled. The increase of the semiconducting content in the nanotube network from 70% (as-grown) to 90% (separated), with the same channel geometry, leads to a decreased off-current, indicated by the reduced conductance at $V_{GS} = 0$ V. The conductance increases with increasing network density, as

(a)

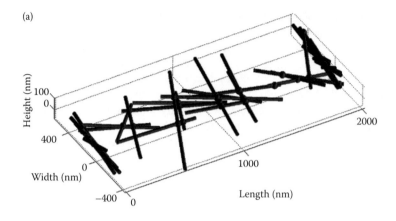

(b)

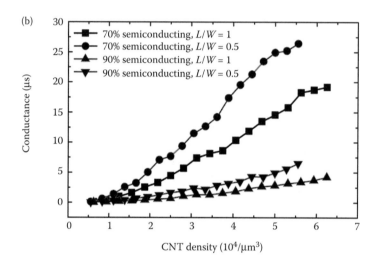

FIGURE 42.4 (a) An example of a randomly generated 3D CNT network. Metallic nanotubes were randomly placed with defined content in the nanotube network. (b) Comparison of conductance at different nanotube densities for random CNT networks with 70% and 90% semiconducting contents with channel geometries: $L/W = 0.5$ and 1. $V_{DS} = -5$ V and $V_{GS} = 0$ V. (From Q. Gong et al., *11th Proc. IEEE-Nano*, Portland, pp. 378–381. Copyright 2011 IEEE, with permission.)

more nanotubes contribute to the formation of percolation paths between the electrodes. Decreasing the channel length significantly increases the probability of forming percolation paths between electrodes. For an optimum design, the channel length should be short enough to allow the formation of semiconducting percolating paths with a small amount of nanotubes in order to keep the metallic network sparse, while keeping a minimum length in order to minimize the influence of metallic paths.

42.3 OUTLOOK AND CHALLENGES

Single-walled CNTs are promising novel materials with superior electronic, mechanical, and optical properties. Solution-processable random CNT networks are easy to manipulate. They can be produced at ambient pressure and room temperature with a low cost and high yield. However, the performance of random CNT network transistors is so far not competitive with traditional silicon-based devices. The potential application of random carbon networks may first be realized in novel devices such as chemical biosensors, because the cylindrical nanometer scale structure of CNTs can be readily attached by molecules. The chemical inertia, the stability in the ambient environment, and the sensitivity of CNTs enable reliable and sensitive sensing processes. Another possibility may be in ultra-high-frequency applications due to the ballistic transport inside CNTs. Recently, random CNT network transistors have been integrated in flexible stretchable circuits, with potential applications in electronic textiles and artificial skin.

There are several challenges for future applications of CNT devices. Purer and more homogeneous nanotube samples with narrow chirality and diameter distributions, high purity of electronic type (semiconducting vs. metallic), and precise doping are needed. Scalable precision methods are needed for oriented nanotube network deposition with controlled density and alignment. Finally, efficient gate dielectric and gate metal, and low resistance contact are demanded for high-performance electronics.

ACKNOWLEDGMENT

This work was sponsored by the German Research Foundation (DFG) through the TUM International Graduate School of Science and Engineering (IGSSE) and the TUM Institute for Advanced Studies (IAS).

REFERENCES

1. S. Iijima and T. Ichihashi, Single-shell carbon nanotubes of 1-nm diameter, *Nature*, 363, 603–605, 1993.
2. P. Avouris, Z. Chen, and V. Perebeinos, Carbon-based electronics, *Nat. Nanotechnol.*, 2, 605–615, 2007.
3. M. Treacy, T. Ebbesen, and J. Gibson, Exceptionally high Young's modulus observed for individual carbon nanotubes, *Nature*, 381, 678–680, 1996.
4. Z. Wu, Z. Chen, X. Du, J. Logan, J. Sippel, M. Nikolou, K. Kamaras, J. Reynolds, D. Tanner, A. Hebard, and A. Rinzler, Transparent, conductive carbon nanotube films, *Science*, 305, 1273–1276, 2004.
5. J. Wildöer, L. Venema, A. Rinzler, R. Smalley, and C. Dekker, Electronic structure of atomically resolved carbon nanotubes, *Nature*, 391, 59–62, 1998.
6. K. Yanagi, Y. Miyata, T. Tanaka, S. Fujii, D. Nishide, and H. Kataura, Colors of carbon nanotubes, *Diam. Relat. Mater.*, 18, 935–939, 2009.
7. L. Huang, H. Zhang, B. Wu, Y. Liu, D. Wei, J. Chen, Y. Xue, G. Yu, H. Kajiura, and Y. Li, A generalized method for evaluating the metallic-to-semiconducting ratio of separated single-walled carbon nanotubes by UV-vis-NIR characterization, *J. Phys. Chem. C.*, 114, 12095–12098, 2010.
8. P. Collins, M. Arnold, and P. Avouris, Engineering carbon nanotubes and nanotube circuits using electrical breakdown, *Science*, 292, 706–709, 2001.
9. M. Arnold, A. Green, J. Hulvat, S. Stupp, and M. Hersam, Sorting carbon nanotubes by electronic structure using density differentiation, *Nat. Nanotechnol.*, 1, 60–65, 2006.
10. A. Javey, J. Guo, Q. Wang, M. Lundstrom, and H. Dai, Ballistic carbon nanotube field-effect transistors, *Nature*, 424, 654–657, 2003.
11. T. Dürkop, S. Getty, E. Cobas, and M. Fuhrer, Extraordinary mobility in semiconducting carbon nanotubes, *Nano Lett.*, 4, 35–39, 2004.
12. C. Rutherglen, D. Jain, and P. Burke, Nanotube electronics for radiofrequency applications, *Nat. Nanotechnol.*, 4, 811–819, 2009.
13. L. Nougaret, H. Happy, G. Dambrine, V. Derycke, J. Bourgoin, A. Green, and M. Hersam, 80 GHz field-effect transistors produced using high purity semiconducting single-walled carbon nanotubes, *Appl. Phys. Lett.*, 94, 243505, 2009.
14. C. Wang, A. Badmaev, A. Jooyaie, M. Bao, K. Wang, K. Galatsis, and C. Zhou, Radio frequency and linearity performance of transistors using high-purity semiconducting carbon nanotubes, *ACS Nano*, 5, 4169–4176, 2011.
15. J. Novak, E. Snow, E. Houser, D. Park, J. Stepnouski, and R. McGill, Nerve agent detection using networks of single-walled carbon nanotubes, *Appl. Phys. Lett.*, 83, 4026–4028, 2003.
16. J. Li, Y. Lu, Q. Ye, M. Cinke, J. Han, and M. Meyyappan, Carbon nanotube sensors for gas and organic vapor detection, *Nano Lett.*, 3, 929–933, 2003.
17. G. Gruner, Carbon nanotube transistors for biosensing applications, *Anal. Bioanal. Chem.*, 384, 322–335, 2006.

18. Q. Cao, H. Kim, N. Pimparkar, J. Kulkarni, C. Wang, M. Shim, K. Roy, M. Alam, and J. Rogers, Medium-scale carbon nanotube thin-film integrated circuits on flexible plastic substrates, *Nature*, 454, 495–500, 2008.
19. M. Jung, J. Kim, J. Noh, C. Lim, G. Lee, J. Kim, H. Kang et al., All-printed and roll-to-roll-printable 13.56-MHz-operated 1-bit RF tag on plastic foils, *IEEE T. Electron. Dev.*, 57, 571–580, 2010.
20. S. Kim, S. Kim, J. Park, S. Ju, and S. Mohammadi, Fully transparent pixel circuits driven by random network carbon nanotube transistor circuitry, *ACS Nano*, 4, 2994–2998, 2010.
21. T. Takahashi, K. Takei, A. Gillies, R. Fearing, and A. Javey, Carbon nanotube active-matrix backplanes for conformal electronics and sensors, *Nano Lett.*, 11, 5408–5413, 2011.
22. D. Sun, M. Timmermans, Y. Tian, A. Nasibulin, E. Kauppinen, S. Kishimoto, T. Mizutani, and Y. Ohno, Flexible high-performance carbon nanotube integrated circuits, *Nat. Nanotechnol.*, 6, 156–161, 2011.
23. H. Dai, Nanotube growth and characterization, *Topics Appl. Phys.*, 80, 29–53, 2001.
24. C. Journet, W. Maser, P. Bernier, A. Loiseau, M. de la Chapelle, S. Lefrant, P. Deniard, R. Lee, and J. Fisher, Large-scale production of single-walled carbon nanotubes by the electric-arc technique, *Nature*, 388, 756–758, 1997.
25. C. Scott, S. Arepalli, P. Nikolaev, and R. Smalley, Growth mechanisms for single-wall carbon nanotubes in a laser-ablation process, *Appl. Phys. A*, 72, 573–580, 2001.
26. S. Bachilo, L. Balzano, J. Herrera, F. Pompeo, D. Resasco, and R. Weisman, Narrow (n,m)-distribution of single-walled carbon nanotubes grown using a solid supported catalyst, *J. Am. Chem. Soc.*, 125, 11186–11187, 2003.
27. J. Bahr, E. Mickelson, M. Bronikowski, R. Smalley, and J. Tour, Dissolution of small diameter single-wall carbon nanotubes in organic solvents, *Chem. Commun.*, 2, 193–194, 2001.
28. R. Haggenmueller, S. Rahatekar, J. Fagan, J. Chun, M. Becker, R. Naik, T. Krauss et al., Comparison of the quality of aqueous dispersions of single wall carbon nanotubes using surfactants and biomolecules, *Langmuir*, 24, 5070–5078, 2008.
29. D. Jain, N. Rouhi, C. Rutherglen, C. Densmore, S. Doorn, and P. Burke, Effects of source, surfactant, and deposition process on electronic properties of nanotube arrays, *J. Nanomater.*, 2011, 174268, 2010.
30. R. Krupke, S. Malik, H. Weber, O. Hampe, M. Kappes, and H. Löhneysen, Patterning and visualizing self-assembled monolayers with low-energy electrons, *Nano Lett.*, 2, 1161–1164, 2002.
31. M. Vosgueritchian, M. LeMieux, D. Dodge, and Z. Bao, Effect of surface chemistry on electronic properties of carbon nanotube network thin film transistors, *ACS Nano*, 4, 6137–6145, 2010.
32. F. Krebs, Fabrication and processing of polymer solar cells: A review of printing and coating techniques, *Sol. Energ. Mat. Sol. C.*, 93, 394–412, 2009.
33. Q. Gong, E. Albert, B. Fabel, A. Abdellah, M. Chan-Park, P. Lugli, and G. Scarpa, Solution-processable random carbon nanotube networks for thin-film transistors, in *Proc. 11th IEEE-Nano*, Portland, pp. 378–381, 2011.
34. M. Roberts, M. LeMieux, A. Sokolov, and Z. Bao, Self-sorted nanotube networks on polymer dielectrics for low-voltage thin-film transistors, *Nano Lett.*, 9, 2526–2531, 2009.
35. E. Bekyarova, M. Itkis, N. Cabrera, B. Zhao, A. Yu, J. Gao, and R. Haddon, Electronic properties of single-walled carbon nanotube networks, *J. Am. Chem. Soc.*, 127, 5990–5995, 2005.
36. W. Wong and A. Salleo, *Flexible Electronics: Materials and Applications*. New York, NY: Springer, 2009, Chapter 10.
37. P. Chen, Y. Fu, R. Aminirad, C. Wang, J. Zhang, K. Wang, K. Galatsis, and C. Zhou, Fully printed separated carbon nanotube thin film transistor circuits and its application in organic light emitting diode control, *Nano Lett.*, 11, 5301–5308, 2011.
38. Y. Zhou, L. Hu, and G. Grüner, A method of printing carbon nanotube thin films, *Appl. Phys. Lett.*, 88, 123109, 2006.
39. S. Jang, H. Jang, Y. Lee, D. Suh, S. Baik, B. Hong, and J. Ahn, Flexible, transparent single-walled carbon nanotube transistors with graphene electrodes, *Nanotechnology*, 21, 425201, 2010.
40. Z. Chen, J. Appenzeller, J. Knoch, Y. Lin, and P. Avouris, The role of metal-nanotube contact in the performance of carbon nanotube field-effect transistors, *Nano Lett.*, 5, 1497–1502, 2005.
41. D. Schroder, *Semiconductor Material and Device Characterization*. New York, NY: John Wiley & Sons, 1998, Chapter 8.
42. F. Schwierz, Graphene transistors, *Nat. Nanotechnol.*, 5, 487–496, 2010.
43. Q. Cao, M. Xia, C. Kocabas, M. Shim, and J. Rogers, Gate capacitance coupling of single-walled carbon nanotube thin-film transistors, *Appl. Phys. Lett.*, 90, 023516, 2007.
44. M. O'Connell, *Carbon Nanotubes: Properties and Applications*. New York, NY: Taylor & Francis Group, 2006, Chapter 4.

45. A. Behnam and A. Ural, Computational study of geometry-dependent resistivity scaling in single-walled carbon nanotube films, *Phys. Rev. B*, 75, 125432, 2007.
46. J. Hicks, A. Behnam, and A. Ural, Resistivity in percolation networks of one dimensional elements with a length distribution, *Phys. Rev. E*, 79, 012102, 2009.
47. L. Hu, D. Hecht, and G. Gruener, Carbon nanotube thin films: Fabrication, properties, and applications, *Chem. Rev.*, 110, 5790–5844, 2010.
48. D. Jack, C. Yeh, Z. Liang, S. Li, J. Park, and J. Fielding, Electrical conductivity modeling and experimental study of densely packed SWCNT networks, *Nanotechnology*, 21, 195703, 2010.
49. R. Tenent, T. Barnes, J. Bergeson, A. Ferguson, B. To, L. Gedvilas, M. Heben, and J. Blackburn, Ultrasmooth, large-Area, high-uniformity, conductive transparent single-walled-carbon-nanotube films for photovoltaics produced by ultrasonic spraying, *Adv. Mater.*, 21, 3210–3216, 2009.

43 Analysis of Yield Improvement Techniques for CNFET-Based Logic Gates

Rehman Ashraf, Malgorzata Chrzanowska-Jeske, and Siva G. Narendra

CONTENTS

43.1 INTRODUCTION

Carbon nanotubes are hollow cylinders in which carbon atoms are arranged in the honeycomb lattice [1]. They were first demonstrated by Bethune [2] and Iijima [3] in 1993. Carbon nanotubes have found numerous applications in nanoelectronics because of their superior electronic properties.

Single-walled carbon nanotube field-effect transistors (CNFETs) are promising candidates for future integrated circuits [1,4] because of their excellent properties, like the long scattering mean free path (MFP) >1 µm [5], resulting in near ballistic transport [6], high carrier mobilities ($10^3 \sim 10^4$ cm²/Vs) in semiconducting carbon nanotubes (CNTs) [7], and the easy integration of high-k dielectric material such as HfO_2 [8], or ZrO_2 [9], that improve gate electrostatics. Owing to the aforementioned properties, CNFETs have the potential to deliver higher performance and lower power as compared to FETs built in silicon technology [10,11].

The 1-D structure of metallic carbon nanotubes allows the electrons to travel without scattering for long distances. The mean free path of metallic CNTs is estimated to be 1000 nm [1], 25 times longer than the mean free path of copper interconnects that is equal to 40 nm at room temperature. Moreover, the current carrying capacity of metallic carbon nanotubes is almost 10^{10} (A/cm²) [12] and is several orders of magnitude larger than the current carrying capacity of copper interconnects. These potential advantages of metallic carbon nanotubes make them a suitable candidate for future interconnects as well as vertical vias.

On-chip capacitors are required in certain analog circuits and for decoupling purposes in digital circuits. Current integrated circuit technology uses metal–insulator–metal (MIM) and metal oxide semiconductor (MOS) capacitors as decoupling capacitors. However, the major problem with this approach is the small capacitance per unit area. Owing to of their low resistivity at nanoscale dimensions, carbon nanotubes have a great potential to be used as integrated capacitors in future integrated circuits [13]. Researchers have shown that the use of CNT-based integrated capacitors significantly increases the capacitance per unit area as well as the quality factors as compared to the capacitors fabricated with MIM and MOS technology [14,15].

Carbon nanotubes can also be used as on-chip inductors [16,17] because of their smaller footprint, higher drive current, and smaller curvatures. Recent research works have shown promising results for the use of CNTs as passive inductors in low-noise amplifiers (LNA) [18].

In addition to the excellent electronics properties, carbon nanotubes have excellent mechanical properties. Their mechanical strength makes them potential candidates for being used in flexible electronics. Various groups have reported the fabrication of individual CNFETs [19] and CNFET-based circuits [20] on flexible substrates with performance ranging from 40 MHz to 6 GHz.

The focus of this chapter is on the application of carbon nanotubes as a channel material for building CNFETs because of their potential of providing large performance advantage over Si complementary metal oxide semiconductor (CMOS) devices. Figure 43.1 shows a cross section and top view of a CNFET with an array of four single-wall (SW) CNTs used as a channel material. Carbon nanotubes are grown by chemical synthesis, and therefore it is very difficult to obtain the precise control on the exact positioning and chirality of CNTs during their growth. In this work, we have analyzed the impact of the unwanted growth of metallic tubes on the performance, power, and yield of CNFET-based circuits. Moreover, we propose solutions to build robust CNFET-based circuits with reduced variability in the performance and power in the presence of fabrication imperfections. It was reported that under ideal conditions, CNFETs are 13 times faster than a p-channel metal oxide semiconductor (PMOS) transistor and 6 times faster than an n-channel metal oxide semiconductor (NMOS) transistor [21] built in 32 nm technology node. To harness these performance advantages, the challenges listed below need to be overcome:

1. Variation in the diameter of tubes
2. Misalignment of CNTs during the fabrication of CNTs
3. The unwanted growth of metallic tubes

These carbon nanotube fabrication imperfections adversely impact the performance, power, and yield of circuits designed with carbon nanotube-based transistors. The variation in the diameter of tubes results in a variation in the drive current of CNFET, and therefore impacts the performance of CNFET-based circuits. The lack of precise control on the positioning of CNTs during the fabrication of CNFETs can result in a misalignment of the tubes [22]. Significant progress has been made in the fabrication of aligned CNTs, and less than 0.5% of CNTs fabricated on the single-crystal quartz substrate are misaligned [23]. The misaligned tubes can cause either a short between the output and the supply rail, or an incorrect logic function. Recently, efficient layout techniques have been proposed to tackle this problem [24]. The presence of metallic tubes creates an ohmic short between the source and

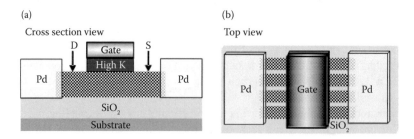

FIGURE 43.1 (a) Cross section of CNFET. (b) Top view of CNFET with an array of four CNTs. (R. Ashraf et al., Analysis of yield improvement techniques for CNFET-based logic gates, *2011 11th IEEE Conference on Nanotechnology (IEEE-NANO)*, August 15–18, pp. 724–729. © (2011) IEEE. With permission.)

drain of a CNFET and thus has a detrimental impact on the delay, static power, and functional yield of logic gates. Current CNT synthesis techniques produce between 4% and 40% [25,26] metallic tubes.

In this chapter, our focus is on analyzing the impact of diameter variation and the presence of metallic tubes on the functional yield of CNFET-based logic gates. The percentage of metallic tubes falls roughly into two ranges: a range for which robust CNT-based circuits can be implemented with circuit-level techniques, and a range for which extra processing techniques are required to remove the metallic tubes. We will analyze and identify these ranges.

43.2 BACKGROUND

As mentioned in the previous section, the major challenges faced by the CNFET technology are variation in the diameter of tubes, misalignment of CNTs during the fabrication of CNFETs, and the unwanted growth of metallic tubes. The diameters of typically fabricated CNTs show a Gaussian distribution [27] and vary between 1 and 2 nm [28]. The diameter of the tubes impacts the bandgap of CNTs, which then impacts the drive current of CNFETs. Tubes with a smaller diameter have a large bandgap and therefore smaller *ON* and *OFF* currents, whereas tubes with a larger diameter have higher *ON* and *OFF* currents. Therefore, a variation in the diameter of tubes leads to a variation in the performance (delay) of CNFETs. To reduce this variation and increase the drive current, researchers have proposed to create multiple parallel transport paths [23] by using an array of densely packed CNTs as a channel material.

Monte Carlo simulations are performed to obtain the sigma-to-mean I_{ON} and I_{OFF} currents as a function of the number of parallel CNTs (N_{tur}) in a transistor with the assumption that all the tubes are present and semiconducting. A sample size (n) of 10,000 transistors is used for Monte Carlo simulations. Simulation results showed that the σ/μ variation in the *ON* and *OFF* currents decreases by increasing the number of tubes in the transistor. The maximum variation obtained in the *ON* and *OFF* currents is almost 10%, and 3.5 times and this is obtained for $N_{tur} = 1$. Similarly, the minimum variation obtained in the *ON* and *OFF* currents is almost 2% and 0.5 times and this is obtained for $N_{tur} = 32$. This decrease in variations while increasing the number of parallel CNTs is due to the statistical averaging of currents among the multiple tubes of the transistors [29].

Although the grown CNTs are well aligned, a small percentage of CNTs can still end up being misaligned. The misalignment of tubes can cause incorrect logic functionality of gates. Recently, new layout techniques to design circuits such that misaligned tubes can be easily etched out have been proposed [24,30]. These techniques ensure the correct functionality of logic gates in the presence of misaligned tubes. The proposed layouts introduce small overheads of the area and delay. In this work, we assume that the misalignment of tubes is corrected by proper layout designs of CNFET-based gates, and therefore the tube misalignment problem is eliminated and the array of CNTs is perfectly aligned.

To use CNTs as the channel material, only semiconducting CNTs are required. However, as was mentioned earlier, depending on the chirality, an SW-CNT can be either metallic or semiconducting. At present, there is no CNT synthesis technique that can produce 100% semiconducting tubes. Various existing metallic tube removal techniques were proposed but none of them can eliminate metallic tubes completely. Therefore, we have to consider the presence of metallic tubes in tube arrays used as channels. In the case of a metallic tube being present, the gate terminal has no control over the channel due to an ohmic short between the source and drain. Therefore, the presence of metallic tubes in complementary CNFETs circuits has a detrimental impact on yield, static current, and, delay, and yield of CNT-based circuits. In the next section, we develop the methodology to analyze the impact of diameter variation and the presence of metallic tubes on the CNT-based circuits.

43.3 METHODOLOGY

The variations in diameter and spacing between tubes that result from the synthesis of CNTs mainly impact the performance and power consumption of CNT-based circuits. The presence of metallic tubes, however, creates shorts between the source and drain of transistors and reduces yield as well as leading to increased delay and static power of CNFET-based gates. To develop modeling solutions for analyzing the delay and power of logic gates implemented with CNFETs, we first assume that all tubes used as CNFET channels are semiconducting. Next, we relax that assumption and develop a methodology to statistically analyze the impact of the presence of metallic tubes on the delay and power through Monte Carlo simulation.

43.3.1 SEMICONDUCTING CNTs

We analyze the delay and power of logic gates implemented with CNFETs by assuming all the CNTs are semiconducting and are of a fixed diameter of 1.5 nm. For simulation purposes, a 32 nm technology node and supply voltage of 0.9 V, projected by International Technology Roadmap for Semiconductors (ITRS) guidelines [31], are used. The energy and delay of CNFET-based logic gates are compared with the energy and delay of logic gates implemented in 32 nm Si-CMOS. To simulate the behavior of CNFET-based circuits, we used an HSpice-compatible model developed in Stanford [32]. For the computation of current versus voltages of CNFET-based circuits, the model considers the practical device nonidealities, such as quantum confinement effects, acoustic/optical phonon scattering, elastic scattering, resistance of the source and drain, the resistance of Schottky barrier, and parasitic gate capacitance. The circuit-compatible model allows simulating CNFET-based circuits with multiple parallel tubes as transistor channels and with a large range of tube diameters. The current versus voltage characteristics obtained from the circuit-compatible model [32] are in close agreement with the experimental CNFET data [33]. For Si-CMOS devices, we used the Spice-compatible model of 32 nm CMOS using BSIM4 predictive technology models [34]. The results show that logic gates implemented with CNFETs have a delay almost 5 times lower and an energy 5.5 times lower as compared to gates implemented in 32 nm silicon CMOS.

43.3.2 METALLIC CNTs ARE PRESENT

Currently, there is no CNT synthesis technique that can guarantee 100% semiconducting tubes. The presence of metallic tubes has a significant impact on the static current and the delay of CNT-based circuits. The complementary CNFET-based gates have pull-up and pull-down networks and the delay of the gate depends upon the current flowing in the *ON* network as well as the *OFF* network. The consequence of the presence of metallic tubes results is a large increase in the *OFF* current, which in turn increases the delay of the gate as shown in Equation 43.1:

$$D_g \propto \frac{C_L V_{DD}}{(I_{ON} - I_{OFF})} \tag{43.1}$$

Here, D_g is the delay of the logic gate, I_{ON} and I_{OFF} are the currents flowing in the *ON* and *OFF* networks, respectively, and C_L is the load capacitance of the gate. We use Monte Carlo simulations to generate and study the statistical distribution of I_{ON} and I_{OFF}. We assumed Gaussian distribution of the tube diameter variation [27] represented by distribution μ and 3σ of 1.5 and 0.5 nm, respectively. We obtain the *ON* and *OFF* currents for semiconducting tubes in a CNFET using the circuit-compatible model developed in Ref. [32].

The *ON* current (I_{onm}) of a metallic CNT in a CNFET is obtained from Equation 43.2 in which x_m is the ratio of *ON* currents of a metallic to a semiconducting tube. Since the metallic tube is always conducting, we can assume that the I_{offm} (*OFF* current of a metallic tube would be equal to I_{onm}. The increase in the *OFF* current due to the presence of metallic tubes also results in an increase in the static power.

$$I_{onm} = x_m I_{ons} \tag{43.2}$$

$$I_{offm} = x_{onm} \tag{43.3}$$

43.3.3 Metallic CNTs Are Being Removed

When a large percentage of metallic tubes is present, extra processing techniques must be used to remove these unwanted metallic tubes. The removal of tubes, however, increases the delay of CNFET-based gates, and results in large variability in the performance and power of CNFET-based devices. Furthermore, in the worst case, all tubes, if all are metallic, can be removed from a single transistor and an open-circuit gate is created. Two scalable processing techniques to remove unwanted metallic tubes from CNFET transistors were proposed in the literature: selective chemical etching (SCE) [35] and VLSI-compatible metallic carbon nanotube removal (VMR) [36]. Both these extra processing techniques remove almost all the metallic tubes, but at the same time, they also remove some of the needed semiconducting tubes. The removal of tubes produces a large variation in the performance of CNT-based circuits and in the worst case, as was mentioned earlier, it can create open-circuit gates. Moreover, a large static power can be expected if the removal process is not perfect and some metallic tubes remain.

43.3.4 Functional Yield

We consider a gate/circuit, in the presence of fabrication imperfection, to be functional if its delay and power are smaller than the maximum allowable delay and static power constraints we have defined. In this chapter, fabrication imperfections, include the presence of metallic tubes, and variations of tube diameters. The functional yield (Y_f) of logic gates is obtained as a function of the drive strength of the gates, percentage of metallic tubes, and percentage of tubes removed if the tube removal process is applied. The functional yield is then defined as the ratio of a number of functional gates to the total number of gates.

43.4 RESULTS

Monte Carlo simulations are used to generate the functional yield of CNFET-based logic gates, that is, INV and NAND, in the presence of CNT diameter variations, and for different percentages of metallic tubes initially present. It has been noticed that the correlation among CNTs used as transistor channels has a strong impact on the functional yield when metallic tubes are present, and when tubes are being removed. In this chapter, we analyze the impact of correlation among tubes on the functional yield of CNFET-based gates/circuits.

The normalized delay versus normalized static power distribution for the parallel tube (PT) NAND gate generated by Monte Carlo simulations is shown in Figure 43.2. The delay and power data are normalized with respect to only the semiconducting tube results. The number of tubes in the gate (N_{tug}) is 48. All CNFETs in the NAND gate transistors are implemented with uncorrelated tubes. A sample size (n) of 10,000 gates is used for Monte Carlo simulations. From the figure, it can be observed that for 10% metallic tubes, an extremely low functional yield is obtained. To increase the yield in the presence of metallic tubes, two new CNFET transistor configurations are proposed, that is, transistor stacking (TrS) and tube stacking (TuS). These configurations exploit the correlation among CNTs to increase the functional yield of logic gates in the presence of metallic tubes. Figure 43.3a shows the TrS configuration where two transistors with uncorrelated (different) parallel tubes are stacked through a common intermediate node that connects all paths between the power and the output. In Figure 43.3b, TuS configuration is shown in which each stacked parallel path from the output to the power is isolated from each other by not having a shared intermediate node. These stacking configurations, especially TuS, help to reduce the probability of an ohmic short between the power and the output. The main objective is to reduce the statistical probability of an ohmic short between the power and the output. In a TrS configuration, two transistors implemented with uncorrelated tubes are stacked on top of each other, and in a TuS configuration, each pair of uncorrelated tubes is stacked on top of each other without shared intermediate contact. To maintain

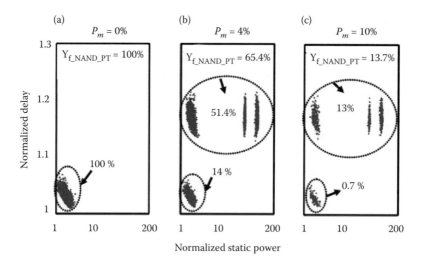

FIGURE 43.2 Monte Carlo simulation for a Parallel Tube (*PT*) NAND gate with $N_{tug} = 48$, showing normalized delay vs. static power for (a) absence of metallic tubes – $P_m = 0\%$, (b) $P_m = 4\%$ metallic tubes, and (c) $P_m = 10\%$ metallic tubes with the delay constraint of 1.3× and the static power constraint of 200×. (R. Ashraf et al., Analysis of yield improvement techniques for CNFET-based logic gates, *2011 11th IEEE Conference on Nanotechnology (IEEE-NANO)*, August 15–18, pp. 724–729. © (2011) IEEE. With permission.)

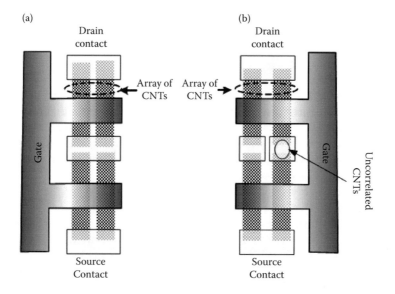

FIGURE 43.3 Tube configurations for (a) Transistor Stacking (*TrS*) and (b) Tube Stacking (*TuS*) CNFET. Both the configurations have the same number of channels to preserve the iso-input capacitance. (R. Ashraf et al., Analysis of yield improvement techniques for CNFET-based logic gates, *2011 11th IEEE Conference on Nanotechnology (IEEE-NANO)*, August 15–18, pp. 724–729. © (2011) IEEE. With permission)

iso-input capacitance, the total number of tubes is kept the same in both stacking configurations as it is in the *PT* configuration. The stacking configurations can result in up to 4 times performance penalty because the number of parallel channels in the stacking configurations is half of the *PT* configuration and there is also a decrease in drive strength due to the stacking of tubes or the stacking of transistors. However, the probability of an ohmic short between the drain and source of a transistor in the *TuS* configuration is lower than for the *TrS* configuration. The fabrication of *TuS* configuration, however, requires more precise control in terms of tube alignment and positioning of contacts. The probability of stack effect-based leakage reduction is higher in *TuS* since there is no node sharing as in *TrS*.

Table 43.1 summarizes the normalized mean delay, static power, and yield results generated by Monte Carlo simulations for an inverter and a two-input NAND gate with different transistor configurations. The percentage of metallic tubes, P_m, is assumed to be 4% and 10%. From the table, it can be observed that circuit-level techniques can be used to obtain acceptable functional yield if the percentage of metallic tubes is less than 5%. When the percentage of metallic tubes is relatively small below 5%, hybrid CNFET circuits built with mixtures of *PT* and *TrS* transistors achieve an acceptable yield.

When the percentage of metallic tubes is larger than 5%, the proposed circuit-level techniques are not sufficient. Extra processing steps to remove the metallic tubes are necessary. Both SCE and VMR techniques have trade-offs, as the removal process is not perfect and removes semiconducting tubes in addition to removing unwanted metallic tubes. Therefore, stochastic removal of tubes results in a large variation in the performance of CNFET-based gates, and in the worst case, it can create open-circuit gates. We have developed a Monte Carlo simulation engine to estimate the impact of the removal of tubes from driving gates on the performance and power of CNFET-based logic circuits.

In the analysis presented in Ref. [37], the authors assumed the fanout of the logic gates to be constant. However, this scenario is only applicable while driving the internal or external interconnect buses. In most other cases, gates will be driving other gates through local interconnects, and tubes removed from both the driving gates and fanout gates will impact the performance of the gates. If we consider the fanout to be constant, then it will result in the underestimation of the functional yield of logic gates. Therefore, in this chapter, we also discuss the effect of gate fanout on the functional yield of logic gates. It will be shown that fanout influence is especially important when extra processing steps are used to remove the finite numbers of tubes.

It has been shown previously that when metallic tubes are present, the use of highly uncorrelated (different) tubes among different transistors reduces the probability of an ohmic short, and increases the functional yield of logic gates. Now, if the extra processing steps such as SCE and VMR are used to remove the metallic tubes, the Monte Carlo results show that either of these techniques removes more than 99.9% of metallic tubes. The trade-off of using these removal techniques is large performance variation due to removal of the metallic tubes and also due to a side effect of undesired removal of semiconducting tubes. It is observed that when the tubes are removed, the use of highly uncorrelated tubes (different tubes) among different CNFETs results in a large variation in performance, and hence low functional yield. On the other hand, when the tube removal process removes almost all the metallic tubes, then use of highly correlated tubes results in less variation in

---

TABLE 43.1

Normalized Mean Delay, Normalized Static Power and Yield Comparisons for Inverter and 2-Input NAND Gate When Percentage of Metallic Tubes is 4% and 10%

		Inverter				NAND			
P_m	Configuration	N_{tug}	D_μ	SP_μ	$Y_f(\%)$	N_{tug}	D_μ	SP_μ	$Y_f(\%)$
4%	Parallel Tube	16	1.1	60.7	99.4	48	1.1	36.5	65.4
	Transistor Stacking		4.1	5.1	99.7		4.1	0.6	81.7
10%	Parallel Tube		1.1	99.6	66.0		1.2	55.3	13.6
	Transistor Stacking		4.1	0.7	77.3		4.1	0.7	35.1

Source: R. Ashraf et al., Analysis of yield improvement techniques for CNFET-based logic gates, *2011 11th IEEE Conference on Nanotechnology (IEEE-NANO)*, 15–18 August, pp. 724–729. © (2011) IEEE. With permission.

Note: N_{tug} is the number of tubes in the gate.

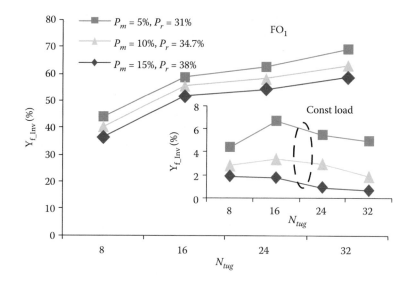

FIGURE 43.4 Summary of Monte Carlo yield simulation for inverters with a delay constraint of 1.3×, and a static power constraint of 10×. The yield is shown as a function of N_{tug}, P_m and P_r assuming a fanout of one (FO$_1$) and when tubes are removed from both the driving and fanout gates (realistic scenario). The inset graph shows the MC yield simulation for inverters with the same delay and static power constraints and with the constant load. (R. Ashraf et al., Analysis of yield improvement techniques for CNFET-based logic gates, *2011 11th IEEE Conference on Nanotechnology (IEEE-NANO)*, August 15–18, pp. 724–729. © (2011) IEEE. With permission.)

performance and higher functional yields of CNFET-based gates. Therefore, when a large fraction of tubes needs to be removed, we expect transistors to be implemented with highly correlated tubes.

Figure 43.4 shows a summary of the Monte Carlo yield simulation for inverters (with a delay constraint of 1.3 times and a static power constraint of 10 times) as a function of the number of tubes in the gate (N_{tug}), the percentage of metallic tubes (P_m) prior to the application of the tube removal process, and the percentage of metallic and semiconducting tubes removed (P_r). We consider the fanout of one, FO$_1$, and assume that tubes are removed from both the driving gate and the fanout (realistic scenario as discussed in the previous paragraph). A sample size (n) of 10,000 gates was used for all Monte Carlo simulations. The inset graph in Figure 43.4 shows the Monte Carlo yield simulated for inverters when no tubes are removed from the fanout gates. Please observe that extremely low yields are obtained when we consider the fanout to be constant.

From the data discussion in the previous paragraph, it can be concluded that the removal of tubes creates two main problems: (1) open-circuit transistors/gates when all tubes are removed from a transistor and (2) low functional yields because of the finite number of tubes removed from the drive and fanout gates. To design robust circuits when finite numbers of tubes are removed from the gates by extra processing steps, we propose tube-level redundancy (TLR) to reduce the probability of open-circuit gates and to improve a functional yield of logic gates.

Our first objective is to find the minimum number of tubes (N_{turmin}) required in a transistor prior to the tube removal process for less than a 0.001% probability of open-circuit CNFETs being created after the tube removal step. N_{turmin} can be calculated as shown in Equation 43.4.

$$N_{turmin} = \log\left(\frac{10^{-5}}{P_r}\right) \tag{43.4}$$

Table 43.2 shows the N_{turmin} required for a negligible probability of open-circuit transistors for different percentages of metallic tubes. Numbers in Table 43.2 are calculated by assuming that the

TABLE 43.2

Minimum Number of CNTs Required in a CNFET to Produce 0.001% Probability of Open Circuit Transistors

P_m	0%	5%	10%	15%	20%	25%	30%
N_{turmin}	1	8	9	10	11	12	13

Source: R. Ashraf et al., Analysis of yield improvement techniques for CNFET-based logic gates, *2011 11th IEEE Conference on Nanotechnology (IEEE-NANO)*, August 15–18, pp. 724–729. © (2011) IEEE. With permission.

SCE technique is applied to remove metallic tubes and the cutoff diameters for metallic and semi-conducting tubes are $D_{CS} = 1.4$ nm and $D_{CM} = 2$ nm, respectively. The same methodology can also be applied to the VMR technique.

An efficient TLR technique that allows us to obtain acceptable levels of yield without sacrificing too much area and power is proposed in this chapter. It is based on adding the redundant tubes with the objective of obtaining the same mean number of tubes in the CNFET after tube removal as are required by the design prior to the tube removal process. Table 43.3 shows the efficient redundancy estimation technique to increase the functional yield of gates (when finite numbers of tubes are removed) with a minimal impact on the area and energy. For example, as can be noticed in Table 43.3, if for $P_m = 10\%$ the number of tubes required in a CNFET, prior to tube removal, is 8, then after the tube removal process the mean number of tubes remaining is 5. However, if 13 tubes are present in the CNFET prior to the tube removal process, then the mean number of tubes after the tube removal process is 8 and is equal to the required number of tubes.

Figure 43.5 shows the Monte Carlo simulation results for NAND gates with the number of tubes, $N_{tug} = 48$. Figure 43.5b shows the functional yield results for $P_m = 10\%$ and $P_r = 35\%$ of the tubes

TABLE 43.3

Original CNTs in a CNFET before Tube Removal (*BTR*) and after Tube Removal (*ATR*) for Different Percentage of Metallic Tubes

P_m (%)	Original		Redundancy	
	CNTs BTR	μ CNTs ATR	CNTs BTR	μ CNTs ATR
5	8	5	12	8
10		5	13	
15		5	13	
5	16	11	24	16
10		10	25	
15		9	26	
5	32	22	47	32
10		20	49	
15		19	52	

Source: R. Ashraf et al., Analysis of yield improvement techniques for CNFET-based logic gates, *2011 11th IEEE Conference on Nanotechnology (IEEE-NANO)*, August 15–18, pp. 724–729. © (2011) IEEE. With permission.

Note: Number of CNTs required in a CNFET *BTR* that will produce the same mean CNTs in a CNFET after tube removal as are initially required by design *BTR*.

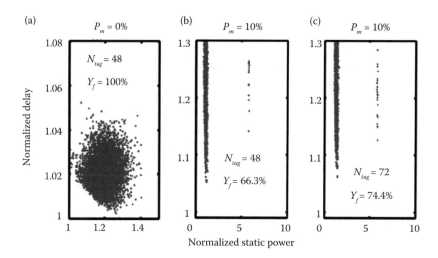

FIGURE 43.5 Monte Carlo simulation for NAND gates showing normalized delay vs. static power for (a) $N_{tug} = 48$, 0% metallic and no tube removal, (b) $N_{tug} = 48$, 10% metallic and 35% tubes are removed, (c) $N_{tugr} = 72$, 10% metallic and 35% tubes are removed. The yields are 100%, 66%, and 74%, respectively. (R. Ashraf et al., Analysis of yield improvement techniques for CNFET-based logic gates, *2011 11th IEEE Conference on Nanotechnology (IEEE-NANO)*, August 15–18, pp. 724–729. © (2011) IEEE. With permission.)

removed by the tube removal process. In such a case, only 66% of the gates are functional. By the redundancy-based methodology explained in the previous paragraph, 72 tubes, $N_{tugr} = 72$, are required in the gate to yield the mean number of 48 tubes after the tube removal process. This added redundancy of 50% increased the functional yield from 66% to 74%. By using this redundancy technique, the increase in area is 50% and the mean energy of the gate is increased by 17%.

Notice that the yield of 72% is still considered low, but it is anticipated that it would improve in large circuits, where we should consider multiple gates that are cascaded to form multistage logic networks. Depending on the logic depth of a logic network, a certain amount of statistical averaging in delay variation is expected and observed. Thus, less than 100% yield at the gate level might be sufficient to obtain an acceptable yield at the system level.

43.5 CONCLUSION

In this chapter, we explored the potential of CNTs to be used as a channel material in the future field-effect transistors for application in integrated circuits. We also showed that despite the detrimental impact of the unwanted metallic tubes on the performance, power, and yield of CNFET-based circuits, possible solutions to build robust CNFET-based circuits with an acceptable performance and a reasonable functional yield can be found. For small percentages of metallic tubes, circuit-level techniques such as *TrS* and *TuS* configurations can be used to build robust CNT-based circuits. For larger percentages of metallic tubes, robust CNT-based circuits, with acceptable functional yield, can be designed using the proposed TLR approach and fabrication by using extra processing techniques for metallic tube removal.

REFERENCES

1. P. L. McEuen, M. S. Fuhrer, and P. Hongkun, Single-walled carbon nanotube electronics, *IEEE Trans. Nanotechnol.*, 1(1), 78–85, Mar. 2002.
2. D. S. Bethune, C. H. Klang, M. S. de Vries, G. Gorman, R. Savoy, J. Vazquez, and R. Beyers, Cobalt-catalysed growth of carbon nanotubes with single-atomic-layer walls, *Nature*, 363(6430), 605–607, Jun. 1993.

3. S. Iijima and T. Ichihashi, Single-shell carbon nanotubes of 1-nm diameter, *Nature*, 363(6430), 603–605, Jun. 1993.
4. A. Raychowdhury, A. Keshavarzi, J. Kurtin, V. De, and K. Roy, Carbon Nanotube field-effect transistors for high-performance digital circuits-DC analysis and modeling toward optimum transistor structure, *IEEE Trans. Electron Devices*, 53(11), 2711–2717, Nov. 2006.
5. A. Javey, J. Guo, D. B. Farmer, Q. Wang, E. Yenilmez, R. G. Gordon, M. Lundstrom, and H. Dai, Self-aligned ballistic molecular transistors and electrically parallel nanotube arrays, *Nano Lett.*, 4(7), 1319–1322, Jul. 2004.
6. Z. Yao, C. L. Kane, and C. Dekker, High-field electrical transport in single-wall carbon nanotubes, *Phys. Rev. Lett.*, 84(13), 2941–2944, Mar. 2000.
7. T. Durkop, S. A. Getty, E. Cobas, and M. S. Fuhrer, Extraordinary mobility in semiconducting carbon nanotubes, *Nano Lett.*, 4(1), 35–39, 2004.
8. A. Javey, J. Guo, D. B. Farmer, Q. Wang, D. Wang, R. G. Gordon, M. Lundstrom, and H. Dai, Carbon nanotube field-effect transistors with integrated ohmic contacts and high-k gate dielectrics, *Nano Lett.*, 4(3), 447–450, Mar. 2004.
9. A. Javey, H. Kim, M. Brink, Q. Wang, A. Ural, J. Guo, P. McIntyre, P. McEuen, M. Lundstrom, and H. Dai, High-K dielectrics for advanced carbon-nanotube transistors and logic gates, *Nat. Mater.*, 1(4), 241–246, Dec. 2002.
10. J. Guo, S. Datta, M. Lundstrom, M. Brink, P. McEuen, A. Javey, H. Dai, H. Kim, and P. McIntyre, Assessment of silicon MOS and carbon nanotube FET performance limits using a general theory of Ballistic transistors, in *Proc. Int. Electron Devices Meet.*, 2002, pp. 711–714.
11. J. Guo, A. Javey, H. Dai, and M. Lundstrom, Performance analysis and design optimization of near Ballistic carbon nanotube field-effect transistors, in *Proc. Int. Electron Devices Meet.*, 2004, pp. 703–706.
12. B. Q. Wei, R. Vajtai, and P. M. Ajayan, Reliability and current carrying capacity of carbon nanotubes, *Appl. Phys. Lett.*, 79(8), 1172–1174, 2001.
13. A. Nieuwoudt and Y. Massoud, Evaluating the impact of resistance in carbon nanotube bundles for VLSI interconnect using diameter-dependent modeling techniques, *IEEE Trans. Electron Devices*, 53(10), 2460–2466, 2006.
14. M. Budnik, A. Raychowdhury, A. Bansal, and K. Roy, A high density, carbon nanotube capacitor for decoupling applications, in *Proc. ACM/IEEE Design Automat. Conf.*, 2006, pp. 935–938.
15. A. Nieuwoudt and Y. Massoud, High density integrated capacitors using multi-walled carbon nanotubes, in *IEEE Conf. Nanotechnol.*, 2007, pp. 387–390.
16. K. Tsubaki, H. Shioya, J. Ono, Y. Nakajima, T. Hanajiri, and H. Yamaguchi, Large magnetic field induced by carbon nanotube current proposal of carbon nanotube inductors, in *Device Res. Conf. Digest*, 2005, vol. 1, pp. 119–120.
17. K. Tsubaki, Y. Nakajima, T. Hanajiri, and H. Yamaguchi, Proposal of carbon nanotube inductors, *J. Phys. Conf. Ser.*, 38, 49–52, May 2006.
18. A. Nieuwoudt and Y. Massoud, Carbon nanotube bundle-based low loss integrated inductors, in *IEEE Conf. Nanotechnol.*, 2007, pp. 714–718.
19. N. Chimot, V. Derycke, M. F. Goffman, J. P. Bourgoin, H. Happy, and G. Dambrine, Gigahertz frequency flexible carbon nanotube transistors, *Appl. Phys. Lett.*, 91(15), 153111–3, Oct. 2007.
20. Q. Cao, H. Kim, N. Pimparkar, J. P. Kulkarni, C. Wang, M. Shim, K. Roy, M. A. Alam, and J. A. Rogers, Medium-scale carbon nanotube thin-film integrated circuits on flexible plastic substrates, *Nature*, 454(7203), 495–500, Jul. 2008.
21. J. Deng, N. Patil, K. Ryu, A. Badmaev, C. Zhou, S. Mitra, and H.-S. P. Wong, Carbon nanotube transistor circuits: Circuit-level performance benchmarking and design options for living with imperfections, in *Proc. Int. Solid State Circuits Conf.*, 2007, pp. 70–588.
22. C. Kocabas, M. Shim, and J. A. Rogers, Spatially selective guided growth of high-coverage arrays and random networks of single-walled carbon nanotubes and their integration into electronic devices, *J. Am. Chem. Soc.*, 128(14), 4540–4541, Apr. 2006.
23. S. J. Kang, C. Kocabas, T. Ozel, M. Shim, N. Pimparkar, M. A. Alam, S. V. Rotkin, and J. A. Rogers, High-performance electronics using dense, perfectly aligned arrays of single-walled carbon nanotubes, *Nat. Nanotechnol.*, 2(4), 230–236, Apr. 2007.
24. Y. Li, D. Mann, M. Rolandi, W. Kim, A. Ural, S. Hung, A. Javey et al., Preferential growth of semiconducting single-walled carbon nanotubes by a plasma enhanced CVD method, *Nano Lett.*, 4(2), 317–321, 2004.
25. L. Qu, F. Du, and L. Dai, Preferential syntheses of semiconducting vertically aligned single-walled carbon nanotubes for direct use in FETs, *Nano Lett.*, 8(9), 2682–2687, Sep. 2008.

26. Y. C. Tseng, K. Phoa, D. Carlton, and J. Bokor, Effect of diameter variation in a large set of carbon nanotube transistors, *Nano Lett.*, 6(7), 1364–1368, Jul. 2006.
27. P. Avouris, J. Appenzeller, R. Martel, and S. J. Wind, Carbon nanotube electronics, *Proc. IEEE*, 91(11), 1772–1783, Nov. 2003.
28. A. Raychowdhury, V. K. De, J. Kurtin, S. Y. Borkar, K. Roy, and A. Keshavarzi, Variation tolerance in a multichannel carbon-nanotube transistor for high-speed digital circuits, *IEEE Trans. Electron Devices*, 56(3), 383–392, Mar. 2009.
29. N. Patil, J. Deng, H.-S. P. Wong, and S. Mitra, Automated design of Misaligned-carbon-nanotube-immune circuits, in *Proc. ACM/IEEE Design Automat. Conf.*, San Diego, CA, 2007, pp. 958–961.
30. S. Bobba, J. Zhang, A. Pullini, D. Atienza, H.-S. P. Wong, and G. De Micheli, Design of compact imperfection-immune CNFET layouts for standard-cell-based logic synthesis, in *Proc. Design, Automat. Test Eur.*, 2009.
31. http://www.itrs.net/Links/2009ITRS/Home2009.htm/, Emerging Research Devices.
32. J. Deng and H.-S. P. Wong, A circuit-compatible SPICE model for enhancement mode carbon nanotube field effect transistors, in *Int. Conf. Simulation Semiconductor Processes Devices, SISPAD*, 2006, pp. 166–169.
33. I. Amlani, J. Lewis, K. Lee, R. Zhang, J. Deng, and H.-S. P. Wong, First demonstration of AC gain from a single-walled carbon nanotube common-source amplifier, in *Proc. Int. Electron Devices Meet*, 2006, pp. 1–4.
34. W. Zhao and Y. Cao, New generation of predictive technology model for sub-45 nm design exploration, in *IEEE Trans. Electron Devices*, 2006, pp. 585–590.
35. G. Zhang, P. Qi, X. Wang, Y. Lu, X. Li, R. Tu, S. Bangsaruntip, D. Mann, L. Zhang, and H. Dai, Selective etching of metallic carbon nanotubes by gas-phase reaction, *Science*, 314(5801), 974–977, Nov. 2006.
36. N. Patil, A. Lin, J. Zhang, Hai Wei, K. Anderson, H.-S. P. Wong, and S. Mitra, VMR: VLSI-compatible metallic carbon nanotube removal for imperfection-immune cascaded multi-stage digital logic circuits using carbon nanotube FETs, in *Proc. Int. Electron Devices Meet*, Baltimore, MD, USA, 2009, pp. 1–4.
37. R. Ashraf, N. Rajeev, C. Malgorzata, and N. Siva G., Yield enhancement by tube redundancy in CNFET-based circuits, in *IEEE Int. Conf. Electron. Circuits Syst.*, 2010.
38. R. Ashraf, M. Chrzanowska-Jeske, and S.G. Narendra, Analysis of yield improvement techniques for CNFET-based logic gates, *2011 11th IEEE Conference on Nanotechnology (IEEE-NANO)*, August 15–18, 2011, pp. 724–729.

44 Low-Power and Metallic-CNT-Tolerant CNTFET SRAM Design

Zhe Zhang and José G. Delgado-Frias

CONTENTS

44.1 INTRODUCTION

For the past four decades, CMOS scaling has offered improved performance from one technology node to the next. However, as device scaling moves beyond the 32 nm node, significant technological challenges will be faced. Currently, two of the main challenges are the considerable increase of standby power dissipation and the increasing variability in device characteristics, which in turn affect the circuit and system reliability. The aforementioned challenges will become more prominent as CMOS scaling approaches atomic and quantum-mechanical physics boundaries [1].

Efforts to extend silicon scaling through innovations in materials and device structures continue. Fin field effect transistors (FinFETs) (with a number of demonstrated designs [2,3]) are expected to continue CMOS scaling [4]. One of the most important features of FinFETs is that, by having independent front and back gates, these gates can be biased differently to control the current and the device threshold voltage. This ability to control the threshold voltage variations offers a temporary means to manage the challenge of standby power dissipation. Gate lengths of 10 nm and below will be achievable with FinFETs. Variability in other device characteristics could be a hurdle. The inability to reduce gate insulator thickness will be one of the issues that would possibly prevent further channel length reductions.

The International Technology Roadmap for Semiconductors (ITRS) has identified seven postsilicon innovations that are likely candidates of the postsilicon CMOS era, with carbon-based nanoelectronics being recommended as the "Beyond CMOS" technology for accelerated development [2]. The electrical properties of carbon nanotube (CNT) offer the potential for molecular-scale electronics.

They conduct as either metal or semiconductor, depending on their chirality. Semiconducting tubes have an energy bandgap that is inversely proportional to the nanotube diameter [5]. A single-walled semiconducting CNT has a typical diameter of 1.4 nm giving an energy bandgap in the range of 0.5–0.65 eV [6]. These devices have unique electron 1-D transport properties. This, in turn, reduces phase space for scattering of the carriers and opens up the possibility of ballistic transport [7]. Power dissipation is low because of the reduction of carrier scattering. Owing to their superior transport properties, low voltage bias, and improved current density, semiconducting CNTs have potential application in future nanoelectronic systems. In addition to the electrical properties, CNTFETs are less sensitive to many process parameter variations compared to conventional MOSFETs [8]. Kim et al. [9] present two memory cells with six and eight transistors. It should be pointed out that the 8T cell presented in Ref. [9] is different than the one presented here. Kim et al. compare the performance of these circuits by evaluating them under CMOS (MOSFETs), FinFET, and CNTFET technologies at the 32 nm technology node. Their study shows that the CNTFET-based cells significantly outperform the MOSFET- and FinFET-based memory cells in the metrics of dynamic and leakage power, write and read delays, as well as in static noise margins (SNMs).

Memory modules are widely used in most digital/computer systems. Memory is accessed frequently in most systems; very high-performance systems do this every clock cycle. In microprocessors, the clock network power consumption due to memory devices (i.e., cache memory, register, and pipeline registers) accounts for 51% of the total power [7].

In this chapter, we present the design of static random access memory (SRAM) cells in CNTFET technology. This technology has great potential to replace bulk CMOS in the near future. In addition, it has a significant reduction in power requirements. This chapter has been organized as follows. A brief description of both standard 6T SRAM and a decoupled read and write 8T SRAM is presented in Section 44.2. The performance of CNTFET memory cells without metallic CNTs is presented in Section 44.3. The performance and parameter variation influences on CNTFET memory cells are presented in Section 44.4. Section 44.5 presents the metallic-CNT-tolerant technology and its influence on our memory design. Section 44.6 evaluates and optimizes the proposed cell design for both performance and functionality. Memory module and spare column schemes are discussed in Section 44.7 to further improve the reliability of large-sized memory. The influence of technology scaling is discussed in Section 44.8. Finally, in Section 44.9, some concluding remarks are presented.

44.2   SRAM CELLS

SRAM is a major component of digital systems such as microprocessors, reconfigurable hardware, and field programmable gate arrays, just to name a few. Fast memory access times and design for density have been two of the most important target design criteria for many years. However, with device scaling to achieve even faster designs, power supply voltages and device threshold voltages have scaled as well, leading to degradation of standby power.

44.2.1   Six-Transistor SRAM Cell

The six-transistor (6T) static memory cell shown in Figure 44.1 has been widely accepted as the standard memory cell. It is designed to achieve fast read times with the inclusion of sense amplifiers. The standard 6T cell requires that a logic value and its inverse be placed on the bit lines during a write operation. The word line (WL) is raised to logic 1 and the logic levels on the bit lines passed into the cross-coupled inverter pair. Device sizing for a CMOS-based cell is driven primarily by area and functional operation constraints; sizing must be carefully performed to enable the correct logic values to be transferred into the cell. Reading from the memory cell entails precharging the bit lines and then asserting logic 1 on the word line (Word-line). The complexity of this cell is in arriving at the appropriate device sizes for proper functionality. The transistors of the cross-coupled inverter must be sized such that the effort to overwrite a previously stored value does not impact the pulse width of the word line.

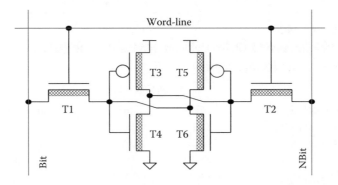

FIGURE 44.1 6T SRAM cell. (Z. Zhang and J.G. Delgado-Frias, Low power and metallic CNT tolerant CNTFET SRAM design, *2011 11th IEEE Conference on Nanotechnology (IEEE-NANO)*, Portland, OR, August 15–18, 2011, pp. 1177–1182, © (2011) IEEE. With permission.)

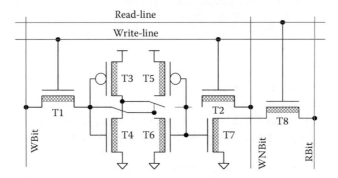

FIGURE 44.2 8T SRAM cell. (Z. Zhang and J.G. Delgado-Frias, Low power and metallic CNT tolerant CNTFET SRAM design, *2011 11th IEEE Conference on Nanotechnology (IEEE-NANO)*, Portland, OR, August 15–18, 2011, pp. 1177–1182, © (2011) IEEE. With permission.)

44.2.2 EIGHT-TRANSISTOR SRAM CELL

An eight-transistor (8T) memory cell is shown in Figure 44.2. This 8T SRAM cell has a similar structure as 6T with two additional transistors that decouple the read and write operations. The read operation is performed by setting the read-word line to logic 1; the additional transistors (T7 and T8) discharge the RBit line that has been precharged before the Read-line is set. The 8T basic cell provides a more orthogonal design; the read and write operations are performed by different transistors. Since the read operation does not affect the contents of the cell (the two back-to-back inverters), the worst-case SNM is simply that for two cross-coupled inverters [10]. For a 6T cell, on the other hand, the worst-case SNM occurs in the read condition.

44.3 PERFORMANCE OF CNTFET SRAM CELLS

Simulations have been performed using HSPICE and CNTFETs parameter models by Stanford University's Nanoelectronics Group [11–13]. The simulations assume memory arrays with 256 cells to evaluate performance with significant capacitive loading on the bit lines. The nominal chirality of the CNTs is (19, 0) and the pitch is 20 nm. The number of CNTs per transistor for the 6T SRAM cell is T1 (3), T2 (3), T3 (3), T4 (3), T5 (1), and T6 (1). For the 8T cell, the number of CNTs per transistor is T1 (6), T2 (6), T3 (1), T4 (1), T5 (1), T6 (1), T7 (12), and T8 (12). Table 44.1 shows the performance of the two SRAM cells in terms of access delays, energy, and power. From Table 44.1, it can be observed that the 8T cell has a lower write delay than the 6T cell. This is attributed to

TABLE 44.1

Performance of SRAM Cells on Read and Write Delay, Energy, and Static Power

Parameters	SRAM Cell		Comparison[a]
	6T	8T	
Write delay (ps)	3.06	2.46	−19.5%
Read delay (ps)	29.23	8.96	−69.4%
Max delay (ps)	29.23	8.96	−69.4%
Energy (fJ)	1.62	1.83	13.2%
Static power (nW)	0.185	0.110	−40.5%

Source: Z. Zhang and J.G. Delgado-Frias, Low power and metallic CNT tolerant CNTFET SRAM design, *2011 11th IEEE Conference on Nanotechnology (IEEE-NANO)*, Portland, OR, August 15–18, 2011, pp. 1177–1182, © (2011), IEEE. With permission

[a] Percentage comparison calculation: 100(8T value–6T value)/6T value.

having two ports (WBit and WNBit buses) that write simultaneously to the cell and smaller inverter transistors. The 8T cell read delay is significantly smaller than the 6T read delay (69.4%). Since the 6T cell has to discharge either the Bit or NBit bus, this makes it difficult to increase the current driving capability of the n-type transistors without greatly affecting the write delay. The energy and dynamic power for the 8T cell are slightly higher; these metrics are measured when the R-line needs to be precharged. If a 1 was read previously, this line is already at a high; this is not the case with 6T, which needs to precharge a line at every read. When a read occurs, the precharged bit lines affect either the stored bit or its inverse (the cross-coupled inverter inputs). Although this voltage change is not enough to change the state of the memory cell, the circuit consumes power to retain the proper voltage.

In a large memory array, static power becomes a dominant factor in power consumption. Static power has a multiplicative effect; all cells in the array are drawing this power. The 8T cell reduces this static power by 40% in simulations with nominal parameter values.

44.4 PERFORMANCE OF CNTFET SRAM CELLS UNDER PARAMETER VARIATIONS

Paul et al. [8] report that the process parameter that, when varying, affects CNTFETs the most is the diameter. When varied, the effective channel length, oxide thickness, dopant fluctuation, and so on show weak influence on the gate capacitance and drive current of CNTFETs. The device width is another parameter that influences the gate capacitance and drive current. The width of a CNTFET is given by the number of tubes per transistor. This number has been adjusted to achieve the proper operation of each memory cell. The device widths, that is, the number of CNTs per transistor, are not varied any further.

The impact of the power supply voltage, pitch, and temperature variations on 6T and 8T SRAM cells' delay, energy, and leakage power is reported in Refs. [14,15]. That study showed that a supply voltage reduction of 44.4% (from 0.9 to 0.5 V) decreased energy by 71.2% and 69.5% and increased delay by 65.4% and 168.6% for the 6T and 8T cells, respectively [14]. Temperature has no major impact on delay and energy. However, temperature has a great effect on leakage power by an order of magnitude. The static power goes from 1.1 and 0.76 nW (at 27°C) to 11.5 and 8.06 nW (at 100°C) for 6T and 8T SRAM cells, respectively.

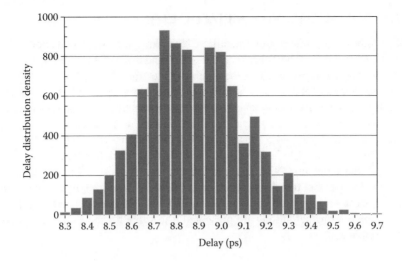

FIGURE 44.3 8T write delay distribution density with random diameter variation. (Z. Zhang and J.G. Delgado-Frias, Low power and metallic CNT tolerant CNTFET SRAM design, *2011 11th IEEE Conference on Nanotechnology (IEEE-NANO)*, Portland, OR, August 15–18, 2011, pp. 1177–1182, © (2011) IEEE. With permission.)

CNT diameter variations have a large impact on the current that a CNTFET can drive [8], affecting the SRAM cell's read and write delays. If the diameter of all the CNTFETs is set to be 10% smaller or 10% larger than the nominal diameter, the read delay is increased by 32.1% or decreased by 13.8%, respectively [15]. A smaller diameter than the nominal CNT diameter has a larger impact on delay. Figure 44.3 shows the write delay of RAM cells with randomly distributed diameters of ±10% of nominal; each memory cell has CNTs that have different diameters. This, in turn, is a more realistic simulation of diameter variations in a memory cell. Since smaller diameters have a larger effect on delay, the distribution is asymmetric with a larger tail toward longer delays. Thus, it is necessary to pay particular attention to the configuration of the circuit/transistor that might be most impacted by diameter variations.

Figure 44.4 shows how the static power of the cell is affected by diameter variations. A 16 × 16 SRAM array with nominal diameter leaks current of 0.763 nW with a 0.9 V power supply. The ±10% variation could cause static power change from −55.8% to 95.8%.

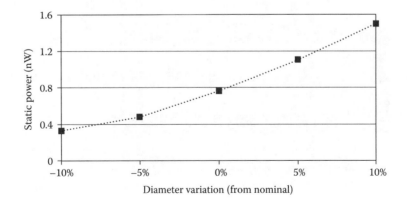

FIGURE 44.4 Diameter variation impact on static power. (Z. Zhang and J.G. Delgado-Frias, Low power and metallic CNT tolerant CNTFET SRAM design, *2011 11th IEEE Conference on Nanotechnology (IEEE-NANO)*, Portland, OR, August 15–18, 2011, pp. 1177–1182, © (2011) IEEE. With permission.)

44.5 CNTFET SRAM CELL WITH METALLIC CNTS

Current CNT synthesis techniques grow metallic CNTs along with semiconductor CNTs. Metallic CNTs essentially form resistive shorts between the source and drain of a CNTFET. It is expected that new techniques will be developed to reduce the number of metallic CNTs to less than 5%. For instance, Zhang et al. developed a novel process of selective etching that removes a large number of metallic CNTs [16]. In this study, we use a technique to tolerate these metallic CNTs that yield CNTFETs.

An array of series and parallel individual CNTFETs is used to tolerate metallic CNTs to obtain a semiconductor CNTFET. Figure 44.5 shows how a CNTFET transistor can be built using this array [17]. Each CNTFET has two CNTs. The 3×2 array of transistors has three transistors in series with uncorrelated (independent) CNTs; this array also has two transistors in parallel with correlated (identical) CNTs. The parallel transistors share common nodes, resulting in a compact layout.

Based on this structure, the probability of having a CNTFET is calculated [17]. Before getting the probability, some terms/definitions are presented (the value for the example in Figure 44.5 is given within parentheses).

P_{semi} = probability that the CNT is a semiconductor
n = number of CNTs per CNTFET ($n = 2$)
j = number of series CNTFETs ($j = 3$)
k = number of parallel CNTFETs ($k = 2$)

The probability of having an n-type CNTFET (where all the tubes are semiconductor CNTs) is

$$P_{CNTFET} = (P_{semi})^n \tag{44.1}$$

Having j CNTFETs in series with uncorrelated (independent) CNTs yields the series probability (a CNT can be either a semiconductor or metallic CNT; a semiconductor series is produced when at least one semiconductor CNT is present in each CNTFET).

$$P_{series} = 1 - (1 - P_{CNTFET})^j \tag{44.2}$$

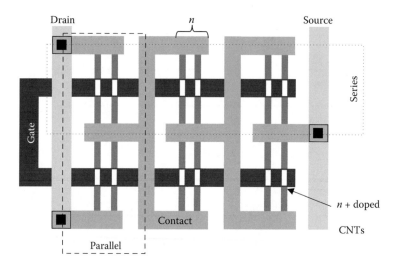

FIGURE 44.5 A transistor with series–parallel CNTFETs. (Z. Zhang and J.G. Delgado-Frias, Low power and metallic CNT tolerant CNTFET SRAM design, *2011 11th IEEE Conference on Nanotechnology (IEEE-NANO)*, Portland, OR, August 15–18, 2011, pp. 1177–1182, © (2011) IEEE. With permission.)

Transistors in series greatly increase the probability of having a semiconductor CNTFET. Series CNTs, however, decrease the ON current (I_{ON}). Parallel CNTFETs that use correlated (identical) CNTs help to increase I_{ON}. Having parallel transistors with correlated CNTs does not affect the overall probability. Thus, the probability of having a transistor including the P_{semi} term is

$$P_{transistor} = 1 - (1 - P_{semi}^n)^j \tag{44.3}$$

Using series and parallel CNTFETs, we propose an approach where the same CNTs are used to form different transistors. In the memory cell, the CNTs run in the vertical orientation. CNTs in the same vertical line are identical; portions of some CNTs are etched to form different transistors. Using the proposed approach results in identical transistor pairs: T1–T2 (write), T3–T5 (inv_p), T4–T6 (inv_n), and T7–T8 (read). Transistors T1, T2, T7, and T8 have both series and parallel individual transistors. Thus, the memory cell's probability is

$$P_{Mem\_cell} = P_{write} \times P_{inv\_p} \times P_{inv\_n} \times P_{read} \tag{44.4}$$

where P_{write} is the probability of having T1 (or T2 since they are identical) as a semiconductor transistor; thus, Equation 44.3 can be used. In a similar fashion, P_{inv_p}, P_{inv_n}, and P_{read} are the probabilities of having T3 (or T5), T4 (or T6), and T7 (or T8) as semiconductor transistors, respectively. The number of series transistors is the same for all the transistors. The number of CNTs per CNTFET is set to 1 ($n = 1$). It should be mentioned that, depending on the required ON current (I_{ON}), the number of parallel transistors could vary in each transistor of the memory cell. Using $P_{transistor}$ expression from Equation 44.3, the memory cell probability becomes (with $n = 1$)

$$P_{Mem\_cell} = (P_{transistor})^4 = \left[1 - (1 - P_{semi})^j\right]^4 \tag{44.5}$$

The proposed scheme reduced the number of terms in the P_{Mem_cell} expression from eight individual probabilities to just four. Table 44.2 shows the probability of having a functional 6T and 8T SRAM cells, given a P_{semi} of 0.9 and 0.95. In this table, the number of CNTFETs in series (j) is increased from 1 to 6.

The memory cell probability depends on one transistor per pair. Parallel transistors provide additional paths to increase I_{ON}. For the 8T SRAM cell, the back-to-back inverters (T3, T4, T5, and T6)

TABLE 44.2

6T and 8T Memory Cell Probability P_{mem_cell} (with $n = 1$)

j	$P_{semi} = 0.9$		$P_{semi} = 0.95$	
	6T	8T	6T	8T
1	0.729	0.6561	0.857375	0.814506
2	0.970299	0.960596	0.992519	0.990037
3	0.997003	0.996006	0.999625	0.9995
4	0.9997	0.9996	0.999981	0.999975
5	0.99997	0.99996	0.999999	0.999999
6	0.999997	0.999996	1	1

Source: Z. Zhang and J.G. Delgado-Frias, Low power and metallic CNT tolerant CNTFET SRAM design, *2011 11th IEEE Conference on Nanotechnology (IEEE-NANO)*, Portland, OR, August 15–18, 2011, pp. 1177–1182, © (2011) IEEE. With permission

TABLE 44.3
j (Series) and k (Parallel) CNTFETs for SRAM Cells

Cell	j (Serial CNTFETs)	k (Parallel CNTFETs per Transistor)							
		T1	T2	T3	T4	T5	T6	T7	T8
6T	1	3	3	2	3	2	3	–	–
	2	3	3	2	3	2	3	–	–
8T	1	3	3	1	1	1	1	3	3
	2	3	3	1	1	1	1	3	3
	3	3	3	1	1	1	1	3	3
	4	3	3	1	1	1	1	3	3
$8T_{kmin}$	2	1	1	1	1	1	1	1	1
	3	1	1	1	1	1	1	1	1
	4	1	1	1	1	1	1	1	1

only hold the stored bit; thus, the inverter transistors need not have parallel transistors. Table 44.3 shows the number of parallel transistors (k) along with the series transistors (j). The 8T cell has two sets of values for k for each value of j. The first set is optimized to obtain a short delay. The second set (labeled as $8T_{k.min}$) is set to have the minimum number of parallel transistors; this, in turn, leads to smaller SRAM cells.

The 6T cell structure is chosen to have appropriate read and write operations according to Ref. [18]. The 8T cell is designed to have the same number (16) of CNTs with the 6T cell in the $j = 1$ case. The read performance for 6T and 8T will be close since they both have three CNTs to discharge the bit lines while reading; the 8T cell will have a shorter write delay because its inverter pair has the minimum size.

We are considering two cases in the optimization of the 8T SRAM cell: having parallel transistors (larger k) to obtain short read delays, and having the minimum number of parallel transistors (k_{min}) to minimize the required CNTs per SRAM cell. With the j and k parameters specified in Table 44.3, simulations of the 6T and 8T SRAM cells were performed. Table 44.4 shows the delay, energy, static power, and SNM of the different 6T, 8T, and 8T with k_{min} memory cell configurations. Figures 44.6 and 44.7 show the information in a graphical manner. As j is increased, both the write and read delays increase. The required energy to write (0 or 1) increases much faster for the 6T cell as j increases; this is due to the need to have strong n-type transistors for the read operation. The 8T cells do not require energy to read a "1"; if reading a "1" becomes dominant, more energy is saved. It should be pointed out that the static power drops drastically as j increases. This is due to having transistor in series or the stack effect [19], which also applies to the CNT-based circuits.

It should be pointed out that the 8T cell requires no parallel transistors for the inverters since they are only needed to hold the stored bit but not to drive a large load. Thus, the static power due to drain source leakage remains low for the 8T cell. From Table 44.4, it can be observed that static power of the 6T cell for $j = 2$ is 1.74 times higher than the 8T cell with the same j, and the SNM of the 6T cell is approximately 100 mV below all the 8T cells.

Having a minimum value of k (i.e., minimum number of parallel CNTFETs per transistor) has an impact on the write delay and a much greater impact on the read delay. This is shown in Figure 44.8. Write to a cell is done by external drivers while read operations are driven by the cells that need to discharge the bit line.

To reduce delays in a particular read delay, it is necessary to increase the number of CNTs (k variable) in the cells. Table 44.5 shows both the 6T and 8T cell structures with k specified in parentheses for each cell. To reduce the delay, larger k values are required to increase the driving current. The 6T cell structure limits the flexibility for the design trade-off between speed and stability. By

TABLE 44.4
6T, 8T, and 8T$_{kmin}$ Performance

Parameter (32 nm)		6T SRAM Cell		8T SRAM Cell				8T SRAM Cell with k_{min}		
		$j=1$	$j=2$	$j=1$	$j=2$	$j=3$	$j=4$	$j=2$	$j=3$	$j=4$
Write (ps)		2.6951	8.322	1.8659	5.0871	9.7773	15.931	10.74	21.117	34.939
Read (ps)		14.578	29.019	14.462	28.334	42.912	57.764	84.439	125.64	168.23
	read 0	0.62853	0.64466	0.61007	0.61869	0.63121	0.62664	0.59437	0.58843	0.53192
	read 1	0.62853	0.64443	–	–	–	–	–	–	–
Energy (fJ)	write 0	0.057456	0.10441	0.03206	0.058499	0.088537	0.11781	0.059377	0.092112	0.11464
	write 1	0.056628	0.10362	0.020351	0.032933	0.045965	0.06034	0.052887	0.072097	0.09121
Static power (pW)		183.12	73.122	102.74	41.984	28.09	20.279	39.193	26.765	17.922
SNM (mV)		235.54	229.75	327.4	328.63	329.66	331.27	328.63	329.61	331.27
Number of CNTs		16	32	16	32	48	64	16	24	32

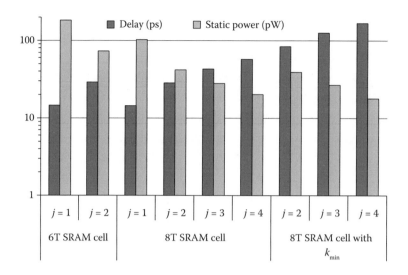

FIGURE 44.6 Delay and static power comparison of various cell structures for the 32 nm technology node.

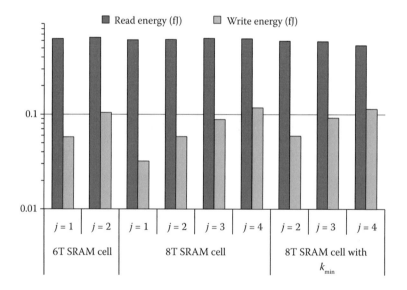

FIGURE 44.7 Read and write energy comparison of various cell structures for the 32 nm technology node.

maintaining the same SNM, we increase the k values to (12 12 8 12 8 12) for (T1 T2 T3 T4 T5 T6). As is shown in the second column for 6T in Table 44.5, the read delay is reduced to 3.98 ps; the write delay, on the other hand, is increased to 5.9678 ps, which is even longer than the read delay. This is due to the larger capacitive loads on the internal inverter pair, which means higher energy is required to change the stored value. The static power increases by four times due to the number of parallel CNTs in the internal inverters. On the other hand, reducing the read delay on the 8T cell due to its orthogonal design requires an increase in the size of transistors T7 and T8. Thus, with $k = $ (3 3 1 1 1 1 12 12), a similar read delay to the improved 6T cell can be achieved. The 8T write delay is slightly increased (2.7572 ps) due to the larger read transistor pair. The new 8T cell has a similar static power as the original 6T cell. The decoupled read and write characteristic of this cell structure ensures that its inverters can stay with the minimum size while improving the dynamic

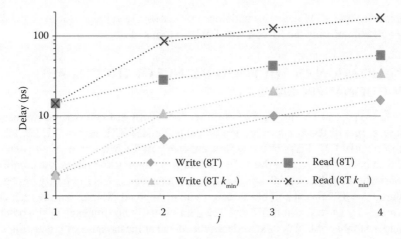

FIGURE 44.8 Read and write delays of 8T and 8T with k_{min} SRAM cells.

TABLE 44.5

Performance of the SRAM Cell with Different Cell Sizes ($j = 1$, Larger k)

Parameter (32 nm) $j = 1$	6T		8T	
	(3 3 2 3 2 3)	(12 12 8 12 8 12)	(3 3 1 1 1 1 3 3)	(3 3 1 1 1 1 12 12)
Write delay (ps)	2.6951	5.9678	1.8659	2.7572
Read delay (ps)	14.578	3.9801	14.462	4.0616
Static power (pW)	183.12	750.4	102.74	184.1
Number of CNTs	16	64	16	34

read performance. It should be noted that the improved 8T cell outperforms the improved 6T cell with about half the total CNT counts.

Reducing the power supply would slow down the circuit but has the potential of saving energy. Table 44.6 presents the data for the 6T and 8T cells in this chapter with different two power supplies (0.9 and 0.6 V). It can be observed that for both cells the worst-case delay is increased from 14.5 to 22.3 ps (53.7%), but the energy during the read or write activities are reduced by around

TABLE 44.6

Low-Power Performance for SRAM Cell in 32 nm Technology

Parameter (32 nm)		6T (3 3 2 3 2 3)		8T (3 3 1 1 1 1 3 3)	
($j = 1$)		0.9 V	0.6 V	0.9 V	0.6 V
Write (ps)		2.6951	5.5219	1.8659	3.57
Read (ps)		14.578	22.345	14.462	22.226
Energy	read 0	0.62853	0.27726	0.61007	0.27534
(fJ)	read 1	0.62853	0.27726	–	–
	write 0	0.05745	0.02661	0.03206	0.01347
	write 1	0.05662	0.02616	0.02035	0.00841
Static power (pW)		183.12	123.76	102.74	73.97
SNM (mV)		235.54	121.75	327.4	250.33

55.9%. As expected, the SNM degrades with lower power supply voltage. However, the 8T cell with $V_{DD} = 0.6$ V still outperforms the 6T cell with 0.9 V in terms of SNM.

44.6 OPTIMIZATION OF THE PERFORMANCE OF THE CELL AND FUNCTIONAL PROBABILITY

In this section, we present an approach to optimize the 8T SRAM cell design; we use $j = 3$ as an example. Several sets of the k values are simulated using HSPICE; we use k1, k2, k3, and k4 as the number of parallel CNTFETs in the write transistor pair T1–T2, inv_n pair T4–T6, inv_p pair T3–T5, and read pair T7–T8, respectively. In the following figures and tables, the combinations are denoted in parentheses as (k1, k2, k3, k4). A (19, 0) semiconducting CNT chirality has been selected as the standard. The 8T SRAM cell has a read–write orthogonal design, and the read delay of the cell is determined by k4 (transistors T7 and T8) with virtually no influence of the other transistor sizes. The larger number of CNTs in k4 reduces this delay at the expense of larger cells.

To evaluate the performance of the SRAM cell with metallic CNT tolerance, 20,000 simulations based on Monte Carlo randomization have been performed on three different cell sizes; cell size is determined by both j and k parameters; in this particular example, j is constant. The cell sizes are the minimum-k (1,1,1,1), (2,1,1,4), and (3,1,1,8). Table 44.7 shows the average write delay of the different combinations. It can be observed that write delay is influenced by k1. As k1 is made larger (or stronger), the write delay decreases. Having metallic CNTs may render some SRAM cells as nonfunctional; these cells may have short circuits. When all the CNTs that form a transistor are metallic, we refer to this problem as a short transistor; these are reported in the third column of the table. Another problem is called write failure; this is due to a combination of having stronger inverters and weaker write transistors. With $j = 3$, a transistor that has one or two metallic CNTs is able to deliver higher current and thus is stronger than the CNTFET, which has no metallic CNTs. As k1 is made larger than 1, the write failure problem is greatly reduced as shown in Table 44.7. Since the read delay is the dominant delay for this cell, the write delay is not a factor in the cell functionality.

The write and read delay distributions for the last two cell structures (2,1,1,4) and (3,1,1,8) in our study are shown in Figures 44.9 and 44.10, respectively. Most of the delays are centered at their standard delay values with a long tail in each case. The read delay could be reduced by increasing the transistor array size. However, this comes at the expense of additional hardware (i.e., more CNTs). Figure 44.11 illustrates the distribution of SNM of the two cells. In both cases, the SNM values are above 280 mV and centered at 330 mV. The high correlation between the two curves suggests that the size of either write or read transistor pairs has very little influence on the SNM.

TABLE 44.7

Cell Structure, Write Delay, Cell Failure: Short Transistor and Write Fail

Cell Structure[a]	Write Delay (ps)	Short Transistor (%)	Write Fail (%)	Total Nonfunctional (%)
(1,1,1,1)	21.117	0.45	2.59	96.96
(2,1,1,4)	14.886	0.41	0	99.59
(3,1,1,8)	13.555	0.43	0	99.57

[a] Number of CNTs in transistor pair (k1, k2, k3, k4).
Transistors included in transistor pair: k1 [T1,T2], k2 [T4,T6], k3 [T3,T5], k4 [T7,T8].

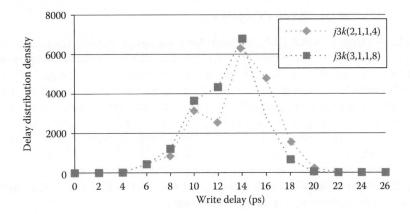

FIGURE 44.9 Write delay distribution density.

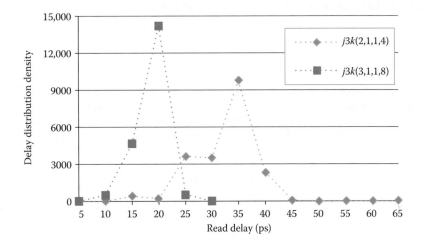

FIGURE 44.10 Read delay distribution density.

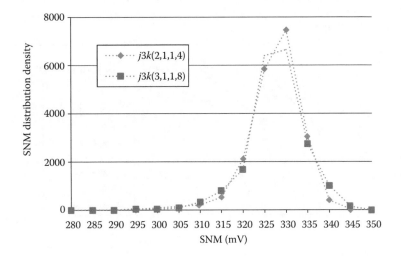

FIGURE 44.11 SNM distribution density.

44.7 MEMORY WITH SPARE COLUMNS

Building a large memory out of SRAM cells in the presence of metallic CNTs requires a fault-tolerant scheme to overcome faulty memory cells. For this study, we have chosen spare columns to deal with faults. This scheme was chosen because it is simple and achieves the goal of having functional memory. Table 44.8 shows the probability of having a functional 1024×32 bit memory. The parameter setting is as follows: P_{semi} is 95%, the number of transistors in series (j) is 3 and 4, and k is specified in Table 44.3. The large memory is divided into smaller modules that go from a single module of 1024 entries to a module of 16 entries. From this table, it can be seen that a small value of j (i.e., 3) would require a large number of spare columns to yield a functional memory.

As the modules have fewer entries, the number of columns needed decreases. It can be observed that 64- or 32-entry modules (for $j = 4$) yield a good probability of a functional memory. Figure 44.12 shows the same information in a graphical manner.

TABLE 44.8

Probability of Functional Memory with Spare Columns

j	Mod Size	Spare Columns							
		1	2	3	4	5	6	7	8
3	1024	0.0000	0.0000	0.0000	0.0001	0.0004	0.0012	0.0028	0.0059
	512	0.0000	0.0001	0.0008	0.0044	0.0163	0.0454	0.1010	0.1876
	256	0.0000	0.0018	0.0208	0.0994	0.2665	0.4859	0.6883	0.8330
	128	0.0005	0.0305	0.2263	0.5610	0.8157	0.9370	0.9812	0.9949
	64	0.0053	0.2178	0.6814	0.9203	0.9844	0.9974	0.9996	0.9999
	32	0.0391	0.5854	0.9306	0.9920	0.9992	0.9999	1	1
	16	0.1574	0.8510	0.9890	0.9994	1	1	1	1
4	1024	0.7973	0.946	0.9886	0.998	0.9997	1	1	1
	512	0.8726	0.9817	0.998	0.9998	1	1	1	1
	256	0.9267	0.9947	0.9997	1	1	1	1	1
	128	0.9603	0.9986	1	1	1	1	1	1
	64	0.9793	1	1	1	1	1	1	1
	32	0.9894	1	1	1	1	1	1	1
	16	0.9947	1	1	1	1	1	1	1

Source: Z. Zhang and J.G. Delgado-Frias, Low power and metallic CNT tolerant CNTFET SRAM design, *2011 11th IEEE Conference on Nanotechnology (IEEE-NANO)*, Portland, OR, August 15–18, 2011, pp. 1177–1182, © (2011) IEEE. With permission.

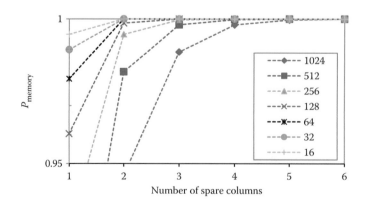

FIGURE 44.12 P_{memory} as a function of the number of spare columns and module size for $j = 4$.

44.8 TECHNOLOGY SCALING

So far, we have discussed the performances and evaluated the functionality of a CNT-based SRAM cell at the 32 nm technology node. In this section, we show the performance of the CNTFET memory cells at 22 and 16 nm technologies. The process parameters for 22 and 16 nm N-type CNTFET are listed in Table 44.9. The 32 nm CNTFET parameters are from the values included in the Stanford University CNTFET HSPICE Model [11], and the 16 nm CNTFET values are compiled from a study by Yanan Sun et al. [20]. The supply voltage drops from 0.9 to 0.7 V as the technology is scaled down, and dielectric materials with high k values are used to maintain the same driving current and reduce leakage current.

Tables 44.10 and 44.11 present the performances of the 6T and 8T cells under 22 and 16 nm technology. It can be observed that all the delays and power dissipations decrease as the transistor size is scaled down. The 22 nm 6T and 8T cells take only 85% of the time, and the 32 nm cells need to read 0, while dissipating about 76% energy because of the reduced capacitive loads and lower power supply. The improvement for 16 nm cells is 75% in read delay and 60% in read energy. The static power for 22 nm cells is only about 40~60% of the amount for 32 nm. The 16 nm cells dissipate only 9–19% of the static power compared to the 32 nm counterparts. The only parameter that is degraded is the SNM. For the 8T cells, the lower power supply can still hold the SNM at about 97% for 22 nm and 90% for 16 nm of the 32 nm SNM. The 6T cells are more vulnerable to noise; their values drop around 20% and 40% for 22 and 16 nm technologies, respectively.

44.9 CONCLUDING REMARKS

In this chapter, we have presented the design of a CNTFET-based memory system that has a simple metallic CNT tolerance. Some features of the 8T CNTFET SRAM and memory modules include:

- *CNTFET SRAM cell tolerant to metallic CNT.* Metallic CNT tolerance is accomplished by having a series of CNTs to form a transistor. With $j = 3$, the probability of a nonfunctional cell can be as low as 99.59%; the worst-case delay and SNM are as high as 30 ps and 350 mV, respectively.

TABLE 44.9
Process Parameters for 32, 22, and 16 nm CNTFET

Process Parameters	32 nm [11,20]	22 nm	16 nm [20]
	CNTFET Technology Node		
Supply voltage (V_{DD})	0.9 V	0.8 V	0.7 V
Physical channel length (L_g)	32 nm	22 nm	16 nm
Length of doped CNT source/drain (L_s/L_d)	32 nm	22 nm	16 nm
Dielectric constant of bottom gate (K_{sub})	SiO_2 (4)	SiO_2 (4)	SiO_2 (4)
Oxide thickness	10 µm	10 µm	10 µm
Top gate dielectric material (High K_{ox})	HfO_2 (16)	ZrO_2 (20)	ZrO_2 (25)
Thickness of top gate dielectric (T_{ox})	4 nm	3.3 nm	3 nm
Flatband voltage of CNTFET (V_{fbn}/V_{fbp})	0	0	0
Fermi level of n+ doped source/drain CNT region (E_{fo})	0.6 eV (~0.8%) (uniformly distributed)	0.6 eV (~0.8%) (uniformly distributed)	0.6 eV (~0.8%) (uniformly distributed)
Mean free path: intrinsic CNT	200 nm	200 nm	200 nm
Mean free path: doped CNT	15 nm	15 nm	15 nm
Interconnect capacitance	0.213 fF/µm	0.25 fF/µm	0.3 fF/µm

TABLE 44.10
6T, 8T, and 8T$_{kmin}$ Performance at 22 nm

Parameter (22 nm)		6T SRAM Cell		8T SRAM Cell				8T SRAM Cell with k_{min}		
		$j=1$	$j=2$	$j=1$	$j=2$	$j=3$	$j=4$	$j=2$	$j=3$	$j=4$
Write (ps)		2.05	5.9116	1.5477	3.9496	7.6749	12.413	7.9042	16.127	26.264
Read (ps)		12.349	24.304	12.324	24.246	36.74	49.514	74.163	110.11	144.17
Energy (fJ)	read 0	0.4817	0.49131	0.47692	0.47778	0.48453	0.48024	0.46282	0.46205	0.4324
	read 1	0.48167	0.49124	–	–	–	–	–	–	–
	write 0	0.037251	0.06578	0.022344	0.040001	0.060277	0.079972	0.038564	0.05928	0.073256
	write 1	0.037256	0.06485	0.013143	0.021582	0.030123	0.039633	0.033928	0.046917	0.060109
Static power (pW)		97.038	37.718	47.031	18.533	16.615	10.981	15.848	13.7	8.9216
SNM (mV)		188.92	185.2	318.8	319.46	319.78	320.28	319.19	319.58	320.21

TABLE 44.11
6T, 8T, and 8T$_k$ min Performance at 16 nm

Parameter (16 nm)		6T SRAM Cell		8T SRAM Cell				8T SRAM Cell with k$_{min}$		
		$j=1$	$j=2$	$j=1$	$j=2$	$j=3$	$j=4$	$j=2$	$j=3$	$j=4$
Write (ps)		1.8367	4.6262	1.2459	3.1501	6.1322	11.044	6.9501	13.106	21.414
Read (ps)		10.323	21.218	10.964	20.835	31.996	42.21	69.37	102.45	130.96
Energy (fJ)	Read 0	0.36228	0.36947	0.36097	0.36164	0.36694	0.36326	0.35208	0.35388	0.34322
	Read 1	0.36228	0.36947	–	–	–	–	–	–	–
	Write 0	0.021725	0.040878	0.015458	0.028895	0.04598	0.057024	0.023486	0.033566	0.04535
	Write 1	0.02232	0.040618	0.0085746	0.014929	0.020295	0.027001	0.019108	0.026855	0.03473
Static power (pW)		25.66	6.6842	15.673	6.6091	5.3967	3.5201	5.6824	4.4403	3.0512
SNM (mV)		141.62	142.86	295.85	295.84	295.04	296.22	295.83	294.98	295.32

- *Low static power.* Leakage current is one of the major contributors in the power consumption in SRAM arrays. The 32 nm 8T CNTFET SRAM cell's static power stays between 20.79 and 102.74 pW with different j and k configurations. The 6T static power with $j = 2$ is 1.47 times higher than the 8T cell with $j = 2$. The leakage current drops drastically as j increases.
- *High SNM.* The 8T cell's SNM is higher than the 6T's SNM. For all the configuration of the 8T, the SNM range is between 327 and 331 mV while the 6T's range is between 229 and 235 mV. With the power supply lowered from 0.9 to 0.6 V, the 8T cell's SNM drops by 23.5%, while the 6T cell's SNM drops by 48.3%.
- *Small cell size.* The delay can be improved by having more parallel CNTs; the 8T cell achieves similar read and write delays using about half the number of CNTs in the 6T cell.
- *Low k (parallel FETs) values.* The 8T cells can tolerate low values of k; this, in turn, impacts the read delay.
- *Spare columns to tolerate faulty SRAM cells.* By means of a simple fault-tolerant scheme, the proposed SRAM cell can be used to form a large memory system.
- *Orthogonal design space for read and write.* Decoupled write and read operations provide an orthogonal design space that helps to balance speed and area. 6T cells, on the other hand, require a great deal of design effort to ensure proper read and write operations.
- *Read delay of the 8T cell without metallic nanotubes tends to cluster around 8.9 ps with random ±10% diameter variations.* A Monte Carlo simulation shows a delay distribution with a mean of 8.91 ps. Owing to the nonlinear effect of diameter variation on current, the delay distribution extends to the longer delays.
- *Low energy (dynamic power) for writing.* While performing a write into an 8T cell, smaller gate capacitance is engaged in restoring the cell's value due to the minimum-sized inverter pair.
- *Robustness to technology scaling.* SRMA cells at 22 and 16 nm technologies reduce the delay by 15% and 25%, respectively. Both dynamic and static power consumptions are also reduced by up to 24% and 60% for 22 nm, and 43% and 91% for 16 nm. 8T cells maintain a higher SNM than 6T cells as technology scales down.

As new techniques to grow CNTs reduce the number of metallic CNTs, the proposed memory system will require lower j (fewer CNTFETs in series) and fewer spare columns.

ACKNOWLEDGMENT

The research reported in this chapter has been partially supported by WSU's Boeing Centennial Endowed Chair.

REFERENCES

1. T.-C. Chen, Overcoming research challenges for CMOS scaling: Industry directions, *International Conference on Solid-State and IC Technology*, Shanghai, China, pp. 4–7, October 23–26, 2006.
2. T. Cakici, K. Kim, and K. Roy, FinFET based SRAM design for low standby power applications, *8th International Symposium on Quality Electronic Design*, San Jose, CA, pp. 127–132, March 26–28, 2007.
3. Y. B. Kim, Y.-B. Kim, and F. Lombardi, New SRAM cell design for low power and high reliability using 32 nm independent gate FinFET technology, *IEEE International Workshop on Design and Test of Nano Devices, Circuits and Systems*, Cambridge, MA, pp. 25–28, 2008.
4. International Technology Roadmap for Semiconductors, http://www.itrs.net/.
5. A. Lundstrom, A top-down look at bottom-up electronics, *2003 Symposium on VLSI Circuits, Digest of Technical Papers*, Kyoto, Japan, pp. 5–8, June 12–14, 2003.
6. P. H.-S. Wong, Field effect transistors-from silicon MOSFETs to carbon nanotube FETs, *23rd International Conference on Microelectronics,* Nis, Yugoslavia, pp. 103–107, May 12–15, 2002.

7. P. Avouris, Z. Appenzeller, R. Martel, and S. Wind, Carbon nanotube electronics, *Proceedings of the IEEE*, 91(11), 1772–1784, November 2003.
8. B. C. Paul, S. Fujita, M. Okajima, T. H. Lee, P. H.-S, Wong, and Y. Nishi, Impact of a process variation on nanowire and nanotube device performance, *IEEE Transactions on Electron Devices*, 54(9), 2369–2376, September 2007.
9. Y. B. Kim, Y.-B. Kim, F. Lombardi, and Y. J. Lee, A low power 8T SRAM cell design technique for CNTFET, *IEEE International SoC Design Conference (ISOCC)*, Pusan, South Korea, pp. 176–179, November 24–25, 2008.
10. L. Chang, R. K. Montoye, Y. Nakamura, K. A. Batson, R. J. Eickemeyer, R. H. Dennard, W. Haensch, and D. Jamsek, An 8T-SRAM for variability tolerance and low-voltage operation in high-performance caches, *IEEE Journal of Solid-State Circuits*, 43(4), 956–963, April 2008.
11. CNTFET Models. http://nano.stanford.edu/models.php.
12. J. Deng and H.-S. P. Wong, A compact SPICE model for carbon nanotube field effect transistors including non-idealities and its application—Part II: Full device model and circuits performance benchmarking, *IEEE Transactions on Electron Devices*, 54(12), 3195–3205, December 2007.
13. J. Deng and H.-S. P. Wong, A compact SPICE model for carbon nanotube field effect transistors including non-idealities and its application—Part I: Model of intrinsic channel region, *IEEE Transactions on Electron Devices*, 54(12), 3186–3194, December 2007.
14. J. G. Delgado-Frias, Z. Zhang, and M. A. Turi, Low power SRAM cell design for FinFET and CNTFET technologies, *International Conference on Green Computing*, Chicago, IL, pp. 547–553, 2010.
15. Z. Zhang, Y. Liu, J. Nyathi, and J. G. Delgado-Frias, Performance of CNFET SRAM cells under diameter variation corners, *MWSCAS 2009. 52nd Midwest Symposium on Circuits and Systems*, Cancun, Mexico, pp. 547–550, August 2–5, 2009.
16. G. Zhang, P. Qi, X. Wang, Y. Lu, X. Li, R. Tu, S. Bangsaruntip, D. Mann, L. Zhang, and H. Dai, Selective etching of metallic carbon nanotubes by gas-phase reaction, *Science*, 314, 974–977, 2006.
17. A. Lin, N. Patil, H. Wei, S. Mitra, and H.-S. P. Wong, A metallic-CNT-tolerant carbon nanotube technology using asymmetrically-correlated CNTs (ACCNT), *2009 Symposium on VLSI Technology*, Honolulu, HI, pp. 182–183, 2009.
18. S. Lin, Y.-B. Kim, F. Lombardi, and Y. J. Lee, A new SRAM cell design using CNTFETs, *SoC Design Conference, 2008. ISOCC '08. International*, 01, Busan, South Korea, pp. I-168–I-171, November 24–25, 2008.
19. C. Zhanping, M. Johnson, W. Liqiong, and W. Roy, Estimation of standby leakage power in CMOS circuit considering accurate modeling of transistor stacks, *Proceedings of the 1998 International Symposium on Low Power Electronics and Design, 1998*, Monterey, CA. pp. 239–244, August 10–12, 1998.
20. Y. Sun and V. Kursun, N-type carbon-nanotube MOSFET device profile optimization for very large scale integration, *Transactions on Electrical and Electronic Materials*, 12(2), 43–50, April 25, 2011.
21. Z. Zhang and J. G. Delgado-Frias, Low power and metallic CNT tolerant CNTFET SRAM design, *2011 11th IEEE Conference on Nanotechnology (IEEE-NANO)*, Portland, OR, pp. 1177–1182, August 15–18, 2011.

Section XIV

Nano-Redundant Systems

45 Optimized Built-In Self-Test Technique for CAEN-Based Nanofabric Systems

Maciej Zawodniok and Sambhav Kundaikar

CONTENTS

45.1 INTRODUCTION

Nanotechnology enables future advancements in integrated circuitry's miniaturization, energy and cost efficiency, and capabilities. However, a popular, chemically assembled electronic nanotechnology (CAEN) has a high rate of defects that negates these benefits of the nanofabric. To address this challenge, we propose a testing technique that maximizes the yield from a nanofabric while minimizing the testing overhead. In contrast, traditional testing techniques, for example, the ones employed in field programable gate array (FPGA) applications, assume a low defect rate and fail to achieve high effectiveness when applied to testing nanofabrics. In this chapter, we propose a novel approach to testing a nanofabric that includes new testing configurations, test-set optimization methodology, and design of customized configuration, which provide a reduction in testing time while enhancing the utilization of the nanofabric. Part of the proposed scheme is a recovery procedure that further increases the utilization of nanoblocks at the expense of testing time. The proposed procedure tests all the components in parallel and identifies the defective nanoblocks in a nanofabric. A defect map is generated to aid logic function implementation in a nanofabric. The proposed technique results in less number of test configurations compared to other proposed methods and a significant reduction in the test time.

45.1.1 BENEFITS AND CHALLENGES OF NANOTECHNOLOGY

Moore's law predicted that the number of transistors that could be placed on a chip doubles every 2 years. The complementary metal oxide semiconductor (CMOS) technology has been able to keep up with Moore's law as a result of scaling. However, CMOS technology today faces a number of challenges such as leakage currents, process variation, costs, and reliability issues, which may put an end to scaling. Therefore, new technologies are needed to replace CMOS in the future to continue Moore's law advancement.

CAEN has shown promising potential for the future of nanoscale design [1–3]. A basic building block of a nanofabric is a nanoblock, which is an interconnected 2D array of nanoscale wires that can be electronically configured as logic networks, memory units, and signal-routing cells [4,5]. These nanofabric architectures are reconfigurable in nature and can achieve a density of 10^{10}–10^{12} gates per cm^2, which is significantly higher than that of CMOS-based devices. However, owing to the high defect rate in these nanofabrics, new testing strategies need to be devised to effectively test and diagnose the nanofabric within a reasonable time.

The low-cost manufacturing process of self-assembly and self-alignment is associated with a high defect rate approaching 10%, thus lowering the yield and increasing the manufacturing costs. It is impractical to throw away a fabricated chip once it is diagnosed to be defective. Thus, a defect-tolerant approach is required such that partially defective blocks can be utilized. The ability to identify and diagnose the faulty sites on a chip and develop techniques to avoid these faulty areas is known as defect tolerance [4,6–11]. Also, the testing techniques needed for nanofabrics are complicated due to the large density, high defect rate, and large size of the nanoblocks. Hence, traditional FPGA testing approaches [12,13] are not suitable for the nanofabric systems.

In this chapter, we discuss a testing methodology for nanofabric systems that configures the components as block under tests (BUT) and comparators (C). This method allows us to handle high defect densities and large-size nanofabrics. Our method uses a set of test architectures and BUT configurations to test the nanofabric. Also, we explain the design of each of the BUT configurations for the targeted faults. An external tester is used to apply test patterns to the BUTs. The nanoblocks are tested in parallel, and hence the testing time does not depend on the size of the nanofabric. Further, we introduce an optimization technique, which can increase both the testing speed and the usability (yield) of the nanoblock.

45.2 BACKGROUND AND RELATED WORKS

First, the basic concept of the nanofabric-based digital circuits is presented. Then an example application of the self-test procedure is discussed in the context of a reconfigurable digital logic design.

The nanofabric is a regular architecture consisting of a two-dimensional array of interconnected nanoblocks and switchblocks as shown in Figure 45.1. A nanoblock and a switchblock are two main components in a nanofabric. A nanoblock in a nanofabric is similar to a configurable logic block (CLB) in FPGA. A nanoblock is based on a molecular logic array (MLA). There is a reconfigurable switch at each intersection of MLA in series with the diode. Diode-resistor logic is used to perform logical operations. The region between a set of nanoblocks is known as a switchblock that connects wire segments of adjacent nanoblocks. The configuration of the switchblock determines the direction of data flow between the nanoblocks. The detailed explanation of the nanofabric architecture, nanoblocks, and switchblocks can be found in Ref. [14].

The defect tolerance methodology is presented in Ref. [2] such that components of a nanofabric are configured to test circuits to infer the defect status of individual components. Only stuck-at and stuck-open faults are targeted during the test and the proposed technique does not provide high recovery. In contrast, the proposed approach considers additional faults, including nanowire bridging and cross-point bias. Application-dependent testing of FPGA has also been proposed in Refs. [11,15]. A defective FPGA may pass the test for a specific application. This is a one-time configuration; in other words, the FPGA will no longer be reconfigurable. This reduces yield loss and leads to manufacturing cost savings.

In Ref. [4], testing is performed by configuring the blocks and switches as linear feedback shift registers (LFSR). If the final bit stream generated by LFSR is correct, all the components are assumed to be defect-free. Otherwise, there is at least one faulty component in LFSR. To diagnose, the components in LFSR and other components are used to configure a new LFSR. If the new LFSR is faulty, the component at the intersection of faulty LFSRs is considered defective. A defect database is created after completing the test and diagnosis. In contrast to the proposed methodology, the LFSR-based approach targets only stuck-at and stuck-open faults, thus reducing the recovery rate.

The CAEN built-in self-test (BIST) approach presented in Ref. [9] configures a nanoblock as a tester to test its neighboring nanoblocks. Test patterns are fed to both the tester and the nanoblock under test (BUT) from an external source. A defect-free BUT generates output patterns that are identical to the input patterns. The tester compares the input test patterns and the output patterns from the BUT to see if the BUT is defective. However, CAEN-BIST is performed in a wavelike manner in which a set of nanoblocks in the same diagonal tests another set of nanoblocks until the entire nanofabric has been tested. Therefore, the complexity and testing time depend on the size of the nanofabric under test.

Another BIST approach was proposed in Ref. [16] where the nanoblocks can be configured as test pattern generators (TPGs), block under test (BUTs), or output response analyzers (ORAs). These blocks, along with the corresponding switchblocks, comprise a TG (test group). In a TG, the TPG generates the testing patterns for a BUT and ORAs examine the BUT output response. A total of $4k + 6$ configurations are needed to test for the stuck-at, stuck-open, bridging, and defective crosspoints.

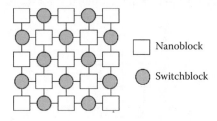

Nanoblock

Switchblock

FIGURE 45.1 Nanofabric architecture. (S. Kundaikar and M. Zawodniok, Optimized built-in self-test technique for CAEN-based nanofabric systems, *Proceedings of 11th IEEE Conference on Nanotechnology (IEEE-NANO'11)*, Portland, OR, pp. 1717–1722, © (2011) IEEE. With permission.)

In the BIST procedure discussed in Ref. [17], each nanoblock is configured either as a pattern generator (PG) or as a response generator (RG). A test group is created using a set of PGs, RGs, and switchblock(s) between the two. The nanoblock configured as a PG tests itself and generates the test pattern for RG. An external device is needed to program the nanoblocks and read the RGs' responses. In the test configurations, stuck-at, stuck-open, forward-biased and reverse-biased diode, and AND and OR bridging faults are targeted. If the size of the nanoblock is $k \times k$, it is estimated that $8k + 5$ configurations are needed to provide 100% fault coverage. This number of configurations is still large for a large value of k.

45.2.1 NOVEL LOGIC MAPPING TECHNIQUE

Once the nanofabric is manufactured, it needs to be tested thoroughly to identify the faulty blocks. The testing methodology used will test for certain types and numbers of faults. Once the faulty areas in the chip have been identified, the next step is to map logic onto the nanofabrics by avoiding the faulty blocks. The reconfigurability and functional redundancy within the nanofabric and nanoblocks enable the development of robust, defect-, and fault-tolerant design mechanisms. When a defective resource or component is identified in the chip using test and diagnosis, the postfabrication configuration and logic mapping can counter the effects of those defects. Logic mapping onto nanoblocks refers to the programming of the various cross points of the nanoblock to implement different gates and connecting them together to obtain the required logic function. A nanoblock consists of two groups of parallel nanowires, which are perpendicular to each other in the same plane and programmable cross points at the junction of these nanowires [1]. These cross points can be programmed to behave as diodes by the application of suitable voltage levels. A nanoblock can implement AND/OR gate functionalities by programming the different cross points and by configuring pull-up and/or pull-down resistors. Figure 45.2 illustrates the example implementation of a simple logic by programming the cross points in a nanoblock and the use of pull-up/pull-down resistors. Note that not all cross points are employed in this realization.

The process of logic mapping could be simplified if there were standard configurations of nanoblocks available, which could implement specific functionality. This would reduce the effort needed to design custom configurations for each nanoblock. By configuring a nanoblock, we mean programming the various cross points within the block to implement logic functionality.

We propose a novel logic mapping technique, which would use the result of the testing technique presented in this work. Here, the types of faults targeted are stuck-at faults, bridging, and cross-point faults. It is assumed that each nanoblock implements only AND/OR logic and is tested to make sure it can successfully implement such a configuration. This requires only a certain set of cross points to be fault-free to successfully implement the function. This results in a substantial increase in the yield of the nanofabric. Also, the nanoblocks are assumed to be implementing only AND/OR gates, and the nanoblock configurations are fixed. A $k \times k$ nanoblock implementing a $(k \times 1)$ input AND/OR gate is shown in Figure 45.3. The F_1 output implements the AND function for inputs $A_1, A_2, \ldots, A_{k-1}$, while F2 implements the OR function for the $B_1, B_2, \ldots, B_{k-1}$ inputs. For such

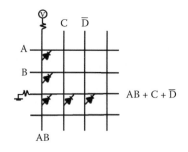

FIGURE 45.2 Simple logic implementation.

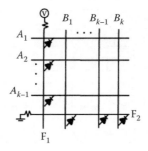

FIGURE 45.3   AND/OR implementation.

a configuration, the effective cross points (required for function realization) are known *a priori*. Hence, the customized testing, such as the one in Ref. [17], provides a map of useful nanoblocks that can be utilized even if partial defects are present. Consequently, the nanoblocks marked as fault-free can implement AND/OR logic using standard configurations, such as the one in Figure 45.3.

The main advantage of such an implementation is that we need not know the location of defects inside the nanoblock. Also, no effort would be needed in designing the nanoblock configurations for each nanoblock since we are using standard AND/OR implementation. This implementation method requires the function to be in the sum of products (SOP) or product of sums (POS) form. Once the function is available in the required form, we need to determine the number of input blocks required to implement the function and also map the functionality to each of the blocks used, that is, whether the block will implement AND logic or OR logic. We need not worry about configuring the blocks internally since we are using standard configurations. Once functionality has been mapped to the nanoblocks, we need to connect them and route the result to the output of the nanofabric. Here, we illustrate this method using three examples.

Consider that we have a simple function $Z_1 = AB + C + DE$. We need a total of two AND and two OR gates to implement this function. Consider that we have a fault-free nanofabric consisting of 36 nanoblocks and that each nanoblock is 3×3. The implementation of this function is shown in Figure 45.4. The circles represent switchblocks, which are used to connect the output of one block to the input of the next one. The rectangles represent the nanoblocks that are configured to implement either an AND or OR gate or are just used for routing purposes. As can be seen, the mapping procedure is relatively simpler in the absence of defective blocks. However, in the presence of defects, we need to make sure that we avoid the defective blocks and route the result to the output of the nanofabric.

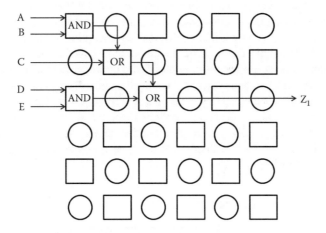

FIGURE 45.4   Logic mapping using AND/OR blocks.

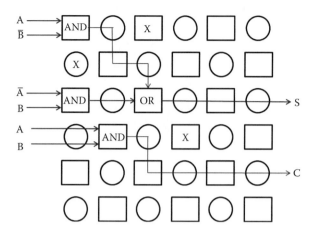

FIGURE 45.5   Half-adder implementation in the presence of defects.

Consider a similar nanofabric with a few defective blocks as shown in Figure 45.5. Here, the defective blocks have been identified by the testing procedure and indicate that they cannot be used to implement any logic. However, in some cases, they may be used to transfer signals from one block to the next. This means that they can be used for routing purposes. Let us implement a half-adder with inputs A, B and outputs sum (S) and carry (C) in the presence of faulty blocks. As can be seen, we need to go around the faulty blocks and find a fault-free path. Thus, in the presence of faulty blocks, we need to use a lot of nanoblocks just for routing around the faulty blocks.

Next, the details of the self-testing approach are presented and discussed.

45.3   TESTING APPROACH

45.3.1   Overview

In this section, a testing approach is presented that efficiently and cost effectively identifies the faulty components. The entire nanofabric is set up based on test architectures to perform a self-test among the nanoblocks. In concert with carefully selected configurations for the nanoblocks under test, the proposed approach minimizes the testing time while ensuring all fault testability. The analysis of the tests results in a defect map that indicates the faulty nanoblocks.

Moreover, we propose two methods that increase the nanofabric yield by identifying partially defective nanoblocks that still can correctly implement the AND/OR logical functions. The proposed *optimization* approach selects the minimal subset from the entire proposed list of configurations to test for the desired functionality only. Additionally, this test set optimization reduces the overall testing time. The second method, the *customization* method, redesigns the test configurations such that only the required cross points are tested. In contrast, the optimization technique only reduces the number of unnecessary configurations while still testing the unnecessary cross points. The drawback of the customization method is the need for creating a new set of test configurations for each desired logical function. The increased overhead may result in a longer and more complex testing process.

First, a set of test architectures is presented in Section 45.3.2. Next, the nanoblocks configurations are introduced and discussed in Section 45.3.3. The optimization and customization techniques are proposed in Section 45.4.

45.3.2   Test Architectures and Procedure

A test architecture (TA) defines the manner in which the nanoblocks and switchblocks are configured and connected. Each of the nanoblocks is configured either as a block under test (BUT) or as a comparator (C) [18] and the switch blocks are used to connect the outputs of BUTs to comparators. A nanoblock configured as a BUT in one TA is configured as a comparator in the other TA. Hence, half of the total number of nanoblocks is tested in each TA. In contrast to the traditional approach utilized in an FPGA-type application, the proposed TAs are constrained by the fabric topology. In a typical nanofabric, nanoblocks have outputs on two sides called the east (right) and the south (bottom). To test the nanoblock functionality in both directions, a total of four TAs are required, as shown in Figure 45.6. The "B" block refers to a nanoblock that is configured as a BUT, and "C" refers to a nanoblock configured as a comparator.

The proposed approach defines two pairs of complementary TAs: TA-1 for outputs on the east side, and TA-2 for outputs on the south side. TA-1b is the complement of TA-1a since a BUT in TA-1a is a comparator in TA-1b and vice versa. Similarly, TA-2b is the complement of TA-2a. The selection of the particular test sets (TA-1a and TA-1b) or (TA-2a and TA-2b) is dictated by the test configuration as discussed in the Section 45.3.3.

The test patterns are applied simultaneously using an external tester. Each comparator compares outputs from two BUTs and each BUT output is connected to two comparators. The output of the comparator will be successful if the BUTs being compared and the comparator itself are defect-free. If any of the BUTs or the comparator itself is faulty, it is an unsuccessful comparison. Since the defect rate is of the order of 10–15%, it is assumed that the probability of two defective BUTs being compared by the same comparator is very low. It is assumed that the comparator generates a "0" for a successful comparison and a "1" for an unsuccessful comparison. This helps generate an intermediate defect map called the partial defect map and, in turn, the final defect map. A test is run for each type of fault to be targeted to create partial defect maps for corresponding faults. Combining all the partial defect maps gives the final defect map, which the compiler can use to configure the nanofabric by avoiding defective blocks.

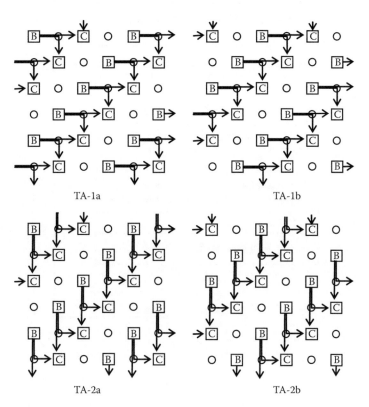

FIGURE 45.6 Test architectures.

45.3.3 TESTING PROCEDURE

A block is declared fault-free only if it does not manifest any of the faults targeted. Initially the final defect map is set to NULL. Next, the BUT configuration is optimized/customized for the targeted fault. A raw defect map is a defect map for a particular fault. For each targeted fault, the raw defect map is initially set to null. The first test architecture is then selected and the corresponding BUT configuration and test patterns are applied. The comparator outputs are read out and the possible faulty blocks are marked as suspects. The complementary test architecture is then applied and the procedure is repeated. The comparator outputs are read out and a decision is made whether the marked blocks are actually faulty or not. Figure 45.7 gives the flowchart for this testing procedure.

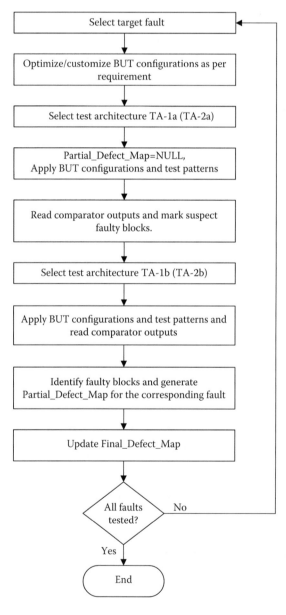

FIGURE 45.7 Testing procedure.

45.3.4 Algorithm for Locating Defects and Example Defect Detection

The comparator detects inconsistencies between the outputs of two BUTs, which indicate the presence of a fault. To uniquely identify the defective nanoblocks, we need to analyze both (1) outputs of adjacent comparators and (2) test results in the complementary architecture. Once the defect is located, the raw defect map is updated accordingly. The faulty block identification algorithm is presented in Table 45.1. In general, the faulty block is identified when it is marked as faulty by two corresponding comparators. The complementary test architecture is used to identify faults in the comparators.

45.3.5 Example of a Faulty Block Identification

Figure 45.8 illustrates a section of a nanofabric for two test architectures with the blocks numbered 1 through 18. The two complementary architectures are needed to ensure that each of the blocks is tested.

Consider the following assumptions:

- The fault targeted is F1 and the test architectures used are TA-1a and TA-1b.
- The size of the nanofabric is 6×6.
- Blocks B4 and B5 have fault F1.

It can be observed that block B5 is tested in TA-1a and block B4 is tested in TA-1b. Consider the first test architecture TA-1a. The blocks C2, C4, C6, C7, C9, C11, C13, C16, and C18 are configured

---

TABLE 45.1

Pseudocode of the Faculty Block Detection Algorithm

1. Assign a score of zero (0) to all the nanoblocks.
2. For each TA:
   a. For each comparator:
      i. If the comparator gives false output, increment score "+1" for each of the corresponding BUTs
   b. For each BUT:
      i. If total score is equal to two (2), then the block is marked as defective
   c. Create the raw defect map.

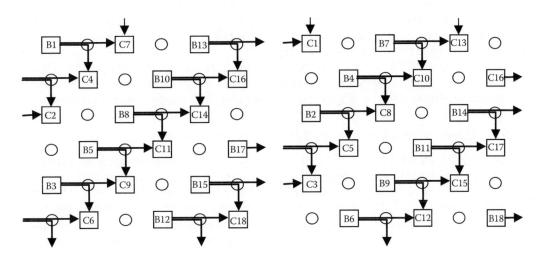

FIGURE 45.8   Subset of a nanofabric in TA-1a and TA-1b.

as comparators and compare the outputs of the BUTs. Initially, all the nanoblocks are assigned a score of "0."

Owing to the presence of the faulty block B5, the following comparators would generate a "1" at the output:

- C9: Blocks B3 or B5 have fault F1.
  - Increment B3 and B5 by "1." B3 = 1, B5 = 1.
  - C11: Blocks B5 or B8 have fault F1
- Increment B5 and B8 by "1." B5 = 2, B8 = 1.

The remaining comparisons are successful. Analyzing the above results, the following is obtained:

- B3 = 1
- B5 = 2
- B8 = 1
- Rest of the BUTs = 0

Therefore, eliminating all the blocks with a score of "1" and "0" and only retaining blocks with a score of "2" as faulty blocks, block B5 is diagnosed as faulty and the defect map is updated accordingly. Similarly, the faulty block B4 is identified using test architecture TA-1b. Since the nanoblock size is very large, efficient data structures based on Bloom filters have been proposed for the storage of the defect map [19].

45.3.6 Design of New BUT Configurations and Test Patterns and Fault Coverage Analysis

To test the BUTs for defects, we need to configure them internally and apply appropriate test patterns. Configuring a BUT refers to programming of the various cross points in the nanoblock and applying the desired voltages to the nanowires. The BUT configuration and the test pattern target specific fault types. In turn, the particular configuration dictates which pair of complementary test architecture has to be used based on the direction of inputs and outputs. In each configuration, one or more faults can be targeted.

In general, the faults can be divided into two categories: nanowire faults and cross-point faults. The former affects the entire vertical or horizontal wire in nanoblocks, thus rendering it useless. The latter affects only individual cross points and can potentially have no effect on the correctness of a logic function implemented using the remaining good cross points inside the particular block. The configurations are grouped in two sets: (a) *Set I* that targets the nanowire faults and includes configurations C1 through C4, and (b) *Set II* that targets the cross-point faults and includes configurations C5 and C6. The following faults are targeted in the next subsection: stuck-at and stuck-open faults, connection faults, and bridging faults using a set of test architectures, BUT configurations, and test patterns.

45.3.7 Configuration C1: Stuck-at-0 and Stuck-Open Faults

The BUT configuration C1 is shown in Figure 45.9a. All junctions are programmed as diodes and all inputs are connected to a "1." Hence, all outputs should be a "1" for a fault-free BUT. However, if any of the lines has a stuck-at-0 and/or stuck-open fault, the corresponding output will be a "0." Test architectures TA-1a and TA-1b will be used in conjunction with this BUT configuration since the inputs are on the west and outputs on the south of the nanoblock. Consider, for instance, that a vertical line has a stuck-at-0 fault as shown in Figure 45.5b and hence is permanently tied to a 0. As a result, the output for the vertical line 2 will be a "0," while for all other defect-free lines the output will be a "1" (Figure 45.9).

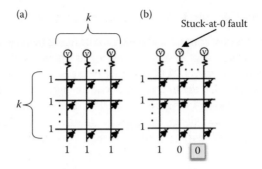

FIGURE 45.9 (a) BUT configurations and test patterns for C1. (b) Example fault detection.

45.3.8 CONFIGURATION C2: STUCK-AT-1 FAULTS

The BUT configuration C2 is shown in Figure 45.10a. All the junctions are programmed to behave as diodes and all inputs are connected to "0." Hence, for a defect-free BUT, all the outputs should be "0." However, if any of the lines has a stuck-at-1 fault, the corresponding output will be "1." Test architectures TA-2a and TA-2b will be used in conjunction with this BUT configuration since the inputs are on the north and outputs on the east of the nanoblock. Consider, for instance, that a vertical line has a stuck-at-1 fault as shown in Figure 45.10b and hence is permanently tied to 1. As a result, the outputs will be "0."

45.3.9 CONFIGURATION C3: AND/OR BRIDGING FAULTS (H/H)

The bridging faults between adjacent horizontal wires can be detected by using the configuration shown in Figure 45.11a. The BUT is configured in such a way that the programmed diodes are forward-biased. Consider that there is AND bridging between the second and third horizontal wire as shown in Figure 45.11b, and as a result the second horizontal wire is pulled down to a "0"; the output will be inverted. Test architecture TA-1a and TA-1b will be used in conjunction with this BUT configuration.

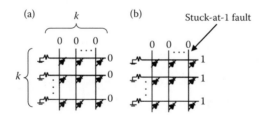

FIGURE 45.10 (a) BUT configurations and test patterns for C2. (b) Example fault detection.

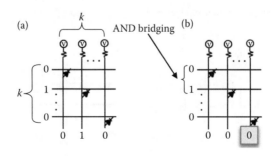

FIGURE 45.11 (a) BUT configurations and test patterns for C3. (b) Example fault detection.

45.3.10 Configuration C4: AND/OR Bridging Faults (v/v)

This is similar to the previous configuration, and tests for bridging faults between the vertical wires. The configuration is as shown in Figure 45.12a. The inputs applied and the expected outputs are as shown. However, if there is any AND/OR bridging between the vertical wires, then the output changes. Consider, for example, that the first vertical wire changes to a "1" due to OR bridging with its neighboring wires. This is shown in Figure 45.12b. The diode on that wire will now be reverse-biased and hence act as an open switch and the corresponding horizontal output will become a "1." Test architecture TA-2a and TA-2b will be used in conjunction with this BUT configuration.

45.3.11 Configuration C5: Reverse-Biased Diode and v/h Bridging Faults

All junctions are programmed to behave as diodes. Horizontal wires are connected to a "1" while the vertical wires are connected to a "0" as shown in Figure 45.13a. The output can be taken either from the east side or from the south side. Hence, either of the two test architecture sets can be used. Such a configuration reverse biases all the junction diodes and the outputs are all 1's if taken on the east side and all 0's if taken on the south side. In the figure shown, outputs are taken on the east side. If any of the reversed-biased diode is defective and has a small resistance, or there is bridging between the horizontal and vertical wires, the output on the east side will be pulled down to a "0." This configuration thus detects defective reversed-biased diodes and bridging faults between the vertical and horizontal wires. Consider the example shown in Figure 45.13b. The circled diode is faulty and offers a small resistance in the reverse-biased state. Hence, the corresponding horizontal output changes to a 0.

45.3.12 Configuration C6: Forward-Biased Diode Faults

For a $k \times k$ nanoblock, to detect forward-biased diode faults, we need a total of k configurations to test all the cross points. Figure 45.14a shows these configurations. Here, the vertical wires are connected to V_{dd} and only one junction is programmed as a diode on each of the horizontal and vertical wires in each configuration. Thus, in the first configuration, we program the cross points in a diagonal fashion. The different subconfigurations are obtained by shifting the programmed cross points one position to the right. The outputs are "0" for a defect-free BUT since the diodes

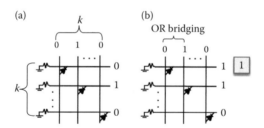

FIGURE 45.12   (a) BUT configurations and test patterns. (b) Example fault detection.

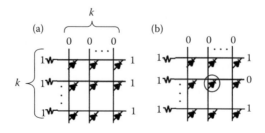

FIGURE 45.13   (a) BUT configurations and test patterns for C5. (b) Example fault detection.

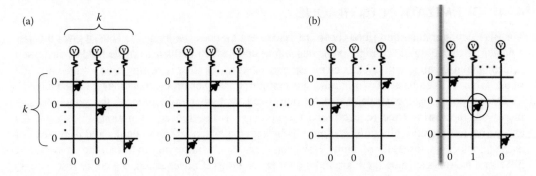

FIGURE 45.14 (a) BUT configurations and test patterns for C6. (b) Example fault detection.

are forward-biased and transmit a "0" to the output. If any of the forward-biased diodes is defective as shown in Figure 45.14b, the corresponding vertical wire output will be pulled up to a "1." Test architecture TA-1a and TA-1b will be used in conjunction with this BUT configuration.

45.3.13 Fault Coverage and Number of Configurations

Table 45.2 summarizes the various faults covered using these configurations, and the BUT configuration test architectures to be used for each of the faults. A total of $2k + 10$ configurations are needed to the test the entire nanofabric for the given faults. This number is reasonably small compared to other BIST techniques and also provides sufficient fault coverage. Also, it is independent of the number of nanoblocks in a nanofabric. These configurations can be modified as per the optimization technique that is presented in Section 45.4. The optimization technique reduces the number of configurations and/or the number of cross points programmed, thereby reducing the testing time.

TABLE 45.2
Fault Coverage and Configurations

| | Configuration | | | | | |
| | Set I | | | | Set II | |
Fault Model	C1	C2	C3	C4	C5	C6
Stuck-at-0	x					
Stuck-at-1		x				
open-line	x					
AND bridging (h/h)			x			
AND bridging (v/v)				x		
OR bridging (h/h)			x			
OR bridging (v/v)				x		
OR bridging (v/h)					x	
Cross points (forward)						x
Cross points (reverse)					x	
Configurations/test patterns	1	1	1	1	1	k
Test architecture	TA-1	TA-2	TA-1	TA-2	TA-1/TA-2	TA-1
Elimination of redundant configurations	N	N	N	N	N	Y
Customization for effective cross-point testing	N	N	N	N	Y	Y

45.4 OPTIMIZATION TECHNIQUE

The BUT configurations described above are used to test for one or more faults. Thus, it gives the user a flexibility to choose the configurations depending upon which faults are targeted. Also, each fault detecting configuration will have its own cost associated with it, that is, testing time, complexity, and so on. Table 45.2 can be used to minimize the configurations to be used depending on these factors.

The proposed testing strategy can be optimized such that utilization of the nanoblocks is increased, the testing time reduced, and the number of required configurations decreased. An exhaustive test includes all configurations C1 through C6, where all cross points and nanowires of each nanoblock are tested for all faults. However, it may not be necessary to test all the cross points/ NWs in a nanofabric since there are redundant cross points. Consequently, the entire test process includes unnecessary checks and it is possible to test only a subset of cross points and NWs, which are required to implement a logic function. For example, the optimization scheme would select a subset of tests that target only the needed cross points out of all k^2 ones. First, we tabulate the list of cross points/NWs tested in each configuration. Next, we remove columns that correspond to the cross points not required by the targeted logic function. Then, using Petrick's method of reduction, we can obtain the minimal test set required for the targeted faults. In fact, Petrick's method will list all possible combinations of the tests that are required. Consequently, it is straightforward to assign various weights to performance metrics such as testing time and configuration time, thus enabling a more flexible optimization.

A typical logic implementation in nanoblocks realizes AND and OR gates only [20], since these are the basic gates employed when implementing any logic function. Note that nanoblocks require an external component to implement a NOT logic function, thus complementing the minimal and complete set of gates. Using a $k \times k$ nanoblock, we can implement an AND/OR gate with $(k − 1)$ inputs.

Figure 45.2 shows an example of four-input AND/OR gate implemented using 4×4 nanoblock. It can be observed that all NWs need to be fault-free. However, only five out of the total sixteen cross points are programmed and take part in implementing the logic function. We will refer to these cross points as *effective cross-points*. Similarly, for a 4×4 nanoblock, to implement a three-input AND/OR gate, there are six effective cross points. In general, for a $k \times k$ nanoblock, we can implement a $(k − 1)$ input AND/OR gate, using only $2(k − 1)$ effective cross points while still needing to test all NWs. The BUT configurations and test patterns described in Section 45.4.2 have been split into two sets—set I that tests the nanowire faults and set II that tests the cross-point faults. Thus, it is possible to optimize the BUT configurations C5 and C6, which belong to set II, in such a way so as to test only the effective cross points.

The performance will be evaluated using the utilization metric, which is a probability that a nanoblock is successfully used to implement a logic function. Considering the effective cross points, the utilization of nanoblocks can be improved since faults in unused cross points will neither be tested nor interfere with correct function implementation. The increase in utilization is due to the fact that we can still use a defective nanoblock to implement a logic function as long as the effective cross points are defect-free.

Consider configuration C5 where all the cross points are tested. This needs to be modified to test only the required cross points. The optimized configuration C5 is shown in Figure 45.15. The entire test targets only the effective cross points based on the assumption that nanoblocks implement only the AND and OR gates. For this, we need to program $(k − 1)$ cross points on one vertical line and $(k − 1)$ cross points on one horizontal line. The number of cross points programmed (and tested) is thus reduced from k^2 to $2(k − 1)$. This reduces a lot of time overhead associated with programming of the cross points and also increases the utilization of the nanoblock.

Now consider the configuration C6 shown in Figure 45.14a where all the cross points are tested. We will demonstrate the optimization using a 3×3 nanoblock for simplicity. Three subconfigurations are needed to test all the cross points (for a 3×3 nanoblock). We describe two approaches to optimizing C6 and note that each has its own pros and cons.

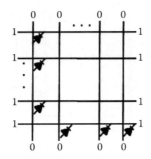

FIGURE 45.15 Customized configuration C5.

45.4.1 Elimination of Redundant Configurations

It is possible to eliminate the efforts required in designing new configurations to test only the effective cross points by selecting a subset from the defined subconfigurations of C6. Since C6 needs a total of k subconfigurations to test all the cross points, we can lay out a table that lists the different cross points tested in each of the configuration of C6, and then from this table select only the set of configurations that are needed to test the effective cross points. This is demonstrated below in Table 45.3.

Table 45.3 shows the cross points tested in each of the subconfigurations of C6 for a 3×3 nanoblock. Here, the assumption is that the nanoblocks are used to implement AND and OR gates only. Hence, from Figure 45.3, we see that there are only four effective cross points—cp11, cp21, cp32, and cp33. From Table 45.3, it can be seen that these four cross points can be tested in two configurations C61 and C63 and the configuration C62 is redundant and can be avoided. Thus, we need only two configurations instead of three as a result of optimization. This can be extended to a $k \times k$ nanoblock, wherein the total number of configurations required is now $(k-1)$ instead of k. Moreover, since the number of cross points to be programmed is reduced, this leads to a considerable reduction in testing time.

45.4.2 Customized Configurations

The advantage of the above approach is that there is no need to design new configurations. We can simply select the desired configurations with the help of a table. This is simpler and less time consuming. However, it can be seen in Table 45.3 that, in addition to the effective cross points, some additional cross points are also tested (six cross points in the above case instead of four), which are not required. This reduces the utilization slightly as opposed to the maximum utilization that can be obtained. To have maximum utilization, we can design custom configurations wherein we test only the cross points, which are necessary. Figure 45.16 shows the customized configurations for C6 for a 3×3 nanoblock.

TABLE 45.3

Junction Cross Points Tested in Each Configuration

Configuration	Junction Cross Points Tested								
	cp11	cp12	cp13	cp21	cp22	cp23	cp31	cp32	cp33
C61	x				x				x
C62		x				x	x		
C63			x	x				x	

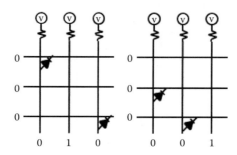

FIGURE 45.16 Customized configuration C6.

Here, again, we need just two configurations but we are testing only the effective cross points. This gives the maximum utilization with a slightly reduced testing time than the first approach. However, this demands that we design new customized configurations and not reuse existing ones.

45.4.3 Recovery Procedure

Once the entire testing phase has been completed and the faulty nanoblocks identified, we can have an additional recovery procedure to diagnose the faulty nanoblocks and further increase the utilization of nanoblocks. This is optional and may be carried out only if utilization is a major concern since this will increase the total testing time.

A defective nanoblock implies that it cannot implement the AND and OR logic within the same nanoblock. However, since the cross points used to implement the AND and OR logic are distinct, we can still have nanoblocks among the defective ones, which can implement either of the two functions but not both. We can have an additional testing phase for the defective nanoblocks to test for the AND or OR implementation. The BUT configurations need to be modified accordingly since we are testing a nanoblock only for one implementation; only the corresponding cross points need to be tested. Thus, at the end of these test phases, we have three sets of usable nanoblocks: (i) those that can be used to implement both the AND and OR functions, (ii) those that can implement only the OR function, and (iii) those capable of implementing only the AND function.

45.5 ANALYSIS AND DISCUSSION

45.5.1 Number of Configurations/Test Patterns

The BIST approaches developed by the authors in Refs. [16,17] have been discussed in Section 45.2. It can be seen from Table 45.1 that the total number of configurations/test patterns required to test the entire nanofabric is $2k + 10$. After using the optimization technique, it results in $2(k-1) + 10$. From the analysis of BUT configurations in Section 45.3.3, it can be deduced that only configuration C6 depends on the value of k while the rest of them are independent of k.

45.5.2 Nanoblock Utilization

We defined the term "nanoblock utilization" as the probability that a nanoblock can be successfully utilized to implement a logic function in the presence of defects. We will derive the expression for nanoblock utilization for the optimized technique and compare it with that of a nonoptimized approach.

Assume a nanoblock of size $k \times k$. Let p be the probability of a faulty cross point. For a nonoptimized testing approach, a nanoblock is deemed as faulty if at least one of its cross points is found to be faulty. Hence, the condition for a nanoblock to be utilized successfully is that all its cross points have to be fault-free.

Utilization (nonoptimized) = probability {all cross points are fault-free}

$$U1 = (1 - p)^{k^2} \tag{45.1}$$

For the optimized technique, as described in Section 45.5.1, there are $2(k-1)$ effective cross points for a $k \times k$ nanofabric. For the nanoblock to be successfully utilized to implement the logic function (AND/OR), these $2(k - 1)$, cross points need to be fault-free. Hence, the utilization in this case is given by:

Utilization (optimized) = probability$\{2(k - 1)$ cross points are fault-free$\}$

$$U2 = (1 - p)^{2(k-1)} \tag{45.2}$$

45.5.3 Testing Time

The total testing time for the nanofabric can be split into a number of components. Let t_{config} denote the total time required to configure the BUT, that is, to program the various cross points and set up power connections. Thus, t_{config} would not be the same for all configurations since the number of cross points programmed is not the same for all the configurations.

Let t_p denote the time required to program a single cross point and t_s be the time required to set up the V_{dd}/GND connections to the wires. Let t_{test} denote the time required to apply a test pattern to the BUT, wait for the comparator outputs to be generated, read the comparator outputs, and generate the partial defect map. Here, it is assumed that $t_{test} \gg t_{config}$. Using these different components along with the procedure described in Ref. [2], the total time required for testing a nanofabric can be calculated for a $k \times k$ nanoblock.

Let T_i denote the testing time for configuration Ci (as described in Section 45.4.2), T_i^* the testing time for the configuration Ci after using the optimization technique, and T_i^{**} the testing time after using the customization approach. The total testing time is given by the sum of the testing times for each of the individual configurations. Summary of improvement in testing times is shown in Table 45.4.

$$T_1 = T_2 = T_5 = 2 * \left\{ (k^2 * t_p) + t_s + t_{test} \right\} \tag{45.3}$$

$$T_3 = T_4 = 2 * \left\{ (k * t_p) + t_s + t_{test} \right\} \tag{45.4}$$

$$T_6 = 2 * \left\{ \left[(k * t_p) + t_{test} \right] * k + t_s \right\} \tag{45.5}$$

$$T_1^* = T_2^* = 2 * \left\{ (k^2 * t_p) + t_s + t_{test} \right\} \tag{45.6}$$

$$T_3^* = T_4^* = 2 * \left\{ (k * t_p) + t_s + t_{test} \right\} \tag{45.7}$$

$$T_5^* = 2 * \left\{ (2(k - 1) * t_p) + t_s + t_{test} \right\} \tag{45.8}$$

$$T_6^* = 2 * \left\{ \left[(k * t_p) + t_{test} \right] * (k - 1) + t_s \right\} \tag{45.9}$$

TABLE 45.4

Summary Comparison of the Testing Time

		Change due to Optimization	Change due to Customization
Wire testing time	C1	$T_1{}^* - T_1 = 0$	$T_1{}^{**} - T_1 = 0$
	C2	$T_2{}^* - T_2 = 0$	$T_2{}^{**} - T_2 = 0$
	C3	$T^{3*} - T_3 = 0$	$T_3{}^{**} - T_3 = 0$
	C4	$T_4{}^* - T_4 = 0$	$T_4{}^{**} - T_4 = 0$
Cross point testing time	C5	$T_5{}^* - T_5 = -2t_p[k-1]^2$	$T_5{}^{**} - T_5 = -2t_p[k-1]^2$
	C6	$T_6{}^* - T_6 = -2[kt_p + t_{\text{test}}]$	$T_6{}^{**} - T_6 = -2k[(k+2)t_p - t_{\text{test}}]$
Total difference		$-2[t_p(k^2 + k - 1) - t_{\text{test}}]$	$-2[t_p(2k^2 - k + 1) - kt_{\text{test}}]$

$$T_5^{**} = 2 * \left\{ (2(k-1) * t_p) + t_s + t_{\text{test}} \right\} \tag{45.10}$$

$$T_6^{**} = 2 * \left\{ \left[(k-1) * t_p + t_{\text{test}} \right] * (k-1) + t_s \right\} \tag{45.11}$$

45.5.4 Recovery Procedure

The recovery procedure is discussed in Section 45.4.3. Here, the nanoblock would be usable if it can implement either an AND or OR logic but not both. Though the functionality of the nanoblock is reduced, it is still usable. Hence, we would expect an increase in the utilization of the nanoblock.

If we assume the parameters defined in Section 45.5.2, then the utilization after using the recovery procedure would be given as

$$U_3 = (1 - p)^{k-1} \tag{45.12}$$

45.6 RESULTS

Figure 45.17 shows the comparison between the proposed method and the approaches discussed in Refs. [16,17] with respect to the number of configurations required for testing a $k \times k$ nanofabric. The proposed method reduces the number of required configurations over the schemes in Refs. [16,17] since configurations without the effective cross points are eliminated from the test process. Consequently, the testing time for the proposed scheme reduces as shown in the analysis in Section 45.5.

Figure 45.18 shows the plots of (1) and (2) for a value of k between 3 and 10. The optimization technique improves the utilization of the nanoblock since partially defective blocks are not marked as faulty ones, provided the desired logic gates could be correctly implemented.

Figure 45.19 shows the total testing time for a nanoblock size varying between 3 and 10. The improvement in testing time increases with nanoblock size (k) since the fraction of effective cross points to the total number of cross points decreases with the nanoblock size. The number of effective cross points grows linearly with the nanoblock size, while the total number of cross points increases exponentially (k^2).

Figure 45.20 shows the plot of (45.12) for a nanoblock size between 3 and 10. The utilization of the nanoblocks improves since the customized configurations target only the effective cross points instead of fewer configurations. Consequently, only defects related to the effective cross points are detected and will mark the block as faulty. This is in contrast to the only optimization technique where unnecessary configurations were removed from the test while still testing the redundant cross points in the remaining configurations. However, since we are testing for partial

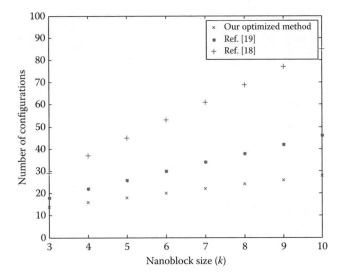

FIGURE 45.17   Number of configurations required in comparison with the approaches discussed in Refs. [18] and [19].

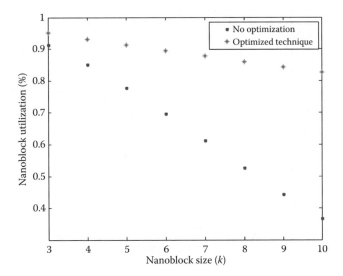

FIGURE 45.18   Improvement in the utilization of the nanoblock.

functionality, we need to carry out additional tests. This results in an increase in the number of tests and testing time. Figure 45.21 shows the increase in testing time as a result of using the recovery procedure. Hence, the improvement in utilization is obtained at the expense of increased testing time. As a result, the recovery procedure should be used only in cases where utilization is more important than testing time.

45.7   CONCLUSION

The novel testing procedure presented, has been shown to outperform the existing testing approaches. The proposed configurations and test patterns provide 100% fault coverage for stuck-at, stuck-open, connection, and bridging faults. The parallel testing architecture does not increase testing time with

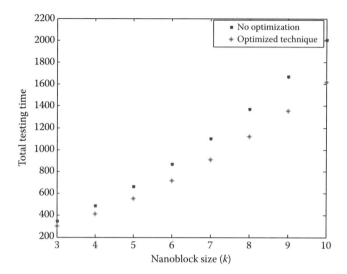

FIGURE 45.19   Total testing time.

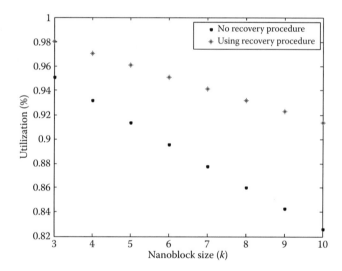

FIGURE 45.20   Improvement in utilization using recovery procedure.

nanofabric size, thus making it suitable for the dense architectures. Also, it requires only $2k + 10$ configurations, which corresponds to 23% reduction over the method discussed in Ref. [16] and 55% reduction over the method proposed in Ref. [17] for a nanoblock size of $k = 5$. Moreover, using the optimization technique described here, the nanoblock usability is increased from 36% to 76% for nanoblock size $k = 5$. A significant reduction in the testing time of the nanofabric is also observed. For example, for $k = 5$, the testing time is reduced by 29%. Moreover, the proposed recovery procedure can tailor testing to a particular functionality (either an AND gate or an OR gate). This further increases the utilization of nanoblocks by 10%. In summary, the proposed technique is simple, efficient, quick, and outperforms the existing approaches. Further work will include theoretical analysis to demonstrate the optimality of the proposed approach, and later the development of a new, fault-tolerant design methodology.

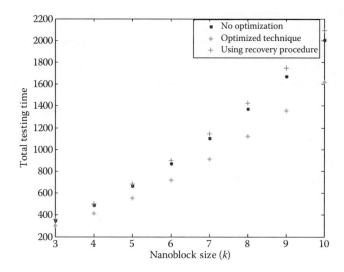

FIGURE 45.21 Increase in testing time as a result of using recovery procedure.

REFERENCES

1. S.C. Goldstein and M. Budiu, NanoFabric: Spatial computing using molecular electronics, *Proceedings of the International Symposium in Computer Architecture*, Victoria, Espírito Santo, Brazil, pp. 178–189, 2001.
2. M. Mishra and S.C. Goldstein, Scalable defect tolerance for molecular electronics, *Workshop on Non-Silicon Computing*, Cambridge, MA, 2002.
3. S.C. Goldstein and D. Rosewater, What makes a good molecular scale computer device? School of Computer Science, Carneige Mellon University, Tech. Rep. CMU-CS-02-181, September 2002.
4. M. Mishra and S.C. Goldstein, Defect tolerance at the end of the roadmap, *Proceedings of the International Test Conference (ITC'03)*, pp. 1201–1210, 2003.
5. D. Akinwande, S. Yasuda, B. Paul, S. Fujita, G. Close, and H.-S.P. Wong, Monolithic integration of CMOS VLSI and carbon nanotubes for hybrid nanotechnology applications, *IEEE Transactions on Nanotechnology*, pp. 636–639, 2008.
6. R.M.P. Rad and M. Tehranipoor, A reconfiguration-based defect tolerance method for nanoscale devices, *21st IEEE International Symposium on Defect and Fault Tolerance in VLSI Systems, DFT '06*, Washington DC, pp. 107–118, 2006.
7. D. Jianwei, W. Lei, and J. Faquir, Analysis of defect tolerance in molecular crossbar electronics, *IEEE Transactions on Very Large Scale Integration (VLSI) Systems*, pp. 529–540, 2009.
8. A.A. Al-Yamani, S. Ramsundar, and D.K. Pradhan, A defect tolerance scheme for nanotechnology circuits, *IEEE Transactions on Circuits and Systems I: Regular Papers*, pp. 2402–2409, 2007.
9. J.G. Brown and R.D.S. Blanton, CAEN-BIST: Testing the nanofabric, *Proceedings of the International Test Conference*, Charlotte, NC, pp. 462–471, 2004.
10. C. Stroud, S. Konala, P. Chen, and M. Abramivici, Built-in self-test of logic blocks in FPGAs (finally, a free lunch: BIST WithoutOverhead), *Proceedings of the IEEE VLSI Test Symposium (VTS'96)*, Princeton, NJ, pp. 387–392, 1996.
11. M.B. Tahoori, E.J. McCluskey, M. Renovel, and P. Faure, A multi-configuration strategy for an application dependent testing of FPGAs, *Proceedings of IEEE VLSI Test Symposium (VTS'04)*, Napa Valey, CA, pp. 154–159, 2004.
12. C. Metra, G. Mojoli, S. Pastore, D. Salvi, and G. Sechi, Novel technique for testing FPGAs, *Proceedings of Design, Automation and Test in Europe (DATE'98)*, Paris, France, pp. 89–94, 1998.
13. S.J. Wang and T.M. Tsai, Test and diagnosis of faulty logic blocks in FPGAs, *IEE Proceedings: Computers and Digital Techniques*, Kwangju, South Korea, pp. 100–106, 1999.
14. S. Kundaikar and M. Zawodniok, Optimized built-in self-test technique for CAEN-based nanofabric systems, *Proceedings of 11th IEEE Conference on Nanotechnology (IEEE-NANO'11)*, Portland, OR, pp. 1717–1722, 2011.

15. M. Tahoori, Application-dependent diagnosis of FPGAs, *Proceedings of the International Test Conference (ITC'04)*, Charlotte, NC, pp. 645–654, 2004.

16. Z. Wang and K. Chakrabarty, Built-in self-test of molecular electronics-based nanofabrics, *European Test Symposium*, Tallinn, Estonia, pp. 168–173, 2005.

17. M. Tehranipoor, Defect tolerance for molecular electronics-based NanoFabrics using built-in self-test procedure, *20th IEEE International Symposium on Defect and Fault Tolerance in VLSI Systems (DFT)*, Monterey, CA, 2005.

18. M.V. Joshi and W.K. Al-Assadi, A BIST approach for configurable nanofabric arrays, *8th IEEE Conference on Nanotechnology*, Arlington, TX, pp. 695–698, 2008.

19. G. Wang, On the use of Bloom filters for defect maps in nanocomputing, *Proceedings of ICCAD 2006*, San Jose, CA pp. 743–746, 2006.

20. M. Abramovici, E. Lee, and C. Stroud, BIST-based diagnostics for FPGA logic blocks, *Proceedings of the International Test Conference*, Washington, DC, pp. 539–547, 1997.

46 Adaptive Fault-Tolerant Architecture for Unreliable Device Technologies

Nivard Aymerich, Sorin Cotofana, and Antonio Rubio

CONTENTS

46.1 INTRODUCTION

Computer architecture constitutes one of the key and strategic application fields for new, emerging devices at nanoscale dimensions, potentially getting benefit from the expected high-component density and speed. However, these future technologies are expected to suffer from a reduced device quality, exhibiting a high level of process and environmental variations as well as performance degradation due to the high stress of materials [1–4]. This clearly indicates that if we are to make use of those novel devices, we have to rely on fault-tolerant architectures. Currently, most of the redundancy-based fault-tolerant architectures make use of majority gates (MAJ) [5,6] and require high area overhead. An alternative to majority gate voting is the averaging cell (AVG) [7–9], which can exhibit higher reliability at a lower cost. The underlying principle of the AVG is to average several input replicas in order to compute the most probable output value. This approach is quite effective in case the AVG inputs are subject to independent drifts with the same/similar magnitude. In practice, however, input deviations can be nonhomogeneous, in which case the balanced average cannot provide a response that minimizes the output error probability.

To alleviate this problem, we propose the adaptive average cell (AD–AVG) [10], which is able to better cope with nonhomogeneous input drifts. This technique uses the principle of nonparity voting and gives more weight to the inputs that are known to be more reliable to the detriment of the less reliable ones. This unbalanced voting is performed by adjusting the relative magnitudes of each average weight.

In Section 46.2, we present the AVG architecture and its main features, and we obtain the equation to calculate the output error probability. In Section 46.3, we introduce the idea of unbalanced voting with the AD–AVG technique and demonstrate that it is possible to optimize the values of the average weights in order to minimize the output error probability. We also find the analytic expression of the optimal weights. In Section 46.4, we present the results of Monte Carlo simulations of the AVG structure with balanced and optimal unbalanced weights and compare the behavior of both approaches in the presence of the heterogeneous variability levels in the input replicas. Our experiments indicate that the proposed method of optimal unbalanced weights is more robust against degradation effects and external aggressions, and that for the same reliability level it requires a lower redundancy level (less area overhead). In Section 46.5, we propose a technology implementation for the AD–AVG structure based on switching resistance crossbars and explore its behavior by means of Monte Carlo simulations. Finally, we present the conclusions.

46.2 THE AVERAGING CELL

The averaging cell (AVG), widely known for its application in the four-layer reliable hardware architecture (4LRA) [9], stems from the perceptron, the McCulloch–Pitts neuron model [11,12]. Associated with fault-tolerant techniques based on redundancy, the AVG graphically depicted in Figure 46.1 can calculate the most probable value of a binary variable from a set of error-prone physical replicas. While the MAJ-based voting technique operates in the digital domain, the AVG performs a weighted average of the replicated inputs in the analog domain. There are R physical replicas of an ideal variable y:

$$y_i = y + \eta_i \quad i = 1,...,R \tag{46.1}$$

Each of the input replicas y_i has associated an independent drift η_i that modifies its ideal value y. As a consequence, input signals y_i are observed in the system as continuous voltage levels, where 0 and V_{cc} stand for ideal logical values "0" and "1," respectively. The output of the AVG $\hat{y}$ is an estimation of y according to (46.2) and (46.3).

$$y' = W(y_1,...,y_R) = \frac{1}{\sum_{i=1}^{R} k_i} \sum_{i=1}^{R} k_i y_i \tag{46.2}$$

$$\hat{y} = T(y') = \begin{cases} V_{cc} & \text{if } y' \geq V_{cc}/2 \\ 0 & \text{if } y' < V_{cc}/2 \end{cases} \tag{46.3}$$

To simplify the mathematical formulation, we use the normalized weights $c_i = k_i/\Sigma_{j=1}^{R} k_j$ instead of weights k_i, $i = 1, ..., R$. We model the drift magnitudes η_i as Gaussian random variables with a null mean and a standard deviation σ_i, $\eta_i \sim N(0, \sigma_i)$.

When y' is processed by the threshold operation $T(y')$, an error is produced if and only if the deviation in the weighted average $\epsilon = y' - y$ reaches $V_{cc}/2$ or $-V_{cc}/2$, depending on the ideal logic value y. Since this deviation parameter ϵ can be expressed as a linear combination of normally distributed variables η_i, by the properties of the normal distribution, the probability density function (PDF) $f_\epsilon(\epsilon)$ can be described as a normal distribution with parameters:

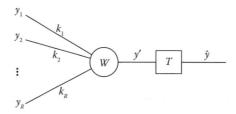

FIGURE 46.1 Averaging cell (AVG) architecture. (N. Aymerich et al. Adaptive fault-tolerant architecture for unreliable device technologies, *11th IEEE International Conference on Nanotechnology*, Portland, Oregon, pp. 1441–1444 © (2011) IEEE. With permission.)

$$\mu_\epsilon = E\left\{\sum_{i=1}^{R} c_i y_i\right\} - y = 0 \tag{46.4}$$

$$\sigma_\epsilon^2 = \sigma_{y'}^2 + 0 = \sum_{i=1}^{R} c_i^2 \sigma_i^2 \tag{46.5}$$

Thus, using the complementary Gauss error function, we can formulate analytically the output error probability as Equation 46.6. Figure 46.2 depicts the relationship between $\sigma_{y'}$ and P_e and its monotonically increasing behavior. Thus, given a requirement of output error probability P_e, there is a maximum admissible output standard deviation $\sigma_{y'}$. For example, an output error probability of $P_e < 2 \times 10^{-4}$ implies having an output standard deviation lower than $\sigma_{y'} < 0.14\,\mathrm{V}$.

$$P_e = \int_{V_{cc}/2}^{\infty} f_\epsilon(\epsilon)d\epsilon = \frac{1}{2} \times erfc\left(\frac{V_{cc}/2}{\sqrt{2\sigma_{y'}^2}}\right) \tag{46.6}$$

A previous AVG-based study assumes homogeneous input drifts [8], a case in which a balanced weight set produces a weighted average y' with a minimum standard deviation $\sigma_{y'}$, and therefore the output error probability P_e is minimized. However, when the input drifts lose homogeneity due to aging and variability effects, the output standard deviation increases and so does the output error probability P_e. Under these conditions, the error probability P_e becomes highly dependent on the level of heterogeneity. In the next sections, we make use of the fact that the replica weights can be different and introduce a method to compute them in such a way that P_e is minimized even under severe nonhomogeneous aging and variability conditions.

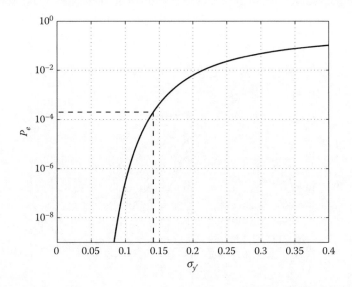

FIGURE 46.2 Output error probability P_e against the output standard deviation $\sigma_{y'}$. The relationship is monotonically increasing; thus, $P_e < 2 \times 10^{-4}$ implies having an output standard deviation lower than $\sigma_{y'} < 0.14\,\mathrm{V}$. (N. Aymerich, S. Cotofana, and A. Rubio. Adaptive fault-tolerant architecture for unreliable device technologies, *11th IEEE International Conference on Nanotechnology*, Portland, Oregon, pp. 1441–1444 © (2011) IEEE. With permission.)

46.3 THE ADAPTIVE AVERAGING CELL

The AVG provides robustness when all the inputs are under the same aggression factors, in which case balanced weights provide the best drift compensation. However, in practice, this may not be the case, as some replicas may have a larger drift with respect to others. As a consequence, owing to this decompensation, the balanced weights approach becomes suboptimal. In this section, while preserving the AVG architecture, we consider the nonhomogeneity of the aggression factors and degradation effects. We propose to adjust the AVG weighting scheme according to the following principle: assign greater weight to the less degraded and more reliable inputs, and lower weight to the ones that are more prone to be unreliable. Inductively speaking, such an approach should improve the overall reliability.

Before the calculation of the optimal set of weights, we demonstrate that it always exists regardless of the environmental variability conditions. To show this, we analyze the sensitivity of the error probability P_e against the variation of one specific weight c_j. We show that there is always a minimum in the error probability and thus an optimal value for the weight c_j. For the analysis, we assume different levels of variability in the input j modeled with the parameter σ_j ranging from 0 to 0.4 V. The rest of the inputs are considered all together with a fixed contribution to the variability modeled by the standard deviation parameter $\sigma_{y'} \mid_{\{c_j=0\}} = 0.2$ V. In these conditions, taking into account that increasing the weight c_j implies decreasing the rest of the weights, according to the restriction $\sum_{i=1}^{R} c_i = 1$, the variance of the weighted average can be expressed as $\sigma_{y'}^2 = c_j^2 \sigma_j^2 + (1 - c_j)^2 \sigma_{y'}^2 \mid_{\{c_j=0\}}$. Using this result and substituting into Equation 46.6, we can analyze the influence of c_j on the error probability P_e. Figure 46.3 depicts the error probability against the weight c_j for different levels of input variability. One can observe the different locations of the P_e minimum and the relation between optimal weights and different levels of variability.

The following comments can be extracted from Figure 46.3:

- There is always one and only one value of weight c_j in the range from 0 to 1 that minimizes the error probability P_e in each possible variability environment.
- The optimal value of weight c_j is never exactly equal to 0. Even for large levels of deviation in the jth input with respect to the others, it is useful to have a contribution from the input j.

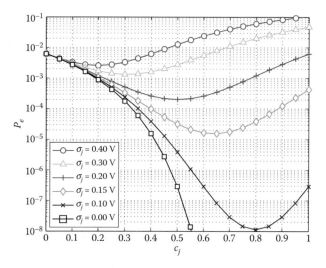

FIGURE 46.3 Variation of the error probability P_e against the weight c_j. Standard deviation in the input j ranges from $\sigma_j = 0$V to $\sigma_j = 0.4$V and the standard deviation of the weighted average due to the rest of the inputs is $\sigma_{y'}^2 \mid_{\{c_j=0\}} = 0.2$ V. (N. Aymerich, S. Cotofana, and A. Rubio. Adaptive fault-tolerant architecture for unreliable device technologies, *11th IEEE International Conference on Nanotechnology*, Portland, Oregon, pp. 1441–1444 © (2011) IEEE. With permission.)

- The optimal value of weight c_j is never exactly equal to 1. Even for small levels of deviation in the jth input with respect to the others, it is useful to have a contribution from the rest of the inputs.
- Based on one's intuition, in order to minimize the output error probability P_e, higher weights (close to 1) should be assigned to less deviating inputs and vice versa.
- When the deviating effect in the input j (σ_j) is equal to the deviating effect of the rest of the inputs ($\sigma_{y'}^2 \mid_{\{c_j=0\}}$), the optimal value for the weight c_j is one-half. Note that in Figure 46.3 the curve with minimum at $c_j = 0.5$ ($\sigma_j = \sigma_{y'} \mid_{\{c_j=0.5\}}$).

The set of weights that minimizes the output error probability P_e can be found analytically by minimizing the standard deviation of the weighted average $\sigma_{y'}$; this magnitude is directly related to P_e as Equation 46.6 evidences. In order to perform this minimization considering all the weights c_i simultaneously, we make use of the Lagrange multipliers as there is an additional restriction to be held for the weights. The target function to minimize is the standard deviation of the weighted average $\sigma_{y'}$, or equivalently the variance $\sigma_{y'}^2$, and the variables are the magnitudes of the averaging weights c_j. The condition to be held is the restriction of normalized weights ($\Sigma_{i=1}^R c_i = 1$). Therefore, the target function is

$$F\left(c_1, c_2, ..., c_R, \lambda\right) = \sigma_{y'}^2 - \lambda\left(\sum_{i=1}^R c_i = 1\right) \tag{46.7}$$

Differentiating with respect to the normalized weights c_j, recall Equation 46.5, and Lagrange multiplier λ and equating to zero to find the minimum, we obtain the following equations:

$$\frac{d(F)}{dc_j} = 2c_j^{opt}\sigma_j^2 - \lambda = 0 \quad j = 1,...,R \tag{46.8}$$

$$\frac{d(F)}{d\lambda} = 1 - \sum_{i=1}^R c_i^{opt} = 0 \tag{46.9}$$

Equation 46.8 shows that the optimal weights are inversely proportional to the input variances $c_j^{opt} = A/\sigma_j^2$ and Equation 46.9 expresses the condition of normalized weights. Combining both conditions, we deduce the value of A:

$$A = \frac{1}{\sum_{i=1}^R 1/\sigma_i^2} \tag{46.10}$$

Now, we can calculate the explicit formula for the optimal weights. It depends only on the input variances σ_j^2:

$$c_j^{opt} = \frac{1}{\sum_{i=1}^R \sigma_j^2/\sigma_i^2} \quad j = 1,...,R \tag{46.11}$$

This is the optimal configuration of weights that minimizes the error probability P_e. Depending on the input drifts distribution, each weight should be tuned according to Equation 46.11 in order to achieve the lowest possible output error probability P_e. We can observe how the optimal weights are small when the input variance is large and vice versa. We can also calculate the value of the standard deviation $\sigma_{y'}$ when we apply optimal weights; this value is the minimum of $\sigma_{y'}$ and coincides with A ($\sigma_{y'min} = A$). So it is also possible to express each optimal weight in terms of its input variance and the minimum variance of the weighted average:

$$c_j^{opt} = \frac{\sigma_{y'\,min}^2}{\sigma_j^2} \quad j = 1,...,R \tag{46.12}$$

The optimal configuration has all the weights directly proportional to the constant $\sigma_{y'\,min}^2$ and inversely proportional to its input variance σ_i^2. A further comment can be made regarding the analytical expression of the optimal weights: the particular case in which one or more inputs have null variability $\sigma_i = 0$ has to be treated separately. First, we have to say that in practice this situation will never happen because there is always at least a small contribution of noise that affects all the input signals. However, if this case occurs, then the minimization of the output error probability would be straightforward: assigning the maximum weight to the input with null variability $c_i = 1$, we would obtain a null output error probability $P_e = 0$, see Equations 46.5 and 46.6.

46.4 COMPARISON: BALANCED VERSUS UNBALANCED AVG

To assess the implications of our proposal, we carried out a reliability analysis for the AVG with balanced and optimal unbalanced weights. Given a nonuniform input drift scenario, we calculated the percentage of circuits that achieve an output error probability P_e lower than 2×10^{-4} (equivalently $\sigma_{y'} < 0.14$ V). We used this condition as criteria to define the yield of the AVG circuit. In the simulation, in order to reproduce realistic heterogeneous environments, the per replica drift variances were generated with a chi-square distribution $\sigma_i^2 \sim \chi_1^2$. We modeled the effect of degradation using different mean values for the drift distribution σ_m. In the experiment, we used two levels: the first one corresponding to less degraded environments ($\sigma_m = 0.6$ V) and the second one reproducing more degraded environments ($\sigma_m = 0.8$ V). Using Monte Carlo computation, we estimated for both approaches (balanced and unbalanced weights) the yield of circuits. Figure 46.4 presents the results

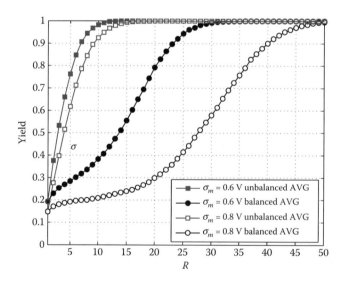

FIGURE 46.4 Percentage of AVG circuits that accomplish $P_e < 2 \times 10^{-4}$ against the redundancy level R at different levels of heterogeneous variability. Circle markers correspond to the conventional AVG and square markers correspond to the optimal unbalanced AVG approach. (N. Aymerich, S. Cotofana, and A. Rubio. Adaptive fault-tolerant architecture for unreliable device technologies, *11th IEEE International Conference on Nanotechnology*, Portland, Oregon, pp. 1441–1444 © (2011) IEEE. With permission.)

against the redundancy level R. One can observe in the figure that the AVG with unbalanced weights can deliver the same yield with a much lower redundancy factor R. For example, if we require a 90% yield in a technology with a level of heterogeneity that can be modeled with a chi-square distribution with σ_m between 0.6 and 0.8 V, the use of optimal unbalanced weights can save from 70% to 77% of redundancy level R with respect to the balanced approach.

Thus, we demonstrated that if we knew the distribution of deviations among the replicas, we could design better devices at lower cost. The capability of adjusting the average weights in the AVG structure provides a way to counteract the nonhomogeneous variability effects of degradation and external aggressions.

46.5 AD–AVG TECHNOLOGY IMPLEMENTATION

In this section, we propose a realistic technology implementation for the presented AD–AVG structure. This implementation is based on the use of switching resistance crossbars. In order to efficiently modify the configuration of weights as required by the structure and follow the changes of the input variability levels to maximize the AD–AVG reliability, we need a technology implementation with high reconfiguration capabilities. In this sense, the use of crossbars of resistive switching devices provides the required framework, such as memristor crossbars [14,15].

Figure 46.5 depicts the AD–AVG structure implementation using a switching resistance crossbar, a threshold block, a variability monitor, and a set of weight drivers. The complete architecture can be decomposed into three layers. The first one corresponds to the input layer and receives the input signals from the replicas. The second layer performs the averaging operation by means of the resistive switching crossbar configured taking into account the information provided by the variability monitor. The third one corresponds to the decision layer; it restores the binary output value by means of a threshold function.

In the following, we analyze the practical repercussions of implementing the proposed AD–AVG structure by means of resistive switching crossbars. The basic idea of implementing different averaging weights for each input is to connect or disconnect more or less devices in the corresponding input line. Figure 46.6 reproduces a detailed view of the resistive switching crossbar layout used to implement the averaging weights. Each of the inputs is connected to a horizontal metal line that can be connected or disconnected to N vertical lines. There is one resistive switching device situated in each crossing point of the crossbar to connect or disconnect the lines. In Figure 46.6, black dots correspond to the connected devices. The state of each device can be easily controlled with the weight drivers applying specific configuration voltages to each vertical and horizontal line.

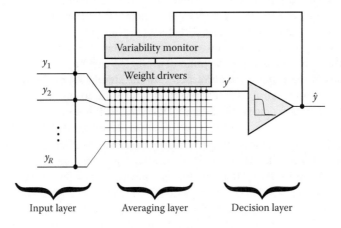

FIGURE 46.5 Adaptive averaging cell (AD–AVG) implementation in switching resistive crossbar technology.

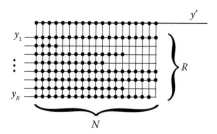

FIGURE 46.6 Crossbar layout view of the AD–AVG to configure the averaging weights.

Figure 46.6 shows that the output line, corresponding to the signal y', is connected to all the vertical lines, whereas the horizontal lines have a different number of connections. Using this feature, we can set up a network of interconnects that averages the input replicas with specific averaging weights. The resistive switching devices exhibit a different resistance depending on the state (R_{ON} and R_{OFF}). For example, in the case of memristors, the values of resistances are: $R_{ON} \approx 1\,M\Omega$ and $R_{OFF} \approx 1\,G\Omega$. Analyzing the equivalent circuit, we can calculate the analytic expression of signal y' in terms of the inputs y_i and the crossbar configuration of connections n_i, $i = 1,...,R$, see Equation 46.13. The n_i values correspond to the number of devices connected in the input line i. This result allows us to deduce the particular value of the averaging weights in terms of parameters n_i.

$$y' = \frac{1}{RNR_{ON} + \sum_{i=1}^{R} n_i\left(R_{OFF} - R_{ON}\right)} \sum_{i=1}^{R}\left[NR_{ON} + n_i\left(R_{OFF} - R_{ON}\right)\right] \times y_i \qquad (46.13)$$

In the following, we perform an experiment to prove the behavior of the proposed implementation. We simulate a 2-input AD–AVG implemented in a switching resistance crossbar with $N = 20$ vertical lines. We analyze the impact of the connecting and disconnecting devices in the structure against different levels of variability in the inputs. To perform this experiment, we run 1000 Monte Carlo simulations of the structure and find the AD–AVG yield for each possible configuration (from 0 to N devices connected in both input lines). Figure 46.7 reproduces the simulation result of the yield against the connected devices and for a homogeneous input variability level with $\sigma = 0.15$ V. We confirm that the yield is maximized when n_1 is equal to n_2, as expected. The homogeneity in the variability levels implies a symmetric optimal configuration of weights. However, we also observe that the larger the number of connected devices, the wider is the region with a high yield. This is an interesting result as it means that if we have difficulties in obtaining the exact value of variability in the inputs, we can look for configurations with a large number of connected devices to reduce the impact of nonperfect determination of the optimal averaging weights.

The second part of the experiment corresponds to the simulation of the same structure with a nonhomogeneous environment of variability with $\sigma_1 = 0.13$ V and $\sigma_2 = 0.17$ V. Figure 46.8 shows the simulation result for the yield of the 2-input AD–AVG. In this case, we observe that the maximum yield condition is not $n_1 = n_2$. As the input variability levels are different, the optimal weights are not balanced. Using the result (46.12), we can calculate $c1/c2$ as σ_2^2/σ_1^2; this yields the condition $(c_1/c_2) \cong 1.7$. Applying now the result (46.13), we get the condition required for the optimal averaging weights: $n_2 \cong -0.0082 + 0.6n_1$. If we assign the maximum possible value to $n_1 = 20$, then the optimal n_2 value is $n_2 = 12$, which is coherent with the simulation in Figure 46.8.

With this result, we prove that by using the proposed implementation for the AD–AVG it is easy to determine and configure the number of connections that we need in each input line to maximize the reliability. We just need a good estimation of the input levels of variability. Our simulation results correspond to a particular case of the 2-input AD–AVG structure, but the mechanism can

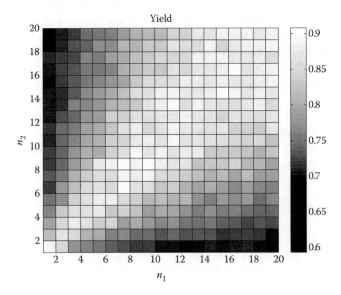

FIGURE 46.7 A 2-input AD–AVG yield against the connected devices in the input line 1 (n_1) and line 2 (n_2) for a homogeneous level of variability in the inputs with $\sigma = 0.15$ V.

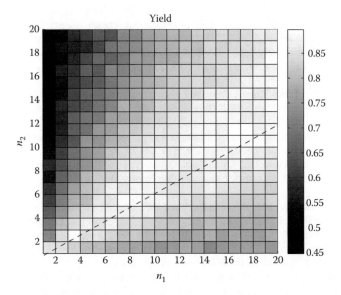

FIGURE 46.8 A 2-input AD–AVG yield against the connected devices in the input line 1 (n_1) and line 2 (n_2) for a nonhomogeneous variability level in the inputs with $\sigma_1 = 0.13$V and $\sigma_2 = 0.17$V. The dashed line corresponds to the optimal condition of the parameters n_1 and n_2 to maximize the yield ($n_2 \cong -0.0082 + 0.6 n_1$).

be easily extended to structures with more input replicas by distributing the crossbar connections among a larger number of input replicas.

46.6 CONCLUSIONS

In this section, we introduce an adaptive averaging cell (AD–AVG) structure tailored for realistic environments with nonhomogeneous variability. We demonstrate the efficiency of using adjustable weights in the averaging cell structure taking into account the heterogeneous variability levels among the replicas. We provide a method to calculate the optimal weights to minimize the output error probability and find that these values are inversely proportional to the input variances. The simulation of the AD–AVG technique indicates that it potentially results in substantial reliability improvement, at a lower cost than other equivalent state-of-the-art methods. A 70% reduction of the redundancy overhead has been found for realistic, nonhomogeneous scenarios with a chi-square distribution of input drift variances. We also propose a technology implementation based on the use of switching resistance crossbars and simulate its behavior, proving that it is possible to establish an adaptive mechanism to maximize the AD–AVG reliability according to the particular nonhomogeneous variability conditions.

REFERENCES

1. M. Mishra and S. Goldstein, Scalable defect tolerance for molecular electronics, in *First Workshop on Non-Silicon Computing*. Citeseer.
2. S. Borkar, T. Karnik, S. Narendra, J. Tschanz, A. Keshavarzi, and V. De, Parameter variations and impact on circuits and microarchitecture, in *Proceedings of the 40th Annual Design Automation Conference*. ACM, 2003, pp. 338–342.
3. S. Borkar, Electronics beyond nano-scale CMOS, in *Proceedings of the 43rd Annual Design Automation Conference*. ACM, 2006, pp. 807–808.
4. S. Borkar, Designing reliable systems from unreliable components: The challenges of transistor variability and degradation, *IEEE Micro*, 25(6), 10–16, 2005.
5. M. Stanisavljevic, A. Schmid, and Y. Leblebici, Optimization of nanoelectronic systems' reliability under massive defect density using cascaded R-fold modular redundancy, *Nanotechnology*, 19, 465202, 2008.
6. M. Stanisavljevic, A. Schmid, and Y. Leblebici, Optimization of nanoelectronic systems reliability under massive defect density using distributed R-fold modular Redundancy (DRMR), in *Defect and Fault Tolerance in VLSI Systems*, 2009. DFT'09. *24th IEEE International Symposium on. IEEE*, 2009, pp. 340–348.
7. A. Schmid and Y. Leblebici, Robust circuit and system design methodologies for nanometer-scale devices and single-electron transistors, *IEEE Transactions on Very Large Scale Integration(VLSI) Systems*, 12(11), 1156–1166, 2004.
8. F. Martorell, S. Cotofana, and A. Rubio, An analysis of internal parameter variations effects on nanoscaled gates, *IEEE Transactions on Nanotechnology*, 7(1), 24–33, 2008.
9. M. Stanisavljevic, A. Schmid, and Y. Leblebici, Optimization of the averaging reliability technique using low redundancy factors for nanoscale technologies, *Nanotechnology, IEEE Transactions on*, 8(3), 379–390, 2009.
10. N. Aymerich, S. Cotofana, and A. Rubio. Adaptive fault-tolerant architecture for unreliable technologies with heterogeneous variability, *IEEE Transactions on Nanotechnology*, 11(4), 818–829, July 2012.
11. W. S. McCulloch and W. Pitts, A logical calculus of the ideas immanent in nervous activity, *Bulletin of Mathematical Biophysics 5*, 115–133 (Reprinted in *Neurocomputing Foundations of Research* Edited by J.A. Anderson and E. Rosenfeld, the MIT Press, 1988), 1943.
12. W. Pitts and W. S. McCulloch, How we know universals: The perception of auditory and visual forms, *Bulletin of Mathematical Biophysics 9*, 127–147 (Reprinted in *Neurocomputing Foundations of Research* Edited by J.A. Anderson and E. Rosenfeld, the MIT Press, 1988), 1947.
13. N. Aymerich, S. Cotofana, and A. Rubio. Adaptive fault-tolerant architecture for unreliable device technologies, *11th IEEE International Conference on Nanotechnology*, Portland, Oregon, pp. 1441–1444, 2011.
14. G. Snider, Computing with hysteretic resistor crossbars, *Applied Physics A*, 80(6), 1165–1172, Mar 2005.
15. J. Borghetti, Z. Li, J. Straznicky, X. Li, D. Ohlberg, W. Wu, D. Stewart, and R. Williams, A hybrid nano-memristor/transistor logic circuit capable of self-programming, *Proceedings of the National Academy of Sciences*, 106(6), 1699–1703, Feb 2009.

Section XV

Nanowire Fabrication

47 Growth and Characterization of GaAs Nanowires Grown on Si Substrates

Jung-Hyun Kang, Qiang Gao, Hark Hoe Tan,
Hannah J. Joyce, Yong Kim, Yanan Guo, Hongyi Xu, Jin Zou,
Melodie A. Fickenscher, Leigh M. Smith, Howard E. Jackson,
Jan M. Yarrison-Rice, and Chennupati Jagadish

CONTENTS

47.1 INTRODUCTION

Recently, nanowires (NWs) have come into the spotlight for future integrated optoelectronic devices. III–V compound semiconductor NWs are unique, in that they can be grown epitaxially on Si substrates, which allow the integration of III–V-based optoelectronics with established Si microelectronics technologies [1–3]. That is to say, well-aligned NW morphology with vertical growth is a rudimentary necessity in the applications of III–V compound semiconductor NWs on Si substrates. However, several problems must be overcome to obtain the successful integration of III–V compound semiconductors with Si [4,5], including lattice and thermal expansion mismatch, and the formation of antiphase domains. For the case of Si, in particular, the existence of a native oxide layer at the interface between NWs and substrates is a major factor preventing the epitaxial growth of NWs [1]. To achieve nanoscale optoelectronic devices, it is also crucial to fabricate NWs with a high-quality crystal structure and controllable composition. For optoelectronic device applications, NW defects such as twin defects, phase polytypism, and stacking faults may give rise to the deterioration of optical and carrier transport properties and are therefore undesirable [6]. Overcoming these issues, namely, the interface defects between NWs and substrates and crystal structure defects, is the first priority to obtain perfect III–V compound semiconductor NWs on Si substrates for nanoelectronic device applications.

47.2 GROWTH OF III–V NANOWIRES ON SILICON SUBSTRATES

47.2.1 MORPHOLOGY OF III–V NANOWIRES GROWN ON SI (111) SUBSTRATES

Generally, III–V NWs grown on Si substrates without any special treatments have messy and randomly oriented morphology as shown by scanning electron microscopy (SEM) images in Figure 47.1. Here, GaAs NWs (Figure 47.1a), InP NWs (Figure 47.1b), and GaP NWs (Figure 47.1c) were grown on HF-etched Si (111) substrates using Au catalysts by metal–organic chemical vapor deposition (MOCVD). This messy morphology has resulted from the nonepitaxial growth of the NWs on the Si substrates, possibly due to the native oxide layer and interface defects.

In terms of the general planar growth of III–V semiconductor materials on Si, a high density of defects is the result due to their mismatch in lattice parameters and thermal expansion coefficients, and naturally, NWs are no exception. Owing to these mismatches, lattice expansion or contraction can take place by stressing its bonds in the growth direction, which will result in the formation of crystal defects at the interface between III–V and Si, once the so-called pseudomorphic growth regime is exceeded [7]. "Antiphase domain formation" arises during the growth of polar materials (III–V semiconductors) on nonpolar materials (Si). In terms of GaAs NWs on Si, for instance, silicon has a diamond crystal structure with homogeneous two-fcc sublattices (only Si atoms). However, GaAs has a zinc blende (ZB) structure, where the Ga and As atoms occupy a sublattice each. Therefore, the interface between the GaAs epilayer and Si substrates will have a defective boundary with the antiphase domains consisting of Ga–Ga and As–As bonding. In addition, the existence of an amorphous native oxide layer at the Si surface is a decisive factor to prevent the epitaxial growth of high-quality material on silicon [8].

Even though there are a few reports of epitaxial growth of straight and vertical III–V NWs on Si substrates [8–10], it is still a major challenge to prepare the surface properties of Si suitable for NW nucleation, such as substrate etching and cleaning, baking, and specific growth temperature. A simpler and easily reproducible approach is highly desirable to achieve epitaxial growth of III–V NWs on Si.

47.2.2 CRYSTAL STRUCTURE OF GAAS NANOWIRES GROWN ON SI (111) SUBSTRATES

A typical morphology of GaAs NWs grown directly on Si was shown in Figure 47.2a, which has a tapered shape and a serrated sidewall. The transmission electron microscopy (TEM) image

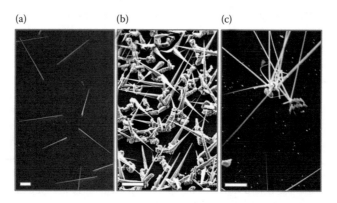

   (a)   (b)   (c)

FIGURE 47.1   SEM images of (a) GaAs NWs, (b) InP NWs, and (c) GaP NWs on Si (111) substrate; scale bars are 2 μm.

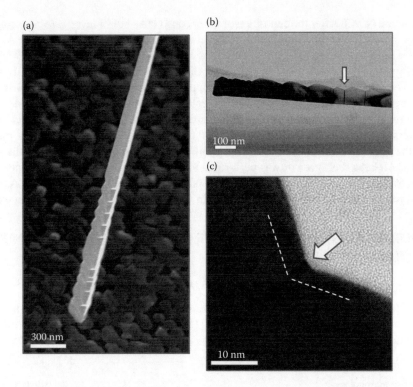

FIGURE 47.2 (a) A SEM image (45° tilted) of single GaAs NW on Si (111) substrates, (b) a TEM image of the NW showing sawtooth-like facets with many segments, and (c) a HR-TEM image illustrating a twin defect at one of the segments (arrow).

(Figure 47.2b) shows a high density of stacking faults (the dark bands) in this NW and a sawtooth-like facet. The high-resolution (HR) TEM image of Figure 47.2c clearly shows the ZB structure with a thin twin plane. In fact, this type of twin defects is commonly associated with the sidewall faceting behavior of ZB III–V compound NWs [11–13]. Generally, faceting takes place in association with the high density of twins at high growth temperatures [12,14].

To achieve efficient nanoscale optoelectronic devices, an overall improvement is necessary in the morphology and crystal structure.

47.3 EXPERIMENTAL DETAILS

In this study, the GaAs layers and NWs were grown on Si substrates by horizontal-flow low-pressure MOCVD. Au particles of 50 nm were used as catalysts via the vapor-liquid-solid (VLS) mechanism for NW growth and the substrates used here were the Si substrates having (111) orientation with 4°-miscut toward the (112) direction. Trimethylgallium (TMGa) and arsine (AsH_3) were used as the group III and V precursor source materials, respectively, and the reactor pressure was fixed at 100 mbar.

To improve the morphology of the NWs, buffer layer growth was introduced first. The Si substrates were etched with 4.8% diluted hydrofluoric (HF) solution to remove the native oxide layer prior to buffer growth. Three types of buffer layer structures were investigated. For a single buffer layer structure, only one GaAs layer was grown on Si substrates at 400°C for 1 h with a low V/III ratio of 15.4 (structure No. 1). This single buffer layer also acts as the initial layer in the double buffer layer structure, where the subsequent GaAs buffer layer was grown at 700°C for 10 min with a higher V/III ratio of 154.3 without growth interruption during the heating up step from 400°C to

700°C (structure No. 2). After the deposition of the second GaAs buffer layer, *selected* samples were annealed *in situ* at 750°C for 15 min under AsH₃ atmosphere to prevent desorption of As from the GaAs buffer layers (structure No. 3).

For NW growth, the Si substrates coated with a GaAs buffer were treated by poly-L-lysine (PLL) the solution (to attach Au particles) and then the Au particles were dispersed onto the surface. After all treatments were finished, the sample was annealed again at 600°C for 10 min under the AsH₃ ambient to remove contamination and form the eutectic alloys. Three different GaAs NWs were grown at temperatures of 375 ~ 450°C for 6 or 30 min with different precursor flow rates. More details of the growth conditions of each sample will be given later as required. After the growth of GaAs NWs, AlGaAs shell growth was performed at 650°C for 20 min with an Al vapor concentration of 26% and the same flows of TMGa and AsH₃ as described above. The AlGaAs shell was finally covered by a thin GaAs cap grown for 5 min at the same temperature to prevent oxidation.

47.4 MORPHOLOGY IMPROVEMENT OF GaAs NANOWIRES BY THIN BUFFER LAYERS

47.4.1 GROWTH OF THIN GaAs BUFFER LAYERS ON SI (111) SUBSTRATES

Figures 47.3 shows the illustrations, SEM, and TEM images of the three different types of GaAs buffer layers. The single buffer layer grown on Si (111) substrates at a low temperature (400°C) with a low V/III ratio (15.4) as shown in Figure 47.3a has a rough surface as seen in the SEM image. We

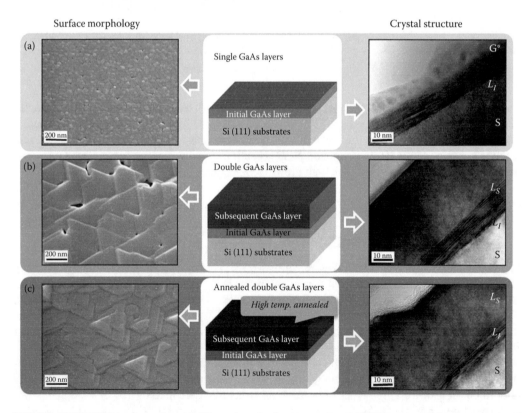

FIGURE 47.3 SEM images, pictorial illustrations, and cross-sectional TEM images of the three types of GaAs buffer layers grown on Si (111) substrates: (a) a single buffer layer, (b) double buffer layers, and (c) annealed double buffer layers, respectively (S: Si substrate, L_I: the initial/single GaAs buffer layer, L_S: the subsequent GaAs buffer layer, and G*: amorphous glue used during sample preparation).

speculate that, during growth, islands initially build up due to the lattice mismatch between the Si substrate and the growing GaAs material [15,16] until finally the layer forms a polycrystalline structure made up of the coalescence of these small islands. The HR-TEM image of the single buffer layer illustrates that abundant structural defects exist in this thin buffer layer (~12 nm), such as dislocations and stacking faults, with surface irregularity. The poor crystalline quality of this single buffer layer indicates that this layer is unsuitable for the subsequent growth of NWs.

The double GaAs buffer layer structure consists of an initial layer that is the same as a single buffer layer and the subsequent layer grown at a high temperature (700°C) with a high V/III ratio (154). The SEM image in Figure 47.3b shows the surface with triangular terraces and the occasional presence of voids. The triangular symmetry of the crystals is related to the ZB structure of GaAs grown on the (111) surface. The TEM image taken from the [110] zone axis shows that the upper region of the GaAs layer is almost free of lattice defects and superior to the single buffer layer.

In situ annealing at high temperature (750°C) effectively improves the surface morphology of the double buffer layer structure as seen in the SEM image of Figure 47.3c. The triangular terraces have become smoother and the surface voids have also disappeared due to recrystallization. The TEM image also shows a dislocation-free upper GaAs layer and the lattice defects in the initial layer are also reduced after annealing.

47.4.2 MORPHOLOGY OF GaAs NANOWIRES GROWN ON Si (111) SUBSTRATES COATED WITH BUFFER LAYERS

SEM images in Figure 47.4 show the morphology of the GaAs NWs grown on the Si (111) substrates directly and coated with various GaAs buffer layers explained earlier. The growth conditions for all

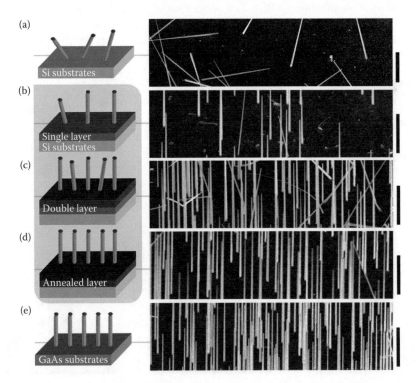

FIGURE 47.4 Schematic illustrations with SEM images of GaAs NWs grown on Si (111) substrates (a) without any buffer layer, (b) with a single GaAs buffer layer, (c) with a double GaAs buffer layer, and (d) with an annealed double GaAs buffer layer. (e) The morphology of GaAs NWs grown on GaAs (111)B substrates is shown in (e) for comparison; scale bars are 2 μm.

NWs are the same: growth temperature, growth time, and V/III ratio were 450°C, 30 min, and 46.3, respectively. In all cases, except the direct growth of GaAs NWs on Si substrates, most GaAs NWs grow vertically along the [111] direction, indicating the epitaxial relationship of NWs with the buffer layers and substrates. A typical morphology of GaAs NWs grown directly on an Si (111) substrate is shown in Figure 47.4a, where a very low density of NWs was observed and found to exhibit a random growth direction. Although HF etching was used to remove the native oxide on the Si surface prior to the application of Au particles, it was likely that surface oxidation reoccurred during the DI water rinsing step. This oxide layer most likely resulted in lower adhesion of the Au particles, and hence the low density of NWs despite the use of PLL. The morphology of GaAs NWs can be significantly improved by covering the Si surface with a thin GaAs buffer layer. Further improvement in NW density and quality can be achieved by optimizing the quality of the buffer layers. Figures 47.4b and 47.4d show the morphology of GaAs NWs following deposition on the three buffer layers described above and, as a result, on comparing with GaAs NWs grown on the Si (111) substrates directly (Figure 47.4a), the morphology and density of GaAs NWs were dramatically improved. In addition, the NWs grown on the annealed double buffer layers are similar to those grown on GaAs (111)B substrates (Figure 47.4e).

47.5 IMPROVEMENT OF CRYSTAL STRUCTURE OF III–V NANOWIRES

To obtain defect-free NWs, optimization of growth conditions in terms of growth temperature and a V/III ratio, together with a precursor flow rate was investigated in this study.

Figure 47.5 shows the illustrations of three different types of GaAs NWs grown for this study. Here, all NWs were grown on Si (111) substrates coated with annealed double GaAs buffer layers as discussed above (Figure 47.3b). The first set of GaAs NWs was grown at 450°C for 30 min with a V/III ratio of 46.3 as reference (NW_N). The second set of GaAs NWs was grown via the two-temperature procedure (NW_T) [14], where an initial nucleation temperature of 450°C was employed for a minute before growth was continued for 30 min at a reduced growth temperature of 375°C. The

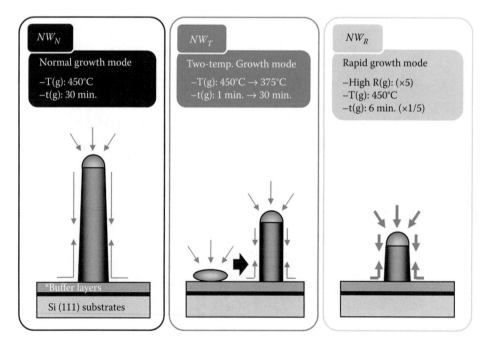

FIGURE 47.5 Illustration of the three types of GaAs NWs investigated: NW_N grown by the normal procedure as a reference, NW_T grown via a two-temperature procedure, and NW_R grown via a rapid growth rate procedure. T(g): growth temperature, t(g): growth time, and R(g): growth rate.

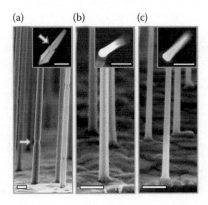

FIGURE 47.6   SEM images of (a) NW_N, (b) NW_T, and (c) NW_R, respectively. The samples were tilted 10° from the normal surface. Arrows in the image (a) indicate surface grooves on the nanowires: all scale bars are 1 µm.

V/III ratio was maintained constant at 46.3. The last set of GaAs NWs was grown with a rapid growth rate (NW_R) at 450°C for 6 min with the TMGa flow rate increased by a factor of 5 while maintaining the V/III ratio at 46.3. The growth time was reduced to have a comparable length as NW_N.

In Figure 47.6a, the reference GaAs NWs (NW_N) are seen to be straight and vertical with serrated sidewalls (indicated by arrows) and substantial tapering, the latter two features being undesirable for device applications. These surface grooves and rotated body segments are related to the sidewall faceting behavior [11,14]. As shown in the inset image, the NW base appears to be a truncated triangular shape due to different {112}A and {112}B sidewall growth rates [12]. In contrast to the morphology of NWs grown at 450°C, GaAs NWs grown by the two-temperature procedure (NW_T) in Figure 47.6b show much improved morphology with little tapering and smooth sidewalls. Also, the base of the NW appears to be almost hexagonal in shape (inset image). It should be noted that without the nucleation step at 450°C, all NWs grown (on GaAs substrate) at 375°C were kinked [14]. Similar to NW_T, GaAs NWs with a high growth rate (NW_R) (Figure 47.6c) also exhibit straight and vertical orientation with a smooth surface morphology. Significantly less tapering in comparison to NW_N is noted, as has been reported for similar growths on GaAs (111)B substrates previously [17]. NW_R also has a near hexagonal-shaped bottom as shown by the inset image.

High-magnification TEM images of these three NWs are shown in Figures 47.7a through 47.7c. All NWs have a ZB crystal structure that is consistent with previous reports [14,17]. NW_N in Figure 47.7a, as discussed in Section 47.2.2, has some stacking faults and a serrated sidewall [11,13,18]. The HR-TEM image (inset) shows a thin twin defect across two ZB crystal segments (A–B–A type). This is in contrast with Figures 47.7b and 47.7c, where no such defects are present and the sidewalls are smooth. In the case of NW_T, it is believed that the low growth temperature results in a twin-free ZB crystal structure due to the low supersaturation growth condition while the availability of high concentration of Ga and As adatoms for NW_R is believed to reduce the Au–GaAs interfacial tension [17] and thus hinder the formation of twin planes [19].

47.6   OPTICAL PROPERTIES OF NANOWIRES FOR DEVICE APPLICATIONS

Photoluminescence (PL) is routinely used to characterize the optical properties of GaAs NWs. Bare GaAs NWs, however, emit only very weak PL, even at low temperatures [14,20,21], which is due to the high density of surface states that act as nonradiative recombination centers. These surface states not only quench GaAs NW PL but also adversely affect optoelectronic device performance. These surface states can be removed by cladding the GaAs NWs with a shell of a wider band-gap AlGaAs layer, creating a core–shell structure. Figure 47.8 shows a schematic image of core–shell (CS-) GaAs NWs with an AlGaAs/GaAs shell and SEM images for morphology of CS-NW_N, CS-NW_T, and CS-NW_R,

(a)

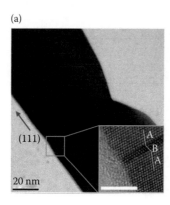

(b)

(c)

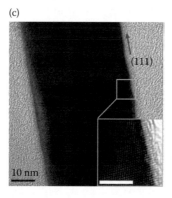

FIGURE 47.7 High-magnification TEM images and HR-TEM images (inset) of (a) NW_N, (b) NW_T, and (c) NW_R, respectively. All scale bars in the inset images are 5 nm.

respectively. Here, the thin GaAs cap has a role to prevent the oxidation of the AlGaAs shell. The shell spatially separates the electronic states of the GaAs core from the surface and acts as a carrier reservoir for the inner core [22]. Thus, GaAs NWs capped by the AlGaAs shell have much a higher PL efficiency.

In Figure 47.9a, normalized micro-PL spectra from the three types of core–shell NW (CS-NW) samples are shown. Regular ripples in the PL spectra may be considered as artifacts arising from the interference filter used to remove the 1.65 eV laser line while broader structures come from the AlGaAs shell due to nonuniformities of the Al composition owing to changes in the Al incorporation rate with diameter [23,24]. All PL peaks are near 1.518 eV (±0.006 eV) and show a similarly strong intensity (albeit normalized here), which is attributed to free exciton recombination in bulk GaAs [20,25]. Figure 47.9b plots time-resolved (TR-) PL spectra at the free exciton recombination energy for each of the GaAs/AlGaAs CS-NWs. CS-NW_N, grown with standard core-NWs (NW_N), shows a short lifetime due to crystal defects inside the NW core (stacking faults and twin defects) while the two temperature growth, CS-NW_T, exhibits the longest lifetime reflecting a high degree of crystalline quality [14,26]. Interestingly, CS-NW_R grown with a rapid growth rate displayed a very short lifetime, which is also the resolution limit of our TR-PL measurement system, possibly due to the point defects as nonradiative recombination centers of gallium and arsenic-related defects such as vacancies, interstitials, and antisites.

47.7 SUMMARY

For the integration of III–V device applications with Si-based electronics, well-aligned, straight morphology, and pure crystal structure without any structural defects, such as twin defects or

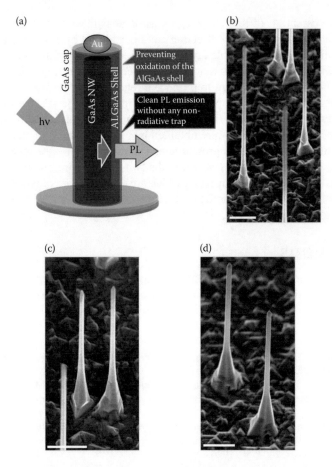

FIGURE 47.8 (a) Schematic image of core–shell (CS-) GaAs NWs with a AlGaAs/GaAs shell. (b)–(d) SEM images for morphology of CS-NW_N, CS-NW_T, and CS-NW_R; all scale bars are 1 μm.

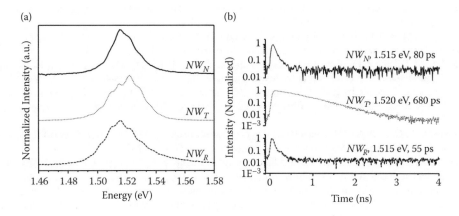

FIGURE 47.9 Normalized (a) micro- and (b) time-resolved (TR)-photoluminescence (PL) emission peaks of CS-NW_N, CS-NW_T, and CS-NW_R.

stacking faults, are essential. In this chapter, it was demonstrated that straight, epitaxial GaAs NWs on Si substrates along the <111> direction can be achieved by first covering the Si surface with a thin GaAs buffer layer. A double-layer structure, consisting of an initial layer grown at a low temperature with a low V/III flow ratio and a subsequent layer grown at a high temperature with a high V/III flow ratio, was shown to produce a relatively flat surface with good crystalline quality; *in situ* annealing at high temperature further improved the surface morphology of the double buffer layers, and subsequently the yield of high-quality NWs. In addition, it was also demonstrated that defect-free GaAs NWs were obtained by using specific growth techniques, such as the two-temperature growth and rapid growth rate. The rapid growth rate method resulted in the highest yield of straight and vertical NWs while the two-temperature growth method resulted in the longest exciton lifetime and least tapering, which are desirable for optoelectronic device applications. These results show promise for the future integration of III–V NW optoelectronic devices with Si-based microelectronic devices.

ACKNOWLEDGMENTS

The Australian Research Council is gratefully acknowledged for the financial support. This work was performed in part using the ACT Node of the Australian National Fabrication Facility (ANFF).

REFERENCES

1. N. Dasgupta and A. Dasgupta, *Semiconductor Devices—Modelling and Technology*, Prentice Hall of India, New Delhi, 2004, pp. 56–57.
2. R. Chau, S. Datta, M. Doczy, B. Doyle, B. Jin, J. Kavalieros, A. Majumdar, M. Metz, and M. Radosavljevic, Benchmarking nanotechnology for high-performance and low-power logic transistor applications, *IEEE Trans. Nanotechnol.*, 2005, 4, 153–158.
3. H. S. P. Wong, Beyond the conventional transistor, *J. IBM Res. Dev.*, 2002, 46, 133–168.
4. V. A. Sverdlov, T. J. Walls, and K. K. Likharev, Nanoscale silicon MOSFETs: A theoretical study, *IEEE Trans. Electron Devices*, 2003, 50, 1926–1933.
5. A. A. Bergh and P. J. Dean, *Light-Emitting Diodes*, Clarendon, Oxford, 1976.
6. C. Jeong, S. W. Lee, J. M. Seo, and J. M. Myoung, The effect of stacking fault formation on optical properties in vertically aligned ZnO nanowires, *Nanotechnology*, 2007, 18, 305701.
7. T. M. Chang and E. A. Carter, Structures and growth mechanisms for heteroepitaxial fcc (111) thin metal films, *Phys. Chem.*, 1995, 99, 7637–7648.
8. T. Mårtensson, C. P. T. Svensson, B. A. Wacaser, M. W. Larsson, W. Seifert, K. Deppert, A. Gustafsson, L. R. Wallenberg, and L. Samuelson, Epitaxial III-V nanowires on silicon, *Nano Lett.*, 2004, 4, 1987–1999.
9. X. Y. Bao, C. Soci, D. Susac, J. Bratvold, D. P. R. Aplin, W. Wei, C. Y. Chen, S. A. Dayeh, K. L. Kavanagh, and D. Wang, Heteroepitaxial growth of vertical GaAs nanowires on Si (111) substrates by metal-organic chemical vapour deposition, *Nano Lett.*, 2008, 8, 3755–3760.
10. L. Roest, M. A. Verheijen, O. Wunnicke, S. Serafin, H. Wondergem, and E. P. A. M. Bakkers, Position-controlled epitaxial III–V nanowires on silicon, *Nanotechnology*, 2006, 17, S271–S275.
11. J. Zou, M. Paladugu, H. Wang, G. J. Auchterlonie, Y. N. Guo, Y. Kim, Q. Gao, H. J. Joyce, H. H. Tan, and C. Jagadish, Growth mechanism of truncated triangular III–V nanowires, *Small*, 2007, 3, 389–393.
12. J. Johansson, L. S. Karlsson, C. P. T. Svensson, T. Mårtensson, B. A. Wacaser, K. Deppert, L. Samuelson, and W. Seifert, Structural properties of <111> B -oriented III–V nanowires, *Nat. Mater.*, 2006, 5, 574–580.
13. J. Bauer, V. Gottschalch, H. Paetzelt, G. Wagner, B. Fuhrmann, and H. S. Leipner, MOVPE growth and real structure of vertical-aligned GaAs nanowires, *J. Cryst. Growth*, 2007, 298, 625–630.
14. H. J. Joyce, Q. Gao, H. H. Tan, C. Jagadish, Y. Kim, Z. Zhang, Y. N. Guo, and J. Zou, Twin-free uniform epitaxial GaAs nanowires grown by a two-temperature process, *Nano Lett.*, 2007, 7, 921–926.
15. K Kitahara, M. Ozeki, and K. Nakajima, Reflection high-energy electron diffraction of heteroepitaxy in chemical vapour deposition reactor: Atomic-layer epitaxy of GaAs, AlAs and GaP on Si, *Jpn. J. Appl. Phys.*, 1993, 32, 1051–1055.
16. Y. B. Bolkhovityanov and O. P. Pchelyakov, GaAs epitaxy on Si substrates: Modern status of research and engineering, *Physics-Uspekhi*, 2008, 51, 431–456.

17. H. J. Joyce, Q. Gao, H. H. Tan, C. Jagadish, Y. Kim, M. A. Fickenscher, S. Perera et al., Unexpected benefits of rapid growth rate for III – V nanowires, *Nano Lett.*, 2009, 9, 695–701.

18. K. A. Dick, P. Caroff, J. Bolinsson, M. E. Messing, J. Johansson, K. Deppert, L. R. Wallenberg, and L. Samuelson, Control of III–V nanowire crystal structure by growth parameter tuning, *Semicond. Sci. Technol.*, 2010, 25, 204009.

19. Q. Li, X. Gong, C. Wang, J. Wang, K. Ip, and S. Hark, Size-dependent periodically twinned ZnSe nanowires, *Adv. Mater.*, 2004, 16, 1436–1440.

20. L. V. Titova, T. B. Hoang, H. E. Jackson, L. M. Smith, J. M. Yarrison-Rice, Y. Kim, H. J. Joyce, H. H. Tan, and C. Jagadish, Temperature dependence of photoluminescence from single core-shell GaAs-AlGaAs nanowires, *Appl. Phys. Lett.*, 2006, 89, 173126.

21. H. J. Joyce, Q. Gao, H. H. Tan, C. Jagadish, Y. Kim, J. Zou, L. M. Smith et al., III–V semiconductor nanowires for optoelectronic device applications,. *Prog. Quantum Electron.*, 2011, 35, 23–75.

22. F. Jabeen, S. Rubini, V. Grillo, L. Felisari, and F. Martelli, Room temperature luminescent InGaAs/GaAs core–shell nanowires, *Appl. Phys. Lett.*, 2008, 93, 083117.

23. J. Noborisaka, J. Motohisa, S. Hara, and T. Fukui, Fabrication and characterization of freestanding GaAs/AlGaAs core-shell nanowires and AlGaAs nanotubes by using selective-area metalorganic vapour phase epitaxy, *Appl. Phys. Lett.*, 2005, 87, 093109.

24. F. Bailon-Sominatic, J. J. Ibanez, R. B. Jaculbia, R. A. Loberternos, M. J. Defensor, A. A. Salvador, and A. S. Sominac, Low temperature photoluminescence and Raman phonon modes of Au-catalyzed MBE-grown GaAs-AlGaAs core-shell nanowires grown on a pre-patterned Si (111) substrate, *J. Cryst. Growth*, 2011, 314, 268–273.

25. V. Swaminathan, D. L. V. Haren, J. L. Zilko, P. Y. Lu, and N. E. Schumaker, Characterization of GaAs films grown by metalorganic chemical vapour deposition, *J. Appl. Phys.*, 1985, 57, 5349–5353.

26. P. Parkinson, H. J. Joyce, Q. Gao, H. H. Tan, X. Zhang, J. Zou, C. Jagadish, L. M. Herz, and M. B. Johnston, Carrier lifetime and mobility enhancement in nearly defect-free core-shell nanowires measured using time-resolved terahertz spectroscopy, *Nano Lett.*, 2009, 9, 3349–3353.

48 Synthesis and Characterization of n- and p-Doped Tin Oxide Nanowires for Gas Sensing Applications

Hoang A. Tran and Shankar B. Rananavare

CONTENTS

48.1 INTRODUCTION

1D nanostructure materials have attracted great attention in recent years[1–3] as they provide interesting possibilities for resistance-less transport due to quantum effects. Another remarkable property of such 1D semiconductor materials is their sensitivity to ambient conditions such as humidity, oxygen level, and so on. Their resistance can be modulated by adsorbed gases capable of electron/hole injection/withdrawal from conduction and valence bands. Such an electron transfer creates a carrier depletion layer near the solid–air interface increasing the device impedance. This increase/decrease in resistance provides a convenient response with respect to analyte concentration. Furthermore, the sensor devices can be made more specific for a given type of gas through the use of dopants that selectively interact with the analyte gas. Additionally, doping can further amplify the device sensitivity and its dynamic range by providing an increase in the carrier concentration.

The most common method for the synthesis of 1D nanowires (NWs)/nanotubes (NTs) is based on the vapor–liquid–solid (VLS) mechanism that can be carried out in a chemical vapor deposition (CVD) reactor. VLS is considered to be one of the best methods for producing single-crystal NWs and NTs in relatively large quantities.[4,5] An adaptation of the VLS method involving solution–liquid–solid phases is developed by Buhro et al.[6] A similar solvothermal method uses solvents under very specific conditions of pressure and temperature to increase the solubility of solids and speeds up reactions[7] that lead to the formation of NWs. Solution-state methods involving templating by a combination of surfactants that preferentially cap the sidewalls of growing NWs have also been developed for the synthesis of metallic NWs of gold, silver, and so on.[8,9]

This chapter focuses on solution-state syntheses of doped SnO_2 NWs. A key benefit of tin oxide over other contemporary semiconductors such as silicon or gallium arsenide and even carbon nanotubes (CNTs) is that it is a chemically stable material that is corrosion resistant. Thus, this wide bandgap semiconductor (3.56 eV) finds wide-ranging applications in nanoelectronics as a transparent conductor,[10–13] in chemistry as heterogeneous catalysts,[14–19] and also as a solid-state gas sensor.[20–25] In a series of papers, we have illustrated the synthesis,[26] characterization,[26] and applications[22,27] of nanoparticles (NPs) of SnO_2. Using compressed powders of n-doped and p-doped SnO_2 NPs, we have been able to fabricate homo-junction diode sensors capable of detecting the 100 ppb level of a toxic oxidizing gas, chlorine. One key limitation of these sensors arises from poor electron transport across the NP powders. This is a result of the interparticle depletion barrier and the effects of coulomb blockade at each physical contact. Practically, this limits the dynamic range of devices confining them to low concentration (below 20 ppm) of the analyte. A simple method to minimize these interparticle electron/hole hops is through the replacement of NPs with aligned NWs. The resistance for transport along the single-crystal nanowire is lower due to the favorable quantum effects in 1D systems.[28] Studies of electron transport through CNTs and SiNWs indicate extraordinarily high mobility in these 1D quantum fluids.[29,30]

Our goal here is to present synthetic methods for synthesizing *doped* NWs for tin oxide and illustrate their application in gas sensing. Current methods so far for the synthesis of 1D nanostructures of undoped SnO_2 include laser ablation,[31,32] molten salt,[33,34] solvothermal and hydrothermal methods,[35,36] thermal evaporation,[37,38] carbothermal reduction,[39,40] and rapid oxidation.[41] Herein, we describe the synthesis of doped SnO_2 NWs adopting and modifying the strategy of Yang and Lee.[33] It has enabled us to synthesize n-type, antimony-doped SnO_2 NWs. We introduce the first synthesis of a p-doped SnO_2 1D nanostructure using lithium ion as a dopant. The latter is developed based on our current understanding of p-doped SnO_2 NPs.[26,42,43] We also present a comparison of the structural and optical properties of SnO_2 NPs and SnO_2 NWs. Both n- and p-type NWs display a characteristic red shift in their photoluminescence (PL) spectra. Surface plasmons observed in these systems imply high carrier concentrations. These corrosion-resistant materials are useful in fabricating ultrasensitive gas detectors and transparent electronics, and we demonstrate the application to toxic gas Cl_2.

48.2   EXPERIMENTAL PROCEDURE

48.2.1   Materials

The following chemicals were used without further purification: $SnCl_4 \cdot 5H_2O$ (>98%, Fischer Scientific), 1,10-phenanthroline monohydrate (ACS reagent, puriss. p.a., ≥99.5%, Sigma Aldrich), $NaBH_4$ (98%, Sigma Aldrich), LiCl, KCl, and NaCl (>95%, Fischer Scientific).

48.2.2   Material Characterizations

SEM data were collected on an FEI Sirion FEG SEM operating at 5 KV with a working distance of 5 mm. TEM data were collected using Holey carbon TEM grids on a FEI Tecnai F-20 microscope operating at 200 kV. The optical PL of the doped and undoped SnO_2 NWs was characterized on a Shimadzu spectrofluorophotometer RF-5301PC, using a xenon light source and a 3 nm bandwidth. As-prepared samples including NPs and NWs were dispersed in water for the PL measurements. Fourier transform infrared (FTIR) spectra were acquired on a Perkin-Elmer RXI instrument. For infrared as well as UV–visible–NIR (Shimadzu UV 3600) spectroscopic investigations, samples consisted of solid powders sandwiched between quartz plates. This protocol is different from the commonly used, solvent-dispersed studies of these materials.

48.2.3   Synthetic Methods

n-Doped tin oxide NWs: The precursor Sn NPs were synthesized by reducing a thoroughly mixed solution of 50 mL of 0.05 M $SnCl_4 \cdot 5H_2O$ and 0.5 g of 1,10-phenanthroline with 100 mL of 0.1 M

NaBH$_4$ aqueous solution. The reducing solution was introduced dropwise to synthesize phenanthroline-capped Sn NPs.[44] NPs were separated from the reaction mixture by centrifuging (10,000 rpm for 15 min) after 2 h of reaction time. The precipitate was dried at 50°C for 2 h on a hot plate. 0.2 g of 1,10-phenanthroline-capped Sn NPs powder was mixed with a mixture of 0.4 g of NaCl and 0.6 g of KCl, grounded into a fine powder, and heated at 750°C for 2 h in a furnace. The molten mixture was then slowly cooled to room temperature. The solidified product was washed several times with water, to remove KCl and NaCl. Salt removal was tested with 0.05 mM AgNO$_3$ solution for any residual Cl$^-$ anion. The wet powder was dried in an oven at 100°C overnight to remove water. For n-doping, varying amounts of antimony chloride (1–5% at.wt. in relation to Sn) were added to the initial solution of SnCl$_4$·5H$_2$O and 1,10-phenanthroline. The rest of the synthetic procedure was identical to the synthesis of undoped NWs as described above. Note that the actual amount of dopant inserted in NPs could be different from the concentration of dopant used during synthesis (see below).

A slightly modified synthetic method was developed to produce much higher quantities of undoped SnO$_2$ and n-doped SnO$_2$ NWs. 5 g of SnCl$_4$·5H$_2$O in was dissolved in 150 mL (antimony chloride was added for the synthesis of n-doped nanowires) water. The solution was stirred for 15 min before the continuous dropwise addition of 50 mL of 4% NaBH$_4$ (aqueous) solution. The mixture turned black, indicating the formation of oxide-coated Sn NPs. The precipitate was collected by filtering and was left to dry in air overnight. 2 g of the above precursor was mixed with 0.2 g of 1,10-phenanthroline, 4 g of NaCl, and 6 g of KCl, ground into a fine powder, and kept at 750°C for 2 h.

p-Doped tin oxide NWs: For p-doping, LiCl replaced antimony chloride as a dopant. The amounts of lithium chloride added to the initial reaction mixture ranged from 1 to 5 wt%. The precursor Sn NPs coated with a thin layer of tin oxide were synthesized in a similar manner to that in which LiCl was replaced as the doping agent. 0.2 g of the above as-synthesized precursors was mixed with 0.2 g of 1,10-phenanthroline, 0.31 g of LiCl, and 0.6 g of KCl, ground into a fine powder, and kept at 480°C (which is near the eutectic temperature and composition for the LiCl/KCl mixture) for 2 h. The choice of this temperature is critical, as at a higher temperature Li$_2$SnO$_3$ begins to form and phase separate from the SnO$_2$ lattice. Furthermore, the presence of LiCl in liquid form enables a higher degree of p-doping in this synthetic method. The molten mixture was then slowly cooled to room temperature with a cooling rate of 4°C/min. The product was collected and purified as in the case of undoped NWs.

48.2.4 Cl$_2$ Sensor Fabrication and Characterization

To fabricate the sensor, a simple powder compression technique was employed. For the chlorine sensor, 2 g of Sb-doped SnO$_2$ was mixed with 0.5 mL of 2% ethyl silicate in acidified ethanol. The paste was compressed using a pressure of 5463 kg/cm^2 to yield a 13 mm (diameter) pellet with a thickness of 800 μm. The pellet was heated at 450°C for 30 min. It was then mounted onto a glass substrate and copper contacts attached using conductive silver epoxy.

The sensor performance was tested in a home-built test chamber at room temperature. The resistance of the sensor material was monitored by a Fluke 8840A multimeter, which was interfaced to a National Instrument's GPIB-PCII interface on a personal computer. During the measurements, there was a continuous flow of diluted test gas (10 L/min) into the test chamber (volume 1.2 L) (illustrated below). The sensor response time was measured by analyzing resistance versus time data. To refresh the sensor, a flow of fresh air was maintained in the chamber and the sensor temperature was adjusted. The sample temperature could be maintained between 25°C and 120°C by varying the resistive heater voltage (0–5 V). A Cr–Al thermocouple was placed on the surface of the sensor to measure the sensor operating temperature. The output of the thermocouple was fed to a temperature indicator Mastech (MAS-345) multimeter. A schematic view of the sensor testing system is shown in Figure 48.1.

The sensor response was defined as ($R_{Chlorine} - R_{Air}$)/R_{Air} air. Typically, the measurements were repeated three times and the sensor measurements were reproducible over a period of a week at Cl$_2$ concentrations (<1 ppm (μg of Cl$_2$ per g of carrier gas)), while a steady drift in resistance was observed at high chlorine concentrations, perhaps due to the formation of AgCl at the electrode contact points.

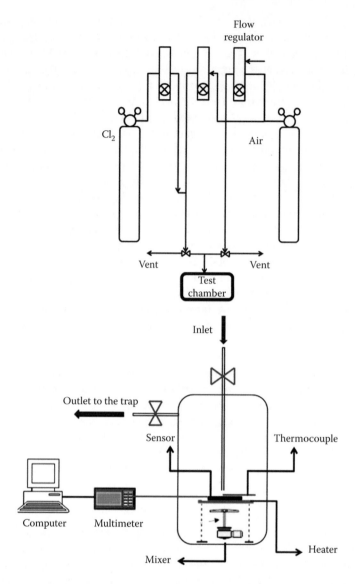

FIGURE 48.1 Outline of the dynamic sensor testing system.

48.3 RESULTS AND DISCUSSION

The bulk-scale synthesized powders of NWs exhibit distinct colors (Figure 48.2). The undoped NWs are pale yellow; p-doped NWs are deeper yellow; while the n-doped NWs are dark blue. This may simply reflect the higher carrier concentration in n-doped systems.

The morphology of the as-prepared undoped and Sb-doped SnO_2 NWs was characterized by SEM; see Figure 48.3. The majority of the undoped NWs showed an average NW diameter ranging from 20 to 50 nm and lengths up to a few microns. This NW structure was mixed with a small portion of whisker-like nanorod structure. In n-doped NWs, the size distribution was much broader due to the incorporation and morphological effects of dopant atoms (see below). Also, note that the smaller-diameter NWs tended to fuse with one another. Since such structures are rarely observed in the nucleation and growth of NWs mediated by the VLS mechanism, the mechanism of growth of these NWs is likely to be governed by the surface energy of NW walls as controlled by the capping

FIGURE 48.2   Bulk-scale synthesized (a) undoped, (b) n-doped, and (c) p-doped SnO$_2$ nanowires.

agent phenanthroline. TEM and the high-resolution TEM of n-doped NWs, shown in Figure 48.3 along with the electron diffraction as inset, indicated a single crystalline nature of the synthesized NWs. The crystal structure of these 20 nm diameter NWs was a characteristic cassiterite type (details are not shown here), which is commonly observed upon calcination at 500°C in SnO$_2$ NPs synthesized by the sol–gel method.[26] Also, it can be seen that the nanowire growth mechanism is neither VLS (no spherical catalyst droplet on the tip) nor solution–liquid–solid (SLS). The NWs showed a very straight and smooth surface morphology suggesting a 1D Ostwald ripening mechanism[34] in which NPs deposit onto, dissolve into each other, and extrude to form a 1D nanostructure.

Room-temperature PL spectra of undoped SnO$_2$ NPs and NWs appear in Figure 48.4. The excitation wavelength used in this study was 310 nm corresponding to energy greater than the bulk bandgap energy for SnO$_2$. Two different PL characteristics were observed for NPs and NWs. Both nanostructures gave a strong UV emission at 360 nm, corresponding to the bandgap of SnO$_2$ (3.56 eV), which partially overlapped with the Raman scattering of water. The latter peak appears at a constant frequency offset from the excitation wavelength.

Bulk SnO$_2$ is an indirect bandgap semiconductor, so it is not expected to have a strong PL corresponding to its bandgap energy, as it would violate the principle of momentum conservation.

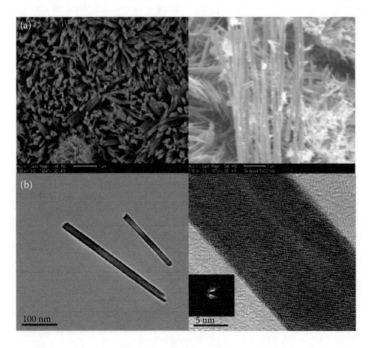

FIGURE 48.3   (a) SEM images of undoped (left) and Sb-doped NWs (right). (b) TEM of n-doped SnO$_2$ NWs with low and high magnifications. (H.A. Tran and S.B. Rananavare, Synthesis and characterization of N- and P-doped tin oxide nanowires. *2011 11th IEEE Conference on Nanotechnology (IEEE-NANO),* Aug. 15–18, 144–149, © (2011) IEEE. With permission.)

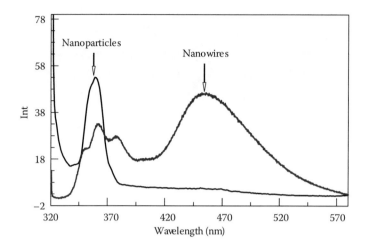

FIGURE 48.4 PL of undoped NWs and NPs (excitation wavelength 310 nm). (H.A. Tran and S.B. Rananavare, Synthesis and characterization of N- and P-doped tin oxide nanowires. *2011 11th IEEE Conference on Nanotechnology (IEEE-NANO)*, Aug. 15–18, 144–149, © (2011) IEEE. With permission.)

Observation of PL in these nanogeometries (i.e., in both NPs and NWs) implies a PL characteristic of a direct bandgap-type semiconductor. Similar effects are well known in silicon NPs; porous silicon and silicon NWs[45] are not still well understood. Interestingly, the NWs have an additional broad emission peak near 460 nm (2.7 eV). A precise understanding of this PL peak does not exist currently, although explanations in terms of defects, including vacancies of oxygen, dopant segregation, and lattice disorders inside the lattice of SnO_2[34,46–48] or near surface, can be advanced.

The presence of a dopant, especially in high concentrations, can lead to an impurity band near the conduction (in the case of an n-type dopant) or valence (in the case of p-type dopant) bands. The resulting bandgap narrowing could result in a red shift in the PL. As shown in Figure 48.5, under the same excitation wavelength of 280 nm, both undoped and n-doped SnO_2 NWs gave two UV emission peaks, one at 310 nm, which is due to the solvent Raman scattering, and the other

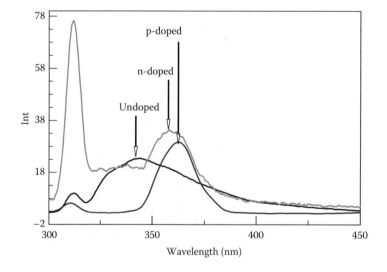

FIGURE 48.5 PL of undoped, 5% n-doped, and 5% p-doped SnO_2 NWs. (H.A. Tran and S.B. Rananavare, Synthesis and characterization of N- and P-doped tin oxide nanowires. *2011 11th IEEE Conference on Nanotechnology (IEEE-NANO)*, Aug. 15–18, 144–149, © (2011) IEEE. With permission.)

broad peak at 360 nm wavelength for n- and p-doped nanowires, but for undoped nanowires, it is at 340 nm. For the latter peak, the observed red shift of 20 nm is consistent with a dopant-induced bandgap narrowing concept.

At this time, we do not have an adequate explanation for the disappearance of the 450 nm peak observed in undoped NWs when excited with a 310 nm wavelength (Figure 48.4). The currently accepted interpretation of the 450 nm peak, ascribed to oxygen vacancies, cannot explain the disappearance of this peak when excited by a 280 nm wavelength.

Figure 48.6 shows the *I–V* characteristics of n-type NWs as a function of dopant concentration. To carry out these measurements, powders of NWs were compressed into the pellets[22] and silver epoxy contacts were used. Good ohmic contacts were observed due to degenerate doping. As the impurity concentration is increased, the conductivity significantly increases and achieves its maximum value at 5% dopant concentration. However, at higher dopant concentration, a decrease in the conductivity was observed. Nanoscopic phase separation of the Sb_2O_5 (insulating amorphous) layer from SnO_2 that wraps around NWs can explain this anomalous conductivity. This phenomenon can also decrease the carrier mobility.

Figure 48.7 presents SEM images of p-doped NWs. Broad size distributions for length and diameter were observed. The length of wires varied from 200 nm up to a few μm while the diameter of a single wire ranged from 50 nm to hundreds of nanometers. Like n-doped NWs, whisker-like nanorods with smaller diameters and shorter lengths were also observed along with the tendency of NWs to fuse. Thus, preformed NPs transformed into NWs through recrystallization from alkali halide salts exhibited similar characteristics regardless of the nature (n or p type) of the dopant. The morphology of the NW nanostructure was influenced by a variety of other factors, such as precursors, growth temperature, and the moisture content of the ground mixture before recrystallization. The presynthesized powder, when dried to get rid of the moisture, gave a better yield of wires and more uniform dimensional distributions. Also, the NWs formed at temperatures higher than 450°C gave longer wires (>1 μm in length) with more uniform length distribution as shown in Figure 48.7. The effect of different synthetic conditions is shown in Table 48.1. Replacing coated Sn NP precursors by SnO NPs[49] did not produce NWs at 450°C. However, recrystallization at 750°C did produce NWs with low yield.

Single crystalline SnO-coated Sn NP precursors and NWs were observed in TEM (Figure 48.8a). NPs showed a spherical shape with a mean diameter around 10 nm, as well as a tendency to agglomerate due to the absence of an interfacial capping agent.

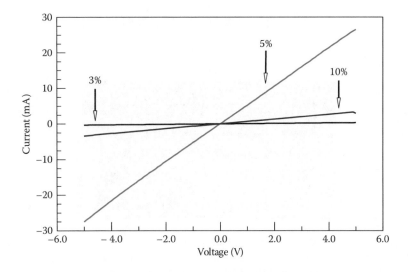

FIGURE 48.6 *I–V* curves of n-doped NWs at different dopant concentrations.

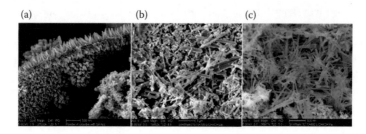

FIGURE 48.7   SEM image of p-doped SnO_2 NWs synthesized at (a) 480°C, (b) 600°C, and (c) 700°C.

TABLE 48.1

Characteristics of p-Doped SnO_2 Nanowires (NW) Synthesized from Sn Nanoparticles under Varying Synthetic Conditions

Temperature (°C)	In Stagnant Air	N_2 Flow (120 sccm)	Air Flow (120 sccm)	Phenanthroline
480	No NWs	200–500 nm size (length) NWs	No NWs, but NPs observed	200 nm–1 μm size NWs
600 or 700	2–5 μm size NWs and mixed with NPs	Not tested	Not tested	2–5 μm long NWs

The agglomeration of NPs was perhaps responsible for the observed fused NWs during the growth process. HR-TEM images of p-doped SnO_2 NWs shown in Figure 48.8b indicated their single crystalline nature, which was further confirmed by electron diffraction shown in the inset of Figure 48.8b. The TEM data showed that an interplanar spacing is about 0.34 nm, which corresponds to the (110) plane of the cassiterite crystal structure of SnO_2.

Tin oxide doping was also investigated through studies of surface plasmons as it provides a convenient means of determining the carrier concentration by contact-free measurement. Specifically, the surface plasmon resonance frequency is given by

$$\omega_{sp} = \sqrt{\frac{\omega_{p^2}}{1 + 2\varepsilon_m} - \gamma^2} \qquad \omega_p = \sqrt{\frac{N_{he^2}}{\varepsilon_0 m_h}}$$

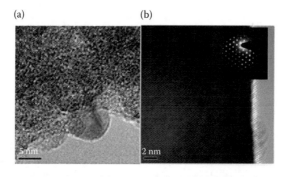

FIGURE 48.8   (a) TEM of SnO_{2-x}-coated Sn NP precursors and (b) TEM of p-doped SnO_2 NWs with inset showing selective area electron diffraction. (H.A. Tran and S.B. Rananavare, Synthesis and characterization of N- and P-doped tin oxide nanowires. *2011 11th IEEE Conference on Nanotechnology (IEEE-NANO),* Aug. 15–18, 144–149, © (2011) IEEE. With permission.)

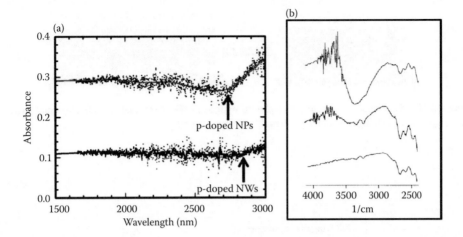

FIGURE 48.9 (a) Optical absorption spectra for p-doped NP and NWs. The surface plasmon peaks are just beyond the range of the instrument. Smooth lines are drawn to guide the eye. (b) FTIR spectra of p-doped NPs (top), NWs (middle), and quartz substrate (bottom). Note that the y-axis scale is transmittance and the spectra are shifted vertically for better display. The surface plasmon wavelength for NPs is about 3400 nm and the corresponding carrier density is 1.6×10^{20} cm^{-3}. The carrier density for the p-doped NWs is similar. (H.A. Tran and S.B. Rananavare, Synthesis and characterization of N- and P-doped tin oxide nanowires. *2011 11th IEEE Conference on Nanotechnology (IEEE-NANO)*, Aug. 15–18, 144–149, © (2011) IEEE. With permission.)

Where ω_{SP} is the observed surface plasmon resonance peak frequency and γ is its width; ε_m and ε_0 are the dielectric constant and dielectric permittivity of the space, respectively; and m_h and N_h are the hole mass and concentration in the SnO$_2$ semiconductor, respectively; and e is the electronic charge. For Sb-doped tin oxide NPs, the surface plasmons in the near-IR to mid-IR range were observed and carrier densities in the range of 10^{20}–10^{22} cm^{-3} were reported[50] for 5–20% dopant concentration.[*] Our studies of n-doped NPs and NWs gave surface plasmon peaks near 2300 nm (not shown) corresponding to the carrier density of 1×10^{22} cm^{-3}.

UV–visible–NIR studies of p-doped NPs and NWs, shown in Figure 48.9, suggested a shift in the surface plasmon resonance peak to the mid-IR region reflecting a significant reduction in the carrier density for p-type NWs and NPs. Note that even an undoped SnO$_2$ tends to be an n-type semiconductor due to the nonstoichiometric nature of the oxide, that is, it is SnO$_{2-x}$ and not SnO$_2$, where x is small but positive. The loss of oxygen deposits excess electrons in the conduction band and positively charged immobile oxygen vacancies in the crystalline lattice. Thus, to realize a p-type semiconductor, the excess electrons have to be first neutralized, that is, the Fermi level has to be pushed toward the valence band requiring significantly higher p-type dopant concentration[26,42] to insert the equivalent hole carrier density. To verify if the surface plasmon peak indeed appears in the mid-IR region, we collected IR spectra from the powders of NP and NWs. The data, shown in Figure 48.9b, exhibited a strong peak near the 3400 nm region, that is, a clear continuation of the incipient peaks seen in the NIR spectra as seen in UV–visible spectra shown in Figure 48.9a. The corresponding carrier concentration was 1.6×10^{21} cm^{-3} for p-doped NPs.

Interestingly, the plasmon peak position for the p-doped NWs barely shifted from its location in p-type NPs, yet the width of the peak was clearly broader, implying a shorter collisional lifetime of the carriers. This could arise from the relatively high surface area to volume ratios for NWs than

---

[*] These concentrations correspond to the concentration of dopants used during sol–gel synthesis, which are not necessarily the actual dopant concentrations in the final NPs. Nanoscopic phase separation, compound formation, and so on can significantly reduce the actual dopant concentrations in the semiconductor lattice.

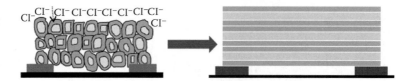

FIGURE 48.10 Schematic illustration of depletion regions in NPs (left) and NWs (right) upon exposure to oxidizing gas such as chlorine in the SnO_2-based resistive sensor.

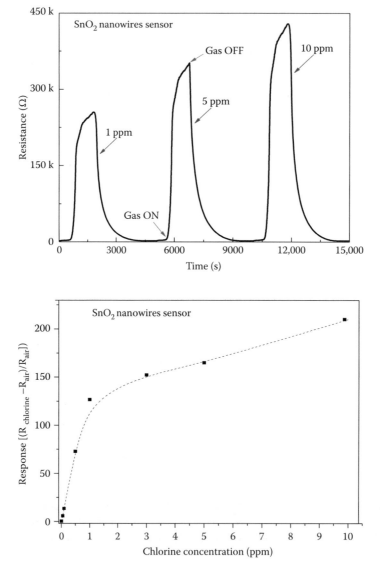

FIGURE 48.11 Time-dependent resistance change as a function of chlorine concentration and overall response concentration curve.

NPs. The failure to shift the peak to lower wavelengths, that is, lower concentration of inserted p-type dopants (even when synthesized in LiCl solvent), could imply a radial concentration gradient of the dopant or even the formation $Li_2SnO_3^{27}$ on the NW outer shell. Further NMR studies[42] are currently underway to explore the location of the inserted Li in the NWs.

48.3.1 Sensor Application of n-Doped NWs: Chlorine Detection

The success of using n-doped SnO_2 NPs as a smart sensor based on noncorrodible SnO_2 NPs to selectively detect chlorine at room temperature[22] was the motivation for the current synthesis of n-doped nanowires. The basic mechanism of the increase in the resistance of the sensor is due to the creation of the depletion layer and an associated barrier for electron tunneling. Increasing the concentration of chlorine in the gas phase results in a proportionate increase in the chloride ion concentration at the gas–solid interface, which in turn increases the depletion region and hence the resistance of the sensor. This is shown schematically in Figure 48.10.

Replacing n-doped NPs with n-doped nanowires should lead to a faster response by reducing the number of interparticle electron hops between the sensing electrodes (Figure 48.10, right).

The data in Figure 48.11 show a monotonic response with Cl_2 concentration. It should be noted that the switching speeds near room temperature are about 4–5 min. These n-doped NW-based sensors are capable of pushing the chlorine detection limit to the ppb level. By analogy, an increase in impedance is expected for reducing gases with p-type SnO_2 nanowires. Currently, our ongoing studies of H_2 gas detection have shown a relatively poor sensitivity for its detection using p-type nanowires/nanoparticles. We ascribe these effects to the mid-bandgap location of the H_2 redox level. The sensitivity of these devices can be improved by using palladium (Pd) nanoparticles. Pd can absorb and presumably form hydrides in the presence of hydrogen gas. Interestingly, Pd nanoparticle composites of n-type nanowires were sensitive to H_2 gas. The results of these extensive studies will be reported elsewhere.

48.4 CONCLUSION

The first synthesis of single crystalline n-doped and p-doped SnO_2 NWs is reported. These structures exhibit two strong emission bands in the fluorescence spectra and display surface plasmon peaks in the near- to mid-IR regions, implying a high carrier concentration. The availability of both n- and p-type NWs opens up new opportunities for crafting transparent nanoelectronic devices, sensors, and so on. Preliminary studies of the sensor properties of n-type NW-based devices have yielded ppb-level sensitivity for chlorine gas detection.

ACKNOWLEDGMENTS

We are indebted to Professors Carl C. Wamser and Andrea M. Goforth for allowing access to UV–visible–NIR and fluorescence spectrometers, respectively. Partial financial support from Intel Corporation NIH and ONAMI is gratefully acknowledged.

REFERENCES

1. A. P. Alivisatos, *Science* **1996**, *271*, 933–937.
2. C. Dekker, *Physics Today* **1999**, *52*, 22–28.
3. J. Hu, T. W. Odom, and C. M. Lieber, *ChemInform* **1999**, *32*, 435–445.
4. R. S. Wagner and W. C. Ellis, *Applied Physics Letters* **1964**, *4*, 89–90.
5. X. Duan and C. M. Lieber, *Advanced Materials* **2000**, *12*, 298–302.
6. T. J. Trentler, K. M. Hickman, S. C. Goel, A. M. Viano, P. C. Gibbons and W. E. Buhro, *Science* **1995**, *270*, 1791–1794.

7. J. R. Heath and F. K. LeGoues, *Chemical Physics Letters* **1993**, *208*, 263–268.
8. F. Kim, K. Sohn, J. Wu and J. Huang, *Journal of the American Chemical Society* **2008**, *130*, 14442–14443.
9. G. Yan, L. Wang and L. Zhang, *Reviews on Advanced Materials Science* **2010**, *24*, 10–25.
10. I. Hamberg and C. G. Granqvist, *Journal of Applied Physics* **1986**, *60*, R123–R159.
11. J. F. Wager, *Science* **2003**, *300*, 1245–1246.
12. R. L. Hoffman, B. J. Norris and J. F. Wager, *Applied Physics Letters* **2003**, *82*, 733–735.
13. R. E. Presley and et al., *Journal of Physics D: Applied Physics* **2004**, *37*, 2810.
14. G. E. Batley, A. Ekstrom and D. A. Johnson, *Journal of Catalysis* **1974**, *34*, 368–375.
15. G. Croft and M. J. Fuller, *Nature* **1977**, *269*, 585–586.
16. G. C. Bond, L. R. Molloy and M. J. Fuller, *Journal of the Chemical Society, Chemical Communications* **1975**, *19*, 796–797.
17. P. G. Harrison, C. Bailey and W. Azelee, *Journal of Catalysis* **1999**, *186*, 147–159.
18. F. Solymosi and J. Kiss, *Journal of Catalysis* **1978**, *54*, 42–51.
19. T. Matsui, T. Okanishi, K. Fujiwara, K. Tsutsui, R. Kikuchi, T. Takeguchi and K. Eguchi, *Science and Technology of Advanced Materials* **2006**, *7*, 524–530.
20. C. Cané, I. Gràcia, A. Götz, L. Fonseca, E. Lora-Tamayo, M. C. Horrillo, I. Sayago, J. I. Robla, J. Rodrigo and J. Gutiérrez, *Sensors and Actuators B: Chemical* **2000**, *65*, 244–246.
21. F. Allegretti, N. Buttá, L. Cinquegrani and S. Pizzini, *Sensors and Actuators B: Chemical* **1993**, *10*, 191–195.
22. A. Chaparadza and S. B. Rananavare, *Nanotechnology* **2008**, *19*, 245501.
23. N. Buttà, L. Cinquegrani, E. Mugno, A. Tagliente and S. Pizzini, *Sensors and Actuators B: Chemical* **1992**, *6*, 253–256.
24. J. Tamaki, T. Maekawa, N. Miura and N. Yamazoe, *Sensors and Actuators B: Chemical* **1992**, *9*, 197–203.
25. J. M. Wu, *Nanotechnology* **2010**, *21*, 235501.
26. A. Chaparadza, S. B. Rananavare and V. Shutthanandan, *Materials Chemistry and Physics* **2007**, *102*, 176–180.
27. J. C. Chan, N. A. Hannah, S. B. Rananavare, L. Yeager, L. Dinescu, A. Saraswat, P. Iyer and J. P. Coleman, *The Japan Society of Applied Physics* **2006** *45*, L1300–L1303.
28. F. D. M. Haldane, *Journal of Physics C: Solid State Physics* **1981**, *14*, 2585.
29. T. Darkop, S. A. Getty, E. Cobas and M. S. Fuhrer, *Nano Letters* **2003**, *4*, 35–39.
30. O. Gunawan, L. Sekaric, A. Majumdar, M. Rooks, J. Appenzeller, J. W. Sleight, S. Guha and W. Haensch, *Nano Letters* **2008**, *8*, 1566–1571.
31. J. Q. Hu, Y. Bando, Q. L. Liu and D. Golberg, *Advanced Functional Materials* **2003**, *13*, 493–496.
32. Z. Liu, D. Zhang, S. Han, C. Li, T. Tang, W. Jin, X. Liu, B. Lei and C. Zhou, *Advanced Materials* **2003**, *15*, 1754–1757.
33. Y. Wang and J. Y. Lee, *The Journal of Physical Chemistry B* **2004**, *108*, 17832–17837.
34. W. Wang, J. Niu and L. Ao, *Journal of Crystal Growth* **2008**, *310*, 351–355.
35. C. Yu, J. C. Yu, F. Wang, H. Wen and Y. Tang, *Crystal Engineering Communication* **2010**, *12*, 341–343.
36. Z. Guifu et al., *Nanotechnology* **2006**, *17*, S313.
37. Z. W. Pan, Z. R. Dai and Z. L. Wang, *Science* **2001**, *291*, 1947–1949.
38. H. Kim, J. Lee and C. Lee, *Journal of Materials Science: Materials in Electronics* **2009**, *20*, 99–104.
39. J. X. Wang, D. F. Liu, X. Q. Yan, H. J. Yuan, L. J. Ci, Z. P. Zhou, Y. Gao et al., *Solid State Communications* **2004**, *130*, 89–94.
40. P. Nguyen, H. T. Ng, J. Kong, A. M. Cassell, R. Quinn, J. Li, J. Han, M. McNeil and M. Meyyappan, *Nano Letters* **2003**, *3*, 925–928.
41. X. L. Ma, Y. Li and Y. L. Zhu, *Chemical Physics Letters* **2003**, *376*, 794–798.
42. A. Chaparadza and S. B. Rananavare, *Nanotechnology* **2010**, *21*, 035708.
43. B.-M. Mohammad-Mehdi and S.-S. Mehrdad, *Semiconductor Science and Technology* **2004**, *19*, 764.
44. Y. Wang, J. Y. Lee and T. C. Deivaraj, *The Journal of Physical Chemistry B* **2004**, *108*, 13589–13593.
45. J. C. Chan, H. Tran, J. W. Pattison and S. B. Rananavare, *Solid-State Electronics* **2010**, *54*, 1185–1191.
46. T. W. Kim, D. U. Lee and Y. S. Yoon, *Journal of Applied Physics* **2000**, *88*, 3759–3761.
47. C. M. Liu, X. T. Zu, Q. M. Wei and L. M. Wang, *Journal of Physics D: Applied Physics* **2006**, *39*, 2494.
48. B. Wang, Y. H. Yang, C. X. Wang, N. S. Xu and G. W. Yang, *Journal of Applied Physics* **2005**, *98*, 124303–124304.
49. S. Majumdar, S. Chakraborty, P. S. Devi and A. Sen, *Materials Letters* **2008**, *62*, 1249–1251.
50. T. Nutz, U. z. Felde and M. Haase, *The Journal of Chemical Physics* **1999**, *110*, 12142–12150.
51. H.A. Tran and S.B. Rananavare, *11th IEEE Conference on Nanotechnology (IEEE-NANO)*, Aug. 15–18, **2011**, 144–149.

49 Cu Silicide Nanowires
Fabrication, Characterization, and Application to Li-Ion Batteries

Poh Keong Ng, Reza Shahbazian-Yassar,
and Carmen Maria Lilley

CONTENTS

49.1 INTRODUCTION

Although batteries are inherently simple in concept, surprisingly, their development has progressed much more slowly than other areas of electronics [1]. The slow progress is due to the lack of suitable electrode materials and electrolytes, together with difficulties in mastering the interfaces between them. Traditional lithium-ion batteries employ carbonaceous anodes with a capacity of 372 mAhg^{-1}. To obtain a substantial improvement in the specific capacity of Li-ion cells, it is essential to replace carbonaceous anodes with anodes having greater capacity. The most attractive candidate to replace carbonaceous anodes is silicon (Si), which has the highest known capacity, in excess of 4000 mAhg^{-1} [2,3]. Two major drawbacks have hindered the application of Si structures as anodes for Li-ion batteries. One is related to its electrical conductivity, which is much lower than graphite anodes. Consequently, during the charging process, the Li-ions cannot penetrate deep into the active Si anodes. As such, it is highly desirable to dope Si with other elements and improve its electrical transport. Among these dopants, alloying with copper (Cu) is favorable due to the fact that the current collector is also made of Cu [4]. In addition to the electrical conductivity drawback, the mechanical stresses associated with silithiation can be problematic. It is observed that upon driving Li-ions into Si, a volume expansion on the order of 300–400% [5,6] occurs due to the formation of various phases like $Li_{12}Si_7$, Li_7Si_3, $Li_{13}Si_4$, and $Li_{22}Si_5$ [7]. This leads to mechanical stresses large enough to fracture and pulverize Si into powder after the first few cycles of charging/discharging and eventually capacity fade during cycling [8]. To address this issue, it is suggested that a nanowire morphology will facilitate the lateral expansion of Si and enhance their fracture resistance [9]. In particular, single crystalline nanowires have the potential to exhibit ideal material characteristics for the design of nanotechnology-based fuel cells. For example, single crystalline nanowires may have a lower electrical resistivity and a higher tolerance to failure as compared to polycrystalline nanowires [10]. Recently, the fabrication of single crystalline freestanding Cu_3Si nanowires was reported in Ref. [11] by the annealing of Cu/Ge bilayer films on a SiO_2/Si substrate. However, obtaining the high-density coverage of single crystalline nanowires needed for anode materials remains a technical challenge. Self-assembly of copper silicide (Cu-Si) by e-beam evaporation will produce single crystalline and low-defect-density nanowires, which are ideal test systems to study the fundamental properties and integration into Li-ion batteries. In addition, Cu-Si alloy nanowires can be synthesized by the code-position of Si and Cu at high temperatures using oblique angle deposition (OAD). In this technique, two target sources of Si and Cu will be evaporated using high-energy ions and the evaporated atoms will be deposited on a substrate in the form of nanowires and nanorods [4], which are freestanding as well. Unlike the technique described in Ref. [11], the OAD method produces a far larger amount of Cu-Si nanowires. This chapter will outline research by the authors on the material properties of Cu-Si fabricated via self-assembly and OAD techniques. In addition, Cu-Si nanowires fabricated with the OAD method and integrated into Li-ion batteries will also be briefly discussed.

49.2 SYNTHESIS METHODS FOR CU SILICIDE NANOWIRES

49.2.1 Single Crystalline Cu Silicide

The self-assembly of silver (Ag) nanowires on miscut Si substrates, a technique pioneered by Roos et al. [12], has attracted considerable interest recently [13,14]. This same method has been applied to synthesize Cu-Si nanowires [15–17] for fabricating Cu-Si nanowires on Si(001). The e-beam evaporation method was chosen here for Cu-Si nanowire fabrication because it can produce low-defect-density crystalline nanowires. Fabrication of Cu-Si nanowires using the e-beam evaporation approach is given in Ref. [15,16]. However, Cu and Si are very diffusive even at a relatively low temperature of 200°C [18]. Therefore, the preparation of Cu nanowires on Si at a high temperature will likely result in significant intermixing. Because there are limited publications on self-assembled Cu-Si nanowires, little material characterization data are available. The following section of this chapter presents a high-resolution analysis of Cu-Si nanowires grown by the e-beam evaporation method. The aim of the discussion presented here is to provide insight on the growth characteristics and chemical composition of Cu-Si nanowires fabricated by the self-assembly method described in Ref. [16].

Cu-Si nanowires were fabricated by self-assembly on both flat and vicinal Si. The general fabrication process is as follows: Small dies (10×3 mm) of Si, n-type (Ph-doped) and (001)-oriented, with a 4° miscut (or <1° for flat Si) were cleansed with acetone and isopropyl-alcohol (IPA) organic solvents to remove organic matter before loading it into an e-beam evaporation chamber. The Si substrate was degassed for 12 h at 600°C to remove unwanted hydrocarbons. The Si substrate was then flash-heated five times to 1200°C for 30 s to remove the native oxide on the Si surface. Cu pellets of 99.99% purity were then e-beam-evaporated and deposited onto the Si substrate in an ultrahigh vacuum (UHV) environment ($<1 \times 10^{-8}$ mbar) while annealing the substrate at approximately 600°C. During material deposition, both Cu-Si islands and nanowires randomly formed on the Si substrate. Upon deposition, *in situ* analysis was performed with a scanning electron microscope (SEM) to confirm the formation of nanowires and to measure their geometries. The e-beam evaporation experiments were conducted with an Omicron UHV nanoprobe system located at the Center for Nanoscale Materials (CNM) at the Argonne National Laboratory (ANL), Illinois, USA. Sample preparations for the transmission electron microscope (TEM) and scanning transmission electron microscope (STEM) were done at CNM in ANL. Finally, the TEM and STEM analyses were performed at the Research Resources Center (RRC) of the University of Illinois at Chicago (UIC). The results from the characterization of the self-assembled nanowires are discussed in Section 49.3 of this chapter.

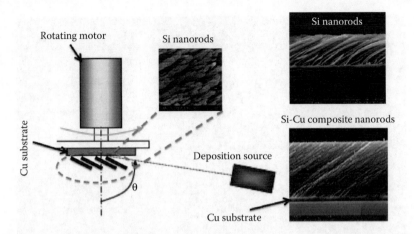

FIGURE 49.1 The fabrication of Si and Cu-Si alloy nanostructures using the approach of oblique angle deposition. Deposition sources of Si and Cu can be selected to synthesize the composite nanowires.

49.2.2 Oblique Angle Codeposition of Cu-Si

The fabrication of the Cu_3Si nanowires can be conducted by an oblique angle (co)deposition technique [19–21] in a two-source electron-beam deposition system, where two quartz crystal microbalances (QCMs) can be installed to monitor the near-normal deposition thickness and rate of each source independently [4]. The OAD method is simply a physical vapor deposition (VPD) technique in which flux arrives at a large oblique incidence angle (>80°) from the substrate normal (while the substrate is rotating), as shown schematically in Figure 49.1. This results in the formation of isolated nanorods by the self-shadowing effect during growth. Either pure Si or Cu-Si nanowires can be codeposited on the Cu film-coated glass substrates at a vapor incident angle of α with respect to the substrate normal. PVD methods such as sputtering and e-beam evaporation are clean and repeatable, have high deposition rates, and can be grown over large areas with good adhesion to the substrate. OAD provides a convenient method to grow Cu-Si nanorods by using the additional Cu deposition source [4].

49.3 CHARACTERIZATION OF Cu SILICIDE NANOWIRES

49.3.1 Self-Assembled Nanowires

Since the self-assembly of Cu-Si nanowires has not been widely studied, the crystalline structure was studied to identify the type of Cu-Si that is formed. We have also investigated the influence of the surface preparation on the growth mode. Cu-Si nanowires grown on clean and oxidized Si surfaces were compared. The nanowires (or nanoislands) were randomly formed on the Si substrate and located using an *in situ* SEM and are shown in Figure 49.2. Figure 49.2a shows a low-magnification view after a typical fabrication process. Similar to experiments on Ag nanowires described in Refs. [12–14], the length rather than the width increases with deposition time. Some of the nanoislands were observed to have pyramidal shapes with truncated tops, whereas the rest of the nanoislands had undefined shapes, as seen in Figure 49.3.

49.3.2 Growth Characteristics of Self-Assembled Nanowires

To investigate the structural properties of the nanowires in three dimensions, cross-sectional SEM images were obtained. These cross sections were prepared by FIB ion-milling. Two randomly chosen Cu-Si nanowires were ion-milled to reveal their cross-sectional shapes, as shown in Figure 49.4. The cross-sectional *ex situ* SEM images show that the two Cu-Si nanowires do not have

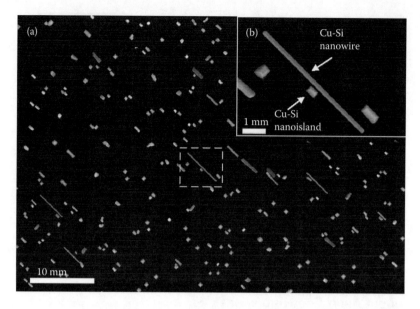

FIGURE 49.2   An *in situ* SEM image of Cu-Si nanowires and nanoislands with (a) a low-magnification view and the dashed box is shown with (b) a high-magnification view.

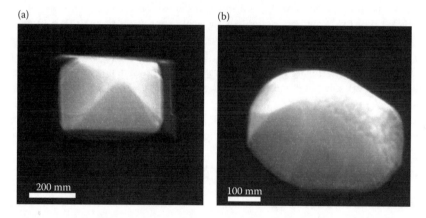

FIGURE 49.3   *Ex situ* SEM images taken with a 52° tilt using the FEI Nova 600 Dual Beam system. (a) A Cu-Si nanopyramid with a truncated top. (b) An irregularly shaped nanoisland. (P. K. Ng et al. Self assembled Cu nanowires on vicinal Si(001) by the E-beam evaporation method, Presented at the *11th IEEE International Conference on Nanotechnology (IEEE-NANO)*, Portland, OR. © (2011) IEEE. With permission.)

well-defined facets. The cross-sectional dimensions for the nanowires in Figure 49.4a and b are the following: widths are 210 and 295 nm, respectively; thicknesses are 93 and 100 nm, respectively; and height of the nanowires measured between the peak of the nanowire and the surface of Si substrate are 50 and 50 nm, respectively. Also, the Cu-rich material extends into the substrate, indicating severe interdiffusion (indicated by dashed lines). To show the differences between Ag and Cu-Si more clearly, we also fabricated and analyzed Ag nanowires. The resulting SEM image is shown in Figure 49.5. As found in Refs. [12–14], the Ag nanowire has a well-defined trapezoidal cross-sectional shape and is more uniform than the two Cu-Si nanowires observed in Figure 49.4. Furthermore, the Ag does not significantly extend into the Si substrate. Rather, some Si "wetting" toward the Ag nanowire seems to have taken place. In contrast, little or no Si wetting toward Cu-Si nanowires is seen at the edges of the two Cu-Si nanowires in Figure 49.4.

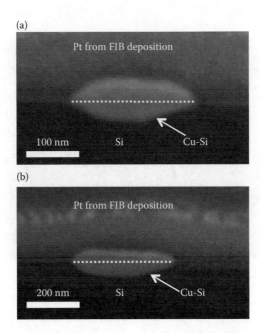

FIGURE 49.4 Cross-sectional view of the *ex situ* SEM images, (a) and (b), of two different Cu–Si nanowires tilted at 52° prepared by using an FIB. Although both nanowires in (a) and (b) were fabricated on the same Si substrate, the two nanowires do not exhibit similar cross section shape and geometry. Each Cu–Si nanowire in (a) and (b) had undefined and irregular facets. The dashed lines portray the imaginary Si surface's plane. (P. K. Ng et al. Self assembled Cu nanowires on vicinal Si(001) by the E-beam evaporation method, Presented at the *11th IEEE International Conference on Nanotechnology (IEEE-NANO)*, Portland, OR. © (2011) IEEE. With permission.)

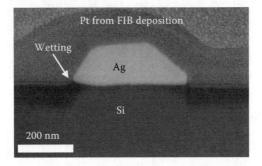

FIGURE 49.5 Cross-sectional view of the *ex situ* SEM image of a Ag nanowire tilted at 52° prepared by using an FIB. (P. K. Ng et al. Self assembled Cu nanowires on vicinal Si(001) by the E-beam evaporation method, Presented at the *11th IEEE International Conference on Nanotechnology (IEEE-NANO)*, Portland, OR, P. K. Ng et al. High resolution analysis of self assembled Cu nanowires on vicinal Si(001), Presented at the *11th IEEE International Conference on Nanotechnology (IEEE-NANO)*, Portland, OR. © (2011) IEEE. With permission.)

The different degrees of interdiffusion might be explained by Ag-Si [22] and Cu-Si phase diagrams [18]. Ag and Si do not form silver silicide (Ag-Si) even at the deposition temperature of 600 °C. In contrast, Cu-Si diffusion occurs well below 600°C. Indeed, Cu-Si interdiffusion has been confirmed with the STEM experiments discussed below. In another set of experiments, we have investigated the effect of the Si surface quality on the growth of Cu-Si nanowires. The e-beam evaporation procedure was exactly the same as before. The only difference was that the Si surface was not flash-treated, and thus the native oxide was present during the e-beam evaporation process. The resulting morphology is shown in Figure 49.6a. The fuzzy appearance of the image is caused by

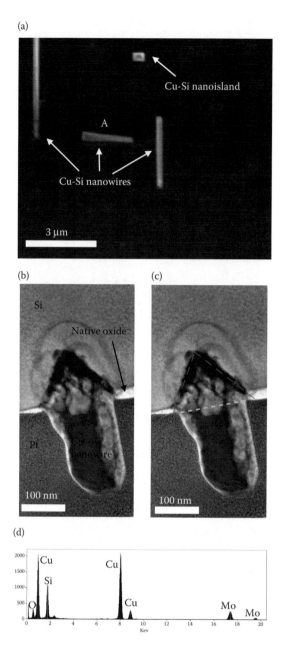

FIGURE 49.6 Cu-Si nanowires on an Si substrate with native oxide. (a) An *ex situ* SEM view with a 52° tilt, (b,c) a cross-sectional TEM view, and (d) XEDS analysis showing the presence of O, Si, and Cu in the nanowire. The element Mo is from the TEM grid. (P. K. Ng et al. Self assembled Cu nanowires on vicinal Si(001) by the E-beam evaporation method, Presented at the *11th IEEE International Conference on Nanotechnology (IEEE-NANO)*, Portland, OR, © (2011) IEEE. With permission.)

the poor electrical conductivity of SiO, which leads to charging and affects the SEM performance. A lower density of Cu-Si nanowires was observed than without a native oxide. In addition, the nanowire labeled with "A" in Figure 49.6a shows that some Cu-Si nanowires exhibit an inhomogeneous cross section. Obviously, the left side of the nanowire "A" is wider than the right side. Again, details of the cross section were obtained from the cross-sectional TEM shown in Figure 49.6b. The width of the nanowire amounts to 150 nm and the height to 220 nm. It is not known what causes the different contrast in the figures.

Figure 49.6c shows the outline of the nanowire in dashed lines. The shape is very different from the Cu-Si nanowires grown on clean Si. The oxide material on the Si surface obviously modifies the growth mechanism, possibly through a higher surface diffusion barrier. The horizontal dashed line indicates the imaginary surface plane of the Si substrate. It is obvious that the Cu nanowire interdiffuses with the Si substrate even though there is an oxide barrier. Although this mechanism is currently not understood, it is believed that the Cu atoms can penetrate through the thin and porous layer of the native oxide at the annealing temperature of 600°C. Nevertheless, details of the growth mechanisms require further studies. The preliminary x-ray energy-dispersive spectrometer (XEDS) experiments shown in Figure 49.6d reveal that the material is composed of Cu, Si, and O. Further research is also needed to determine the exact material composition and phase.

49.3.3 HIGH-RESOLUTION ANALYSIS OF SELF-ASSEMBLED CU SILICIDE NANOWIRES

TEM diffraction pattern tests were performed on nanowires to determine their crystalline structures. Two Cu-Si nanowires from the same substrate are labeled as "Cu-Si wire 1" and "Cu-Si wire 2," as shown in Figures 49.7a and 49.8a. In the right panels of these figures are the local diffraction patterns. They reveal that wire 1 possesses a single crystalline structure. In contrast, wire 2 is composed of two grains with a low-angle boundary between them. The two low-angle Cu grains appear to have a projected rotational mismatch of 3° based on the projection of the Si[110] zone

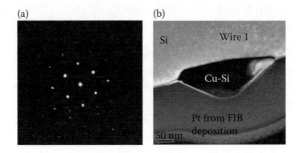

FIGURE 49.7 Single crystalline, self-assembled Cu-Si nanowires with (a) their diffraction pattern obtained from TEM analysis and (b) a cross-sectional view prepared by FIB. (P. K. Ng et al. High resolution analysis of self assembled Cu nanowires on vicinal Si(001), Presented at the *11th IEEE International Conference on Nanotechnology (IEEE-NANO)*, Portland, OR. © (2011) IEEE.)

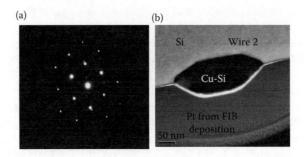

FIGURE 49.8 Low-angle grain boundary, self-assembled Cu-Si nanowires with (a) their diffraction pattern obtained from TEM analysis and (b) a cross-sectional view prepared by FIB. (P. K. Ng et al. High resolution analysis of self assembled Cu nanowires on vicinal Si(001), Presented at the *11th IEEE International Conference on Nanotechnology (IEEE-NANO)*, Portland, OR. © (2011) IEEE.)

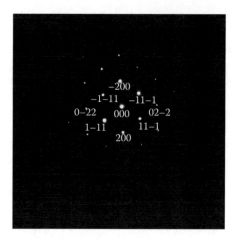

FIGURE 49.9 Indexed Si[110] zone pattern. (P. K. Ng et al. High resolution analysis of self assembled Cu nanowires on vicinal Si(001), Presented at the *11th IEEE International Conference on Nanotechnology (IEEE-NANO)*, Portland, OR. © (2011) IEEE. With permission.)

axis, which is shown in Figure 49.9. It is important to note that this measured angle is derived from the Si[110] zone projection. It is possible that the two low-angle grains may actually be slightly rotated along another arbitrary axis that is close to the Si[110] axis. The reason for the formation of the two different structures is presently unknown. However, the hexagonal patterns obtained from the two diffraction patterns strongly indicate an epitaxial relationship between the Cu-Si nanowires and the Si substrate.

To study the chemical composition of the wires, two nanowires were selected as shown in Figure 49.10a. The nanowire with a triangular top is labeled as "A" while the other nanowire with a flat top is labeled as "B." An FIB-TEM analysis was performed on these two nanowires. The TEM images of nanowires "A" and "B" are shown in Figure 49.10b and c. The dashed lines in Figure 49.10b and c illustrate the original surface plane of the Si substrate. The reason for this observation may be Cu-Si diffusion during the 600°C annealing process since diffusion occurs at this temperature [18].

STEM energy z-contrast images were obtained for the two nanowires, as shown in Figure 49.11a and b. The numeric labels in the figures are the locations of the XEDS spectra that were measured. Elemental analysis data are shown in Figure 49.11c. From these XEDS data, we find discrete Cu:Si ratios of roughly 50:50, 30:70, and 70:30. Indeed, Cu_3Si is a stable phase [18]. However, the other Cu-Si phases with ratios 50:50 and 30:70 are not known to be stable after annealing to 600°C [18]. Future studies are needed to understand the existence of unstable Cu-Si phases in the Cu-Si nanowires.

In summary, an in-depth discussion is presented here on the synthesis and material characteristics of self-assembled Cu-Si nanowires. The self-assembly method may not be the ideal method for integrating Cu-Si nanowires to Li-ion nanowires. However, the methods do enable the synthesis of ideal Cu-Si nanowires (i.e., single crystalline or with a few defects and contaminant free) that can be characterized for electrical and defect tolerance that is necessary for fundamental research as Li-ion anode materials. Furthermore, *in situ* electrical characterization can be performed after the self-assembling Cu-Si nanowires in the UHV environment without exposure to the ambient impurities, which are currently being studied. In the following section, the synthesis of Cu-Si nanowires as a Li-ion anode will be discussed. Although the synthesis techniques are different, the characterization of the material properties is complementary to study the fundamental properties of Cu-Si nanowires and select the ideal crystalline structure, doping and Cu-Si phase properties, and dimensions.

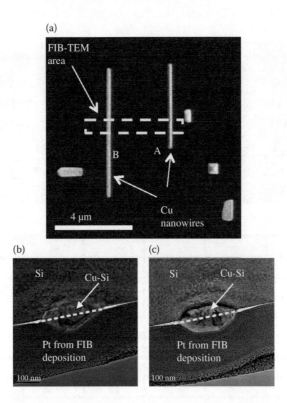

FIGURE 49.10 (a) SEM image of self-assembled Cu-Si nanowires that appeared to have an "A" triangular top surface and a "B" regular flat top surface. Cross-sectional view of the FIB-TEM area in (a) for (b) a regular top "B" wire and (c) a triangular top "A" wire. (P. K. Ng et al. High resolution analysis of self assembled Cu nanowires on vicinal Si(001), Presented at the *11th IEEE International Conference on Nanotechnology (IEEE-NANO)*, Portland, OR. © (2011) IEEE. With permission.)

49.3.4 CHARACTERIZATION OF (CO)DEPOSITED CU SILICIDE NANOWIRES

Both the pure Si and Cu-Si composite nanorods were fabricated by an oblique angle (co)deposition technique in a custom-designed, two-source electron-beam deposition system, where two QCMs were installed to monitor the near-normal deposition thickness and rate of each source independently [23,24]. During the anode nanostructure fabrication, first, a layer of a 200-nm-thick Cu film was coated onto the 10×30 mm glass substrates as the current collector of battery anodes at a deposition rate of $r_{Cu} = 0.2$ nm/s. Then, Cu-Si alloy nanowires were deposited on the Cu film–coated glass substrates at a vapor incident angle of $\alpha = 88°$ and both Cu and Si sources were coevaporated at $r_{Cu} = 0.5$ nm/s and $r_{Si} = 0.4$ nm/s. The mass of the active Si on an area of 10×30 mm was estimated to be approximately 3.7×10^{-5} g for the Cu-Si alloy nanowires.

The codeposited samples consist of arrays of well-aligned and tilted Si and Cu-Si nanowires (Figure 49.1). The x-ray diffraction (XRD) results in Figure 49.12 show that the as-deposited Si nanowires are amorphous (Figure 49.12a); however, when codeposited with Cu, a polycrystalline orthorhombic Cu$_3$Si phase forms, as revealed by the peak Cu$_3$Si (012) at $2\theta \approx 44.57°$ in Figure 49.12. In the TEM images, the amorphous Si matrix and the high-contrast Cu$_3$Si nanoparticles can be seen. This indicates that the Cu-Si nanowires are composed of the polycrystalline Cu$_3$Si nanoparticles embedded in the amorphous Si nanowire matrix. The nanoparticles are estimated to have diameters of ~3–5 nm.

(a)

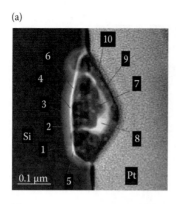

(b)

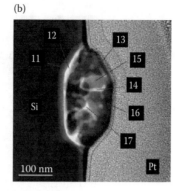

(c)

Location	Si	Cu	Location	Si	Cu
1	69.51	30.49	9	52.72	47.28
2	71.8	28.2	10	64.84	35.16
3	70.86	29.14	11	73.7	26.3
4	28.76	71.24	12	31.4	68.6
5	32.98	67.02	13	68.54	31.46
6	32.62	67.38	14	26.85	73.15
7	60.65	39.35	15	43.25	56.75
8	29.97	70.03	16	55.56	44.44
			17	26.54	73.46

FIGURE 49.11 STEM z-contrast data for (a) wire "A" and (b) wire "B," where the numeric labeled locations of the spot analyses were performed. (c) EDX data where the atomic percentages of Si and Cu are given for the locations in (a) and (b). (P. K. Ng et al. High resolution analysis of self assembled Cu nanowires on vicinal Si(001), Presented at the *11th IEEE International Conference on Nanotechnology (IEEE-NANO)*, Portland, OR. © (2011) IEEE. With permission.)

49.4   CU SILICIDES AS LI-ION ANODE MATERIALS

To study the lithiation behavior of Cu-Si nanowires, the nanowires fabricated via OAD were subjected to a charging process inside a TEM [25,26]. The applied bias voltage was set to be −2 V on a Cu-Si nanowire to drive Li-ions into the nanowire. Figure 49.13 shows the structural evolution of a Cu-Si nanowire as a function of charging time. As expected, the diameter of a nanowire increases during Li-ion intercalations in the anode nanowire. From Figure 49.13a and b, one can observe the change of diameter from 173 to 201 nm upon 15 min of charging. In addition, the formation of high-contrast nanoparticles was detected, which can be related to various Li_xSi alloys. According to Figure 49.13c, the diameter of the Cu-Si alloy nanowire increases much faster during the lithiation process in comparison to the pure Si nanowires. This indicates that alloying Si with Cu and the Cu_3Si crystals can improve the rate of lithiation.

The lithiation performance of single crystals of Cu_3Si nanowires is expected to be superior to that of the amorphous Cu-Si alloys. The electrical conductivity of Cu_3Si single crystals is greater than the low electrical conductivity of amorphous Cu-Si alloys. Prior results have shown that the metallic Cu_3Si nanowires are excellent conductors with a resistivity of less than 30 μΩ·cm [11]. In addition to the better electronic conductivity, the single crystalline nature of the nanowires will be ideal to prevent localized lithiation and delithiation, which may result in the premature fracture

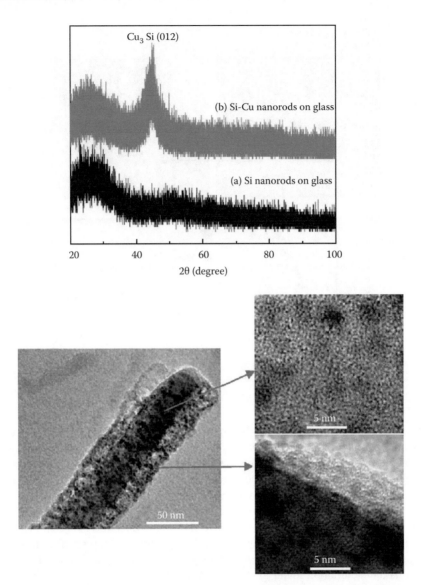

FIGURE 49.12 (Top) XRD patterns of the deposited (a) Si and (b) Cu-Si composite nanowires. (Bottom) The TEM images of the Cu-Si nanowire indicate the presence of crystalline Cu_3Si embedded in an amorphous matrix. (From *Journal of Power Sources*, 196, M. Au et al., Silicon and silicon–copper composite nanorods for anodes of Li-ion rechargeable batteries, 9640–9647, Copyright 2011, with permission from Elsevier.)

of the nanowires. In fact, our previous results [25] show that lithiation can be surface diffusion dominant and the selective locations on the surface can be the predominant place for the lithiation process. The single crystalline structure of the Cu_3Si nanowires can promote a uniform diffusion path for the Li-ions to transfer into the crystal structure and interact with the host atoms.

49.5 SUMMARY AND CONCLUSIONS

In summary, we have briefly discussed the fabrication methods for single-crystal and Cu-alloyed amorphous Si nanowires. Although the synthesis technique for single-crystal nanowires cannot be directly integrated into anode material for Li-ion batteries, these nanowires enable fundamental

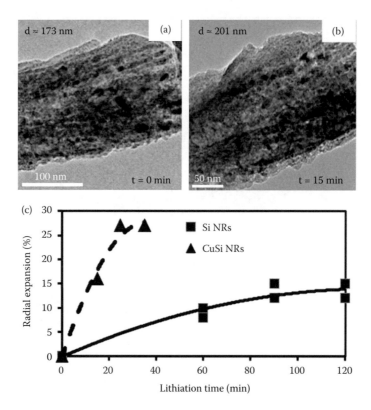

FIGURE 49.13 (a) A Cu-Si nanowire just before the start of the lithiation process. The diameter is around 173 nm. (b) The same Cu-Si nanowire after 15 min of the lithiation process and the diameter increased to 201 nm. (c) The graph compares the radial expansion of Cu-Si and pure Si nanowires as a function of lithiation time. The Cu-Si alloy nanowire shows a higher rate of mechanical straining, which indicates that the rate of charging is faster for a Cu-Si nanowire.

studies of the properties needed for Cu-Si nanowire anode materials. It is expected that single crystalline Cu-Si nanowires will provide lower electrical resistivity and better surface Li ion diffusion than the alloyed Si nanowires due to their uniformity in crystalline structure. This will prevent localized lithiation/delithiation found in the OAD Cu-Si nanowires that may result in the formation of cracks and fractures of the nanowires. Further work is needed on synthesizing freestanding single crystalline Cu-Si nanowires for *in situ* TEM lithiation/delithiation studies and on improving the crystalline structure of the OAD nanowires.

ACKNOWLEDGMENTS

C. M. Lilley acknowledges the support from the National Science Foundation (Award No. 846814, Division of Civil, Mechanical and Manufacturing Innovation). The use of the Center for Nanoscale Materials was supported by the U.S. Department of Energy, Office of Science, Office of Basic Energy Sciences, under Contract No. DE-AC02-06CH11357. C.M.L. would also like to acknowledge the technical assistance from Dr. Ke-Bin Low of the Research Resource Center at the University of Illinois at Chicago in the TEM characterization of the self-assembled Cu-Si nanowires and Dr. Alexandra Joshi-Imre from the Center for Nanoscale Materials for her assistance in the TEM sample preparation. In addition, C.M.L. thanks Mr Brandon Fisher for his technical assistance on the Omicron system at the Center for Nanoscale Materials. Finally, C.M.L. acknowledges the contribution to the growth mechanics analysis by Dr. Matthias Bode at the University of Wuerzburg.

R. S. Yassar acknowledges the support from the National Science Foundation (Award No. 0820884, Division of Materials Research) and the American Chemical Society—Petroleum Research Fund (Award No. 51458-ND10). R.S.Y. also recognizes the contributions from Dr. Ming Au at the Savannah River National Laboratory for the preparation of Cu-Si nanostructures and the XRD characterization. Also, thanks to Dr. Hessam Ghassemi at Drexel University for the *in situ* microscopy of the Cu-Si alloy nanostructures.

REFERENCES

1. M. Armand and J. M. Tarascon, Building better batteries, *Nature,* 451, 652–657, 2008.
2. M. Green, E. Fielder, B. Scrosati, M. Wachtler, and J. S. Moreno, Structured silicon anodes for lithium battery applications, *Electrochemical and Solid State Letters,* 6, A75–A79, 2003.
3. U. Kasavajjula, C. Wang, and A. J. Appleby, Nano- and bulk-silicon-based insertion anodes for lithium-ion secondary cells, *Journal of Power Sources,* 163, 1003–1039, 2007.
4. M. Au, Y. He, Y. Zhao, H. Ghassemi, R. S. Yassar, B. Garcia-Diaz, and T. Adams, Silicon and silicon–copper composite nanorods for anodes of Li-ion rechargeable batteries, *Journal of Power Sources,* 196, 9640–9647, 2011.
5. L. Baggetto, R. A. H. Niessen, F. Roozeboom, and P. H. L. Notten, High energy density all-solid-state batteries: A challenging concept towards 3D integration, *Advanced Functional Materials,* 18, 1057–1066, 2008.
6. J. Yang, M. Winter, and J. O. Besenhard, Small particle size multiphase Li-alloy anodes for lithium-ion batteries, *Solid State Ionics,* 90, 281–287, 1996.
7. H. Föll, H. Hartz, E. Ossei-Wusu, J. Carstensen, and O. Riemenschneider, Si nanowire arrays as anodes in Li ion batteries, *Physica Status Solidi (RRL)—Rapid Research Letters,* 4, 4–6, 2010.
8. J. P. Maranchi, A. F. Hepp, A. G. Evans, N. T. Nuhfer, and P. N. Kumta, Interfacial properties of the a-Si/Cu: Active–inactive thin-film anode system for lithium-ion batteries, *Journal of the Electrochemical Society,* 153, A1246–A1253, 2006.
9. C. K. Chan, H. Peng, G. Liu, K. McIlwrath, X. F. Zhang, R. A. Huggins, and Y. Cui, High-performance lithium battery anodes using silicon nanowires, *Nature Nanotechnology,* 3, 31–35, 2008.
10. H. Qiaojian, C. M. Lilley, R. Divan, and M. Bode, Electrical failure analysis of Au nanowires, *IEEE Transactions on Nanotechnology, I,* 7, 688–692, 2008.
11. S. J. Jung, T. Lutz, A. P. Bell, E. K. McCarthy, and J. J. Boland, Free-standing, single-crystal Cu$_3$Si nanowires, *Crystal Growth & Design,* 12(6), 3076–3081, 2012.
12. K. R. Roos et al., High temperature self-assembly of Ag nanowires on vicinal Si(001), *Journal of Physics: Condensed Matter,* 17, S1407, 2005.
13. Q. Huang, C. M. Lilley, and M. Bode, Surface scattering effect on the electrical resistivity of single crystalline silver nanowires self-assembled on vicinal Si (001), *Applied Physics Letters,* 95, 103112-3, 2009.
14. B. Stahlmecke, F. J. M. z. Heringdorf, L. I. Chelaru, M. H.-v. Hoegen, G. Dumpich, and K. R. Roos, Electromigration in self-organized single-crystalline silver nanowires, *Applied Physics Letters,* 88, 053122-3, 2006.
15. P. K. Ng, B. Fisher, K. B. Low, A. Joshi-Imre, M. Bode, and C. M. Lilley, Comparison between bulk and nanoscale copper-silicide: Experimental studies on the crystallography, chemical, and oxidation of copper-silicide nanowires on Si(001), *Journal of Applied Physics,* 111, 104301-7, 2012.
16. P. K. Ng, B. Fisher, M. Bode, and C. M. Lilley, Self assembled Cu nanowires on vicinal Si(001) by the E-beam evaporation method, Presented at the *11th IEEE International Conference on Nanotechnology (IEEE-NANO),* Portland, OR, 2011.
17. P. K. Ng, B. Fisher, K. B. Low, A. Joshi-Imre, M. Bode, and C. M. Lilley, High resolution analysis of self assembled Cu nanowires on vicinal Si(001), Presented at the *11th IEEE International Conference on Nanotechnology (IEEE-NANO),* Portland, OR, 2011.
18. H. Okamoto, Cu-Si (copper-silicon), *Journal of Phase Equilibria,* 23, 281–282, 2002.
19. R. Teki, N. Koratkar, T. Karabacak, and T.-M. Lu, Enhanced photoemission from nanostructured surface topologies, *Applied Physics Letters,* 89, 193116-3, 2006.
20. Y. P. Zhao, D. X. Ye, G. C. Wang, and T. M. Lu, Novel nano-column and nano-flower arrays by glancing angle deposition, *Nano Letters,* 2, 351–354, 2002.
21. K. Robbie, J. C. Sit, and M. J. Brett, Advanced techniques for glancing angle deposition, *Journal of Vacuum Science & Technology B: Microelectronics and Nanometer Structures,* 16, 1115–1122, 1998.

22. Ag-Si, *Journal of Phase Equilibria,* 10, 673–674, 1989.
23. Y. He, Z. Zhang, C. Hoffmann, and Y. Zhao, Embedding Ag nanoparticles into MgF$_2$ nanorod arrays, *Advanced Functional Materials,* 18, 1676–1684, 2008.
24. Y. He, J. Fan, and Y. Zhao, The role of differently distributed vanadium nanocatalyst in the hydrogen storage of magnesium nanostructures, *International Journal of Hydrogen Energy,* 35, 4162–4170, 2010.
25. H. Ghassemi, M. Au, N. Chen, P. A. Heiden, and R. S. Yassar, *In situ* electrochemical lithiation/delithiation observation of individual amorphous Si nanorods, *ACS Nano,* 5, 7805–7811, 2011.
26. H. Ghassemi, M. Au, N. Chen, P. A. Heiden, and R. S. Yassar, Real-time observation of lithium fibers growth inside a nanoscale lithium-ion battery, *Applied Physics Letters,* 99, 123113-3, 2011.

50 High-Aspect-Ratio Metallic Nanowires by Pulsed Electrodeposition

Matthias Graf, Alexander Eychmüller, and Klaus-Jürgen Wolter

CONTENTS

50.1 INTRODUCTION

Since the miniaturization and the performance increase of electronic components nowadays are ongoing parallel trends in the microelectronics industry, the introduction of nanometer (nm)-scale materials emerges even in the field of packaging, and interconnections within the back-end-of-line (BEoL) processes are also subjects of these processes [1]. Their potential as thermally, mechanically, or electrically enhancing additive to established packaging materials reduces their potential to a passive role. The use of carbon nanotubes (CNTs) inside Cu-based through-silicon vias (TSVs) [2,3] or packaging polymers [4,5] are prominent examples. On the other hand, the use of one-dimensionally elongated nanoparticles featuring a high aspect ratio (AR, the ratio of particle length to diameter), so-called nanowires (NWs) gains more and more attention when it comes to an active, that is, interconnecting incorporation of nanoparticles into packaging [6–8]. Their mostly bottom-up-based fabrication has already been realized by versatile approaches [9]. In general, one-dimensional growth can be achieved either by energetically favored growth based on high crystal structure anisotropy of the growing material or by spatial confinement during growth achieved by a template with parallel pores of a defined geometry. In the latter case, this geometry determines the later geometry of the NWs, that is, their diameter, length, and order. Diameters in the range of a few nanometers up to several hundred nanometers have already been achieved using different template materials. Among mesoporous silica [10], track-etched polycarbonate [11], or self-organizing block copolymers [12], anodically oxidized alumina (AAO) [13–15] enables beneficial features for template-assisted NW synthesis. AAO offers a wide range of variable pore geometries (pore diameters from 25 to 500 nm, interpore distances from 50 to 400 nm, pore depths from a few nanometers to micrometers, depending on the thickness of the Al used) while enabling once tremendous pore densities up to 10^{11} pores/cm², through-going and nearly defect-free pores, and a very high degree of hexagonal pore order. Pores of AAO templates have been successfully filled with metals [16–18], metallic alloys [19], or (conductive) polymers [20], demonstrating the high material versatility for AAO-assisted NW synthesis. Different architectures, such as core–shell structures [21], multisegmented NWs [22], or nanoporous NWs [23], have also been obtained by filling the pores of AAO templates.

Similar to the manifoldness in materials used for NW synthesis are the methods for pore filling. Among, for example, chemical deposition [24] or monomer injection followed by polymerization [25], electrodeposition of the desired NW species is the most common method to fabricate metal-based NWs. As depicted in Figure 50.1e, electrochemical deposition can be once potential driven or otherwise controlled by the applied current, which forces the deposition reaction. In this sense, the first mode applies a thermodynamic force to the system, whereas the current-driven mode—the most common for electrochemical deposition—enables kinetic control simply by the current used for deposition. Since—as a consequence of the ongoing deposition reaction—the properties (i.e., the resistance, R) of the system electrolyte–surface change drastically with time, one of the two parameters current, i, or potential, E, has to change as the other is kept constant by a galvanostat or potentiostat, respectively. When depositing under constant current conditions (amperostatic), the cell potential can reach regions in which the NW deposition is not favored thermodynamically and side reactions (e.g., the evolution of gases and the decomposition of the deposited species) occur and disturb the deposition reaction. When applying a thermodynamic force (in the case of potential-driven deposition) to the system, the growth does not show linear behavior anymore but more regularity with time. From the kinetic point of view, a reductive deposition reaction proceeds much faster than the diffusion of ions to the pore bottom through thin and long pores. Consequently, the deposition is highly controlled by diffusion. To enable the diffusion process to level the concentration of the electrolyte at the electrochemical double layer (EDL) of the cathode (pore bottom) to the concentration inside the bulk of the above-standing electrolyte, a deposition pause after the deposition enables the diffusion to take place. In this pause (which is much longer than the deposition period itself), no reaction apart from the diffusion shall be allowed. Such pulsed deposition processes are known to lead to much more homogenous deposits with a smaller grain size [26]. In this chapter, we focus on the pulsed electrodeposition of Ag into the pores of AAO with the objective of gaining higher NW homogeneity, faster NW growth, and an evaluation of growth kinetics. In the concluding section, we substantiate the need for higher ARs within the use of NW arrays for an anisotropically conductive and adhesive NW film acting as a fine pitch interconnecting layer, for example, for vertical three-dimensional or flip-chip interconnection applications.

50.2 EXPERIMENTAL SECTION

50.2.1 TEMPLATE PREPARATION

AAO templates used for these investigations were fabricated as reported elsewhere [14] and schematized in Figure 50.1. High-purity (99.999%) Al foils were purchased from ABCR, Germany. Al pretreatment included ultrasonic degreasing twice, alternating in ethanol and acetone, followed by a 2 h vacuum annealing step ($p < 10^{-3}$ mbar, $T = 500°C$), and an electropolish (48 V, $T = 20°C$, ethanol–HClO$_4$–methylcellulose 1:4:1 per volume). Anodic oxidation was carried out in a two-step process (40 V DC, 2°C under stirred oxalic acid). First, anodization was conducted for 4 h, followed by a stripping step ($T = 60°C$, stirring in 5 wt.-% H$_3$PO$_4$ and 1.8 wt.-% CrO$_3$, 2 h) to dissolve the first porous layer of Al$_2$O$_3$. The second anodization was carried out under the same conditions as the first, but for 10 h (Figure 50.1b), resulting in a pore depth of approximately 20 μm. Then

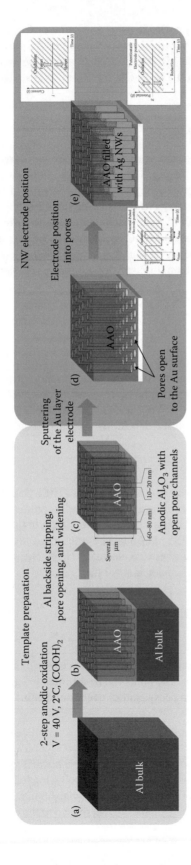

FIGURE 50.1 Schematic procedure of the AAO template-assisted electrodeposition of Ag NWs: (a) Starting from high-purity bulk Al; (b) continuing over a two-step anodization and generation of a regularly ordered pore structure; (c) stripping the remaining carrier Al layer and opening/widening the pores; (d) after sputtering a continuous thin layer of Au onto one surface acting as the cathode layer while performing electrodeposition inside the pores under different deposition.

the carrying Al was dissolved from the backside using a mixture of $CuCl_2$ and concentric HCl for 15 min (Figure 50.1c). A closed thin layer (thickness ~200 nm) of Au was sputtered onto the front side of the gained membrane (Figure 50.1d). For electrodeposition, the pores on the backside were opened by dipping the membrane into 0.3 M NaOH for 5 min. Pores were finally widened using 5 wt.-% H_3PO_4 for 15 min.

50.2.2 ELECTRODEPOSITION OF AG NWS

To evaluate the deposition potential, cyclic voltammetry (CV) was carried out with the Ag(I)-containing electrolyte (2.3 M KSCN, 0.1 M Ag_2SO_4, pH ~6) onto a flat Au surface. Electrodeposition was carried out according to Riveros et al. [16]. For sufficient wetting of the AAO pores with the electrolyte, substrates were exposed to the electrolyte solution for at least 30 min prior to the deposition. All depositions were done using a PC-controlled potentiostat by Princeton Applied Research. Potentiostatic deposition was carried out under DC conditions for 30 min at room temperature and a cathodic deposition potential of −250 mV versus standard calomel electrode (SCE) as reference and a flat Pt counterelectrode (CE). Potential-pulsed deposition was carried out for different pulse frequencies, f, as defined by

$$f = \frac{t_{\text{pulse}}}{t_{\text{pulse}} + t_{\text{pause}}} \tag{50.1}$$

Thereby, the total time for one pulse and pause ($t_{\text{pulse}} + t_{\text{pause}}$) was kept constant at 1 s throughout all experiments. For growth homogenization, the corresponding potentials (all vs. SCE) were varied between $E_{\text{pulse}} = -250$ mV with $E_{\text{pause}} = 0$ mV and $E_{\text{pulse}} = -320$ mV with $E_{\text{pause}} = -220$ mV. For the evaluation of NW growth kinetics, pulsed deposition was carried out for 5, 10, 20, and 30 min.

50.2.3 CHARACTERIZATION

Scanning electron microscopy (SEM) was carried out with a ZEISS Supra 40VP equipped with a backscatter electron-sensitive detector enabling a higher material contrast. Electrochemical impedance spectroscopy (EIS) was performed using a Gamry potentiostat in the frequency range of 1 Hz to 1 MHz with −10 mV cathodic bias, a 135 nm Au layer evaporated onto Al as a working electrode (WE), a Pt sheet as CE, and SCE as a reference electrode (RE). Nyquist plot's semicircle fit in Figure 50.3h was performed by parabolic mathematical fits.

50.3 RESULTS AND DISCUSSION

50.3.1 AAO TEMPLATES

As shown in Figure 50.2a, the pores of AAO can be estimated to have a diameter of around 60–70 nm and an interpore distance of 100–110 nm. The pores can be estimated to grow in depth with a velocity of about 2 µm/h of anodization, giving easy control over pore depth.

50.3.2 CYCLIC VOLTAMMETRY

From the cathodic current occurring in the cyclovoltammogram of Figure 50.2b, the reductive potential region (i.e., $i < 0$ mA) of the used electrolyte can be estimated to lie between −220 and −350 mV followed by cathodic currents related to other decomposition reactions inside the electrolyte. In the anodic potential region, a strong anodic current related to the oxidative dissolution of the deposited Ag can be observed.

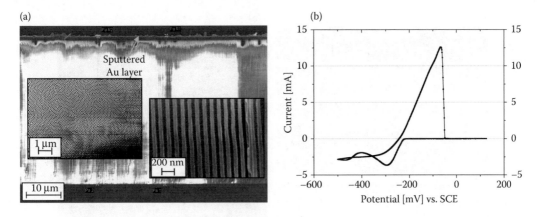

FIGURE 50.2 (a) SEM picture of an empty AAO template (second anodization 20 h) with a thin Au layer visible on top. Left inset: Top view onto hexagonal pore arrangement after second anodization. Right inset: Magnified picture of parallel pore channels after pore widening of 30 min. (b) Cyclovoltammogram of a 0.1 M Ag_2SO_4 + 2.3 M KSCN electrolyte on Au surface (starting from 0 mV into cathodic region with 50 mV/s).

50.3.3 NW DEPOSITION BY DC PLATING

According to CV results, DC plating was performed at $E = -250$ mV (vs. SCE) since the highest cathodic current appears at this point. Results are shown in Figure 50.3a and c. Ag NW growth could be obtained with nearly 100% lateral filling degree in AAO. The current measured during deposition shows a slight decrease with time after a long initiation period of approximately 200 s of seed growth on the pore bottoms. The slight current decrease with ongoing deposition is related to the decreasing concentration of the electrolyte in the EDL region according to Faraday's law. Theoretically, this is compensated weakly by the slowly decreasing length of the diffusion path through the pores since they are filled with time and the deposition region shifts toward the AAO upside. Apart from that, these NWs are observed to be rather inhomogeneous in length with a relative standard deviation of 13% so that we expected pulsed plating as an option to level the NW length.

50.3.4 NW DEPOSITION BY PULSED PLATING

In pulse plating experiments, the two most important parameters are pulse frequency, f, and pulse/pause potentials. Apart from that, we have investigated the influence of Ag^+ concentration on the homogeneity of the NW precipitate. Figure 50.3b shows the results of a deposition with $f = 0.01$ and pulse potentials same as in the DC deposition experiment (i.e., $E_{pulse} = -250$ mV, $E_{pause} = 0$ mV). It is obvious that the NWs are very inhomogeneous in structure, length, and filling. This is mostly related to the long deposition pauses. When switching to 0 mV, the system crosses the oxidative potential region (compare Figure 50.3) so that strong dissolution effects occur. To avoid the dissolution during pulse pauses, we chose a pause potential that causes no significant effect (i.e., no pause current) in the system. SEM pictures of samples where a pause potential of $E_{pause} = -220$ mV and a deposition potential $E_{pulse} = -320$ mV were applied for deposition are shown in Figure 50.3d–f. No traces of dissolution can be observed in these figures. Apart from that, since the filling degree is related to seed growth that is higher with increased seed concentration, the strong influence of Ag(I) concentration on pore filling can be observed when comparing Figure 50.3d with Figure 50.3e. Figure 50.3g shows the time evolution of the deposition current during pulse plating with $E_{pause} = -220$ mV and $E_{pulse} = -320$ mV. The inset explains the attributed curves that evolve due to the high resolution of measurement. Three conclusions can be made from these curves: (1) The pause current of nearly 0 mA during pulsed plating proves that there is actually no significant process occurring, which is

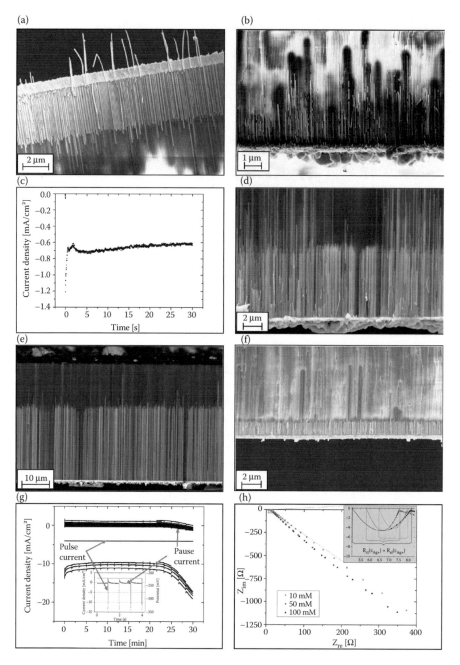

FIGURE 50.3 Results of Ag electrodeposition into pores of AAO: (a) NWs obtained by 30 min DC plating at $E = -250$ mV with corresponding time-current density developing (c); NWs obtained by pulse plating for 30 min with $f = 0.01$, $c = 0.05$ mM, and $E_{pulse} = -250$ mV, $E_{pause} = 0$ mV (b), with $f = 0.05$, $c = 0.05$ mM, $E_{pulse} = -320$ mV, and $E_{pause} = -220$ mV (d), with $f = 0.05$, $c = 0.1$ mM, $E_{pulse} = -320$ mV, and $E_{pause} = -220$ mV (e; inset shows material composition obtained by energy dispersive x-ray spectroscopy [EDS]) with corresponding time-current density development (g; inset shows a magnification of the different deposition curves) and with $f = 0.05$, $c = 0.1$ mM and, $E_{pulse} = -320$ mV, $E_{pause} = -220$ mV after 5 min (f). (h) EIS results (Nyquist plot) of the Ag_2SO_4/2,3 M KSCN electrolyte at a flat Au electrode in dependence of Ag^+ concentration with a magnified inset indicating RC semicircle behavior at high frequencies (fitted by parabolic function, distance between roots represents the ohmic resistance for charge transfer at the Au surface) without Warburg impedance/constant phase element.

connected to a Faraday conversion during deposition pauses. (2) Pulse currents are about two orders of magnitude higher compared to DC plating, resulting in much longer NWs than obtained by DC plating in 30 min. To evaluate the growth rate during pulsed plating, we performed electrodeposition under $f = 0.05$, $c = 0.1$ mM, $E_{pulse} = -320$ mV, and $E_{pause} = -220$ mV for 5, 10, 20, and 30 min. (3) The pulse current curve shows an increase in the deposition current starting at approximately 22 min. This might be attributed to the fast increasing NW length, which reduces the diffusion path through the pores and leads to a faster deposition. This stands in agreement with the nonlinear growth behavior indicated in Figure 50.5. Figure 50.3f shows small NWs with an average length of 1.42 ± 0.07 µm and very high regularity obtained after 5 min of electrodeposition. Figure 50.3h shows a graph of time-dependent NW length evolution. It is obvious that NW growth by pulsed plating is not linear with time, which we expected and explained before. Apart from that, an increase in standard deviation (the absolute standard deviation is indicated by the error bars) with time can be observed, but compared with the relative standard deviation (DC: 13.0%, pulsed plating: 6.2%; each after 30 min), a considerable decrease in length inhomogeneity can be observed by optimized pulse plating.

50.3.5 THE NEED FOR HIGH-ASPECT-RATIO NWs: PATHWAY TO PACKAGING AT NANOSCALE

The obtained NW arrays being embedded in AAO are planned to be applied as vertically interconnecting layers inside flip-chip or three-dimensionally stacked architectures as described before [6] and depicted in Figure 50.4. As experienced, a very thin layer of AAO is mechanically very brittle. Consequently, to realize such a described film with vertically oriented NWs, a sufficient film thickness is important for later lamination processability in chip-to-chip (C2C), chip-to-wafer (C2W), or wafer-to-wafer (W2W) assembly processes. Since the NWs shall interconnect both sides of the film, their length is determined by the lowest applicable film thickness, which is assumed here to be in the range of 10–20 µm. This refers to NW ARs of 140–280 (assuming NW diameters of 70 nm). To gain these values by simple DC plating, very long deposition times have to be accepted. With pulse plating, these ARs can be realized in much shorter times and higher regularity. Apart from that, the NWs in the film are also expected to act as a mechanically enhancing agent, that is, they provide higher film flexibility as they are longer.

The NW film is expected to show various beneficial features when applied as a chip-interconnecting film. Among them, the reduction of the effective signal transporting cross section from a currently established solder microball to numerous ($\sim 10^3$) thin metallic NWs enables the elimination of the frequency-limiting skin effect (since the skin thickness is more than the NW diameter, NWs do not suffer a reduction of transmission cross section). Apart from that, the laterally highly resolved

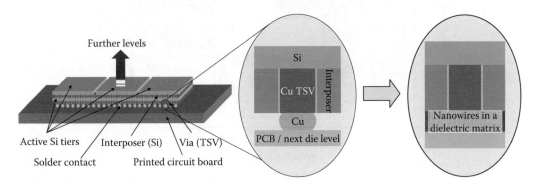

FIGURE 50.4 Application scheme for high-aspect-ratio metallic NWs embedded in an AAO matrix. The filled membrane shall be applied within vertical packaging either between two active Si chips in 3D chip stacks or—as depicted—for direct vertical interconnection from a chip/TSV interposer to the printed circuit board (PCB).

NWs fit present as well as future demands of contact geometries and provide electrically contacted neighbored to noncontacted NWs, which can (because of their metallic nature) act as thermally trapping devices and so contribute to the reduction of thermal stress in densely packed chip stacks.

50.3.6 Deposition Mechanism during Pulsed Electroplating into Pores

From the kinetic NW growth curve in Figure 50.5, three distinct observations can be made: First, the growth can be significantly accelerated and pushed toward higher length homogeneity by applying pulsed electrodeposition. Second, the filling of nanoscale pores by pulsed electrodeposition is no linear process since the NW length obviously does not increase linearly with time and is better fitted by a parabolic function. Third, the speed of NW growth seems to be dependent on pore depth since the 30 min point (40 μm membrane thickness) is not well fitted by the curve made up by 20 μm membrane points. To understand the reasons for all the three observations made, we hereby formulate a qualitative deposition mechanism.

When observing the time-current transients of the one pulse–pause sequence, one notices in both a current decay from a very high peak value to a much lower end value. When fitting the much longer pause behavior mathematically, one would suggest a single exponential decay that is corresponding to the typical discharging behavior of a capacitor. This capacitor can be the EDL between an Au electrode and an electrolyte, which is discharged when positive charge carriers are reduced at the electrode surface. EIS measurements (see Figure 50.3h) on the system of a blank Au electrode facing the same electrolyte as used in the pulsed pore electrodeposition experiments described above showed that the system cannot be described by this simple model, but rather by a more complex system such as suggested in Figure 50.6 involving the intrinsic resistance of the electrolyte itself, the charge transfer resistance (depending on the Ag^+ concentration in the electrolyte), and a supposedly Warburg, or more general, a constant phase element impedance that can be caused by a diffusive barrier in front of the blank electrode [27]. This is obvious since there is no typical semicircle behavior but a continuous increase in phase shift between the potential and the current. Diffusively caused Warburg impedance would then cause a slope with an angle α of 45° between real part impedance,

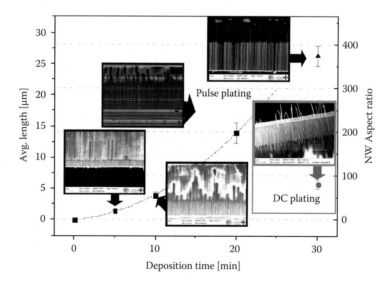

FIGURE 50.5 Kinetic growth curve of Ag NWs grown in AAO templates by constant potential electroplating (DC plating) and pulsed potential electroplating with SEM pictures, which were the basis for NW length measurement. The electrolyte was of the same composition in all cases. AAO template membranes were 20 μm in thickness for 5, 10, and 20 min and 40 μm for pulse and DC plating samples after 30 min each. Error bars represent the relative deviation in length with respect to the measured average length.

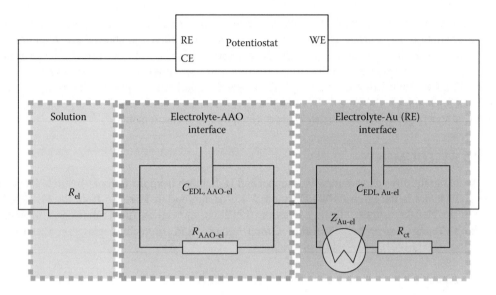

FIGURE 50.6 Equivalent electrical circuit diagram corresponding to AAO pore deposition with the following elements: R_{el}, the intrinsic electrical resistance of the electrolyte, R_{AAO-el} and $C_{EDL, AAO-el}$, the capacitance of the EDL between the AAO surface and the electrolyte, $C_{EDL, Au-el}$, the EDL capacitance of the double layer between the Au cathode and the electrolyte, Z_{Au-el}, the constant phase element's impedance of the Au cathode (as observed by EIS) and R_{ct}, the electrical resistance of the cathodic charge transfer process at the Au electrode; RE, the reference electrode (SCE), CE, the counterelectrode (Pt), and WE, the working electrode.

Z_{re}, and imaginary part impedance, Z_{im}. Since the observed slope is higher, this additive impedance can more generally be characterized by the model of a constant phase element.

By adding a membrane with thin pores in front of this electrode, convective acceleration of the diffusion is avoided actively, depending on the pore length, so that with longer pores, the diffusion path for the electrolyte increases. In contrast, it decreases as long as the pores are being filled, and the cathode surface (e.g., the top of the NW, the place where the growth proceeds) is more or less slowly shifted to the membrane's top, which so reduces the diffusion length again. In addition to this, Chien et al. predict that a second EDL is added in a series circuit to the Au-electrolyte, resulting in a proposed equivalent electrical circuit as depicted in Figure 50.6 [28]. Further investigations on the particular assembly of the AAO-EDL to the Au-EDL will follow. We now cannot surely predict whether the equivalent circuitry represents the complete sum of all electrical elements involved in the electrochemical deposition process. Also, a possible influence of the pore geometry cannot be evaluated, yet. Since the range of influences of surface effects, such as superviscosity causing very high friction coefficients at the nanoscale, is—to the best of our knowledge—not further characterized for the system of AAO pores, we cannot predict a possible threshold of pore width/length where pore diffusion is significantly hindering the electrolyte from diffusing to the electrode [29].

50.4 CONCLUSION

In this chapter, we have demonstrated the feasibility of generating high-aspect-ratio metallic NWs by enhanced pulse plating of Ag into highly ordered pores of AAO. We have shown once a significant increase in NW length and AR (DC: 56, pulsed plating: 373), respectively, within the same time of deposition (30 min) and a decrease of length inhomogeneity when performing pulsed electrodeposition. The parameters being evaluated and expected to have the most significant influence on the deposition process were pulse frequency, Ag(I) concentration, as well as pulse and pause potential. When choosing a pause potential too high, strong dissolution effects during the pauses cause very high inhomogeneity of the NWs. The effect of faster and denser NW growth with higher Ag⁺ concentration was proven. NW growth kinetics were recorded and showed a continuous but nonlinear behavior in growth rate. In future, the obtained NW arrays shall be used inside an anisotropically conductive interconnecting film featuring vertically aligned NWs that have to feature higher ARs for better film processability. We sketched a qualitative model for the nanopore-confined electrochemical NW growth featuring electrical elements (such as a Warburg impedance at the Au cathode), which will be further characterized by EIS measurements.

ACKNOWLEDGMENTS

The authors gratefully acknowledge V. Haehnel and H. Schlörb from the Leibniz-Institute for Solid State and Materials Research (IFW) Dresden, J. Katzmann and T. Härtling from the Fraunhofer Institute for Non-Destructive Testing Dresden (IZFP-D), as well as the Technische Universität Dresden's Young Researchers Training Group "Nano- and Biotechnologies for Electronics Packaging" (DFG 1401/1) for technical guidance.

REFERENCES

1. Roadmap by ITRS, Issue 13: Assembly & Packaging, pp. 56–58, **2009**.
2. Y. Chai, K. Zhang, M. Zhang, P. C. H. Chan, and M. M. F. Yuen, Carbon nanotube/copper composites for via filling and thermal management, *IEEE Proc. Elec. Components Tech. Conf. (ECTC)*, **2007**.
3. M. Fayolle et al., Integration of dense CNTs in vias on 200 mm diameter wafers: Study of post CNT growth processes, *IEEE Proc. Int. Interconn. Tech. Conf. (IITC)*, **2011**.
4. M. Heimann, B. Boehme, S. Scheffler, M. Wirts-Ruetters, and K.-J. Wolter, CNTs—A comparable study of CNT-filled adhesives with common materials, *IEEE Proc. Elec. Components Tech. Conf. (ECTC)*, pp. 1871–1878, **2009**.
5. B. Li, Y.-C. Zhang, Z.-M. Li, S.-N. Li, and X.-N. Zhang, Easy fabrication and resistivity-temperature-behaviour of an anisotropically conductive carbon nanotube-polymer composite, *J. Phys. Chem. B*, 114(2), 689–696, **2010**.
6. M. Graf et al., Nanowire filled polymer films for 3D system integration, *IEEE Proc. Int. Interconn. Tech. Conf. (IITC)* **2011**, pp. 1–3.
7. K. J. Ziegler et al., Conductive films of ordered nanowire arrays, *J. Mater. Chem.*, 14, 585–589, **2004**.
8. R.-J. Lin et al., Design of nanowire anisotropic conductive film for fine pitch flip chip interconnection, *IEEE Proc. 6th Electr. Pack. Tech. Conf. (EPTC)*, pp. 120–125, **2004**.
9. Y. Xia et al., One-dimensional nanostructures: Synthesis, characterization, and applications, *Adv. Mater.*, 15(5), 353–389, **2003**.
10. F. Schüth, Non-siliceous mesostructured and mesoporous materials, *Chem. Mater.*, 13, 3184–3195, **2001**.
11. C. Schonenberger et al., Template synthesis of nanowires in polycarbonate membranes: Electrochemistry and morphology, *J. Phys. Chem. B*, 101, 5497–5505, **1997**.
12. M. Stamm, S. Minko, I. Tokarev, A. Fahmi, and D. Usov, Nanostructures and functionalities in polymer thin films, *Macromol. Symp.*, 214, 73–83, **2004**.
13. C. R. Martin, Nanomaterials—A membrane-based synthetic approach, *Science*, 266, 1961–1966, **1994**.
14. O. Jessensky, F. Müller, and U. Gösele, Self-organized formation of hexagonal pore arrays in anodic alumina, *Appl. Phys. Lett.*, 72(10), 1173–1175, **1998**.
15. A. P. Li, F. Müller, A. Birner, K. Nielsch, and U. Gösele, Hexagonal pore arrays with 50–420 nm interpore distance formed by self-organization in anodic alumina, *J. Appl. Phys.*, 84(11), 6023–6026, **1998**.
16. G. Riveros et al., Silver nanowire arrays electrochemically grown into nanoporous anodic alumina templates, *Nanotechnology*, 17(2), 561–570, **2006**.
17. H. Cao, L. Wang, Y. Qiu, and L. Zhang, Synthesis and properties of aligned copper nanowires, *Nanotechnology*, 17(6), 1736–1739, **2006**.
18. A. J. Yin, J. Li, W. Jian, A. J. Bennett, and J. M. Xu, Fabrication of highly ordered metallic nanowire arrays by electrodeposition, *Appl. Phys. Lett.*, 79(7), 1039–1041, **2001**.
19. V. Haehnel, C. Mickel, S. Fahler, L. Schultz, and H. Schlörb, Structure, microstructure, and magnetism of electrodeposited Fe₇₀Pd₃₀ nanowires, *J. Phys. Chem. C*, 114(45), 19278–19283, **2010**.

20. D. J. Shirale, M. A. Bangar, W. Chen, N. V. Myung, and A. Mulxhandani, Effect of aspect ratio (length:Diameter) on a single polypyrrole nanowire FET device, *J. Phys. Chem. C*, 114(31), 13375–13380, **2010**.

21. P. R. Evans, W. R. Hendren, R. Atkinson, and R. J. Pollard, Nickel-coated gold-core nanorods by template assisted electrodeposition, *J. Electrochem. Soc.*, 154(9), K79–K82, **2007**.

22. F. Gao and Z. Gu, Nano-soldering of magnetically aligned three-dimensional nanowire networks, *Nanotechnology*, 21(11), 115604–115610, **2010**.

23. C. Ji and P. C. Searson, Fabrication of nanoporous gold nanowires, *Appl. Phys. Lett.*, 81(23), 4437–4439, **2002**.

24. Z. Hu, T. Xu, R. Liu, and H. Li, Template preparation of high density, and large-area Ag nanowire arrays by acetaldehyde reduction, *Mater. Sci. Eng. A*, 371236–240, **2004**.

25. H. Masuda and K. Fukada, Ordered metal nanohole arrays made by a two-step replication of honeycomb structures of anodic alumina, *Science*, 268, 1466–1468, **1995**.

26. N. Kanani, *Electroplating—Basic Principles, Processes and Practice*, 1st ed., Elsevier, Amsterdam, pp. 117–124, **2004**.

27. E. Barsoukov, *Impedance Spectroscopy—Theory, Experiment and Applications*, 2nd ed., Wiley Interscience, Hoboken, NJ, pp. 85–88, **2005**.

28. M.-C. Chien, G.-J. Wang, and W.-C. Yu, Modelling ion diffusion current in nanochannel using infinitesimal distribution resistor-capacitor circuits, *Jpn. J. Appl. Phys.*, 46(11), 7436–7440, **2007**.

29. S. Guriyanova, V. G. Mairanovsky, and E. Bonaccurso, Superviscosity and electroviscous effects at an electrode/aqueous electrolyte interface: An atomic force microscope study, *J. Coll. Interface Sci.*, 360(2), 800–804, **2011**.

Section XVI

Nanowire Applications

51 Zinc Oxide Nanowires for Biosensing Applications

Anurag Gupta, Bruce C. Kim, Dawen Li, Eugene Edwards, Christina Brantley, and Paul Ruffin

CONTENTS

51.1 INTRODUCTION

One-dimensional (1D) and quasi-1D systems, such as nanowires, nanobelts, and nanotubes, have been the focus of intensive research due to their unique physical and chemical properties. These novel and interesting characteristics have been shown to stem primarily from their nanometer size resulting in quantum confinement of their electronic wavefunctions. Pertaining to nanometer dimensions, these quasisystems manifest additional surface-related properties, which have opened new frontiers of research in sensing [1–3], energy [4], catalysis [5], photovoltaics [6], and bio-nanotechnology [7].

Recently, considerable research effort has been directed toward zinc oxide (ZnO) nanostructures, due to their superior electronic, optical, piezoelectric, and thermal properties, which have provided a new perspective in nanoelectronics, optoelectronics, and sensor development [8–10]. ZnO is a wide bandgap (3.37 eV) semiconductor, which has been particularly interesting to the research community due to its large exciton binding energy (60 meV) that facilitates room-temperature lasing action based on the exciton recombination. Owing to such excellent optical properties, ZnO nanostructures have versatile applications in short-wavelength-based optoelectronic devices, such as light emitting diodes (LEDs), laser diodes, nanocatalysis, OLED-based displays, and optoelectronic chemical sensors in particular [11–18]. The aforementioned applications require ZnO nanostructures in different morphologies, such as nanowires, nanorods, and nanobelts, which must be subjected to postsynthesis procedures to achieve the desired functionality. High-aspect-ratio morphologies of ZnO are desirable due to their large surface-to-volume ratio, which facilitates enhanced surface interactions. Therefore, it is rational to expect surface-related effects to be crucial for any electronic or optoelectronic devices based on ZnO nanowires. The technological relevance of surface functionalized ZnO nanowires has already been demonstrated in sensing devices, dye-sensitized solar cells (DSSCs) for efficient light harvesting, and other hybrid photovoltaic devices [19–21]. The polar nature of ZnO and the ability to maintain high-symmetry *c*-axis orientation during growth has also facilitated the growth of densely packaged nanowire arrays, which are promising candidates for nonlinear optical sensing devices. Furthermore, at the nanowire surface, there is a loss of symmetry, which is expected to increase the nonlinear optical response of these nanowires as compared to the bulk. The functionalization of these nanowires with additional organic or inorganic compounds should therefore allow for tailoring of the surface for its nonlinear signal, thereby enabling the design of efficient nanoscale optical sensors.

Although these quasi-1D ZnO nanowires cover a breadth of applications, their deployment in the fabrication of different types of chemical and biological sensors is pertinent to the present-day scenario. Novel sensors for detecting hazardous compounds need to be developed due to the rapidly evolving military, DHS (Department of Homeland Security), and NIH (National Institutes of Health) requirements. The potential of ZnO nanowires for chemical gas sensing, biological moiety detection, and environmental probes has already been demonstrated [22–25]. ZnO nanowires operating as a single nanowire probe [26] or in field effect transistor (FET) [27] mode have provided sensitive detection up to ppb ranges. The advantages of using ZnO nanowires as sensing elements are manifold. First, rapid progress made in synthesis methods for ZnO nanostructures and the ability to perform controlled synthesis make them a cost-effective choice for developing sensing platforms. Second, the ZnO nanowire surface is resistant to atmospheric oxidation due to the inherent oxide structure of the material, which provides robustness and high chemical stability. Third, the surface of the nanowire can be tailored for high selectivity toward specific bioanalytes through surface functionalization techniques. Finally, the nanowires in array morphology provide high permeability for analyte molecules, which decreases the detection time and enhances sensitivity [28,29]. Furthermore, the diameter of these nanostructures is comparable to the biomolecules being detected, which effectively makes them an excellent transducer for producing signals to interface with macroscopic instruments. Owing to the aforementioned properties and advantages, ZnO nanowires have emerged as excellent candidates for fabricating biosensor devices with a wide spectrum of applicability.

51.2 BACKGROUND

ZnO nanowires not only possess a large number of surface sites, owing to their nanoscale morphology, but they also provide a favorable microenvironment for retaining the bioactivity of the functionalized enzyme or antigens on the nanowire surface. The high isoelectric point (IEP) (~9.5) of these nanowires helps in immobilizing antigens with low IEP. Furthermore, the high electron mobility concomitant with the biocompatible nature of ZnO nanowires has led to a considerable thrust in the development of mediator-less implantable biosensors. Despite the potential of ZnO nanowires for biosensing applications, the literature on the subject is far from comprehensive. However, the research is catching up in the light of new challenges and development in existing biotechnology and therapeutics. Lee and coworkers demonstrated the use of the ZnO/Si surface acoustic wave (SAW) device for detecting prostate-specific antigen (PSA) antibody–antigen immunoreaction as a function of PSA concentration [30]. The immobilization of PSA on the SAW device leads to a shift in resonance frequency, which can be related to the amount of PSA detected. A detection limit of 2–10,000 ng/mL was reported. Cai and coworkers focused on the development of a pH nanosensor based on ZnO [31]. The large surface-to-volume ratio was observed to enhance the diffusion time of the analyte, concomitantly increasing the response time of the sensor. In another recent study by Ahmad and coworkers, the pH sensitivity and biochemical sensing capability of ZnO nanostructures were reported. Pt-ZnO nanospheres on a glass carbon electrode were utilized for estimating the amount of cholesterol in a solution by Ahmad and coworkers [32]. A sensitivity of 1886.4 mA M^{-1} cm^{-2} was reported. Glucose is yet another compound whose concentration in blood plays a crucial role in diseases, such as diabetes and endocrine disorder. There is compelling empirical evidence that suggests that ZnO nanowires can be appropriately tailored to provide sensitive and selective detection of glucose in blood serum. In a recent study, ZnO 1D nanostructures have been investigated as a suitable candidate for glucose sensing with a limit of detection of 1 μM [33]. Good anti-interference ability and prolonged shelf life were also observed. Gu and coworkers developed ZnO nanowire-based H$_2$O$_2$

sensors by alternate immobilization of poly(sodium 4-styrenesulfonate) (PSS) and horseradish peroxidase on a nanowire surface [34]. A wide linear range and a low detection limit up to 1.9 µM were deduced. In another recent study, Ibupoto and coworkers utilized ZnO nanorods immobilized with penicillinase enzyme through N-5-azido-2-nitrobenzoyloxysuccinimide (ANB-NOS) cross-linking molecules, and estimated the detection limit of penicillin at 100 µM [35].

The literature surveyed above only serves to corroborate the argument that much more exhaustive research needs to be conducted to obtain complete benefits of ZnO nanowire-based biosensors. This could be ascertained through a detailed investigation of the surface chemistry, mechanistic studies, and biochemistry of the receptors to be immobilized on the nanowire surface. In the forthcoming sections, we would outline a generic design approach for developing a ZnO nanowire-based biosensor. Through the oleic acid–ZnO nanowire model system, steps for successful synthesis, characterization, and surface functionalization will be outlined. Subsequently, the discussion would be culminated in a prospective application for developing a selective and sensitive *p*-nitrophenol biosensor.

51.3 GENERIC DESIGN APPROACH AND POTENTIAL APPLICATIONS

As outlined in the previous sections, to effectively engineer the surface of ZnO nanowires for biological sensing, an in-depth understanding of the surface phenomenon and interactions is crucial. Therefore, we have reported the immobilization of a common organic compound, oleic acid, on a ZnO nanowire surface. We synthesized ZnO nanowires in a customized chemical vapor deposition (CVD) furnace and established the morphology and composition of the synthesized nanowires through scanning electron microscopy (SEM), transmission electron microscopy (TEM), and x-ray diffraction (XRD). Thereafter, the ZnO nanowires were modified with oleic acid and the nature of bonding and surface orientation of oleic acid molecules on nanowire surface was determined with surface-sensitive characterization techniques of Raman and Fourier transform-infrared (FT-IR) spectroscopies. Photoluminescence (PL) measurements on modified and unmodified ZnO nanowires showed changes in peak heights based on which a prospective mechanism of sensing oleic acid on a ZnO nanowire surface is proposed. Oleic acid is a surfactant that has been shown to attract different moieties present in biologically important molecules [36]. It is believed that modifying the ZnO nanowire surface with oleic acid can improve the detection limits of these moieties to the molecular level. Hence, we chose this model system to demonstrate the capability of covalent functionalization of the ZnO nanowire surface with an organic moiety, as pertinent to previous biosensor research. The most crucial factor in designing a sensitive and selective ZnO-based biosensor is the appropriate choice of an antigen/receptor, which is capable of altering either optical or electrical characteristics of the ZnO nanowire postgrafting. This inorganic–organic heterostructure must be capable of producing a measurable response to an analyte stimulus to ascertain sensitive detection. Now, let us focus on the ZnO nanowire–oleic acid model system to highlight significant steps in a ZnO nanowire-based biosensor fabrication. The schematic of the idea is presented in Figure 51.1.

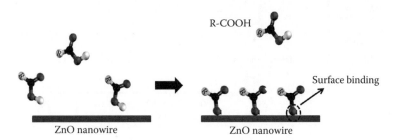

FIGURE 51.1 Schematic representation of nanowire-based detection. (A. Gupta et al., Zinc oxide nanowires for biosensing applications, *Proc. 11th IEEE International Conference on Nanotechnology*, August 15–18, Portland, Oregon, pp. 1615–1618. © (2011) IEEE. With permission.)

51.3.1 Synthesis of ZnO Nanowires

The ZnO nanowires were synthesized on the ZnO (0001) substrate using the vapor liquid solid (VLS) mechanism widely published in the literature [37,38]. In brief, ZnO substrates were coated with gold (4 nm) by sputtering (Shirley Sputtering System), which serves as a template for the growth of ZnO nanowires during the VLS process. Precursor powders of ZnO (99.9%, from J. T. Baker) and graphite (99%, from Alfa Aesar) were homogenously mixed in a 1:1 ratio and introduced in a customized CVD furnace at 950°C. Mixed gas (2% O_2 + Ar) was utilized as a carrier medium. Using an optimal flow rate and a growth time of 30 min, dense nanowire growth was observed.

51.3.2 Characterization and Surface Studies

The morphology and crystal structure of the synthesized nanowires were determined by FE-SEM (JEOL 7000), x-ray diffraction (Bruker-AXS), and TEM (Technai). Figure 51.2a and b shows the SEM images of ZnO nanowires on the ZnO substrate along with the XRD spectrum of the nanowires, as synthesized by the CVD process. It can be observed that the nanowires are vertically aligned with respect to the substrate and are laterally branched close to the tip. While the vertical alignment is due to the absence of lattice mismatch between the nanowires and the substrate, hierarchical growth could be due to the high supersaturation of Zn vapors in the furnace as suggested by Zhang and coworkers [39]. Figure 51.2c corresponds to the XRD spectrum of the as-synthesized ZnO nanowires, which indicates preferential growth along the [0001] direction as the (001) peak is sufficiently high in intensity than other peaks, consistent with the documented literature. In addition, the

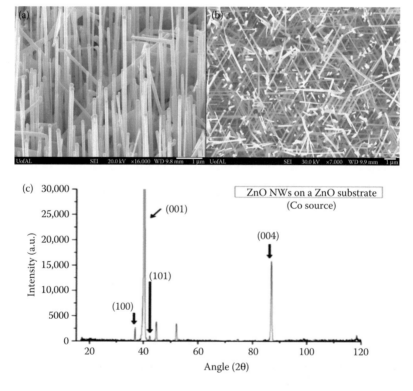

FIGURE 51.2 (a) Tilted and (b) top view of ZnO nanowires synthesized on a ZnO substrate. (c) XRD spectrum of nanowires showing preferential c-axis growth. (A. Gupta et al., Zinc oxide nano-wires for biosensing applications, *Proc. 11th IEEE International Conference on Nanotechnology*, August 15–18, Portland, Oregon, pp. 1615–1618. © (2011) IEEE. With permission.)

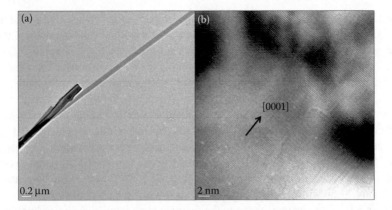

FIGURE 51.3 **(a)** TEM image of the as-synthesized ZnO NW; **(b)** HR-TEM image showing the [0001] growth direction of an NW. (A. Gupta et al., Zinc oxide nano-wires for biosensing applications, *Proc. 11th IEEE International Conference on Nanotechnology*, August 15–18, Portland, Oregon, pp. 1615–1618. © (2011) IEEE. With permission.)

hexagonal wurtzite structure of the synthesized nanowires can be confirmed by the superimposition of standard spectra (ICDD PDF# 01-089-0510) over the obtained spectrum. This observation can be further supported by the TEM analysis of the nanowires as presented in Figure 51.3.

After characterization, the ZnO nanowires as-synthesized on the ZnO substrate were dipped in oleic acid solution (1% v/v in hexane) for 15 min to functionalize the ZnO nanowire surface. Thereafter, the specimens were copiously rinsed in pure hexane to remove residual oleic acid. It is essential to remove unadsorbed oleic acid to prevent erroneous signal generation during subsequent characterization. The nature of bonding and orientation of oleic acid on a nanowire surface were determined by comparing unmodified and modified ZnO nanowires through Raman (JobinYvon, HR800 UV) and Fourier transform–infrared spectroscopy. Figure 51.4a shows the Raman spectrum of the as-synthesized ZnO nanowires. The characteristic ZnO peaks were observed at 439

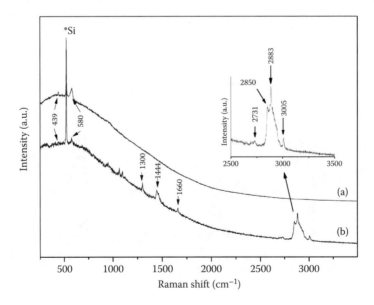

FIGURE 51.4 (a) Raman spectrum of as-synthesized ZnO nanowires. (b) Raman spectrum of modified ZnO nanowires. (A. Gupta et al., Zinc oxide nano-wires for biosensing applications, *Proc. 11th IEEE International Conference on Nanotechnology*, August 15–18, Portland, Oregon, pp. 1615–1618. © (2011) IEEE. With permission.)

(E_2 mode) and 580 cm^{-1} ($A_{1(LO)}$ mode), which serve as a reference to further modification procedures [40]. Figure 51.4b is the Raman spectrum of the oleic acid-modified ZnO nanowires. It can be seen that the spectrum has peaks corresponding to oleic acid in addition to the characteristic ZnO peaks. This observation is suggestive of possible modification of the nanowire surface by oleic acid, but the nature of carboxylate bonding is ambiguous. Accordingly, complementary evidence was obtained through FT-IR. Figure 51.5a is the FT-IR spectrum of 1% oleic acid (v/v) in hexane. Peaks at 937 and 1464 cm^{-1} correspond to the out-of-plane and in-plane OH deformations, respectively. The broad peak at 1284 cm^{-1} can be attributed to the C–O stretch, while the intense peak at 1710 cm^{-1} is C=O. Small peaks at 2570 and 2673 cm^{-1} indicate the oleic acid dimer formation. The group of peaks from 2825 to 2950 cm^{-1} is from CH$_2$ and CH$_3$ asymmetric and symmetric vibration stretches and are typical of long-chain organic compounds. The weak peak at 3005 cm^{-1} is the C–H stretch of (*cis*)-alkene [41]. Figure 51.5b shows the modified ZnO nanowire spectrum. It can be observed that the changes in the spectrum are primarily due to the features associated with the carboxylic acid group, leading to bound carboxylate. The absence of peaks at 2570 and 2673 cm^{-1} rules out any possibility of oleic acid dimerization. It can also be observed that the C=O band at 1710 cm^{-1}, C–O stretch at 1284 cm^{-1}, and the out-of-plane O–H deformation mode are no longer present. Instead, new features indicative of the carboxylate species appear. Peaks at 1407 and 1589 cm^{-1} can be attributed to COO– symmetric and asymmetric stretching, respectively. The remaining unidentified peaks are characteristic of the long hydrocarbon chain as mentioned before. All peaks beyond 2800 cm^{-1}, associated with CH$_3$ and CH$_2$ as well as C=C modes beyond 2800 cm^{-1}, remain in the same position pre- and postmodification of ZnO nanowires with oleic acid.

After the immobilization of oleic acid molecules on the ZnO nanowires surface, we tested the PL behavior of unmodified and modified ZnO nanowires. Figure 51.6a shows the PL spectrum of as-synthesized nanowires. Weak peaks at 510 and 536 nm can be attributed to the defect-related emission, and prominent peaks at 360 and 380 nm are due to the onset of excitonic band-edge-related emission. When the nanowire surface sites are occupied with oleic acid molecules, two significant changes occur in the spectrum. In Figure 51.6b, the defect-related peaks disappear, which could be due to the occupation of surface sites by oxygen from oleic acid, which otherwise would have contributed to visible luminescence. Furthermore, the band-edge-related emission peaks merge and display a slight red-shift that can be attributed to the surface adsorption of oleic acid molecules. Both these effects could be due to the surface interaction between the ZnO nanowire and oxygen

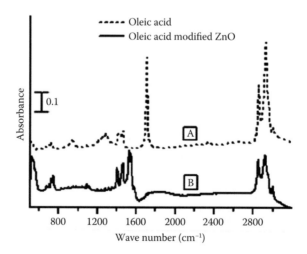

FIGURE 51.5 FT-IR spectra of (a) oleic acid 1% (v/v) in hexane. (b) Oleic acid-modified ZnO nanowires. (A. Gupta et al., Zinc oxide nano-wires for biosensing applications, *Proc. 11th IEEE International Conference on Nanotechnology*, August 15–18, Portland, Oregon, pp. 1615–1618. © (2011) IEEE. With permission.)

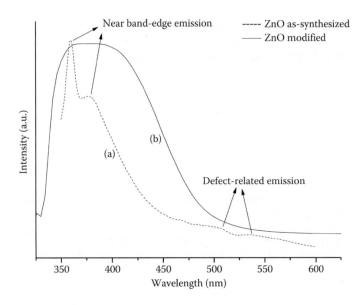

FIGURE 51.6 (a) PL spectrum of as-synthesized ZnO NWs. (b) PL spectrum of modified ZnO NWs. (A. Gupta et al., Zinc oxide nano-wires for biosensing applications, *Proc. 11th IEEE International Conference on Nanotechnology*, August 15–18, Portland, Oregon, pp. 1615–1618. © (2011) IEEE. With permission.)

atoms from oleic acid. Although the role of oxygen in ZnO PL is still debated, it can nonetheless serve as a functional tool for quantification and characterization. For developing biosensors based on ZnO nanowires, emission characteristics, both band-edge and defect-related, can together serve as a fingerprint to identify different biological moieties.

51.3.3 Development of the ZnO Nanowire-Based *p*-Nitrophenol Sensor

In the previous section, a generic approach for synthesizing a ZnO nanowire-based biosensor was highlighted. It is possible to extend this approach for sensing desired analytes by performing the functionalization of the nanowire surface through the appropriate receptor. In one of our recent studies, we studied the surface chemistry of the 1-pyrenebutyric acid (PyBA) receptor for developing the *p*-nitrophenol sensor [42]. Nitrophenols are a class of organic compounds that are generated as degradation products of many organophosphorus compounds (OPs). OP compounds are key components of the modern agricultural industry and find application in the manufacture of herbicides, insecticides, and so on. Although, these compounds do not get accumulated in biological systems, they exhibit high toxicity and are purported to have carcinogenic behavior on human subjects [43]. To develop the aforementioned ZnO nanowire-based sensor, the receptor was immobilized onto the ZnO nanowire surface through covalent grafting. Subsequently, vapor phase detection of *p*-nitrophenol was envisaged up to 20 ppb. We believe that the synergistic utilization of optical and electrical characteristics of the ZnO nanowire–PyBA system could lead to an optoelectronic device platform for selective and selective determination of *p*-nitrophenol.

51.4 CONCLUSIONS

Based on the above findings and previous empirical evidence, it can be deduced that ZnO nanowires coupled with a suitable receptor or antigen has great potential for detecting a wide spectrum of biologically important compounds. To achieve this, however, rigorous research related to all fundamental aspects of biochemistry, surface phenomenon, and functionalization is required. ZnO nanostructures have the required capabilities to provide a sensitive and selective detection platform, and the imminent needs of biodetection, imaging, and therapeutics are expected to provide the much-needed thrust to the ZnO-based nanosensor technology research.

ACKNOWLEDGMENT

Approved for public release; distribution unlimited. Review completed by the AMRDEC Public Affairs Office June 13, 2012; FN5869.

REFERENCES

1. T. Asefa, C. T. Duncan, K. K. Sharma, Recent advances in nanostructured chemosensors and biosensors, *Analyst*, **10**, 1980–1990, 2009.
2. S. Börner, R. Orghici, S. R. Waldvogel, U. Willer, W. Schade, Evanescent field sensors and the implementation of waveguiding nanostructures, *Appl. Opt.*, **48**, B183–B189, 2009.
3. Y. Cui, Q. Wei, H. Park, C. M. Lieber, Nanowire nanosensors for highly-sensitive, selective and integrated detection of biological and chemical species, *Science*, **293**, 1289–1292, 2001.
4. C. B. Murray, D. J. Norris, M. G. Bawendi, Synthesis and characterization of nearly monodisperse CdE (E = sulfur, selenium, tellurium) semiconductor nanocrystallites, *J. Am. Chem. Soc.*, **115**, 8706–8715, 1993.
5. M. C. Daniel, D. Astruc, Gold nanoparticles: Assembly, supramolecular chemistry, quantum-size related properties, and applications towards biology, catalysis and nanotechnology, *Chem. Rev.*, **104**, 293–346, 2004.
6. B. Tian, X. Zheng, T. J. Kempa, Y. Fang, N. Yu, G. Yu, J. Huang, C. M. Lieber, Coaxial silicon nanowires as solar cells and nanoelectronic power sources, *Nature*, **449**, 885–889, 2007.
7. C. M. Niemeyer, Nanoparticles, proteins, and nucleic acids: Biotechnology meets materials science, *Angew. Chem. Int. Ed.*, **40**, 4128–4158, 2001.
8. D. H. Cobden, Nanowires begin to shine, *Nature*, **409**, 32–33, 2001.
9. Y. Cui, C. M. Lieber, Functional nanoscale electronic devices assembled using silicon nanowire building blocks, *Science*, **291**, 85–853, 2001.
10. R. J. Tonucci, B. L. Justus, A. J. Campillo, C. F. Ford, Nanochannel array glass, *Science*, **258**, 783–785, 1992.
11. E. Comini, G. Faglia, G. Sberveglieri, Z. Pan, Z. L. Wang, Stable and high-sensitive gas sensors based on semiconducting oxide nanobelts, *Appl. Phys. Lett.*, **81**, 1869–1871, 2002.
12. Q. Wan, K. Yu, T. H. Wang, C. L. Lin, Low-field electron emission from tetrapod-like ZnO nanostructures synthesized by rapid evaporation, *Appl. Phys. Lett.*, **83**, 2253–2255, 2003.
13. B. Y. Oh, M. C. Jeong, T. H. Moon, W. Lee, J. M. Myoung, J. Y. Hwang, D. S. Seo, Transparent conductive Al-doped ZnO films for liquid crystal displays, *J. Appl. Phys.*, **99**, 124505–124509, 2006.
14. K. Nomura, H. Ohta, K. Ueda, T. Kamiya, M. Hirano, H. Hosono, Thin-film transistor fabricated in single-crystalline transparent oxide semiconductor, *Science*, **300**, 1269–1272, 2003.
15. T. Yoshida, H. Minoura, Electrochemical self-assembly of dye-modified zinc oxide thin films, *Adv. Mater.*, **12**, 1219–1222, 2000.
16. K. Önenkamp, R. C. Word, C. Schlegel, Vertical nanowire light-emitting diode, *Appl. Phys. Lett.*, **85**, 6004–6006, 2004.
17. H. Nanto, T. Minami, S. Takata, Zinc-oxide thin-film ammonia gas sensors with high sensitivity and excellent selectivity, *J. Appl. Phys.*, **60**, 482–484, 1986.
18. M. H. Sarvari, H. Sharghi, Zinc oxide (ZnO) as a new, highly efficient, and reusable catalyst for acylation of alcohols, phenols and amines under solvent free conditions, *Tetrahedron*, **61**, 10903–10907, 2005.
19. M. Law, L. E. Greene, J. C. Johnson, R. Saykally, P. D. Yang, Nanowire dye-sensitized solar cells, *Nat. Mater.*, **4**, 455–459, 2005.
20. D. C. Olson, Y. J. Lee, M. S. White, N. Kopidakis, S. E. Shaheen, D. S. Giney, J. A. Voigt, J. W. P. Hsu, Effect of polymer processing on the performance of poly(3-hexylthiophene)/ZnO nanorod photovoltaic devices, *J. Phys. Chem. C*, **111**, 16640–16645, 2007.
21. P. Ravirajan, A. M. Peiro, M. K. Nazeeruddin, M. Graetzel, D. D. C. Bradley, J. R. Durrant, J. Nelson, Hybrid polymer/zinc oxide photovoltaic devices with vertically oriented ZnO nanorods and an amphiphilic molecular interface layer, *J. Phys. Chem. B*, **110**, 7635–7639, 2006.

22. P. C. Chen, S. Sukcharoenchoke, K. Ryu, A. Gomez, A. Badmaev, C. Wang, C. Zhou, 2,4,6-Trinitrotoluene (TNT) chemical sensing based on aligned single-walled carbon nanotubes and ZnO nanowires, *Adv. Mater.*, **22**, 1900–1904, 2010.

23. S. Das, J. P. Kar, J. H. Choi, T. I. Lee, K. J. Moon, J. M. Myoung, Fabrication and characterization of ZnO single nanowire-based hydrogen sensor, *J. Phys. Chem. C*, **114**, 1689–1693, 2010.

24. Y. Gui, C. Xie, J. Xu, G. Wang, Detection and discrimination of low concentration explosives using MOS nanoparticle sensors, *J. Hazard. Mater.*, **164**, 1030–1035, 2009.

25. A. Choi, K. Kim, H. I. Jung, S. Y. Lee, ZnO nanowire biosensors for detection of biomolecular interactions in enhancement mode, *Sens. Actuat. B Chem.*, **148**, 577–582, 2010.

26. L. Liao, H. B. Lu, J. C. Li, C. Liu, D. J. Fu, Y. L. Liu, The sensitivity of gas sensor based on single ZnO nanowire modulated by helium ion radiation, *Appl. Phys. Lett.*, **91**, 173110-3, 2007.

27. Z. Fang, J. G. Lu, Gate-refreshable nanowire chemical sensors, *Appl. Phys. Lett.*, **86**, 12, 2005.

28. F. Patolsky, B. P. Timko, G. Yu, Y. Fang, A. B. Greytak, G. Zheng, C. M. Lieber, Detection, stimulation, and inhibition of neuronal signals with high-density nanowire transistor arrays, *Science*, **313**, 1100–1104, 2006.

29. N. Kakati, S. H. Jee, S. H. Kim, H. K. Lee, Y. S. Yoon, Sensitivity enhancement of ZnO nanorod gas sensors with surface modification by an InSb thin film, *Jpn. J. Appl. Phys.*, **48**, 10500, 2–5, 2009.

30. D. S. Lee, Y. Q. Fu, S. Maeng, J. Luo, N. M. Park, S. H. Kim, M. Y. Jung, W. I. Milne, *ZnO Surface Acoustic Wave Biosensor*, Inter. Elec. Devices. Meeting, Washington DC, 2007.

31. X. Cai, N. Klauke, A. Glidle, P. Cobbold, G. L. Smith, J. M. Cooper, Ultra-low-volume, real-time measurements of lactate from the single heart cell using microsystems technology, *Anal. Chem.*, **74**, 908–914, 2002.

32. M. Ahmad, C. Pan, L. Gan, Z. Nawaz, J. Zhu, Highly sensitive amperometric cholesterol biosensor based on pt-incorporated fullerene-like ZnO nanospheres, *J. Phys. Chem. C*, **114**, 243–250, 2010.

33. M. Ahmad, C. Pan, Z. Luo, J. Zhu, A single ZnO nanofiber-based highly sensitive amperometric glucose biosensor, *J. Phys. Chem. C*, **114**, 9308–9313, 2010.

34. B. X. Gu, C. X. Xu, G. P. Zhu, S. Q. Liu, L. Y. Chen, M. L. Wang, J. J. Zhu, Layer by layer immobilized horseradish peroxidase on zinc oxide nanorods for biosensing, *J. Phys. Chem. B*, **113**, 6553–6557, 2009.

35. Z. H. Ibupoto, S. M. U. Ali, K. Khun, C. O. Chey, O. Nur, M. Willander, ZnO nanorods based enzymatic biosensor for selective determination of penicillin, *Biosensors*, **1**, 153–163, 2011.

36. I. Mahmood, C. Guo, H. Xia, J. Ma, Y. Jiang, H. Liu, Lipase Immobilization on oleic acid – pluronic (L-64) block copolymer coated magnetic nanoparticles, for hydrolysis at the oil/water interface, *Ind. Eng Chem. Res.*, **47**, 6379–6385, 2008.

37. P. C. Chang, Z. Fan, J. G. Lu, J. Hong, W. Y. Tseng, ZnO nanowires synthesized by vapor trapping CVD method, *Chem. Mat.*, **16**, 5133–5137, 2004.

38. C. Y. Lee, T. Y. Tseng, S. Y. Li, P. Lin, Growth of zinc oxide nanowires on Si (100), *Tamkang J. Sci. Eng.*, **6**, 127–132, 2003.

39. Z. Zhang, S. J. Wang, T. Yu, T. Wu, Controlling the growth mechanism of ZnO nanowires by selecting catalyst, *J. Phys. Chem. C*, **111**, 17500–17505, 2007.

40. Y. Huang, M. Liu, Z. Li, Y. Zeng, S. Liu, Raman spectroscopy study of ZnO-based ceramic films fabricated by novel sol–gel process, *Mater. Sci. Eng. B*, **97**, 111–116, 2003.

41. L. J. Bellamy, *The Infrared Spectra of Complex Molecules*, Halsted, UK, pp. 233–237, 1975.

42. A. Gupta, B. C. Kim, E. Edwards, C. Brantley, P. Ruffins, Synthesis and functionalization study of hierarchical ZnO nanowires for potential nitroaromatic sensing applications, *App. Phys. A: Mater. Sci. Process*, **107**, 709–714, 2012.

43. R. C. Gupta, *Toxicology of Organophosphate and Carbamate Compounds*, Elsevier Academic Press, USA, 2005.

44. A. Gupta, B. C. Kim, D. Li, E. Edwards, C. Brantley, P. Ruffin, Zinc oxide nanowires for biosensing applications, *Proc. 11th IEEE International Conference on Nanotechnology*, August 15-18, 2011, Portland, Oregon, pp. 1615–1618.

52 Aqueous Synthesis of n-/p-type ZnO Nanorods on Porous Silicon for the Application of p–n Junction Device

Eunkyung Park, Jungwoo Lee, Taehee Park, Jongtaek Lee, Donghwan Lee, and Whikun Yi

CONTENTS

52.1 INTRODUCTION

One-dimensional (1D) materials have stimulated much attention due to their promising potential in extensive applications [1]. The recent research in this field focuses on investigating the dependence of electrical transport and optical and mechanical properties on size and dimensionality [2,3]. Among them, semiconducting zinc oxide (ZnO), which has a wide bandgap of 3.37 eV and a large exciton binding energy of 60 meV, has attracted much attention due to its promising potential as a "future material" [4]. At ambient pressure and temperature, ZnO crystallizes in the wurtzite structures, as shown in Figure 52.1. This is a hexagonal lattice and is characterized by two interconnecting sublattices of Zn^{2+} and O^{2-}, such that each Zn ion is surrounded by tetrahedral O ions and vice versa. This tetrahedral coordination gives rise to polar symmetry along the hexagonal axis. This polarity is responsible for a number of properties of ZnO, including its piezoelectricity and spontaneous polarization, and is also a key factor in crystal growth, etching, and defect generation. Many groups have focused on novel nanostructures with different shapes ranging from nanowires to nanobelts and even nanosprings. ZnO nanorods or nanowires are generally synthesized by four main techniques: chemical vapor deposition [5], metal–organic vapor deposition [6], aqueous solution method (equal to chemical bath deposition (CBD)) [7], and electrodeposition [8].

In spite of the many studies on the ZnO synthesis on various substrates [9,10] or porous templates [11,12], ZnO nanorod films grown on porous silicon (PS) have been little studied. PS is one

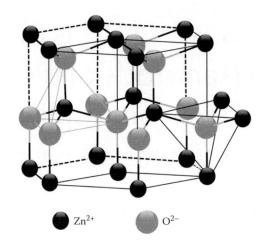

Zn²⁺ ... O²⁻

FIGURE 52.1   Wurtzite structure model of ZnO.

of the most important Si-based materials because its open structure and large surface area, combined with unique optical and electrical properties, make it a challenging position for templates [13,14]. PS is a very promising material due to its excellent mechanical and thermal properties, obvious compatibility with silicon-based microelectronics, and its low cost. The large surface area within a small volume, combined with controllable pore sizes, convenient surface chemistry, and the ability to modulate its refractive index as a function of depth, makes PS a suitable dielectric material for the formation of a multilayer.

Alternatively, as-grown ZnO is typically n-type, and different hypotheses have been proposed to explain n-type conductivity [15]. The achievement of reproducible p-type ZnO still remains one of the significant obstacles to wide applications of ZnO in a p–n junction device. Thus, the n- or p-type ZnO grown on n-type PS using the CBD method offers a cheap route for an application to various p–n junction devices. We adopted the atomic layer deposition (ALD) technique to make seed layer for p-type ZnO nanorods. ALD is a chemical gas-phase, thin-film deposition method based on alternate saturative surface reactions. In ALD, the source vapor is pulsed into the reactor alternately, one at a time, separated by purging or evacuation periods. Each precursor exposure step saturates the surface with a monomolecular layer of that precursor. This results in a unique self-limiting growth mechanism with a number of advantageous features, such as good conformality and uniformity, and accurate film thickness control.

In this chapter, ZnO nanorod films are synthesized on a PS template using the CBD technique. We demonstrate that the conductivity type (n- or p-type) of ZnO nanorods can be changed by controlling the seed layer preparation. Two types of seed layers are prepared by ALD with/without nitrogen (N) doping.

52.2   EXPERIMENT

The PS layer was made by the electrochemical anodization method on an n-type Si (100) wafer. The Si substrate has a resistivity of 1–20 ohm·cm. Al (100 nm) was deposited on the back side of the substrate to form an electrode as an anode and the Pt plate was served as a cathode (counterelectrode). Electrochemical anodization was carried out for 20 min in a Teflon cell containing HF:ethanol:deionized (DI) water = 1:2:1 volume ratio with 50 mA/cm² current densities. A 50 W halogen lamp was used for the back side illumination for derivation of effective holes during electrochemical anodization. The resultant PS layer has a thickness of about 20–30 μm. After the anodization process, the substrate was rinsed with DI water and dried in a nitrogen atmosphere.

ZnO seed layers were prepared on the PS surface from diethylzinc (DEZn, Zn 52.0%, Aldrich) and H_2O precursors, and nitrogen (N) doping was achieved by NH_3 gas. Typical sequences for growing with the N-doped and non-N-doped ZnO seed layers were [DEZn/Ar/H_2O/Ar/NH_3/Ar] 2/10/2/10/3/15 sec and [DEZn/Ar/H_2O/Ar] 1/5/2/7 sec, respectively. The process was repeated for the formation of the ZnO layer up to 20 nm in depth inside the PS. The role of Ar was to remove the residual diethylzinc inside the reaction chamber so as not to react with the next H_2O precursor. The working pressure was 0.3 Torr and Ar gas was used as a carrier gas and purging gas with a flow rate of 50 sccm. The temperature of the PS substrate was kept at 423.15 K.

The synthesis of the ZnO nanorod films was carried out by the CBD method. Plastic bottles were filled with a mixture a of 0.02 M equimolar aqueous solution of zinc nitrate hexahydrate (($Zn(NO)_3)_2 \cdot 6H_2O$) and hexamethylenetetramine (HMT; $C_6H_{12}N_4$). Subsequently, the ZnO seed layer-grown PS substrates were placed in the bottles and heated at a constant temperature of 343.15 K for 24 h. As-grown ZnO nanorod films were washed with DI water to eliminate residual salts, dried in air, and annealed at 673.15 K to enhance the crystallinity. An annealing process was essential to remove the residual organic material and improve the crystallinity of the ZnO nanorods. The crystalline character and doping state of the samples were examined via x-ray diffraction (XRD) and photoluminescence (PL) spectroscopy, respectively. Au electrodes were fabricated using the magnetron sputtering method on each side of the sample to measure the electrical properties. Two samples (n-type ZnO or p-type ZnO on n-type PS) were investigated throughout the measurement of the forward and reverse bias I–V curve.

52.3　RESULTS AND DISCUSSION

52.3.1　Effect of Concentration of Zn Solution on the Formation of ZnO Nanorods

We confirmed the effect of concentration of Zn solution (zinc nitrate and hexamethylenetetraamine equimolecular solution) on the formation of ZnO nanorods on an Si wafer. Figure 52.2 shows FE-SEM images of the product resulting from the aqueous chemical growth of ZnO obtained by a further lowering of the overall concentration of the precursors to the low millimolar range. A seed layer of ZnO with 20 nm thickness was deposited by using RF sputtering. Crystalline nanorods, about 100–400 nm in diameter and up to 1–2 μm long, were identified at various concentrations of the Zn solution after 40 h of aging. When the reaction time was constant, the diameter of the ZnO nanorod was decreased and adjacent nanorods were coalesced as the concentration of the solution increased.

52.3.2　XRD and PL Spectra of ZnO Nanorods

The scanning electron microscopy (SEM) images of the ZnO nanorods prepared with different seed layers are shown in Figure 52.3. The geometric parameters for growing rods such as growth time and solution composition were the same for two samples as described in the experimental section. We can observe that the N-undoped seed layer yields smaller nanorods (Figure 52.3a and b) and the N-doped seed layer by NH_3 gas yields larger nanorods (Figure 52.3c and d). N-doping would affect the seed layer density.

Figure 52.4 shows the XRD patterns of the N-undoped and N-doped ZnO nanorod films on an n-type PS substrate. The XRD was operated at the condition of 40 kV and 100 mA with Cu Kα line radiation. The major peaks of the samples are identified and compared with JCPDS file 36-1451. This reveals that two samples are composed of hexagonal ZnO with lattice constants $a = 0.325$ nm and $c = 0.522$ nm, and have a hexagonal wurtzite crystalline ZnO structure. The (002) peak of the N-doped ZnO nanorods implies better alignment than undoped ZnO nanorods. The higher intensity of the (002) peak for the sample indicates that the ZnO nanorods are perpendicularly oriented to the substrate. To evaluate the mean crystallite size of two samples, Scheerer's formula is adopted [16]. The mean crystallite size of two samples corresponds to 36, 41, and 32 nm for the (110), (002), and

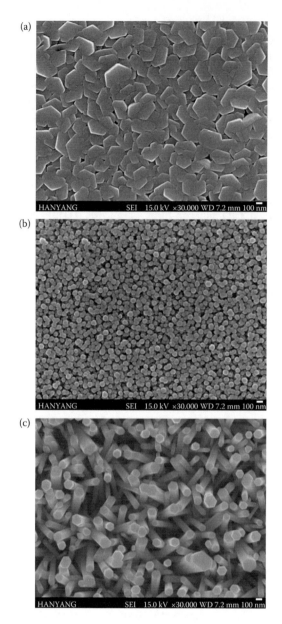

FIGURE 52.2 SEM images showing differences of ZnO nanorod's formation grown on Si wafers by dipping in ZnO solutions with a concentration of (a) 0.1 M, (b) 0.05 M, and (c) 0.02 M.

(101) directions, respectively, which means that the crystallite growth is slightly faster along the c axis direction.

Figure 52.5 shows the room-temperature PL spectra of the N-undoped and N-doped ZnO nanorods grown on the n-type PS substrate. The peak (380–415 nm) emerging at the visible range of the N-undoped sample results from the near-band-edge emission. Blue light emission appears at 460 nm, which is predicted from intrinsic defects such as zinc and oxygen interstitials produced through the annealing process. The effect of N-doping is confirmed from the result of the PL spectra. Compared to the N-undoped ZnO nanorods, N-doped ZnO nanorods show the red shift phenomenon. N-doping leads to the increase of internal hole concentration inside the ZnO nanorods; thus, the bandgap of the N-doped ZnO nanorods becomes narrower.

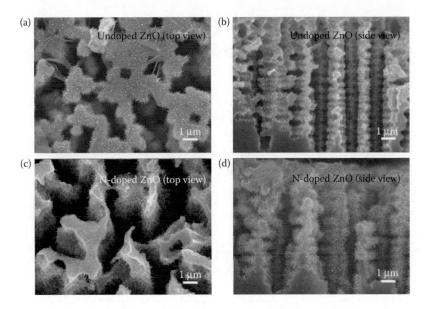

FIGURE 52.3 SEM images of the ZnO nanorod: (a) top and (b) side view images of the N-undoped ZnO nanorods on PS (n-type ZnO/n-type PS), (c) top and (d) side view images of the N-doped ZnO nanorods on PS (p-type ZnO/n-type PS).

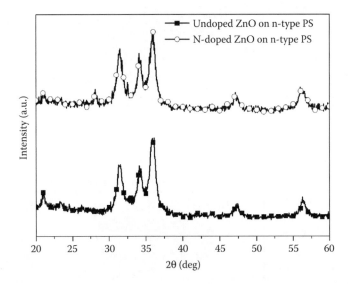

FIGURE 52.4 XRD patterns of the undoped and N-doped ZnO nanorod films.

52.3.3 *I–V* CHARACTERISTICS OF THE UNDOPED (N-TYPE) AND N-DOPED (P-TYPE) NANORODS GROWN ON N-TYPE PS

To study whether N-doping effectively changes the carrier type of the N-doped ZnO nanorods, we compare the *I–V* characteristics of the N-undoped and N-doped ZnO nanorods grown on n-type PS. The *I–V* characteristics of the N-undoped and N-doped ZnO nanorods on n-type PS are measured at room temperature. Figure 52.6 clearly shows that the N-undoped ZnO nanorods (n-type ZnO nanorods) do not form a rectifying diode with the n-type PS. However, the N-doped ZnO nanorods

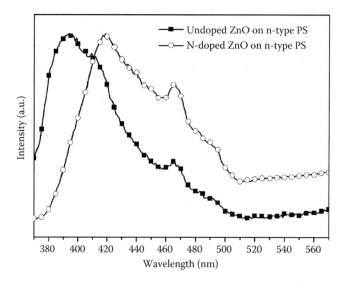

FIGURE 52.5   Room-temperature PL spectra of the N-undoped and N-doped ZnO nanorods on an n-type PS.

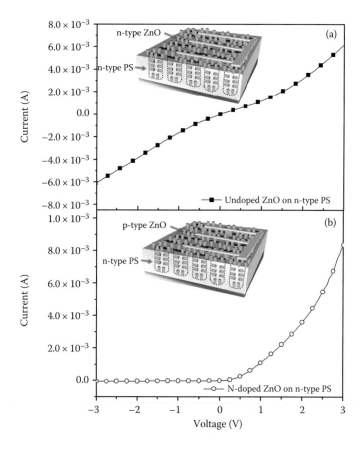

FIGURE 52.6   *I–V* characteristics of (a) undoped (n-type) and (b) N-doped ZnO nanorods (p-type) grown on an n-type PS.

(p-type ZnO nanorods) form a good rectifying diode with an n-type PS. These results clearly indicate that the N-doping process achieved via NH_3 gas during the seed layer formation changes the carrier type of the ZnO nanorods from n-type to p-type semiconducting materials.

52.4 CONCLUSION

Well-directed ZnO nanorods are grown on n-type PS substrates via CBD. N-type ZnO nanorods are changed to p-type by the N-doping process achieved by NH_3 gas during the formation of seed layers. The N-doped ZnO nanorod (p-type ZnO nanorod) films can form a good p–n heterojunction with the n-type PS substrate.

ACKNOWLEDGMENT

This study was supported by the National Research Foundation of Korea funded by the Ministry of Education, Science, and Technology (2011-0028850, 2011-0027329, and 2011-0003056).

REFERENCES

1. Y. N. Xia, P. D. Yang, Y. G. Sun, Y. Y, Wu, B. Mayers, B. Gates, Y. D. Yin, F. Kim and H. Yan, One-dimensional nanostructures: Synthesis, characterization and application, *Adv. Mater.*, 15, 353–389, 2003.
2. M. S. Gudiksen, L. J. Lauhon, D. C. Smith and C. M. Lieber, Growth of nanowire superlattice structures for nanoscale photonics and electronics, *Nature*, 415, 617–620, 2002.
3. Y. Haung, X. F. Daun, Y. Cui, L. J. Lauhon, K. H. Kim and C. M. Libber, Logic gates and computation from assembled nanowire building blocks, *Science*, 294, 1313–1317, 2001.
4. C. W. Bunn, The lattice-dimensions of zinc oxide, *Proc. Phys. Soc.*, 47, 835–842, 1935.
5. R. S. Wagner and W. C. Eliss, Vapor-liquid-solid mechanism of single crystal growth, *Appl. Phys. Lett.*, 4, 89–90, 1964.
6. W. Lee, M. C. Heong and J. M. Myoung, Evolution of the morphology and optical properties of ZnO nanowires during catalyst-free growth by thermal evaporation, *Nanotechnology*, 15, 1441–1445, 2004.
7. A. Dricil, G. Djeteli, G. Tchangbedji, H. Ferouiche, K. Jondo, K. Napo, J. C. Bernede, S. Ouro-Djobo and M. Gbagba, Structured ZnO thin films grown by chemical bath deposition for photovoltaic applications, *Phys. Stat. Sol. A*, 201, 1528–1536, 2004.
8. M. A. Verges, A. Mifsud and C. J. Serna, Formation of rod-like zinc oxide microcrystals in homogeneous solutions, *J. Chem. Soc. Farad. Trans.*, 86, 959–963, 1990.
9. H. J. Ko, Y. F. Chen, Z. Zhu, T. Yao, I. Kobayashi and H. Uchiki, Photoluminescence properties of ZnO epilayers grown on CaF_2.(111) by plasma assisted molecular beam epitaxy, *Appl. Phys. Lett.*, 76, 1905–1907, 2000.
10. A. Ohtomo, K. Tamura, K. Saikusa, T. Takahashi, T. Makino, Y. Segawa, H. Koinuma and M. Kawasaki, Single crystalline ZnO films grown on lattice-matched ScAlMgO4(0001) substrate, *Appl. Phys. Lett.*, 75, 2635–2637, 1999.
11. W. H. Zhang, J. L. Shi, L. Z. Wang and D. S. Yan, Preparation and characterization of ZnO clusters inside mesoporous silica, *Chem. Mater.*, 12, 1408–1413, 2000.
12. Y. Li, G. W. Meng, L. D. Zhang and F. Phillipp, Ordered semiconductor ZnO nanowire arrays and their photoluminescence properties, *Appl. Phys. Lett.*, 76, 2011–2013, 2000.
13. D. C. Look, Progress in ZnO materials and devices, *J. Electron. Mater.*, 35, 1299–1305, 2006.
14. Q. Ahsanulhaq, A. Umar and Y. B. Hahn, Growth of aligned ZnO nanorods and nanopencils on ZnO/Si in aqueous solution: Growth mechanism and structural and optical properties, *Nanotechnology*, 18, 115603-1–110503-7, 2007.
15. M. Caglar, S. Ilican and Y. Caglar, Influence of dopant concentration on the optical properties of ZnO: In films by sol-gel method, *Thin Solid Films*, 517, 5023–5028, 2009.
16. B. D. Cullity, *Elements of X-Ray Diffraction* (Addison-Wesley, Reading, MA, 1978), p. 10.

53 High Surface-Enhanced Raman Scattering (SERS) as an Analytical Tool Using Silver Nanoparticles on GaN Nanowires

Nitzan Dar, Wen-Jing Wang, Kuo-Hao Lee, Yung-Tang Nien, and In-Gann Chen

CONTENTS

53.1 INTRODUCTION

Surface-enhancement Raman spectroscopy (SERS) is a technique to observe analytes such as molecules, polymers, or biological materials at low concentration per surface area by high amplification of the Raman signal [1–3]. Every molecule has its unique Raman spectrum, which makes SERS a powerful tool for detection [4]. SERS effect was discovered by Fleischmann et al. [5]. He reported the first high-quality Raman spectra of pyridine adsorbed on an electrochemically roughened surface of a silver electrode. However, understanding the new phenomena came rather through other researchers. High-enhancement SERS substrate was discovered independently by Craighton et al. [6] and Van Duyne et al. [7] in 1977. In 1980, theoretical explanations of the SERS effect were given by Gersten and Nitzan [8].

Most of the SERS enhancement occurs due to the electromagnetic contribution which causes the Raman signals to be enhanced by a factor of 10^6–10^8. An electromagnetic effect occurs when the absorption of light creates an enhanced local electric field at the location of the adsorbed species due to the surface plasmon resonance. The local electric field decays within 10 nm of the metallic source [8]. Therefore, the analyte should be close to the metal enhancer but not necessarily in contact with it. The electromagnetic effect gives an enhancement of the 4th power of the surrounding electric field that was created by plasmon resonance [9]. Another source of the SERS effect arises from the chemical contribution [10,11], which further increases the SERS signal. A chemical effect may occur due to resonance Raman if the target analyte absorbs the laser wavelength significantly [4]. An additional source for the chemical effect occurs from a metal to a molecule charge transfer [12]. In contrast to the electromagnetic effect, the chemical effect causes at most 10^2 enhancements [13]. Moreover, the analyte must be adsorbed on the SERS enhancer.

Every molecule has its unique Raman spectrum, which makes SERS a powerful tool for detection. The SERS technique was used to detect a variety of analytes from pollutants and explosives to biological matter such as cancer cells and viruses. SERS as an analytical tool is extensively reviewed in the application chapter.

Because porous GaN was used as a SERS template with a deposited Ag or Au film [14], we assumed that GaN nanowires (GaN NWs) can also be used as a template for SERS. They have already been used for many nanoscale applications [15] due to the wide bandgap of GaN (3.4 eV at room temperature) [16], low refractive index [17], and chemical and thermal stability [18].

This chapter shows high SERS amplification of a composite nanostructure made of Ag NPs as SERS enhancers adsorbed on GaN NWs. GaN NWs were synthesized by vapor–liquid–solid chemical vapor deposition (CVD) on a Pt nanofilm-coated Si substrate. Ag NPs were adsorbed on the GaN NWs by electron beam physical vapor deposition. Rhodamine-6G (R6G) was applied to the surface at different concentrations (10^{-3}–10^{-12} M) to measure the detection limit and to derive the enhancement factor. Relationships between the percentage of Ag coverage and SERS intensity are also discussed.

53.2 EXPERIMENTAL SECTION

Ga was reacted with ammonia in a CVD furnace to produce GaN NWs on an Si substrate. A detailed description of the CVD synthesis was reported elsewhere [19]. Briefly, GaN NWs were grown by reacting metallic Ga with NH_3 at 950°C, 0.347 bar, and 0.5 L/min flow rate for 30 min in a horizontal quartz-tube furnace with a diameter of 80 mm and length of 120 cm; a platinum-coated 10-nm-thick Si(111) substrate was placed 2 cm downstream from the Ga source. GaN NWs were checked by x-ray diffraction (Rigaku, D-Max-IV scanned from 30° to 70°, scanning rate 2°/min) and a scanning electron microscope (SEM) (Zeiss, EVO50). Ag was deposited by an E-beam machine (ULVAC VT1-10CE) at a deposition rate of 3 Å/s. Two series of experiments were conducted at a different calibration of the E-beam machine. After silver deposition, samples were examined again by SEM and XRD. Energy-dispersive x-ray spectroscopy (EDS) (EDAX, TSC) was used to show Ag coverage on the GaN NWs. Transmission electron microscope (TEM) (JEOL, JEM-2100F) measurements were conducted to show the morphology of the Ag-deposited particles and to calculate the percentage of Ag coverage. A volume of 5 μL of Rhodamine-6G (R6G) (Sigma Aldrich) in ethanol solution was dropped on the GaN NWs-coated surface. Then the drop was evaporated in an oven at 80 °C. Raman (Jobin Yvon, Labram HR) laser measurements at 532 and 633 nm were carried out after calibration with a silicon scattering peak at 520 cm^{-1}; laser spot diameter was around 1 μm. Five SERS measurements in different spots of the ensemble were carried out for 1 s each to ensure reproducibility of the results. An SERS signal was given as the average signal of these five measurements. Graphs are obtained after reduction of cosmic rays, polynomial fittings, and then finally by smoothing. R6G were checked at concentrations of 10^{-3}, 10^{-6}, 10^{-9}, and 10^{-12} M, all of which were prepared from a stock solution of 50 mL 10^{-2} M R6G. GaN NWs with 1 M R6G were also checked to calculate the enhancement factor.

53.3 RESULTS AND DISCUSSION

53.3.1 TEM AND RAMAN RESULTS

Size and morphology of the Ag NPs are the main factors which determine the intensity of the SERS signal [20]. Therefore, we analyzed the samples by TEM, XRD, and EDX. A clear XRD pattern of cubic Ag (38, 44, 64, 77, 81°, and wurtzite hexagonal GaN NWs (32°, 34°, 36°, 48°, 57°, 63°, 69°, and 70°) was observed. Also the broad XRD peaks mentioned earlier indicate a nanoscale structure. Successful Ag deposition was also supported by the EDX spectra. EDX results show 13% ± 2.5 at% of Ag, 42 ± 1.3 at% of Ga, and 44 ± 2.1 at% of N were also detected, indicating the presence of GaN NWs. Figure 53.1 compares TEM images of E-beam physical vapor deposition Ag NPs on GaN NWs at different deposition times. Figure 53.1a shows isolated oval and spherical NPs (major axis 2–25 nm) deposited on the NWs. Figure 53.1b shows both isolated NPs (major axis 15–45 nm long) and smaller aggregates (40–300 nm long). In Figures 53.1c and d, there is a different pattern from Figure 53.1a; more NPs cover the GaN NWs and some of them were aggregated to form nanorod-like bodies longer than 100 nm. Ag NPs size distribution was similar to other studies as shown in Table 53.1.

We used R6G, investigated by Hildebrandt et al. [21] as an enhancement reagent, as an SERS probe analyte. R6G is an established SERS probe molecule with a known peak assignment [22]. The

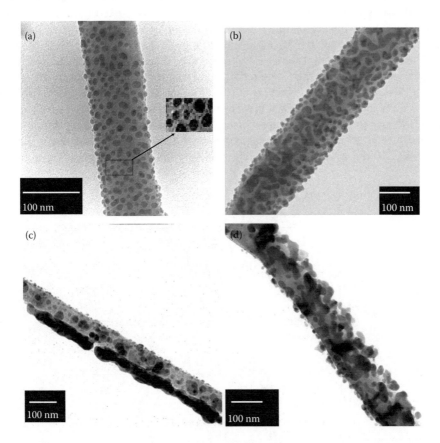

FIGURE 53.1 TEM images of E-beam-deposited Ag NPs grown at different deposition times: (a) 3 s, (b) 15 s, (c) 30 s, (d) 60 s (location of the hot spots shown in the inset of (a)). (D. Nitzan et al., High surface-enhanced Raman scattering (SERS) sensitivity of R6G by fabrication of silver nanoparticles over GaN nanowires, *2011 11th IEEE International Conference on Nanotechnology*, Portland, OR, August 15–18, pp. 297–300, © (2011) IEEE. With permission.)

TABLE 53.1

Comparison of the SERS Detection Limit of R6G by Various Research Groups

Research group	Detection Limit (M)	Repetition No. of Scans	Ag NPs Size Distribution (nm)
Leng et al. [23]	10^{-7}	No data	30–100
Galopin et al. [24]	10^{-12}	No data	4–40
Deng et al. [25]	10^{-7}	No data	a:6–14; b:80–220
Shao et al. [26]	10^{-16}	20 Spots (R6G)	5–20
Zhou el al. [27]	10^{-5}	40 Spots (4-MPY)	a:50–60; b:80–90 c:120
Our paper	10^{-12}	5 Spots (R6G)	a:2–25; b: >100

key peaks of the R6G spectrum are 1124, 1189, 1310, 1360, 1508, 1577, and 1649 cm^{-1}. To compare the SERS effect for R6G at 532 and 633 nm, Raman laser measurement with R6G for both laser wavelengths was carried out. We showed that using a 532 nm laser gives a 2–4 times higher SERS effect, depending on the peak. This effect is probably due to the resonance of the R6G absorption at 532 nm [21]. Therefore, we chose to continue the study with the 532 nm Raman laser. In this study, the highest SERS intensity was achieved by growing Ag NPs by E-beam for only 3 s. When we deposited Ag NPs by E-beam for longer deposition times than 3 s, detection limit was 10^{-9} M for 30 and 60 s or even 10^{-6} M R6G for 180 and 600 s deposition time. Figure 53.2a–c shows the results from 3-s E-beam deposition time; all R6G peaks are observed at high intensity. Measuring the R6G at a concentration of 10^{-12} M (Figure 53.2d) showed the four highest peaks of R6G (1360, 1508, 1577, and 1649 cm^{-1}). This lowest concentration to be detected, that is, detection limit, is comparable to the results of Leng [23], Galopin [24], and Deng [25] but lower than Shao's results [26] (Table 53.1). However, the Raman wavelength was different, and the amount of R6G and the measured time were higher. Moreover, GaN NWs can be used in some conditions, which are not suitable for Si NWs due to better oxidation resistance. Furthermore, reaction temperature was higher and the experimental session was longer, implying that Si NWs are a less favorable SERS substrate than GaN NWs. To ensure reproducibility of the SERS signals, any given result is an average of five

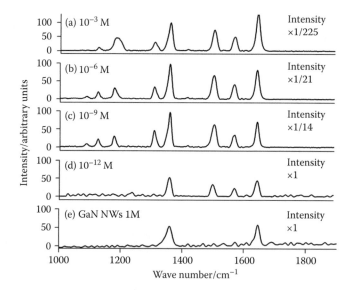

FIGURE 53.2 SERS enhancement of R6G at 3 s deposition time and at different concentrations. (a) 10^{-3} M, (b) 10^{-6} M, (c) 10^{-9} M, (d) 10^{-12} M, and (e) Raman spectrum of GaN with 1M R6G reference. Intensity is normalized for clear comparison.

measurements of different spots on the substrate. However, studies carried out by Shao et al. [26] have repletion of the same results. Also, studies carried out by Zhou et al. [27], even though they measured the R6G signal, showed repeatability but not for the R6G (Table 53.1).

53.3.2 RELATIONSHIP BETWEEN AG COVERAGE AND SERS INTENSITY

To investigate the relationships between adsorbed Ag morphology, silver NPs' coverage, and SERS intensity and magnification, silver% coverage was calculated as shown in Equation 53.1.

$$\text{Ag \% coverage} = \frac{\text{Area covered by Ag} * 100}{\text{Surface area of GaN NWs}} \tag{53.1}$$

When we investigated the relationships between % coverage and deposition time, we found that at short deposition times, there was a linear relationship between deposition time and % coverage, reaching a plateau when the deposition time was over a minute (% coverage becomes constant). SERS intensity of the 1360 cm^{-1} peak was lower when the % of coverage was higher (Figure 53.3a). Therefore, SERS intensity decreased slightly with increased deposition time. We hypothesized that not all particles increase the SERS intensity equally; the aggregated bigger particles do not contribute to the SERS amplification as much as the smaller nano-sized ones. To test this hypothesis, we calculated the % coverage of only Ag NPs on the GaN NWs in respect of their SERS intensity (Figure 53.3b). We found a linear relationship between the coverage of Ag NPs and SERS intensity, results which supported the hypothesis. Theoretical predictions of SERS intensity versus particle size, studied by Boyack et al. [28], are in agreement with our results. The "hot-spot" effect, which causes high SERS intensity, occurs in a site between two NPs, where the distance between them is less than 5 nm due to the high-induced electric field in the vicinity of the NPs [29]. An example of a plausible "hot-spot" site was shown in the inset of Figure 53.2a.

There is a trade-off between % coverage and SERS amplification. On the one hand, too low a % coverage (i.e., too few NPs) decreases the enhancement factor. On the other hand, too high a coverage is characterized by a few larger NPs, an island growth pattern which decreases the enhancement factor. Also, too high a % coverage is rather characterized by an island growth pattern with a few NPs, which decreases the SERS intensity. Wang [30] et al. investigated the SERS intensity of

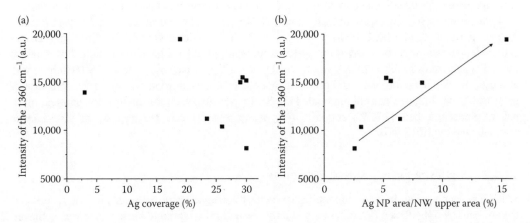

FIGURE 53.3 (a) Relationship between total Ag% coverage and SERS of the 1360 cm^{-1} peak of 10^{-3} M R6G. (b) Relationship between Ag NPs' % coverage and SERS intensity of the 1360 cm^{-1} peak of 10^{-3} M R6G. (D. Nitzan et al., High surface-enhanced Raman scattering (SERS) sensitivity of R6G by fabrication of silver nanoparticles over GaN nanowires, *2011 11th IEEE International Conference on Nanotechnology*, Portland, OR, August 15–18, pp. 297–300, © (2011) IEEE. With permission.)

4-mercaptopyridine absorbed on a silicon nanostructure with Ag NPs. He found a similar relationship between SERS intensity and coverage. To conclude, isolated spherical Ag NPs improve the SERS effect, and the higher their planar density, the better the SERS amplification. However, % NP coverage is the way to measure the density of the Ag NPs, which causes the hot-spot effect.

53.3.3 GaN Contribution to the SERS Effect

Figure 53.2e shows that the SERS intensity of 1 M R6G on GaN NWs without Ag NPS is very low. However, the amount of R6G which adsorbed on the GaN NWs is unknown. The effect of GaN on the SERS intensity is still unclear since GaN cannot be an ordinary Raman enhancer. In general, Raman enhancers are conductive materials, such as silver or gold, which create plasmon resonance through absorption of light. It is suggested that tuning the plasmon resonance maxima due to its dielectric medium properties such as refractive index and dielectric constant [31] may enhance the R6G signal. Also, Ag NPs-coated GaN can transfer the surface plasmon energy due to the waveguide effect [32].

53.4 APPLICATIONS

53.4.1 SERS as a Biosensing Tool

As already written, SERS technique is feasible for the analytical detection of various substrates such as biological materials and for environmental analysis. For using SERS as a biosensing tool, SERS was used to examine the structure, conformation, or charge transfer of biochemicals [33]. The protein redox structure of cytochrome C has been studied by SERS on Ag electrodes by various research groups [34]. Using the recognition chemistry principles of biotin–avidin, researchers have detected 10^{-7} M of dyed label biotin [35]. Detection of the DNA, which is interesting and important for basic research and technology, has been studied by Vo-Dinh and coworkers [36]. They have immobilized a single DNA strand on a nitrocellulose fiber, followed by exposure of a SERS-sensitive dye-labeled hybrid DNA that yields a strong SERS signal. Washing the unbound labeled DNA and measuring the remaining hybrid DNA removed and transferred on a SERS substrate helps detect the initial DNA concentration. Using similar procedures, HIV genes [37] and cancer markers [38] can also be detected. Detection of DNA without markers was carried out by using silver nanoparticles on silicon nanowires [36]. An influenza virus was also detected using a focused ion beam Au nanostructure growth technique [39]. Detection of bacteria can be indirect [40] using Ag nanoparticles synthesized by silver mirror reaction for the detection of dipicolinic acid. Direct pathogen detection was also carried out by Reinhard et al. [41]. They have made nanocluster arrays for SERS substrates to discriminate among different species of bacteria. The SERS technique was also used for the detection of natural products that occur in the environment and in the detection of polysaccharides [42] and natural pigments [43]. SERS was also used for the *in vivo* chemical mapping of plants and seeds [44]. Water quality (i.e., presence of bacteria, eukaryotes, and viruses) was also detected by SERS [45].

53.4.2 SERS for Environmental Detection

SERS has been applied in environmental science for trace analysis of pollutants in water, soil, and the atmosphere [46]. Trace detection of pollutants in natural water and waste water was achieved by SERS by the addition of a small volume of the contaminant onto the metallic SERS enhancer. Trace amounts of pesticides, industrial waste, and other contaminants were detected by SERS in natural water [47,48]. Detection of pollutants in air is more challenging due to the difficulties in the suspended particles in air making contact with the metal enhancer. Contaminants which have functional groups with high affinity to gold or silver enhancers can be naturally adsorbed onto these surfaces. These surfaces were produced from colloids [49,50] and by nanosphere lithography [51]. However, most pollutants are not readily adsorbed on metallic enhancers [46]. To overcome this problem, hybrid materials with hydrophobic interfaces were made [52,53]. They were used to detect polycyclic aromatic hydrocarbons, well known as human carcinogens. In a similar way, mechanical trapping was used for the detection of dioxin and by SERS in aqueous solutions [54].

For the detection of contaminants in the soil by SERS, soil lixiviates are analyzed. SERS was applied for the detection of organo-phosphorous compounds [55]. Humic acids, the major constituent of organic matter in soil, can be detected by SERS using colloidal silver [56]. The behavior of humic acids in natural colloids for understanding their fate as well as the soil structure was studied by [57] using the deposition of humic acid on silver island films. Humic acid reduces gold to make gold colloids [58]. This makes it an excellent substrate for contaminant SERS detection in the soil [59]. Real-time studies with remote monitoring of contaminants in the environment were carried out by Lim et al. [60]. Inorganic ions such as perchlorate, sulfate, nitrate, cyanide, and sulfocyanide were also detected by SERS [61]. Metallic inorganic ions were also detected by complexation of metallic ions to organic ligands [62]. However, the detection of metals that do not have a Raman vibration signal still need to be developed for the actual use as a tool for field measurements [46].

The overall feasibility of the SERS technique for environmental detection has been proved. It is a mature technique with a wide range in fields and applications. It can be used both for solids and solutions, as well as in the lab and for field assays, whenever a portable Raman spectrophotometer is available. However, the real implementation of the SERS technique in the marketplace and routine analysis require valid quantitative analysis and reproducibility, which have not been reported so far. The SERS system should be implemented for molecular and ionic species, regardless of their size composition and aggregation state [46].

53.4.3 SERS as an Analytical Tool

To assess the feasibility of SERS as an analytical tool, important factors should be taken into account. These factors are: uniformity of the substrates, reproducibility of the results, SERS enhancement, and cost [63]. Although virtually every molecule has its unique spectrum, it is quite difficult to interpret signals when a mixture of molecules are adsorbed on the substrate [3]; therefore, a separation technique should be applied to obtain a few kinds of molecules on the substrate that make the signal meaningful. For trace analysis, high enhancement can be obtained for many analytes; sometimes even a single molecule can be detected [64]. The SERS acquisition time is usually very short; it takes just seconds; also, the operation of SERS is easy, which makes it a good choice for industrial purposes. However, for quantitative analysis, reproducibility is still a problem due to the hot-spot effect [24]. Sometimes laser-induced photodesorption as well as photochemistry may occur [65]. Unless the former cases occur, SERS is usually a nondistractive analysis [3]. However, the SERS intensity is linearly related to the molecule's concentration [3], also due to the hot-spot effect. Substrates can also adsorb molecules which interfere with the SERS signal [66]. It is a problem especially when the signal is weak due to the low concentration of the analyte or when impurities are adsorbed that are much stronger than the analyte. To solve that problem to get mainly the analyte signal without interference, vacuum or plasma cleaning can be introduced [67]. However, if the substrate is reexposed to air, impurities will readsorb on the substrate [68].

Our study shows both high enhancement factor and good reproducibility of the SERS signal ensemble made of GaN NWs partially coated with AgNPs. The SERS substrate has good enough uniformity. The E-beam growing method to prepare SERS substrates was recently used for cancer cell detection [38] and for the analysis of biomolecules [69]. Proven as a good SERS substrate for the R6G dye, apparently, it can be used for the detection of other molecules and substrates at low concentration and high reproducibility.

53.5   CONCLUSION

An ensemble of GaN NWs partially coated with Ag NPs was shown to be an effective SERS template. R6G, which was chosen as a reagent for SERS detection, was detected at concentrations as low as 10^{-12} M. We have shown that SERS intensity has a linear relationship with the % of coverage of Ag NPs. Also, Ag NPs show better enhancement than larger particles. The SERS technique has been applied for many kinds of substances such as SERS as a biosensing tool for DNA, viruses, bacteria, and cells. SERS was also applied for environmental detection in air, water, and soil. Other applications of SERS are in forensic science and explosive detection. However, for the actual use of SERS as a commercial tool, research should overcome some drawbacks, especially reproducibility of the substrate, detection limit (for trace detection), and substrate contamination.

REFERENCES

1. Kneipp, K., M. Moskovits, and H. Kneipp, *Surface-Enhanced Raman Scattering: Physics and Applications*. Vol. 103. 2006: Springer Verlag, pp. 1–17.
2. Stiles, P.L. et al., Surface-enhanced Raman spectroscopy. *Annual Review of Analytical Chemistry*, 2008. **1**: 601–626.
3. Aroca, R. and E. Corporation, *Surface Enhanced Vibrational Spectroscopy*. 2006: Wiley Online Library, pp. 73–176.
4. Kneipp, K. et al., Ultrasensitive chemical analysis by Raman spectroscopy. *Chemical Reviews*, 1999. **99**: 2957–2976.
5. Fleischmann, M, P.J. Hendra, and A.J. McQuillan, Raman spectra of pyridine adsorbed at a silver electrode. *Chemical Physics Letters*, 1974. **26**(2): 163–166.
6. Albrecht, M.G. and J.A. Creighton, Anomalously intense Raman spectra of pyridine at a silver electrode. *Journal of the American Chemical Society*, 1977. **99**(15): 5215–5217.
7. Jeanmaire, D.L. and R.P. Vanduyne, Surface Raman spectro electrochemistry 1: Heterocyclic, aromatic and aliphatic-amines adsorbed on anodized silver electrode. *Journal of Electroanalytical Chemistry*, 1977. **84**(1): 1–20.
8. Gersten, J. and A. Nitzan, Electromagnetic theory of enhanced Raman scattering by molecules adsorbed on rough surfaces. *Journal of Chemical Physics*, 1980. **73**(7): 3023–3037.
9. Williamson, T.L. et al., Porous GaN as a template to produce surface-enhanced Raman scattering-active surfaces. *The Journal of Physical Chemistry B*, 2005. **109**(43): 20186–20191.
10. Moskovits, M., Surface-enhanced spectroscopy. *Reviews of Modern Physics*, 1985. **57**(3): 783–826.
11. Persson, B.N.J., K. Zhao, and Z. Zhang, Chemical contribution to surface-enhanced Raman scattering. *Physical Review Letters*, 2006. **96**(20): 207401.
12. Moskovits, M., Surface-enhanced Raman spectroscopy: A brief retrospective. *Journal of Raman Spectroscopy*, 2005. **36**(6–7): 485–496.
13. Lombardi, J.R. et al., Charge-transfer theory of surface enhanced Raman spectroscopy: Herzberg–Teller contributions. *The Journal of Chemical Physics*, 1986. **84**: 4174.
14. Persson, B.N.J., On the theory of surface enhanced Raman scattering. *Chemical Physics Letters*, 1981. **82**(3): 561–565.
15. Chattopadhyay, S. et al., One-dimensional group III-nitrides: Growth, properties, and applications in nanosensing and nano-optoelectronics. *Critical Reviews in Solid State and Materials Sciences*, 2009. **34**(3–4): 224–279.
16. Bloom, S. et al., Band-structure and reflectivity of GaN. *Physica Status Solidi B-Basic Research*, 1974. **66**(1): 161–168.
17. Goldberger, J. et al., Single-crystal gallium nitride nanotubes. *Nature*, 2003. **422**(6932): 599–602.
18. Kang, B. et al., Capacitance pressure sensor based on GaN high-electron-mobility transistor-on-Si membrane. *Applied Physics Letters*, 2005. **86**(25): 253502–253502-3.
19. Lee, K.-H. et al., The effect of nanoscale protrusions on field-emission properties for GaN nanowires. *Journal of the Electrochemical Society*, 2007. **154**(10): K87–K91.
20. Stamplecoskie, K.G. et al., Optimal size of silver nanoparticles for surface-enhanced Raman spectroscopy. *Journal of Physical Chemistry C*, 2011. **115**(5): 1403–1409.
21. Hildebrandt, P. and M. Stockburger, Surface-enhanced resonance Raman spectroscopy of rhodamine 6G adsorbed on colloidal silver. *Journal of Physical Chemistry*, 1984. **88**(24): 5935–5944.

22. Nie, S. and S.R. Emory, Probing single molecules and single nanoparticles by surface-enhanced Raman scattering. *Science*, 1997. **275**(5303): 1102.

23. Leng, W. et al., Silver nanocrystal-modified silicon nanowires as substrates for surface-enhanced Raman and hyper-Raman scattering. *Analytical Chemistry*, 2006. **78**(17): 6279–6282.

24. Galopin, E. et al., Silicon nanowires coated with silver nanostructures as ultrasensitive interfaces for surface-enhanced Raman spectroscopy. *ACS Applied Materials & Interfaces*, 2009. **1**(7): 1396–1403.

25. Deng, S. et al., An effective surface-enhanced Raman scattering template based on a Ag nanocluster–ZnO nanowire array. *Nanotechnology*, 2009. **20**: 175705.

26. Shao, M.-W. et al., Ag-modified silicon nanowires substrate for ultrasensitive surface-enhanced Raman spectroscopy. *Applied Physics Letters*, 2008. **93**(23): 233118(1–3).

27. Zhou, J. et al., In situ nucleation and growth of silver nanoparticles in membrane materials: A controllable roughened SERS substrate with high reproducibility. *Journal of Raman Spectroscopy*, 2009. **40**(1): 31–37.

28. Boyack, R. and E.C. Le Ru, Investigation of particle shape and size effects in SERS using T-matrix calculations. *Physical Chemistry Chemical Physics*, 2009. **11**(34): 7398–7405.

29. Hao, E. and G.C. Schatz, Electromagnetic fields around silver nanoparticles and dimers. *Journal of Chemical Physics*, 2004. **120**(1): 357–366.

30. Wang, Z. and L. Rothberg, Silver nanoparticle coverage dependence of surface-enhanced Raman scattering. *Applied Physics B: Lasers and Optics*, 2006. **84**(1): 289–293.

31. Miller, M.M. and A.A. Lazarides, Sensitivity of metal nanoparticle surface plasmon resonance to the dielectric environment. *The Journal of Physical Chemistry B*, 2005. **109**(46): 21556–21565.

32. Hutchison, J.A. et al., Subdiffraction limited, remote excitation of surface enhanced Raman scattering. *Nano Letters*, 2009. **9**(3): 995–1001.

33. Picorel, R. et al., Surface-enhanced resonance Raman scattering spectroscopy of photosystem II pigment-protein complexes. *The Journal of Physical Chemistry*, 1994. **98**(23): 6017–6022.

34. Feng, J.J. et al., Gated electron transfer of yeast Iso-1 cytochrome c on self-assembled monolayer-coated electrodes. *The Journal of Physical Chemistry B*, 2008. **112**(47): 15202–15211.

35. Pieczonka, N.P.W., P.J.G. Goulet, and R.F. Aroca, Chemically selective sensing through layer-by-layer incorporation of biorecognition into thin film substrates for surface-enhanced resonance Raman scattering. *Journal of the American Chemical Society*, 2006. **128**(39): 12626–12627.

36. Culha, M. et al., Surface-enhanced Raman scattering substrate based on a self-assembled monolayer for use in gene diagnostics. *Analytical Chemistry*, 2003. **75**(22): 6196–6201.

37. Isola, N.R., D.L. Stokes, and T. Vo-Dinh, Surface-enhanced Raman gene probe for HIV detection. *Analytical Chemistry*, 1998. **70**(7): 1352–1356.

38. Liu, Y. et al., Fabrication of silver ordered nanoarrays SERS-active substrates and their applications in bladder cancer cells detection. *Spectroscopy and Spectral Analysis*, 2012. **32**(2): 386–390.

39. Fan, M., G.F.S. Andrade, and A.G. Brolo, A review on the fabrication of substrates for surface enhanced Raman spectroscopy and their applications in analytical chemistry. *Analytica Chimica Acta*, 2011. **693**(1–2): 7–25.

40. Daniels, J.K. and G. Chumanov, Nanoparticle-mirror sandwich substrates for surface-enhanced Raman scattering. *The Journal of Physical Chemistry B*, 2005. **109**(38): 17936–17942.

41. Yan, B. et al., Engineered SERS substrates with multiscale signal enhancement: Nanoparticle cluster arrays. *ACS Nano*, 2009. **3**(5): 1190–1202.

42. Shafer-Peltier, K.E. et al., Toward a glucose biosensor based on surface-enhanced Raman scattering. *Journal of the American Chemical Society*, 2003. **125**(2): 588–593.

43. Schulte, F. et al., Characterization of pollen carotenoids with in situ and high-performance thin-layer chromatography supported resonant Raman spectroscopy. *Analytical Chemistry*, 2009. **81**(20): 8426–8433.

44. Rösch, P., J. Popp, and W. Kiefer, Raman and surface enhanced Raman spectroscopic investigation on Lamiaceae plants. *Journal of Molecular Structure*, 1999. **480–481**(0): 121–124.

45 Shanmukh, S. et al., Rapid and sensitive detection of respiratory virus molecular signatures using a silver nanorod array SERS substrate. *Nano Letters*, 2006. **6**(11): 2630–2636.

46. Alvarez-Puebla, R. and L. Liz-Marzan, Environmental applications of plasmon assisted Raman scattering. *Energy & Environmental Science*, 2010. **3**(8): 1011–1017.

47. Weißenbacher, N. et al., Continuous surface enhanced Raman spectroscopy for the detection of trace organic pollutants in aqueous systems. *Journal of Molecular Structure*, 1997. **410**: 539–542.

48. Taurozzi, J.S. and V.V. Tarabara, Silver nanoparticle arrays on track etch membrane support as flow-through optical sensors for water quality control. *Environmental Engineering Science*, 2007. **24**(1): 122–137.

49. Ayora, M. et al., Detection of atmospheric contaminants in aerosols by surface-enhanced Raman spectrometry. *Analytica Chimica Acta*, 1997. **355**(1): 15–21.

50. Fernandez-Lopez, C. et al., Highly controlled silica coating of PEG-capped metal nanoparticles and preparation of SERS-encoded particles. *Langmuir*, 2009. **25**(24): 13894–13899.

51. Biggs, K.B. et al., Surface-enhanced Raman spectroscopy of benzenethiol adsorbed from the gas phase onto silver film over nanosphere surfaces: Determination of the sticking probability and detection limit time†. *The Journal of Physical Chemistry A*, 2009. **113**(16): 4581–4586.

52. Jones, C.L., K.C. Bantz, and C.L. Haynes, Partition layer-modified substrates for reversible surface-enhanced Raman scattering detection of polycyclic aromatic hydrocarbons. *Analytical and Bioanalytical Chemistry*, 2009. **394**(1): 303–311.

53. Guerrini, L. et al., Sensing polycyclic aromatic hydrocarbons with dithiocarbamate-functionalized Ag nanoparticles by surface-enhanced Raman scattering. *Analytical Chemistry*, 2009. **81**(3): 953–960.

54. Aldeanueva-Potel, P. et al., Recyclable molecular trapping and SERS detection in silver-loaded agarose gels with dynamic hot spots. *Analytical Chemistry*, 2009. **81**(22): 9233–9238.

55. Alak, A.M. and T. Vo-Dinh, Surface-enhanced Raman spectrometry of organo phosphorus chemical agents. *Analytical Chemistry*, 1987. **59**(17): 2149–2153.

56. Sánchez-Cortés, S. et al., pH-dependent adsorption of fractionated peat humic substances on different silver colloids studied by surface-enhanced Raman spectroscopy. *Journal of Colloid and Interface Science*, 1998. **198**(2): 308–318.

57. Alvarez-Puebla, R.A., J.J. Garrido, and R.F. Aroca, Surface-enhanced vibrational microspectroscopy of fulvic acid micelles. *Analytical Chemistry*, 2004. **76**(23): 7118–7125.

58. Machesky, M.L., W.O. Andrade, and A.W. Rose, Interactions of gold (III) chloride and elemental gold with peat-derived humic substances. *Chemical Geology*, 1992. **102**(1–4): 53–71.

59. Baigorri, R. et al., Optical enhancing properties of anisotropic gold nanoplates prepared with different fractions of a natural humic substance. *Chemistry of Materials*, 2008. **20**(4): 1516–1521.

60. Lim, C. et al., Optofluidic platforms based on surface-enhanced Raman scattering. *Analyst*, 2010. **135**(5): 837–844.

61. Gu, B., C. Ruan, and W. Wang, Perchlorate detection at nanomolar concentrations by surface-enhanced Raman scattering. *Applied Spectroscopy*, 2009. **63**(1): 98–102.

62. Bhandari, D. et al., Characterization and detection of uranyl ion sorption on silver surfaces using surface enhanced Raman spectroscopy. *Analytical Chemistry*, 2009. **81**(19): 8061–8067.

63. Tripp, R.A., R.A. Dluhy, and Y. Zhao, Novel nanostructures for SERS biosensing. *Nano Today*, 2008. **3**(3–4): 31–37.

64. Kneipp, K. et al., Single molecule detection using surface-enhanced Raman scattering (SERS). *Physical Review Letters*, 1997. **78**(9): 1667–1670.

65. Zhdanov, V.P. and B. Kasemo, Specifics of substrate-mediated photo-induced chemical processes on supported nm-sized metal particles. *Journal of Physics: Condensed Matter*, 2004. **16**: 7131.

66. Lin, X.M. et al., Surface-enhanced Raman spectroscopy: Substrate-related issues. *Analytical and Bioanalytical Chemistry*, 2009. **394**(7): 1729–1745.

67. Norrod, K.L. and K.L. Rowlen, Removal of carbonaceous contamination from SERS-active silver by self-assembly of decanethiol. *Analytical Chemistry*, 1998. **70**(19): 4218–4221.

68. Taylor, C.E., S.D. Garvey, and J.E. Pemberton, Carbon contamination at silver surfaces: Surface preparation procedures evaluated by Raman spectroscopy and X-ray photoelectron spectroscopy. *Analytical Chemistry*, 1996. **68**(14): 2401–2408.

69. David, C. et al., SERS detection of biomolecules using lithographed nanoparticles towards a reproducible SERS biosensor. *Nanotechnology*, 2010. **21**: 475501.

70. D., Nitzan, W.-J. Wang, K.-H. Lee, and I.-G. Chen, High surface-enhanced Raman scattering (SERS) sensitivity of R6G by fabrication of silver nanoparticles over GaN nanowires, *2011 11th IEEE International Conference on Nanotechnology*, Portland, OR, August 15–18, 2011, pp. 297–300.

Section XVII

Nanowire Transistors

54 High-Speed and Transparent Nanocrystalline ZnO Thin Film Transistors

Burhan Bayraktaroglu and Kevin Leedy

CONTENTS

54.1 INTRODUCTION

Thin film transistors (TFTs) made from amorphous or organic semiconductors are commonly used in the control circuits of large-area display electronics such as flat-panel TV screens. They can also be used in applications requiring flexible or nonplanar surfaces where the use of regular single crystal electronics is problematic. The usefulness of TFTs, however, has not been extended to high-performance applications due to significantly inferior electronic properties of thin films compared to their single crystal counterparts. For example, the field-effect electron mobility of amorphous Si is typically 0.1 cm²/V.s, whereas the mobility values in single crystal Si can be in excess of 1000 cm²/V.s [1]. Metal-oxide semiconductors based on Zn, In, Ga, and Hf have shown great promise in solving these problems for the current and next-generation flat-panel display electronics [2,3]. However, most metal-oxide semiconductor thin films are also amorphous and have limitations in current density and switching on/off ratios. Although they offer higher electron mobilities than amorphous Si films and therefore show promise in higher-speed circuit applications [4], they are not yet suitable for high-performance digital or analog circuits beyond display control electronics.

The bonding in ZnO and similar oxide semiconductors is strongly ionic and the states near the conduction band minimum arise almost completely from cation (zinc) *s*-orbitals. This is in strong contrast to covalent or near-covalent semiconductors like Si or GaAs where the *sp*-bonding is sensitive to both the bond angle and bond length disorder that easily generates localized states (traps). In ZnO, the nearly spherical and relatively large, empty zinc cation orbitals that form the conduction band result in electron transport that is largely unaffected by the bond angle or bond length disorder, and the disorder does not result in localized states [5]. It is anticipated that radiation-induced damage will have minimal effect on electronic device operation due to this insensitivity to lattice disorder [6]. Initial results show that ZnO TFTs and circuits have remarkable tolerance to high-dose gamma ray and neutron radiation and that damages caused by irradiation can be removed by low-temperature annealing [7].

Because ZnO is also a wide bandgap semiconductor, it can be used for both transparent contact layers and the channel semiconductor in transparent electronics applications. We have shown that high-performance transparent transistors can be fabricated using only nc-ZnO films and no indium-containing layers.

In this report, a comprehensive study of the nc-ZnO thin film growth conditions was undertaken to understand the factors influencing the film properties as they relate to the transistor operation. The influence of the gate dielectric material on the physical properties of ZnO films and the transistor performance was identified. Device design issues related to the high-frequency (microwave) operation were identified and approaches used to mitigate these issues were described.

Ordered nc-ZnO thin films composed of closely packed nanocolumns offer a unique solution to improving TFT performance to levels comparable to single crystal semiconductors while maintaining their thin film properties. High-speed transistors were fabricated using 1.2–2 μm gate length devices with cut-off frequencies as high as 10 GHz [8].

Thin films of spontaneously ordered and closely packed nanocolumns of ZnO were used to fabricate high-speed and transparent TFTs. The use of nanocrystalline ZnO (nc-ZnO) helps to achieve the intrinsic electronic properties of single crystals while providing substrate agnostic thin films that can be grown on nonplanar surfaces of rigid or flexible substrates. We have developed low-temperature pulsed laser deposition (PLD) techniques for both doped and undoped nc-ZnO films and demonstrated microwave transistor operation (f_{max} = 10 GHz). Unlike in amorphous TFTs, the operation of nc-ZnO transistors relies on field-effect charge control at vertical grain boundaries between nanocolumns and can produce very high on/off ratios ($>10^{12}$), very high current densities, and near-ideal subthreshold voltage swings (~74 mV/decade). Using a combination of doped and undoped nc-ZnO films, we have fabricated the first indium-free transparent TFTs with excellent performance and transparency.

54.2 THIN FILM DEPOSITION AND DEVICE FABRICATION

Nc-ZnO films were deposited in a Neocera Pioneer 180 pulsed laser deposition system with a KrFexcimer laser (Lambda Physik COMPex Pro 110, λ = 248 nm, 10 ns pulse duration). The following processing conditions were used to deposit undoped ZnO films for active layers of transistors: laser energy density of 2.6 J/cm², laser repetition rate of 30 Hz, deposition temperature of 25–400°C, oxygen partial pressure of 1–100 mTorr during the deposition, and substrate-to-target distance of 9.5 cm. The target was a 50 mm diameter by 6 mm thick sintered ZnO ceramic disk (99.999%). The deposition parameters for conductive transparent ZnO films (used as contact layers in transparent transistors) were the same, except that Ar gas was used during deposition instead of oxygen [9] and the source material was high-purity ZnO with 3% Ga_2O_3 used as an n-type dopant. Various gate insulators, including SiO_2, Al_2O_3, and HfO_2, were used in this study. SiO_2 films were grown in a PlasmaTherm 790 plasma-enhanced chemical vapor deposition (PECVD) system at 250°C. Al_2O_3 and HfO_2 films were prepared in a Cambridge Nanotech Fiji F200 atomic layer deposition (ALD) system at 150–250°C.

The ZnO crystal structure was determined by using a PANalytical X'Pert Pro MRD x-ray diffractometer. Film morphologies were analyzed with an FEI DB235 scanning electron microscope (SEM) and a JEOL 4000EX transmission electron microscope (TEM) operating at 400 kV. Surface roughness was measured with a Veeco Dimension 3000 atomic force microscope (AFM). Dielectric film thickness and refractive index were measured with a Horiba JobinYvon UVISEL spectroscopic ellipsometer. Optical transmission measurements were made with a Varian Cary 5000 spectrophotometer.

Devices for low-frequency characterization were fabricated on p-type Si wafers or flexible polyimide films. The gate lengths, L_G, were varied from 2 to 25 μm. High-speed devices were fabricated

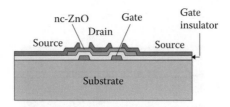

FIGURE 54.1 Cross-sectional schematic of the baseline nc-ZnO thin film transistor. (B. Bayraktaroglu and K. Leedy, *2011 11th IEEE International Conference on Nanotechnology*, August 15–18, Portland, Oregon, pp. 1450–1455. © (2011) IEEE. With permission.)

on high-resistivity Si substrates (>2000 ohm.cm) to minimize capacitive parasitics. Gate lengths for these devices varied from 1.2 to 3 μm.

Transparent devices were fabricated on quartz or glass substrates. A bottom-gate configuration was used for all transistors, where the gate contact was fabricated first directly on the substrate. As shown in Figure 54.1, the gate contact is covered with the gate insulator and the nc-ZnO film is produced conformally over the gate insulator. The Ni/Au and Ti/Au contacts were used as the gate and the source/drain contacts for the nontransparent devices, whereas the Ga-doped ZnO (GZO) films were used as contacts for the transparent devices. Other device-processing details were reported elsewhere [3].

The source and drain contacts overlap the gate contact by 0.5–3 μm. To reduce the parasitic capacitance effects due to this overlap of gate and source–drain contacts for high-frequency devices, a thicker gate insulator was used in these regions while maintaining a thinner gate insulator under the channel region. In this way, high transconductance was maintained while reducing the effects of parasitic elements. The gate length is defined as the distance between the source and drain contacts. No intentional surface passivation was applied to the free surface of the ZnO film.

54.3 ZnO THIN FILMS

54.3.1 Thin Film Analysis

Cross-sectional TEM images of ZnO films deposited conformally over the gate contact indicate that the film is composed of densely compacted, highly faulted columnar-shaped grains that predominantly extend through the thickness of the film, as shown in Figure 54.2. The nanocolumnar grain diameters were in the 20–50 nm range depending on the film growth parameters such as substrate temperature and oxygen pressure, but self-ordering of nanocolumns was observed with films grown on other types of gate insulators also, as shown in Figure 54.3.

X-ray diffraction scans of the ZnO films deposited on the SiO_2, Al_2O_3, and HfO_2 gate dielectrics exhibited a highly textured c-axis orientation with only the ZnO (002) peak present, consistent with other studies of PLD ZnO films [10–13]. The RMS surface roughness of ZnO films deposited on SiO_2 and HfO_2 was 0.65 and 0.94 nm, respectively, as measured by AFM. AFM images in Figure 54.4 also show slightly larger ZnO grains grown on HfO_2 compared to SiO_2 in agreement with the cross-sectional TEM results. The measured surface roughness of the underlying dielectric material was approximately 0.2 nm, regardless of the film composition, deposition technique, or deposition temperature. This smooth surface was necessary for the ordered ZnO grain growth and produced a well-defined dielectric–ZnO interface, which was essential for device performance.

54.3.2 Influence of Growth Temperature

At deposition temperatures of <75°C, the grains exhibited diffraction contrast differences and, in some areas, nearly equiaxed grain structures. As the film growth temperature increased from 100°C to 400°C, films displayed an increasingly uniform columnar grain structure and consistent

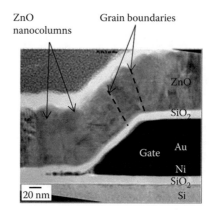

FIGURE 54.2 Cross-sectional TEM image of a ZnO film over the gate metal showing continuous, closely packed nanocolumnar structures over nonplanar surfaces. (B. Bayraktaroglu and K. Leedy, *2011 11th IEEE International Conference on Nanotechnology*, August 15–18, Portland, Oregon, pp. 1450–1455. © (2011) IEEE. With permission.)

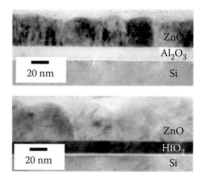

FIGURE 54.3 Cross-sectional TEM images of a ZnO thin film deposited at 200°C on Al_2O_3/Si and HfO_2/Si. (B. Bayraktaroglu and K. Leedy, *2011 11th IEEE International Conference on Nanotechnology*, August 15–18, Portland, Oregon, pp. 1450–1455. © (2011) IEEE. With permission.)

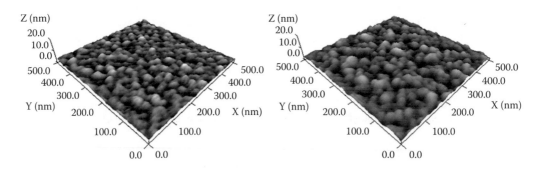

FIGURE 54.4 AFM images of ZnO thin films deposited at 200°C on (a) 20 nm SiO_2/Si and (b) 30 nm HfO_2/Si. (B. Bayraktaroglu and K. Leedy, *2011 11th IEEE International Conference on Nanotechnology*, August 15–18, Portland, Oregon, pp. 1450–1455. © (2011) IEEE. With permission.)

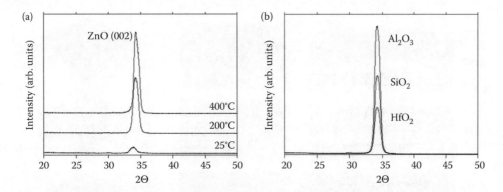

FIGURE 54.5 X-ray diffraction patterns of (a) ZnO thin films deposited on SiO_2/Si at 25°C, 200°C, and 400°C and (b) ZnO thin films deposited at 200°C on SiO_2/Si, Al_2O_3/Si, and HfO_2/Si.

diffraction contrast. With increasing deposition temperature up to 400°C, the intensities of the ZnO (002) peak increased and full-width at half-maximum values of the ZnO (002) decreased, as shown in Figure 54.5a, indicating improved film crystallinity. The 2Θ positions were less than the 34.421° 2Θ (002) peak from the JCPDS #36-1451 powder diffraction file indicating strained lattice structures. Progressively higher ZnO (002) peak intensities were obtained from ZnO films deposited on HfO_2, SiO_2, and Al_2O_3, respectively, as shown in Figure 54.5b. The corresponding 2Θ positions were 34.21°, 34.21°, and 34.17° for ZnO films on HfO_2, SiO_2, and Al_2O_3. Consistent with the diminished x-ray diffraction intensity, larger ZnO grains with a rougher surface texture were found on films grown on HfO_2 (see Figure 54.4). Further details on XRD results are given elsewhere [14].

A summary of the TEM, SEM, and AFM characteristics of films grown at different temperatures is shown in Figure 54.6. Within the temperature range studied, the surface morphologies for all ZnO films showed no trend in grain size with deposition temperature. AFM images with the 500 nm × 500 nm collection areas in this figure also show that the films were predominantly smooth.

RMS roughness values ranged from 0.65 to 1.65 nm. Although the lowest roughness occurred in a film deposited at 200°C, no trend in roughness as a function of deposition temperature was observed. The RMS roughness is similar to other reported values of ZnO deposited by PLD [11,12]. Grain size calculations based on grain boundary intercepts in a scanning probe image processing software indicated 25–35 nm ZnO grains, again with no trend observed as a function of deposition temperature.

54.4 THIN FILM TRANSISTORS

54.4.1 DC CHARACTERISTICS

The typical common-source $I–V$ characteristics of nc-ZnO TFTs fabricated on the HfO_2, SiO_2, and Al_2O_3 gate insulators are shown in Figure 54.7. The ZnO films were grown at 200°C and the devices were fabricated in the same fabrication batch for a direct comparison of results. All devices showed hysteresis-free operation and positive threshold voltages of 1.5 ± 0.6 V. As expected, higher dielectric constants of Al_2O_3 and HfO_2 compared to SiO_2 resulted in higher drain currents for the same gate voltage (i.e., higher transconductance) and flatter current characteristics in the saturation region.

The influence of the gate insulator on the device transconductance is shown in Figure 54.8. The same reciprocal dependence of transconductance on the gate length was obtained for these insulators in the gate length range of $L_G = 2–25$ μm. This indicates that the number of charge states at the insulator–ZnO interface is low and such charges do not heavily influence the device operation.

The low interface state density values can also be inferred from the near-ideal subthreshold voltage swing, S, values typically obtained with the nc-ZnO TFTs. From the transfer characteristics shown in Figure 54.9, we determine the S-values of 74 and 107 mV/decade for devices with the SiO_2 and HfO_2

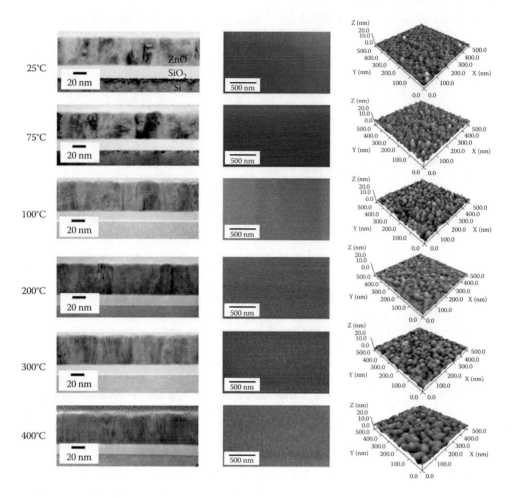

FIGURE 54.6 Cross-sectional TEM images, surface SEM images, and AFM images of ZnO thin film deposited on 20 nm SiO₂/Si at 25°C, 75°C, 100°C, 200°C, 300°C, and 400°C.

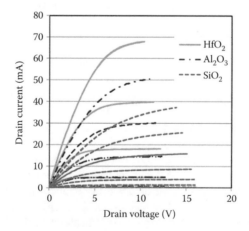

FIGURE 54.7 *I–V* characteristics of nc-ZnO TFTs with various gate insulators. Gate insulator thickness = 30 nm, $L_G = 5$ μm, $W_G = 400$ μm, $V_G = 2$ V/step. (B. Bayraktaroglu and K. Leedy, *2011 11th IEEE International Conference on Nanotechnology*, August 15–18, Portland, Oregon, pp. 1450–1455. © (2011) IEEE. With permission.)

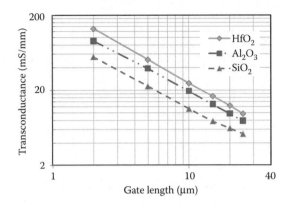

FIGURE 54.8 Transconductance dependence on gate length for three different gate insulators. $V_D = 12$ V, $V_G = 10$ V. (B. Bayraktaroglu and K. Leedy, *2011 11th IEEE International Conference on Nanotechnology*, August 15–18, Portland, Oregon, pp. 1450–1455. © (2011) IEEE. With permission.)

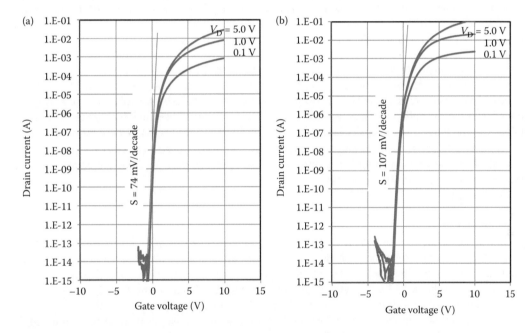

FIGURE 54.9 Transfer characteristics of nc-ZnO TFTs with (a) SiO_2 and (b) HfO_2 gate insulators for various drain voltage values. ZnO films were grown at 200°C. $L_G = 2$ μm, and $W_G = 400$ μm.

gate insulators. These S-values correspond to the total surface state density values of 2.51×10^{11} and 5.38×10^{12} cm^{-2} using the expression developed in Ref. [15]. These transfer characteristics also show very high on/off ratios of about 10^{12}, which is typical for the nc-ZnO TFTs fabricated on other substrates [16]. The maximum field-effect mobility for the nc-ZnO TFT with HfO_2 gate insulators was 63 cm^2/V · s, as shown in Figure 54.10. Field-effect mobility values of as high as 110 cm^2/V · s were obtained with SiO_2 gate insulator devices due to exceptionally low interface state densities [16,17].

We interpret these record performance values for nc-ZnO TFTs deposited in O_2 to be a result of high crystal quality preserved within nanocolumns as well as the presence of acceptor-like charges at the grain boundaries due to Zn-vacancies [9]. Transistors show exceptionally low leakage currents in the OFF state due to back-to-back depletion layers formed at the vertical grain boundaries, which effectively block horizontal current conduction between the source and the drain contacts.

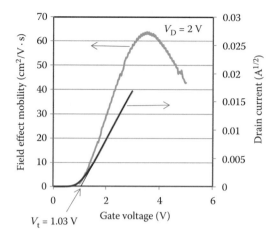

FIGURE 54.10 Field-effect electron mobility and drain current as a function of gate voltage for an nc-ZnO TFT fabricated on a polyimide substrate at low temperatures. $L_G = 5\ \mu m$, $W_G = 400\ \mu m$. (B. Bayraktaroglu and K. Leedy, *2011 11th IEEE International Conference on Nanotechnology*, August 15–18, Portland, Oregon, pp. 1450–1455. © (2011) IEEE. With permission.)

In the ON state, the vertical depletion layers collapse under the influence of the gate field and large current densities are made possible between nanocrystals. This is similar to high current conduction in microcrystalline ZnO-based varistors where back-to-back depletion layers at grain boundaries rapidly collapse at high electric fields [18]. A combination of low leakage current in the OFF state and large current density in the ON state produces the very high ON/OFF ratios observed.

Other factors contributing to large current density operation are the high electron mobility and low interface state density described earlier. Such low interface state densities are consistent with the measured high field-effect mobility.

54.4.2 HIGH-SPEED TRANSISTORS

TFTs are mostly used in circuits that do not require high-speed operation [1] because of electron mobility limitations. However, with the demonstrated high electron mobility and the anticipated high electron velocity [19], ZnO-based TFTs have a greater potential for high-speed circuit applications than amorphous TFTs. Recently, devices capable of small-signal amplification at microwave frequencies were demonstrated using relatively long gate lengths of 1.2–3 μm [8,20].

In the design of high-speed transistors, several parasitic device elements must be minimized in addition to shrinking the gate length. The most prominent parasitic elements that degrade the high-frequency operation of a TFT are the gate resistance and the capacitance between the gate and the source/drain contacts. The location of the parasitic capacitances is illustrated in Figure 54.11.

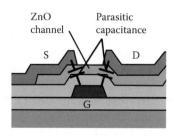

FIGURE 54.11 The parasitic capacitance between the source/drain and gate contacts that limit the high-frequency performance of the bottom-gate-type thin film transistors.

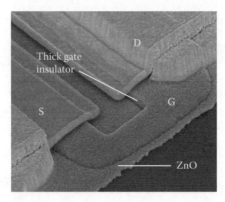

FIGURE 54.12   SEM picture showing the use of a thick gate insulator to reduce parasitic capacitance in high-speed TFTs.

The gate resistance can be lowered by the use of thick, high-conductivity metals such as Au or Cu. As shown in the cross-sectional drawing of Figure 54.2, a thick gate metal produces nonplanar topography for the bottom-gate-type transistors. It is important that the gate insulator and the ZnO thin film are produced conformably over the gate contact.

The parasitic capacitance resulting from the overlap of source/drain contacts over the gate can be reduced in several ways. The first is to minimize the overlap by careful alignment of these contacts and reduction of the actual gate length to as close to the source–drain spacing as possible (i.e., self-aligned contacts). The second is to employ a thicker gate insulator in the regions where the contacts overlap and a thinner gate insulator under the actual channel region. This approach requires the use of two different thickness gate insulators over the gate contact, as shown in Figure 54.12. Again, such an approach results in nonplanar surfaces over which the ZnO thin film is produced. However, the nc-ZnO films fabricated by PLD have shown that such topological variations can be readily overcome.

The high-speed potential of nc-ZnO TFTs was examined in a series of designs fabricated on high-resistivity Si substrates (>2000 ohm.cm). The gate contact was Ni/Au (5 nm/120 nm) to ensure low gate resistance. The gate insulator was PECVD SiO_2 and the ZnO films were grown at 400°C. Devices with two different gate lengths ($L_G = 1.2$ and 2.1 μm) and two different gate widths ($W_G = 50$ and 100 μm) were used to examine device scalability. From the measured s-parameters, the maximum available gain (MAG) and current gain ($|h_{21}|^2$) values were determined as shown in Figure 54.13. The current gain $|h_{21}|^2$ values for both devices showed −6 dB/octave slopes with cut-off frequencies, f_T, of 2.45 and 1.1 GHz for devices with $L_G = 1.2$ and 2.1 μm, respectively. MAG values had slopes of −3 dB/octave for stability factor $K < 1$, and −6 dB/octave for $K > 1$. Power gain cut-off frequencies, f_{max}, of 7.45 and 3.02 GHz were achieved for the two size devices.

The cut-off frequency dependence on device size and bias conditions is shown in Figure 54.14. It is clear that the small-signal microwave performances of both size devices (i.e., $W_G = 50$ and 100 μm) at the same bias conditions are nearly identical. This is an indication of good device scalability and indicates that the microwave performance is not degraded due to the increased gate resistance of longer gate fingers on larger devices. It also shows that the microwave performance has a weak dependence on gate bias and therefore on drain current. Such device characteristics, which are desirable for linear circuit applications, can be attributed to fully depleted channels characteristic of nc-ZnO TFTs.

Reduced parasitic capacitance with the use of double gate insulator thicknesses resulted in overall improvement in the high-speed performance of devices, as shown in Figure 54.15. Device designs with varying degrees of overlap between the source/drain contacts and the gate contact were examined for devices with $L_G = 1.2$ μm and $W_G = 2 \times 50$ μm. All three device designs (with 1.5, 1.0,

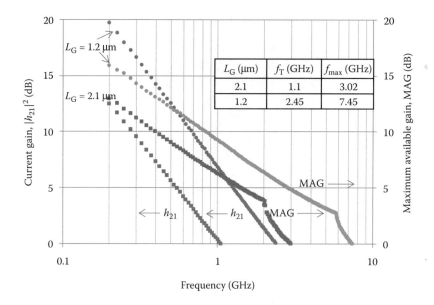

FIGURE 54.13 Small-signal microwave characteristics of ZnO TFTs with different gate lengths. The gate width was 100 μm for both devices.

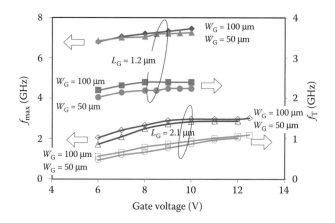

FIGURE 54.14 Microwave performance dependence on device size and gate voltage. Bias conditions were $V_D = 13$ V for the $L_G = 1.2$ μm device and $V_D = 16$ V for the $L_G = 2.1$ μm device.

and 0.5 μm of gate electrode overlap with source/drain contacts) showed an identical current gain cut-off frequency of 2.9 GHz. The highest power cut-off frequency was $f_{max} = 10$ GHz for devices with a 0.5 μm overlap. To our knowledge, this is the highest-frequency operation obtained with any TFT using any material technology. Based on the small-signal device models extracted from the measured s-parameters, we expect $f_{max} > 40$ GHz for devices with $L_G = 0.5$ μm.

54.4.3 Transparent Transistors

ZnO thin films are suitable for transparent thin film transistor (TTFT) applications because the wide bandgap properties (3.4 eV) of both doped and undoped layers can provide excellent optical transparency. The semiconducting properties of undoped layers make them suitable for active layers, as shown earlier, while the doped films can be considered for contact layers. An all-ZnO

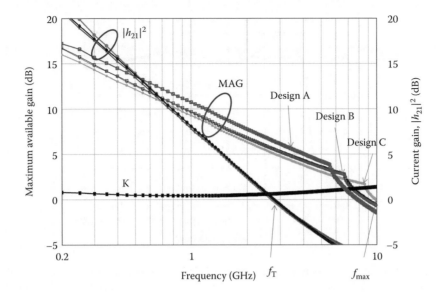

FIGURE 54.15 Small-signal microwave characteristics of nc-ZnO TFTs. TFTs with different gate contacts overlap with S/D contacts. Devices were biased at $V_G = 6$ V, $V_D = 11$ V. (B. Bayraktaroglu and K. Leedy, *2011 11th IEEE International Conference on Nanotechnology*, August 15–18, Portland, Oregon, pp. 1450–1455. © (2011) IEEE. With permission.)

approach is important for this application because of the high cost and the worldwide shortage of indium caused by the rapidly increasing use of ubiquitous touch screen devices with ITO (indium-tin oxide) transparent conductive oxide (TCO) films [21].

The resistivity, transparency, and stability of contact layers made of Al- or Ga-doped ZnO films were shown to be competitive with indium-based TCOs [22–25]. However, the integration of doped and undoped ZnO films in the same device faces some technical challenges, including low-temperature processing for contact layers, autodoping between layers, and the lack of selective etching between doped and undoped ZnO films.

We have developed indium-free transparent transistors by replacing the metal contacts of the nontransparent TFTs described earlier with Ga-doped ZnO films grown at low temperatures and in an Ar atmosphere [26]. These Ar-grown TCOs do not require postgrowth annealing to achieve low resistivity and therefore comply with the low process temperature budget of ≤200°C. To address the difficulty of selectively etching the contact layer over the channel (both ZnO), we employed a thin Ti etch-stop layer between the source/drain contacts and the transistor channel layer as shown in Figure 54.16. During processing, the exposed parts of this etch-stop layer oxidized and became a nonconducting TiO_x surface passivation layer.

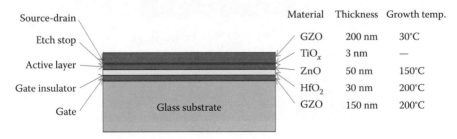

	Material	Thickness	Growth temp.
Source-drain	GZO	200 nm	30°C
Etch stop	TiO_x	3 nm	—
Active layer	ZnO	50 nm	150°C
Gate insulator	HfO_2	30 nm	200°C
Gate	GZO	150 nm	200°C

FIGURE 54.16 The entire thin film stack of indium-free TTFT grown on a transparent substrate at successively lower growth temperatures. (B. Bayraktaroglu and K. Leedy, *2011 11th IEEE International Conference on Nanotechnology*, August 15–18, Portland, Oregon, pp. 1450–1455. © (2011) IEEE. With permission.)

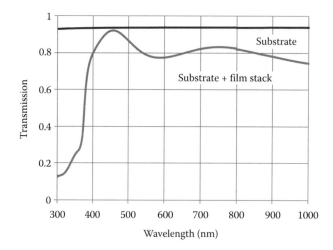

FIGURE 54.17 Optical transmission rate for the entire TTFT films and the glass substrate. (B. Bayraktaroglu and K. Leedy, *2011 11th IEEE International Conference on Nanotechnology*, August 15–18, Portland, Oregon, pp. 1450–1455. © (2011) IEEE. With permission.)

The entire film stack was grown on glass or quartz substrates before device fabrication and had about 90% transparency (not including the substrate) in the visible spectrum, as shown in Figure 54.17. The growth temperature of layers was reduced as the layers were added to minimize the impact of temperature on the properties of the previously grown layers. The gate insulator was 30-nm-thick HfO_2 grown by the ALD technique. Mesa etching techniques were used to define the S/D contacts, device active area, and gate contacts in successive process steps. Figure 54.18 shows a cross-sectional drawing of the transistor and a photograph of several devices with different gate lengths ranging from 5 to 20 µm.

The linear portion of the *I–V* characteristics of a device with $L_G = 5$ µm and $W_G = 70$ µm, shown in Figure 54.19, indicates that the GZO layers make good ohmic contacts with the active layer and there is no evidence of current blocking by the TiO_x etch-stop layer. Also, devices exhibit excellent

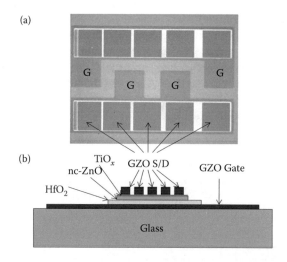

FIGURE 54.18 (a) A photograph and (b) a cross-sectional drawing of transparent nc-ZnO TFTs. A thin TiO_x layer is employed as an etch-stop layer to facilitate the selective etching of S/D GZO layers. (B. Bayraktaroglu and K. Leedy, *2011 11th IEEE International Conference on Nanotechnology*, August 15–18, Portland, Oregon, pp. 1450–1455. © (2011) IEEE. With permission.)

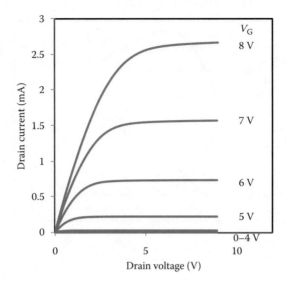

FIGURE 54.19 Current–voltage characteristics of the nc-ZnO TTFT. $L_G = 5\ \mu m$, $W_G = 70\ \mu m$, $V_G = 1\ V/$ step. (B. Bayraktaroglu and K. Leedy, *2011 11th IEEE International Conference on Nanotechnology*, August 15–18, Portland, Oregon, pp. 1450–1455. © (2011) IEEE. With permission.)

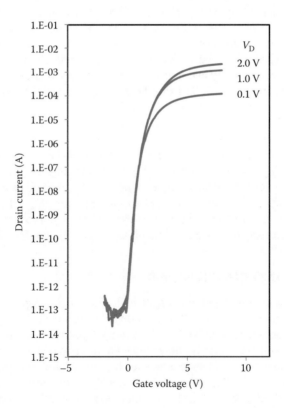

FIGURE 54.20 Transfer characteristics of the indium-free nc-ZnO TTFT. (B. Bayraktaroglu and K. Leedy, *2011 11th IEEE International Conference on Nanotechnology*, August 15–18, Portland, Oregon, pp. 1450–1455. © (2011) IEEE. With permission.)

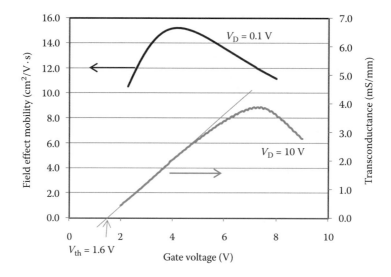

FIGURE 54.21   Field-effect mobility and transconductance of the indium-free nc-ZnO TTFT as a function of gate voltage. (B. Bayraktaroglu and K. Leedy, *2011 11th IEEE International Conference on Nanotechnology*, August 15–18, Portland, Oregon, pp. 1450–1455. © (2011) IEEE. With permission.)

saturation characteristics for this gate length dimension. The current density was lower than the non-transparent devices fabricated on Si substrates due to the thermal limitations of the glass substrate.

The transfer characteristics shown in Figure 54.20 exhibit an on/off ratio of 5×10^{10} and a sub-threshold voltage swing of $S < 200$ mV/decade. These values are comparable to the values obtained with nontransparent devices fabricated within the same temperature budget.

The field-effect electron mobility and transconductance values obtained from the transfer characteristics are shown in Figure 54.21. The threshold voltage, $V_{th} = 1.6$ V, was slightly higher than the value obtained with nontransparent devices (i.e., $V_{th} = 1.03$ V). On the other hand, the maximum field-effect mobility, 15 cm^2/V · s, was lower than the value obtained with nontransparent devices. The measured S value corresponds to an increase in the interface state density to 4.61×10^{12} cm^{-2}, which can explain the corresponding decrease in mobility. A higher interface state density can be caused by the ZnO–HfO$_2$ interface roughness [27] due to the rougher gate contact layer surfaces compared to the metal gates used in nontransparent devices. Improvements in the surface morphology of the doped nc-ZnO gate contact layer are expected to bring closer the performance metrics of the transparent and nontransparent transistors. Nevertheless, to our knowledge, the performance values reported here represent the highest values obtained with indium-free TTFTs.

54.5   SUMMARY AND CONCLUSIONS

PLD-grown ZnO thin films were used to fabricate high-speed and transparent thin film transistors. TEM and x-ray examination showed that films consisted of closely packed nanocolumns of ZnO with a predominantly (002) orientation and 20–50 nm diameter. Films maintained their ordered structure conformably over nonplanar surfaces. We have studied doped and undoped films grown on various gate insulators and at low substrate temperatures for high-speed and transparent thin film transistor applications. Excellent switching characteristics (on/off ratios greater than 10^{12}), subthreshold voltage swing ($S = 74$ mV/decade), and field-effect mobilities (63 cm^2/V · s) were obtained with low-temperature grown films. We attributed these superior performances of nc-ZnO TFTs to the high crystal quality preserved within nanocrystals and the presence of acceptor-like states at the vertical grain boundaries due to Zn-vacancies. High-speed performance was demonstrated by devices fabricated on Si substrates with the achievement of a current gain cut-off frequency of 2.9 GHz and a power gain cut-off frequency of 10 GHz. Doped and undoped nc-ZnO films were integrated on glass substrates using etch-stop layers to fabricate the first all-ZnO TTFTs with record performance. Indium-free TTFTs are important to address the cost and availability issues related to indium.

ACKNOWLEDGMENTS

This work was supported in part by AFOSR under LRIR No. 07SN03COR (Program Manager: Dr. Jim Hwang). The authors thank D. Agresta, J. Brown, and D. Tomich for optical transmission, AFM, and XRD analyses, respectively.

REFERENCES

1. C. R. Kagan and P. Andry, *Thin Film Transistors*, Marcel Dekker Publishing, New York, 2003.
2. K. Nomura, H. Ohta, A. Takagi, T. Kamiya, M. Hirano, and H. Hosono, Room temperature fabrication of transparent flexible thin-film transistors using amorphous oxide semiconductors, *Nature*, 432, 488–492, 2004.
3. T. Kamiya, K. Nomura and H. Hosono, Present status of amorphous In–Ga–Zn–O thin-film transistors, *Sci. Technol. Adv. Mater.*, 11, 044305, 2010.
4. D. A. Mourey, D. A. Zhao, J. Sun, and T. N. Jackson, Fast PEALD ZnO thin-film transistor circuits, *IEEE Trans. Elect. Dev.*, 57, 530–4, 2010.
5. H. Hosono, Ionic amorphous oxide semiconductors: Material design, carrier transport, and device application, *J. Non-crystalline Solids*, 352, 851–858, 2006.
6. D. Zhao, D. A. Mourey, and T. N. Jackson, Gamma-ray irradiation of ZnO thin film transistors and circuits, *2010 Device Research Conference Technical Digest*, pp. 241–242, 2010.
7. B. Bayraktaroglu, K. Leedy, Y. V. Li, D. Zhao, I. Ramirez, H. H. Fok, and T. N. Jackson, Radiation tolerance and recovery characteristics of ZnO thin film transistors, *GOMACTech-12 Conference Technical Digest*, pp. 193–196, 2012.
8. B. Bayraktaroglu, K. Leedy, and R. Neidhard, Nanocrystalline ZnO microwave thin film transistors, *Proc. SPIE*, 7679, 767904-1, 2010.
9. R. C. Scott, K. Leedy, B. Bayraktaroglu, D. C. Look, and Y-H. Zhang, Effects of Ar vs. O_2 ambient on pulsed-laser-deposited Ga-doped ZnO, *J. Cryst. Growth*, 324, 110–114, 2011.
10. L. Bentes, R. Ayouchi, C. Santos, R. Schwarz, P. Sanguino, O. Conde, M. Peres, T. Monteiro, and O. Teodoro, ZnO films grown by laser ablation with and without oxygen CVD, *Superlattices Microstruct.*, 42, 152–157, 2007.
11. S. Amirhaghi, V. Craciun, D. Craciun, J. Elders, and I. W. Boyd, Low temperature growth of highly transparent c-axis oriented ZnO thin films by pulsed laser deposition, *Microelectron. Eng.*, 25, 321–326, 1994.
12. L. Han, F. Mei, C. Liu, C. Pedro, and E. Alves, Comparison of ZnO thin films grown by pulsed laser deposition on sapphire and Si substrates, *Physica E*, 40, 699–704, 2008.
13. C-F. Yu, C-W. Sung, S-H. Chen, and S-J. Sun, Relationship between the photoluminescence and conductivity of undoped ZnO thin films grown with various oxygen pressures, *Appl. Surf. Sci.*, 256, 792–796, 2009.
14. B. Bayraktaroglu, K. Leedy, and R. Neidhard, ZnO thin film transistors for RF applications, *Mater. Res. Soc. Symp. Proc.*, 1201, H09–07, 2010.
15. A. Roland, J. Richard, J. P. Kleider, and D. Mencaraglia, Electrical properties of amorphous silicon transistors and MIS-devices: Comparative study of top nitride and bottom nitride configurations, *J. Electrochem. Soc.*, 140, 3679–3683, 1993.
16. B. Bayraktaroglu, K. Leedy, and R. Neidhard, Microwave ZnO thin-film transistors, *IEEE Electron Dev. Lett.*, 29, 1024–1026, 2008.
17. S.-J. Chang, M. Bawedin, B. Bayraktaroglu, J.-H. Lee, S. Cristoloveanu, Low–temperature properties of ZnO on insulator MOSFETs, *IEEE Int. SOI Conf.*, Tempe, AZ, 2011.
18. D. R. Clarke, Varistor Ceramics, *J. Am. Ceram. Soc.*, 82, 485–502, 1999.
19. J. Albrecht, P. P. Ruden, S. Limpijumnong, W. R. Lambrecht and K. F. Brennan, High field electron transport properties of bulk ZnO, *J. Appl. Phys.*, 86, 6864–6867, 1999.
20. B. Bayraktaroglu, K. Leedy, and R. Neidhard, High-frequency ZnO thin film transistors on Si substrates, *IEEE Electron Dev. Lett.*, 30, 946–948, 2009.
21. B. O'Neill, Indium market forces, a commercial perspective, *35th IEEE Photovoltaic Specialists Conference (PVSC) Proceedings*, DOI: 10.1109/PVSC.2010.5616842, 2010.

22. H. Agura, A. Suzuki, T. Matsushita, T. Aoki, and M. Okuda, Low resistivity transparent conducting Al-doped ZnO films prepared by pulsed laser deposition, *Thin Solid Films*, 445, 263–267, 2003.

23. B. Bayraktaroglu, K. Leedy and R. Bedford, High temperature stability of postgrowth annealed transparent and conductive ZnO:Al films, *Appl. Phys. Lett.*, 93, 022104, 2008.

24. T. Minami, S. Ida, and T. Miyata, High rate deposition of transparent conducting oxide thin films by vacuum arc plasma evaporation, *Thin Solid Films,* 416, 92–96, 2002.

25. R. C. Scott, K. Leedy, B. Bayraktaroglu, D. Look, D. Smith, D. Ding, X. Lu, and Y-H. Zhang, Influence of substrate temperature and post-deposition annealing on material properties of Ga-doped ZnO prepared by pulsed laser deposition, *J. Electronic Mat.*, 40, 419–428, 2010.

26. R. C. Scott, K. Leedy, B. Bayraktaroglu, D. Look, and Y-H. Zhang, Highly conductive ZnO grown by pulsed laser deposition in pure Ar, *Appl. Phys. Lett.*, 97, 072113, 2010.

27. K. Okamura and H. Hahn, Carrier transport in nanocrystalline field-effect transistors: Impact of interface roughness and geometrical carrier trap, *Appl. Phys. Lett.*, 97, 153114, 2010.

28. B. Bayraktaroglu and K. Leedy, Ordered nanocrystalline ZnO films for high speed and transparent thin film transistors, *2011 11th IEEE International Conference on Nanotechnology*, August 15–18, Portland, Oregon, pp. 1450–1455.

55 First-Principle Study of Energy-Band Control by Cross-Sectional Morphology in [110]-Si Nanowires

Shinya Kyogoku, Jun-Ichi Iwata, and Atsushi Oshiyama

CONTENTS

55.1 INTRODUCTION

Si nanowire field-effect transistors (SiNW FETs) are expected to be boosters in postscaling semiconductor technology. Bangsaruntip et al. have indeed achieved excellent control of SiNW sizes utilizing conventional complementary metal oxide semiconductor (CMOS) fabrication processes and then observed clear scaling of short-channel effects versus NW sizes with the fixed gate length [1]. Reducing the dimensions of SiNWs is shown to improve the short-channel control. They have fabricated the [110]-SiNW FETs with channel dimensions of 5.0 nm × 6.3 nm and even smaller, 2.0 nm × 3.3 nm. In such small dimensions, quantum confinement becomes prominent. In addition, SiNW FETs with various sizes and shapes, including circular, elliptic, and rectangular, have been fabricated [2–6]. The quantum effects depend on crystallographic directions of the SiNW axes and also on the cross-sectional morphology of the SiNWs, resulting in substantial modification of the energy-band structures and the transport characteristics of SiNW FETs.

Hydrogenated SiNWs that are formed by treatments of pristine or oxidized SiNWs with hydrogen fluorides offer a stage where salient properties of SiNWs are investigated. Electronic structures of such hydrogenated SiNWs have been indeed studied by density functional theory (DFT) [7–13], although the NWs considered so far are much thinner than those experimentally synthesized. Ng et al. and Leu et al. have investigated band structures of SiNWs oriented along the [100], [110], [111], and [112] directions and shown variation of the bandgaps as a function of the diameter [9,10]. Ng et al. have further explored the effects of cross-sectional shapes on bandgaps in [110]-SiNWs with a diameter of about 1 nm and argued that the effects are minor [9]. Other DFT calculations provide different results, however. Singh et al. and Sorokin et al. have found that energy-band structures of hydrogenated SiNWs with diameters less than 2.0 nm depend strongly on the cross-sectional morphology [11,12]. Even though the SiNWs are oriented along the same direction and the dimensions of the wires are close to each other, they have direct or indirect gaps depending on the morphology. Hence, at the present stage, the relation between the cross-sectional morphology and the electronic structure is under debate even in thin NWs. The relation in the thicker SiNWs experimentally available has not been discussed in the past. Clarifying the underlying physics that is decisive to determine the electronic structure is in high demand.

Furthermore, Scheel et al. have calculated the density of states (DOS) of [110]-SiNW [13], and found that the conduction-band minimum (CBM) at the zone center Γ contributes only small DOS, whereas the second minimum that is located at some point between the Γ and the zone boundary X provides a peak of DOS peculiar to a one-dimensional system. This aspect is extremely important to discuss the characteristics of SiNW FETs.

Hence, it is imperative to clarify the electronic structures of SiNWs, focusing not only on the bandgap but also on the relative locations of several energy minima and its morphology dependence. We have performed extensive electronic structure calculations based on the DFT for [110]- and [100]-SiNWs with various cross-sectional shapes.

On the other hand, sidewalls of the SiNWs calculated by the earlier-cited papers are of ideal structures without roughness, namely, there are few calculations based on DFT about a local cross-sectional morphology. In the experiment, however, SiNWs have subnanometer roughness by oxidation or etching in fabrication processes [1]. Therefore, we have also investigated the effects of sidewall roughness by changing amounts of the roughness and a pattern of the roughness.

We have found that [110]-SINWs have lighter electron effective masses compared to [100]-SiNWs. In this chapter, restricting ourselves to the [110]-SiNWs, we report on DFT calculations that clarify the relationship between structural morphology and energy bands.

55.2 METHOD

Calculations have been performed in the local density approximation (LDA) in DFT [14,15]. Norm-conserving pseudopotentials generated by using the Troullier–Martins scheme are adopted to describe the electron–ion interaction [16,17]. We use our newly developed real-space finite-difference scheme [18] that allows us to perform LDA calculations for systems with tens of thousands of atoms on a few tens-of-Tera FLOPS computers. The grid spacing in the real-space calculations is taken to be 0.034 nm, corresponding to the cutoff energy of 260 eV, which is calculated by

$$E_{\text{cut}} = \left(\frac{\pi}{d}\right)^2, \tag{55.1}$$

where d is the grid spacing in real space. The sixth-order finite difference is adopted for the kinetic energy operator. We use a supercell model in which [110]-SiNWs are arranged periodically. To avoid interactions among the SiNWs, they are separated from each other by at least 10+. We take five k points along the [110] direction in the Brillouin zone (BZ) integration. These calculational parameters assure the accuracy required in this work. Sidewalls of the wires are terminated by hydrogen atoms, reflecting certain experimental situations on the one hand and focusing on essential features of the energy bands on the other. The starting geometry for each SiNW is constructed by cleaving the bulk Si. Geometry optimization has been performed until the maximum force becomes less than 0.5 eV/nm.

The sidewall roughness of SiNWs is introduced as a random fluctuation of the radius of the nanowire around an average radius. In cylindrical coordinates, the radius of the rough wire is expressed as follows:

$$R(z,\theta) = R_0 + \Delta(z,\theta), \tag{55.2}$$

where R_0 is the average radius. The distribution of the random fluctuation is characterized by assuming the following autocorrelation function [19,20]:

$$(\Delta(r)\Delta(r + r')) = \Delta_m^2 e^{-r'/L_m} \tag{55.3}$$

where Δ_m is the root mean square of the radius fluctuation and L_m is the correlation length. In our calculation, we examined four values from 0.2 to 0.8 nm, which seem to be realistic. In addition, we examined three-pattern SiNWs with sidewall roughness in 0.4 nm of the root mean square. We set 0.54 and 4.3 nm for L_m and R_0, respectively.

55.3 RESULTS AND DISCUSSION

55.3.1 EFFECTS OF CROSS-SECTIONAL SHAPE

We have explored various [110]-SiNWs with the dimensions ranging from 3 to 6 nm. Figure 55.1 shows some examples of cross-sectional views of obtained geometry-optimized [110]-SiNWs, representing typical cross-sectional shapes. The dimensions of these wires are close to each other as shown. Both Figure 55.1b and d show the elliptic cross-sectional shapes, but their extended directions are not equivalent to each other. We call the ellipses that have the long axes along the [001] and the [1$\bar{1}$0] directions and the [001]-ellipse and the [1$\bar{1}$0]-ellipse, respectively. Similarly, the dumbbells that are shown in Figure 55.1c and e are called the [001]-dumbbell and [1$\bar{1}$0]-dumbbell.

Figure 55.2 shows the calculated conduction-band structure near the gap for each SiNW shown in Figure 55.1. These band structures qualitatively agree with the results in the past for the thinner [110]-SiNWs [9,12,13]. Table 55.1 shows the calculated bandgaps and effective masses. In all cases, the conduction-band minima are located at the zone center Γ, whereas the second minima are located at approximately 85% from the Γ to the X, which we call Δ_0. The energy at each minimum is shifted from the bulk conduction-band bottom by E_{shift} due to the quantum confinement. In Figure 55.2, the difference ΔE_{shift} between the first and the second minima is also shown. The Γ-point minimum originates from the two valleys along [00 ± 1] of the conduction bands in bulk Si, whereas the second minimum comes from another two valleys along the [010] and the [100] directions, respectively. Hence, all the minima are doubly degenerate.

This situation is clearly recognized in Figure 55.3. In the plane perpendicular to the [110] direction, two primitive vectors, a_1 and a_2, are introduced. Together with the primitive vector along the [110] direction, a unit cell is defined and the corresponding BZ is shown in Figure 55.3b. The six conduction-band valleys in an original BZ corresponding to the diamond structure are shown in this new BZ. The [00 ± 1] valleys are marked in dark gray, whereas the [±100] and [0 ± 10] are marked in light gray. In Figure 55.3c and d, we show the energy contours of the conduction band near the valleys on the planes perpendicular to the wire axis. We indeed observe the corresponding conduction-band

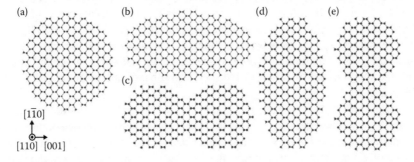

FIGURE 55.1 Cross-sectional views of various [110]-SiNWs. Large and small balls represent Si and H atoms, respectively: (a) the circle whose diameter is 4.3 nm, (b) the [001]-ellipse, (c) the [001]-dumbbell, (d) the [1$\bar{1}$0]-ellipse, and (e) the [1$\bar{1}$0]-dumbbell. (S. Kyogoku et al., First-principle study of energy-band control by cross-sectional morphology in [110]-Si nanowires. *2011 11th IEEE Conference on Nanotechnology (IEEE-NANO)*, Portland, OR, pp. 1322–1326, © (2011) IEEE. With permission.)

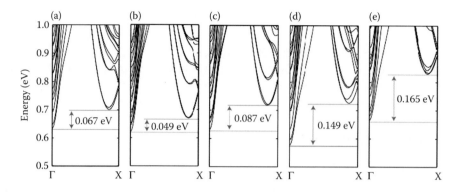

FIGURE 55.2 Conduction-band structures of various SiNWs shown in Figure 55.1: (a) the circle, (b) the [001]-ellipse, (c) the [001]-dumbbell, (d) the [1 $\overline{1}$0]-ellipse, and (e) the [1 $\overline{1}$0]-dumbbell. The values of ΔE_{shift} are also shown. The origin of the energy is at the valence-band top. The bandgap of the bulk Si is 0.49 eV in the present LDA calculations. (S. Kyogoku et al., First-principle study of energy-band control by cross-sectional morphology in [110]-Si nanowires. *2011 11th IEEE Conference on Nanotechnology (IEEE-NANO)*, Portland, OR, pp. 1322–1326, © (2011) IEEE. With permission.)

TABLE 55.1

Calculated Bandgap in Increases and Effective Masses of [110] SiNWs

	Bandgap (eV)	m_Γ^* (m_e)	$m_{X'}^*$ (m_e)
Circle	0.633	0.16	0.60
[001]-dumbbell	0.621	0.15	0.63
[001l-ellipse	0.632	0.14	0.61
[1 $\overline{1}$0]-ellipse	0.578	0.15	0.63
[1 $\overline{1}$0]-dumbbell	0.666	0.17	0.59

valleys on these contour lines. The energy bands become minimum at the wave number projected on the [110] direction of each valley, thus leading to the behavior shown in Figure 55.2.

Our results clearly show that the bandgap, the effective mass, and the ΔE_{shift} depend on the cross-sectional shapes. The bandgap of the [1 $\overline{1}$0]-ellipse is the narrowest, and that of the [1 $\overline{1}$0]-dumbbell is the widest among all the cross-sectional shapes considered here. The difference between the two is 88 meV. All the effective masses at Γ, m_Γ^*, are smaller than the transverse (lighter) mass of the Si bulk, $m_t = 0.19$. This modification of the effective mass from the bulk value can be explained in terms of the nonparabolicity and anisotropy near the conduction-band valleys of Si [21].

The effective mass at Δ_0, $m_{\Delta 0}^*$, is between the longitudinal mass m_l (=0.95) and m_t for any cross-sectional shaped SiNW. This finding is a consequence of the wire-axis direction being obliquely crossed with the longitudinal direction of the conduction-band valleys along [100] and [010] as shown in Figure 55.3b.

As for the ΔE_{shift}, the [001]-ellipse and the dumbbell provide smaller values than the [1 $\overline{1}$0]-ellipse and the dumbbell. In addition, the ΔE_{shift} of the [001]-ellipse is smaller than that of the circle. Energy shifts of the dumbbell-shaped NWs are larger than those of the elliptic-shaped NWs, irrespective of the extended direction. This is attributed to the fact that the standing waves in the perpendicular plane in the dumbbell-shaped NWs have the maximum amplitude at two points even for the lowest-energy state. Our results unequivocally show that the [001]-ellipse SiNW has benefits in terms of the number of channels under the ballistic regime of FETs. The difference between the largest and the smallest ΔE_{shift} is 116 meV, which is larger than the difference of the bandgap, indicative of the importance of cross-sectional morphology in the characteristics of NW-FETs.

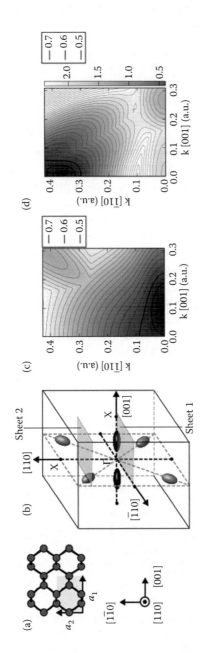

FIGURE 55.3 (a) Two primitive vectors a_1 and a_2 in the plane perpendicular to the wire-axis direction; (b) corresponding first Brillouin zone and the six conduction-band valleys. The planes named by sheet 1 and sheet 2 cut across the center of each conduction-band valley. Energy contour plots on the planes named by sheet 1 and sheet 2 are shown in (c) and (d), respectively, in eV units. (S. Kyogoku et al., First-principle study of energy-band control by cross-sectional morphology in [110]-Si nanowires. *2011 11th IEEE Conference on Nanotechnology (IEEE-NANO)*, Portland, OR, pp. 1322–1326, © (2011) IEEE. With permission.)

55.3.2 Origins of Dependence on Cross-Sectional Shapes

Our findings explained above come from the fact that the wave numbers in the plane perpendicular to the wire axis become discrete, depending on the cross-sectional shapes, and that the energy minima along the wire axis at these discrete wave numbers are different from each other. To elucidate this point, we consider an SiNW that has a rectangular cross section with dimensions of $L_1 = n_1|a_1| \times L_2 = n_2|a_2|$. Then the wave numbers in the first BZ allowed in the perpendicular plane become

$$k_1 = \frac{\pi}{n_1|a_1|}, \frac{2\pi}{n_1|a_1|}, \cdots \frac{\pi}{|a_1|},$$

$$k_2 = \frac{\pi}{n_2|a_2|}, \frac{2\pi}{n_2|a_2|}, \cdots \frac{\pi}{|a_2|}. \tag{55.4}$$

To mimic the circle- and the ellipse-shaped NWs examined earlier, we consider the three-type rectangular SiNWs: (1) a rectangular shape extended along [1$\bar{1}$0], (2) a nearly square shape, and (3) a rectangular shape extended along [001]. We call them the [1$\bar{1}$0]-rectangular, the square, and the [001]-rectangular, respectively. Figure 55.4 shows the allowed wave numbers on the perpendicular plane for the three rectangular-shaped NWs introduced earlier. Some of the energy contours shown by thick lines in Figure 55.3c and d are reproduced in Figure 55.4.

In the [1$\bar{1}$0]-rectangular NW (Figure 55.4a), the confinement along the [001] direction is stronger than that along the [1$\bar{1}$0] direction. Hence, the discrete wave numbers are sparsely distributed along the [001] direction compared with the [1$\bar{1}$0] direction. Consequently, the allowed wave vector marked by the square is close to CBM in the [00 ± 1] valley, whereas the allowed wave vector marked by the circle is less close to the CBM in the [100] and [010] valleys. The former corresponds to the Γ-point minimum and the latter to the Δ_0-point minimum in the dispersion along [110]. Therefore, we obtain relatively large ΔE_{shift} in this case.

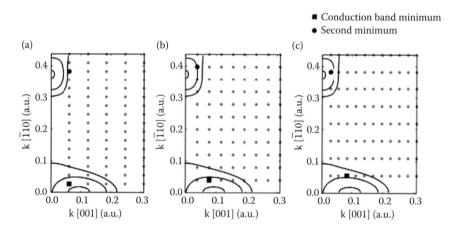

FIGURE 55.4 Allowed wave numbers on the plane perpendicular to the [110] wire axes for the three rectangular NWs. The side lengths of the three rectangular NWs are (a) 5.3 nm × 3.2 nm, (b) 3.8 nm × 4.3 nm, and (c) 2.3 nm × 7.0 nm. Gray diamonds and black squares and circles show the allowed wave numbers. Energy contours near the conduction-band minima shown by thick lines in Figure 55.3c and d are also shown. The allowed wave numbers that are the closest to the conduction-band minimum along [001] and along [1$\bar{1}$0] are marked by a black square and a black circle, respectively. Values of k are in units of the corresponding BZ. (S. Kyogoku et al., First-principle study of energy-band control by cross-sectional morphology in [110]-Si nanowires. *2011 11th IEEE Conference on Nanotechnology (IEEE-NANO)*, Portland, OR, pp. 1322–1326, © (2011) IEEE. With permission.)

On the contrary, in the [001]-rectangular NW (Figure 55.4c), the confinement along the [1$\overline{1}$0] direction is stronger than that along the [001] direction. Hence, the discrete wave numbers are densely distributed along the [001] direction compared with the [1$\overline{1}$0] direction. Consequently, the allowed wave vector marked by the square is far from CBM in the [00 ± 1] valley, whereas the allowed wave vector marked by the circle is close to the CBM in the [100] and [010] valleys. Therefore, we obtain relatively small ΔE_{shift} in this case.

These results obtained in the rectangular-shaped NW models are consistent with the calculated ΔE_{shift} for real SiNWs shown in Figure 55.2. We thus argue that the cross-sectional shape is an important factor in determining ΔE_{shift} and the properties of SiNW FETs.

55.3.3 Effects of Sidewall Roughness

We showed the results of ideal-shaped NWs explained in Section 55.3.1. In the experiment, however, SiNWs have a subnanometer roughness by the oxidation or etching in fabrication processes [1], and it is difficult to eliminate roughness in the sidewall of SiNWs in their fabrication processes. Therefore, it is important to clarify the effects of sidewall roughness. These effects can be regarded as the effects of the cross-sectional morphology. If we call the effects explained in Section 55.3.1 as the global-shape effects, we can call these effects by the sidewall roughness effects as the local-shape effects.

To clarify these effects of sidewall roughness on DOS, we have explored [110]-SiNWs with sidewall roughness. The roughness is introduced as a radius fluctuation ranging from 0.2 to 0.8 nm. To examine the pure effects of the amounts of roughness, we calculate these SiNWs with sidewall roughness generated by a single set of random numbers in Equation 55.3. Figure 55.5 shows the cross-sectional views and side views from [111] of various [110]-SiNWs with sidewall roughness. All the averaged diameters are 4.3 nm. Moreover, to examine the effects of a roughness pattern, we

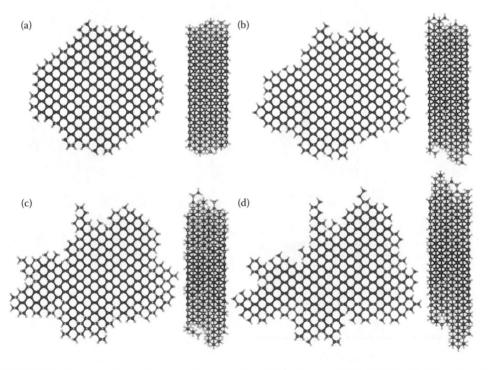

FIGURE 55.5 Cross-sectional views and side views from [111] of various [110]-SiNWs with sidewall roughness. The root mean squares of the radius fluctuation, Δ_m, are (a) 0.2 nm, (b) 0.4 nm (pattern 1), (c) 0.6 nm, and (d) 0.8 nm.

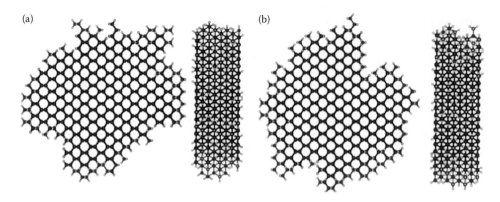

FIGURE 55.6 Cross-sectional views and side views from [111] of two-pattern rough SiNWs. The root mean square of the radius fluctuation, Δ_m, is 0.4 nm. (a) Pattern 2 and (b) pattern 3.

calculate other two-pattern rough SiNWs that are generated by different sets of random numbers in Equation 55.3. Figure 55.6 shows the cross-sectional views and side views from [111] of such [110]-SiNWs with sidewall roughness. Indeed, these cross-sectional shapes are different from those shown in Figure 55.5b. All the averaged diameters are also 4.3 nm and the root mean squares of the radius fluctuations, Δ_m, are 0.4 nm in this case. As in Section 55.3.1, we have focused not only on the bandgap itself but also on the relative locations of several energy minima and DOS near the CBM.

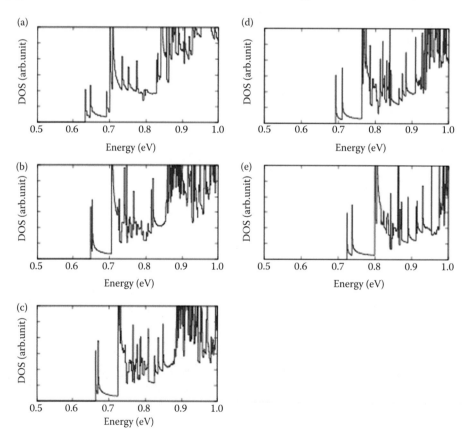

FIGURE 55.7 Densities of states with Δ_m of (a) 0.0 nm, which is shown in Figure 55.1a, (b) 0.2 nm, (c) 0.4 nm, (d) 0.6 nm, and (e) 0.8 nm.

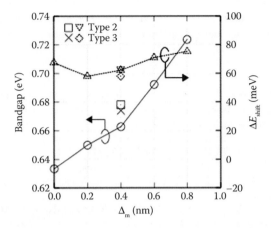

FIGURE 55.8 The bandgap and ΔE_{shift} versus the amounts of roughness. The values of three patterns of roughness in 0.4 nm are also shown.

Figure 55.7 shows DOS of the SiNWs with sidewall roughness, along with the ideal SiNWs. Shapes of the DOS show a little change with or without the roughness. In addition, this tendency does not change by the amount of roughness. However, the energy shift of each peak increases with an increase in the roughness. Figure 55.8 shows variations of a bandgap and a ΔE_{shift} of SiNWs with sidewall roughness. We found that the bandgap increases with an increase in the amount of roughness. This is attributed to the fact that the wave function confinement is narrower than that in those ideal structures due to sidewall roughness. In contrast, there is little change in the ΔE_{shift} versus the amount of roughness. This indicates that the ΔE_{shift} is unaffected by change of a local cross-sectional shape near the sidewall. According to the results of pattern 2 and pattern 3 in 0.4 nm of the root mean fluctuation, we can find that these behaviors are unchanged by the pattern of sidewall roughness. The reason why ΔE_{shift} is unchanged is given by the discussion in Section 55.3.2. The allowed discrete wave numbers only depend on the global cross-sectional shapes of the SiNWs. Hence, ΔE_{shift} does not depend on the local cross-sectional shape of the SiNWs.

55.4 CONCLUSIONS

We have performed extensive electronic structure calculations based on the LDA in DFT for [110]-SiNWs with various cross-sectional shapes with the diameter ranging from 3 to 6 nm. We have found that the quantum effects substantially modify the energy bands, and that the global cross-sectional shapes play important roles in the modification. We have found that the [110]-SiNWs with the elliptic cross section extending in the [001] direction has the largest number of channels within 50 meV from the conduction-band bottom, thus being a good candidate for FET applications.

We have analyzed our LDA results by using simple rectangular-shaped NW structures, and found that the substantial effects of the cross-sectional morphology come from the particular arrangement of the allowed wave numbers in the lateral BZ, and that the particular arrangement is peculiar to each cross-sectional shape. The relative positions of the allowed wave numbers to the conduction-band bottom are crucial in determining the number of channels. Moreover, we have performed the DOS of SiNWs with sidewall roughness with an average diameter of 4.3 nm. We have found that sidewall roughness does not modify the shape of DOS, but increases the energy shift. Therefore, the bandgap increases as the amounts of roughness increase. However, we have found that the relative locations of several energy minima near the CBM do not depend on the amounts and the pattern of sidewall roughness. In conclusion, we have clarified the significance of the global-shape effects and the insignificance of the local-shape effects that imply the sidewall roughness for the number of channels near the CBM. The knowledge obtained here is applicable to other crystallographic directions, opening a possibility to the design and fabrication of optimum SiNWs for FETs.

REFERENCES

1. S. Bangsaruntip, G. M. Cohen, A. Majumdar, Y. Zhang, S. U. Engelmann, N. C. M. Fuller, L. M. Gignac et al., High performance and highly uniform gate-all-around silicon nanowire MOSFETs with wire size dependent scaling, in *IEDM Tech. Dig.*, 297, 2009.
2. S. D. Suk, K. H. Yeo, K. H. Cho, M. Li, Y. Y. Yeoh, K.-H. Hong, S.-H. Kim et al., Investigation of nanowire size dependency on TSNWFET, in *IEDM Tech. Dig.*, 717, 2005.
3. K. H. Yeo, S. D. Suk, M. Li, Y. Y. Yeoh, K. H. Cho, K.-H. Hong, S.-K. Yun et al., Gate-all-around (GAA) twin silicon nanowire MOSFET (TSNWFET) with 15 nm length gate and 4 nm radius nanowires, in *IEDM Tech. Dig.*, 539, 2006.
4. M. Li, K. H. Yeo, S. D. Suk, Y. Y. Yeoh, D.-W. Kim, T. Y. Chung, K. S. Oh, and W.-S. Lee, Sub-10 nm gate-all-around CMOS nanowire transistors on bulk Si substrate, in *VLSI Tech. Dig.*, 123, 2009.
5. H. S. Wong, L.-H. Tan, L. Chan, G.-Q. Lo, G. Samudra, and Y.-C. Yeo, Gate-all-around uantum-wire field-effect transistor with Dopant Segregation at Metal-Semiconductor-Metal heterostucture, in *VLSI Tech. Dig.*, 92, 2009.
6. C. Dupré, A. Hubert, S. Becu, M. Jublot, V. Maffini-Alvaro, C. Vizioz, F. Aussenac et al., 15 nm-diameter 3D stacked nanowires with independent gates operation: ΦFET, in *IEDM Tech. Dig.*, 749, 2008.
7. X. Zhao, C. M. Wei, L. Yang, and M. Y. Chou, Quantum Confinement and Electronic Properties of Silicon Nanowires, *Phys. Rev. Lett.* 92, 236805, 2004.
8. T. Vo, A. J. Williamson, and G. Galli, First principles simulations of the structural and electronic properties of silicon nanowires, *Phys. Rev. B* 74, 045116, 2006.
9. M. -F. Ng, L. Zhou, S. -W. Yang, L. Y. Sim, V. B. C. Tan, and P. Wu, Theoretical investigation of silicon nanowires: Methodology, geometry, surface modification, and electrical conductivity using a multiscale approach, *Phys. Rev. B* 76, 155435, 2007.
10. P. W. Leu, B. Shan, and K. Cho, Surface chemical control of the electronic structure of silicon nanowires: Density functional calculations, *Phys. Rev. B* 73, 195320, 2006.
11. A. K. Singh, V. Kumar, R. Note, and Y. Kawazoe, Effects of Morphology and Doping on the Electronic and Structural Properties of Hydrogenated Silicon Nanowires, *Nano Lett.* 6, 920, 2006.
12. P. B. Sorokin, P. V. Avramov, A. G. Kvashnin, D. G. Kvashnin, S. G. Ovchinnikov, and A. S. Fedorov, Density functional study of 110 -oriented thin silicon nanowires, *Phys. Rev. B* 77, 235417, 2008.
13. H. Scheel, S. Reich, and C. Thomsen, Electronic band structure of high-index silicon nanowires, Phys. Status Solidi B, 242, 2474, 2005.
14. J. P. Perdew and A. Zunger, Self-interaction correction to density-functional approximations for many-electron systems, *Phys. Rev. B* 23, 5048, 1981.
15. P. Hohenberg and W. Kohn, Inhomogeneous Electron Gas, *Phys. Rev.* 136, B864, 1964; W. Kohn and L. J. Sham, ibid. 140, A1133, 1965.
16. N. Troullier and J. L. Martins, Efficient pseudopotentials for plane-wave calculations, *Phys. Rev. B* 43, 1993, 1991.
17. L. Kleinman and D. M. Bylander, Efficacious Form for Model Pseudopotentials, *Phys. Rev. Lett.* 48, 1425, 1982.
18. J.-I. Iwata, D. Takahashi, A. Oshiyama, B. Boku, K.Shiraishi, S. Okada, and K. Yabana, J., A massively-parallel electronic-structure calculations based on real-space density functional theory, *Comput. Phys.* 229, 2339 (2010).
19. J. Wang, E. Polizzi, A. Ghosh, S. Datta, and M. Lundstorm, Theoretical investigation of surface roughness scattering in silicon nanowire transistors, *Appl. Phys. Lett.* 87, 043101, 2005.
20. A. Lherbier, M.P. Persson, Y.-M. Niquet, F. Triozon, and S. Roche, Quantum transport length scales in silicon-based semiconducting nanowires: Surface roughness effects, *Phys. Rev. B* 77, 085301, 2008.
21. N. Neophytou, A. Paul, M. S. Lundstrom, and G. Klimeck, Bandstructure Effects in Silicon Nanowire Electron Transport, *IEEE Trans. Electron Devices* 55, 1286, 2008.
22. S. Kyogoku, J.-I. Iwata, and A. Oshiyama, First-principle study of energy-band control by cross-sectional morphology in [110]-Si nanowires, *2011 11th IEEE Conference on Nanotechnology (IEEE-NANO)*, Portland, OR, pp. 1322–1326, 2011.

56 Interplay of Self-Heating and Short-Range Coulomb Interactions due to Traps in a 10 nm Channel Length Nanowire Transistor

Arif Hossain, Dragica Vasileska, Katerina Raleva, and Stephen M. Goodnick

CONTENTS

56.1 INTRODUCTION

Random telegraph noise fluctuations (RTF) manifest themselves as fluctuations in the transistor threshold voltage and drive (ON) current. RTF is caused by random trapping and detrapping of charges lying at the inversion channel of the device close to the oxide–semiconductor interface [1]. Traditionally, RTF were important only in analog design at low frequencies [2]. However, as complementary metal-oxide-semiconductor (CMOS) is scaling into the sub-100 nm regime, the effect of RTF as well as its variability is no longer negligible, even in digital design [3]. In fact, we have illustrated in past work that the presence of a single trap at the source end of the channel in a nanowire transistor can significantly degrade the on current [4]. In these simulations, we have utilized a 3D Monte Carlo device simulator in which the short-range portion of the Coulomb interaction was accounted for by a real-space molecular dynamics (MD) model, the details of which can be found in Ref. [5]. The model accounts for both the short-range and the long-range components of the Coulomb interaction and has been applied in many other studies [6].

The purpose of this chapter is to present the results of our current investigations of the influence of the negatively charged trap/impurity on the magnitude of the current for the case when, in addition to the short-range Coulomb interactions, the self-heating effects are incorporated in the theoretical model.

56.2 SIMULATION RESULTS

The nanowire field effect transistor (FET) simulated in this work has a gate oxide that is 0.8 nm thick and a BOX that is 10 nm thick. The dimensions of the silicon nanowire are 10 nm channel length, 7 nm channel thickness, and 10 nm channel width. For the thermal conductivity that

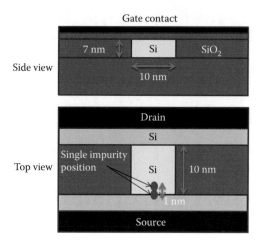

FIGURE 56.1 Schematic of the simulated nanowire MOSFET device, where the position of the trap/impurity atom is indicated.

appears in the acoustic phonon temperature energy balance solvers, we have taken the value of 13 W/m/K from measurements by Li Shi [7] that approximately correspond to a wire with a cross section of 7×10 nm^2. A schematic of the device structure, in which we also indicate the position of the trap, is shown in Figure 56.1.

The incorporation of self-heating effects in the existing model is achieved by self-consistently solving a multiband 3D Monte Carlo/MD coupled to a 3D Poisson equation solver, which is then self-consistently coupled with 3D energy balance equations for the acoustic and optical phonons [8–10]. A schematic of the complete program flowchart is given in Figure 56.2. Briefly, after specifying the structure and relevant input parameters, an equilibrium 3D Poisson equation is solved

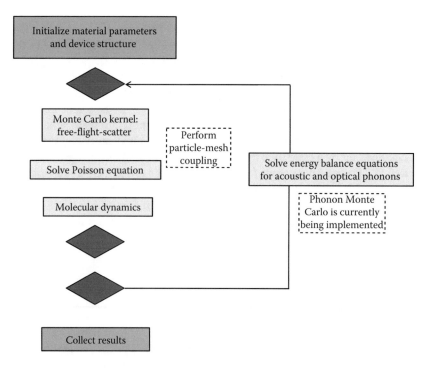

FIGURE 56.2 Flowchart of the Monte Carlo–molecular dynamics–3D Poisson–3D thermal solver.

given the applied gate bias. Then a charge assignment is performed and the Monte Carlo free-flight scatter procedure is initiated. During the carrier free-flights, carriers are accelerated according to the electric fields that result from the solution of the 3D Poisson equation (long-range portion of the Coulomb interaction) + electric fields that arise from particle–particle–particle–mesh coupling (short-range portion of the Coulomb interaction). The ensemble Monte Carlo (EMC)–Poisson–MD routine is run for 10 ps, after which the energy balance equations are solved for the optical and the acoustic phonon bath temperature profiles. It is necessary to incorporate separately the acoustic and the optical phonon baths due to the different timescales involved in the energy transfer processes involved, as schematically depicted in Figure 56.3. The information for the local acoustic and optical phonon temperatures is used in the Monte Carlo free-flight scatter routine, in the choice of the scattering mechanism (scattering tables). When the EMC–MD–Poisson sequence is complete, the information about the local electron density, electron temperature, and the electron drift velocity is inputted into the energy balance solver to calculate the corresponding lattice temperature and optical phonon temperature profiles.

When solving the energy balance equations for the acoustic and optical phonons, boundary conditions on the lattice temperature must be established. Recalling that there is an analogy between the electrical and thermal variables, from Ohm's law for electrical conduction and Fourier's law for heat conduction, one immediately sees that electrostatic potential is analogous to lattice temperature and electrical current is analogous to heat flux. We know that when solving the Poisson equation for electrostatic potential, one has to define at least one node on the Poisson mesh with Dirichlet boundary conditions to connect to the outside world. Hence, in the lattice temperature mesh, at least one node has to have Dirichlet boundary conditions. In all the simulations presented in this chapter, the bottom (substrate) electrode is taken to be at lattice temperature $T = 300$ K and the top gate is also assumed to be at temperature $T = 300$ K. For all the other boundaries, Neumann boundary conditions are assumed.

In Table 56.1, we show simulation results for the case when a negative trap is placed right at the source end of the channel or 1 nm inside the channel toward the gate. From the results presented in Table 56.1, it is evident that larger current degradation is observed for the case when the trap is located 1 nm into the channel, namely, when the trap is at the source end of the channel, the current degradation is 0.47% and when the trap is located 1 nm in the channel, the current degradation is 2.12%, a factor of five higher. The explanation to this behavior is as follows: when the trap is located at the source end of the channel, screening from the source electrons is very effective in reducing the strength of the trap's Coulomb potential. Therefore, the current that flows through the structure is reduced by a negligible amount. More current means more self-heating effects, which can be observed from the lattice and optical temperature plots shown in Figure 56.4. In other

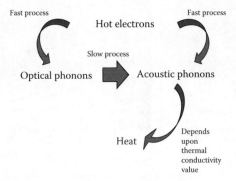

FIGURE 56.3 Most likely energy loss paths for energetic particles in a semiconductor device. (D. Vasileska, A. Hossain, and S. M. Goodnick, Self-heating and short-range coulomb interactions due to charging of traps in nanowire transistors, in *Proceedings of the 2011 IEEE Nano Conference on Nanotechnology*, Portland, Oregon, pp. 1110–1113. © (2011) IEEE. With permission.)

TABLE 56.1

Single Negative Impurity Impact on Current Degradation

Gummel Cycle	No Impurity Case (μA/μm)	Source Edge Case (μA/μm)	1 nm Toward Drain (μA/μm)
1	4154	4122	4075
3	4068	4030	3974
5	4052	4035	3968
Degradation	N/A	0.47%	2.12%

Screening of the source charges
reduce the impact of the negative trap

Maximal impact

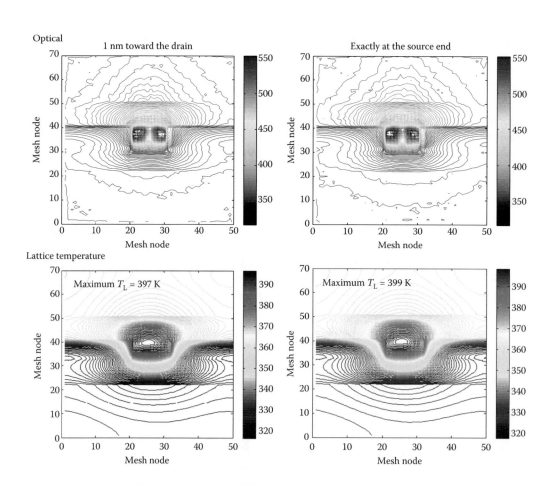

FIGURE 56.4 Optical (top panel) and lattice (bottom panel) temperature profiles for the case of a negative impurity placed 1 nm into the channel (left panels) and a negative impurity placed at the source edge of the channel (right panels). Applied bias is $V_{GS} = V_{DS} = 1.2$ V.

words, maximum lattice temperature is 399 K when the trap is located right at the source end and is reduced to 397 K when the trap is moved 1 nm into the channel.

56.3 CONCLUSIONS

In summary, we have presented simulation results for the degradation of the ON current in the presence of a negatively charged impurity placed in the middle of the source end injection barrier. We find that the interplay of self-heating and Coulomb effects is very complicated in nanowire transistors and both have to be accounted for to get proper estimates of the ON current.

REFERENCES

1. M. J. Kirton and M. J. Uren, Noise in solid-state microstructures: A new perspective on individual defects, interface states and low frequency (1/f) noise, *Advances in Physics*, 38, 367–468, 1989.
2. K. S. Ralls, W. J. Skocpol, L. D. Jackel, R. E. Howard, L. A. Fetter, R.W. Epworth, and D. M. Tennat, Discrete resistance switching in submicrometer silicon inversion layers: Individual interface traps and low-frequency (1/f) noise, *Physical Review Letters*, 52(3), 228–231, 1984.
3. H. Kurata, K. Otsuga, A. Kotabe, S. Kajiyama, T. Osabe, Y. Sasago, S. Narumi, K. Tokami, S. Kamohara, and O. Tsuchiya, The impact of random telegraph signals on the scaling of multilevel flash memories, in *Proceedings of the Symposium on VLSI Circuits*, 125–126, 2006.
4. D. Vasileska and S. S. Ahmed, Narrow-width SOI devices: The role of quantum mechanical size quantization effect and the unintentional doping on the device operation, *IEEE Transactions on Electron Devices*, 52, 227236, 2005.
5. W. J. Gross, D. Vasileska, and D. K. Ferry, A novel approach for introducing the electron-electron and electron-impurity interactions in particle-based simulations, *IEEE Electron Device Letters*, 20, 463–465, 1999.
6. Z. Aksamija and I. Knezevic, Anisotropy and boundary scattering in the lattice thermal conductivity of silicon nanomembranes, *Physical Review B*, 82, 045319, 2010.
7. D. Li, Y. Wu, P. Kim, L. Shi, P. Yang, and A. Majumdar, Thermal conductivity of individual silicon nanowires, *Applied Physics Letters*, 83, 2934–2936, 2003.
8. D. Vasileska, K. Raleva, and S. M. Goodnick, Self-heating effects in nano-scale FD SOI devices: The role of the substrate, boundary conditions at various interfaces and the dielectric material type for the BOX, *IEEE Transactions on Electron Devices*, 56(12), 3064–3071, 2009.
9. D. Vasileska, K. Raleva, and S. M. Goodnick, Modeling heating effects in nanoscale devices: The present and the future, *Journal of Computational Electronics*, 7(2), 66–93, 2008.
10. D. Vasileska, A. Hossain, and S. M. Goodnick, The role of the source and drain contacts on self-heating effect in nanowire transistors, *ECS Transactions*, 31, 83–91, 2010.
11. D. Vasileska, A. Hossain, and S. M. Goodnick, Self-heating and short-range coulomb interactions due to charging of traps in nanowire transistors, in *Proceedings of the 2011 IEEE Nano Conference on Nanotechnology*, Portland, Oregon, pp. 1110–1113, 2011.

57 Impact of Phonon Scattering in an Si GAA Nanowire FET with a Single Donor in the Channel

Antonio Martinez, Manuel Aldegunde, and Karol Kalna

CONTENTS

57.1 INTRODUCTION

Nanowire field effect transistors (NFETs), FinFETs, and other 3D architectures are progressively replacing the bulk metal-oxide semiconductor field-effect transistor (MOSFET) architecture in a complementary metal-oxide semiconductor (CMOS) Si technology. This process has been started by Intel's introduction of TriGate (a variant of FinFET) MOSFET architecture into high-volume production of the 22 nm technology node [1]. The superior electrical integrity of FinFETs and nanowires resulting in very low leakage current at the off-state, as compared to much leakier bulk architectures, is the main reason for this trend [2]. However, when the transistor cross section and channel lengths reach sub-10 nm dimensions, quantum effects such as confinement, gate tunneling, and source/drain (S/D) tunneling will dominate over classical electrostatic alternately affecting the electrostatic integrity of 3D device architectures. Therefore, we deploy nonequilibrium Green's function (NEGF) [3] simulations to study the effect of a single dopant unintentionally present in the channel of an NFET [4]. The probability of occurrence of this unintentional dopant will also increase in devices with sub-10 nm dimensions.

The NEGF simulations include a phonon scattering considering the electron interactions with acoustic and nonpolar optical phonons dominant in silicon [5], unlike previous studies investigating the effect of the single dopant [6,7]. Even though phonon scattering has already been incorporated into NEGF simulations of nanowire transistors [8–10], its effect on the single dopant in a nanowire has not been studied. Furthermore, the separate impact of elastic and inelastic phonon scatterings has not been looked at. Therefore, we study the effect of phonon scattering on the Si NFET with the single dopant close to the source. The donor is introduced as a charge distribution and the potential is calculated self-consistently within a Hartree approximation. The asymmetrical location of the donor in the middle of the channel allows the electrons near the source to tunnel into the donor potential and then be heated up to the top of the potential barrier via inelastic phonon processes.

57.2 NONEQUILIBRIUM GREEN'S FUNCTION MODEL

The NEGF formalism within one-electron approximation reduces a many-body Schrödinger equation to a solution of the following equations [3,11]:

$$(g_0^{-1} - U)g = 1 + \sigma g \tag{57.1}$$

$$(g_0^{-1} - U)g^< = \sigma g^< + \sigma^< g_a \tag{57.2}$$

where g_0 is the retarded Green's function for a single-electron part of a Hamiltonian with the external field U turned off, and σ is the retarded self-energy including the coupling to leads and other interactions (i.e., the electron–phonon interactions). The g and g_a are the retarded and advanced Green's functions, respectively. In Equation 57.2, $g^<$ and $\sigma^<$ are "the-lesser-than" pieces of the complex contour path of Green's function (i.e., the first leg is always in earlier time than the second leg in the complex contour [3]). The Hamiltonian of NEGF is written in the effective mass approximation and the effective mass itself is extracted from tight-binding calculations [12] ($m_l = 1.07$, $m_t = 0.3$ in units of electron mass), which accounts for a bandstructure change due to strong confinement occurring in the nanowires. In other words, Equation 57.1 is equivalent to the Schrödinger equation in single-electron approximation and Equation 57.2 is a quantum generalization of the semiclassical Boltzmann equation. The contacts are assumed to be in equilibrium and at room temperature. We employ a recursive algorithm [13] that allows us to calculate only the diagonal and first off-diagonal blocks of the less-than Green's function $g^<$. $g^<$ is then required for the computation of current and electron density in the NEGF formalism [3].

57.3 ELECTRON–ELECTRON INTERACTION MODEL BEYOND THE HARTREE APPROXIMATION

The incorporation of an electron–electron interaction into a single-electron model is a cumbersome task [3]. One of the most practical ways based on a set of theorems had been introduced by Kohn and Sham [14] who proposed their famous one-electron Schrödinger equation in the following form:

$$\left[-\frac{\hbar^2 \nabla^2}{2m(r)} + V(r) + v_{XC}(r) \right] \psi_i(r) = E_i \psi_i(r) \tag{57.3}$$

where $-\hbar^2 \nabla^2 / 2m(r)$ is a kinetic operator and $m(r)$ is the position-dependent mass. $V(r)$ is the Coulomb potential given by

$$V(r) = v(r) + \int \frac{e^2 \rho(r')}{|r - r'|} \, dr' \tag{57.4}$$

where $v(r)$ is the electrostatic, external potential, $\rho(r)$ is the ground-state density, and r represents x, y, z coordinates in a real space. Finally, $v_{XC}(r)$ is an exchange–correlation (XC) potential [14] generally defined as the difference between a full, many-body kinetic plus potential energy and one-body kinetic plus potential energy. The XC potential is impossible to express for large, many-body systems like NFETs, but there are various physical approximations like the local density approximation (LDA) or the generalized gradient approximation (GGA) that can serve. The accuracy of these approximations is quite strongly dependent on the nature of the physical system and physical quantities one would like to compute. Therefore, these approximations are typically suitable for particular calculations performed in solid-state physics (energy band calculations of metals, semiconductors, insulators, ferromagnetic materials, calculations of material permittivity, etc.). The analytical parametrization introduced by Hedin and Lundqvist [15] has proven to be the most practical with sufficient accuracy for a large number of solid-state problems related to semiconductors. In this parametrization, the XC potential is split into two parts, the first being the exchange plus correlation hole, and the second being the correlation energy correction to chemical potential, as follows [15]:

$$v_{XC}(x,y,z) = -\frac{e^2}{4\pi^2 \varepsilon_{sc} \varepsilon_0} \left[3\pi^2 \, n(x,y,z) \right]^{1/3} \times \left[1 + 0.7734 \, X \, \ln\left(1 + \frac{1}{X}\right) \right] \tag{57.5}$$

where $X = r_s/21$. This XC potential parametrization is valid for a wide range of r_s, while r_s is defined via a local carrier density $n(r)$ as

$$r_s = \left[4\pi \, b^3 n(x,y,z)/3 \right]^{-1/3}$$

and

$$b = \frac{4\pi \, \varepsilon_{sc} \, \hbar^2}{m(r)e^2}.$$

where ε_{sc} is the permittivity of a semiconductor and ε_0 is the permittivity of a vacuum. In silicon, r_s values span a range of 2.6–0.26, corresponding to electron concentrations of 10^{18}–10^{21} cm^{-3}. The XC potential can then be relatively easily included into self-consistent simulations of carrier transport in semiconductor devices by adding it to the electrostatic potential $v(x,y,z)$ [16,17] since the carrier density $n(r)$ is often known locally. In our NEGF simulation approach, the electron density $n(x,y,z)$ entering the XC potential is obtained from self-consistent simulations iteratively solving the NEGF and Poisson equations. The XC potential is plotted in Figure 57.1 as a function of the carrier density showing the contribution due to the exchange only (LDA) and the exchange–correlation (Hedin–Lunqvist) parts represented by the first and second terms in the last square brackets of Equation 57.5, respectively. The XC potential induces a 20–80 meV shift in the electron self-energy at the carrier density typical in Si NFETs. This shift is more significant in the access regions to the

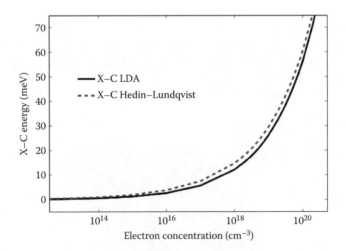

FIGURE 57.1 The XC potential approximation of Ref. [15] compared with the standard local density approximation (LDA). The difference between the two potentials is smaller than 6 meV in the range of relevant electron densities in Si transistors. (A. Martinez, K. Kalna, and M. Aldegunde, Impact of phonon scattering in a Si GAA nanowire FET with a single donor in the channel, *Proc. 11th IEEE International Conference on Nanotechnology*, August 15–18, 2011, Portland, Oregon, pp. 1348–1351. © (2011) IEEE. With permission [18].)

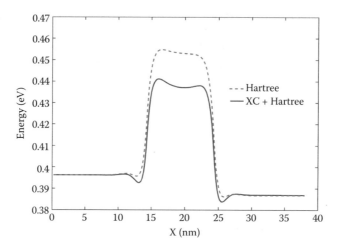

FIGURE 57.2 Potential energy across the channel of Si NFET at equilibrium including and excluding the XC potential at $V_D = 0.01$ V. (A. Martinez, K. Kalna, and M. Aldegunde, Impact of phonon scattering in a Si GAA nanowire FET with a single donor in the channel, *Proc. 11th IEEE International Conference on Nanotechnology*, August 15–18, 2011, Portland, Oregon, pp. 1348–1351. © (2011) IEEE. With permission [18].)

channel, where the carrier density approaches 10^{20} cm^{-3}, inducing a lowering of the S/D barrier and thus increasing the on-current as indicated in Figure 57.2. Figure 57.2 demonstrates that the potential profile given by conduction band edge substantially decreases the potential barrier between the source and the drain. Note also that the potential in the source and the drain practically do not change. The XC potential-induced self-energy shift is of the order of threshold voltage shift induced by statistical sources of fluctuations [2,4] such as random dopants and surface roughness and thus should be considered as an accurate description of carrier transport.

57.4 SIMULATED DEVICE

The gate-all-around (GAA) nanowire transistor considered in this chapter is schematically shown in Figure 57.3. It has 14 nm long S/D regions and a 10 nm channel length. It has a square cross section of 2.2×2.2 nm^2 coated with a 0.8 nm SiO$_2$ dielectric layer providing a confinement potential of 3.1 eV. The channel is undoped and the S/D regions have an *n*-type doping concentration of 10^{20} cm^{-3}. A metal gate is wrapped all around the nanowire on the outer surface of the oxide and is assumed to have a work function of 4.5 eV.

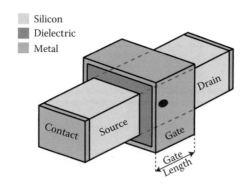

FIGURE 57.3 Schematic view of the considered Si NFET with the indicated position of the single dopant in the channel (a dark dot).

We would like to stress that phonon scattering is considered in the whole device, including the S/D regions and not just in the channel as in many previous studies [6,8]. This approach makes the device simulations much more realistic since the electron interaction with phonons plays a considerable role in the heavily doped S/D regions despite the frequent interactions with ionized impurities.

57.5   EFFECT OF SINGLE DOPANT ON PERFORMANCE OF NFET

Figure 57.4 shows the I_D–V_G characteristics of a pristine nanowire transistor both in the presence and absence of phonon scattering (all the simulations are done at $V_D = 10$ mV). To isolate the effect of inelastic scattering, we also show the I_D–V_G characteristics in which we consider all the scattering mechanisms as elastic. An expected result is that elastic scattering produces a lower current than inelastic scattering. This is partially a consequence of the impossibility of electrons being heated in the source and cooled in the drain because the emission or absorption of phonons is prohibited. In the case of inelastic scattering, the electrons, cooled in the drain, have a small chance to return to the source and therefore to contribute to an increase in the current, as opposed to a more realistic simulation in which we consider phonon scattering to be elastic. Figure 57.5 shows the corresponding cases but with the impurity located 3 nm from the source/channel interface.

The overall effect of phonon scattering is less severe, as shown in Figure 57.6. Figure 57.6 shows the percentage decrease in the ballistic current due to phonon scattering in a pristine device and a device with a donor in the channel. A case of the NFET with a donor in the channel considering only the elastic phonon scattering is shown for comparison. Figure 57.6 also shows that, in a pristine wire, the scattering increases as the gate bias increases but an impurity in the channel behaves unexpectedly. The energy states induced by the single dopant into the nanowire channel can completely remove the effect of phonon scattering and increase the current to its ballistic value, as seen in Figure 57.6 at $V_G = 0.6$ V. The resonant level here has a major impact on the current and the optical phonon scattering maximizes the absorption of carriers in the source.

Finally, our results predict a decrease in the on-current of more than 70% for the pristine wire due to phonon scattering. The current reduction is slightly smaller if the device has a donor positioned in

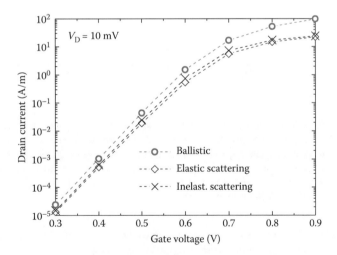

FIGURE 57.4 I_D–V_G characteristics of the 10 nm gate length pristine Si nanowire. The ballistic NEGF simulations are compared with the simulations, which include elastic and inelastic phonon scattering. (A. Martinez, K. Kalna, and M. Aldegunde, Impact of phonon scattering in a Si GAA nanowire FET with a single donor in the channel, *Proc. 11th IEEE International Conference on Nanotechnology*, August 15–18, 2011, Portland, Oregon, pp. 1348–1351. © (2011) IEEE. With permission [18].)

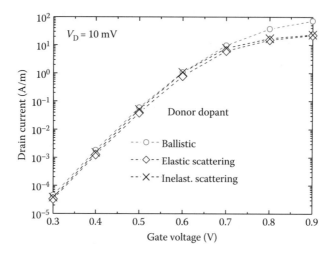

FIGURE 57.5 I_D–V_G characteristics of the same Si nanowire simulated with ballistic NEGF and NEGF, including elastic and inelastic phonon scattering. (A. Martinez, K. Kalna, and M. Aldegunde, Impact of phonon scattering in a Si GAA nanowire FET with a single donor in the channel, *Proc. 11th IEEE International Conference on Nanotechnology*, August 15–18, 2011, Portland, Oregon, pp. 1348–1351. © (2011) IEEE. With permission [18].)

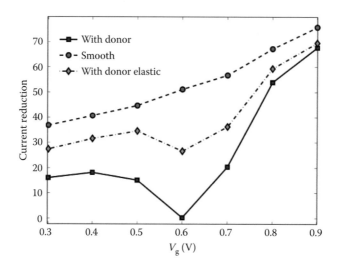

FIGURE 57.6 Current reduction in the NFET at $V_D = 0.01$ V due to the phonon scattering against the ballistic transport model for a pristine nanowire (Smooth), with a single dopant including only elastic scattering (with donor Elastic), and with a single dopant including all scatterings (with donor). (A. Martinez, K. Kalna, and M. Aldegunde, Impact of phonon scattering in a Si GAA nanowire FET with a single donor in the channel, *Proc. 11th IEEE International Conference on Nanotechnology*, August 15–18, 2011, Portland, Oregon, pp. 1348–1351. © (2011) IEEE. With permission [18].)

the middle of the channel. This result is partially due to the increase in electron tunneling from the source to the drain in the presence of a single donor in the channel.

Figure 57.7 shows the electron density at $V_G = 0.6$ V, illustrating the sharp peak of the density along the channel induced by the single dopant. The peak of the density is reflected in the local density of states (LDoS) and in the current spectra along the axis of the nanowire at $V_G = 0.6$ V shown in Figures 57.8 and 57.9. The LDoS in Figure 57.8 indicates the resonant level of the impurity potential as well as the induced resonances into the density of states. The current spectra in Figure 57.9 illustrate clearly

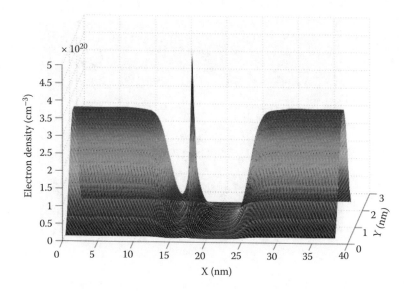

FIGURE 57.7   Electron density along the nanowire at $V_G = 0.6$ V and $V_D = 0.01$ V. The source and the drain are uniformly *n*-type doped up to 10^{20} cm^{-3} and the single dopant is placed in the middle of the channel, 3 nm from the source. (A. Martinez, K. Kalna, and M. Aldegunde, Impact of phonon scattering in a Si GAA nanowire FET with a single donor in the channel, *Proc. 11th IEEE International Conference on Nanotechnology*, August 15–18, 2011, Portland, Oregon, pp. 1348–1351. © (2011) IEEE. With permission [18].)

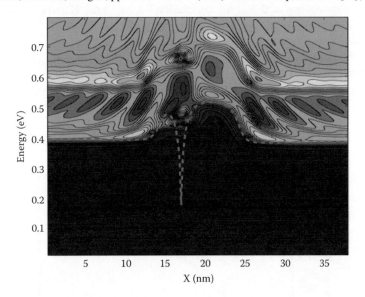

FIGURE 57.8   Local density of states (LDoS) along the axis of the nanowire at $V_G = 0.6$ V and $V_D = 0.01$ V.

two conduction channels created by the split due to the single dopant in the channel. At this gate bias of $V_G = 0.6$ V, the contribution of the resonant state to the total current is 7%, but at $V_G = 0.7$ V, the resonant state contribution increases to about 30%. The heating and the cooling of carriers, indicated by the red and blue colors for the high and low energy of carriers, can also be seen in Figure 57.9. The heated carriers are accumulated just above the wide barrier near the drain created by the single dopant. One can also see how they then quickly cool down when entering the drain region.

Furthermore, we go beyond the standard Hartree–Fock approximation and simulate the impact of XC potential introduced in the previous section using the NEGF approach from Silvaco's ATLAS

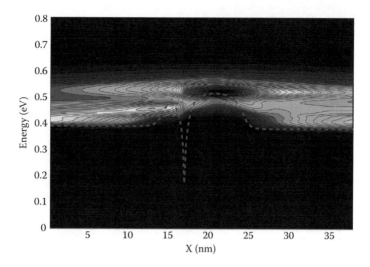

FIGURE 57.9 The current spectra along the axis of the nanowire at $V_G = 0.6$ V and $V_D = 0.01$ V. (A. Martinez, K. Kalna, and M. Aldegunde, Impact of phonon scattering in a Si GAA nanowire FET with a single donor in the channel, *Proc. 11th IEEE International Conference on Nanotechnology*, August 15–18, 2011, Portland, Oregon, pp. 1348–1351. © (2011) IEEE. With permission [18].)

toolbox [19]. The exchange takes into account that electrons are indistinguishable particles reducing thus the ground-state energy, and that the correlation improves the Hartree–Fock approximation accounting for the many-body effects. The resulting effective potential will then be lower than the potential excluding the exchange–correlation, and the density in the channel increases as observed mostly close to the threshold voltage. Figure 57.10 shows a comparison of I_D–V_G characteristics at a low drain voltage of 0.01 V. Note first that our in-house NEGF simulator gives I_D–V_G characteristics that are in excellent agreement with those obtained from Silvaco's NEGF simulations in a ballistic regime. The lowering of the current due to the presence of the single dopant is also illustrated, assuming the same ballistic transport regime. When the XC potential is included in ballistic simulations, the current is not changed in the subthreshold region due to a low carrier density in the

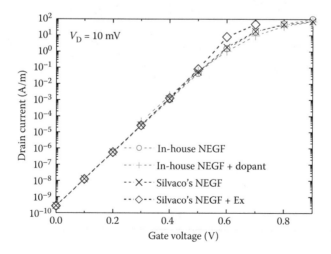

FIGURE 57.10 I_D–V_G characteristics at $V_D = 0.01$ V of the 10 nm gate length Si nanowire simulated with our ballistic NEGF (in-house) w/o dopant compared with Silvaco's ATLAS NEGF in a ballistic regime and also including exchange and correlation (Ex).

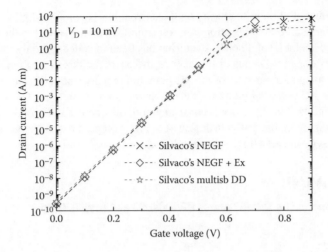

FIGURE 57.11 I_D–V_G characteristics at $V_D = 0.01$ V of the pristine Si nanowire comparing the ATLAS NEGF simulations in a ballistic regime with/without the exchange and correlation (Ex) to multi-subband drift-diffusion simulations (multisb DD). (A. Martinez, K. Kalna, and M. Aldegunde, Impact of phonon scattering in a Si GAA nanowire FET with a single donor in the channel, *Proc. 11th IEEE International Conference on Nanotechnology*, August 15–18, 2011, Portland, Oregon, pp. 1348–1351. © (2011) IEEE. With permission [18].)

channel. However, when the gate bias drives the NFET into on-state, the current starts to increase more swiftly when the XC potential is included compared to the result when it is excluded.

Finally, Figure 57.11 shows a comparison of the results obtained from drift-diffusion (DD) multi-subband simulations [19]. The results confirm the same electrostatic behavior by exhibiting the same subthreshold slope. However, the on-current obtained from the DD multi-subband simulations of Silvaco's ATLAS toolbox is lower. This is because the carrier scattering in the DD is taken into account through a mobility model, while the NEGF simulations of Silvaco ATLAS were carried out in a ballistic regime. The ballistic simulation will thus predict a much larger current since this model assumes that electrons do not scatter and thus cannot lose their kinetic energy (gained from the electric field present in the channel). Furthermore, the DD multi-subband Silvaco simulations show a very similar on-current with a difference of only 10% when compared to the on-current obtained from the corresponding in-house NEGF simulations with the phonon scattering included as presented in Figure 57.10. We are not able to draw any conclusions on this close agreement as to speculate that this is due to a decisive effect of the electron density along the nanowire channel on the current while the electron dynamics (electron drift velocity/mobility) plays a less important role in determining this current at $V_D = 0.01$ V.

57.6 CONCLUSIONS

3D NEGF simulations including dissipative processes due to elastic and inelastic phonon scattering have been carried out to investigate its effect on device characteristics of a 10 nm gate length Si NFET with a cross section of 2.2×2.2 nm^2. Compared to previous studies of NFETs, we have investigated the effect of dissipative processes in the nanowire with and without a single dopant in the channel, also taking into account the impact of heavily doped S/D access regions. The role of the elastic and inelastic phonon scatterings in electron transport has been compared with the results from ballistic NEGF simulations. We have found that the presence of the dopant in the channel close to the source region reduces the impact of phonon scattering on the drain current.

This phonon scattering reduction is the largest at the gate voltage at which the resonant level of the impurity has the maximum impact on the drain current. The contribution of the electron tunneling current, enhanced by the resonant level of impurity, is the largest relative to the total current. In other words, the effect of phonon scattering on the current has become negligible at the gate bias at which a relatively large amount of electrons are transmitted from the source to the drain through the impurity-induced resonant level. Phonon scattering has been included not only into the channel but also into the heavily doped S/D, thus distinguishing these simulations as much more realistic.

We have also evaluated the impact of the XC potential in a pristine NFET. The results show that the inclusion of the XC potential increases the on-current more than twice but has little impact on the subthreshold regime. This behavior is consistent with the decrease of overall potential energy in the channel of the nanowire due to the inclusion of the XC potential and with the fact that the overall potential in the heavily doped S/D remains unaltered.

ACKNOWLEDGMENT

This work was supported by EPSRC grants EP/I004084 and EP/D070236.

REFERENCES

1. http://www.intel.com/content/www/xa/en/silicon-innovations/intel-22 nm-technology.html, 2011.
2. D. Hisamoto, W.-C. Lee, J. Kedzierski, E. Anderson, H. Takeuchi, K. Asano, T.-J. King, J. Bokor, and C. Hu, A folded-channel MOSFET for deep-sub-tenth micron era. *Int. Electron Devices Meeting Tech. Dig.*, pp. 1032–1034, 1998.
3. S. Datta, *Electronic Transport in Mesoscopic Systems*, Cambridge University Press, Cambridge, UK, 1995.
4. A. Martinez, M. Bescond, J. R. Barker, and A. Svizhenko, M. P. Anantram, C. Millar, and A. Asenov, A self-consistent full 3-D real-space NEGF simulator for studying nonperturbative effects in nano-MOSFETS, *IEEE Trans. Electron Devices* **54**, 2213–2222 (2007).
5. C. Jacoboni and P. Lugli, *The Monte Carlo Method for Semiconductor Device Simulations*, Springer, Wien-N.Y. 1989.
6. M. Bescond, M. Lannoo, L. Raymond, and F. Michelini, Single donor induced negative differential resistance in silicon n-type nanowire metal-oxide-semiconductor transistors, *J. Appl. Phys.* 107, 093703-1-6, 2010.
7. A. Martinez, N. Seoane, M. Aldegunde, A. R. Brown, J. R. Barker, and A. Asenov, 3-D nonequilibrium Green's function simulation of nonperturbative scattering from discrete dopants in the source and drain of a silicon nanowire transistor, *IEEE Trans. Nanotech.* 8, 603–610, 2009.
8. S. Jin, Y.-J. Park, and H. S. Min, A three-dimensional simulation of quantum transport in silicon nanowire transistor in the presence of electron-phonon interactions, *J. Appl. Phys.* 99, 123719-1-123719-10, 2006.
9. A. Svizhenko and M. P. Anantram, Role of scattering in nanotransistors, *IEEE Trans. Electron Devices* 50, 1459–1466, 2003.
10. M. Luisier and G. Klimeck, Atomistic full-band simulations of silicon nanowire transistors: Effects of electron-phonon scattering, *Phys. Rev. B* 80, 155430, 2009.
11. D. C. Langreth and J. W. Wilkins, Theory of spin resonance in dilute magnetic alloys, *Phys. Rev.* B 6, 3189–3227, 1972, Equations 57.1 and 58.2 correspond to Equations A15–A16.
12. K. Nehari, N. Cavassilas, J. L. Autran, M. Bescond, D. Munteanu, and M. Lannoo, Influence of band structure on electron ballistic transport in silicon nanowire MOSFET's: An atomistic study, *Solid-St. Electron.* 50, 716–721, 2006.
13. A. Svizhenko, M. P. Anantram, T. R. Govindan, B. Biegel, and R. Venugopal, Two-dimensional quantum mechanical modeling of nanotransistors, *J. Appl. Phys.* 91, 2343–2354, 2002.
14. W. Kohn and J. Sham, *Phys. Rev.* 140, A1133–1138, 1965; J. Sham and W. Kohn, *Phys. Rev.* 145, 561–567, 1966.
15. L. Hedin and B. I. Lundqvist, Explicit local exchange-correlation potentials, *J. Phys. C: Solid St. Phys.* 4, 2064–2083, 1971.
16. H. Iwata, T. Matsuda, and T. Ohzone, Influence of image and exchange-correlation effects on electron transport in nanoscale DG MOSFETs, *IEEE Trans. Electron Devices* 52, 1596–1602, 2005.
17. D. Vasileska, D. K. Schroder, and D. K. Ferry, Scaled silicon MOSFETs: degradation of the total gate capacitance, *IEEE Trans. Electron Devices* 44, 584–1587, 1997.
18. A. Martinez, K. Kalna, and M. Aldegunde, Impact of phonon scattering in a Si GAA nanowire FET with a single donor in the channel, *Proc. 11th IEEE International Conference on Nanotechnology*, August 15–18, 2011, Portland, Oregon, pp. 1348–1351.
19. Silvaco Int., *Atlas User's Manual* [online]. Available: www.silvaco.com.

58 Modeling and Minimizing Variations of Gate-All-Around Multiple-Channel Nanowire TFTs

Po-Chun Huang, Lu-An Chen, C. C. Chen,
and Jeng-Tzong Sheu

CONTENTS

58.1 INTRODUCTION

Although polysilicon thin-film transistors (poly-Si TFTs) are attracting much attention for their use in active-matrix liquid crystal displays [1,2], fine-grain structures in the channel can affect their carrier transport and device performance. Poly-Si consists of a number of single-crystal grains. Between poly-Si grains, there exists a high defect density region called the grain boundary with a typical value of defect density ca. 10^{12} cm^{-2}. Such a large number of randomly oriented grain boundaries usually cause large variations in the device's electrical characteristics, including the threshold voltage and subthreshold swing (SS) [3]. Hence, several methods, including excimer laser annealing [4] and metal-induced lateral crystallization [5], have been developed to increase the grain size and minimize the grain boundary defects and thereby improve the electrical characteristics of poly-Si TFTs. Recently, a multiple-gate structure was reported in which the additional electric field enhanced the control over the channel surface potentials and the device performance, providing improved immunity to short channel effects [6,7]. In addition, TFT characteristics are dramatically improved upon reducing the channel width, particularly when using a nanowire (NW) as a channel [8,9]. Poly-Si TFTs featuring multiple NW channels exhibit improved performance relative to that of traditional planar devices because of the increment of effective channel width and the reduced grain boundary trap density in the channel region [10,11]. When the channel width and the poly-Si grain size are of the same order of magnitude, the random distribution of the grain boundary causes an even greater electrical fluctuation including threshold voltage and SS [9,12,13]. Therefore, uniformity of the grain boundary is critical when scaling down device dimensions.

In a previous study [7], we found that a gate-all-around (GAA) poly-Si NW TFT exhibited excellent gate controllability because the GAA structure enhanced the electric field. In this study, we investigated the performance of GAA NW TFTs featuring multiple gate configurations; in

particular, we compared the tri-gate and GAA structures and studied the impact of multiple channels to further reduce the fluctuations in device performance. Indeed, the variations in threshold
voltage and SS were effectively modulated. Although the Poisson area scatter model was proposed
to estimate the impact of statistical grain boundary distribution on TFT characteristics [14]. To the
best of our knowledge, this model has never been adopted to model poly-Si GAA TFTs featuring
multiple NW channels, especially for devices with a grain size and channel diameter in the same
order of magnitude. Finally, we found that NH_3 plasma treatment further minimized the fluctuations
and improved the performance of these GAA NW TFTs.

58.2 EXPERIMENT

The 1-μm-long multiple channels of the poly-Si NWs were fabricated using a spacer patterning technique [7]. At first, a 75-nm-thick undoped amorphous-Si (α-Si) layer was deposited onto a 200-nm-
thick bottom oxide (BOX) at 550°C using low-pressure chemical vapor deposition (LPCVD). In
sequence, solid-phase crystallization (SPC) was performed at 600°C for 24 h in a nitrogen ambient
to turn the α-Si into a polycrystalline Si structure. The key parameter of the spacer patterning
technique was the thickness of followed deposition of the tetraethylorthosilicate (TEOS) oxide and
nitride spacer. The TEOS oxide thickness defined the height of the nitride spacer. After definition of the TEOS pattern by reactive ion etching (RIE), the nitride was deposited and then etched
using RIE. The thickness of the nitride film corresponded to the width of the spacer. In the study,
the thicknesses of the TEOS and nitride layers were chosen to be 70 and 60 nm, respectively, for
these processing conditions. Using the nitride spacer as a hard mask, 1-μm-long multiple channels
of poly-Si NW having a channel width (W_{ch}) of 44 nm were formed. By varying the thicknesses
of the TEOS and nitride layers, nanoscale structures were readily defined without the need for
advanced lithography techniques. After wet etching in a diluted HF solution (1:100) to remove the
70–80 nm BOX layer, the poly-Si NW was released from the BOX substrate. Next, the sequential
conformal deposition (LPCVD) of dielectric of oxide-nitride-oxide (ONO) stacks (TEOS/nitride/
TEOS = 10.5 nm/5.5 nm/15.3 nm) and 200-nm-thick *in situ* N^+ poly-Si was performed to surround
the channels. After gate pattern transformation, the etching of dielectric on the source and drain surface was performed. The self-aligned phosphorus ion implantation was then completed at a dose of
5×10^{15} cm^{-2} for the formation of ohmic contact. The TEOS passivation, thermal annealing (600°C,
6 h), standard metal contact formation, and sintering processes were performed to provide the final
structure. To study the impact of the grain boundary defects, the samples were subjected to NH_3
plasma treatment in a parallel-plate plasma reactor at a power density of 0.7 W/cm² at 300°C for 1 h.

58.3 RESULTS AND DISCUSSION

Figure 58.1 presents transmission electron microscopy (TEM) images of NW channels possessing
a GAA structure and a tri-gate structure surrounded by ONO layers. These devices had a nominal
channel length (L) of 1 μm (effective channel length ~ 0.75 μm), a channel width (W_{ch}) of 44 nm,
and a channel thickness (T_{ch}) of 62 nm. The ONO stack (O/N/O = 10.5 nm/5.5 nm/15.3 nm) was
conformably deposited around the channel. Such a stack layer can improve the future integrating
memory function when using a system-on-panel (SOP) approach [15]. To prevent the effects of non-
uniformity of our fabricated NWs, especially during the film deposition and etching processes, the
devices under investigation were selected from the same vicinity of the chip.

Figure 58.2 presents the transfer curves of the GAA and tri-gate NW TFTs. Relative to the tri-gate
device, the GAA NW TFT, even when it had not been subjected to the treatment process, suppressed
the short channel effects (SCEs), providing a smaller drain induce barrier lowering (DIBL), a lower
leakage current, and a steeper SS, because the GAA devices provided additional electric field strength
at the corner and bottom regions, leading to improvements in the values of the drain current and V_{th} [the
gate voltage required to yield a normalized drain current of I_d ($W = L$) = 10 nA at $V_d = 0.1$ V]. These

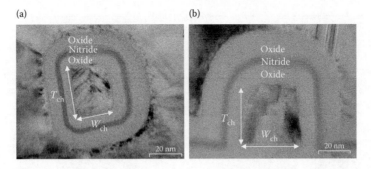

FIGURE 58.1 TEM images of (a) GAA and (b) tri-gate devices having a channel thickness of 44 nm and a channel width of 62 nm. The ONO stack (O/N/O = 10.5 nm/5.5 nm/15.3 nm) was prepared through conformal deposition using LPCVD. (Reprinted from *Microelectronic Engineering*, 91, P. C. Huang, L. A. Chen, C. C. Chen, and J. T. Sheu, Minimizing variation in the electrical characteristics of gate-all-around thin film transistors through the use of multiple-channel nanowire and NH_3 plasma treatment, 54–58, (2012), with permission from Elsevier.)

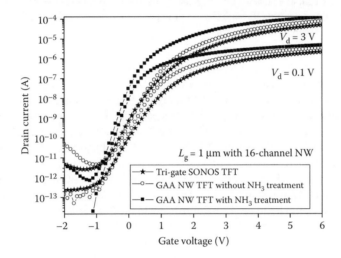

FIGURE 58.2 A comparison of transfer curves for GAA and tri-gate NW TFTs. After NH_3 treatment, the GAA NW TFT exhibited superior performance. The GAA NW TFT that had been subjected to NH_3 plasma treatment exhibited improved device characteristics, including a lower value of V_{th} (0.18 V), a steeper SS (184 mV/dec), a very high on/off current ratio (1.64 × 108), and a small DIBL (58 mV/V). (Reprinted from *Microelectronic Engineering*, 91, P. C. Huang, L. A. Chen, C. C. Chen, and J. T. Sheu, Minimizing variation in the electrical characteristics of gate-all-around thin film transistors through the use of multiple-channel nanowire and NH_3 plasma treatment, 54–58, (2012), with permission from Elsevier.)

improvements were also as a result of the channel potential being tightly controlled by the surrounding electric field. Moreover, the large number of trap states in the poly-Si channel region also played a critical role influencing the device performance and the SCEs. As a result, the poly-Si TFT exhibited a large leakage current due to the large field emission current near the drain side and grain boundary trap states in the channel [16]. Thus, after treatment process, the N and H atoms passivated the traps on the channel surface and/or in the channel area [7]. The passivation species diffused mainly through the gate oxide into the channel from the channel edge. These passivation radicals also further reduced the grain boundary potential barrier; hence, the off-state leakage current was minimized and the on-state current was enhanced. As a result, the proposed device exhibited a higher on/off current ratio (>10^8) than that of the untreated samples (>10^7). Therefore, the GAA NW TFT that had been subjected to NH_3 plasma treatment exhibited improved device characteristics, including a lower value of V_{th} (0.18 V), a steeper SS (184 mV/dec), a very high on/off current ratio (1.64 × 10^8), and a small DIBL (58 mV/V).

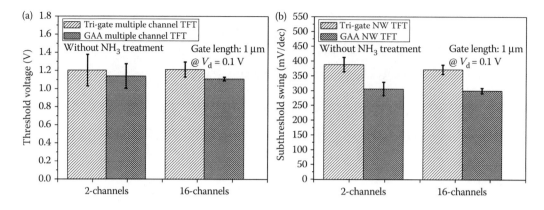

FIGURE 58.3 Mean values and standard deviations of the (a) threshold voltage and the (b) subthreshold swing of the GAA and tri-gate NW TFTs featuring 2 and 16 channels. (Reprinted from *Microelectronic Engineering*, 91, P. C. Huang, L. A. Chen, C. C. Chen, and J. T. Sheu, Minimizing variation in the electrical characteristics of gate-all-around thin film transistors through the use of multiple-channel nanowire and NH₃ plasma treatment, 54–58, (2012), with permission from Elsevier.)

Figure 58.3a displays the mean values and the standard deviations of V_{th} for the different gate configurations possessing different numbers of channels. For the GAA NW TFTs possessing 2 and 16 channels, the mean values of V_{th} were 1.14 and 1.11 V, respectively; for the corresponding tri-gate devices, these values were 1.20 and 1.21 V, respectively. Notably, because the electric field surrounded the entire channel, the GAA NW TFT exhibited a smaller mean value of V_{th} relative to that of its tri-gate counterpart. In addition, the GAA NW TFT exhibited a smaller standard deviation of its value of V_{th}; such relative stability is critical for larger-glass active-matrix liquid crystal displays (AMLCD) applications. Furthermore, the grain boundaries of the SPC poly-Si were randomly distributed in the channel region, resulting in a variation of its electrical characteristics. These grain boundaries trap charges and build up potential barriers to the flowing carriers, resulting in additional scattering that leads to device degradation [17]. After SPC, the grain size in the channel region was approximately 25–40 nm in our experiment; relative to the NW channel having a value of W_{ch} of 44 nm, we expected the device to suffer from fluctuating electrical characteristics, mainly due to these random distribution grain boundaries [12]. Our results reveal that the gate structure and the number of channels both influence the device characteristics. The standard deviations of the electrical characteristics of both the tri-gate and GAA NW TFTs were reduced effectively upon increasing the number of channels. The standard deviation of the value of V_{th} of the tri-gate TFT decreased from 174 to 82 mV; that of the GAA NW TFT decreased from 135 to 21 mV. Interestingly, the improvement of variation was greater when increasing the number of channels from 2 to 16 than it was when changing the gate configuration from a tri-gate to a surrounding gate. Figure 58.3b reveals similar results for the SS. The GAA NW TFT with 16 channels had the smallest average value (300 mV/dec) and standard deviation (9.38 mV/dec) of SS among all of the TFTs. Thus, the GAA structure improved the mean values of V_{th} and SS, and increasing the number of channels can effectively minimize the electrical variations.

Figure 58.4 displays the transfer characteristics of devices with channel numbers of 2, 4, 8, and 16. The transfer characteristics resulted from the measurements of 15 devices. Device-to-device variation was minimized by increase of channel numbers. On the basis of a previous study [14], the Poisson area scatter distribution was adopted to characterize the grain size variation relating to the properties of poly-Si GAA TFTs featuring multiple-channel NWs. An average grain area in the channel is assumed with an area of A_g. The active area of a transistor is defined by $W \times L$. k is a Poisson random variable with an expected value of the number of grains, λ, within the active region. λ is defined by $\lambda = (W \times L)/A_g$. For multiple channels, the channel width is defined as $W = n \times W_{eff}$,

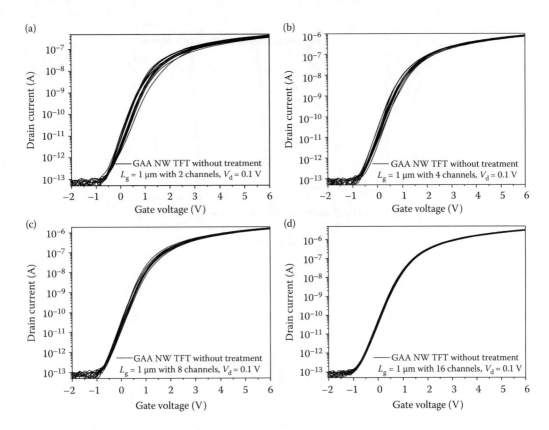

FIGURE 58.4 Transfer characteristics of GAA NW TFTs for channel numbers of (a) 2, (b) 4, (c) 8, and (d) 16. The uniformity of electrical characteristics has been effectively improved by increase of channel numbers. (Reprinted from *Microelectronic Engineering*, 91, P. C. Huang, L. A. Chen, C. C. Chen, and J. T. Sheu, Minimizing variation in the electrical characteristics of gate-all-around thin film transistors through the use of multiple-channel nanowire and NH_3 plasma treatment, 54–58, (2012), with permission from Elsevier.)

where n is the channel number and W_{eff} is the effective channels width of an NW channel. The Poisson distribution probability $P(k)$ with exactly k hits in an area $W \times L$ is defined as

$$P(k) = \frac{e^{-\lambda} \times \lambda^k}{k!}. \tag{58.1}$$

The average grain size within a specific transistor is also a random variable and is given by

$$L_{g,TFT} = \sqrt{\frac{W \times L_{eff}}{k}}. \tag{58.2}$$

The threshold voltage is related to the grain size through the model [14,17]:

$$V_T = V_{FB} + \sqrt{\frac{8kTN_{trt}t_{OX}}{C_{OX}L_{g,TFT}}}\sqrt{\frac{\varepsilon_{Si}}{\varepsilon_{OX}}}. \tag{58.3}$$

Figure 58.5 exhibits the cumulative probability distributions of threshold voltage of a 2-channel poly-Si GAA TFT with different grain sizes. The threshold voltage is the minimum voltage when the grain boundary traps are filled. The significant conduction occurred only when all these traps were

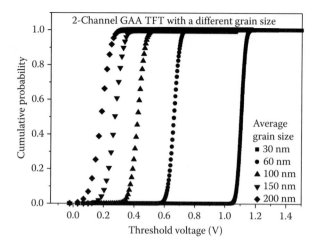

FIGURE 58.5 Cumulative probability distributions of threshold voltage for a 2-channel GAA TFT for different grain sizes; the effective channel length L_{eff} is 0.75 μm. (P. C. Huang, L. A. Chen, C. C. Chen, and J. T. Sheu, Modeling and Minimizing Variations of Gate-All-Around Multiple-channel Nanowire TFTs, *IEEE NANO 2011 Conference*, August 15–18, 2011, Portland, Oregon. © (2011) IEEE. With permission.)

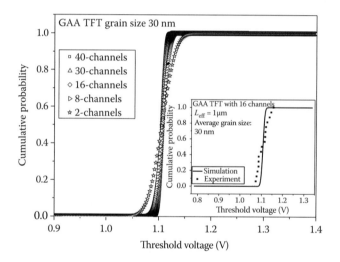

FIGURE 58.6 Cumulative probability distributions of threshold voltage for GAA NW TFTs featuring different channel numbers; grain size= 30 nm, L_{eff} is 0.75 μm. Inset: Comparison of cumulative distributions of threshold voltage distribution in the experimental measurement data and the calculated model with 16 channels and effective channel length is 0.75 μm. (P. C. Huang, L. A. Chen, C. C. Chen, and J. T. Sheu, Modeling and Minimizing Variations of Gate-All-Around Multiple-channel Nanowire TFTs, *IEEE NANO 2011 Conference*, August 15–18, 2011, Portland, Oregon. © (2011) IEEE. With permission.)

passivated. As a result, the larger the grain size the less the V_{th}, which presumably result from the fewer trapped levels that need to be filled. The larger the grain size, the smaller the V_{th}. However, a device with a larger grain size exhibits a larger electrical variation. Figure 58.6 presents the threshold voltage distribution related to the different channel numbers with an average grain size of 30 nm. The spread of V_{th} decreases as the channel number increases. This result exhibited that the statistics variation caused by grain size distribution has been balanced by the increasing channel numbers. Also, a comparison of magnitude of threshold voltage distribution from the experimental data and that calculated from the model shows a reasonable agreement as shown in the inset of Figure 58.6.

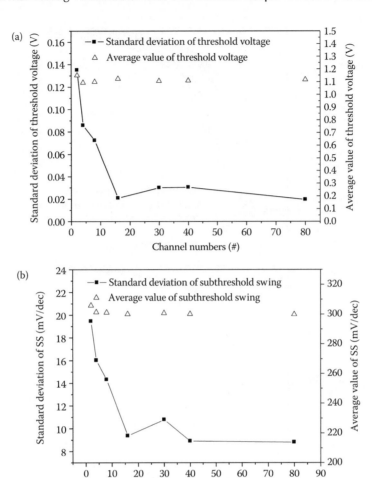

FIGURE 58.7 The electrical fluctuation has been minimized after devices featuring 16 channels. (a) The mean and standard deviation of the threshold voltage and (b) the mean value and standard deviation of subthreshold swing. The statistics were obtained from measurements of 15 GAA NW TFTs from three different chips, at $V_d = 0.1$ V. (Reprinted from *Microelectronic Engineering*, 91, P. C. Huang, L. A. Chen, C. C. Chen, and J. T. Sheu, Minimizing variation in the electrical characteristics of gate-all-around thin film transistors through the use of multiple-channel nanowire and NH_3 plasma treatment, 54–58, (2012), with permission from Elsevier.)

Figure 58.7 shows the standard deviation and average value of threshold voltage and SS of devices with respect to the nanowire channel number. A device with 16 channels reaches a local minimum in both standard deviation of threshold voltage and SS (about 21 mV and 9.38 mV/dec, respectively). Figure 58.7a also shows that the average threshold voltage is around 1.1 V, and that extending the channel width by increasing the channel numbers did not affect the mobility for the constant current extraction method of V_{th}. Figure 58.7b exhibits a similar result of the SS, where the average value is maintained at 300 mV/dec with an increase in the channel number. However, the variation was reduced by increasing the channel number and reached a minimum value after utilizing 16 channels in our proposed device. Many crystallization technologies have been demonstrated to enlarge the grain size and to effectively reduce the impact of the grain boundary in the channel for device performance improvement [4,5]. However, it inherently brings a statistical variation induced by the grain boundary [18]. In this study, a multiple channel design provides a promising alternative to further improve the uniformity of electrical characteristics. Moreover, increasing the number of channels not only decreased the variation of the electrical properties but also improved the on/off

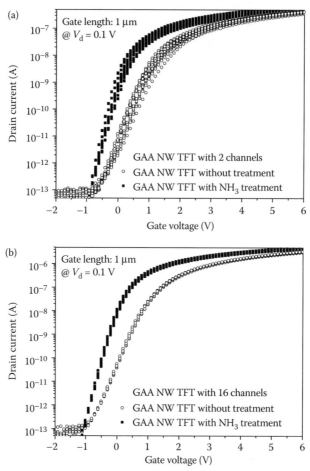

FIGURE 58.8 Transfer characteristics of GAA NW TFTs with (a) 2 and (b) 16 channels, with and without NH$_3$ plasma treatment. $I-V$ curves resulted from the measurements of 15 devices from three chips. (Reprinted from *Microelectronic Engineering*, 91, P. C. Huang, L. A. Chen, C. C. Chen, and J. T. Sheu, Minimizing variation in the electrical characteristics of gate-all-around thin film transistors through the use of multiple-channel nanowire and NH$_3$ plasma treatment, 54–58, (2012), with permission from Elsevier.)

current ratio (as shown in Figure 58.4), presumably due to the increase in the number of corners. However, the off-state leakage current remained tightly controlled by the surrounding electric field.

Figure 58.8 displays the transfer characteristics of GAA NW TFTs prepared with and without NH$_3$ plasma treatment. The curves were then collected from different chips of the same batch. Five devices were characterized from each chip. The multiple-channel structure results in higher treatment efficiency relative to that of a traditional planar device [7]. According to the results shown in Figure 58.2, we observed that the threshold voltage and SS are all improved after NH$_3$ plasma treatment, which reduced the trap states in the channel region. Moreover, the plasma passivation treatment also effectively reduces the electrical fluctuation. The 2-channel devices exhibited further minimization in standard deviation in their values of V_{th} (106 mV) and SS (15.88 mV/dec) compared to the untreated device (V_{th} is 135 mV and SS is 19.44 mV/dec) as shown in Figure 58.8a. Figure 58.8b shows that the treated 16-channel GAA NW TFT exhibits a very small value of standard deviation in its value of V_{th} and SS (30 mV and 11.4 mV/dec, respectively). In spite of that, NH$_3$ plasma-treated 16-channel devices present a slight deterioration in electrical uniformity compared to the untreated ones; however, the device performance still shows obvious improvement.

58.4 CONCLUSION

We have characterized GAA NW TFTs featuring multiple channels to minimize the variation of their electrical characteristics. The gate structure and the number of channels both improved the device performance relative to that of tri-gate TFTs. Although the different gate configurations affected both the mean values and standard deviations of the performance metrics, increasing the number of channels decreased the grain boundary variation, effectively minimizing the electrical variations. To further improve the variations in device performance, devices were subjected to NH_3 plasma treatment to reduce the trap density of states in the grain boundary. The proposed device of GAA NW TFT with 16 channels combined with plasma treatment exhibited a low threshold voltage (0.18 V), a steep SS (184 mV/dec), a high on/off current ratio (1.64×10^8), a small DIBL (58 mV/V), and displayed a minimized standard deviation of V_{th} (30 mV) and SS (11.4 mV). The kink effect was also suppressed. Finally, we also utilize the Poisson area scatter model to characterize the electric variations of poly-Si NW TFTs with different grain size and channel numbers. The result shows a good agreement between the numerical simulation and experiment data. We suspect that such GAA multiple-channel TFTs would be suitable devices for applications in low-voltage circuit operations.

ACKNOWLEDGMENT

We acknowledge the National Science Council, Taiwan, for financial support and the National Nano Device Laboratories for use of their facilities.

REFERENCES

1. K. Yoneda, R. Yokoyama, and T. Yamada, Development trends of LTPS TFT LCDs for mobile applications, in *VLSI Circuits, 2001. Digest of Technical Papers. 2001 Symposium on*, Kyoto, Japan, 2001, pp. 85–90.
2. T. Aoyama, K. Ogawa, Y. Mochizuki, and N. Konishi, Inverse staggered poly-Si and amorphous Si double structure TFT's for LCD panels with peripheral driver circuits integration, *IEEE Transactions on Electron Devices*, 43, 701–705, 1996.
3. Y. Nakajima, Y. Kida, M. Murase, Y. Toyoshima, and Y. Maki, Latest development of system-on-glass display with low temperature poly-Si TFT, *Proc. SID Int. Symp. Dig. Tech. Papers*, pp. 864–867, 2004.
4. A. Baiano, R. Ishihara, J. van der Cingel and K. Beenakker, Strained single-grain silicon n- and p-channel thin-film transistors by excimer laser, *IEEE Electron Device Letters*, 31, 308–310, 2010.
5. C.-W. Chang, S.-F. Chen, C.-L. Chang, C.-K. Deng, J.-J. Huang, and T.-F. Lei, High-performance nanowire TFTs with metal-induced lateral crystallized poly-Si channels, *IEEE Electron Device Letters*, 29, 474–476, 2008.
6. M. Im, J.-W. Han, H. Lee, L.-E. Yu, S. Kim, C.-H. Kim, S. C. Jeon et al., Multiple-gate CMOS thin-film transistor with polysilicon nanowire, *IEEE Electron Device Letters*, 29, 102–105, 2008.
7. J.-T. Sheu, P.-C. Huang, T.-S. Sheu, C.-C. Chen, and L.-A. Chen, Characteristics of gate-all-around twin poly-Si nanowire thin-film transistors, *IEEE Electron Device Letters*, 30, 139–141, 2009.
8. H.-W. Zan, T.-C. Chang, P.-S. Shih, D.-Z. Peng, T.-Y. Huang, and C.-Y. Chang, Analysis of narrow width effects in polycrystalline silicon thin film transistors, *Japanese Journal of Applied Physics, Part 1: Regular Papers and Short Notes and Review Papers*, 42, 28–32, 2003.
9. N. Yamauchi, J. J. J. Hajjar, and R. Reif, Polysilicon thin-film transistors with channel length and width comparable to or smaller than the grain size of the thin film, *IEEE Transactions on Electron Devices*, 38, 55–60, 1991.
10. Y.-C. Wu, T.-C. Chang, C.-Y. Chang, C.-S. Chen, C.-H. Tu, P.-T. Liu, H.-W. Zan et al., High-performance polycrystalline silicon thin-film transistor with multiple nanowire channels and lightly doped drain structure, *Applied Physics Letters*, 84, 3822–3824, 2004.
11. T.-C. Liao, S. W. Tu, M. H. Yu, W. K. Lin, C. C. Liu, K. J. Chang, Y. H. Tai, and H. C. Cheng, Novel gate-all-around poly-Si TFTs with multiple nanowire channels, *IEEE Electron Device Letters*, 29, 889–891, 2008.
12. T. Noguchi, A. J. Tang, J. A. Tsai, and R. Reif, Comparison of effects between large-area-beam ELA and SPC on TFT characteristics, *IEEE Transactions on Electron Devices*, 43, 1454–1458, 1996.

13. H. H. Hsu, H. C. Lin, L. Chan, and T. Y. Huang, Threshold-voltage fluctuation of double-gated poly-Si nanowire field-effect transistor, *IEEE Electron Device Letters,* 30, 243–245, 2009.
14. A. W. Wang and K. C. Saraswat, A strategy for modeling of variations due to grain size in polycrystalline thin-film transistors, *IEEE Transactions on Electron Devices,* 47, 1035–1043, 2000.
15. H. C. Lin, T. W. Lin, H. H. Hsu, C. D. Lin, and T. Y. Huang, Trigated poly-Si nanowire SONOS devices for flat-panel applications, *IEEE Transactions on Nanotechnology,* 9, 386–391, 2010.
16. G. A. Bhat, J. Zhonghe, H. S. Kwok, and M. Wong, Effects of longitudinal grain boundaries on the performance of MILC-TFTs, *IEEE Electron Device Letters,* 20, 97–99, 1999.
17. C. A. Dimitriadis and D. H. Tassis, On the threshold voltage and channel conductance of polycrystalline silicon thin-film transistors, *Journal of Applied Physics,* 79, 4431–4437, 1996.
18. L. Jing, K. Kunhyuk, and K. Roy, Variation estimation and compensation technique in scaled LTPS TFT circuits for low-power low-cost applications, *IEEE Transactions on Computer-Aided Design of Integrated Circuits and Systems,* 28, 46–59, 2009.
19. P. C. Huang, L. A. Chen, C. C. Chen, and J. T. Sheu, Minimizing variation in the electrical characteristics of gate-all-around thin film transistors through the use of multiple-channel nanowire and NH_3 plasma treatment, *Microelectronic Engineering,* 91, 54–58, 2012.
20. P. C. Huang, L. A. Chen, C. C. Chen, J. T. Sheu, Modeling and minimizing variations of gate-all-around multiple-channel nanowire TFTs, *Conference on IEEE NANO 2011,* Portland, Oregon, August 15–18, 2011.

59 Characterization of Gate-All-Around Si-Nanowire Field-Effect Transistor

Extraction of Series Resistance and Capacitance–Voltage Behavior

Yoon-Ha Jeong, Sang-Hyun Lee, Ye-Ram Kim, Rock-Hyun Baek, Dong-Won Kim, Jeong-Soo Lee, and Dae Mann Kim

CONTENTS

59.1 INTRODUCTION

The silicon nanowire field-effect transistor (NWFET) has been extensively studied because of its potential for further scaling down of the complementary metal-oxide semiconductor (CMOS) technology. In particular, it has been reported that the NWFET with gate-all-around (GAA) structure shows excellent gate controllability by which low leakage current and effective suppression in a short channel effect are guaranteed [1–5]. Furthermore, as an effective channel length (L_{EFF}) and channel diameter (d_{NW}) shrink down to a sub-20 nm regime, the quantum confinement effect in volume-channel inversion and less surface scattering of carriers in such a short nanowire channel enables the NWFET to operate in a quasi-ballistic transport regime [6–8]. Owing to this enhanced electrostatic nature, the NWFET is expected to exhibit superb electrical characteristics, including large current drivability and high mobility, some of which have already been observed [1–5,9]. The NWFET, however, in spite of its numerous advantages, has a few technical issues such as parasitic resistance and capacitance components due to the structural particularity [10–12]. It is thus required to exactly extract such parasitic components to exactly estimate the electrical properties and optimize the structure of the NWFET, providing feedback to the fabrication process.

59.2 EXTRACTION AND CHARACTERIZATION OF R_{SD} IN NWFET

The series resistance (R_{SD}) placed at the source/drain region becomes an important parameter as the devices become smaller and smaller. Most parameters of MOSFET have been scaled as devices have shrunk, but R_{SD} could not be decreased because controlling the junction structure in a nanoscale device is a challenging problem [13]. The channel resistance (R_{CH}), for instance, has become correspondingly small as the channel length has been shortened, but R_{SD} has been maintained and thus contributes a comparatively larger fraction to total device resistance ($R_{TOT} = R_{SD} + R_{CH}$) as illustrated in Figure 59.1. As a result, the advanced CMOS process has focused on minimizing R_{SD}. R_{SD} plays an especially crucial role in the degradation of NWFET performance such as drain current (I_D) and transconductance (g_m) due to the complicated junction structure from a one-dimensional nanowire channel to a three-dimensional source/drain area [14]. Accurate extraction of R_{SD}, therefore, is an important issue for the characterization of NWFET [15,16].

59.2.1 EXTRACTION PROCEDURE OF R_{SD} IN NWFET

To extract R_{SD}, some groups of researchers used the channel resistance method (CRM) [3,17]. CRM is a very simple technique to extract R_{SD}, but it is invalid for very short channel devices. R_{TOT} versus gate length curves for different gate voltage often fail to intersect at one point, which is the main key to extract using CRM. Another researcher, T. Tanaka, proposed the YΦ method, which can extract R_{SD} and other parameters such as the first and second mobility attenuation factors (θ_1 and θ_2) [18]. However, it needs several devices with different effective channel lengths (L_{EFF}) to extract a single R_{SD} value. The method assumes that several devices have the same R_{SD}, and thus we cannot examine the individual R_{SD} of each device with the same geometry parameter.

Here, we use the Y function technique based on the Y parameter in NWFET to extract R_{SD} [11]. The Y function technique is an accurate method to extract for very short channel devices and has a simple process. Also, it needs only a single device to obtain the R_{SD} value, not several devices with different effective channel lengths. The extraction method is valid only when the linearity of the Y function is observed in a measured device. The I_D and Y function in strong inversion can be generally expressed as

$$I_D = \mu_{EFF}\, C_{OX}\, \frac{W}{L_{EFF}}(V_G - V_{TH})\, V_D' \tag{59.1}$$

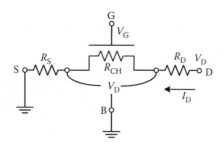

FIGURE 59.1 Equivalent circuit diagram of MOSFET with the source/drain series resistance (R_{SD}) and the channel resistance (R_{CH}).

$$Y \equiv \frac{I_D}{\sqrt{g_m}} = \sqrt{\frac{U_0 V_D'}{1 - \theta_2 (V_G - V_{TH})^2}} \cdot (V_G - V_{TH}) \quad (59.2)$$

where U_0 is $\mu_0 C_{ox}(W_{CIR}/L_{EFF})$ with μ_0 denoting low field mobility, C_{OX} is the gate oxide capacitance per unit area, W_{CIR} is the circumference of the nanowire, L_{EFF} is the effective gate length, and V_D' is the voltage drop at the intrinsic channel ($V_D' \equiv V_D - I_D R_{SD}$). If the θ_2 effect is negligible when the device is operating, the Y function is able to be simply modified as

$$Y \equiv \frac{I_D}{\sqrt{g_m}} \approx \sqrt{U_0 V_D'} \cdot (V_G - V_{TH}) \quad (59.3)$$

In a strong inversion regime, V_{GEFF} can be substituted for $V_G - V_{TH}$, and thus the Y function versus the overdrive voltage curve behaves linearly. Fortunately, in NWFET, the linearity of the Y function, indicating little θ_2, is observed and it makes the extraction of the R_{SD} process much easier and more accurate. Figure 59.2 shows that the Y function of NWFET is almost linear compared with that of the planar MOSFET resulting from the highly suppressed surface roughness scattering due to the volume inversion effect [19].

As shown in Figure 59.3, the Y function curves of NWFETs with different channel circumferences have linearity indicating that the calculation of parameters such as threshold voltage and low field mobility is accurate for all measured devices with different channel circumferences. The procedure for extracting R_{SD} using the Y function technique is listed as follows:

Step 1. In the first step, the initial value U_0 and V_{TH} are extracted from the Y function versus the V_G curve in a strong inversion regime, assuming that V_D' is equal to V_D. The slope of the Y function curve gives U_0, and the x-intercept point of the curve gives V_{TH} (Equation 59.3).

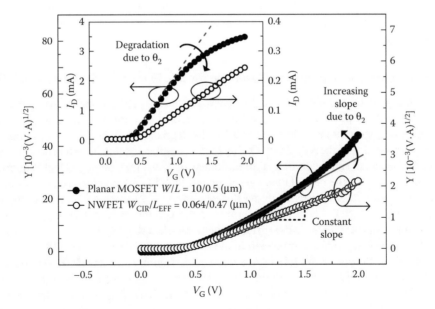

FIGURE 59.2 Measured Y parameter in planar MOSFET and NWFET. The inset shows measured I_D in planar MOSFET and NWFET. Lines represent linearly regressed lines.

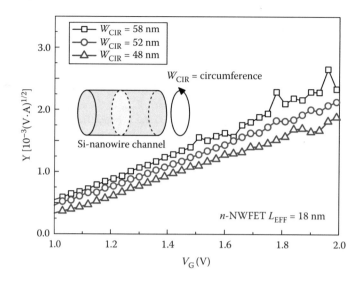

FIGURE 59.3 Y parameter versus V_G curves of NWFETs with different channel circumferences (W_{CIR}). Linear dependence on V_G was observed in devices with all W_{CIR}s.

Step 2. By using Equation 59.1, U_0 and V_{TH} extracted in the previous step are adjusted to the measured I_D and g_m for a V_G range close to V_{TH} and below the V_G point at $g_{m.max}$, where the difference between V_D and V_D' is quite small.

Step 3. Once U_0 and V_{TH} are extracted, R_{SD} can be fitted to I_D and g_m for the entire range, including the strong inversion regime.

Step 4. If the root-mean-square error between the measured I_D and g_m and the simulated I_D and g_m curve is not the minimal in the subthreshold and strong inversion region, steps 2, 3, and 4 should be performed again after replacing V_D by V_D' to extract precise values of U_0, V_{TH}, and R_{SD}. This iteration flow ends when the root-mean-square error is smallest among the error values resulting from step 4. The details of the extraction process are exhibited in Ref. [20].

59.2.2 Modeling and Comparison Result of the R_{SD} Extracting Procedure in NWFET

Figure 59.4 shows that the simulated I_D and g_m curves with extracted parameters using the Y function technique such as U_0, mobility attenuation factors, and R_{SD} are well fitted to the measured curves. The most important factor in Figure 59.4 is the degradation in the I_D curve with increasing V_G, which is mainly due to the R_{SD} effect. Owing to the complicated junction structure in NWFET, from a one-dimensional source/drain to a three-dimensional channel, the R_{SD} effect is dominant in I_D degradation. Also, I_D, the degradation caused by the mobility attenuation effect, is almost negligible, and this result has consistency in the linearity of the Y function curve. Again, the volume inversion effect in NWFET, where the carrier moves through the channel forming away from the oxide-to-channel interface, reduces the surface roughness scattering [8,21].

Figure 59.5 shows R_{SD} values normalized in the n- and p-type NWFETs with 48 and 58 nm circumference (W_{CIR}). Because NWFETs have a large dispersion in electrical characteristics, including R_{SD}, the average values of R_{SD} with the same geometry parameter must be examined. In both the 48 and 58 nm W_{CIR} devices, the averaged value of normalized R_{SD} in n-type devices is larger than that in p-type devices. Boron atoms that are implanted into the source/drain region in p-type MOSFET is more diffusive than arsenic, which is used in n-type MOSFET [2,3], and thus it makes the channel-to-source/drain junction, where the R_{SD} value is essentially determined, more ohmic.

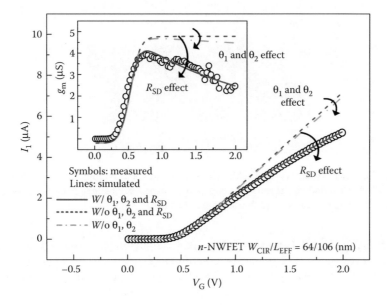

FIGURE 59.4 Measured I_D of NWFET (open symbols), simulated I_D without the effect of mobility attenuation factors (θ_1 and θ_2) and R_{SD} (dashed line), simulated I_D without the effect of R_{SD} or after considering the effects of θ_1 and θ_2 only (dash-dotted line), and simulated I_D with the effects of θ_1, θ_2, and R_{SD} (solid line).

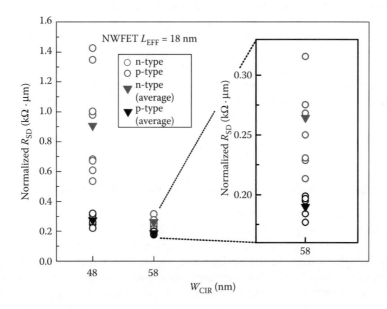

FIGURE 59.5 Extracted R_{SD} values of several NWFETs with different circumferences and doping types.

59.3 INVESTIGATION ON THE *C–V* CHARACTERISTICS AND EFFECT OF THE BOTTOM PARASITIC TRANSISTOR

59.3.1 DEVICE STRUCTURE AND *C–V* MEASUREMENT

The capacitance–voltage (*C–V*) characterization, including gate-to-channel capacitance, includes the essential information of a transistor such as threshold voltage, flatband voltage, inversion

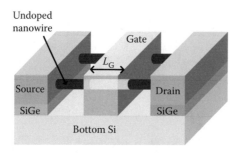

FIGURE 59.6 Schematic of the NWFET with twin Si nanowires. The bottom Si is a p-type substrate and the gate material is TiN, surrounding the intrinsic (undoped) Si nanowire channels. The SiGe layer is embedded under the source/drain region to suspend the nanowire channels.

charge, and mobility [22,23]. However, $C–V$ characterization in the NWFET has technical issues that need to be investigated in depth due to structural properties. The undoped Si nanowire channel is surrounded by gate material and suspended by the embedded SiGe layer under the source/drain region (refer to Figure 59.6). The structure is built on the Si substrate from which the undoped nanowire channel is electrically floated by a thick dielectric. However, the bottom Si bulk acts as a parasitic transistor with different V_{TH} and mobility from an intrinsic nanowire channel transistor [24] by a gate signal delivered through the SiGe layer. Since the two types of transistors that share the gate and source/drain contribute differently to the charge supply, which determines the capacitance behavior, an extensive study on each capacitance component of the NWFET is needed.

In this study, the array structure that consists of a 100×100 single nanowire capacitor (NWCAP) with $W_{CIR}/L_G = 64/250$ nm was used for experiments as the single device has an extremely scaled active area and is hard to be precisely measured. The detailed fabrication process is reported in Refs. [2,9]. To obtain each capacitance component, various basic configurations were set up; C_{GSDB} represents the capacitance measured with all terminals grounded, C_{GB} represents the capacitance measured with the source/drain floated and body grounded, and C_{GS} denotes the capacitance measured with the body floated and source/drain grounded, respectively. All capacitance values are normalized by the number of nanowires in the array.

59.3.2 Gate Signal Response of the Intrinsic Si Nanowire Channel

In this section, the capacitance generation mechanism from component to component is discussed, based mainly upon the descriptions reported in Refs. [12,25].

The total capacitance modulated by a gate signal consists of two components: capacitance from electrons that are injected from the source/drain into the intrinsic Si nanowire channel and capacitance from electrons and holes that are generated at the bottom Si interface. As shown in Figure 59.7, C_{GB} shows a substantial reduction in its value in the strong inversion region (positive V_G for an n-type NWFET) compared to C_{GSDB} because the electron injection from the source/drain is eliminated. In such a gate bias region, electrons in the C_{GB} configuration are supplied from the generation in the bottom p-type Si substrate, in which the electrons are minor carriers, showing a frequency-dependent behavior. On the other hand, in the accumulation region (negative V_G for an n-type NWFET), C_{GSDB} and C_{GB} have almost the same values, indicating that C_{GB} by "fast" hole charging/discharging is most responsible for C_{GSDB}. This interpretation is conceptually illustrated in Figure 59.8. The effect of charging/discharging from the bottom Si only represents the C_{GB} component.

C_{GSD} can be understood in a similar way. Contrary to the C_{GB}, C_{GSD} represents the capacitance component whose charge is supplied solely by the injection from the source/drain into the nanowire channel. In the strong inversion region, C_{GSD} is independent of the frequency as a sufficient number

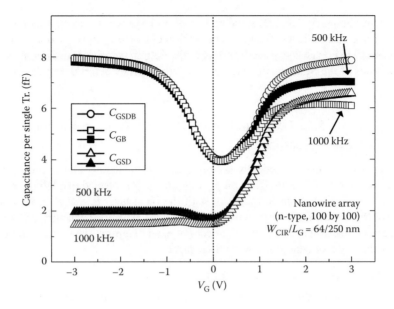

FIGURE 59.7 *C–V* curves of the n-type NWFET (n-NWFET) with several biasing configurations. Only 1000 kHz data are presented for C_{GSDB} as no frequency dependence was observed in all V_G range, indicating that a majority of the carriers (electrons or holes) can be supplied from either source/drain or bottom Si or both at any V_G regime. All capacitance values were normalized and represent the value of the single transistor.

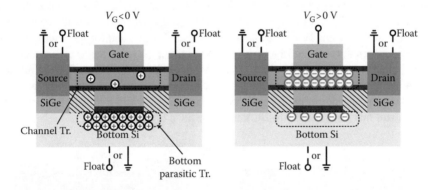

FIGURE 59.8 Illustrative charge distribution in the n-type NWFET with negative and positive V_G conditions, respectively. With a corresponding combination of grounded or floated source/drain and body, the charge supply of C_{GSDB}, C_{GB}, and C_{GSD} can be estimated.

of electrons are injected from the source/drain to follow the gate signal (Figure 59.7). The reason why C_{GSD} shows slightly smaller values than C_{GSDB} is that C_{GSDB} contains a "slow" supply of electrons from the bottom Si. It is important to note, however, that the charging/discharging speed of electrons in the Si channel supplied by the source/drain is much faster than electron generation in the bottom Si. Furthermore, steep V_G dependence in C_{GSD} means effective channel inversion and charge supply of the nanowire channel transistor [12]. In the accumulation region, C_{GSD} is much smaller than C_{GSDB} or C_{GB}, resulting from the dominant hole injection from the bottom Si over the hole injection from the source/drain region. It also indicates that the intrinsic nanowire channel has a better off-state gate controllability, thereby suppressing leakage current. This interpretation is illustratively described in Figure 59.8. The effect of charging/discharging from an undoped nanowire channel only represents the C_{GSD}.

As depicted in Figures 59.7 and 59.8, the bottom Si bulk plays a role as another parasitic bottom transistor and has an insignificant effect on the electrical characteristics in the operation region [24]. Still, it is necessary to precisely extract the effect of the bottom parasitic transistor to investigate the intrinsic properties of the Si nanowire channel explicitly, thereby determining the inversion status of the device.

59.3.3 EFFECT OF THE SiGe LAYER AND BOTTOM PARASITIC TRANSISTOR IN THE Si SUBSTRATE ON *C–V* CHARACTERISTICS

The SiGe layer is embedded to basically sustain a nanowire channel and improve the electrical characteristics of the p-type NWFET by compressive stress. It is needed to qualitatively investigate the parasitic effect of the bottom Si transistor with the SiGe layer to estimate the exact intrinsic performance of the device. Based on the results in Ref. [26], parasitic capacitance is analyzed with C_{GSD} measurement results with the help of the energy band diagrams. The simulation result is also added to show the effect of conditions in the SiGe layer.

59.3.3.1 Roles of the SiGe Layer in NWFET

In the NWFET structure investigated, owing to the application of the SiGe layer under the source/drain region as a compressive stressor to the nanowire channel, p-type NWFET exhibits greatly enhanced device performance with a slightly degraded n-type NWFET [2]. It was reported that the compressive stress increases as the thickness of SiGe increases, enhancing the hole mobility due to the quantum-confinement-induced suppression in surface scattering [27]. As illustrated in Figure 59.9a, the gate trimming structure of the single devices comprises two different cross sections: A–A′ in Figure 59.9b with the nanowire channel region and B–B′ in Figure 59.9c without the nanowire channel region. In particular, the unexpected electrical leakage path delivering the gate signal to the bottom Si area is formed along B–B′. The contact between the gate and the source/drain causes the SiGe layer to be inverted, resulting in a turn-on of the bottom Si substrate as a parasitic transistor. The parasitic bottom transistor affects the C_{GSD} even though the body is floated. It is thus important to examine the generation mechanism of the parasitic bottom transistor and its capacitance behavior to explicitly extract properties of the intrinsic nanowire transistor and optimize the structure of the nanowire transistor fabricated on the Si Bulk with the SiGe layer embedded.

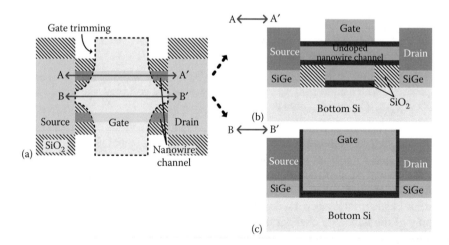

FIGURE 59.9 (a) Top view of the NWFET having two nanowire channels. Owing to the gate trimming process, the device has two different cross sections. (b) A–A′ with the nanowire channel between the source and the drain and (c) B–B′ without the channel. The contact between the gate and the source/drain along B–B′ forms an undesired leakage path delivering the gate signal through the SiGe layer.

59.3.3.2 Analysis of the Turning-On Effect of the Parasitic Bottom Transistor

Figure 59.10 shows the C_{GSD} of the n- and p-NWFET with the parasitic bottom transistor and the nanowire channel transistor depicted, respectively. The effect of the parasitic bottom transistor can be understood using charging/discharging behavior, considering the structural effect in the NWFET with energy band diagrams in Figures 59.11, 59.12, and 59.13, which describe the spatial charge placement in the gate, undoped nanowire channel, and bottom Si substrate at each bias condition. Near $V_G = 0.2$ V, the nanowire channel transistor is turned on so that the electron conduction starts between the source and the drain (Figure 59.11), and this is corresponding to the early steep increase in C_{GSD} in Figure 59.10. The SiGe layer is not inverted enough to provide the electron conduction path through which the gate, source/drain, and bottom Si are connected (Figure 59.12a). Considering the work function of the TiN metal gate, band bending in the SiGe layer and bottom Si substrate is not sufficient yet (Figure 59.12b). However, as V_G increases further and reaches 0.9 V, the SiGe is turned on to act as an electron channel, supplying a number of electrons to the bottom

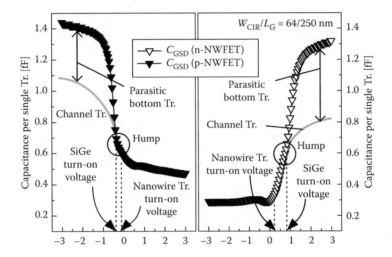

FIGURE 59.10 C_{GSD} versus V_G curves of the n-type and p-type NWFET. The hump in the curve corresponds to the SiGe turning-on point and the channel transistor (in thick line) is conceptually drawn to explicitly express the effect of the bottom parasitic transistor.

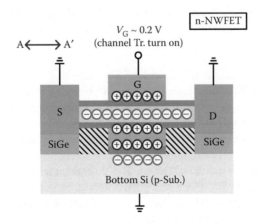

FIGURE 59.11 Charge distribution of the *n*-type NWFET along A–A' in the nanowire channel and bottom Si substrate at $V_G = 0.2$ V, which is the Si nanowire channel transistor turn-on point.

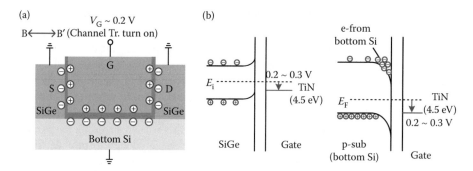

FIGURE 59.12 (a) Charge distribution of the n-type NWFET along B–B′ in the gate and bottom Si substrate at $V_G = 0.2$ V. (b) The SiGe layer and bottom parasitic transistor are not sufficiently biased to be inverted.

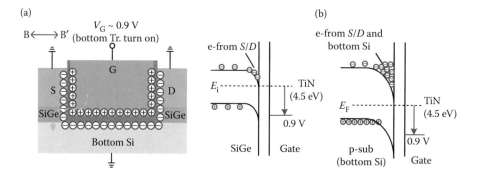

FIGURE 59.13 (a) Charge distribution of the n-type NWFET along B–B′ in the gate and bottom Si substrate at $V_G = 0.9$ V. (b) The SiGe layer and the bottom parasitic transistor are sufficiently biased and strongly inverted. SiGe acts as a channel, a conduction path between the source/drain, gate, and bottom Si transistor so that the electrons from the source/drain are injected to the bottom Si interface area.

Si substrate (Figure 59.13a). The energy band bending in such a bias condition is enough for both the SiGe and the bottom Si region to be strongly inverted (Figure 59.13b). This is responsible for a hump in the C_{GSD}–V_G curve in Figure 59.10 (around SiGe turn-on voltage) even though the body is floated. The effect of the parasitic bottom transistor in p-NWFET can be explained in a similar way. The difference from n-NWFET is the value of the turn-on voltage of the SiGe layer. The built-in potential applies as a barrier to an electron in the n-NWFET but helps a hole to be inverted.

59.4 CONCLUSION

The structure of the NWFET includes parasitic components such as series resistance R_{SD} and capacitance in the bottom parasitic transistor. It is important to precisely characterize the parasitic components to evaluate the intrinsic performance of the device.

The Y function technique is known to be helpful in dealing with devices such as the NWFET, which has a large sample-to-sample variation in its electrical characteristics, because this technique enables the R_{SD} to be extracted from the individual device. The Y function technique is valid only if a parameter Y is linearly dependent on V_G, which was proved to be true in the NWFETs due to the volume inversion effect and thus suppressed the surface roughness scattering. Furthermore, the parameter Y was also observed to have linearity in devices having different channel circumference W_{CIR} (or effective width). The R_{SD} extraction procedure, where low field mobility, threshold voltage, and effective drain voltage considering the voltage drop across the R_{SD} are iteratively corrected, was proposed as the

Y-parameter-based modeling methodology. It was shown that simulated data are successfully fitted to the measured I_D and g_m data. Most of the degradation in the I_D is from the R_{SD} due to the complicated junction structure and quantum effect in the nanowire channel. Also, R_{SD} values were extracted with respect to the doping type. It was found that the R_{SD} of the p-type NWFET are much smaller than that of the n-type NWFET since, with the high diffusivity of boron in the p-well, the channel-to-source/drain junction in the p-type NWFET becomes more ohmic than in the n-type NWFET.

Using various biasing configurations, capacitance components in the NWFET were obtained. The value of each capacitance component and its frequency are dependent on which charging/discharging source, charge injection from the source/drain into the channel, or generation from the bottom Si transistor contributes to the electron/hole supply. It was concluded that in spite of the floated body condition, the bottom parasitic transistor turns on as the V_G increases because the SiGe layer is inverted and acts as a leakage path delivering the gate signal to the bottom parasitic transistor. The parasitic transistor affects the inversion status of the NWFET with the intrinsic nanowire channel transistor.

REFERENCES

1. S. D. Suk, S.-Y. Lee, S.-M, Kim, E.-J. Yoon, M.-S. Kim, M. Li, C. W. Oh et al., High performance 5 nm radius twin silicon nanowire MOSFET (TSNWFET): Fabrication on bulk Si wafer, characteristics and reliability, in *Proc. IEDM Tech. Dig.*, 2005, pp. 717–720.
2. K. H. Yeo, S. D. Suk, M. Li, Y.-Y. Yeoh, K. H. Cho, K.-H. Hong, S. K. Yun et al., Gate-all-around (GAA) twin silicon nanowire MOSFET (TSNWFET) with 15 nm length gate and 4 nm radius nanowires, in *Proc. IEDM Tech. Dig.*, 2006, pp. 1–4.
3. S. D. Suk, M. Li, Y. Y. Yeoh, K. H. Yeo, K. H. Cho, I. K. Ku, H. Cho et al., Investigation of nanowire size dependency on TSNWFET, in *Proc. IEDM Tech. Dig.*, 2007, pp. 891–894.
4. S. D. Suk, K. H. Yeo, K. H. Cho, M. Li, Y. Y. Yeoh, K.-H. Hong, S.-H. Kim et al., Gate-all-around twin silicon nanowire SONOS memory, in *Proc. VLSI Symp. Tech. Dig.*, 2007, pp. 142–143.
5. S. D. Suk, Y. Y. Yeoh, M. Li, K. H. Yeo, S.-H. Kim, D.-W. Kim, D. Park and W.-S. Lee, TSNWFET for SRAM cell application: Performance variation and process dependency, in *Proc. VLSI Symp. Tech. Dig.*, 2008, pp. 38–39.
6. K. H. Cho, S. D. Suk, Y. Y. Yeoh, M. Li, K. H. Yeo, D.-W. Kim, S. W. Hwang, D. Park, and B.-I. Ryu, Observation of single electron tunneling and ballistic transport in twin silicon nanowire MOSFETs (TSNWFETs) fabricated by top-down CMOS process, in *Proc. IEDM Tech. Dig.*, 2006, pp. 1–4.
7. K. Uchida, J. Koga, R. Ohba, T. Numata, and S. Takagi, Experimental evidences of quantum-mechanical effects on low-field mobility, gate-channel capacitance and threshold voltage of ultrathin body SOI MOSFETs, in *Proc. IEDM Tech. Dig.*, 2001, pp. 633–636.
8. J. Chen, T. Saraya, K. Miyaji, K. Shimizu, and T. Hiramoto, Experimental study of mobility in [110]- and [100]-directed multiple silicon nanowire GAAMOSFETs on (100) SOI, in *Proc. VLSI Symp. Tech. Dig.*, 2008, pp. 32–33.
9. M. Li, K. H. Yeo, S. D. Suk, Y. Y. Yeoh, D.-W. Kim, T. Y. Chung, K. S. Oh, and W.-S. Lee, Sub-10 nm gate-all-around CMOS nanowire transistors on bulk Si Substrate, in *Proc. VLSI Symp. Tech. Dig.*, 2009, pp. 94–95.
10. F. Léonard and A. A. Talin, Size-dependent effects on electrical contacts to nanotubes and nanowires, *Phys. Rev. Lett.*, 97, 026804-1-02684-4, Jul. 2006.
11. R.-H. Baek, C. K. Baek, S. W. Jung, Y. Y. Yeoh, D.-W. Kim, J.-S. Lee, D. M. Kim, and Y.-H. Jeong, Characteristics of the series resistances extracted from Si-nanowire FETs using the Y-function technique, *IEEE Trans. Nanotechnol.*, 9(2), 212–217, Mar. 2010.
12. R.-H. Baek, C.-K. Baek, S.-H. Lee, S. D. Suk, M. Li, Y. Y. Yeoh, K. H. Yeo et al., C-V characteristics in undoped gate-all-around nanowire FET array, *IEEE Electron Device Lett.*, 32(2), 116–118, Feb. 2011.
13. K. K. Ng and W. T. Lynch, The impact of intrinsic series resistance on MOSFET scaling, *IEEE Trans. Electron Devices*, 34(3), 503–511, Mar. 1987.
14. J. Hu, Y. Liu, C. Z. Ning, R. Dutton, and S.-M. Kang, Fringing field effects on electrical resistivity of semiconductor nanowire-metal contacts, *Appl. Phys. Lett.*, 92, 083503-1-083503-3, Feb. 2008.
15. S. D. Kim, C.-M. Park, and J. C. S. Woo, Advanced model and analysis for series resistance in sub-100 nm CMOS including poly-depletion and overlap doping gradient effect, in *Proc. IEDM Tech. Dig.*, 2000, pp. 723–726.

16. S. Thompson, P. Packan, T. Ghani, M. Stettler, M. Alavi, I. Post, S. Tyagi, S. Ahmed, S. Yang, and M. Bohr, Source/drain extension scaling for 0.1 μm and below channel length MOSFETs, in *Proc. VLSI Symp. Tech. Dig.,* 1998, pp. 132–133.

17. J. Kim, S. Yang, J. Lee, S. D. Suk, K. Seo, D. Park, B.-G. Park, J. D. Lee, and H. Shin, Investigation of mobility in twin silicon nanowire MOSFETs (TSNWFETs), in *Proc. Int. Conf. Solid-State Integr. Circuit Technol. (ICSICT),* 2008, pp. 50–52.

18. T. Tanaka, Novel parameter extraction method for low field drain current of nano-scaled MOSFETs, in *Proc. Int. Conf. Microelectron. Test Struct. (ICMTS),* 2007, pp. 265–267.

19. R.-H. Baek, C.-K. Baek, S.-W. Jung, Y. Y. Yeoh, D.-W. Kim, J.-S. Lee, D. M. Kim, and Y.-H. Jeong, Comparison of series resistance and mobility degradation extracted from n- and p-type Si-nanowire field effect transistors using Y-function technique, *Jpn. J. Appl. Phys.,* 49(4), 04DN06–04DN06-5, Apr. 2010.

20. R.-H. Baek, C.-K. Baek, S.-W. Jung, Y. Y. Yeoh, D.-W. Kim, J.-S. Lee, D. M. Kim, and Y.-H. Jeong, Series resistance behavior extracted from silicon nanowire transistors using the Y-function technique, *Int. Conf. Solid State Dev. Mater. (SSDM),* 2009, pp. 1108–1109.

21. Y.-R. Kim, S.-H. Lee, C.-K. Baek, R.-H. Baek, K.-H. Yeo, D.-W. Kim, J.-S. Lee, and Y.-H. Jeong, Reliable extraction of series resistance in silicon nanowire FETs using Y-function technique, in *Proc. Int. Conf. Nanotechnol. Mater. Devices (NMDC),* 2011, pp. 262–265.

22. S. Severi, L. Pantisano, E. Augendre, E. San Andrés, P. Eyben, and K. De Meyer, A reliable metric for mobility extraction of short-channel MOSFETs, *IEEE Trans. Electron Devices,* 54(10), 2690–2698, Oct. 2007.

23. I. Ferain, L. Pantisano, B. J. O'Sullivan, R. Singanamalla, N. Collaert, M. Jurczak, and K. De Meyer, Methodology for flatband voltage measurement in fully depleted floating-body FinFETs, *IEEE Trans. Electron Devices,* 55(7), 1657–1663, Jul. 2008.

24. S. D. Suk, K. H. Yeo, K. H. Cho, M. Li, Y. Y. Yeoh, S.-Y. Lee, S. M. Kim et al., High-performance twin silicon nanowire MOSFET (TSNWFET) on bulk Si wafer, *IEEE Trans. Nanotechnol.,* 7(2), 181–184, Mar. 2008.

25. Y.-H. Jeong, R.-H. Baek, C.-K. Baek, K. H. Yeo, D.-W. Kim, J. Y. Chung, and D. M. Kim, Comparative study of C-V characteristics in Si-NWFET and MOSFET, in *Proc. Int. Conf. Nanotechnol. Mater. Devices (NMDC),* 2010, pp. 26–29.

26. R.-H. Baek, M.-D. Ko, S.-H. Lee, C.-K. Baek, K. H. Yeo, D.-W. Kim, J.-S. Lee, and Y.-H. Jeong, Analysis of parasitic bottom capacitance in n- and p-type Si-nanowire field effect transistors on bulk, in *Proc. Int. Conf. Nanotechnol.,* 2011, pp. 139–143.

27. M. Li, K. H. Yeo, Y. Y. Yeoh, S. D. Suk, K. H. Cho, D.-W. Kim, D. Park, and W.-S. Lee, Experimental investigation on superior PMOS performance of uniaxial strained <110> silicon nanowire channel by embedded SiGe source/drain, in *Proc. IEDM Tech. Dig.,* 2007, pp. 899–902.

Section XVIII

Nanomagnetic Logic

60 Nonvolatile Logic-in-Memory Architecture

An Integration between Nanomagnetic Logic and Magnetoresistive RAM

Jayita Das, Syed M. Alam, and Sanjukta Bhanja

CONTENTS

60.1 INTRODUCTION

The use of magnetism for data storage dates back to the early days of computers when magnetic core memories were popular. Magnetic core memories were built of tiny magnetic toroids (commonly called cores). The memories exhibited hysteresis characteristics that enabled them to store 1 and 0 along two stable states of magnetization [1–3]. The magnetic core memories were rewritable, non-volatile, and radiation hard. However, they were bulky and slow. On the other hand, the semiconductor memories, static RAM (SRAM), and dynamic RAM (DRAM) that replaced the magnetic core memories are fast, rewritable, and compact. However, they are volatile [4] and require constant voltage supply. In addition, the DRAM requires periodic refreshing to retain data. With the introduction of semiconductor memories, both data storage and logic computation became electronic in nature. But the storage and computation continued to remain in separate regions inside a computer. Any logic operations inside the central processing unit (CPU) required data to be fetched from memory, which in turn posed a bottleneck to high-speed computing. To resolve this issue, researchers started focusing on ways to embed logic and memory into a single plane and various logic-in-memory architectures were developed. Various devices were also studied as potential candidates for logic-in-memory applications. Further discussion on these previous studies in logic-in-memory is scheduled for Section 60.5.1.

The latest trend in data storage has moved back to the magnetic domain. The modern magnetoresistive memories, called magnetoresistive RAMs (MRAMs), use magnetization to store binary information. MRAMs are nonvolatile and radiation hard, and can sustain high temperatures. They are built using magnetic tunnel junctions (MTJs) that have correlated electric and magnetic properties. MTJs are primarily composed of two single-domain ferromagnetic layers that are vertically stacked and separated by a nonmagnetic barrier layer. One of the layers has a fixed magnetization while the other is free to rotate in a plane. The layers are called the fixed and free layers respectively, in accordance with their magnetic properties. In MTJs, a logic 1 or a logic 0 is represented with the help of relative magnetization between the free and fixed layers (see Figure 60.1). The 1 or 0 in an MTJ is written with the help of external magnetic fields or a spin-transfer torque (STT) current. In Toggle MRAMs [5], external magnetic fields generated by current-carrying wires are used for writing. In the more recent STT-MRAMs, STT current is used to write 1 or 0 into MTJs. The STT current-induced writing is low power in comparison to external field-driven writing and will be the only writing mechanism considered in this chapter. The value once written into MTJ remains stored through magnetic remanence and there is no leakage of magnetism with time. This makes the MRAMs truly nonvolatile memories.

Apart from STT current-induced magnetization switching, the electrical resistance of MTJs is also closely interlinked to their magnetic state. The electrical resistances (R_0 and R_1) of the logic 0 and 1 states of MTJs are well distinguished from each other and their difference is used to define the tunnel magnetoresistance (TMR) of MTJs (see Equation 60.1).

$$\text{TMR} = \frac{R_1 - R_0}{R_0} \quad \text{where } R_1 > R_0 \tag{60.1}$$

The electrical resistance of MTJs is used to read the bits in MRAM. MTJs are also integrable to CMOS. In MRAM, every single MTJ is integrated to an NMOS access transistor [6] (see Figure 60.2). The combined MTJ-CMOS unit is then used to define a bit in memory. In MRAM, three layers of metal are used to define the source, bit, and word lines. The bit and source lines are used to apply the desired potential across MTJs, whereas the word line is used to select the access transistors of the MTJs. Together, the three metal lines are used to select a bit in memory for read or write.

The placement of MTJs inside MRAM is like a regular 2D array with a separation d between any two MTJs (see Figure 60.3). The bit and source lines are parallel to the rows of MTJs while the word lines are parallel to the columns of MTJs. For example, to select a bit in the (i,j)th location in MRAM, i and j being the row and column numbers, respectively, the jth word line is first raised high. An appropriate potential is next applied across the ith bit and source line pair. The separation d determines the density of the memory. The lower the d, the denser is the memory. However, lowering d beyond a limit d_m increases the tendency of the interaction between the free layers of MTJs. This interaction is undesirable in memory where the bits are independent of each other. The interaction would however play a decisive role if used in logic. Designed at this junction of logic and memory behavior of MTJs is the logic-in-memory architecture [7] that is described in this chapter.

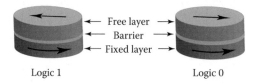

Logic 1   Logic 0

FIGURE 60.1 Logic states in MRAM.

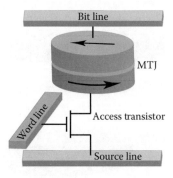

FIGURE 60.2 MTJ with an access transistor. (From G. Sun et al., A novel architecture of the 3D stacked MRAM l2 cache for cmps, in *IEEE 15th International Symposium on High Performance Computer Architecture, 2009, HPCA 2009*, pp. 239–249, Feb. 2009.)

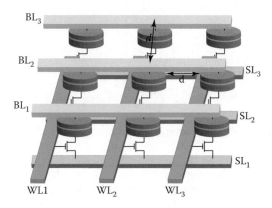

FIGURE 60.3 MRAM layout.

The main motivation behind the logic-in-memory design has remained the same: to reduce data traffic between memory and the CPU. The goal of the architecture described in this chapter is to compute some simplistic logic within memory while performing the complex operations in the CPU. The architecture is regular like MRAM in its MTJ placements. However, it remains unique from MRAM in its CMOS integration and layout of metal lines. In this architecture, the clock signal acts as a classifier between logic and memory operations. With a clock, the MTJs interact among their free layers and compute logic. Without a clock, the MTJs retain their contents and behave as memory. In the next few sections, we will closely focus on the operating principles and the key specifications of the architecture.

60.2 OPERATING PRINCIPLES OF LOGIC INSIDE THE ARCHITECTURE

The working principles of nanomagnetic logic (NML) [8] govern the logic execution inside the architecture. In NML, single-domain single-layer nanomagnets are used to build logic. The magnetizations in the nanomagnets are used to represent 1 and 0 in logic (see Figure 60.4). Both the logic states are along the easy axes of the nanomagnets, which are also the energy minimum states for the nanomagnets (see Figure 60.5). The magnetic interaction between the nanomagnets, together with a clocking signal, is used to compute logic. External magnetic fields generated by current carrying wires are used as a clock in NML [9–11]. When clocked, the magnetization in the nanomagnets orient along the saddle points or the energy maximum state (see Figure 60.6). This state is unstable

FIGURE 60.4 Bit representation in single-layer nanomagnets.

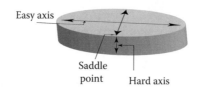

FIGURE 60.5 Easy axis, hard axis, and saddle point in a nanomagnet.

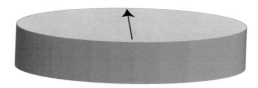

FIGURE 60.6 A clocked single-layer nanomagnet in NML.

and once the clock is released, the nanomagnets settle in accordance with the net magnetic fields acting on them. The magnetic interaction or coupling between the nanomagnets are of two different types: ferromagnetic and antiferromagnetic. When ferromagnetically coupled, the nanomagnets have the same magnetization. When antiferromagnetically coupled, the nanomagnets have opposite magnetizations (see Figure 60.7).

In this architecture, the logic computation takes place in the free layers of MTJs. The free layers behave like the single-domain nanomagnets in NML. The dipolar magnetic coupling between the free layers of MTJs takes responsibility for logic computation. Two ferromagnetically coupled MTJs will have their free layers magnetized in the same direction while two antiferromagnetically coupled MTJs will have their free layers magnetized in opposite directions (see Figure 60.8). However, like NML, these couplings play an active role in logic computation only when they are accompanied by a suitable clock signal.

Like NML [12], the clock in this architecture has dual functions: (i) to synchronize events during the course of logic execution and (ii) to help the magnetization of the free layer overcome the potential barrier separating the logic 0 from the logic 1 state and vice versa. When clocked, the MTJs are in an energy maximum state [13]. When the clock is released, the MTJs settle into one of their energy minimum states (i.e., logic 0 or logic 1), depending on the resultant coupling force

Ferromagnetic coupling   Antiferromagnetic
                                 coupling

FIGURE 60.7 Dipolar magnetic coupling between single-layer nanomagnets.

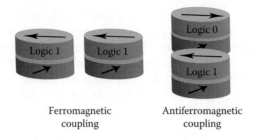

FIGURE 60.8 Dipolar magnetic coupling between free layers of MTJs.

acting from neighboring MTJs. However, the implementation of the clock in this architecture is uniquely different from that of NML. It uses a train of STT current pulses to clock the MTJs [14]. The STT current-driven clock used in this architecture is low-power compared to the external field-driven clock practiced in NML. The STT clock also requires the fixed layers of MTJs to have a 45° magnetization polarization (see bottom layer in Figure 60.8). The magnetization of the free layer of MTJs remains in plane. The other two major operations in logic are writing inputs and reading outputs. In this architecture, both the operations are carried out electrically using current. Further details of the clocking, writing, and reading techniques followed in the architecture are discussed under Section 60.4.

60.3 SPECIFICATIONS OF THE LOGIC-IN-MEMORY ARCHITECTURE

A section of the architecture is shown in Figure 60.9. The MTJs rest above a CMOS plane that houses access transistors for selected MTJs in the architecture. Ideally, a designer would prefer to have an access transistor with every MTJ in the architecture to gain user control over every logic cell. However, certain technology constraints have prohibited this ideal integration between access transistors and MTJs in this architecture as we will discuss below.

For successful logic computation, MTJs need to be 20 nm apart [15]. The planar surface area chosen for this architecture is 100×50 nm^2. The values are derived from the single-domain behavior of MTJs and stability at room temperature. On the other hand, the metal pitch of the CMOS technology to be used determines the minimum separation between the two access transistors. With the latest available 22 nm CMOS technology node, the metal pitch requirement for the layer 1 is 64 nm [16]. This is again the minimum metal pitch available from any present CMOS technology. This contention between the CMOS access transistor spacing and the MTJ spacing puts a restriction

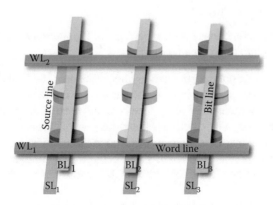

FIGURE 60.9 Regular placement of MTJs in the logic-in-memory architecture. Only the dark gray MTJs have an access transistor integrated underneath. For simplicity, access transistors are not shown in this figure.

over CMOS integration with every MTJ in the architecture. The integration between CMOS and MTJs therefore needs to be optimized. As an optimum solution to this problem, in this architecture, an access transistor is integrated with every alternate MTJ that is located either in a column or in a row (see Figure 60.9). The outcome is that we now have two sets of MTJs in the architecture: one with an access transistor (Figure 60.9: X_{11}, X_{13}, X_{31}, X_{33}) and the other without an access transistor (Figure 60.9: X_{12}, X_{21}, X_{22}, X_{23}, X_{32}). The rule of their placements in the architecture sets the salient features for the architecture [14].

60.3.1 SALIENT FEATURES OF THE ARCHITECTURE

1. An access transistor for every 2×2 MTJ array: Figure 60.9 shows the integration of access transistors with MTJs in the architecture. All the access transistors in the architecture are NMOS without any exception.
2. A source and bit line for every row and a word line for every alternate column: A source and bit line pair, housed in metal layers 1 and 2, run parallel to the rows of cells in the architecture. The purpose of the source and bit lines in the architecture is to supply a suitable potential across MTJs during writing, clocking, and reading. The bit lines are always aligned with the row axes of the architecture. They are connected to the free layers of the MTJs through vias (see Figure 60.10). The connection between the source lines and MTJs vary depending on the presence or absence of access transistors. For MTJs without access transistors, the source lines are directly connected to the fixed layers of the MTJs through vias. For MTJs with access transistors, the source lines are connected to either the source or drain terminals of the access transistors. The second terminal (drain or source) of the access transistor then connects to the fixed layer of the MTJ, thereby completing the electrical path between the source and bit line through the MTJ (see Figures 60.10 and 60.11).

   The word line is housed in metal layer 3 and runs orthogonal to the first two metal layers. The word line is used to connect the gates of access transistors that are all integrated to MTJs in a single column (see Figure 60.10). Since the access transistors are present with MTJs that lie in alternate columns, the word lines are also present along alternate columns of MTJs (see Figure 60.9). The metal pitch calculated from the layout (Figure 60.11) for the source, bit, and word lines are 70, 120, and 140 nm, respectively. The metal 1 pitch in 22 nm CMOS technology is 64 nm [16], while the metal 2 and 3 pitches are each >64 nm. The architecture therefore satisfies 22 nm CMOS pitch requirements.

60.3.2 CELL TYPES IN THE ARCHITECTURE

Depending on their assigned functions inside a logic, the MTJs (alternately addressed as cells) in the architecture are classified into three broad categories:

1. *Input cells:* These cells only participate as inputs of logic. Only those MTJs that have access transistors are selected as input cells. During the entire course of the logic operation, these cells are written only once. Their values remain unchanged throughout the course of logic execution. These cells are therefore never clocked.
2. *Output cells:* The final output of logic is produced in these MTJs. Again, only those MTJs that have access transistors are selected. These cells participate in logic and are therefore always clocked. In addition, their contents are also read whenever the output of the logic needs to be known.
3. *Logic cells:* These cells form the body of the logic and are present in between the input and output cells. MTJs with and without access transistors belong to this group of cells. These cells are always clocked whenever logic is computed.

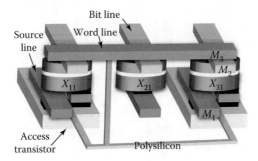

FIGURE 60.10 A column of MTJs inside the architecture. (J. Das, S.M. Alam, and S. Bhanja, Ultra-low power hybrid CMOS-magnetic logic architecture, *IEEE Transactions on Circuits and Systems I: Regular Papers*, 59(9), pp. 2008–2016 © (2012) IEEE.)

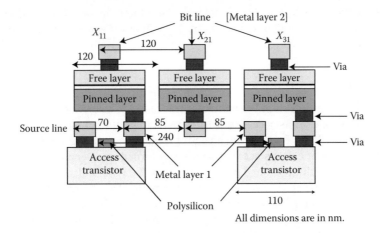

FIGURE 60.11 Cross section of the column of MTJs shown in Figure 60.10. (J. Das, S. M. Alam, and S. Bhanja, Low power CMOS-magnetic nano-logic with increased bit controllability, in *11th IEEE Conference on Nanotechnology IEEE-NANO*, pp. 1261–1266 © (2012) IEEE.)

Figure 60.12 shows an example layout of a two-input *AND* gate built in this architecture. Inputs A and B are written into input cells X_1 and X_2. The result $A \cdot B$ is produced in X_3. The cell on the left of X_3 has a fixed magnetization of 1. However, with the given architecture specifications, X_3 cannot have an access transistor; neither can any MTJ (X_4 and X_5) in its row. Therefore, none of X_3, X_4, and X_5 can qualify as an output cell. X_3 now becomes a logic cell and its value needs to be propagated to the nearest MTJ that fulfills the criteria of an output cell. An MTJ in the row of X_1 or X_2 can be a possible candidate. A second condition of the output cell is that it should be located at least one cell apart from the input cells X_1 or X_2. This ensures that the inputs do not have any direct influence on the output. X_6 fulfills the requirement for an output cell. X_4 and X_5 are then used to propagate the value from X_3 to X_6. X_4 and X_5 also become logic cells.

60.4 BASIC OPERATIONS FOR LOGIC COMPUTATION

The three basic operations to be carried out in any logic are

1. Writing inputs
2. Clocking cells
3. Reading outputs

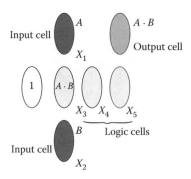

FIGURE 60.12 Two-input AND.

In the architecture described in this chapter, all these three operations are current driven and vary in magnitude and the duration of the current. We will now focus on how each of these operations is carried out in the architecture.

60.4.1 WRITING INPUTS

Every input cell in the architecture has an access transistor. Whenever an input, 1 or 0, needs to be written into an input cell, its access transistor needs to be turned on to complete the path for the current through the MTJ. By applying an active high voltage on the word line corresponding to the column of the input cell, its access transistor is turned on. Depending on the input value to be written, the current direction and magnitude through the MTJ also change. For example, to write logic 1, current needs to flow from the bit to the source line. Current from a source to a bit line is required to write a logic 0. During writing 1 and 0, the terminals of the access transistor switch roles between the source and the drain, depending on the direction of the current. The typical current magnitudes calculated for writing 1 and 0 into the input cells are 280 and 216 μA, respectively, with approximate durations of 20 ps [17]. Once all the inputs are written, the next step in logic is to compute. The most important operation during logic computation is clocking, which is discussed in the next section.

60.4.2 CLOCKING CELLS

As mentioned in Section 60.2, the logic cells are clocked using a train of STT current pulses. Whenever an MTJ is clocked, the magnetization of its free layer is held in a stationary state along the direction of the saddle points (see Figure 60.13, Phase III). To gain a deeper insight into the STT clock, we need to focus on the modified Landau–Lifshitz–Gilbert (LLG) equation (Equation 60.2) that describes the magnetodynamics behavior of an MTJ free layer with STT current.

$$\frac{d\mathbf{m}}{dt} = -\gamma M_s \mathbf{m} \times \left(\mathbf{h_{eff}} - \frac{\alpha}{\gamma M_s}\frac{d\mathbf{m}}{dt} - \frac{J_e G}{J_p}\mathbf{e}_p \times \mathbf{m} \right) \tag{60.2}$$

Phase I   Phase II   Phase III

FIGURE 60.13 Different phases of STT-clocking in MTJ.

In the above equation, **m** and $\mathbf{e}_p$ are unit vectors along the magnetization of the free and fixed layers, respectively. M_s is the saturation magnetization of the free layer. $\mathbf{h}_{\text{eff}}$ is the effective normalized magnetization field on the free layer that is a sum of the crystalline and shape anisotropic fields, the demagnetization field, the exchange field, and the coupling from the underlying fixed layer. γ is the gyromagnetic ratio and α is the damping coefficient. Parameters G and J_p are defined in Equation 60.3. The other parameters in Equation 60.3 are P—the spin-polarizing factor; $\hat{s}_1$ and $\hat{s}_2$—the unit vectors along the global spin orientation of the fixed and free layers of MTJ; e—the electron charge; $\hbar$—the reduced Planck's constant; and d—the thickness of the free layer. J_e is the applied STT current. When an MTJ is STT clocked, the current magnitude $|J_e|$ (Equation 60.4, [13]) cancels the magnetic precession of the free layer, reducing the left-hand side of Equation 60.2 to zero. In Equation 60.4, H_d is the coupling between the fixed and free layers. It is also worth mentioning here that the MTJ fixed layer polarization of 45° that is mentioned in Section 60.2 is imposed by this STT clocking.

$$G = \left[-4 + (1+P)^3 \frac{(3+\hat{s}_1 \cdot \hat{s}_2)}{4P^{3/2}} \right]^{-1} \qquad J_p = \mu_0 \cdot M_s^2 \frac{|e|d}{\hbar} \tag{60.3}$$

$$J_e = \left(\frac{\mu_0 \cdot M_s \cdot |e| \cdot d \cdot H_d}{\hbar \cdot G} \right) \tag{60.4}$$

The STT clocking in the logic takes place in three phases: (i) an initial writing of 1 into the MTJs, (ii) an intermediate half precession sweep to the saddle point, and (iii) the final stationary magnetization state along the saddle point (see Figure 60.13). Each of the three phases requires three different currents. The current magnitudes for the first and second phases remain the same as the STT writing of logic 1 and 0 into the MTJs. The current estimate for the final phase is 170 μA [13].

60.4.3 READING OUTPUTS

Once a logic is computed, its output needs to be read. The TMR of MTJs described by Equation 60.1 provides an electrical base for determining the logic output. By reading the electrical resistance of an output cell, its values can be determined. In our logic-in-memory architecture, the output is determined through a differential read approach using complementary output values [14]. In logic, there exists a dependency among bits. The dependency between two adjacent logic bits X_{i-1} and X_i in a row can be probabilistically expressed in the ideal case as $p(X_i = 1 | X_{i-1} = 0) = p(X_i = 0 | X_{i-1} = 1) = 1$. Here, X_i is influenced by only its left neighbor X_{i-1} through antiferromagnetic coupling. The differential read discussed in this section is developed on this property of logic. Figure 60.14 shows the read circuit. S_a and $\overline{S}_a$ are the complementary output cells from the logic that are placed in the two arms of the sense amplifier. M_{1a} and M_{2a} are their respective access transistors. (At this point, ignore cells S_b, $\overline{S}_b$ and their access transistors M_{1b}, M_{2b} in Figure 60.14. Their functions will be explained later in the section.)

Brief description of the circuit operation: The read circuit operates in two consecutive phases: the precharge phase followed by the sense phase. In the precharge phase, signals ϕ_2 and ϕ_3 are pulled low to (i) turn off transistors M_3 and M_4 and (ii) simultaneously precharge nodes X and Y to V_{DD}. The potential at nodes X and Y are equalized by setting E_q to an active high potential. In the sense phase, a low voltage signal V_{read} is applied on ϕ_2. This is used to provide an equal bias on the output MTJs S_a and $\overline{S}_a$. Simultaneously, ϕ_3 is raised high and E_q is pulled low to turn off transistors M_7, M_8, and M_9. The difference in electrical resistances between S_a and $\overline{S}_a$ starts to develop a differential voltage

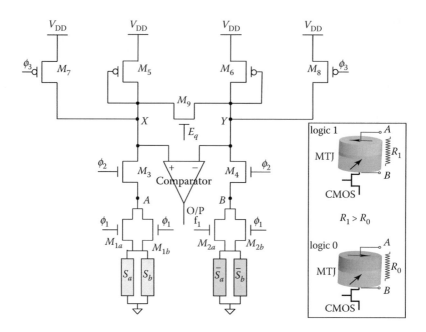

FIGURE 60.14   Nondestructive variability tolerant differential read for logic-in-memory architecture. (J. Das, S.M. Alam, and S. Bhanja, Ultra-low power hybrid CMOS-magnetic logic architecture, *IEEE Transactions on Circuits and Systems I: Regular Papers*, 59(9), pp. 2008–2016. © (2012) IEEE.)

across X and Y that is finally read by the comparator. A high or low at the comparator output decides in favor of $S_a = 1$ or $S_a = 0$.

The read operation described above is low-power. It requires an average current of 31 μA to read a 0 and 28 μA to read a 1 from an output cell. It is nondestructive and provides a near 50% improvement in sense margin over the conventional MRAM read using reference values.

The bit dependency in logic also produces equal values among alternate cells in a row of logic. This inherent property of logic is used to further modify the read circuit to make it variability tolerant. Instead of comparing a single bit (S_a) against its complement ($\overline{S_a}$), the sense amplifier now compares a pair of similar-valued bits (S_a, S_b) against their complements ($\overline{S_a}, \overline{S_b}$) (see Figure 60.14). Bit dependency in logic makes these bits readily available for read. Apart from an increased tolerance to variability, the modification also increases the sense margin of the read circuit [18].

60.5   PERFORMANCE ANALYSIS OF THE ARCHITECTURE

Before we conclude our discussion on our logic-in-memory architecture, we would present before our readers two brief comparative studies with our present architecture: (a) comparison with operating principles of existing logic-in-memory architectures and (b) comparison of logic principles and their performance with traditional NML.

60.5.1   COMPARISON WITH EXISTING LOGIC-IN-MEMORY ARCHITECTURES

The motivation behind all logic-in-memory architectures is to reduce the communication bottleneck between memory and CPU. Research has evolved over different devices for logic-in-memory architectures. One of the early works in logic-in-memory uses cellular arrays, a two-dimensional iterative configuration of identical cells that contain both logic and memory elements [19]. The logic and memory elements are both fabricated in CMOS technology. Each cell in the array is a combination of gates and flip-flops. Computation and storage location inside a cell, however, remain distinct, thereby increasing the hardware complexity and associated costs. Additional costs involve invoking special design techniques to map a logic design onto the cellular arrays [20]. The multiple-valued floating gate MOS transistor used for logic-in-memory VLSI [21] compiles storage and switching functions in a multiple-valued input and binary output combinational logic. In this design concept, the stored data are distributed in word-parallel and digit-parallel manners throughout the logic plane. A third logic-in-memory design uses ferroelectric capacitors [22]. Ferroelectric capacitors in combination with MOS transistors are used to develop a complementary ferroelectric-capacitor (CFC) logic gate with both logic and memory functions. However, within a gate, the logic and memory elements still remain separate. A stored input vector is distributed into CFC logic gates and a series parallel combination of CFC logic gates is used to realize any logic. The final logic-in-memory architecture that we would discuss here is TMR based. In the TMR-based logic-in-memory [23], a literal generator, a wired sum circuit, and a threshold detector are combined together. The TMR-based devices are distributed in a logic circuit plane along with CMOS transistors. The devices are used for *storing primary inputs*, in particular those inputs that change least frequently. The resistances of these devices are then used to compute logic.

From the above, we can observe the following trends in all of the previous logic-in-memory architectures:

* Embedding memory elements within the logic plane
* Selective storing of inputs from an entire logic operation
* Keeping the logic and memory elements separate inside a logic-in-memory cell

The principle of our logic-in-memory architecture is different from the above-mentioned trend. In our architecture, the logic elements themselves behave as memory. The clock acts as a classifier between their participation in logic and behavior as memory. The logic operations are carried to the memory plane to perform some simplistic operations within the memory. The more complex logic operations will however remain with the CPU. The entire operation is nonvolatile. Information from various stages of operation can be retrieved as and when required without reexecuting any logic. It is worth mentioning at this point about the zero standby power from our logic-in-memory design.

60.5.2 Comparison with NML

Table 60.1 compiles a comparison between logic operations in traditional NML and our logic-in-memory architecture. The following are evident from the table.

Logic built with multilayer STT-MTJs are more favorable in terms of

1. Overall power consumption, which is the sum of the power consumed from writing inputs, clocking cells, and reading outputs
2. User control over cells in the logic
3. CMOS compatibility
4. Electrical interface of input and output signals

Other benefits of the logic-in-memory architecture include provision for logic decomposition that significantly reduces the cell count and overall power consumption of the logic. Further details of it are available in Ref. [24] and are beyond the scope of this chapter. A layout constraint in NML that arises whenever a bit and its complement are required in a single column of cells can be solved in this architecture. This would further optimize the cell count and power in logic [25]. In this chapter, we have only considered Boolean logic computations using dipolar magnetic coupling. Non-Boolean computing using magnetic logic is also an active research area and is discussed in Chapter 62. One of its major applications is in computer vision.

TABLE 60.1

Comparison of Logic Behavior and Logic Performance between Traditional NML and Discussed Logic-in-Memory Architecture

	Nanomagnetic Logic	Logic-in-Memory Architecture
Logic Characteristics		
Cell types	Single layer nanomagnets	Multilayer magnetic tunnel junctions [14]
Cell dimensions and spacing	100×50 nm^2, 20 nm [26]	100×50 nm^2, 20 nm [15]
Computation style	Dipolar magnetic coupling: ferro and antiferro [8]	Dipolar magnetic coupling: ferro and antiferro
CMOS compatibility	None	Yes, access transistor with MTJs
User control	Restricted to entire rows and columns, not to individual cells	Up to one in 2×2 cells [14]
Signal Types and Operating Styles		
Inputs	Magnetic, external fields or through explicit neighbor magnets [27]	Electric, STT current [14]
Clock	Magnetic, external fields [28]	Electric, STT current [13]
Outputs	Magnetic, magnetic force microscopy [27]	Electric, TMR based [14]
Current and Timing		
Input current (writing 1)	>2.29 mA, 3 ns [28,29]	280 µA, 20 ps [14]
Input current (writing 0)	>2.29 mA, 3 ns [28,29]	216 µA, 20 ps [14]
Clocking current	2.29 mA, 3 ns [28]	170 µA, 20 ps [13]
Output current	—	30 µA, 4 ns [14]
Overall Performance		
Standby power	0	0
Logic volatility	Nonvolatile [30]	Nonvolatile [31]

REFERENCES

1. H. Ringler and B. Homburg von der Hohe, *Magnetic Core Memories*, Jun. 15 1965. US Patent 3188721.
2. R. Dadamo J., D. Hill, and W. Henessey M., *Magnetic Core Memories*, Oct. 5 1965. US Patent 3210745.
3. W. Bartik J. and K. Gruensfelder, *Magnetic Core Memory*, Feb. 16 1965. US Patent 3170147.
4. R. J. Baker, *CMOS Circuit Design, Layout, and Simulation.* John Wiley & Sons, Inc., Hoboken, NJ, 3rd ed., 2010.
5. J. Nahas, T. Andre, B. Garni, C. Subramanian, H. Lin, S. Alam, K. Papworth, and W. Martino, A 180 kbit embeddable MRAM memory module, *IEEE Journal of Solid-State Circuits*, 43, 1826–1834, 2008.
6. G. Sun, X. Dong, Y. Xie, J. Li, and Y. Chen, A novel architecture of the 3D stacked MRAM l2 cache for cmps, in *IEEE 15th International Symposium on High Performance Computer Architecture, 2009. HPCA 2009*, pp. 239–249, Feb. 2009.
7. J. Das, S. M. Alam, and S. Bhanja, Low power CMOS-magnetic nano-logic with increased bit control-lability, in *11th IEEE Conference on Nanotechnology, IEEE-NANO*, pp. 1261–1266, Aug. 2011.
8. A. Imre, G. Csaba, L. Ji, A. Orlov, G. H. Bernstein, and W. Porod, Majority logic gate for magnetic quantum-dot cellular automata, *Science*, 311(5758), 205–208, 2006.
9. M. Niemier, M. Alam, X. Hu, G. Bernstein, W. Porod, M. Putney, and J. DeAngelis, Clocking structures and power analysis for nanomagnet-based logic devices, in *Proceedings of the 2007 ISLPED*, pp. 26–31, ACM, New York, USA, 2007.
10. A. Kumari and S. Bhanja, Magnetic cellular automata MCA arrays under spatially varying field, in *Nanotechnology Materials and Devices Conference, 2009. NMDC '09. IEEE*, pp. 50–53, Jun. 2009.

11. A. Kumari and S. Bhanja, Landauer clocking for magnetic cellular automata MCA arrays, *IEEE Transactions on Very Large Scale Integration Systems*, 19, 714–717, 2011.

12. M. Alam, M. Siddiq, G. Bernstein, M. Niemier, W. Porod, and X. Hu, On-chip clocking for nanomagnet logic devices, *IEEE Transactions on Nanotechnology*, 9(3), 348–351, 2010.

13. J. Das, S. Alam, and S. Bhanja, Low power magnetic quantum cellular automata realization using magnetic multi-layer structures, *IEEE JETCAS*, 1, 267–276, 2011.

14. J. Das, S. Alam, and S. Bhanja, Ultra-low power hybrid CMOS-magnetic logic architecture, *IEEE Transactions on Circuits and Systems I: Regular Papers*, 59(9), 2008–2016, September 2012.

15. D. K. Karunaratne and S. Bhanja, Study of single layer and multilayer nano-magnetic logic architectures, *Journal of Applied Physics*, 111, 07A928–07A928–3, 2012.

16. International technology roadmap for semiconductor, 2009. Available at: http://www.itrs.net/Links/2009ITRS/Home2009.htm

17. A. D. Kent, B. Ozyilmaz, and E. del Barco, Spin-transfer-induced precessional magnetization reversal, *Applied Physics Letters*, 84, 3897–3899, 2004.

18. J. Das, S. M. Alam, and S. Bhanja, Non-destructive variability tolerant differential read for non-volatile logic, *2012 IEEE 55th International Midwest Symposium on Circuits and Systems (MWSCAS)*, 5–8 August 2012, pp.178–181

19. W. Kautz, Cellular logic-in-memory arrays, *IEEE Transactions on Computers*, C-18, 719–727, 1969.

20. R. C. Minnick, Cutpoint cellular logic, *IEEE Transactions on Electronic Computers*, EC-13(6), 685–698, 1964.

21. T. Hanyu, K. Teranihi, and M. Kameyama, Multiple-valued floating-gate-MOS pass logic and its application to logic-in-memory VLSI, in *Proceedings of the 1998 28th IEEE International Symposium on Multiple-Valued Logic*, pp. 270–275, May 1998.

22. H. Kimura, T. Hanyu, M. Kameyama, Y. Fujimori, T. Nakamura, and H. Takasu, Complementary ferroelectric-capacitor logic for low-power logic-in-memory VLSI, *IEEE Journal of Solid-State Circuits*, 39, 919–926, 2004.

23. A. Mochizuki, H. Kimura, M. Ibuki, and T. Hanyu, TMR-based logic-in-memory circuit for low-power VLSI, *IEICE Transactions on Fundamentals of Electronics, Communications and Computer Sciences*, E88-A, 1408–1415, 2005.

24. J. Das, S. M. Alam, and S. Bhanja, A novel design concept for high density hybrid CMOS-nanomagnetic circuits, *2012 12th IEEE Conference on Nanotechnology (IEEE-NANO)*, 20–23 August 2012, pp. 1–6.

25. J. Das, S. M. Alam, and S. Bhanja, Addressing the layout constraint problem when cascading logic gates in nanomagnetic logic, *2012 12th IEEE Conference on Nanotechnology (IEEE-NANO)*, 20–23 August 2012, pp. 1–4.

26. J. Pulecio, P. Pendru, A. Kumari, and S. Bhanja, Magnetic cellular automata wire architectures, *IEEE Transactions on Nanotechnology*, 10, 1243–1248, 2011.

27. S. Bhanja and J. Pulecio, A review of magnetic cellular automata systems, in *2011 IEEE International Symposium on Circuits and Systems ISCAS*, pp. 2373–2376, 2011.

28. A. Dingler, M. T. Niemier, X. S. Hu, and E. Lent, Performance and energy impact of locally controlled NML circuits, *Journal of Emerging Technologies in Computing Systems*, 7, 2:1–2:24, 2011.

29. R. P. Cowburn and M. E. Welland, Room temperature magnetic quantum cellular automata, *Science*, 287(5457), 1466–1468, 2000.

30. G. Csaba, *Computing with Field-Coupled Nanomagnets*. PhD thesis, University of Notre Dame, 2003.

31. W. Zhao, E. Belhaire, C. Chappert, and P. Mazoyer, Spin transfer torque STT-MRAM–based runtime reconfiguration FPGA circuit, *ACM Transactions on Embedded Computing Systems*, 9, 14:1–14:16, 2009.

61 Implementation of a Nanomagnet Full Adder Circuit

Edit Varga, György Csaba, G. H. Bernstein, and Wolfgang Porod

CONTENTS

61.1 INTRODUCTION

The development and characterization of a nanomagnet full adder circuit are demonstrated in this chapter. To our knowledge, this is the most complex magnetic circuit attempted to date. The circuit comprises 53 magnets that perform as three majority gates plus internal interconnections, with only one internal error. The inputs of the gate can be set independently, allowing an exhaustive test of the magnetic operation. We introduce two other different designs of a full adder circuit built from nanomagnets. We use previously demonstrated and tested nanomagnet logic (NML) components, such as inverters, wires [1], programmable inputs [2], shape-engineered magnets [3], and fanouts [4], to realize a larger-scale circuit.

NML is a novel nanoscale computing paradigm where circuits are constructed from sub-100 nm, single-domain magnets. Logic signals are propagated in magnetic "wires" built from nanomagnets placed side by side. The emanating magnetic field of a magnet provides the interaction. NML technology is receiving increasing attention due to its special benefits, such as low power dissipation with high integration density of functional devices, operation over a wide range of temperatures, and nonvolatility [5].

The basic unit of NML is the nanomagnet, as shown in the schematic of Figure 61.1, as a rectangle. It is well known that a symmetrically shaped magnet has an energy barrier between two stable states, as shown in Figure 61.1, with the energy (E) versus magnetization (H) landscape as shown. Consider a symmetric magnet that is subjected to a strong hard-axis field (indicated by the thick blue arrow pointing from left to right in Figure 61.1). For a large hard-axis external field, referred to as the "clocking field," the magnetization is pointing in the direction of the field, and favors neither the up nor the down direction, a condition referred to as "nulled." When the field is removed, the nanomagnet relaxes into one of the two energetically equivalent ground states, that is, pointing either up or down. Even a

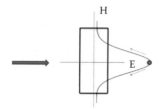

FIGURE 61.1 Energy landscape and clocking process of a symmetric, rectangular-shaped nanomagnet. The thick, horizontal, blue arrow indicates the strong hard-axis clocking field. As the field is removed, the symmetric nanomagnet relaxes into one of the two energetically equivalent ground states.

small biasing field along the easy axis can influence which magnet ground state the magnetization will select. In Figure 61.1, the blue curve shows the potential landscape of the nanomagnet immediately after the removal of the nulling field. The presence of the energy barrier requires that an external field stronger than the nulling field is required to reevaluate the magnet, that is, set the magnet so that a new logic value, either up or down, can be written to it. It relaxes into a newly ordered state, in accordance with any present biasing field. The magnet retains its new state without an externally applied field, since the size of the magnet is assumed to be larger than the superparamagnetic limit.

61.2 NML COMPONENTS OF THE FULL ADDER

All NML circuits are built from closely spaced nanomagnets, and are constructed in such a way that the magnets are located on a grid-based array, that is, they form horizontal and vertical lines in a plane. Various basic structures are necessary to build a complex NML circuit. NML devices are interconnected by magnetic "wires," which in the case of NML refers to a line of nanomagnets along which the signal is propagated. The basic logic element of the currently existing NML library is the majority gate in which one magnet reflects the total magnetic forces of the surrounding magnets. Signal distribution, that is, fanout, is performed by splitting the input signal and passing on as three output signals. Above all, shape engineering helps to design programmable inputs as well as to reduce the overall footprint. All these components are described in the following sections.

61.2.1 Nanomagnet Wires

Nanomagnets comprising NML wires can be coupled in one of two ways. One of these is the vertically aligned wire with ferromagnetically coupled (FC) magnets, and the other is the horizontally aligned wire built from antiferromagnetically coupled (AFC) magnets. Figure 61.2a is a scanning electron micrograph (SEM) of a typical vertically aligned wire. The one horizontally aligned magnet on the top is the "driver" magnet; it initializes the entire wire through FC. Figures 61.2b and 61.2c are magnetic force microscope (MFM) images showing the magnetic state of all the magnets in the wire.

One horizontally aligned, five-magnet-long wire is shown in Figure 61.3a. Here, the information propagates from the horizontal driver magnet through the AFC magnets, that is, from left to right. Two MFM images (Figures 61.3b and 61.3c) show the magnetization of the same wire for two possible states of the driver magnet.

61.2.2 Programmability and Majority Gate

Most NML designs use magnets with a rounded-edge rectangular shape, as shown in the previous figures. Magnets with different aspect ratios can enhance logic functionality, as summarized in Ref. [2]. The switching field required to reverse the magnetization coercivity along the long axis increases with its length. This phenomenon has been used in the programmable majority gates [3] shown in Figure 61.4. The majority gate is the basic logic structure of NML because influence by fixing one of the

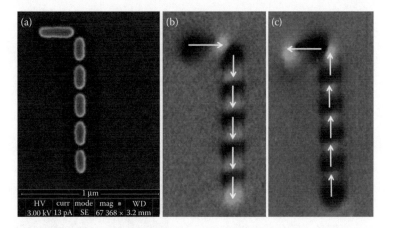

FIGURE 61.2 (a) SEM image of a five-nanomagnet-long vertical wire with a horizontally aligned driver magnet. (b) MFM image of the same wire for one magnetization state of the driver magnet and (c) for the other. The information propagates from the driver magnet toward the bottom of the wire by ferromagnetic coupling.

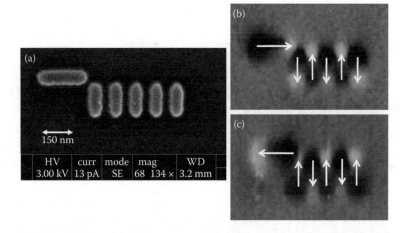

FIGURE 61.3 (a) SEM image of a five-nanomagnet-long horizontal wire with a horizontally aligned driver magnet. (b) MFM image of the same wire for one magnetization state of the driver magnet and (c) for the other. The horizontal driver magnet initializes the entire wire, thereby defining the magnetization state of all the magnets. The information propagates from left to right, that is, from the region of the strongest influence to the region of the weakest influence.

inputs, it can take on the roles of the AND and OR gates [1]. The three horizontal driver magnets have different aspect ratios, enabling programmability of the logic gate. Programmability is achieved using external fields with various strengths, such that the longest driver magnet is set at the highest field, and subsequent lower fields set shorter driver magnets. The driver magnets set the "input" magnets through FC. The center magnet is the "compute" magnet, which responds to the total magnetic force of all three input magnets in a majority fashion. The compute magnet influences the "output" magnet through AFC. All eight input combinations are shown in Figure 61.5 for the same gate.

61.2.3 FANOUT

In addition to the NML structures discussed above, nanomagnet fanout is also necessary to construct large circuits such as an NML full adder. The aspect-ratio-dependent switching phenomenon

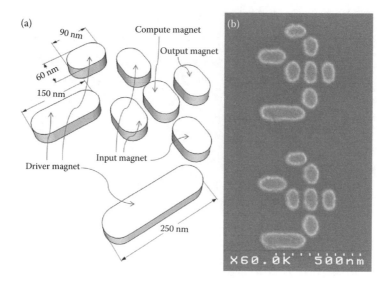

FIGURE 61.4 (a) Schematic of the majority gate with various shapes of the driver magnets providing signals to the input bits. (b) SEM of the fabricated permalloy structure.

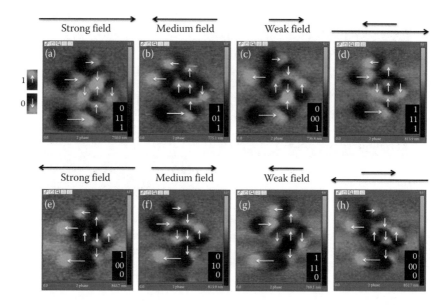

FIGURE 61.5 MFM images of the programmable majority gate for all input combinations. The black insets show the three input values along the output, majority value. Note that because the middle input is AFC to the compute magnet, its physical orientation results in the opposite logic value compared with the other two inputs, that is, down is logic "1."

has been exploited for the design of the fanout structures [4] as well to provide a stable input for magnetic circuits. A fanout structure is shown in Figure 61.6 with the schematic (Figure 61.6a) and the SEM image (Figure 61.6b). Two MFM images summarize the magnetization of the same structure shown in the SEM according to the two possible magnetization states of the vertical driver magnet.

Elongation of the driver magnet allows it to be set at a higher external field while the rest of the fanout circuit is nulled at a lower field without causing the driver to change its state.

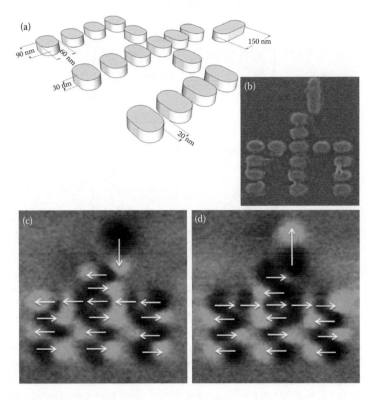

FIGURE 61.6 (a) Schematic, (b) SEM, and (c), (d) MFM images for both possible driver magnetization states of the fanout circuit.

61.2.4 SHAPE ENGINEERING: ASYMMETRIC NANOMAGNETS

Shapes other than oval have been investigated in Refs. [6,7], where the nanomagnets are asymmetric, having a slanted edge, as shown in Figure 61.7 on the left side. Simulations based on the single-domain model show that the asymmetry of the magnet shifts the entire energy landscape of the magnet, whereby the energy minima are shifted as well. The thin line along the diagonal of the

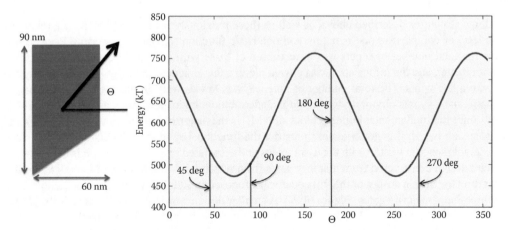

FIGURE 61.7 Schematic of the slant magnet (left) and simulation result based on the single-domain model for the slant nanomagnet energy landscape. The energy minima are shifted from the long geometrical axis, that is, from the 90° and 270° magnetization direction.

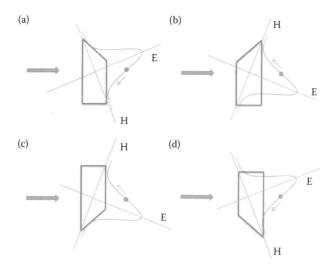

FIGURE 61.8 Energy curves of the slant magnets for different slant locations. The external magnetic field is applied from left to right shown by horizontal, thick, blue arrows. The thin line along the diagonal of the magnet corresponds to the effective easy axis. The energy maximum is perpendicular to this axis.

magnet in Figure 61.8 corresponds to the effective easy axis. The energy maximum is perpendicular to this axis. The horizontally applied field is not precisely in the direction of the energy maximum, so the resulting energy of the magnet is to one side of the maximum and falls toward the appropriate ground state when the field is removed. As a result of the tilted easy axis, the nanomagnet takes on a preferred magnetization direction, as summarized in Figure 61.8.

The schematic lists all possible slant orientations and the corresponding overlaid energy diagrams, energy (E) versus magnetization (H). The magnets in (a) and (d) relax into the downward pointing magnetization, and in (b) and (c) relax into the upward pointing direction. This specific shape can be exploited to reduce the device footprint as shown below.

61.3 NML FULL ADDER

61.3.1 Design

The logic structures described above, as well as those previously published [8,9], have a relatively basic level of complexity; one structure had one basic function (AND/OR/majority). Prior to this study, the total number of inputs was a maximum of three with one single compute nanomagnet (majority gate), and the information was passed along a few-magnet-long magnetic wire, or fanned out with a few wires. The total number of magnets was low as well. In the case of the full adder, the basic, most simple circuit design would include multiple logic gates, such as the AND and OR gates. Since the fundamental building block of NML is the majority gate, upon which the AND and OR gates are based, it is not efficient to restrict the functionality to the AND/OR functions since considerable space is wasted with the extra inputs and associated magnets. Rather, it is better to find a circuit design constructed from majority gates from which it is straightforward to realize the full adder directly. Such a design of the full adder was proposed in Ref. [10] proving the operation using five three-input majority gates (Figure 61.9a). A smaller design was introduced later in Refs. [11,12] where the circuit is composed of only three three-input majority gates (Figure 61.9b).

More complicated designs were published that required one three-input majority gate and one five-input majority gate [13]. Since the three-input nanomagnet majority gate was tested successfully in Refs. [1,5], we chose them to construct the full adder of Figure 61.9b instead of the one with five inputs. Figure 61.10 shows a possible nanomagnet-based layout with three three-input majority

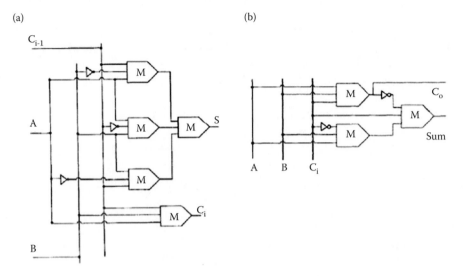

FIGURE 61.9 (a) An early full adder design constructed from five majority gates (E. Varga et al., Experimental demonstration of non-majority, nanomagnet logic gates, *IEEE DRC (Device Research Conference) Technical Digest*, pp. 87–88, © (2010) IEEE. With permission.) and (b) the smaller full adder design proposed in several publications (W. Wang, K. Walus and G. A. Jullien, Quantum-dot cellular automata adders, *3rd IEEE Conference on Nanotechnology*, 1–2, 461–464, © (2003) IEEE. H. Cho and E. E. Swartzlander, Adder designs and analyses for quantum-dot cellular automata, *IEEE Transactions on Nanotechnology*, 6(3), 374–383, © (2007) IEEE. With permission.). Only three majority voting gates are necessary for the full adder implementation. Since the AFC wire comprises a series of inverters, depending on the total length being an even or odd number of magnets, the inverter function is included in the wire length.

gates (the cross point of the horizontal and the vertical nanomagnet wires) built in [14]. Each input is applied at multiple positions in the circuits of Figure 61.10, making it possible to implement the layout without wire crossings.

61.3.2 Functionality

The long magnets on the left side of the circuit in Figure 61.10 are the programmable drivers providing the data for testing the adder. A particular B_{ext} applied field can switch only magnets with a switching field (coercivity) $B_{sw} < B_{ext}$. Higher aspect ratios yield higher switching fields. Therefore, a globally applied magnetic field can switch only magnets smaller than a certain length. Figure 61.11 shows the calculated [15] switching field of several different-sized magnets. The thicknesses of the magnets are 20 and 30 nm, their width is fixed at 60 nm, and the coercivity is plotted as the function of their length. The field is applied at a 45° angle to the long axis of the magnets. Exploiting the aspect ratio as a design variable, the driver magnets can be set separately, resulting in a programmable circuit. The relation between the switching field and shape has already been experimentally demonstrated [5].

61.4 FABRICATION AND CHARACTERIZATION

The magnetic structure of Figure 61.10 was fabricated with electron-beam lithography, evaporation, and a lift-off of Supermalloy. The smaller magnets are 90 nm and the drivers are 220, 340, and 440 nm long with 60 nm width. The thickness of the Supermalloy magnets is 30 nm. An SEM of a fabricated structure is shown in Figure 61.12a. The device was tested by MFM (Figure 61.12b) after the application of external fields.

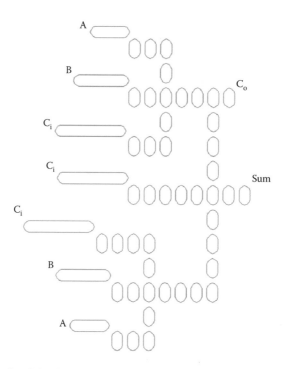

FIGURE 61.10   Schematic of the full adder including three three-input majority gates, and several high-aspect-ratio magnets to act as drivers providing the input data for the circuit. The majority gates are the intersections of horizontal and vertical wires at which the center magnet is influenced by three input magnets (top, left, and bottom neighbors) and influences an output magnet (right neighbor). (E. Varga et al., Implementation of a nanomagnetic full adder circuit, *11th International Conference on Nanotechnology (IEEE Nano)*, August 15–19, pp. 1244–1247, © (2011) IEEE. With permission.)

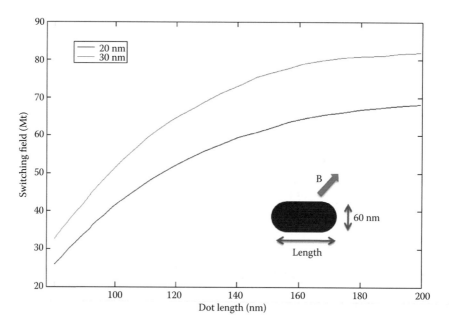

FIGURE 61.11   Simulated switching field versus dot (magnet) length. The thickness and width of nanomagnets are fixed. The monotonically increasing curvature suggests that the aspect ratio change can be exploited to design separately programmable inputs for the NML circuit.

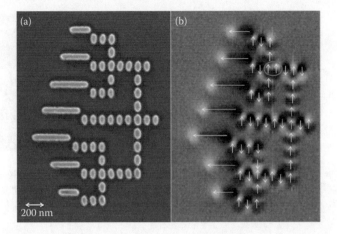

FIGURE 61.12 (a) SEM and (b) MFM images of a full adder constructed from 53 nanomagnets. Nearly perfect nanomagnet ordering is shown. Of the 53 magnets, only a single coupling failed (circled). (E. Varga et al., Implementation of a Nanomagnetic Full Adder Circuit, *IEEE International Conference on Nanotechnology,* Portland, OR, August 15–19, pp. 1244–1247, © (2011) IEEE. With permission.)

A homogeneous 200 mT field was applied first to set all the drivers pointing in the same direction, as shown in Figure 61.12b. Here, we do not magnetize the drivers separately; our goal is to test the operation of the entire structure for one set of input bits. The 200 mT magnetization was followed by a rotating field continuously decaying from 40 to 0 mT; this field demagnetizes the magnets to their computational ground state. More details about the choice of these field values are given in Ref. [4]. The MFM image (Figure 61.12b) shows a nearly perfect ordering of the nanomagnets, according to their neighbors, as required to correctly perform the logic operation. Of the 53 magnets, only a single coupling failed during the testing (at the output of the topmost majority gate—circled). This result is encouraging, as 52 of the 53 magnets fabricated in this complex geometry are correctly coupled to their neighbors, and all three majority gates functioned correctly. To the best of our knowledge, this circuit is the most complex NML circuit fabricated so far, and the first one that includes several logic gates. The error occurred at the circled position for three independent demagnetization events, suggesting that this error is most likely due to a design or fabrication error (an odd-shaped or pinned magnet). To achieve a truly error-free operation, the design and the fabrication parameters need to be optimized. There are several ways to design more compact NML full adders, as discussed below.

61.5 PROPOSED DESIGNS OF FULL ADDER

The tested full adder structure has a relatively high number of magnets, which increases the probability of errors during the switching process. One potential optimization process is to reduce the footprint of the structure and the number of magnets, as shown in Figure 61.13a. Here, there are 6 driver magnets and 25 small magnets. This design is already much smaller (altogether 31 magnets) than the previously tested one with 53 magnets (Figure 61.8).

The reduced footprint structures were fabricated successfully, as shown in Figure 61.13b. However, in this structure, the relatively long vertical and horizontal lines reveal a design flaw: the horizontal and vertical magnetic wires relax to their ground state at different external fields. This is due to the fact that for a horizontal applied field, the ferromagnetic coupling superposes to the applied field, so the horizontal (AFC) wires relax later than the vertical (FC) wires. This is why the magnet denoted with a star in Figure 61.11b fails to work correctly in 75% of the cases. An example is shown in Figure 61.14. The MFM image clearly displays two errors; the top one is the most common error for this layout.

OOMMF [16] simulations were performed to better understand the error formation mechanism. The results are summarized in Figure 61.15.

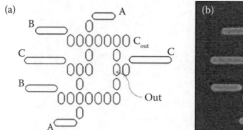

FIGURE 61.13 The reduced footprint NML full adder design. (a) Schematic and (b) SEM image of the Permalloy structure. The (*) symbol is explained in the text. (E. Varga et al., Implementation of a Nanomagnetic Full Adder Circuit, *IEEE International Conference on Nanotechnology*, Portland, OR, August 15–19, pp. 1244–1247, © (2011) IEEE. With permission.)

As the external field is decreased from the maximum value, the long horizontal wires start to switch to their antiferromagnetically ordered ground state at around the 60 mT field, while the vertical wires start to order at around 200 mT. This is a significant difference, and may cause the information to flow in the wrong direction. If a long horizontal wire precedes a long vertical wire segment in the signal flow, then the magnets of the vertical segment may fall into a random state before the effect of the drivers can reach them.

To avoid the above-mentioned issues coming from the different behaviors of the horizontal and vertical wires, we propose a new design (Figure 61.16). This design has a smaller footprint than any previous designs. The horizontal and vertical wires have decreased lengths, and the lengths of the driver magnets have been adjusted according to the simulation results shown in Figure 61.11.

The footprint can be reduced further by using asymmetric, slant magnets as inputs to the adder. The behavior of this magnet is summarized in Figure 61.8. As discussed above, the energy is not the highest when a device is magnetized along the horizontal direction, as is the case for the symmetric magnets (Figure 61.1). The design constructed from the slant and regular oval-shaped magnets is shown in Figure 61.17. The inputs are provided by the seven slant magnets, and the number of the rounded-edge magnets is only 14 (21 magnets altogether). The full adder has four different designs

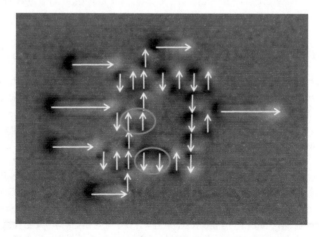

FIGURE 61.14 MFM image of the second full adder design (shown in Figure 61.13). Two errors (circled) are presented here; the top one is common and appears in 75% of the gates. (E. Varga et al., Implementation of a nanomagnetic full adder circuit, *11th International Conference on Nanotechnology (IEEE Nano)*, pp. 1244–1247, © (2011) IEEE. With permission.)

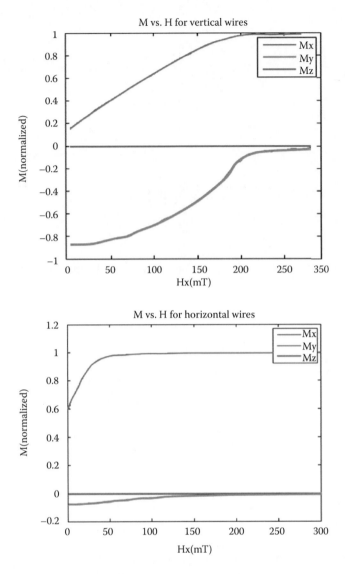

FIGURE 61.15 Simulation results for vertical (top) and horizontal (bottom) magnetic wires. The M versus H graph shows smooth switching behavior for the vertical line, though not for the horizontal one. Since the field is reduced from high to low values, the horizontal wire switches after the vertical wire. This difference has to be taken into account during the design of any NML structure. (E. Varga et al., Implementation of a nanomagnetic full adder circuit, *11th International Conference on Nanotechnology (IEEE Nano)*, pp. 1244–1247, © (2011) IEEE. With permission.)

(Figures 61.17a through 61.17d) for testing all possible input combinations. Our goal is to demonstrate the correct operation of the gate for all logic cases. This can be completed by using fixed inputs, as shown in Figure 61.17 with slant input magnets.

61.6 SUMMARY AND CONCLUSIONS

In this chapter, we have presented the realization of complex NML circuits. A full adder design containing 53 magnets has been fabricated and tested with only one coupling error throughout the entire

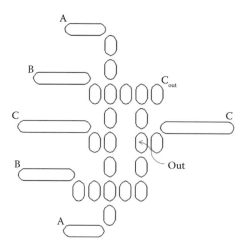

FIGURE 61.16   The proposed, new full adder design with decreased length of the magnetic wires.

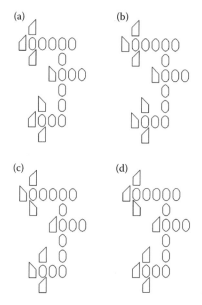

FIGURE 61.17   A reduced footprint design of the full adder. The inputs are provided by the slant magnets. Four different designs are shown (a through d) for testing all possible input combinations.

circuit. We concluded that these errors were most likely due to the fabrication and suboptimal design. It is challenging to design and fabricate high-yield NML circuits of such complexity. The low number of errors and the position of their occurrences suggest that improvement to no errors is possible.

To realize complex functions in NML, one does not necessarily need to achieve a perfect ordering of large nanomagnet arrays. Large circuits can be split into smaller, individually clocked units, which then order and compute sequentially [17]. However, using large units significantly reduces the overhead of the clocking circuitry. With careful design and optimized technology, the realization of complex NML circuit units seems feasible.

REFERENCES

1. A. Imre, G. Csaba, L. Ji, A. Orlov, G. H. Bernstein, and W. Porod, Majority logic gate for magnetic quantum-dot cellular automata, *Science*, 311(5758), 205–208, 2006.
2. E. Varga, M. T. Niemier, G. H. Bernstein, W. Porod, and X. S. Hu, Programmable nanomagnet-logic majority gate, *IEEE Device Research Conference (DRC), Notre Dame, IN, Technical Digest*, pp. 85–86, 2010.
3. E. Varga, M. T. Niemier, G .H. Bernstein, X. S. Hu, and W. Porod, Programmable nanomagnet-logic majority gate, *IEEE DRC (Device Research Conference) Technical Digest*, pp. 85–86, 2010.
4. E. Varga, A. Orlov, M. T. Niemier, X. S. Hu, G. H. Bernstein, and W. Porod, Experimental demonstration of fanout for nanomagnet logic, *IEEE Transactions on Nanotechnology*, 9(6), 668–670, 2010.
5. G. Csaba, W. Porod, and A. I. Csurgay, A computing architecture composed of field-coupled single domain nanomagnets clocked by magnetic field, *International Journal of Circuit Theory and Applications*, 31, 67–82, 2003.
6. M. Niemier, E. Varga, G. H. Bernstein, W. Porod, M. T. Alam, A. Dingler, A. Orlov, and X. Sharon Hu, Shape engineering for controlled switching with nanomagnet logic, *IEEE Transactions on Nanotechnology*, 11(2), 220–230, 2012.
7. E. Varga, S. Kurtz, M. T. Niemier, W. Porod, G. H. Bernstein, and X. S. Hu, Two input, non-majority magnetic logic gates: experimental demonstration and future prospects, *Journal of Physics: Condensed Matter*, 23(5), 053202, 2011.
8. E. Varga, M. Siddiq, M. T. Niemier, M. T. Alam, G. H. Bernstein, W. Porod, X. S. Hu, and A. Orlov, Experimental demonstration of non-majority, nanomagnet logic gates, *IEEE DRC (Device Research Conference) Technical Digest*, pp. 87–88, 2010.
9. E. Varga, M. T. Niemier, G. H. Bernstein, W. Porod, and X. S. Hu, Non-volatile and reprogrammable MQCA-based majority gates, *IEEE DRC (Device Research Conference) Tech. Digest*, pp 1–2, 2009.
10. P. D. Tougaw and C. S. Lent, Logical devices implemented using quantum cellular-automata, *Journal of Applied Physics*, 75(3), 1818–1825, 1994.
11. W. Wang, K. Walus, and G. A. Jullien, Quantum-dot cellular automata adders, *3rd IEEE Conference on Nanotechnology*, 1–2, 461–464, 2003.
12. H. Cho and E. E. Swartzlander, Adder designs and analyses for quantum-dot cellular automata, *IEEE Transactions on Nanotechnology*, 6(3), 374–383, 2007.
13. K. Navi, S. Sayedsalehi, R. Farazkish, and M. R Azghadi, Five-input majority gate, a new device for quantum-dot cellular automata, *Journal of Computational and Theoretical Nanoscience*, 7(8), 1546–1553, 2010.
14. E. Varga, G. Csaba, G. H. Bernstein, and W. Porod, Implementation of a nanomagnetic full adder circuit, *11th International Conference on Nanotechnology (IEEE Nano)*, August 15–19, pp. 1244–1247, 2011.
15. G. Csaba, M. Becherer, and W. Porod, Development of CAD tools for nanomagnetic logic devices, *International Journal of Circuit Theory and Applications*, Published online in Wiley Online Library (wileyonlinelibrary.com). DOI: 10.1002/cta.1811, 2012.
16. M. Donahue and D. Porter, OOMMF User's Guide, Version 1.0, Interagency Report NISTIR 6367, http://math.nist.gov/oommf.
17. M. T. Alam, M. J. Siddiq, G. H. Bernstein, M. T. Niemier, W. Porod, and X. S. Hu, On-chip clocking for nanomagnet logic devices, *IEEE Transactions on Nanotechnology*, 9, 348–351, 2010.

62 Investigations on Nanomagnetic Logic by Experiment-Based Compact Modeling

Stephan Breitkreutz, Josef Kiermaier, Irina Eichwald,
Xueming Ju, György Csaba, Doris Schmitt-Landsiedel,
and Markus Becherer

CONTENTS

62.1 INTRODUCTION

Nanomagnetic logic (NML) is an emerging technology using field-coupled nanomagnets that combine logic and memory functionality in a single device [1]. The magnetization state of nonvolatile, bistable nanomagnets is used as a logic state. Digital computation is achieved by magneto-static field-coupling of the nanomagnets. High-density integration due to a scalability of the nanomagnets down to a sub-50 nm size and potential low power consumption for switching of the nanomagnets are key features of NML.

NML with perpendicular magnetic anisotropy (PMA) uses nanomagnets made from, for example, Co/Pt or Co/Ni multilayers and has been proposed in 2002 [2,3]. In contrast to NML with in-plane nanomagnets made of permalloy [4,5], the switching behavior of the out-of-plane nanomagnets is defined by the interface and crystalline anisotropy allowing for a user-defined dot geometry. Additionally, the switching field of the out-of-plane nanomagnets can be controlled by partial focused ion beam (FIB) irradiation [6].

The reversal process of an as-grown nanomagnet is governed by domain wall (DW) nucleation at the weakest spot in the nanomagnet [7]. By applying an external field in the easy-axis direction, the DW nucleates at this weakest link and, after depinning, propagates through the entire dot and

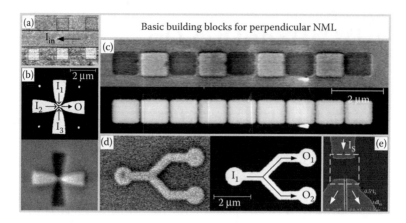

FIGURE 62.1 Basic building blocks for nanomagnetic logic with perpendicular magnetic anisotropy. (a) MFM image of a current wire as electrical input, (b) MFM images of a majority gate for logic operation, (c) MFM images of an inverter chain for signal inversion and transmission, (d) MFM images of a fanout to split magnetic signals, and (e) SEM image of a Hall sensor as electrical output.

completely reverses it [8,9]. The perpendicular magnetic anisotropy of the utilized Co/Pt nano-magnets is achieved by the interface and crystalline anisotropy of the Co and Pt layers [10]. FIB irradiation intermixes the single layers and reduces the switching field of the thin film [11]. Partial FIB irradiation locally reduces the anisotropy in a nanomagnet and therefore creates an artificial nucleation center (ANC) at a user-defined position. The switching behavior of the nanomagnet can be precisely controlled by the size and dose of partial FIB irradiation [6]. For the realization of logic operations, the DW nucleation at the ANC can be supported or constrained by the coupling fields of the surrounding nanomagnets, which superpose with the clocking field and therefore support or inhibit the switching of the dot [12]. The basic building blocks for NML with PMA, such as a major-ity gate, an inverter chain, or a fanout structure, have already been demonstrated [6,13,14].

In a majority gate with PMA (Figure 62.1b), the superposing coupling fields of the input dots I_1, I_2, and I_3 stabilize or destabilize the magnetization state of the output O during clocking and therefore force or inhibit its switching. Owing to the antiparallel orientation of the coupling fields from a nanomagnet, compared to its own magnetization state, the output O is always aligned anti-parallel compared to the input majority [13]. In an inverter chain (Figure 62.1c), magnetic signals are transmitted over two dots per clocking cycle. During clocking, consecutive nanodots with the same magnetization state as their prior neighbors are destabilized by its coupling field and therefore forced to switch to the antiparallel state [14]. A fanout structure (Figure 62.1d) can be realized by cloning a propagating domain wall from the input I_1 to two (or more) output wires O_1 and O_2 [6].

Perpendicular NML uses alternating, homogeneous magnetic fields in an easy-axis direction for clocking [14], which can be generated by either external or integrated coils. On-chip inductors operating up to 100 MHz have been shown in Ref. [15], but they have to be modified for the imple-mentation of NML. Alternative clocking concepts using the magnetic stray field of a propagating DW as a clocking field are currently under investigation [16].

Integration in standard CMOS circuits can be achieved by electrical input and output structures. Current wires generating oersted fields (Figure 62.1a) as input structures or Hall sensors in the split-current design (Figure 62.1e) to read out the magnetization state of a nanodot are standard I/O devices for magnetic systems [17,18]. Increased sensitivity can be achieved by GMR and MTJ struc-tures. Those devices are standard CMOS-compatible structures to set or read out the magnetization state of sub-100 nm nanodots with perpendicular magnetic anisotropy [19,20].

Micromagnetic simulations like object-oriented micromagnetic framework OOMMF [21] are widely used for simulations and investigations on magnetic structures. Such micromagnetic modeling tools are highly accurate but also very time intensive and therefore not capable of simulating large NML circuits. Hence, simplified physical dot models that accurately mimic the behavior of the nanomagnets are needed. In this chapter, an experiment-based compact model for NML with PMA is presented. Compact modeling of the switching and the interaction of the nanomagnets allows for fast and accurate predictions on the behavior of complex NML circuits [22].

62.2 DOT MODEL

Figure 62.2 shows the simplified physical model for the nanomagnets. The switching behavior of the single-domain nanomagnets is described by the switching field H_c (Figure 62.2a). Once the applied external field H_{ext} reaches the switching field, a DW is nucleated at the weakest spot and propagates through the entire dot to switch its magnetization M.

Interaction of the nanomagnets is achieved by their magnetic stray field, often named as the coupling field k, which superposes with the external field H_{ext} on a neighboring nanomagnet (62.2b). Hence, nanomagnets with parallel magnetization destabilize each other and the switching field H_c is reduced by the coupling field k to $H_{p \to ap}$. Once one of the dots switches, the nanomagnets are in the stabilizing, antiparallel magnetization state. Hence, the switching field H_c of the second dot to switch it in the parallel state is increased by k to $H_{ap \to p}$.

An applied, sinusoidal clocking field with adequate amplitude H_{clock} will force parallel magnetized dots in the antiparallel state without affecting the magnetization of antiparallel ordered dots. Owing to the influence of fabrication variations, the switching field is not constant from dot to dot and is

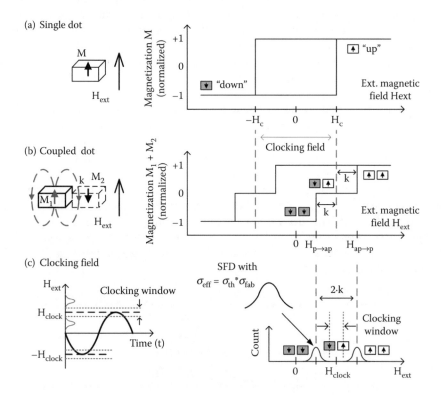

FIGURE 62.2 Simplified model of field-coupled nanomagnets. The switching behavior of the single-domain nanomagnets is described by the switching field H_c (a). The coupling field k of a neighboring nanomagnet superposes with the external field H_{ext} and therefore reduces the switching field to the antiparallel state to $H_{p \to ap}$ by the strength of the coupling field k (b). The switching field distribution of the nanomagnets caused by fabrication variations and thermal noise reduces the clocking window (c).

described by a switching field distribution (SFD) [23]. Additionally, the nanomagnets underlie the influence of thermal noise, which causes an additional SFD for each dot [24]. Both SFDs convolute to an effective SFD with standard deviation σ_{eff}, which narrows the clocking window from the former two times the coupling strength [14]. For the correct operation of the perpendicular NML circuits, the coupling fields of the nanomagnets have to overcome the effective SFD to keep up the clocking window.

62.3 EXPERIMENTS

62.3.1 Fabrication

The $Pt_{5\ nm}[Co_{0.4\ nm}Pt_{1.0\ nm}]_{\times 8}Pt_{3\ nm}$ multilayer stack is magnetron sputtered on a thermally oxidized Si <100> wafer. The 5 nm Pt seed layer prefers the (111) growth that is required to generate the perpendicular anisotropy of the Co layers [25]. The 3 nm Pt top layer prevents the Co layers from oxidation. The nanomagnets are fabricated by FIB lithography using a PMMA photoresist and an evaporated Ti hard mask. After ion beam etching, the nanomagnets are partially irradiated using a 50 kV Ga⁺ FIB system. In the following, the switching behavior of the nanomagnets will be analyzed by magneto-optical microscopy.

62.3.2 Switching Behavior

The switching field H_c of a single magnet varies over time due to thermal noise and may be described by an Arrhenius-type model [24,26]. Figure 62.3 shows the thermally induced SFD of a 1 μm·1 μm dot. In a first approximation, the influence of thermal noise can be modeled by a Gaussian distribution with mean $m_n = H_c$ and standard deviation (SD) σ_n. Comparisons between measurements and the following simulations show that modeling of thermal noise by a Gaussian distribution is justified. For our sample, we measured a thermally induced SFD with an SD of $\sigma_n = 3$ mT.

Owing to fabrication variations, the switching field H_c also varies from dot to dot [23]. The origin of those variations can be traced back to edge roughness, edge oxidation, and grain formations occurring during the sputter process [7]. By partial FIB irradiation of the nanomagnets, the intrinsic switching field distribution (SFD) is rendered ineffective and replaced by the distribution caused by the ANCs [6]. Figure 62.4 shows the measurement results for the switching field H_c, depending on the irradiation dose for a 10 nm·10 nm and 20 nm·20 nm irradiation area. The coercivity H_c remains constant at approximately $m_{asgrown} \approx 400$ mT below a certain dose, depending on the irradiation size. For higher doses, the coercivity drops below the DW nucleation and depinning limit at the ANC and then remains constant again at roughly $m_{ANC} \approx 100$ mT. Fortunately, the absolute SFD of the ANCs is considerable smaller than the intrinsic SFD. The SD is reduced from $\sigma_{asgrown} = 25$ mT to $\sigma_{ANC} = 6.5$ mT for the irradiated dots at high doses.

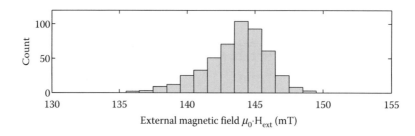

FIGURE 62.3 Thermally induced switching field distribution of a single nanomagnet with 1 μm·1 μm size.

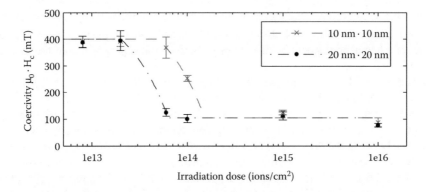

FIGURE 62.4 Switching field H_c of Co/Pt nanomagnets depending on the dose and size of the partial irradiation. The intrinsic switching field distribution of the nanomagnets is replaced by the distribution of the artificial nucleation centers. Both the mean and standard deviation of the coercivity are significantly reduced.

Summing up, the switching field H_c of a nanomagnet can be generally described by

$$H_c = m_{asgrown} + z_{Hc} \cdot \sigma_{asgrown} + z_n(t) \cdot \sigma_n \tag{62.1}$$

with normally distributed random variables z_{Hc}, $Z_n(t) \in 1(0, 1)$. As discussed before, partial FIB irradiation creates artificial nucleation centers and replaces the intrinsic SFD by one of the partially irradiated nanomagnets with ANC:

$$H_{c,pi} = m_{ANC} + z_{Hc} \cdot \sigma_{ANC} + z_n(t) \cdot \sigma_n. \tag{62.2}$$

Previous experiments showed that the switching field $H_{c,pi}$ has to be lower than the intrinsic switching field H_c by two times the coupling field k to achieve directed signal flow in a wire [27]:

$$H_{c,pi} \leq H_c - 2 \cdot k. \tag{62.3}$$

62.3.3 COUPLING FIELD

The interaction between two dots is described by the coupling field k caused by the magnetic stray field of the nanomagnets. In a first approximation, it can be calculated by finite element simulations using the point-dipole approximation [2].

Here, the nanomagnet is divided into small finite elements of roughly 10 nm·10 nm·10 nm size. The dipole field $\vec{B}(m,r)$ of each element is then calculated by

$$\vec{B}(\vec{m},\vec{r}) = \frac{\mu_0}{4\pi r^3}\left(3(\vec{m} \cdot \vec{e_r})\vec{e_r} - \vec{m}\right) + \frac{2\mu_0}{3}\vec{m} \cdot \delta^3(\vec{r}) \tag{62.4}$$

with $\vec{r}$ being the vector to the point of interest, r the absolute value of $\vec{r}$, $\vec{e_r}$ the unit vector in the direction of $\vec{r}$, $\vec{m}$ the dipole moment of each element, and $\delta^3(\vec{r})$ the three-dimensional delta function. The dipole moment $\vec{m}$ can be calculated by

$$\vec{m} = M_s \cdot (d_x \cdot d_y \cdot d_z) \cdot \vec{e_m} \tag{62.5}$$

with M_s being the saturation magnetization of the utilized multilayer, $d_x \cdot d_y \cdot d_z$ the size of each element, and $\vec{e}_m$ the unit vector with the direction of the nanomagnet magnetization. The coupling field of the complete nanomagnet can then be calculated by the superposition of the dipole field of each element.

Figure 62.5 shows the simulation results of the coupling field k of a 400 nm·400 nm dot with eight multilayers (ML) of 0.4 nm Co each. The coupling field drops with $1/d^n$ over the distance d from the border of the nanomagnet with $1 < n < 3$, depending on the geometry of the nanomagnet (squared dot: $n \approx 1$). The strength of the coupling field on the subsequent nanomagnet depends not on the distance between the dots but on the effective distance to its DW nucleation center, which is considerably larger [14]. In an experiment with coupled 800 nm·800 nm dots, we observed a coupling field of $k = 4.5 \pm 1.1$ mT for an interdot distance of $d = 100$ nm. The results coincide with the simulation results, where the SD σ_k can be traced back to variations on the distance between the nanomagnet and the ANC of the subsequent dot. Hence, an additional standard deviation σ_k due to fabrication variations in the shape and edge roughness is taken into account.

Summing up again, the coupling field k of the nanomagnets can be modeled by

$$k = m_k + z_k \cdot \sigma_k \qquad (62.6)$$

where m_k is calculated using the point-dipole approximation described above and $Z_k \in N\,(0, 1)$.

The coupling field may be enhanced by increasing the number of multilayers, but a reduction in the natural domain size and, therefore, the maximal size of the nanomagnets has to be taken into account. Another way to enhance the coupling field is to adapt the shape of the nanomagnets that is not limited due to the interface anisotropy of the Co/Pt multilayer film. The stray field of a nanomagnet with an optimized geometry is concentrated at one spot, where the ANC of the subsequent nanomagnet can be located. Figure 62.6 shows the simulation results for the coupling fields of an optimized inverter structure (a) and a majority gate (b). The stray field of the fork-like inverter structure is concentrated right in the gap between both arms, in which the ANC of the following dot (marked as dashed line) is placed. Measurements on such inverter structures show coupling fields in the 15 mT range, which is sufficient for the realization of complex NML circuits. Both the measurements and simulations on the majority gate resulted in coupling fields of ≈ 5 mT for each input magnet.

Further simulations show that a misalignment of the ANC would degrade the functionality of the majority gate due to the rapid decay of the coupling fields over a distance. A misalignment by several nanometers in one direction would increase the influence of one input and, at the same time, decrease the influence of the others [13]. A variance on the alignment of the ANC due to fabrication variations is considered during the calculation of the coupling fields of each nanomagnet and results in correlated inequalities of the coupling fields in a majority gate.

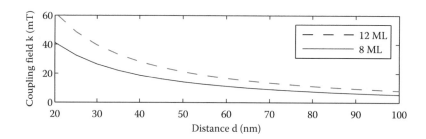

FIGURE 62.5   Simulation results for the coupling field k depending on the distance d from the edge of a 400 nm · 400 nm nanomagnet and the number of multilayers (ML) of the utilized Co/Pt film.

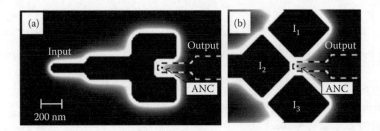

FIGURE 62.6   Simulation results for the coupling fields of an optimized structure for an inverter structure (a) and a majority gate (b). The nanomagnets are marked as a black area; the white color indicates field strengths of at least 20 mT. The shape of the input magnets concentrates the coupling fields on the position of the ANC of the subsequent nanomagnet (indicated as a dashed line).

62.4   NANOMAGNETIC COMPACT MODEL

The analyzed switching behavior of the nanomagnets and the coupling field calculated by the point-dipole approximation are now implemented in the compact models. Figure 62.7 shows the simulation model and the corresponding technological implementation of an inverter and a majority gate.

In our model, the coupling fields of each nanomagnet on the subsequent dot are calculated with respect to the exact geometry of the nanomagnets and the location of the ANC. Each dot can be described to have the switching field $H_{c,pi}$ caused by the ANC. Once the irradiated part of the dot switches, the whole dot gets reversed by a propagating domain wall.

Figure 62.8 shows the block diagram of the implemented model of an output dot in a majority gate. The magnetization states M_1, M_2, and M_3 of the input dots I_1, I_2, and I_3 with $M_i = \pm 1$ are multiplied by the strength of their coupling field k. Owing to the antiparallel impact of the coupling field compared to the input's own magnetization state, the coupling fields are subtracted from the external clocking field H_{ext}. The hysteresis of the nanomagnet is modeled by the switching field $H_{c,pi}$. Once the sum of the coupling and clocking fields reaches the switching field, the nanomagnet switches its magnetization state $M_{out} = \pm 1$. The switching delay models the time needed to propagate the DW through the entire dot.

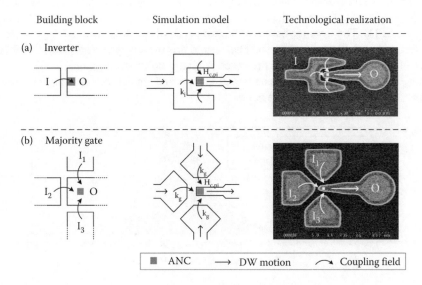

FIGURE 62.7   Compact models used for the simulation and technological realization of an inverter (a) and a majority gate (b).

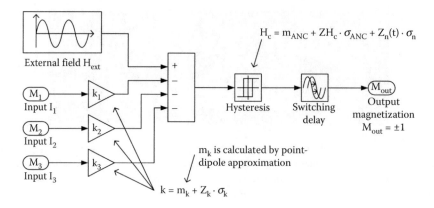

FIGURE 62.8 Block diagram of the implemented model of an output dot in a majority gate.

62.5 SIMULATIONS

62.5.1 MODEL VERIFICATION

The presented compact models are now compared to experiments to validate the accuracy of the model. First, the simulation of the thermally induced SFD of a single dot is compared to the experimental results (Figure 62.9). In both experiments and simulations, 500 hysteresis curves of a single, uncoupled dot are performed and analyzed in 1 mT steps to extract the switching events of each curve. The simulation has been performed with a time-dependent, Gaussian distributed switching field $H_{c,pi}$ with $\sigma_n = 3$ mT. The simulation results are in very good agreement with the experiment and therefore confirm the assumption of a Gaussian distributed thermal noise of the switching field.

To verify the simulated interaction between the dots, a coupled pair with an input dot D_1 and a partially irradiated dot D_2 are both measured and simulated. The size of the dots is 1 μm·1 μm with an interdot distance of d = 100 nm. Depending on the magnetization of the input dot D_1, its coupling field supports or counteracts the switching of the nanomagnet D_2. Hence, the thermally induced SFD of D_2 is shifted depending on the input dot magnetization. Figure 62.10 shows the corresponding SFDs of both the measurement and simulation results. 200 hysteresis curves for each of the input configurations $D_1\downarrow$ and $D_1\uparrow$ were both measured and simulated and afterward analyzed in 1 mT steps to identify the SFDs. The distance between the two peaks of the SFDs is $\mu_0 \cdot \Delta H_{c,mean} = 2 \cdot k = 10$ mT for both measurements and simulations. The results are in very good agreement, verifying the calculation of the coupling field by the presented point-dipole approximation.

Summing up, the comparison between measurements and simulations demonstrates the validity and accuracy of the presented compact models.

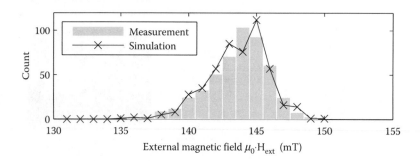

FIGURE 62.9 Comparison between measurement and simulation of the thermally induced switching field distribution of a single nanomagnet.

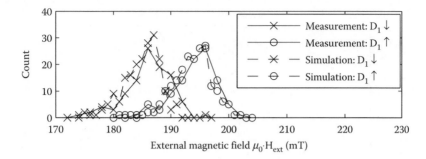

FIGURE 62.10 Comparison between measurement and simulation of the switching behavior of a partially irradiated, coupled nanomagnet. Owing to the influence of the coupling field of its neighbor D_1, the thermally induced switching field distribution of D_2 is shifted to the left ($D_1\downarrow$) and the right ($D_1\uparrow$).

62.5.2 NANOMAGNETIC 1-BIT FULL ADDER

For investigations on a complex NML circuit, simulations were performed on a 1-bit full adder (Figure 62.11). The structure only consists of 12 nanomagnets arranged in three majority gates (G_1, G_2, and G_3) and four inverters, demonstrating the great benefit of the majority decision. The input magnets are A (with $A = A_1 = A_2$), B (with $B = B_1 = B_2$), and the carry-in C_{in}. The sum S and the carry-out C_{out} are the output magnets. The width of the nanomagnets is 200 nm and the gap between the nanomagnets is d = 50 nm.

The structure is clocked using a homogeneous, external clocking field. The input magnets are not partially irradiated and therefore set prior to the simulation. With random initial states of all nanomagnets and with respect to the longest path, the correct logic computation of the output magnets (complete antiparallel ordering of the nanomagnets) is achieved during five clocking cycles in the worst case scenario.

For simulations on the error rate e of the full adder, the total amount of Cobalt t_{Co} of the utilized multilayer and the fabrication-dependent standard deviation of the switching field σ_{ANC} are varied. The standard deviation of the thermal noise has been set to $\sigma_n = 3$ mT, the SD of the coupling field to $\sigma_k = 0.5$ mT, and the SD of the ANC alignment to $\sigma_{align} = 10$ nm. Monte Carlo simulations with a total of 35,000 simulations were performed to investigate the parameter-dependent error rate of the full adder. An error is counted if the nanomagnets of the full adder are not in the correct order after a maximum of five clocking cycles. Simulation results are shown in Figure 62.12.

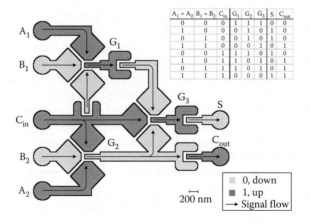

FIGURE 62.11 Simulation model of a 1-bit full adder in perpendicular NML.

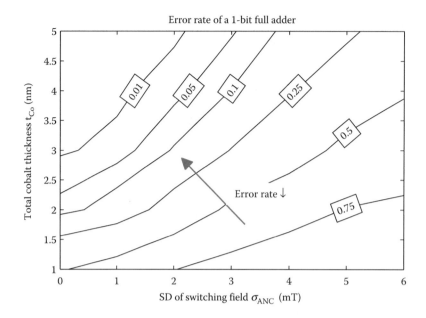

FIGURE 62.12   Error rate e of a 1-bit full adder depending on the total amount of cobalt t_{Co} of the utilized multilayer and the fabrication-dependent standard deviation of the switching field σ_{ANC}.

As a result, the error rate strongly depends on the switching field variations and the total amount of cobalt, which corresponds to an almost linear increase of the coupling field. This shows the importance of small fabrication variations and at the same time strong dipolar coupling for accurate computing devices and systems realized with perpendicular NML.

Higher coupling fields can be achieved by small distances between the field-coupled nanomagnets in the order of 20–30 nm. Additionally, the total amount of cobalt in the utilized multilayer system can be increased to enhance the magnetic moment. Smaller variations can be reached by optimization on partial irradiation. A reduction of both the mean and SD of the nucleation and depinning fields at the ANC may be achieved by a smooth anisotropy gradient from the ANC to the unirradiated area of the nanomagnet [28].

62.6   CONCLUSION

Nanomagnetic logic with perpendicular magnetic anisotropy uses field-coupled nanomagnets and homogeneous clocking fields for the implementation of logic circuits. A partially focused ion beam irradiation is used to create artificial nucleation centers to realize directed signal flow and logic computation in an array of field-coupled nanomagnets. Coupling fields are superposed with the external clocking field and therefore enforce or inhibit the switching of the nanomagnets.

Experiments show that the switching field $H_{c,pi}$ of partially irradiated nanomagnets can be described by a Gaussian distribution with $m_{ANC} \pm \sigma_{ANC}$, which strongly depends on the size and dose of the partial irradiation. Additionally, a time-dependent Gaussian distribution (σ_{ANC}) caused by thermal noise has to be taken into account. The coupling field k between the nanomagnets can be calculated using the point-dipole approximation considering the saturation magnetization M_s of the utilized multilayer, the geometry of the nanomagnets, and the location of the ANC of the subsequent dot.

Comparisons to experiments on single and coupled nanomagnets demonstrate the accuracy of the described model of the nanomagnets. Further simulations on the error rate of a 1-bit full adder demonstrate the great benefit of the compact models, which allow for fast and accurate investigations on the behavior of complex NML circuits.

Potential applications for NML are memory intensive, low-power applications, where robustness, synchronicity, and parallel pipelined computing is more important than extremely high speed. Owing to the compatibility of NML to CMOS technology, hybrid CMOS/NML circuits utilizing I/O structures like GMRs or MTJs are feasible. The scalability of the nanomagnets has been proven to be less than 50 nm. Hence, a programmable NAND/NOR gate would fit in an area of 200 nm·200 nm, which is comparable to CMOS in a 22 nm technology. The speed of NML circuits is limited by the propagation speed of the domain walls (10–100 m/s), the size of the nanomagnets, and the frequency of the clocking field (up to 100 MHz with integrated coils). Regarding the power consumption during the clocking of perpendicular NML, low DW nucleation and depinning fields at the ANCs as well as small DW propagation fields in the unirradiated area of the nanomagnets are essential. In patterned Co/Pt nanomagnets, the field needed for the DW propagation is significantly lower than that needed for the DW nucleation and depinning [29]. Measurement results of Ref. [28] show that by an anisotropy gradient the pinning of the DW may be decreased and the coercivity can be reduced below 10 mT. Co/Ni multilayers are a promising candidate to further reduce both the DW nucleation and propagation fields to the low mT range [30].

REFERENCES

1. The International Technology Roadmap for Semiconductors: Emerging Research Devices (ERD), 2011. [Online]. Available at: http://www.itrs.net/Links/2011ITRS/2011Chapters/2011ERD.pdf.
2. G. Csaba, A. Imre, G. H. Bernstein, W. Porod, and V. Metlushko, Nanocomputing by field-coupled nanomagnets, *IEEE Transactions on Nanotechnology*, 1(4), 209–213, 2002.
3. M. Becherer, G. Csaba, R. Emling, W. Porod, P. Lugli, and D. Schmitt-Landsiedel, Field-coupled nanomagnets for interconnect-free nonvolatile computing, *Digest Technical Papers IEEE International Solid-State Circuits Conference (ISSCC)*, pp. 474–475, 2009.
4. A. Imre, G. Csaba, L. Ji, A. Orlov, G. Bernstein, and W. Porod, Majority logic gate for magnetic quantum-dot cellular automata, *Science*, 311, 205–208, 2006.
5. E. Varga, G. Csaba, G. H. Bernstein, and W. Porod, Implementation of a nanomagnetic full adder circuit, *Proceedings of the 11th IEEE Conference on Nanotechnology (IEEE-NANO)*, pp. 1244–1247, 2011.
6. S. Breitkreutz, J. Kiermaier, V. K. Sankar, G. Csaba, D. Schmitt-Landsiedel, and M. Becherer, Controlled reversal of Co/Pt dots for nanomagnetic logic applications, *Journal of Applied Physics*, 111, 07A715, 2012.
7. J. W. Lau, R. D. McMichael, S. H. Chung, J. O. Rantschler, V. Parekh, and D. Litvinov, Microstructural origin of switching field distribution in patterned Co/Pd multilayer nanodots, *Applied Physics Letters*, 92, 012506, 2008.
8. J. Pommier, P. Meyer, G. Pénissard, J. Ferré, P. Bruno, and D. Renard, Magnetization reversal in ultrathin ferromagnetic films with perpendicular anisotropy: Domain observations, *Physical Review Letters*, 65(16), 2054–2058, 1990.
9. G. Hu, T. Thomson, C. T. Rettner, and B. D. Terris, Rotation and wall propagation in multidomain Co/Pd islands, *IEEE Transactions on Magnetics*, 41, 3589–3591, 2005.
10. W. B. Zeper, H. W. van Kesteren, B. A. J. Jacobs, J. H. M. Spruit, and P. F. Carcia, Hysteresis, microstructure, and magneto-optical recording in Co/Pt and Co/Pd multilayers, *Journal of Applied Physics*, 70, 2264–2271, 1991.
11. C. Vieu, J. Gierak, H. Launois, T. Aign, P. Meyer, J. P. Jamet, J. Ferré et al., Modifications of magnetic properties of Pt/Co/Pt thin layers by focused gallium ion beam irradiation, *Journal of Applied Physics*, 91(5), 3103–3110, 2002.
12. S. Breitkreutz, J. Kiermaier, X. Ju, G. Csaba, D. Schmitt-Landsiedel, and M. Becherer, Nanomagnetic logic: Demonstration of directed signal flow for field-coupled computing devices, *IEEE Proceedings of the 41st European Solid-State Device Research Conference ESSDERC*, pp. 323–326, 2011.
13. S. Breitkreutz, J. Kiermaier, I. Eichwald, X. Ju, G. Csaba, D. Schmitt-Landsiedel, and M. Becherer, Majority gate for nanomagnetic logic with perpendicular magnetic anisotropy *IEEE Transactions on Magnetics*, 48, 4336–4339, 2012.
14. I. Eichwald, J. Kiermaier, S. Breitkreutz, G. Csaba, D. Schmitt-Landsiedel, and M. Becherer, Nanomagnetic logic: Error-free, directed signal transmission by an inverter chain *IEEE Transactions on Magnetics*, 48, 4332–4335, 2012.

15. D. Gardner, G. Schrom, F. Paillet, B. Jamieson, T. Karnik, and S. Borkar, Review of on-chip inductor structures with magnetic films, *IEEE Transactions on Magnetics*, 45(10), 4760–4766, 2009.
16. G. Csaba, J. Kiermaier, M. Becherer, S. Breitkreutz, X. Ju, P. Lugli, D. Schmitt-Landsiedel, and W. Porod, Clocking magnetic field-coupled devices by domain walls, *Journal of Applied Physics*, 111, 07E337, 2012.
17. J. Kiermaier, S. Breitkreutz, G. Csaba, D. Schmitt-Landsiedel, and M. Becherer, Electrical input structures for nanomagnetic logic devices, *Journal of Applied Physics*, 111, 07E341, 2012.
18. J. Kiermaier, S. Breitkreutz, X. Ju, G. Csaba, D. Schmitt-Landsiedel, and M. Becherer, Field-coupled computing: Investigating the properties of ferromagnetic nanodots, *Solid-State Electronics*, 65–66, 240–245, 2011.
19. D. Ravelosona, S. Mangin, Y. Lemaho, J. A. Katine, B. D. Terris, and E. E. Fullerton, Domain wall creation in nanostructures driven by a spin-polarized current, *Physical Review Letters*, 96, 186604, 2006.
20. S. Mangin, D. Ravelosona, J. A. Katine, M. J. Carey, B. D. Terris, and E. E. Fullerton, Current-induced magnetization reversal in nanopillars with perpendicular anisotropy, *Nature Materials*, 5, 210–215, 2006.
21. http://math.nist.gov/oommf/.
22. S. Breitkreutz, J. Kiermaier, C. Yilmaz, X. Ju, G. Csaba, M. Becherer, and D. Schmitt-Landsiedel, Nanomagnetic logic: Investigations on field-coupled computing devices by experiment-based compact modeling, *Proceedings of the 11th IEEE International Conference on Nanotechnology (IEEE-NANO)*, pp. 1248–1251, 2011.
23. T. Thomson, G. Hu, and B. D. Terris, Intrinsic distribution of magnetic anisotropy in thin films probed by patterned nanostructures, *Physical Review Letters*, 96(25), 257204, 2006.
24. J. B. C. Engelen, M. Delalande, A. J. le Febre, T. Bolhuis, T. Shimatsu, N. Kikuchi, L. Abelmann, and J. C. Lodder, Thermally induced switching field distribution of a single CoPt dot in a large array, *Nanotechnology*, 21, 035703, 2010.
25. C. L. Canedy, X. W. Li, and Gang Xiao, Large magnetic moment enhancement and extraordinary Hall effect in Co/Pt superlattices, *Physical Review B*, 62, 508–519, 2000.
26. H.-T. Wang, S. T. Chui, A. Oriade, and J. Shi, Temperature dependence of the fluctuation of the switching field in small magnetic structures, *Physical Review*, 69, 064417, 2004.
27. S. Breitkreutz, J. Kiermaier, C. Yilmaz, X. Ju, G. Csaba, D. Schmitt-Landsiedel, and M. Becherer, Nanomagnetic logic: Compact modeling of field-coupled computing devices for system investigations, *Journal of Computational Electronics*, 10, 352–359, 2011.
28. J. H. Franken, M. Hoeijmakers, R. Lavrijsen, J. T. Kohlhepp, H. J. M. Swagten, B. Koopmans, E. van Veldhoven, and D. J. Maas, Precise control of domain wall injection and pinning using helium and gallium focused ion beams, *Journal of Applied Physics*, 109, 07D504, 2011.
29. M. Delalande, J. de Vries, L. Abelmann, and J. C. Lodder, Measurement of the nucleation and domain depinning field in a single Co/Pt multilayer dot by anomalous Hall effect, *Journal of Magnetism and Magnetic Materials*, 324, 1277–1280, 2012.
30. D. Stanescu, D. Ravelosona, V. Mathet, C. Chappert, Y. Samson, C. Beigne, N. Vernier, J. Ferre, J. Gierak, E. Bouhris, and E. E. Fullerton, Tailoring magnetism in CoNi films with perpendicular anisotropy by ion irradiation, *Journal of Applied Physics*, 103, 07B529, 2008.

63 Parallel Energy Minimizing Computation via Dipolar Coupled Single Domain Nanomagnets

Javier Pulecio, Sanjukta Bhanja, and Sudeep Sarkar

CONTENTS

63.1 INTRODUCTION

At the onset of the twenty-first century, we are presented with stimulating prospects to develop technologies capitalizing on the reduction of feature sizes to produce unique behaviors. The term nanotechnology is often used loosely, but a well-accepted definition is technology that advantageously uses phenomena occurring at critical features of approximately 1–100 nm [1,2]. Particularly for current computational technologies, this creates challenges to overcome, but also provides opportunities to explore novel paradigms and unconventional implementations of logic and memories.

Traditional complementary metal-oxide-semiconductor (CMOS) devices began to feel the concluding "squeeze" of Moore's law when the transistor's critical dimension, the gate length, approached sub-100 nm around 2002 known as the 90 nm node. To mitigate various detrimental effects in microprocessor units, several new strategies were introduced, such as high-k gate dielectrics, strained silicon, and more recently, the use of a three-dimensional Tri-gate structure which can be found commercially in Intels 22 nm node Ivy Bridge processors [3,4]. By 2013, the doubling of device density will cease to trend every 2 years and move to a 3-year cycle [5]. With optimistic scaling predictions of CMOS gate lengths to approximately 6 nm and the aforementioned conclusion of Moore's law, alternative logic paradigms are on the forefront of exploration.

There have been recent proposals for the use of nanomagnets to directly solve quadratic minimization problems, especially those arising in computer vision applications. This is unlike proposals for using nanomagnets to represent binary states. A collection of nanomagnets, when driven to their ground states, can be seen to optimize a quadratic energy function that is determined by their relative placement. By controlling the relative placement of nanomagnets, we can change the energy function being minimized. In this work, we experimentally demonstrate this capability by fabricating and testing an example of a quadratic optimization problem that accomplishes line grouping.

63.2 FIELD-COUPLED COMPUTATIONAL TECHNOLOGIES

There has been a significant growth of interest in field-coupled computing as a radically different computing paradigm. The customary way for charge-based switches, such as a CMOS transistor, to propagate information is through the displacement of charge over a physical connection. The dilemma with this architecture is that once it reaches a critical dimension, classical mechanics makes way for effects like quantum tunneling and the device is no longer operational. Field-based computing is distinctive in that devices are no longer physically connected but still communicate through force fields and may also rely on effects such as quantum tunneling. One such architecture that has been proposed is cellular automata [6,7]. Cellular automata architectures are based on cells where electrons tunnel from one redox site, abstracted in Figure 63.1, to another and interact with neighboring cells through Coulombic forces (quantum dot cellular automata) [8–11].

Another possibility is to use magnetic coupling. This is particularly attractive since this form of computing can be implemented at room temperatures [12], unlike those based on electronic charge interactions, which require very low temperatures. Bandhyopadhay [13] has advocated the use of local spin coupling for computing, a middle ground between quantum computing and spintronic transistors. The architecture is essentially a cellular automata architecture, but with spin coupling energies for logic computing. So far, suggestions for field-coupled computing have been made for Boolean logic-based computing [7,12]. Indeed, most work on nanologic seeks to replicate traditional computing involving logic and arithmetic operations [14–16]. Other applications that have been proposed with QCAs are signal processing [17], permutation matrices [18], interconnection networks [19], and fast Fourier transforms [20]. *However, the energy minimization aspects of CAs have not been directly harnessed* [21]. Since nanodevices are expected to have high error rates, both fabrication related and during operations, it makes sense to consider *error-tolerant applications* where the cost of failure of not finding *the* optimal solution is not high; even solutions that are close to optimal

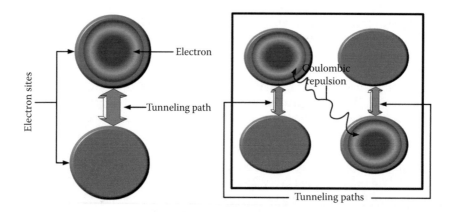

FIGURE 63.1 An abstraction of a quantum-dot cellular automata cell. The electrons tunnel from one site to another and configure themselves in the lowest possible energy state. Similarly, placing cells sufficiently close to one another would induce inter-cell coupling via Coulombic interactions, allowing for more complex functionality.

ones suffice in practice. One such context is in quadratic optimization that arises in computer vision. Energy minimization plays a central role in computer vision algorithms. Nanomagnets offer a tantalizing alternative to traditional form of digital computing for solving quadratic energy minimization problems, drastically reducing the computational time required. Collection of nanomagnets, when driven to their ground states, can be seen to optimize a quadratic energy function that is determined by their relative placement. By controlling the relative placement of magnets, we can change the energy function being minimized. It is somewhat like analog computing from the past, except that instead of solving differential equations, we solve minimization problems.

63.3 APPLICATION: DIGITAL MEDIA OBJECT DETECTION

Imagine an energy minimization co-processor based on nanomagnets that are heterogeneously integrated with CMOS. In the long term, these vision computing circuits can be integrated with camera circuitry to design cognitive cameras, capable of higher level reasoning. The "complier" for this form of computing would transform a given energy minimization problem into a set of equivalent coordinates for a magnet collection. These magnets would be "selected" from a regular grid of nanomagnets by driving the noncomputing magnets into noninteracting vortex states. The array would then be clocked to its ground state. The final states would be read off as the solution to the problem. The vision problem would then use these magnetic measurements as the solution. We will essentially be solving an optimization problem with each input-and-read-out cycle as compared to orders of magnitude more clock cycles that would be needed in a Boolean logic circuit. The current work is toward this long-term goal. We experimentally demonstrate the viability of using single domain nanomagnetic coupling for function minimization computing. Unlike logic and arithmetic computing tasks that demand exact computations, vision problems can work with near-optimal solutions. These vision problems place high demand on computational resources (on Boolean logic-based computing platforms). There have been proposals for using regular arrays of quantum dots [22,23] and nanomagnets [24] for low-level vision, mainly segmentation where the input and the output are both regular grid of pixels. In this chapter, we consider quadratic energy forms that arise in contexts involving extended image features rather than individual pixels. These problems have high-computational complexity and are not amenable to single-instruction-multiple-data (SIMD)-type hardware solutions. To solve the vision problems that we target, VLSI implementation using traditional logic would require complex multiple-instruction-multiple-data (MIMD) architectures as opposed to just SIMD architectures, which are prevalent in the design of vision chips [25]. For instance, a recent proposal for an object recognition chip uses both SIMD and MIMD components [26].

63.3.1 Image Processing

The detection of significant features in an image is a computationally expensive process. The term perceptual organization is used to describe the act of recognizing important features of an image. It is commonly broken down into three steps:

- *Segmentation-extraction* of features from an image such as edges, textures, pixel intensity, and so on.
- *Grouping-relating* low-level segments of an image into larger perceptual groups such as surfaces, background, foreground, and so on.
- *Recognition-matching* groupings identified as significant to a known model such as a building, human, and so on.

Although each step has its own associated complexities, grouping is specifically important when reducing computational intensity. The grouping of low-level segments can be accomplished through a quadratic energy maximization process [24]. Traditional Boolean architectures found in super-computers normally reduce problems such as the optimization of quadratic problems into finding exact solutions of arithmetic problems and logical operations. This can be very demanding on a Boolean system and is not necessarily required for the grouping of visual objects, and magnetic nano-systems present a unique way to accomplish such a task. For instance, in Figure 63.3a, the task is to find the visually salient or important edges, such as the 3 parallel lines in the center of the image. The grouping of these lines does not require exact computational accuracy, meaning that solutions that are near the optimal result are acceptable [27,28].

Let there be N straight lines in the image that we would like to group. With each straight line, let us associate a variable, x_i, taking on values 1 or 0, denoting whether it is significant or not significant, respectively. Every pair of straight lines can be associated with an affinity value capturing its saliency (or perceptual importance). Various functional forms have been proposed in the computer vision literature for this affinity function. They are all designed to capture perceptual organization of the straight lines. For instance, if two straight lines are parallel to each other, they will have higher affinity than any other random arrangements. The justification for this is that it is highly unlikely for lines to be parallel to each other by chance. There must be an underlying reason, and they very likely belong to some object in the scene. One mathematical form that captures the pairwise saliency of the ith and jth line segments in the image is the following:

$$a_{ij} = \sqrt{l_i l_j} \, \exp^{-\frac{o_{ij}}{\max l_i, l_j}} \exp^{-\frac{d_{min}}{\max l_i, l_j}} \cos^2(2\theta_{ij}) \tag{63.1}$$

where l_i and l_j are the lengths of the two segments, θ_{ij}, is the angle between them, o_{ij} is the overlap with each other, and d_{min} is the minimum distance between them. As one can see from the expression, the affinity between two straight lines will be high for longer segments, segments that are parallel to each other, or continuous to each other, or at right angles to each other. Given these pairwise affinity values, the vision problem is to find the values of x_i for each segment such that the following measure is maximized:

$$\sum_i \sum_{j \neq i} a_{ij} x_i x_j + \left(k - \sum_i x_i \right) \tag{63.2}$$

The first term is the total of the pairwise affinities among the segments with $x_i = 1$ and the second term tries to enforce that we have k segments with $x_i = 1$. This is a hard problem to solve. Traditional Boolean logic-based approaches would reduce this binary quadratic problem into finding exact solutions based on arithmetic and logical operations. This is very demanding on a Boolean system and is not necessarily required for the grouping of visual objects, and magnetic nanosystems present a unique way to accomplish such a task in a direct manner.

63.3.2 Magnetic Hamiltonian

The basic unit of computation for magnetic logic is a nanomagnet with dimensions and materials such that it exhibits single domain behavior, that is, it can be modeled as one overall magnetic state. The material composition and the geometry (shape and size) of the nanomagnet determine the overall magnetic behavior. For instance, for disk-shaped magnets that are thin (say 20 nm) and with a diameter of 100 nm, all the magnetic vectors are aligned perpendicular to the z-direction, in the xy-plane (single domain). The vectors are either all aligned in one direction, resulting in one overall effective magnetic vector direction, or aligned in a circular fashion, resulting in a vortex state. For a vortex state, there is only a small, effective magnetic vector in the z-direction at the center of the magnet, but the overall magnetic effect in the xy-direction is zero. Let the flat nanoscale disks be of height h, radius r, and magnetization M_0 ordered in an array in the xy-plane. Bennett and Xu [29] showed that a disk of uniform magnetization can be approximated well by a point dipole with moment $\pi r^2 h M_0$ that is oriented in the plane forming an angle ϕ with the x-axis and with $m(z) = 0$. The magnetization vector of the ith magnet can be represented by $\mathbf{m}_i$. The total Hamiltonian of an arrangement of magnets is given by

$$\mathcal{H} = \sum_i \mathbf{m}_i^\mathsf{T} \mathbf{D}_i \mathbf{m}_i + \mu_0 \sum_i \mathbf{m}_i^\mathsf{T} \mathbf{h}_{\text{ext}} + \sum_i \sum_{j \neq i} \mathbf{m}_i^\mathsf{T} \mathbf{C}_{ij} \mathbf{m}_j \tag{63.3}$$

where $\mathbf{D}_i$ is the demagnetization tensor of the ith magnet capturing the shape anisotropy, $\mathbf{C}_{ij}$ is the interaction matrix between the ith and jth magnets, and $\mathbf{h}_{\text{ext}}$ is the external field. A diagonal matrix with a value of 1/2 along the diagonal can approximate the demagnetization tensor for a thin disk, so the first energy term is a constant.

$$\mathcal{H} = D_0 + \mu_0 \sum_i \mathbf{m}_i^\mathsf{T} \mathbf{h}_{\text{ext}} + \sum_i \sum_{j \neq i} \mathbf{m}_i^\mathsf{T} \mathbf{C}_{ij} \mathbf{m}_j \tag{63.4}$$

For dipole-to-dipole interaction approximation, the coupling term (the third term) is given by

$$\mathbf{C}_{ij} = \frac{\mu_0 |M|}{4 \pi d_{ij}^3} \left(3 \mathbf{e}_{ij} \mathbf{e}_{ij}^\mathsf{T} - \mathbf{I} \right) \tag{63.5}$$

$$= c_{ij} \left(3 \mathbf{e}_{ij} \mathbf{e}_{ij}^\mathsf{T} - \mathbf{I} \right) \tag{63.6}$$

where $\mathbf{e}_{ij}$ is the *unit* vector line joining the centers of the two dipoles, d_{ij} is the distance between the centers, and $\mathbf{I}$ is the identity matrix [30]. The term $|M|$ is the product of the magnetic moment magnitudes of the two magnets and is constant for our magnets. The interaction term between the ith and jth magnets will be dependent on the relative placement of the magnets. For any particular magnet, if the interaction was low with the other magnets, then it would be easy to change its magnetization vector with a low external field. Conversely, if the interaction with other magnets were high, then it would be hard to change its magnetization using a low external field.

63.4 CORRESPONDENCE BETWEEN VISION AND MAGNETS

Notice the correspondence between Equations 63.2 and 63.4. Each line segment in the image corresponds to a magnet. The magnetizations, $\mathbf{m}_i$, in Equation 63.4 correspond to the saliencies, x_i, in Equation 63.2. The pairwise coupling constants, c_{ij}, (Equation 63.6) correspond to the affinities, a_{ij}. By engineering the distances between the magnets, if we can modify the coupling constants to match the respective affinities, then the minimum state of the arrangement will give us an approximate solution to the original problem. Figure 63.2 shows some examples of this placement for two straight lines. For the general case of N straight lines, we rely on the body of work in statistics called multidimensional scaling (MDS) [32].

The objective is to find a configuration of points, representing the low-level features, in a 2D space such that the distance between the ith and jth points, d_{ij}, will be proportional to affinity between the corresponding lines, a_{ij}. If magnetic cells are placed at these point coordinates, then the pairwise interaction between them will be proportional to the given energies, that is, $c_{ij} \propto a_{ij}$. For this process, we look into the rich areas of graph embedding onto planes [33,34] and multidimensional scaling [32]. The affinity matrix can be considered to represent the adjacency matrix of a weighted graph.

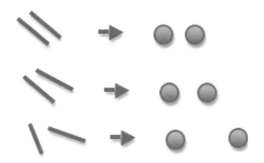

FIGURE 63.2 Correspondence between nanomagnets (MFC) and vision problem formulation. On the left are sample arrangements of pairs of straight lines and on the right are the corresponding nanomagnet placements whose interactions match the affinities between the lines. (J. Pulecio, S. Bhanja, and S. Sarkar, An experimental demonstration of the viability of energy minimizing computing using nano-magnets, *Proc. 11th IEEE International Conference on Nanotechnology*, Portland, OR, August 15–18, © (2011) IEEE. With permission.)

The problem then is to embed the nodes of the graph in the plane in such a way as to preserve an edges weight at a Euclidean distance between them. If we allow for distortions of weights of the graphs, this is indeed possible [34,35]. We have developed an approach based on multidimensional scaling.

Let the matrix Λ be constructed out of given affinities such that: $\Lambda_{rs} = 1/(A_{ij})^2$. We desire to find the coordinate of each point in a 2D space, which we denote by the matrix of the coordinate vector, $\mathbf{X}_{MDS} = [\mathbf{x}_1, \ldots, \mathbf{x}_N]$, such that

$$(\mathbf{x}_i - \mathbf{x}_j)^{\mathrm{T}}(\mathbf{x}_i - \mathbf{x}_j) = c\Lambda_{rs} \tag{63.7}$$

$$\mathbf{X}_{MDS}^{\mathrm{T}}\mathbf{X}_{MDS} = -c\frac{1}{2}\mathbf{H}\Lambda\mathbf{H}, \quad \text{where } \mathbf{H} = \left(\mathbf{I} - \frac{1}{N}\vec{1}\vec{1}^{\mathrm{T}}\right) \tag{63.8}$$

with $\mathbf{I}$ as the identity matrix and $\vec{1}$ as the vector of ones. This operator $\mathbf{H}$ is referred to as the centering operator. These coordinates $\mathbf{X}$ can be arrived at by the classical MDS scheme [32]. The solution is based on the singular value decomposition of the centered distance matrix $\frac{1}{2}\mathbf{H}\Lambda\mathbf{H} = \mathbf{V}_{MDS}\Delta_{MDS}\mathbf{V}_{MDS}^{T}$ where $\mathbf{V}_{MDS}$, Δ_{MDS} are the eigenvectors and eigenvalues, respectively. Assuming that the centered distance matrix represents the inner product distances of a Euclidean distance matrix, the coordinates are given by

$$\mathbf{X}_{MDS} = \left(\mathbf{V}_{MDS}\Delta_{MDS}^{1/2}\right)^{\mathrm{T}} \tag{63.9}$$

Note that we have dropped the constant of proportionality, c, since the energy minimizing solutions are invariant to scaling of the original function. Our nanomagnet selection solution is given by the first two rows of $\mathbf{X}_{MDS}$; each column of this matrix gives us the coordinates of the corresponding nanomagnet to consider.

The computational overhead of this synthesis step is linear in the number of the image features. This replaces the complexity of the software solution to the minimization problem.

63.5 FABRICATION PROCESS

An Si wafer was coated with PMMA (poly/methyl methacrylate) via a Laurell Technologies WS-400A-8NPP/Lite Spin Processor. A single thin layer of 950 molecular weight PMMA in anisole was spun with a resulting thickness of approximately 120 nm. Afterward, it was baked in an oven which provided even heat over the entire wafer to evaporate any residual solvent. The magnetic field-based computing (MFC) systems were designed using DesignCAD2000 NT. The most effective line spacing, exposure doses, points, and focus were determined by using the diagnostic wheel pattern. The sample was then loaded into a JOEL 840 m retrofitted with the NPGS lithography system and a beam blanker for pattern exposure. Subsequently, the sample was unloaded for development in MIBK:Isopropanol 1:3, after which a thin Permalloy film was deposited via a Varian Model 980-2462 electron beam evaporator. A vacuum of about 2 μTorr was achieved and evaporation was conducted at a fast rate to reduce contamination. Liftoff was accomplished by placing the sample coated with Permalloy in a heated ultrasonic acetone bath for approximately 15 min.

63.6 READ-OUT SCHEME

Figure 63.3a shows an example of a vision problem where we have to find the visually salient subset of straight lines—the three parallel lines in the middle. Figure 63.3b shows the placement of magnets that can be used to solve these problems. The magnetostatic coupling between the magnets matches the pairwise affinities of the lines. Each line is represented by a magnet. The white circles represent magnetic nano-disks, enumerated from the left to the right. The disks were made of permalloy with the thickness much less than the lateral dimensions to force in-plane single domain magnetic dipole moments. The experiment would proceed by applying an external magnetizing field along a particular direction. Afterward, the field would be removed and the magnetic cells would be allowed to settle into a ground state as shown in Figure 63.3c. The natural tendency for magnetic energy is to be minimized, in this case primarily due to magneto-static coupling. It would lead magnets 1–3, to arrange themselves in a ferromagnetic fashion. It would also be possible for

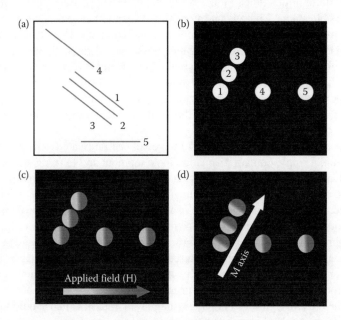

FIGURE 63.3 Read-out scheme for validation. (a) The vision problem: find the visually salient lines. (b) Equivalent magnet arrangement with a similar Hamiltonian as the vision problem. The disks correspond to the lines. (c) Magnetic initialization of the MFC system into a possible ground state of the system after the external field is removed. (d) If the magnetization is taken along axis M, the magnetization component of nanodisks 1, 2, and 3 will be larger than those of 4 and 5. (J. Pulecio, S. Bhanja, and S. Sarkar, An experimental demonstration of the viability of energy minimizing computing using nano-magnets, *Proc. 11th IEEE International Conference on Nanotechnology*, Portland, OR, August 15–18, © (2011) IEEE. With permission.)

magnet 4 to experience a degree of coupling with magnets 1–3 which could be demonstrated via an antiferromagnetic coupling. The coupling exhibited among magnets 1–3 is expected since the inter-spacing distance is smaller than that of magnets 4 and 5. Our experimental measurements do indeed demonstrate this to be the case.

If the magnetization vector of each magnet is taken along axis M, as shown in Figure 63.3d, the magnetization of magnets 1–3 would be greater than those of magnets 4 and 5. As mentioned previously, this magnetic interaction can be modeled via a quadratic term and is computationally intensive but occurs naturally in the physical world, at least on the order of nanoseconds. The magnetostatic quadratic term is inversely proportional to the quadratic edge affinity, which determines the salient features of the image. So by setting appropriate affinity thresholds, the edges corresponding to magnets 1–3 would be regarded as a significant grouping, while magnets 4 and 5 would not.

63.7 RESULTS

The size, spacing, and placement all correlated to the features of an image; therefore, proper fabrication of a design was imperative. We were able to achieve the accuracy of these parameters for the system shown in Figure 63.3, although on a larger scale, there were instances where some common irregularities associated with the electron beam fabrication process occurred. It is important to note how all these parameters affect the ability of a field-based system to couple and therefore extract salient features. The magnetic nanodisks used throughout the experiments were made of permalloy with negligible magnetocrystalline anisotropy. The circular shape combined with the absence of magnetocrystalline anisotropy allows for a nano-disk to have an easy axis of magnetization of 360° in-plane. For the proof of concept, the thickness was reduced to try to induce an in-plane single domain magnetic moment. The in-plane moments would couple with neighbors through magneto-static interactions. Figure 63.4a shows MFM phase graphs with corresponding images of the possible states. There is an interesting dynamic between the formation of a vortex or single domain moment in nanodisks, which is a formulation of material, diameter, neighboring elements, and thickness. Indeed, a complete implementation of magnetic field computation could benefit from the ability to switch between the single domain and vortex states, particularly because the vortex state exhibits no in-plane stray field at rest. This would essentially remove any participation of a nanodisk (in the vortex state) from the computational process. For the purpose of these experiments, the goal was to create in-plane dipole interactions through single domain magnetic moments.

Figure 63.5a and b shows AFM and MFM images of the fabricated cells scanned along three directions. To ensure a single domain magnetic dipole moment and to mitigate vortex states, engineering of the shape anisotropy should be such that the out-of-plane magnetization component of the vortex core should produce higher demagnetization energy than that of an in-plane single domain state. The average dimensions of the nanomagnetic disks were approximately 130 nm in diameter and 20 nm thick and are referred to as magnets from the left to the right. As desired, the three leftmost magnets (1, 2, 3) during all the three different scan directions were in a single domain state, with their magnetic configurations unaltered. This is due to the ferromagnetic dipole–dipole coupling between nearby neighbors that reduces the susceptibility of the nanomagnets to the stray field emanating from the magnetically coated proximal probe.

When analyzing the two right-most magnets (magnets 4 and 5), which should be in a decoupled single domain state due to the distance in between their nearest neighbors, a pinwheel magnetization is observed. This was due to the sample–probe interaction which caused the nanodisks to flip their magnetization during data acquisition, signifying a weak interaction with the neighboring elements. Even though the probe tip altered the magnetic state of the nanodisks, the single domain moment can still be extrapolated via the presence of strong dipoles. By retracing the scanning process of the tip with a scan angle of 0, as shown in Figure 63.5b by the arrow, where the slow scan progresses from the bottom to the top and the fast scan from the left to the right, the single

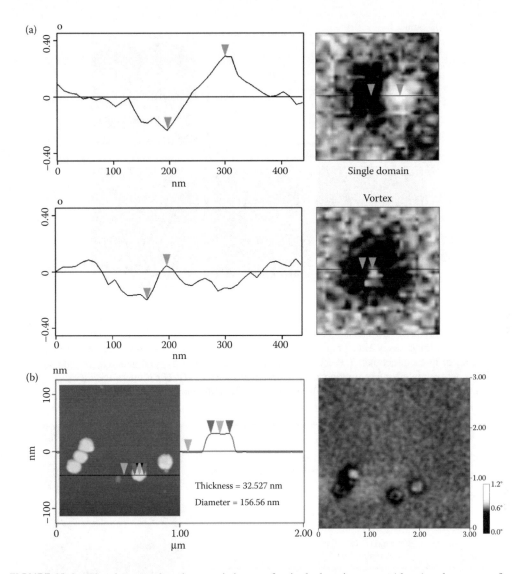

FIGURE 63.4 The phase graph and magnetic image of a single domain magnet (above) and vortex configuration (below) taken via MFM (a). The out-of-plane curl at the pole ends for the single domain state and the out-of-plane central core for the vortex state are detected by the proximal probe. Critical dimensions of the diameter and thickness are shown in (b) where both the vortex and single domain states are present.

domain moment can be followed as it flips throughout the scan. The stray field from the scanning tip was sufficiently strong to flip the magnetization of magnet 4 at least 3 times during the scan because it was decoupled from the neighboring elements. A similar process altered the magnetization of magnet 5 and the different flipping behavior is explained by a preferred magnetization axis in the vertical direction and is detailed in Figure 63.6b. This becomes evident once the scan angle is 90 and the slow scan is in the vertical direction. Magnet 5 flipped once during the scan, in a similar fashion to magnet 4, with a scan angle of 0. The development of structural minor and major axes in the elliptical magnets results in the introduction of shape anisotropy in the nanoelements. This alters the energy landscape of each element, and hence the different behavior, as the proximal probe traverses each element along their respective hard (minor) and easy (major) axes of magnetization.

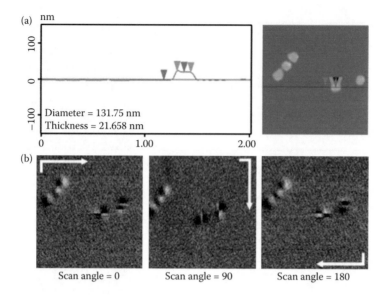

FIGURE 63.5 Fabricated nanomagnets for the vision problem. (a) AFM height amplitude image and the fabricated layout of the MFC system. As can be seen in the graph, the diameter of the measured nanodisk is 131.75 nm and the thickness is 21.658 nm. (b) Three images show MFM scans of the fabricated arrangement, scanned along different directions. (J. Pulecio, S. Bhanja, and S. Sarkar, An experimental demonstration of the viability of energy minimizing computing using nano-magnets, *Proc. 11th IEEE International Conference on Nanotechnology*, Portland, OR, August 15–18, © (2011) IEEE. With permission.)

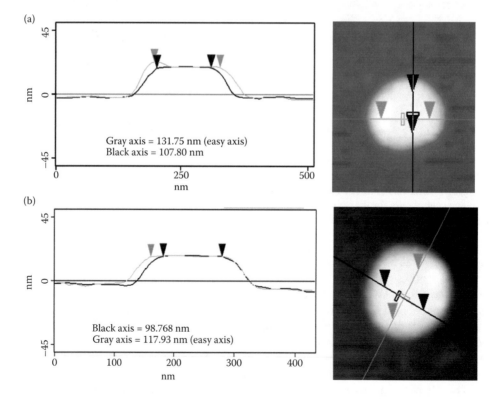

FIGURE 63.6 AFM images of major and minor axes of magnet 4 (a) and magnet 5 (b) are depicted. The major axes of the magnets create an easy axis of magnetization. This is significant when interpreting sample–tip interactions.

63.8 CONCLUSION

Magnetic field-based computing (MFC) has the potential to offer a unique solution for the quadratic minimization problem. This is unlike other implementations of magnetic logic in that it does not force the magnetostatic coupled elements to perform Boolean operations; instead, the natural tendencies of magnets to reach the lowest possible ground state for the entire system are exploited. The time necessary for the magnetic field-based systems to settle into an energy minimum is related to the relaxation time of the constituting magnetic elements (in this case on the order of nanoseconds). Excitingly, the computational time should not scale with the size/complexity of the image to process since the magnetic interactions between nanoelements are massively parallel, meaning that the relative computation time for a small system such as Figure 63.5 should be similar to a larger system such as Figure 63.7. MFC should also provide the associated benefits of other nanomagnetic logic such as radiation hardness, scalability, and low-power operation.

Here, we presented an image processing application of a quadratic problem where the grouping of low-level segments into relevant features was demonstrated. A proof-of-concept device, which consisted of five edge segments from an image, was mapped to nanomagnetic disks through a placement algorithm. The nanomagnets were fabricated and strictly adhered to placement and dimensions that correlated magnetostatic interactions to the edge affinity energies of an image. The significance of creating a nanodisk to specific dimensions was displayed by the undesired formation of preferred

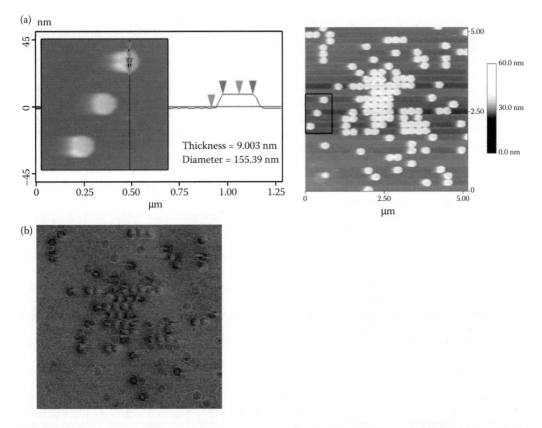

FIGURE 63.7 A more intricate system of nanomagnets for the vision problem. (a) AFM height amplitude image and the fabricated layout of the MFC system. As can be seen in the graph, the diameter of the measured nanodisk is 155.39 nm and the thickness is 9.0 nm. (b) A corresponding MFM image of the system. Note the complicated magnet structure throughout the entire system. Both the single domain and vortex states are present.

axes of magnetization and the formation of magnetic vortices. The strong magnetostatic interaction revealed by nanodisks 1–3 was established via the ferromagnetic coupling exhibited by the nanomagnets. Nanodisks 4–5 were strongly influenced by the MFM tip and caused the magnet moments of the magnets to flip during data acquisition. The switching of nanodisks 4–5, and the lack of variation in magnetic dipole moments of nanodisks 1–3 due to the magneto-static coupling, was determined to be a satisfactory way of qualitatively determining strong field interactions of neighboring magnets. The strong coupling exhibited between nanomagnets 1–3, which are directly mapped to the specific edges of an image, was determined to be a salient feature for segment grouping. Larger systems, such as those described in Figure 63.7, have also been fabricated and are currently being evaluated to investigate the effectiveness of MFC for perceptual organization in computer vision.

ACKNOWLEDGMENTS

This work was supported in part by the National Science Foundation under Grant CCF 0829838.

REFERENCES

1. National Nanotechnology Initiative. Standards for Nanotechnology. http://www.nano.gov/you/standards. Accessed: 17/05/2012.
2. International Organization for Standardization. TC 229 Nanotechnologies. http://www.iso.org/iso/iso_technical_committee?commid=381983, 2005. Accessed: 17/05/2012.
3. Intel's 45 nm CMOS technology. *Intel Technology Journal,* 12(45), 2008.
4. K. J. Kuhn. CMOS scaling for the 22 nm node and beyond: Device physics and technology. In *Proceedings of 2011 International Symposium on VLSI Technology, Systems and Applications,* pp. 1–2. IEEE, April 2011.
5. International Technology Roadmap for Semiconductors. Executive Summary. http://www.itrs.net/Links/2011ITRS/Home2011.htm, 2011. Accessed: 18/05/2012.
6. G. Csaba, A. Imre, G. Bernstein, W. Porod, and V. Metlushko, Nanocomputing by field-coupled nanomagnets, *IEEE Transactions on Nanotechnology,* 1(4), 209–213, Dec. 2002.
7. A. Imre, G. Csaba, L. Ji, A. Orlov, G. Bernstein, and W. Porod, Majority logic gate for magnetic quantum-dot cellular automata, *Science,* 311(5758), 205–208, Jan. 2006.
8. C. Lent and P. Tougaw, A device architecture for computing with quantum dots, in *Proceedings of the IEEE,* 85(4), 541–557, April 1997.
9. J. C. Lusth, C. B. Hanna, and J. C. Diaz-Velez, Eliminating non-logical states from linear quantum-dot cellular automata, *Mircoelectronics Journal,* 32, 81–84, 2001.
10. R. Kummamuru, J. Timler, G. Toth, C. Lent, R. Ramasubramaniam, A. Orlov, G. Bernstein, and G. Snider, Power gain in a quantum-dot cellular automata latch, *Applied Physics Letters,* 81, 1332–1334, August 2002.
11. P. D. Tougaw and C. S. Lent, Dynamic behavior of quantum cellular automata, *Journal of Applied Physics,* 80, 4722–4736, Oct. 1996.
12. R. Cowburn and M. Welland, Room temperature magnetic quantum cellular automata, *Science,* 287(5457), 1466–1468, 2000.
13. S. Bandyopadhyay, Power dissipation in spintronic devices: A general perspective, *Journal of Nanoscience & Nanotechnologies,* 7(1), 168–80, 2007.
14. J. Pulecio and S. Bhanja, Magnetic cellular automata coplanar cross wire systems, *Journal of Applied Physics,* 107(3), 034308–034308-5, Feb. 2010.
15. M. T. Alam, S. Kurtz, M. T. Niemier, S. X. Hu, G. H. Bernstein, and W. Porod, Magnetic logic based on coupled nanomagnets: Clocking structures and power analysis, in. *8th IEEE Conference on Nanotechnology, 2008. NANO '08,* Arlington, TX, pp. 637–637, 2008.
16. D. Allwood, G. Xiong, C. Faulkner, D. Atkinson, D. Petit, and R. Cowburn, Magnetic domain-wall logic, *Science,* 309(5741), 1688–1692, 2005.
17. A. Csurgay, W. Porod, and C. Lent, Signal processing with near-neighbor-coupled time-varying quantum-dot arrays, *IEEE Transactions on Circuits and Systems,* 47, 1212–1223, Aug. 2000.
18. A. Fijany, N. Toomarian, and M. Spotnitz, Implementing permutation matrices by use of quantum dots, *Tech. Rep.,* Jet Propulsion Laboratory, California, Oct. 2001.

19. A. Fijany, N. Toomarian, K. Modarress, and M. Spotnitz, Compact interconnection networks based on quantum dots, *Tech. Rep.*, Jet Propulsion Laboratory, California, Jan. 2003.
20. A. Fijany, N. Toomarian, K. Modarress, and M. Spotnitz, Hybrid vlsi/qca architecture for computing FFTs, *Tech. Rep.*, Jet Propulsion Laboratory, California, Apr. 2003.
21. T. Cole and J. C. Lusth, Quantum-dot cellular automata, *Progress in Quantum Electronics,* 25, 165–189, 2001.
22. W. H. Lee and P. Mazumder, Color image processing with quantum dot structure on a multi-peak resonant tunneling diode, in *7th IEEE Conference on, Nanotechnology, 2007. IEEE-NANO 2007.* pp. 1161–1165, Aug. 2007.
23. K. Karahaliloglu, S. Balkir, S. Pramanik, and S. Bandyopadhyay, A quantum dot image processor, *IEEE Transactions on Electron Devices,* 50, 1610–1616, 2003.
24. S. Sarkar and S. Bhanja, Direct quadratic minimization using magnetic field-based computing, in *2008 IEEE International Workshop on, Design and Test of Nano Devices, Circuits and Systems*, Cambridge, MA, pp. 31–34, Sept. 2008.
25. A. Moini, *Vision Chips.* Springer, Netherlands, 2000.
26. J. Kim, M. Kim, S. Lee, J. Oh, S. Oh, and H. Yoo, Real-time object recognition with neuro-fuzzy controlled workload-aware task pipelining, *Micro, IEEE,* 29(6), 28–43, 2009.
27. W. Grimson, *Object Recognition by Computer: The Role of Geometric Constraints.* MIT Press, Cambridge, MA, 1991.
28. D. Clemens and D. Jacobs, Model group indexing for recognition, in *1991. Proceedings CVPR'91. IEEE Computer Society Conference on, Computer Vision and Pattern Recognition,* Maui, HI, pp. 4–9, IEEE, 1991.
29. A. Bennett and J. Xu, Simulating collective magnetic dynamics in nanodisk arrays, *Applied Physics Letters,* 82, 2503, 2003.
30. M. Levitt, *Spin Dynamics: Basics of Nuclear Magnetic Resonance.* Wiley, Chichester, UK, 2001.
31. J. Pulecio, S. Bhanja, and S. Sarkar, An experimental demonstration of the viability of energy minimizing computing using nano-magnets, *Proc. 11th IEEE International Conference on Nanotechnology,* Portland, OR, August 15–18, 2011.
32. M. Cox and T. Cox, Multidimensional scaling, in *Handbook of Data Visualization*, C. -H. Chen, W. Hardie, and A. Unwin, (eds.) Springer-Verlag, Berlin, Heidelberg, pp. 315–347, 2008.
33. P. Indyk and J. Matousek, Low-distortion embedding of finite metric space, *Handbook of Discrete and Computational Geometry,* p. 177, 2004.
34. J. Nievergelt and N. Deo, Metric graphs elastically embeddable in the plane, *Information Processing Letters,* 55(6), 309–315, 1995.
35. J. Bourgain, On Lipschitz embedding of finite metric spaces in Hilbert space, *Israel Journal of Mathematics,* 52(1), 46–52, 1985.

Section XIX

Spintronics

64 On Physical Limits and Challenges of Graphene Nanoribbons as Interconnects for All-Spin Logic

Shaloo Rakheja and Azad Naeemi

CONTENTS

64.1 INTRODUCTION

The semiconducting material silicon is at the heart of the current complementary metal–oxide–semiconductor (CMOS) technology, which today has developed into a \$270-billion market [1]. Over the last four decades as the minimum feature size (MFS) on the microprocessor has shrunk from a few microns to tens of nanometers, the productivity of the silicon technology has roughly increased by a factor of billion [1]. Dr. G. Moore from Intel first pointed out this exponential growth in the semiconductor industry; his observation later became the celebrated Moore's law. It is, indeed, Moore's law that has driven the economics of the semiconductor industry by setting targets for research and development over the past few decades. However, as we move into an era of sub-10 nm technology nodes, it is natural to ask if Moore's law will hold forever. One of the main limits of dimensional scaling is the power barrier of the CMOS technology. It is now well established that the fundamental limit of the energy dissipation of a single binary transition in a CMOS switch is $k_B T \ln 2$, where k_B is the Boltzmann constant, and T is the temperature of the system [2–4]. Using materials other than silicon to implement field-effect transistors (FETs) might provide one-time performance gains but will eventually be plagued by the same fundamental limits governing silicon-based FETs [5].

In the CMOS technology, electron charge is used as the state variable. Hence, the power dissipation in the CMOS technology comes from charging and discharging nodal capacitances at the output node. The output-node capacitance is a sum of the parasitic capacitance of the driver, the load capacitance, and the wire capacitance. In modern microprocessors, more than 50% of the on-chip capacitance is

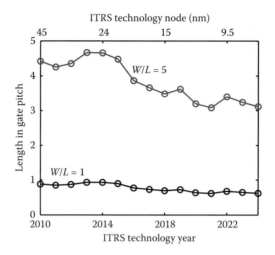

FIGURE 64.1   The length of the interconnect whose energy dissipation is equal to that of a CMOS inverter for two sizes of the inverter: (i) $W/L = 1$ (minimum-sized inverter) and (ii) $W/L = 5$ (5× the minimum-sized inverter). (Taken from S. Rakheja and A. Naeemi, Performance, energy-per-bit, and circuit size limits of post-CMOS logic circuits–Modeling, analysis and comparison with CMOS logic, in *International Interconnect Technology Conference (IITC)*, Dresden, 2011. Copyright © IEEE 2012, with permission.)

associated with wires; roughly half of the wiring capacitance is due to the short, local wires [6]. To understand the issue of interconnect capacitance, we plot the length of the interconnect that consumes energy equal to a CMOS inverter with various sizes in Figure 64.1 as a function of the technology year from the International Technology Roadmap for Semiconductors (ITRS). As can be seen from this figure, interconnects as short as one gate pitch consume energy equal to that of a minimum-sized inverter in any given technology node beyond 45 nm. Hence, ignoring interconnect capacitance can severely underestimate the amount of energy dissipation in a circuit even with short wires.

Another challenge associated with interconnects is their growing resistivity to dimensional scaling. Owing to the enhanced electron scatterings with the grain boundaries and sidewalls in the interconnect, the material resistivity of Cu has gone up consistently with dimensional scaling. The data from the ITRS show an enhancement of Cu resistivity from 4.08 to 14.06 $\mu\Omega$ cm as the technology scales from 45 nm in the year 2010 to 7.5 nm by the year 2024 [8]. This increase in resistivity has enhanced the disparity in the delays of the transistors and on-chip wires. There are two metrics that capture the interconnect challenge in gigascale integration very effectively. The signal-drive distance (SDD) is the distance that a signal can travel through the interconnect in one-gate delay without using a buffer. The other metric, clock locality (CL), is the number of gate delays in a one-clock cycle and dictates how far a signal can reach on the chip without the use of a synchronization register. Owing to the degradation in the speed of interconnects, both SDD and CL show no improvement or rather a degradation with dimensional scaling as plotted in Figure 64.2.

To sustain Moore's Law beyond the 2024 technology node, less power-hungry computational systems are required. In particular, systems that use state variables other than electronic charge may introduce a paradigm shift and open up a new scaling path [5]. An overview of the potentials and limitations of various novel state variables for beyond-CMOS computation is given in Ref. [9]. In this chapter, we consider only electron spin as the state variable for post-CMOS computation, particularly because it is one of the most studied state variables and has the potential to offer nonvolatile operation with enhanced functionality [10]. Most of the current research is focused on spin-based devices [11–13], while not much research has been conducted so far in interconnects for spin devices. The quintessential purpose of interconnects is communication of information between various devices on the microprocessor. Hence, an analysis of the interconnection aspects early on

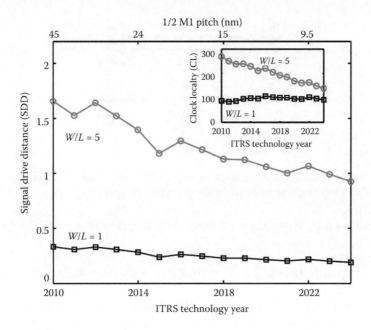

FIGURE 64.2 The signal-drive distance normalized to the gate pitch versus ITRS technology year for two inverter sizes: *W/L* = 1 (minimum-sized inverter) and *W/L* = 5 (5× the minimum-sized inverter). The inset plot shows clock locality normalized to the gate pitch versus the ITRS technology year.

is necessary to quantify the advantages, limitations, and opportunities of the spintronic logic. A preliminary analysis of the interconnect limits in spintronic technology, conducted in Refs. [14,15], shows that interconnects will continue to be a major bottleneck even in spintronic technology.

The remainder of the chapter is organized as follows. First, the reference CMOS circuit that is used for comparison with the spintronic circuit is introduced. The closed-form models for the delay and the energy dissipation of the CMOS circuit are explained. The second part of the chapter introduces the all-spin logic (ASL) circuit in which graphene nanoribbons (GNRs) will be used as the channel/interconnect material of choice. Without delving into the rich physics of graphene, only the physical models of transport parameters relevant for the work in this chapter will be discussed. This part of the chapter also includes a discussion on the nanomagnets in the ASL circuit that serve as digital capacitors in the spin domain. The nanomagnets add to the overall delay and the energy consumption of ASL circuits; hence, the treatment in this chapter will be incomplete without considering the nanomagnet overhead in the spin circuit. The fourth part of the chapter is devoted to a comparison of the electrical and spintronic circuits at the 2024 technology node, which corresponds to an MFS of 7.5 nm as per the ITRS. Finally, the conclusions and outlook are provided in the last part of the chapter.

64.2 CMOS CIRCUIT

The reference CMOS circuit used for further analysis is shown in Figure 64.3. The interconnect in the CMOS circuit is implemented with Cu/low-κ and has a length *L*. The interconnect is driven by a CMOS driver with channel width to the length ratio denoted by *W/L*. The CMOS load is sized equally as the driver. The CMOS driver/load has a p-FET width double that of the n-FET to match the ON resistance of the devices. The equivalent circuit diagram is also shown in Figure 64.3, where the interconnect is represented by a distributed resistance–capacitance ($r_w c_w$) network. The source resistance and capacitance are denoted as R_s and C_s, respectively. The load capacitance is denoted as C_L.

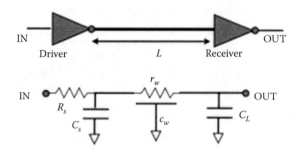

FIGURE 64.3 The schematic of the CMOS system with a CMOS driver, an interconnect, and a CMOS load (top). The equivalent circuit representation of the CMOS system (bottom).

Using the Elmore delay formula, the net delay of the circuit in Figure 64.3 can be expressed as [16]

$$t_{CMOS} = 0.69R_s(C_s + C_L) + 0.69(R_s c_w + r_w C_L)L + 0.38r_w c_w L^2. \qquad (64.1)$$

The various symbols used in Equation 64.1 and their meanings are given in Table 64.1. To evaluate the per-unit-length resistance of the interconnects, the impact of line-edge roughness and scatterings (collectively termed as "size effects") is also considered. The grain-boundary reflectivity for Cu wires is taken to be 0.3, while the sidewall specularity parameter is selected as 0.2;[*] the line-edge roughness is taken to be 40% of the interconnect width. From Table 64.1, it can be seen that the resistance per unit length of the Cu interconnect in the presence of size effects is 8× more than that of bulk Cu with the same aspect ratio and at the same technology node.

The energy dissipation of the CMOS circuit is given by the charging/discharging of all capacitances at the output node. Mathematically, this is expressed as

---

TABLE 64.1

Values of the Resistance and the Capacitance of Devices and Interconnects to Evaluate the Delay of the CMOS Circuit at the 2024 Technology Node

Symbol	Meaning	Value
W	Width = 1/2 M1 pitch	7.5 nm
V_{DD}	Supply voltage	0.6 V
I_{DSAT}	Saturation current of minimum-sized n-FET	2170 μA/μm
R_s	On-resistance of minimum-sized inverter—$V_{DD}/(I_{DSAT} \times W)$	37 kΩ
C_s	Parasitic capacitance of minimum-sized inverter	6.3 aF
C_L	Load capacitance of minimum-sized inverter	6.3 aF
c_w	Per-unit-length capacitance of local-level M1	1.2 pF/cm
AR	Aspect ratio of interconnect = Height/Width	2.1
H	Height of the interconnect	15 nm
r_w	Per-unit-length resistance of local-level M1	1.21×10^7 Ω/cm

Source: *International Technology Roadmap for Semiconductors (ITRS)*, 2009. Semiconductor Industry Assoc., 2009.

---

[*] The interface between Cu and Ta is diffusive due to a mismatch in the Fermi surfaces of the two materials. Hence, electrons in narrow and thin Cu wires suffer from diffuse scatterings at the sidewalls rendering a low value to the sidewall specularity parameter. From most experiments, the sidewall specularity was extracted between 0 and 0.2 [17].

$$E_{COMS} = \frac{1}{2}(C_s + C_L + c_w L)v_{DD}^2. \tag{64.2}$$

The energy dissipation of the CMOS circuit increases linearly with the interconnect length. Hence, the energy consumption in the interconnect quickly surpasses that of the source and the load particularly when the devices are small.

Only short, local interconnects up to 100 gate pitches are considered in this analysis. The gate pitch at the 7.5 nm node is 140 nm evaluated from the ITRS by assuming an average of four transistors per gate.

64.3 ASL CIRCUIT

The ASL circuit belongs to the category of spin circuits in which the input and output are in the electrical domain, while the processing within the circuit happens in the spin domain. There are a few other notable spin-based circuits that also fall into the same category: the magnetic-tunnel junctions (MTJs) [18], the Datta–Das spin modulator [12], and the spin-FET [13]. The important feature that distinguishes all the above-mentioned devices from those in which the I/Os are also in the spin domain, such as the spin-wave bus (SWB) [19,–21] and the magnetic quantum cellular automata (MQCA) [22], is that the former category of devices are amenable to scaling. In essence, when the footprint of the ASL circuit is reduced, the requirement on the electrical current needed to switch the magnetization of the nanomagnet devices also goes down.

While there have been several modifications to the originally proposed ASL circuit, we present a prototype in Figure 64.4 that we will use for further analysis.

In this circuit, an electrical current I_{elec} is used to pump spin-polarized electrons through the nanomagnet and a highly resistive tunnel barrier into the interconnect. A pure spin-diffusion current flow is set up in the interconnect; as it flows through the nonmagnetic interconnect, the amplitude of the spin-diffusion current goes down exponentially because of the finite spin flipping processes occurring within the interconnect. Finally, only a small fraction of the initially injected spin current reaches the nanomagnet receiver; depending upon the polarity of the spin current relative to that of the nanomagnet magnetization, the nanomagnet magnetization may undergo spin precession and may finally change its orientation by 180°. This is the commonly known spin-torque effect also utilized in implementing the spin-transfer torque (STT) memories. The net delay of the ASL circuit has two components: nanomagnet switching time constant (τ_{mag}) and the diffusion time constant of the interconnect (τ_{DIFF}). These individual components are discussed in detail in the next subsections. The main component of the energy dissipation of the ASL circuit is due to the ohmic heat generated along the path of the current flow. In the ASL circuit, the electrical current flows only between the terminals labeled C1T and C2T in Figure 64.4, while in the interconnect connecting the transmitter and the receiver terminals, there is no net addition of charge. The resistance along the path of

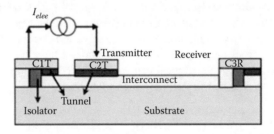

FIGURE 64.4 A representation of the ASL device proposed in Ref. [11]. An electric current pumps spin-polarized electrons through a highly resistive barrier into the interconnect. The receiver has an ohmic contact with the interconnect.

the current flow is the sum of the contact resistances and the interconnect resistance sandwiched between the C1T and C2T terminals. Since the ASL circuit employs highly resistive tunnel barriers to enhance the efficiency of spin injection, the net resistance is dominated by the contacts. The energy dissipation of the ASL circuit is mathematically given as

$$E_{ASL} = I_{elec}^2 R_{tx} \Delta t, \tag{64.3a}$$

$$I_{elec} > \varsigma \frac{I_{spin,thres}}{\eta}, \tag{64.3b}$$

where R_{tx} is the net resistance in the path of the current flow, and Δt is the pulse width of the electric current set to a value greater than the delay of the spin circuit. The inequality in Equation 64.3b must be satisfied to ensure that the amplitude of the spin current reaching the receiver terminal is more than the threshold current needed to toggle it. The factor ς is called the overdrive at the receiver nanomagnet, and it is given as the ratio of the spin current actually available at the receiver to toggle it and its spin threshold current. The factor η is called the spin injection and transport efficiency (SITE), and it includes losses in the spin signal occurring while (i) injection through the tunnel barrier and (ii) the flow of the spin current through the interconnect because of the spontaneous spin flipping processes present in the nonmagnetic interconnect.

64.3.1 NANOMAGNET DYNAMICS

Spin torque refers to the phenomenon in which a pure spin current can rotate the magnetization of a nanomagnet. The component of the spin angular momentum transverse to the nanomagnet's moment is absorbed by the nanomagnet. If the nanomagnet responds as a single domain, the magnetic moment of the nanomagnet may begin to rotate. To understand the dynamics of the nanomagnet, the effect of various torques acting on the nanomagnet must be considered. The various torques include the torques due to an applied magnetic field, magnetic anisotropies, and the intrinsic damping in the nanomagnet.

In this study, the nanomagnet is assumed to be an elliptical body shown in Figure 64.5. The magnetization is assumed to lie in the y–z plane. The shape anisotropy energy of the nanomagnet is given as

$$E_{SHA} = \frac{\mu_0}{2} M_S^2 \Omega (N_{d,zz} \cos^2 \theta(t) + N_{d,yy} \sin^2 \theta(t)), \tag{64.4}$$

where μ_0 is the free-space permeability, M_s is the saturation magnetization of the nanomagnet, Ω is the nanomagnet volume, $\theta(t)$ is the angle subtended by the magnetization along the easy axis (EA) at time t, and $N_{d,zz}$ and $N_{d,yy}$ are the demagnetization coefficients along the z- and y-axes, respectively. For $\theta = 0$ and π, E_{SHA} is minimized. Hence, the axis corresponding to $\theta = 0$ and π (z-axis in Figure 64.5) is called the EA. The axis perpendicular to the EA is called the hard axis (HA).

The dynamics of the nanomagnet under the presence of various torques due to the shape anisotropy, spin torque, and external magnetic field can be obtained by solving the Landau–Lifshitz–Gilbert (LLG) equation [24,25]. An analytical expression of the switching time of the nanomagnet under a pure spin torque as obtained in Ref. [26] is presented here:

$$\tau_{mag} = -\frac{1 + \alpha^2}{2\gamma\alpha B_0} \mu_0 M_S \Omega \frac{m}{m^2 - 1} \left(m \ln\left(\frac{1 - m\cos v}{1 + m\cos v} \right) - \ln\left(\frac{1 - \cos\gamma}{1 + \cos\gamma} \right) \right),$$

$$B_0 = \frac{\mu_0}{2} M_S^2 \Omega (N_{d,zz} - N_{d,yy}), \tag{64.5}$$

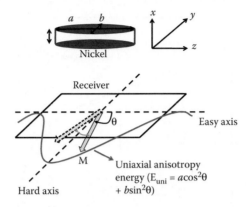

FIGURE 64.5 The top figure shows the elliptical nanomagnet with the coordinate axes. The bottom figure is the shape anisotropy energy landscape of the nanomagnet. (Taken from S. Rakheja and A. Naeemi, Interconnect analysis in spin-torque devices: Performance modeling, optimal repeater insertion, and circuit-size limits, presented at the *International Symposium on Quality Electronic Design*, Santa Clara, 2012. Copyright © IEEE 2012, with permission.)

where α is the dimensionless phenomenological Gilbert damping constant; γ is the gyromagnetic ratio for electrons; $m = 2\alpha B_0/s$, where $s = h/(4\pi e)\eta I_{elec}$ is the spin angular momentum deposition per unit time, and the angle v is the initial deviation of the nanomagnet magnetization from its EA. Typically, thermal fluctuations will make the magnetization deviate slightly from EA such that v is never perfectly zero or π. However, if v were exactly zero or π, the spin torque could never rotate the magnet. That is, for angle $v = 0/\pi$, $\tau_{mag} \to \infty$, and as such these angles are also referred to as "stagnation points" in the energy landscape of the nanomagnet. The material and design parameters chosen for the nanomagnet are given in Table 64.2. For the size of the nanomagnet considered in this analysis, the thermal stability of the nanomagnet is $3k_B T$, which corresponds to a spin threshold current of 0.75 μA.

Using the parameters in the Table 64.2, the switching time of the nanomagnet as a function of the initial orientation v is plotted in Figure 64.6. The nanomagnet switching time is large for a small value of v; this means that it is difficult for a pure spin current to induce a torque on the nanomagnet if the nanomagnet is aligned close to the EA. However, as the nanomagnet magnetization moves

TABLE 64.2
Material and Design Parameters for the Nanomagnet (Nickel)

Parameter	Value
Major axis dimension of the nanomagnet	16.5 nm
Minor axis dimension of the nanomagnet	10 nm
Thickness of the nanomagnet	5 nm
Saturation magnetization, M_s	3.93×10^5 A/m
Damping coefficient, α	0.01
Uniaxial shape anisotropy	$3\,k_B T$
Threshold current of the receiver, $I_{spin,thres}$	0.75 μA

Source: Taken from S. Rakheja and A. Naeemi, Interconnect analysis in spin-torque devices: Performance modeling, optimal repeater insertion, and circuit-size limits, presented at the *International Symposium on Quality Electronic Design*, Santa Clara, 2012. Copyright © IEEE 2012, with permission.

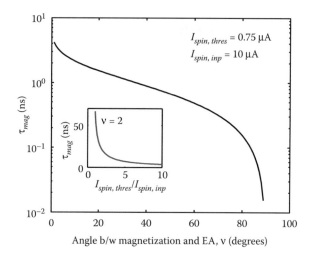

FIGURE 64.6 Switching time of the nanomagnet versus the angle ν. The inset plot shows the impact of the input spin current on nanomagnet switching time.

closer to the HA, it becomes easier for the spin current to rotate the nanomagnet magnetization. Hence, two specific kinds of switching schemes have been discussed in ASL. The first, called the full-spin torque (FST) switching, utilizes only pure spin currents to rotate the nanomagnet. For FST switching, the initial orientation of the nanomagnet is ν → 0. In a different scheme, one can use Bennett clocking to align the nanomagnet along its HA (ν → π/2); hence, the input electrical current is now required to only tip the nanomagnet toward one of its stable states. This is called the mixed-mode (MM) switching. There are, of course, issues related to the thermal stability of the nanomagnet and energy dissipation in the clocking scheme required for the MM switching of the nanomagnet. However, in principle, MM switching would be much faster compared to the FST switching. A novel proposal to use the nanomagnet in the MM switching mode is discussed in Refs. [26,27] in which a voltage-generated stress in a multiferroic magnet consisting of a magnetostrictive layer (nickel) and a piezoelectric layer is used to rotate the nanomagnet to ν ~ π/2. In another MM scheme in Ref. [11], it is discussed that applying a voltage to a fixed magnetic layer in contact with the output magnetic layer through the spacer layer can help accumulate spins in the direction of the fixed layer; these spins can then apply a spin torque on the output magnet and rotate it to its HA. However, the details and issues related to these approaches are beyond the scope of this chapter.

The inset plot of Figure 64.6 shows the impact of increasing the input spin current to induce spin-torque-assisted switching of the nanomagnet. The horizontal x-axis in the inset plot is the factor $1/\zeta$ as in Equation 64.3b. As the input current increases, the overdrive of the spin current available at the nanomagnet increases, making it easier for the spin current to switch the nanomagnet magnetization. However, the improvement in the nanomagnet delay saturates with an increase in the input spin current. An optimal ζ can be determined such that the energy dissipation of the nanomagnet is minimized. As ζ increases, the input electrical current increases, which increases the I^2R heating the circuit. However, because the nanomagnet toggles faster with an increase in ζ, the landscape of the energy dissipated in the nanomagnet versus ζ will exhibit a minimum energy point.

64.3.2 INTERCONNECT DYNAMICS

The interconnect in the ASL circuit can be implemented using a variety of materials. Metals like Cu and Al have an established technology and, therefore, from a fabrication perspective, would be a good choice. In addition, their resistivity values could be comparable to that of the injecting nanomagnet. This can help to overcome the well-known "conductivity mismatch" issue in spin-based devices [28,29]. However, the spin-relaxation length in metals is usually limited to only a few 100s of nanometers [30,31]. Semiconductors such as Si and GaAs may also be good choices for implementing interconnects in the ASL circuit. However, appreciable spin injection in semiconductors could be an issue [32]. Carbon-based material graphene has a very rich physics that provides it superior electrical and spintronic characteristics desirable of interconnects in the ASL circuit. Graphene, owing to its large electron mean free path (MFP), may support a fast spin-current transport through it. Further, owing to its low atomic number, graphene also has a reduced spin–orbit coupling (SOC) lending it long spin-relaxation lengths [33]. In this work, the focus will be on narrow patterned GNRs as the interconnect material of choice in the ASL circuit.

Graphene can support a quasi-ballistic transport of electron spins through it. In the steady state, the flux of the electron spins through the interconnect is given as [34]

$$J_s = \frac{D}{L} \frac{s(0)}{1 + \frac{D}{L} \frac{1}{v_f}}, \tag{64.6}$$

where $s(0)$ is the electron spin concentration in 1/cm at the beginning of the interconnect, D is the diffusivity of carriers in the interconnect in cm²/s, L is the interconnect length in cm, and v_f is the Fermi velocity of carriers in cm/s. The steady-state carrier concentration in the interconnect is given by a linear profile as

$$s(z) = s(0)\left(1 - \frac{z}{L}\right) + \frac{s(L)}{L}. \tag{64.7}$$

The time constant of spin diffusion through the interconnect can be given as

$$\tau_{DIFF} = \frac{\displaystyle\int_0^L s(z)dz}{J_s}. \tag{64.8}$$

Using Equations 64.6 through 64.8, τ_{DIFF} can be simplified as

$$\tau_{DIFF} = \frac{L^2}{2D} + \frac{L}{v_f}. \tag{64.9}$$

The first component in Equation 64.9 is the diffusive time constant and the second component is the ballistic time constant. From Equation 64.9, it can also be seen that the speed of the spin-diffusion interconnects is governed by the material parameters: diffusion coefficient, D, and Fermi velocity, v_f. The delay of spin interconnects increases quadratically with their length much like the intrinsic delay of electrical interconnects.

Next, the physical models of the electron diffusion coefficient in bulk and 1D graphene are presented.

64.3.2.1 Diffusion Coefficient in GNRs

In 2D graphene, the electron diffusion coefficient, D, is related to the electron MFP, λ, according to $D = 1/2\lambda v_f$. The electron MFP in 2D graphene is obtained by taking contributions to electron scatterings from the intrinsic phonons in graphene (acoustic and optical), and extrinsic sources arising from the substrate underneath the graphene sheet. For a suspended graphene sheet, the electron MFP can be as high as 1–1.2 μm as obtained experimentally in [35,36]. However, when deposited or grown epitaxially on a substrate, the electron MFP in graphene is limited to only a few hundred nanometers or even lower [37]. Hence, the substrate can easily limit the best-possible diffusion coefficient in graphene. Table 64.3 shows the impurity concentration and remote-phonon-limited MFP for various substrates for an electron concentration of 10^{12} cm^{-2} in the graphene sheet.

Using the data from Table 64.3, the electron diffusion coefficient in 2D graphene is obtained as a function of carrier concentration and is plotted in Figure 64.7 for various substrates at RT. The best-case diffusion coefficient for graphene is obtained in the case of h-BN substrate because of the nearly perfect lattice matching of h-BN with graphene, which can significantly reduce the surface roughness of h-BN. For an electron concentration, N_s, in the range $(1–5) \times 10^{12}$ cm^{-2}, the diffusion coefficient of electrons in graphene on h-BN lies between 1000 and 1200 cm^2/s. This corresponds to an electron MFP of 250–300 nm. For graphene SiO_2, the diffusion coefficient is 408 cm^2/s for $N_s = 10^{12}$ cm^{-2}; this corresponds to an electron MFP of 102 nm.

However, once graphene is patterned into narrow ribbons for use as interconnects and devices, there occurs a quantization of electron energies along the width of the graphene. The energy-dispersion relation of GNRs is given as [41]

$$E = hv_f \sqrt{k_\parallel^2 + k_\perp^2}, \tag{64.10}$$

where $k_\parallel$ is the wave vector along the length of the ribbon and $k_\perp$ is the wave vector along the ribbon width and is quantized according to

$$k_\perp = \frac{\pi}{W}|m + \beta|, \tag{64.11}$$

where W is the ribbon width, and $\beta = 0$ for metallic ribbons and 1/3 for semiconducting ribbons. Experiments have confirmed that all narrow ribbons are semiconducting and that the orientation effects are all washed out for narrow ribbons [42]. Hence, we choose $\beta = 1/3$ for all further analysis. The low-field 1D conductivity according to Landauer's formalism is given as

$$\sigma_{1D} = -\frac{2e^2}{h} \sum_m \int_{E_{sub,m}}^{\infty} \lambda_m(E) \frac{\partial f_{FD}(E)}{\partial E} \, dE, \tag{64.12}$$

TABLE 64.3

Concentration of Charged Impurities and the Remote-Phonon-Limited MFP at RT for an Electron Concentration, $N_s = 10^{12}$ cm^{-2}

Substrate	Charged-Impurity Concentration (cm^{-2})	Remote-Phonon-Limited MFP at RT for $N_s = 10^{12}$ cm^{-2} (nm)
SiO_2	$(1.3–1.8) \times 10^{11}$	195
SiC	$(0.6–1) \times 10^{11}$	575
$SrTiO_3$ (STO)	$(0.7–1.3) \times 10^{11}$	53
HfO_2	—	35
h-BN	$<7 \times 10^{10}$	920

Source: Data taken from F. Giannazzo et al., *Nano Letters*, 11, 4612–4618, 2011; C. Dean, et al., *Nature Nanotechnology*, 5, 722–726, 2010; and V. Perebeinos and P. Avouris, *Physical Review B*, 81, 2010.

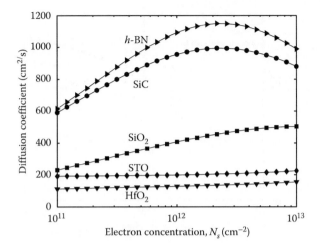

FIGURE 64.7 Diffusion coefficient in graphene on various substrates versus electron concentration at RT.

where e is the electron charge, h is the Planck's constant, $f_{FD}(E)$ is the Fermi–Dirac statistics, and $\lambda_m(E)$ is the electron MFP in the mth conduction channel in the GNR. The summation in Equation 64.12 runs over all the conduction channels in the GNR. The electron diffusion coefficient is then obtained using Equation 64.12 in conjunction with Einstein's relationship and is mathematically expressed as

$$D = \frac{\sigma}{e^2 \partial n_{1D} / \partial E_f}, \tag{64.13}$$

where n_{1D} is the 1D carrier concentration in the GNR and is obtained by the convolution of the density of states with the Fermi–Dirac statistics in the energy space. The electron MFP, $\lambda_m(E)$, is obtained by using Mattheissen's rule for the edge-scattering limited MFP, $\lambda_m^{edge}(E)$, and the substrate-limited MFP, λ_{sub}. Patterning graphene into narrow ribbons renders its edges rough and can introduce additional scattering of electrons. Indeed, experiments have shown that a 2.5 nm wide GNR deposited on SiO_2 has an electron mobility as low as 200 cm²/Vs at RT, which corresponds to an electron MFP of 12 nm [43]. This value of electron MFP is significantly lower than a value of ~100 nm for a wide graphene flake on SiO_2.

The MFP associated with the diffusive scatterings at the edges is a function of the scattering probability at the edges (P_{GNR}) and the average distance that the electrons move along the length of the ribbon before hitting one of the edges. Mathematically, the edge-scattering MFP is given as [41]

$$\lambda_m^{edge}(E) = \frac{1}{P_{GNR}} \frac{k_\parallel}{k_\perp} W = \frac{W}{P_{GNR}} \sqrt{\left(\frac{E}{E_{sub,m}}\right)^2 - 1}, \tag{64.14}$$

where $E_{sub,m}$ is the minimum of the mth sub-band energy. The effective MFP of electrons in GNR on various substrates is plotted as a function of the GNR width for $P_{GNR} = 0.2$ in Figure 64.8. Two values for the Fermi-energy shift, E_f, in graphene have been considered. Experiments have shown that because of a work-function difference in the substrate and graphene sheet, the Fermi energy in the graphene sheet is usually positive. Values of $E_f = 0.2$ eV and 0.4 eV have been experimentally obtained [44,45]. For all GNR widths, the effective MFP increases with an increase in the Fermi energy and substrate-limited MFP. The electron MFP for a suspended wide graphene sheet measured

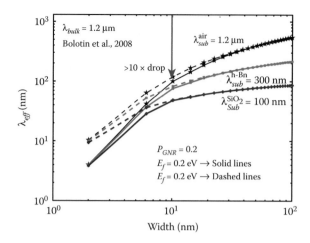

FIGURE 64.8 The effective MFP of electrons in GNR as a function of its width. The substrate-limited MFP for h-BN is taken to be 300 nm, while that for SiO_2 is 100 nm.

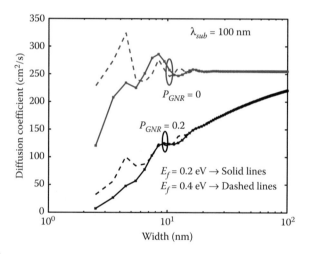

FIGURE 64.9 Electron diffusion coefficient in GNR versus the GNR width for the SiO_2 substrate. Various values of the Fermi-energy shift, E_f, and the edge-scattering coefficient, P_{GNR}, have been selected for simulation.

by Bolotin and coworkers in 2008 was 1.2 μm for an electron concentration of 2×10^{11} cm^{-2} [35]. Even when the substrate-limited MFP is as high as 1.2 μm,* there is more than a 10× drop in the effective electron MFP of a 10-nm-wide GNR with a 20% edge-scattering probability. For sub-10 nm GNR widths, the effective electron MFP will be limited to only a few tens of nm, particularly for substrates like SiO_2 for which $\lambda_{sub} < 100$ nm.

The electron diffusion coefficient in the GNR on SiO_2 as a function of the GNR width is shown in Figure 64.9 for various values of P_{GNR} and E_f. The substrate is chosen to be SiO_2 as most experiments in graphene spin valves have been conducted for graphene deposited on SiO_2; other substrates have been far less explored for potential applications in spintronics. The diffusion coefficient improves with an increase in the GNR width and a reduction in edge roughness (characterized by a lower value of P_{GNR}). At a GNR width of 7.5 nm and a Fermi-energy shift $E_f = 0.2$

---

* A substrate with $\lambda_{sub} = 1.2$ μm is referred to as an "ideal" or "air" substrate.

eV, the electron diffusion coefficient is 280 cm²/s in the absence of edge scatterings ($P_{GNR} = 0$). However, for $P_{GNR} = 0.2$, the electron diffusion coefficient drops to 105 cm²/s if all other parameters are the same.

64.3.2.2 Spin-Relaxation Length in Graphene

Unlike electrical current, spin current is not a conserved quantity. That is to say, the amplitude of the spin current decays in an exponential fashion as it moves through the interconnect. This behavior of the spin current is captured through the spin-relaxation time (τ_s) and the spin-relaxation length (L_s); these parameters are related through the diffusion coefficient as $L_s = (D\tau_s)^{0.5}$. The principal cause of spin relaxation in materials is the presence of a finite atomic SOC in materials that leads to an interaction between the spin and the orbital angular momentum of electrons. The atomic SOC rises sharply with the atomic number ($\sim Z^4$); hence, the atomic SOC in graphene is rather weak due to its low atomic number ($Z = 6$). However, the presence of ripples in graphene creates an additional SOC, which is at least 20× in strength compared to the atomic SOC [33,46,47]. In addition, extrinsic sources such as remote polar phonons and impurities in the substrate on which graphene is grown or deposited can also enhance SOC and lead to a faster spin relaxation.

In Table 64.4, various SOCs in graphene from theoretical estimates are provided. From theoretical studies in Ref. [33], it was found that the spin-relaxation length in pristine graphene could be as high as 15–20 μm. Further, when $(K_f\lambda)^2 \gg 1$, the spin-relaxation time is insensitive to the momentum-relaxation time in graphene. Here, K_f is the Fermi wave-vector. This is the characteristic D'yakonov–Perel' (DP) spin-relaxation mechanism. In the other extreme, when $(K_f\lambda)^2 \ll 1$, the spin-relaxation time is proportional to the momentum-relaxation time. This is commonly referred to as the Elliott–Yafet (EY) spin-relaxation mechanism [48]. Hence, depending upon the product of the Fermi wave vector and the electron MFP in graphene, the spin-relaxation mechanism in graphene could change from EY to DP. Recent experiments on spin relaxation in graphene show a limited spin-relaxation length of 2–3 μm at RT [49–53]. Further, the nature of the spin-relaxation mechanism in Ref. [54] is EY-type, while for the experiment in Ref. [55], the nature of spin-relaxation mechanism is DP. One of the reasons for the controversial experimental results and theoretical estimates is that in the experimental samples there is an appreciable concentration of adatoms. These adatoms hybridize with the carbon atoms of graphene and modify the otherwise sp² bonding of graphene to sp³ bonding. The strength of the SOC associated with these adatoms could be as high as a few meV [56]. Further, the nature of the spin relaxation due to adatoms could also change from EY to DP, depending upon environmental factors such as the electron concentration, adatom concentration, and ambient temperature. For a thorough theoretical investigation of the spin relaxation mediated by the random Rashba field (RRF) of adatoms in graphene, see Ref. [57].

TABLE 64.4

Various SOCs in Graphene

Coupling Constant	Value (eV)	Spin-Relaxation Time, τ_s	Reference
Δ_{so}^{int} (atomic)	8.6×10^{-7}	18 μs	[33]
Δ_{so}^{curv} (curvature)	1.7×10^{-5}	0.28 μs	[33]
Δ_{so}^{sub}(substrate SiO₂) ($N_{imp} = 4 \times 10^{11}$ cm⁻²)	7.24×10^{-8}	0.6 ms	[56]
Δ_{so}^{sub}(substrate SiO₂) ($N_{imp} = 4 \times 10^{12}$ cm⁻²)	2.17×10^{-6}	6 μs	[56]

Note: N_{imp} denotes the impurity concentration. The curvature-induced coupling depends on the ripple radius, which has been assumed to be between 50 and 100 nm. The value of substrate-induced τ_s is quoted for a Fermi energy of 0.1 eV. All values are at RT.

In the case of GNRs, there is an additional SOC introduced by the edge states. However, for the case where $(K_f\lambda)^2 \gg 1$, the spin-relaxation length will not be much different from that in the bulk assuming that the effective SOC is equal to the ripple-induced SOC; for this regime, the spin-relaxation length will only be slightly sensitive to the edge-scattering coefficient.

In this analysis, simulations are conducted for the two values of spin-relaxation length: (i) $L_s = 15$ μm (best-case) and (ii) $L_s = 2$ μm (experimental).

64.3.2.3 Spin Injection and Transport Efficiency

Spin injection and transport efficiency (SITE) in the ASL circuit gives the amount of spin current reaching the receiver per unit of input electrical current in the system. The mathematical details to obtain SITE in the ASL circuit will not be presented. Interested readers can refer to Ref. [29]. Briefly stated, the method consists of solving the spin quasi-chemical potential in various regions of the circuit and applying appropriate boundary conditions. At the interfaces, the spin current is balanced, while a discontinuity in the spin quasi-chemical potential is created at the interface in the presence of tunnel barriers. The spin quasi-chemical potential equation is a second-order ODE given as

$$\frac{d^2\mu_s}{dx^2} = \frac{\mu_s}{L_S^2},\tag{64.15}$$

where μ_s is the difference in the chemical potential of the up-spin and down-spin carriers at any given location in the ASL circuit. The spin current for a nanomagnet is the sum of the current generated due to spatial variation in μ_s and also due to an applied bias. For a nonmagnetic interconnect, the spin current arises only because of the spatial variation in μ_s (also termed as "spin accumulation"), while an applied bias is insufficient by itself to generate a spin current in a nonmagnetic interconnect.

In Figure 64.10, SITE is plotted as a function of the interconnect length for a 7.5-nm-wide GNR for $L_s = 2$ μm and 15 μm for various values of P_{GNR}. It can be seen from this figure that SITE degrades rapidly with an increase in the interconnect length, particularly for a short L_s and

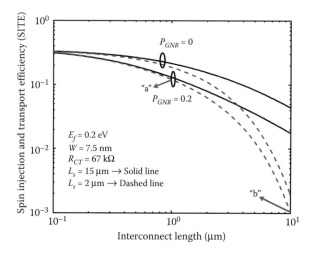

FIGURE 64.10 SITE for an ASL circuit with a 7.5-nm-wide GNR interconnect versus the interconnect length. R_{CT} denotes the total contact resistance at the transmitter. Two values of the spin-relaxation length in the GNR are considered. A 2 μm value of L_s is from experiments, while 15 μm is from theoretical estimates (best-case value).

a greater edge roughness. For example, for $P_{GNR} = 0.2$ and $L_s = 2$ μm, SITE is 0.135 for $L = 1$ μm (point "a" in Figure 64.10), and SITE degrades to 1.1×10^{-3} for $L = 10$ μm (point "b" in Figure 64.10). For $L_s = 15$ μm, the degradation in SITE is from 0.144 to 1.8×10^{-2} as L increases from 1 to 10 μm for $P_{GNR} = 0.2$. This degradation in SITE will necessitate pumping in more electrical current in the transmitter of the ASL so that an appreciable spin current reaches the receiver. This translates to a rapid increase in the energy dissipation of the ASL circuit as will be shown in the next section.

64.4 COMPARISON OF ELECTRICAL AND SPINTRONIC CIRCUITS

In Figure 64.11, the delays of the spintronic and CMOS systems are plotted as a function of the interconnect length at the 7.5 nm technology node. For the CMOS system, the delay is computed using Equation 64.1. For the spin system, the delay is computed using Equations 64.5 and 64.9 to account for both the nanomagnet switching and the interconnect switching. To be able to compute the delay of the nanomagnet, the overdrive at the nanomagnet must be known. As illustrated in Figure 64.6, the nanomagnet becomes particularly sluggish if the spin current available to switch the nanomagnet is not sufficiently higher than its threshold current. For the simulation of the delay, an overdrive of 10× is assumed at all interconnect lengths. That is, $\zeta = 10$. However, the input current can be optimized to yield a minimum energy dissipation of the ASL circuit for the given interconnect dimensions.

From Figure 64.11, it can be seen that even without considering the overhead of the nanomagnet switching, the delay of the ASL circuit governed only by the native interconnect delay increases more rapidly with the interconnect length when compared with the delay of the CMOS system. In fact, even short, local GNR spin interconnects longer than one or two gate pitches at the 7.5 nm interconnect technology node have a delay higher than that of the CMOS system. Further, the nanomagnet switching time for FST mode is 9.7 ns for a 10× overdrive; hence, in the FST mode, the dominant time constant is that of the receiver nanomagnet. The ASL circuit is orders of magnitude slower than the CMOS circuit. The switching time of the receiver nanomagnet can be lowered significantly by utilizing the MM switching mode. However, in this case, as the overall time constant

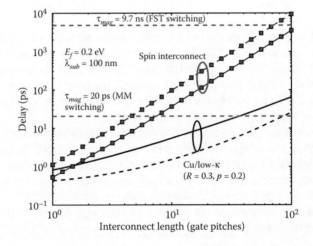

FIGURE 64.11 Delay versus interconnect length for both the ASL and CMOS circuits. For the CMOS circuit, R denotes the grain-boundary reflectivity, while p denotes the sidewall specularity. No line-edge roughness has been considered. The dashed horizontal lines correspond to the intrinsic delay of nanomagnet switching in both the FST and MM switching modes.

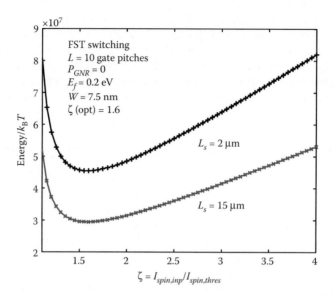

FIGURE 64.12 Energy of the ASL circuit versus the overdrive at the receiver nanomagnet. The optimal value of the overdrive in the FST switching mode is 1.6, independent of the interconnect length and the spin-relaxation length in the interconnect.

is dominated by the spin interconnect, the ASL circuit delay increases with the interconnect length, and it is still much slower compared to the CMOS circuit.

As discussed in Section 64.3.1, the value of the overdrive at the receiver nanomagnet can be optimized to minimize the energy dissipation of the ASL circuit. In Figure 64.12, the total energy dissipation of the ASL circuit with the FST mode of nanomagnet switching is shown as a function of the overdrive factor, ζ. This optimal energy point depends only upon the nanomagnet dynamics. For the material and design parameters chosen, the optimal $\zeta = 1.6$ for the FST mode of switching. In the MM switching mode, the delay of the nanomagnet decreases very gradually with an increase in ζ, while the Joule heating term I^2R grows quadratically with ζ. Hence, in the MM switching mode, the optimal value of ζ that minimizes the energy dissipation is very close to unity.

In Figure 64.13, the energy dissipation of the ASL circuit is plotted as a function of the interconnect length using the optimal value of ζ. The total transmitter-side resistance is taken to be 67 kΩ.[*] It can be seen that for the simulation parameters selected, the energy dissipation of the ASL circuit is higher than that of the CMOS circuit. Further, owing to the rapid increase in the energy dissipation of the ASL circuit with interconnect length, the disparity in the energy dissipation of ASL and CMOS circuits grows with an increase in length.

The simulation results presented here conclusively show that interconnects will continue to be a major challenge even in spintronics technology. Hence, spin circuits that require shorter interconnects either by scaling the footprint of devices more aggressively or by using smart architectures with fewer gates to implement a function will be important to make spintronic technology competitive with respect to the charge-based technology. In addition, parallelism that can mask the delay of interconnects will also be favorable to the spintronic technology. It is quite likely that short, local interconnects are implemented as spin interconnects while intermediate and global-level interconnects are electrical in future spin circuits.

---

[*] This corresponds to approximately 1.9 conduction channels at RT for a 7.5 nm wide GNR with a Fermi-energy shift of 0.2 eV. The transmission coefficient of electrons from the contacts into the graphene channel is taken to be 0.1 for the transmitter side.

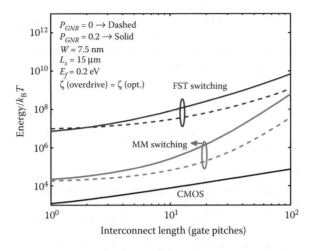

FIGURE 64.13   Energy dissipation versus the interconnect length for both the ASL and CMOS circuits. Both the FST and MM switching modes are considered for the ASL circuit.

64.5   CONCLUSIONS AND OUTLOOK

In this chapter, the limits imposed by interconnects in the ASL circuit are quantified. While for CMOS circuits it is well established that interconnects are one of the grandest challenges in gigas-cale integration, relatively little research has been conducted in quantifying interconnect limits in spin-based circuits.

The net delay of the ASL circuit is given by the sum of the delays of the nanomagnet and the interconnect. For the nanomagnet switching, a full-spin torque-assisted switching is much more sluggish when compared with a MM switching utilizing Bennett clocking to align the nanomagnet along its HA that utilizes spin current to only tip the nanomagnet toward one of its stable states. The interconnect delay in the ASL circuit is shown to increase quadratically with the interconnect length and varies as $1/D$, where D is the spin diffusion coefficient in the interconnect. Owing to the finite spin-relaxation length in the interconnect material, the amount of spin current reaching the receiver is substantially less than the ideal case. This makes the nanomagnet sluggish particularly for interconnects longer than the spin-relaxation length. The energy dissipation of the ASL circuit is given by the Joule heating occurring along the path of the current flow, which in the case of the ASL circuit is limited to the transmitter side.

Owing to its unique and rich physics, graphene has turned out to be an interesting material for future devices in both the electrical and spin-based domains. This chapter provides physical models of the electron diffusion coefficient in narrow graphene ribbons as a function of the interconnect dimensions and the edge roughness. The spin-relaxation length in graphene could be as high as 15–20 μm if only ripple-induced SOC is considered. However, in most experiments, a limited spin-relaxation length of 2 μm is measured, illustrating the limits imposed by metallic adatoms on the spin transport through a graphene channel. Using both the optimistic and realistic values of the spin-relaxation length in graphene, SITE of the ASL circuit is obtained. It is shown that SITE degrades rapidly with an increase in the interconnect length and with an increase in the edge rough-ness of the GNR. SITE also drops drastically for shorter spin-relaxation lengths in the interconnect.

It is shown that both the delay and the energy dissipation of the ASL circuit do not compare well with those of its CMOS counterparts. However, by designing spin-based circuits with smarter architectures that favor shorter interconnects, it may be possible to achieve faster spin circuits while dissipating relatively low energy. Parallelism of computation may also play a significant role in lowering the overall switching time constant of the ASL circuit, thereby improving both its speed and energy dissipation.

In its quest to sustain Moore's law for sub-10 nm nodes, the current research in nanoelectronics is rich with exploratory ideas for the next-generation switch. With this accelerated progress, one can certainly be hopeful for a new technology to emerge in the near future that would either augment or replace the current CMOS technology and establish a new scaling path for computation.

REFERENCES

1. M. S. Bakir and J. D. Meindl, *Integrated Interconnect Technologies for 3D Nanoelectronic Systems (Integrated Microsystems)*, 1st ed. Boston, MA: Artech House Publishers, 2008.
2. V. V. Zhirnov et al., Limits to binary logic switch scaling—A Gedanken model, *Proceedings of the IEEE*, 91, 1934–1939, 2003.
3. R. K. Cavin et al., Energy barriers, demons, and minimum energy operations of electronic devices, *Fluctuation and Noise Letters*, 5, 29–38, 2005.
4. R. Cavin et al., Research directions and challenges in nanoelectronics, *Journal of Nanoparticle Research*, 8, 841–858, 2006.
5. G. Bourianoff, The future of nanocomputing, *Computer*, 36, 44–53, 2003.
6. N. Magen et al., Interconnect-power dissipation in a microprocessor, presented at the *Proc. Int. Workshop System Level Interconnect Prediction*, Paris, France, 2004.
7. S. Rakheja and A. Naeemi, Performance, energy-per-bit, and circuit size limits of post-CMOS logic circuits–Modeling, analysis and comparison with CMOS logic, in *International Interconnect Technology Conference (IITC)*, Dresden, 2011.
8. *International Technology Roadmap for Semiconductors (ITRS)*, 2009. Semiconductor Industry Assoc., 2009.
9. K. Galatsis et al., Alternate state variables for emerging nanoelectronic devices, *IEEE Trans. Nanotechnology*, 8, 66–75, 2009.
10. D. D. Awschalom and M. E. Flatte, Challenges for semiconductor spintronics, *Nature Physics*, 3, 153–159, 2007.
11. B. Behin-Aein et al., Proposal for an all-spin logic device with built in memory, *Nature Nanotechnology*, 5, 266–270, 2010.
12. S. Datta and B. Das, Electronic analog of the electro-optic modulator, *Applied Physics Letters*, 56, 665–667, 1990.
13. S. Sughara and J. Nitta, Spin transistor electrons: An overview and outlook, *Proceedings of the IEEE*, 98, 2124–2154, 2010.
14. S. Rakheja and A. Naeemi, Interconnects for novel state variables: Performance modeling and device and circuit implications, *IEEE Transactions on Electron Devices*, 57, 2711–2718, 2010.
15. S. Rakheja and A. Naeemi, Modeling interconnects for post-CMOS devices and comparison with copper interconnects, *IEEE Transactions on Electron Devices*, 58, 1319–1328, 2011.
16. H. B. Bakoglu, *Circuits, Interconnections and Packaging for VLSI*, 1 ed., Addison-Wesley, 1990.
17. S. M. Rossnagel and T. S. Kuan, Alteration of Cu conductivity in the size effect regime, *Journal of Vacuum Science & Technology B: Microelectronics and Nanometer Structures*, 22, 240–247, 2004.
18. J.-G. Zhu and C. Park, Magnetic tunnel junctions, *Materials Today*, 9, 36–45, 2006.
19. A. Khitun et al., Feasibility study of logic circuits with a spin wave bus, *Nanotechnology*, 18, 465202 (9 pages), 2007.
20. A. Khitun and K. L. Wang, Nano scale computational architectures with spin wave bus, *Superlattices and Microstructures*, 38, 184–200, 2005.
21. A. Khitun et al., Efficiency of spin-wave bus for information transmission, *Electron Devices, IEEE Transactions on*, 54, 3418–3421, 2007.
22. A. Imre et al., Majority logic gate for magnetic quantum-dot cellular automata, *Science*, 311, 205–208, 2006.
23. S. Rakheja and A. Naeemi, Interconnect analysis in spin-torque devices: Performance modeling, optimal repeater insertion, and circuit-size limits, presented at the *International Symposium on Quality Electronic Design*, Santa Clara, 2012.
24. J. Z. Sun et al., A three-terminal spin-torque-driven magnetic switch, *Applied Physics Letters*, 95, 083506, 2009.
25. J. Z. Sun, Spin-current interaction with a monodomain magnetic body: A model study, *Physical Review B*, 62, 570–578, 2000.

26. K. Roy et al., Hybrid spintronics and straintronics: A magnetic technology for ultra low power energy computing and signal processing, *Applied Physics Letters*, 99, 063108 (3 pages), 2011.
27. M. Salehi-Fashmi et al., Ultra low-power straintronics with multiferroic nanomagnets: Magnetization dynamics, universal logic gates and associated energy dissipation, presented at the *APS March Meeting*, Boston, MA, 2012.
28. J. Fabian et al., Semiconductor spintronics, *Acta Physica Slovaca*, 57, 565–907, 2007.
29. I. Zutic et al., Spintronics: Fundamentals and applications, *Reviews of Modern Physics*, 76, 323–410, 2004.
30. F. J. Jedema et al., Electrical detection of spin precession in a metallic mesoscopic spin valve, *Nature*, 46, 713–716, 2002.
31. F. J. Jedema et al., Spin injection and spin accumulation in all-metal mesoscopic spin valves, *Physical Review B*, 2003, 085319 (16 pages), 2003.
32. S. P. Dash et al., Electrical creation of spin polarization in silicon at room temperature, *Nature*, 462, 491–494, 2009.
33. D. Hernando et al., Spin–orbit coupling in curved graphene, fullerenes, and nanotube caps, *Physical Review B*, 74, 155426 (15 Pages), 2006.
34. M. Lundstrom, *Fundamentals of Carrier Transport*, 2nd ed. Cambridge, UK: Cambridge University Press, 2000.
35. K. I. Bolotin et al., Ultrahigh electron mobility in suspended graphene, *Solid State Communications*, 146, 351–355, 2008.
36. X. Du et al., Approaching ballistic transport in suspended graphene, *Nature Nanotechnology*, 3, 491–495, 2008.
37. T. Shimizu et al., Large intrinsic energy bandgaps in annealed nanotube-derived graphene nanoribbons, *Nature Nanotechnology*, 6, 6, 2011.
38. F. Giannazzo et al., Mapping the density of scattering centers limiting the electron mean free path in graphene, *Nano Letters*, 11, 4612–4618, 2011.
39. C. Dean et al., Boron nitride substrate for high-quality graphene electronics, *Nature Nanotechnology*, 5, 722–726, 2010.
40. V. Perebeinos and P. Avouris, Inelastic scattering and current saturation in graphene, *Physical Review B*, 81, 195442 (8 pages), 2010.
41. A. Naeemi and J. D. Meindl, Compact physics-based circuit models for graphene nanoribbon interconnects, *IEEE Transactions on Electron Devices*, 56, 1822–1833, 2009.
42. D. Queriloz et al., Suppression of the orientation effects on bandgap in graphene nanoribbons in the presence of edge disorder, *Applied Physics Letters*, 92, 042108, 2008.
43. X. Wang et al., Room-temperature all-semiconducting sub-10-nm graphene nanoribbon field-effect transistors, *Physical Review Letters*, 100, 206803–206804, 2008.
44. S. Y. Zhou et al., Substrate-induced bandgap opening in epitaxial graphene, *Nature Materials*, 6, 770–775, 2007.
45. C. Berger et al., Electronic confinement and coherence in patterned epitaxial graphene, *Science*, 312, 1191–1196, 2006.
46. D. Huertas-Hernando et al., Spin–orbit-mediated spin relaxation in graphene, *Physical Review Letters*, 103, 146801 (4 pages), 2009.
47. D. Heurtas-Hernando et al., Spin relaxation times in disordered graphene, *The European Physical Journal-Special Topics*, 148, 178–181, 2007.
48. H. Ochoa, A. H. Castro Neto, and F. Guinea, Elliot-Yafet mechanism in graphene, *Applied Physics Letters*, 108 (20), 206808, 2012.
49. W. Han et al., Enhanced spin injection into graphene with MgO tunnel barriers, *APS March Meeting*, 35.006, 2010.
50. W. Han et al., Electrical detection of spin precession in single layer graphene spin valves with transparent contacts, *Applied Physics Letters*, 94, 22109–22109-3, 2009.
51. M. Shiraishi, M. Ohishi, R. Nouchi, N. Mitoma, T. Nozaki, T. Shinjo, and Y. Suzuki, Robustness of spin polarization in graphene-based spin valves, *Advanced Functional Materials*, 19(23), 3711–3716, 2009.
52. M. Popinciuc et al., Electronic spin transport in graphene field effect transistors, *Physical Review B*, 80, 214427, 2009.
53. N. Tombros et al., Electronic spin transport and spin precession in single graphene layers at room temperature, *Nature*, 448, 571–574, 2007.
54. C. Jozsa et al., Linear scaling between momentum and spin scattering in graphene, *Physical Review B*, 80, 241403 (4 pages), 2009.

55. W. Han and R. Kawakami, Spin relaxation in single-layer and bilayer graphene, *Physical Review Letters*, 107, 047207 (4 pages), 2011.

56. C. Ertler et al., Electron spin relaxation in graphene: The role of the substrate, *Physical Review B*, 80, 4, 2009.

57. P. Zhang and M. W. Wu, Electron spin relaxation in graphene with random Rashba field: Comparison of the D'yakonov–Perel' and Elliott–Yafet-like mechanisms, *New Journal of Physics*, 14, 033015 (22 pages) 2012.

65 Influence of Impurity and Dangling Bond Scattering on the Conductance Anomalies of Side-Gated Quantum Point Contacts

J. Wan, J. Charles, M. Cahay, P. P. Das, N. Bhandari, and R. S. Newrock

CONTENTS

65.1 INTRODUCTION

For more than a decade, there have been many experiments reporting anomalies that appear at noninteger values of the quantized conductance G_0 in the ballistic conductance regime of quantum point contacts (QPCs) based on GaAs. These include the observation of an anomalous plateau at $G \cong 0.5\, G_0$ [1–4] and the well-known "0.7 structure" [5]. The majority of the theoretical models link them to spontaneous spin polarization in the QPC [6–8]. Recently, we used a nonequilibrium Green's function (NEGF) approach to study in detail the ballistic conductance of asymmetrically biased, side-gated QPCs in the presence of lateral spin–orbit coupling (LSOC) and strong electron–electron interactions. We performed simulations for a wide range of QPC dimensions and gate bias voltages [9].

Various conductance anomalies appeared below the first quantized conductance plateau ($G_0 = 2e^2/h$); these occur due to spontaneous spin polarization in the narrowest portion of the QPC. We have found that the number of conductance anomalies increases with the aspect ratio (length/width) of the QPC constriction. These anomalies are fingerprints of spin textures in the narrow portion of the QPC [9].

Since the early 1990s, there has been a considerable amount of work studying the influence of impurity scattering on the quantized conductance plateaus of QPCs [10–16]. We recently showed that asymmetric LSOC, resulting from the lateral in-plane electric field of the confining potential of a side-gated QPC, can be used to create a strongly spin-polarized current by purely electrical means [17] *in the absence of* an applied magnetic field. Using an NEGF analysis of a small model

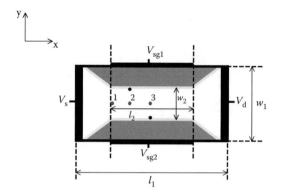

FIGURE 65.1 Schematic of the QPC configuration used in the numerical simulations. The width and length of the narrow portion of the QPC are equal to w_2 and l_2, respectively. In the simulations, we used w_2, l_2, w_1, and $l_1 = 16, 32, 48,$ and 64 nm, respectively. The circles in the middle of the narrow portion of the QPC represent impurities with coordinates $y_1 = w_1/2$ and $x_1 = 16, 24,$ and 32 nm, respectively. The two circles on the side walls are dangling bonds located one-fourth (top side wall) and one-half of the way (bottom side wall) from the left entrance to the narrow portion of the QPC.

QPC [18], three ingredients were found to be essential for generating a strong spin polarization: an asymmetric lateral confinement, an LSOC induced by the lateral confining potential of the QPC, and a strong electron–electron (e–e) interaction.

In this chapter, we investigate the influence of impurities and dangling bonds on the conductance anomalies in GaAs QPCs created by side gates, in the presence of LSOC. We consider the effects of attractive and repulsive scatterers in the narrow portion of the QPC and dangling bonds on its side walls, as depicted in Figure 65.1.

One of the main ingredients needed to generate spin polarization in the central portion of a QPC is an asymmetric potential profile in the channel. It is expected that off-center impurities can lead to such an asymmetry even for a symmetric bias voltage on the two side gates. In this case, our simulations predict that conductance anomalies can be observed in an otherwise perfectly symmetric QPC due to unwanted impurities in the channel.

65.2 NUMERICAL SIMULATIONS

The conductance through a QPC was calculated using an NEGF approach, described in detail earlier [2,9,19]. At the interface between the rectangular region of size $w_2 \times l_2$ and a vacuum, the conduction band discontinuities at the bottom and the top interface were modeled using a smooth conduction band change over a small distance d from $\Delta E_c(y)$ at the trench/channel interface down to zero in the channel. This gradual change in $\Delta E_c(y)$ is responsible for the LSOC that triggers the spin polarization of the QPC in the presence of an asymmetry in V_{sg1} and V_{sg2}. We modeled $\Delta E_c(y)$ for the bottom side wall in Figure 65.1 as

$$\Delta E_c(y) = \frac{\Delta E_c}{2}\left[1 + \cos\frac{\pi}{d}\left(y - \frac{w_1 - w_2}{2}\right)\right] \tag{65.1}$$

and, for the top side wall, we used

$$\Delta E_c(y) = \frac{\Delta E_c}{2}\left[1 + \cos\frac{\pi}{d}\left(\frac{w_1 + w_2}{2} - y\right)\right] \tag{65.2}$$

where d represents the distance (in nm) over which the conduction band profile varies from its value inside the quantum wire to that of the vacuum region. In our simulations, we used $d = 1.6$ nm. A similar grading was also used along the walls going from the wider part of the channel to the central constriction of the QPC (Figure 65.1).

In our simulations, we model the e–e interaction V_{int} (r,r') using the following *nonlocal* 2D screened potential [20]:

$$V_{int}(r,r') = \frac{e^2}{4\pi\varepsilon_0\varepsilon_r}\left\{\frac{1}{|r-r'|} - \frac{\pi}{2\lambda}\left(H_0\left(\frac{|r-r'|}{\lambda}\right) - N_0\left(\frac{|r-r'|}{\lambda}\right)\right)\right\} \tag{65.3}$$

where λ is the screening length, $H_0(x)$ is the Struve function, and $N_0(x)$ is a Bessel function of the second kind. The LSOC parameter β (see appendix) and screening length λ were set equal to 5 Å^2 and 5 nm, respectively.

For an impurity or dangling bond located at location (x_1, y_1), we model the potential energy in the two dimensional electron gas (2DEG) as

$$U_{impurity}(x,y) = \frac{q^2}{4\pi\varepsilon_0\varepsilon_r\sqrt{(x - x_1)^2 + (y - y_1)^2 + \Delta^2}} \tag{65.4}$$

where $\Delta = q/4\pi\varepsilon_0\varepsilon_r U_0$ and U_0 is the maximum strength of the impurity potential.

For an attractive impurity, we use the above expression with a negative sign. We use $\varepsilon_r = 12.9$, the relative dielectric constant of GaAs. In all simulations, the source potential $V_s = 0$ V, and the potential at the drain contact $V_d = 0.1$ mV. An asymmetry in the QPC potential confinement is introduced by taking $V_{sg1} = 0.2$ V $+ V_{sweep}$ and $V_{sg2} = -0.2$ V $+ V_{sweep}$ and the conductance of the constriction was studied as a function of the sweeping (or common mode) potential, V_{sweep}. The conductance of the QPC was then calculated using the NEGF with a nonuniform grid configuration that contained more grid points at the interface of the QPC with a vacuum. All calculations were performed at a temperature $T = 4.2$ K. The screening length λ in Equation 65.1 was set to 5 nm and assumed to be independent of the gate potentials. Hereafter, we also calculate the spin conductance polarization $\alpha = [G_\uparrow - G_\downarrow]/[G_\uparrow + G_\downarrow]$, where $G_\uparrow$ and $G_\downarrow$ are the conductances due to the majority and minority spin bands, respectively. We study how the maximum of alpha is affected by the strength, polarity, and locations of the impurities and dangling bonds.

65.3 RESULTS

65.3.1 Effect of Impurity Scattering

Figure 65.2 is a plot of the conductance as a function of V_{sweep} for a QPC containing an impurity at location 3 in Figure 65.1. Shown for comparison is the conductance with no impurity in the channel. The conductance anomalies are highly sensitive to the strength and type of impurity (attractive and repulsive). All the curves show a conductance anomaly around 0.5 G_0 followed by a negative differential resistance region (NDR) and a second anomaly somewhere between 0.5 and 1 G_0. The peak-to-valley ratio of the NDR after the 0.5 G_0 increases with the strength of the impurity potential for the attractive impurity potential. The opposite trend is observed for a repulsive impurity. Figure 65.2 shows that there is a substantial shift of the conductance along the common signal V_{sweep} for an impurity that is either attractive or repulsive.

In QPC experiments, the charge state of an impurity is often affected by sample handling, such as temperature cycling, when the sample is brought back to room temperature between low temperature measurements. This typically leads to markedly different conductance traces for

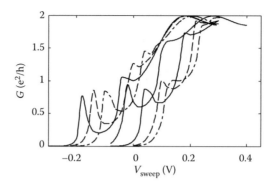

FIGURE 65.2 Conductance as a function of V_{sweep} for a QPC containing an impurity (either attractive or repulsive) at location 3 in Figure 65.1. The three leftmost curves correspond to an attractive impurity with scattering strength (U_0) equal to (from left to right) −30, −20, and −10 meV, respectively. The three rightmost curves correspond to a repulsive impurity with scattering strength (U_0) equal to (from left to right) 10, 20, and 30 meV, respectively. The middle curve corresponds to the case of no impurity in the channel.

identical bias conditions, and recently, we observed this phenomenon while studying the conductance of InAs-based QPCs in the presence of LSOC, while varying the asymmetric bias applied between the two side gates [21]. Thermal cycling is expected to change the charge state of either the remote impurities used in the modulation doping to form the 2DEG or in the charge state of dangling bonds formed on the side walls of the QPC during etching, or even due to a single impurity located directly in the path of the current flow. The sensitivity of the conductance anomalies to the impurity location (points 1, 2, and 3 in Figure 65.1) in the central portion of the QPC is illustrated in Figure 65.3.

The maximum value of α was found to be located near the first maximum in the conductance plots in Figures 65.2 and 65.3. In Figure 65.1, α_{max} changes by 6% (from 0.989 to 0.927) when U_0 varies from −30 to 30 meV. The change in α_{max} is only 4% (from 0.976 to 0.948) when U_0 changes from −20 to 20 meV, and less than 1% (from 0.976 to 0.97) when U_0 changes from −10 to 10 meV. This dependence of α_{max} on the type and strength of an impurity, even though by just a few percent, could lead to a substantial reduction in the ON/OFF conductance ratio of a spin valve built of two

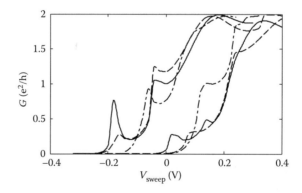

FIGURE 65.3 Conductance as a function of V_{sweep} for a QPC containing an impurity in the central portion of the channel at locations 1, 2, and 3 in Figure 65.1. The strength of the scattering potential is −30 meV for the three leftmost curves (attractive impurity) and 30 meV for the three rightmost curves (repulsive impurity). The full, dashed, and dot-dashed curves correspond to an impurity at locations 1, 2, and 3, respectively, in Figure 65.1.

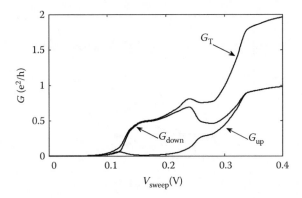

FIGURE 65.4 Conductance as a function of V_{sweep} for a QPC containing two off-center repulsive impurities located at $(x_1, y_1) = (16\ nm, 28\ nm)$ and $(x_2, y_2) = (32\ nm, 28\ nm)$. We used $V_{sg1} = -0.2\ V + V_{sweep}$ and $V_{sg2} = 0.2\ V + V_{sweep}$.

QPCs in series. Interestingly, the value of α_{max} is larger (smaller) in the presence of an attractive (repulsive) impurity compared to the case of no impurity in the channel, for which $\alpha_{max} = 0.973$. To reach a larger value of α_{max}, a tunable repulsive impurity potential could be generated through the use of a negatively biased scanning tunneling microscope (STM) tip on top of the narrow portion of the QPC.

In Figure 65.4, we plot the conductance versus V_{sweep} for a QPC containing two repulsive impurities located at $(x_1, y_1) = (16\ nm, 28\ nm)$ and $(x_2, y_2) = (32\ nm, 28\ nm)$, that is, slightly off-center. In this case, a reverse polarity was used, that is, $V_{sg1} = -0.2\ V + V_{sweep}$ and $V_{sg2} = 0.2\ V + V_{sweep}$, the overall potential energy in the narrow portion of the QPC is larger closer to gate 1 and near the conductance anomaly, the spin-down electrons are the majority carriers. We found $\alpha_{max} = -0.956$, still a rather large value despite the two impurities in the channel.

Figure 65.5 illustrates that even for the case of asymmetric bias on the two gates $(V_{sg1} = V_{sg2} = 0.0\ V + V_{sweep})$, a conductance anomaly is found with the same two off-center repulsive impurities as in the previous figure. In this case, the potential energy maximum is also closer to gate 1 in the narrow portion of the QPC and the spin-down electrons are the carriers in the channel near the conductance anomaly. This simulation shows that even a slight asymmetry due to unwanted impurities can lead to spin polarization in an otherwise perfectly symmetric channel.

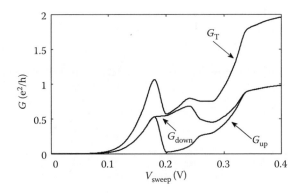

FIGURE 65.5 Conductance as a function of V_{sweep} for a QPC containing two repulsive impurities impurities located at $(x_1, y_1) = (16\ nm, 28\ nm)$ and $(x_2, y_2) = (32\ nm, 28\ nm)$ for the case of a symmetrical bias condition, that is, $V_{sg1} = V_{sg2} = 0.0\ V + V_{sweep}$.

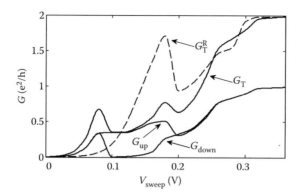

FIGURE 65.6 Conductance as a function of V_{sweep} for a QPC containing two dangling bonds shown as dark dots in Figure 65.1. The strength of the dangling bond potential U_0 was set to 200 meV. The contributions from the up-spin and down-spin electrons to the total conductance G_T are shown for the case $V_{sg1} = -0.2$ V $+ V_{sweep}$ and $V_{sg2} = 0.2$ V $+ V_{sweep}$. The dashed curve is the conductance G_T^R, that is, for the case $V_{sg1} = 0.2$ V $+ V_{sweep}$ and $V_{sg2} = -0.2$ V $+ V_{sweep}$.

65.3.2 EFFECT OF DANGLING BOND SCATTERING

In Figure 65.1, the black circles represent the two dangling bonds considered in our simulations. The first dangling bond has coordinates $(x_1, y_1) = ((l_1 - l_2/2) + (l_2/4), (w_1 - w_2/2) - (d/2))$, that is, it is located one-fourth of the way from the left side in the narrow portion of the QPC and in the middle of the top side wall interface modeled with Equation 65.1. The second dangling bond has coordinates $(x_2, y_2) = ((l_1 - l_2)/2) ((w_1 - w_2)/2) (d/2)$, (that is, it is located in the middle of the narrow portion of the QPC on the bottom side wall interface modeled with Equation 65.2.

Figure 65.6 shows a plot of the total conductance $G_T = G\uparrow + G\downarrow$ versus V_{sweep} for the case of two dangling bonds. The strength of a dangling bond potential U_0 was set equal to 200 meV. The full curve for G_{tot} corresponds to the case where $V_{sg1} = -0.2$ V $+ V_{sweep}$ and $V_{sg2} = 0.2$ V $+ V_{sweep}$. Also shown in Figure 65.6 are the contributions from $G\uparrow$ and $G\downarrow$, which differ substantially for V_{sweep} in the range from 0.02 to 0.1 V. Figure 65.6 shows conductance anomalies around 0.3 G_0 and 0.36 G_0 and also a shoulder in the plot around 0.75 G_0. The dashed curve in Figure 65.6 is a plot of G_T^R

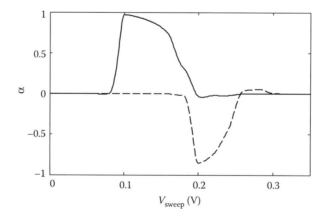

FIGURE 65.7 Spin polarization of the conductance α as a function of V_{sweep} calculated using the results in Figure 65.6. The full and dashed curves are α for the case of $V_{sg1} = -0.2$ V $+ V_{sweep}$ and $V_{sg2} = 0.2$ V $+ V_{sweep}$ and $V_{sg1} = 0.2$ V $+ V_{sweep}$ and $V_{sg2} = -0.2$ V $+ V_{sweep}$, respectively.

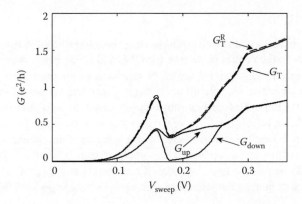

FIGURE 65.8   Same as Figure 65.6 for the same dangling bonds but with an impurity potential $U_0 = 100$ meV.

r reverse asymmetry, that is, with $V_{sg1} = 0.2$ V $+ V_{sweep}$ and $V_{sg2} = -0.2$ V $+ V_{sweep}$. The difference between the two curves for G_T is due to the asymmetry in the potential energy profile in the narrow portion of the QPC prior to the application of the different biases on the side gates.

The spin conductance polarization α for the two biasing cases $V_{sg1} = \pm 0.2$ V $+ V_{sweep}$ and $V_{sg2} = \pm 0.2$ V $+ V_{sweep}$ is shown in Figure 65.7. When $V_{sg1} = -0.2$ V $+ V_{sweep}$ and $V_{sg2} = 0.2$ V $+ V_{sweep}$, α reaches a maximum of 0.97 for $V_{sweep} = 0.1$ V. For this value of V_{sweep}, $G_T = 0.47$ G_0. When $V_{sg1} = 0.2$ V $+ V_{sweep}$ and $V_{sg2} = -0.2$ V $+ V_{sweep}$, α is negative because the role of the majority and minority spin bands are flipped compared to the previous case and α reaches a minimum of -0.85 for $V_{sweep} = 0.2$ V. For this value of V_{sweep}, $G_T^R = 0.47$ G_0. The conductance curve versus V_{sweep} is different for the two polarities of the bias asymmetry between the gates. Despite the presence of the dangling bonds, α is quite large, indicating a substantial amount of spin polarization in the QPC channel.

The simulations were repeated for a lower value of the dangling bond impurity strength $U_0 = 100$ meV. The plots of the conductance and spin polarization α versus V_{sweep} are shown in Figures 65.8 and 65.9, respectively. In this case, there is hardly any difference between the two conductance curves for the two biasing configurations, cases $V_{sg1} = \pm 0.2$ V $+ V_{sweep}$ and $V_{sg2} = \pm 2$ V $+ V_{sweep}$. The plot of α versus V_{sweep} for the two biasing configurations is the reverse of one another, indicating that spins of opposite polarities are the major carriers when the biasing conditions are flipped; α reaches a maximum of 0.89 for $V_{sweep} = 0.18$ V at which $G_T = 0.17$ G_0. For the opposite polarity, α reaches an absolute minimum of -0.90 for $V_{sweep} = 0.18$ V at which G_T^R 0.175 G_0. These simulations show that a hysteresis in the conductance curve, when flipping the polarity of the bias asymmetry $\Delta V_G = V_{G1} - V_{G2}$, can only be observed if the dangling bond scattering potential is sufficiently large.

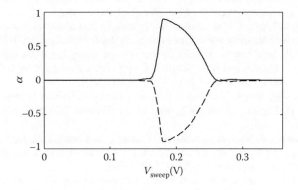

FIGURE 65.9   Same as Figure 65.7 for the same dangling bonds but with an impurity potential $U_0 = 100$ meV.

65.4 CONCLUSIONS

In this chapter, the NEGF formalism was used to study how the spin polarization is affected by the presence of impurities in the central portion of the QPC and by dangling bonds on the side walls. It is found that the number, location, and shape of the conductance anomalies occurring below the first quantized conductance plateau ($G_0 = 2e^2/h$) are strongly dependent on the nature (attractive or repulsive) and locations of the impurities and dangling bonds. Our simulations show that a spin conductance polarization over 90% can be achieved despite the presence of impurities and dangling bonds and that this maximum is not necessarily reached where the conductance of the channel is equal to 0.5 G_0. For QPCs with impurities off-center, a conductance anomaly appears below the first integer step even for the case of symmetric bias on the two side gates. These results are of practical importance if QPCs in series are to be used to fabricate all-electrical spin valves with a large ON/OFF conductance ratio.

REFERENCES

1. W.K. Hew et al., Spin incoherent transport in quantum wires, *Phys. Rev. Lett.* 101, 036801, 2008.
2. R. Crook et al., Conductance quantization at a half-integer plateau in a symmetric GaAs quantum wire, *Science* 312, 1359, 2006.
3. D.J. Reilly et al., Density-dependent spin polarization in ultra-low-disorder quantum wires, *Phys. Rev. Lett.* 89, 246801, 2002.
4. S. Kim, Y. Hashimoto, Y. Iye, and S. Katsumoto, Evidence of spin-filtering in quantum constrictions with spin-orbit interaction, arXiv: 1102.4648v1, 2011.
5. A.P. Micolich, What lurks below the last plateau: Experimental studies of the $0.7 \times 2e^2/h$ conductance anomaly in one-dimensional systems, *J. Phys. Cond. Matter* 23, 443201, 2011.
6. P. Jaksch, I. Yakimenko, and K.-F. Berggren, From quantum point contacts to quantum wires: Density-functional calculations with exchange and correlation effects, *Phys. Rev. B* 74, 235320, 2006.
7. K.J. Thomas et al., Possible spin polarization in a one-dimensional electron gas, *Phys. Rev. Lett.* 77, 135, 1996.
8. D.J. Reilly, Phenomenological model for the 0.7 conductance feature in quantum wires, *Phys. Rev. B* 72, 033309, 2005.
9. J. Wan, M. Cahay, P. Debray, and R.S. Newrock, Spin texture in quantum point contacts in the presence of lateral spin orbit scattering, *J. Nanoelectron. Optoelectron.* 6, 95, 2011.
10. J. Faist, P. Gueret, and H. Rothinzen, Observation of impurity effects on conductance quantization, *Superlattices Microstructures* 7, 349, 1990.
11. C.S. Chu and R.S. Sorbello, Effect of impurities on the quantized conductance of narrow channels, *Phys. Rev. B* 40, 5941, 1989.
12. E. Tekman and S. Ciraci, Ballistic transport through a quantum point contact: Elastic scattering by impurities, *Phys. Rev. B* 42, 9098, 1990.
13. J.A Nixon, J.H. Davies, and H.U. Baranger, Breakdown of quantized conductance in point contacts calculated using realistic potentials, *Phys. Rev. B* 43, 12638, 1991.
14. A. Grincwajg, G. Edwards, and D.K. Ferry, Quasi-ballistic scattering in quantum point contact systems, *J. Phys. Condens. Matter* 9, 673, 1997.
15. M.J. Laughton, J.R. Barker, J.A. Nixon, and J.H. Davies, Modal analysis of transport through quantum point contacts using realistic potentials, *Phys. Rev. B* 44, 1150, 1991.
16. J.C. Chen, Y. Lin, K.T. Lin, T. Ueda, and S. Komiyama, Effects of impurity scattering on the quantized conductance of a quasi one-dimensional quantum wire, *Appl. Phys. Lett.* 94, 012105, 2009.
17. P. Debray et al., All-electric quantum point contact spin-polarizer, *Nat. Nanotechnol.*, 4, 759–764, 2009.
18. J. Wan, M. Cahay, P. Debray, and R.S. Newrock, On the physical origin of the 0.5 plateau in the conductance of quantum point contacts, *Phys. Rev. B* 80, 155440, 2009.
19. S. Datta, *Electron Transport in Mesoscopic Systems*, Cambridge University Press, New York, 1995.
20. G. Giuliani and G. Vignale, *Quantum Theory of the Electron Liquid*, Cambridge University Press, Cambridge, 2005.
21. P.P. Das, N. Bhandari, J. Wan, J. Charles, M. Cahay, K.B. Chetry, R.S. Newrock, and S.T. Herbert, Influence of surface scattering on the anomalous plateaus in an asymmetrically biased InAs/InAlAs quantum point contact, *Nanotechnology* 23, 215201, 2011.
22. J. Wan, M. Cahay, P. P. Das, and R. S. Newrock, Influence of impurity scattering on the conductance anomalies of quantum point contacts with lateral spin-orbit coupling, *Proc. 11th IEEE International Conference on Nanotechnology*, August 15–18, Portland, Oregon, pp. 1395–1398, 2011.

66 Electric Field-Controlled Spin Interactions in Quantum Dot Molecules

Kushal C. Wijesundara and Eric A. Stinaff

CONTENTS

66.1 INTRODUCTION

Device miniaturization with advanced fabrication techniques has revolutionized the semiconductor industry. This, along with the property of intrinsic carrier spin, may ultimately result in a functional unit utilizing quantum mechanical effects at the fundamental device operating limit. Potential platforms to implement quantum information processing has been demonstrated in many systems, including ion traps [1], nuclear magnetic spin resonances [2], microwave resonators [3], and photonic materials [4]. Among these systems, semiconductor quantum dot molecules (QDMs) are excellent candidates due to 3-D confinement, discrete energy levels, optical access, controllable coupling, and the wealth of semiconductor technology and techniques to draw upon. Spins in these QDMs are considered as a candidate to provide quantum bits (qubits) as they can be initialized, manipulated, and measured through established spectroscopic techniques, for example, via recombination of neutral exciton or charged exciton states (neutral or charged electron–hole pairs). QDMs may also help to realize potential next-generation quantum computing schemes [5] through the generation of entangled photon pairs via neutral biexcitons (exciton pair) [6].

One of the main challenges in implementing quantum information processing is identifying and controlling the spin states that are affected by the fine structure splitting in exciton states [7]. For example, the failure to demonstrate as an ideal source of entangled photons arises due to the inherent electron and hole exchange interaction, related to the structural anisotropy within the QDMs. As biexciton states in QDMs have proven to be useful for demonstrating conditional dynamics, biexciton–exciton cascade paths are useful to identify optical transitions and especially to account for exchange splitting. The anisotropic electron–hole exchange interaction that is typically present in QDMs mixes the pure exciton states, and splits the resulting states, giving distinguishable which-path information.

The main focus of this chapter is to present some of the effects due to the electron and hole exchange interaction in QDMs and to demonstrate control of the exchange splitting energy. In the devices discussed here, the electron and hole exchange interaction can be varied using an applied electric field. Furthermore, with precise device engineering, one can tune the carrier wave function [8], which may lead to spintronics-based applications. The formation of other charged exciton states can be readily identified through carrier tunneling [9,10] via variations in luminescence intensity. By far, the most attractive feature that differentiates the QDMs from single quantum dots (QDs) is the strong electric field dependence of the indirect exciton states where the electron and hole reside in different QDs within the QDM. Through polarization-dependent photoluminescence (PL) spectra, we could clearly identify the neutral exciton states (direct and indirect configuration) and singly charged exciton states (positive trion) and the spin-dependent interactions in QDMs were investigated. Overall, the electron and hole exchange interaction was controlled either by additional charges that resulted in singly charged states or by spatial variation in the carrier wave function in the InAs/GaAs QDM system.

66.2 FUNDAMENTALS IN SPIN INTERACTIONS

66.2.1 OPTICAL ORIENTATION

One of the primary experimental techniques mentioned in the introduction is that of optical orientation, which involves the transfer of angular momentum from the light to the crystal. In charge-tunable QDs, many efforts have been undertaken to investigate the optical properties of excitonic states through polarization-resolved measurements. Here, we utilize similar techniques to better understand spin effects and probe the carrier spin states in QDMs. In principle, the optical orientation approach demonstrates the angular momentum conservation between light and the crystal nanostructure. With resonant or nonresonant circular polarized laser excitation corresponding to either the right or left circularly polarized light (σ^+ or σ^-), photons of angular momentum of ± 1 (units of $\hbar$) are attained. When the QDMs are excited with such polarized excitation, it conserves the angular momentum by producing photo-generated electron–hole pairs while conserving the total spin. In the direct band gap semiconductor QDMs, electron states in the conduction band are twofold degenerate with a total angular momentum of $J = 1/2$. The valence band hole states with the associated p-type Bloch wave character demonstrate a fourfold degeneracy that gives rise to a total angular momentum of $J = 3/2$ as illustrated in Figure 66.1. The angular momentum projections corresponding to the twofold degenerate heavy holes $m_j^{hh} = \pm 3/2$ and the twofold degenerate light holes $m_j^{lh} = \pm 1/2$ provide the subbands related to the heavy and light hole states. Furthermore, owing to the compressive strain corresponding to the lattice mismatch between the InAs and GaAs materials and the structural confinement, the light hole subband is shifted in energy and the heavy holes are predominant in the low energy states. Thus, the recombination of excitons comprising

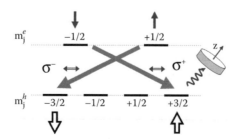

FIGURE 66.1 Optical orientation process. Defined spin orientations of the electrons and holes are attained with circular polarized light and identified in agreement with the optical selection rules and device characteristics. Detecting the orientation of the emitted light also reveals the associated spin states of the electron and hole pair.

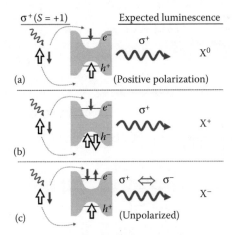

FIGURE 66.2 Expected luminescence from circularly polarized σ^+ excitation corresponding to a (a) neutral exciton, (b) positively charged exciton, and (c) negatively charged exciton with associated spin configurations.

electrons and heavy holes, in agreement with optical transition selection rules, gives rise to the circularly polarized light of σ^+ or σ^- as depicted in Figure 66.1.

Measuring the luminescence polarization, the spin states of different excitons can be investigated through optical pumping as discussed above. To gain insight into the spin-dependent interactions, we consider elementary charged excitons. Excitation into the quasicontinuum wetting layer with circularly polarized light would essentially generate electron and hole pairs with an electron spin of $-(1/2)$ and heavy hole spin of $+(3/2)$ that conserves the angular momentum of $+1$. Some of the carriers would then relax and become trapped inside the QDs. However, in the relaxation process, a spin of $-(1/2)$ electrons and unpolarized holes can prevail due to the relatively fast hole spin relaxation rate arising from the spin–orbit interaction within the valence band [11] of the confined structure. Therefore, if the electron predominantly remains in a spin of $-(1/2)$ configuration, neutral exciton (X^0) luminescence would be expected to be equivalent to the initial circularly polarized excitation (σ^+) as shown in Figure 66.2a.

With an initial ground-state heavy hole, polarized excitation can create two paired holes in the positively charged exciton [positive trion (X^+)] configuration (Figure 66.2b). The expected PL luminescence can be determined from the spin-down electron that results in the same helicity as the excitation. Optical excitation of a QDM with an initial isolated electron will result in a negatively charged exciton [negative trion (X^-)] state as depicted in Figure 66.2c where the PL emission can be predicted from the hole spin state. However, owing to the efficient hole relaxation and paired electron spins, the expected luminescence will be unpolarized. Beyond the expected luminescence from basic exciton configurations, the experimental measurements have provided further insight into the polarization effects, including polarization signatures associated with different charge states in single QDs that have been widely discussed in the literature [12]. To understand these observed results of exciton luminescence, the electron and hole exchange interactions that arise due to the structural anisotropy of the crystal is discussed in the next section.

66.2.2 EXCHANGE INTERACTION AND EXCITON FINE STRUCTURE

To better understand the origin of the expected polarized luminescence from exciton states, fundamentals of the electron–hole exchange interaction are explored. Qualitatively, the electron and hole exchange interaction arises due to the anisotropy that is present in the QDM structure. This lifts the heavy hole and light hole degeneracy along with the inversion asymmetries in the QD potential [13]. The exchange Hamiltonian can be presented along with the group symmetry considerations as [14,15]

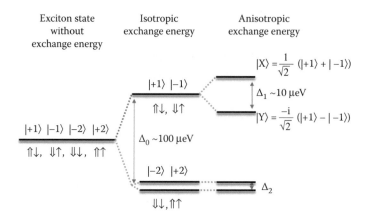

FIGURE 66.3 Schematic representation of the exciton fine structure due to exchange interactions.

$$H_{EX} = \underbrace{2\Delta_0 I_z S_z}_{H_0} + \underbrace{\Delta_1(I_x S_x - I_y S_y)}_{H_1} + \underbrace{\Delta_2(I_x S_x + I_y S_y)}_{H_2} \qquad (66.1)$$

where $S(S_x, S_y, S_z)$ and $I(I_x, I_y, I_z)$ denote the total electron and hole spins, respectively, and the corresponding exchange splitting strengths are represented by Δ. The isotropic part of the exchange interaction denoted by the first term, H_0 in the Hamiltonian, splits the fourfold degenerate exciton ground state by ~100 µeV into two doublets with total spins of ±1 and ±2, respectively, as schematically illustrated in Figure 66.3. The doublet with a total spin of ±1 is termed as a bright exciton as it couples with the light field conserving the angular momentum. The dark exciton (total spin of ±2) results in nonradiative processes but is accessible through mixing with bright excitons [16]. The second term in the Hamiltonian (H_1) denotes the anisotropic exchange energy corresponding to the bright excitons that split the doublet Δ_1 due to a lowering of the confinement symmetry (piezoelectric effects [17,18]). This is opposed to the already-lifted dark state doublet [19,20] Δ_2, (Hamiltonian H_2). Thus, the circularly polarized excitation (σ^+) mixes the bright states into linearly polarized states (π^x, π^y), corresponding to the anisotropic exchange interaction, and splits them into a linearly polarized doublet (~10 µeV).

When the exchange interaction effects are considered, further insight can be obtained for the exciton states presented in Figure 66.2. The unpaired electron and hole in the neutral exciton experiences a strong anisotropic exchange energy that wipes out the circular polarization memory, resulting in a linearly polarized doublet. With two holes and an electron, the hole spins pair in the positive trion and the exchange energy is diminished, resulting in a positively polarized luminescence. Similarly, for the negative trion with two paired electrons and a hole, the degree of polarization mainly depends on the spin configuration of the hole state as the exchange energy is eliminated due to the total spin zero electrons. As such, owing to the anisotropy, the expected polarized luminescence from exciton states are in clear agreement with the reported experimental observations [12].

66.3 POLARIZATION-RESOLVED PHOTOLUMINESCENCE

For the present study, the samples consist of InAs QDs on a GaAs matrix, which has been epitaxially grown through the use of the Stransky–Krastanov technique [21]. Quantum mechanical coupling is achieved via two InAs QD layers and a GaAs barrier between the two dots. Quantization energies of the individual layers of QDs are controlled by the indium flush technique during the growth [22]. Electrical controllability is attained by embedding them in a Schottky diode structure. Samples were kept in a closed cycle cryostat operated at ~10 K. The PL was attained by dispersing through a 0.75 m spectrometer. The excitation was provided using a mode locked Ti:sapphire laser

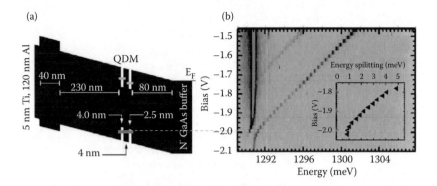

FIGURE 66.4 Band edge diagram with applied electric field to the QDM. (a) Schematic view of the QDM embedded in Schottky diode structure along with hole level resonance attained with applied electric field. (b) Photoluminescence spectra showing anti-crossing signature corresponding to molecular state of the exciton and the inset provides evidence of having minimum energy splitting between the two excitonic states at the same applied electric field. (From K. C. Wijesundara et al., *IEEE Nanotechnol.* 11, 887 © (2011) IEEE. With permission.)

operated in the continues wave (CW) mode. The schematic structure of the QDM embedded in an n+ Schottky diode along with the PL spectra is illustrated in Figure 66.4. With an applied electric field along the growth direction (z direction), we observe hole-level resonance at high-enough forward biases because the device itself has a relatively larger bottom dot compared to the top dot. For polarization-resolved PL measurements, linear polarizers and liquid crystal retarders were introduced to the basic PL experiment. Initially, linear polarized light is created by sending the CW laser beam through the linear polarizer with energy above the quasicontinuum wetting layer of the QDM. Next, the desired circular polarized light necessary to excite the QDMs is generated using liquid crystal retarders. By varying the applied voltage given to the liquid crystal retarder, its effective birefringence and hence retardance can be varied, which is used to generate a phase shift of $\lambda/4$ or $3\lambda/4$ to create σ^+ or σ^- circularly polarized light. After the recombination process, the emitted light is sent through an analyzer system. This process measures both σ^+ and σ^- helicities and evaluates the degree of circular polarization of the emitted light that is dispersed by a three-stage spectrometer and detected through an LN_2-cooled CCD. To gain insight into the polarization signatures through the measurement of the rate at which the unpaired photo-generated charge carrier spin flip occurs, the degree of polarization is determined from Equation 66.2.

$$P = \frac{(I^+ - I^-)}{(I^+ + I^-)} \tag{66.2}$$

Here, I^+ (I^-) represents the intensity of σ^+ (σ^-) polarization under σ^+ polarized excitation. By varying the applied electric field given to the Schottky photodiode, polarization signatures associated with different exciton states that arise from the embedded QDMs can be monitored. Spin states of the excitons are denoted by $^{\uparrow_B, \downarrow_T}_{\Downarrow_B, \Uparrow_T} X^q$, with spin-up or-down electrons ($\uparrow, \downarrow$), holes ($\Uparrow, \Downarrow$) in the bottom (B) or top (T) dot along with charge (q) throughout this chapter.

66.4 ELECTRIC FIELD-TUNABLE EXCHANGE INTERACTION

As discussed in Section 66.2.2, the polarization-resolved PL results of neutral and singly charged exciton states arise as a consequence of anisotropic exchange interaction [12]. These results further exemplify that the polarized emission is determined by the unpaired carrier spin configurations. As a result, one may alleviate the anisotropic exchange interaction effects through the exclusion of the overall spin. A unique feature of the QDM structure is its ability to spatially separate the electron

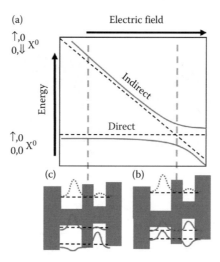

FIGURE 66.5 (a) Schematic representation of direct and indirect excitons as a function of an applied electric field along with relevant spin configurations. Band edge diagrams shown in (b) and (c) represent the QDMs in a Schottky diode structure with a larger bottom dot (B) and a smaller top dot (T). The variation in the band edge diagram at high electric fields (b) shows hole level resonances along with electron wave functions (dotted curves) and symmetric (direct) and anti-symmetric (indirect) hole wave functions (solid curve). At low electric fields (c), overlap between the electron and the anti-symmetric hole wave function is decreased and more atomic-like behavior can be observed.

and the hole as a function of an applied electric field as schematically depicted in Figure 66.5. This may allow for effective control of the overall exchange interaction.

In the n-doped QDM system studied here, by varying the electric field, we can attain hole-level resonances, where the tunnel coupling of hole states results in the formation of molecular states that can be observed as unique anticrossing signatures. At high electric fields, the hole levels become resonant, leading to the enhancement of the symmetric and antisymmetric molecular hole wave functions (Figure 66.5b). Away from the anticrossing region, the holes can be predominantly localized on either the top or the bottom dot, leading to indirect or direct excitonic states as schematically illustrated in Figure 66.5c. Thus, varying the electric field leads to the tuning of the excitonic emission from the interdot to the intradot. This spatial separation of the electron and the hole provides a method to control the overall exchange interaction.

Next, polarization-resolved PL measurements were performed on neutral exciton states to determine the degree of circular polarization according to Equation 66.2. Using a basic PL spectral map, the different excitonic state spectral lines that are inherent to the QDM emission as depicted in Figure 66.6a were recognized, especially inside the squared region, where the indirect and direct transitions associated with the neutral exciton state were predominant. Figure 66.6b represents the polarization-dependent PL spectra for the neutral exciton, and on the top, the polarization memory analysis from each point of the spectra denoted by dashed lines is depicted in Figure 66.6c. From the analysis of the polarization-dependent spectra, it is evident that the degree of circular polarization memory corresponding to the direct transition demonstrates relatively lower polarization percentage values due to the strong exchange interaction in its direct configuration. At the anticrossing region ~53.96 kV/cm, the degree of circular polarization displays similar polarization percentage values for both indirect and direct transitions as evident by Figure 66.6c. In the n-doped QDMs, molecular symmetric and antisymmetric hole wave functions corresponding to the hole-level molecular state observed at ~53.96 kV/cm have comparable amplitude in both dots. Therefore, direct configuration of the neutral exciton demonstrated an increased polarization percentage, whereas the polarization memory associated with the indirect configuration exhibits a reduced value. As the applied electric

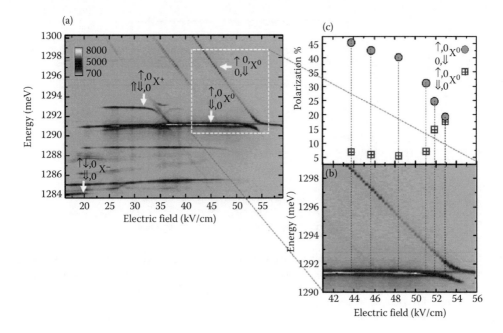

FIGURE 66.6 (a) Photoluminescence spectra from the 4 nm QDM structure. Spectral lines associated with exciton states are clearly identified, including the inter-dot and intra-dot exciton states along with the anti-crossing signature highlighted in the squared region. (b) Polarization-resolved PL spectra for the neutral exciton state near the anti-crossing region are acquired through excitation to the quasi-continuum wetting layer with σ^+ polarized excitation and detected for both the σ^+ and σ^- polarizations. (c) The degree of circular polarization as a function of an applied electric field is evaluated from Equation 66.2, corresponding to the direct (open squares) and indirect (solid circles) exciton states.

field is reduced, the wave function amplitude of the hole is shifted, which involves a change in the overlap of the electron–hole wave functions. This change results in reduced exchange interaction effects, which cause the indirect exciton transition to show an increased degree of circular polarization, as depicted in Figure 66.6. Since the degree of circular polarization is influenced by the anisotropic exchange interaction, by experimentally tuning the electric field, this is evidence that one can control the overall electron–hole exchange interaction [23,24].

Electric field-tunable polarization-resolved measurements were extended to similar QDM structures with different barrier heights of 4 and 2 nm, respectively. Experimental parameters including laser power were kept constant while exciting into the quasicontinuum of the wetting layer. For the two different barrier QDM structures, hole-level anticrossings were observed at different applied electric fields due to structural variations and different tunneling strengths [25]. As such, hole-level anticrossing signatures were identified at 43.48 and 53.96 kV/cm, respectively, for the 2 and 4 nm barrier QDM structures. To signify comparable results on the polarization-resolved PL, measurements were presented relative to the anticrossing point of the two QDMs. The difference in the degree of circular polarization between the indirect and direct exciton states of the neutral exciton as a function of the relative electric field is depicted in Figure 66.7 for the 2 nm (circles) and 4 nm (squares) barrier samples. Overall, the difference in the degree of polarization increase as a function of the relative electric field with increasing barrier separation between the top and the bottom dot in the QDM structures is evident by the plot. For a given relative electric field, as the barrier between the top and the bottom dot is reduced, the overlap between the electron and the hole wave function increases due to the higher tunneling strengths. This results in a relatively higher exchange interaction for the lower-barrier QDMs. These barrier-dependent, polarization-resolved results further reveal the electric field-controllable electron–hole exchange interaction effects.

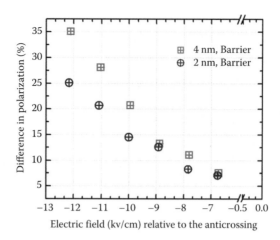

Electric field (kv/cm) relative to the anticrossing

FIGURE 66.7 Difference in the circular polarization percentage as a function of the relative electric field, relative to the interdot and intradot neutral exciton states. At low relative electric fields, a higher degree of circular polarization memory can be observed with the 4 nm barrier (squares) as opposed to the 2 nm barrier (circles) QDM structure. This common trend in the degree of circular polarization is a consequence of the reduced wave function overlap and the electron-hole exchange interaction with increase in the barrier separation.

66.5 SPIN EFFECTS OF CHARGED EXCITON STATES

When the electric field is further reduced relative to the neutral exciton anticrossing, luminescence intensity variations are observed along with additional spectral lines associated with other charge states as depicted in Figure 66.6a. Below 40 kV/cm, a singly charged positive trion luminescence is observed due to charge carrier tunneling. The key to the detailed polarization signatures is the identification of an "X"-shaped pattern within the singly charged positive trion state, which contains both crossing and anticrossing signatures [8]. This characteristic pattern can be identified from the higher energy lines of the positive trion and by determining the four recombinations between the initial and final states as illustrated in Figure 66.8. The spin fine structure is observed as a result of exchange interactions and the detailed spectral pattern of the positive trion arises as a combined result of tunneling, exchange interaction, and the Pauli exclusion principle. The origin of the spin states associated with the fine structure splitting in the optical spectra arises from the kinetic hole–hole and electron–hole exchange interactions as discussed in the literature [16].

In subsequent experiments, polarization-dependent PL was utilized as a tool to probe the spin fine structure via detailed polarization signatures in singly charged positive trion states. Initially, the characteristic "X" pattern was identified as depicted in Figure 66.9a within the squared region where the spectra have been plotted relative to the intradot neutral exciton energy and anticrossing electric field. In the zoomed region of the trion state, different spectral lines were identified from associated spin states as illustrated in Figure 66.9b, which is in agreement with the results discussed in the literature [16].

From the spin configuration associated with the positive trion state "X" patterned region, two spin states were identified as X_H^+, X_L^+, where the subscripts H and L correspond to the relatively high and low energy states [24,26]. Polarization-resolved PL measurements were obtained in the fine structures of the singly charged positive trion. The degree of circular polarization is evaluated according to Equation 66.2 and the observed circular polarization percentage results for the spin states of X_H^+, X_L^+ are shown in Figure 66.10, where the higher energy state revealed relatively larger circular polarization memory results as opposed to the low energy spin state.

From the above results, the lower energy state X_L^+ with the spin configuration of $_{\Downarrow,\Uparrow}^{\uparrow,0}X_L^+$ can be thought of as a neutral exciton with a spectator hole in the top dot; hence, the hole spins are not

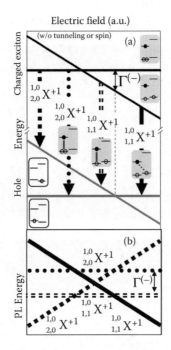

FIGURE 66.8 Origin of the "X" pattern in the singly charged exciton state. (a) Four possible recombination paths from the initial charged exciton to the ground state hole level. (b) Basic singly charged exciton spectral recombination signature without tunneling or spin effects.

paired. This results in relatively higher anisotropic exchange energies similar to intradot exciton states, which give rise to a low degree of circular polarization. From the higher energy state spin configuration of $_{\uparrow\downarrow,0}^{\uparrow,0}X_H^+$, it is evident that the hole spin singlet states are formed and the degree of polarization depends on the unpaired electron, which results in relatively higher circular polarization percentages.

From the isotropic exchange interaction, the energy splitting associated with the spin fine structure of the singly charged positive trion state can be identified. Related spin states are represented by $X_{HD}^+ \equiv {}_{\downarrow,\downarrow}^{\uparrow,0}X_{HD}^+$ and $X_{LD}^+ \equiv {}_{\uparrow,\downarrow}^{\uparrow,0}X_{LD}^+$, respectively, for the lower and higher energy configurations of the spin fine structure doublet. The two spectral lines of the spin fine structure doublet are observed due to the high electric field-dependent indirect transitions that arise from the associated spin configurations of the fine structure doublet and from the allowed recombination path via the electron and the hole in the top dot.

In the spin fine structure doublet, the electron and the top dot hole states have opposite spin configurations, in agreement with the optical selection rules, which gives rise to the observed spectral lines as depicted in Figure 66.11, whereas the energy splitting of the spin fine structure doublet is essentially determined from the isotropic exchange energy between the bottom dot electron and the hole states. However, bottom dot hole states exhibit either singlet (antiparallel) or triplet (parallel) configurations with the top dot hole. Therefore, even without an applied magnetic field (which is normally used to mix the states and create dark exciton to be optically active), bright and dark exciton splitting can be measured via the top dot hole state using optical techniques.

Next, the degree of circular polarization was evaluated for the spin states $X_{HD}^+ \equiv {}_{\downarrow,\downarrow}^{\uparrow,0}X_{HD}^+$ and $X_{LD}^+ \equiv {}_{\uparrow,\downarrow}^{\uparrow,0}X_{LD}^+$. From the circular polarization percentage results shown in Figure 66.11b, an overall increase in the polarization memory can be observed for the fine structure doublet as a function of a relative electric field, relative to the intradot anticrossing. Within the fine structure doublet, an applied electric field causes the top dot hole to spatially separate and it results in reduced exchange

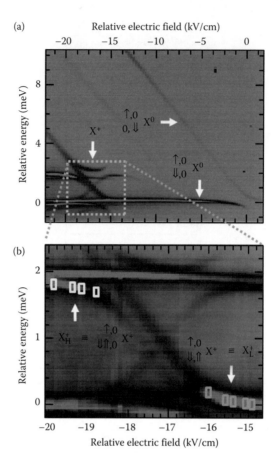

FIGURE 66.9 (a) Identification of the singly charged trion state. (b) Spin fine structure of the positive trion along with spin states of X_H^+ and X_L^+.

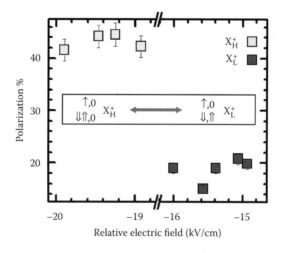

FIGURE 66.10 The degree of circular polarization for the two spin states X_H^+ and X_L^+.

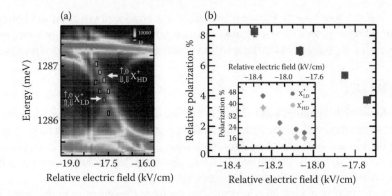

FIGURE 66.11 (a) X_{HD}^+, X_{LD}^+ fine structure doublet of the positive trion which arises due to the isotropic exchange interaction along with a corresponding degree of polarization memory results (b).

interaction energy. Therefore, the observed trend in circular polarization memory for the doublet is analogous to the interdot neutral exciton state as discussed in Section 66.4 with a resident hole in the bottom dot. The inset in Figure 66.11b also indicates a relative increase in the degree of circular polarization memory for the fine structure doublet as a function of an applied electric field [24,26]. Moreover, this trend can be prominently observed through the difference in polarization memory for the doublet as depicted in the main plot of Figure 66.11b.

From the above plot, it is evident that the circular polarization percentage is higher for the X_{LD}^+ state as compared to the higher-energy component X_{HD}^+. These observed results may arise due to the mixing with the dark states, which causes an asymmetry in polarization memory.

To further the effects of the spatial separation of the hole states in the spin fine structure doublet, barrier-dependent polarization-resolved measurements were performed. The observed degree of circular polarization memory results for the 2, 4, and 6 nm barriers between the top and the bottom dots of the QDM structure is depicted in Figure 66.12. As the barrier separation is increased, the hole in the top dot becomes more spatially separated from the bottom dot electron and hole, and this tends to reduce its overall effect to the electron–hole exchange interaction. This creates a more intradot neutral exciton-like configuration that depicts a relatively higher electron–hole exchange interaction energy at low relative electric fields for the spin fine structure doublet. This is indeed a singly charged positive trion with two holes in separate dots. Moreover, with relatively low applied electric fields, the degree of circular polarization percentage is reduced for the higher barrier QDM structures as evident from Figure 66.12.

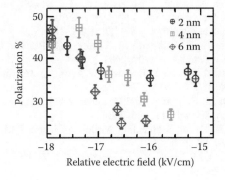

FIGURE 66.12 Barrier dependence of the degree of circular polarization on X_{LD}^+ as a function of the relative electric field.

When the top dot hole becomes more localized, it has a reduced effect on the bottom dot electron–hole pair. Therefore, both barrier separation and applied electric field can be tuned to effectively control the overall spin interactions in the spin fine structure of the singly charged positive trion state.

66.6 SUMMARY

The primary focus of this chapter was to identify and control the electron–hole exchange interaction in QDM systems through multiple techniques, including applied electric field and barrier width control between the top and bottom dots within the QDM. The spin-dependent interactions in QDMs were investigated through polarization-resolved PL spectra, as they directly couple to the spin states of the carriers while conserving the angular momentum.

The degree of circular polarization was used to measure the anisotropic exchange interaction, which causes pure states to be mixed and causes a shift in degeneracy. A possible technique was introduced to control the exchange energy by simply varying the applied electric field. In indirect neutral exciton states, the electron and hole were spatially separated via an applied field, which resulted in a reduction in the electron and hole wave function overlap. Because of this, the effects of the exchange splitting energy were reduced, providing a possible tool for spin manipulation.

An optically resolved spectral doublet was identified on the singly charged positive trion state due to the isotropic part of the exchange interaction. Observed results on the trion doublet were illustrated with the associated spin states. Interestingly, the spectator hole in the top dot of the singly charged state could be spatially separated with the applied field, again providing a degree of control over the exchange interactions. Barrier-dependent polarization-resolved spectra further confirmed the control of the wave function overlap in the singly charged trion doublet. Control of the spin interactions of these QDM device structures may be useful in quantum information processing schemes. Nonetheless, these techniques may also open up new avenues for nanotechnological applications in applied optical and spintronic-based devices.

ACKNOWLEDGMENTS

The authors would like to thank Dan Gammon and Allan Bracker for helpful discussions. This work was supported by the Ohio University CMSS program and NSF grant number DMR-1005525.

REFERENCES

1. J. I. Cirac and P. Zoller, *Phys. Rev. Lett.* **74**, 4091, 1995.
2. L. M. K. Vandersypen, M. Steffen, G. Breyta, C. S. Yannoni, M. H. Sherwood, and I. L. Chuang, *Nature* **414**, 883, 2001.
3. Q. A. Turchette, C. J. Hood, W. Lange, H. Mabuchi, and H. J. Kimble, *Phys. Rev. Lett.* **75**, 4710, 1995.
4. M. Woldeyohannes and S. John, *Phys. Rev. A* **60**, 5046, 1999.
5. D. Loss and D. P. DiVincenzo, *Phys. Rev. A* **57**, 120, 1998.
6. M. Scheibner, I. V. Ponomarev, E. A. Stinaff, M. F. Doty, A. S. Bracker, C. S. Hellberg, T. L. Reinecke, and D. Gammon, *Phys. Rev. Lett.* **99**, 197402, 2007.
7. N. Akopian, N. H. Lindner, E. Poem, Y. Berlatzky, J. Avron, D. Gershoni, B. D. Gerardot, and P. M. Petroff, *Phys. Rev. Lett.* **96**, 130501, 2006.
8. E. A. Stinaff, M. Scheibner, A. S. Bracker, I. V. Ponomarev, V. L. Korenev, M. E. Ware, M. F. Doty, T. L. Reinecke, and D. Gammon, *Science* **311**, 636, 2006.
9. D. Haft, R. J. Warburton, K. Karrai, S. Huant, G. Medeiros-Ribeiro, J. M. Garcia, W. Schoenfeld, and P. M. Petroff, *Appl. Phys. Lett.* **78**, 2946, 2001.
10. M. Ediger, P. A. Dalgarno, J. M. Smith, B. D. Gerardot, R. J. Warburton, K. Karrai, and P. M. Petroff, *Appl. Phys. Lett.* **86**, 211909, 2005.
11. T. C. Damen, L. Via, J. E. Cunningham, J. Shah, and L. J. Sham, *Phys. Rev. Lett.* **67**, 3432, 1991.
12. A. S. Bracker, E. A. Stinaff, D. Gammon, M. E. Ware, J. G. Tischler, A. Shabaev, Al. L. Efros et al., *Phys. Rev. Lett.* **94**, 047402, 2005.

13. D. Gammon, E. S. Snow, B. V. Shanabrook, D. S. Katzer, and D. Park, *Phys. Rev. Lett.* **76**, 3005, 1996.

14. B. Urbaszek, R. J. Warburton, K. Karrai, B. D. Gerardot, P. M. Petroff, and J. M. Garcia, *Phys. Rev. Lett.* **90**, 247403, 2003.

15. E. L. Ivchenko and G. E. Pikus, *Superlattices and Other Heterostructures*, 2nd ed. Springer, Berlin, 1995.

16. M. Scheibner, M. F. Doty, I. V. Ponomarev, A. S. Bracker, E. A. Stinaff, V. L. Korenev, T. L. Reinecke, and D. Gammon, *Phys. Rev. B* **75**, 245318, 2007.

17. Udo W Pohl et al., Size-tunable exchange interaction in InAs/GaAs quantum dots, In R. Haug, *Advances in Solid State Physics*, 45. Springer-Verlag, Berlin, 2007.

18. M. Grundmann, O. Stier, and D. Bimberg, *Phys. Rev. B* **52**, 11969, 1995.

19. G. Bacher, Optical spectroscopy on epitaxially grown II–VI single quantum dots, In P. Michler, *Single Quantum Dots: Fundamentals, Applications, and New Concepts,* 147, Springer, Berlin, 2003.

20. V.D. Kulakovskii, G. Bacher, R. Weigand, T. Kümmell, A. Forchel, E. Borovitskaya, K. Leonardi, and D. Hommel, *Phys. Rev. Lett.* **82**, 1780, 1999.

21. I. Stranski and L. Krastanow, Stizungsberichte d. mathem.-naturw. *Kl., Abt. IIb* **146**, 797, 1938.

22. Z. R. Wasilewski, S. Fafard, and J. P. McCaffrey, *J. Cryst. Growth* **201–202**, 1131, 1999.

23. K. C. Wijesundara, M. Garrido, S. Ramanathan, E. A. Stinaff, M. Scheibner, A. S. Bracker, and D. Gammon, *Mater. Res. Soc. Proc.* **1117E**, 1117-J04-08.R1, 2009.

24. K. C. Wijesundara, *Ultrafast Exciton Dynamics and Optical Control in Semiconductor Quantum Dots*, PhD Dissertation, Ohio University, 2012.

25. M. Scheibner, M. Yakes, A. S. Bracker, I. V. Ponomarev, M. F. Doty, C. S. Hellberg, L. J. Whitman, T. L. Reinecke, and D. Gammon, *Nat. Phys.* **4**, 291, 2008.

26. K. C. Wijesundara, A. Bracker, D. Gammon, and E. A. Stinaff, *IEEE Nanotechnol.* **11**, 887, 2011.

67 Material Issues for Efficient Spin-Transfer Torque RAMs

Kamaram Munira, William A. Soffa, and Avik W. Ghosh

CONTENTS

67.1 INTRODUCTION

Owing to the physical and electrical scaling challenges, the MOSFET-based memory industry (Flash, DRAM, SRAM, etc.) is predicted to hit the end of the road in the near future. Not only do MOSFET-based memories have scalability issues, but also they face increased power leakage and endurance problems [1–3]. A lot of effort is accordingly being invested into alternate forms of information storage to sustain Moore's law for memory over the next decades.

A possible road forward is to store information by using the spin degree of freedom of electrons, instead of charge-based storage that most MOSFET-based memory technologies follow. Spin-transfer torque random access memory (STT-RAM) is one of the frontrunners in spin-based memory technology today [4,5]. STT-RAM devices are based on magnetic tunnel junctions (MTJs) that can store a bit of information (Figure 67.1) [6]. A typical MTJ consists of two ellipsoidal ferromagnetic films separated by an insulating spacer film. The magnetization of one ferromagnetic layer (the fixed layer) is pinned by exchange coupling with an antiferromagnetic layer, while the magnetization of the second (free) layer can change freely. Data in an MTJ can be retained in the relative spin orientations of the fixed and free ferromagnetic layers (bit '0' or low resistance for parallel (P) and bit '1' or high resistance for antiparallel (AP)). The stored data can thereafter be sensed by measuring the tunneling magnetoresistance (TMR) [7]. Data are written by switching the magnetization of the free layer by direct transfer of the spin angular momentum from the torque due to a current injected across the insulator, spin polarized by the fixed layer [8,9]. Figure 67.2 shows the simulation results from a comprehensive STT-RAM model, combining non-equilibrium Green's function with macrospin dynamics [10,11] with parameters extracted from a published, in-plane rectangular 70×250 nm CoFeB/MgO/CoFeB (2 nm) [12] MTJ. AP to P switching was observed to

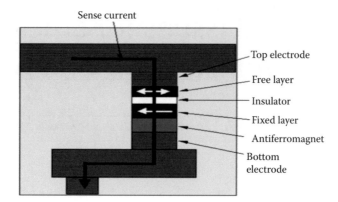

FIGURE 67.1 Magnetic tunnel junction—two ferromagnetic layers separated by an insulating spacer layer. The magnetization of one ferromagnetic layer (the fixed layer) in the MTJ is pinned by exchange coupling with an antiferromagnetic layer, while the magnetization of the second (free) layer can change by a spin-transfer torque.

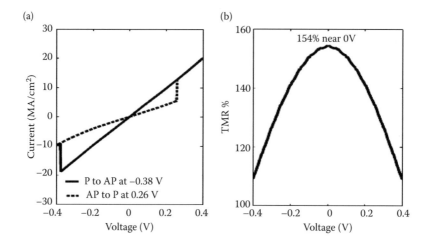

FIGURE 67.2 (a) STT-RAM simulation result with parameters extracted from the published CoFeB/MgO/CoFeB MTJ [1], [2] experiment. In our simulations, the AP to P switching was observed to take place at 0.26 V and P to AP switching at −0.38 V given within a 10 ns pulse, matching the experimental switching voltages quite well. (b) Tunneling magnetoresistance is 154% near 0 V, in agreement with the experiments.

take place in 10 ns at 0.26 V and P to AP switching at 0.38 V, consistent with the experimental data. The tunneling magnetoresistance is 154% near 0 V.

With the help of the comprehensive model, we can study how AP to P and P to AP switching happens in an STT-RAM cell. AP to P switching occurs in the free layer when a positive voltage is applied to the free layer relative to the fixed layer (Figure 67.3). A majority of the electrons with respect to the fixed layer tunnel through the barrier and build up at the free layer, as shown in Figure 67.3b, adding angular momentum and exerting a torque. After enough majority spin is accumulated to exert a critical torque, the free layer starts switching from the antiparallel to the parallel configuration at 2.6 ns. By 5 ns, the magnetization of the free layer has completed its switching from the antiparallel to the parallel mode. Conversely, for P to AP switching (Figure 67.4), a negative voltage is applied on the free layer. At 0 ns, the free layer is parallel but slightly noncollinear to the fixed layer (at room temperature, thermal torques ensure slight deviation from strict collinearity). At 6 ns, a majority of the electrons with respect to the fixed layer tunnel through the barrier, while the antiparallel minority electrons build up at the free layer, reducing its angular momentum. The

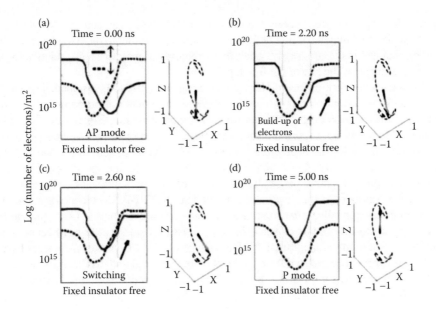

FIGURE 67.3 AP to P switching using CoFeB/MgO/CoFeB MTJ parameters at 0.4 V applied to the free layer. The starting angle is taken to be root-mean-square of the initial thermal distribution, θ_{rms}. (a) Time = 0 ns: The free layer is almost antiparallel to the fixed layer, (b) Time = 2.2 ns: A majority of the electrons with respect to the fixed layer tunnel through the barrier and build up at the free layer. The accumulated majority of electrons exerts a torque on the free layer, which causes its magnetization to start switching. (c) Time = 2.6 ns: The free layer continues switching from the antiparallel to the parallel configuration. (d) Time = 5 ns: The final magnetization of the free layer is parallel to the fixed layer.

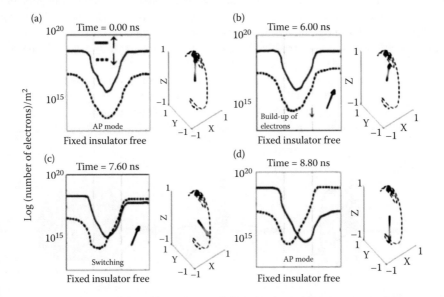

FIGURE 67.4 P to AP switching at −0.4 V applied to the free layer. (a) Time = 0 ns: The free-layer is almost parallel to the fixed layer. (b) Time = 6 ns: The magnetization of the free layer is non-collinear with respect to the fixed layer magnetization. A majority of the electrons with respect to the fixed layer tunnel through the barrier, while the residual antiparallel minority electrons build up at the free layer. The accumulated down spin electrons create a torque on the free layer, causing it to start switching from the up to down. (c) Time = 7.6 ns: The free layer continues switching from the parallel to the antiparallel configuration. (d) Time = 8 ns: The final magnetization of the free layer is antiparallel to the fixed layer.

accumulated downspin electrons create a torque on the free layer, causing it to switch from up to down. By 8 ns, the final magnetization of the free layer is antiparallel to the fixed layer. In short, the switching mechanism is the progressive addition or removal of majority spins relative to the fixed layer, creating a torque at the site of the free layer.

The critical current densities for spin torque switching in STT-RAMs is nonetheless too high for commercial applications, prompting intense investigation into material properties of the free layer [13]. The energy dissipation during the spin-transfer torque switching is given by $I^2R\tau$, where I is the current used to induce spin-transfer torque switching, R is the resistance of the MTJ, and τ is the total delay time it takes for the magnetization of the free layer to switch. In this work, for a given spin-polarized current density, J, of 2 MA/cm^2, we study the switching speed τ of different classes of magnetic materials, in-plane, perpendicular, and partially perpendicular (Figure 67.5) at 0 K temperature. We consider the free layer at the 45 nm feature STT-RAM technology, and employ a single-domain macrospin solver that solves the Landau–Lifshitz–Gilbert (LLG) equation [14]. For enough stability against thermally driven switching, to retain data for at least 10 years, Δ needs to be greater than 75 at room temperature [15]. The voltage needed to generate a current of such magnitude, and thus the resistance R, will be dependent on the material and insulator properties, such as the effective masses, insulator thickness, band-offset, and contact polarizations. Extracting the MTJ resistance corresponding to a given I will require the solution of the quantum transport problem using the non-equilibrium Green's function formalism (NEGF) [10] or, in a simpler incarnation, the Simmons model for tunneling across a barrier, modified to include its crucial magnetization-dependent prefactors [11,16]. While such a coupled transport-macromagnetic study is ultimately needed for overall energy efficiency studies, the aim of this chapter is to identify materials for fast switching in STT-RAM within 10 ns, with acceptable thermal stability ($\Delta > 75$).

A proper model for spin torque-induced switching should include thermal effects, which provide the initial torque to nudge the magnetization away from stagnation points along the energy landscape (i.e., away from strict parallel or antiparallel configurations), while at the same time hindering the motion of the magnetization on its journey past stagnation. In Ref. [11], we include an average over the thermal distribution of the initial angle evaluated using a Fokker–Planck equation. In this chapter, we simplify this treatment by directly replacing the initial angle with the root mean squared value θ_{rms} at 300 K, set by the equipartition theorem. While thermal agitation during the course of the reversal adds some uncertainty to the final delay time, the difference is likely to be small [17]. Therefore, we will disregard thermal effect during the course of reversal in this work.

In Section 67.2, the macrospin simulator that solves the LLG equation is introduced, and benchmarks with experimental results are shown. Section 67.3 outlines the critical switching currents and

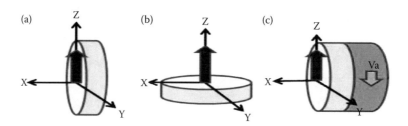

FIGURE 67.5 Definition of the magnetization of the free ferromagnetic layer in a spherical coordinate system. Magnetization for (a) In-plane material: the easy axis is along the major axis, Z, of the ellipsoidal plane,. (b) Perpendicular material: the easy axis is perpendicular to the disk. (c) Partially-perpendicular: an in-plane disk is capped with a vanadium layer that reduces the demagnetization anisotropy of the system by antiferromagnetic exchange coupling. In all three cases, magnetization prefers to lie near $\theta = 0$ (parallel case) or $\theta = \pi$ (antiparallel case). (K. Munira, W. A. Soffa, and A. W. Ghosh, Comparative material issues for fast reliable switching in STT-RAMs, *2011 11th IEEE Conference on Nanotechnology (IEEE-NANO)*, pp. 1403–1408, © (2011) IEEE. With permission.)

thermal stabilities for the in-plane, perpendicular, and partially perpendicular ferromagnetic free layers. Section 67.4 describes the variation in switching speeds across the material classes for a given switching current density, providing a path for the STT-RAM industry for fast, reliable switching.

67.2 MICROMAGNETIC SOLVER AND BENCHMARKS WITH EXPERIMENTAL DATA

The magnetization dynamics of the free layer is modeled by solving the modified LLG equation including spin torque [14]

$$\frac{d\hat{m}}{dt} - \alpha\left(\hat{m} \times \frac{d\hat{m}}{dt}\right) = -\gamma\mu_0\hat{m} \times \vec{H}_{\text{eff}} + \vec{T} \tag{67.1}$$

$$\vec{T} = \frac{\gamma\hbar\eta I}{2qM_S\Omega}\hat{m} \times (\hat{m}_p \times \hat{m}) \tag{67.2}$$

where $\hat{m}$ and $\hat{m}_p$ are the unit magnetization vectors of the free layer and the spin-polarized current density, respectively, α is the Gilbert damping coefficient, M_S is the saturation magnetization, γ is the gyromagnetic ratio, and $\vec{H}_{\text{eff}}$ is the effective magnetic field that includes the effects of magnetocrystalline and shape anisotropy. The spin-transfer torque is introduced by the current I of polarization η. The initial angle for all materials is set at the root mean square θ of the free layer nanomagnets under thermal perturbation, calculated by the equipartition theorem (Figure 67.6), $\sin^2(\theta_{\text{rms}}) = 1/2\Delta$ [18]. Note that we are including only the current-induced spin torque in our treatment. The "field-like" direct exchange torque between the layers can also play a role in switching, although it is typically significantly weaker than the purely current-driven term.

The current values entering the above equation are benchmarked separately by fitting NEGF simulation results with multiple STT-RAM experiments. In experiments on in-plane rectangular 70×250 nm CoFeB/MgO/CoFeB (2 nm) [12], AP to P switching was observed to take place at 0.26 V and P to AP switching at -0.38 V with a spin-polarized current density $\eta J = 28.5$ MA/cm^2, given a 10 ns pulse. Using H_K of 39.78 KA/m, M_S of 1050 KA/m, and Gilbert damping of 0.02 in our LLG model, our NEGF simulation reproduces these results. Indeed, with $\eta I = 28.5$ MA/cm^2,

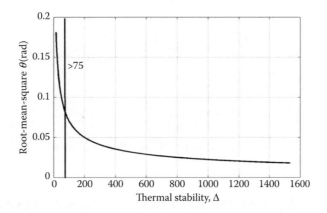

FIGURE 67.6 Thermal stability, Δ, vs. root-mean-square, θ_{rms}(rad) of the free layer nanomagnet. Δ needs to be greater than 75 for commercial applications. Root-mean-square θ_{rms}(rad) identifies the initial angle that we use in our simulations. (K. Munira, W. A. Soffa, and A. W. Ghosh, Comparative material issues for fast reliable switching in STT-RAMs, *2011 11th IEEE Conference on Nanotechnology (IEEE-NANO)*, pp. 1403–1408, © (2011) IEEE. With permission.)

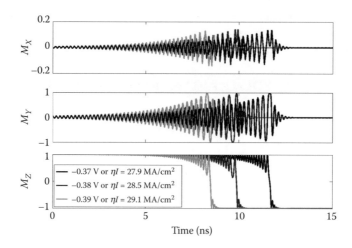

FIGURE 67.7 Parallel to antiparallel switching takes place in an in-plane CoFeB/MgO/CoFeB magnetic tunnel junction [1] at −0.38 V, given a 10 ns pulse of spin current, with a spin current density, $\eta J = -28.5$ MA/cm². A current pulse of at least 10 ns is needed for switching at −0.38 V. Voltages greater than −0.38 V will not be able to switch the free ferromagnet within 10ns as shown in the figure. H_K of 39.78 KA/m, M_S of 1050 KA/m, and Gilbert damping of 0.02 were used. The magnetization of the free ferromagnet switches from $\theta \approx 0$ to $\theta \approx \pi$. $M_Z = \cos \theta$: The simulation results agree quantitatively with the experiments [1]. (K. Munira, W. A. Soffa, and A. W. Ghosh, Comparative material issues for fast reliable switching in STT-RAMs, *2011 11th IEEE Conference on Nanotechnology (IEEE-NANO)*, pp. 1403–1408, © (2011) IEEE. With permission.)

we find that a current pulse of at least 10 ns is needed for switching. For AP to P switching, voltages greater than −0.38 V are unable to switch the free ferromagnet within 10 ns as seen in Figure 67.7. Similarly, for P to AP switching, negative voltages weaker than 0.26 V are unable to switch within 10 ns (Figure 67.8). Our NEGF currents that agree with the experimental values are thereafter incorporated into the simplified LLG solver above, through the I term that enters the torque.

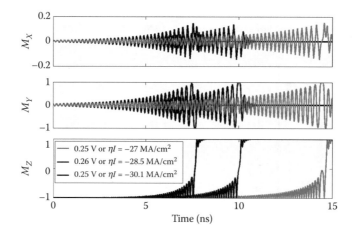

FIGURE 67.8 Antiparallel to parallel switching takes place in an in-plane CoFeB/MgO/CoFeB magnetic tunnel junction [1] at 0.26 V, given a 10ns pulse of spin current, with a spin current density $\eta J = 28.5$MA/cm². A current pulse of at least 10 ns is needed for switching at 0.26 V. Voltages less than 0.26 V will not be able to switch the free ferromagnet within 10 ns as shown in the figure. The magnetization of the free ferromagnet switches from $\theta \approx \pi$ to $\theta \approx 0$. The simulation results agree quantitatively with the experiments [1]. (K. Munira, W. A. Soffa, and A. W. Ghosh, Comparative material issues for fast reliable switching in STT-RAMs, *2011 11th IEEE Conference on Nanotechnology (IEEE-NANO)*, pp. 1403–1408, © (2011) IEEE. With permission.)

67.3 CRITICAL SWITCHING CURRENTS AND THERMAL STABILITIES FOR VARIOUS CLASSES OF FREE LAYER MAGNETIC MATERIALS

67.3.1 IN-PLANE

Figure 67.5 defines our choice of axes and angular conventions that we will adopt for the different material classes. For in-plane materials, the easy axis is aligned along the major axis of the ellipsoidal plane (Figure 67.5a). For a 45 nm feature technology, the free layer ellipsoidal disk has a minor axis of 45 nm, a major axis of 90 nm, and a thickness of 2 nm [15]. The potential energy density for an in-plane system is defined by its magnetocrystalline anisotropy and demagnetization field, $U = K \sin^2\theta + \mu_0 M_S^2 \sin^2\theta \cos^2\phi/2$. The critical current for switching is

$$I_c = \frac{2q}{\eta\hbar} \alpha \, \Omega\mu_0 H_K M_S \left[1 + \frac{M_S}{2H_K} \right] \tag{67.3}$$

H_K is the magnetocrystalline anisotropy field, M_S is the saturation magnetization, and Ω is the total volume of the free layer. The thermal stability is

$$\Delta = \frac{\mu_0 H_K M_S \Omega}{2K_B T} \tag{67.4}$$

The switching time required at zero temperature [17] for the initial angle, θ_{rms}, and applied current, I, is

$$\tau^{-1} = \frac{\mu_0 \alpha\gamma}{ln(\pi/2\theta_{rms})} \left(H_K + \frac{M_S}{2} \right) \left[\frac{I}{I_C} - 1 \right] \tag{67.5}$$

67.3.2 PERPENDICULAR

The easy axis is perpendicular to the plane (Figure 67.5b). For a 45 nm feature technology, the free layer circular disk has a diameter of 90 nm and a thickness of 2 nm [15]. The potential energy density for the system is defined by its magnetocrystalline anisotropy and demagnetization field, $U = K \sin^2\theta + \mu_0 M_S^2 \cos^2\theta/2$. The critical current for switching [19] is

$$I_c = \frac{2q}{\eta\hbar} \alpha \, \Omega\mu_0 H_K M_S \left[1 - \frac{M_S}{H_K} \right] \tag{67.6}$$

The decreased barrier reduces the critical current but also has an adverse effect on thermal stability

$$\Delta = \frac{\mu_0 (H_K M_S - M_S^2)\Omega}{2K_B T} \tag{67.7}$$

Materials with higher saturation magnetization will have less thermal stability, making the range of reliable materials narrower. Most perpendicular materials require epitaxial growth at elevated *in situ* temperatures, making them harder to integrate with CMOS processes than in-plane MTJ materials. They also typically have higher damping constants α than in-plane MTJ free layer materials [20–23]. The switching time required at zero temperature for initial angle, θ_{rms}, and applied current, I, is

$$\tau^{-1} = \frac{\mu_0 \alpha\gamma}{ln(\pi/2\theta_{rms})} (H_K - M_S) \left[\frac{I}{I_C} - 1 \right] \tag{67.8}$$

67.3.3 PARTIALLY PERPENDICULAR

An in-plane material is capped antiferromagnetically with a vanadium cap (Figure 67.5c) that reduces the demagnetization field of the disk [24,25]. The term "partially perpendicular" is used because the overall switching barrier is being reduced as in perpendicular materials (by the demagnetization field in the later case), but the magnetization still lies in the plane of the elliptical disk. The potential energy density for the system is defined by its magnetocrystalline anisotropy, demagnetization field, and exchange coupling with the capping layer, $U = K \sin^2\theta + (\mu_0 M_S^2/2 - K_i/t_{\text{freelayer}})\sin^2\theta \cos^2\phi$, where K_i (J/m²) is the interfacial energy between the free layer and capping and $t_{\text{freelayer}}$ (nm) is the thickness of the free layer. In Ref. [24], for a 2 nm CoFeB free layer capped with vanadium, we see an 85% decrease in the demagnetization field. The critical current for switching is

$$I_c = \frac{2q}{\eta\hbar}\alpha\Omega\mu_0 H_K M_S\left[1 + \frac{M_S}{2H_K} - \frac{H_\perp}{H_K}\right]$$

(67.9)

where $H_\perp$ is the reduced demagnetization field. Thermal stability is the same as the in-plane materials. The switching time required at zero temperature for initial angle, θ_{rms}, and applied current, I, is

$$\tau^{-1} = \frac{\mu_0\alpha\gamma}{\ln(\pi/2\theta_{\text{rms}})}\left(H_K + \frac{M_S}{2} - \frac{H_\perp}{H_K}\right)\left[\frac{I}{I_C} - 1\right]$$

(67.10)

67.4 SWITCHING SPEEDS OF DIFFERENT CLASSES OF FERROMAGNETIC MATERIALS

The previous section lists the critical currents that correspond to a destabilization of the initial magnetization from their respective easy axes toward the harder axes. To actually accomplish the switching in a given time, we will need to inject a substantially larger current. The switching time depends on the interactive dynamics between the various effective fields along the trajectory of the switching magnetization. Figures 67.9 through 67.11 show the switching delays of the in-plane, perpendicular, and partially perpendicular materials with varying M_S, H_K, and α, given a current density of 2 MA/cm². The phase plots designate the switching times, while those enclosed by the white boundaries identify the range of suitable materials that successfully switch within 10 ns (using Equations 67.5, 67.8, and 67.10) with a $\Delta > 75$ (using Equations 67.4 and 67.7). Table 67.1 lists the set of industrially relevant STT-RAM free layer materials that we study. Numerically calculated H_K and M_S combinations that switch within 10 ns and solutions of Equations 67.5, 67.8, and 67.10 for the in-plane, perpendicular, and partially perpendicular materials are compared in Figure 67.12. The numerical and analytical results are in agreement.

We can rationalize the shapes of the phases and white polygons designating efficient switching by outlining the various constraints they need to satisfy. In particular, we find that in-plane materials with low M_S and low H_K tend to switch faster, as their switching energy barrier to the spin current is proportional to $H_K M_S + M_S^2/2$. Perpendicular materials with low M_S and low H_K also switch faster, their barriers being proportional to $H_K M_S - M_S^2$. Even though a higher M_S also cuts down the energy barrier through the negative term, materials with high H_K do not help switching, as the dominant energy term is proportional to $H_K M_S$.

67.4.1 LOW VERSUS HIGH H_K IN-PLANE MATERIALS

For a given M_S, we find that materials with high H_K switch almost as fast as low-H_K materials. This seems counterintuitive, as a higher H_K yields a higher switching barrier Δ and a correspondingly low initial angle θ_{rms}. The explanation lies in the effective magnetic fields during the first half (initial

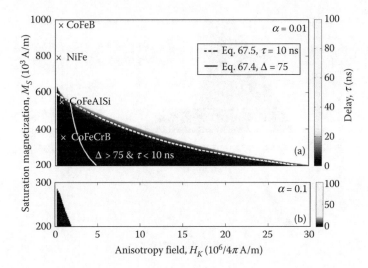

FIGURE 67.9 Numerically calculated switching delay (ns) for in-plane materials with varying saturation magnetization (M_S), anisotropy field (H_K), and Gilbert damping (a) $\alpha = 0.01$ and (b) $\alpha = 0.1$. A current density of 2 MA/cm² is applied. Materials that switch faster, <10 ns with thermal stability greater than 75, are preferred and are indicated by the area enclosed by the white boundaries. The white solid line indicates the boundary where Δ is 75 (using Equation 67.4, which defines a rectangular hyperbola), while the dashed line indicates the boundary where the switching time is 10ns (using Equation 67.5, another rectangular hyperbola). Compared to materials with $\alpha = 0.01$, materials with $\alpha = 0.1$ rarely switch. At $\alpha = 0.01$ and low magnetization, materials with low and high H_K switch at the same speed, even though the energy barrier that the spin current needs to overcome is proportional to $H_K M_S + M_S^2 = 2$. The explanation for this is outlined in Figure 67.12. (K. Munira, W. A. Soffa, and A. W. Ghosh, Comparative material issues for fast reliable switching in STT-RAMs, *2011 11th IEEE Conference on Nanotechnology (IEEE-NANO)*, pp. 1403–1408, © (2011) IEEE. With permission.)

configuration to equator) versus the second half (equator to flipped configuration) of the switching process. Figure 67.13 shows simulations with the saturation magnetization set at 200 KA/m and H_K set to 159.15 KA/m (top) and 1591.5 KA/m (bottom). To isolate the field dynamics, the same spin current of 2 MA/cm² is applied to the two systems, and the starting θ angle is set to 0.129 rad in each case. At 159.15 KA/m, the magnetization takes 0.5 ns to come to the equator and then another 0.4 ns to reach the south pole. At 1591.5 KA/m, the magnetization takes 1.03 ns to come to the equator, but only 0.22 ns to come to the south pole. What is noteworthy is the unequal times taken for the two switching steps. The asymmetry arises because while the high anisotropy field hinders the magnetization from moving from the north pole to the equator, it helps while moving from the equator to the south pole. Therefore, the large amount of time taken to travel to the equator at high H_K is compensated by the fast switching from the equator to the north pole.

In summary, in-plane materials with high H_K and low M_S would be good candidates for the free layer. The high H_K promotes high thermal stability, and does not compromise on switching speed because it actually helps with the switching process from the hard axis to the south pole.

67.4.2 IN-PLANE VERSUS PERPENDICULAR MATERIALS

Figure 67.14 shows the difference between in-plane and perpendicular magnetization switching at 2 MA/cm², $H_K = 159.15$ KA/m, $M_S = 500$ KA/m, and $\alpha = 0.01$. Recall that for in-plane magnetization switching, both the anisotropy and demagnetization fields hinder the switching from the north pole to the equator, and both fields help the switching from the equator to the south pole. For perpendicular magnetization switching on the other hand, the anisotropy field opposes while the demagnetization field helps the switching from the north pole to the equator. However, after passing the equator, the

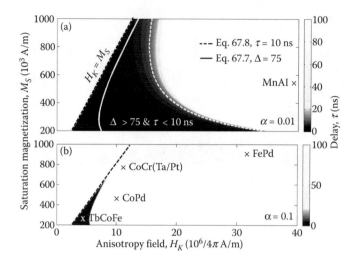

FIGURE 67.10   Numerically calculated switching delay (ns) for perpendicular materials with varying saturation magnetization (M_S), anisotropy field (H_K), and Gilbert damping: (a) $\alpha = 0.01$ and (b) $\alpha = 0.1$. A current density of 2 MA/cm^2 is applied. Overall, perpendicular materials switch faster than in-plane materials, as the energy barrier that the spin current needs to overcome is proportional to $H_K M_S = M_S^2$. However, the decreased barrier has an adverse effect on thermal stability, $\Delta (= \mu_0(H_K M_S = M_S^2) = 2K_B T)$. For increased M_S, thermal stability is reduced, making the range of reliable materials narrower. Most perpendicular materials have $\alpha = 0.1$. Materials that switch within 10 ns and Δ greater than 75 are preferred and are indicated by the enclosed area by the white boundaries. The white solid line indicates the boundary where Δ is 75 (using Equation 67.7, which defines a rectangular hyperbola added to a linear term), while the dashed line indicates the boundary where the switching time is 10 ns (using Equation 67.8). Even though higher M_S cuts down the energy barrier, materials with high H_K do not switch as the dominant energy barrier term is proportional to $H_K M_S$. The shape of the colored region is bounded to the left by a linear separatrix that corresponds to maintaining a perpendicular magnetization, $H_K M_S > M_S^2$. (K. Munira, W. A. Soffa, and A. W. Ghosh, Comparative material issues for fast reliable switching in STT-RAMs, *2011 11th IEEE Conference on Nanotechnology (IEEE-NANO)*, pp. 1403–1408, © (2011) IEEE. With permission.)

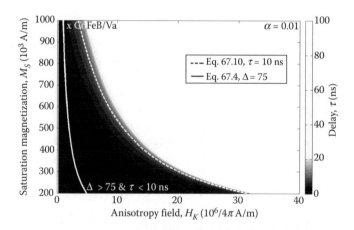

FIGURE 67.11   Numerically calculated switching delay (ns) for partially-perpendicular materials with varying saturation magnetization (M_S), anisotropy field (H_K), and $\alpha = 0.01$. The demagnetization is reduced by 85% by capping. A current density of 2MA/cm^2 is applied. A wider range of materials (indicated by the enclosed area by the white boundaries using Equations 67.4 and 67.10) switch within 10 ns with a thermal stability greater than 75, as compared to in-plane materials with $\alpha = 0.01$, generating thereby a more suitable class of free layer materials. (K. Munira, W. A. Soffa, and A. W. Ghosh, Comparative material issues for fast reliable switching in STT-RAMs, *2011 11th IEEE Conference on Nanotechnology (IEEE-NANO)*, pp. 1403–1408, © (2011) IEEE. With permission.)

TABLE 67.1

A List of Ferromagnetic Materials That Are Being Investigated for the Free Layer in STT-TAM

Material	Anisotropy Field, H_k ($10^3/4\pi$ KA/m)	Saturation Magnetization, M_S (KA/m)	α	∥ or ⊥
TbCoFe [21]	1.2	139	0.1	⊥
CoFeB [12], [27]	0.5	1050	0.01	∥
FePd [22]	33	1100	0.1	⊥
FePt [22]	116	1140	0.1	⊥
$Co_2FeAl_{0.5}Si_{0.5}$ [28]	0.3	560	0.01	∥
MnAl [22]	61	560	0.01	⊥
CoPd [23], [24]	10	450	0.1	⊥
NiFe [29], [30]	0.25	800	0.01	∥
CoPt [31]	91	900	0.1	⊥
CoFeGe [32]	5	350	0.1	⊥
CoFeCrB [32]	0.1	350	0.01	∥
CoCrPt [33]	14.2	800	0.1	⊥
CoCrTa [33]	12.2	800	0.1	⊥

Note: ∥ stands for in-plane materials while ⊥ for perpendicular materials.

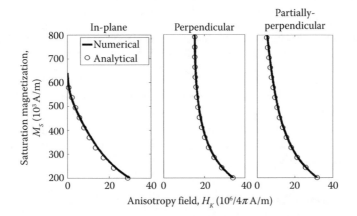

FIGURE 67.12 Numerically calculated H_K and M_S combinations that switch within 10ns, overlaid with solutions of Equations 67.5, 67.8, and 67.10 for in-plane, perpendicular and partially-perpendicular materials. The numerical and analytical results are in excellent agreement.

anisotropy helps in moving to the south pole, while the demagnetization field opposes it. Thus, the times taken for the magnetization to travel from the north pole to the equator and then from the equator to the south pole are comparable.

For comparable Gilbert damping α, a wider range of perpendicular materials switch than in-plane materials, the energy barrier being proportional to $H_K M_S - M_S^2$ (Figure 67.10a). The steep curve to the right of the colored region at 100 ns coalesces ultimately with the linear boundary to the left that corresponds to $H_K M_S > M_S^2$, in other words, maintaining a perpendicular magnetization. The white polygon includes the additional inequality $\Delta > 75$, which gives a rectangular hyperbola

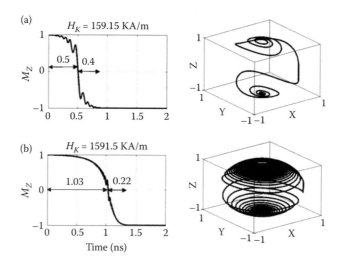

FIGURE 67.13   For in-plane materials with $\alpha = 0.01$ and low magnetization, materials with low and high H_K switch at almost the same speed, even though the energy barrier that the spin current needs to overcome is proportional to $H_K M_S + M_S^2/2$. The above figure shows two cases where the saturation magnetization is fixed at (200 KA/m), and the H_K of (a) 159.15 KA/m and (b) 1591.5 KA/m are used for the LLG simulations. At 159.15 KA/m, the magnetization takes 0.5 ns to come to the equator and then another 0.4 ns to come to the south pole. At a larger H_K of 1591.5 KA/m Oe, the magnetization takes 1.03 ns to come to the equator but only 0.22 ns to come to the south pole. The asymmetry arises because the high magnetocrystalline anisotropy field hinders the magnetization while moving from the north pole to the equator, but assists it while moving from the equator to the south pole. Thus, the longer time taken to reach the equator at higher H_K is compensated for by the faster switching from the equator to the south pole. (K. Munira, W. A. Soffa, and A. W. Ghosh, Comparative material issues for fast reliable switching in STT-RAMs, *2011 11th IEEE Conference on Nanotechnology (IEEE-NANO)*, pp. 1403–1408, © (2011) IEEE. With permission.)

on the $M_S H_K$ plot. Unfortunately, most perpendicular materials known today have high Gilbert damping, making them undesirable for STT-RAM use (Figure 67.10b).

Perpendicular materials with a low damping constant will be good candidates for STT-RAM free layers. Their switching speeds will be comparable to in-plane materials, but there will be a greater probability of switching.

67.4.3   IN-PLANE VERSUS PARTIALLY PERPENDICULAR MATERIALS

A wider range of partially perpendicular materials switch within 10 ns when compared to in-plane materials, making partially perpendicular materials ultimately a more energy-efficient class of materials. Their dampings at $\alpha = 0.01$ are small, yet their barriers are reduced by an overall negative term, allowing a wide range of fast switching events. Figure 67.15 shows the difference between in-plane and partially perpendicular magnetization switching at 2 MA/cm², $H_K = 1591.5$ KA/m, $M_S = 500$ KA/m, and $\alpha = 0.01$. The decrease in the demagnetization field causes the magnetization to switch three times faster. Capping in-plane materials with a Va layer should be further investigated as their fast switching speed shows great promises for low-energy switching.

67.5   CONCLUSION

A wide variety of materials, cataloged into three magnetic classes, are explored for their switching speeds at a given thermal stability and switching current density. To be considered a suitable candidate for the free layer in STT-RAM, the material needs to conform to three requirements: (1) high

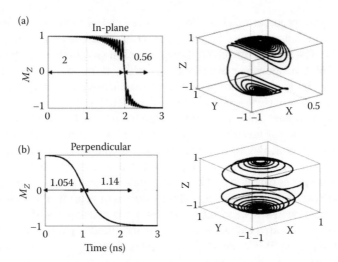

FIGURE 67.14 Difference between (a) in-plane and (b) perpendicular magnetization switching at 2 MA/cm², H_K = 159.15 KA/m, M_S = 500 KA/m, and α = 0.01. For in-plane magnetizations, both the magnetocrystalline anisotropy and demagnetization fields hinder the magnetization during the north pole to the equator transition. After passing the equator, both fields help in moving to the south pole. For perpendicular magnetizations, the anisotropy field opposes, while the demagnetization field helps the magnetization move from the north pole to the equator. After passing the equator, the anisotropy helps while the demagnetization field opposes the subsequent equator to south pole switching. Thus, the times taken for the north pole to equator and equator to south pole transitions are comparable. (K. Munira, W. A. Soffa, and A. W. Ghosh, Comparative material issues for fast reliable switching in STT-RAMs, *2011 11th IEEE Conference on Nanotechnology (IEEE-NANO)*, pp. 1403–1408, © (2011) IEEE. With permission.)

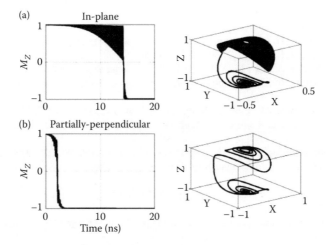

FIGURE 67.15 Difference between (a) in-plane and (b) partially-perpendicular magnetization switching at 2MA/cm², H_K = 1591.5 KA/m, M_S = 500 KA/m, and α = 0.01. The decrease in the demagnetization field causes the magnetization to switch 3 times faster for the latter case. (K. Munira, W. A. Soffa, and A. W. Ghosh, Comparative material issues for fast reliable switching in STT-RAMs, *2011 11th IEEE Conference on Nanotechnology (IEEE-NANO)*, pp. 1403–1408, © (2011) IEEE. With permission.)

thermal stability to prevent soft error, (2) ability to switch with low applied spin current, and (3) high switching speed, as the energy consumed during the write process is proportional to the total time it takes to switch. For faster and more reliable energy-efficient switching in the thermally stable free layer of an STT-RAM, the following material paths should be taken: (a) in-plane materials with high H_K and low M_S (high H_K increases the speed of switching, especially during the second half, and has greater thermal stability); (b) perpendicular materials with low damping—reaching comparable switching speeds with in-plane but with greater probability to switch; and (c) antiferromagnetically capped partially perpendicular materials—capping with a Va layer decreases the demagnetization field, which enables the switching to occur faster without compromising the thermal stability. It is worth emphasizing at this stage that the study presented here is entirely based on the magnetic properties of the contacts, as captured by LLG. A separate material phase space will correspond to the electronic properties and ultimately the voltage required to generate the drive current, namely, the contact and barrier effective masses, tunnel barrier height and width, and the polarization in the contacts [11]. These can be extracted and studied using the NEGF equation, and alternately using a modified Simmons equation. While the LLG study yields the critical current density and switching speeds, the Simmons equation would, in addition, provide the switching voltages, that is, the tunnel barrier resistances required to accomplish these switching events, thereby providing the total energy dissipated during the switching event.

ACKNOWLEDGMENTS

K. Munira thanks E. Chen of Grandis Inc. for discussions on a variety of switching materials, A. Natarajarathinam and W. H. Butler of the University of Alabama for discussions on perpendicular and partially perpendicular materials, and A. Nigam and Mircea Stan of the University of Virginia for discussions regarding the macromagnetic solver.

REFERENCES

1. G. Atwood, Future directions and challenges for ETox flash memory scaling, *IEEE Trans. Device Mater. Reliab.*, 4(3), 301–305, 2004.
2. R. Baumann, Soft errors in advanced computer systems, *IEEE Des. Test Comput.*, 22(3), 258–266, 2005.
3. D. Frank, R. Dennard, E. Nowak, P. Solomon, Y. Taur, and H.-S. P. Wong, Device scaling limits of Si MOSFETs and their application dependencies, *Proc. IEEE*, 89(3), 259–288, 2001.
4. V. Zhirnov, J. Hutchby, G. I. Bourianoff, and J. Brewer, Emerging research memory and logic technologies, *IEEE Circuits Devices Mag.*, 21(3), 47–51, 2005.
5. S. A. Wolf, D. D. Awschalom, R. A. Buhrman, J. M. Daughton, S. von Molnr, M. L. Roukes, A. Y. Chtchelkanova, and D. M. Treger, Spintronics: A spin-based electronics vision for the future, *Science*, 294(5546), 1488–1495, 2001.
6. E. Hirota, H. Sakakima, and K. Inomata, Magnetic random access memory (MRAM). In: *Giant Magneto-Resistance Devices*, G. Ertl, R. Gomer, H. Luth, and D. Mills (Eds.), Springer, Berlin, Germany, pp. 135–136, 2002.
7. M. Julliere, Tunneling between ferromagnetic films, *Phys. Lett. A*, 54(3), 225–226, 1975.
8. J. C. Slonczewski, Current-driven excitation of magnetic multilayers, *J. Magn. Magn. Mater.*, 159(1–2), L1–L7, 1996.
9. L. Berger, Emission of spin waves by a magnetic multilayer traversed by a current, *Phys. Rev. B*, 54(13), 9353–9358, 1996.
10. K. Munira and A. Ghosh, Energy-efficient switching in STT-RAM using higher polarized ferromagnetic layers, in preparation.
11. K. Munira, W. H. Butler, and A. Ghosh, A quasi-analytical model for energy-delay-reliability tradeoff studies during write operations in perpendicular STT-RAM cell, *IEEE Trans. Electron Devices*, 59, 1–6, 2012.
12. H. Kubota, A. Fukushima, K. Yakushiji, T. Nagahama, S. Yuasa, K. Ando, H. Maehara, Y. Nagamine, K. Tsunekawa, D. D. Djayaprawira, N. Watanabe, and Y. Suzuki, Quantitative measurement of voltage dependence of spin-transfer torque in MgO-based magnetic tunnel junctions, *Nat. Phys.*, 4(1), 37–41, 2001.

13. Y. Huai, D. Apalkov, Z. Diao, Y. Ding, A. Panchula, M. Pakala, L.-C. Wang, and E. Chen, Structure, materials and shape optimization of magnetic tunnel junction devices: Spin-transfer switching current reduction for future magnetoresistive random access memory application, *Jpn. J. Appl. Phys.*, 45(5A), 3835–3841, 2006.

14. J. Z. Sun, Spin-current interaction with a monodomain magnetic body: A model study, *Phys. Rev. B*, 62(1), 570–578, 2000.

15. E. Chen, D. Apalkov, Z. Diao, A. Driskill-Smith, D. Druist, D. Lottis, V. Nikitin, X. Tang, S. Watts, S. Wang, et al., Advances and future prospects of spin-transfer torque random access memory, *IEEE Trans. Magn.*, 46(6), 1873—1878, 2010.

16. J. Simmons, Generalized formula for the electric tunnel effect between similar electrodes separated by a thin insulating film, *J. Appl. Phys.*, 34, 1793–1803, 1963.

17. J. Sun, Spin angular momentum transfer in current-perpendicular nanomagnetic junctions, *IBM J. Res. Dev.*, 50(1), 81–100, 2006.

18. K. Munira, W. A. Soffa, and A.W. Ghosh, Comparative material issues for fast reliable switching in STT-RAMs, *2011 11th IEEE Conference on Nanotechnology (IEEE-NANO)*, pp. 1403–1408, 2011.

19. J. Mallinson, Magnetization fluctuation noise, In: *Magneto-Resistive and Spin Valve Heads: Fundamentals and Application*, A. Press, Ed. Academic Press, San Diego, CA, p. 161, 2002.

20. S. Mangin, D. Ravelosona, J. Katine, M. Carey, B. Terris, and E. Fullerton, Current-induced magnetization reversal in nanopillars with perpendicular anisotropy, *Nat. Mater.*, 5, 210–215, 2006.

21. M. Nakayama, T. Kai, N. Shimomura, M. Amano, E. Kitagawa, T. Nagase, M. Yoshikawa, T. Kishi, S. Ikegawa, and H. Yoda, Spin transfer switching in TbCoFe/CoFeB/MgO/CoFeB/TbCoFe magnetic tunnel junctions with perpendicular magnetic anisotropy, *J. Appl. Phys.*, 103(7), 07A710–07A713, 2008.

22. T. Klemmer, D. Hoydick, H. Okumura, B. Zhang, and W. A. Soffa, Magnetic hardening and coercivity mechanisms in 110 ordered FePd ferromagnets, *Scr. Metall. Mater.*, 33(10–11), 1793–1805, 1995, *Proceedings of an Acta Metallurgica Meeting on Novel Magnetic Structures and Properties.*

23. S. Hashimoto, Y. Ochiai, and K. Aso, Perpendicular magnetic anisotropy in sputtered CoPd alloy films, *Jpn. J. Appl. Phys.*, 28(Part 1, 9), 1596–1599, 1989.

24. D. Smith, V. Parekh, C. E, S. Zhang, W. Donner, T. R. Lee, S. Khizroev, and D. Litvinov, Magnetization reversal and magnetic anisotropy in patterned Co/Pd multilayer thin films, *J. Appl. Phys.*, 103(2), 023920, 2008.

25. A. Natarajarathinam, Z. Tadisina, and S. Gupta, Partial perpendicular anisotropy of CoFeB with Vanadium capping, in *AVS 57th International Symposium & Exhibition, Magnetic Interfaces and Nanostructures*, 2010.

26. S. Ikeda, K. Miura, H. Yamamoto, K. Mizunuma, H. Gan, M. Endo, S. Kanai, J. Hayakawa, F. Matsukura, and H. Ohno, A perpendicular-anisotropy CoFeB-MgO magnetic tunnel junction, *Nat. Mater.*, 9(1476–1122), 721–724, 2010.

27. Y. Huai, M. Pakala, Z. Diao, D. Apalkov, Y. Ding, and A. Panchula, Spin-transfer switching in mgo magnetic tunnel junction nanostructures, *J. Magn. Magn. Mater.*, 304(1), pp. 88–92, 2006.

28. H. Sukegawa, S. Kasai, T. Furubayashi, S. Mitani, and K. Inomata, Spin-transfer switching in an epitaxial spin-valve nanopillar with a full-heusler Co[sub 2]FeaL[sub 0.5]Si[sub 0.5] alloy, *Appl. Phys. Lett.*, 96(4), p. 042508, 2010.

29. R. C. Hall, Single crystal anisotropy and magnetostriction constants of several ferromagnetic materials including alloys of NiFe, SiFe, ALFe, CoNi, and CoFe, *J. Appl. Phys.*, 30(6), pp. 816–819, 1959.

30. N. Nishimura, T. Hirai, A. Koganei, T. Ikeda, K. Okano, Y. Sekiguchi, and Y. Osada, Magnetic tunnel junction device with perpendicular magnetization films for high-density magnetic random access memory, *J. Appl. Phys.*, 91(8), pp. 5246–5249, 2002.

31. H. Shima, K. Oikawa, A. Fujita, K. Fukamichi, K. Ishida, S. Nakamura, and T. Nojima, Magnetocrystalline anisotropy energy in 110-type copt single crystals, *J. Magn. Magn. Mater.*, 290–291(Part 1), pp. 566–569, 2005, *Proceedings of the Joint European Magnetic Symposia (JEMS'04).*

32. M. Ding, Y. Cui, J. Lu, T. Mewes, and J. Poon, Amorphous Gd-Fe-Co as prospective material for perpendicular STT-MRAM, in *APS March Meeting,* 2011.

33. B. B. Lal, S. S. Malhotra, and M. A. Russak, Magnetic recording medium having a CoCr alloy interlayer of a low saturation, U.S. Patent 5 922442, Jul 13, 1999.

Section XX

Nanodevice Modeling

68 Atomic-Scale Modeling of Nanoscale Devices

Anders Blom and Kurt Stokbro

CONTENTS

68.1 INTRODUCTION

Even though the active components of semiconductor devices have been shrunk down radically over the past decade, to the point where certain feature sizes are in the range of 30–40 nm, the motion and behavior of electrons are still reasonably well described by semiclassical models—Ohm's law coupled with macroscopic electrostatic models, drift–diffusion equations, and so on. The number of carriers is still large enough that quantum fluctuations are averaged out statistically, and the materials involved can, for the most part, be characterized by bulk material parameters. In this domain, technology computer aided design (TCAD) models are an essential tool to model and predict the physical properties of device components.

However, as the downscaling of the gate length, oxide thickness, and other crucial device parameters continues, quantum effects that are not captured by the semiclassical models start to play a dominating role. Furthermore, the device functionality of, for example, tunnel field-effect transistors (TFET) is a consequence not just of the properties of a single material but rather of the interfaces formed between different semiconductors and/or metals. The smaller the active regions are made, the larger the effects of individual atomic defects, vacancies, dislocations, and grain boundaries also become, and these can significantly change the properties of the Schottky barrier height of an interface, or the ability of a thin dielectric layer to block the leakage current. This can have devastating consequences for the device variability during the fabrication process, and for reliability under operating conditions. Adding to this the veritable zoo of "exotic" elements (including structures like graphene) that are currently part of the palette of options for device engineers, it becomes increasingly clear that new models need to be incorporated into the design workflow used for developing novel transistors, memory elements, radiofrequency resonators, and so on.

From a physical perspective, there are two fundamental factors to consider. First of all, atomic-scale defects are discrete in nature, as opposed to the continuum material description used in TCAD models. Moreover, the properties of an interface between two or more materials, or, for instance, a thin dielectric layer embedded between two other materials, cannot simply be deduced from the combined properties of the different materials. In fact, the most advanced gate stacks today contain just a few atomic layers of high-κ dielectric materials like HfO$_2$, and it is highly questionable whether these layers can be described reliably by the same parameters that apply to bulk HfO$_2$. In the case of nanowires, the influences of surface adsorbants, interface roughness, or diameter variation (Figure 68.1) are also effects that require a detailed investigation on the atomic scale.

Second, as the length scales approach the mean-free path of the material, electrons will not travel diffusively across the barrier anymore. Instead, the charge transport through the barrier becomes a tunneling process, and the electrons no longer behave as particles, but must be described as wave functions. In the ultimate limit, where scattering can be neglected, we thus have a ballistic, coherent tunneling process, and the corresponding transmission probability and current can only be computed quantum-mechanically; Ohm's law is no longer valid, and we cannot define uniquely the conductivity of the material, as it becomes bias-dependent with a nonlinear current–voltage relation. The tunneling current is furthermore strongly influenced by the presence of atomic-scale defects, which act as elastic scattering centers. Inelastic scattering may, of course, also be present in the quantum-mechanical picture, especially when the mean-free path is comparable to the tunneling barrier thickness.

To capture both sets of effects, it is necessary to consider the system as a whole, with all the different materials interacting and influencing each other in a way that can only be described with quantum-mechanical atomistic models.

In addition, with the introduction of new materials such as low-κ dielectrics with low thermal conductivity, the need arises for combined thermal, mechanical, optical, and electrical modeling, as stressed in the International Technology Roadmap for Semiconductors (ITRS) [1], since these properties are no longer necessarily independent of each other, especially as more and more functionality becomes integrated on a single chip. The mechanical and thermal properties can to a large extent still be treated classically, but atomistically—or rather, the large computational cost for quantum-mechanical models leaves us no other solution for the moment. An important stumbling block here, however, is the fact that very few reliable potentials exist for most of the materials of interest, and new material combinations are introduced all the time. A methodology that allows for the derivation of new potentials for novel materials is therefore necessary to facilitate this effort, and in this process, first-principles modeling will play an important role for the fitting of the potentials.

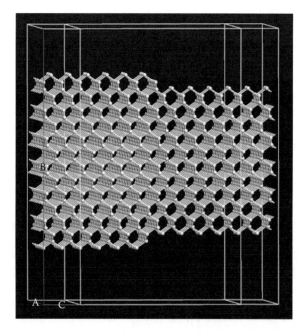

FIGURE 68.1 Model for a step-like change in diameter in a Si (110) nanowire.

It thus appears evident that there is a strong need for a new generation of device simulation tools, which can incorporate methodologies that are capable of integrating different methods, deal with combinations of metals, semiconductors, as well as organic materials, and that can compute both electronic and thermal transport properties, as well as mechanical and optical properties, in realistic nanoscale device geometries, taking atomistic features into account. Such tools will be an essential enabling technology for the nodes ahead in the ITRS, which mentions "atomistic" or "atomic-scale" modeling at least 15 times in the latest section on Modeling & Simulation [1]. This applies not only to the far-out nodes at 10 or 5 nm but also the more imminent ones.

We will review here some aspects of the state-of-the-art modeling of currents from a quantum-mechanical atomistic perspective. Particular focus will be placed on recent developments of relevance for more traditional (yet nanoscale) semiconductor materials and device types, although the principles apply very well also to more exotic device designs based on molecular junctions, nanotubes, graphene, and so on.

68.2 ELECTRICAL CURRENTS ON THE NANOSCALE

It is interesting to note that the two effects discussed above, that is, the discrete nature of sources of scattering and the atomistic effects involved in the combination of different materials to form an interface on the one hand, and the wave nature of electrons on the other, can individually be treated with simpler models. Using effective mass theory, either in a simple single-band description or with more sophisticated models like kp, the quantum-mechanical electronic structure of quantum wells and other heterojunctions can be computed while still assuming that the materials are homogeneous, and thus can be described with a set of material parameters, typically derived from bulk properties. Conversely, classical molecular mechanics simulations are used extensively to model growth processes and other properties of collections of atoms (usually very many)—provided that relevant potentials are available. Complications arise, however, when we combine the quantum effects with the atomistic features, and a more complete description will be needed.

In the classical picture of electrical conductivity, electrons are accelerated by an electric field caused by the voltage bias applied across the material, but owing to a series of scattering events— occurring with a frequency determined by the mean-free path of the material, which is typically of the order of 50–100 nm—the conductivity remains finite. At the nanoscale—that is, in the limit where the feature sizes are significantly smaller than the mean-free path of the material—a completely different picture emerges. The electronic current must now be viewed as the propagation of a quantum-mechanical wave function. Concepts like resistivity or conductivity no longer have any meaning since they arise via scattering processes that now do not have time to occur as the electron traverses the device. Instead, we must speak of conductance as a measure of the "quality" by which the material (or, rather, the device as a whole) can conduct electricity. Conductance is a central concept in the Landauer–Büttiker formalism [2], which is the standard approach for describing coherent ballistic or tunneling currents in a complex structure containing several leads. In this picture, the current in a two-terminal device is obtained by integrating the transmission probability $T(E)$ (the summed transmission probability of all available channels or modes) over all energies E, weighted with the difference in the Fermi distributions of both the leads. Ignoring the details for the present discussion, in the zero-temperature limit, the expression for the current reduces to

$$I = \frac{2e^2}{h} \int_{-V_{\text{bias}}/2}^{V_{\text{bias}}/2} T(E)dE, \tag{68.1}$$

and from this expression one can derive the zero-bias conductance from the transmission at the Fermi level E_f as $G = G_0 T(E_f)$, where $G_0 = 2e^2/h = 77.48 \ \mu\Omega^{-1}$ is the quantum conductance unit.

The difficulty in this formalism lies in actually computing the transmission probabilities for a real atomic-scale system, taking into account the relevant boundary conditions and self-consistent electrostatic response, with an accurate account of the electronic energy levels. This will be the topic of the following sections.

As a side comment, we note that the transmission probabilities are typically not greatly affected by the electron temperatures of the leads. One can therefore often obtain an accurate estimate of the temperature dependence of the thermionic emission current without having to recompute the transmission spectrum for each temperature. It is, however, almost always necessary to compute $T(E)$ for each bias, that is, the linear-response assumption that $T(E)$ is independent of the bias rarely holds, even for small values of the bias.

68.3 ATOMIC-SCALE DEVICE STRUCTURES

Modeling of nanoscale structures such as nanowires, nanotubes, graphene devices, molecular junctions, and ultrathin dielectric junctions has evolved over the last decade into a position where several thousand atoms can now be simulated with atomistic, quantum-mechanical methods, using methods ranging from density-functional theory (DFT) to tight-binding models.

With the introduction of nonequilibrium techniques, it has furthermore become possible to compute not only the electronic structure properties of various materials, but also the transport properties of device-like structures under an applied finite bias [3–5]. In these models, the system geometry is generally made up of two conducting electrodes, between which a semiconducting or insulating material is inserted. The methodology is very general (Figure 68.2); the electrodes, which need not be identical, may consist of nanotubes, bulk metals, nanowires, a graphene nanoribbon, a doped semiconductor, or any other (preferably conducting) material. The central part, or scattering region, on the other hand, can contain any atomic configuration imaginable; it may be a molecule, another material to form a sandwich junction, a differently shaped piece of a graphene nanoribbon, and so on. It is also possible to introduce additional electrodes in the system, either atomistic ones to allow true multiterminal transport to be studied [6] or purely electrostatic electrodes [5] that introduce a way to modulate the electron density in the central region of the device to simulate the gating effect in a transistor.

This makes it possible to study the contact resistance or Schottky barrier of a single metal–semiconductor junction or grain boundaries, the leakage current through an ultrathin dielectric layer, field-effect transistors made of nanowires or graphene nanoribbons, rectification and negative differential resistance in molecular diodes and switches, the change in conductance of a functionalized nanotube for use in sensor applications, and so on. If spin is considered, the spin current or tunnel magneto-resistance can also be calculated, to just mention a few of the many application areas. Since the central region is treated purely atomistically, effects of impurities like vacancies, dopants, and dislocations are automatically incorporated by just modifying the structure accordingly.

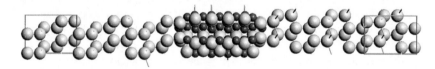

FIGURE 68.2 A device model of a thin slice of Si_3N_4 embedded in Si (111). The Si electrodes are marked by the unit cell "boxes," which are repeated infinitely to the left and right, respectively. The atoms in between the electrodes form the central region, in which the entire electronic structure must be computed self-consistently. We note that there is also Si in the central region—this is necessary to provide a smooth transition from bulk-like Si near the electrodes to the Si layers closest to the Si_3N_4, where the electronic structure is different due to interactions with the sandwiched material, and where the Si atoms even rearrange themselves geometrically to best fit the interface structure. The electron transport takes place between the electrodes, and the entire system is periodic in the plane transverse to the transport direction.

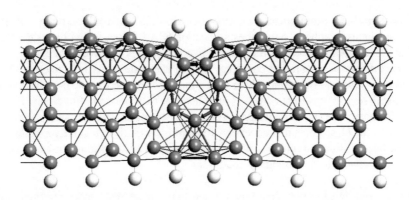

FIGURE 68.3 Transmission pathway at the Fermi level for a graphene nanoribbon with a Stone–Wales defect on one of the edges [7]. This visual way of representing the results provides an intuitive way to understand the electron transport mechanism.

As noted in the previous section, the fundamental quantity to be computed is the transmission spectrum, which is a function of the energy of the incoming electrons from the electrode. The transmission spectrum can then be integrated to give the current at the given bias, but it can also be decomposed and projected in different ways, resulting in supporting quantities that can assist in the analysis and understanding of the transport mechanisms.

Examples of such quantities are transmission eigenvalues and eigenstates, transmission pathways (Figure 68.3), current density, and the real-space density of states. All of these provide a visual image of how the electrons travel through the structure, although one has to keep in mind that the final current depends on the collective behavior of all states, integrated over wave vectors and energy.

Other quantities that provide insight are the nonequilibrium density of states (atom and orbital projected) and the complex band structures of the electrodes. The regular band structure is of course of relevance too, but the complex band structure also gives information about states in the band gap, in the case of semiconductor electrodes.

This deep-level understanding is one of the key takeaways from modeling and simulation. Although one can often achieve a good qualitative agreement between experiments and computed results, the possibility to dig deeper into the mechanism is only available *in silico*. This can provide far more insight about the system than can be obtained in an experiment—insight that can be used to guide new experiments in directions of most likely success, or give rise to entirely new ideas not tried before. Moreover, even if the calculations do not take all possible environmental effects into account, the simulated results provide a limit for the performance of an ideal device, thus giving an indication of the feasibility of a suggested device design. Used in this way, modeling does not replace experiments and measurements but supports and guides them.

68.4 QUANTUM-MECHANICAL DEVICE MODELING

68.4.1 BASIC INGREDIENTS

This section will review the essential points of the models needed to compute the transmission spectrum accurately and reliably from a quantum-mechanical perspective. In summary, the crucial ingredients are

- *Atomistic models*—we need to describe the detailed scattering potential landscape of the central device region, with individual atomic impurities, and so on; this was already discussed above.

- *Quantum models*—required to accurately describe the electronic structure of ultrathin films, molecules, graphene, nanotubes; in Section 68.4.1 below, we will consider the models that are available, as well as their respective advantages and disadvantages.
- *Boundary conditions*—a major point that we will elaborate more on in Section 68.4.2.
- *Nonlinear response*—in toy models or very simple systems, it may be possible to compute the transmission spectrum at zero bias and then just extrapolate to finite bias. Experience shows, however, that in the majority of realistic systems, the transmission spectrum itself depends on the bias. Thus, it is necessary to employ methods that can treat systems in the nonequilibrium situation where a finite bias is applied across the central region. This is also closely related to the boundary conditions.

We can also point out the two single most important factors that determine the quality of the results:

- An accurate band structure (including the HOMO-LUMO gap of molecules) of the electrodes and the scattering region (see Section 68.4.2); this primarily determines the zero-bias properties of the transport system.
- A correct description of the voltage drop that occurs across the central region (see Section 68.4.3) allows for an accurate computation of the tunneling current at finite bias.

One should pay attention to the fact that the electronic structure and transport properties are not two separate quantities. They are deeply coupled since there is information extracted from the electronic structure that goes into the electrostatic model used to compute the voltage drop, and the potential distribution in turn influences the occupation of the electronic states. We therefore end up with a rather complex nonlinear problem with a higher complexity than the usual self-consistent iterative scheme used in the coupled Schrödinger/Poisson equation solvers, since the total number of carriers is not conserved in the calculation—charge can move in and out of the electrodes that are infinite reservoirs for electrons. A powerful approach for computing the occupations of the states in the central region in this situation (with or without a bias applied) is the nonequilibrium Green's function (NEGF) formalism, which has become the standard tool for this task [3] due to its computational efficiency and numerical stability, although in principle one can also use scattering states, transfer matrices, and other methods.

For the electronic structure, however, there are a few choices, which we will discuss in the following section.

68.4.2 ELECTRONIC STRUCTURE METHODS FOR ATOMIC-SCALE DEVICE MODELING

Ab initio methods, in which we also count DFT, have the advantage of being able to describe any combination of elements without the need for predefined parameters. This is clearly of importance in cases where new materials are being developed, and in particular when materials are being combined in new ways, for example, at interfaces, where the bonding character changes from bulk to molecular. It also makes it possible to combine organic and inorganic materials.

A common concern about DFT is the issue related to obtaining a correct description of the band gap of semiconductors. For some applications, this is not really an issue—geometrical properties and total energies are often well represented by DFT. The traditionally inaccurate estimation of the band gap does however raise some questions for the electronic transport properties in some cases. Recent developments of models that allow the inclusion of the on-site Hubbard terms can however "cure" the band gap in many situations, and alleviate the problems of interface states ending up in the conduction band, rather than in the band gap, without incurring any real increase in computational effort. One can also try to use novel exchange-correlation functionals like meta-generalized gradient approximation (GGA) [8], although it may introduce nontrivial effects for, for example,

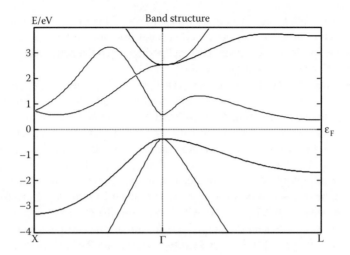

FIGURE 68.4 Band structure of germanium, computed with DFT and the TB09 meta-GGA functional in Atomistix ToolKit. Unlike for LDA or GGA, we here obtain band gaps at both the L and Γ points that agree with experiments to within 5%, and also a very reasonable effective mass for the lowest conduction band at the Γ point. Interestingly, at the L point, the conduction band effective mass is accurately estimated by basic GGA, even if the band gap comes out completely wrong.

layered systems that contain both metals and dielectric layers; even if the band gap of the insulator can be adjusted (cf. Figure 68.4), the metal may suffer from distortion of its band structure and a shift of the Fermi level. Thus, the Hubbard model may be preferable for device systems since it treats each chemical element individually, and can be fitted to accurate band structures—perhaps produced with meta-GGA. Screened exchange or functionals of similar nature like HSE06 may offer a third and, in many ways, better alternative, however, at the cost of a noticeable increase in the computational time.

The Slater–Koster tight-binding models are not in the same way transferable, meaning that the parameters usually have to be tailored for each specific problem. However, they do offer treatment of far larger systems than DFT [4] and give a correct band gap essentially by construction. In particular, tight-binding models often only take into account interactions between adjacent atoms, and combined with routines for operating on sparse matrices; this allows for efficient handling of even hundreds of thousands of atoms. At that scale, one is actually approaching small but realistic models of entire transistors. Also, for example, for III–V materials composed of random alloys like AlGaAs, it is necessary to use large supercells to provide statistical accuracy in terms of the material composition fluctuations; methods based on special quasirandom structures (SQS) or virtual crystal approaches attempt to provide some systematic approach to this problem, but have not been tried extensively for transport modeling yet.

An intermediate position is taken by semiempirical models such as the extended Hückel theory (EHT) [5] and DFTB [9], where a smaller number of empirical parameters are fitted to experimental results or very accurate calculations. The computational demands are smaller than for DFT, while offering a more transferable approach than tight-binding, not least thanks to the self-consistency included in the model. DFTB has the additional advantage of incorporating a repulsive potential, and thus this formalism can also take ionic movement into account, to enable geometry optimizations, molecular dynamics simulations, and calculations of reaction barriers using nudged elastic bands (NEB). As long as the geometry remains within the space covered reasonably well by the parameterization, this method can be very accurate.

In summary, as long as the model is able to correctly describe the band structure and other electronic structure properties of the system, there is no inherent additional accuracy in more complex models like DFT. Their advantage is a higher transferability; there is no need to consider different parameters for different problems. If, on the other hand, we have suitable parameters for a broad range of systems, the lower computational cost and perhaps even the better band gap description of parameterized models make them very attractive.

The main limitation for the use of semiempirical models is however the poor availability of parameters for material combinations outside certain well-established areas; in particular, there are almost no metal/semiconductor parameter sets. Precisely the same issue holds for classical potentials, which could enable even process simulations with millions of atoms for semiconductor materials. A task for the future is therefore to develop methods by which tight-binding parameters and classical potentials can be fitted to new materials and material combinations in an efficient manner, so that one can take advantage of the higher calculation speed of the simpler methods without relying on generic parameters, which may not apply to the actual system. An interesting development here are bond order potentials like reaxFF, which can describe chemical processes where bonds are formed and broken, and which have parameters for almost the entire periodic table, at least for certain types of materials [10]. Another good example of this are the EHT parameters tabulated by Cerda [11], which among others contain an excellent set of parameters for sp^2-bonded carbon, that is, graphene and nanotubes. If, on the other hand, one attempts to use the "standard" carbon EHT parameters (by Hoffmann or Müller) for graphene, the results are completely disastrous, due to the fact that these parameters apply to sp^3-bonded atoms. Fitting semiempirical parameters is a huge task by itself, but one of the fundamental prerequisites is an accurate DFT code that can easily be integrated with automated fitting schemes, to reduce manual labor. Thus, having a platform where different codes can be accessed through a common interface helps a lot, not just for the fitting but also for making it possible to quickly compare the accuracy of different models.

Our software package, Atomistix ToolKit® (ATK) [7], is developed with this in mind. It comprises a wide range of electronic structure models (essentially all those mentioned above), which can all be combined with an NEGF framework to compute transport properties of all possible device geometries discussed in this work.

68.4.3 ELECTROSTATICS AND BOUNDARY CONDITIONS

There is a multitude of quantum-mechanical software packages available for computing the electronic structure of atomic-scale systems. Generally, they apply to one of two types of boundary conditions: periodic structures that are infinite in three or fewer dimensions (examples include three-dimensional bulk crystals, two-dimensional graphene sheets, and one-dimensional nanotubes or nanowires), or finite structures (clusters, molecules). In a transport system, however, neither applies: although the electrodes are periodic, and are computed as such during the transport calculation, the central region is finite. Thus, a method that can combine these two geometry types is required. In addition, we must also consider the possibility of applying a potential difference between the source and drain electrodes to drive a current through the structure, as well as electrostatic gates that can modulate the current. This breaks any periodicity that might otherwise be present in the system.

Therefore, we need to use open boundary conditions in the transport direction, such that charge can flow between the source and the drain. This is where the NEGF method comes in, as it can describe a nonequilibrium distribution of electrons in the central region. As mentioned above, the entire problem must be solved self-consistently, not only for the electronic structure—as is inherent at least in DFT—but also for the charge redistribution that takes place between the electrodes, acting as reservoirs of electrons, and the central region.

The problem is highly nonlinear, and convergence can be hard to achieve unless care is taken in the mixing of algorithms and the precise way in which matrix elements are computed close to the edges of the central region. This is one of the main points where different implementations of the NEGF transport method differ the most. In ATK, we have made sure to include all possible contributions to the Hamiltonian matrix elements of the central region, also those that arise from interactions with atoms in the electrodes, the so-called spill-in terms. We have also implemented a double-contour algorithm that allows for bound states to be present in the bias window. This is crucial for the physical correctness of the model, and the ability to apply a realistically large bias to the structures.

For the electrostatic potential, we employ Dirichlet boundary conditions on the surfaces connecting the electrodes and the central region, which allow for inclusion of the finite bias directly in the boundary conditions. This requires a multigrid method, which is considerably more time-consuming than the usual fast Fourier transform (FFT) methods that can be used to solve the Poisson equation under periodic boundary conditions. For systems with periodic boundary conditions in the transverse plane, it is possible to combine FFT methods in these directions with a multigrid method in the transport direction. This makes it easy to study interface systems using a small transverse unit cell, although for 1D or 2D systems one needs to introduce vacuum padding to avoid spurious electrostatic and other interactions between the repeated copies of the system.

If, additionally, we also add one or more electrostatic gates to the device (in practice, this means a metallic region plus a dielectric layer that can screen the gate potential), the periodicity in the transverse plane is also broken, and we need to employ a full multigrid method in all directions. Recently, capabilities for including metallic and dielectric regions have been incorporated into quantum transport calculations (see for example Figure 68.5), enabling calculations of realistic transistor characteristics [4,5].

It should be noted that in these simulations, the gate electrodes and dielectric screening regions are not described atomistically, but as continuum regions without structure. This allows for a great flexibility in the design of the gates, and one can consider cylindrical wrap gates, double-gates, or single back gates without changing the methodology. We also have here a multiscale simulation, where part of the system is described on one complexity level (the device is a discrete atomic system) and another on a different one (continuum regions). This paves the way for more advanced quantum device simulators that can treat realistically large structures without the need to necessarily describe the entire device atomistically but only the relevant parts that carry the current.

A complication that possibly arises in such simulations is, however, that to mimic a realistic device geometry, it is often necessary to place the electrostatic gates at a distance from the active device region that far exceeds the dimensions of the atomic-scale features themselves. Put shortly, the simulation cell will contain a lot of inactive space (vacuum), which becomes rather costly at the atomic level, not least in terms of memory. A solution here is to introduce real-space grids based on finite-element methods (FEM). This is rather straightforward to do for methods like tight-binding where the primary real-space grid is the external potential in the Poisson equation, but decidedly harder to do for DFT where we also have to evaluate the real-space density and the

FIGURE 68.5 The self-consistent electrostatic potential of a z-shaped graphene transistor, constructed from fusing two metallic zigzag ribbons (the electrodes) and an armchair segment (the central region). The results are show for a gate potential of −1 V, at a source/drain bias of 1 V (symmetrically applied). Note how the electrostatic potential from the gate is screened by the graphene layer [5].

exchange–correlation functional based on it [12]. It is also not clear if FEM grids actually require less memory since the description of each FEM vertex requires considerably more memory than the definition of a point in a simple, regular space grid. This is at least true as long as the total simulation volume and atom count are of the order where DFT is still reasonably manageable (less than 2000 atoms or so). For tight-binding calculations of truly large-scale structures with thousands of atoms, the balance shifts, however.

There are also efforts being made for handling multiterminal devices [6], but this is still an area for further research, as there are significant challenges in obtaining stable convergence and also for the computational demands in terms of time and memory. In general, such simulations are currently only tenable on supercomputer resources.

68.5 SUMMARY AND OUTLOOK

It is hopefully evident from the presentation above that there is a strong need for a new generation of device simulation tools, which can incorporate methodologies that are capable of integrating different methods, deal with combinations of metals, semiconductors as well as organic materials, and which can compute both electronic and thermal transport properties in realistic nanoscale device geometries, as well as mechanical and optical properties. Whether or not this type of modeling software will be an extension of existing TCAD tools (similar to the downscaling of the devices themselves) or part of a separate bottom-up approach remains to be seen. It is far from trivial to just add another layer of "quantum effects" that somehow correct the classical model to account for effects that occur on the nanoscale, since these effects come in so many flavors and with a huge amount of parameters. What is clear, however, is that these tools will be an essential enabling technology for the nodes ahead in the ITRS, which mentions "atomistic" or "atomic-scale" modeling at least 15 times in the latest section on Modeling & Simulation [1]. This applies not only to the far-out nodes at 10 or 5 nm but also the more imminent ones. With a deeper fundamental understanding of the processes, materials, and operation behavior of the novel devices comes improved reliability and scalability, while development speed can increase and uncertainties and risks can be reduced.

In general, the methods outlined above, whether referring to the way the geometry is optimized or the method that is used for the electronic structure calculation, have their own set of advantages and specific areas of applicability. No single method—or even code—can hope to solve all the problems. What is required is thus a simulation framework that comprises several methods. The real power, however, comes from not only having these different methods coexisting, but also making it possible to combine them, so that one can leverage the respective advantages of each methodology. This will enable true multiscale modeling that can span many orders of length and time scales. In this way, the modeling tool can deliver reliable results in a timely manner to the developer or engineer working not only with standard structures but also novel device ideas, even based on exotic combinations of materials.

It is our desire and goal to create such a framework with our simulation platform, ATK. In order for this effort to be successful in delivering a tool that is useful for solving real problems at hand, a close dialog is however needed with relevant industrial and academic partners. With this chapter, we would like to invite anyone interested to thus discuss the foundation of "nano TCAD."

REFERENCES

1. The International Technology Roadmap for Semiconductors (ITRS). http://www.itrs.net/.
2. S. Datta, *Electronic Transport in Mesoscopic Systems*, Cambridge University Press, Cambridge, 1997.
3. M. Brandbyge, J.-L. Mozos, P. Ordejon, J. Taylor, and K. Stokbro, *Phys. Rev. B.* **65**, 165401, 2002.
4. G. Klimeck. F. Oyafuso, T.B. Boykin, C.R. Bowen, and P. von Allmen, *Comput. Model. Eng. Sci.* **3**, 601, 2002.

  5. K. Stokbro, D.-E. Petersen, S. Smidstrup, M. Ipsen, A. Blom, and K. Kaasbjerg, *Phys. Rev. B*. **82**, 075420, 2010.

  6. K.K. Saha, W. Lu, J. Bernholc, and V. Meunier, *Phys. Rev. B*. **81**, 125420, 2010.

  7. Atomistix ToolKit (ATK), http://www.quantumwise.com.

  8. F. Tran and P. Blaha, *Phys. Rev. Lett*. **102**, 226401, 2009.

  9. M. Elstner et al., *Phys. Rev. B*. **58**, 7260, 1998.

 10. A.C.T. van Duin, S. Dasgupta, F. Lorant, and W. A. Goddard III, *J. Phys. Chem. A*. **105**, 9396, 2001.

 11. http://www.icmm.csic.es/jcerda/EHT_TB/TB/Periodic_Table.html

 12. J. Avery, PhD Thesis, Copenhagen University, February 2011.

Index

A

O

OAD, *see* Oblique angle deposition (OAD)
Object-oriented micromagnetic framework (OOMMF), 780
Oblique angle deposition (OAD), 628
O/C ratio, *see* Oxygen/carbon atomic ratio (O/C ratio)
Octadecylphosphonic acid (ODPA), 42
Octadecyltrichlorosilane (OTS), 43
ODPA, *see* Octadecylphosphonic acid (ODPA)
Office of Naval Research (ONR), 406
Oleic acid, 657
    FT-IR spectra, 660
ON current degradation simulation results, 711
    energy loss paths, 713
    incorporation of self-heating effects, 712
    Monte Carlo free-flight scatter, 713
    Monte Carlo–molecular dynamics–3D Poisson–3D
        thermal solver, 712
    simulated nanowire MOSFET device, 711
    single negative impurity impact, 714
    solving energy balance equations, 713
On-chip capacitors, 535
1D, *see* One-dimension (1D)
One-dimension (1D), 27
    nanostructure materials, 615
    Ostwald ripening mechanism, 619
One electron at a time transport, 144
One-electron Schrödinger equation, 718
ONO, *see* Oxide-nitride-oxide (ONO)
ONR, *see* Office of Naval Research (ONR)
OOMMF, *see* Object-oriented micromagnetic
        framework (OOMMF)
OPs, *see* Organophosphorus compounds (OPs)
Optoelectronic device, 437
OR type primitive, 270, 271
OR-2 gate, *see* Two-input OR gate (OR-2 gate)
OR-3 gate, *see* Three-input OR gate (OR-3 gate)
ORAs, *see* Output response analyzers (ORAs)
Organic semiconductors, 110; *see also* Humidity sensing
    capacitive sensors, 111
    humidity measurement and control, 110
    laser ablation, 111, 115
    nanoparticle generation, 111
    phthalocyanines, 111
    sensor materials, 110
    sensors based on, 111
    synthesis and characterization of, 109, 110
Organophosphorus compounds (OPs), 661
Orthodox theory, 188
OTF, *see* Oxide thickness fluctuation (OTF)
OTS, *see* Octadecyltrichlorosilane (OTS)
Output response analyzers (ORAs), 571
Output-node capacitance, 807
Overdrive at receiver nanomagnet, 812
Oxidation processes, 450
Oxide thickness fluctuation (OTF), 30
Oxide-nitride-oxide (ONO), 728
Oxygen/carbon atomic ratio (O/C ratio), 451
Oxynitrides, 82

P

P, *see* Parallel (P)
Palladium nanoparticles (Pd nanoparticles), 625

PANI, *see* Polyaniline (PANI)
Parallel (P), 849
    plate model, 527
Parallel energy minimizing computation, 791, 801–802
    correspondence between vision and magnets, 795–796
    digital media object detection, 793
    energy minimization co-processor, 793
    fabricated nanomagnets, 800
    fabrication process, 796–797
    field-coupled computational technologies, 792–793
    image processing, 793–794
    intricate system of nanomagnets, 801
    magnetic Hamiltonian, 794–795
    read-out scheme, 797–798
    single domain magnet, 798, 799
Parallel tube (PT), 539
Parasitic current paths, 316
Parasitic leakage paths, 300
Partial bias schemes, 310
Particle-based device simulation, 16–17
Particle–mesh coupling (PM coupling), 16
Particle–particle–particle–mesh coupling method, 17;
        see also Corrected Coulomb approach; Fast
        multipole method
Particulate capacitors, 85; *see also* Etched-foil
        Nanocapacitors
    high *K* ferroelectrics, 87
    nanocapacitor with copper nano-electrode, 88
    stable dielectric, 86
    Tantalum capacitors, 86, 87, 88
    three-dimensional trench structure, 86
    valve metal challenges, 87
Patterned graphene printing, 384; *see also* Graphene;
        Highly oriented pyrolytic graphite (HOPG)
    direct printing, 387–388
    FLG pattern on PDMS, 388
    graphene nanowire printing, 388
    MLG pattern on glass slide, 387
    process for, 386
    thermal tape method, 384–387
Pattern generator (PG), 572
Patterning processes, 520
PC, *see* Propylene carbonate (PC)
PCLO, *see* Probability of correct logical output (PCLO)
PDF, *see* Probability density function (PDF)
PDMS, *see* Polydimethylsiloxane (PDMS)
Pd nanoparticles, *see* Palladium nanoparticles
        (Pd nanoparticles)
PDP, *see* Power-delay product (PDP)
PECVD, *see* Plasma-enhanced chemical vapor
        deposition (PECVD)
PEDT, *see* Polyethylenedioxythiophene (PEDT)
PEDT-based capacitors, 83
PEO, *see* Polyethylene oxide (PEO)
Percolation-based analytical model, 18; *see also*
        Analytical models
Perfluorophenylazide (PFPA), 410
Performance metrics, 528
Perpendicular magnetic anisotropy (PMA), 779
PET, *see* Positron emission tomography (PET)
Petrick's method, 582
PFPA, *see* Perfluorophenylazide (PFPA)
PG, *see* Pattern generator (PG); Polarity gate (PG)
Phase-change memory, 300

For Product Safety Concerns and Information please contact our
EU representative GPSR@taylorandfrancis.com Taylor & Francis
Verlag GmbH, Kaufingerstraße 24, 80331 München, Germany